Albert Overhauser

Anomalous Effects in Simple Metals

Related Titles

Friebolin, H.

Basic One- and Two-Dimensional NMR Spectroscopy

2005

ISBN: 978-3-527-31233-7

Wöhrle, D., Pomogailo, A. D. (eds.)

Metal Complexes and Metals in Macromolecules
Synthesis, Structure and Properties

2003

ISBN: 978-3-527-30499-8

Grant, D. M., Harris, R. K. (eds.)

Encyclopedia of Nuclear Magnetic Resonance
9 Volumes

2002

ISBN: 978-0-470-84784-8

Salibi, N., Brown, M. A. (eds.)

Clinical MR Spectroscopy
First Principles

1998

ISBN: 978-0-471-18280-1

Albert Overhauser

Anomalous Effects in Simple Metals

WILEY-VCH Verlag GmbH & Co. KGaA

The Author

Prof. Albert Overhauser
Department of Physics
Purdue University
525, Northwestern Avenue
West Lafayette, IN 47907
USA

Library of Congress Card No.: applied for

British Library Cataloguing-in-Publication Data: A catalogue record for this book is available from the British Library.

Bibliographic information published by the Deutsche Nationalbibliothek
The Deutsche Nationalbibliothek lists this publication in the Deutsche Nationalbibliografie; detailed bibliographic data are available on the Internet at http://dnb.d-nb.de.

Typesetting le-tex publishing services GmbH, Leipzig
Printing and Binding Strauss GmbH, Mörlenbach
Cover Design Adam Design, Weinheim

Printed in the Federal Republic of Germany
Printed on acid-free paper

ISBN 978-3-527-40859-7

Foreword

Comments by M. S. Dresselhaus and G. Dresselhaus

This volume on the physics of simple metals features a collection of articles constituting the seminal contributions that Albert Warner Overhauser made to this field with a view toward its future as derived from his own research. He was attracted to simple metals like potassium at an early time (1951) because simple metals allowed him to study the effect of interacting electrons which are responsible for the many interesting and fundamental phenomena exhibited by these simple metals. This research area is now called "emergent phenomena" which address the question of "how do complex phenomena emerge from simple ingredients". This topic remains at the forefront of condensed matter physics, as cited in the present decadal study by the US National Research Council Condensed Matter and Materials Physics 2010 Committee entitled "The Science of the World Around Us".

Over his active career of about 55 years, Al Overhauser has written extensively (about 65 papers published in prestigious journals) on the subject of the properties of the electronic structure of the simplest metals, namely potassium and other alkali metals. This is a subject that is usually covered rather briefly in every elementary condensed matter physics course for both undergraduates and graduate students because it is so fundamental. Overhauser has clearly demonstrated why these materials have such a fundamental importance for our understanding of condensed matter systems. Furthermore Overhauser's research papers have become important for advancing our understanding of the "many body" aspect of all metallic systems which so strongly depend on the interactions between electrons and with their associated spins. This book on an apparently simple topic clearly points out the basic interactions which are necessary to understand this important area of physics.

Now, let us say a few words about Albert Warner Overhauser's physics career. He was nominally the first Ph.D. student of Charles Kittel at the University of California (U.C.) Berkeley, who in the early 1950's became a tenured full professor faculty acquisition at U.C. Berkeley. Kittel had visited U.C. Berkeley from Bell Labs in 1950, the year before Kittel's permanent appointment to U.C. Berkeley. Shortly before Kittel returned to Bell Labs from his visiting appointment at U.C. Berkekry,

Anomalous Effects in Simple Metals. Albert Overhauser

ISBN: 978-3-527-40859-7

Kittel met Al Overhauser who was looking for a thesis topic. Kittel then suggested the research which in later years led to the discovery of the “Overhauser Effect”. Soon after this encounter between Kittel and Overhauser, Kittel left Berkeley to return to Bell Labs. During Kittel's absence, Al Overhauser worked independently on this research project and quickly reached the point of writing a classic paper on the subject which was soon published in the Physical Review. [“Paramagnetic relaxation in metals”, Phys. Rev. 89, 689 (1953)]. When Kittel returned to Berkeley after winding up his affairs at Bell Labs, he started looking into the Status of his new U.C. Berkeley research group. He then noticed that Overhauser had no stipend for the Fall term. When Kittel discussed this issue with Overhauser, Al informed Kittel that he didn't need support as a graduate student because he had already finished his thesis. Overhauser then informed Kittel that now he needed a job instead. At that point Kittel contacted his friend Professor Fred Seitz at the University of Illinois who arranged for a post-doc position for Overhauser at the University of Illinois, which was then the Mecca of Condensed Matter Physics.

Because of his absence from the Berkeley campus, Kittel didn't really fully understand the work of his student, but since Overhauser was known at Berkeley to be a brilliant student, Kittel thought the thesis work was important. So when Gene Dresselhaus joined the Kittel research group in the fall on 1953, Gene's first assignment was to check over Overhauser's thesis. Reading Overhauser's thesis was educational firstly, in becoming calibrated on the great creative work expected of a new entrant to the field of theoretical condensed matter physics when working with Professor Charles Kittel. Secondly, looking for mistakes in Al Overhauser's published work was not a good use of research time.

After Overhauser's postdoc at Illinois, where he discovered the fundamental importance of the Overhauser effect, Al accepted a Cornell professorship at Cornell University. The transition from a postdoc to a Cornell faculty position was unusual even at that time. It was at Cornell that Gene Dresselhaus, starting in 1956, got well acquainted with the Overhauser research program and family. Later in 1958, when Millie Dresselhaus arrived at Cornell, she joined the group of Overhauser friends and admirers. This group was very happy together for only a short time. The group was soon broken up when Overhauser left for Ford Research Laboratories in 1958. It was while Overhauser was at Ford that he started his long time creative work on Charge Density Waves and Spin Density waves in potassium and other simple systems.

Without Al's presence at Cornell, the Cornell job lost its attraction for the Dresselhaus duo and they shortly left in 1960 to establish their own careers at MIT.

Gene Dresselhaus
Millie Dresselhaus

Contents

Anomalous Effects in Simple Metals. Albert Overhauser

ISBN: 978-3-527-40859-7

Part I Introduction and Overview

1
The Simplest Metal: Potassium

The five alkali metals (Li, Na, K, Rb, Cs) are monovalent, so their conduction-electron momentum states occupy one half of the Brillouin zone. The periodic potentials which create the energy gaps at the (twelve) Brillouin-zone faces are small and therefore, the Fermi surface of each metal is nearly spherical.

The noble metals (Cu, Ag, Au) are also monovalent; but the periodic potentials which create energy gaps at the eight hexagonal faces of their zone are strong. Consequently the Fermi surface is distended along the $\{1, 1, 1\}$ directions until it is terminated by the energy gaps at the hexagonal faces. Such a Fermi surface is multiply connected; and this leads to a variety of conduction-electron orbits in the presence of a large magnetic field.

The anticipated electric, magnetic, optical, and thermal properties of a simple metal (possessing free-electron-like conduction electrons and, therefore, a spherical Fermi surface) have been elaborated in many monographs and textbooks published during the last seventy years. One would expect that most theoretical predictions would agree with experimental behavior found in alkali metals. The surprise is that such agreement is not found! The purpose of this volume is to document the many phenomena that have violated expectations during the last forty years and to collect in one place the research of the author and his collaborators which has led to a unified synthesis of alkali metal peculiarities.

Most of the experimental studies have focused on potassium. Many of the phenomena must be studied at low-temperatures so that the electron mean-free-path, λ, can be long. That is: $\lambda \gg r_c$, where r_c is the radius of a cyclotron orbit for an electron traveling at the Fermi velocity. Li and Na are disqualified because they undergo a martensitic transformation from their (room temperature) b.c.c. structure to close-packed alternatives near 78 K and 35 K, respectively. See [R38], i.e., reprint no. 38. Such transformations change a good single-crystal sample into a polycrystalline jumble. Long λ's are then impossible. Although Rb and Cs do not suffer similarly, they are quite difficult to work with on account of their environmental chemical reactivity and mechanical softness.

Potassium is then the preeminent, simple metal of the periodic table. Its role in metal physics is analogous to that of hydrogen in atomic physics. The conduction-electron effective mass (near E_F) is $m^*/m = 1.25$. Furthermore, de Haas-van Alphen studies [Ref. 16, R5] indicate that the Fermi surface is spherical to within a few parts per thousand. Nevertheless, thirty phenomena, summarized in the content of this volume, show that the foregoing remark is strikingly inadequate. The research reprints, [R1], [R2], ..., [R65], presented in Part II, document (in a personal chronology) the ultimate reconciliation of the many anomalous phenom-

Anomalous Effects in Simple Metals. Albert Overhauser

ISBN: 978-3-527-40859-7

ena within a unified panorama. Interim summaries are: [R19], [R34], [R35], and [R45].

The fundamental influence that alters the anticipated behavior of conduction electrons in potassium is a spontaneous, collective breach of translation symmetry. For example, the electronic ground state might incorporate a spin-density-wave (SDW) or charge-density-wave (CDW) superstructure. Such a (time independent) modulation modifies the topology of the Fermi surface, and many important electron orbits are severely altered. Astonishing properties ensue.

2
SDW and CDW Instabilities

SDW or CDW broken symmetries of an (otherwise) homogeneous Fermi sea of conduction electrons exhibit a sinusoidal modulation of up-spin number density, $\rho^+(\mathbf{r})$, and of down-spin number density, $\rho^-(\mathbf{r})$:

$$\rho^+(\mathbf{r}) = \frac{1}{2}\rho_0\left[1 + p\cos\left(\mathbf{Q}\cdot\mathbf{r} + \phi\right)\right] ,$$
$$\rho^-(\mathbf{r}) = \frac{1}{2}\rho_0\left[1 + p\cos\left(\mathbf{Q}\cdot\mathbf{r} - \phi\right)\right] \tag{2.1}$$

ρ_0 is the (mean) total density, p is the modulation amplitude, and $\mathbf{Q}$ is the wave vector of the spatial oscillation. If $\phi = 0$, Eq. (2.1) describes a CDW, and the spin density is zero everywhere. If $\phi = \frac{1}{2}\pi$, Eq. (2.1) describes an SDW, and the total electron density is ρ_0 everywhere. If $0 < \phi < \frac{1}{2}\pi$, Eq. (2.1) describes a mixed SDW-CDW, a phenomenon which might be expected in a metal that is elastically anharmonic [R37].

The physical model of a simple metal requires, of course, electrical neutrality. One takes for granted that the charge density of the positive-ion background is homogeneous and equal to $\rho_0 e$. This simplified model is often called "jellium". The "rigid-jellium" model does not allow any spatial modulation of the positively-charged jelly. The "deformable-jellium" model postulates a jelly that has zero rigidity; so it can be (sinusoidally) modulated without input of elastic energy. Nevertheless, all Coulomb energies arising from spatial modulations of the conduction electrons and the jellium background must always be recognized. For example in rigid jellium, incipience of a large Coulomb repulsion prevents a CDW broken symmetry. In contrast, it is easy to show that an SDW instability occurs in a one-dimensional, rigid metal because exchange terms of the repulsive interactions are negative [R1]. The Fermi occupation span, $2k_f$, is the optimum wave vector magnitude, Q, of the SDW modulation.

In 3d, it seems remarkable that one can prove within the (Hartree-Fock) mean-field approximation, that SDW instabilities always arise [R3]. However, dynamic electron-correlation corrections tend to suppress SDWs [R6]. Nevertheless SDW modulations can be coaxed into existence by the presence of magnetic ions in solid solution. The most well-known example is that of dilute, random alloys of Mn in Cu [R2], where long-range antiferromagnetic order has been verified at low T, even if the Mn concentration is only 1/2%. Many low-temperature anomalies in the electric, magnetic, and thermal properties of Cu-Mn alloys can be explained quantitatively [R2]. The exchange interactions, $-g\mathbf{s}\cdot\mathbf{S}_i$, between the local spin density, $\mathbf{s}(\mathbf{r})$, of the SDW and the 3d moments ($S = 5/2$) of the Mn^{2+} ions (when polarized by the SDW at low-T) provide the negative energy which brings forth the

Anomalous Effects in Simple Metals. Albert Overhauser

ISBN: 978-3-527-40859-7

SDW sinusoidal spin density, $p\rho_0 \sin(\mathbf{Q} \cdot \mathbf{r})$. The elastic rigidity of the Cu lattice prevents a CDW instability (for reasons discussed above).

Although dynamic electron-correlation corrections tend to suppress SDWs, they promote CDWs. This dichotomy is easily understood [R6]. The dominant term of the Fermi-liquid correlation energy arises from the virtual scattering of opposite-spin electrons. Most such terms are negative (on account of the Pauli exclusion principle). The matrix elements for such excitations are reduced in magnitude for an SDW since the opposite-spin, electron modulations are out-of-phase. For a CDW, the opposite-spin modulations are in-phase (spatially). Consequently the virtual excitations are enhanced, and so is the magnitude of the (negative) correlation energy.

A CDW instability should be expected in a "deformable-jellium" metal. Both exchange and correlation's incremental increases (in magnitude) support a CDW state. Furthermore, any Coulomb energy increase that might be anticipated (on account of the conduction-electron modulation) is reduced to zero by a compensating, elasticity-free modulation of the positive "jelly".

The foregoing analysis suggests that metals which can be well approximated by the "deformable-jellium" model are likely to exhibit a genuine CDW broken symmetry. The bulk moduli of the alkali metals are very small. For example, potassium's bulk modulus is 43 times smaller than that of Cu. Potassium may be expected to behave according to the "deformable-jellium" model. Accordingly one should seriously entertain a CDW broken symmetry for K, and explore the influence of the CDW on observable properties. Such a program is described below and has generated the 65 publications (reproduced in Part II), that emerged during forty years of study.

3
The CDW Wavevector Q and Q-domains

When a CDW is present a conduction-electron wave function is no longer a pure, plane wave, $|\mathbf{k}>$. Instead,

$$\psi_{\mathbf{k}}(CDW) \cong |\mathbf{k}\rangle + \left[\langle \mathbf{k}+\mathbf{Q}|A_{ex}|\mathbf{k}\rangle/(E_{\mathbf{k}} - E_{\mathbf{k}+\mathbf{Q}})\right] |\mathbf{k}+\mathbf{Q}\rangle . \qquad (3.1)$$

The admixture of $|\mathbf{k}+\mathbf{Q}>$ can occur only if $|\mathbf{k}+\mathbf{Q}>$ is above the Fermi energy (and therefore unoccupied). This requirement, together with energy denominators as small as possible, leads to:

$$|\mathbf{Q}| \cong 2k_F , \qquad (3.2)$$

the diameter of the Fermi sphere. A_{ex} is the exchange operator of the electron-electron interaction. An electrostatic potential does not occur since (for the "deformable-jellium" model) an energy-free spatial modulation of the positive "jelly" arises (passively) to cancel any electric field. The augmentative feedback of the exchange operator, A_{ex}, which drives the instability, is a maximum for $|\mathbf{Q}| = 2k_F$, [R9].

One should appreciate that the CDW broken symmetry considered in this book is driven by electron-electron interactions! This behavior is quite distinct when compared to a Peierls instability, which develops from a Hamiltonian having no electron-electron interactions. Instead, only electrons which couple to a static, sinusoidal lattice modulation are postulated.

The direction of **Q** is of crucial interest for most of the phenomena recounted below. The wavevector of the acoustic phonon mode which screens the electric field of the CDW (in b.c.c. K) cannot be **Q** because **Q** lies outside of the Brillouin zone. The (longitudinal acoustic) mode involved should then be [Section R24.3]:

$$\mathbf{Q}' = \mathbf{G}_{110} - \mathbf{Q} , \qquad (3.3)$$

since $\mathbf{G}_{110}$ is the reciprocal lattice vector having a magnitude close to that of **Q**. One would expect **Q** to be parallel to $\mathbf{G}_{110}$ so that $|\mathbf{Q}'|$ could be optimally small. However, the large elastic anisotropy of *K* causes **Q** to tilt several degrees away from the [1,1,0] axis [R24]. As a consequence of the (original) cubic symmetry of the lattice, there will be four (energy-equivalent) tilts for each of the six [110] axes. Accordingly, there are 24 possible **Q** axes in a single crystal!

In the next chapter it will become evident that two (or more) CDWs are not simultaneously present. However, a good single crystal is usually divided into **Q**-domains (similar to magnetic domains in a ferromagnet). Obviously physical properties will depend on the orientational texture of the **Q**-domains. Invention of techniques to control **Q** orientation has been (and remains) a challenge [Section R19.3.8].

Anomalous Effects in Simple Metals. Albert Overhauser

ISBN: 978-3-527-40859-7

4
Optical Anomalies

The optical absorption spectrum of an alkali metal is expected to have two contributions. The first is the Drude intraband absorption, which is proportional to the electrical resistivity and inversely proportional to the square of the photon energy. The second arises from interband transitions, which (in K) have a threshold at 1.3 eV and a maximum near 2.0 eV [Figure R19.7]. However, Mayer and El Naby [Ref. 1, R1] found a very large absorption band with a threshold at 0.62 eV and a maximum near 0.8 eV. This anomaly was observed at $T = -183°$C, $-80°$C, 20°C, and (in the liquid) at 85°C.

This strange absorption was reproduced by Harms [Ref. 5, R11]; see Figure R11.1. Furthermore, B. Hietel informed me that he had reproduced all of the observations of Mayer–El Naby (on K) before studying Na. Hietel and Mayer [Ref. 10, R43] found a similar anomalous absorption band (in Na) with a 1.2 eV threshold and a peak near 1.6 eV. It is important to note that all three of the K measurements mentioned above were made on bulk-metal surfaces. Several other workers failed to observe the Mayer–El Naby absorption; but their studies [Refs. 2,3,4, R11] employed thin evaporated films of K. The significance of this variance will become clear below.

An early attempt to explain the Mayer–El Naby anomaly postulated the presence of a (static) SDW [R4]. The only free parameter was the SDW energy gap, G [Figure R4.1], created by the SDW's (sinusoidal) exchange potential, [Eq. R4.1]. The Mayer–El Naby absorption was then explained both in shape and magnitude by letting $G = 0.62$ eV, the observed threshold. However, Hopfield pointed out [Ref. 26, R11] that matrix elements for optical transitions across (pure) SDW energy gaps must be zero. A new contribution to the matrix element, which exactly cancels the direct optical one, arises from a time dependent (exchange potential) oscillation of the SDW phase caused by the photon's electric field. Such cancellation is guaranteed since the electron-photon interaction commutes with the electron-gas Hamiltonian.

Hopfield's theorem does not apply to an optical absorption anomaly created by a CDW because the CDW phase is "locked" to the lattice by the spatial modulation of the positive-ion background. Therefore the absorption spectrum, calculated (mistakenly) for an SDW broken symmetry, applies instead to K with a CDW. It is important to realize that the CDW absorption is proportional to $\cos^2\theta$, where θ is the angle between the CDW $\mathbf{Q}$ and the polarization vector of the photon inside the metal [Eq. R11.3]. If the CDW $\mathbf{Q}$-domains have no orientational texture, $\langle\cos^2\theta\rangle_{av} = 1/3$; and the (macroscopic) absorption will be isotropic.

The failure to observe the Mayer–El Naby absorption in K films deposited on microscope slides is easily understood. A low-energy electron-diffraction study [Ref. 33, R11] of such films has shown that a [110] crystallographic direction is

Anomalous Effects in Simple Metals. Albert Overhauser

ISBN: 978-3-527-40859-7

perpendicular to the surface. This epitaxial effect should carry over to the direction of **Q**, which is nearly parallel to a [110]; this allows the planes of optimum charge density to be parallel to the surface. The surface energy may then be minimized by adjusting the CDW phase. Since the polarization vector of a photon inside the metal is nearly parallel to the surface, $\cos^2\theta \approx 0$. Accordingly, the CDW optical anomaly will be suppressed. This disappearance shows that K is optically anisotropic and that its (presumed) cubic symmetry is broken. It also shows that each **Q** domain has only one CDW since a second one would have its **Q** nearly parallel to one of the five remaining $\{110\}$ axes. For a random distribution among these remaining possibilities, $\langle\cos^2\theta\rangle_{av} \sim 2/5$.

The spectral shape of the Mayer–El Naby absorption depends dramatically on the Fermi-surface distortion caused by the CDW potential, $-G\cos(\mathbf{Q}\cdot\mathbf{r})$, [Eq. R45.18]. This subject will be discussed below, in Chapter 8. (The distortion leads to a conduction-electron, spin-resonance splitting.)

An enormous optical absorption peak was discovered (in K) at 8 eV (in the vacuum ultraviolet) by Whang *et al.* [Ref. 1, R44] in 1972; see Figure R44.1. This striking absorption band went unexplained for thirteen years. It does *not* depend on the CDW broken symmetry. (Similar absorption bands were also found in Rb and Cs.) The explanation involves the dynamic vibration of the eight M-shell electrons of the K ions. The Coulomb potential of the M-shell lattice acquires an oscillatory, time modulation which contributes a large interband matrix element at the photon frequency [R44]. The agreement of the observed spectrum with calculation is outstanding [Figure R44.3], and proves that V_{110}, the (110) pseudopotential of K, is −0.2 eV.

5
Phase Excitations of an Incommensurate CDW

An incommensurate CDW is one for which the wave vector **Q** is not a simple rational fraction of a reciprocal lattice vector. The electron charge density can be chosen:

$$\rho(\mathbf{r}) = \rho_o \left[1 + p \cos\left(\mathbf{Q} \cdot \mathbf{r} + \varphi\right)\right] . \tag{5.1}$$

An incommensurate relationship implies that the CDW energy is independent of the phase, φ. This continuous symmetry leads to a number of interesting phenomena when $\varphi(t)$ is allowed to depend on time. For example, with an electric field parallel to **Q** [Eq. R14.32]:

$$\varphi(t) = -\mathbf{Q} \cdot \left(\mathbf{D}_o t + \frac{1}{2}\mathbf{a}t^2\right) , \tag{5.2}$$

where $\mathbf{D}_o$ is the CDW drift velocity (along **Q**) at $t = 0$; and **a** is the CDW acceleration caused by the electric field. Impurities in the metal can pin the CDW; so Eq. (5.2) applies only if the field is strong enough to depin the CDW. For potassium the acceleration, **a**, of the CDW is smaller than that of the conduction electrons by a large factor [Section R14.6].

Naturally, the drift velocity does not increase without limit; one can show that electron scattering leads to a frictional force on a drifting CDW [R17]. The steady-state drift velocity of the CDW does *not* equal the electron drift velocity. At low temperatures, where electronic transport is limited by impurity scattering, the CDW drift velocity is $\sim$ 22% of the electron drift velocity (parallel to **Q**). The contribution of CDW drift to the conductivity tensor is, of course, anisotropic [Eq. R14.18]. The influence of a magnetic field on CDW drift dynamics has also been studied [R15].

A more interesting consequence of the broken symmetry is the occurrence of oscillatory phase excitations [Section R10.5]. (These are new "Goldstone bosons" caused by the incommensurate broken symmetry.) The displacement of ions from their bcc lattice sites, **L**, is for a mode, **q**, [Eq. R10.22]:

$$\mathbf{u}(\mathbf{L}) = \mathbf{A} \sin\left[\mathbf{Q} \cdot \mathbf{L} + \varphi_\mathbf{q} \sin\left(\mathbf{q} \cdot \mathbf{L} - \omega_\mathbf{q} t\right)\right] . \tag{5.3}$$

Here **A** ($\sim$ 0.03Å) is the amplitude of the CDW lattice distortion and **q** is the wave vector of a phase modulation mode, i.e., a "phason". Consider a local region near $L = 0$ (and let $t = 0$). Then Eq.(5.3) becomes,

$$\mathbf{u}(\mathbf{L}) \cong \mathbf{A} \sin\left[\left(\mathbf{Q} + \phi_\mathbf{q}\mathbf{q}\right) \cdot \mathbf{L}\right] . \tag{5.4}$$

The factor, $(\mathbf{Q} + \varphi_\mathbf{q}\mathbf{q})$, shows that phase modulation corresponds to a small, local change in direction and magnitude of **Q**. Application of Newton's laws within the

Anomalous Effects in Simple Metals. Albert Overhauser

ISBN: 978-3-527-40859-7

CDW energy valley [Figure R24.2], (centered at the optimum **Q**) [R10], [R24] leads to

$$\omega_{\mathbf{q}}^2 = c_\ell^2 q_\ell^2 + c_{t1}^2 q_{t1}^2 + c_{t2}^2 q_{t2}^2 \,. \tag{5.5}$$

q_ℓ is the component of **q** parallel to **Q**, and q_{t1}, q_{t2} are the components perpendicular to **Q**.

The phason frequencies are linear in |**q**|. Phason velocities are comparable to phonon velocities [R25]. Each mode is a linear combination of two (old) phonon modes near **Q** [Figure R28.1], [Figure R55.2]. The orthogonal linear combination describes a sinusoidal modulation of the CDW amplitude, A. (Accordingly the total number of vibrational modes in the crystal remains unchanged.) Phason excitations and phason-electron interactions play an important role in several of the phenomena described below.

6
Neutron Diffraction Satellites

A CDW induces displacements of the K ions from their ideal, bcc lattice sites, **L**, [Eq. R10.1]:

$$\mathbf{u}(\mathbf{L}) = \mathbf{A}\sin(\mathbf{Q}\cdot\mathbf{L}) \ . \tag{6.1}$$

Without the CDW, (elastic) x-ray or neutron diffraction reflections occur only at the reciprocal lattice vectors, $\mathbf{G}_{hk\ell}$. When the displacements, Eq.(6.1), are present, each Bragg reflection will have many small satellite reflections [Figure R53.1]:

$$\mathbf{G}_{hk\ell} \pm \mathbf{Q}_j \ . \tag{6.2}$$

The index ($j = 1, 2, \ldots, 24$) is inserted since a large single-crystal sample will usually be divided into 24 **Q** domains (as described in Chapter 3). The pattern of 48 satellites near each $G_{hk\ell}$ is presented in R51.

The expected intensities of CDW satellites are very small. $\sim 10^{-5}$ times the intensity of a G_{110} Bragg peak, even at liquid He temperature [R58]. The penetration of x-rays in K is so small that, neutrons are the probe of choice. The first such experiment [R49] confirmed both the anticipated intensity and the nearness of **Q** to (110). (In $2\pi/a$ units.)

$$\mathbf{Q}_1 = (0.995, 0.975, 0.015) \ , \tag{6.3}$$

which is tilted 0.85° from [110]. (The 23 other $\mathbf{Q}_j$ axes are related to $\mathbf{Q}_1$ by rotations consistent with the underlying cubic symmetry.)

Unfortunately the satellite reflections are so weak that it is difficult to prove their authenticity. (There are also higher-order satellites, but these are much weaker than the first-order ones considered here [R36].) It may be possible to explain apparent small peaks as double-scattering artifacts if there is sufficient diffuse scattering from surface oxides on the K sample or from crystalline disorder in the cryogenic capsule [R53]. Elimination of such alternatives requires, for example, a consistent determination of **Q** in separate experiments using different neutron energies. Another strategy is to measure the "satellite" locations near both (110) and (220) and to confirm that both sets require the same **Q**. The difficulty of such verification is compounded by anisotropic, thermal-diffuse phonon scattering near the reciprocal-lattice points. Diffraction near (110) and (220) [R55] did indeed agree with the CDW hypothesis. However, this confirmation should be regarded as tentative until it can be repeated with higher momentum and energy resolution of the neutron beams.

CDW satellites are surrounded by an anisotropic cloud of thermal-diffuse phason scattering [R10], [R28]. This (inelastic) scattering creates an extra temperature

Anomalous Effects in Simple Metals. Albert Overhauser

ISBN: 978-3-527-40859-7

factor for CDW satellites (analogous to the Debye-Waller factor associated with phonon scattering) [Section R10.6]. The phason temperature factor is so extreme that studies of CDW satellites (in K) must be carried out at liquid He temperatures.

7
Phason Phenomena

The simplest model of the phason spectrum is a miniature Debye model, [Figure R45.15]. The phason wave vectors, $\mathbf{q}$, are taken to lie within a sphere (centered at $\mathbf{Q}$) of radius q_ϕ. If $\bar{c}_\phi$ is the mean phason velocity, then the phason "Debye" temperature, θ_ϕ, is defined by:

$$k_B \theta_\phi = \hbar \bar{c}_\phi q_\phi \ . \qquad (7.1)$$

q_ϕ is the approximate distance, in Figure R55.2, from $\mathbf{Q}$ (where $\omega_q = 0$) to the point where ω_q is a maximum. Since q_ϕ is much smaller than the radius, q_D, of the (ordinary) Debye sphere, the phason contribution to the heat capacity will create a low-temperature anomaly, [R22], [R62]. Such an anomalous peak was first observed in metallic Rb (by Lien and Phillips [Ref. 1, R62]) near 0.6° K [Figure R62.3]. (Rb also has a CDW.) A similar anomaly was observed in potassium by Amarasecara and Keesom [Ref. 82, R45] near 0.8 °K. The optimum fit was obtained with $\theta_\phi = 6$ °K; and $(q_\theta/q_D)^3 = 2\times10^{-5}$, the ratio of the phason-sphere volume to the Debye-sphere volume.

A second phenomenon that can be explained by the existence of phasons is the behavior of the electrical resistivity between 0.4 and 1.6 °K, which was measured with precision by Rowlands *et al.* [Ref. 1, R23]. The standard theory contains four contributions [Eq. R23.1]: The residual resistivity (caused by impurities and other lattice imperfections) is independent of T. The second term (associated with electron scattering by phonons) is proportional to T^5. The third term arises from "umklapp" scattering by phonons:

$$\mathbf{k}' = \mathbf{k} + \mathbf{q}_p + \mathbf{G} \ , \qquad (7.2)$$

where $\mathbf{q}_\mathrm{p}$ is a phonon wavevector and $\mathbf{G}$ is a reciprocal-lattice vector. This resistivity contribution varies exponentially with $1/T$, and becomes negligible below 1.5 °K. The fourth term, proportional to T^2, is caused by electron-electron scattering. Only umklapp events need to be considered, i.e.,

$$\mathbf{k}_1' + \mathbf{k}_2' = \mathbf{k}_1 + \mathbf{k}_2 + \mathbf{G} \ ; \qquad (7.3)$$

since (without $\mathbf{G}$) total electron momentum (and current) would be conserved, and the resulting contribution to the resistivity would be negligible.

The temperature dependence of the resistivity between 0.4 and 1.3 °K cannot be explained by the four terms just enumerated [R23], [R27]. However a good fit becomes possible if one includes electron-phason, CDW-umklapp scattering. That is:

$$\mathbf{k}' = \mathbf{k} + \mathbf{q}_\phi \pm \mathbf{Q} \ . \qquad (7.4)$$

Anomalous Effects in Simple Metals. Albert Overhauser

ISBN: 978-3-527-40859-7

Such a process is visualized in Figure R20.6. The temperature dependence of this process involves a sum of Bloch-Grüneisen functions [Eq. R27.74a], and provides an excellent fit to the data [Figure R27.1]. (The closest pure power-law is $T^{1.5}$ [Figure R27.2].) The magnitude of this mechanism varies with the **Q** domain distribution (and therefore is sample dependent). It is largest for domains having their **Q** axis parallel to the current. The best fit for the K data of Figure R27.1 used the value, $\theta_\varphi = 3.25\,°\mathrm{K}$, somewhat smaller than the value (6 °K) obtained from the heat capacity anomaly (which should be sample independent).

The most dramatic evidence for the phason spectrum is direct observation by point-contact spectroscopy [Section R45.4.13]. This technique, developed by Yanson *et al.* [Ref. 84, R45], measures the spectral density of phonons (and phasons) by electron injection from a very sharp point to a larger surface, where the electron can be scattered back into the point by a phonon (or phason). This enhances the junction resistance, and provides a spectrum of the vibrational modes versus junction voltage. The point-contact spectrum for K [Ref. 85, R45] was measured at 1.2 °K. It has an unexpected peak between 0 and 1 meV, [Figure R45.16], where the spectral density of phonons is nil. Ashraf and Swihart [Ref. 86, R45] calculated the spectral density caused by phasons, and found excellent quantitative agreement [Figure R45.16] for the anomalous peak.

Finally, the phasons play an important role in nuclear magnetic resonance. In a metal the Pauli paramagnetism causes an NMR frequency shift,

$$\omega = \omega_d(1 + K_o)\,, \tag{7.5}$$

where ω_d is the NMR frequency in a diamagnetic salt. The Knight shift, $K_o \sim 0.26\%$, arises from the Fermi hyperfine coupling to the electron-spin polarization. The CDW will cause K_o to vary in proportion to the local conduction-electron charge density:

$$K_o'(\mathbf{r}) = K_o\left[1 + p\cos(\mathbf{Q}\cdot\mathbf{r})\right]\,. \tag{7.6}$$

In a 6T magnetic field, $K_o = 156\,Oe$. It follows, with $p \sim 0.11$ [Eq. R26.19], that the width of the NMR line would be 34 Oe [Figure R47.1]. Follstaedt and Slichter [Ref. 16, R47] measured the ^{39}K linewidth at 1.5 °K in a 6T field. The observed width, $0.215\,Oe$, agrees with the value expected if a CDW were *not* present.

Phase fluctuations can motionally narrow the Knight-shift broadening caused by the CDW [R47], and this can explain the null result found by Follstaedt and Slichter. The theory of motional narrowing (in general) is due to Pines and Slichter [Ref. 18, R47]. The narrowing caused by phasons is very dependent on temperature and, of course, is sensitive to the phason frequency spectrum. The NMR line in K should approach the width it would have without a CDW at temperatures above 40 mK [Figures R50.2, R50.3]. It is clear that NMR studies below 40 mK, where the rise in line width is predicted to be extremely rapid, should be very interesting.

The influence of phasons in diffraction experiments was discussed in the previous chapter. They cause a Debye-Waller-like reduction in the strength of CDW satellites [R16], [R28]; and they also cause an anisotropic cloud of diffuse scattering surrounding each satellite [Figure R49.1]. The theory of electron-phason scattering indicates that phasons are usually underdamped [R18].

8
Fermi-Surface Distortion and the Spin-Resonance Splitting

The CDW potential, $G\cos(\mathbf{Q}\cdot\mathbf{r})$, causes the Fermi surface to be distorted from its (otherwise) spherical shape. It is of considerable interest to know whether the CDW energy-gap planes cut the Fermi surface, make critical contact, or miss the surface entirely. Figure R7.1 illustrates critical contact. Figure R6.1 shows that the electronic density of states at E_F has a sharp maximum if $|\mathbf{Q}|$ is chosen to provide critical contact. The correlation energy (which arises mainly from virtual-pair excitations near the Fermi surface) will then be optimized. Critical contact occurs when

$$Q \approx 2k_F[1 + (G/4E_f)] , \tag{8.1}$$

which is 7% larger than $2k_F$; i.e. $Q \approx 1.33 \times (2\pi/a)$.

The Mayer–El Naby optical anomaly also indicates that critical contact obtains. The theoretical curve of Figure R4.2 was calculated on that basis. If Q were smaller than (8.1), the CDW optical absorption would jump discontinuously from zero to a finite value at $\hbar\omega = G$. If Q were larger, the CDW absorption would increase linearly (from 0) with the excess of $\hbar\omega$ above threshold. For critical contact, the initial rise is proportional to $(\hbar\omega - G)^{1/2}$.

The foregoing remarks took into account only the main CDW energy gap, $G = 0.6\,\text{eV}$. There are many small, higher-order gaps [R36] which create absorption bands in the far infrared [R43]. Nevertheless the Mayer–El Naby absorption shows that the Fermi surface has a "lemon" shape, illustrated in Figure R7.1 with some exaggeration. This lemon-shaped anisotropy leads, as we now show, to a splitting of the conduction-electron spin resonance.

Electron spin resonance was studied by Walsh, Jr. *et al.* [Ref. 1, R7] in very pure potassium plates (~ 0.2 mm thick), which had been squeezed under oil between parafilm sheets. The resonance line width was 0.13 G at $T = 1.3\,^\circ\text{K}$ in a field, $H = 4200$ G. A g-factor shift, $\Delta g_o = -0.0025$ (caused by spin-orbit coupling) is found from the 5.3 G shift of the resonance relative to that for a free electron. The theory for g shifts in metals is due to Yafet [Ref. 10, R7]. If θ is the angle between $\mathbf{H}$ and the wave vector $\mathbf{k}$,

$$\Delta g(\mathbf{k}) \cong -\mu k^2 \sin^2\theta . \tag{8.2}$$

Since an electron scatters $\sim 10^3$ times in a spin-relaxation time, the observed Δg will be the average of (8.2) over the Fermi surface. It is clear that Δg will depend on the shape of the Fermi surface. If the Fermi surface is spherical,

$$\Delta g_o = -\frac{2}{3}\mu k_F^2 . \tag{8.3}$$

Anomalous Effects in Simple Metals. Albert Overhauser

ISBN: 978-3-527-40859-7

For the present purpose the coefficient, μ, is taken from the experimental value of (8.3), and the result agrees reasonably with Yafet's theory.

Since the CDW causes a lemon-shaped distortion of the Fermi surface [Figure R7.1], Δg will be modified and, of course, it will depend on the angle between **Q** and **H** [R7]. For **Q** parallel or perpendicular to **H**:

$$\begin{aligned} \Delta g_{\parallel} &= \Delta g_{o}(1-2\alpha)\,, \\ \Delta g_{\perp} &= \Delta g_{o}(1+\alpha)\,, \end{aligned} \tag{8.4}$$

where $\alpha \equiv G/8E_F \approx 0.035$, using potassium's CDW energy gap, $G = 0.6\,\text{eV}$. (Only terms linear in α have been retained; $E_F = 2.12\,\text{eV}$.) The fractional difference between $\Delta g_{\perp}$ and $\Delta g_{\parallel}$ is 3α [Eq. R7.26]. The resulting spread in any possible resonance field is 0.5 G (at 12 GHz).

If **H** is parallel to the potassium plate, electrons will be guided by the field and will sample many crystal grains. The consequence will be a single resonance line, appropriate to an average g-factor in the interval limited by (8.4). Such indeed was observed [Ref. 1, R7]. However, when **H** was tilted out of the sample plane, the resonance split into two components. The maximum splitting (at a 9° tilt) was $\sim$ 0.5 G. The reason two (or more) resonances become possible can be seen from Figure R7.2. A tilted magnetic field can confine an electron to an individual crystallite. Without knowledge of the **Q** directions of crystal grains, one cannot predict the number and locations of the resonances. Both lines disappeared with increasing angle of tilt (which permits rapid diffusion out of the microwave skin depth).

Without the anisotropic Fermi-surface distortion caused by the CDW broken symmetry, $\alpha = 0$; and any splitting of the spin resonance would be impossible. Experiments by Dunifer and Phillips near $H = 40\,\text{kG}$ [Ref. 24, R64] showed that the resonance splitting is proportional to H, and that it can have as many as five well-resolved components.

9
Magnetoresistivity and the Induced Torque Technique

Experimental studies of galvanomagnetic effects in the alkali metals have revealed many phenomena which prove that their Fermi surface intersects CDW energy-gap planes and becomes (thereby) multiply connected. The importance of magnetoresistance, for example, stems from transport theorems due to Lifshitz *et al.* [Ref. 6, R8], who showed that the resistivity of a metal with a simply-connected Fermi surface must be independent of H, especially when $\omega_c\tau \gg 1$. (ω_c is the cyclotron frequency and τ is the electronic relaxation time.) Contrary to such expectation, the magnetoresistance of the alkali metals (at liquid He temperatures) increases linearly with H, even to very high fields (e.g. 110 kG, $\omega_c\tau \sim 275$) without any sign of saturation [Refs. 1,2,4, R8].

Concern that such anomalous behavior might result from distorted current paths near voltage or current probes led to the development of probeless techniques. The induced-torque method, of Lass and Pippard [Ref. 15, R61], has been employed frequently. A spherical sample is suspended vertically (y-axis) from a movement that can measure the torque (exerted by the suspension); and a horizontal magnetic field $\mathbf{H}$ is rotated slowly (Ω rad/sec) in the xz plane. The induced currents create a magnetic moment perpendicular to $\mathbf{H}$. The resulting (y component) torque is [Eq. R61.4],

$$N_y = \frac{2\pi R^5 \Omega n^2 e^2 \rho_o}{15}\left[\frac{(\omega_c\tau)^2}{1+(\frac{1}{2}\omega_c\tau)^2}\right]. \tag{9.1}$$

R is the radius of the spherical sample, and n is the number of conduction electrons per cm^3. ρ_o is the resistivity which, for a metal having cubic symmetry, is a scalar. (The c^2 in the denominator of [Eq. R63.1] is a misprint and should be deleted.)

If a metal has (only) a spherical Fermi surface, the field dependence of the induced torque, N_y, should arise only from the factor in square brackets. However this factor saturates at the value, 4, when $H > 4$ kG, i.e. $\omega_c\tau >\sim 10$. Any significant increase of N_y with magnetic field must then be attributed to a field dependence of $\rho_o(H)$. The observed linear increase of N_y with magnetic field [Ref. 16, R31] shows that (for high fields) $\rho_o(H)$ increases linearly with H, apparently without limit. A.M. Simpson [Ref. 11, R31] has followed such behavior to high fields, e.g., $\omega_c\tau \sim 150$.

The most significant question is how a term, linear in H, for the magnetoresistivity can arise. The answer is "from open orbits". Energy gaps arising from the CDW cut through the Fermi surface. Five possible open orbits (for a given $\mathbf{Q}$-domain) are shown in Figure R30.2. Since even a single crystal can have as many as 24 $\mathbf{Q}$ do-

Anomalous Effects in Simple Metals. Albert Overhauser

ISBN: 978-3-527-40859-7

mains (Chapter 3), there can be up to 120 open-orbit axes. Effective-medium theory [R29] can then be used to compute the macroscopic conductivity tensor.

The existence of an open-orbit can be modeled by incorporating a small cylindrical Fermi surface (in addition to the usual spherical one). If $\hat{w}$ is the cylinder axis in **k** space, the open-orbit path in real space is helical with a central axis parallel to $\hat{w}$ (provided $\hat{w}$ and **H** are not exactly perpendicular). If θ is the angle between **H** and the transverse component of $\hat{w}$ [Figure R29.1], both in the xz plane, the transverse magnetoresistance acquires a peak at $\theta = 90°$ [Figure R29.2], [Figure R30.2]. The height of the peak is proportional to H^2, but its width is proportional to $1/H$ [Section R32.2]. Accordingly, the area (under such a peak) is linear in H. If the **Q**s of the **Q** domains are randomly oriented, the observed torque should be isotropic (independent of θ) when H is small (and the peak widths are broad).

The 120 peaks (in a 180° rotation of θ) are not resolved below $H \sim 20k\,G$. It follows that the macroscopic magnetoresistance arises from the peak area, and therefore is proportional to H, contrary to theoretical expectation (which requires zero magnetoresistance if there is only a spherical Fermi surface). The evolution of potassium's magnetoresistance from low fields to high fields is described in the following chapter.

10
Induced Torque Anisotropy

Without a CDW broken symmetry, the induced torque exerted by a potassium sphere would be isotropic (versus magnet angle, θ, in the horizontal plane of rotation), since the crystal structure would then be cubic. It is remarkable that the torque anisotropy is not only large, but is qualitatively different depending on whether H is small, medium, or large. The corresponding mechanisms that lead to anisotropic torques in these three magnetic field regimes are, respectively, CDW umklapp scattering, anisotropic Hall effects, and open-orbit magnetoresistance peaks. The anisotropy for small and medium fields will be manifested only if the crystal has just a few CDW **Q**–domains and these also have a textured orientation of their **Q**s lying near the rotation plane.

For small H, say $H < 4\,\mathrm{kG}$, the induced torque usually exhibits two sinusoidal oscillations in a 360° rotation of H. Schaefer and Marcus [Ref. 2, R12] observed this anisotropy in 193 of 200 runs on seventy different single-crystal spheres. Similar behavior was found by Holroyd and Datars [Ref. 4, R20], who noticed also that the anisotropy would appear in a sample (without two-fold anisotropy) if a little oil is placed on the surface. Presumably thermal stress (occurring when the oil freezes on cooling to liquid-He temperature) stimulates the growth of large **Q**-domains, which then exhibit their resistivity anisotropy.

Analysis of the torque anisotropy is based on the theory of Visscher and Falicov [Ref. 7, R39], who calculated the induced torque [Eq. 39.2] for a sphere with a general resistivity tensor. The residual-resistivity derived from the data (mentioned above) requires the resistivity, ρ, parallel to **Q** to be about five times ρ perpendicular to **Q** [R12]. Such an anisotropy arises from CDW umklapp scattering of (isotropic) impurity potentials [R20] on account of the CDW admixtures, Eq. 3.1, in the wavefunctions. The resistivity anisotropy increases (with increasing T) when phason umklapp scattering becomes important [R27].

The induced-torque anisotropy undergoes a profound change when H exceeds about 4 kG. This "medium-field" range extends to 30 kG (where the "high-field" range begins). Above $H \sim 4\,\mathrm{kG}$, the induced torque transforms from two sinusoidal oscillations into a four-peak pattern (in a 360° rotation). This behavior was discovered by Schaefer and Marcus [Ref. 1, R39] and replicated by Holroyd and Datars [Ref. 2, R39]. The four-peak phenomena arise from anisotropic Hall coefficients [R39]. Such anisotropy has been experimentally confirmed in potassium slabs by Chimente and Maxfield during their study of helicon resonances [Ref. 19, R61].

The torque maxima grow approximately as H^2 [Figure R61.7], and can be 30 times larger than the minima. The torque minima are evenly spaced: $90^\circ, 90^\circ, 90^\circ$, 90°. However the torque maxima are staggered: $75^\circ, 105^\circ, 75^\circ, 105^\circ$. This stag-

Anomalous Effects in Simple Metals. Albert Overhauser

ISBN: 978-3-527-40859-7

gering is correctly predicted by the theory (without need for an adjustable parameter). The experimental and theoretical torque families (for integral values of H between $1 kG$ and 23 kG) are virtually identical [Figures R61.2, R61.6]. CDW umklapp scattering was omitted to optimize the simplicity of the theoretical torque, N_y [Eq. R61.15]. Umklapp scattering, when included, causes the heights of the torque minima to be staggered [Figures R39.4, R39.5].

The origin of the Hall-effect anisotropy is the Bragg-reflection-like response of a cyclotron orbit as it encounters small heterodyne gaps (created by periodic potentials, $\mathbf{Q}'$ and $2\mathbf{Q}'$, where $\mathbf{Q}' \equiv \mathbf{G}_{110} - \mathbf{Q}$), [Figure R61.5].

Finally, the high-field region is extremely spectacular. Open-orbit torque peaks, described in Chapter 9, spring up and become sharp. Figure R30.1 displays typical data of Coulter and Datars [Ref. 1, R30]. The magnetic field at which the open-orbit peaks begin to be resolved depends on the **Q**-domain size, as illustrated in Figure R30.4 and Figure R31.10. Small **Q**-domain size leads to an (equivalent) shorter relaxation time, τ [Eq. R30.5] which, in turn, broadens the magnetoresistance peaks and delays (to higher H) their ultimate emergence in $H_y(\theta)$.

Theoretical induced-torque spectra are shown in Figures R31.6–31.12 for several magnetic fields. At 80 kG the torque curves have 20–30 resolved peaks, similar to the data [Figure R31.2] (in a 180° scan of θ). Figures R31.11, R31.12 show anticipated spectra for $H = 240$ kG. Without CDW energy gaps that intersect the Fermi surface, all of the induced-torque curves would have to be horizontal straight lines. The success of the CDW broken symmetry in accounting simultaneously for the striking torque spectra observed with small, medium, and high magnetic fields is decisive.

11
Microwave Transmission Through K Slabs in a Perpendicular Field H

Dunifer *et al.* [Ref. 5, R57] have studied 79 GHz transmission through $\sim 0.1mm$ thick potassium slabs at 1.3 °K (with a large magnetic field, $H \sim 34\,000$ G, perpendicular to the slab). The amplitude of their transmitted signal vs. ω_c/ω is shown in Figure R57.1, and Figure R59.1. ($\omega_c \equiv eH/m^* c$). Five different phenomena in the transmitted, linearly-polarized, microwave signal were studied in fifteen samples. Only the conduction-electron spin resonance (CESR) can be explained by a simple (spherical Fermi surface) model [Ref. 7, R57]. The other four are completely anomalous unless one takes into account the CDW broken symmetry.

A Gantmakher–Kaner (GK) oscillation occurs whenever the time, L/v_F, to traverse a slab (of thickness L) equals an integer number of cyclotron periods. Accordingly the oscillations are expected to be periodic in H, and indeed they are. However, the transmitted signal, computed for a spherical Fermi surface [Figure R57.4] is incorrect in all other respects: The observed signal is too small by two orders of magnitude. The computed signal has no cyclotron resonance structure at, $\omega_c/\omega = 1, 1/2$, or $1/3$, as is observed in Figure R59.1; and it has no high-frequency, Landau-level oscillations [Figure R57.1]. Finally the observed five-fold growth of the GK oscillations with increasing H cannot be accounted for unless the CDW minigaps are recognized (Figure R59.4). The minigap and heterodyne-gap distortions of the Fermi surface are shown (exaggerated) in Figure R57.5, but are accurately depicted in Figure R54.3.

The ω_c/ω values of the "high-frequency oscillations" [Figure R56.1] are catalogued in Table R56.1. A quantitative study of these values was made to optimize the fit to the function:

$$\Delta y \sim \cos(\lambda H^p + \varphi) . \tag{11.1}$$

The mean square deviations of the function (11.1), fitted to the data, are shown in Figure R56.3 versus any chosen periodicity exponent, p. The optimum p is precisely,

$$p = -1 , \tag{11.2}$$

which is the exponent appropriate for Landau-level oscillations. Eq.(11.1) then becomes:

$$\Delta y \sim \cos(\frac{2\pi F}{H} + \varphi) , \tag{11.3}$$

where F is the deHaas-van Alphen frequency [Eq. R56.6]. These oscillations arise from a cylindrical Fermi surface with cross-sectional area 69 times smaller than

Anomalous Effects in Simple Metals. Albert Overhauser

ISBN: 978-3-527-40859-7

πk_F^2 . The fractional number of conduction electrons within this cylinder is $\eta \sim 4 \times 10^{-4}$. See Figure R56.2 and Figure R59.2. This cylindrical Fermi surface accounts for the cyclotron resonance transmission peak [Figure R59.5], its subharmonics, and for the small GK oscillations. Landau-level oscillations from the smallest cylinder in Figure R59.2 are also shown in Figure R59.6 near $\omega_c/\omega = 0.6$.

Another phenomenon which arises from the Fermi-surface cylinder in the minigrap region is a sharp cyclotron resonance in the surface impedance when the magnetic field is perpendicular to the surface. Data due to Baraff *et al.* [Ref. 5, R58] are shown in Figure R48.1 and Figure R58.1. The theoretical curve for potassium without a CDW is also shown. There is then no structure at all near cyclotron resonance, as was first emphasized by Chambers [Ref. 4, R48]. The surprising behavior is fully explained when the CDW modifications of the Fermi surface are included [R58].

12
Angle-Resolved Photoemission

In Chapter 4, we observed that evaporated films of potassium (or Na) are found (by low energy electron diffraction) to have a [110] crystal direction perpendicular to the surface. A consequence is that one would also expect the CDW **Q** vector to be nearly perpendicular to the surface, since the direction of **Q** is almost parallel to a [110] direction [R24]. These features led to the prediction that the CDW structure in Na or K could be detected with angle-resolved photoemission [R21].

The method involves measuring the kinetic energy, K_ν, of a photoemitted electron which is travelling (in vacuum) perpendicular to the surface. Since the photoexcitation (inside the metal) is essentially a vertical transition in the Brillouin zone, the initial energy, $E_i(k_z)$, relative to the Fermi energy, is [Eq. R21.1], [Figure R21.1]:

$$E_i(k_z) = K_\nu + \varphi - \hbar\omega \,, \tag{12.1}$$

where φ is the work function, and k_z is the component of **k** along [110]. The initial state must (of course) lie below E_F. Therefore if one plots E_i versus photon energy, $\hbar\omega$, using a nearly free electron model, there is a gap in the curve near $\hbar\omega = 35ev$ (for Na), [Figure R46.2]. (In potassium the gap is near $\hbar\omega = 25ev$). The reason for the expected gap is that (for a spherical Fermi surface) k_F is about 12% smaller than the distance to the center of the Brillouin zone face. See Figure R46.1. The gap would disappear if the Fermi surface were sufficiently distorted to extend all the way to the zone boundary. Energy band calculations of Ham, [Ref. 9, R21], imply that distortion of the Fermi surface (caused by the bcc lattice potential) is much too small to close the gap at $E_i = E_F$.

The initial electron energy, E_i, in sodium versus photon energy, $\hbar\omega$, was first measured by Jensen and Plummer [Ref. 6, R46] in 1985. Surprisingly the 6 ev gap (near $\hbar\omega = 35ev$) was filled in by a nearly flat "bridge", (Figure 3 of Jensen and Plummer) just below E_F. These (originally) unexpected photoexcitations can be explained by the presence of the CDW potential (having wave vector **Q**), which is large enough to distort the Fermi surface so that it intersects the CDW energy gaps at $k_F \cong \pm Q/2$ [Figure R46.3]. The CDW-induced photoexcitations are easily understood from Figure R21.2. (They are no longer vertical transitions in the bcc Brillouin zone.) The resulting plot of E_i versus $\hbar\omega$, which includes CDW minigaps and concomitant plane-wave mixing [Eq. R46.3] is shown in Figure R46.5. The nearly flat "bridge" across the "expected" gap near 35 ev is explained.

Jensen and Plummer remark (in their last sentence on page 1914) that "very recent data from K(110) show similar behavior." The K data were published five years later: Itchkawitz *et al.*, Phys. Rev. B41, 8075 (1990). Ma and Shung, Phys. Rev.

Anomalous Effects in Simple Metals. Albert Overhauser

ISBN: 978-3-527-40859-7

B50, 5004 (1994), conclude that a CDW in K is essential in order to explain the sharp photoemission peaks having E_i near E_F.

13
Concluding Remarks

The sixty five reprints included in this book cover a research span of forty years. Not all topics have been addressed in the preceding sections for the sake of brevity. However Part III lists thirty anomalous properties of metallic potassium and identifies the reprints that relate each anomaly to the CDW broken symmetry.

Needless to say there have been important advances in understanding potassium's broken symmetry. Originally a SDW ground state [R4] was entertained since SDW energy gaps could account for several magneto-transport anomalies [R8], [R5]. Until Hopfield's theorem (Chapter 4) was published in 1965, it was (mistakenly) believed that transitions across SDW energy gap could also explain the Mayer–El Naby optical anomaly. Fortunately this mistake was esasily rectified by recognizing that the broken symmetry in potassium ought to be a CDW [R6]. The accompanying periodic lattice distortion effectively pins the CDW (exchange and correlation) potential to the lattice. Optical transitions across CDW energy gaps are then allowed, and explain the shape and anisotropy of the anomalous absorption [R11].

Magnetoflicker noise [R64] is a phenomenon which was postulated in 1974 in an attempt to explain the four-peaked induced torque patterns described in Chapter 10. The idea was that fluctuations in the direction of **Q** would create noise and also allow **Q**'s mean direction to switch during rotation of **H**. However, Hockett and Lazarus [Ref. on page 355; R19] found no excess noise in *K*. This observation rendered unlikely the creation of extra torque peaks [Figure R63.3] by switching the direction of **Q**. Ten years passed before the discovery that the four-peaked torque patterns could arise from an anisotropic Hall effect [R39]. After yet another eighteen years, a microscopic mechanism for such anisotropy was discovered [R61]. One can appreciate that the absence of magnetoflicker noise played a significant historical role in the progress of understanding *K*.

The theoretical foundation of the CDW broken symmetry in alkali metals is, of course, the SDW instability theorem (in the Hartree–Fock approximation), [R3]. The initial proof of this theorem was rather difficult. Giuliani and Vignale (Quantum Theory of the Electron Liquid, Cambridge University Press, 2005, Sect. 2.6.3, page 96) have formulated an alternative proof which is easier to follow. That the alkali metals manifest a CDW instability, rather than an SDW, can be attributed to the (many-electron) correlation corrections (to Hartree–Fock) which enhance the instability on account of the elastic "softness" of the positive ion lattice (which cancels most of the Coulomb energy that otherwise would arise from the electronic charge modulation), [R6].

Anomalous Effects in Simple Metals. Albert Overhauser

ISBN: 978-3-527-40859-7

All thirty experimental anomalies listed in Part III can be understood as the result of a CDW broken symmetry. This success implies that the CDW structure of potassium is a fact.

Part II Reprints of SDW or CDW Phenomena in Simple Metals

The sixty-five reprints listed below cover the research on broken translation symmetry due to the author, together with his coauthors, during the period from 1960 to 2002. The purpose of this compendium is to collect together in one volume the theoretical treatments that enable one to understand the extraordinary, and in many cases spectacular, effects that arise from the (static) CDW in the simplest metal, potassium. Part~III lists thirty such unexpected phenomena and identifies the relevant reprints from the list below. This collection also provides some insight into the historical development, which in some cases was painfully slow. For example, the four-peaked induced-torque anisotropy discovered in 1971, was not fully understood until 2002.

Reprint 1 Giant Spin Density Waves[1)]

*A.W. Overhauser**

* Scientific Laboratory, Ford Motor Company, 525, Northwestern Avenue, Dearborn, Michigan 47907, USA

Received 4 April 1960

The observations made in this paper call into serious question many details of the modern electron theory of metals. It will be shown below that, almost certainly, the Hartree–Fock ground state of a Fermi gas with Coulomb interactions is not the familiar Fermi sphere of occupied momentum states, but rather a state in which there are large static spin density waves, and in which large energy gaps exist in the single-particle excitation spectrum.

In order to emphasize the essential physical simplicity of the new low-energy states, we shall treat first a one-dimensional model. (Only translational freedom will be restricted to one dimension; the ordinary spin degrees of freedom will be retained.) The kinetic energy operator will be the usual one; and we shall assume that the repulsive interactions are delta functions:

$$V_{ij} = \gamma\delta\left(z_i - z_j\right) . \tag{1.1}$$

The normal state of such a gas – N electrons in a box of length L has all (plane wave) states occupied for $|k| \leq k_0 = \pi N/2L$. The total kinetic energy is $\frac{1}{3} N E_F$, where $E_F = \hbar^2 k_0^2/2m$; and the expectation value of the interactions (direct plus exchange) is $\gamma N^2/4L$. It is of interest to compare the energy of the normal state with that of the ferromagnetic state (all spins parallel):

$$W_{\text{ferro}} - W_{\text{normal}} = N E_F\left(1 - n^{-1}\right), \tag{1.2}$$

where

$$n = (N/L)/\left(2m\gamma/\pi^2\hbar^2\right), \tag{1.3}$$

a dimensionless quantity proportional to the electron density, N/L. The critical density, at which a transition between the normal and ferromagnetic states would occur, corresponds to $n = 1$. We shall prove, however, that the normal state is never the Hartree–Fock ground state, and that the ferromagnetic state is stable only for $n \leq \frac{3}{4}$.

We shall "begin" by writing down the self-consistent Hartree–Fock potential for the solutions of interest.

$$U(z) = 2gE_F\left(\sigma_x \cos qz + \sigma_y \sin qz\right), \tag{1.4}$$

where g and q are parameters to be determined later, and σ_x and σ_y are the usual Pauli matrices. The z direction need not be perpendicular to the (spin) x–y plane, but the spiral exchange potential (1.4) is perhaps easiest to visualize for that special case. This potential has the remarkable property that its only nonzero matrix elements are between the states (k, α) and $(k + q, \beta)$, where α and β are the spin-up and spin-down spin functions. The *exact* single-particle solutions for this potential are

$$\phi_k^+ = \left[ge^{ikz}\alpha + \left(E_k^+ - \omega_k\right)e^{i(k+q)z}\beta\right]/L^{1/2}\left[g^2 + \left(E_k^+ - \omega_k\right)^2\right]^{1/2}, \tag{1.5a}$$

1) Phys. Rev. Letters 4, 9 (1960)

Anomalous Effects in Simple Metals. Albert Overhauser

ISBN: 978-3-527-40859-7

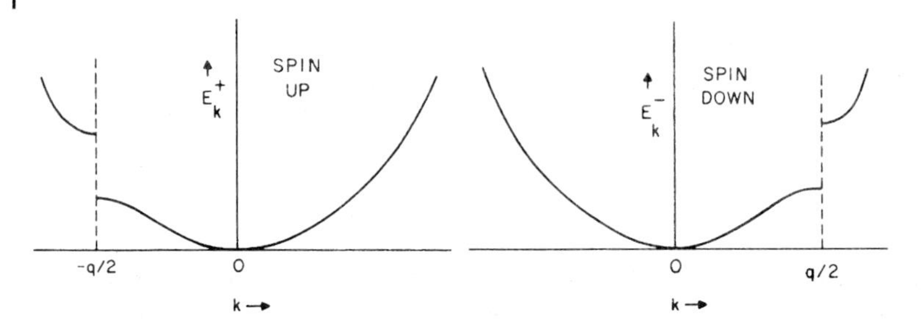

Figure 1.1 Single-particle energy level spectrum for an electron gas with a giant spiral spin density wave.

for spin-up states, and

$$\phi_k^- = \left[g e^{ikz} \beta + \left(E_k^- - \omega_k \right) e^{i(k-q)z} \alpha \right] / L^{1/2} \left[g^2 + \left(E_k^- - \omega_k \right)^2 \right]^{1/2} , \tag{1.5b}$$

for spin-down states. The free-particle energies are $2E_F \omega_k$, where

$$\omega_k = \frac{1}{2} (k/k_0)^2 , \tag{1.6}$$

and the perturbed single-particle energies, $2E_F E_k$, are given by

$$E_k^+ = \frac{1}{2} \left(\omega_k + \omega_{k-q} \right) \pm \left[\frac{1}{4} \left(\omega_k - \omega_{k+q} \right)^2 + g^2 \right]^{1/2} , \tag{1.7}$$

together with a similar expression for E_k^- having q replaced by $-q$. The single-particle energy spectrum is shown in Figure 1.1. Note the unusual feature that the energy gap, $4gE_F$, occurs on only one side of $k = 0$ for a given spin. Near the energy gaps the real spin directions spiral in a plane perpendicular to the axis of spin quantization.

The N-electron wave function will be a single Slater determinant of wave functions (1.5). For any given q, the lowest energy is achieved if the occupied states satisfy $-\frac{1}{2}q \le k \le 2k_0 - \frac{1}{2}q$ for spin up, and $-2k_0 + \frac{1}{2}q \le k < \frac{1}{2}q$ for spin down. The amplitude g must then be determined so that the Hartree–Fock equations are satisfied. Such a procedure yields the required value of g immediately. An alternative way is to calculate the expectation value of the total energy, and subsequently to minimize that energy with respect to g. A somewhat long, but straightforward, calculation gives the following result for the kinetic energy of the N-electron wave function, relative to that for the normal state:

$$\Delta T = N E_F (1-b)^2 + N E_F \left[2b - \left(4b^2 + g^2 \right)^{1/2} + \left(g^2/2b \right) \ln S \right] , \tag{1.8}$$

where

$$S = \left[2b + \left(4b^2 + g^2 \right)^{1/2} \right] / g , \tag{1.9}$$

and $b = q/2k_0$. Since the total electron density of the new state is spatially uniform (as it is for the normal state), the only new contribution to the potential energy is an (algebraic) decrease of the total exchange energy, J. After another lengthy calculation, one finds

$$\Delta J = -\left(N E_F / 4n \right) \left[(g/b) \ln S \right]^2 . \tag{1.10}$$

An appreciable number of additional algebraic steps allows one to conclude that the sum of (1.8) and (1.10) is a minimum when

$$g = 2b / \sinh(2nb) . \tag{1.11}$$

If this value is inserted into (1.8) and (1.10), the total energy of the new N-electron wave function relative to the normal state becomes

$$W = NE_F \left[1 + b^2 - 2b \coth(2nb)\right] . \tag{1.12}$$

This result is negative definite for $b = 1$, the value which corresponds to the same k-state occupation as the normal state. Consequently, the normal state is always a highly excited state. In fact, within the field of variation employed here, for $b = 1$, the normal state is an energy maximum.

The value (1.11) may be inserted into the wave functions (1.5), and the exchange potential operator computed according to its basic definition. One obtains a constant plus (1.4), the coefficient of the latter being in agreement with (1.11). Therefore, an N-electron state employing the functions (1.5) provides an exact solution of the Hartree–Fock equations for arbitrary b. One can show easily that the value of b which, in turn, minimizes (1.12) is less than unity. In fact, as n decreases (decreasing density or increasing interaction strength), $b \to 0$, at which value (1.12) and (1.2) become equal. This occurs at $n = \frac{3}{4}$. It should not be necessary to emphasize that the new lowest energy state has not been proved to be the Hartree–Fock ground state. However, we shall refer to it as the lowest (known) state. This lowest state has a spiral antiferromagnetic structure. The wavelength of the spiral is small, $\sim \pi/k_0$, for high density and gradually becomes larger, approaching ∞, as the density is reduced. Consequently the transition from spiral antiferromagnetism to ferromagnetism is a gradual one.

States above the energy gap can be occupied (e.g., by thermal excitations), but the spin directions of these states are out of phase with the spin density wave. As a result, the self-consistent amplitude of the spin density wave will be reduced, together with its contribution to the total energy. At a sufficiently high temperature – some fraction of the Fermi temperature – a second order phase transition will occur, and the normal state, with excitations, will (at last) become the state of minimum free energy. One can also construct nonspiral spin density wave states by employing an exchange potential $\propto \sigma_z \cos 2k_0 z$, instead of (1.4). These states, too, are always much lower than the normal state, but by an amount less than half that of the lowest spiral spin density wave.[2)]

The physical reason why the spin density wave states are always lower than the normal state is, of course, the increase in magnitude of the exchange energy resulting from the *local* augmented parallelism of spins. The opposing term is of course the increase in kinetic energy. But the states most highly perturbed – those near the energy gap – are mixed with states of almost equal kinetic energy. Consequently this increase is sufficiently small to allow the exchange energy to dominate. In fact, in the limit of small g (considered again as a variational parameter), the ratio of exchange energy increase to kinetic energy increase approaches infinity (although only logarithmically). This observation makes it easy to prove that the normal state of a Fermi gas is unstable with respect to spin density wave formation for rather general repulsive interactions. Indeed, if the Fourier transform $V(k)$ of the interaction is large for small k, as it is for Coulomb interactions, the instability is enhanced, since the highly deformed single-particle states are close together in k space. (One must observe here that the fractional transfer of wave function amplitude to states across the Fermi sea is compensated by transfers in the reverse direction.)

2) J. des Cloizeaux, J. phys. radium **20**, 606 (1959) and **20**, 751 (1959), has employed oscillating (nonspiral) spin density waves to describe the electronic structure of antiferromagnetic transition-metal oxides, an approach originally suggested by Slater [J.C. Slater, Phys. Rev. **82**, 538 (1951)]. Here, the wave vector $\vec{q}$ is required to be half of a reciprocal lattice vector, and the state is energetically stable only if the Coulomb interactions are sufficiently large.

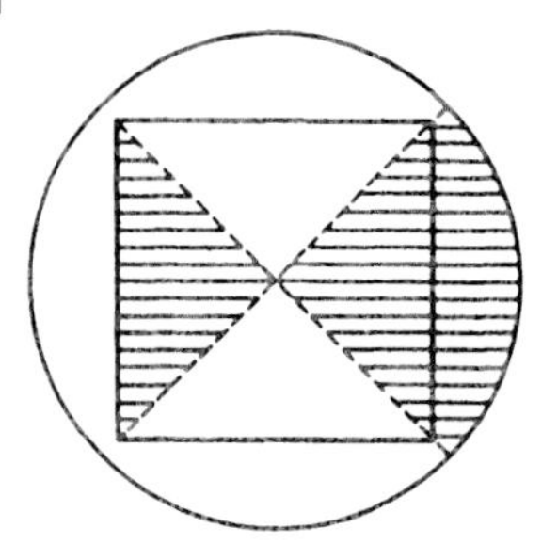

Figure 1.2 Fermi surface for an electron gas with large repulsive interactions. The shaded region – a pyramidal cone – indicates the occupied states of a given spin for one (of three) groups of electrons. A large energy gap exists at the surface of the cube.

The theory for a three-dimensional electron gas is almost identical to that given above for one dimension. It is necessary, however, to divide the electrons into at least three groups, and to allow each group to be deformed by its own spin density wave. For this reason the problem remains essentially one-dimensional. The occupied $\vec{k}$ states of each group will lie within a pyramidal cone (whose central axis is parallel to the wave vector of its spin density wave). For a given spin direction of a given group, the Fermi surface will be a plane on the energy gap side, and (approximately) a spherical polygon on the opposite side, as illustrated by the shaded region in Figure 1.2. For strong interactions the complete Fermi surface will consist of a cube, at the surface of which is a large energy gap, and a sphere, at which $E(\vec{k})$ is continuous. Additional kinetic energy is required, of course, to occupy the unperturbed states in this way, but this will be more than compensated by the lower energy resulting from the spin density wave deformations. For weak interactions the inner Fermi surface will be a many-faced polyhedron, each pair of opposite faces arising from a spin density wave, and its volume will almost equal that of the sphere.

The foregoing description provides a means for "putting together" a three-dimensional electron gas in which the main effects of giant spin density waves are incorporated. The description is accurate only if the (off-diagonal) oscillatory part of the exchange potential between electrons of one group and those of another is neglected. Such effects, if included, will cause additional perturbation of the single-particle wave functions, will lower the total energy even further, and will prevent (fortunately) the boundary planes between different groups from having any sharp physical significance. Also, additional energy gaps in the spectrum of each group will be introduced by these perturbations. Consequently, the spherical part of the Fermi surface will lose some of its continuity. Such refinements need not be elaborated here, however.

A quantitative estimate of the energy gap at the surface of the cube for a typical metal with Coulomb interactions yields −10 ev. (This estimate involves the formulation and solution of a complex nonlinear integral equation, and will be discussed elsewhere.) There seems to be little doubt that giant spin density waves should have numerous and profound consequences with regard to the properties of metals. Fortunately there remains some semblance of a Fermi sphere, but its states are no longer doubly degenerate. On the other hand, one can think of many experiments which might have revealed the existence of giant spin density waves, but have not: neutrons should suffer coherent magnetic diffraction in any metal, Langevin paramagnetism should not occur in metals or alloys, nuclear resonance lines should not be observed at their expected frequencies, if at all, etc. A possible escape from such enigmas is provided by the existence of collective excited states, of at least three varieties, relative to the spin density waves. Besides providing additional transport mechanisms, the excitation of collective modes (spin waves on the giant spin wave) will cause rapid fluctuations in the local spin density directions. It is by no means obvious that such fluctuations will be sufficiently rapid to avoid the unfortunate consequences just mentioned. Correlation energy differences are probably far too small to invert the new lowest state and the normal state in approximations beyond the Hartree–Fock scheme. Further research is necessary before one can decide whether or not a real paradox exists.

Reprint 2 Mechanism of Antiferromagnetism in Dilute Alloys[1)]

A.W. Overhauser *

* Scientific Laboratory, Ford Motor Company, 525, Northwestern Avenue, Dearborn, Michigan 47907, USA

Received 2 October 1959

A mechanism for the antiferromagnetic ordering of a dilute paramagnetic solute in a metal is proposed and discussed in relation to the phenomena that occur in copper–manganese alloys. Long-range antiferromagnetic order results from a static spin-density wave in the electron gas of the metal. This new state of the gas is dynamically self-sustaining as a result of the exchange potentials arising from the spin-density distribution. The paramagnetic solute atoms are then oriented by their exchange interaction with the spin-density wave. The resulting interaction energy more than compensates the increase in energy associated with the formation of the spin-density wave. The theory predicts correctly the magnitude and concentration-dependence of the critical temperature, the anomalous low-temperature specific heat and the anomalous electrical and magnetic properties of the alloys.

2.1 Introduction

Recent experimental studies [1–4] of dilute alloys of manganese in copper have revealed several interesting and unusual properties. Among these is a low-temperature maximum in the magnetic susceptibility – behavior that is characteristic of an antiferromagnetic transition. On the other hand, the temperature-dependence of the reciprocal susceptibility at higher temperatures has a positive intercept on the temperature axis (when extrapolated) – behavior that is characteristic of a ferromagnetic transition. Low-temperature specific-heat measurements [5] on the alloys show an excess specific heat (compared to pure copper) in the temperature region of the susceptibility maximum. Very likely a cooperative phenomenon of some sort is occurring. The transition region is broad, however, presumably as a result of in-homogeneities associated with a random distribution of the paramagnetic solute. The transition temperature is approximately proportional to the solute concentration, being $\sim 25°$K for 1 atomic per cent manganese. Small magnetic remanences, corresponding to $\sim$1 per cent magnetic alignment of the manganese, can be induced. It is difficult to decide experimentally whether the low-temperature order is basically antiferromagnetic or whether for example, it consists of small ferromagnetic domains coupled antiferromagnetically [6]. Both the entropy associated with the excess specific heat and the susceptibility above the transition temperature indicate that spin state of the manganese is $S = 2$.

Finding a mechanism of long-range order in a dilute alloy is not easy at first sight. Sato *et al.* [7] have recently pointed out that long-range order can occur only if the concentration of magnetic solute is greater than 10 or 15 atomic per cent, assuming only short-range interactions are operative. They have emphasized that the familiar molecular-field approximation is completely erroneous when the product of coordination number and solute concentration is not larger than unity. Imagining a superlattice structure associated with a randomly distributed species is an even more formidable task. It seems essential to postulate a long-range interaction, or its equivalent, for an approach to the problem in copper–manganese alloys.

1) J. Phys. Chem. Solids. Pergamon, 13 pp71–80 (1960), Great Britain

Anomalous Effects in Simple Metals. Albert Overhauser

ISBN: 978-3-527-40859-7

Hart [8] has shown that interactions between the spins of solute atoms in a metal are always screened by the dynamical readjustment of the conduction electrons. Yosida [9] has exhibited this behavior in detail by means of perturbation theory. Although the resulting interaction is not restricted to nearest neighbors, the long-range part of it oscillates in sign and diminishes rapidly in magnitude with distance. Consequently this pseudo-exchange interaction seems to have meager potentiality for producing cooperative phenomena in a dilute alloy.

Lack of an alternative suggests that the phenomena of interest result from the exchange interaction between the spins of the paramagnetic solute and those of the conduction electrons. The exchange Hamiltonian can be approximated:

$$H' = -(G/N)\sum_{i,j} s_i \cdot S_j \delta(r_i - R_j) , \tag{2.1}$$

where i labels the spin and position operators of the conduction electrons and j labels the spin and lattice site of the paramagnetic solute atoms. Employment of the delta function in (2.1) is equivalent to neglecting the dependence of the exchange integrals on the wave vectors of the conduction electrons. The magnitude of the coefficient G can be estimated from the spectroscopic-term values associated with the $s-d$ exchange interaction of the $(3d)^4 4s$ or $(3d)^6 4s$ manganese configurations, which yield a value

$$G \sim 0.4\,\text{eV} . \tag{2.2}$$

The number of atoms per unit volume is N. The Hamiltonian, (2.1), has been used by Yosida in deriving the effective interaction [9] between pairs of manganese spins, and in calculating the contribution of spin-disorder scattering to the resistivity of the alloys [10]. In the present work the interaction, (2.1), will be used only as the origin of an effective field for orienting the spins of the solute atoms. As will be shown in the following sections, long-range antiferromagnetic order is caused by a static spin-density wave in the conduction-electron gas:

$$s(R) = bN\epsilon \cos q \cdot R , \tag{2.3}$$

where b is a (dimensionless) amplitude, ϵ is a unit polarization vector, and $s(R)$ is the electron spin density (in units of $\hbar$), given by:

$$s(R) = \left\langle \sum_i s_i \delta(r_i - R) \right\rangle , \tag{2.4}$$

the expectation value of the total spin-density operator. The effective field H_j acting on the solute spin S_j, as defined by the interaction energy,

$$W = -H_j \cdot S_j , \tag{2.5}$$

is given by:

$$H_j = Gb\epsilon \cos q \cdot R_j . \tag{2.6}$$

The wave vector q of the spin-density wave will be unrelated to interatomic distances of the lattice. On a statistical basis, half of the solute spins will be oriented parallel to ϵ and the other half antiparallel.

The spin-density wave will constitute a collective, deformed state of the electron gas that is an eigenfunction of the Hartree–Fock equations with an energy greater than that of the ground state. Slater [11] has pointed out that exchange potentials arising from a spin-density wave are attractive, and therefore tend to dynamically maintain the spin-density wave. This problem is formulated below, and complete dynamic stability of a spin-density wave is found to be possible for a particular magnitude of the wave vector $\boldsymbol{q}$, and then only if the mean exchange interaction between conduction electrons is sufficiently large. The spin-density wave can be thought of most easily by considering it to be the sum of two charge-density waves, one of electrons with spin parallel to $\boldsymbol{\epsilon}$ and the other of antiparallel spin, which are equal in magnitude but opposite in phase. Therefore, the total charge density of the gas remains uniform in space.

The higher energy of a spin-density wave precludes its formation in a pure metal. However, in the alloys under consideration, and at sufficiently low temperatures, the thermodynamic orientation of the solute spins in the effective field of the spin-density wave provides an interaction energy which more than compensates the higher energy of the spin-density wave. Therefore, below a critical temperature, the amplitude b of the wave will increase from zero and approach a maximum at $0\,^\circ$K.

Several appealing features of this mechanism can be appreciated immediately. The spin-density wave, together with its tendency to orient solute spins antiferromagnetically, vanishes above the critical temperature. Consequently, the residual, short-range solute–solute interactions of the direct overlap or pseudo-exchange type remain and can account for a positive paramagnetic Curie point, although they could never produce long-range ferromagnetic order. This mechanism provides, therefore, an easy explanation of how an antiferromagnet can be paramagnetically "ferromagnetic".

Since the wave vector of the spin-density wave is not an integral multiple of a reciprocal lattice vector, the solute spins will see differing effective fields, of magnitude varying from 0 to Gb, depending on the location of each. Such a distribution of effective fields explains why the low-temperature specific heat arising from the cooperative interaction is linearly proportional to T far enough below the transition temperature [5]. The reason is that each incremental increase in temperature allows a new group of solute spins to reach the temperature at which they become thermally disoriented, and the integrated heat capacity of such a group is proportional to the temperature at which disorientation occurs.

One of the most unusual properties of copper–manganese alloys is the behavior of the remanent magnetization [4]. If a sample is cooled through the transition region in the absence of an applied field, little or no remanence occurs. A saturation remanence corresponding to about 1 atomic per cent alignment of the manganese spins can be induced by temporarily applying a field greater than 15 kOe. The remarkable feature is that the saturation remanence, once produced, can be reversed by fields of only a few kOe. Such behavior can be understood within the framework of the present model. It is necessary, of course, to have a domain structure; otherwise remanence could not occur on a macroscopic basis, since for a given domain the numbers of parallel and antiparallel solute spins can be expected to differ only by the square root of their number, according to statistical fluctuations. The smaller the domain size, the greater the relative unbalanced moment. The inhomogeneous distribution of solute atoms, to which the lack of sharpness of the transition region has already been attributed, should provide also a spontaneous means of domain formation.

Consider then an individual domain. Only the magnitude of the wave vector of the spin-density wave is fixed by the condition of self-consistency. Its direction and phase are free to adjust so as to minimize the free energy. In the presence of an applied field such adjustment will maximize the unbalanced moment parallel to the field. Presumably 15 kOe is sufficient to overcome all barriers which inhibit such adjustment, and which serve also to lock in the unbalance. Furthermore the axis of spin quantization assumed for the spin-density wave plays no role in the energetic or dynamic stability of the wave. Consequently all spin directions are free to rotate coherently, together with any unbalanced moment. Presumably the few-kOe field necessary for reversal provides opposition to magnetic dipole–dipole interactions or crystalline anisotropy arising from spin-orbit interactions, which have not been introduced into the model for the sake of simplicity. The foregoing discussion is admittedly conjectural, but it illustrates the ease with which the experimental phenomena can be interpreted.

One further enigma deserves discussion. Both copper–cobalt and copper–manganese alloys are paramagnetic, but only the latter undergo a cooperative transition at low temperatures [4]. Why is there a difference? A somewhat speculative explanation can be given in terms of the present model. As has already been mentioned, dynamic stability of a spin-density wave is possible only if the mean exchange interaction between conduction electrons of the metal is sufficiently large. On this basis alone, copper–cobalt and copper–manganese alloys should be similar. But the cooperative mechanism requires a further condition, specific to the solute: the paramagnetic orbital of the solute must be localized in a region small compared to the wavelength of the spin-density wave. (This wavelength is expected to be of the order of a lattice constant.) The physical origin of such a condition is clear. If the paramagnetic orbital, although localized, were to extend over several half-wavelengths of the spin-density wave, the orbital's exchange interaction with the wave would average to zero, and the effective fields necessary for long-range order would vanish. Since cobalt is two atomic numbers closer to copper than manganese it is to be expected that *d*-hole orbitals of cobalt in copper are less localized than those of manganese in copper. (Were the solute atomic number two units higher than cobalt, a *d*-hole orbital would be completely non-localized.) Although copper–cobalt alloys are paramagnetic, they show marked deviations from a Curie–Weiss law with regard to both field-dependence and concentration-dependence of the susceptibility [4]. This behavior, occurring even in alloys with ~ 1 atomic per cent cobalt, is indicative of long-range solute–solute interactions. In view of the screening action proved by Hart [8], these interactions probably arise by direct overlap of extensive orbitals. If such is indeed the case, the present theory explains also the basic difference between copper–cobalt and copper–manganese alloys.

2.2 Dynamics of a Spin-Density Wave

The free-electron model of a metal will be used in treating the quantum mechanics of a spin-density wave. (Such an approximation should not entail any essential compromise.) The objective is to find solutions of the Hartree–Fock equations for which the expectation value, (2.4), of the total spin-density operator is given by (2.3). The Hartree–Fock Hamiltonian for the problem is:

$$H = H_0 + A\,, \tag{2.7}$$

where H_0 is the electronic kinetic energy and A is the exchange operator, defined as follows:

$$A\psi(r_1) \equiv -\sum_k \left[\int \phi_k^\dagger(r_2)\psi(r_2)V(r_{12})d^3r_2\right]\phi_k(r_1)\,. \tag{2.8}$$

Here, the set of (two-component) functions $\phi_k(r)$ are the occupied one-particle states of the electron gas, $V(r_{12})$ is the electron–electron interaction and ψ is any one-electron wave function in Hilbert space. It must be appreciated that A is a linear Hermitian operator under these general assumptions. It is not necessary to include the Hartree term in (2.7), since the deformations of the electron gas under consideration do not alter the total charge density, which is zero.

The usual solutions of (2.7) are plane waves. The solutions of interest here will be slightly perturbed plane waves. The amplitude b of the spin-density wave will be very small compared to unity. Consequently, it should be adequate to represent the perturbed plane waves as power series in b (retaining, of course, only the first-order correction terms). The following matrix elements of the exchange operator between plane waves may be defined:

$$\left.\begin{aligned} g^{\pm}(k) &\equiv \{\exp[i(k \pm q)\cdot r]\alpha|A|\exp(ik\cdot r)\alpha\} \\ h^{\pm}(k) &\equiv \{\exp[i(k \pm q)\cdot r]\beta|A|\exp(ik\cdot r)\beta\} \end{aligned}\right\} \tag{2.9}$$

where α and β are spin-up and spin-down spin functions relative to an arbitrary direction ϵ of spin quantization. It is easy to see from (2.8) that there are no matrix elements of A between waves of opposite spin states if the occupied states, ϕ_k, have unmixed spin states, as will be assumed. (Such an assumption is therefore a consistent one.) The spin-up perturbed states can be written, for example:

$$\begin{aligned} \phi_k(r) &= \exp(ik\cdot r)\alpha \\ &\cdot\left(1 + \frac{g^{+}(k)}{E_k - E_{k+q}}\exp(iq\cdot r) + \frac{g^{-}(k)}{E_k - E_{k-q}}\exp(-iq\cdot r)\right), \end{aligned} \tag{2.10}$$

where E_k is the single-particle kinetic energy. If these wave functions are used for the occupied states appearing in the exchange operator, one can evaluate the matrix elements (2.9) and thereby obtain a pair of coupled integral equations for $g^{\pm}(k)$. They are:

$$\left.\begin{aligned} g^{+}(K) &= -\sum_k\left[\frac{J(K-k)}{E_k - E_{k+q}}g^{+}(k) + \frac{J(K-k+q)}{E_k - E_{k-q}}g^{-}(k)^*\right] \\ g^{-}(K) &= -\sum_k\left[\frac{J(K-k-q)}{E_k - E_{k+q}}g^{+}(k)^* + \frac{J(K-k)}{E_k - E_{k-q}}g^{-}(k)\right] \end{aligned}\right\} \tag{2.11}$$

where $J(q)$ is the exchange integral associated with the electron–electron interaction:

$$J(q) \equiv \Omega^{-1}\int V(r)\exp(iq\cdot r)d^3r\,, \tag{2.12}$$

Ω being the volume in which the wave functions are normalized. An identical set of equations applies to $h^{\pm}(k)$, but the requirement that the charge-density deviations of spin-up electrons cancel those of spin-down entails that the solutions of interest obey

$$h^{\pm}(k) = -g^{\pm}(k)\,, \tag{2.13}$$

a relation that can always be satisfied by the solutions of (2.11) as a consequence of their partial linearity. (A solution multiplied by a *real* constant is also a solution.)

The factors in the integral equations have the following important symmetry properties:

$$\left.\begin{aligned} J(q) &= J(-q) \\ E_k &= E_{-k} \end{aligned}\right\} \tag{2.14}$$

As a consequence, it is relatively easy to see that if

$$f = [g^{+}(K), g^{-}(K)] \tag{2.15}$$

is a solution of (2.11), then:

$$f' = Tf \equiv [g^{-}(-K)^*, g^{+}(-K)^*] \tag{2.16}$$

is also a solution which is either $\pm f$ or is linearly independent of f. In the latter case it follows that the functions

$$F = \left[g^{+}(K) + g^{-}(-K)^{*}, g^{-}(K) + g^{+}(-K)^{*}\right] \tag{2.17}$$

and

$$F' = \left[g^{+}(K) - g^{-}(-K)^{*}, g^{-}(K) - g^{+}(-K)^{*}\right] \tag{2.18}$$

are also solutions, and which, when considered as basis vectors, span the same space as f and f'. But, $TF = F$ and $TF' = -F'$. Therefore it has been proved that basis vectors which span the space of solutions of (2.11) can *all* be chosen to satisfy:

$$f = \pm T f \tag{2.19}$$

Basis vectors obeying the + sign in (2.19) can then be written:

$$f = \left[g(K), g(-K)^{*}\right], \tag{2.20}$$

where $g(K)$ is the solution of the integral equation:

$$g(K) = -\sum_{k}\left[J(K-k) + J(K+k+q)\right]\frac{g(k)}{E_k - E_{k+q}}\,; \tag{2.21}$$

and basis vectors obeying the − sign in (2.19) can be written:

$$f = \left[g(K), -g(-K)^{*}\right], \tag{2.22}$$

where $g(K)$ satisfies:

$$g(K) = -\sum_{k}\left[J(K-k) - J(K+k+q)\right]\frac{g(k)}{E_k - E_{k+q}}\,. \tag{2.23}$$

(Since the kernels of (2.21) and (2.23) are real, the basic functions of their solutions can be chosen real without loss of generality.) Each solution of (2.21) or (2.23) generates two linearly independent solutions of (2.11). For example, if $g(K)$ is a solution of (2.21), the two corresponding solutions of (2.11) are (2.20) and

$$f = \left[ig(K), -ig(-K)^{*}\right]. \tag{2.24}$$

Spin-density waves corresponding to (2.20) and (2.24) have the same structure, but are shifted in real space by one-quarter wavelength. The phase of a spin-density wave can be varied continuously, therefore, by appropriately varying the linear combination of these two degenerate solutions.

It has been proved that the space of solutions of the coupled integral equations, (2.11), is identical to the space generated by the solutions of the uncoupled equations (2.21) and (2.23). It has not been proved that solutions other than $g(K) \equiv 0$ exist. It should be noted that the wave vector $\boldsymbol{q}$ is a parameter in the integral equation, and its magnitude may play the role of an eigenvalue parameter. Treating equation (2.21) or (2.23) exactly is a very difficult task because of the complexity of the kernels. It seems necessary to make simplifying approximations in order to gain insight about the solutions. A very effective approximation is to neglect the variation of $J(q)$ by replacing it with a constant. This is perhaps a better approximation than it may seem at first sight, since much of the variation of $J(q)$ is averaged out by the integrations over the planes of constant-energy denominator. For a coulomb interaction

$$J(q) = 4\pi e^2/\Omega q^2\,. \tag{2.25}$$

The wave-number difference q appearing in the denominator of (2.25) will generally have a magnitude of the order of the radius k_0 of the Fermi sphere. Consequently, an average exchange integral, $\bar{J}$, can be defined:

$$\bar{J} = 4\pi e^2/\Omega k_0^2 \xi , \tag{2.26}$$

where ξ is the required numerical factor, of unit order of magnitude, in order that (2.26) be the appropriate average value of J as defined by (2.21). The total exchange energy per electron of a Fermi gas is:

$$-J = -9\pi n e^2/4k_0^2 , \tag{2.27}$$

where n is the number of electrons per cm^3. It will be convenient to express $\bar{J}$ in terms of J:

$$\bar{J} = 16J/9n\Omega \xi . \tag{2.28}$$

If the constant $\bar{J}$ is inserted into (2.23), only the trivial solution, $g(K) \equiv 0$, results. If $J(q)$ is replaced by $\bar{J}$ in (2.21), a solution, $g(K) = g$, an arbitrary constant, is obtained provided that the following equation can be satisfied for some value of q:

$$1 = -2\bar{J} \sum_k \left(E_k - E_{k+q}\right)^{-1} . \tag{2.29}$$

If no root of this equation exists, only the trivial solution ($g = 0$) is possible, and a spin-density wave cannot occur. The integrand in (2.29) has singularities, but one can prove that the principal value of the integral is the correct interpretation. The result is,

$$\sum_k \left(E_k - E_{k+q}\right)^{-1} = -\left(3\pi\Omega/8E_F\right) p\left(q/2k_0\right) , \tag{2.30}$$

where E_F is the Fermi energy and $p(x)$ is given by:

$$p(x) = \frac{1}{2} + \frac{1-x^2}{4x} \log\left|\frac{1+x}{1-x}\right| . \tag{2.31}$$

If the foregoing equations are combined, the relation determining q is:

$$1 = (4J/3\xi E_F)p(q/2k_0) . \tag{2.32}$$

Since $p(0) = 1$ and is monotonic, decreasing for increasing x, (2.32) has a solution if

$$J > (3\xi/4)E_F . \tag{2.33}$$

This inequality is not too stringent, since $J \sim E_F$ for most metals. Dynamic stability of a static spin-density wave, at least in some metals, seems to be an assured possibility.

The amplitude b (see (2.3)) of the spin-density wave can be computed directly from the operator in (2.4) and the wave functions (2.10). One finds

$$b = -(9\xi\eta/8J)g , \tag{2.34}$$

where $\eta = n/N$, the number of conduction electrons per atom, and where use has been made of (2.30) and (2.32). The $-$ sign in (2.34) reflects the attractive character of exchange interactions. Wherever the spin density is positive, the exchange potential is negative (as is necessary for a dynamically stable situation).

The remaining task of this section is to calculate the energy increase associated with a spin-density wave. This can be accomplished by summing the kinetic-energy change of each occupied state together with one-half the energy change arising from the exchange potential. Employing the wave functions (2.10) and taking care to normalize, then to order g^2 one finds the kinetic energy increase per cm^3 to be,

$$\Delta T = (9\xi n/8J)g^2 \ . \tag{2.35}$$

Similarly, the total coulomb energy change turns out to be the negative of (2.35). Consequently, in the Hartree–Fock approximation, the energy of a spin-density wave is zero. It is necessary to include correlation-energy changes in order to obtain an energy increase. The magnitude of the correlation-energy change can be estimated by means of the following relatively crude argument. Consider a small region in the Fermi gas for which the number of + spins is $M(1+\epsilon)$. The exchange energy is proportional to the number of pairs of parallel spins, which is:

$$(1/2)M^2(1+\epsilon)^2 + (1/2)M^2(1-\epsilon)^2 = M^2(1+\epsilon^2) \ . \tag{2.36}$$

The correlation energy is proportional (mainly) to the number of pairs of antiparallel spins, which is:

$$M(1+\epsilon)M(1-\epsilon) = M^2(1-\epsilon^2) \ . \tag{2.37}$$

From (2.36) and (2.37) it is clear that the fractional (algebraic) decrease in the exchange energy should roughly equal the fractional increase in the correlation energy, arising from a small spin-density deviation. Therefore, if we let $-C$ be the correlation energy per electron of the Fermi gas, the energy W per cm^3 of a spin-density wave is approximately C/J times the kinetic-energy increase (2.35). With the help of (2.34), this energy density can be written:

$$W(b) = 8nCb^2/9\xi\eta^2 \ . \tag{2.38}$$

For typical metals the correlation energy per electron, C, is ~ 1 eV [12].

Equation (2.38) is the only result of the present section that is needed for treating the thermodynamics of the antiferromagnetic phase. However, the main burden of the foregoing development has been to give theoretical justification to static spin-density waves as a mechanism of long-range anti-ferromagnetic order.

2.3 Thermodynamics of the Antiferromagnetic Phase

The magnetic contributions to the free energy of a dilute alloy can be evaluated immediately by considering the distribution of effective fields, (2.6), experienced by the paramagnetic solute spins as a result of their interaction, (2.5), with the spin-density wave. If the orbitals of the solute spins are sufficiently localized, compared to a wavelength of the spin-density wave, the probability distribution for the (absolute) magnitude of the effective field H is:

$$P(|H|) = (2/\pi)\left(G^2b^2 - H^2\right)^{-1/2} , \tag{2.39}$$

for $|H|$ smaller than the maximum value, Gb. The partition function of a spin S_j in a field H_j is:

$$Z_j = \sinh\left\{\beta\left[S_j + (1/2)\right]H_j\right\} / \sinh(1/2)\beta H_j \ , \tag{2.40}$$

where $\beta = 1/kT$. The free energy of such a spin

$$A_j = -(\log Z_j)/\beta \ . \tag{2.41}$$

The magnetic free-energy density of the dilute alloy can be expressed as the average of (2.41), weighted by the probability distribution (2.39), and added to the self-energy of the spin-density wave:

$$A = W(b) - \frac{2cN}{\pi\beta} \int_0^{Gb} (G^2 b^2 - H^2)^{-1/2} \times \log\left[\frac{\sinh\beta[S + (1/2)]H}{\sinh(1/2)\beta H}\right] dH \,, \tag{2.42}$$

where c is the atomic fraction of solute spins. The amplitude b of the spin-density wave must be chosen so as to minimize (2.42). $b(T)$ will have a maximum value $b(0)$ at $T = 0$, and will become zero at the transition temperature T_c. By appropriate power-series expansions of (2.42) near $T = 0$ and T_c, it is easy to derive the following results: near $T = 0$, including terms up to order T^2 (2.42) becomes:

$$A = \frac{8nCb^2}{9\xi\eta^2} - \frac{2cNSGb}{\pi} - \frac{2\pi ScN}{3(2S-1)Gb} k^2 T^2 \,. \tag{2.43}$$

The value of b which minimizes this free energy is, to order T^2

$$b(T) = \frac{9\xi\eta SG}{8\pi C} c - \frac{8\pi^3 C}{27\xi\eta S(2S+1)G^3 c} k^2 T^2 \,. \tag{2.44}$$

The low-temperature free energy is determined by inserting (2.44) into (2.43):

$$A = -\frac{9n\xi S^2 G^2 c^2}{8\pi^2 C} - \frac{16\pi^2 NC}{27\xi\eta(2S+1)G^2} k^2 T^2 \,. \tag{2.45}$$

Since the specific heat (at constant volume) is given by:

$$C_V = -T\left(\frac{\partial^2 A}{\partial T^2}\right)_V \,, \tag{2.46}$$

the low-temperature magnetic specific heat is

$$C_{VM} = \gamma_M T \,, \tag{2.47}$$

where,

$$\gamma_M = \frac{32\pi^2 NCk^2}{27\xi\eta(2S+1)G^2} \,. \tag{2.48}$$

This is a most remarkable result, because γ_M is *independent* of solute concentration. The experimental measurements of Zimmerman [5] are in complete accord with this conclusion. He finds

$$\gamma_M \sim 4 \times 10^{-3}\ \text{J/mole, deg}^2 \,, \tag{2.49}$$

for samples of 1/6, 1/2, 1, 2, and 4 atomic per cent manganese in copper. The magnitude of (2.49), as well as the lack of concentration-dependence, is in excellent agreement with (2.48). (One should appreciate that the maximum temperature below which (2.47) is valid is proportional to the solute concentration.)

By appropriate power-series expansions near T_c, the following results can be established:

$$kT_c = \frac{3\xi\eta S(S+1)G^2}{32C} c \,. \tag{2.50}$$

The critical temperature is proportional to the solute concentration. By combining (2.48) and (2.50), all of the less-known parameters can be eliminated:

$$\gamma_M T_c = \frac{\pi^2}{9} \cdot \frac{S(S+1)}{2S+1} cNk \,. \tag{2.51}$$

If the experimental value (2.49) for γ_M (together with $S = 2$) is used in this expression, the indicated critical temperature of a 1 atomic per cent manganese alloy is $T_c = 27°$K. The temperature variation of b, the spin-density wave amplitude, near T_c is given by:

$$b(T) \sim (T_c - T)^{1/2} \,, \tag{2.52}$$

and $b = 0$ for $T > T_c$, as would be expected. The variation of b with T throughout the antiferromagnetic phase can, with an accuracy of 2 per cent or better, be taken to be:

$$b(T) = b(0)\left[1 - (T/T_c)^2\right]^{1/2} \,, \tag{2.53}$$

where $b(0)$ is given by the first term in (2.44).

The magnetic contribution to the specific heat just below T_c is:

$$C_{VM}(T_c) = \left[\frac{30(2S+1)}{\pi^2(2S^2+2S+1)}\right] \gamma_M T_c \,. \tag{2.54}$$

Since the coefficient in brackets has a value 1·17 (for $S = 2$), the variation of the specific heat throughout the antiferromagnetic region deviates very little from the linear behavior given by equation (2.47). The magnetic specific heat rises slightly above the linear low as T_c is approached, and then drops discontinuously to 0 above T_c. This sudden drop is not very sharp experimentally, especially in the less-dilute alloys. Such broadening is probably associated with inhomogeneities, and one may expect a high-temperature tail in the magnetic specific heat as a result of solute–solute pseudo-exchange interactions, particularly in the more concentrated alloys. The temperature at which C_{VM}/T drops to half its low-temperature value should provide a relatively reliable measure of T_c. The experimental results for 1/6 and 1/2 atomic per cent manganese alloys, so interpreted, yield:

$$T_c = 26 \pm 4°\text{K/atomic per cent} \,, \tag{2.55}$$

a value in good agreement with (2.51), and with the onset of the anomalous resistivity decrease [3]. If $\xi, \eta \sim 1$, and $C \sim 1\,\text{eV}$, then the $s-d$ exchange interaction parameter G which yields agreement between (2.50) and (2.55) is:

$$G = 0.6\,\text{eV} \,. \tag{2.56}$$

This value also gives agreement between the theoretical and experimental magnitudes of the anomolous resistivity decrease [10], and compares well with the estimate, (2.2), derived from spectroscopic data of free ions.

2.4 Concluding Remarks

The applications of the foregoing theory to the electrical and magnetic properties of copper–manganese alloys will be elaborated in a subsequent paper. The static spin-density wave mechanism of antiferromagnetic order may obtain in systems other than dilute paramagnetic alloys. For example, it may provide a mechanism of nuclear antiferromagnetism, the hyperfine interaction playing the role of the $s-d$ exchange interaction for coupling localized spins to the spin-density wave. The mechanism may also operate in a pure metal, such as a-manganese, which is antiferromagnetic and which has an anomalously large linear contribution to the low-temperature specific heat.

Acknowledgements

The author is grateful to Dr. A. Arrott for many helpful discussions regarding magnetic structures and behavior.

A Appendix

In presenting the idea of a static spin density wave, the author has encountered frequently several questions related to the theoretical framework within which the concepts associated with such excited states have been elaborated. It seems worthwhile to discuss in greater detail some of these "objections".

A.1 Objection 1

The criterion, equation (2.33), for the occurrence of static spin density waves as collective excited states of a Fermi gas is similar in form to that for free electron ferromagnetism, ($J > 1 \cdot 36 E_F$ in the Hartree–Fock approximation). If one replaces the Coulomb interaction by a delta function, the resulting criteria are in fact numerically identical. Consequently, it would appear that spin density waves can occur only if the Fermi gas were ferromagnetically unstable, which fact would preclude their formation in the first place (at least for the proposed applications).

A.2 Reply to Objection 1

The factor ξ in equation (2.33), and hence the precise criterion for a spin density wave, can be determined only by solving the integral equation (2.21) exactly. The value of ξ is related to the average value of the exchange integral (2.25) according to the averaging process defined by the integral operator in (2.21). The best *a priori* estimate of ξ, perhaps, is the value required for the approximation (2.26) to yield the correct total exchange energy, which is known exactly and which defines thereby *an* average exchange integral. Such an estimate of ξ yields 4/9; so that the condition for static spin density waves would be, $J > E_F/3$. This criterion is weaker than that for ferromagnetism by at least a factor of 4, and would allow a Fermi gas to have static spin density waves as long as the electron concentration were between 1 and 64 times the limiting density compatible with ferromagnetism. This result provides, therefore, an adequate "margin of safety" and justifies the concept of static spin density waves as a working hypothesis, subject of course to subsequent experimental or theoretical verification. Lack of precision in the latter respect cannot be considered serious, however, since by analogy, one should remember that the entire electron theory of metals rests on the theoretically *unproved* assumption (among others) that the Hartree–Fock ground state of an ideal Fermi gas is the familiar Fermi sphere of occupied momentum states.

A.3 Objection 2

The proposed mechanism of antiferromagnetism is similar to the mechanism of ferromagnetism suggested by Zener. However, Yosida has shown (with simplifying approximations) that the energy contribution resulting from first order excitation of ferromagnetic excited states of the Fermi gas is cancelled by second order contributions of the $s-d$ exchange interaction. Similarly, one might expect such second order contributions to cancel as well the antiferromagnetic interaction resulting from first order excitation of a static spin density wave.

A.4 Reply to Objection 2

No such cancellation occurs in this case. The totality of intermediate states employed by Yosida in his second order calculation exhausts all of Hilbert space. No further second order contributions remain to be evaluated. (Of course, in so far as intermediate states employed in a perturbation calculation should approximate eigenstates, the recognition of collective modes – e.g. plasma waves, spin density waves, etc. – as better "eigenstates" than the single particle excitations, of which they are linear combinations, will suggest improvements to such computations. But the improvements would involve replacement rather than addition of terms.) The spin density wave mechanism of antiferromagnetism assumes and incorporates Yosida's work as a foundation. The second order $s-d$ exchange interaction between solute spins is presumed to be always present (and to be responsible for the positive paramagnetic Curie temperature of Cu–Mn alloys). A localized spin which interacts with a spin density wave is not just the "bare" solute ion spin-orbital, but that together with its localized conduction electron, spin density readjustment, as envisioned by Yosida. It is incorrect to think that spin density waves give rise to a solute–solute interaction of infinite range. Rather, the spin density wave must be considered as having independent existence, and as interacting individually with localized solute spins; the very *tendency* towards long range antiferromagnetic order vanishes above the Néel point, where the spin density wave amplitude is zero.

A.5 Objection 3

It is difficult to understand why a single spin density wave having a unique wave vector magnitude, q_0, should be involved so pre-eminently instead of a cooperative participation by many spin density waves with a distribution in q; and it seems odd that the allowed amplitudes of the spin density wave form a continuous distribution rather than a discrete set like the harmonic oscillator levels of an ordinary collective mode.

A.6 Reply to Objection 3

The special character of a static spin density wave is that it is *static*. The uniqueness of the eigenvalue q_0 is associated with the employment of stationary state perturbation theory in deriving the integral equations (2.11). There are, to be sure, non-static spin density waves, which oscillate harmonically in time, and these can be constructed by employing time dependent perturbation theory in a straightforward fashion. One finds that the frequency ω of such waves varies with q according to $(q_0 - q)^{1/2}$. (These collective modes will be quantized and will yield a negligible contribution, $\sim Nk(kT/E_F)^2$, to the specific heat.) The molecular fields arising from such waves will oscillate in time, and that is why they are unimportant for long range magnetic order. It seems likely, however, that waves with q near q_0 may play a role in the description of antiferromagnetic domains. The Hamiltonian of a collective mode can be written, $P^2 + \omega^2 Q^2$, where P and Q are appropriate generalized coordinates. It is well known that a Hamiltonian of this type leads to discrete energy levels, but only as long as $\omega \neq 0$. When $\omega = 0$, only the P^2 term remains, and the allowed energies form a continuous spectrum, analogous to that of free particles.

A.7 Objection 4

Since solutions of the integral equation (2.11) obey a superposition theorem, it seems that a spiral spin density wave (such as a cos qz polarized in x plus a sin qz polarized in y) could be constructed and would lead to an antiferromagnetic state of lower energy, since the spin density would have the same magnitude everywhere, changing only its direction with position.

A.8 Reply to Objection 4

The integral equations satisfy a superposition principle, but only *after* a specific axis of spin quantization has been chosen. One can superpose two spin density waves only if they have the same axis of spin quantization. To construct a spiral spin density wave as suggested, one would require conduction electron spin states simultaneously quantized along two perpendicular axes – an obvious impossibility.

References

1 Owen, J., Browne, M., Knight, W.D., and Kittel, C. (1956) *Phys. Rev.*, **102**, 1501.
2 Owen, J., Browne, M., Arp, V., and Kip, A.F. (1957) *J. Phys. Chem. Solids*, **2**, 85.
3 Schmitt, R.W. and Jacobs, I.S. (1957) *J. Phys. Chem. Solids*, **3**, 324.
4 Jacobs, I.S. and Schmitt, R.W. (1959) *Phys. Rev.*, **113**, 459.
5 Zimmerman, J.E. To be published.
6 Gorter, C.J., van den Berg, G.J., and de Nobel, J. (1956) *Canad. J. Phys.*, **34**, 1281.
7 Sato, H., Arrott, A., and Kikuchi, R. (1959) *J. Phys. Chem. Solids*, **10**, 19.
8 Hart, E.W. (1957) *Phys. Rev.*, **106**, 467.
9 Yosida, K. (1957) *Phys. Rev.*, **106**, 893.
10 Yosida, K. (1957) *Phys. Rev.*, **107**, 396.
11 Slater, J.C. (1951) *Phys. Rev.*, **82**, 539.
12 Seitz, F. (1948) *Modern Theory of Solids*, Chap. X, McGraw-Hill, New York.

Reprint 3 Spin Density Waves in an Electron Gas[1]

A.W. Overhauser*

* Scientific Laboratory, Ford Motor Company, 525, Northwestern Avenue, Dearborn, Michigan 47907, USA

Received 11 June 1962

It is shown rigorously that the paramagnetic state of an electron gas is never the Hartree–Fock ground state, even in the high-density – or weak-interaction – limit. The paramagnetic state is always unstable with respect to formation of a static spin density wave. The instability occurs for spin-density waves having a wave vector $Q \approx k_F$, the diameter of the Fermi sphere. It follows that the (Hartree–Fock) spin susceptibility of the paramagnetic state is not a monotonic decreasing function with increasing Q, but rather a function with a singularity near $Q = 2k_F$. Rather convincing experimental evidence that the antiferromagnetic ground state of chromium is a large-amplitude spin density wave state is summarized. A number of consequences of such states are discussed, including the problem of detecting them by neutron diffraction.

3.1 Introduction

An idealized electron gas – namely, N electrons confined to a box of volume V and having a uniform, rigid background of positive charge to guarantee electrical neutrality – has been the subject of much theoretical study during the past thirty years. The elementary theory is the Hartree–Fock (HF) approximation, which defines the best N-electron wave functions that can be written as a single Slater determinant of one-electron wave functions. HF states have almost always been employed as the N-electron basis functions for further refinements of the theory. In any event, the nature of the HF ground state is of some interest.

Traditionally, the belief has been that an electron gas can have either a paramagnetic or ferromagnetic HF ground state, depending on the electron density, $n = N/V$. The paramagnetic state is a Slater determinant of plane waves, having occupied all one-electron states with wave vector $\vec{k}$ lying within a Fermi sphere of radius k_F for both spin states. The HF energy per electron of the paramagnetic state is

$$(W/N)_p = (3\hbar^2 k_F^2/10m) - (3e^2 k_F/4\pi)\,, \tag{3.1}$$

the sum of the kinetic energy and the exchange energy. The relation between n and k_F is

$$n = k_F^3/3\pi^2\,. \tag{3.2}$$

The ferromagnetic state for an electron gas of identical density is a Slater determinant having all electron states within a sphere of radius $2^{1/3} k_F$ occupied, but only for spin up, say. The HF energy per electron of the ferromagnetic state is

$$(W/N)_f = 2^{2/3}\left(3\hbar^2 k_F^2/10m\right) - 2^{1/3}\left(3e^2 k_F/4\pi\right)\,. \tag{3.3}$$

A comparison of (3.1) and (3.3) indicates that the ferromagnetic state has lower energy than the paramagnetic state when

$$k_F < 5me^2/(2^{1/3}+1)2\pi\hbar^2 \approx 0.66 \times 10^8\ \text{cm}^{-1}\,. \tag{3.4}$$

1) Phys. Rev. 128, 3 (1962)

Anomalous Effects in Simple Metals. Albert Overhauser

ISBN: 978-3-527-40859-7

Consequently, a high-density electron gas would be paramagnetic and a low-density one would be ferromagnetic. This criterion was first derived by Bloch [1]. The ferromagnetic state considered above is completely polarized. Partially polarized states must also be compared, but one can easily show that they always have higher energy than the smaller of (3.1) and (3.3). If the electron density is extremely low, Wigner [2] has shown that a HF state with electrons localized near lattice positions would have lower energy than (3.3). The system, then, could no longer be described as an electron gas.

In Section 3.3 it will be shown, contrary to general belief, that the paramagnetic state is never the HF ground state. Specifically, the paramagnetic state is always unstable with respect to formation of a static spin-density wave (SDW). In fact, the instability is sufficiently great that a general proof valid even in the high-density – or weak-interaction – limit can be constructed. A low-density electron gas will still be ferromagnetic (in the HF approximation), but the critical density will be smaller than that corresponding to (3.4). It seems safe to presume that the HF ground state is indeed a SDW state (except for the low-density ferromagnetic range) although, strictly speaking, all that has been proved is that is has lower energy than the historical HF "ground state".

The SDW instability of an electron gas in the HF approximation has been pointed out earlier [3] by the writer, although the demonstration given there was, for the sake of simplicity, limited to a one-dimensional gas. The original surmise had arisen from a study of a three-dimensional electron gas, similar to that presented here in Section 3.3. Several authors [4, 5] have suggested that such HF instabilities cannot occur in the three-dimensional case, either generally or for weak interactions. However, their arguments have neglected the wave-vector dependence of exchange interactions characteristic of Coulomb forces. It is true that SDW instabilities are less general in three dimensions that in one, e.g., they are never ground states for δ-function interactions in three dimensions.[2)] So considerable care is required to establish the nature of the ground state in any particular case. It is necessary to emphasize that a consideration of only small deviations from the paramagnetic state is not sufficient for establishing the stability of that state. For example, the critical electron density corresponding to (3.4) is 35 % larger than the critical density that would have been obtained if only small ferromagnetic deviations from the paramagnetic state were tested.

Of course, the ultimate question of physical interest is the nature of the true ground state of an electron gas. Until methods are developed which do not depend on a prejudgment of the final outcome, recourse to experimental behavior can serve as a guide, even though real metals do not conform to the idealized model. Correlation corrections will tend to suppress SDW's just as they tend to suppress ferromagnetism. Therefore it is not surprising that almost all metals appear to have paramagnetic ground states. But there are some exceptions. Considerable experimental evidence indicates that the antiferromagnetic ground state of chromium is just a large amplitude SDW state. This is discussed in Section 3.8. The recently discovered anomalies in the magnetic susceptibilities of vanadium [6] and molybdenum [7] may also arise from a SDW transition. The susceptibility maximum observed near 100 °K in palladium [8] is another candidate. A SDW of small amplitude would have very minor physical manifestation, so the (remote) possibility of a more frequent occurrence cannot be easily ruled out on the basis of existing experimental evidence.

2) This question has been investigated partially by Yoshimori (see [5]) who studied the instability resulting from δ-function interactions, but without allowance for the repopulation of k space or large SDW deformations. The present writer has extended this study by evaluating the HF energy for both large and small SDW's, with k space always repopulated in such a way that the Fermi surface is a surface of constant HF (one-electron) energy. A similar study has been carried out independently by Conyers Herring (private communication). SDW ground states do not occur for δ-function interactions, whatever their strength.

In the following section the nature of a SDW in the HF approximation will be elaborated. It will be assumed that the reader is familiar with HF theory. The usual derivation of the HF equations does not require that the one-electron functions be separable into products of pure space and pure spin functions, although such a factorization has almost always been imposed. This unnecessary constraint must be dropped in order to allow a treatment of spiral SDW's.

3.2 Nature of a Spin Density Wave

Exchange interactions arising from repulsive forces always favor the parallel alignment of spins. Partial or complete ferromagnetic polarization of an electron gas increases the magnitude of the exchange energy. However the cost, in terms of increased kinetic energy, is very high. Ferromagnetic polarization introduces a long-range parallelism of spins as well as a short range, but it is the latter that is more important with respect to the exchange energy. It is not unreasonable, then, to consider states of an electron gas which have a net fractional spin polarization $\vec{P}(\vec{r})$ at every point, but with the direction of $\vec{P}$ varying continuously with position. A spiral SDW is the simplest example:

$$\vec{P} = P(\vec{x}\cos Qz + \vec{y}\sin Qz)\,. \tag{3.5}$$

The axis $\vec{Q}$ of the SDW, taken here to be the $\vec{z}$ direction, need not be perpendicular to the plane of polarization defined by the unit vectors $\vec{x}$ and $\vec{y}$. No generality is lost, however, by considering this special case. It should be noted that (3.5) describes a static polarization, the phase and amplitude of the wave being independent of time.

As will be shown below, a spin polarization of the form (3.5) leads to an off-diagonal contribution A'' to the one electron exchange potential $A \equiv A' + A''$,

$$A = A' - g\vec{\sigma}\cdot(\vec{x}\cos Qz + \vec{y}\sin Qz)\,, \tag{3.6}$$

where $\vec{\sigma}$ is the Pauli spin operator. A' is the diagonal part (in the pure plane-wave representation) of the one-electron exchange energy. The amplitude g of the off-diagonal contribution will, in general, be dependent on the wave vector of the electron considered. If the spin operators are substituted explicitly,

$$A'' = -g\begin{pmatrix} 0 & e^{-iQz} \\ e^{iQz} & 0 \end{pmatrix}, \tag{3.7}$$

The operator A'' connects plane-wave states in pairs: $\vec{k}, \alpha \rightleftarrows (\vec{k} + \vec{Q}), \beta$, where α and β are the spin-up and spin-down spin functions. Consequently, the HF equation,

$$\left[(p^2/2m) + A\right]\psi = E_\psi\,, \tag{3.8}$$

can be solved formally. The eigenvalues are those of a two-dimensional secular equation:

$$E_k = \frac{1}{2}(\epsilon_k + \epsilon_{k+Q}) \pm \left[\frac{1}{4}(\epsilon_k - \epsilon_{k+Q})^2 + g^2\right]^{1/2}. \tag{3.9}$$

The one-electron energy ϵ_k is the free electron energy plus the diagonal part A' of the one-electron exchange energy; see (3.15)a,b). The two branches of the eigenvalue spectrum are shown in Figure 3.1. The exact wave functions for the lower branch are

$$\phi_k = \left\{\alpha\cos\theta\exp(i\vec{k}\cdot\vec{r}) + \beta\sin\theta\exp\left[i(\vec{k}+\vec{Q})\cdot\vec{r}\right]\right\}/V^{1/2}\,, \tag{3.10}$$

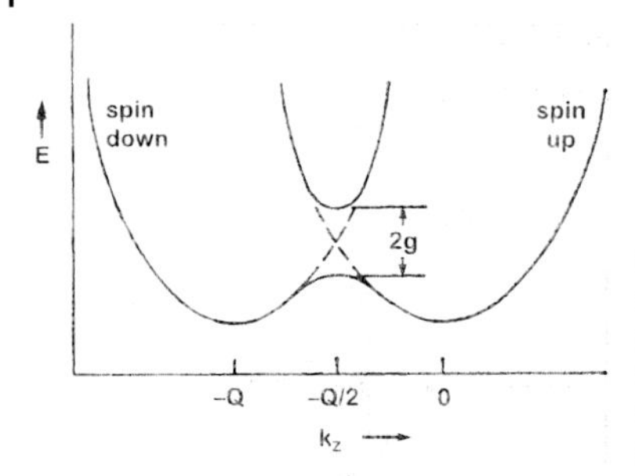

Figure 3.1 Single-particle energy level spectrum for a spiral spin density wave. The spin-down branch has been displaced Q to the left of the spin-up branch in order that the unperturbed states coupled by the SDW exchange potential have the same abscissa. Only lower branch states will be occupied in a SDW ground state.

where

$$\cos\theta(\vec{k}) \equiv g/\left[g^2 + (\epsilon_k - E_k)^2\right]^{1/2} . \tag{3.11}$$

The algebraic signs have been defined so that $0 \leq \theta \leq \frac{1}{2}\pi$. The wave functions for the upper branch can be obtained from (3.10) by replacing θ with $\theta + \frac{1}{2}\pi$. However, states of the upper branch will not be occupied in the ground state. They are, of course, orthogonal to the lower branch. They become occupied at finite temperatures, and cause a decrease in the SDW amplitude.

The square modulus of (3.10) is a constant, $1/V$, because the two terms have orthogonal spin functions. Therefore a charge density wave does not accompany a SDW. That is why a Coulomb potential term need not appear in the HF equation (3.8).

The exchange operator A that arises in the HF scheme is defined by the following operator equation, valid for any two-component wave function $\psi(\vec{r})$

$$A\psi(\vec{r}_1) = -\sum_k \left[\int (e^2/r_{12})\varphi_k^\dagger(\vec{r}_2)\psi(\vec{r}_2)d^3r_2\right]\varphi_k(\vec{r}_1) . \tag{3.12}$$

For the present problem the φ_k are the occupied states (3.10). The matrix elements of A between the pure plane wave states, $\vec{k}'$, α and $(\vec{k}' + \vec{Q})$, β can be evaluated explicitly. By definition they are,

$$-g(\vec{k}') \equiv V^{-1}\int \alpha^\dagger \exp\left[-i\vec{k}'\cdot\vec{r}\right] A \times \beta \exp\left[i(\vec{k}' + \vec{Q})\cdot\vec{r}\right] d^3r . \tag{3.13}$$

The indicated coordinate integrations of (3.12) and (3.13) can be carried out, and there remains only the integrations in $\vec{k}$ space for the occupied states φ_k:

$$g(\vec{k}') = \int \frac{4\pi e^2}{|\vec{k}' - \vec{k}|^2}\sin\theta\cos\theta\frac{d^3k}{8\pi^3} . \tag{3.14}$$

This is an integral equation for $g(\vec{k})$ since $\theta(\vec{k})$ is a function of $g(\vec{k})$, as given by (3.11). Actually, the dependence is quite intricate since θ also depends through ϵ_k on the diagonal elements A' of the exchange operator (3.12). For a spin-up plane wave,

$$\epsilon_{k'} = \frac{\hbar^2|\vec{k}'|^2}{2m} - \int \frac{4\pi e^2}{|\vec{k}' - \vec{k}|^2}\cos^2\theta\frac{d^3k}{8\pi^3} , \tag{3.15a}$$

and for a spin-down plane wave,

$$\epsilon_{k'+Q} = \frac{\hbar^2|\vec{k}' + \vec{Q}|^2}{2m} - \int \frac{4\pi e^2}{|\vec{k}' - \vec{k}|^2}\sin^2\theta\frac{d^3k}{8\pi^3} . \tag{3.15b}$$

An exact solution of the HF equations reduces therefore to a solution of the coupled integral equations (3.14) and (3.15), together with the algebraic relations (3.9) and (3.11). Needless to say, explicit nontrivial solutions cannot be written down easily, although it follows from the demonstration of this paper that such solutions always exist.

The expectation value W of the Hamiltonian for a single Slater determinant of the functions (3.10) can be evaluated in a straightforward manner:

$$\frac{W}{V} = \frac{\hbar^2}{2m}\int\left(k^2\cos^2\theta + |\vec{k}+\vec{Q}|^2\sin^2\theta\right)\frac{d^3k}{8\pi^3} - \frac{1}{2}\iint\frac{4\pi e^2}{|\vec{k}'-\vec{k}|^2}\cos^2(\theta'-\theta)\frac{d^3k}{8\pi^3}\frac{d^3k'}{8\pi^3} \tag{3.16}$$

The first term is the kinetic energy and the second is the exchange energy. The only difference between the exchange integral for two pure plane-wave states and two states such as (3.10) is the additional factor $\cos^2(\theta'-\theta)$ for the latter. (Of course, the pure plane-wave case has a similar factor, which is either 1 or 0, depending on whether the pair has parallel or antiparallel spin.) If $\theta' = \theta$, the two spins will be parallel at every point, even though their direction of polarization varies from point to point. This direction lies on the surface of a cone having semivertical angle 2θ. The azimuthal angle of the polarization is Qz. The latter component is responsible for the net local spin polarization of the entire electron gas, as given by (3.5). The fractional amplitude P is readily found to be

$$P = \int \sin 2\theta\,(d^3k/8\pi^3 n)\,. \tag{3.17}$$

The energy (3.16) is a functional of the parameter $\theta(\vec{k})$, and the problem at hand is to find that $\theta(\vec{k})$ which minimizes W. The integral equation (3.14) defines the solution of this variational problem. The region in k space which is taken to be occupied can also be varied, provided its total volume is $8\pi^3 n$. Of course, each choice requires a separate variational calculation.

The paramagnetic state of an electron gas is a trivial solution of (3.14) although, as will be shown in Section 3.3, it is a saddle point of W rather than an absolute minimum. The occupied region of k space for the paramagnetic state (according to the convention of Figure 3.1) is two spheres of radius k_F, one centered at $k_z = 0$ and the other at $k_z = -2k_F$. (They touch at the point $k_z = -k_F$, $k_x = k_y = 0$.) The solution is $\theta = 0$ for states in the sphere centered at $k_z = 0$, and $\theta = \frac{1}{2}\pi$ for states in the other sphere, as illustrated by the dashed curve of Figure 3.2. A nonzero, conical spin polarization arises whenever θ takes on values between 0 and $\frac{1}{2}\pi$. The qualitative behavior of $\theta(\vec{k})$ for a SDW state is indicated by the continuous curve in Figure 3.2.

The integral equation can be solved easily if the wave-vector dependence of the exchange integral is neglected. This is not only an instructive exercise but the result will be used in the general proof of the following section. Let $4\pi e^2/|\vec{k}'-\vec{k}|^2$ be replaced by the constant γ. The $\vec{k}'$ dependence vanishes from the right-hand side of (3.14), so g is then a constant, independent of $\vec{k}$. Using (3.9) and (3.11) one finds that (3.14) can be written,

$$g = \gamma\int\frac{g}{2\left[\mu^2\left(k_z+\frac{1}{2}Q\right)^2+g^2\right]^{1/2}}\frac{d^3k}{8\pi^3}\,, \tag{3.18}$$

where

$$\mu \equiv (-\partial\epsilon_k/\partial k_z)_{k_z=-\frac{1}{2}Q}\,. \tag{3.19}$$

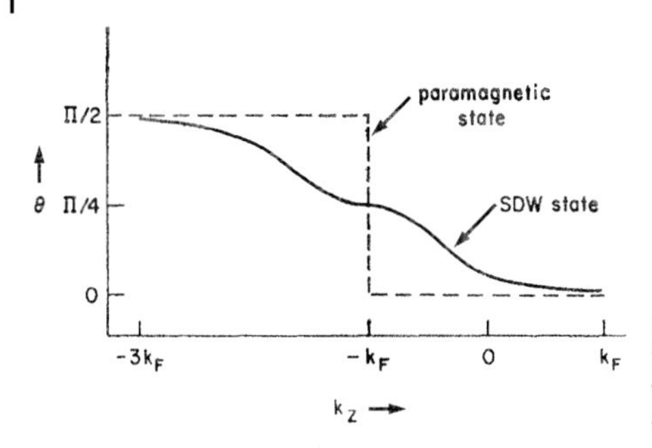

Figure 3.2 Variation of the parameter θ with k_z for the low energy branch of Figure 3.1. The SDW wave vector Q is taken equal to the diameter $2k_F$ of the Fermi sphere.

The volume of integration must now be selected. For reasons that will become clear later, it will be taken to be a circular cylinder of radius R and length L, centered in the space of Figure 3.1 at $k_z = -\frac{1}{2}Q$, and with the cylinder axis parallel to the z axis. The integration in (3.18) can now be carried out and the resulting equation solved explicitly for g.

$$g = \mu L/2 \sinh \beta \ , \tag{3.20}$$

where

$$\beta = 8\pi^2 \mu/\gamma R^2 \ . \tag{3.21}$$

The only other solution of (3.18) is the trivial one, $g = 0$.

The quantity of major interest is the energy difference between the solution (3.20) and the trivial one. This can be evaluated directly from (3.16), after replacing the exchange integral by γ, as before.

$$\Delta W/V = -\left(\mu L^2 R^2/32\pi^2\right)(\coth \beta - 1) \ . \tag{3.22}$$

This result is negative definite, so that the SDW solution always has lower energy than the trivial solution. It must be appreciated, of course, that (3.22) is the SDW energy relative to that of a pure plane wave state for which the occupied region in k space is a cylinder, and not the pair of touching spheres that would minimize the kinetic energy. The foregoing calculation can be employed to show that SDW states exist whenever the occupied region of k space (as defined by Figure 3.1) has a finite area of intersection with the plane $k_z = -\frac{1}{2}Q$. This conclusion means that exact solutions of the HF equations having SDW character always exist, but it does not imply that the HF ground state has that character. For it is possible that the energy of repopulation associated with the achievement of a finite area of intersection is larger than the subsequent energy decrease resulting from the SDW deformation. Such is the case for short-range interactions in the weak-coupling limit.[3] But it is not the case for Coulomb interactions. No matter how weak the coupling nor how high the density, the paramagnetic state of an electron gas is unstable in the HF approximation.

3.3 General Proof of the SDW Instability

The method employed will be dependent on the well-known variational theorem. It will be shown that an N-electron determinantal wave function having SDW character and lower

3) This can be shown, for example, by the argument of Kohn and Nettel (see [4]), which is not valid, it must be pointed out, for Coulomb interactions. This result may be surmised easily from the R dependence of (3.21), (3.22), and (3.24).

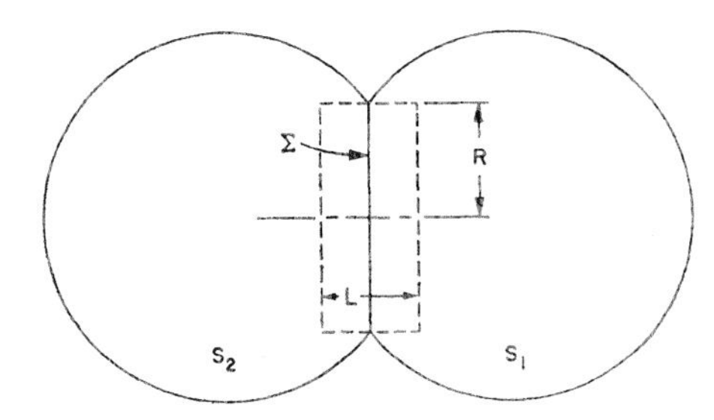

Figure 3.3 Truncated spheres, S_1 and S_2, for spin-up and spin-down electrons, respectively. S_2 is displaced Q to the left of S_1, as in Figure 3.1, so that the truncated faces Σ coincide. Electron states allowed to participate in a SDW deformation, during proof of the instability theorem, lie within the circular cylinder of length L and radius R.

energy than the paramagnetic state can always be constructed. The wave function will by no means be close to the optimum one energetically, but it will provide a brief, yet rigorous demonstration of the existence of the instability.

Let the Fermi surface be truncated, as shown in Figure 3.3, so that the circular interface Σ has radius R, assumed small compared to k_F. The wave vector $\vec{Q}$ of the SDW that will arise will be slightly smaller than $2k_F$:

$$Q = 2k_F - (R^2/k_F) , \tag{3.23}$$

to the lowest power in R. The radius of each sphere must also be increased slightly to make up for the volume, $\pi R^4/2k_F$ of the two truncated spherical segments. This repopulation requires a kinetic energy increase ΔT given, to the lowest power of R, by

$$\Delta T/V = (\hbar^2/96\pi^2 m k_F) R^6 . \tag{3.24}$$

The R^6 dependence arises because the number of displaced electrons is proportional to R^4 and their energy increase to R^2.

The foregoing repopulation also causes an increase in the exchange energy of the electron gas. The exchange potential A' of an electron, arising from its interaction with all the others, is

$$A'(k) = -\frac{e^2 k_F}{\pi}\left[1 + \frac{k_F^2 - k^2}{2k_F k}\ln\left|\frac{k_F + k}{k_F - k}\right|\right], \tag{3.25}$$

which can be derived by evaluating the integral in (3.15), employing the paramagnetic state solution for θ. The maximum exchange energy increase for a single electron involved in the repopulation is, for small R,

$$(\Delta A')_{\max} = (e^2 R^2/\pi k_F)\ln(2k_F/R) . \tag{3.26}$$

A simple integration determines ΔJ, the total exchange energy increase resulting from the repopulation:

$$\frac{\Delta J}{V} = \left[\frac{-7}{12} + \ln\frac{2k_F}{R}\right]\frac{e^2 R^6}{96\pi^3 k_F^2} , \tag{3.27}$$

which is about a factor of 6 smaller than the product of (3.26) and the number of displaced electrons per unit volume.

Consider now those states lying within a circular cylinder of radius R and length L, as shown in Figure 3.3. The cylinder lies entirely within the truncated spheres S and S'. Those states outside the cylinder will be kept undeformed, whereas those inside will be allowed to generate a SDW, as described in the preceding section. States having the same k_z will be taken to have equal deformations, θ. Therefore, when the interaction between two groups of states located at k_z and k'_z is computed, it will be useful to know the average value M of the exchange integral:

$$M \equiv \langle 4\pi e^2/|\vec{k}' - \vec{k}|^2 \rangle_{\text{av}} . \quad (3.28)$$

The average is over all $\vec{k}'$ and $\vec{k}$ lying, respectively, on two circular planes inside the cylinder, separated by $\zeta = k'_z - k_z$. Only elementary integrations enter the computation of this average. The result is

$$M = (4\pi e^2/R^2)\, f(u) , \quad (3.29)$$

where $u = \zeta^2/R^2$ and

$$f(u) = 2\ln\left[(u + (u^2 + 4u)^{1/2})/2u\right] + \frac{1}{2}\left[(u^2 + 4u)^{1/2} - u\right] - 1 . \quad (3.30)$$

This is a monotonic decreasing function with increasing u, varying as $\ln(1/u)$ for small u and as $1/u$ for large u. The average (3.29) will now be written as the sum of two contributions,

$$M = \gamma + D , \quad (3.31)$$

where

$$\gamma = (4\pi e^2/R^2)\, f(L^2/R^2) . \quad (3.32)$$

The constant γ is the smallest average interaction between two circular planes of the cylinder, corresponding to the average for the two faces at opposite ends. Consequently, the remainder D is positive for all other pairs.

The one-electron wave functions for the states inside the cylinder will be taken to be those defined by the solution (3.20) of the preceding section, with γ given by (3.32). The parameter β becomes essentially independent of R if we fix the shape of the cylinder as follows:

$$L = \lambda R , \quad (3.33)$$

where λ is a constant. Then,

$$\beta = 2\pi\mu/e^2\, f(\lambda^2) . \quad (3.34)$$

It follows from (3.22) that the SDW lowers the energy by a term proportional to R^4 whereas the repopulation energy, (3.24) + (3.27), increases it by a term essentially proportional to R^6. Since the value of R is at our disposal, the net energy change can always be made negative.

The foregoing argument indicates the basic strategy of the proof, but two difficult points remain. The energy (3.22) includes, to be sure, the entire kinetic-energy increase arising from the deformation, but only the exchange-energy decrease arising from the constant term γ of (3.31). Naturally, the total energy (3.16) includes as well the exchange-energy difference ΔJ_D arising from the remainder D in (3.31), summed over all pairs of states within the cylinder. Furthermore, the entire exchange interaction between states inside the cylinder and those outside has yet to be considered. The latter contribution is the more troublesome one.

The contribution ΔJ_D is readily shown to be negative by considering an infinite sequence of steps by which the trivial solution in the cylinder is changed to the nontrivial one. One begins, say, by changing the θ values for $k_z > -\frac{1}{2}Q$ from $\theta = 0$ to their final ones for each infinitesimal interval $(\Delta k_z)_i$ in sequence, beginning at the interface Σ. Each resulting energy increment is a change in an energy W_i which involves a sum of interactions of the interval i with intervals $i + p$ and $i - p$, equally spaced on opposite sides of the interval i:

$$W_i = -\frac{1}{2} \sum_p \left[\cos^2(\theta_i - 0) + \cos^2(\theta_i - \theta_{i-p})\right] D_p \ . \tag{3.35}$$

The interval $i + p$ (to the right of i) still has its $\theta = 0$. The important point to note is that the function in square brackets has its minimum magnitude as a function of θ_i at $\theta_i = 0$ and $\theta_i = \theta_{i-p} \leq \frac{1}{2}\pi$. Therefore, since $\theta(k_z)$ is a monotonic decreasing function of k_z, the change in θ_i from 0 to its final value – intermediate between 0 and θ_{i-p} – must increase the magnitude of the exchange interaction. D_p in (3.35), is the interaction between the interval i and $i \pm p$ and is positive, being just the product of $D(\zeta)$ by the numbers of states for the two intervals of each pair. If the interval $i + p$ lies outside of the cylinder, the first term of (3.35) must be omitted, leaving the increase of the second term completely – instead of only partially – uncancelled. A similar sequence of steps must be carried out for $k_z < -\frac{1}{2}Q$, and by a corresponding argument each step increases the magnitude of the exchange interaction. Therefore,

$$\Delta J_D < 0 \ . \tag{3.36}$$

Consequently, the remainder D of the exchange interaction (3.31), summed throughout the cylinder, necessarily enhances the instability under consideration.

Finally, the exchange interaction between states inside the cylinder and those outside must be accounted for. This should be done at the outset by including an exchange potential $\bar{A}'(\vec{k})$ in the single-particle energy $\vec{\epsilon}_k$. The parameter that enters the theory in this regard is μ, defined in (3.19):

$$\mu = (\hbar^2 Q/2m) + (-\partial \bar{A}'/\partial k_z)_{k_z = -\frac{1}{2}Q} \ . \tag{3.37}$$

$\bar{A}'$ differs from A', given by (3.25), in that the contribution of the half-cylinder inside the truncated sphere must not be included in the integration of (3.15). When this contribution is subtracted from (3.25), the well-known singular slope of the potential at the Fermi surface is removed for electron states on the interface Σ. This subtraction is accomplished most directly by employing (3.29) and (3.30) in the integrations, which then yield a value for μ appropriately averaged over the circular cross section of the cylinder:

$$\mu = \hbar^2 Q/2m + (e^2/\pi)\ln(4k_F/L) \ . \tag{3.38}$$

Actually, the magnitude of the slope at the interface Σ is a trifle smaller than (3.38). However, employment of the value (3.38) will prevent the exchange potential from being underestimated anywhere in the cylinder. Direct verification of this result is straight forward, but tedious. It can be seen intuitively rather easily, since $\frac{1}{2}L$ times the second term of (3.38) is equal to the difference in A' for states at the interface Σ and states $\frac{1}{2}L$ nearer the center of the sphere. (The contribution of the half-cylinder to this energy difference is zero, by symmetry. And the curvature of $\bar{A}'$ is such that the average energy difference $\bar{A}'(k_z) - \bar{A}'(-\frac{1}{2}Q)$ is smaller in magnitude than $|k_z + \frac{1}{2}Q|$ times the second term of (3.38), for all states inside the cylinder.)

Substitution of (3.38) into (3.34) together with the use of (3.33) determines the parameter β,

$$\beta = (\pi\hbar^2 Q/me^2 f) + (2/f)\ln(4k_F/\lambda R) \ . \tag{3.39}$$

The deformation energy (3.22) with μ and β given by (3.38) and (3.39) now includes adequately the exchange interactions under consideration. Since β depends slightly on R, (3.22) no longer varies precisely as R^4. The proof of the instability requires only that ΔW be proportional to a smaller power than R^6 for small R. When $R \to 0$, β becomes large, and

$$(\coth \beta - 1) \to 2e^{-2\beta} . \tag{3.40}$$

The deformation energy ΔW then varies essentially as follows:

$$-\Delta W \propto R^{4+(4/f)} , \tag{3.41}$$

in the small R limit. Since $f(\lambda^2)$ can be chosen arbitrarily large by selecting $\lambda = L/R$ sufficiently small, the exponent appearing in (3.41) is easily made smaller than six. For example, a cylindrical shape corresponding to $\lambda = 0.2$, for which $f = 2.6$, is adequate.

The validity of the foregoing demonstration does not depend on the magnitude of e^2 or k_F. Consequently, the SDW instability exists in the HF approximation even in the weak-coupling or high-density limit.[4] To be sure, the magnitude of the instability and the optimum value of R become very small in this limit. The HF ground state would then not correspond to a single SDW, but would probably contain a large number, $\sim (kF/R)^2$, of SDW's, having $\vec{Q}$'s appropriately oriented throughout the available solid angle. The total instability would thereby be increased by a similar numerical factor.

The magnitude of the instability derived above is, of course, only a lower limit. Allowing the exchange potential to perturb states outside the cylinder will lower the energy of the SDW state even more, and will increase the amplitude of the exchange potential, thereby causing still larger deformations, etc. A further enhancement will arise when k space is repopulated so that the Fermi surface is a surface of constant energy E_k. Numerical calculations that include such features are planned, and should provide a smaller upper bound than heretofore for the energy of the HF ground state.

3.4 Linear Spin Density Waves

The occupied region of k space for a right-handed spiral SDW, corresponding to (3.5), is illustrated in Figure 3.4(a), where the spin-up and spin-down distributions are indicated separately. The states that are most strongly perturbed by the SDW exchange potential (3.7) are those located near S and S' of Figure 3.4, for spin-up and spin-down electrons, respectively. The energy gap occurs at the plane $k_z = -\frac{1}{2}Q$ for spin-up electrons and at the plane $k_z = \frac{1}{2}Q$ for spin-down electrons. If the self-consistent amplitude g of the SDW exchange potential is comparable to the Fermi energy, all of the occupied states will be strongly perturbed, and the fractional spin polarization P will be large. In this circumstance one would not anticipate further instabilities from additional SDW formation since most of the available exchange energy decrease has been achieved. A second SDW would necessarily "interfere" with the first one.

However, if the amplitude of the first SDW is small compared to the Fermi energy, the states near T and T' of Figure 3.4 will hardly be perturbed at all, and can participate fully in a left-handed spiral SDW instability, say:

$$\vec{P} = P(\vec{x} \cos Qz - \vec{y} \sin Qz) . \tag{3.42}$$

4) The instability of an electron gas has also been studied recently by [9]. They succeeded in finding only a low-density instability.

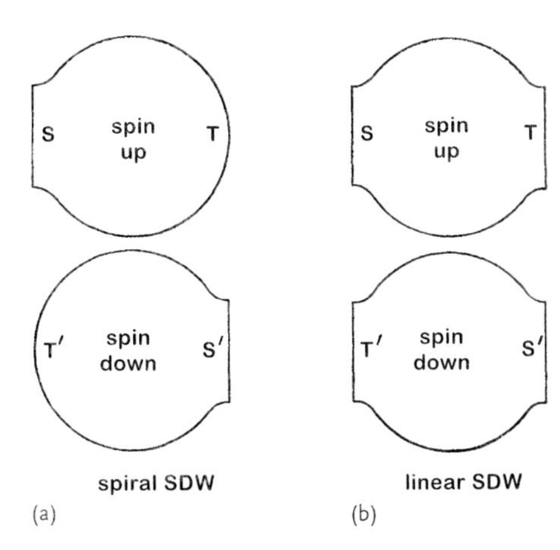

Figure 3.4 Occupied regions of k space (a) for a right-handed spiral SDW, and (b) for a linear SDW. A right-handed spiral exchange potential perturbs strongly the electron states near S and S', whereas a left-handed spiral perturbs the states near T and T'.

Consequently, the energy decrease resulting from SDW formation will be almost doubled. There will be some interference, since the contribution of states near T and T' to the right-handed spiral will be reduced, and the contribution of states near S and S' to the left-handed spiral will be similarly moderated. The occupied region of k space for two colinear spirals is illustrated in Figure 3.4(b). The total polarization is the sum of (3.5) and (3.42).

$$\vec{P}_{\text{tot}} = P_0 \vec{\varepsilon} \cos Qz \,, \tag{3.43}$$

where $P_0 = 2P$. A polarization wave of this form will be referred to as a linear SDW. It should be remembered that the theory of a SDW is insensitive to the direction of spin polarization, so that the unit vector $\vec{\varepsilon}$ of (3.43) can have any orientation with respect to $\vec{Q}$. (For the particular example employed above, it would have been the $\vec{x}$ direction.) Of course, in a real metal dipolar interaction, spin-orbit coupling, and other possible mechanisms could operate to select say a transverse or longitudinal polarization preferentially.

The orientation of $\vec{Q}$ is arbitrary for an idealized, isotropic electron gas. Very likely, in a real metal one or more symmetry directions would be selected. From (3.21) and (3.22) it can be seen that a high density of states at the Fermi surface is the most important factor influencing the magnitude of a SDW instability. Therefore, one would anticipate that the $\vec{Q}$ directions would intersect the Fermi surface where the local density of states is a maximum. For example, if a portion of the Fermi surface makes a close approach to a Brillouin zone boundary, the local density of states will be large. A SDW instability with $\vec{Q}$ parallel to, but somewhat smaller, say, than the reciprocal lattice vector associated with the zone boundary might be anticipated. The Fermi surface would then be truncated by a SDW energy gap lying inside of the zone boundary. Topologically, the Fermi surface would be equivalent to one slightly larger in enclosed volume and intersecting instead the zone boundary.

The multiple SDW instability that permits a linear SDW (which can usually be regarded as two colinear spirals) may also allow the simultaneous existence of several linear SDW's. The amplitude of each could not be too large, of course, For example, the ground state of a cubic metal could contain three linear SDW's with $\vec{Q}$'s parallel to each of the three cubic axes, or possibly four linear SDW's with $\vec{Q}$'s parallel to each of the four [111] axes, etc. It must be emphasized that the direction of the polarization vector $\vec{\varepsilon}$ of each linear SDW can be different. This is possible because each SDW arises essentially from a different group of electron states in k space. For example, each linear SDW could be longitudinally polarized. To illustrate the point, the total spin polarization density for the case of a three linear SDW

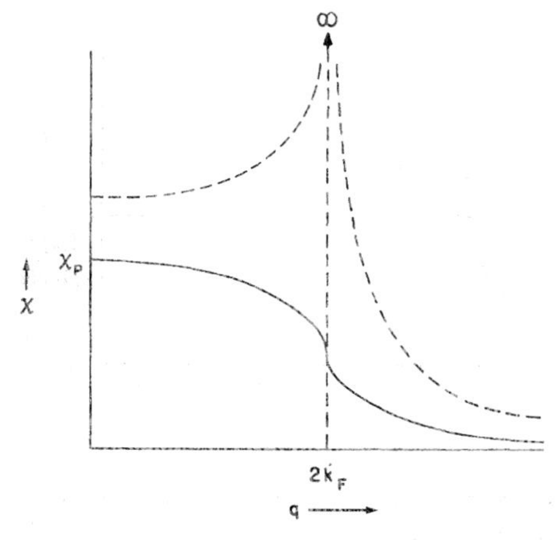

Figure 3.5 Schematic illustration of the wave-vector-dependent spin susceptibility. The solid curve is that of a noninteracting electron gas, with intercept χ_P, at $q = 0$, being the well-known static Pauli susceptibility. The dashed curve is that of an electron gas, with Coulomb interactions, in the Hartree–Fock approximation.

ground state could be

$$\vec{P} = P_0(\vec{x} \cos Qx + \vec{y} \cos Qy + \vec{z} \cos Qz) . \tag{3.44}$$

A cyclic permutation of the polarization vectors in (3.44) would correspond to three linear SDW's, each having transverse polarization. The antiferromagnetic ground state of chromium is similar to that given by (3.44), as will be discussed in Section 3.8.

3.5 Spin Susceptibility of the Paramagnetic State

The wave-vector-dependent spin susceptibility of a noninteracting electron gas is

$$\chi(q) = \chi_P \left[\frac{1}{2} + \frac{4k_F^2 - q^2}{8k_F q} \ln \left| \frac{2k_F + q}{2k_F - q} \right| \right] , \tag{3.45}$$

where $\chi_P \equiv \chi(0)$ is the well-known Pauli susceptibility. Equation (3.45) is a monotonic decreasing function with increasing q, as shown schematically by the solid curve in Figure 3.5. The spin susceptibility of an interacting electron gas has been studied by Wolff [10], who found that the HF spin susceptibility remained a monotonic decreasing function of q in an approximate treatment employing δ-function interactions. However, the SDW instability derived in Section 3.3 shows that the HF spin susceptibility of the paramagnetic state cannot be a monotonic function for the case of Coulomb interactions. The existence of the instability requires that $\chi(q)$ have a singularity near $q = 2k_F$, as shown schematically by the dashed curve of Figure 3.5. Whether or not there is a susceptibility minimum between the intercept,

$$\chi(0) = \chi_P \left[1 - (me^2/\pi\hbar^2 k_F)\right]^{-1} , \tag{3.46}$$

and the singularity near $2k_F$ very likely depends on the electron density.

The spin susceptibility (for $q \neq 0$) is ordinarily evaluated by perturbation theory[5] without allowing for the repopulation of k space that will occur as a consequence of the self-consistent

5) The reader should appreciate that the Hartree–Fock perturbation theory (with exchange) is required. This formalism has been given by [11], and has been used explicitly by [12, 13] in a collective electron treatment of a ferromagnetic Bloch wall.

potential. This omission is justified because the repopulated volume is proportional to the square of the potential and cannot contribute a first order term to the spin polarization. It will have been noted, however, that the proof of the instability given in Section 3.3 employed a repopulation of k space. The question arises, then, whether the instability would have occurred without including repopulation. The answer is yes. The repopulated volume occuring in the proof is proportional to R^4, and from (3.20) and (3.33) the exchange potential is proportional essentially to R. Therefore, if the repopulated volume were depleted, at most an additional energy term of order R^5 would be needed. But by an appropriate choice of λ, (3.33), the SDW deformation energy can be made to vary with a power of R less than five, as shown in (3.41). Consequently, the instability would still have occurred, despite a neglect of repopulation, but it would have been less severe.

It would seem to be purely academic to discuss the spin susceptibility of an unstable state. The motivation is, of course, the fact that the HF instability discussed here is probably not indicative of the character of the true ground state for most metals. That would require the true susceptibility peak, as in Figure 3.5, to have at most a finite height for typical metals. The important point remains, though, that a susceptibility maximum near $q = 2k_F$ is a likely consequence of Coulomb interactions, and has not been pointed out previously. The occurrence of such a maximum has interesting implications with regard to the magnetic properties of metals and alloys, as is discussed briefly below.

The present paper is concerned primarily with the possible existence of the SDW ground states. These may indeed be rare; but the existence of SDW excited states is more general. One may select an arbitrary wave vector $\vec{q}$, and then repopulate k space so that the Fermi surface has a finite area in common with the planes, $\vec{k} = \pm\frac{1}{2}\vec{q}$. It follows that the repopulated plane wave state is unstable with respect to SDW deformations, for any repulsive interaction, however weak. (The proof is essentially equivalent to the one-dimensional case [3].) Therefore, exact SDW solutions of the HF equations always exist. And they encompass a continuous range of SDW polarization amplitudes, $0 \leq P < 1$, for all wave vectors $q > 0$. The HF equations of a SDW excited state should of course include appropriately the effective potential of any interactions that stabilize the excitation, when such are present.

SDW excited states have been employed previously[6] by the writer to explain the apparent antiferromagnetic properties of paramagnetic alloys. The HF theory of a SDW was formulated there within the framework of perturbation theory, which does not adequately approximate the exact HF formulation given here in Section 3.2. However, the associated treatment of the interaction of localized spins with SDW excited states, and the long-range antiferromagnetic order that can occur at low temperature remains essentially unaffected. A possible misgiving with regard to such an antiferromagnetic mechanism might arise from the incorrect view that $\chi(q)$ is a monotonic decreasing function of q. The excitation energy of a SDW state, with a given fractional polarization P, is inversely proportional to $\chi(q)$. The lowest energy excitation (for a fixed P) would then correspond to $q = 0$, and would favor ferromagnetism rather than antiferromagnetism. This argument is incomplete, as pointed out below, since energy contributions from SDW excitations having wave vectors differing from $\vec{q}$ by a reciprocal lattice vector must also be included. In any case, the observation that $\chi(q)$ need not have its maximum at $q = 0$ obviates even the plausibility of such a misgiving.

6) [14]. The reader should note that an alternative explanation that ignores the possibility of long-range order, proposed by [15], when evaluated quantitatively, does not in fact account for the experimental heat capacity of Cu–Mn alloys at the observed Mn concentrations.

The effective exchange interaction between a localized spin $\vec{S}_j$ and, say, a linear SDW such as (3.43) would be

$$H_j = -G(q)\vec{S}_j \cdot \vec{\varepsilon} P_0 \cos(\vec{q} \cdot \vec{R}_j) , \qquad (3.47)$$

where $\vec{R}_j$ is the position of the localized spin $\vec{S}_j$, and $G(q)$ is an appropriate exchange interaction constant between the spin and a SDW of wave vector $\vec{q}$. $G(q)$ will depend on q, since the exchange interaction should have a "form factor", associated of course with the finite size of the localized orbital. It is the interaction energy (3.47), depending linearly on P_0, which can overcome the SDW excitation energy:

$$W(q) = n^2\mu^2 P_0^2/4\chi(q) , \qquad (3.48)$$

where μ is the Bohr magneton, $W(q)$ is the excitation energy per unit volume of a linear SDW with maximum fractional polarization P_0, $\chi(q)$ is the spin susceptibility per unit volume. At 0 °K all spins $\vec{S}_j$ will be oriented by the interaction (3.47), causing a net energy decrease per cc arising from the sum of (3.47) and (3.48):

$$\Delta W = -(2SN_p/\pi n\mu)^2 G^2(q)\chi(q) , \qquad (3.49)$$

where N_p is the concentration of localized spins. Ignore, now, for the sake of simplicity all other SDW excitations. The wave vector of the long-range order would then be expected to be that value of q which maximizes $G^2\chi$. If this occurs for $q \neq 0$, antiferromagnetic order is possible. If the maximum occurs at $q = 0$, a ferromagnetic or, possibly, a superparamagnetic [16] state may result. The striking difference between the properties of dilute Cu–Mn alloys [17, 18] and dilute Cu–Co alloys [19] may possibly be explained on this basis – the result of significantly different form factors $G(q)$ for the two species of paramagnetic solute.

A considerable amount of additional effort is needed to explore details of SDW mechanisms in magnetic ordering. The foregoing remarks are intended only to illustrate the elementary physical basis of the method. The point of view is that the natural coordinates for elaborating the theory are the spin directions $\vec{S}_j$ and the amplitudes P_q of the SDW excitations for all $\vec{q}$. To order P^2 the latter coordinates can be eliminated, their physical effect replaced by equivalent interactions between all pairs of spins.[7] But the temptation to truncate such indirect interactions for pragmatic reasons – considering say only pairs that are relatively near – is always present and can lead to spurious conclusions. Indeed, for an ordered ground state described by a wave vector $\vec{Q}$ (including the ferromagnetic case, $Q = 0$), the amplitudes $P_{\vec{q}}$ are identically zero for all $\vec{q}$ excepting $\vec{Q}$ itself and those which differ from $\vec{Q}$ by a reciprocal lattice vector (times 2π). This result can be recovered in the indirect interaction model only if the interactions are summed exactly for all pairs, regardless of range. In the SDW description just one – or at most a few – SDW's play a role in an ordered ground state, or in ordered thermally excited states treated by a molecular field or random phase approximation.

Errors that are more difficult to avoid when the indirect interaction model is employed occur in cases when a single SDW is highly excited – e.g., the helical spin structures occurring in the rare earth metals. Terms of higher order than P^2 and the repopulation of k space, for example, become important and will play perhaps a crucial role in such phenomena as order–order transitions, the temperature variation of the helical pitch, resistivity anomalies, etc. Such nonlinear phenomena arise naturally in the theory of an individual SDW. Consequently, this approach is the appropriate and generally valid one.

7) These calculations cannot be considered quantitative until they are carried out with a selfconsistent field method including exchange interactions, and until appropriate form factors $G(q)$ are also incorporated [20, 21].

3.6 Detection of SDW's by Neutron Diffraction

Direct observation of the Fourier components of magnetization density by neutron diffraction is the standard technique for determining antiferromagnetic order. Detection of SDW ground states in metals, when present, may nevertheless be quite difficult. Very likely the fractional polarization of a SDW is of the order of one percent, so that scattering amplitudes would be a hundred times smaller than typical. A further difficulty is that the wave vector $\vec{Q}$ of a SDW is determined by the diameter of the Fermi surface, not by the lattice parameter. Consequently, the neutron-scattering vector $\vec{K}$ will generally be unpredictable and will correspond to an unknown point in reciprocal space, both in magnitude and direction.

Consider first the case of a spiral SDW in an ideal electron gas, with the circle of polarization perpendicular to $\vec{Q}$, as in (3.5). The allowed scattering vectors are,

$$\vec{K} = \pm\vec{Q} \,. \tag{3.50}$$

It has been pointed out that if the incident neutrons are unpolarized, each of the two scattered beams will be completely polarized [22], provided the diffraction sample contains only right-handed (or left-handed) spirals. Since there is no magnetostatic energy tending to cause domain formation in an antiferromagnet, and since SDW's are a conduction electron phenomenon with consequent long range coherence, it is likely that subdivision into domains will not occur in good specimens. This possibility requires that particular caution must be used in searching for SDW reflections. For suppose there is a single SDW in, for example, an otherwise cubic crystal. All equivalent directions in reciprocal space must be scanned since $\vec{Q}$ would be parallel to just one of them.

Consider next a linear SDW in an ideal electron gas, such as (3.43). The allowed reflections will be the two given by (3.50), but only if the polarization vector $\vec{\varepsilon}$ has a component perpendicular to $\vec{K}$. For if $\vec{\varepsilon}$ and $\vec{K}$ are parallel, the magnetic induction B will be identically zero, and there will be no magnetic interaction with neutrons. A longitudinally polarized linear SDW in an ideal electron gas is invisible to neutrons.

In real metals the wave functions of conduction electrons have, of course, a periodic modulation within each unit cell of the crystal, e.g.,

$$\psi_{\vec{k}} = u_{\vec{k}}(\vec{r})e^{i\vec{k}\cdot\vec{r}_a} \,. \tag{3.51}$$

If we neglect the $\vec{k}$ dependence of the Bloch periodic part, $u_{\vec{k}}(\vec{r})$, the one-electron functions for a SDW deformation will be,

$$\varphi_{\vec{k}} = u(\vec{r})\left\{\alpha\cos\theta\exp(i\vec{k}\cdot\vec{r}) + \beta\sin\theta\exp\left[i(\vec{k}+\vec{Q})\cdot\vec{r}\right]\right\} \tag{3.52}$$

which is a slight generalization of the wave function (3.10). The conjecture that has been made here is that the spin polarization within a unit cell follows the phase of the SDW, as in the empty lattice case, but that the amplitude is modulated by the additional multiplicative factor $|u(\vec{r})|^2$ For a linear SDW the magnetization density $\vec{M}(\vec{r})$ would be,

$$\vec{M}(\vec{r}) = n\mu|u(\vec{r})|^2\vec{\varepsilon}P_0\cos(\vec{Q}\cdot\vec{r}) \,, \tag{3.53}$$

with $u(\vec{r})$ normalized in a unit volume. The conduction electron density of a unit cell can be Fourier analyzed in the usual way:

$$|u(\vec{r})|^2 = \sum_G f_G\exp(2\pi i\vec{G}\cdot\vec{r}) \tag{3.54}$$

where $\{\vec{G}\}$ are the reciprocal lattice vectors and $\{f_G\}$ are the Fourier coefficients, or form factors. It follows from (3.53) and (3.54) that the Fourier components of the magnetization density, and therefore the allowed scattering vectors, are

$$\vec{K} = 2\pi\vec{G} \pm \vec{Q} \,. \tag{3.55}$$

In other words, there will be two magnetic reflections associated with each reciprocal lattice vector $\vec{G}$, and we shall refer to them as the satellites of $\vec{G}$. The satellites of a particular $\vec{G}$ might lie closer to another $\vec{G}$ since $\vec{Q}$ can be large, but it should be easy to make the proper identification. Such assignments would be clear from relative intensity measurements, since from (3.53) and (3.54), the relative intensity is determined by f_G^2 the form factor of $\vec{G}$ rather than the form factor of the scattering vector $\vec{K}$.

The form factor of valence electrons ordinarily decreases rapidly in magnitude with increasing wave vector. Consequently, the only intense reflections from a conduction electron SDW will be the satellites of the origin ($G = 0$). The other satellites should be present in principle, but one would expect them to be extremely weak. A longitudinally polarized linear SDW in a conduction band should not be completely invisible. The satellites of the origin would have zero intensity for the same reason given previously. But the weak satellites of the other reciprocal points, for which $\vec{\varepsilon}$ is not parallel to $\vec{K}$, would theoretically be present.

The foregoing model was based on a "flexible spin" hypothesis, in which it was assumed that the spin polarization density conforms smoothly to the phase of the SDW within each unit cell. This model is probably an adequate one for SDW's arising entirely from conduction electrons. We must now consider the other extreme; a rigid spin hypothesis – in which it is assumed that the cellular part of the wave function is so completely centralized by the ion core potential that the direction and amplitude of the spin polarization density within a unit cell is determined by the phase of the SDW at the center of the cell. Undoubtedly the "truth" will always be somewhere between the two extremes. But the latter extreme may be an appropriate model for a *d*-band SDW, or for a SDW in which oriented localized moments (as in the rare-earth metals) are the major contribution to the magnetization density.

Let $u(\vec{s})$ be redefined so that it is nonzero only in the unit cell centered at the origin of coordinate space. In the rigid-spin model, the magnetization density of a linear SDW will be given by

$$\vec{M}(\vec{r}) = \mu\vec{\varepsilon}\,P_0 \sum_L |u(\vec{r} - \vec{L}|^2 \cos(\vec{Q}\cdot\vec{L}) \,, \tag{3.56}$$

where $\{\vec{L}\}$ are the lattice sites. If f_K is the Fourier transform of $|u(\vec{s})|^2$, then the Fourier coefficient $\vec{M}_K$ of $\vec{M}(\vec{r})$ is,

$$\vec{M}_K = \mu\vec{\varepsilon}\,P_0\,f_K \sum_L \exp(2\pi i\vec{K}\cdot\vec{L})\cos(\vec{Q}\cdot\vec{L}) \,. \tag{3.57}$$

The sum over $\vec{L}$ is nonzero only if $\vec{K} \pm \vec{Q}$ is a reciprocal lattice vector. Consequently, the Fourier components of (3.56) are those given previously by (3.55), but the amplitudes are proportional to f_K, the form factor of the scattering vector, rather than the form factor of the reciprocal lattice vector of which $\vec{K}$ is a satellite. For the rigid-spin model experimental determination of the wave vector $\vec{Q}$ is ambiguous to within any reciprocal lattice vector. However, a slight admixture of the flexible-spin model, which is required if $\vec{M}(\vec{r})$ is to be a continuous function, together with reasonable assumptions about the shape of f_K should allow $\vec{Q}$ to be determined uniquely. This will be illustrated in Section 3.8 for the case of chromium, which approximates the rigid-spin model.

It is evident that a systematic search for SDW neutron reflections would be extremely tedious, even with neutron beams of the highest available flux. Such experiments, however, would be of significant interest, at least for metals where weak antiferromagnetism is suspected. A search should not be confined to scans of lines in reciprocal space that pass through the origin. This would allow longitudinally polarized linear SDW's to go undetected. It is necessary to scan also lines in reciprocal space that pass through (nonzero) reciprocal lattice points. Furthermore, the experiments must, in general, be carried out at low temperatures since SDW amplitudes will approach zero at a critical temperature T_c (and remain zero at higher temperatures) as discussed in the following section.

Inelastic neutron (or x-ray diffuse) scattering may reveal indirectly the presence of SDW states. The phonon spectrum will be altered as a result of the energy gaps in the electron energy level spectrum caused by the periodic SDW exchange potential. Changes in the phonon frequencies are expected, of course, because the adiabatic readjustment of the electron gas during a lattice vibration plays an important role in determining its frequency. Kohn [23] has pointed out how such effects may be expected to leave an image of the Fermi surface in the phonon spectrum, solely from the sharpness of the Fermi surface in the paramagnetic state. Energy gaps across planes tangent to the Fermi surface would be expected to alter appreciably the frequencies of the modes with wave vector approximately equal (in magnitude and direction) to the wave vectors of the SDW's present, give or take a reciprocal lattice vector. The effect may show up as an additional kink or cusp (or both) in the phonon frequency spectrum. This phenomenon should differ from the one proposed by Kohn in that it is restricted to specific wave vector directions and should disappear abruptly at the critical temperature. A search for this effect by, say, temperature diffuse x-ray scattering in single crystal chromium is of considerable interest.

3.7 Temperature Dependence of SDW Parameters

The formulation of the HF theory of a static SDW given in Section 3.2 is for an electron gas at $0\,^\circ$K, since it was tacitly assumed that the states in k space were either occupied or empty depending on their location with respect to the Fermi surface. At finite temperature occupation of higher energy one-electron levels by thermal excitation will occur, so the integral equation (3.14) must be appropriately generalized. The details of such a generalization are obvious, and it is our intention here to discuss only the important features that will result.

Occupation of one-electron levels of the upper energy branch of Figure 3.1, together with a corresponding depletion of the lower branch, will cause the amplitude of a SDW to be smaller than its $0\,^\circ$K value. This arises from the fact that the spin polarization of a state of the upper branch is opposite (at every position) to the corresponding state of the lower branch. Consequently, the polarization contribution of upper branch states will tend to cancel that arising from the remaining lower branch electrons. Moreover, the exchange potential will be similarly reduced, so that the fractional polarization component of each state will be smaller, thereby causing a further reduction of the SDW amplitude, etc. Such effects will be greater the higher the temperature. The SDW amplitude $P(T)$ will therefore be a mono tonic decreasing function with increasing T. It is reasonable to conjecture that P will approach zero continuously, giving rise to a second order transition at a critical temperature T_c. Above T_c a nontrivial solution of the generalized integral equation will not exist.

It is an interesting coincidence that the structure of the integral equation (3.14) is similar to the integral equation that occurs in the Bardeen, Cooper, and Schrieffer (BCS) theory of superconductivity [24]. The similarity is perceived readily by comparing the integral equa-

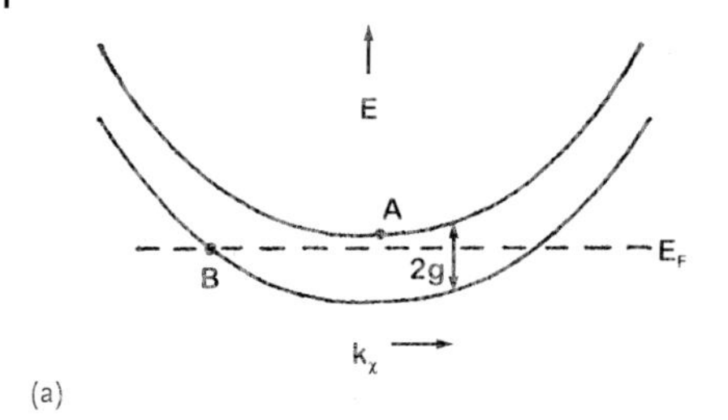

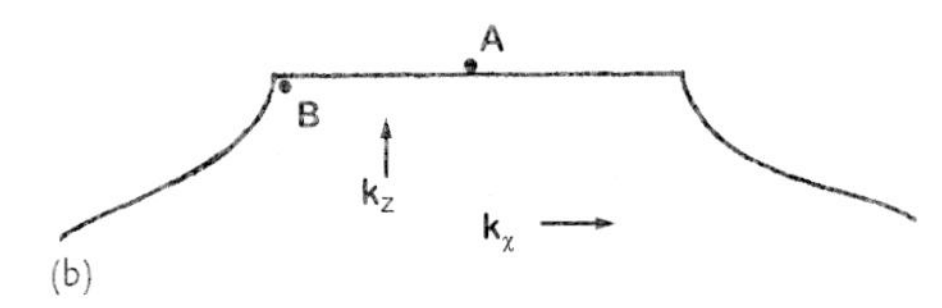

Figure 3.6 Single-particle energy level spectrum, (a) on the plane in k space coincident with the SDW energy gap, and (b) the occupied region in k space near the plane of the gap. Observe the neck in the Fermi surface caused by the SDW exchange potential.

tions that result in both theories when the relevant interactions are replaced by constants; compare (3.18) with Eq. (2.38) of reference [24]. They are essentially identical, differing only in the shape of the integration volume. This similarity will also prevail with the generalized equations for finite T. On the basis of this analogy, one might guess that the SDW amplitude $P(T)$, as well as the SDW energy gap $2g$, will vary with T/T_c in a manner quite similar to the superconducting energy gap. Indeed, Swihart has shown [25, 26] that the temperature variation of such parameters in the theory of superconductivity is quite insensitive to the form and strength of the interaction that appears in the kernel of the integral equation. Consequently, the similarity may be semiquantitative as well as qualitative. This conclusion is of course tentative since the SDW theory is complicated by several features not present in the superconductivity case. Nevertheless, the comparison for the case of Cr is quite satisfactory, as is shown in Section 3.8.

One may anticipate on the basis of the foregoing analogy that the SDW energy gap $2g$ at $0\,°$K is 3 or 4 times kT_c as it is in superconductors. It is probable that the ratio should be somewhat larger for SDW's since $2g$ is not the minimum energy gap, as shown in Figure 3.6(a), but the energy difference between states on opposite sides of the gap whose wave vectors have identical projections in the plane of the gap. The true thermal energy gap is the energy difference between points A and B of Figure 3.6, and is necessarily smaller than $2g$. This minimum thermal gap may indeed be quite small because a SDW instability can be greater the smaller the magnitude of $\vec{Q}$, provided that the states above the gap (near A) remain unoccupied. This feature can be seen from the deformation energy (3.22), which depends exponentially on μ the gradient of ϵ_k at the Fermi surface. Therefore, in an electron gas for which μ is an increasing function of ϵ_k the SDW energy gaps will tend to "clamp down" tightly on the Fermi surface. On the other hand, if the Fermi surface lies beyond an inflection point in ϵ_k, this tendency to clamp down would not be so great.

The magnitude Q of the SDW wave vector will also be a function of temperature, as can be seen easily. When the energy gap decreases with increasing T, the periodic potential will be less able to sustain the repopulation of k space, which takes place predominantly near the gap at points such as B in Figure 3.6. As a consequence the energy gaps will move apart with increasing T in order to accommodate the constant total number of electrons while keeping the fraction above the gap at a minimum. Also, the clamp-down effect discussed above will diminish with decreasing SDW amplitude, causing a further change in the re-population and an additional increase of Q with T. Consequently, a detailed calculation should predict that Q

will increase monotonically with increasing T. The fractional change of Q between $0\,^\circ$K and T_c will of course be larger the larger the fractional polarization P of the SDW at $0\,^\circ$K.

At T_c the magnitude of $\vec{Q}$ should equal the diameter of the Fermi surface (in the direction of $\vec{Q}$) that obtains for the paramagnetic state. This conclusion depends however on an assumption that only one unfilled energy band is participating appreciably in the SDW deformation. For if more than one band plays a significant role, it is difficult to anticipate the value of $\vec{Q}$ for which $\chi(\vec{Q}, T)$ would first become singular as T is lowered through T_c. The reason that the Fermi surface diameter is the critical value for a single band is associated with the fact that the maximum in the Fourier transform, V_q, of the Coulomb interaction occurs at $q = 0$. (And one expects this feature to remain even after correlation screening is appropriately accounted for.) Therefore, instabilities are maximized by keeping the highly deformed states as close together in k space as possible. This is almost self-evident from the structure of the integral equation (3.14). If Q is made smaller, states above the gap must be occupied, and their effect would cancel out the enhanced part of the exchange interaction between pairs of states just below the gap. The burden for enhancing exchange interactions would then fall upon pairs of states that have a larger average separation in k space, with a consequently smaller average value of V_q. In particular, for $Q = 0$ (a ferromagnetic deformation), V_q must be averaged equally over all pairs of points around the entire Fermi surface, and this would lead to a relatively small average value. The foregoing considerations are, of course, qualitative in nature. A delicate instability might require a quantitative account of all factors before $Q(T_c)$ could be predicted reliably.

3.8 Antiferromagnetism of Chromium

The antiferromagnetic state of chromium metal has recently received extensive study by single-crystal neutron diffraction techniques [27–30]. The wave vectors of the magnetization waves are parallel to the [100] directions of the bcc lattice. The wavelength is incommensurate with the lattice constant and varies continuously with temperature, as expected for a SDW state. The magnitude of the magnetization wave(s) is about a half Bohr magneton per atom. Several interpretations of this structure have been discussed [8] [27, 31], assuming that Cr atoms have a localized moment (with a spin degree of freedom) in the metal. It is now apparent, however, that localized moments do not occur in Cr.

The excess entropy near the antiferromagnetic transition temperature ($T_c = 311\,^\circ$K) would be about $\frac{1}{2} R \ln 2 \approx 0.7\,\text{cal/deg}$, if the magnetization arose from localized moments. Such a large anomaly would be easily observed. Its apparent absence [33] had previously indicated that a collective electron mechanism is operating [34–36]. The heat capacity anomaly associated with the transition has now been observed [37] and the integrated (molar) entropy is

$$\Delta S = 0.0044\,\text{cal/deg}\,. \tag{3.58}$$

This very small value can be satisfactorily explained on the basis of a SDW mechanism.

The absence of localized moments has also been confirmed by the nonoccurrence of paramagnetic neutron scattering above the critical temperature [38]. The only sources of entropy, then, are the phonons and the electron gas. Changes in both contributions occur at the transition, but we shall consider only the latter one since the phenomenon is primarily electronic. The Sommerfeld constant γ has been measured [39] for Cr. Consequently, the electronic en-

8) The experiment proposed in this paper was based on the presumed existence of localized moments, which are now known not to occur in Cr. A field cooling effect is not anticipated on the basis of the current interpretation [32].

tropy at the transition temperature is,

$$S = \int_0^{311} \gamma T(dT/T) \approx 0.10\,\text{cal/deg}\,. \tag{3.59}$$

Therefore, the entropy (3.58) associated with the antiferromagnetic transition is only 4% of the total Sommerfeld entropy at T_c. On the basis of the SDW model the entropy loss that occurs as a result of lowering the temperature below T_c arises from the truncation of the Fermi surface by the energy gaps of the SDW exchange potential. (Electron states adjacent to the gap will, for the most part, be completely occupied or completely empty and therefore will not contribute significantly to the entropy.) From the experimental data one would conclude that about 4% of the Fermi surface of Cr is truncated by energy gaps of magnetic origin in the antiferromagnetic state.

Now, the fraction of the Fermi surface that is truncated by energy gaps can be estimated independently in several ways. That portion of the Fermi surface truncated by energy gaps will not contribute to the electrical conductivity, since the normal component of electron velocity at an energy gap is zero. Therefore, SDW energy gaps should cause an abrupt resistivity increase as a result of the decrease in the effective number of current carriers. This conclusion assumes that the phase of the SDW(s) is locked to the lattice – a phenomenon that could result from paramagnetic impurities, inhomogeneous strains, etc. The electrical resistivity of Cr does indeed undergo an abrupt increase [40, 41] below T_c and the magnitude of the increase is about 5%. This estimate of the fractional truncation is in satisfactory agreement with the heat-capacity result. The comparison should be only semiquantitative, since truncation of the Fermi surface will tend to reduce the density of final states available to electron scattering processes. Consequently, the resistivity increase may be partly cancelled by decreased scattering rates. But changes in phonon frequencies, which do occur [40, 41], will enhance the thermal scattering. An accurate comparison would have to include all such effects, including the phonon contribution to the heat-capacity anomaly as well as another contribution from spin-wave excitations of the "giant" SDW.

A more interesting estimate of the fractional truncation derives from the coincidental mathematical similarity, discussed in Section 3.7, between the BCS theory and SDW theory. The temperature dependence of the fractional polarization P should compare with the energy gap parameter of the BCS theory. That it does so very well is shown in Figure 3.7, for which the relative values of P were determined from the neutron diffraction intensity measurements of Shirane and Takei [30]. Therefore, it is reasonable to suppose that the ratio of energy gap to kT_c is about 3.5 for Cr. This would imply an energy gap of about 0.1 eV.

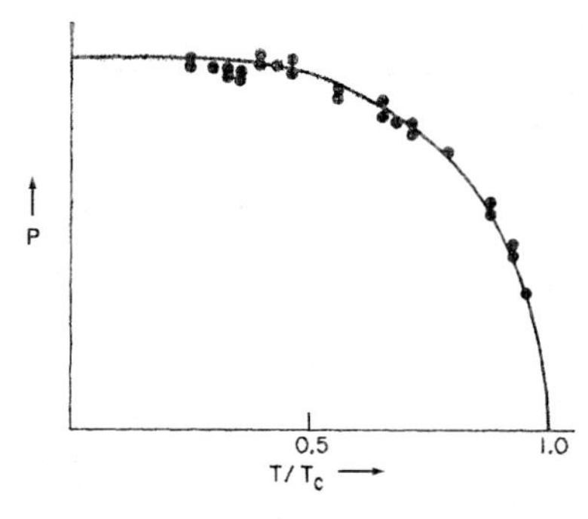

Figure 3.7 Temperature dependence of the SDW amplitude P for a chromium single crystal. The experimental points are taken from Shirane and Takei (see reference [29]). The small break near $T/T_c = 0.4$ occurs at the spin-flip transition. The solid curve is the theoretical temperature dependence of the superconducting energy gap, after the BCS theory.

The largest possible area of each truncated face can now be estimated by setting the energy difference between points A and B of Figure 3.6 to zero. If R_0 is the radius of the truncated face,

$$\hbar^2 R_0^2/2m^* = 0.1\,\text{eV}\,, \tag{3.60}$$

where m^* is the effective mass. The fractional area of the Fermi surface truncated by energy gaps is

$$t = p\,R_0^2/4k_F^2\,, \tag{3.61}$$

where p is the number of truncated faces. The diameter $2k_F$ of the Fermi surface is determined directly from the neutron diffraction scattering vector $\vec{Q}$, according to the argument given in Section 3.8. The observations indicate that for the [100] directions,

$$k_F \approx 1.1 \times 10^8\,\text{cm}^{-1}\,. \tag{3.62}$$

The effective mass can be estimated from the measured electronic heat capacity, but one must first know the total area of the Fermi surface. We have already assumed it to be spherical in writing (3.60) and (3.61). However, calculations by Lomer [43] indicate that the shape is somewhat fluted, and that the energy bands are essentially of d character. His estimate of the Fermi surface radius in the [100] directions is in agreement with (3.62). But he also finds that the relevant d band is doubly degenerate in that direction. There is an electron band and a hole band of approximately equal size, and their Fermi surfaces touch where they intersect the [100] axes. Taking both surfaces into account, one obtains from the Sommerfeld constant γ,

$$m^* \approx 1.5m\,. \tag{3.63}$$

The maximum radius of each truncated face, from (3.60), is therefore

$$R_0 \approx 0.2 \times 10^8\,\text{cm}^{-1}\,. \tag{3.64}$$

The number p of truncated faces is presumably six, since the linear SDW instabilities in each of the [100] directions are equal by symmetry. The polarization density that would result is given (crudely) by (3.44). From (3.61) the maximum fraction truncated by energy gaps is

$$t \approx 5\%\,. \tag{3.65}$$

This estimate is in excellent agreement with those derived from the entropy argument and the resistivity argument.

The SDW energy gaps will cause necks, Figure 3.6(b), in the Fermi surface, and these should give rise to de Haas-van Alphen oscillations, magnetoacoustic oscillations, etc. Two de Haas-van Alphen periods have been observed [44] with orbits in (100) planes, confirming both the double degeneracy in the [100] directions pointed out by Lomer and the estimated size (3.64) of the necks.

Chromium undergoes an order–order transition at low temperature, which has been reported at about 155 °K by some workers [28, 29] and near 110 °K by others [30, 45, 46]. Below this critical temperature the neutron reflections located on the cubic axes disappear, proving [45] that the magnetization waves are longitudinally polarized in the low temperature phase. Intensity studies [30, 45] of the satellites of the (110) reciprocal lattice points indicate that the order–order transition is merely a change from longitudinal to transverse polarization of the linear SDW's. The absence of third harmonics [30] of the magnetization waves indicates that the magnetization density follows a cosine modulation, in conformity with the SDW model.

The polarization density given by (3.44), which describes qualitatively the magnetic ground state of Cr, would be reasonably accurate if the flexible-spin model were a good approximation here. However, the neutron-diffraction intensities that obtain in Cr can be interpreted only if the rigid-spin model is the appropriate approximation, as discussed below. This conclusion indicates that the electron energy bands which contribute to the polarization are d bands, as would be anticipated from Lomer's work on the location of the Fermi surface.

The neutron reflections that lie on the [100] axis in reciprocal space have approximately the coordinates,

$$Q_1 \approx (0.96, 0, 0) , \quad Q_2 \approx (1.04, 0, 0) . \tag{3.66}$$

One of them is the satellite of the origin $(0, 0, 0)$ and the other is the satellite of the $(2, 0, 0)$ reciprocal lattice point. If the flexible-spin model applies, one of them would have an intensity corresponding to the form factor for zero scattering angle, as discussed in Section 3.6. The other would have an intensity corresponding to the form factor (squared) of a $(2, 0, 0)$ scattering angle. Consequently, the intensities would differ by an order of magnitude or more. Actually, the intensity of Q_1 is only about 30% larger than the intensity of Q_2 [28–30]. Therefore, one is forced to conclude that the rigid-spin model is the better approximation in Cr. And one should compare the observed intensities with that expected, say, for d-electron form factors in the rigid-spin approximation. The form factors for Mn^{++} d electrons have been measured [47] and, if applicable here, could account for only a 10% intensity difference between Q_1 and Q_2. In view of the fact that a slight admixture of the flexible-spin model is necessary in order that the magnetization density be a continuous function of position, which would add appreciable scattering intensity only to the satellites of the origin, the larger intensity difference observed can be easily interpreted. One would surmise, then, that the SDW wave vector $\vec{Q}$ is to be identified with Q_1 and not Q_2. (This conclusion is not absolutely certain, since a less centralized d-electron density distribution in the atomic cell, calling for a more rapidly falling form factor vs scattering angle, could reverse the assignment, assuming that loss in centralization is possible without compromising the rigid-spin hypothesis.)

The Cr^{53} nuclear resonance in Cr metal has recently been observed [48] above the critical temperature T_c. The resonance line broadens as the temperature is reduced near T_c and disappears gradually below T_c. This behavior is explained by the fact that the SDW wave length is incommensurate with the lattice. Consequently, the hyperfine fields at the nuclei arising from the magnetization density take on a wide, continuous spectrum of values. This type of behavior need not be characteristic of all SDW antiferromagnets, however, A small amplitude SDW state may have very low energy spin wave excitations, which would allow the hyperfine field direction to change rapidly at finite temperatures. Observation of the nuclear resonance in this case would be impaired only if the nuclear relaxation rate were comparable to or greater than the spin wave excitation frequencies. Otherwise the effect of the hyperfine fields would tend to a null time-averaged value.

Magnetomechanical damping at low frequencies ($\sim$1 cps) has been observed [49] in polycrystalline Cr wire by the torsion technique. The damping increases markedly with decreasing temperature below T_c, but the measurements were not extended below 200 °K. A possible mechanism for this damping, suggested by de Morton [49], is antiferromagnetic domain wall motion. An alternative mechanism arises from the transverse polarization of the linear SDW's. Since each of the three SDW's can have two transverse polarization directions, there are eight energetically equivalent states which can become inequivalent as a result of elastic strain, and so give rise to anelastic transitions. (The wide variation of the spin-flip transition temperature from sample to sample, cited previously, indicates that there is a significant interaction between the SDW polarization modes and crystal imperfections, e.g., elastic strains.) This suggested mechanism can easily be subjected to experimental test. Below the spin-flip transition temperature there is just one magnetic state, since the SDW's are then longitudinally polarized. Therefore, it is of considerable interest to determine whether or not the large magnetomechanical damping found by de Morton disappears below the spin-flip transition temperature.

SDW theory provides a consistent and satisfactory interpretation of the magnetic phenomena in Cr. On the other hand it is not at all clear, from an *a priori* basis, why Cr should be unique in possessing large amplitude SDW's. The rapid variation of the transition temperature with small concentrations of alloying elements [40–42] suggests that it may be associated with a fortuitously favorable band configuration.

3.9 Accidental Ferrimagnetism

Metals with a SDW ground state are multiply periodic structures. There is no pure translation operation which leaves the system invariant, provided the SDW wave vectors are incommensurate with the reciprocal lattice vectors (times 2π). In the absence of magnetocrystalline interactions of sufficient strength, a commensurate relationship between the two types of wave vector would be accidental. Such a coincidence, however, could cause a bulk magnetization of the material to occur, analogous to that in ferrimagnetic materials.

A SDW is commensurate with the lattice if its wave vector $\vec{Q}$ satisfies

$$\vec{Q} = 2\pi \vec{G}_0 / p \,, \tag{3.67}$$

for some reciprocal lattice vector $\vec{G}$ and integer p. We shall take $\vec{G}_0$ to be the smallest vector satisfying this relation. Suppose for the sake of simplicity that $p = 1$, and that the rigid-spin model applies. Then the moments in each primitive unit cell will all have the same phase, as defined by the SDW. Consequently, a net moment will arise. (Of course, if there are several identical atoms per primitive cell, the moment would be zero if $\vec{G}_0$ is a reciprocal lattice vector having zero x-ray structure factor. Also the moment could be zero for a monatomic Bravais lattice if each lattice site were at the node of a linear SDW. But such a situation seems unlikely energetically.) It is relatively easy to see that similar situations can prevail for $p > 1$, especially if p is prime. For example, in thulium [50], which has localized moments oriented by a linear SDW, $p = 7$. Alternately, four hexagonal layers are polarized up and three down. It is not our purpose, here, to envision all conceivable structures and special requirements, but rather to cite a few examples and to discuss qualitatively some of the relevant features.

One would not, in general, anticipate a SDW instability in an electron gas with $Q = 2\pi \vec{G}$, since the periodic potential of the lattice will cause energy gaps in the conduction electron spectrum across the same planes in k space as the exchange potential of the commensurate SDW. It should be recalled that the SDW instability arises as a result of the approximate degeneracy between filled levels on one side and empty levels on the other side of the Fermi distribution. This near degeneracy would be destroyed by the periodic crystal potential, excepting the chance situation where the latter potential is very small. This might occur at a superlattice $\vec{G}$ of an ordered alloy, if the two species of atoms are, say, of equal valence. A different magnetic response of the two species could then cause a net moment per superlattice cell.

The remarkable ferromagnetism of ordered Sc–In alloys [51] near the Sc_3In composition may be an example of the foregoing phenomenon. The ferromagnetic state is observed only for In concentrations between 23.8 and 24.2 at.%. This very narrow range is hard to understand if the ferromagnetic mechanism is an ordinary alignment of local moments depending, though, on long-range lattice order, expecially since it does not occur at the stoichiometric composition. The proposed SDW mechanism would require that the ordered alloys are antiferromagnetic over a wide composition range, but that the wave vector $\vec{Q}$ happens to be coincident with the superlattice $\vec{G}$ only at 24% indium (presuming that the Fermi surface location varies with composition). This suggestion can, in principle, be checked by neutron diffraction. A somewhat similar phenomenon has been observed in $ZrZn_2$ [51, 52].

Magnetocrystalline interactions that could force commensurateness of a SDW for $\vec{Q}$ near to, but not quite satisfying (3.67) is, of course, possible. It is perhaps surprising that Cr, with large amplitude linear SDW's, appears insensitive to the nearness of the $(1, 0, 0)$ point in reciprocal space, even though $\vec{Q}$ is short of that point by only four percent. Similarly, the wave vectors of the spiral spin configurations in dysprosium and holmium [53], which have a significant temperature variation, show no evidence of locking when they pass through submultiples of the reciprocal lattice vectors. Locking does occur for the linear spin wave configurations in erbium [54] and thulium [50]. For both cases, $p = 7$. But the energetic cause is easy to understand in these two cases. The magnetic interaction responsible for the ordered state is the $s - f$ exchange interaction between a linear SDW in the conduction electrons and the localized f electrons. This interaction is negligible for atoms at nodes of the linear SDW; and there will always be a significant fraction of lattice sites at or very near such nodes for an incommensurate linear SDW. This loss of interaction energy is avoided by a commensurate SDW wave vector. The phase of the SDW can adjust so that no lattice site is near a node. Consequently, a locking mechanism is provided. This cannot occur for a spiral spin wave configuration, since a spiral SDW has no nodes. It is interesting that the locking in erbium stops when a spiral component appears, as might be anticipated, since the (perpendicular) spiral component eliminates the nodes, even though the linear component remains.

Acknowledgements

Throughout the course of this research the author has received stimulation, enlightenment, and encouragement from Dr. Anthony Arrott. It is a pleasure to acknowledge his contributions and to thank him for his continued interest.

References

1 Bloch, F. (1929) *Z. Physik*, 57, 545.
2 Wigner, E.P. (1938) *Trans. Faraday Soc.*, 34, 678.
3 Overhauser, A.W. (1960) *Phys. Rev. Letters*, 4, 462.
4 Kohn, W. and Nettel, S.J. (1960) *Phys. Rev. Letters*, 5, 8.
5 Yoshimori, A. (1961) *Phys. Rev.*, **124**, 326.
6 Burger, J. and Taylor, M.A. (1961) *Phys. Rev. Letters*, 6, 185.
7 Kojima, R.S., Tebble, R.S., and Williams, D.E.G. (1961) *Proc. Roy. Soc.* (London), **A260**, 237.
8 Hoare, F.E. and Matthews, J.C. (1952) *Proc. Roy. Soc.* (London), **A212**, 137.
9 Iwamoto, F. and Sawada, K. (1962) *Phys. Rev.*, **126**, 887.
10 Wolff, P.A. (1960) *Phys. Rev.*, **120**, 814.
11 Peng, H.W. (1941) *Proc. Roy. Soc.* (London), **A178**, 499.

12 Herring, C. (1952) *Phys. Rev.*, **85**, 1003.
13 Herring, C. (1952) *Phys. Rev.*, **87**, 60.
14 Overhauser, A.W. (1960) *J. Phys. Chem. Solids*, **13**, 71.
15 Marshall, W. (1961) *Phys. Rev.*, **118**, 1519.
16 Bean, C.P. and Livingston, J.D. (1959) *J. Appl. Phys.*, **30**, 120S.
17 Zimmermann, J.E. and Hoare, F.E. (1960) *J. Phys. Chem. Solids*, **17**, 52.
18 Crane, L.T. and Zimmermann, J.E. (1961) *J. Phys. Chem. Solids*, **21**, 310.
19 Crane, L.T. and Zimmermann, J.E. (1961) *Phys. Rev.*, **123**, 113.
20 Ruderman, M.A. and Kittel, C. (1954) *Phys. Rev.*, **96**, 99.
21 Yoshida, K. (1957) *Phys. Rev.*, **106** 893.
22 Overhauser, A.W. (1962) *Bull. Am. Phys. Soc.*, **7**, 241.
23 Kohn, W. (1959) *Phys. Rev. Lett.*, **2**, 393.
24 Bardeen, J., Cooper, L.N., and Schrieffer, J.R. (1957) *Phys. Rev.*, **108**, 1175.
25 Swihart, J.C. (1962) *IBM J. Research Develop.*, **6**, 14.
26 Swihart, J.C. (1959) *Phys. Rev.*, **116**, 346.
27 Corliss, L.M., Hastings, J.M., and Weiss, R.J. (1959) *Phys. Rev. Lett.*, **3**, 211.
28 Bykov, V.N., Golovkin, V.S., Ageev, N.V., and Levdik, V.A. (1959) *Doklady Akad. Nauk S.S.S.R.*, **128**, 1153; [translation: Soviet Phys. – Doklady **4**, 1070 (1960)].
29 Bacon, G.E. (1961) *Acta Cryst.*, **14**, 823.
30 Shirane, G. and Takei, W.J. (1961) *Proceedings of the International Conference on Magnetism and Crystallography*, Kyoto, Japan, September; [(1962) *J. Phys. Soc. Japan*, **17**, Suppl. **BIII**, 35].
31 Kaplan, T.A. (1959) *Phys. Rev.*, **116**, 888.
32 Overhauser, A.W. and Arrott, A.*Phys. Rev. Lett.*, **4**, 226 (1960).
33 Goldman, J.E. (1953) *Revs. Modern Phys.*, **25**, 113.
34 Gorter, C.J. (1953) *Revs. Modern Phys.*, **25**, 113.
35 Lidiard, A.B. (1953) *Proc. Phys. Soc.* (London), **A66**, 1188.
36 Slater, J.C. and Koster, G.F. (1954) *Phys. Rev.*, **94**, 1498.
37 Beaumont, R.H., Chihara, H., and Morrison, J.A. (1960) *Phil. Mag.*, **5**, 188 .
38 Wilkinson, M.K. (to be published).
39 Rayne, J.A. and Kemp, W.R.G. (1956) *Phil. Mag.*, **1**, 918.
40 Pursey, H. (1958) *J. Inst. Metals*, **86**, 362.
41 Arajs, S., Colvin, R.V., and Marcinkowski, M.J. (1962) *J. Less-Common Metals*, **4**, 46.
42 Marcinkowski, M.J. and Lipsitt, H.A. (1961) *J. Appl. Phys.*, **32**, 1238.
43 Lomer, W.M. (1962) *Proc. Phys. Soc.* (London), **80**, 489. The writer is very grateful to Dr. Lomer for an extensive discussion of his work prior to publication. This work is based to a large extent on the calculations of J.H, Wood, *Phys. Rev.*, **126**, 517 (1962).
44 Shoenberg, D. (private communication).
45 Hastings, J.M. (1960) *Bull. Am. Phys. Soc.*, **5**, 455.
46 Wilkinson, M.K., Wollan, E.O., and Koehler, W.C. (1960) *Bull. Am. Phys. Soc.*, **5**, 456.
47 Corliss, L., Elliot, N., and Hastings, J. (1956) *Phys. Rev.*, **104**, 924.
48 Barnes, R.G. and Graham, T.P. (1962) *Phys. Rev. Lett.*, **8**, 248.
49 de Morton, M.E. (1961) *Phil. Mag.*, **6**, 825.
50 Koehler, W.C., Cable, J.W., Wollan, E.O., and Wilkinson, M.K. (1962) *J. Appl. Phys.*, **33**, 1124.
51 Matthias, B.T., Clogston, A.M., Williams, H.J., Corenzwit, E., and Sherwood, R.C. (1961) *Phys. Rev. Lett.*, **7**, 7.
52 Matthias, B.T. and Bozorth, R.M. (1958) *Phys. Rev.* **109**, 604.
53 Wilkinson, M.K., Koehler, W.C., Wollan, E.O., and Cable, J.W. (1961) *J. Appl. Phys.*, **32**, 20S, 48S.
54 Cable, J.W., Wollan, E.O., Koehler, W.C., and Wilkinson, M.K. (1961) *J. Appl. Phys.*, **32**, 49S.

Reprint 4 Spin-Density-Wave Antiferromagnetism in Potassium[1)]

*A.W. Overhauser**

* Scientific Laboratory, Ford Motor Company, 525, Northwestern Avenue, Dearborn, Michigan 47907, USA

Received 8 July 1964

It is shown that the recently discussed [1] optical-absorption threshold in metallic K, found by El Naby [2] at $\hbar\omega = 0.62\,\mathrm{eV}$, provides striking evidence that the electronic state of the metal is a giant spin-density-wave state. The optical absorption associated with the spin-density-wave energy gap is calculated and found to be in remarkable quantitative agreement with the observed spectrum. Direct observation of the spin-density wave by neutron diffraction is feasible, since its amplitude is estimated to be 0.09 Bohr magneton per atom.

In the Hartree–Fock (HF) approximation the normal state of every metal is unstable with respect to spin-density-wave (SDW) formation [3]. However, without an exact solution of the many-body problem, the presence of an SDW can be surmised only from experimental evidence, direct (as in [3] Cr) or indirect (as in Cu, Ag, and [4] Au). Low electron density favors their formation, so the alkali metals may be expected to exhibit them in a more pronounced way compared to most other metals.

Within the HF scheme SDW energy gaps should truncate the Fermi surface, thereby altering the Fermi surface topology by neck formation [3]. However, correlation energy considerations lead to the conclusion that for a simple metal an SDW energy gap should touch the Fermi surface only at a single point, as shown in Figure 4.1. The argument is as follows: The correlation energy arises from virtual electronic excitations near the Fermi surface. As long as E vs K is continuous at the Fermi surface, low-energy virtual excitations, arising from electron interactions, can occur easily and contribute to the binding energy. If the wave vector $\vec{Q}$ of the SDW is small enough so that a finite area of Fermi surface is truncated by the energy gap, the virtual excitations will be locally inhibited and the correlation energy reduced in magnitude. The net effect is a repulsion between the Fermi surface and the energy gaps, which allows the total energy to be a minimum when the gaps just touch the Fermi surface. (One needs to remember that the HF energy leads to a small attractive interaction between the Fermi surface and the energy gaps [3].)

We will now compute the optical absorption associated with a single linear SDW. The periodic exchange potential causing the energy gaps is

$$A(\vec{r}) = G\vec{u}\cdot\vec{\sigma}\cos\vec{Q}\cdot\vec{r}\,, \tag{4.1}$$

where $\vec{\sigma}$ is the Pauli spin operator and $\vec{u}$ is the SDW polarization vector. The one-electron energy E_K obtained by including (4.1) in the Schrödinger equation is the familiar result of weak-binding theory. In the vicinity of the gap at A (Figure 4.1),

$$E_K = \epsilon_K + \mu z \pm \left(\mu^2 z^2 + G^2/4\right)^{1/2}\,, \tag{4.2}$$

where z is the perpendicular distance in K space from the gap, $\epsilon_K = \hbar^2K^2/2m$, and $\mu = \hbar^2 Q/2m$. The space parts of the wave functions above and below the gap, respectively (for spin parallel to $\vec{u}$), are

$$\varphi_K = \cos\theta\,\exp i\vec{K}\cdot\vec{r} + \sin\theta\,\exp i(\vec{K}+\vec{Q})\cdot\vec{r}\,, \tag{4.3a}$$

1) Phys. Rev. Letters 13, 6 (1964)

Anomalous Effects in Simple Metals. Albert Overhauser

ISBN: 978-3-527-40859-7

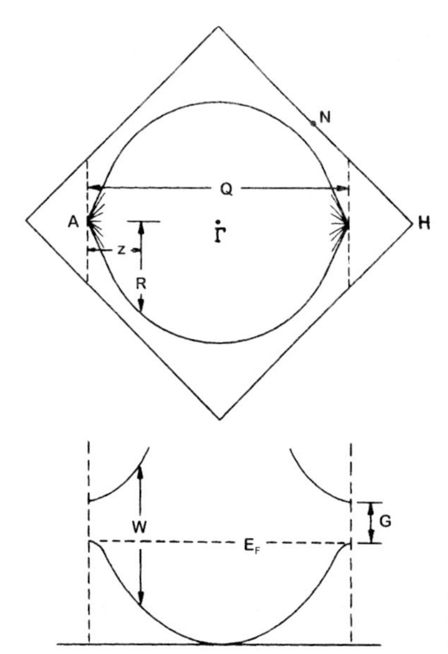

Figure 4.1 Brillouin zone, Fermi surface, and electron energy spectrum of potassium with a single linear spin-density-wave state. The wave vector of the spin-density wave is assumed to be parallel to a cubic axis.

$$\varphi_{K'} = \sin\theta \exp i\vec{K}' \cdot \vec{r} - \cos\theta \exp i(\vec{K}' + \vec{Q}) \cdot \vec{r}\,, \tag{4.3b}$$

where

$$\sin\theta \cos\theta = G/2W\,, \tag{4.4}$$

and

$$W = 2(\mu^2 z^2 + G^2/4)^{1/2} \tag{4.5}$$

is the vertical transition energy. The radius R of the Fermi surface at z is found with the help of (4.2), (4.5), and the condition $R = 0$ at $z = 0$:

$$\frac{R^2}{Q^2} = \frac{W - G}{2\mu Q}\left(1 - \frac{W + G}{2\mu Q}\right). \tag{4.6}$$

Consider now a light wave traveling in the ζ direction of a centimeter cube of metal. The vector potential may be taken to be

$$\vec{A} = \vec{\nu} \exp\left[(in - k)(\omega\zeta/c) - i\omega t\right], \tag{4.7}$$

where n and k are the optical constants, and $\vec{\nu}$ is the polarization vector. The energy flux at the front surface, $\zeta = 0$, is accordingly

$$P = (c/4\pi)\text{Re}\left(\vec{E}^* \times \vec{H}\right) = n\omega^2/4\pi c\,. \tag{4.8}$$

We need only compute the SDW band-to-band transition rate caused by the perturbation, $V = (e/mc)\vec{A} \cdot \vec{p}$, and set the result equal to P divided by $W = \hbar\omega$:

$$\frac{P}{W} = 4\sum_{K',K}(2\pi/\hbar)|V_{K'K}|^2\delta\left(E_K - E_{K'} - \hbar\omega\right), \tag{4.9}$$

where the factor 4 takes account of the spin degeneracy and the two gaps at opposite sides of the Fermi surface. Employing (4.3), one finds

$$|V_{KK'}|^2 = \left(e\hbar\vec{v}\cdot\vec{Q}\sin\theta\cos\theta/mc\right)^2 \times\left[(K'_\zeta - K_\zeta + \omega n/c)^2 + (\omega k/c)^2\right]^{-1}, \tag{4.10}$$

the $\vec{K}$ components transverse to ζ being conserved. The double sum in (4.9) can be evaluated:

$$\frac{P}{W} = \frac{c}{2\pi^2 Wk}\left(\frac{e\hbar QG\cos\alpha}{2mcW}\right)^2 \pi R^2\frac{dz}{dW}, \tag{4.11}$$

where α is the angle between $\vec{v}$ and $\vec{Q}$. With the help of (4.5), (4.6), and (4.8), the imaginary part, $2nk$, of the dielectric constant is found:

$$2nk = \frac{G^2e^2Q}{3W^3}\left(\frac{W-G}{W+G}\right)^{1/2}\left(1-\frac{W+G}{2\mu Q}\right), \tag{4.12}$$

where $\cos^2\alpha$ has been replaced by its average value, $\frac{1}{3}$.

It is emphasized that the final result (4.12) has no adjustable parameter other than the threshold energy G. Equation (4.6) can be used to find Q, which is slightly larger than the diameter, $2k_F = 1.24(2\pi/a)$, of the undeformed Fermi sphere:

$$Q = 1.33(2\pi/a), \tag{4.13}$$

where a is the lattice constant. The theoretical absorption coefficient, $2nk/\lambda$, is shown in Figure 4.2 for $G = 0.62$ eV. The experimental data of El Naby [2], after subtracting the Drude background [5], are also shown. The remarkable agreement, both as to shape and absolute magnitude, indicates that potassium has a single linear SDW. The peculiar absorption shape is characteristic of interband transitions at an energy gap which touches the Fermi surface at a single point. (A truncated face of finite area yields a singular absorption behavior above threshold, whereas a gap which does not touch at all causes a linear rise above threshold.) The absorption anomaly cannot be ascribed to the Fermi surface touching the 12 zone faces (at N) since that would require a sixfold larger magnitude.

Possible difficulties in observing SDW's by neutron diffraction have been discussed [3]. However, it is anticipated that direct observation in this case should be easy. The fractional spin polarization of the SDW can be computed using the observed G. The amplitude is

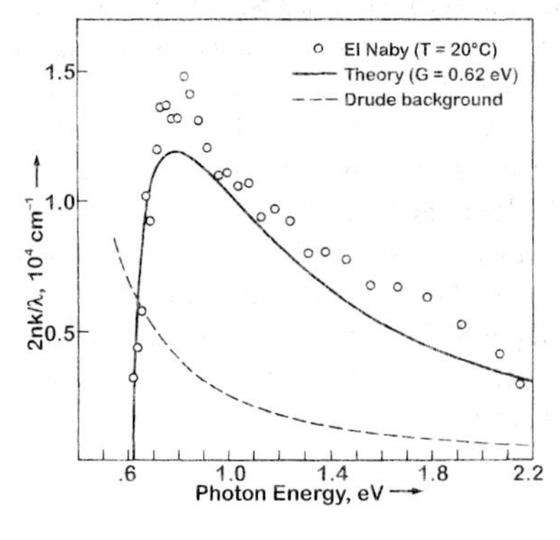

Figure 4.2 Optical-absorption spectrum of potassium. The theoretical curve, according to (4.12) of the text, has no adjustable parameter other than the threshold energy G. The experimental values shown are those obtained after subtracting the Drude background.

found to be 9%, which is adequately large provided the SDW has transverse polarization. This should be the case for a simple metal like K, since the magnetostatic energy favors transverse polarization, and the situation is not complicated by multiple SDW interactions. The direction of $\vec{Q}$ is probably either the cubic axis or the body diagonal. Consequently, a single SDW domain should exhibit a slight tetragonal (or rhombohedral) distortion. However, the optical properties of a domain should be highly anisotropic, since the anomalous absorption is proportional to $\cos^2 \alpha$.

Near a point of contact with the energy gaps, the Fermi surface has the shape of a circular cone with semivertical angle β given by

$$\tan \beta = \left[(\mu Q/G) - 1 \right]^{1/2} . \tag{4.14}$$

For potassium, $\beta = 75°$. Several anomalies [6] in the cyclotron resonance of K may possibly arise from the conical shape of the Fermi surface at such points.

References

1. Cohen, M.H. and Phillips, J.C. (1964) *Phys. Rev. Lett.*, **12**, 662.
2. El Naby, M.H. (1963) *Z. Physik*, **174**, 269.
3. Overhauser, A.W. (1962) *Phys. Rev.*, **128**, 1437.
4. Van Zandt, L.L. and Overhauser, A.W. (1963) *Bull. Am. Phys. Soc.*, **8**, 337; and to be published; L.L. Van Zandt, thesis, Harvard University, 1964 (unpublished).
5. Mayer, H. and El Naby, M.H. (1963) *Z. Physik*, **174**, 289.
6. Grimes, C.C. and Kip, A.F. (1963) *Phys. Rev.*, **132**, 1991.

Reprint 5 Helicon Propagation in Metals Near the Cyclotron Edge[1)]

A.W. Overhauser and Sergio Rodriguez***

* Scientific Laboratory, Ford Motor Company, 525, Northwestern Avenue, Dearborn, Michigan 47907, USA
** Department of Physics, Purdue University, Lafayette, Indiana 47906, USA

Received 12 July 1965

The theory of Doppler-shifted cyclotron resonance with helicon waves is elaborated by calculating the magnetic-field dependence of the surface impedance near the cyclotron absorption (Kjeldaas) edge. It is found that the location of the edge, defined as that field for which the field derivative of the surface reactance is a minimum, deviates significantly from the threshold field defined by the Doppler-shifted frequency criterion. The deviations are large when $\omega_c\tau < 50$. The theory is compared with the experimental work of Taylor on Na and K. The data agree well with the free-electron model for Na, but there is a large discrepancy (770 G for 10-Mc/sec helicons) in K. The surface impedance is also calculated for a metal with a spin-density-wave ground state, in order to test the hypothesis of such a ground state, previously made to explain other anomalies in K. The edge field derived for a spin-density-wave state appropriate to K is 730 G less than that required by the free-electron model, and agrees with the published data.

5.1 Introduction

Helicons [1, 2] are circularly polarized electromagnetic waves propagating in an electron gas in the presence of a magnetic field. A helicon has low damping if the cyclotron period of the electrons in the magnetic field is short compared to the average time between two successive collisions of an electron, i.e., if $\omega_c\tau \gg 1$ where ω_c is the electron cyclotron frequency and τ the electron relaxation time. Now, if the frequency of the helicon co is such that $\omega \ll \omega_c \ll \omega_p^2/\omega$ where ω_p is the plasma frequency of the electron gas, the complex wave number q of the helicon is given by

$$q^2 = (q_1 + iq_2)^2 = (4\pi\omega ne/Bc)\, f_\pm \,. \tag{5.1}$$

In (5.1), B is the applied magnetic field, n the concentration of electrons, and $-e$ the charge on the electron. We have assumed here that the wave propagates parallel to the direction of the magnetic field $\mathbf{B}$ which we take parallel to the z axis of a Cartesian coordinate system. The factor $f_\pm$ is equal to ± 1 in the limit of long wavelengths $\left[|q| \ll (\omega_c/v_F)\right]$, but differs from unity if $|q| \gtrsim (\omega_c/v_F)$. The analytical expression for $f_\pm$ is[2)]

$$f_\pm = \frac{3}{4}\int_0^\pi d\theta \sin^3\theta \left[(qv_F/\omega_c)\cos\theta \pm 1 + (i/\omega_c\tau)\right]^{-1} . \tag{5.2}$$

In the relations given above the upper and lower signs correspond to left-circularly-polarized and right-circularly-polarized waves, respectively. Of these, of course, the right-circularly-

1) Phys. Rev. 141, 1 (1966)
2) See for example, [3].
3) This is true for the model considered in the text, namely, a free-electron gas embedded in a uniform background of positive charge of the same density. In a metal in which the conduction is due to holes the right-circularly-polarized wave is undamped.

Anomalous Effects in Simple Metals. Albert Overhauser

ISBN: 978-3-527-40859-7

polarized wave has a large attenuation and need not be considered further.[3] Typical experimental conditions are, for example, $\nu = (\omega/2\pi) = 10\,\text{Mc/sec}$, $B = 10\,\text{kG}$, and $n = 10^{22}\,\text{cm}^{-3}$. Then $|q| \approx 4 \times 10^3\,\text{cm}^{-1}$. In this case, the phase velocity of the helicon is approximately $(\omega/|q|) \approx 10^4\,\text{cm/sec}$. This velocity is several orders of magnitude smaller than the Fermi velocity v_F of the electrons. Thus, an electron whose component of velocity parallel to the direction of $\boldsymbol{B}$ is v_z experiences a periodic electric field of frequency

$$\omega_a = \omega \pm q_1 v_z \approx \pm q_1 v_z \,. \tag{5.3}$$

Clearly, a resonant absorption of energy occurs when $\omega_c = \omega_a$, i.e., if electrons exist having a component v_z of the velocity satisfying the condition,

$$\omega_c = q_1 v_z \,. \tag{5.4}$$

If the radiofrequency ω is kept fixed and the magnetic field is decreased from a value such that $\omega_c > q_1 v_F$ to values which reverse the inequality, then one expects an absorption edge[4] (called the Kjeldaas edge) at $\omega_c = q_1 v_F$. This resonance is also called the Doppler-shifted cyclotron resonance [6]. The Kjeldaas resonance has been observed by Kirsh [7] in bismuth using conditions appropriate for propagation of Alfvén waves [8] and by Taylor *et al.*, [9] and Taylor [10, 11] using helicon waves in sodium, potassium, and indium.

It is of the utmost importance to distinguish the theoretical and experimental definitions of the edge. The "theoretical" edge B_e is

$$B_e \equiv \text{that field for which } \omega_c = q_1 v_{\max} \,, \tag{5.5}$$

where $v_{\max}$ is the maximum value of the z component of electron velocity on the Fermi surface. The experimental definition of the edge is arbitrary. Because the most salient feature of Taylor's data [10, 11] is the sharp minimum in dX/dB, where X is the reactive part of the surface impedance Z, we define the experimental edge B_X to be

$$B_X \equiv \text{that field for which } dX/dB \text{ is a minimum} \,. \tag{5.6}$$

Taylor made the conjecture that $B_X = B_e$, an assumption which we shall demonstrate to be generally invalid.

For the free-electron model B_e is easily calculated. Equation (5.1) together with $v_{\max} = v_F$ yield

$$B_e = (8\hbar^2 c/3e)^{1/3} k_F^{5/3} \nu^{1/3} f_e^{1/3} \,, \tag{5.7}$$

where k_F is the Fermi wave number, and

$$f_e \equiv \left\{ \text{Re}\left[f_+(B_e) \right]^{1/2} \right\}^2 . \tag{5.8}$$

Taylor has evaluated approximately the dependence of f_e on the collision time. $f_e \approx 1.5$ for $\omega_c \tau > 50$, and decreases monotonically from this asymptotic value with decreasing $\omega_c \tau$. We have verified this behavior by a more precise calculation. For potassium (taking the lattice constant [12] $a = 5.225$ Å) we find, for $\nu = 10\,\text{Mc/sec}$,

$$B_e(\omega_c \tau = \infty) = 18.58\,\text{kG} \,. \tag{5.9}$$

The calculation of the "experimental" edge B_X is considerably more difficult than that of B_e. The details are given in Section 5.2. The conclusions may be summarized here:

4) See [4] for a discussion of the edge in connection with the attenuation of ultrasonic shear waves propagating parallel to an applied magnetic field. For helicons see [5].

5) We have defined for convenience the parameter $\omega_c(18)\tau$ to mean the value of $\omega_c\tau$ when the applied magnetic field is 18 kG, the approximate position of the Kjeldaas edge in potassium. Thus $\omega_c(18)\tau$ is merely a measure of τ and is independent of B.

(a) $B_X \approx B_e$ for large $\omega_c(18)\tau$ (in excess of about 40).[5)]

(b) B_X *increases* monotonically with *decreasing* τ. This behavior is the reverse of that found for B_e.

(c) For the case of potassium at 10 Mc/sec, and $\omega_c(18)\tau \sim 25$, we find B_X to be about 18.73 kG. This is to be compared with the edge observed by Taylor in material of comparable collision time, which was 17.96 kG. There is therefore a discrepancy of 770 G, which seems significant. In Table I of Taylor's paper he found what appeared to be agreement between experiment and theory for the free-electron model. It is clear that this apparent agreement was a spurious consequence of the conjecture which identified B_X with B_e. It should be appreciated that both of these quantities are independent of the electron effective mass.

The existence of this large discrepancy has been verified independently by T. Kushida (private communication). The magnetic field near the Kjeldaas edge was accurately calibrated by Cs NMR. Furthermore, the shape of the Cs NMR was used to precisely determine the pure resistance and reactance modes; and this was verified by an independent experiment using (instead of an impedance bridge) a Pound box, which automatically selects the pure resistance mode.

(d) For the case of sodium the agreement found by Taylor between experiment and the free-electron model is pertinent because the specimen collision times, $\omega_c(25)\tau \sim 100$, were sufficiently long to justify the use of conclusion (a), above.

In order to explore a possible reason for the large discrepancy between experiment and theory for the case of potassium, we have computed B_X, assuming a spin-density-wave ground state [13]. The details are given in Section 3.3. The calculated value of B_X at 10 Mc/sec is 18.0 kG, assuming, as above, $\omega_c(18)\tau = 25$. The spin-density-wave energy gap was taken to be 0.62 eV, the value required [13] to explain the optical anomalies [14, 15] in K. The agreement between this calculated value and Taylor's experimentally observed edge is an interesting coincidence.

5.2 The Surface Impedance

In this section we calculate the surface impedance of a metal in the presence of a magnetic field $\boldsymbol{B}$ perpendicular to the surface and of a radiofrequency electromagnetic field of frequency $\omega \ll \omega_c$. The calculation is carried out for the free-electron model. The numerical applications that we give are for the case of potassium with the parameters given in Section 5.1. The surface impedance of the material for a left-circularly-polarized wave, assuming that the electrons are reflected specularly from the surface of the metal, is given by the relation

$$Z = R - iX = (8i\omega/c^2)\int_0^\infty dq\left[q^2 + \frac{4\pi i\omega}{c^2}(\sigma_{xx} - i\sigma_{xy})\right]^{-1}, \tag{5.10}$$

In (5.10) σ_{xx} and σ_{xy} are components of the magneto-conductivity tensor with respect to a Cartesian coordinate system whose z axis is parallel to $\boldsymbol{B}$.

We now show how to calculate σ_{ij} $(i, j = x, y, z)$ for an electron gas with an arbitrary dispersion formula $E(\boldsymbol{k})$. This calculation is considerably more general than is required for our purposes. However we give it here to provide a basis for the analysis of Section 5.3 and because of its intrinsic interest. The procedure we use is that due to Eckstein [16]. We consider the electrons within a metal in the presence of the dc magnetic field $\boldsymbol{B}$ and of an alternating electric field $\boldsymbol{\varepsilon}$ that varies as $\exp(i\omega t - i\boldsymbol{q}\cdot\boldsymbol{r})$. The Boltzmann transport equation for the electron distribution function $f = f(\boldsymbol{k}, \boldsymbol{r}, t)$ is

$$\frac{\partial f}{\partial t} + \boldsymbol{v} \cdot \nabla f - \frac{e}{\hbar}\left(\boldsymbol{\varepsilon} + \frac{1}{c}\boldsymbol{v} \times \boldsymbol{B}\right) \cdot \nabla_k f = -(f - f_0)/\tau \,, \tag{5.11}$$

where $\boldsymbol{v}$ is the velocity of the electron having wave vector $\boldsymbol{k}$ and $f_0 = f_0(E(\boldsymbol{k}))$ is the distribution function in thermal equilibrium. Equation (5.11) is linearized in $\boldsymbol{\varepsilon}$ the usual way and its solution is found to be

$$f = f_0 + f_1 \,, \tag{5.12}$$

where

$$f_1 = \int_{-\infty}^{u} du' e\boldsymbol{\varepsilon} \cdot \boldsymbol{v}' \frac{d f_0}{d E} \exp\left[\left(\frac{1}{\tau} + i\omega\right) \times (u' - u) + i\boldsymbol{q} \cdot (\boldsymbol{R} - \boldsymbol{R}')\right]. \tag{5.13}$$

In writing (5.13) we found it convenient to express $\boldsymbol{k}$ in terms of the new variables E, k_z, and u where u is a quantity having the dimension of a time and which describes the position of the electron on the orbit defined by the intersection of the surface $E(\boldsymbol{k}) = E$ and the surface k_z equals a constant when there is no radio-frequency field present. Thus u is the time parameter in the equation $\hbar(d\boldsymbol{k}/du) = -(e/c)\boldsymbol{v} \times \boldsymbol{B}$ which governs the motion of the electron in $\boldsymbol{k}$ space in the presence of $\boldsymbol{B}$ alone. The function $\boldsymbol{R}(u)$ is the position in real space of an electron of energy E and component k_z of the wave vector at the time u under these conditions. Thus, with a suitable choice of origin $\boldsymbol{k} = -(e/\hbar c)\boldsymbol{R}(u) \times \boldsymbol{B}$, so that $\boldsymbol{R}(u)$ describes an orbit in space which is identical to that described by $\boldsymbol{k}$ but rotated by 90° in the clockwise direction about an axis parallel to $\boldsymbol{B}$ and amplified by the factor $\hbar c/eB$. Of course, the component of $\boldsymbol{R}$ parallel to $\boldsymbol{B}$ is not determined by these considerations. In (5.13) $\boldsymbol{R}$ is the function $\boldsymbol{R}(E, k_z, u)$ while $\boldsymbol{R}'$ is the same function evaluated for the value u' of u, i.e., $\boldsymbol{R}' = \boldsymbol{R}(E, k_z, u')$. In the same manner $\boldsymbol{v}' = \boldsymbol{v}(E, k_z, u')$. We can always separate $\boldsymbol{R}$ into two terms $\boldsymbol{R}_s$ and $\boldsymbol{R}_p$ so that $\boldsymbol{R} = \boldsymbol{R}_s + \boldsymbol{R}_p$. Here $\boldsymbol{R}_p$ is a periodic function of $\boldsymbol{k}$ while $\boldsymbol{R}_s = \boldsymbol{v}_s(E, k_z)u$ increases linearly with the time u. Now $\boldsymbol{v}_s(E, k_z)$ is the average velocity of an electron on the orbit defined by E and k_z. In the free-electron model $\boldsymbol{v}_s$ is the component of the velocity parallel to the applied magnetic field but in the more general case, to which our discussion is applicable, $\boldsymbol{v}_s$ can have another component when the orbit defined by E and k_z in $\boldsymbol{k}$ space corresponds to an open orbit. Making use of the facts that the volume element in $\boldsymbol{k}$ space $d\boldsymbol{k} = (eB/\hbar^2 c)dE dk_z du$ and that the electrical current density is $\boldsymbol{j} = [(-2e)/(2\pi)^3]\int \boldsymbol{v} f_1 d\boldsymbol{k}$, we obtain

$$\boldsymbol{j} = \boldsymbol{\sigma} \cdot \boldsymbol{\varepsilon} \,, \tag{5.14}$$

where

$$\boldsymbol{\sigma} = \frac{2e^3 B}{(2\pi)^3 \hbar^2 c} \int dE \left(-\frac{d f_0}{d E}\right) dk_z T(E, k_z) \sum_{n=-\infty}^{\infty} V_n V_n^* \times \left[\frac{1}{\tau} + i\left(\omega - \frac{2\pi n}{T(E, k_z)} - \boldsymbol{q} \cdot \boldsymbol{v}_s\right)\right]^{-1} . \tag{5.15}$$

In this expression $T(E, k_z)$ is the period of the motion in $\boldsymbol{k}$ space of an electron having energy E and a component of its wave vector equal to k_z along the direction of $\boldsymbol{B}$. The quantities V_n are the Fourier components of $\boldsymbol{v}(E, k_z, u)\exp[i\boldsymbol{q} \cdot \boldsymbol{R}_p(E, k_z, u)]$ considered as periodic function of u, i.e., we have

$$\boldsymbol{v}(E, k_z, u)\exp\left[i\boldsymbol{q} \cdot \boldsymbol{R}_p(E, k_z, u)\right] = \sum_{n=-\infty}^{\infty} V_n(E, k_z)\exp\left[\frac{2\pi i n u}{T(E, k_z)}\right], \tag{5.16}$$

so that

$$V_n(E, k_z) = \frac{1}{T}\int_0^T du\boldsymbol{v}(E, k_z, u)$$
$$\times \exp\left[i\boldsymbol{q}\cdot\boldsymbol{R}_p(E, k_z, u) - (2\pi i n u/T)\right]. \tag{5.17}$$

The application of these results to the case of interest to us is particularly simple. We wish to consider the components of the conductivity tensor $\boldsymbol{\sigma}$ in the particular situation in which the Fermi surface possesses an axis of rotational symmetry about $\boldsymbol{B}$. In such a case we obtain

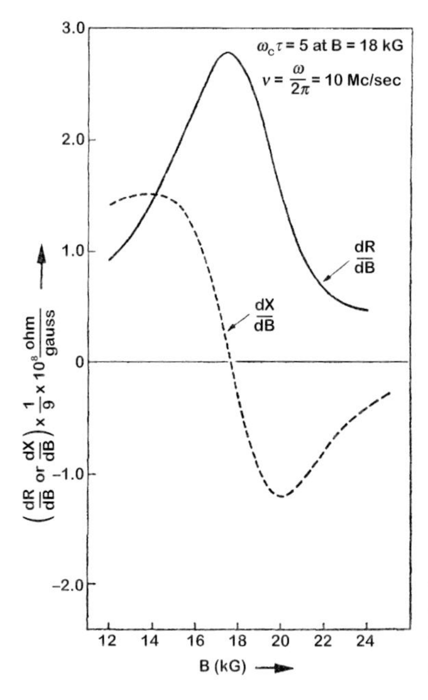

Figure 5.1 Derivatives of the surface resistance and surface reactance with respect to an applied magnetic field B for a potassium slab with the field B perpendicular to the surface of the slab. The model used is the free-electron gas for a frequency of 10 Mc/sec and $\omega_c 818)\tau = 5$.

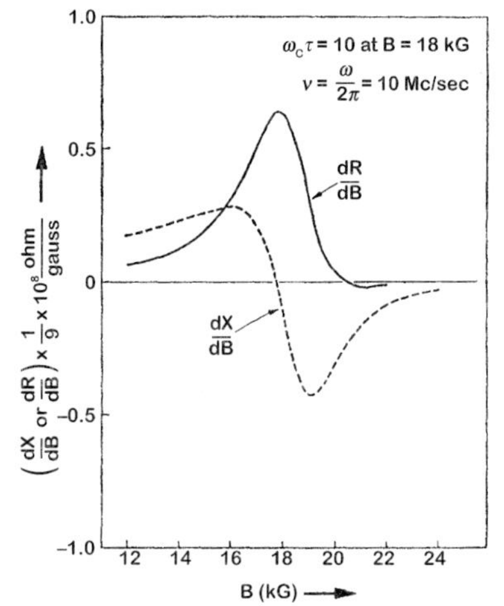

Figure 5.2 dR/dB and dX/dB for potassium with $\omega_c(18)\tau = 10$.

$V_{nx} \pm i V_{ny} = i v_\perp \delta_{n,n+1}$, where $v_\perp$ is the component of the electron velocity on a plane perpendicular to $\boldsymbol{B}$. Then we obtain

$$\sigma_{xx} - i\sigma_{xy} = \frac{e^2}{4\pi^2\hbar^2} \int dk_z \frac{m_\tau v_\perp^2}{1 + i(\omega\tau - \omega_c\tau - q v_z \tau)}, \tag{5.18}$$

where m is the cyclotron effective mass. For the free-electron model we use $E(\boldsymbol{k}) = \hbar^2 k^2/2m$, where m is the mass of the electron and assume that τ is a constant over the Fermi surface. The latter assumption is not really necessary as long as $\omega_c\bar{\tau} \gg 1$, where $\bar{\tau}$ is an appropriate average collision time but we make it for the sake of simplicity. In this case $\sigma_+ = \sigma_{xx} - i\sigma_{xy}$ can be evaluated in terms of elementary functions. The calculation of Z has been carried out numerically and the results are displayed in Figures 5.1–5.5. We notice that B_X increases monotonically with decreasing τ. We can understand this phys-

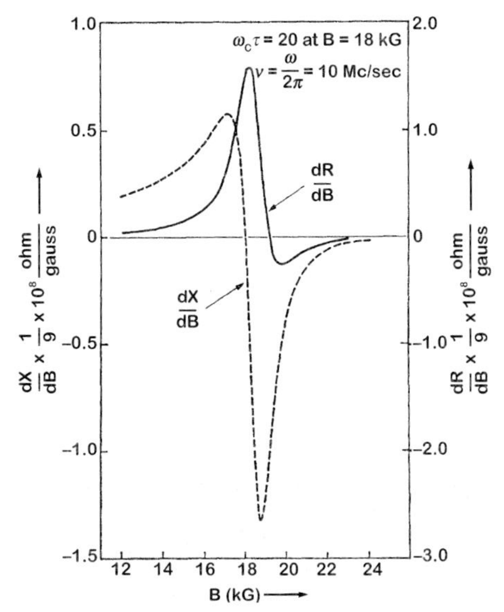

Figure 5.3 dR/dB and dX/dB for potassium with $\omega_c(18)\tau = 20$.

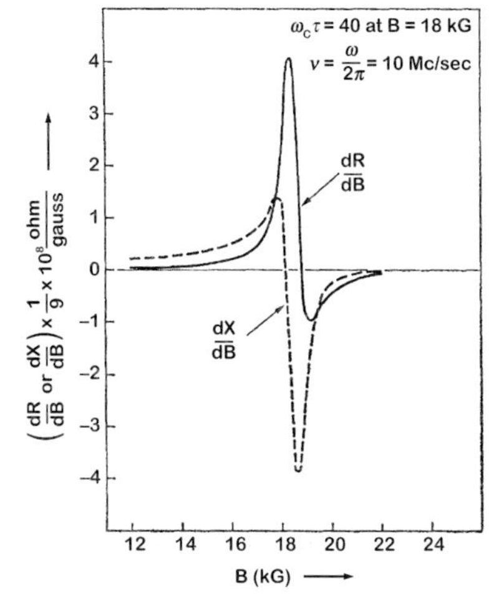

Figure 5.4 dR/dB and dX/dB for potassium with $\omega_c(18)\tau = 40$.

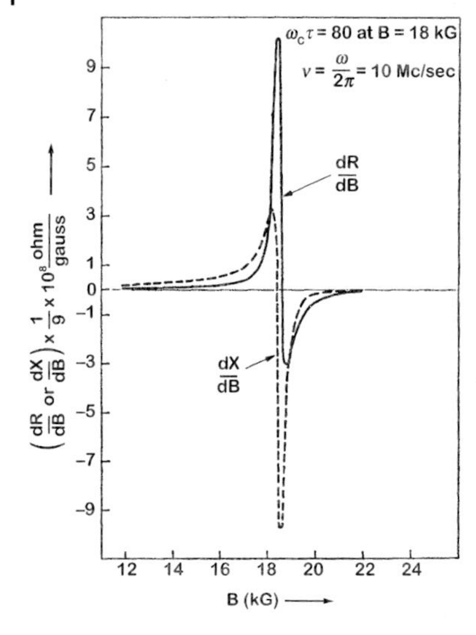

Figure 5.5 dR/dB and dX/dB for potassium with $\omega_c(18)\tau = 80$.

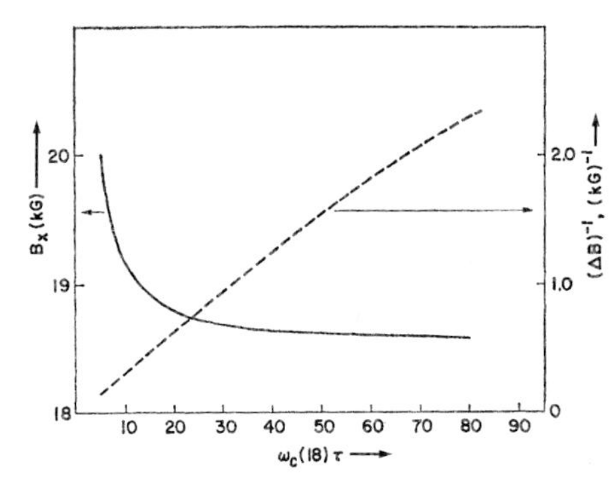

Figure 5.6 Graph of B_X and $(\Delta B)^{-1}$ versus $\omega_c(18)\tau$ corresponding to a frequency of 10 Mc/sec for a free-electron gas with a density of electrons equal to that of potassium.

ically as follows. For the case in which $\omega_c\tau$ is not very much larger than unity the helicon is attenuated and is not a single sinusoidal wave in space. If a Fourier analysis of the wave is made one would find a Lorentzian distribution of wave numbers q and therefore the condition for the absorption edge is reached at higher magnetic fields for lower collision times.

In Figure 5.6 we have displayed the variation of B_X and of $(\Delta B)^{-1}$ with τ. ΔB is defined as

$$\Delta B \equiv \text{Difference between the values of } B \text{ for which a minimum and a maximum of } dR/dB \text{ occur} . \tag{5.19}$$

ΔB is a measure of the width of the resonance line and we see that it is approximately proportional to τ^{-1} as one would expect. For large values of τ, $(\Delta B)^{-1}$ deviates from a linear relation as might be anticipated, since even for $\tau = \infty$ the line should have a finite width. It would be desirable to carry out experiments to verify the τ dependence of B_X.

5.3 Helicon Propagation in a Spin-Density Wave Metal

The anomalous optical reflectivity of potassium [14, 15] gives support to the hypothesis [13] that the electronic ground state of potassium possesses a spin-density wave [17] (SDW). In the SDW model the one-electron states can be labeled by a wave vector $\boldsymbol{k}$ and have an energy dispersion relation of the form

$$E(\boldsymbol{k}) = \epsilon(\boldsymbol{k}) + \mu\left(\frac{1}{2}Q - k_z\right) - \left[\mu^2\left(\frac{1}{2}Q - k_z\right)^2 + \frac{1}{4}G^2\right]^{1/2}, \tag{5.20}$$

which is valid for $k_z > 0$ in the vicinity of $k_z = Q/2$. In this relation k_z is the component of $\boldsymbol{k}$ along the direction of the wave vector $\boldsymbol{Q}$ of the SDW. The quantity G is the energy gap at $k_z = Q/2$, $\mu = \hbar^2 Q/2m$, and $\epsilon(\boldsymbol{k}) = \hbar^2 k^2/2m$. In virtue of time-inversion symmetry we must have $E(-\boldsymbol{k}) = E(\boldsymbol{k})$, a relation that defines $E(\boldsymbol{k})$ for $k_z < 0$. On the other hand, de Haas–van Alphen studies by Shoenberg and Stiles [18] and cyclotron resonance experiments by Grimes and Kip [19] have shown that the Fermi surface of potassium is a sphere to within a few parts in a thousand. It is possible to reconcile these two results by assuming that the wave vector $\boldsymbol{Q}$ of the SDW orients itself parallel to a sufficiently strong dc magnetic field $\boldsymbol{B}$. If this is correct, then the extremal cross sections of the Fermi surface by planes perpendicular to $\boldsymbol{B}$ are circles. The area of these circles is, however, slightly less than their values for the free-electron model. This assumption would explain why in a SDW model a rotation of the specimen with respect to the magnetic field would give rise to a configuration in which an extremal cross section of the Fermi surface is always a circle.

We consider now the position of the Kjeldaas edge in the SDW model. Here we expect the edge to occur at considerably lower magnetic fields. In fact the velocity v_z in this case is not linear in k_z but behaves in the fashion displayed schematically in Figure 5.7. The position of the Kjeldaas edge is at the value of the magnetic field such that $\omega_c = q_1 v_{\max}$. Now, in this case the dispersion relation (5.1) is the same as for the free-electron gas provided we set $f_\pm$ equal to

$$f_\pm = \frac{eBm}{4\pi^2 nc\hbar^2 q}\int \frac{v_\perp^2\, dk_z}{v_z \pm \omega_c/q + i/q\tau}. \tag{5.21}$$

In (5.21) $v_\perp$ is the magnitude of the component of velocity of an electron perpendicular to $\boldsymbol{B}$. Now

$$v_\perp^2 = \frac{2E_F}{m} - \frac{\hbar^2 k_z^2}{m^2} - \frac{2\mu}{m}\left(\frac{Q}{2} - k_z\right) + \frac{2}{m}\left[\mu^2\left(\frac{Q}{2} - k_z\right) + \frac{G^2}{4}\right]^{1/2}, \tag{5.22}$$

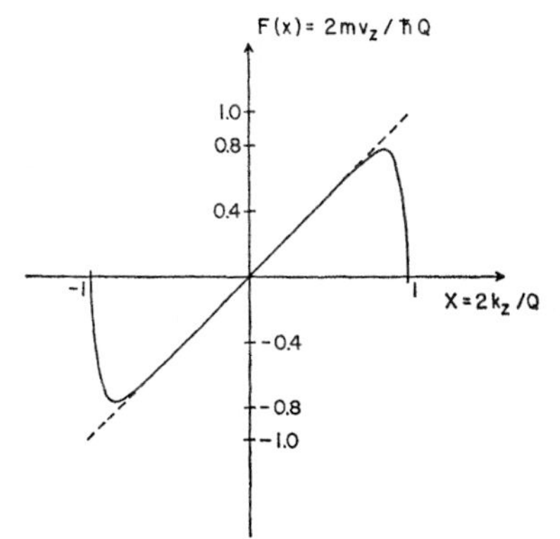

Figure 5.7 Schematic diagram showing v_z versus k_z for a SDW model of a metal.

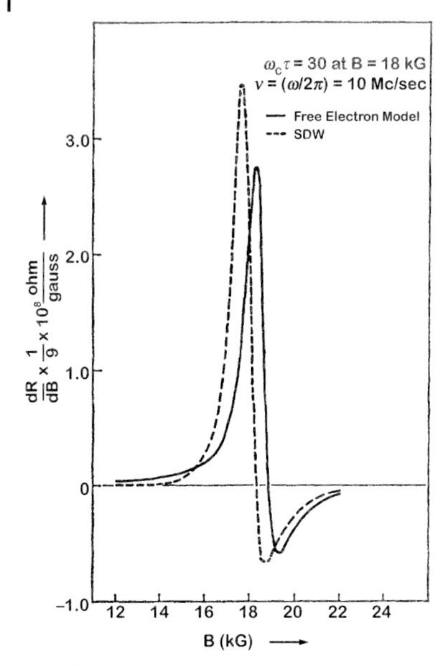

Figure 5.8 dR/dB for potassium with $\omega_c(18)\tau = 30$ for both the free-electron and SDW models.

for $k_z > 0$. In this equation E_F is the Fermi energy. Equation (5.21) has been obtained using (5.18) in Section 5.2. The value of $v_{\max}$ is

$$v_{\max} = (\hbar Q/2m)\left[1 - (G/\mu Q)^{2/3}\right]^{3/2} . \tag{5.23}$$

For potassium, if we choose $G = 0.62\,\text{eV}$ [and $Q = 1.33(2\pi/a)$ where a is the lattice constant] to obtain agreement with the optical data [14, 15] we obtain $v_{\max} = 0.714 \times 10^8$ cm/sec which is to be compared with the Fermi velocity $v_F = 0.864 \times 10^8$ cm/sec in the free-electron model. Now, the dielectric constant of the electron gas in a SDW state appropriate to helicon propagation is ϵ_+ where[6)]

$$\epsilon_+ = (e^2 Q^2/2\pi\omega\hbar q)\, U_+ , \tag{5.24}$$

and

$$U_+ = \int_{-1}^{1} \frac{(2E_F/\mu Q) - \frac{1}{2}x^2 - (1-|x|) + \{(1-|x|)^2 + (G/\mu Q)^2\}^{1/2}}{F(x) + (\hbar\omega_c/\mu q) + (i\hbar/\mu q\tau)} . \tag{5.25}$$

In (5.25) the function $F(x)$ is the ratio of the velocity v_z to $(\hbar Q/2m)$ as a function of $x = 2k_z/Q$.

The surface impedance of potassium has been calculated using a SDW model in the same way as for the free-electron model. We have displayed the results for the derivative of the real and imaginary parts of the surface impedance $Z = R - iX$ with respect to the applied magnetic field B in Figures 5.8 and 5.9 assuming that $\omega_c\tau = 30$ at $B = 18$ kG. The curves for dR/dB and dX/dB are compared with the results for the free-electron gas for the same value of $\omega_c(18)\tau$. We see that the minimum in (dX/dB) occurs at $B = 18.68$ kG for the free-electron model and at $B = 18.0$ kG for a model with a SDW ground state. We see, then, that if potassium had a spin-density-wave ground state, B_X at $\nu = 10$ Mc/sec would occur

6) The dielectric constant ϵ_+ is related to $\sigma_{xx} - i\sigma_{xy}$ by the relation $\sigma_{xx} - i\sigma_{xy} = (i\omega/4\pi)(\epsilon_+ - 1) \approx (i\omega\epsilon_+/4\pi)$.

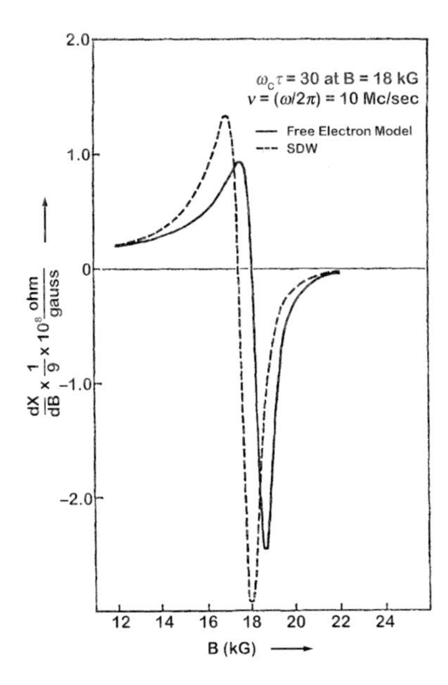

Figure 5.9 dX/dB for potassium with $\omega_c(18)\tau = 30$ for both the free-electron and SDW models.

at a magnetic field of about 18.0 kG. It is interesting to notice that this agrees with Taylor's observations, especially since the calculation employed no adjustable parameter. Numerical methods that would guarantee at least four figure accuracy were used throughout the calculations.

It is surprising that B_X for the SDW model is only 4% less than B_X for the free-electron model. One might have expected a larger deviation because v_{max} for the SDW model is 17% less than v_F. In fact, such a large shift is predicted for the location of the Kjeldaas edge of acoustic waves [20]. Evidently the helicon dispersion relation near the Kjeldaas edge is more sensitive to the nature of the electronic model than is the case for acoustic waves.

Acknowledgements

The authors are grateful to T. Kushida for providing accurate experimental confirmation of the discrepancy with free-electron theory of the helicon absorption edge in potassium. We also wish to thank P.R. Antoniewicz and K.M. Brown for their assistance with the numerical calculations, and J.C. McGroddy, J.L. Stanford, and E.A. Stern for communicating their results of similar, independent research prior to publication.

References

1 Aigrain, P. (1961) *Proceedings of the International Conference on Semiconductor Physics, Prague, 1960*, Czechoslovak Academy of Sciences, Prague, p. 224.
2 Bowers, R., Legendy, C., and Rose, F. (1961) *Phys. Rev. Lett.*, **7**, 339.
3 Quinn, J.J. and Rodriguez, S. (1964) *Phys. Rev.*, **133**, A1589.
4 Kjeldaas, T. Jr., (1959) *Phys. Rev.*, **113**, 1473.
5 Stern, E.A. (1963) *Phys. Rev. Lett.*, **10**, 91.
6 Miller, P.B. and Haering, R.R. (1962) *Phys. Rev.*, **128**, 126.

7 Kirsh, J. (1964) *Phys. Rev.*, **133**, A1390.

8 Buchsbaum, S.J. and Gait, J.K. (1961) *Phys. Fluids*, **4**, 1514.

9 Taylor, M.T., Merrill, T.R., and Bowers, R. (1964) *Phys. Letters*. **6**, 159.

10 Taylor, M.T. (1964) *Phys. Rev. Lett.*, **12**, 497.

11 Taylor, M.T. (1965) *Phys. Rev.*, **137**, A1145.

12 Barrett, C.S. (1956) *Acta Cryst.*, **9**, 671.

13 Overhauser, A.W. (1964) *Phys. Rev. Lett.*, **13**, 190.

14 El Naby, M.H. (1963) *Z. Physik*, **174**, 269.

15 Mayer, H. and El Naby, M.H. (1963) *Z. Physik*, **174**, 280, 289.

16 Eckstein, S.G. (1964) *Bull. Am. Phys. Soc*, 9, 550, and private communication.

17 Overhauser, A.W. (1962) *Phys. Rev.*, **128**, 1437.

18 Shoenberg, D. and Stiles, P. (1964) *Proc. Roy. Soc.* (London), **A281**, 62.

19 Grimes, C.C. and Kip, A.F. (1963) *Phys. Rev.*, **132**, 1991.

20 Alig, R.C., Quinn, J.J., and Rodriguez, S. (1965) *Phys. Rev. Lett.*, **14**, 981.

Reprint 6 Exchange and Correlation Instabilities of Simple Metals[1)]

A.W. Overhauser *

* Scientific Laboratory, Ford Motor Company, 525, Northwestern Avenue, Dearborn, Michigan 47907, USA

Received 25 September 1967

The influence of electron-electron correlation on exchange instabilities of a metal is examined. The employment of screened interactions does not constitute a proper treatment. Correlation effects suppress ferromagnetic instabilities, as is well known, but they need not suppress instabilities of the spin-density-wave type. On the contrary, it is shown that correlation enhances exchange instability of the charge-density-wave type. For either type, the wave vector of such a state adjusts so that the Fermi surface makes critical contact with the energy gaps introduced by the instability. This circumstance optimizes the correlation energy. The observed conjunction of the long-period-superlattice periodicity with the Fermi surface in order-disorder alloys is probably an example of this phenomenon. It is suggested that charge-density-wave ground states are likely in simple metals having weak Born–Mayer ion–ion interactions, such as the alkali metals. The intensity of Bragg reflection satellites caused by a concomitant positive-ion modulation is computed.

6.1 Introduction

Exchange interactions among itinerant electrons tend to cause magnetic instabilities, an effect first discussed by Bloch [1]. The extensive bibliography which has accumulated on this topic is summarized and enlarged upon by Herring [2]. The prevailing opinion is that correlation corrections to the Hartree–Fock approximation always cancel substantially the effects of exchange. This was shown originally for ferromagnetic instabilities by Wigner [3, 4].

The influence of electron-electron correlation on the stability of a spin-density-wave (SDW) state has not been adequately investigated. The question is critical because there are always SDW states of lower Hartree–Fock energy than the normal state for all electron densities [5]. One purpose of this paper is to show that modifications of the electronic density of states $N(E)$ by, say, SDW energy gaps augment the correlation energy and enhance the instability.

This effect is pertinent to a general exchange-instability wave, which we now define. Consider a (supposed) electronic ground state for which the spin-up and spin-down electron densities are

$$\begin{aligned} \rho^{+}(\mathbf{r}) &= \frac{1}{2}\rho_0\left[1 + p\cos(\mathbf{Q}\cdot\mathbf{r} + \varphi)\right], \\ \rho^{-}(\mathbf{r}) &= \frac{1}{2}\rho_0\left[1 + p\cos(\mathbf{Q}\cdot\mathbf{r} - \varphi)\right]. \end{aligned} \tag{6.1}$$

The mean electron density is p_0, and the fractional modulation is p. We shall refer to a state for which the phase φ is 0 as a charge-density wave (CDW). The three possible types of

1) Phys. Rev. 167, 3 (1968)

Anomalous Effects in Simple Metals. Albert Overhauser

ISBN: 978-3-527-40859-7

exchange-instability wave are

$$\begin{aligned} &\varphi = 0\,, &&\text{pure CDW}\,;\\ &\varphi = \tfrac{1}{2}\pi\,, &&\text{pure SDW}\,;\\ &0 < \varphi < \tfrac{1}{2}\pi\,, &&\text{mixed CDW-SDW}\,. \end{aligned}$$

At first sight, the possibility of a CDW ground state seems remote. A large Coulomb energy, the volume integral of $\mathcal{E}^2/8\pi$, is the obvious reason. However, our primary concern is not with an ideal electron gas having a rigid background of neutralizing positive charge; it is with a real metal, where the positive ions are more or less free to adjust their positions to minimize their local potential energy. For example, the Born–Mayer ion–ion interactions are known to be extremely weak in the alkali metals. Consequently, the equilibrium positions of the positive ions could be displaced from their ideal cubic sites, to cancel most of the Coulomb energy mentioned above.

The existence of an exchange-instability wave, described by (6.1), requires a nonconstant potential $V(\mathbf{r})$ in the one-electron Hamiltonian. It will have the form

$$V(\mathbf{r}) = A\sigma_z \sin \mathbf{Q}\cdot\mathbf{r} - C\cos\mathbf{Q}\cdot\mathbf{r}\,, \tag{6.2}$$

where A and C are the coefficients of the exchange and Coulomb contributions. Both will be proportional to the fractional modulation p. σ_z is the usual Pauli matrix. A spin-up electron will accordingly experience a potential

$$V(\mathbf{r}) = -G\cos(\mathbf{Q}\cdot\mathbf{r} + \varphi)\,. \tag{6.3}$$

A spin-down electron will experience a similar potential, but with the sign of φ reversed. The relationship between (6.2) and (6.3) is

$$A = G\sin\varphi\,, \quad C = G\cos\varphi\,.$$

The periodic potential (6.3) will introduce energy gaps of magnitude G in the one-electron energy spectrum $E(\boldsymbol{k})$. These will occur on planes perpendicular to the wave vector $\mathbf{Q}$, a distance $\frac{1}{2}Q$ from the origin of k space.

The effect of a CDW or SDW on the correlation energy cannot be determined with precision. It is convenient to take the second-order perturbation correction to the Hartree–Fock energy E_0 as the basis for discussion. In this crude approximation, the correlation energy is

$$W_c = -\sum_i |\langle 0|U|i\rangle|^2/(E_i - E_0)\,. \tag{6.4}$$

The matrix elements are those an an electron-electron interaction. (See the Appendix for an explicit interpretation of $\langle 0|U|i\rangle$.) The excited states $\{i\}$ are those for which two electrons have been removed from below the Fermi energy and placed in states above the Fermi energy, but with no change in total momentum. A CDW or SDW changes the numerical value of W_c by altering both the matrix elements and energy denominators of (6.4). We shall consider first the latter effect, since this has heretofore been entirely neglected. It is obvious that electron excitations of low energy play a proportionately greater role in the sum than those of high energy. Consequently, the electronic density of states $N(E)$near the Fermi energy E_F is of paramount interest.

$N(E)$ for electrons interacting with a CDW-SDW potential is shown in Figure 6.1. There are striking deviations from the $\sqrt{E}$ dependence appropriate to an energy spectrum without such a potential. Point A of Figure 6.1 corresponds to the constant-energy surface (in k space) which makes critical contact with the energy gaps introduced by the potential (6.3). (We shall show below that point A will coincide with E_F, but this may be ignored for the present.) Point B of Figure 6.1 corresponds to the constant-energy surface which just begins to include states on the high-energy side of the gap G. $N(E)$ is rigorously horizontal between A and B.This may be seen by considering a two-dimensional electron gas, with

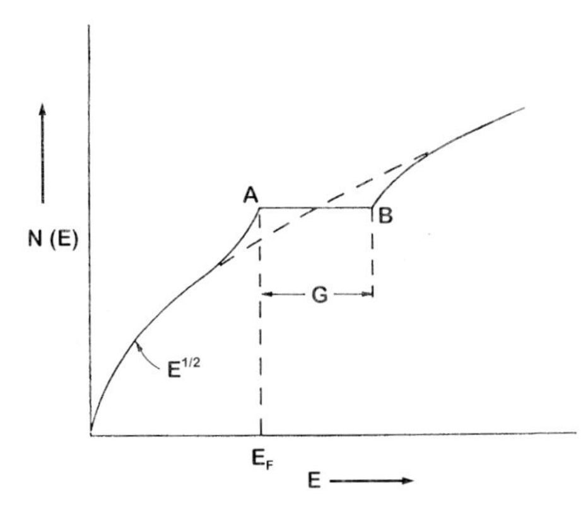

Figure 6.1 Electronic density of states versus energy for a free-electron gas. The parabolic curve corresponds to the normal, paramagnetic state. The solid curve, passing through points A and B, applies to a collective-electron deformation of the CDW-SDW type, having energy gap G.

$E(k) = \hbar^2 k^2/2m$. For this case, the (unit-area) density of states is

$$N_2(E) = m/2\pi\hbar^2 \,, \tag{6.5}$$

a constant. Consider now the three-dimensional electron gas to be sliced up perpendicular to $\mathbf{Q}$ into a large number of thin circular segments. Because (6.5) is constant, the increase in $N(E)$ with increasing E occurs only because new segments are continually being added. However, as soon as E reaches the value for which its surface just touches the energy-gap plane (critical contact), new segments are no longer added with increasing E. The surfaces of constant energy acquire necks at each energy gap, and $N(E)$ remains constant. Only when E exceeds its critical contact value by G, allowing further addition of new segments, does $N(E)$ resume its increase.

The location of A or B in Figure 6.1 depends on the magnitude of Q. If Q is large compared to the diameter $2k_F$ of the Fermi surface, the energy at which critical contact occurs will exceed E_F, and $N(E_F)$ will not differ appreciably from its normal value $N_0(E_F)$. Consider the change of $N(E_F)$ as Q is gradually reduced (with G held constant). The variation is shown in Figure 6.2. There is a cusplike maximum in $N(E_F)$ at point A, corresponding to critical contact. At this point,

$$N(E_F) \approx N_0(E_F)\left[1 + (G/4E_F)\right]. \tag{6.6}$$

This incremental increase in $N(E_F)$ can be deduced from Figure 6.1, to terms linear in G. It is just the slope of $\sqrt{E}$ (at E_F) times $\frac{1}{2}G$. The value of Q corresponding to critical contact also depends on G, because the constant-energy surfaces are appreciably distorted by the potential (6.3). The equation of the surface in critical contact has been derived previously [6]. It requires

$$Q \approx 2k_F\left[1 + (G/4E_F)\right]. \tag{6.7}$$

Consequently, the wave vector of a CDW or SDW in critical contact with the Fermi surface is slightly larger than $2k_F$.

Since the correlation energy W_c is negative, a larger magnitude for W_c is associated with greater stability. Obviously, modifications of the one-electron $E(\boldsymbol{k})$ which increase the proportion of low-energy virtual excitations contributing to (6.4), increase this magnitude and reduce the total energy. It follows that the upward cusp at A in Figure 6.2 will correspond to a downward cusp in the total energy. Therefore, if the ground state contains a CDW or SDW, the optimum Q will be that given by (6.7) for critical contact.

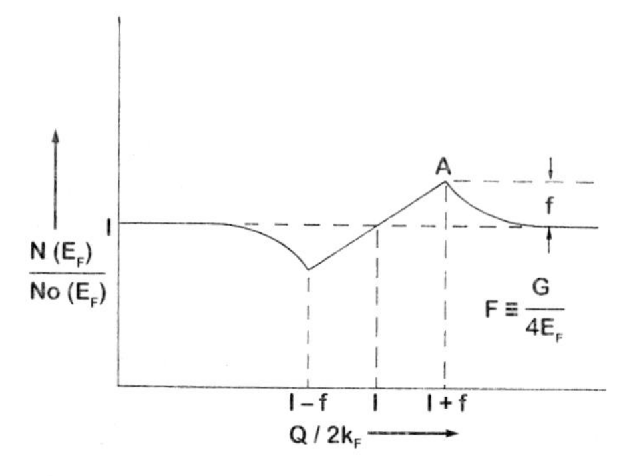

Figure 6.2 Electronic density of states at the Fermi energy versus wave vector Q of an exchange-instability wave, for a given energy gap G. The optimum Q [maximum $N(E_F)$] is slightly larger than $2k_F$ and corresponds to critical contact of the Fermi surface with the energy-gap planes in k space.

The foregoing argument assumes that the cusplike behavior of $N(E_F)$ provides the sharpest variation to the Q dependence of the total energy. In a Hartree–Fock approximation, the optimum Q will be smaller than $2k_F$, since the Hartree–Fock energy decreases smoothly with decreasing Q near $Q \sim 2k_F$. This has been called the "clamp-down" effect [5], which describes the tendency of SDW energy gaps to clamp down on the Fermi surface, causing neck formation. This effect is still present in a Hartree–Fock treatment using screened interactions.[2] However, such calculations with realistically screened interactions do not show SDW-type instabilities [2, 7, 8]. It follows that if a CDW or SDW were the ground state of a typical metal, the correlation-energy stabilization arising from the enhanced $N(E_F)$ has dominated, and should likewise dominate the determination of Q.

The fallacy of all calculations employing screened interactions in attempts to discuss delicate questions such as stability is apparent. The appropriate screened interaction, if such is ever possible, must certainly depend on the character of the deformation (i.e., Q) and the amplitude of the deformation p. Even for ferromagnetic deformations ($Q = 0$), the remaining dependence on p must be included [9]. To ignore the deformation parameters in U, as almost everyone does, is to beg the question. Unfortunately, it is difficult to envision how to determine U. The alternative is to compute deformed one-electron wave functions and energy spectra for employment in an expression such as (6.4). But the computational task seems prohibitive.

Consider the $N(E_F)$ variation for a ferromagnetic polarization of conduction electrons. This is easily derived:

$$N(E_F) = \frac{1}{2} N_0(E_F) \left[(1 + P)^{1/3} + (1 - P)^{1/3} \right], \tag{6.8}$$

where P is the fractional polarization. This variation is shown in Figure 6.3, along with that derived above for a CDW-SDW state. In contrast with the latter case, a ferromagnetic polarization decreases the density of states at the Fermi surface. Thus the correlation energy is reduced in magnitude, inhibiting the deformation. The $N(E_F)$ decrease is a contributing factor to the suppression of a ferromagnetic instability by correlation corrections. An argument

2) These authors ignored the correlation-energy contribution leading to critical contact. Consequently, their conclusions regarding neck formation are invalid. Another serious error in their work is a claim to have shown that finite-amplitude SDW instabilities could occur before differential ones. For the finite case, their calculation allowed repopulation of k space, as is required, in order that the Fermi surface be one of constant energy. This should also have been done in the differential case, since the factorized interaction empolyed for a fixed Q would require a nonspherical Fermi surface even in the zero-amplitude limit. Their apparent failure to do this prevents a meaningful comparison of the two cases [7].

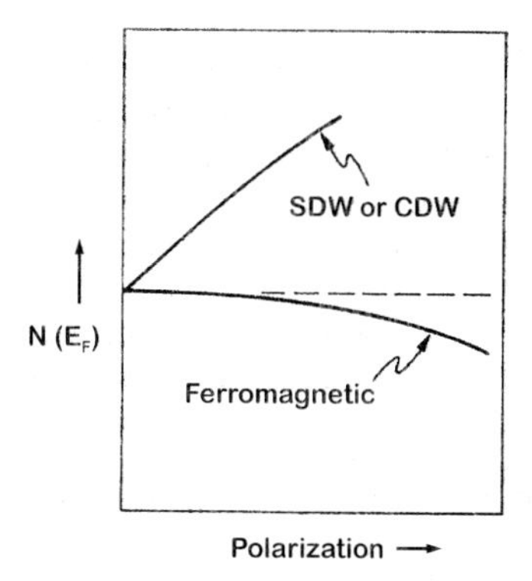

Figure 6.3 Electronic density ot states at the Fermi energy versus fractional polarization for ferromagnetic, SDW, and CDW deformations.

is frequently advanced that SDW instabilities will be suppressed by correlation corrections analogous to the ferromagnetic case. We emphasize that the deformations have opposite behavior in a very crucial aspect, so no analogy exists.

We have shown that the two common arguments for discounting the likelihood of anomalous conduction-electron ground states are incorrect.[3] This by no means establishes the contrary. The energy differences between paramagnetic, CDW and SDW states are probably much smaller than the accuracy of many-body calculations. Experimental evidence[4] will possibly provide the only information on such questions for the near future.

The negative cusp in correlation energy that occurs when the Fermi surface is near critical contact with an energy gap should be a general phenomenon. Sato and Toth[5] have established experimentally that the periodicity of the long-period superlattice in CuAu and other alloys is governed by such a critical-contact requirement. The generality of their results is hard to explain without a cusplike term in the energy versus super lattice periodicity. The effect we have described provides a promising explanation.

6.2 Matrix-Element Contributions to the Correlation Energy

In Section 6.1, we have discussed qualitatively the effect of energy-level shifts on the correlation energy (6.4). Here we shall estimate the effect of matrix-element changes for virtual excitation of electron–hole pairs. We shall find that these are also important.

To achieve perspective, consider again a half-filled conduction band with fractional ferromagnetic polarization P. The number of parallel-spin pairs per unit volume is

$$n_P = \frac{1}{8}\left[(1+P)^2 + (1-P)^2\right]\rho_0^2 = \frac{1}{4}\rho_0^2(1+P^2) ,$$

3) Herring [2](p. 114) presents a third argument, but because of a logical error, this is also incorrect. He invokes a lemma, proved on p. 15, that electron correlations eliminate any energy contributions from repulsive δ-function interactions. He then argues: Real interactions can be divided into high-q and low-q components. The latter determine the plasma frequency, which is independent of magnetic deformation. Appeal is then made to the lemma to show that the high-q components can have negligible effect on magnetic stability. However, we observe that proof of the lemma does not derive from the high-q components of a δ function. It depends on the fact that a δ function is a precisely phased synthesis of low-q and high-q components. A δ function with its low-q parts deleted would have no such lemma.

4) In a subsequent paper, the author will summarize the extensive, yet still inconclusive, evidence that alkali metals have anomalous electronic ground states.

5) [10–12] have attempted to explain the behavior by treating the superlattice potential in perturbation theory. They found relatively smooth and shallow minima in energy versus Q for superlattice Q's near the required values. However, they neglected several Q-dependent contributions to the energy which will probably shift the minima and spoil agreement.

where ρ_0 is the electron density. Since the exchange energy is proportional to n_P (if all exchange integrals are taken equal),

$$W_{\text{ex}}(P)/W_{\text{ex}}(0) \sim 1 + P^2 \,. \tag{6.9}$$

The increase of $|W_{\text{ex}}|$ with P^2 is the major contribution favoring magnetic instability. As is well known, the correlation energy opposes such a tendency. This is easily understood by considering Equation (6.4). The dominant contributions to the sum over virtual states are those from electron–hole-pair excitations having antiparallel spin. (The parallel-spin pairs have smaller matrix elements because of cancellation by an exchange term.) If wave-vector conservation is ignored, the total number of antiparallel-spin electron–hole excitations in a half-filled band is

$$n_A = \left(\frac{1}{2}\rho_0\right)^4 (1+P)(1-P)(1-P)(1+P) \,.$$

Since the correlation energy is crudely this number multiplied by an average matrix element and divided by an average-energy denominator,

$$W_c(P)/W_c(0) \sim (1-P^2)^2 \sim 1 - 2P^2 \,. \tag{6.10}$$

The last approximation is limited to small P. The decrease of $|W_c|$ with P^2 only partially cancels the increase of $|W_{\text{ex}}|$, since generally $|W_c(0)|$ is three or four times smaller than $|W_{\text{ex}}(0)|$.

The foregoing analyses of exchange- and correlation-energy changes fail when exchange-instability waves [Equation (6.1)] are considered. In this case, the numbers of spin-up and spin-down electrons remain the same when the amplitude p of the fractional modulation deviates from zero. It is clear, however, that variations similar to (6.9) and (6.10) must still occur. These are to be found instead in matrix elements, as we now show. Suppose, for the sake of simplicity, that all of the modulated wave functions for occupied states can be written

$$\psi_k = e^{i\mathbf{k}\cdot\mathbf{r}}\left(1 + \frac{1}{2}p\cos\mathbf{Q}\cdot\mathbf{r}\right) \Big/ \left(1 + \frac{1}{8}p^2\right)^{1/2} \,. \tag{6.11}$$

The exchange energy associated with two of these functions is, by a tedious evaluation,

$$\langle \boldsymbol{k}, \boldsymbol{k}'|U|\boldsymbol{k}', \boldsymbol{k}\rangle \cong U(\boldsymbol{k}'-\boldsymbol{k}) + \frac{1}{4}p^2\left[U(\boldsymbol{k}'-\boldsymbol{k}+\boldsymbol{Q}) + U(\boldsymbol{k}'-\boldsymbol{k}-\boldsymbol{Q})\right], \tag{6.12}$$

with p^4 and higher terms neglected. $U(\boldsymbol{k})$ is the Fourier transform of $U(\mathbf{r})$. If we were to neglect the $\boldsymbol{k}$ dependence of $U(\boldsymbol{k})$, we would have

$$\langle \boldsymbol{k}, \boldsymbol{k}'|U|\boldsymbol{k}', \boldsymbol{k}\rangle \approx U(\boldsymbol{k}'-\boldsymbol{k})\left(1 + \frac{1}{2}p^2\right).$$

This agrees with (6.9), since $\frac{1}{2}p^2$ is the mean-square fractional polarization for SDW deformations and correspond to P^2 in the ferromagnetic case.

The correlation energy involves the matrix element for processes in which two electrons in occupied states $\boldsymbol{k}, \boldsymbol{k}'$ (having opposite spin) are virtually excited to empty states $\boldsymbol{k}+\boldsymbol{q}$, $\boldsymbol{k}'-\boldsymbol{q}$. It is important to know whether these latter states are above or below the energy gap. Assume first that they are both below, which is the predominant case. Then both $\psi_{\boldsymbol{k}}$ and $\psi_{\boldsymbol{k}+\boldsymbol{q}}$ can be taken to have the form (6.11). For an SDW deformation, $\psi_{\boldsymbol{k}'}$ and $\psi_{\boldsymbol{k}'+\boldsymbol{q}}$ will have the form

$$\psi_{k'} = e^{i\mathbf{k}'\cdot\mathbf{r}}\left(1 - \frac{1}{2}p\cos\mathbf{Q}\cdot\mathbf{r}\right) \Big/ \left(1 + \frac{1}{8}p^2\right)^{1/2} , \tag{6.13}$$

since the down-spin modulation is 180° out of phase with the spin-up modulation. The matrix element of the virtual excitation can be evaluated directly (all but 19 of the 81 terms which arise are 0:

$$\langle k' - q, k + q | U | k', k\rangle \cong U(q) - \frac{1}{4}p^2 \left[U(q + Q) + U(q - Q)\right], \tag{6.14}$$

where p^4 and higher terms have again been neglected. Were we to neglect the q dependence of $U(q)$,

$$|\langle k' - q, k + q | U | k', k\rangle|^2 \approx U(q)^2 \left[1 - 2\left(\frac{1}{2}p^2\right)\right],$$

which agrees with (6.10). (It is the behavior of the square-matrix element that must be compared in this case.) Naturally, it is not correct to assume that all ψ_k have the same p [or, for that matter, that the $\exp(i\mathbf{Q} \cdot \mathbf{r})$ and $\exp(-i\mathbf{Q} \cdot \mathbf{r})$ components of the modulation have equal coefficients], as we have done for the sake of simplicity. Matrix elements of general validity, corresponding to (6.12) and (6.14), can with patience be written down, but that is not necessary for the present purpose; as long as $U(q)$ is positive, all of the correction terms that would arise have the same sign.

We have shown that the usual suppression of a ferromagnetic instability by correlation energy persists in the SDW case, but manifests itself as a decrease in magnitude of the matrix elements appearing in Equation (6.4). In Section 6.1, we showed that this decrease is partly compensated by an increase resulting from changes in $N(E)$, provided Q has the value required for critical contact. Such cancellation cannot occur for a ferromagnetic polarization.

Consider now the behavior of the correlation energy for a CDW deformation. The only difference from the SDW case above is that the modulation of the spin-up and spin-down wave functions are now in phase. Since the most numerous excited states are again those for which both virtually excited electrons are below the energy gap, all four states which determine the matrix element can be taken to have the form (6.11). We then find

$$\langle k' - q, k + q | U | k', k\rangle \cong U(q) + \frac{1}{4}p^2 \left[U(q + Q) + U(q + Q)\right]. \tag{6.15}$$

The correction terms are all positive. This result is important because it shows that correlation energy enhances a CDW instability, opposite to the behavior in the magnetic case. This is in sharp contrast to the exchange-energy trend, which is always the same. (Equation (6.12) applies generally, irrespective of the phase φ in Equation (6.1).)

The foregoing conclusions are easy to anticipate physically. The correlation energy is a consequence of virtual scattering of pairs of electrons having (predominantly) antiparallel spin. Wave-function deformations which partially localize both probability densities in identical regions magnify this mechanism. Deformations which localize them in contiguous but separate regions diminish it.

Contemporary literature in the theory of metals is often characterized by a total neglect of exchange and correlation effects, presumably based on a naive hope they thay always almost cancel. Many workers pretend that correlation effects are somehow adequately accounted for by carrying out Hartree–Fock calculations with screened interactions.[6] We have shown that such an artifice can be qualitatively wrong. Phonon spectra and electron-phonon interaction are examples of phenomena that are sensitive, and probably need reinvestigation. If a modified-interaction artifice must be employed for pragmatic reasons, "antiscreening" will

6) Most workers seem content to screen the Fock term without also screening the Hartree term.

sometimes be the appropriate choice. The alternatives depend on wave vector, as we now show.

When $Q \geq 2k_F$, which is the case we have been considering, almost all low-lying excited states are below the energy gap. This can be understood from Figure 6.1. The contribution to $N(E)$ from states below the gap corresponds to the area of the curve below an extended horizontal line running through points A and B. The only excited states that are above the gap (so the spatial phase of their modulation is reversed) are those above the line, to the right of point B. Such states of low excitation energy are obviously very few in number. This is no longer the case if Q is small, allowing energy gaps to cut through the occupied region of k space. In this case, virtual excitations for which one or both excited states lie above the gap must be evaluated. If only one excited state is above,

$$\langle \boldsymbol{k}' - \boldsymbol{q}, \boldsymbol{k} + \boldsymbol{q} | U | \boldsymbol{k}', \boldsymbol{k} \rangle \cong U(q) \left(1 - \frac{1}{4} p^2 \right), \tag{6.16}$$

instead of (6.14) or (6.15). If both excited states are above (and both initial states below),

$$\langle \boldsymbol{k}' - \boldsymbol{q}, \boldsymbol{k} + \boldsymbol{q} | U | \boldsymbol{k}', \boldsymbol{k} \rangle \cong U(q) \left(1 - \frac{1}{2} p^2 \right). \tag{6.17}$$

The results given by (6.16) and (6.17) are obtained for both SDW and CDW modulations. Consequently, for such excitations, the correlation energy is diminished by the modulation. Numerous other cases must also be evaluated, since the locations of all four states relative to the energy gaps are pertinent.

The general trend is indicated. Exchange and correlation reinforce one another for charge modulations with Q sufficiently large. The value of Q at which virtual transitions having this property tend to predominate is likely near $Q = k_F$, for which the probabilities of low-energy excitations being above and below the gap are equal.

The ability of exchange and correlation to reinforce one another rather than cancel when spin-up and spin-down modulations are in phase has been emphasized previously in a different context [13].

6.3 Parallel-Spin Correlation and Umklapp Correlation

We were primarily concerned, in the previous sections, with the variations in antiparallel-spin correlation caused by an exchange-instability wave. Parallel-spin correlation (in excess of the exchange energy that arises directly from the Pauli exclusion principle) is considerably smaller in magnitude. This can be seen by the following argument. Consider the excited states $\{|i\rangle\}$ that enter Equation (6.4). They are enumerated by three wave vectors $\boldsymbol{k}, \boldsymbol{k}'$, and $\boldsymbol{q}$. $\boldsymbol{k}$ and $\boldsymbol{k}'$ are the one-electron states emptied by excitation to $\boldsymbol{k}+\boldsymbol{q}, \boldsymbol{k}'-\boldsymbol{q}$. There are $\frac{1}{4}\rho_0^2$ initial pairs of parallel spin, half for spin-up pairs and half for spin-down. This is the same number of $\boldsymbol{k}, \boldsymbol{k}'$ pairs having antiparallel spin. It seems at first sight that the total number of excited states $|i\rangle$ would be the same. However, for parallel-spin pairs, the final state $\boldsymbol{k} + \boldsymbol{q}, \boldsymbol{k}' - \boldsymbol{q}$ is the same as $\boldsymbol{k}' - \boldsymbol{q}, \boldsymbol{k} + \boldsymbol{q}$. Consequently, the number of distinguishable excited states is fewer by a factor of 2.

Furthermore, square-matrix elements of parallel-spin excitations are smaller on the average than their anti-parallel-spin counterparts, since direct and exchange terms have opposite sign:

$$|\langle 0 | U | i \rangle^2 = \left[U(q) - U(\boldsymbol{k}' - \boldsymbol{k} - \boldsymbol{q}) \right]^2 .$$

Suppose, for example, that the probability distribution for $|U(q)|$ were uniform in an interval $(0, U_m)$. For anti-parallel-spin pairs,

$$\langle U^2 \rangle_{\text{av}} = U_m^2 \int_0^1 x^2 dx = \frac{1}{3} U_m^2 .$$

For parallel-spin pairs,

$$\langle U^2 \rangle_{\text{av}} \approx U_m^2 \int_0^1 \int_0^1 (x-y)^2 dx dy = \frac{1}{6} U_m^2 ,$$

which is smaller by a factor of 2. Combining all factors? we conclude that parallel-spin correlation is approximately 20% of the total correlation energy. This estimate depends on the probability distribution assumed above, but is not very sensitive to reasonable modifications.

Our present interest is the parallel-spin correlation-energy changes introduced by ferromagnetic, SDW, and CDW deformations. A crude argument for the ferromagnetic case, similar to that employed in obtaining Equation (6.10), leads again to Equation (6.10). This result is understandable. For a half-filled band with $P = 1$, no virtual excitations within the band are possible, so each contribution to the correlation energy must decrease as $P \to 1$.

Matrix-element changes for SDW and CDW parallel-spin excitations are identical, and are given by Equation (6.15). Both the direct and exchange terms, say, U and U', are increased in magnitude to $U + p^2 u$, $U' + p^2 u'$. The square-matrix element to order p^2 is

$$(U - U')^2 + 2p^2(u - u')(U - U') .$$

Because the wave vector $\mathbf{Q}$ appearing in Equation (6.15) is large, the relative magnitudes of U and u and of U' and u' will be uncorrected. Consequently, there will be no predictable trend in the sign of the p^2 correction term above. For this case, we conclude that the over-all effect of the density modulation via matrix elements is negligible. However, the Fermi-energy density-of-states effect, discussed in Section 6.1, plays a proportionate role here also, enhancing the correlation energy of CDW-SDW states attributable to parallel-spin excitations.

For the paramagnetic and ferromagnetic states, the one-electron states are eigenfunctions of momentum, and the virtual excitations conserve momentum. This is not the case for CDW and SDW ground states. We still catalog the one-electron wave functions with wave-vector labels, but they are no longer eigenfunctions of momentum. The most important virtual excitations are those which conserve wave-vector label; and these are the only ones we have discussed until now. These excitations can still be shown to conserve expectation value of momentum; but one must introduce the entire metal center-of-mass coordinate and include its recoil. The recoil energy is of course completely negligible and does not affect the correlation energy.

Modulation of the one-electron wave function by the CDW-SDW potential introduces an entirely new class of virtual excitations that contribute to the correlation energy, namely, those for which

$$(\boldsymbol{k}, \boldsymbol{k}') \to (\boldsymbol{k} + \boldsymbol{q}, \boldsymbol{k}' - \boldsymbol{q} \pm \mathbf{Q}) .$$

The matrix elements of these umklapp excitations will be proportional to p. They necessarily enhance the correlation energy of an exchange-instability wave by an amount proportional to p^2 We shall refer to this contribution as umklapp correlation. It favors CDW or SDW instability. Which type benefits more is hard to assess. There are also virtual excitations for which wave-vector label is nonconserved by $\pm 2\mathbf{Q}$, $\pm 3\mathbf{Q}$, and $\pm 4\mathbf{Q}$. The matrix elements are of higher order in p, so that their contributions can be neglected.

6.4 Charge-Density-Wave Instabilities

The pertinent contributions to the exchange instability of an electron gas are summarized in Table 6.1 for ferromagnetic, SDW, and CDW deformations. There is always an increase in total kinetic energy, together with a countervailing increase in magnitude of the exchange energy. The SDW-instability theorem [5] proves that the latter always dominates the former for CDW and SDW deformations. Only for a low-density electron gas can this occur in the ferromagnetic case. The effect of correlation energy differs for all three cases. It suppresses a ferromagnetic instability and enhances a CDW instability. The net effect in the SDW case is uncertain because the correlation contributions have opposite sign. The degree of cancellation cannot be determined without an elaborate numerical calculation.

Perhaps the most surprising conclusion of the present analysis is the strongly indicated possibility that CDW instabilities can occur in a simple metal. The (unfavorable) electric-field energy $\mathcal{E}^2/8\pi$ will be cancelled in large part by a displacement of the positive ions from their normal equilibrium sites. The fractional charge modulation of the electrons is

$$p = (3G/4E_F)\left\{1 + \left[(1-u^2)/2u\right]\ln\left[(u+1)/(u-1)\right]\right\}, \tag{6.18}$$

where

$$u \equiv Q/2k_F \approx 1 + (G/4E_F) .$$

The last equality is equivalent to (6.7). The result (6.18) is obtained by perturbation theory and differs insignificantly from an exact calculation. Suppose now, that the ions of a monovalent metal have normal lattice sites $\{\boldsymbol{L}\}$. If the displacement $s(\boldsymbol{L})$ of the ions from these sites is

$$s(\boldsymbol{L}) = (p\mathbf{Q}/Q^2)\sin\mathbf{Q}\cdot\boldsymbol{L} , \tag{6.19}$$

the positive-ion charge modulation will just cancel the electronic CDW. The electrostatic potential arising from this modulation must of course be included in the one-electron potential (6.2).

Naturally the ion displacements will not be quite so large as that given by (6.19). The Born–Mayer ion–ion interactions will resist a static phononlike modulation of the lattice periodicity. However, it is well known that in alkali metals this interaction is very weak and, for example, contributes little to the bulk modulus. For this reason alkali metals are perhaps the best candidates for the occurrence of CDW instabilities. The Coulomb interaction between ions,

Table 6.1 Energy contributions to ferromagnetic, SDW, and CDW instabilities. F indicates a favorable and U an unfavorable contribution.

Energy contribution	Ferro	SDW	CDW
Kinetic	U	U	U
Exchange	F	F	F
$\mathcal{E}^2/8\pi$	...	...	U
Correlation, $N(E_F)$	U	F	F
Correlation, $\lvert\langle 0\lvert U\rvert i\rangle\rvert^2$	U	U	F
Correlation, ‖ spin ‖	U	F	F
Correlation, umklapp	...	F	F

dielectrically screened by the conduction electrons, will also be altered by the lattice modulation. This change, which excludes contributions from Fourier components $\pm\mathbf{Q}$, will also be small compared to the $\mathcal{E}^2/8\pi$ energy that can be eliminated by the lattice modulation (6.19). The importance of the aforementioned ion–ion interactions, though small, lies in their dependence on the orientation of $\mathbf{Q}$ relative to the crystal axes. Determination of the optimum direction of $\mathbf{Q}$ is an important problem.

Modulation of the positive-ion lattice, given approximately by (6.19), will necessarily accompany a CDW ground state of the conduction-electron system. This will give rise to satellite Bragg reflections of neutrons or x rays. Observation of satellites would provide unambiguous evidence for such a state. However, the expected intensities are very weak, as we now show.

Consider a monatomic Bravais lattice of point ions. The positive-ion density is

$$\rho(\mathbf{r}) = \sum_L \delta[\mathbf{r} - L] ,$$

where $\delta[\mathbf{r}]$ is a δ function. Suppose the lattice positions are modulated according to (6.19). Then

$$\rho(\mathbf{r}) = \sum_L \delta[\mathbf{r} - \mathbf{L} - (p\mathbf{Q}/Q^2)\sin\mathbf{Q}\cdot L] . \tag{6.20}$$

Bragg reflections are found by taking the Fourier transform of (6.20):

$$\rho(K) = \int \rho(\mathbf{r})\exp(iK\cdot\mathbf{r})d^3r .$$

Integration yields

$$\rho(K) = \sum_L \exp\left\{iK\cdot\left[L + (p\mathbf{Q}/Q^2)\sin\mathbf{Q}\cdot L\right]\right\} .$$

The second term of the exponential can be expanded in a power series in p, which allows the sum over $\{L\}$ to be performed. One obtains a nonzero result only for scattering vectors,

$$K = 2\pi G, 2\pi G + \mathbf{Q}, 2\pi G - \mathbf{Q} ,$$

where $\{G\}$ are the reciprocal lattice vectors. If the Bragg-reflection intensity of the unmodulated lattice is I_0, the intensities of the allowed reflections for the modulated lattice are

$$I(2\pi G) = \left[1 - \frac{1}{2}p^2(2\pi G\cdot\mathbf{Q}/Q)^2\right] I_0 , \tag{6.21a}$$

$$I(2\pi G \pm \mathbf{Q}) = \frac{1}{4}p^2\left[(2\pi G \pm \mathbf{Q})\cdot\mathbf{Q}/Q^2\right]^2 I_0 . \tag{6.21b}$$

The satellite intensities (6.21b) are weak if experimental data are used to determine p. For potassium, the instability-wave energy gap is 0.62 eV [6], and $E_F = 2.1$ eV. Equation (6.18) predicts $p = 0.17$; so that $I(2\pi G \pm \mathbf{Q})/I_0 \sim 0.007$.

This relatively small intensity could be obtained only if the sample were a "single-**Q**" crystal. A "poly-**Q**" crystal, containing modulation domains of at most 24 different (but equivalent) orientations, would have individual satellite intensities $-3\times10^{-4} I_0$. Obviously these would be quite difficult to observe, even if the favored orientations of $\mathbf{Q}$ were known in advance.

In the absence of direct observation by Bragg reflection, indirect evidence for electronic ground states of the SDW-CDW type can occur in varied phenomena. Anomalous behavior, otherwise unexplained, has been reported in the alkali metals. Quantititive interpretation is possible in five areas: (a) optical absorption [6], (b) magnetoresistance,[7] (c) de Haas–van Alphen periodicity [15],[4] (d) helicon wave absorption,[8] and (e) positron annihilation [17]. Only the optical absorption depends sensitively on the phase angle φ of the CDW-SDW admixture [(6.1)] The theoretical treatment of this effect [6] assumed implicitly that the phase of the exchange-instability wave (regarded at the time as a pure SDW) was pinned to the positive-ion lattice. It was emphasized subsequently [18] that the absorption could not occur without pinning. The present investigation began as a theoretical study of SDW pinning, by incorporation of a CDW component. The surprisingly favorable correlation energy of the CDW component, shown in Section 6.2, leads to the conclusion that exchange-instability waves in simple metals are more likely of the pure-CDW type.

The foregoing conjecture solves a difficulty pointed out in the original work on SDW instabilities [19]. Large nuclear hyperfine fields ($\sim$ 500 kG in Cs) should be manifest, at least in a time-averaged, vestigial way. For the alkali metals, such evidence has never been found in nuclear-magnetic-resonance (NMR) experiments nor in the Mössbauer effect [20].

A Appendix

A well-known difficulty of Equation (6.4) is that W_c diverges when the matrix elements of e^2/r are employed. This is caused by the small-momentum-transfer singularity of Rutherford scattering, and arises from the long-range tail of the Coulomb interaction. Collective-screening effects in a metal prevent the dominance of virtual scattering by small-angle events. This restraint can be incorporated into the perturbed-ground-state wave function ψ by letting

$$\psi \cong |0\rangle - \sum_i |i\rangle\langle i| U_{sc}|0\rangle/(E_i - E_0)\,, \tag{A1}$$

where $|0\rangle$ is the ground-state Slater determinant, and $\{|i\rangle\}$ are those of higher energy having two electrons excited above the Fermi surface. U_{sc} is some appropriate screened Coulomb interaction between all electron pairs. It is clear, however, that U_{sc} cannot be used to calculate the energy. Only the exact Hamiltonian $T + U$ can be employed. In other words, we interpret (A1) as a variational wave function. Accordingly, the energy is

$$E = \langle\psi|T + U|\psi\rangle/\langle\psi|\psi\rangle\,. \tag{A2}$$

Naturally, the question arises about the choice of U_{sc}. From the variational point of view, this is discretionary. One might employ, for example,

$$U_{sc} = e^2 \sum_{ij} \exp(-\mu r_{ij})/r_{ij}\,, \tag{A3}$$

and determine the optimum value of μ by minimizing (A2). The results of such a calculation would provide an interesting comparison with other estimates of the correlation energy.[9]

In the event that $\langle\psi|\psi\rangle$ does not deviate too much from unity, (A2) can be expanded in powers of U and U_{sc}. To second order, we obtain

$$E \approx E_0 - \sum_i \left[\langle 0|U|i\rangle\langle i|U_{sc}|0\rangle + \langle 0|U_{sc}|i\rangle\langle i|U|0\rangle - |\langle 0|U_{sc}|i\rangle|^2\right]/(E_i - E_0)\,, \tag{A4}$$

7) The theory of this effect will be published by J.R. Reitz and the author [14].

8) Conflicting evidence from ultrasonic absorption is vitiated by severe elastic stress of specimens [16].

9) A computation similar to that proposed here has already been carried out and gives reasonable values for the correlation energy [21].

where $E_0 \equiv \langle 0|T + U|0\rangle$. By way of illustration, suppose (A3) were used for U_{sc}. Then the correction term of (A4) becomes

$$W_c \approx -(4\pi e^2)^2 \sum_i \left(q_i^2 + 2\mu^2\right) / q_i^2 \left(q_i^2 + \mu^2\right)^2 (E_i - E_0) \,, \tag{A5}$$

where $\hbar q_i$ is the momentum transfer involved in the transition to $|i\rangle$. Comparison of Equations (6.4) and (A5) provides an explicit (though somewhat arbitrary) interpretation of the matrix element appearing in (6.4):

$$\langle 0|U|i\rangle = 4\pi e^2 \left(q^2 + 2\mu^2\right)^{1/2} / q \left(q^2 + \mu^2\right) \,. \tag{A6}$$

The advantage of this identification is that long-range screening effects, as they effect small-angle virtual excitations, are incorporated by wave-function modification, as it should be done, and by keeping the exact Hamiltonian intact. The sum appearing in Equation (6.4) will of course converge.

Ideally, the energy of a CDW-SDW state should be computed by an *ab initio* repetition of the perturbation-variation scheme described above. The reason is that the optimum parameter μ of the screened interaction for the CDW-SDW state may differ from the value μ_0 that minimizes the energy of the paramagnetic state. The resulting correction to the CDW-SDW-state energy, however, is fourth order in p, the fractional modulation of the exchange-instability wave, as we now show. The total energy, to relevant order in p and $\mu - \mu_0$, is

$$W(p, \mu) = A + B(\mu - \mu_0)^2 + p^2 \left[C + D(\mu - \mu_0)\right] , \tag{A7}$$

where A, B, C, and D are appropriate parameters. This is optimized with respect to μ by setting $\partial W / \partial\mu = 0$. Accordingly,

$$\mu = \mu_0 - D p^2 / 2B \,.$$

Inserting this result back into (A7), we obtain

$$W(p) = A + C p^2 - (D^2/4B) p^4 \,,$$

which differs from $W(p, \mu_0)$ only by a term in p^4. Consequently, in calculating correlation-energy changes associated with CDW-SDW deformations, we may employ the effective matrix elements, e.g., (A6), derived for the paramagnetic state.

Finally, it should be appreciated that the approximate expression [Equation (6.4)] for the correlation energy is not a correction term in a standard Rayleigh–Schrödinger perturbation scheme. Rather, it is the term caused by first-order configuration-interaction corrections to a Hartree–Fock state. Brillouin's theorem [22] shows that the only configurations connected to a Hartree–Fock state in first order are those determinantal functions differing from the Hartree–Fock state by just two orbitals. Consequently, the only term in the Hamiltonian which can contribute to such matrix elements is the electron–electron interaction. One might suspect that this involves some double counting of the interaction, since the Hartree–Fock potential includes already some effects of the interaction. This is not so. The potential terms in the Hartree–Fock equations are one-electron operators; so that even if they were to be subtracted from U, they could not contribute to the configuration mixing allowed by Brillouin's theorem.

References

1 Bloch, F. (1929) *Z. Physik*, **57**, 545.
2 Herring, C. (1966) in *Magnetism*, (eds. T. Rado and H. Shul), Academic Press Inc., New York, Vol. VI.
3 Wigner, E.P. (1934) *Phys. Rev.*, **46**, 1002.
4 Wigner, E.P. (1938) *Trans. Faraday Soc.*, **34**, 678.
5 Overhauser, A.W. (1962) *Phys. Rev.*, **128**, 1437.
6 Overhauser, A.W. (1964) *Phys. Rev. Lett.*, **13**, 190.
7 Penn, D.R. and Cohen, M.H. (1967) *Phys. Rev.*, **155**, 468.
8 Hamann, D.R. and Overhauser, A.W. (1966) *Phys. Rev.*, **143**, 183.
9 Misawa, S. (1965) *Phys Rev.*, **140**, A1645.
10 Sato, H. and Toth, R.S. (1961) *Phys. Rev.*, **124**, 1844.
11 Sato, H. and Toth, R.S. (1962) *Phys. Rev.*, **127**, 469.
12 Tachiki, M. and Teramoto K. (1966) *J. Phys. Chem. Solids*, **27**, 335.
13 Overhauser, A.W. (1967) *Phys. Rev.*, **156**, 884.
14 Penz, P.A. and Bowers, R. (1967) *Solid State Commun.*, **5**, 341.
15 Shoenberg, D. and Stiles, P.J. (1964) *Proc. Roy. Soc.* (London), **A281**, 62.
16 Overhauser, A.W. and Rodriguez, S. (1966) *Phys. Rev.*, **141**, 431.
17 Gustafson, D.R. and Barnes, G.T. (1967) *Phys. Rev. Lett.*, **18**, 1.
18 Hopfield, J.J. (1965) *Phys. Rev.*, **139**, A419.
19 Overhauser, A.W. (1960) *Phys. Rev. Lett.*, **4**, 462.
20 Boyle, A.J.F. and Perlow, G.J. (1966) *Phys. Rev.*, **151**, 211.
21 Macke, W. (1950) *Z. Naturforsch.*, **5a**, 192.
22 Slater, J.C. (1963) *Quantum Theory of Molecule and Solids*, McGraw-Hill Book Co., New York, Vol. I, p. 259.

Reprint 7 Splitting of Conduction-Electron Spin Resonance in Potassium[1]

A.W. Overhauser and A.M. de Graaf**

* Scientific Laboratory, Ford Motor Company, 525, Northwestern Avenue, Dearborn, Michigan 47907, USA

Received 27 November 1967

The 0.5-G splitting of the conduction-electron spin resonance in potassium (at 4200 G) observed by Walsh, Rupp, and Schmidt can be quantitatively explained providing the conduction electrons are in a charge-density-wave (CDW) ground state. The Fermi-surface distortion caused by the CDW energy gap leads to an anisotropic conduction-electron g factor depending on the angle between **H** and the wave vector **Q** of the CDW. The extremal values of g, corresponding to $\mathbf{Q} \perp \mathbf{H}$ and $\mathbf{Q} \parallel \mathbf{H}$, differ by $(3V/8E_F)\Delta g$, where $\Delta g = -0.0025$ is the observed g shift. V is the observed threshold energy of the Mayer–El Naby optical-absorption anomaly, and E_F is the Fermi energy. The predicted maximum splitting is 0.56 G. Interpretation of the data requires the sample to have a macroscopic domain structure, caused by thermal stress and plastic flow when the potassium-Parafilm sandwich is cooled to He temperature. The orientation of **Q** in stress-free regions should be parallel to **H**. In regions of high stress, **Q** is presumed perpendicular to the surface, and therefore approximately perpendicular to **H**.

7.1 Introduction

Two years ago, Walsh, Rupp, and Schmidt [1] observed conduction-electron spin resonance (CESR) in extremely pure potassium. Linewidths as narrow as 0.13 G were obtained at a resonance field of 4200 G and a temperature of 1.3 °K. The observed g shift, caused by spin–orbit coupling, was $\Delta g = -0.025(4)$, and represents an experimental shift of 5.3 G. A very puzzling feature of their result was that the CESR signal split into two well-resolved components as the magnetic field was tilted away from an initial orientation parallel to the surface of the potassium. The splitting reached a maximum value of about 0.5 G at a tilt angle of 9°. At larger tilt angles the signal intensity diminished, and one component vanished rapidly. The splitting has been observed in a number of specimens [2] having sufficient purity to yield comparably narrow lines.

The possibility that the splitting is caused by an artifact arising from excitation of both faces of a thin sample has been ruled out by Lampe and Platzman [3], who computed the surface impedance spectra for a variety of configurations. They found line-shape variations but no splitting. The possibility that it is caused by new collective modes (paramagnetic spin waves), resulting from conduction-electron exchange interactions, was ruled out by Platzman and Wolff.[2] They indeed found that such modes exist theoretically and provide a quantitative explanation of the sidebands reported by Schultz and Dunifer [6]. The 0.5-G splitting, however, remained a mystery.[2]

An important characteristic of CESR is that individual electrons undergo several thousand quantum transitions among levels at the Fermi energy during a spin relaxation time T_2. Con-

1) Phys. Rev. 168, 3 (1968)

2) [4]. For a theory of short-wavelength paramagnetic spin waves, see [5].

Anomalous Effects in Simple Metals. Albert Overhauser

ISBN: 978-3-527-40859-7

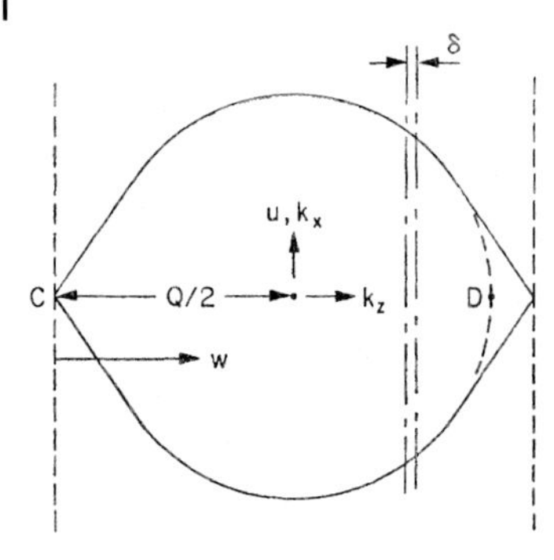

Figure 7.1 The lemon-shaped Fermi surface of a CDW ground state. The vertical dashed lines through the conical points, e.g., point C, are the planes in *k* space where the energy gaps caused by the CDW potential occur. The conical-point distortion is exaggerated by about a factor of 3.

sequently the resonance field is determined by the average g factor (the average over all levels that contribute to the paramagnetism). Since an averaging process necessarily results in a unique number, a single resonance must always be anticipated. This is strikingly confirmed by resonance experiments on dilute paramagnetic alloys [7, 8]. Paramagnetic Mn dissolved in Cu or Ag produces a single line with g factor intermediate between that of the Mn ion and the host metal. Spin exchange between conduction-band levels and localized paramagnetic levels is sufficiently rapid to merge the contributions into a single, sharp resonance line.

We conclude therefore that a split CESR indicates that the metal sample is heterogeneous. That is, we suggest that a K sheet (~ 0.02 cm thick) sandwiched between layers of Parafilm divides into contiguous regions having either of two possible g factors. This hypothesis requires that there are two distinguishable variants of K, and that both occur in samples where CESR splitting was observed. Had this occurred in Na, an immediate explanation would have been possible. Below 35° K, Na partially transforms martensitically [9] from bcc to hcp structure. Conceivably the two structures would have slightly different g factors. However, there is no similar transformation in K. It must also be appreciated that the splitting, though only 0.5 G, is relatively large, $\sim 10\%$ of the spin–orbit shift, which is the true measure of such an effect.

The purpose of this paper is to show that the foregoing hypothesis is a natural one provided the electronic ground state of the metal is a spin-density-wave (SDW) or charge-density-wave (CDW) state. For alkali metals a CDW state is preferred for theoretical [10] and experimental reasons. In this case, as with other alkali-metal anomalies that have been explained quantitatively [11–14] by assuming the ground state has an exchange instability wave, the interpretation is the same for either a CDW or SDW. The theoretical characteristic of either state is that each electron experiences a selfconsistent periodic potential,

$$U(\mathbf{r}) = V \cos \mathbf{Q} \cdot \mathbf{r} \,. \tag{7.1}$$

The wave vector $\mathbf{Q}$ is unrelated to the lattice periodicity, and for a simple metal, has a length just right for critical contact between the Fermi surface and the two energy gaps of magnitude V. This distorts the Fermi surface into a lemon shape as shown in Figure 7.1.

An immediate consequence of the distorted Fermi surface and modified electron wave functions, arising from (7.1), is that the average g factor depends on the angle between $\mathbf{Q}$ and the magnetic field $\mathbf{H}$. The principal axes of the g-factor tensor will of course correspond to $\mathbf{Q} \parallel \mathbf{H}$ and $\mathbf{Q} \perp \mathbf{H}$, and will be associated with the extremal values of g. In Section 7.2 we show that

$$g(\mathbf{Q} \perp \mathbf{H}) - g(\mathbf{Q} \parallel \mathbf{H}) \cong (3V/8E_F)\Delta g \,, \tag{7.2}$$

where E_F is the Fermi energy. Derivation of this result does not require an accurate theory of the conduction-electron g factor. All one needs is the simple result, valid for alkali metals [15], that the g shift of a nearly-free-electron Bloch state $\mathbf{k}$ is

$$\Delta g(\mathbf{k}) \cong -\mu k^2 \sin^2\theta \;, \tag{7.3}$$

where θ is the angle between $\mathbf{k}$ and $\mathbf{H}$. For a nearly spherical Fermi surface,

$$\Delta g = \langle \Delta g(\mathbf{k}) \rangle_{\text{av}} = -\frac{2}{3}\mu k_F^2 \;. \tag{7.4}$$

We need not compute the coefficient μ since the experimental Δg can be used in Equation (7.2).

The CDW energy gap V in K is 0.6 eV, the threshold energy of the Mayer–El Naby optical-absorption anomaly [16]. $E_F = 2.12$ eV. Consequently Equation (7.2) predicts that the maximum splitting is 10.6% of the observed Δg, which implies a maximum observed splitting of 0.56 G.

In Section 7.3 we indicate why a thin K sheet subjected to severe thermal stress on cooling would likely divide into contiguous patches having $\mathbf{Q}$ either parallel to $\mathbf{H}$ or perpendicular to the surface (and therefore perpendicular to $\mathbf{H}$). The remarkable quantitative agreement between the predicted maximum splitting and that observed provides significant additional evidence for a CDW state in K.

7.2 Anisotropy of g

In order to compute the conduction-electron g factor, starting from Equation (7.3), we must know the eigenfunctions and energy spectrum $E(\mathbf{k})$ of the Schrödinger equation having the potential-energy term (7.1). Compact solutions of this – the Mathieu equation – cannot be written down, so we make use of the following artifice which has sufficient accuracy. The perturbation (7.1) can be divided into two parts, one which leads to the energy gap for $k_z = -\frac{1}{2}Q$, the gap on the left in Figure 7.1, and the other which leads to the gap for $k_z = \frac{1}{2}Q$. Each part must be treated with great accuracy because wave functions and k-space occupation are significantly altered near the gaps. The simplifying feature of the problem is that the physical consequences of the two parts are additive. Accordingly, we shall compute the anisotropy of g arising from that part of the perturbation associated with the left-hand gap, and multiply the result by 2.

The part of (7.1) which we shall treat causes the unperturbed state $\mathbf{k}$ to be mixed with $\mathbf{k}+\mathbf{Q}$ only. That is, we solve the secular equation,

$$\begin{pmatrix} (p^2/2m) - E & \frac{1}{2}V\exp(-i\mathbf{Q}\cdot\mathbf{r}) \\ \frac{1}{2}V\exp(i\mathbf{Q}\cdot\mathbf{r}) & (p^2/2m) - E \end{pmatrix} = 0 \;, \tag{7.5}$$

within the space defined by the basis states $\mathbf{k}$ and $\mathbf{k}+\mathbf{Q}$. The energy for the solution below the gap is

$$E_{\mathbf{k}} = \frac{1}{2}(\epsilon_{\mathbf{k}} + \epsilon_{\mathbf{k}+\mathbf{Q}}) - \frac{1}{2}\left[(\epsilon_{\mathbf{k}} - \epsilon_{\mathbf{k}+\mathbf{Q}})^2 + V^2\right]^{1/2} , \tag{7.6}$$

where $\epsilon_{\mathbf{k}} = \hbar^2 k^2/2m$, the unperturbed energy. The corresponding eigenfunction is

$$\psi_{\mathbf{k}} = \cos\varphi\exp(i\mathbf{k}\cdot\mathbf{r}) - \sin\varphi\exp[i(\mathbf{k}+\mathbf{Q})\cdot\mathbf{r}] \;, \tag{7.7}$$

where

$$\cos\varphi(\mathbf{k}) = V/[V^2 + 4(\epsilon_{\mathbf{k}} - E_{\mathbf{k}})^2]^{1/2} . \tag{7.8}$$

It is of interest that Equation (7.6) is the exact solution of the spiral SDW problem [17, 18]. [For that case, the two terms of Equation (7.7) would have opposite spin states.]

The equation describing the Fermi surface is obtained by setting

$$E_{\mathbf{k}} = E_F = \left[\hbar^2\left(\frac{1}{2}Q\right)^2/2m\right] - \frac{1}{2}V . \tag{7.9}$$

The last equality is imposed because E_F must be the energy of the point of critical contact [10], point C of Figure 7.1. We choose $\mathbf{Q}$ in the z direction and introduce dimensionless variables:

$$u \equiv k_x/Q , \quad v \equiv k_y/Q , \quad w \equiv \left(k_z + \frac{1}{2}Q\right)/Q , \tag{7.10}$$

so point C is (0,0,0). The equation of the Fermi surface in the (u, w) plane is

$$u = \left[(w^2 + \alpha^2)^{1/2} - \alpha - w^2\right]^{1/2} , \tag{7.11}$$

where $\alpha \equiv mV/\hbar^2 Q^2$. Note that $u = 0$ at $w = 0$, which verifies our choice for E_F in (7.9). This surface has a peculiar shape. Near point C it is a circular cone. But near point D of Figure 7.1, the other intersection with the w axis, it is nearly spherical, as indicated by the dashed curve. The coordinate of point D is obtained by setting (7.11) to zero.

$$w(\mathrm{D}) \equiv \gamma = (1 - 2\alpha)^{1/2} . \tag{7.12}$$

The radius $r_0 Q$ of the Fermi surface at $w = \frac{1}{2}$, the origin in k space, can be written as a power series in α. To first order, Equation (7.11) is

$$r_0 \cong \frac{1}{2}(1 - 2\alpha) . \tag{7.13}$$

This radius must equal the free-electron-sphere radius to order α,[3] since the effect of the perturbation on states near the belly of the lemon is of order α^2. Consequently,

$$Q/k_F \cong 1/r_0 \cong 2(1 + 2\alpha) . \tag{7.14}$$

Together with the definition of α given above, this implies

$$Q \cong 2k_F\left[1 + (V/4E_F)\right]. \tag{7.15}$$

This result then provides an expression for α depending only on V/E_F.

$$\alpha \cong (V/8E_F)/\left[1 + (V/4E_F)\right]^2 . \tag{7.16}$$

The geometry of the semideformed Fermi surface has now been completely specified.

The g shift for the case $\mathbf{Q} \parallel \mathbf{H}$ can now be computed from (7.3):

$$\Delta g(\mathbf{Q} \parallel \mathbf{H}) = \mu\langle k_x^2 + k_y^2\rangle_{\mathrm{av}} . \tag{7.17}$$

3) The reader should not concede this point, however, without verifying that the volume enclosed by the surface, Equation (7.11), differs from that of a sphere of radius r_0 only by terms of order $\geq \alpha^2$.

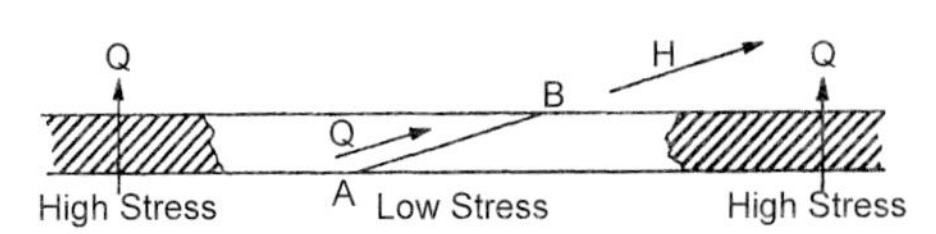

Figure 7.2 Schematic conception of the stress-induced Q domains in a thin sheet of K, sandwiched between layers of Parafilm. For large $\omega_c \tau$ spatial diffusion of electrons is very directional, and is parallel to H. For example, an electron originally at point A will, upon diffusion to the other side, arrive near point B. Subsequent diffusion back will again be along the line from B to A.

The density-of-states average over the Fermi surface is the one required. For the case $\mathbf{Q} \perp \mathbf{H}$, with $\mathbf{H}$ in the y direction,

$$\Delta g(\mathbf{Q} \perp \mathbf{H}) = -\mu \langle k_x^2 + k_z^2 \cos^2 \varphi + (k_z + Q)^2 \sin^2 \varphi \rangle_{\text{av}} . \tag{7.18}$$

The reason why (7.18) is more complex than (7.17) is that the wave vector $\mathbf{k}$ in Equation (7.3) is not the wave-vector label of the state, but the actual eigenfunction component wave vector. Since (7.7) has two components, $\mathbf{k}$ and $\mathbf{k} + \mathbf{Q}$, they both must be included in proportion to their probabilities. There is no contribution to (7.18) from the interference term of the two components. The reason for this depends on the fact that $\mathbf{Q}$ is incommensurate with the reciprocal lattice. For the alkali metals the conduction-electron g shift can be considered to originate from the corelike p functions admixed into the (otherwise) pure plane waves by the process of core orthogonalization. If one were to examine the g-shift contribution attributable to a particular atomic cell of the metal, there would be an interference term dependent on the relative phase of the $\mathbf{k}$ and $\mathbf{k}+\mathbf{Q}$ components at that cell. However, the crystal sum of the interference term vanishes because $\mathbf{Q}$ is not a simple rational fraction of a reciprocal-lattice vector.

The axial symmetry of the Fermi surface about the z axis makes the density-of-states average a very simple operation. Consider a slice of k space perpendicular to the z axis and having thickness δ, as shown in Figure 7.1. The number of quantum states of energy $\leq E$ within the slice is

$$Z = \pi t^2 \delta / 8\pi^3 = (m\delta / 4\pi^2 \hbar^2) \left[E - E_0(K) \right] . \tag{7.19}$$

Here t is the radius of the energy surface E at $k_z = K$, and $E_0(K)$ is the z component of energy, Equation (7.6) evaluated at $(0, 0, K)$. The density of states N in the slice is

$$N = (\partial Z / \partial E)_K = m\delta / 4\pi^2 \hbar^2 . \tag{7.20}$$

Since this is constant, the density-of-states average of an axially symmetric function F is merely

$$\langle F \rangle_{\text{av}} = \gamma^{-1} \int_0^{\gamma} F(w) dw , \tag{7.21}$$

where $F(w)$ is the value of F at $k_z = \left(w - \frac{1}{2}\right) Q$ on the Fermi surface.

With Equations (7.10), (7.17), (7.18), and (7.21) in view we define two subsidiary functions:

$$A(\alpha) \equiv \int_0^{\gamma} u^2 dw ,$$

$$B(\alpha) \equiv \int_0^{\gamma} \left[\left(w - \frac{1}{2} \right)^2 \cos^2 \varphi + \left(w + \frac{1}{2} \right)^2 \sin^2 \varphi \right] dw . \tag{7.22}$$

Furthermore, we define the fractional g-factor anisotropy f of the CDW state:

$$f(\alpha) \equiv \left[g(\mathbf{Q}\perp\mathbf{H}) - g(\mathbf{Q} \parallel \mathbf{H})\right]/\Delta g \,, \tag{7.23}$$

where Δg is the g shift for $\mathbf{Q} \parallel \mathbf{H}$. Inspection of (7.17), (7.18), and (7.22) allows us to conclude

$$f(\alpha) = 2\left\{\left[\frac{1}{2}A(\alpha) + B(\alpha)\right] - A(\alpha)\right\}/A(\alpha) \,. \tag{7.24}$$

The factor 2 accounts for the anisotropy introduced by the other energy gap, at $k_z = \frac{1}{2}Q$, as anticipated in the first paragraph of this section. The factor $\frac{1}{2}$ inside the square brackets follows from the obvious relation $\langle k_x^2\rangle_{av} = \frac{1}{2}\langle k_x^2 + k_y^2\rangle_{av}$. The integral $A(\alpha)$ is easily performed after using Equation (7.11). The integral $B(\alpha)$ is likewise elementary after the functional dependence of $\cos^2\varphi$ on w is found. From (7.6), (7.8), and (7.10),

$$\cos^2\varphi = \frac{1}{2}\left[1 + w(w^2 + \alpha^2)^{-1/2}\right]. \tag{7.25}$$

The final result, with the help of (7.12), is to terms linear in α

$$f(\alpha) \cong 3\alpha \,. \tag{7.26}$$

Use of Equation (7.16) to order V/E_F completes the derivation of (7.2), $f \cong 3V/8E_F$.

The question remains as to what fraction of f is caused by repopulation of k space and what fraction is caused by admixture of $\mathbf{k} \pm \mathbf{Q}$ components into $\psi_{\mathbf{k}}$. This is easily answered by setting $\varphi \equiv 0$ in (7.22) and repeating the balance of the calculation. One obtains $3V/8E_F$ again. Thus, to lowest order, the entire g-shift anisotropy is caused by repopulation of k space. The $\mathbf{k} \pm \mathbf{Q}$ admixture contribution is of order $(V/E_F)^2$.

The tedious analysis of this section was of course necessary. However, with the understanding that inevitably is acquired one can surmise the final result in a single glance: The fractional increase in $\int k_z^2 dk_z$ is just three times the conical-point distortion, $p = V/4E_F$. The factor 3 comes from the binomial coefficient of $(1 + p)^3$, associated with the upper limit of integration in k_z. This increase is diluted by 50% due to the invariance of $\int k_x^2 dk_z$. Consequently $f = \frac{3}{2}p$.

7.3 Stress-Induced Q Domains

In Section 7.1 we concluded that the only possible explanation of CESR splitting in K was a macroscopic division of the sample into two species having different g factors. Our purpose here is to elaborate the required physical characteristics of such a domain structure and to show how its origin is easily understood on the basis of a CDW ground state. This will not require any novel or new subsidiary assumptions.

K has a very large thermal expansion coefficient. Its lattice constant [9] at 5, 78, and 293° K is 5.225, 5.247, and 5.344 Å, respectively. Consequently, the linear expansion between 5° and 78° K alone is 0.4%. The elastic moduli are known [19]. It follows that a negative pressure of 4.5 kg/mm^2 would be required to prevent the contraction below 78° K. The polycrystalline yield stress has been measured[4] throughout this temperature range and is only 0.1 kg/mm^2 Consequently, a sheet of K cooled to He temperatures between layers of another material to which it adheres will undergo severe thermal stress and plastic deformation. It is reasonable

4) See Figure 5 of [20].

to assume that the stress state of such a sample will be macroscopically heterogeneous. Some regions where a great deal of slip has occurred will be relatively stress free. Others will be in planar tension.

We believe that in relatively stress-free regions **Q** will tend to align parallel to **H**. This is not a new assumption, since it is also required [21] for the explanation of other anomalous phenomena. However, we also assume that severe stress prevents alignment parallel to **H**. The striking stress dependence [22] of the high-field magnetoresistance establishes elastic stress as a crucial variable affecting electronic properties. The directional effect of stress on orientation of **Q** can possibly be surmised from the conflicting data on the optical anomaly. Hodgson [23] failed to observe it. An important difference between his experiment and that of Mayer and El Naby [16] is that the former was carried out by reflection from an evaporated metal layer on a glass interface, whereas the latter was by reflection from a bulk-metal vacuum interface. The theory [11–14] of the anomaly based on a CDW or SDW model predicts that it should be observed only if **Q** has a component parallel to the reflecting surface. Hodgson's failure to see the anomaly can therefore be explained if **Q** were perpendicular to the surface of his evaporated layer. Such a layer is probably under severe planar tension because of the large differential expansion between K and glass. We therefore postulate that the similarly stressed patches of the CESR samples also have **Q** nearly perpendicular to the specimen surface. Since **H** is approximately parallel to the surface, these patches will have **Q** nearly perpendicular to **H**.

The dependence of **Q** orientation on **H** and uniaxial stress that is needed to explain the CESR splitting is identical to the behavior that has been postulated previously to explain other anomalies in K. It should also be appreciated that similar behavior has been established experimentally in other systems. For example, the ability of magnetic field[5)] or uniaxial stress [27] to orient the SDW **Q** of Cr has been conclusively demonstrated by neutron diffraction.

A schematic illustration of stress-induced **Q** domains in a thin K sheet is shown in Figure 7.2. The linear size of the domains in the plane of the sheet must be about 1 or 2 mm, as shown below. To understand the behavior of CESR as **H** is tilted slightly from an orientation parallel to the surface, one must recall the salient properties of conduction-electron diffusion in a magnetic field [28]. The diffusion constant parallel to **H** is $D_{11} = \frac{1}{3} v_F^2 \tau$, where v_F is the Fermi velocity and τ the conductivity relaxation time. Diffusion perpendicular to **H** depends on $\omega_c \tau$, where ω_c is the cyclotron frequency. For $\omega_c \tau \gg 1$, $D_\perp \approx D_{11}/(\omega_c \tau)^2$. The relevant quantity for CESR is the spin diffusion length $L = (D T_2)^{1/2}$, which is the average distance an electron will diffuse before the phase coherence of its spin polarization is lost. T_2 is the transverse spin relaxation time. For the extremely pure specimens of Walsh, Rupp, and Schmidt $L_{11} \sim 1$ cm. Since $\omega_c \tau \sim 40$, $L_\perp \sim 0.025$ cm.

Suppose now that **H** is perfectly parallel to the surface of the specimen. The electrons travel in tight helical paths about the field lines. Since their spin diffusion length parallel to **H** is about 1 cm, the volume average g factor of the stress-induced **Q** domains will characterize the CESR. There will be a single resonance line, as observed [1]. Suppose that **H** is now tilted as shown in Figure 7.2. The extensive diffusion in the plane is suppressed. An electron originally at point A of Figure 7.2 will diffuse along the line AB. After reaching point B, it can only diffuse back to A, etc. Such electrons are trapped in the low-stress domain and will contribute a CESR with $g = g(\mathbf{Q} \parallel \mathbf{H})$. Geometrical considerations govern the orientation at which a significant splitting can occur. No splitting should occur if the domain size is too

5) [24, 25]. This behavior was also surmised from susceptibility measurements by [26].

small, i.e., smaller than $L_{\perp}$. Since the observed maximum splitting occurred for a tilt angle of $\sim \frac{1}{6}$ rad, the average domain size was probably $\sim$ 10 times the thickness of the sheet.

It is clear that CESR splitting in K is a phenomenon that is sensitively dependent on the metallurgical preparation of the sample. Even when preparation techniques are reproduced, the precise behavior of the resonance with tilt angle will vary from sample to sample [2]. The ideal specimen – one prepared in a stress-free condition – should not exhibit any splitting.

Since $\Delta g < 0$, Equation (7.2) predicts that $g(\mathbf{Q} \parallel \mathbf{H}) > g(\mathbf{Q} \perp \mathbf{H})$. Therefore the low-field CESR component must be attributed to the stress-free domains, which have **Q** parallel to **H**. One puzzling feature of the data is the rapidity with which this component disappears for tilt angles greater than 9°. One can only speculate about the cause. The stress-free domains are presumably those in which extensive plastic flow has occurred. Consequently slip bands will have caused originally smooth surface to become irregular, providing a greater likelihood for contamination and surface spin relaxation. As tilt angle increases, the low-field component reflects more uniquely the relaxation behavior peculiar to its own domains.

References

1. Walsh, W.M. Jr., Rupp, L.W. Jr., and Schmidt, P.H. (1966) *Phys. Rev.*, **142**, 414.
2. Walsh, W.M. Jr. (private communication).
3. Lampe, M. and Platzman, P.M. (1966) *Phys. Rev.*, **150**, 340.
4. Platzman, P.M. and Wolff, P.A. (1967) *Phys. Rev. Lett.*, **18**, 280.
5. Van Zandt, L.L. (1967) *Phys. Rev.*, **162**, 399.
6. Schultz, S. and Dunifer, G. (1967) *Phys. Rev. Lett.*, **18**, 283.
7. Cowan, D.L. (1967) *Phys. Letters*, **18**, 770.
8. Schultz, S., Shanabarger, M.R., and Platzman, P.M. (1967) *Phys. Letters*, **19**, 749.
9. Barrett, C.S. (1956) *Acta Cryst.*, **9**, 671.
10. Overhauser, A.W. *Phys. Rev.*, (to be published).
11. Overhauser, A.W. (1964) *Phys. Rev. Letters*, **13**, 190.
12. Overhauser, A.W. and Rodriquez, S. (1966) *Phys. Rev.*, **141**, 431.
13. Gustafson, D.R. and Barnes, G.T. (1967) *Phys. Rev. Lett.*, **18**, 1.
14. Reitz, J.R. and Overhauser, A.W. *Phys. Rev.*, (to be published).
15. Yafet, Y. (1952) *Phys. Rev.*, **85**, 478.
16. Mayer, H. and El Naby, M.H. (1963) *Z. Physik*, **174** 289.
17. Overhauser, A.W. (1960) *Phys. Rev. Lett.*, **4**, 462.
18. Overhauser, A.W. (1962) *Phys. Rev.*, **128**, 1437.
19. Marquardt, W.R. and Trivisonno, T. (1965) *J. Phys. Chem. Solids*, **26**, 273.
20. Hull, D. and Rosenberg, H.M. (1959) *Phil. Mag.*, **4**, 303.
21. Overhauser, A.W. (1965) *Bull. Am. Phys. Soc.*, **10**, 339.
22. Penz, P.A. and Bowers, R. (1967) *Solid State Commun.*, **5**, 341.
23. Hodgson, J.N. (1963) *Phys. Letters*, **7**, 300.
24. Arrott, A., Werner, S.A., and Kendrick, H. (1965) *Phys. Rev. Lett.*, **14**, 1024.
25. Arrott, A., Werner, S.A., and Kendrick, H. (1967) *Phys. Rev.*, **155**, 528.
26. Montalvo, R.A. and Marcus, J.A. (1964) *Phys. Letters*, **8**, 151.
27. Bastow, T.J. and Street, R. (1966) *Phys Rev.*, **141**, 510.
28. Gaspari, G.D. (1966) *Phys. Rev.*, **151**, 215.

Reprint 8 Magnetoresistance of Potassium[1)]

John R. Reitz and A.W. Overhauser**

* Scientific Laboratory, Ford Motor Company, 525, Northwestern Avenue, Dearborn, Michigan 47907, USA

Received 3 November 1967

The linear magnetoresistance of single-crystal specimens of potassium, which is observed far into the high-magnetic-field regime, is not in accord with generally accepted ideas concerning the bands structure of potassium. The linear dependence of resistivity on field is, however, in accord with a charge-density-wave model for the metal. The charge-density wave modifies the Fermi surface by introducing many energy gaps that slice the Fermi surface. These heterodyne gaps undergo progressive magnetic breakdown. The model predicts the largest magnetoresistance for crystals in which the magnetic field is oriented along the [100] or [111] directions, and the smallest effect for crystals in which **H** is oriented parallel to [110]; these predictions are in agreement with observation.

8.1 Introduction

One of the most puzzling and unexplained observations on the alkali metals is that of their magnetoresistance: first, that they show mangetoresistance at all, and secondly, that the magnetoresistivity is linear in field to the highest magnetic fields measured [1, 2]. Attempts to explain away the magnetoresistance as due to probe effects have not succeeded since probeless techniques also yield a linear change in resistance with field [3]. Recently, Penz and Bowers [4, 5] used the helicon method to determine the magnetoresistivity of single-crystal specimens of high-purity potassium to fields of 55 kG and of polycrystalline potassium to 110 kG. Although the results show the magnetoresistivity to vary somewhat with crystallographic direction, it again is linear in field over the range studied.

It is well known that a degenerate electron gas with a spherical Fermi surface shows no magnetoresistance, and with a closed (although not necessarily spherical) Fermi surface shows constant magnetoresistance in large magnetic fields. It is generally believed that the alkali metals can be characterized by such a model and that their Fermi surfaces (at least those of sodium and potassium) deviate only very slightly from sphericity [6, 7]. Even if the Fermi surfaces are more complicated than is generally believed, it is difficult to understand the *linear* magnetoresistance; theory predicts [8] that the transverse magnetoresistance should saturate in the high-field regime if all orbits are closed, but should vary as H^2 for certain crystallographic directions if open orbits are present. (The situation is somewhat different in compensated metals, but here again the only predicted behavior at high fields is an H^2 dependence or saturation.) Once open orbits are admitted, an approximately linear magnetoresistance over a restricted range of magnetic fields might be obtained through an averaging over various pieces of the Fermi surface. But to achieve this result, a complicated Fermi surface would be required. Since the Fermi surfaces of the alkali metals are generally thought to be simple (and closed) the resistance versus field ought to saturate for $\omega_c\tau > 1$, whereas experimentally it remains linear to $\omega_c\tau \approx 100$. (Here ω_c is the cyclotron frequency and τ is the relaxation time for scattering of electrons.)

1) Phys. Rev. 171, 3 (1968)

Anomalous Effects in Simple Metals. Albert Overhauser

ISBN: 978-3-527-40859-7

The purpose of this paper is to show that one model of the Fermi surface of potassium does in fact lead to a linear magnetoresistivity over a large range of magnetic fields, although it would eventually saturate at very high fields. The model requires the existence of either a charge-density wave (CDW) or a spin-density wave (SDW), which is orientable by the field. A CDW (or SDW) state has not been established conclusively for any of the alkali metals, but the existence of such a state has been postulated to explain other anomalies which have not been explained on other bases [9–11]. The CDW (or SDW) modifies the Fermi surface by introducing many additional energy gaps which are progressively *broken down* by the magnetic field through the phenomenon of magnetic breakdown [12, 13]. The transverse magnetoresistance shows a linear dependence upon field throughout the breakdown region.

In Section 8.2 we review the experimental situation for single crystal potassium. In Section 8.3 we discuss several model calculations of magnetoresistance in which magnetic breakdown plays a role. Finally, in Section 8.4 we discuss the CDW model of potassium and calculate its magnetoresistance.

Table 8.1 Magnetoresistivity of single-crystal potassium [4, 5].

Sample	[*hkl*] field direction	$\omega_c\tau$ at 55 kG	*S* (%)	10^{-3} RRR
1	100	133	0.34	3.1
2	100	146	0.42	3.4
3	100	143	0.26	3.4
4	100	56	0.54	1.3
5	100	46	0.54	1.1
6	100	64	0.65	1.5
7	100	63	0.43	1.5
8	100	61	0.25	1.4
9	110	143	0.20	3.4
10	110	39	0.18	0.9
11	110	54	0.18	1.3
12	110	61	0.20	1.4
13	110	74	0.26	1.7
14	110	50	0.10	1.2
15	110	60	0.10	1.4
16	111	165	0.55	3.9
17	111	127	0.31	3.0
18	111	136	0.33	3.2
19	123	92	0.55	2.2
20	123	139	0.22	3.3
21	123	166	0.46	3.9
22	123	59	0.22	1.4
23	123	49	0.10	1.2
24	123	100	0.10	2.4

8.2 Single-Crystal Magnetoresistivity of Potassium

Penz and Bowers [4, 5] measured the transverse magnetoresistance of single crystals prepared from high purity potassium with residual resistance ratios (RRR) at 4.2° K in the range 1000 to 4000. These were measured to 55 kG, which for the purity mentioned corresponds to a $\omega_c \tau$ of the order of 100. Penz and Bowers used a probeless technique, namely, the helicon method; with the z direction being defined as that of the magnetic field, this technique measures $\frac{1}{2}(\rho_{xx} + \rho_{yy})$.

Their results are summarized in Table 8.1 in which the data are presented in terms of the parameter S, defined as

$$S = \frac{\rho(H) - \rho(0)}{\rho(0)\omega_c \tau} \times 100\% \,. \tag{8.1}$$

S is essentially the normalized slope of the linear variation in resistivity ρ as a function of field H. One interesting result is that S apparently depends upon the orientation of magnetic field relative to the crystallographic axes. When the field is oriented along [100] or [111] the magnetoresistance (averaged over the samples measured) appears to be two and one half times larger than when the field is oriented along [110]. This is an important point for the present paper, since the model to be discussed in Section 8.4 predicts strong directional dependences. For fields oriented along [123] the situation is less clear since the spread in S values is quite large; however, the average S is less than that for [100] or [111].

The samples used to obtain the data in Table 8.1 were prepared in such a way that they were subjected to a minimum amount of strain. By deliberately straining the crystal Penz was able to increase the slope S by as much as a factor of two. Presumably the spread in S values for various specimens with the same field orientation is due to small residual strains in these crystals.

8.3 Model Calculations of Magnetoresistance in Metals with Magnetic Breakdown

Magnetoresistance calculations for metals with specified Fermi-surface topology are usually made by a semi-classical approach in which the time dependence of the electron's group velocity $\mathbf{v}$ is determined from the equations

$$\hbar\dot{\mathbf{k}} = -(e/c)\mathbf{v} \times \mathbf{H} \,, \tag{8.2}$$

$$\mathbf{v}(\mathbf{k}) = \hbar^{-1}\left[\partial\epsilon(\mathbf{k})/\partial\mathbf{k}\right] . \tag{8.3}$$

The conductivity tensor σ_{ij} as a function of magnetic field may be calculated by the path integral method [14, 15] yielding

$$\sigma_{ij} = -(e^2/4\pi^3)\int\limits_{\text{all }\mathbf{k}} v_i(\mathbf{k})\frac{d f_0}{d\epsilon}d^3k \int\limits_{-\infty}^{t(k)} v_j(s) \times \exp\left[\frac{s-t}{\tau}\right] ds \,. \tag{8.4}$$

Here f_0 is the electron distribution function, $\mathbf{k}$ is the electron's wave number, and $\mathbf{H}$ is the magnetic field. This method does not include quantum oscillatory effects (de Hass–Shubnikov oscillations) but does reproduce the gross features of the field-dependent conductivity.

When magnetic breakdown effects the important, the connectivity of the various electron orbits is changed as a function of field, and Equation (8.4) cannot be used directly. Falicov and Sievert [16] devised a suitable matrix generalization of (8.4) which allowed them to include magnetic breakdown, and, as a result, they were able to investigate a large number of models in which the connectivity of the Fermi surface is modified in a prescribed way by the magnetic field. Thus, for example, they studied the transition from open to closed orbits, the transition from extended orbits to circular closed orbits, and transitions from closed hole-like orbits to closed electronlike orbits. Each model calculation was investigated as a function of the following parameters: ω_c, ω_0, and τ. $\omega_c \equiv eH/mc$ is the cyclotron frequency, τ is the relaxation time, and ω_0 is a frequency related to the probability of magnetic breakdown; in fact, the probability of an energy gap being broken down by the field is $\exp(-\omega_0/\omega_c)$ per passage through the gap region. An alternative definition of ω_0 is the following: at high magnetic fields, when the orbit is almost completely broken down, ω_0 is the probability per unit time of Bragg reflection into the low-field topology by the energy gap under consideration. Thus, the effective scattering time at high magnetic field is

$$\tau_{\text{eff}}^{-1} = \tau^{-1} + \omega_0 \,. \tag{8.5}$$

ω_0 depends upon the magnitude of the energy gap Δ. Falicov and Sievert write $\omega_0 = C\Delta^2/\epsilon_F\hbar$ with C a constant of order unity, and ϵ_F is the Fermi energy. Actually, as has been pointed out by many authors [17–19], the "constant" C depends on the geometry of the orbits, and ω_0 may be written as

$$\omega_0 = \frac{\Delta^2 k_F^2}{8\hbar\epsilon_F \mathbf{K}\cdot(\mathbf{k}\times\mathbf{b})}\,, \tag{8.6}$$

where $\mathbf{K}$ is the wave vector corresponding to the energy gap Δ and $\mathbf{b}$ is a unit vector in the direction of the magnetic field.

Of the various idealized calculations considered by Falicov and Sievert, the only ones that appear to apply to the potassium case (to be discussed in Section 8.4) are transitions from extended closed orbits to circular closed orbits, and transitions from closed (hole) orbits to closed (electron) orbits. Only the latter model calculation seems capable of explaining the *linear* magnetoresistivity of the metal. The results of Falicov and Sieverts calculation for this model are presented in Figure 8.1.

8.4 Fermi Surface of Potassium

We are now in a position to discuss the Fermi surface of potassium and its effect on the magnetoresistance of this metal. The Fermi surface to be discussed derives from a charge-density-wave model with a single charge-density wave of wave vector **Q** which is orientable by magnetic fields of 10 kG or more. The orientation effect is such that **Q** aligns approximately parallel to the field.[2] Stability arguments demand that **Q** be slightly larger than the diameter of the Fermi surface, and calculations[3] for potassium indicate that $Q = 1.33(2\pi/a)$, where a is the lattice constant.

Since $Q > 2k_F$ the charge-density periodicity does not change the connectivity of the Fermi surface. There are, however, other periodicities in the structure; there are, for example, periodic potentials associated with the reciprocal-lattice vectors $\mathbf{G}_{hkl}$. But gaps arising from

2) An SDW model is equally acceptable. The **Q** vector does not have to align exactly parallel to the field, but can align approximately along the field, picking a compromise position determined by crystal strain and magnetic field.

3) The importance of subsidiary periods and their effect on galvanomagnetic properties has been studied previously for the case of Cr [20, 21].

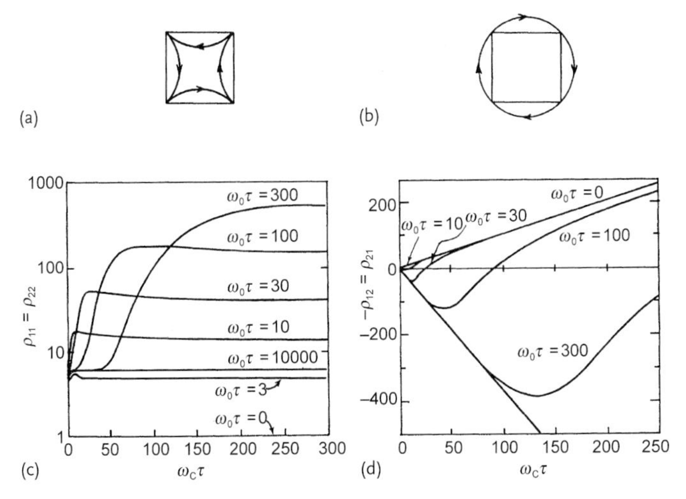

Figure 8.1 Magnetoresistance and Hall resistance for a transition via magnetic breakdown from a closed-hole to a closed-electron orbit (from Falicov and Sievert [16]). (a) The low-field orbit (as $H \to 0$); (b) the high-field orbit (as $H \to \infty$); (c) the transverse magnetoresistance; (d) the Hall resistance.

these do not intersect the Fermi surface either. Further, there are beat periods or heterodyne periods arising from interaction of the charge-density and lattice periods [20, 21]. The most pronounced heterodyne periods are those governed by the relation

$$\mathbf{K} = 2\pi \mathbf{G}_{hkl} \pm \mathbf{Q} \tag{8.7}$$

and these give rise to energy gaps which we denote by Δ. We thus have an energy-band structure which depends on the orientation of the magnetic field.

Figures 8.2 and 8.3 show *all* of the periodicities (8.7) whose **K** is less in magnitude than the diameter of the Fermi surface for magnetic fields (and **Q**) oriented along [100], [110], [123], and [111]. These give rise to energy gaps and an associated zone structure. The magnitude of the energy gaps Δ is unknown; this will be treated as a parameter to be determined from fitting to the magnetoresistivity. For two of the orientations shown there is a group of hole-like orbits at low fields. For **H** ∥ [100] the orbit marked with arrows is such an orbit (solid line represents that part of the orbit on the front surface of the sphere and broken line that part on the rear surface). For **H** ∥ [111], a remapping of the orbit on the midplane shows one of the hole orbits. These hole orbits are converted into electron orbits at high fields as a result of magnetic breakdown. Magnetic breakdown causes other connectivity changes also – small electron orbits being converted into large electron orbits, – but the largest effects and those which contribute to the linear magnetoresistivity are those caused by the change from hole orbits to electron orbits. For **H** strictly parallel to [110] there is no change in connectivity as a function of field. If **H** (and hence **Q**) is misaligned slightly from the [110] direction, there will be a few extended double orbits at low field which can be broken down. This effect does not lead to a linear magnetoresistivity. For **H** ∥ [123] we again have only changes from extended orbits to circular orbits.

We return now to the [100] and [111] orientations, where we have hole orbits at low field. We have a group of orbits and hence a range of ω_0. [For the geometry considered the $\mathbf{K} \cdot (\mathbf{k} \times \mathbf{b})$ element gives rise to a cos Θ factor, where Θ is the angle between the orbit tangent at the zone intersection and **K**.] At large $\omega_c\tau$ we thus have a magnetoconductivity which is

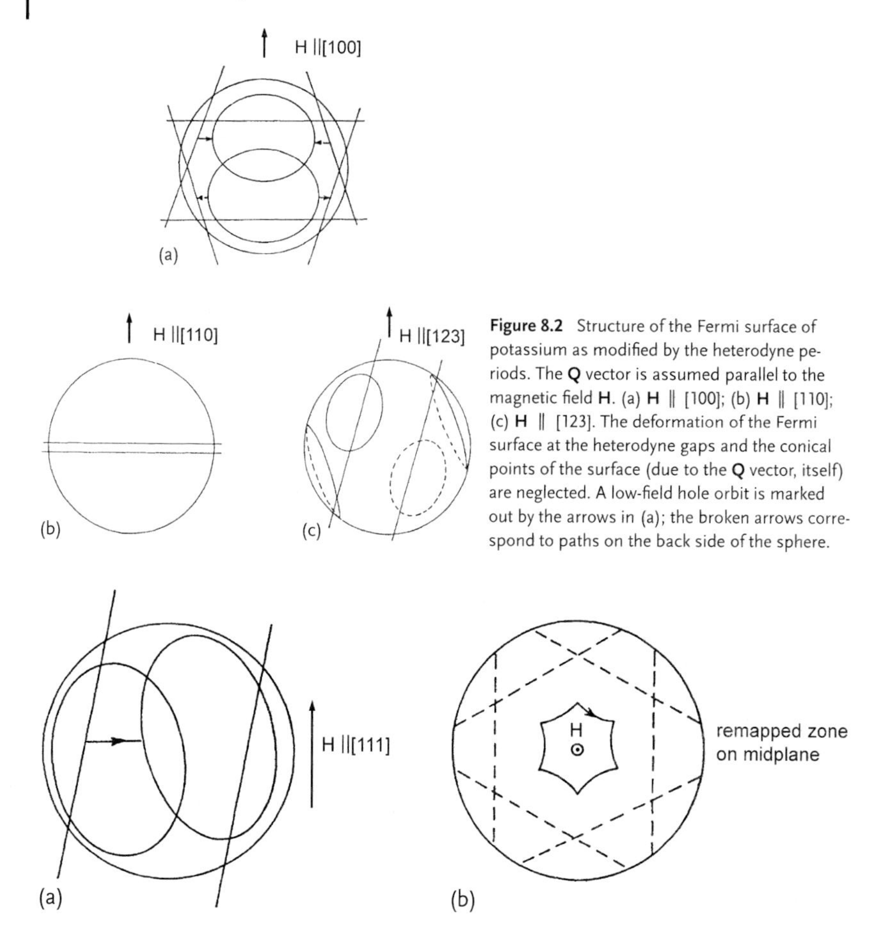

Figure 8.2 Structure of the Fermi surface of potassium as modified by the heterodyne periods. The **Q** vector is assumed parallel to the magnetic field **H**. (a) **H** ∥ [100]; (b) **H** ∥ [110]; (c) **H** ∥ [123]. The deformation of the Fermi surface at the heterodyne gaps and the conical points of the surface (due to the **Q** vector, itself) are neglected. A low-field hole orbit is marked out by the arrows in (a); the broken arrows correspond to paths on the back side of the sphere.

Figure 8.3 Same as Figure 8.2. (a) **H** ∥ [111]. The arrow marks part of a low-field-hole orbit; (b) A cut through the midplane of the Fermi surface shown in (a). Parts of the Fermi surface are remapped inside the zone formed by the heterodyne planes in order to show the hole orbit.

made up predominantly from circular orbits but with a small admixture of orbits which are transforming (via magnetic breakdown) from hole orbits to electron orbits. These orbits are characterized by a range in the parameter, ω_0. The contributions to the conductivity from various parts of the Fermi surface add together.

In order to test this model to see if it produces a linear magnetoresistance, the calculations of Falicov and Sievert (Figure 8.1) were inverted to obtain the components of their conductivity tensor as a function of $\omega_c \tau$. The conductivity of a small group of these orbits, with $\omega_0 \tau$'s ranging from 50 to 150, was added to that of circular orbits without breakdown. The result (transformed back again to resistivity) is an essentially linear magnetoresistance over the range of interest: to $\omega_c \tau = 150$. With breakdown orbits making up about 1% of the total orbits, the increase in resistivity over this range is 60%.

8.5 Conclusions

A charge-density-wave model for potassium in which the Fermi surface is modified by heterodyne gaps and varies with orientation of the magnetic field gives rise to a linear magnetoresistance for single crystals when the magnetic field is oriented either along the [100] or [111] direction. The resistance will eventually saturate at very high magnetic field. The intrinsic energy gap (corresponding to $\cos\Theta = 1$), obtained by fitting the model to Penz and Bower's data, corresponds to $\omega_0\tau \approx 50$; this gives $\Delta = 0.07$ eV. Such a value appears to be reasonable since the energy gap associated with **Q** has been estimated at 0.6 eV.

There is no linear magnetoresistance predicted for **H** $\parallel$ [110] and **H** $\parallel$ [123]; in fact, if the field and **Q** are oriented precisely along [110] in a strain-free specimen, we would predict no magnetoresistance at all. In a strained crystal the charge-density wave vector **Q** will pick a compromise alignment depending on the strain field and the magnetic field. This will produce a different zone structure. Since low-field hole orbits do not appear to be uncommon, orbits of this type may be mixed in, even though they do not appear in the "ideal" orientation. In order to test this possibility we have looked at the band structure resulting from small misalignments of **Q** and **H**. With **H** $\parallel$ [110], but **Q** misaligned 11° (toward [100]), three new pairs of heterodyne gaps intersect the Fermi surface. At a 13° misalignment a well developed hole orbit appears.

The Fermi-surface model described here predicts a number of small low-field orbits which can possible be observed in de Haas–van Alphen studies of the alkali metals. There are, however, reasons why these might not have been seen before: (1) the curvature of the Fermi surface in the vicinity of the extremal orbits may be so large that not enough orbits contribute in phase with one another to be observed; (2) since strain effects are more important at low magnetic fields there is the likelihood of competing zone structures at low field which would again make individual orbits more difficult to observe; and (3) low-field de Haas–van Alphen studies in strain-free alkali metal crystals have perhaps not been adequately investigated. One observation which lends support to the Fermi-surface model proposed here is that of Okumura and Templeton [22] concerning high-field de Haas–van Alphen studies in cesium. They found that the quality of de Haas–van Alphen signals depended upon crystal orientation in the magnetic field, the best signals being obtained when **H** $\parallel$ [110]. For other field directions, e.g., **H** $\parallel$ [111], it was difficult to obtain a signal at all.

Note added in proof. The model discussed in this paper predicts that the Hall coefficient of potassium should decrease with increasing H. This effect arises because the high-field Hall coefficient essentially measures the difference between electron and hole concentrations. P.A. Penz [23] has indeed observed this phenomenon on both single-crystal and polycrystalline material. The largest effect was for a polycrystalline specimen, which had a reduction in Hall coefficient of 7% at $\omega_c\tau = 300$.

References

1 Justi, E. (1948) *Ann. Physik*, **3**, 183.
2 MacDonald, D.K.C. (1956) in *Handbuch der Physik*, (ed. S. Flügge), Springer-Verlag, Berlin, Vol. 14, p. 137.
3 Rose, F.E. (1964) Ph.D. thesis, Cornell University (unpublished).
4 Penz, P.A. and Bowers, R. (1967) *Solid State Commun.*, **5**, 341.
5 Penz, P.A. (1967) Ph.D. thesis, Cornell University (unpublished).
6 Shoenberg, D. and Stiles, P.J. (1964) *Proc. Roy. Soc.* (London), **A281**, 62.
7 Lee, M.J.G. (1966) *Proc. Roy. Soc.* (London), **A295**, 440.
8 Lifshitz, I.M., Azbel, M.Y., and Kaganov, M.I. (1956) *Zh. Eksperim. i Teor. Fiz.*, **31**,

63. [English transl.: (1957) *Soviet Phys. – TETP*, **4**, 41].

9 Overhauser, A.W. (1964) *Phys. Rev. Lett.*, **13**, 190.

10 Gustafson, D.R. and Barnes, G.T. (1967) *Phys. Rev. Lett.*, **18**, 3.

11 Overhauser, A.W. (1968) *Phys. Rev.*, **167**, 691. This paper introduces the concept of the charge-density wave and shows that the earlier discussed anomalies can be explained in terms of either the CDW or the SDW. As is further shown in this paper, the CDW appears to be a more likely candidate for the alkali metals.

12 Cohen, M.H. and Falicov, L.M. (1961) *Phys. Rev. Lett.*, **8**, 231.

13 Blount, E.I. (1962) *Phys. Rev.*, **126**, 1636.

14 Chambers, R.G. (1952) *Proc. Phys. Soc.* (London), **A65**, 458.

15 Chambers, R.G. (1956) *Proc. Phys. Soc.* (London), **A238**, 334.

16 Falicov, L.M. and Sievert, P.R. (1965) *Phys. Rev.*, **138**, A88.

17 Joseph, A.S., Gordon, W.L., Reitz, J.R., and Eck, T.G. (1961) *Phys. Rev. Lett.*, **8**, 334.

18 Harrison, W.A. (1962) *Phys. Rev.*, **126**, 504.

19 Reitz, J.R. (1964) *J. Phys. Chem. Solids*, **25**, 53.

20 Lomer, W.M. (1965) in *Proceedings of the International Conference on Magnetism, Nottingham 1964*, The Institute of Physics and the Physical Society, London, p. 127.

21 Falicov, L.M. and Zuckermann, M.J. (1967) *Phys. Rev.*, **160**, 372.

22 Okumura, K. and Templeton, I.M. (1965) *Proc. Roy. Soc.* (London), **A287**, 89.

23 Penz, P.A. (1968) *Phys. Rev. Letters*, **20**, 725.

Reprint 9 Exchange Potentials in a Nonuniform Electron Gas[1)]

A.W. Overhauser *

* Scientific Research Staff, Ford Motor Company, 525, Northwestern Avenue, Dearborn, Michigan 48121, USA

Received 17 November 1969

Off-diagonal matrix elements of the exchange operator are computed for a degenerate electron gas having a small sinusoidal density modulation. The extreme nonlocal character of exchange is shown explicitly by its wave-vector dependence. The Slater exchange approximation severely underestimates the off-diagonal action of the exact exchange operator (by a numerical factor approaching ∞ for long-wavelength modulations). Such errors are largely compensated by neglect of the correlation potential.

In recent years there has been considerable debate[2)] on how best to approximate the (nonlocal) exchange operator A with a local potential. A frequent choice is the Slater $\rho^{1/3}$ relation

$$A_S = -3e^2(3\rho/8\pi)^{1/3} , \tag{9.1}$$

where $\rho(\vec{r})$ is the electron density. We believe that an explicit display of the properties of the exact A for a simple case indicates the futility of debate.

Consider a degenerate electron gas having a density

$$\rho(\vec{r}) = \rho_0(1 - p\cos\vec{q}\cdot\vec{r}) . \tag{9.2}$$

The mean density is $\rho_0 = k_F^3/3\pi^2$ and the fractional modulation p is assumed small. We shall presume that the modulation is caused by a (total) perturbing potential $V\cos\vec{q}\cdot\vec{r}$. Accordingly, the one electron wave functions are

$$\psi_{\vec{k}} \cong e^{i\vec{k}\cdot\vec{r}}\left[1 + (V/2\Delta_+)e^{i\vec{q}\cdot\vec{r}} + (V/2\Delta_-)e^{-i\vec{q}\cdot\vec{r}}\right], \tag{9.3}$$

where the energy denominators are $\Delta_+(\vec{k}) \equiv E(\vec{k}) - E(\vec{k}+\vec{q})$, etc. We neglect any $\vec{k}$ dependence of V. These wave functions are the ones generally employed, e.g., in the random-phase approximation (RPA), with or without local exchange and correlation corrections. If we take $E(\vec{k}) = \hbar^2k^2/2m$ and sum $|\psi_{\vec{k}}|^2$ over all occupied states, we find the modulation to be

$$p = (3V/2E_F)g(q/2k_F) , \tag{9.4}$$

where

$$g(x) \equiv \frac{1}{2} + \left[(1-x^2)/4x\right]\ln|(1+x)/(1-x)|$$

and $E_F = \hbar^2k_F^2/2m$.

The matrix element of the Slater potential is, from (9.1) and (9.2), in the *pure-momentum* representation

$$\langle\vec{k}+\vec{q}|A_s|\vec{k}\rangle \cong e^2k_Fp/4\pi . \tag{9.5}$$

One should observe that, for a given p, this is independent of both $\vec{k}$ and $\vec{q}$.

1) Phys. Rev. B 2, 3 (1970)
2) For a recent review see [1].

Anomalous Effects in Simple Metals. Albert Overhauser

ISBN: 978-3-527-40859-7

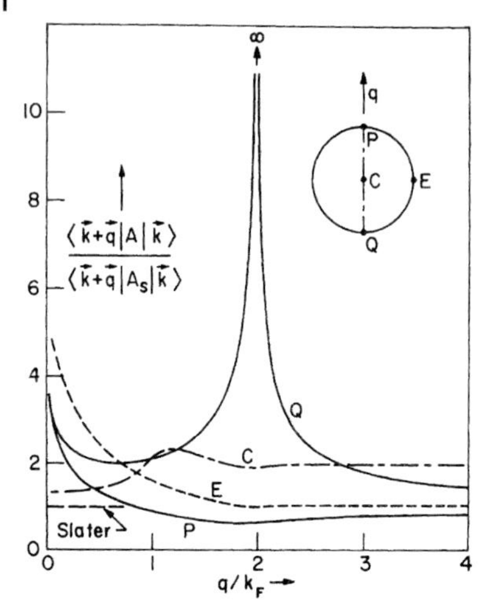

Figure 9.1 Matrix elements of the exchange operator A for an electron gas with a small density modulation $p \cos \vec{q} \cdot \vec{r}$. The ratio of the exact to the Slater values are shown for $\vec{k}$ at the center C, the equator E, and the poles P and Q of the Fermi sphere. The curves remain the same if the sign of $\vec{q}$ is reversed, except that P and Q are interchanged. Matrix elements of A and A_s are in the pure-momentum representation. All curves for $\vec{k}$ on the Fermi surface between P and the equator will have a singularity at $q = 0$. All curves for $\vec{k}$ on the Fermi surface between Q and the equator will have two singularities: at $q = O$ and at $q = 2|k_z|$. From (9.7) it is easy to show analytically that all singularities are logarithmic.

On the other hand, the exact exchange operator is defined by the transformation it effects on a general function $\varphi(\vec{r})$,

$$A\varphi(\vec{r}) \equiv -\sum_{\vec{k}} \left[\int \psi_{\vec{k}}^*(\vec{s})(e^2/|\vec{r}-\vec{s}|)\varphi(\vec{s})d^3s \right] \psi_{\vec{k}}(\vec{r}) \,. \tag{9.6}$$

The summation includes all occupied states k with spin parallel to that of $\varphi(\vec{r})$. We insert (9.3) into (9.6), eliminate V with the help of (9.4), and obtain off-diagonal matrix elements in the *pure-momentum* representation

$$\begin{aligned}&\langle \vec{k}+\vec{q}|A|\vec{k}\rangle \\ &\cong -pE_F/3g \times \sum_{\vec{k}'} \left(\frac{4\pi e^2}{|\vec{k}'-\vec{k}|^2 \Delta_+(\vec{k}')} + \frac{4\pi e^2}{|\vec{k}'-\vec{k}-\vec{q}|^2 \Delta_-(\vec{k}')} \right) .\end{aligned} \tag{9.7}$$

In contrast with (9.5) this depends markedly on both $\vec{k}$ and $\vec{q}$.

The ratio of (9.7) to (9.5) is shown (as a function of q) in Figure 9.1 for several points in $\vec{k}$ space. The striking (logarithmic) singularity at $q = 2k_F$ tor the point Q is the mathematical origin of the spin-density wave-instability theorem.[3] The extreme variation of (9.7) with $\vec{k}$ and $\vec{q}$ indicates that, for wave-function calculations pertaining to real materials, approximate exchange potentials should be judged empirically.

The average of (9.7) over $\vec{k}$ within the Fermi sphere is not particularly relevant. However, in Figure 9.1 it would fall mono tonic ally from the value 2 at $q = 0$ to 1.5 as $q \to \infty$. This latter limit contradicts a previous calculation due to Payne[4], who concluded that the exchange potential falls rapidly to zero for $q > 2k_F$. The oversight in Payne's work is subtle. He optimized (by a variational calculation) the admixture of $\vec{k} \pm \vec{q}$ components in $\{\psi_{\vec{k}}\}$ caused by an applied sinusoidal perturbation $w(\vec{r})$. The resulting modulation in electron density was misinterpreted by failing to isolate (off-diagonal) exchange contributions from a renormalization

3) [2] The singularity at $q = 2k_F$ has also been noted by [3].

4) [4]. The writer is deeply grateful to Dr. A.D. Brailsford for suggesting the correct interpretation of this reference.

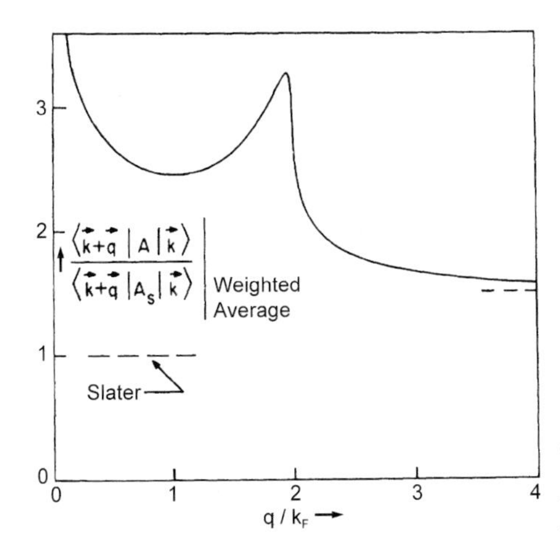

Figure 9.2 Weighted average of the exchange operator for an electron gas with a small density modulation. The weighting function employed was proportional to each electron's contribution to the modulation.

of the $w(\vec{r})$ contributions caused by (diagonal) exchange corrections to the energy differences Δ_+ and Δ_-.[5)]

The foregoing remarks are relevant to an explanation why the exchange potentials for the points *E*, *P*, and *Q* in Figure 9.1 diverge like $\ln(1/q)$ as $q \to 0$. (One might have expected them to approach the Kohn and Sham value $\frac{2}{3}$.) It is well known that the spin susceptibility $\chi(\vec{q})$ is enhanced relative to the Pauli value at $q = 0$. However, if one were to compute $\chi(\vec{q})$ for small (but finite) q by perturbation theory, taking into account only the exchange corrections to $E(\vec{k})$, $\chi(\vec{q})$ would approach 0 as $[\ln(1/q)]^{-1}$. An exchange potential which diverges as $\ln(1/q)$ must be present to compensate this effect. Otherwise the exact (Hartree–Fock) $\chi(\vec{q})$ could not be continuous at $q = 0$.

We observe that the extreme nonlocality of A calls into question the reliability of any local approximation to it, especially in a band calculation where the $\vec{k}$ dependence of the electron wave function and energy is the major question. However, if one insists on replacing A with a local operator, the question arises: What is the most appropriate average (over $\vec{k}$) of $\langle \vec{k} + \vec{q}|A|\vec{k}\rangle$? The answer depends, of course, on what one ultimately intends to calculate. However, a particularly appealing choice is the following: What is the local operator which, when acting in first-order perturbation, gives rise to the same charge-density modulation that A does? One can show that this is given by the average of Equation (9.7) (over occupied $\vec{k}$ states) computed with a $1/\Delta_+(\vec{k})$ weighting function. In other words, it is the average of (9.7) weighted (algebraically) in proportion to each electron's contribution to p. This weighted average shown in Figure 9.2 falls from $\ln(\infty)$ at $q = 0$ to a minimum near $q = k_F$, goes through a sharp maximum near $q = 2k_F$, and approaches 1.5 for large q.

The physical consequences of the large exchange potentials, illustrated in Figures 9.1 and 9.2, will of course be moderated by (compensating) correlation potentials. The work of

5) Since Payne's calculation was in a Hartree–Fock framework, the wave-function modulation attributable to $w(\vec{r})$ should be taken to be $\langle \vec{k} + \vec{q}|w|\vec{k}\rangle/\Delta_+$, where the energy difference Δ_+ includes contributions from exchange. This inclusion reduces the charge modulation attributable to the Hartree term $w(\vec{r})$ so that the exchange potential, derived by requiring that it account for the balance of the charge modulation, is correspondingly larger. This re interpretation of Payne's calculation leads to results in agreement with the present work. We emphasize that the energy denominators which we employ in Equation (9.3) do not include exchange corrections since we intend to take cognizance of the fact that correlation corrections cancel the pathological $\vec{k}$ dependence of the (diagonal) exchange energy.

Kohn and Sham [5] has shown that in the small-q limit the sum of the exchange and correlation potentials is slightly larger than $\frac{2}{3}$ of the Slater potential. However, their work seems to suggest that (for $q \to 0$) the exchange potential is $\frac{2}{3}$ of Equation (9.1) and that the correlation potential is much smaller and of the same sign. The present work indicates that (for $q \to 0$) the exchange potential approaches $\ln(\infty)$. We conclude, then, that the correlation potential approaches $-\ln(\infty)$ in such a way that the sum of exchange and correlation potentials equals the sum given by Kohn and Sham. It is perhaps academic to argue how a sum is divided into parts, if it is only the sum that matters. However, the physical mechanisms which might enter a (future) microscopic theory of the correlation potential may depend on whether the result to be obtained is small and positive or large and negative. A treatment of the nonlocal properties of correlation is an outstanding theoretical challenge.

Inspection of Figures 9.1 and 9.2 shows that A_S severely underestimates the off-diagonal action of the exact exchange operator. The empirical success of the Slater exchange approximation in energy band calculations can be attributed to compensation of this error by neglect of the correlation potential.

References

1 Slater, J.C. (1969) *M.I.T. Solid-State and Molecular Theory Group*, Report No. 71, p. 3 (unpublished).
2 Overhauser, A.W. (1962) *Phys. Rev.*, **128**, 1437.
3 Herring, C. (1966) *Exchange Interactions among Itinerant Electrons* (eds G. Rado and H. Suhl), Academic, New York, Vol. IV, pp. 33, 40.
4 Payne, H. (1967) *Phys. Rev.*, **157**, 515.
5 Kohn, W. and Sham, L.J. (1965) *Phys. Rev.*, **140**, A1133.

Reprint 10 Observability of Charge-Density Waves by Neutron Diffraction[1)]

A.W. Overhauser*

* Scientific Research Staff, Ford Motor Company, 525, Northwestern Avenue, Dearborn, Michigan 48121, USA

Received 7 December 1970

It is shown that atomic displacements $\vec{A}\sin\vec{Q}\cdot\vec{L}$ associated with a charge-density wave (CDW) modulation of a metal crystal cause several changes in the anticipated diffraction properties. Very weak satellite reflections should occur at locations $\pm\vec{Q}$ from the ordinary (cubic) reflections. However it is found that vibrational excitations, corresponding to phase modulation of the CDW, may weaken the satellite intensities below the level of observation. The structure factors of the cubic reflections are reduced from unity to (the Bessel function) $J_0(\vec{K}\cdot\vec{A})$, where $\vec{K}$ is the scattering vector. Measurements of structure factors are difficult to interpret in the alkali metals because of severe primary extinction (mosaic block size ~ 0.5 mm) and unknown anharmonic contributions to the Debye–Waller factor. Nevertheless, an unambiguous test for CDW structure is possible since $\vec{Q}$ (and $\vec{A}$) must be orientable by a large magnetic field at 4 °K. Consequently high-index Bragg reflections should be turned off or on by a magnetic field rotated parallel or perpendicular to $\vec{K}$. Motivation for such an experiment is suggested by a survey of electronic anomalies which have been reported in the alkali metals and which can be explained with the CDW model.

10.1 Introduction

The alkali metals are generally considered to have a body-centered cubic (bcc) crystal structure. In recent years, however, a variety of electronic properties have been found anomalous. In Section 10.7, we present a brief survey and commentary on these anomalies. Many of these can be explained quantitatively if one assumes that the conduction electrons have experienced a charge-density wave (CDW) instability [1]. Such a phenomenon, if it exists, would result from exchange and correlation interactions among electrons.

The purpose of this study is to show how the presence *or absence* of a CDW in potassium, say, can be determined decisively. We adopt the view that evidence from electronic properties, no matter how extensive, is inconclusive.

Electric fields caused by a CDW will displace the positive ions from their ideal bcc equilibrium sites $\{\vec{L}\}$. The displacements $\vec{u}(\vec{L})$ are

$$\vec{u}(\vec{L}) = \vec{A}\sin(\vec{Q}\cdot\vec{L}+\phi)\,, \tag{10.1}$$

and can be regarded as a static modulation of the lattice. The wave vector $\vec{Q}$ of the CDW has a magnitude slightly larger than $2k_F$ the diameter of the Fermi surface [1]. Its direction is unknown. The amplitude $\vec{A}$ is (approximately) parallel to $\vec{Q}$; and its magnitude can be estimated by requiring the fractional modulation in positive-ion density to equal the fractional modulation ($p \sim 0.2$) of the CDW. Accordingly, $\mathrm{div}\vec{u} \sim AQ \sim 0.2$, so

$$A \sim 0.1\ \text{Å}\,. \tag{10.2}$$

1) Phys. Rev. B 3, 10 (1971)

Anomalous Effects in Simple Metals. Albert Overhauser

ISBN: 978-3-527-40859-7

This estimate neglects effects arising from pseudo-atom form factors which, if included, would lead to a larger A. Nevertheless the value (10.2) is so large that it constitutes a severe modification of the crystal structure. If a CDW exists, it must be directly visible in a diffraction experiment.

We confine our attention to neutron diffraction, which can easily be carried out with large specimens. (Alkali metals are so chemically reactive that X-ray experiments cannot easily be proven to exhibit bulk properties.)

10.2 CDW Satellites

The diffraction pattern expected from a mon-atomic crystal is obtained by Fourier analysis of the positive-ion density $\rho(\vec{r})$

$$\rho(\vec{v}) = \sum_{\vec{L}} \sigma(\vec{r} - \vec{L} - \vec{A} \sin \vec{Q} \cdot \vec{L}) \,, \tag{10.3}$$

where $\sigma(\vec{r})$ is the Dirac σ function. We neglect the phase angle ϕ appearing in (10.1). The Fourier amplitude associated with a scattering vector $\vec{K}$ is

$$\rho(\vec{K}) \equiv \int \rho(\vec{r}) e^{i\vec{K}\cdot\vec{r}} d\vec{r} \,. \tag{10.4}$$

After substituting (10.3) into (10.4) and performing the integration, we find

$$\rho(\vec{K}) = \sum_{\vec{L}} e^{i\vec{K}\cdot(\vec{L}+\vec{A}\sin\vec{Q}\cdot\vec{L})} \,. \tag{10.5}$$

The factor involving $\sin \vec{Q} \cdot \vec{L}$ in (10.5) can be transformed with the help of the Jacobi–Anger generating function for Bessel functions:

$$e^{iz\sin\phi} = \sum_{n=-\infty}^{\infty} e^{in\phi} J_n(z) \,. \tag{10.6}$$

Accordingly, we obtain

$$\rho(\vec{K}) = \sum_{\vec{L},n} J_n(\vec{K}\cdot\vec{A}) e^{i(\vec{K}+n\vec{Q})\cdot\vec{L}} \,. \tag{10.7}$$

The sum over $\{\vec{L}\}$ yields zero unless $\vec{K} + n\vec{Q}$ is 2π times a reciprocal-lattice vector $\vec{G}$. Therefore, the allowed Bragg reflections are those with scattering vector

$$\vec{K} = 2\pi\vec{G} - n\vec{Q} \,. \tag{10.8}$$

For these reflections the sum (10.7) is the number of atoms times $F(\vec{K})$

$$F(\vec{K}) = J_n(\vec{K}\cdot\vec{A}) \,. \tag{10.9}$$

This is the structure factor of the reflection $\vec{K}$.

The ordinary bcc Bragg reflections are those given by (10.8) with $n = 0$. The (first-order) CDW satellites are those with $n = \pm 1$. The intensity (for $\vec{K} = \vec{Q}$) *relative* to an ordinary Bragg reflection is

$$F^2 = J_1^2(\vec{Q}\cdot\vec{A}) \sim \frac{1}{4}(\vec{Q}\cdot\vec{A})^2 \sim 10^{-2} \,. \tag{10.10}$$

However, a macroscopic single crystal will probably subdivide into "$\vec{Q}$ domains", each domain having a different orientation of $\vec{Q}$. Depending on the symmetry of the favored $\vec{Q}$ direction, the expected intensity will be reduced by a factor between 3 and 24.

In Section 10.6, we shall show that low-frequency phononlike excitations of the CDW will contribute an enormous Debye–Waller factor to the GDW satellites (but not to the bcc reflections). Consequently, CDW satellites will likely be too weak to be seen directly, although in principle they must be present (if the metal has a CDW structure). A careful search [2] for CDW satellites in potassium revealed no reflections that could be so identified.

10.3 Structure Factors of Cubic Reflections

Every Bragg reflection of a monatomic bcc lattice will have a structure factor $F = 1$. However, if the crystal is modulated by a CDW, these cubic reflections will remain, but will be reduced in intensity. From (10.9), with $n = 0$, their structure factor will be

$$F(\vec{K}) = J_0(\vec{K} \cdot \vec{A}) \tag{10.11}$$

instead of unity. If θ is the angle between the scattering vector $\vec{K}$ and the amplitude $\vec{A}$ of the CDW, the integrated intensity of a bcc reflection will be proportional to $J_0^2(KA\cos\theta)$. The observed intensity will be proportional to the average of this function over all discrete (or continuous) values of θ obtaining in the sample

$$F^2 = \langle J_0^2(KA\cos\theta)\rangle_{\theta\ \mathrm{av}} \,. \tag{10.12}$$

The solid curve in Figure 10.1 shows the decrease in F^2 with KA for an isotropic average over θ. It is clear that the existence of a CDW state could be determined by quantitative measurement of the structure factors of (observable) bcc reflections. Experimental deviations from $F^2 = 1$ for high-index bcc reflections would indicate that a structure is not really cubic.

Measurements of this type can be rather difficult in alkali metals. The observed integrated intensity I_{obs} of a Bragg reflection is

$$I_{\mathrm{obs}} \sim (\gamma + \alpha)F^2 e^{-2W} / \sin 2\theta_{\mathrm{B}} \,. \tag{10.13}$$

γ is the primary extinction factor, α the contribution of thermal-diffuse (phonon) scattering to a Bragg peak, e^{-2W} the Debye–Waller factor, and θ_{B} the Bragg angle. We include no correction for secondary extinction in (10.13), since we have found that it is negligible (< 2%)

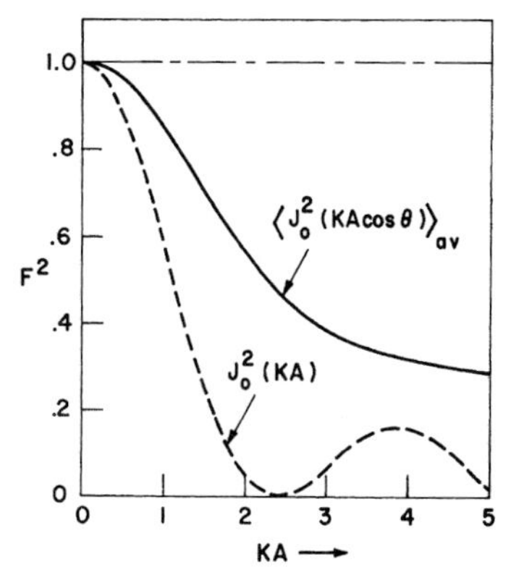

Figure 10.1 Structure factor F^2 vs. scattering vector $\vec{K}$ for Bragg reflections of a metal modulated by a CDW of amplitude $\vec{A}$. Dashed curve applies if $\vec{K}$ is parallel to $\vec{A}$, and the broken curve (at unity) applies if $\vec{K}$ is perpendicular to $\vec{A}$. Solid curve is an isotropic average of F^2 over the angle θ between $\vec{K}$ and $\vec{A}$.

in potassium by transmission experiments through 3 cm of a single crystal rocked through a [110] reflection. (The width of the rocking curve, in the parallel position, was $\sim 0.25°$.)

Primary extinction is severe. We have determined the mosaic block size in potassium by fitting measurements of (10.13) to Zachariasen's theory [3] for y. We found the mosaic block size to be ~ 0.05 cm. Extinction of the [110] reflection was a factor $1/y = 5$. By slight thermal stress we were able to reduce the extinction by a factor of 2; however, it completely recovered in less than an hour at room temperature. This suggests that large mosaic-block size is a characteristic of such material. Ideally one would prefer to measure (10.13) at 4 °K to minimize corrections for thermal diffuse scattering and the Debye–Waller factor. However, even the high-index reflections (in neutron diffraction) will be highly extinguished at 4 °K. Since the Zachariasen theory does not make proper allowance for the dependence of y on θ_B [4], interpretation of high-index intensities would be suspect.

Unfortunately, interpretation of room-temperature measurements of (10.13) is also uncertain. The Debye–Waller factor is then an extreme correction since the appropriate characteristic temperature $\Theta^M = 84°$K is so small. This is known accurately from the lattice-entropy analysis due to Martin [5], which leads to $\Theta^M = 89 \pm 1°$K at 0 °K. A correction for thermal expansion [6]

$$\Theta^M(T)/\Theta^M(0) = (V_0/V_T)^\alpha , \tag{10.14}$$

where V is the molar volume and $\alpha(\approx 1.30)$ the Grüneisen constant, yields the first value given. The Debye–Waller factor for potassium (at 292 °K) in the quasiharmonic approximation is then found to be

$$e^{-2W} = e^{-0.213(h^2+k^2+l^2)} , \tag{10.15}$$

where h, k, l are the Miller indices of the reflection. This factor reduces the [330] by a factor 30 relative to the [110], We have attempted to find F^2 from I_{obs}, corrected for primary extinction [3], thermal diffuse scattering [7, 8], and the foregoing temperature factor (10.15). We found $F^2 \sim 0.5$ for the [330]. This result cannot be taken seriously since Maradudin and Flinn [9] and Willis [10] have shown that there can be anharmonic contributions to the Debye–Waller effect, not included in (10.15), which introduce an extra factor $\exp[-\beta(h^2+k^2+l^2)^2]$. Since little is known about the magnitude (or sign) of β, we are unable to set error limits on our measurements[2)] of F^2.

10.4 Magnetic Field Modulation of $F(\vec{K})$

In Sections 10.1– 10.3, we have shown that it is difficult to prove from nonobservance of CDW satellites or from structure-factor measurements that potassium (for example) does not have a CDW. There remains, however, a diffraction effect required by the CDW model that is necessarily observable and spectacular.

The apparent isotropy of the Fermi surface (to one part per thousand in Na and K) as seen in de Haas–van Alphen measurements [11] rules out the CDW model unless $\vec{Q}$ can be oriented by a magnetic field. Otherwise a large Fermi-surface distortion ($\sim$ 7%) caused by CDW energy gaps would be immediately evident. Consequently, the CDW model is viable

2) We measured the integrated intensities ($\omega/2\theta$ scans) of 70 reflections having $h^2+k^2+l^2 \leq 18$. Counter acceptance angle and scan range (in θ) were 6°. The one unknown parameter in Zachariasen's formula for y was found by requiring $F^2_{\{110\}} = F^2_{\{200\}}$. Our results were insensitive to the coefficient in (10.15) since changes in the coefficient (up to 20%) required a nearly compensating change in primary extinction. Were it not for unknown anharmonic contributions to the temperature factor, mentioned above, our results would require $A \sim 0.4$ Å.

only if it can be taken as axiomatic that the direction of $\vec{Q}$ is (approximately) parallel to that of a sufficiently large magnetic field in a stress-free crystal at 4 °K. Interpretations of spin- and cyclotron-resonance anomalies (Section 10.7) also require this postulate. It is difficult to say how large the field $\vec{H}$ must be, but 15 kG should suffice.

The experiment we propose requires a large stress-free single crystal at 4 °K in a neutron diffractometer and an orientable magnetic field. Since the CDW amplitude $\vec{A}$ is approximately parallel to $\vec{Q}$, and therefore to $\vec{H}$, the structure factors $J_0(\vec{K} \cdot \vec{A})$ of the bcc reflections, Equation (10.11), could be varied at will between $J_0(KA)$ and $J_0(0) = 1$. This would be accomplished merely by rotating $\vec{H}$ from parallel to $\vec{K}$ to perpendicular. If a high-index reflection is chosen, so that $KA \sim 2.4$ (near the root of J_0) the reflection could be virtually turned off and on with the field. Observance of such an effect would obviously establish the existence of a CDW beyond doubt. What is perhaps more important, absence of field modulation would rule out the CDW model unambiguously.

An advantage of this experiment is that no analysis or interpretation of data is required. The Bragg intensities either change or they do not. If they do, the maximum intensity ratio would not be $J_0^2(KA)$, however. When primary extinction is large, $y \sim 1/|F|$, so the intensity ratio would more nearly be $|J_0(KA)|$.

This experiment should be attempted in potassium. (The partial martensitic transformation of sodium below 36 °K may introduce elastic stresses that could hinder alignment of $\vec{Q}$ with $\vec{H}$.) Short-wavelength (~ 0.5 Å) neutrons should be employed so that high-index reflections, i. e., [10, 10, 0], can be reached at angles compatible with pole geometry of the magnet. We also emphasize that the crystal should be free standing, so as to avoid thermal stress (duringcooling to 4 °K), which could inhibit alignment of $\vec{Q}$.

10.5 Phase Modulation of CDW

In this section we shall study the low-lying phononlike excitations of a CDW ground state. This is necessary in order to understand the Debye–Waller temperature factor for CDW satellites, which we treat in Section 10.6.

We must begin by estimating the energy difference between a CDW state and the normal ground state of a simple metal. This is most easily done by analogy with the condensation energy of a superconductor [12],

$$E_s \approx -\frac{1}{2}\rho_F \Delta^2 = -3n\Delta^2/4E_F \,, \tag{10.16}$$

where ρ_F is the electronic density of states at the Fermi surface (energy E_F), Δ is one-half the superconducting energy gap, and n is the electron density. The last equality in (10.16) results if we assume a parabolic band, so $\rho_F = 3n/2E_F$. A simplified interpretation of (10.16) is that the condensation energy can be attributed to the $\rho_F \Delta$ electrons near the Fermi surface which have their energy lowered an average amount $\frac{1}{2}\Delta$ relative to their nonsuperconducting energy spectrum.

We shall assume that an analogous relation also applies to a CDW condensation energy. The major difference is that the CDW energy gap 2Δ touches the Fermi surface only at a point [1]. This is shown by the energy spectrum and Fermi surface in Figure 10.2, The number of electron states contained within the circular cone bounded by the conical point P and the plane w (an energy depth Δ below P) is

$$N_\Delta = n\Delta^2/16E_F^2 \,. \tag{10.17}$$

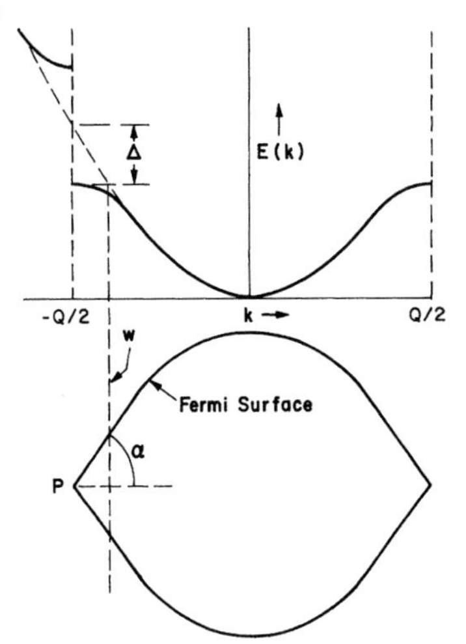

Figure 10.2 One-electron energy spectrum and Fermi surface of a metal with a CDW of wavevector $\vec{Q}$. Energy gap caused by the CDW periodic potential is 2Δ. Distortion of the Fermi surface at the conical point P is exaggerated. Half-angle α of the cone would be $\sim 75°$. Electron wave functions most severely deformed are those between the conical point P and the plane w.

We have used a relation, derived earlier [13], that the half-angle α of the cone is given by $\tan\alpha \approx (2E_F/\Delta)^{1/2}$ Because of the conical shape the average energy shift is $\sim \frac{1}{4}\Delta$ instead of $\frac{1}{2}\Delta$. Accordingly, the condensation energy (including both conical points) is

$$E_0 \approx -n\Delta^3/32E_F^2 \,. \tag{10.18}$$

The difference in form between (10.16) and (10.18) may be verbalized by saying that (for a CDW) only a fraction $\approx \Delta/24E_F$ of the Fermi surface is affected by the CDW energy gaps. With $\Delta \sim 0.3\,\text{eV}$, the condensation energy (10.18) for potassium is $\sim 2\times10^{-4}\,\text{eV}$ per electron. This very, small energy indicates the delicate nature of the instability.

The (supposed) CDW ground state is the minimum point in a four-dimensional potential valley, whose coordinates are the magnitudes of $\vec{Q}$ and $\vec{A}$ and two parameters needed to specify the direction of $\vec{Q}$. If we omit the dependence on $|\vec{A}|$, the CDW energy can be approximated,

$$E \approx E_0[1 - \lambda L(x, y, z) - \mu(\delta Q/Q)^2] \,. \tag{10.19}$$

$\delta Q/Q$ is the fractional deviation of $|\vec{Q}|$ from its equilibrium value, and $L(x, y, z)$ is a function of x, y, z, the direction cosines of $\vec{Q}$. The dimension-less parameters λ and μ are unknown. On account of the isotropy of the Fermi surface we expect $\lambda < 1$, whereas we expect $\mu \gg 1$, since CDW instabilities can arise only for $|\vec{Q}|$ very near $2k_F$. We expect[3] that the easy direction for $\vec{Q}$ is [110]. Thesimplest function having this property and full cubic symmetry is

$$L(x, y, z) = 2(x^4 + y^4 + z^4) + 24x^2y^2z^2 - 1 \,. \tag{10.20}$$

This form was chosen so that $L = 0$ at the minima ([110] directions) and is isotropic for small deviations in direction; also, $L = 1$ at the maxima ([100] directions).

We are now in a position to discuss dynamical modes of excitation that arise from a CDW structure. Since $\vec{Q}/2\pi$ is incommensurate with the (reciprocal) lattice, there is, strictly speaking, no Brillouin zone. The lattice-vibration dynamical matrix will be $3N \times 3N$ in size rather

3) Our reasons for this supposition are not conclusive: A CDW can be regarded as a static phonon. One can show that the lowest-energy longitudinal phonon having wave vector $|\vec{Q}|$ is in a [110] direction.

than 3 × 3. Nevertheless, we should be able to treat long-wavelength phase modulation of the CDW in a manner similar to the quasicontinuum approximation for long-wave length phonons. The physical model is that the direction (and magnitude) of $\vec{Q}$ can vary from its equilibrium value. We suppose that the phase φ in Equation (10.1) can be expanded,

$$\varphi(\vec{L}, t) = \sum_{\vec{q}} \varphi_{\vec{q}} \sin(\vec{q} \cdot \vec{L} - \omega t) . \tag{10.21}$$

$\vec{q}$ is the wave vector of a phase-modulation phonon, or "phason", and $\omega(\vec{q})$ its frequency. For simplicity, consider a single-phason mode to be excited. Then the atomic displacements relative to their bcc sites are

$$\vec{u}(\vec{L}) = \vec{A} \sin[\vec{Q} \cdot \vec{L} + \varphi_{\vec{q}} \sin(\vec{q} \cdot \vec{L} - \omega t)] . \tag{10.22}$$

It is easy to see that phase modulation corresponds to a local change in direction and magnitude of $\vec{Q}$: Consider a point in the lattice where $\vec{q} \cdot \vec{L} \sim 0$. Then (10.22) can be written, for $t = 0$,

$$\vec{u}(\vec{L}) \approx \vec{A} \sin[(\vec{Q} + \varphi_{\vec{q}} \vec{q}) \cdot \vec{L}] . \tag{10.23}$$

We define $q_\perp$ and $q_\parallel$ as the components of $\vec{q}$ perpendicular and parallel to $\vec{Q}$. Then, from (10.19) and (10.20) the potential-energy increase $U_{\vec{q}}$ caused by the phase modulation is, for small $\vec{q}$,

$$U_{\vec{q}} = \frac{1}{2} |E_0| \varphi_{\vec{q}}^2 [6\lambda(q_\perp Q)^2 + \mu(q_\parallel / Q)^2] . \tag{10.24}$$

The factor $\frac{1}{2}$ arises from averaging (10.19) over a full wavelength of the phason.

The kinetic energy $T_{\vec{q}}$ of a phason can also be derived easily. For each ion the kinetic energy is $\frac{1}{2} M(\partial \vec{u}/\partial t)^2$. Accordingly, from (10.22), the energy density is

$$T_{\vec{q}} = \frac{1}{8} n M A^2 \omega^2 \varphi_q^2 , \tag{10.25}$$

where a factor $\frac{1}{4}$ has entered from averaging over all lattice sites. The phason frequency spectrum (in the small q limit) is found by recalling that for harmonic oscillators $T_{\vec{q}}$ and $U_{\vec{q}}$ are equal. By equating (10.24) and (10.25), we obtain

$$\omega^2 = (4|E_0|/n M A^2)[6\lambda(q_\perp/Q)^2 + \mu(q_\parallel/Q)^2] . \tag{10.26}$$

The phase velocity of the phasons cannot be calculated until λ and μ are known. However, if we estimate $\lambda \sim 1$, $\mu \sim 10$, $A \sim 0.1$ Å, and use (10.18) for E_0, we find that phason and phonon velocities are of the same magnitude.

Finally, it is relevant to find the relationship between phasons and phonon modes of the undistorted lattice. This is easily done using (10.22) and standard trigonometric relations. For small amplitudes $\varphi_{\vec{q}} \ll 1$, we find

$$\vec{u}(\vec{L}) \cong \vec{A} \sin \vec{Q} \cdot \vec{L} + \vec{A}\varphi_{\vec{q}} (\cos \vec{Q} \cdot \vec{L}) \sin(\vec{q} \cdot \vec{L} - \omega t) . \tag{10.27}$$

Transforming the last term, we obtain

$$\vec{u}(\vec{L}) = \vec{A} \sin \vec{Q} \cdot \vec{L} + \frac{1}{2} \vec{A} \varphi_{\vec{q}} \sin[(\vec{q} + \vec{Q}) \cdot \vec{L} - \omega t] + \frac{1}{2} \vec{A} \varphi_{\vec{q}} \sin[(\vec{q} - \vec{Q}) \cdot \vec{L} - \omega t] . \tag{10.28}$$

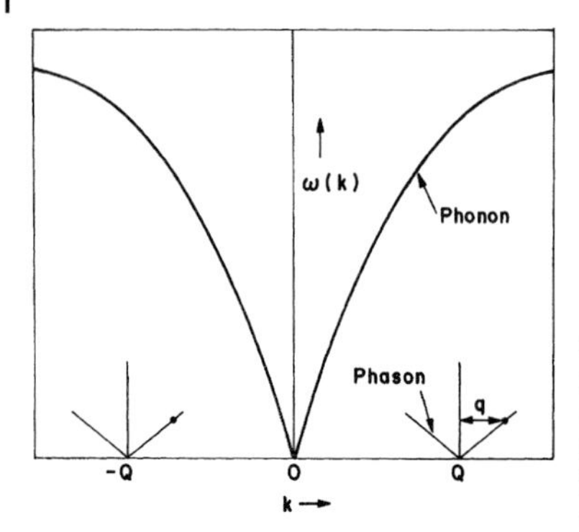

Figure 10.3 Schematic illustration of the vibrational modes in a metal having a CDW. Frequency of the phason branch goes to zero at $\vec{Q}$, the location of the CDW satellite reflections. Such a diagram has only an approximate meaning since, strictly, an incommensurate CDW structure does not have a Brillouin zone.

The last two terms, which represent the phason deviations from the equilibrium CDW state, constitute a coherent superposition of phonon modes having wave vectors $\vec{q}+\vec{Q}$ and $\vec{q}-\vec{Q}$. This means that (in wave-vector space) long-wavelength phason modes are located near the CDW satellite positions, as shown in Figure 10.3.

That the phason frequency $\omega(\vec{q}) \to 0$ as $\vec{q} \to 0$ is related to the fact that $\vec{Q}/2\pi$ is incommensurate with the (reciprocal) lattice. If $\vec{Q}/2\pi$ were commensurate, the phason branch would be a soft optical-phonon branch, with $\omega(0)$ finite, analogous to the soft modes in a ferroelectric near the transformation temperature.

10.6 Debye–Waller Factors for Phasons

In Section 10.5, we have shown that phason modes constitute a separate branch of the phonon spectrum, and that their frequency spectrum goes to zero at the CDW satellite locations, e. g., Figure 10.3. Here we shall investigate the relative contribution of phason and phonon modes to the Debye–Waller temperature factor.

Consider first the effect of phonons on an ordinary Bragg reflection of a bcc crystal. The ionic displacements caused by all the phonons $\{i\}$ having amplitudes and frequencies $\{\vec{a}_i, \omega_i\}$ can be written

$$\vec{u}(\vec{L}) = \sum_i \vec{a}_i \sin(\vec{k}_i \cdot \vec{L} - \omega_i t) . \tag{10.29}$$

One repeats the steps from Equations (10.3) to (10.9) and obtains the structure factor for the bcc reflection $\vec{K}$:

$$F(\vec{K}) = \prod_i J_0(\vec{K} \cdot \vec{a}_i) . \tag{10.30}$$

Since all amplitudes $\vec{a}_i$ are infinitesimal, the product in (10.30) can be evaluated by taking the logarithm of each side and expanding, $J_0(x) \cong 1 - \frac{1}{4}x^2$. One then obtains,

$$F(\vec{K}) = \exp\left(-\sum_i K^2 a_i^2/12\right) . \tag{10.31}$$

We have used the fact that $\langle\cos^2\theta\rangle_{\text{av}} = \frac{1}{3}$ for the angles between $\vec{K}$ and the phonon amplitudes $\vec{a}_i$. The square of (10.31) is just the Debye–Waller factor e^{-2W}.

The kinetic energy of each phonon mode for a sample containing N atoms is

$$T_i = \frac{1}{4} N M \omega_i^2 a_i^2 = \frac{1}{2} k_B T , \tag{10.32}$$

where k_B is the Boltzmann constant and T the temperature. From (10.31) and (10.32), the contribution w_i of each phonon to the exponent in the Debye–Waller factor e^{-2W}

$$w_i = -K^2 k_B T/3NM\omega_i^2 \,. \tag{10.33}$$

This is the standard result, appropriate for sufficiently high temperature. (Our definition is $-2W = \sum_i \omega_i$.)

We next consider the ionic displacements of a phase-modulated CDW, Equation (10.22). Repeating the steps from Equations (10.3) to (10.7), we obtain

$$\rho(K) = \sum_{\vec{L},n} J_n(\vec{K}\cdot\vec{A}) e^{i(\vec{K}+n\vec{Q})\cdot\vec{L}} e^{in\varphi_{\vec{q}}\sin(\vec{q}\cdot\vec{L}-\omega t)} \,. \tag{10.34}$$

Now we apply the Jacobi–Anger generating function (10.6) to the last factor in (10.34),

$$\rho(\vec{K}) = \sum_{\vec{L},n,m} J_n(\vec{K}\cdot\vec{A}) J_m(n\varphi_{\vec{q}}) e^{i(\vec{K}+n\vec{Q}+m\vec{q})\cdot\vec{L}-im\omega t} \,. \tag{10.35}$$

The bcc reflections correspond to $n = m = 0$, for which (10.35) leads again to Equation (10.11) for $F(\vec{K})$. In other words, we have shown that phason excitations do not contribute to the Debye–Waller factor of the bcc reflections.

The (first-order) CDW satellites correspond to $n = 1$, $m = 0$. From (10.35) their intensity will be proportional to

$$F^2 = J_1^2(\vec{K}\cdot\vec{A}) J_0^2(\varphi_{\vec{q}}) \tag{10.36}$$

instead of (10.10). We see that the phason $\vec{q}$ contributes a factor $J_0^2(\varphi_{\vec{q}})$. If we included all phason modes, there would be a similar factor for each one, as in the (infinite) product in (10.30). An exponential Debye–Waller factor would result, analogous to (10.31), with each phason contributing a term to the exponent (in e^{-2W}) of

$$W_{\vec{q}} = -\frac{1}{2}\varphi_{\vec{q}}^2 \,. \tag{10.37}$$

Furthermore, the mean kinetic energy (10.25) of a phason is also $\frac{1}{2}k_B T$. Accordingly, for a sample having N atoms,

$$w_{\vec{q}} = -2k_B T/NMA^2\omega^2 \,. \tag{10.38}$$

Comparing this result with (10.33), we find the relative contribution of phasons and phonons to the exponent of the Debye–Waller factor to be

$$w_{\vec{q}}/w_i = 6\omega_i^2/K^2A^2\omega^2 \,. \tag{10.39}$$

This is an extremely large ratio: For a CDW satellite, $K \sim 1.6 \times 10^8\ \text{cm}^{-1}$ so that with $A \sim 10^{-9}$ cm and the ratio (10.39) is ~ 200.

The foregoing results may be summarized: Phasons do not contribute to the Debye–Waller factor of the cubic reflections. For a CDW satellite, however, each phason contributes to the *exponent* of the temperature factor an amount ~ 200 times larger than that of a phonon having the same frequency. Of course, the complete Debye–Waller factor for satellites will include contributions from all $3N$ modes (ordinary phonons and phasons). Phasons can be regarded as the analog of spin-wave excitations of a magnetic structure. However, they do not represent new degrees of freedom. They are merely a subset of the vibrational degrees of freedom. Their unique properties, derived in this and Section 10.5, justify their separation into a distinct category.

A likely consequence of phason excitations is that intensities of CDW satellites may be reduced below the limits of observability. Furthermore, static deviations in the direction of $\vec{Q}$, caused by imperfections, dislocations, residual stress, etc., may supplement the (dynamic) contribution of phasons to e^{-2W}.

10.7 Survey of Electronic Anomalies

The purpose of this section is to summarize *briefly* the experimental evidence relevant to CDW structure in the alkali metals. We intend this to provide motivation for the field-modulation experiment proposed in Section 10.4. A number of electronic anomalies will be discussed, most of which are major. A unified and quantitative interpretation of these phenomena can be made using the CDW model. However, three additional postulates are required: (i) The direction of $\vec{Q}$ can be oriented approximately parallel to a (sufficiently strong) magnetic field $\vec{H}$ at 4 °K in a stress-free specimen. (ii) The direction of $\vec{Q}$ is subject to elastic stress which, if sufficiently severe, can impede the action of a magnetic field, (iii) The direction of $\vec{Q}$ near a metal surface may orient perpendicular to the surface, especially if the surface is in contact with an oxide layer or substrate to which the metal adheres.

These postulates are reasonable a priori in view of the symmetry of the Fermi surface in alkali metals and the small condensation energy (Section 10.5) of a CDW. There has been no attempt to justify them theoretically. However, similar postulates are known to be valid for the spin-density wave state of chromium; so they are neither novel nor speculative.

10.7.1 Optical Anomalies

The existence of the Mayer–El Naby [14–16] optical-absorption anomaly in K (with a threshold at $\hbar\omega \approx 0.6$ eV) has been called into question by the work of Smith [17], who found no evidence for such absorption. Postulate (iii) allows both sets of data to be consistent with the CDW model: Mayer and El Naby measured the reflectivity at a bulk-metal vacuum interface, whereas Smith measured the reflectivity on evaporated films. The theory [13] of the anomalous absorption predicts the observed intensity (oscillator strength ~ 0.1) and spectral shape of the Mayer–El Naby absorption if $\vec{Q}$ is oriented at random. But if $\vec{Q}$ is perpendicular to the surface, as seems possible in evaporated films, the matrix element (for electronic transitions across a CDW energy gap) becomes zero. This follows from the fact that the electric vector of the light (within the metal) is always parallel to the surface. Similar anomalies have been reported [18] in Na, Rb, and Cs (with thresholds at 1.2, 0.8, and 0.8 eV).

10.7.2 Conduction-Electron Spin Resonances

In ultrapure potassium, Walsh *et al.* [19] found conduction-electron spin resonance (CESR) line-widths as narrow as 0.13 G. They also observed this narrow resonance to split into two well-resolved components when the magnetic field direction was tilted from an initial orientation parallel to the metal surface. The maximum splitting was about 0.5 G. This is a large effect since the resonance shift (caused by the spin–orbit interaction) is only 5 G. This behavior was explained quantitatively [20] by the CDW model. It results from the fact that the conduction-electron g shift depends on the angle between $\vec{Q}$ and $\vec{H}$. The separation into two components is attributed to the highly stressed character of the specimens which, with the help of the three postulates given above, allows a specimen to subdivide into regions having either $\vec{Q}$ parallel to $\vec{H}$ or $\vec{Q}$ (nearly) perpendicular to $\vec{H}$. (The splitting should not occur in a stress-free specimen.) The reported behavior was reproduced in many specimens [21]. Serious attempts [22, 23] to explain it in more conventional ways were unsuccessful.

10.7.3 Doppler-Shifted Cyclotron Resonance

CDW energy gaps which touch the Fermi surface at the conical points P, Figure 10.2, cause the $\vec{Q}$ component of the electron group velocity to approach zero near P. A consequence is that the maximum velocity in the $\vec{Q}$ direction is less than the Fermi velocity $v_F = \hbar k_F/m^*$. For a free-electron-sphere model of a metal, the magnetic field H for doppler-shifted cyclotron resonance (DSCR) is given by $eH/m^*c = qv_F$ where q is the wave vector of a helicon or acoustic wave. (Note that m^* cancels out of this relation.) For a CDW with $\vec{Q}$ aligned parallel to $\vec{H}$ [postutate (i)], DSCR will occur at a smaller field [24–26]. High-precision DSCR with helicons was carried out on free-standing single crystals of Na and K by Penz and Kushida [27]. In K they found the DSCR to be shifted $\sim$ 1 kG below that of the free-electron model, in *quantitative* agreement with the CDW model. There was no comparable shift in Na.[4] In contrast, Libchaber and Grimes [28] have measured the DSCR edge in K, sandwiched between Mylar sheets. (The thermal contraction of K on cooling to 4 °K will introduce the maximum stress permitted by plastic yield.) They found no shift from the free-electron value. These two studies, when juxtaposed, are contradictory if the free-electron-sphere model is supposed, but are consistent with the CDW model: The DSCR shift need not occur in a maximally stressed sample where [according to postulate (ii)] $\vec{Q}$ need not align parallel to $\vec{H}$,

The apparent absence of a DSCR shift with acoustic waves [29, 30] might also be ascribed to thermal stress caused by transducers glued to the K surfaces. Several radio-frequency size-effect studies [31–33] in K, which in principle could reveal CDW distortions of the Fermi surface, were all carried out on K-Mylar or K-Parafilm sandwiches. Stresses introduced on cooling to 4 °K disqualify these experiments as critical tests of the CDW model.

10.7.4 Magnetoresistance

The magnetoresistance of a metal having a simply connected Fermi surface must saturate near $\omega_c\tau \sim 1$ and thereafter become independent of H [34, 35]. However, all workers have found a substantial linear magnetoresistance in Na and K which does not saturate up to $\omega_c\tau \sim 300$. Results obtained by four-terminal [36], helicon-resonance [37], and eddy-current-torque [38] techniques substantially agree. A serious attempt to explain this behavior by anisotropic relaxation[5] fell short by many orders of magnitude. The CDW model explains most of the general features without difficulty. Two mechanisms are involved:

(a) Bragg reflection and magnetic breakdown of electron orbits by heterodyne gaps [40], which arise from ionic displacements having periodicities with wave vector $2\pi\vec{G} \pm \vec{Q}$. The size and k-space location of these gaps is very sensitive to the orientation of $\vec{Q}$ relative to the crystal axes. The mechanism leads to an approximately linear magnetoresistance (which would eventually have to saturate). It explains the observed [37] dependence of the transverse magnetoresistance on orientation of $\vec{H}$ relative to the crystal axes, the wide variation of the linear slope from sample to sample, and the sensitivity to elastic stress;

4) Either Na does not have a CDW or it has, but is highly stressed by its martensitic transformation. Experimental evidence seems to favor the latter interpretation.

5) To explain K data, Young found it necessary to have "hot spots" on the Fermi surface where umklapp scattering takes place at a rate $\sim$ 10 000 times faster than relaxation on the rest of the Fermi surface. The angular size of the hot spot had to be 10^{-5} sr. Even without these extraordinary demands, his suggestion that a hot spot could be caused by *transverse* [110]-phonon umklapp scattering is not admissible since the matrix element for such a process vanishes by symmetry [39].

(b) anisotropy of the resistivity tensor relative to the axis $\vec{Q}$ of the deformed Fermi surface (see Section 10.7.7). The resistivity parallel to $\vec{Q}$ is larger than that perpendicular to $\vec{Q}$. Accordingly, with increasing $\vec{H}$, the longitudinal resistivity should be enhanced (as $\vec{Q}$ domains align parallel to $\vec{H}$) and the transverse resistivity should be reduced (relative to the increase caused by the first mechanism, above). This would explain why the longitudinal magnetoresistance exceeds the transverse [38]. Furthermore, since this mechanism is suppressed by elastic stress, it explains why the longitudinal effect is reduced by stress [41], whereas the transverse is enhanced [37]. Since the two mechanisms need not "track" together with increasing field, dominance of the second mechanism over the first at low field could sometimes occur and explain the initially negative transverse magnetoresistance found in some specimens [36].

10.7.5 Hall Effect

The theory of Lifshitz *et al.* [34, 35] requires that the high-field ($\omega_c\tau > 1$) Hall coefficient of a metal with a simply connected Fermi surface be $R = 1/nec$, independent of H. Penz [42] has observed, however, that $|R|$ decreases monotonically with increasing H in K. The decrease varies from sample to sample and is $\sim 7\%$ at 110 kG. Garland and Bowers [43] verified this by two different methods and found an equally large effect in Na. This behavior can be accounted for semiquantitatively with the CDW model: Bragg reflection of electrons at heterodyne gaps [mechanism (a)] causes a fraction f, say, of the electrons to behave like (positively charged) holes. Accordingly, $|R|$ is proportional to $[n(1-2f)]^{-1}$ instead of to n^{-1}. As magnetic breakdown reduces f (with increasing H), $|R|$ must decrease. The magnitude of the decrease is correctly predicted when f is chosen to fit the observed magneto-resistance [42].

10.7.6 De Haas–van Alphen Effect

The extremely small variation of the de Haas–van Alphen frequency as $\vec{H}$ is rotated relative to the crystal axes of Na [44] and K [11] precludes the CDW model unless postulate (i) is assumed. (The Fermi surface appears spherical to $\sim 0.1\%$.) If $\vec{Q}$ is always parallel to $\vec{H}$ in a 50-kG field, then the observed de Haas–van Alphen frequency F corresponds to the extremal area perpendicular to $\vec{Q}$ of the lemon-shaped Fermi surface, Figure 10.2. Variations of this belly area with $\vec{H}$ orientation should manifest the Fermi-surface anisotropy that would be present without a CDW. However, one might anticipate deviations from perfect alignment when $\vec{H}$ is in a "hard-$\vec{Q}$" direction, cf. Equation (10.19). In such a case, detailed interpretation [45] of anisotropy data in terms of crystalline pseudo-potentials would have to be revised.

Since the volume (in k space) enclosed by the Fermi surface must correspond to one electron per atom, any extra volume associated with the conical-point distortion must be compensated for by a decrease in belly area. Shoenberg and Stiles [11] attempted to compare the observed frequency F with F_0, calculated for a free-electron sphere. They found $(F - F_0)/F_0$ to be 0.6, −0.2, −0.7, and −1.4% for Na, K, Rb, and Cs. The relative accuracy of these numbers was 0.2%, although the absolute error (of magnet calibration) was much larger. Thomas and Turner[6] have measured F precisely in K, which provides a magnet calibration for the prior work. The revised values for $(F - F_0)/F_0$ are 0.4, −0.4, −0.9, and −1.6%. $\{F_0\}$ here are based on the lattice parameters of Barrett [47], and observed anisotropies in F were averaged so that F is the frequency for a Fermi sphere of equal volume.[7] A more precise value for K

6) They found $F = 1.8246 \pm 0.0006) \times 10^8$ G [46].

7) This could not be done for Cs since the crystal orientations were unknown. However Shoenberg and Stiles's data when combined with that of [48] require a belly-area defect of at least −0.8% for Cs.

8) Their 5 °K lattice constant (5.2295 ± 0.0007 Å) requires $F = 1.8283 \times 10^8$ $\vec{G}$ [49].

can be obtained from the Thomas and Turner F together with a new value[8] for the K lattice constant, $(F - F_0)/F_0 = -0.20 \pm 0.04\%$.

We conclude that the (mean) de Haas–van Alphen frequencies in K, Rb, and Cs deviate from the free-electron-sphere model by about five times the experimental uncertainties. If the periodic potential of a CDW is treated as a local potential, the predicted belly-area defects for these three metals can be shown to be −0.8, −1.5, and −1.8%. However, a CDW is caused by exchange and correlation effects [1], and exchange and correlation potentials are known [50] to be *highly* nonlocal. Consequently, Fermi-surface distortions caused by such phenomena cannot be accurately predicted. The CDW model allows discrepancies between F and F_0 of the observed magnitudes, whereas the usual model allows no deviation whatsoever.[9]

10.7.7 Electron–Phonon Interaction

In a recent study, Rice and Sham [51] found significant disagreement between theory and experiment in the angular dependence of electron–phonon scattering at low temperature ($\sim 10^\circ$K) in K. This could be studied because the electrical-resistivity relaxation time τ_1 differs from the ultrasonic-attenuation relaxation time τ_2. If $\sigma(\theta)$ is the cross section for electron scattering through an angle θ, then $1/\tau_1$ is proportional to the solid-angle average of $\sigma(\theta)$ with a $1 - \cos\theta$ weighting function. On the other hand, $1/\tau_2$ depends on the average of $\sigma(\theta)$ with a $\frac{3}{2}\sin^2\theta$ weighting function. Using four different electron–phonon interaction models, Rice and Sham found that τ_2/τ_1 should be ~ 0.9, whereas experimentally this ratio is ~ 2.7. This indicates there is a large unexpected contribution to the electrical resistivity from large-angle scattering. Such a contribution is expected with the CDW model: Electrons near one conical point can be easily scattered to the other conical point by small-q phonons (umklapp scattering caused by CDW energy gaps). Small-q phasons will also contribute since, from Equation (10.28), they have the same $\vec{k}$-conservation rules as umklapp processes. These mechanisms have not been studied quantitatively. It is obvious that the resistivity tensor will be highly anisotropic, with a much larger resistivity parallel to $\vec{Q}$ compared to perpendicular. The degree of anisotropy will be sensitive to both temperature and purity. It is not clear, however, that solution of the transport equation, which would be very complex, would still lead to a low-temperature phonon resistivity $\sim T^5$ as is observed [52]. A realistic calculation of bulk resistivity (or magnetoresistivity) must take account of large-scale heterogeneities caused by $\vec{Q}$ domains.

10.7.8 Positron Annihilation

The CDW periodic potential introduces anisotropy into the conduction-electron momentum distribution. In principle, this can be measured by angular correlation of γ rays arising from positron annihilation. Such an experiment requires a magnetic field orientation of $\vec{Q}$, so that momentum components parallel and perpendicular to $\vec{Q}$ can be compared. The predicted effect is small, 2.4%; the observed effect [53] was $2.3 \pm 0.8\%$.

10.7.9 Other Properties

If CDW's exist, they will influence in some way almost any physical property of the metal. The phenomena we have discussed above are ones which are profoundly affected. Others will be only slightly modified. For example, the electronic specific-heat coefficient should be

9) A suggestion made in [11] that discrepancies of the required magnitude might arise from *negative* pressure associated with thermal contraction is inadmissible. Such (unrelieved) forces would exceed the yield stress by two orders of magnitude.

10) Reference [1], Fig. 1.

enhanced ∼ 10%. It would be difficult to measure this experimentally since band-structure, many-body, and phonon-interaction contributions are not known with sufficient accuracy. However, the electron density of states should rise abruptly to its enhanced value near E_F.[10] Such a rise has been seen [54] in the soft X-ray emission of Na, but this may have another explanation [55]. Nuclear magnetic resonance could be broadened by quadrupole interaction with CDW electric fields or by variations in the Knight shift from crest to trough. (Both of these effects may be diminished by phason excitation.) Anomalous NMR broadening has been reported in Rb and Cs [56], and attributed to pseudodipolar interactions. However, a theoretical study of this latter mechanism [57] indicates that the observed broadening is much too large.

10.8 Conclusion

We have shown that CDW modulation of a cubic metal modifies the diffraction properties of the crystal in three ways: (i) The structure factors of the cubic reflections are reduced to $J_0(\vec{K}\cdot\vec{A})$ from unity; (ii) satellite reflections, at $K = 2\pi\vec{G} \pm \vec{Q}$, are introduced; (iii) intensities of the cubic reflections can be modulated with a magnetic field.

We have carried out structure-factor measurements on K and have found that quantitative interpretation is made difficult either by large primary extinction corrections (mosaic-block size ∼ 0.5 mm) or by unknown anharmonic contributions to the Debye–Waller factor.

Satellite reflections, very small to begin with, are reduced further in intensity by phase-modulation excitations of the CDW. The low-energy frequency spectrum of these modes was calculated. They were shown to have no effect on the intensity of cubic reflections, but contribute extremely to the Debye–Waller factor of CDW satellites. Accordingly, nonobservance of satellites can (perhaps) be rationalized.

Finally, we have emphasized that field modulation of high-index cubic reflections is a decisive experiment. The CDW model requires a spectacularly large effect. Failure to observe it would vitiate the model.

Motivation for carrying out the aforementioned experiment was summarized in a survey of reported alkali-metal anomalies. Published experiments, if all are accepted at face value, provide a spectacle of theoretical failure. None have been explained within the conventional framework – a nearly spherical simply-connected Fermi surface. In contrast, the CDW model provides a unified (and for the most part quantitative) explanation of all of them. But this is entirely circumstantial. If the atoms are displaced from their cubic sites, those displacements must be seen. The field-modulation experiment should successfully separate myth from reality.

Acknowledgements

The author is deeply grateful to the staff of the Nuclear Engineering Department, University of Michigan for their kindness in making their reactor facilities available. He would like to thank L.A. Feldkamp and M. Yessik for guidance during the experimental work. Throughout the course of this research he has received much benefit from the experience and help of A.D. Brailsford.

References

1 Overhauser, A.W. (1968) *Phys. Rev.*, **167**, 691.
2 Atoji, M. and Werner, S.A. (1969) *Solid State Commun.*, **7**, 1681.
3 Zachariasen, W.H. (1967) *Acta Cryst.*, **23**, 558.
4 Cooper, M.J. and Rouse, K.D. (1970) *Acta Cryst. A*, **26**, 214.
5 Martin, D.L. (1965) *Phys. Rev. A*, **139**, 150; Tables VIII and IX. Θ^M is given by $(k_B \Theta^M)^2 = 3h^2/\langle \nu^{-2} \rangle_{av}$.
6 Barron, T.K., Leadbetter, A.J., Morrison, J.A., and Salter, L.S. (1966) *Acta Cryst.*, **20**, 125.
7 Cooper, M.J. and Rouse, K.D. (1968) *Acta Cryst. A*, **24**, 405;
8 Cooper, M.J. and Rouse, K.D. (1969) *Acta Cryst. A*, **25**, 615.
9 Maradudinand, A.A., Flinn, P.A. (1963) *Phys. Rev.*, **129**, 2529.
10 Willis, B.T.M. (1969) *Acta Cryst. A*, **25**, 277.
11 Shoenberg, D. and Stiles, P.J. (1964) *Proc. Roy. Soc. (London) A*, **281**, 62.
12 Kittel, C. (1963) *Quantum Theory of Solids*, John Wiley & Sons, Inc, New York, p. 162.
13 Overhauser, A.W. (1964) *Phys. Rev. Lett.*, **13**, 190; When this was written the periodic potential was assumed to be caused by a spin-density wave rather than a CDW.
14 El Naby, M.H. (1963) *Z. Physik*, **174**, 269.
15 Mayer, H. and El Naby, M.H. (1963) *Z. Physik*, **174**, 280.
16 Mayer, H. and El Naby, M.H. (1963), *Z. Physik*, **174**, 289.
17 Smith, N.V. (1969) *Phys. Rev.*, **183**, 634.
18 Mayer, H. and Hietel, B. (1966) in *Proceedings of the International Colloquium on Optical Properties and Electronic Structure of Metals and Alloys, Paris*, 1965 (ed. F. Abeles), North-Holland, Amsterdam, p. 47.
19 Walsh Jr., W.M., Rupp Jr., L.W., and Schmidt, P.H. (1966) *Phys. Rev.*, **142**, 414.
20 Overhauser, A.W. and de Graaf, A.M. (1968) *Phys. Rev.*, **168**, 763.
21 Walsh Jr., W.M. (private communication).
22 Lampe, M. and Platzman, P.M. (1966) *Phys. Rev.*, **150**, 340.
23 Platzman, P.M. and Wolff, P.A. (1967) *Phys. Rev. Lett.*, **18**, 280, last paragraph.
24 Overhauser, A.W. and Rodriguez, S. (1966) *Phys. Rev.*, **141**, 431.
25 McGroddy, J.C., Stanford, J.L., and Stern, E.A. (1966) *Phys. Rev.*, **141**, 437.
26 Alig, R.C., Quinn, J.J., and Rodriguez, S. (1966) *Phys. Rev.*, **148**, 632.
27 Penz, P.A. and Kushida, T. (1968) *Phys. Rev.*, **176**, 804.
28 Libchaber, A. and Grimes, C.C. (1969) *Phys. Rev.*, **178**, 1145.
29 Thomas, R.L. and Bohm, H.V. (1966) *Phys. Rev. Lett.*, **16**, 587.
30 Greene, M.P., Hoffman, A.R., Houghton, A., and Quinn, J.J. (1967) *Phys. Rev.*, **156**, 798.
31 Koch, J.F. and Wagner, T.K. (1966) *Phys. Rev.*, **151**, 467.
32 Peercy, P.S., Walsh Jr., W.M., Rupp Jr., L.W., and Schmidt, P.H. (1968) *Phys. Rev.*, **171**, 713.
33 Libchaber, A., Adams, G., and Grimes, C.C. (1970) *Phys. Rev. B*, **1**, 361.
34 Lifshitz, I.M., Azbel, M.Ya., and Kaganov, M.I., Eksperim. Zh. (1956) *J. Teor. Fiz.*, **31**, 63 [(1957) *Sov. Phys. JETP*, **4**, 41].
35 Fawcett, E. (1964) *Adv. Phys.*, **13**, 139.
36 Schmidt, R.L.: Cornell Materials Science Center Report No. 1434 (unpublished).
37 Penz, P.A. and Bowers, R. (1968) *Phys. Rev.*, **172**, 991; This paper contains many references to earlier work.
38 Lass, J.S. (1970) *J. Phys. C*, **3**, 1926.
39 Young, R.A. (1968) *Phys. Rev.*, **175**, 813.
40 Reitz, J.R. and Overhauser, A.W. (1968) *Phys. Rev.*, **171**, 749.
41 Jones, B.K. (1969) *Phys. Rev.*, **179**, 637.
42 Penz, P.A. (1968) *Phys. Rev. Lett.*, **20**, 725.
43 Garland, J.C. and Bowers, R. (1969) *Phys. Rev.*, **188**, 1121; Section IIIC; and private communication.
44 Lee, M.J.G. (1966) *Proc. Roy. Soc. (London) A*, **295**, 440.
45 Lee, M.J.G. and Falicov, L.M. (1968) *Proc. Roy. Soc. (London) A*, **304**, 319.

46 Thomas, R.L. and Turner, G. (1968) *Phys. Rev.*, **176**, 768.
47 Barrett, C.S. (1956) *Acta Cryst.*, **9**, 671.
48 Okumura, K. and Templeton, I.M. (1965) *Proc. Roy. Soc.* (London) A, **287**, 89.
49 Werner, S.A., Gürmen, E. and Arrott, A. (1969) *Phys. Rev.*, **186**, 705.
50 Overhauser, A.W. (1970) *Phys. Rev. B*, **2**, 874.
51 Rice, T.M. and Sham, L.J. (1970) *Phys. Rev. B*, **1**, 4546.
52 Garland, J.C. and Bowers, R. (1968) *Phys. Rev Lett.*, **21**, 1007.
53 Gustafsonand, D.R. and Barnes, G.T. (1967) *Phys. Rev. Lett.*, **18**, 3.
54 Crisp, R.S. and Williams, S.E. (1961) *Phil. Mag.*, **6**, 365.
55 Mahan, G.D. (1967) *Phys. Rev.*, **163**, 612.
56 Poitrenaud, J. (1967) *J. Phys. Chem. Solids*, **28**, 161.
57 Mahanti, S.D. and Das, T.P. (1968) *Phys. Rev.*, **170**, 426.

Reprint 11 Questions About the Mayer–El Naby Optical Anomaly in Potassium[1)]

A.W. Overhauser and N.R. Butler**

* Department of Physics, Purdue University, 525, Northwestern Avenue, West Lafayette, Indiana 47907, USA

Received 19 January 1976

The large, optical-absorption anomaly in the reflection spectrum of potassium, discovered by Mayer and El Naby has a threshold near 0.6 eV and maximum near 0.8 eV. The failure of other workers to reproduce this effect has led to a premature conclusion that it was an experimental artifact. The recent discovery by Harms of how to reproduce it in a controlled way reestablishes the need for a satisfactory explanation. Since the anomaly is turned on by the presence of a KOH surface layer, we report transmission measurements on KOH between 0.5 and 1.1 eV. These show that absorption in KOH cannot be the cause. A number of other extrinsic mechanisms are also considered. Finally we entertain the possibility that the absorption is intrinsic to the metal. This would require potassium to be optically anisotropic. If the direction of the optic axis is influenced by the presence of the KOH layer, a quantitative account of the diverse observations is possible. Two critical experiments are proposed which may distinguish intrinsic and extrinsic mechanisms.

11.1 Introduction

The purpose of this paper is to revitalize interest in the problem of the origin of the Mayer–El Naby optical-absorption anomaly [1–3], observed in the near-infrared reflection spectrum of potassium. Study of this enigma has subsided since its discovery in 1963 because other workers [4–6] were unable to reproduce it. The conventional view is that the original anomaly was spurious or, at least, unrelated to the properties of potassium.

Recently, Harms [7], in an extensive study, has shown that the Mayer–El Naby anomaly can be reproduced at will in a controlled way. Freshly prepared (bulk) metal–vacuum surfaces can be obtained which show no trace of the absorption anomaly (see curve *A* of Figure 11.1). Curves *B*, *C*, *D*, and *E* were obtained after successive introduction of small amounts of H_2O vapor into the vacuum chamber, otherwise kept at $\sim 10^{-10}$ Torr. The dashed curve *M* is one obtained by Mayer and El Naby [1–3]. The chemical reaction between potassium and trace amounts of H_2O is

$$\mathrm{K} + \mathrm{H_2O} \rightarrow \mathrm{KOH} + \frac{1}{2}\mathrm{H_2}\,. \tag{11.1}$$

Consequently, we conclude that curves *B*, *C*, *M*, *D*, and *E* are the absorption spectra of potassium surfaces having increasing amounts of KOH on the surface.

The problem, of course, is to account for the magnitude and spectral shape of the absorption shown in Figure 11.1. The absorption threshold is near 0.6 eV, the peak is near 0.8 eV, and the asymmetric high-energy tail extends beyond the 1.3-eV threshold for interband absorption in potassium. When present, the anomalous absorption is an order of magnitude larger than the observed [5, 6] interband absorption. (The data shown were obtained by reflection

1) Phys. Rev. B 14, 8 (1976)

Anomalous Effects in Simple Metals. Albert Overhauser

ISBN: 978-3-527-40859-7

measurements of *p*- and *s*-polarized light having a 75° angle of incidence, and interpreted by employing the Fresnel equations [8].)

11.2 Extrinsic Mechanisms

Since the anomalous loss of reflectivity near 0.8 eV is associated with the presence of KOH on the potassium surface, it is natural to assume that this (apparent) absorption peak has an extrinsic origin, not associated with the fundamental optical properties of bulk potassium. Such a conclusion is not warranted, however, until a physical mechanism can be suggested which accounts for the experimental observations qualitatively. We consider below a number of possibilities.

11.2.1 Absorption in KOH

KOH is a transparent insulator and would not be expected to absorb in the near infrared. The spectral region of interest is between $h\nu = 0.5$ and 1.1 eV. Infrared data for $h\nu < 0.5$ eV can be found [9], but we have not located any for the relevant energy. We therefore report our own observations of the transmission of KOH in the near infrared.

KOH samples 10^{-2} cm thick were grown between glass slides in an argon atmosphere. The transmission of these KOH-glass sandwiches was measured with a Perkin Elmer Model E-1 double-pass spectrometer. The dashed curve *T* of Figure 11.1 is an inverted plot of the transmission data obtained. The transmission is that relative to an empty sandwich. It is obvious that there is no absorption in KOH that could lead to an explanation of the Mayer–El Naby anomaly.

11.2.2 Interference

The observation, shown in Figure 11.1, that the spectral shape of the anomalous absorption remains unchanged as the magnitude of the effect builds up with increasing amounts of KOH on the surface is inconsistent with the supposition that the effect is caused by interference in KOH layers of increasing thickness. A KOH layer of unique thickness, but of increasing area of coverage, would require a reflectivity loss that oscillates with $h\nu$.

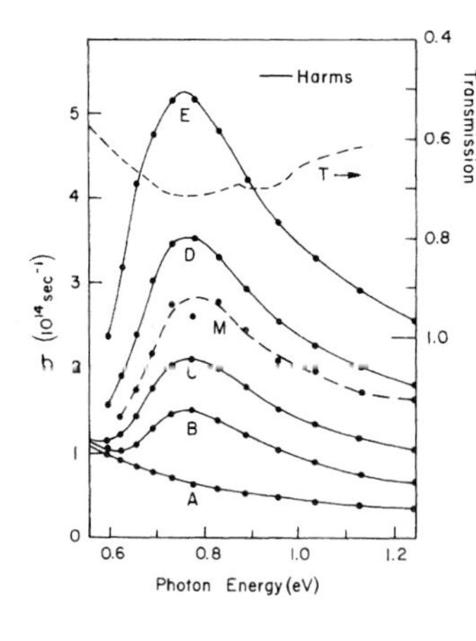

Figure 11.1 Curve *A* is the optical conductivity (absorption) at a clean bulk-potassium–vacuum interface. Curves *B*–*E* were obtained on the same specimen after successive exposure to trace amounts of H_2O. These data are from Harms [7]. The dashed curve *M* is one obtained by Mayer and El Naby [1–3]. The dashed curve *T* is an inverted plot of the transmission through 0.01 cm of KOH sandwiched between glass slides. All data are at room temperature.

11.2.3 Surface Roughness

Attack of a potassium surface by the chemical reaction (11.1) would be expected to roughen the metal surface. This will lead to light scattering [10–12] and a consequent loss in specular reflectivity. The spectral shape of this reflectivity loss depends somewhat on the scale of the surface roughness, but in general, it increases monotonically as $(h\nu)^4$ until an energy -0.7 times the plasma frequency.[2] Since the plasma frequency of potassium corresponds to an energy of 3.8 eV, the spectrum of Figure 11.1 cannot be ascribed to this mechanism. The early work [1–3] of Mayer and El Naby would also preclude this mechanism since they observed the anomaly to be essentially un-changed when the potassium surface became molten. (One would expect the surface roughness to be much smaller for a liquid interface.)

11.2.4 Plasmon Absorption

Surface roughness also allows electromagnetic waves to couple to surface plasmons [13]. The loss in reflectivity caused by this interaction occurs, however, at 2.8 eV in potassium [6] (without a KOH coating). A recent theoretical study [14] has shown that as a dielectric layer is added the frequency of this absorption shifts to lower energy with increasing thickness of the layer. However, for thick layers the frequency approaches $\omega_p/(1+\epsilon)^{1/2}$ where ω_p is the bulk plasma frequency and ϵ is the (optical) dielectric constant. For KOH $\epsilon \sim 2$, so surface plasmon absorption could not occur below ~ 2 eV for potassium coated with KOH.

11.2.5 Scattering by KOH Aggregates

A resonant loss of reflectivity could occur if the KOH formed large ($\sim$ 10 000 Å) clumps on the potassium surface. But one would then expect the spectral shape of such scattering to change continuously as the clumps grow in size, since scattering resonances are geometric in origin. (Even if one were to assume that KOH clumps had a unique size, the sharpness of the observed threshold at 0.6 eV could not be explained.[3] Moreover, a size distribution would be required to reproduce the monotonic behavior for $h\nu > 0.8$ eV so, in addition, one would have to require the size distribution to be independent of accumulated mass of KOH.)

11.2.6 Color Centers in KOH

It is conceivable that KOH formed by reaction (11.1) at a potassium surface acquires a defect structure which would lead to color center absorption. Since excess K is available at the metal surface, the most probable defect would be OH^- vacancies with a trapped electron (for charge compensation), i. e., *F* centers. The energy of the *F*-band absorption in KOH (presuming that it exists) can be estimated from Ivy's law: $h\nu_F$ is a smooth function of the lattice constant. The molecular volume of KOH is intermediate between that of KF and KC1, for which $h\nu_F = 2.8$ and 2.3 eV, respectively. Therefore the *F* band in KOH should be near 2.5 eV. The validity of this argument has been tested [16] for the case of KCN, where the *F* band was found at 2.1 eV, compared to 2.0 eV for KBr. KCN has a slightly smaller lattice constant than KBr. (The forgoing mechanism would also be in difficulty from the standpoint of intensity since extraordinary concentrations of color centers would be required. Furthermore, the striking asymmetry of the anomaly would also be a problem.)

2) See Fig. 4 of [11].
3) Theoretical curves for dielectric spheres are given by [15].

11.2.7 Surface States

The Mayer–El Naby anomaly is 10–20 times stronger than the fundamental interband absorption of potassium. The peak value of the latter [5, 6] corresponds to an optical conductivity of $\sigma \sim 0.3 \times 10^{14}\,\mathrm{s}^{-1}$ The maximum anomalous absorption shown in Figure 11.1 is about $4.5 \times 10^{14}\,\mathrm{s}^{-1}$ The integrated oscillator strength of this peak is about $f \sim 0.1$, a value derived on the basis that every atom within the penetration depth ($\sim$ 260 Å) of the light contributes equally. To attribute such an absorption to surface states would require an active site (having an oscillator strength $f \sim 1$) at each surface atom. Freshly cleaved surfaces of Si and Ge do exhibit [17] an absorption of comparable magnitude. It seems unlikely that a KOH–potassium interface, where surface states would necessarily overlap the conduction band continuum, could have a high density of sharp, occupied surface states and a complimentary narrow band of empty surface states ($\sim$ 0.8 eV higher). Although such hypotheses present serious difficulties, they cannot be ruled out completely.

11.2.8 Absorption by K Particles in KOH

Meessen [18] has suggested that the Mayer–El Naby absorption arises from small (submicron) particles of potassium. In view of the Harms study these could be embedded in the KOH overlayer. For spherical particles the absorption resonance would occur [19] at $\omega_p/(1 + 2\epsilon)^{1/2}$, i. e., at $\sim$ 1.7 eV. One can reduce this energy arbitrarily by assuming the particles are very flat spheroids parallel to the surface. This would seem to be a rather ad hoc and extreme assumption, and one that would be hard to reconcile with the invariant spectral shape shown in Figure 11.1.

11.2.9 Impurity Absorption in Potassium

The surface reaction (11.1) could conceivably provide a source of H or O which, upon diffusion into the metal, might provide sites for optical excitation having the appropriate spectral shape and intensity. However H is insoluble in potassium [20]. A hydride, KH, could occur but it is not stable [21] at ambient pressures of 10^{-10} Torr. Solute oxygen can also be ruled out: Mayer and El Naby also measured the long-wavelength Drude absorption at 90 °K and found that it was consistent with the phonon resistivity of pure potassium ($\sim 2\,\mu\Omega$ cm) at that temperature. Since from Linde's rule [22] the expected residual resistivity of 1-at.% O in a monovalent metal is $\sim 8\,\mu\Omega$ cm, the concentration of O in solid solution could only be 10^{-3} or less. The magnitude of the observed anomalous absorption is too large for such a small concentration, even if the O were to have an absorption (in K) of the required spectral shape and an oscillator strength near unity.

11.2.10 Unknown Mechanism

If the Mayer–El Naby anomaly is to be attributed to an extrinsic mechanism, this one would seem to be more likely than the others discussed above. The structure of the hydrated layer produced by Harms has not been characterized nor has its thickness even been estimated. Under such circumstances it is useless to speculate further. One of the experiments proposed in Section 11.4 should confirm any mechanism that could be attributable to the hydrated layer itself.

11.3 Intrinsic Mechanisms

From the foregoing discussion it seems that, at present, a satisfactory, extrinsic explanation of the Mayer–El Naby anomaly has yet to be found. It is possible that one may eventually be identified and shown experimentally to be the cause. Until such a time it is appropriate to examine the only alternative, namely, that the Mayer–El Naby anomaly arises from the properties of potassium metal itself. Such a postulate leads immediately to three tentative conclusions.

(i) Bulk potassium, although seemingly cubic, must be optically anisotropic. There must be an optic axis $\vec{a}$ such that the anomalous absorption occurs if the polarization vector $\vec{\epsilon}$ of the electromagnetic wave (in the metal) is parallel to $\vec{a}$, whereas if $\vec{\epsilon}$ is perpendicular to $\vec{a}$ absorption does not occur, $\vec{\epsilon}$ parallel to $\vec{a}$ must be the direction for absorption since, when the anomaly is not observed, it is absent for all $\vec{\epsilon}$ lying in a plane (the metal surface).

(ii) The optic axis $\vec{a}$ must be perpendicular to a glass–metal interface or a vacuum–metal interface. This assumption is required to account for the many negative observations.

(iii) Finally, the work of Harms requires that $\vec{a}$ is not perpendicular, and is possibly parallel, to a K–KOH surface. In other words an epitaxial deposit of KOH rotates $\vec{a}$ into the plane of the surface.

A theoretical model must be consistent with the foregoing properties and, in addition, must explain the unusual intensity and asymmetry of the absorption. Most early attempts to explain the Mayer–El Naby result failed to satisfy the properties given above. Only one model [23], which assumes that potassium has a charge-density-wave (CDW) ground state, appears to be consistent with ail of the requirements. In the remainder of this section we will elaborate on this observation.

A CDW introduces an additional periodic potential $V(\vec{r})$ into the one-electron Schrödinger equation for conduction electrons:

$$V(\vec{r}) = G\cos(\vec{Q}\cdot\vec{r}) , \tag{11.2}$$

where $\vec{Q}$, the CDW wave vector, has a magnitude slightly larger than $2k_F$ [23, 24], the diameter of the Fermi surface. This potential introduces new energy gaps of magnitude G into the one-electron energy spectrum $E(\vec{k})$. As a consequence there will arise a new optical absorption mechanism having a threshold at $h\nu = G$. The optical conductivity $\sigma(W)$, $W \equiv h\nu$, caused by electronic transitions from below to above the CDW gap has been calculated [24]:

$$\sigma(W) = \frac{G^2 e^2 Q}{8\pi\hbar W^2}\left(\frac{W-G}{W+G}\right)^{1/2}\left(1-\frac{W+G}{2\mu Q}\right)\cos^2\theta , \tag{11.3}$$

where $\mu = \hbar^2 Q/2m$ and θ is the angle between $\vec{Q}$ and the (electric) polarization vector $\vec{\epsilon}$ of the light. $\sigma(W) = 0$ for $W < G$. This result exhibits a uniaxial absorption, as required. The optic axis is coincident with the wave vector $\vec{Q}$. Multiple-$\vec{Q}$ CDW states are possible and have been observed in layer compounds [25]. However, for the case of potassium, the (apparent) need for a unique optic axis is consistent only with a single-$\vec{Q}$ CDW structure.

One of the remarkable properties of the theoretical $\sigma < (W)$, given by Equation (11.3), is that it explains the magnitude and shape of the Mayer–El Naby anomaly. This is shown in Figure 11.2, where the Drude tail (caused by intraband conductivity) has been subtracted from the experimental points of Mayer and El Naby. The CDW energy gap was taken to be $G = 0.62$ eV and $\cos^2\theta = \frac{2}{3}$ for the theoretical curve. The data of Harms require a larger magnitude by about a factor of 2. A stronger absorption can arise theoretically as a result of collective corrections to the optical matrix element. Exchange and correlation effects lead to such an enhancement [26] when they are large enough to cause a CDW instability.

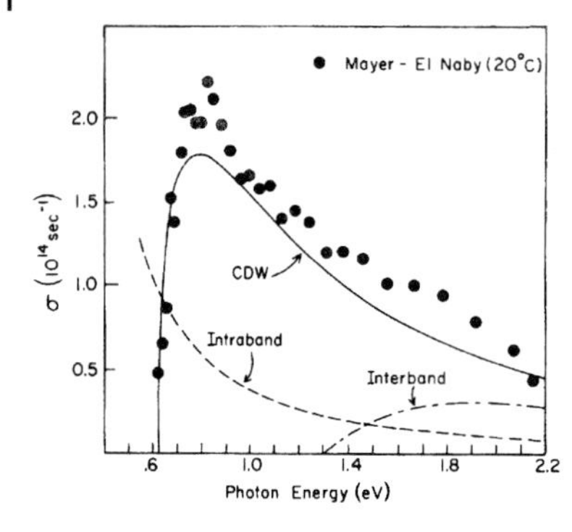

Figure 11.2 Anomalous optical-absorption spectrum of potassium. The intraband conductivity, dashed curve, has been subtracted from the experimental data of Mayer and El Naby before the latter were plotted. The solid curve is the theoretical absorption of a CDW structure given by Equation (11.3) of the text. The dot-dashed curve, starting at 1.3 eV is the fundamental interband absorption caused by the cubic-lattice periodic potential.

[The validity of Equation (11.3) needs to be discussed. It was originally derived for optical transitions across an energy gap created by a spin-density wave (SDW). Hopfield [27] pointed out the matrix elements for transitions across (pure) SDW energy gaps have to be zero. This is true only when the periodic potential causing the energy gap arises exclusively from the conduction electrons. The periodic potential of a CDW, however, arises in part from the lattice modulation which necessarily accompanies the electronic modulation [23]. Consequently, Hopfield's remark does not apply. (Otherwise ordinary interband absorption could never occur.) A self-consistent theory [26] of the optical matrix elements leads to the conclusion that Equation (11.3) is likely a slight underestimate, as observed above.]

A single-$\vec{Q}$ CDW structure is expected to cause a splitting of the conduction-electron spin resonance [28]. This effect has been reported [29] and confirmed as a g-factor splitting [30]. A gap of $G \sim 0.6$ eV is required to fit the observed splitting. If this phenomenon is employed to determine G, then the extraordinary agreement (within a factor of 2) between Equation (11.3) and the observed anomaly can be achieved without an adjustable parameter. Energy threshold, asymmetric shape, magnitude, and uniaxial character are all accounted for.

Orientation of $\vec{Q}$ at various potassium interfaces remains to be discussed. A CDW in the electronic structure must be accompanied by a similar distortion of the positive-ion lattice in order to optimize microscopic charge neutrality [4]. Such ion displacements are equivalent to a static longitudinal phonon. To minimize the energy penalty of this distortion one expects $\vec{Q}$ to have a crystallographic direction for which a longitudinal phonon (having the same $|\vec{Q}|$) has the lowest frequency. For potassium this is a [110] direction.

When soft metals are evaporated on amorphous substrates (e. g., glass), the crystal grains have a preferred texture. The close-packed planes of the lattice lie parallel to the surface [31]. For potassium the normal to a glass–metal interface will be a [110] direction, i. e., one of the allowed directions for $\vec{Q}$. This means that the interfacial energy can be minimized by allowing the phase of the CDW to adjust so that either a maximum or minimum of electron density occurs at the boundary. Since the planes of constant electron density are perpendicular to $\vec{Q}$, optimization of the interfacial energy can occur only if $\vec{Q}$ is parallel to the [110] direction normal to the surface. It must be remembered that when light reflects from a metal surface, the polarization vector $\vec{\epsilon}$ (inside the metal) is parallel to the surface. Accordingly one does not expect a CDW optical anomaly in potassium to be visible at a glass–metal interface or at the vacuum–metal interface of an evaporated film. The CDW model explains therefore why Hodgson [4], Smith [5], and Palmer and Schnatterly [6] could not reporduce the Mayer–El Naby result.

Only Mayer and El Naby [1–3] and Harms [7] have measured the optical constants of potassium at a bulk-metal–vacuum interface. Even if a clean bulk-metal–vacuum interface had, say, a (100) orientation macroscopicaily, one would not expect to see the Mayer–El Naby absorption. Such a surface will quickly regrow into a washboardlike surface, with each facet having a (110)-type orientation, so as to minimize the surface energy. This explains why it is possible to not observe the anomaly on a bulk (and presumably polycrystalline) sample, as shown by curve *A* of Figure 11.1.

The only *ad hoc* assumption that is required to provide a complete (and quantitative) explanation is that a KOH surface layer epitaxially puckers a potassium (110) surface so that $\vec{Q}$ can rotate into, say, the $[1\bar{1}0]$ direction. Or, alternatively, one can assume that a KOH layer stabilizes microscopic facets having other than (110)-type orientations. In either case $\cos^2\theta$, of Eq. (11.3), will acquire a nonzero average value and "turn on" the Mayer–El Naby anomaly.

11.4 Two Critical Experiments

The question whether the Mayer–El Naby anomaly is extrinsic or intrinsic is of considerable importance. An understanding of nature's simplest metal is at stake [32]. We now propose two experiments which may settle the issue. Of course it is of importance first to reproduce the work of Harms under conditions such that the amount of KOH formed on the surface can be measured and correlated with the optical anomaly. Then the following studies seem promising.

(a) Prepare a thin film of K (at least twice as thick as the hydrated layer needed to obtain curve *E* of Figure 11.1) on a transparent substrate. Convert the entire specimen to "KOH" by admitting H_2O. Then measure the transmission coefficient of the completely hydrated sample. An absorption spectrum similar to Figure 11.1 would prove that the anomaly is extrinsic, and is to be associated with electronic transitions in the hydrated layer (or on its surface).

(b) Boutry and Dormont [33] have shown by low-energy electron diffraction that a (100) K film can be epitaxially grown on a clean, cleaved KF substrate. Monin and Boutry [34] have confirmed that K films on optically plane glass are (110). Reflection specimens of both types should be prepared simultaneously and $\sigma(h\nu)$ measured. If both samples show a Mayer–El Naby anomaly, then the conditions of preparation would be at fault. If neither sample shows an anomaly, then the anomaly cannot be intrinsic since the [110] $\vec{Q}$ direction of a CDW would have to have a component parallel to the surface of the (100) specimen. If the (100) specimen were to show an anomaly when the (110) specimen did not, then the CDW mechanism would be definitely established.

Acknowledgements

The authors are most grateful to Dr. Harms for making his results available to us in advance of publication, and to K.L. Kliewer for calling our attention to them. The KOH samples were kindly grown for us by G. Youchunas. We are indebted to a number of colleagues for helpful conversations about various aspects of this research.

Work supported by the NSF under Grant Nos. 41884 and GH 32001A1, and MRL Program No. GH 33574A3.

References

1 Mayer, U. and El Naby, M.H. (1963) *Z. Phys.*, **174**, 269.

2 Mayer, U. and El Naby, M.H. (1963) *Z. Phys.*, **174**, 280 (1963).

3 Mayer, U. and El Naby, M.H. (1963) *Z. Phys.*, **174**, 289 (1963).
4 Hodgson, J.N. (1963) *Phys. Lett.*, **7**, 300.
5 Smith, N.V. (1969) *Phys. Rev.*, **183**, 634.
6 Palmer, R.E. and Schnatterly, S.E. (1971) *Phys. Rev. B*, **4**, 2329.
7 Harms, P. (1972) Dissertation, Technischen Universitat Clausthal, (unpublished). The authors are very grateful to Dr. Harms for permission to reproduce his data, taken from Fig. 24 of his dissertation.
8 Born, M. and Wolf, E. (1959) *Principles of Optics*, Pergamon, London, Chap. XIII.
9 Buchanan, R.A. (1959) *J. Chem. Phys.*, **31**, 870.
10 Kröger, E. and Kretschmann, E. (1970) *Z. Phys.*, **237**, 1.
11 Elson, J.M. and Ritchie, R.H. (1971) *Phys. Rev. B*, **4**, 4129.
12 Maradudin, A.A. and Mills, D.L. (1975) *Phys. Rev. B*, **11**, 1392.
13 Bösenberg, J. and Raether, H. (1967) *Phys. Rev. Lett.*, **18**, 397.
14 Mills, D.L. and Maradudin, A.A. (1975) *Phys. Rev. B*, **12**, 2943.
15 Kerker, M. (1969) *The Scattering of Light*, Academic, New York, p. 112.
16 Susman, S. (1963) Ph.D. thesis, Illinois Institute of Technology, Chicago, 111, (unpublished), Sec. IV, p. 27.
17 Chiarotti, G., Nannarone, S., Pastore, R., and Chiaradia, P. (1971) *Phys. Rev. B*, **4**, 3398.
18 Meessen, A. (1972) *J. Phys. (Paris)*, **33**, 371; see also Marton, J.P. (1971) *Appl. Phys. Lett.*, **18**, 140.
19 Ruppin, R. (1974) *Phys. Rev. B*, **11**, 2871.
20 Barrer, R.M. (1951) *Diffussion In and Through Solids*, Cambridge U.P., Cambridge, England, p. 146.
21 Dushman, S. (1962) *Vacuum Technique*, 2nd edn., Wiley & Sons, Inc, New York, p. 768.
22 Blatt, F.J. (1968) *Physics of Electronic Conduction in Solids*, McGraw-Hill, New York, p. 196.
23 Overhauser, A.W. (1968) *Phys. Rev.*, **167**, 691.
24 Overhauser, A.W. (1964) *Phys. Rev. Lett.*, **13**, 190. The right-hand side of Eq. (12) of this reference should be divided by 2. The error arose from using a complex vector potential, Eq. (7). When a real one is used Eq. (8) is smaller by 2 and Eq. (10) is smaller by 4. Equation (3) of the present paper is obtained from (the corrected) Eq. (12) of this reference by employing the relation $2nk/\lambda = 2\sigma/c$.
25 Wilson, J.A., Di Salvo, F.J., and Mahajan, S. (1975) *Adv. Phys.*, **24**, 117.
26 Overhauser, A.W. (1967) *Phys. Rev.*, **156**, 844.
27 Hopfield, J.J. (1965) *Phys. Rev. A*, **139**, 419.
28 Overhauser, A.W. and de Graaf, A.M. (1968) *Phys. Rev.*, **168**, 763.
29 Walsh Jr., W.M., Rupp Jr., L.W., and Schmidt, P.H. (1966) *Phys. Rev.*, **142**, 414.
30 Dunifer, G.L. (private communication).
31 Jaklevic, R.C. and Lambe, J. (1975) *Phys. Rev. B*, **12**, 4146.
32 Gugan, D. (1972) *Nature Phys. Sci.*, **235**, 61.
33 Boutry, G.A. and Dormant, H. (1969) *Phillips Tech. Rev.*, **30**, 225.
34 Monin, J. and Boutry, G.A. (1974) *Phys. Rev. B*, **9**, 1309.

Reprint 12 Theory of the Residual Resistivity Anomaly in Potassium[1)]

*Marilyn F. Bishop** and *A.W. Overhauser**

* Department of Physics, Purdue University, 525, Northwestern Avenue, West Lafayette, Indiana 47907, USA

Received 12 May 1977

The induced-torque experiments of Holroyd and Datars show evidence for an anomalous anisotropy in the residual resistance of potassium of about five to one. We show that this is consistent with the hypothesis that the conduction electrons are in a static charge-density-wave state. The importance of torque and de Haas–van Alphen experiments on the same specimen is emphasized.

The experiments of Holroyd and Datars [1] have verified the existence of the giant torque anomaly in potassium that was originally observed by Schaefer and Marcus [2]. The results show a variation of the induced torque with magnetic-field direction in spherical samples of potassium, an effect that is sizable even in the limit of zero magnetic field and becomes enormous for high fields. The low-field results imply that the residual resistivity of potassium is highly anisotropic, in contradiction to the simple theory of metals. The purpose of this Letter is to show that this anisotropy can be explained if the conduction electrons in potassium are in a static charge-density-wave state.

In an induced-torque experiment, a magnetic field is rotated with respect to a suspended sample, usually spherical in shape. The time variation of the field induces currents, which, by interacting with the field, exert a torque on the sample. The magnitude of this torque is measured as a function of both the direction and magnitude of the magnetic field. For a simple metal, i. e., a metal with a spherical Fermi surface, the induced torque, for a spherical sample, should be independent of the direction of the magnetic field. De Haas–van Alphen measurements on potassium seem to indicate that its Fermi surface is spherical to within 0.1% [3]. In contradiction, the gigantic anisotropies in the induced-torque experiments suggest that potassium is not so simple and that further study is needed.

In order to facilitate the examination of this problem, we present in Figure 12.1 the data for sample K-10 of Holroyd and Datars [1]. The induced torque is plotted versus direction of magnetic field for axes of rotation along the growth axis (Figure 12.1a) and perpendicular to the growth axis (Figure 12.1b), for field values from 1 to 23 kG. The field was rotated at a speed of 22°/min, and the sample was kept at a temperature of 1.5 K. In order that the sample acquire a precise shape, the potassium was grown in a Kel-*F* mold with a spherical cavity of diameter 1.11 cm, machined to within 10^{-3} in [4]. When the magnetic field was rotated in a plane perpendicular to the growth axis, as in Figure 12.1a, the magnetic-field dependence was essentially isotropic, which is the result expected from a metal with a spherical Fermi surface. In contrast, when the growth axis was in the plane of rotation, as in Figure 12.1b, an anomalous four-peaked anisotropy of 45 to 1 appeared in the torque pattern for high fields. At low fields, this pattern became twofold, the maximum occurring when the field was perpendicular to the growth axis, and the minimum when it was parallel. The observed dependences clearly isolate the growth axis as a unique direction in this specimen.

In order to consider the low-field results in more detail, we plot with circles in Figure 12.2 the data [4] at the minimum, 28° of Figure 12.1b (lower curve), and at 118° (middle curve) in

1) Phys. Rev. Lett. 39, 10 (1977)

Anomalous Effects in Simple Metals. Albert Overhauser

ISBN: 978-3-527-40859-7

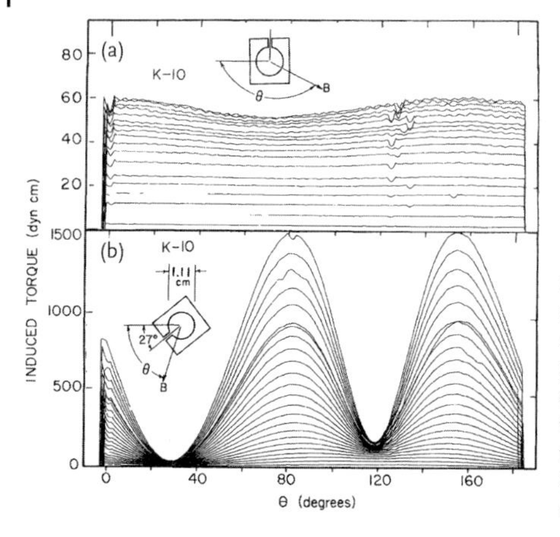

Figure 12.1 Induced-torque vs. magnetic-field direction for field values between 1 and 23 kG for sample K-10 of Holroyd and Datars, which was prepared in a mold. The sample was a sphere of diameter 1.11 cm with the field (a) rotated about the growth axis; (b) rotated in a plane containing the growth axis. The lowest curve in (a) is for 500 G; the second is for 1 kG. Their ratio leads to an x–z average $\omega_c\tau \sim 1.5$ at 1 kG.

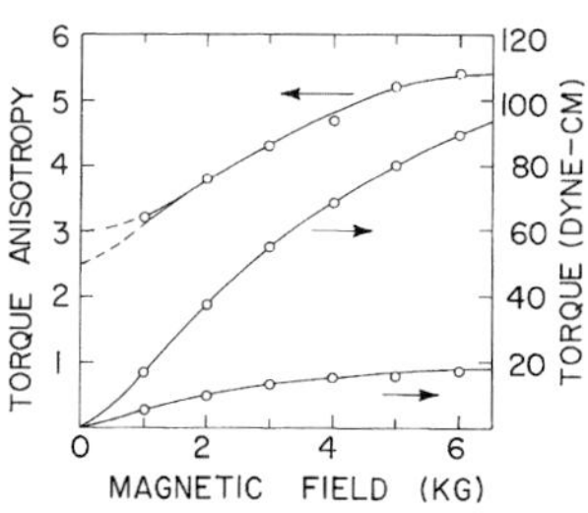

Figure 12.2 Low-magnetic-field values of induced-torque shown in Figure 12.1, with data points plotted as circles. The lower curve is the minimum in induced torque taken at an angle $\theta = 28°$ in Figure 12.1, while the middle curve is taken at $\theta = 118°$, The upper curve is the ratio of these limiting values and is the low-field torque anisotropy. Solid curves are drawn through the experimental points.

the torque pattern versus magnetic field. The upper circles are the ratios for each field value. The solid curves are drawn through the data, while the dashed extensions bracket all reasonable extrapolations of the ratio to zero field. Thus, the zero-field limit of the torque anisotropy is between 2.5 and 3 to 1. In this limit, the residual resistivity tensor $\overleftrightarrow{\rho}$ is diagonal; and with the directional dependence for this sample, we have $\rho_{xx} = \rho_{yy} = \rho_0$ and $\rho_{zz} = \gamma\rho_0$, where z is along the growth axis. The ratio R of the torque with magnetic field in the x direction to that with the field in the z direction for a spherical sample may thus be written as [5]

$$R = \frac{(\gamma + 1)^2 + (\omega_c\tau)^2}{2(\gamma + 1) + (\omega_c\tau)^2} . \tag{12.1}$$

With R at $H = 0$ between 2.5 and 3, γ must be between 4 and 5. This is consistent with the value $\gamma = 4$ that was originally estimated from the data of Schaefer and Marcus [6]. The $\omega_c\tau$ in Eq. (12.1) is that for the x or y direction. The data of Figures 12.1 and 12.2 imply $\omega_c\tau \sim 2.5$ at 1 kG. The corresponding value for relaxation along the z axis would be ~ 0.5.

Lass has suggested that the high-field torque patterns observed by Schaefer and Marcus could be explained with an isotropic resistivity if the sample shapes had been nonspherical by about 10–15% [7]. However, this hypothesis could not possibly explain the data of Holroyd and Datars presented here in Figures 12.1 and 12.2. The low-field anisotropy could be reproduced by Lass's model if the sample were elliptical (i. e., pancake shaped) with the growth-axis dimension reduced to 45% of its reported value, as if the mold had been slightly less than half filled.[2] However, for a magnetic field of 1 kG, the value of the torque at the minimum

2) However, because the Kel-F mold was transparent, both Holroyd and Datars observed that it was completely filled. They also remarked that sample K-10 had a diameter of 1.11 cm (not 1.71), and that the ordinate scale of (their) Figure 5a was incorrect. Magnet rotation rate for this specimen was 22°/min (not 35°/min).

in the angular pattern calculated using this model could never exceed 3.2 dyn cm, regardless of relaxation time. Consequently, the value of 5.4 dyn cm observed by Holroyd and Datars is much too large to be consistent with this explanation.

Finally, we remark that an attempt to extrapolate the anisotropy curve of Figure 12.2 into unity for $H = 0$ would require a 500% magnetoresistance along the z axis at 1 kG. The observed magnetoresistance in K is typically 2% for 1 kG.

Faced with the dilemma of an anisotropic residual resistivity in potassium, we suppose that the conduction electrons are in a static charge-density-wave (CDW) state. This hypothesis has produced successful explanations of other anomalies in potassium [8]. A CDW is characterized by a total self-consistent one-electron potential of the form $G\cos(\vec{Q}\cdot\vec{r})$. This potential mixes plane-wave states $\vec{k}$ with $\vec{k}\pm\vec{Q}$, producing a periodic variation in electron density and energy gaps of magnitude G at $\vec{k}\pm\frac{1}{2}\vec{Q}$. The wave vector $\vec{Q}$ of the CDW spans the Fermi surface so that $Q \sim 2k_F$ where k_F is the Fermi wave vector. Mixing of states $\vec{k}$ with $\vec{k}\pm\vec{Q}$ permits impurity-induced Umklapp scattering along $\vec{Q}$ [6]. Therefore, if an electric field is applied along $\vec{Q}$, the resistivity will be larger than if it is applied perpendicular to $\vec{Q}$. For this reason we take $\vec{Q}$ to be in the z direction, the growth axis of Holroyd and Datars' sample K-10.

In the temperature region of interest, the resistivity of potassium is dominated by impurity scattering. Electrons can scatter from an impurity potential V_c, which for mathematical convenience we take to have the Gaussian form

$$V_c = A\exp\left[-(r/r_s)^2\right] . \tag{12.2}$$

Here r_s is the Wigner–Seitz radius, a value that approximately mimics results of pseudopotential calculations [9]. In addition, scattering can take place from a strain field [10] with potential

$$V_s = B\sum_{\vec{L}\neq 0}\left\{\exp\left[-(\vec{r}-\vec{L}-\vec{u}_{\vec{L}})^2/r_s^2\right]-\exp\left[-(\vec{r}-\vec{L})^2/r_s^2\right]\right\} , \tag{12.3}$$

where $\vec{u}_{\vec{L}} \sim \vec{L}/L^3$ and is the displacement of a neighbor at site $\vec{L}$ from its position in the absence of an impurity [11]. The resistivity along each principal axis was calculated by assuming a uniform displacement δ of the Fermi sphere. δ was determined by balancing impurity-induced momentum transfer with that caused by the applied field.

We calculated the resistivity anisotropy for each of these two potentials as a function of an assumed normalized gap G/E_F of the CDW, where E_F is the Fermi energy; the results are plotted in Figure 12.3. In addition, because the total influence of a defect is actually described by the combined effect of these two potentials, we plot the maximum and minimum values of the anisotropy resulting from any linear combination of the two terms. These are labeled interference (maximum) and interference (minimum) in Figure 12.3. Since the results are ratios of resistivities, parameters such as the absolute magnitude of potentials and the concentration of impurities drop out of the calculations.

Experimental values are shown by the shaded rectangle in Figure 12.3. The range of values of the CDW gap G/E_F (0.29 to 0.36) has been determined by other experiments [8]. It is clear that the calculated anisotropy can be large enough to account for the experimental results if scattering from the strain field is comparable to or greater than that from the central ion. This is certainly reasonable when the valence of an impurity (e. g., Na) is the same as that of the lattice ion or if the impurity introduces a significant strain.

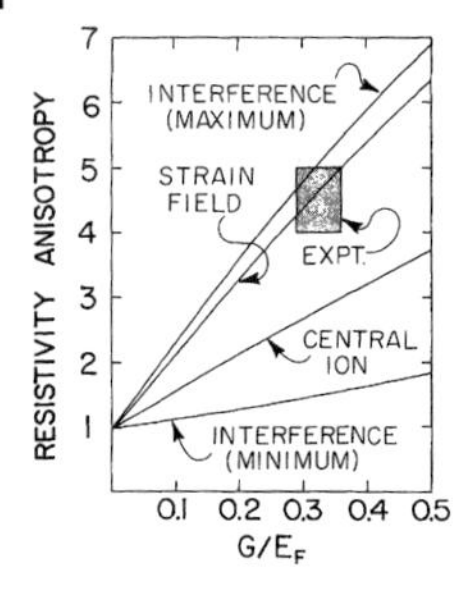

Figure 12.3 Predictions of the resistivity anisotropy vs. charge-density-wave gap G/E_F for the scattering-potential models described in the text. The shaded rectangle represents the range of values determined by experiments.

The high-field torque peaks can be explained only when the multiply connected topology of the Fermi surface is recognized. A CDW structure provides this complication [6], but a quantitative theory is not yet developed.

We conclude that the CDW model for potassium, for reasonable values of the parameters, provides a resolution of the puzzle of the anisotropic residual resistivity that is consistent with experiment. We must note, however, that it is by no means necessary that any single crystal be characterized by a single $\vec{Q}$ direction throughout. It is possible for a $\vec{Q}$-domain structure to exist. This $\vec{Q}$-domain structure, in fact, provides an explanation for the sample dependence of the anisotropics seen in induced-torque experiments [1, 2]. Since the torque anisotropics in sample K-10 of Holroyd and Datars are larger by an order of magnitude than those in other samples, we conjecture that it was probably the only single-$\vec{Q}$ sample that has ever been studied.

A difficulty for the CDW model in potassium stems from the results of the de Haas–van Alphen (dHvA) experiments [3]. However, it is not clear what effect a multiple $\vec{Q}$-domain structure would have on the results of these experiments.[3] Unfortunately, dHvA experiments have never been done on samples that are known to exhibit the large torque anisotropics shown in Figures 12.1 and 12.2, Therefore, with induced-torque measurements as a tool for characterizing samples, it would be extremely interesting to perform dHvA experiments on a sample that exhibits the large torque anisotropics discussed here.

A CDW structure will lead to weak diffraction satellites. Phase excitations [8] may reduce their intensity in K by several orders of magnitude. A search for these satellites at 10^{-7} of the (110) intensity level should be attempted.

Acknowledgements

This work was supported by the National Science Foundation under Grant No. 41884.

References

1 Holroyd, F.W. and Datars, W.R. (1975) *Can. J. Phys.*, **53**, 2517.
2 Schaefer, J.A. and Marcus, J.A. (1971) *Phys. Rev. Lett.*, **27**, 935.
3 Shoenberg, D. and Stiles, P.J. (1964) *Proc. Roy. Soc, Ser. A*, **281**, 62; Lee, M.J.G. and Falicov, L.M. (1968) *Proc. Roy. Soc. Ser. A*, **304**, 319.
4 Holroyd, F.W. and Datars, W.R.: private communication.
5 Visscher, P.B. and Falicov, L.M. (1970) *Phys. Rev. B*, **2**, 1518.

3) We note here that unrecognized $\vec{Q}$-domain structure of the spin-density-wave state of chromium led to initial reports that the dHvA effect had cubic symmetry.

6 Overhauser, A.W. (1971) *Phys. Rev. Lett.*, **27**, 938.
7 Lass, J.S. (1976) *Phys. Rev. B*, **13**, 2247.
8 For a review, see Overhauser, A.W. (1971) *Phys. Rev. B*, **3**, 3173.
9 Rice, T.M. and Sham, L.J. (1970) *Phys. Rev. B*, **1**, 4546.
10 Overhauser, A.W. and Gorman, R.L. (1956) *Phys. Rev.*, **102**, 676.
11 Eshelby, J.D. (1954) *J. Appl. Phys.*, **25**, 255.

Reprint 13 Electromagnetic Generation of Ultrasound in Metals[1)]

N.C. Banik and A.W. Overhauser**

* Department of Physics, Purdue University, 525, Northwestern Avenue, West Lafayette, Indiana 47907, USA

Received 18 May 1977

The electromagnetic generation of transverse acoustic waves in metals in the presence of a static magnetic field normal to the surface is discussed with reference to an isotropic effective mass m* of the conduction electrons. From a semiclassical argument, it is shown that in addition to the direct Lorentz force and the collision-drag force, each lattice ion experiences a Bragg-reflection force proportional to $m/m^* - 1$. In the nonlocal limit, when the ratio m/m^* is greater than unity, this force causes the generated acoustic amplitude as a function of magnetic field to deviate significantly from the monotoriic dependence that is expected from the free-electron theory of metals. However, this force does not provide significant modification to the free-electron theory for predicting the rotation of the plane of polarization, the attenuation coefficient of shear acoustic waves, and the properties of the helicon–phonon interaction.

13.1 Introduction

The free-electron theory for the electromagnetic generation of transverse ultrasound in metals was first studied by Quinn [1] and has since been discussed by a number of other authors [2–5] The mechanism of the process may be summarized as follows. The metal consists of conduction electrons and a lattice of positively charged ions. Electrons move freely in the background of these positive ions except for infrequent collisions with impurities in the lattice, characterized by a constant relaxation time τ. When an electromagnetic wave is incident on the metal surface (Figure 13.1), eddy currents in the skin layer allow the electric field to penetrate. The conduction electrons in the skin layer are accelerated and transfer their excess momentum to the lattice through collisions. The resulting force on the ions is called the collision-drag force. The lattice ions also experience a direct Lorentz force due to the electric field in the skin layer. The Lorentz force due to the ac magnetic field in the electromagnetic wave is negligible. If the conduction electrons are completely free, as is approximately the case in metals like Na or K, these two are the only forces acting on the lattice of the metal.

If these two forces are dynamically unbalanced, they will excite propagating transverse acoustic waves that can be detected at the opposite surface of the sample. In the absence of a static magnetic field this happens only if the electronic mean free path l is larger than the skin depth δ. The collision-drag force is then spatially separated from the direct Lorentz force, and the two produce a shear on the lattice. The shear wave thus generated in the nonlocal limit is polarized parallel to the electric field in the skin layer. The corresponding amplitude of the wave is known as the nonmagnetic-direct-generation (NMDG) amplitude. If, however, $l < \delta$ the two forces locally cancel each other and there is no acoustic generation in this local limit.

If there is a static magnetic field, H_0 present, the electrons experience a Lorentz force,

$$(-e\langle\vec{v}_k\rangle \times \vec{H}_0)/c \equiv (\vec{j}_e \times \vec{H}_0)/nc\ , \tag{13.1}$$

1) Phys. Rev. B 16, 8 (1977)

Anomalous Effects in Simple Metals. Albert Overhauser

ISBN: 978-3-527-40859-7

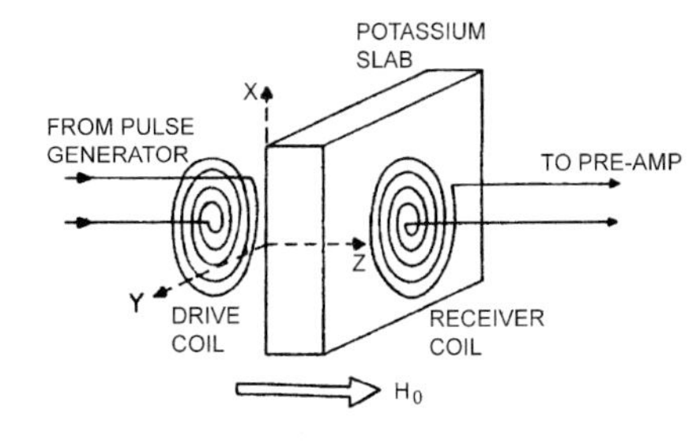

Figure 13.1 Geometry of the coil–coil experimental used by Chimenti *et al.* [7] and definition of the coordinate system (broken lines) used for calculating (Section 13.3) the amplitude of electromagnetic generation.

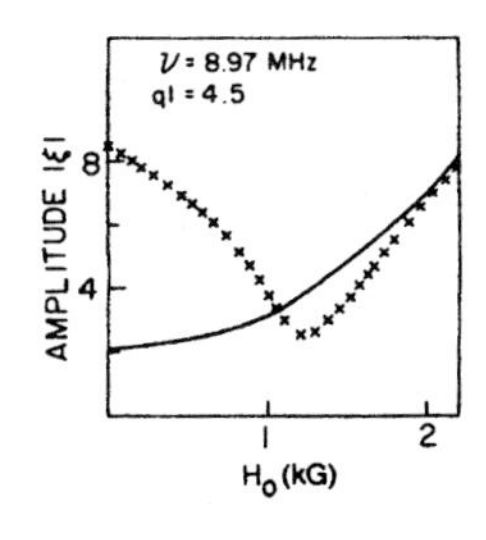

Figure 13.2 Discrepancies between experimental results and free-electron theory for electromagnetic generation of ultrasound in potassium ($\vec{q} \parallel [100]$, $T = 4.2°$K) after Chimenti *et al.* [7]. The generation amplitude has been plotted as a function of static magnetic field. The free-electron theory (solid line) has been normalized at high field.

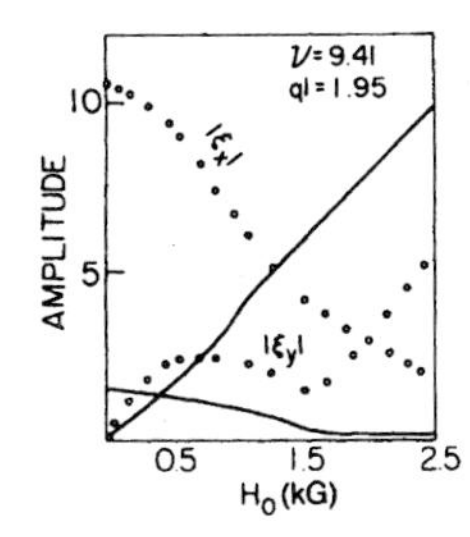

Figure 13.3 Discrepancies between experimental results and free-electron theory in the magnetic field dependence of NMDG and MDG amplitudes in potassium (fast shear wave, $\vec{q} \parallel [110]$, $T = 4.2°$K) after Gaerttner *et al.* [6]. Free-electron theory (solid lines) is for $ql = 1.95$ and has been normalized at high field. Small circles represent experimental points.

where $\vec{j}_e$ is the electronic current density, n is the density of electrons, and c is the velocity of light. The momentum acquired is given up to the lattice in the process of collisions. The result is generation of an acoustic wave with amplitude linear in H_0 and polarized in the direction perpendicular to the electric field. This amplitude has been known as the magnetic-direct-generation (MDG) amplitude. Actually Equation (13.1) represents the total force per lattice ion in the local limit, because only in this limit can one assume that the magnetic Lorentz force on the electron is transmitted directly to the lattice ion.

Quinn's [1] free-electron theory based on the above ideas covers both local and nonlocal conditions and, as reported by Turner *et al.* [2], explains the generation data in potassium samples at large magnetic fields. However, Gaerttner [5] and others [6, 7] in their experiments on potassium and aluminum samples, have found considerable disagreement with the theory in the region of small magnetic fields. The discrepancies are shown in Figures 13.2 and 13.3, and are discussed below.

The magnitude of the generated acoustic-wave amplitude $|\xi|$ as a function of the magnetic field has a large dip near $H_0 \sim 1\,\text{kG}$ (Figure 13.2), compared to the smooth rise predicted by theory. This is probably due to two underlying discrepancies (Figure 13.3): (i) The NMDG component of the amplitude is much larger at $H_0 = 0$ and, with the increase of H_0, falls off much faster than what theory predicts.

(ii) The MDG component is nonlinear for small values of H_0, whereas simple theory predicts only a slight departure from linearity.

In a recent paper Kaner and Falco [8] have introduced another force on the lattice in addition to the magnetic Lorentz force, as shown in Equation (13.1). Although this force has been shown to produce some qualitative features of the experimental data, we do not understand its origin and the degree of its relevance to simple metals. The force does not contain any parameter characterizing a departure from the free-electron model. On the other hand, there is definitely the need of such a parameter in any model, because, as can be seen in [5], the amount of discrepancy between Quinn's theory and experiment is different for different metals.

In the present paper, we introduce, real-metal effects into the theory in terms of an isotropic effective mass m^* of conduction electrons. In real metals there is a nonzero, periodic crystal potential which manifests itself in giving rise to electronic band structures and hence in modifying the dynamics of otherwise free conduction electrons. The introduction of an effective-mass tensor for conduction electrons is a simple way of taking these modified dynamics into account.

For simplicity we assume that the constantenergy surfaces of electrons in wave-vector space are spherical. In this case the electron under the influence of an external force will move in the periodic lattice as If endowed with an effective mass m^* that is different from the free-electron mass m. The fact that the conduction electron does not respond to an external field in the way it should if it were free causes a partial transfer of its momentum to the lattice. This is analogous to the momentum transfer to the lattice by an X-ray photon undergoing a Bragg reflection. The force on the lattice arising from this momentum transfer is what has been called the Bragg-reflection force by Kittel [9]. Thus the Bragg-reflection force is a natural consequence of the existence of a nonzero crystal potential. Although the effective mass m^* is treated as a parameter in the theory, its value should be consistent with the known band structure of the metal in question.

In phenomena concerning acoustic waves in metals, the Bragg-reflection force should play an important role, especially when the direct Lorentz force on the lattice is small. It is, therefore, of interest to obtain an expression for this force and study its effects on the phenomena which might be affected by its presence. For this purpose, we start with Jones and Zener's relation [10],

$$\frac{d(\hbar\vec{k})}{dt} = -e\left(\vec{E} + \frac{\langle\vec{v}_k\rangle \times \vec{H}_0}{c}\right) . \qquad (13.2)$$

Then from a semiciassicai argument, consistent with momentum balance between electrons and ions, we show that each lattice ion experiences a Bragg-reflection force proportional to $m/m^* - 1$. A microscopic derivation of this force is presented in the Appendix. In the case of generation of sound, this force is shown to have substantial effects only at small values of H_0 and in the nonlocal limit. In the local limit or/and at large magnetic fields, we obtain the results of Quinn. We also show that for values of m/m^* appreciably larger than unity, features similar to those found in the data for potassium and aluminum can be produced. However, the value of m/m^* needed to fit the potassium data is intolerably larger than the known band-structure value. In the case of attenuation of shear waves, we find that in the Kjeldaas approximation [11] there is no significant difference in the results for the attenuation coefficient and the rotation of the plane of polarization compared to those obtained from free-electron theory. There is no modification whatsoever in the phenomena of helicon–phonon interaction.

The plan of the paper is as follows. In Section 13.2 we present a simple derivation of the Bragg-reflection force in the presence of external electric and magnetic fields and thus obtain a modified expression for the net force on the lattice ion. The amplitude of the shear acoustic wave is calculated in Section 13.3 following the same method as suggested by Quinn [1]. We also present our results for various values of m/m^* and compare them with the experimental data. Section 13.4 contains a brief study of the effect of the Bragg-reflection force on the phenomena of ultrasonic attenuation and the helicon–phonon interaction. In Section 13.5 we summarize our results and discuss the possibility of extending the theory to metals that have complicated band structures.

13.2 Force on Lattice Ions

In a completely-free-electron model, as we already mentioned in the Introduction, there are two types of forces acting on the lattice ions: (i) the direct Lorentz force and (ii) the collision-drag force. The direct Lorentz force F_{d} for a monovalent metal is given by

$$\vec{F}_{\mathrm{d}} = e[\vec{E} + (\vec{u} \times \vec{H}_0)/c] \tag{13.3}$$

per ion, where $\vec{u}$ is the local velocity of the ion, $\vec{E}$ is the electric field arising from the electromagnetic wave that is incident normally on the metal surface, and H_0 is the static magnetic field applied normal to the surface. In (13.3) we have neglected the Lorentz force due to the magnetic field associated with the electromagnetic wave. The collision-drag force $\vec{F}_{\mathrm{c}}$, in the approximation used by Rodriguez [12] and justified by Holstein [12] from a microscopic viewpoint, can be written as

$$\vec{F}_{\mathrm{c}} = m(\langle \vec{v}_k \rangle - \vec{u})/\tau \,, \tag{13.4}$$

where τ is a constant relaxation time and $\langle \vec{v}_k \rangle$ is the electron velocity averaged over all states $\vec{k}$ in the Fermi sphere and weighted by the zero-temperature Fermi distribution. Although these are the only forces present in the free-electron model, in the present model an additional force on each lattice ion will arise in order to preserve momentum balance.

Consider the energy $\mathcal{E}_k$ versus the wave vector $\vec{k}$ relationship for an electron with an isotropic effective mass m^*,

$$\mathcal{E}_k = \hbar^2 k^2/2m^* \,. \tag{13.5}$$

Then the group velocity of an electron in the state $\vec{k}$ is

$$\vec{v}_k = \hbar \vec{k}/m^* \,. \tag{13.6}$$

Thus the true momentum of the electron is

$$\vec{p}_k = (m/m^*)\hbar \vec{k} \,, \tag{13.7}$$

different from $\hbar \vec{k}$. Also consider the Jones and Zener relation [10],

$$\frac{d(\hbar \vec{k})}{dt} = -e\left(\vec{E} + \frac{\langle \vec{v}_k \rangle \times \vec{H}_0}{c}\right) \equiv -\vec{\Gamma} \,. \tag{13.8}$$

Equation (13.8), together with Equation (13.6) implies that in the presence of external fields, a conduction electron acquires an acceleration $\langle \vec{a}_k \rangle$ which can be written as

$$\langle \vec{a}_k \rangle = -\vec{\Gamma}/m^* \,. \tag{13.9}$$

Since the true mass of an electron is m, the Lorentz force $\vec{F}_L$ on the electron with effective mass m^* is, therefore,

$$\vec{F}_L = (-m/m^*)\vec{\Gamma} \,. \tag{13.10}$$

$\vec{F}_L$, as given by Equation (13.10) represents the true Lorentz force experienced by a conduction electron in.the crystal. Obviously, $\vec{F}_L$ is different from $\hbar\vec{k}$ the Lorentz force on a free electron. As the result of a nonzero crystal potential, created by lattice ions in the metal, there is thus an extra force on the electron. Consequently the lattice ion itself must experience an equal and opposite reaction force $\vec{F}_B$ which is called the Bragg-reflection force and is given by

$$\vec{F}_B = (m/m^* - 1)\vec{\Gamma} \,. \tag{13.11}$$

The Bragg-reflection force $\vec{F}_B$ is a logical consequence of having a nonzero crystal potential. Note that in the limit $m^* \to m$, i. e., in the free-electron limit, $\vec{F}_B \to 0$.

We consider a monovalent metal with n conduction electrons per unit volume. Thus the number of ions per unit volume is also n. The total force per lattice ion can then be written as

$$\begin{aligned} \vec{F}_i &= \vec{F}_d + \vec{F}_c + \vec{F}_B \\ &= e\frac{m}{m^*}\vec{E} - \left(\frac{m}{m^*} - 1\right)\frac{\vec{j}_e \times \vec{H}_0}{nc} - \frac{e\vec{j}_e}{\sigma_0}\frac{m}{m^*} - \frac{m\vec{u}}{\tau} + e\frac{\vec{u} \times \vec{H}_0}{c} \,, \end{aligned} \tag{13.12}$$

with $\sigma_0 = ne^2\tau/m^*$ and $\vec{j}_e = -ne\langle\vec{v}_k\rangle$. From (13.12), one can easily show that in the local limit and in the absence of a magnetic field, there is no net force on the lattice ion. This is a requirement for electrical neutrality of metals and cannot be met rigorously without considering the Bragg-reflection force.

13.3 Generated Sound-wave Amplitude

For the geometry of the experiment, as shown in Figure 13.1, we assume that a plane-polarized electromagnetic (em) wave is incident on the metal surface with the magnetic field vector $\vec{H}$ along the y direction and the propagation vector $\vec{q}$ in the z direction. The surface of incidence is at $z = 0$ while the surface of detection is at $z = \infty$. A static magnetic field $\vec{H}$ is applied normal to the surface and is parallel to $\vec{q}$. We consider a monovalent, cubic metal and assume that $\vec{q}$ is parallel to [100] direction. In this configuration, the two possible shear acoustic modes have equal velocities. Although the degeneracy in the velocity will be lifted due to the presence of the static magnetic field, this effect will be negligibly small in the range of magnetic fields that we shall consider here. Extensions of the following calculations to other orientations of $\vec{q}$ will not yield any new information.

Given $F_i(z)$, the long-range force on the lattice ion, as a function of z, it is straightforward to evaluate the amplitude of generated acoustic waves from the wave equation,

$$\rho\frac{\partial^2\vec{\xi}}{\partial t^2} = \rho s^2\frac{\partial^2\vec{\xi}}{\partial z^2} + \vec{f} \,, \tag{13.13}$$

where $\vec{f}$ is the force on ions per unit volume and is given by

$$\vec{f} = n\vec{F}_i \,. \tag{13.14}$$

Here $\vec{\xi}$ is the displacement of the ion from its equilibrium position, s is the transverse speed of sound, ρ is the density of the metal and n is the number of ions per unit volume and is equal to the density of conduction electrons in the metal, In general, $\vec{\xi}$ is obtained from a self-consistent solution of the acoustic-wave equation (13.13), Maxwell's equations, and the Boltzmann-transport equation. However, self-consistency is not important in the generation of acoustic waves of ultrasonic frequency because the ionic current is small compared to the electronic current and the electromechanical coupling is weak. In contrast, self-consistency becomes important in considering the attenuation of generated acoustic waves, because for this situation the electronic current is comparable to the ionic current. This is not of importance here since the reported experimental data are the actual generated amplitudes after making corrections for attenuation.

With the neglect of self-consistency, it can be shown that for a stress-free surface, [i. e., $(\partial\vec{\xi}/\partial z)_{z=0}$], the amplitude of an acoustic wave propagating towards the opposite end of the surface is given by

$$\vec{\xi}(z,t) = \vec{A}\sin(q_0 z - \omega t) ,$$

with

$$\vec{A} \equiv -\frac{1}{\rho\omega s}\int_0^\infty \cos q_0 z' \vec{f}(z') dZ' . \tag{13.15}$$

The observed acoustic-wave amplitude at infinity, $\xi(\infty)$, is therefore equal to A and we write it as

$$\vec{\xi}(\infty) = \vec{A} . \tag{13.16}$$

The force density $\vec{f}(z)$ depends on the mechanism for electron scattering at the boundary of the metal, hi the following we shall assume that electrons are reflected specularly at the boundary. Influences of diffuse scattering will be considered in a separate paper. The assumption of specular scattering allows one to adopt a simplified, yet mathematically equivalent model to describe the problem of acoustic generation. That is, we may replace the semi-infinite metal by an infinite medium that is symmetrical about the plane $z = 0$. In order to include the effects of an incident electromagnetic wave at the boundary of the semi-infinite metal, we place a source at $z = 0$ in the infinite medium model. In addition, the plane $z = 0$ remains stress-free, as in the semi-infinite case. $\vec{E}(z)$ is continuous across the boundary. However, because of the presence of the source at $z = 0$, components of $d\vec{E}/dz$ parallel to the surface is discontinuous at $z = 0$. The electronic current density $\vec{j}_e(z)$ is then a functional of $\vec{E}(z)$. From Equations (13.12) and (13.14), we see that the force density can also be written as a functional of $\vec{E}(z)$. Thus $\vec{f}(z)$ itself is an even function of ζ and we can write

$$\begin{aligned}\vec{\xi}(\infty) &= -\frac{1}{2\rho\omega s}\int_{-\infty}^{\infty} e^{iq_0 z'} \vec{f}(z') dz' \\ &= -(\pi/\rho\omega s)\vec{f}(q_0) ,\end{aligned} \tag{13.17}$$

with

$$\vec{f}(q_0) = \frac{1}{2\pi}\int_{-\infty}^{\infty} e^{iq_0 z'} \vec{f}(z') dz' ,$$

the Fourier transform of $\vec{f}(z)$. We now evaluate $\vec{f}(z)$ separately for the local and nonlocal limits.

13.3.1 Local Limit

In this case the electric field $\vec{E}(z)$ and the electronic current density $\vec{j}_e(z)$ are related through a local conductivity tensor. We use circular coordinates to write this relation as follows:

$$j_e^{\pm} \equiv j_{ex} \pm i j_{ey} = \sigma^{\pm} E^{\pm} , \tag{13.18a}$$

with

$$\sigma^{\pm} = \sigma_0/(1 \mp i\omega_c \tau) ., \tag{13.18b}$$

Here, σ_0 is the zero-field conductivity as defined in (13.12), τ is a constant relaxation time, and $\omega_c = eH_0/m^* c$. Assuming a time dependence $\sim e^{-i\omega t}$ for the fields, the relevant Maxwell's equations, neglecting the displacement current, can be written

$$\frac{\partial^2 E^{\pm}}{\partial z^2} = -\frac{4\pi i\omega}{c^2} j_e^{\pm} \pm \frac{2\omega}{c} H^{\pm}(0)\delta(z) , \tag{13.19}$$

where $H^{\pm}(0) = \pi i H_y(0)$, the magnitude of the ac magnetic field at the surface and $\delta(z)$ is the Dirac δ function. The second term on the right-hand side represents the source at $z = 0$. As discussed earlier, we have neglected the ionic current. From Equations (13.18) and (13.19), we obtain for the Fourier component of the field,

$$E^{\pm}(q_0) = \frac{\mp(\omega/\pi c)H^{\pm}(0)}{q_0^2 - 4\pi i\sigma^{\pm}/c^2} . \tag{13.20}$$

For ultrasonic frequencies ($\omega \sim 10^7$ Hz) and low magnetic fields ($H_0 \lesssim 20\,\text{kG}$), $q_0^2 \ll |4\pi\omega\sigma^{\pm}/c^2|$, and we obtain

$$E^{\pm} \simeq -cH_y(0)/4\pi^2\sigma^{\pm} \tag{13.21}$$

and

$$j_e^{\pm} \simeq -cH_y(0)/4\pi^2 . \tag{13.22}$$

These results could be obtained from (13.19) by neglecting the left-hand side altogether. This is a consequence of weak electromechanical coupling.

After neglecting the ionic current, we obtain from Equation (13.12) and the definition of $\vec{f}(z)$ as given in (13.14),

$$\vec{f}^{\pm}(z) = \pm i H_0 j_e^{\pm}(z)/c . \tag{13.23}$$

Using Equation (13.22), the Fourier component $\vec{f}^{\pm}(q)$ is given by

$$\vec{f}^{\pm}(q_0) = \mp i H_0 H_y(0)/4\pi^2 . \tag{13.24}$$

Thus from (13.17), the amplitude of the generated acoustic wave can be written

$$\xi^{\pm}(\infty) = \pm i H_0 H_y(0)/4\pi\rho\omega s , \tag{13.25}$$

and transforming back to Cartesian coordinates, final expressions for the x and y components of the generated acoustic-wave amplitude in the local limit are

$$\xi_x = 0 \quad \text{and} \quad \xi_y = H_0 H_y(0)/4\pi\rho\omega\rho . \tag{13.26}$$

The above expressions are exactly what one would have obtained from the free-electron theory of Quinn [1] in the local limit. Thus the Bragg-reflection force does not have any effect on the generated sound-wave amplitude in the local limit.

13.3.2 Nonlocal Limit

When nonlocal conditions are included, the electrons experience spatially varying electric fields between collisions. Thus the conductivity tensor becomes wave-vector dependent and must be obtained from the solution of the Boltzmann-transport equation. For the boundary conditions appropriate for specular scattering, we shall use the results of Kjeldaas [11]:

$$j_e^{\pm}(q) = \sigma_0 G_{\pm}(q) E^{\pm}(q) , \tag{13.27}$$

where $G_{\pm}(q)$ are even functions of q and $G_{-}(H_0) = G_{+}(-H_0)$. Exact expressions for $G_{\pm}(q)$ can be obtained from [11] replacing m by m^*. σ_0 in Equation (13.27) is the zero-field conductivity and has been defined in (13.12). With the above constitutive relation, we proceed as in the local limit and obtain

$$f_{\pm}(q_0) = \mp \frac{C H_{\pm}(0)}{4\pi^2} \left[\frac{im}{e\tau} \left(\frac{1}{G_{\pm}} - 1 \right) - \frac{H_0}{c} \left(\frac{m}{m^*} - 1 \right) \right] \tag{13.28}$$

and

$$\xi^{\pm}(\infty) = \frac{H_{\pm}(0)}{4\pi\rho\omega s} \left[\pm \frac{imc}{e\tau} \left(\frac{1}{G_{\pm}} - 1 \right) - H_0 \left(\frac{m}{m^*} - 1 \right) \right] . \tag{13.29}$$

For a linearly polarized incident electromagnetic wave, $H_{\pm}(0) = \pm i H_y(0)$ and using the symmetry of $G_{\pm}$, namely, Re G_{+} = Re G_{-} and Im $G_{+} = -$Im G_{-}, we finally obtain the following expressions for the components of the acoustic-wave amplitude at infinity:

$$|\xi_x(\infty)| = \frac{C H_y(0)}{4\pi\omega\rho s} \frac{m}{m^*} \frac{(3\pi^2 n)^{1/3} \hbar\omega}{s e q_0 l} \left| \frac{\mathrm{Re}\, G_{+}}{|G_{+}|^2} - 1 \right| \tag{13.30}$$

and

$$|\xi_y(\infty)| = \frac{C H_y(0)}{4\pi\omega\rho s} \left| \frac{m}{m^*} \frac{(3\pi^2 n)^{1/3}_{\hbar\omega}}{s e q_0 l} \frac{\mathrm{Im}\, G_{+}}{|G_{+}|^2} - \frac{H_0}{c} \left(\frac{m}{m^*} - 1 \right) \right| , \tag{13.31}$$

with $l = v_0 \tau$ and v the Fermi velocity. The quantity measured by Chimenti *et al.* [7] is $|\xi(\infty)|$ and is given by

$$|\xi(\infty)| = \left(|\xi_x(\infty)|^2 + |\xi_y(\infty)|^2 \right)^{1/2} . \tag{13.32}$$

On comparing the above expressions with those obtained from the free-electron theory, it is observed that the amplitude $|\xi_x(\infty)|$ is modified by the factor m/m^*. The corresponding term in $|\xi_y(\infty)|$ also contains this factor. In addition $|\xi_y(\infty)|$ contains an extra term proportional to $m/m^* - 1$. This term can enhance or suppress the contribution from the first term depending on whether $m/m^* - 1$ is less than or greater than zero. These features are possible only at low magnetic fields for which $\omega_c \tau / q_0 l < 1$. If $\omega_c \tau / q_0 l \gg 1$, then Im $G_{+}/|G_{+}|^2 \approx \omega_c \tau$, and Re $G_{+}/|G_{+}|^2 \simeq 1$, so that one obtains the results as found in the local limit. Thus the Bragg-reflection force may modify the generated amplitude only at small magnetic fields. The large field amplitudes, however, should behave linearly as in the local case.

Using the parameters appropriate for potassium, we have calculated the amplitudes as given in Equations (13.30)–(13.32) at a frequency of 10 MHz. The results are shown in Figures 13.4–13.8 for various values of m/m^* and $q_0 l$. The results corresponding to $m/m^* - 1$ are the same as those one would have obtained from free-electron theory. The results for

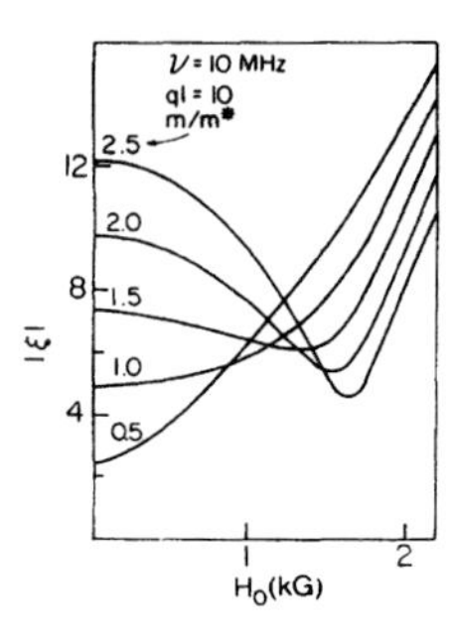

Figure 13.4 Effects of the Bragg-reflection force on the amplitude $|\xi|$ plotted as a function of static magnetic field. Different curves are for different values of m/m^* but with $ql = 10$ for all. Vertical scale is in relative units.

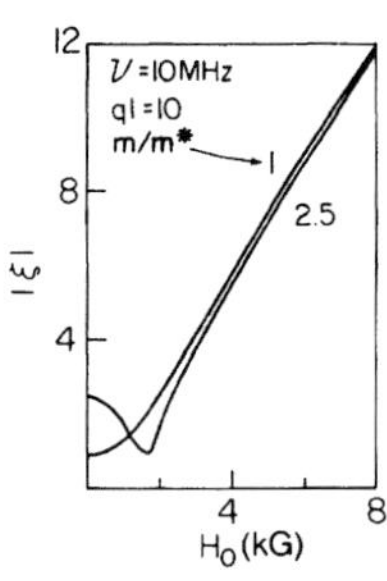

Figure 13.5 Disappearance of the effects of the Bragg-reflection force in the region of higher field. The two curves are for two values of m/m^*, but with $ql = 10$ for both.

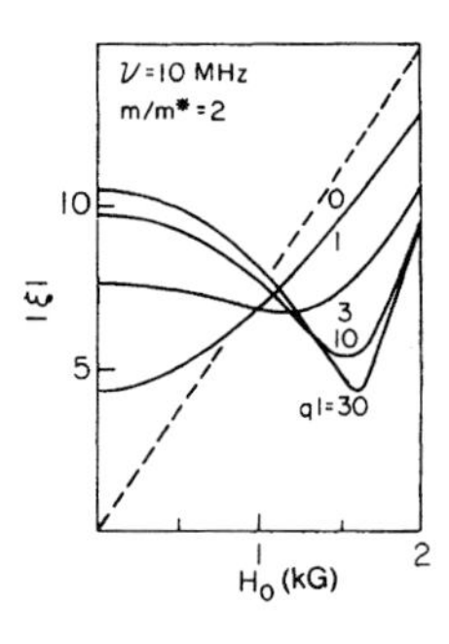

Figure 13.6 Effects of the nonlocality parameter ql on the shape of the dip, appearing in the amplitude vs. magnetic field curve when the Bragg-reflection force is included, $m/m^* = 2$ for all curves. The broken line represents the local limit.

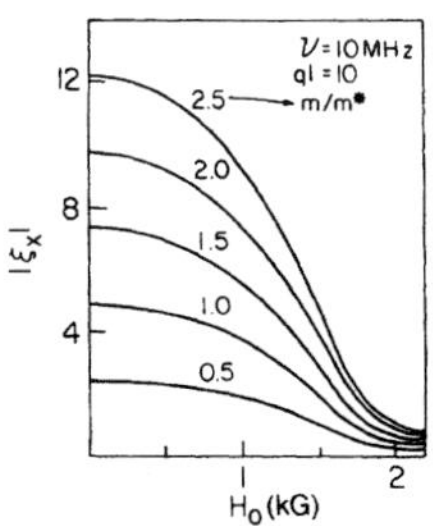

Figure 13.7 Enhancement of the zero-field amplitude and the rapid fall of the NMDG amplitude $|\xi_x|$ as a function of magnetic field after including the Bragg-reflection force. Different curves are for different values of m/m^*, but with $ql = 10$ for all.

$m/m^* > 1$ and $q_0 l > 1$ are encouraging. Here the Bragg reflection force plays a significant role and produces the same features as found in the experimental data. (See Figures 13.1 and 13.2.) When $m/m^* \sim 2$, our theory fits surprisingly well with the experiment. However, the band-structure value for m/m^* in potassium as calculated by Ham [13] is only 0.83. Consequently, if the Fermi surface of potassium is simply connected, this theory does not explain the data. In a subsequent paper we shall show that the anomalous behavior of aluminum

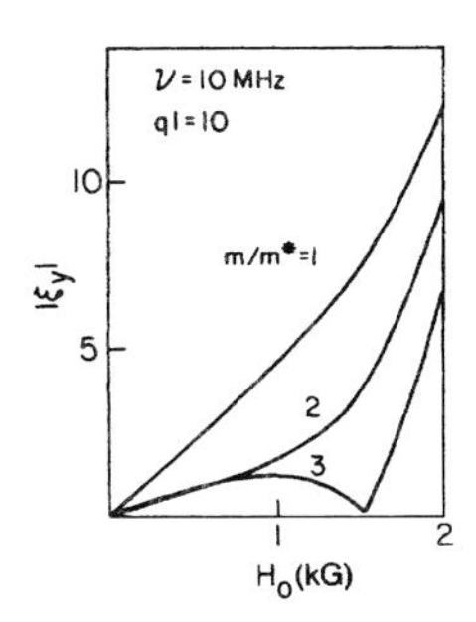

Figure 13.8 Nonmonotonic dependence of the MDG amplitude $|\xi_y|$ on the static magnetic field when the Bragg-reflection force is included (curve with $m/m^* = 3$). $ql = 10$ for all curves.

is explained by this theory even though aluminum is nearly-free-electron-like. The Bragg-reflection force becomes very important when the Fermi surface is not simply connected.

13.4 Ultrasonic Attenuation and the Helicon–Phonon Interaction

In the nonlocal case, the self-consistent force on lattice ions is considerably modified due to the Bragg-reflection force. The dispersion relations for coupled electromagnetic and sound waves will be accordingly modified. One would, therefore, suspect that the phenomena such as the attenuation of sound and the helicon–phonon interaction, which are directly linked with the dispersion relations, might exhibit some new features when the Bragg-reflection force is taken into account. We have investigated this matter and present our findings briefly in Sections 13.4.1 and 13.4.2.

13.4.1 Ultrasonic attenuation

Here we assume that a plane-polarized transverse ultrasonic wave is impressed upon a metal surface at $z = 0$, propagating into the metal along the z direction perpendicular to the surface. As in Sections 13.2 and 13.3, we consider the presence of a static magnetic field normal to the surface of the metal. As a result of the impressed ultrasonic wave, the lattice ions will vibrate about their mean positions producing an electric field that causes the conduction electrons in the metal to accelerate. The electrons in turn, transfer the excess .energy to thermal phonons through collisions with impurities in the lattice. Thus the energy of the ultrasonic phonon is attenuated as it propagates through the metal. In the presence of the static magnetic field there is also a rotation of the plane of polarization of the ultrasonic wave. The plane-polarized ultrasonic wave can be considered to consist of right- and left-circularly-polarized waves of equal amplitudes. These two waves suffer different velocity changes as they propagate through the metal in the presence of the magnetic field. This causes the plane of polarization of the ultrasonic wave to rotate by one-half the phase difference between the right- and left-circularly-polarized waves.

Both the attenuation and the rotation of the plane of polarization can be described in terms of complex wave numbers $q^{\pm}$ of the right- and left-circularly-polarized components of the acoustic wave in the metal. The attenuation coefficient α, defined as the energy of the ultrasonic wave attenuated per unit length of the metal, is then given by

$$\alpha = \operatorname{Im} q^{\pm} + \operatorname{Im} q^{-} . \tag{13.33}$$

The rotation angle Φ of the plane of polarization can be written

$$\Phi = \frac{1}{2}(\operatorname{Re} q^{+} - \operatorname{Re} q^{-}) . \tag{13.34}$$

The quantities $q^{\pm}$ are obtained from the dispersion relation of the acoustic wave. The dispersion relation, in turn, is obtained self-consistently from the elastic wave equation, Maxwell's equations, and the Boltzmann-transport equation. We can, therefore, carry over the notation of Sections 13.2 and 13.3, provided we make the necessary changes in the definitions of variables. For example, $\xi^{\pm}$ will now represent the right- and left-circularly-polarized components of the amplitude of the impressed ultrasonic wave, not of the generated wave. Similarly $E^{\pm}$ will be the self-consistent electric fields instead of the electric fields associated with the incident electromagnetic wave. Since it is the local motion of ions that gives rise to these electric fields, we must not neglect the ionic current here as we have done in previous sections. With these modifications we can write the force on a lattice ion as

$$F_{\pm} = \frac{emE^{\pm}}{m^*} + \left(\frac{im\omega}{\tau} \mp \frac{eH_0}{c}\omega\right)\xi^{\pm} - \left[\frac{m}{ne\tau} \mp \frac{iH_0}{nc}\left(\frac{m}{m^*} - 1\right)\right] j_e^{\pm} \,. \quad (13.35)$$

The constitutive equation, Equation (13.27) is now,

$$j_e^{\pm}(q) = \sigma_0 G_{\pm}[E^{\pm}(q) - J_i^{\pm}/\sigma_0] \,, \quad (13.36)$$

where, $J_i^{\pm} \equiv neu_{\pm}$ is the ionic current density. From Maxwell's equation with the neglect of displacement current, we obtain

$$\frac{d^2E^{\pm}}{dz^2} = -\frac{4\pi i\omega}{c^2}\left(j_e^{\pm} + J_i^{\pm}\right) \,. \quad (13.37)$$

The Fourier component of the self-consistent electric fields can then be written

$$E^{\pm}(q) = \frac{(4\pi i\omega/c^2)J_i^{\pm}(1 - G_{\pm})}{q^2 - 4\pi i\omega\sigma_0 G_{\pm}/c^2} \,. \quad (13.38)$$

For ultrasonic frequencies, $q^2 \ll \omega\sigma_0/c^2$ one can neglect q^2 in the denominator of Equation (13.38). This is Kjeldaas approximation. In this approximation, as can be seen by substituting Equation (13.38) into Equation (13.36), $j_e^{\pm}(q) = -J_i^{\pm}(q)$. In other words, the screening of the ionic current by the electronic current is complete. In this case, the fields may be written

$$E^{\pm}(q) = (ine\omega/\sigma_0)\xi^{\pm}(1/G_{\pm} - 1) \,, \quad (13.39)$$

and the Fourier component of the total force on the lattice ion is given by

$$F_{\pm}(q) = \frac{m}{m^*}e\left(E^{\pm}(q) \mp \frac{H_0\omega}{c}\xi^{\pm}(q)\right) \,. \quad (13.40)$$

If the mass of the lattice ion is M, then from Equations (13.13), (13.39), and (13.40), one obtains the dispersion relations for right- and left-circularly-polarized waves:

$$q_{\pm}^2 = \frac{\omega^2}{s^2}\left[1 + \frac{im}{M\omega\tau}\left(\frac{1}{G_{\pm}} - 1\right) \mp \frac{\Omega_c}{\omega}\frac{m}{m^*}\right] , \quad (13.41)$$

where $\Omega_c \equiv eH_0/Mc$ is the ion cyclotron frequency. Since unity is large compared to other terms in the brackets in Equation (13.41), we may rewrite this relation as

$$q_{\pm} \simeq \frac{\omega^2}{s^2}\left[1 + \frac{im}{2M\omega\tau}\left(\frac{1}{G_{\pm}} - 1\right) \mp \frac{\Omega_c}{2\omega}\frac{m}{m^*}\right] . \quad (13.42)$$

From Equations (13.33) and (13.34), we then obtain the final expressions for the angle of rotation of the plane of polarization Φ and the attenuation coefficient α:

$$\Phi = \frac{m}{m^*}\frac{\Omega_c}{2s}\left(1-\frac{1}{\omega_c\tau}\right)\frac{\operatorname{Im} G+}{|G+|^2} \tag{13.43}$$

and

$$\alpha = \frac{m}{m^*}\left[\frac{\hbar(3\pi^2 n)^{1/3}}{Msl}\left(\frac{\operatorname{Re} G_+}{|G_+|^2}-1\right)\right]. \tag{13.44}$$

Comparing Equations (13.43) and (13.44) with the standard results (e. g. Boyd and Gavenda [14]), we see that the only effect the Bragg-reflection force has on the attenuation of ultrasound in the Kjeldaas approximation is an overall multiplying factor m/m^* for both the rotation of the plane of polarization and the attenuation coefficient *after* replacing m by m^* in the usual free-electron results.

13.4.2 Helicon–Phonon Interaction

When a low-frequency electromagnetic (em) wave with wave vector $\vec{q} \parallel \vec{z}$, is incident on a metal surface, it cannot propagate into the metal. On the other hand, if a constant magnetic field $\vec{H}_0$ is applied parallel to $\vec{q}$ with sufficient magnitude that the electron cyclotron frequency is much larger than the frequency of the em wave, one of the two circularly polarized components of the incident plane-polarized em wave can propagate through the metal without appreciable damping. These propagating em waves are known as helicons. The phase velocity of helicons is very small and can be made to match the velocity of transverse sound waves in the metal. In such a situation, the degeneracy in the velocity of the helicon mode and the phonon mode is lifted due to a coupling between the two modes. This phenomenon is known as the helicon–phonon interaction. The extent of coupling between the two modes is best described by a coupling parameter η which is obtained from the dispersion relation of these coupled modes. The geometries used in Sections 13.2–13.4.1 also apply here, and when the Bragg-reflection force is included, we obtain the dispersion relation for the coupled modes:

$$\left(\omega-\frac{q^2c^2}{4\pi i\sigma_0 G_\pm}\right)\left[\omega^2-q^2s^2+\frac{im\omega}{M\tau}(1-G_-)+\Omega_c\omega+\left(\frac{m}{m^*}-1\right)\Omega_c\omega G_-\right]$$
$$= i\omega^2\left(1-\frac{1}{G_-}\right)\left[\frac{m}{M\tau}(1-G_-)-i\Omega_c G_-\left(\frac{m}{m^*}-1\right)\right]. \tag{13.45}$$

This dispersion relation takes a different form than the standard expression [15] due to the presence of terms containing the factor $m/m^* - 1$, which arises from the Bragg-reflection force. However this is not of importance in the region of interest. Helicons are propagating modes only at comparably large values of the magnetic field. The condition for appreciable coupling between the two modes is that their velocities be approximately equal. This happens, for instance, in potassium for $H_0 \simeq 10^2$ kG at a frequency of 20 MHz. At such high static magnetic fields, G_- is approximately equal to $1/(1+i\omega_c\tau)$ and the dispersion relation reduces to that given in (13.15):

$$\left(\frac{u}{s}-\frac{v_H}{s}\right)\left[\left(\frac{u^2}{s}\right)-1+\frac{\Omega_c}{\omega}\left(\frac{u}{s}\right)^2\right]=\frac{\Omega_c}{\omega}\frac{v_H}{s}, \tag{13.46}$$

where v_H ($\equiv c^2 q\omega_c/\omega_p^2$) is the helicon velocity in the absence of interaction with phonons, ω_p is the electron plasma frequency, and u is the phase velocity of the coupled modes. The coupling parameter η is thus given by

$$\eta = \Omega_c/\omega \,. \tag{13.47}$$

The dispersion relation and the coupling parameter η as given in Equations (13.46), and (13.47) are not altered by the inclusion of the Bragg-reflection force. Therefore, this force has no effect on the helicon–phonon interaction.

13.5 Summary and Concluding Remarks

The Bragg-reflection force was shown to be a natural consequence of the concept of effective mass of the conduction electrons in metals. Though it cannot produce any noticeable effect on the helicon–phonon interaction, it modifies the rotation of the plane of polarization and the attenuation coefficient of the shear ultrasonic wave by a multiplying factor m/m^*. The phenomenon most significantly modified by this force is the electromagnetic generation of ultrasound in metals in the presence of small magnetic fields. Here it is capable of producing a nonmonotonic dependence of the generated amplitude as a function of the magnetic field. These effects will be most important for metals with m/m^* appreciably different from unity.

We have considered only a single isotropic effective mass for all conduction electrons. However one can easily extend the ideas of this paper to the consideration of two or more groups of electrons or holes with different effective masses. The Bragg-reflection force will be obtained by the condition that the total momentum of the ions balance that of the carriers. In this way, metals with complicated band structures can be studied. The Bragg-reflection force is expected to play a significant role in predicting the amplitude of electromagnetic generation of ultrasound in many metals.

Although this study was undertaken in order to understand the anomalous behavior of potassium, we must confess failure if the Fermi surface of potassium is simply connected. On the other hand, if the Fermi surface is multiply connected, as the torque anisotropy data of Holroyd and Datars [16] require, then inclusion of the Bragg-reflection force is necessary. At the present time the only physical mechanism that could lead to a multiply connected Fermi surface in potassium is a charge-density wave instability [17] of the electronic ground state.

A Appendix

For simplicity, we consider the one-dimensional electron gas of a monoatomic linear chain of atoms whose reciprocal lattice vector is $\vec{Q}$. We assume that the average potential energy due to the lattice ions is $2U_1 \cos Qx$ and the unperturbed electronic states are free-particle states. The action of the crystal potential is to cause mixing of the unperturbed state $\vec{k}$ with, for example, $\vec{k}-\vec{Q}$, producing a gap at $k = \frac{1}{2}Q$. Treating this mixing by degenerate perturbation theory, one obtains the energy eigenvalues for states above and below the gap.

$$\mathcal{E}_{\pm}(k) = \frac{1}{2}(\epsilon_k + \epsilon_{k-Q}) \pm \frac{1}{2}\left[(\epsilon_k - \epsilon_{k-Q})^2 + 4U_1^2\right] \,, \tag{A1}$$

where, $\epsilon_k = \hbar^2 k^2/2m$. The corresponding eigenfunctions are

$$\Psi_+ = \cos\theta\, e^{ikx} - \sin\theta\, e^{i(k-Q)x} \,, \tag{A2}$$

$$\Psi_- = \sin\theta\, e^{ikx} - \cos\theta\, e^{i(k-Q)x} \,, \tag{A3}$$

for states above and below the gap, respectively. The coefficients are given by

$$\sin 2\theta = 2U_1/(\mathcal{E}_+ - \mathcal{E}_-) \equiv 2U_1/W \,. \tag{A4}$$

The corresponding effective masses $m^*_\pm$, defined as

$$\frac{1}{m^*_\pm} \equiv \frac{1}{\hbar^2}\frac{\partial^2 \mathcal{E}_\pm}{\partial k^2} \,,$$

can be obtained as

$$m/m^* = 1 \pm 2\hbar^2 Q^2 U_1^2 W^{-3}/m \,. \tag{A5}$$

Under the action of an external electric field $\vec{E}$, the interaction Hamiltonian is $\mathcal{H}_{\text{int}} = eE\hat{x}$, and $\Psi_\pm$ in Equations (A2) and (A3) are the new unperturbed states. Neglecting interband transitions, the action of the electric field has two distinct effects. Firstly it produces a uniform translation of the probability densities in k space. Secondly and most importantly for our purpose, it polarizes the individual electronic wave functions. As an example, we consider the change in the state $\Psi_-(k)$, below the gap, in the presence of the electric field. Treating $\mathcal{H}_{\text{int}}$ by first-order perturbation theory, the polarized state becomes

$$\Psi'_- = \Psi_- + \frac{\langle \Psi_- | eEi\nabla_k | \Psi_+ \rangle \Psi_+}{\mathcal{E}_- - \mathcal{E}_+} \,. \tag{A6}$$

Thus there is an extra charge density $\rho_k(x)$ associated with the matrix element of $eEi\nabla_k$ between the lower and the upper bands:

$$\rho_k(x) = -e\left[|\Psi'_-|^2 - |\Psi_-|^2\right] = e^2 E\left(\frac{2\hbar^2 Q U_1}{m W^3}\right)\sin Qx \,. \tag{A7}$$

This charge density interacts with the crystal potential to produce an additional force F_A on the electron in the state $\vec{k}$, given by

$$F_A = -\frac{1}{e}\frac{\partial U}{\partial x}\rho_k(x) \,, \tag{A8}$$

with $U \equiv 2U_1 \cos Qx$. Averaging this force over one unit cell, we obtain,

$$\langle F_A \rangle = 2eE\hbar^2 Q^2 U_1^2/m W^3 \tag{A9}$$

and from Equation (A5),

$$\langle F_A \rangle = -eE(m/m^* - 1) \,. \tag{A10}$$

The condition of electrical neutrality then leads to the Bragg-reflection force F_B on the lattice ion, which balances the force $\langle F_A \rangle$, and is given by

$$F_B = eE(m/m^* - 1) \,. \tag{A11}$$

From above it is clear that *virtual* interband transitions cause polarization of electronic charge densities which, in turn, produce a Bragg-reflection force on the lattice ion through the crystal potential.

The derivation of the Bragg-reflection force in the presence of a magnetic field is similar to the discussion of this Appendix, and the result is given by Equation (13.11) of Section 13.2 of the text. The expression obtained here in Equation (A11) is the limiting form of Equation (13.11) for zero magnetic field.

Acknowledgements

Supported by the NSF under Grant No. DMR74-03464.

References

1 Quinn, J.J. (1967) *Phys. Lett. A*, **25**, 522.
2 Turner, G., Thomas, R.L., and Hsu, D. (1971) *Phys. Rev. B*, **3**, 3097.
3 Meredith, D.J., Watts-Tobin, R.J., and Dobbs, E.R. (1969) *J. Acoust. Soc. Am.*, **45**, 1393.
4 Alig, R.C. (1969) *Phys. Rev.*, **178**, 1050.
5 Gaerttner, M.R. (1971) Ph.D. thesis, Cornell University, unpublished.
6 Wallace, W.D., Gaerttner, M.R., and Maxfield, B.W. (1971) *Phys. Rev. Lett.*, **27**, 955.
7 Chimenti, D.E., Kukkonen, C.A., and Maxfield, B.W. (1974) *Phys. Rev. B*, **10**, 3228.
8 Kaner, E.A. and Falco, V.L. (1973) *Zh. Eksp. Teor. Fiz.*, **64**, 1016; [Sov. Phys. – JETP **37**, 516 (1973)].
9 Kittel, C. (1954) *Am. J. Phys.*, **22**, 250.
10 Jones, H. and Zener, C. (1934) *Proc. R. Soc. A*, **144**, 101.
11 Kjeldaas Jr., T. (1959) *Phys. Rev.*, **113**, 1473.
12 Rodriguez, S. (1958) *Phys. Rev.*, **112**, 80; Holstein, T. (1959) *ibid.*, **113**, 479.
13 Ham, F.S. (1963) *Phys. Rev.*, **128**, 82.
14 Boyd, J.R. and Gavenda, J.D. (1966) *Phys. Rev.*, **152**, 645.
15 See, for example, Platzman, P.M. and Wolff, P.A. (1973) *Solid State Phys. Suppl.*, **13**, 147.
16 Holroyd, F.W. and Datars, W.R. (1975) *Can. J. Phys.*, **53**, 2517.
17 Overhauser, A.W. (1968) *Phys. Rev.* **167**, 691; (1971) *Phys. Rev. B*, **3**, 3173.

Reprint 14 Dynamics of an Incommensurate Charge-Density Wave[1)]

M.L. Boriack * *and A.W. Overhauser* *

* Department of Physics, Purdue University, 525, Northwestern Avenue, West Lafayette, Indiana 47907, USA

Received 23 June 1977

We present the equation of motion for the drift velocity of a charge-density wave (CDW), the electron drift velocity, and the total electric current in the presence of applied electric and magnetic fields. These equations can be microscopically derived. In this paper, we discuss the electric current and the dynamical effects of an electric field for an incommensurate CDW in a three-dimensional metal. An expression for the effective mass characterizing the CDW acceleration is derived.

14.1 Introduction

The discovery of charge-density waves (CDW) in pseudo-one-dimensional conductors and in transition-metal dichalcogenides has stimulated considerable interest in the properties of CDW systems. One of the most fundamental problems is the effect of CDW's on electrical transport properties. Much of the theoretical effort on CDW transport has been directed toward the one-dimensional conductor TTF-TCNQ. The problem of the conductivity associated with fluctuations into the CDW state above the Peierls critical temperature has been addressed by Patton and Sham [1], Allender, Bray, and Bardeen [2], and Fukuyama, Rice, and Varma [3]. Lee, Rice, and Anderson [4] have studied microscopically the problem of a one-dimensional system containing a CDW which is pinned or fixed in space. Rice [5] has dealt pheno-menologically with the problem of a pinned CDW in one dimension.

Although in some systems, the CDW may be pinned, the existence of phason modes [6] may make it possible to have systems with spatial fluctuations of the CDW's position which are so large that the CDW is not pinned. Thus one of the most fundamental questions which must be answered is that of the electrical transport properties of a system containing an unpinned CDW. It is toward the resolution of this problem that we direct this paper. We begin by presenting a set of equations describing the motion of a CDW and the electron distribution in three-dimensional jellium in the presence of applied electric and magnetic fields. The complete microscopic derivation of these equations will encompass several papers. In this paper we discuss the current and the effects of an applied electric field. The discussion of the magnetic field and scattering processes will be studied in later work.

Because the CDW model has been successful in explaining several of the anomalous properties of potassium, including the Mayer–El Naby optical absorption [7] and the conduction-electron spin resonance [8], and because spatial fluctuations associated with the phason modes are expected to be large if a CDW is assumed to exist in potassium [6], it is of great interest to evaluate the expressions for the current and for the CDW effective mass for values of the various parameters suitable to the CDW model of potassium.

1) Phys. Rev. B 16, 12 (1977)

Anomalous Effects in Simple Metals. Albert Overhauser

ISBN: 978-3-527-40859-7

14.2 Equations of Motion

In this section we write down without motivation a set of equations describing the motion of the CDW and the electron distribution in the presence of an electric field $\vec{\mathcal{E}}$ and a magnetic field $\vec{H}$. These equations are intended to be used as a reference throughout this and future works in which they shall be motivated and microscopically derived.

For the magnetic field $\vec{H}$ in the $\hat{z}$ direction and the CDW wave vector $\vec{Q}$ in the $\hat{x}$ direction, the equation of motion for the CDW drift velocity $\vec{D}$ can be written

$$\frac{dD}{Dt} = (1/m^*)(-e\mathcal{E}_x - eHK_y/c) - (D - \beta K_x)/\tau_D \,, \tag{14.1}$$

where $\vec{D}$ is along $\vec{Q}$. (The symbol D was chosen to bring to mind the word "drift". Nowhere in this paper is the electric displacement discussed.) m^* is the effective mass associated with the acceleration of the CDW, and the electronic charge is $-e$. τ_D is the relaxation time for the CDW velocity $\vec{D}$, and β is another constant which arises due to scattering. $\vec{K}$ is the quasivelocity of the electron distribution and is related to the average electron wave vector $\langle\vec{k}\rangle_{av}$ by

$$\vec{K} \equiv \hbar\langle\vec{k}\rangle_{av}/m \,, \tag{14.2}$$

where m is the electron mass. The βK_x term tries to pull the CDW along with the drifting electrons. By $\langle\vec{k}\rangle_{av}$ we mean the average value of the wave-vector label for occupied states.

It may seem somewhat surprising that a magnetic field can accelerate a CDW. Not only does this arise from a microscopic derivation, but if the electric and magnetic field terms are not as shown in Equation (14.1), it is easy to envision an experiment which violates the second law of thermodynamics. This will be shown in a subsequent paper.

The equations of motion for the components of the electron quasivelocity parallel (K_x) and perpendicular (K_y, K_z) to Q are

$$\begin{aligned} \frac{dK_x}{dt} &= (1/m)(-e\mathcal{E}_x - eHK_y/c) - (K_x - \alpha D)/\tau_x \,, \\ \frac{dK_y}{dt} &= (1/m)(-e\mathcal{E}_y + eHV_x/c) - K_y/\tau_y \,, \\ \frac{dK_z}{dt} &= (1/m)(-e\mathcal{E}_z) - K_z/\tau_z \,. \end{aligned} \tag{14.3}$$

The relaxation times for electron drift parallel (τ_x) and perpendicular (τ_y, τ_z) to $\vec{Q}$ are different due to the distortion of the Fermi surface caused by the CDW. In the case under consideration $\tau_y = \tau_z$. $\vec{V}$ is the mean group velocity of the electrons, i. e.,

$$\vec{V} \equiv \langle\vec{v}_g\rangle_{av} - \langle\vec{p}/m\rangle_{av} \,, \tag{14.4}$$

and is related to $\vec{K}$ and $\vec{D}$ by

$$\begin{aligned} V_x &= (1-\gamma)K_x + \gamma D \,, \\ V_y &= K_y \,, \\ V_z &= K_z \,. \end{aligned} \tag{14.5}$$

The current is given by $\vec{J} = -ne\vec{V}$, where n is the number of electrons per unit volume, α is another constant which arises from scattering, and the term $\alpha\vec{D}$ tries to pull the electrons along with the CDW in a manner analogous to the acoustoelectric effect. γ is discussed in Section 14.4.

In this paper we will discuss Equation (14.5) and the electric field terms in Equations (14.1) and (14.3). The terms arising from the magnetic field and from scattering will be dealt with in future work.

14.3 Jellium Model for a CDW

Our discussion of CDW motion will be given in terms of a three-dimensional jellium model. In the presence of a CDW, the total self-consistent potential in the one-electron Hamiltonian is of the form

$$V(\vec{r}) = G \cos \vec{Q} \cdot \vec{r} . \tag{14.6}$$

This potential produces a density modulation in the electron gas so that the electron density is given by

$$\rho(\vec{r}) = \rho_0(1 - p \cos \vec{Q} \cdot \vec{r}) . \tag{14.7}$$

The mean density is $\rho_0 = k_F^3/3\pi^2$ and the fractional modulation p is given for an unperturbed free electron gas by

$$p = (3G/2\epsilon_F) g(Q/2k_F) , \tag{14.8}$$

where

$$g(x) \equiv \frac{1}{2} + [(1 - x^2)/4x] \ln |(1 + x)/(1 - x)| \tag{14.9}$$

and $\epsilon_F = \hbar^2 k_F^2/2m$.

The density modulation of the electron gas must be compensated by a density modulation of the positively charged background. This requires a local displacement $\vec{u}$ of the background such that

$$\vec{u}(\vec{r}) = (p\vec{Q}/Q^2) \sin \vec{Q} \cdot \vec{r} . \tag{14.10}$$

Throughout our discussion we will take G in the self-consistent potential, Equation (14.6), to be a constant. Since the existence of CDW's depends crucially upon the velocity dependence of G, it is an inconsistent approximation. However, the treatment of velocity dependent effects opens up many intricate questions which would unduly complicate our present discussion and must be postponed to later work.

The one-electron Hamiltonian which incorporates Equation (14.6) is

$$\mathcal{H} = p^2/2m + G \cos \vec{Q} \cdot \vec{r} . \tag{14.11}$$

Compact solutions to this question cannot be written down. However, for our purposes it is sufficiently accurate to divide the potential term into two parts, one which leads to the gap at $\vec{k} = \frac{1}{2}\vec{Q}$ and the other which leads to the gap at $\vec{k} = \frac{1}{2}\vec{Q}$. Each part must be treated accurately near the gap since the wave functions and energy spectrum are significantly altered there. The simplifying feature of this method is that the effects of the two parts of the potential are additive in many cases.

For discussion we take the part of the potential which mixes the plane wave state $\vec{k}$ with $\vec{k} - \vec{Q}$ and produces the gap at $\vec{k} = \frac{1}{2}\vec{Q}$. Treating the mixing by degenerate perturbation theory leads to a secular equation which can be solved for the energies of the states above and below the gap,

$$E_\pm(\vec{k}) = \frac{1}{2}(\epsilon_{\vec{k}} + \epsilon_{\vec{k}-\vec{Q}}) \pm \frac{1}{2}\left[(\epsilon_{\vec{k}} - \epsilon_{\vec{k}-\vec{Q}})^2 + G^2\right]^{1/2} , \tag{14.12}$$

where $\epsilon_k = \hbar^2 k^2/2m$. The corresponding eigen-functions are

$$\varphi_{\vec{k}} = \cos\xi\, e^{i\vec{k}\cdot\vec{r}} - \sin\xi\, e^{i(\vec{k}-\vec{Q})\cdot\vec{r}} \equiv e^{i\vec{k}\cdot\vec{r}} u_{\vec{k}}(\vec{r}) ,$$
$$\psi_{\vec{k}} = \sin\xi\, e^{i\vec{k}\cdot\vec{r}} + \cos\xi\, e^{i(\vec{k}-\vec{Q})\cdot\vec{r}} \equiv e^{i\vec{k}\cdot\vec{r}} w_{\vec{k}}(\vec{r}) , \tag{14.13}$$

for states below and above the gap, respectively. The coefficients obey the relation

$$\sin 2\xi = G/(E_+ - E_-) = G/W . \tag{14.14}$$

Away from the gap these wave functions reduce to those found by nondegenerate perturbation theory.

Due to the "softening" of the energy for wave vectors along $\vec{Q}$, the Fermi surface distorts and takes on a lemon shape [6].

14.4 Current

Equation (14.5) states that the total momentum or current depends upon the difference in CDW and electron drift velocities. This results because the momentum of an electron in a periodic potential is not simply proportional to the wave vector. To derive Equation (14.5) it is most convenient to work in a reference frame in which the CDW is at rest and the electron drift velocity is $\vec{K}'$. The expectation value of the momentum for an electron in a state $\varphi_{\vec{k}}$ is $\hbar\vec{k}$ plus an additional term which is an odd function of $\vec{k}$. The total contribution of this term due to all the electrons clearly depends on the position of the electron distribution in $\vec{k}$ space. The expectation value for the momentum operator $\vec{p} = -i\hbar\vec{\nabla}$ for an electron the state $\varphi_{\vec{k}}$ in Equation (14.13) is

$$\langle\varphi_{\vec{k}}|\vec{p}|\varphi_{\vec{k}}\rangle = \hbar\vec{k} - \hbar\vec{Q}\sin^2\xi . \tag{14.15}$$

There is similar term due to the part of the potential which causes the other gap. We will consider the case where $\vec{Q}$ and the electron quasi-velocity $\vec{k}'$ are in the x direction. To calculate the net group velocity we add up all the contributions of the occupied states for a distribution centered at $mK'_x/\hbar$ in $\vec{k}$ space. To first order in G/E_F we can approximate the Fermi surface by a sphere, and at zero temperature the net contribution to the group velocity comes from a shell of thickness $mK'_x k_x/\hbar k_F$ at the Fermi surface. Accounting for both gaps the total momentum P'_x in the CDW rest frame is

$$P'_x = nmK'_x(1-\gamma) , \tag{14.16}$$

with

$$\gamma = \frac{3Q}{k_F}\int_{-1}^{1} d(\cos\theta)\cos\theta\,\sin^2\xi|_{k=k_F} , \tag{14.17}$$

where θ is the polar angle measured from the x axis. In the laboratory frame $K_x = K'_x + D$ and the average group velocity is

$$V_x = K_x - \gamma(K_x - D) . \tag{14.18}$$

If $\vec{K}$ is perpendicular to z it is easy to see from Equation (14.15) that the periodic potential cannot make the average group velocity unequal to the drift velocity so that we have $V_y = K_y$.

Thus the current along $\vec{Q}$ and perpendicular to $\vec{Q}$ should be different for two reasons: different relaxation times and different relationships between the drift velocity and the current.

It is of interest to calculate γ using the values of the proposed CDW model for potassium [7, 8]. Taking $G/E_F \cong 0.35$ from fitting the Mayer–El Naby optical absorption data and $Q/2k_F \cong 1+G/4E_F$ for $\vec{Q}$ in "critical contact" with the Fermi surface gives a value of $\gamma \cong 0.2$.

14.5 Effects of an Applied Electric Field

It is well known that electrons in a periodic potential (i. e., Bloch electrons) obey the equation of motion

$$\frac{d\vec{k}}{dt} = -e\vec{\mathcal{E}}/\hbar \tag{14.19}$$

in an applied electric field $\mathcal{E}$. However, the electric field has the additional effect of "polarizing" the electron density in a manner analogous to the way an atomic wave function is perturbed by an electric field [9]. This polarization should not be confused with interband transitions which are negligible for our purpose. This effect is present for electrons in any periodic potential and results in an electron density which is out of phase with the original periodicity of the potential. This results in a net force on the lattice and is the microscopic origin of the so-called "Bragg reflection force" [10]. Whereas in many problems the ions are fixed in the lattice, in jellium the ions are free to adjust to the new charge density and thus the CDW can be accelerated by the electric field.

The Hamiltonian of an electron in a periodic potential $V(\vec{r})$ and an applied uniform electric field $\vec{\mathcal{E}}$ is

$$\mathcal{H} = p^2/2m + V(\vec{r}) + e\vec{\mathcal{E}}\cdot\vec{r}\,. \tag{14.20}$$

The spacial periodicity of $V(\vec{r})$ allows the solution in zero electric field to have the Bloch form,

$$\varphi_{n\vec{k}}(\vec{r}) = e^{i\vec{k}\cdot\vec{r}}u_{n\vec{k}}(\vec{r})\,, \tag{14.21}$$

where $u_{n\vec{k}}(\vec{r})$ is periodic in $\vec{r}$ and n is the band index.

In dealing with the term containing the applied electric field, mathematical difficulties arise due to the fact that the coordinate operator $\vec{r}$ is not square integrable with the wave functions in Equation (14.21) over infinite space. To avoid these problems we will use the Bloch or crystal momentum representation. If a general wave function ψ is

$$\psi(\vec{r}) = \frac{1}{8\pi^3}\sum_n \int d^3k\, a_n(\vec{k}) e^{i\vec{k}\cdot\vec{r}} u_{n\vec{k}}(\vec{r}) \tag{14.22}$$

in the coordinate representation, then ψ is

$$\psi(\vec{k}) = [a_0(\vec{k})a_1(\vec{k})a_2(\vec{k})\cdots] \tag{14.23}$$

in the Bloch representation. $\psi(\vec{k})$ is an (infinite) column vector. In this representation the x component of $\vec{r}$, for example, is [11]

$$x = iI\partial/\partial k_x + X_{nn'}(\vec{k})\,, \tag{14.24}$$

where I is an (infinite) unit matrix and

$$X_{nn'}(\vec{k}) \equiv i \int d^3 r u^{\dagger}_{n\vec{k}}(\vec{r}) \partial [u_{n'\vec{k}}(\vec{r})]/\partial k_x \,. \tag{14.25}$$

The Hamiltonian (14.20) with $\vec{\mathcal{E}} = 0$ is, of course, diagonal in this representation with diagonal elements

$$\mathcal{H}_{nn}(\vec{k}) \equiv E_n(\vec{k}) \,, \tag{14.26}$$

which are the energy bands.

From Equation (14.24) for the x operator in the Bloch representation, the two effects of the electric field can be seen. If we consider an electron described by a wave packet confined to one band, say $E_0(\vec{k})$, then

$$\psi(\vec{k}) = [a_0(\vec{k}) 0\, 0 \cdots] \,. \tag{14.27}$$

We assume that $a(\vec{k})$ is nonzero only in a small region of $\vec{k}$ space. The first term on the right in (14.24) leads to an equation of motion for $|a_0(\vec{k})|^2$ which is solved by some general function of the form $F(\vec{k} + e\vec{\mathcal{E}}t/\hbar)$. Thus the wave packet moves in $\vec{k}$ space according to the equation $d\vec{k}/dt = -e\vec{\mathcal{E}}/\hbar$. For a distribution of electrons, this leads to the electric field term in (14.3).

The second term on the right in (14.24) leads to off-diagonal terms in the Hamiltonian matrix $\mathcal{H}_{nn'}$. These off-diagonal terms can be eliminated to first order in $\vec{\mathcal{E}}$ by using a new set of basis functions given by first-order perturbation theory [9]

$$u^{(1)}_{n\vec{k}} = u_{n\vec{k}} + \sum_{n'} u_{n'\vec{k}} \frac{\langle u_{n'\vec{k}} | ie\vec{\mathcal{E}} \cdot \vec{\nabla}_k | u_{n\vec{k}} \rangle}{E_n(\vec{k}) - E'_n(\vec{k})} \,, \tag{14.28}$$

where $\vec{\nabla}_{\vec{k}}$ is the gradient operator in $\vec{k}$ space. Thus it is clear that the electric field polarizes the electron wave functions (as soon as it is turned on) by mixing states with the same $\vec{k}$ in different energy bands. This is analogous to atomic physics where the electric field polarizes the atomic wave functions by mixing the wave functions belonging to various states.

We can now evaluate the change in electron probability density due to the electric field for the CDW state, Equation (14.13). The perturbed wave functions $\varphi^{(1)}_{\vec{k}}$ are

$$\varphi^{(1)}_{\vec{k}} = \phi_{\vec{k}} - \psi_{\vec{k}} \langle w_{\vec{k}} | ie\vec{\mathcal{E}} \cdot \vec{\nabla}_{\vec{k}} | u_{\vec{k}} \rangle \, W \,. \tag{14.29}$$

The change in probability density is given by

$$\left|\varphi^{(1)}_{\vec{k}}\right|^2 - \left|\varphi_{\vec{k}}\right| = (\hbar^2 G e\vec{\mathcal{E}} \cdot \vec{Q}/m W^3) \sin \vec{Q} \cdot \vec{r} \,. \tag{14.30}$$

The corresponding change for a state $\psi_{\vec{k}}$ above the gap is opposite in sign. Note that the part of electron density induced by the electric field is 90° out of phase with that produced by the self-consistent potential $G \cos \vec{Q} \cdot \vec{r}$. It is only the component $\vec{\mathcal{E}}$ of along with $\vec{Q}$ which gives rise to this out-of-phase part of the electron density.

The total electron density induced by the electric field is found by summing the contributions from all states $\varphi_{\vec{k}}$ and $\psi_{\vec{k}}$ weighted by the appropriate Fermi–Dirac occupation probability. For zero temperature only states up to the Fermi energy are filled and the total electron probability density induced by the electric field is

$$\Delta N_\delta = \frac{2\hbar^2 G e\vec{\mathcal{E}} \cdot \vec{Q}}{m} \int \frac{d^3k}{4\pi^3} \frac{1}{W^3} \,, \tag{14.31}$$

where a factor of 4 has been included to account for spin degeneracy and for both parts of the potential. The volume of integration is that enclosed by the Fermi surface. Note that ΔN_δ does not depend on the electron drift or CDW velocities.

14.6 CDW Acceleration and Effective Mass

In an applied electric field and the absence of scattering processes, the total electron density will be phase shifted, i. e., moved in space, with respect to the density of the positive ion background. This phase shift is proportional to the magnitude of the electric field. With the electron and ion densities out of phase, there will be nonzero net forces on the ions; and they will move in such a manner that the local displacement $\vec{u}$ of the background can be described by an accelerating running wave,

$$\vec{u}(\vec{r}) = (p\vec{Q}/Q^2)\sin\left[\vec{Q}\cdot\left(\vec{r} - \vec{D}t - \frac{1}{2}\vec{A}t^2\right)\right], \tag{14.32}$$

where $\vec{D}$ and $\vec{A}$ are the CDW velocity and acceleration respectively. Thus the periodic ion density tries to catch up to that of the electrons only to find the periodic part of the electron density pulled ahead by the electric field like the carrot before the nose of the proverbial donkey. Note that the acceleration $\vec{A}$ of the CDW, i. e., that of the periodic part of the electron and ion densities, is not the same quantity as the rate of change of the electron drift velocity given by Equation (14.19).

The acceleration of the CDW can be determined by making the ions obey Newton's second law of motion, $\vec{F} = M\vec{a}$, where M is the ion mass. Alternatively, the same result can be obtained by requiring that the rate at which the total energy in the system is changing is equal to the rate at which the applied electric field does work on the system. Both methods, of course, give equivalent results; but since the first is tidier, only it will be given here.

It is convenient to work in the laboratory frame of reference and to take $\vec{D} = 0$ at time $t = 0$. The instantaneously accelerating wave of Equation (14.32) becomes

$$u(\vec{r}) \cong (p\vec{Q}/Q^2)[\sin\vec{Q}\cdot\vec{r} - (\vec{Q}\cdot\vec{A}t^2/2)\cos\vec{Q}\cdot\vec{r}]. \tag{14.33}$$

The $\sin\vec{Q}\cdot\vec{r}$ term leads to a term in the one-electron Hamiltonian which is proportional to $\cos\vec{Q}\cdot\vec{r}$. We put this together with the $\cos\vec{Q}\cdot\vec{r}$ parts from the Hartree, exchange and correlation terms to get the static CDW potential in Equation (14.6).

Due to the out-of-phase electron density induced by the electric field [Equation (14.31)], there will be a force $\vec{F}_1$ on a unit volume of the background which is easily found using Poisson's equation. The result is

$$\vec{F}_1 = (4\pi e^2\rho_0\vec{Q}/\epsilon_1 Q^2)\Delta N_\delta\cos\vec{Q}\cdot\vec{r}, \tag{14.34}$$

where ϵ_1 is the electron-gas dielectric function for wave vector $\vec{Q}$ which results from including electron–electron interactions self-consistently. ϵ_1 is' discussed in the Appendix.

The other term in Equation (14.33) is present only when the acceleration of the CDW is nonzero and gives rise to an additional Coulomb term $\mathcal{H}'$ in the one-electron Hamiltonian where

$$\mathcal{H}' = (2\pi e^2\rho_0 p\vec{A}\cdot\vec{Q}/Q^2)t^2\sin\vec{Q}\cdot\vec{r}. \tag{14.35}$$

$\mathcal{H}'$ is explicitly time dependent, and time-dependent perturbation theory can be used to find the perturbed wave functions and the expectation value of the Hamiltonian. This leads to a force $\vec{F}_2$ on a unit volume of the background where

$$\vec{F}_2 = (8\pi e^2 \rho_0 \vec{Q} G/\epsilon_1 Q^2)(\vec{A}\cdot\vec{Q}) \sum_k \frac{1}{W^3} \cos \vec{Q}\cdot\vec{r}\,, \tag{14.36}$$

where ϵ_1 is discussed in the Appendix.

From Equation (14.32) the acceleration of a unit volume of the background at time $t = 0$ is

$$\frac{d^2\vec{u}}{dt^2} = -(p\vec{Q}/Q^2)(\vec{A}\cdot\vec{Q}) \cos \vec{Q}\cdot\vec{r}\,. \tag{14.37}$$

Writing Newton's second law using Equations (14.34), (14.36), and (14.37) results in an equation for the acceleration of the CDW,

$$\vec{F}_1 + \vec{F}_2 = -\rho_0 p M \vec{A}\cdot\vec{Q}/Q^2 \cos \vec{Q}\cdot\vec{r}\,. \tag{14.38}$$

Solving for $\vec{A}\cdot\vec{Q}$ we have,

$$\vec{A}\cdot\vec{Q} = -e\vec{\mathcal{E}}\cdot\vec{Q}/m^*\,, \tag{14.39}$$

where

$$m^*/m = 1 + Mp\epsilon_1/8\pi e^2\hbar^2 G \sum_{\vec{k}} \frac{1}{W^3}\,, \tag{14.40}$$

and m^* is the zero-temperature effective mass characterizing the acceleration of the CDW in an electric field. Note that near the gaps $W \sim 1/G$. The volume in $\vec{k}$ space where $W \sim 1/G$ is of order G^2. Thus $\sum_{\vec{k}}(1/W^3)\sim \sim 1/G$ plus higher-order terms in G. For this reason the second term Equation (14.40) goes to zero linearly with G for small G and m^* approaches m as the CDW amplitude vanishes. Thus the CDW acceleration is the same as that of the electron distribution in this limit.

It is interesting to evaluate m^*/m for the proposed CDW model of potassium [7, 8]. Figure 14.1 shows a plot of m^*/m versus the normalized energy gap G/E_F for a jellium model with M equal to the mass of a potassium ion and ϵ_2 suitable for an electron gas with the density of that in potassium. Using the value $G/E_F = 0.35$, the effective mass m^* would be about 40 000 times the electron mass.

For the large value of m^* which would be required for a CDW in potassium and assuming that the CDW terminal velocity is of the same order of magnitude as the electron drift velocity, Equation (14.5) implies that a transient effect should be seen in the current since the CDW should accelerate about 40 000 times slower than the electrons. For any materials in which the CDW is unpinned, this effect should exist. Since the results derived in this paper assume a spherical Fermi surface, it would be necessary to take account of the actual Fermi surface to derive quantitative expressions for anisotropic materials.

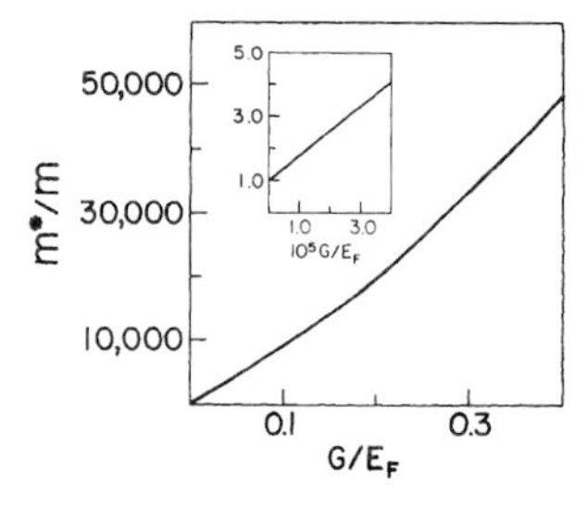

Figure 14.1 Charge-density-wave effective mass m^* versus energy gap G for a jellium model with electron density and ion mass suitable for potassium, m is the free-electron mass and E_F is the Fermi energy. See text for details.

14.7 Conclusion

We have presented a set of equations describing the CDW velocity, the electron drift velocity and the current in the presence of applied electric and magnetic fields for a three-dimensional jellium model of a system with an unpinned CDW. These equations can be microscopically derived. In this paper we discussed the current and the rates of change of the CDW and the electron drift velocities in an applied electric field. We derived an expression for the CDW effective mass in terms of the CDW energy gap G and found that as G goes to zero, the CDW effective mass goes to the electron mass.

A Appendix

he out-of-phase electron density induced by an applied electric field was found in Equation (14.31) to be

$$\Delta N_\delta = \frac{2\hbar^2 G e\vec{\mathcal{E}} \cdot \vec{Q}}{m} \int \frac{d^3k}{4\pi^3} \frac{1}{W^3} . \tag{A1}$$

To calculate the total induced electron density self-consistency it is necessary to include electron–electron interactions. There will be an additional Hartree term due to the change in the electron charge density. Exchange and correlation effects can be accounted for by the observation that to first order in $\vec{\mathcal{E}}$, the effect of the electric field is to change the phase (or spatial position) of the periodic part of the electron density but not to change its amplitude. For a CDW to exist the exchange and correlation potential V_{xc} would of necessity be highly nonlocal or velocity dependent. However, since we have ignored this velocity dependence throughout, for the discussion here we approximate V_{xc} by the Slater potential,

$$V_{xc} = -(e^2 k_{\mathrm{F}}/2\pi\rho_0)\Delta N . \tag{A2}$$

We can then write the total self-consistent perturbation due to the electric field as

$$\mathcal{H}'_{\mathrm{sc}} = e\vec{\mathcal{E}} \cdot \vec{r} + \left(\frac{4\pi e^2}{Q^2} - \frac{e^2 k_{\mathrm{F}}}{2\pi\rho_0} \right) \Delta N_\delta^{\mathrm{sc}} \sin \vec{Q} \cdot \vec{r} \tag{A3}$$

where $\Delta N_\delta^{\mathrm{sc}}$ is the magnitude of the $\sin \vec{Q} \cdot \vec{r}$ part of the self-consistent electron density. The second term in (A3) contains the Hartree and exchange-correlation effects. It follows that

$$\Delta N_\delta^{\mathrm{sc}} = \Delta N_\delta / \epsilon_1 \tag{A4}$$

with

$$\epsilon_1 = 1 + \frac{me^2}{\pi\hbar^2 k_{\mathrm{F}}} \frac{g(x)}{x^2} (1 - 3x^2/2) . \tag{A5}$$

where $g(x)$ is given in Equation (14.9) and $x = Q/2k_{\mathrm{F}}$

A similar treatment of the perturbation (14.35) due to the acceleration of the ions leads to Equation (14.36).

In order that m^*/m can be calculated for the proposed CDW model of potassium [7, 8], we evaluate ϵ_1 for an electron gas with the density of that in potassium. ϵ_1 is a function of G through its dependence on $Q/2k_{\mathrm{F}}$. For $G/E_{\mathrm{F}} \cong 0.35$ we find $\epsilon_1 \cong 0.81$.

Acknowledgements

Research supported in part by the NSF under Grant Nos. Gh 41884 and MRL program DMR-03019A04.

References

1. Patton, B.R. and Sham, L.J. (1973) *Phys. Rev. Lett.*, **31**, 631.
2. Allender, D., Bray, J.W., and Bardeen, J. (1974) *Phys. Rev. B*, **9**, 119.
3. Fukuyama, H., Rice, T.M., and Varma, C.M. (1974) *Phys. Rev. Lett.*, **33**, 305.
4. Lee, P.A., Rice, T.M., and Anderson, P.W. (1974) *Solid State Commun.*, **14**, 703.
5. Rice, M.J. (1975) in *Low Temperature Cooperative Phenomena: The Possibility of High Temperature Superconductivity*, NATO Advanced Study Institute Series Vol. 7, (ed. H.J. Keller), Plenum, New York, p. 23.
6. Overhauser, A.W. (1971) *Phys. Rev. B*, **3**, 3173.
7. Overhauser, A.W. and Butler, N.R. (1976) *Phys. Rev. B*, **14**, 3371. Also see the references listed in [6].
8. Overhauser, A.W. and de Graff, A.M. (1968) *Phys. Rev.*, **168**, 763.
9. Jones, H. and Zener, C. (1934) *Proc. R. Soc. (London) A*, **144**, 101.
10. Banik, N.C. and Overhauser, A.W. (1977) *Phys. Rev. B*, **16**, 3379.
11. Wilson, A.H. (1953) *The Theory of Metals*, 2nd edn., Cambridge U.P., London, p. 48, Eq. (2.82.3).

Reprint 15 Magnetodynamics of Incommensurate Charge-Density Waves[1)]

M.L. Boriack and A.W. Overhauser**

* Department of Physics, Purdue University, 525, Northwestern Avenue, West Lafayette, Indiana 47907, USA

Received 1 August 1977

We present a microscopic derivation of the equations of motion in a magnetic field for the drift velocity of a charge-density wave (CDW) and the electron drift velocity for a three-dimensional system containing an unpinned, incommensurate CDW. We derive an expression for the effective mass of the CDW and show that the effective mass for acceleration by a magnetic field is the same as that for acceleration by an electric field. We examine the effects of the drifting CDW on the magnetotransport properties in the case of a simply-connected Fermi surface. We also discuss the theory of the induced-torque for a spherical sample containing a single CDW which is free to drift.

15.1 Introduction

In a previous paper [1], hereafter referred to as I, we presented a set of equations describing electric transport in a three-dimensional system in the presence of an incommensurate unpinned charge-density wave (CDW). We discussed the dynamic effects of an applied electric field in accelerating the CDW and the electrons and derived an expression for the effective mass of the CDW. Our purpose in this paper is to continue the microscopic derivation of the equations presented in Paper I with a discussion of the dynamic effects of an applied magnetic field.

One of the primary goals of this work has been to discover the effect of a drifting CDW on electrical transport in a magnetic field. There are two types of effects which a CDW can have on magnetotransport: those associated with the drift of the CDW and those associated with the topology of the Fermi surface. This work is limited in that we choose not to include the possible topological effects (e.g., open orbits) in the discussion. Thus, in order to discover those effects associated with the drift of the CDW, we choose a model system in which the Fermi surface is simply-connected. We also choose to ignore any periodicities which would be associated with a real crystal lattice and will discuss the magnetotransport of a CDW in three-dimensional jellium. In particular, we discuss the effect of a drifting CDW on the magnetoresistance and Hall coefficient. The theory of the induced-torque experiment is discussed for a spherical sample of jellium containing a single CDW throughout.

15.2 Equations of Motion

In Paper I, we presented a set of equations describing the drift velocity of the CDW and the electron drift velocity for a three-dimensional jellium model containing a CDW. In this paper we will discuss the effect of an applied uniform magnetic field $\vec{H}$. In order that the goal of our arguments can be kept in mind throughout, we state the resulting equation of motion for the CDW velocity $\vec{D}$ in the absence of scattering for the case where the CDW wave vector

1) Phys. Rev. B 16, 12 (1977)

Anomalous Effects in Simple Metals. Albert Overhauser

ISBN: 978-3-527-40859-7

$\vec{Q}$ is in the $\hat{x}$ direction and $\vec{H}$ is in the $\hat{z}$ direction

$$\frac{dD}{dt} = -\left(\frac{eH}{m^*c}\right) K_y \,, \tag{15.1}$$

where $\vec{D}$ is along $\vec{Q}$. The electronic charge is $-e$. m^* is the effective mass associated with the acceleration of the CDW. $\vec{K}$ is a quasivelocity of the electrons and is related to the average wave vector $\langle \vec{k} \rangle_{\text{av}}$ of the electron distribution by

$$\vec{K} \equiv \hbar \langle \vec{k} \rangle_{\text{av}} / m \,, \tag{15.2}$$

where m is the electron mass.

The equations of motion for the components of the electron quasivelocity parallel (K_x) and perpendicular (K_y, K_z) to $\vec{Q}$ are

$$\frac{dK_x}{dt} = -\left(\frac{eH}{mc}\right) K_y \,,$$
$$\frac{dK_y}{dt} = \left(\frac{eH}{mc}\right) V_x \,, \quad \frac{dK_z}{dt} = 0 \,. \tag{15.3}$$

$\vec{V}$ is the average group velocity for the electron distribution, i. e.,

$$\vec{V} \equiv \langle \vec{V}_g \rangle_{\text{av}} = \langle \vec{p}/m \rangle_{\text{av}} \,, \tag{15.4}$$

and is related to $\vec{K}$ and $\vec{D}$ by

$$V_x = (1-\gamma) K_x + \gamma D \,, \quad V_y = K_y \,, \quad V_z = K_z \,. \tag{15.5}$$

The total electronic current is $\vec{J} = -ne\vec{V}$ where n is the number of electrons per unit volume, γ is a constant discussed in Paper I.

Equation (15.5) was discussed in Paper I. In this paper we will discuss Equations (15.1) and (15.3). It may seem surprising that a CDW could be accelerated by a magnetic field. Not only does this result from a microscopic derivation, but if Equation (15.1) did not have exactly this form, it would be easy to imagine an experiment (as in the Appendix) which violates the second law of thermodynamics.

Since jellium is isotropic, it may seem arbitrary that we have chosen to fix the direction of $\vec{Q}$ and to ignore any torques which might act to change the direction of $\vec{Q}$ in the presence of applied fields and currents. However, in a real metal the direction of $\vec{Q}$ would be some specific crystal direction and thus could not rotate. We therefore impose this property (artificially) while using the jellium model.

15.3 Effects of an Applied Magnetic Field

In the presence of a CDW, the total self-consistent potential in the one-electron Hamiltonian for our jellium model is

$$V(\vec{r}) = G \cos \vec{Q} \cdot \vec{r} \,. \tag{15.6}$$

As in Paper I, we choose to ignore the velocity dependence of G for the sake of simplifying the discussion. This periodic potential leads to a periodic variation in electron density which also has a $\cos \vec{Q} \cdot \vec{r}$ dependence. There is an accompanying modulation of the positive ion background to preserve charge neutrality. In Paper I, it was demonstrated that an applied electric field has two effects on electrons in a periodic potential. First, it causes electrons to move in $\vec{k}$-space according to the equation of motion $d\vec{k}/dt = -e\vec{\mathcal{E}}/\hbar$. Second, the electric field polarizes the electron density in a manner analogous to the polarization of atomic wave functions by an electric field. This polarization of the electrons results in an electron density which is out of phase with that of the positive ion background. Since in jellium the ions are free to move under the nonzero net forces, the CDW will accelerate in the presence of an applied electric field.

The effects of an applied magnetic field are very similar. The magnetic field causes electrons to move in $\vec{k}$ space according to the Jones and Zener equation of motion [2]

$$\frac{d\vec{k}}{dt} = -\left(\frac{e}{\hbar c}\right)\vec{v}_g \times \vec{H} , \tag{15.7}$$

where $\vec{v}_g$ is the group velocity of the electron with wave vector $\vec{k}$. In addition, the magnetic field will, in general, polarize the electron density. The ions will move in response to the nonzero forces acting on them due to the polarized electron density, and the CDW will accelerate.

The one-electron Hamiltonian in the presence of an applied magnetic field and a periodic potential $V(\vec{r})$ is

$$\mathcal{H} = (\vec{p} + e\vec{A}/c)^2/2m + V(\vec{r}) - e\varphi , \tag{15.8}$$

where $\vec{A}$ is the vector potential and φ is the scalar potential. The complication which always arises in problems dealing with a magnetic field is in choosing a gauge to describe the magnetic field $\vec{H}$. Extreme care must always be exercised in the choice of a gauge because the physical significance of a wave function depends on the particular vector potential $\vec{A}(\vec{r})$ that is used [3]. Consider the expression for the quantum mechanical probability current $\vec{j}_p$. In the presence of a nonzero vector potential, $\vec{j}_p$ is given by

$$\begin{aligned}\vec{j}_p = &\frac{\hbar}{2mi}\int d^3r[\Psi^*\vec{\nabla}\Psi - (\vec{\nabla}\Psi^*)\Psi] \\ &- \frac{e}{mc}\int d^3r\Psi^*\vec{A}\Psi .\end{aligned} \tag{15.9}$$

Thus, the current associated with an electron in state Ψ is dependent upon the particular choice of vector potential $\vec{A}$.

In the absence of a magnetic field the group velocity, $\vec{v}_g$ associated with an electron with energy $E(\vec{k})$ is

$$\vec{v}_g = \vec{\nabla}_k E(\vec{k})/\hbar . \tag{15.10}$$

When we turn on a magnetic field, we want our (approximate) wave function to retain this physical property. Otherwise the question we hope to answer by perturbation theory would not be relevant to the physically characterized electron we had in mind [3]. In order that Equation (15.10) be maintained, it is necessary to pick $\vec{A}$ such that the second term in Equation (15.9) is identically equal to zero at all times. This means that we must pick a gauge such that the null point of $\vec{A}(\vec{r})$ travels with the center of the wave packet describing our chosen electron. This is the Jones and Zener gauge [2]

$$\begin{aligned}\vec{A} &= \vec{H} \times \frac{1}{2}(\vec{r} - \vec{v}_g t) , \\ \varphi &= (1/2c)(\vec{H} \times \vec{v}_g) \cdot \vec{r} .\end{aligned} \tag{15.11}$$

The scalar potential φ is required so that the electric field $-\vec{\nabla}\varphi - c^{-1}\partial\vec{A}/\partial t$ is zero. We have chosen without loss of generality that $\langle\Psi|\vec{r}|\Psi\rangle$ at $t = 0$. It is only in the Jones and Zener gauge that Equation (15.7) is valid for Bloch electrons in a magnetic field.

The Hamiltonian, Equation (15.8) with the field on is explicitly time dependent. If we choose $\vec{H}$ in the $\hat{z}$ direction, then for $t = 0$ we have

$$\mathcal{H} = \frac{p^2}{2m} + V(\vec{r}) + \left(\frac{eH}{2mc}\right)(xp_y - yp_x) - e\varphi + \frac{e^2A^2}{2mc^2}\,. \tag{15.12}$$

To discuss the effects of the magnetic field terms for electrons in a periodic potential, we shall use the Bloch or crystal momentum representation as in Paper I. First consider the problem of one electron in a periodic potential $V(\vec{r})$. The periodicity of $V(\vec{r})$ allows the Hamiltonian in zero magnetic field to have eigenfunctions of the Bloch form

$$\Phi_{n\vec{k}}(\vec{r}) = e^{i\vec{k}\cdot\vec{r}}u_{n\vec{k}}(\vec{r})\,, \tag{15.13}$$

where $u_{n\vec{k}}(\vec{r})$ is periodic in $\vec{r}$ and n is the band index. If a general wave function Ψ can be written

$$\Psi(\vec{r}) \equiv \frac{1}{8\pi^3}\sum_n \int d^3k a_n(\vec{k})e^{i\vec{k}\cdot\vec{r}}u_{n\vec{k}}(\vec{r}) \tag{15.14}$$

in the coordinate representation, then Ψ is

$$\Psi(\vec{k}) = [a_0(\vec{k})a_1(\vec{k})a_2(\vec{k})\cdots] \tag{15.15}$$

in the Bloch representation. $\Psi(\vec{k})$ is an (infinite) column vector. In this representation the x component of $\vec{r}$ is [4]

$$x = iI\frac{\partial}{\partial k_x} + X_{nn'}(\vec{k})\,, \tag{15.16}$$

where I is an infinite unit matrix, and

$$X_{nn'}(\vec{k}) \equiv i\int d^3r u^{\dagger}_{n\vec{k}}(\vec{r})\frac{\partial[u_{n'\vec{k}}(\vec{r})]}{\partial k_x}\,. \tag{15.17}$$

Similarly, the operator xp_y in the Bloch representation can be written [4]

$$xp_y = i(P_y)_{nn'}\frac{\partial}{\partial k_x} - \hbar L_{nn'}\,, \tag{15.18}$$

where

$$L_{nn'} \equiv \int d^3r e^{i\vec{k}\cdot\vec{r}}\frac{\partial\Phi^*_{n\vec{k}}}{\partial y}\frac{\partial u_{n'\vec{k}}}{\partial k_x} \tag{15.19}$$

and

$$(P_y)_{nn'} \equiv \int d^3r\Phi^*_{n\vec{k}}\left(-i\hbar\frac{\partial\Phi_{n'\vec{k}}}{\partial y}\right)\,. \tag{15.20}$$

Let us consider an electron described by a wave packet in one band $E_0(\vec{k})$. The wave function describing this electron is

$$\Psi(\vec{k}) = [a_0(\vec{k})0\,0\cdots]\,, \tag{15.21}$$

where we assume $a_0(\vec{k})$ to be nonzero only in a small region of $\vec{k}$ space. We can then find the equation of motion in a magnetic field for the electron described by Equation (15.21) by determining $a_0(\vec{k})$ such that $\Psi(\vec{k})$ satisfies the Schrödinger equation of motion for the Hamiltonian with Equation (15.11) for the vector and scalar potentials. Using Equations (15.16) and (15.18) the terms with $\partial/\partial k_x$ lead to an equation of motion for $|a_0|^2$ which is satisfied by a general function $F(\vec{k} + e\vec{v}_g \times \vec{H}t/\hbar c)$, where $\vec{v}_g$ is the group velocity and is related to the energy by Equation (15.10). Thus, the wave packet moves in k space such that Equation (15.7) is satisfied. When this result is averaged over a distribution of electrons, Equation (15.3) is obtained.

As in Paper I, the additional terms in Equations (15.16) and (15.18) introduce off-diagonal terms into the Hamiltonian matrix $\mathcal{H}_{nn'}$. These can be eliminated to first order by using a new set of basis functions given by first-order perturbation theory

$$u_{n\vec{k}}(\vec{r}) = u_{n\vec{k}}(\vec{r}) + \sum_{n'} \frac{u_{n'\vec{k}}}{E_n - E_{n'}} \left[\langle u_{n'\vec{k}} | \frac{-ie}{2mc} \vec{H} \times \vec{v}_g \cdot \vec{\nabla}_k | u_{n\vec{k}} \rangle \right.$$
$$\left. - \left(\frac{eH}{2mc}\right) \int d^3 r e^{i\vec{k}\cdot\vec{r}} \left(\frac{\partial \Phi^*_{n'\vec{k}}}{\partial_y} \frac{\partial u_{n\vec{k}}}{\partial k_x} - \frac{\partial \Phi^*_{n'\vec{k}}}{\partial x} \frac{\partial u_{n\vec{k}}}{\partial k_y} \right) \right] . \tag{15.22}$$

The second term on the right-hand side in Equation (15.22) comes from the scalar potential φ and the third term ie due to the $\vec{A} \cdot \vec{p}$ term in the Hamiltonian. The A^2 term does not contribute to the out-of-phase part of the electron density, and we shall not consider it further.

As discussed in Paper I, the one-electron states below and above the energy gaps, respectively, in the presence of the CDW potential, Equation (15.6), are

$$\Phi_{\vec{k}} = \cos \xi e^{i\vec{k}\cdot\vec{r}} - \sin \xi e^{i(\vec{k}-\vec{Q})\cdot\vec{r}} \equiv e^{i\vec{k}\cdot\vec{r}} u_{\vec{k}} ,$$
$$\Psi_{\vec{k}} = \sin \xi e^{i\vec{k}\cdot\vec{r}} + \cos \xi e^{i(\vec{k}-\vec{Q})\cdot\vec{r}} \equiv e^{i\vec{k}\cdot\vec{r}} w_{\vec{k}} . \tag{15.23}$$

Φ_k and Ψ_k are the states which result from a perturbation treatment of the part of the potential which produces the energy gap at $\vec{k} = \frac{1}{2}\vec{Q}$. The other part of the potential, which produces the gap at $\vec{k} = -\frac{1}{2}\vec{Q}$ can be treated separately as the effects are additive for our purposes as long as G is not too large compared with the Fermi energy. The energies for the states above and below the gap are

$$E_{\pm} = \frac{1}{2}(\epsilon_{\vec{k}} + \epsilon_{\vec{k}-\vec{Q}}) \pm \frac{1}{2}[(\epsilon_{\vec{k}} - \epsilon_{\vec{k}-\vec{Q}})^2 + G^2]^{1/2} , \tag{15.24}$$

and the coefficients obey the relation

$$\sin 2\xi = G/(E_+ - E_-) \equiv G/W . \tag{15.25}$$

For simplicity let us take $\vec{Q}$ in the $\hat{x}$ direction. In terms of the states in Equation (15.23), the perturbed wave function $\Phi'_{\vec{k}}$ in a magnetic field is

$$\Phi'_{\vec{k}} = \Phi_{\vec{k}} - \frac{\Psi_{\vec{k}}}{W} \left\{ \langle w_{\vec{k}} | \frac{-ieH}{2mc} \left(\frac{V_x \partial}{\partial k_y} - \frac{V_y \partial}{\partial k_x} \right) | u_{\vec{k}} \rangle - \left(\frac{eH}{2mc} \right) \right.$$
$$\left. \times \int d^3 r e^{i\vec{k}\cdot\vec{r}} \left[\frac{\partial \Psi^*_{\vec{k}}}{\partial y} \frac{\partial u_{\vec{k}}}{\partial k_x} - \left(\frac{\partial \Psi^*_{\vec{k}}}{\partial x} \right) \left(\frac{\partial u_{\vec{k}}}{\partial k_y} \right) \right] \right\} . \tag{15.26}$$

The change in electron density Δn_k induced by the magnetic field for a state $\Phi_{\vec{k}}$ below the gap is

$$\Delta n_{\vec{k}} = |\Phi'_{\vec{k}}|^2 - |\Phi_{\vec{k}}|^2 = \frac{\hbar^3 e H G Q k_y}{m^2 c W^3} \sin \vec{Q} \cdot \vec{r} . \tag{15.27}$$

Both terms in the bracketed expression in Equation (15.26) give the same contribution. The corresponding change in electron density for a state above the gap is opposite in sign. Note that the induced electron density is 90° out-of-phase with the $\cos \vec{Q} \cdot \vec{r}$ dependence of the unperturbed electron density. In general, the out-of-phase electron density depends on the product $\vec{H} \cdot \vec{Q} \times \vec{k}$ so that only the components of $\vec{H}$ and $\vec{k}$ perpendicular to $\vec{Q}$ are effective in inducing the out-of-phase electron density.

The total out-of-phase part of the electron density $\Delta N_H \sin \vec{Q} \cdot \vec{r}$ induced by the magnetic field is found by summing up all the contributions such as Equation (15.27) weighted by the appropriate Fermi–Dirac distribution function. At zero temperature only states below the gaps are filled (since we have chosen to discuss the case of a simply-connected Fermi surface). From Equation (15.27) it is easy to see that Δn_k is an odd function of k_y. Thus, unless the electrons have a net quasivelocity in the $\hat{y}$ direction, the total ΔN_H will be zero. Let us take the electrons to have a net quasivelocity K_y in the $\hat{y}$ direction, where K_y is defined in Equation (15.2). The net contribution to ΔN_H comes from a shell of thickness $m K_y k_y / \hbar k_F$ at the Fermi surface. The total out-of-phase electron density is then conveniently written in cylindrical coordinates with k_x the axis of the cylinder,

$$\Delta N_H = \frac{2\pi \hbar^2 e H G Q K_y}{mc} \int \frac{d k_x}{4\pi^3} \frac{\pi k_\perp^2}{W^3} \,. \tag{15.28}$$

$k_\perp$ is the radius of a cross section of the cylindrically symmetric Fermi surface measured from the k_x axis.

Having found ΔN_H a treatment similar to that in Section VI of Paper I leads to the expression for the acceleration dD/dt of the CDW in a magnetic field

$$\frac{dD}{dt} = -\left(\frac{eH}{m_H^* c}\right) K_y \,, \tag{15.29}$$

where

$$\frac{m_H^*}{m} = 1 + M p \epsilon_1 / 8\pi e^2 \hbar^2 G \int \frac{d k_x}{4\pi^3} \frac{\pi k_\perp^2}{W^3} \,. \tag{15.30}$$

p if the fractional modulation of the electron density by the CDW and ϵ_1 is the electron-gas dielectric function for wave vector Q which results from including electron–electron interactions self-consistently and is discussed in Paper I. M is the mass of an ion.

To demonstrate that m^* is the same for electric and magnetic fields we restate the effective mass $m_{\mathcal{E}}^*$ for acceleration by an electric field found in Paper I

$$\frac{m_{\mathcal{E}}^*}{m} = 1 + M p \epsilon_1 / 8\pi e^2 \hbar^2 G \sum_k \frac{1}{W^3} \,, \tag{15.31}$$

where the sum is over states below the Fermi surface. For $\vec{Q}$ in the $\hat{x}$ direction we write the sum as an integral in cylindrical coordinates where the volume element is $2\pi k_\perp d k_\perp d k_x$ for an axially symmetric function. The integration over $k_\perp$ is trivial since W is a function only of k_x. Therefore,

$$\sum_k \frac{1}{W^3} = \int \frac{d k_x}{4\pi^3} \frac{\pi k_\perp^2}{W^3} \,, \tag{15.32}$$

and $m_{\mathcal{E}}^* = m_H^*$. These arguments are easily extended to finite temperatures where states above the Fermi energy are occupied as well.

15.4 Magnetoresistance and Hall Coefficient

In this section, we wish to discuss the Hall coefficient and the magnetoresistance for "jellium" containing an incommensurate CDW. We continue to assume that the Fermi surface is simply connected.

As shown in Figure 15.1, let us imagine the Hall experiment with an applied electric field $\mathcal{E}_{x'}$ along the $\hat{x}'$ direction of the sample. We take $\vec{Q}$ to be along the $\hat{x}$ direction and the magnetic field $\vec{H}$ to point in the $\hat{z}$ direction. The angle between $\vec{Q}$ and $\hat{x}'$ is θ.

From Paper I, the equations describing the drift velocity $\vec{D}$ of the CDW and the electron quasivelocity $\vec{K}$ in the presence of electric and magnetic fields and scattering processes are

$$\begin{aligned}\frac{dD}{dT} &= \frac{1}{m^*}\left(-e\mathcal{E}_x - \frac{eH}{c}K_y\right) - \frac{D-\beta K_x}{\tau_D}\,,\\ \frac{dK_x}{dT} &= \frac{1}{m}\left(-e\mathcal{E}_x - \frac{eH}{c}K_y\right) - \frac{K_x-\alpha D}{\tau_x}\,,\\ \frac{dK_y}{dT} &= \frac{1}{m}\left(-e\mathcal{E}_y + \frac{eH}{c}V_x\right) - \frac{K_y}{\tau_y}\,.\end{aligned} \tag{15.33}$$

τ_D, τ_x and τ_y are relaxation times for the CDW velocity and the components of the electron drift velocity parallel (τ_x) and perpendicular (τ_y) to $\vec{Q}$. α is a constant which arises from scattering because the moving CDW tries to pull along the electrons in a manner analogous to the acoustoelectric effect. Similarly the βK_x term tries to pull the CDW along with the drifting electrons. The discussion of scattering effects and expressions for α, β and the relaxation times will be given in a subsequent paper. $\vec{V}$ is the mean group velocity of the electrons as defined in Equations (15.4) and (15.5), and the total current is $\vec{J} = -ne\vec{V}$, where n is the number of electrons per unit volume.

It is convenient to work in the x–y coordinate system. In terms of the applied electric field $\mathcal{E}_{x'}\hat{x}'$ and the Hall field $\mathcal{E}_{y'}\hat{y}'$, we have $\mathcal{E}_x = \mathcal{E}_{x'}\cos\theta + \mathcal{E}_{y'}\sin\theta$ and $\mathcal{E}_y = \mathcal{E}_{x'}\sin\theta + \mathcal{E}_{y'}\cos\theta$. In equilibrium the equations of motion become, using Equation (15.5),

$$\begin{aligned}&D - \beta K_x + \frac{e\tau_D}{m^*}(\mathcal{E}_{x'}\cos\theta + \mathcal{E}_{y'}\sin\theta) + \frac{eH\tau_D}{m^* c}K_y = 0\,,\\ &-\alpha D + K_x + \left(\frac{e\tau_x}{m}\right)(\mathcal{E}_{x'}\cos\theta + \mathcal{E}_{y'}\sin\theta) + \left(\frac{eH\tau_x}{mc}\right)K_y = 0\,,\\ &-\left(\frac{eH\tau_y\gamma}{mc}\right)D - (1-\gamma)\left(\frac{eH\tau_y}{mc}K_x\right) + \frac{e\tau_y}{m}(-\mathcal{E}_{x'}\sin\theta + \mathcal{E}_{y'}\cos\theta) + K_y = 0\,.\end{aligned} \tag{15.34}$$

In addition, there is the condition that no transverse current can flow, i.e., $V_{y'} = 0$, which becomes

$$\gamma D\sin\theta + (1-\gamma)K_x\sin\theta + K_y\cos\theta = 0\,. \tag{15.35}$$

Equation (15.34) and (15.35) are four equations which are readily solved for the unknowns K_x, K_y, D, and $\mathcal{E}_{y'}$. The current along $\hat{x}'$ is $J_{x'} = -neV_{x'}$ and

$$V_{x'} = (1-\gamma)K_x\cos\theta + \gamma D\cos\theta - K_y\sin\theta\,. \tag{15.36}$$

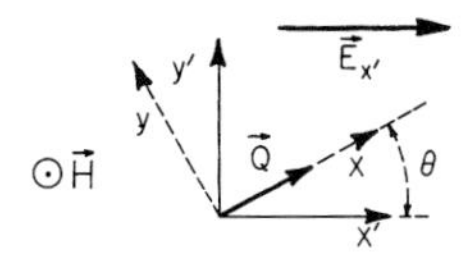

Figure 15.1 Sketch of the hall experiment described in text. The applied electric field $\vec{\mathcal{E}}_{x'}$ is along the $\hat{x}'$ direction. $\vec{Q}$ is along $\hat{x}$ and makes an angle θ with $\hat{x}'$. The magnetic field $\vec{H}$ points along $\hat{z}$ and out of the plane of the figure.

The resulting expression for $J_{x'}$ is

$$J_{x'} = ne^2\tau_y \mathcal{E}_x \Delta_1/m\Delta\,, \tag{15.37}$$

where

$$\Delta = \frac{\tau_y}{m}(1-\alpha\beta)\cos^2\theta + \Delta_1 \sin^2\theta \tag{15.38}$$

and

$$\Delta_1 = (\tau_x/m)(1-\gamma+\beta\gamma) + (\tau_D/m^*)(\alpha+\gamma-\alpha\gamma)\,. \tag{15.39}$$

The transverse electric field $\mathcal{E}_{y'}$ is

$$\mathcal{E}_{y'} = -\frac{\mathcal{E}_{x'}}{\Delta}\left[\left(\Delta_1 - \frac{\tau_y}{m}(1-\alpha\beta)\right)\sin\theta\cos\theta + \omega\tau_y\Delta_1\right], \tag{15.40}$$

where $\omega \equiv eH/mc$. Note that there is a transverse electric field $\mathcal{E}_{y'}$ even in the absence of a magnetic field since the drifting CDW will, in general contribute to the current in the $\hat{y}'$ direction. In zero field the sign of $\mathcal{E}_{y'}$ is dependent on the quadrant in which $\vec{Q}$ lies. Thus, a plot of the transverse electric field versus $\omega\tau$ would give a straight line with an intercept which is positive or negative depending on the quadrant in which $\vec{Q}$ lies. The Hall coefficient R_H is found from Equations (15.37) and (15.40) with the result in all fields

$$R_H = -1/nec\,. \tag{15.41}$$

The magnetoresistance $\rho_{x'x'} = \mathcal{E}_{x'}/J_{x'}$ is

$$\rho_{x'x'} = m\Delta/ne^2\tau_y\Delta_1 \tag{15.42}$$

and is independent of magnetic field for all field strengths.

It is interesting to note the result of ignoring the acceleration of the CDW by the magnetic field or of taking a different effective mass for acceleration by electric and magnetic fields. In this case, the high-field magnetoresistance is linearly dependent on H and changes sign with H and with the quadrant of $\vec{Q}$, which leads to a violation of the second law of thermodynamics. This is discussed in greater detail in the Appendix in the form of a theorem which states that the effective mass for acceleration of a CDW by a magnetic field must be equal to that for acceleration of the CDW by an electric field.

15.5 Theory of the Induced Torque

One of the frequently used tools for studying the topology of the Fermi surface in metals is the induced-torque experiment [5]. It is useful because the theory for a single-crystal sphere predicts that the high-field behavior of the torque gives a dramatic and unambiguous determination of the presence of open orbits. In this section, we discuss the results to be expected for a spherical sample of jellium containing a single CDW which is free to drift.

In an induced-torque experiment the spherical sample is suspended by a rod along the $\hat{y}$ axis in a magnetic field. The magnetic field, which is held constant in strength, rotates in the x–z plane at a constant frequency (typically about 0.01 s^{-1}). The changing magnetic field induces a current in the sample. The $\hat{y}$ component of the torque on the sample, which results from the interaction of the induced current and the magnetic field, is then measured. In general, the torque depends in a fairly complicated way upon the components of the conductivity tensor.

The problem of interest here is the torque to be expected for a spherical sample of jellium containing a single CDW which is free to drift. In the usual derivation of the induced torque, it is assumed that the electric field and the current are related through a local conductivity tensor. However, in case of a sample containing a single CDW, this local relationship cannot be valid since the CDW can have only one drift velocity throughout the sample since planes of constant phase cannot be bunched together. A local relationship between the electric field and the CDW drift velocity implies that the CDW can have a different velocity at each point in the sample. In general, it is by no means necessary that a sample be characterized by a single $\vec{Q}$ throughout. There could be a domain structure with different $\vec{Q}$'s in different domains. The CDW drift velocity in each domain could then be different. At this time we choose to discuss only the case of a sample with a single $\vec{Q}$. In order that planes of constant phase not be bunched together, we take the CDW drift velocity to be a function of the spatial average of the electric and magnetic fields. For the sake of brevity we do not include the details of the solution to the equations of motion, Equation (15.33), of the CDW and the appropriate boundary conditions and Maxwell's equations [5]. If an initial value for $\vec{D}$ is assumed, it is found that $\vec{D}$ decays in time due to a buildup of charge at the surface of the sample. Thus the steady state value for $\vec{D}$ is zero. In addition, any currents associated with the drifting CDW and the electric fields produced by the charge on the surface are uniform (since we take $\vec{D}$ the same everywhere in the sphere). Since there is no torque on a uniform current, the drifting CDW causes no torque on the sample while it has a nonzero velocity. Thus, the induced torque for a spherical sample of jellium with one CDW throughout is given by the usual result of Visscher and Falicov [5].

15.6 Conclusions

In this paper, we have microscopically derived the equations of motion in a magnetic field for the drift velocity of an unpinned incommensurate CDW and the electron drift velocity for a three-dimensional jellium model. The effective mass characterizing the acceleration of the CDW by the magnetic field was shown to be equal to that for the electric-field case previously considered in Paper I. We examined the magnetotransport effects associated with the drift of the CDW for the case of a simply-connected Fermi surface. The magnetoresistance was found to be independent of magnetic-field strength and the Hall coefficient was equal to the free-electron value for all values of the magnetic field. The transverse electric field was found to contain a term which is independent of magnetic field strength and changes sign with the quadrant of the CDW wave vector $\vec{Q}$. Thus, a plot of transverse electric field versus magnetic field would give a straight line with an intercept, the sign of which depends on the quadrant of $\vec{Q}$. We also discussed the theory of the induced torque experiment for a spherical sample of jellium with one CDW throughout which was free to drift and found that the torque did not depend on the CDW drift and that the steady-state CDW drift velocity was zero.

Because the CDW model has been successful at quantitatively explaining several of the highly anomalous properties of potassium including the Mayer–El Naby optical absorption [6], the conduction-electron spin-resonance measurements [7], and the anisotropic residual resistance anomaly [8], one of the motivating factors behind this work was the question whether the CDW model could explain the linear magnetoresistance [9] and the anomaly of the high-field induced-torque anisotropy [10]. It is clear that the results presented in this paper concerning the effects associated with the drift of the CDW in the case of a simply connected Fermi surface cannot offer an explanation of these phenomena. The present discussion has not, however, included a study of the possible topological effects of the Fermi surface on the magnetotransport properties. Although for the discussion here we have taken the Fermi surface to be simply connected in order to examine the effects associated with the drift of the CDW, the actual Fermi surface in jellium in the presence of a CDW is not known and effects associated with the topology of Fermi surface must be investigated. Thus, it remains an open question whether the CDW hypothesis can offer an explanation of the magnetoresistance and high-field induced-torque anisotropy in potassium.

A Appendix

In the text it was stated that the effective mass of a CDW characterizing the acceleration by an electric field must be the same as that for acceleration by a magnetic field. The purpose of this appendix is to present a proof of this theorem based on thermodynamic arguments. In Section 15.3 of the text, a microscopic proof was presented for the case where the CDW potential was taken to be a local non-velocity-dependent function. Because the very existence of CDW's depends crucially on the velocity dependence of electron-electron interactions, the proof on the basis of general thermodynamic arguments is very useful since a microscopic treatment of velocity dependent effects would by highly intricate.

As in Section 15.4, we imagine the Hall experiment shown in Figure 15.1. Let us assume that the effective mass of the CDW for acceleration by an electric field $m_{\mathcal{E}}^*$ is different from that for acceleration by a magnetic field m_H^*.

Then the first of the equations in Equation (15.34) becomes

$$D - \beta K_x + \frac{e\tau_D}{m_{\mathcal{E}}^*}(\mathcal{E}_{x'} \cos\theta + \mathcal{E}_{y'} \sin\theta) + \frac{eH\tau_D}{m_H^* c} K_y = 0 \,. \tag{A1}$$

Solving as in Section 15.4 for the magnetoresistance $\rho_{x'x'}$ obtains

$$\rho_{x'x'} = m(\Delta + \Delta_H)/ne^2\tau_y\Delta_1 \,, \tag{A2}$$

where Δ and Δ_1 are given by Equations (15.38) and (15.39) with m^* replaced by $m_{\mathcal{E}}^*$ and

$$\Delta_H = \omega\tau_D\tau_y(\alpha + \gamma - \alpha\gamma)\sin\theta\cos\theta(1/m_{\mathcal{E}}^* - 1/m_H^*) \tag{A3}$$

where $\omega = eH/mc$.

Thus, for $m_{\mathcal{E}}^* \neq m_H^*$ the magnetoresistance is a linear function of magnetic field. Note, however, that depending on the direction of $\vec{H}$ and the quadrant of angle θ, the high-field magnetoresistance can be negative. This clearly violates the Onsager relations and the second law of thermodynamics. Thus, it must be that $m_H^* = m_{\mathcal{E}}^*$.

Acknowledgements

Research supported in part by the NSF under Grant Nos. Gh 41884 and MRL Program DMR-03018A04.

References

1 Boriack, M.L. and Overhauser, A.W. (unpublished).

2 Jones, H. and Zener, C. (1934) *Proc. R. Soc. Lond. A*, **144**, 101.

3 de Graff, A.M. and Overhauser, A.W. (1969) *Phys. Rev.*, **180**, 701, see especially Ref. 13.

4 Wilson, A.H. (1953) *The Theory of Metals*, 2nd edn., Cambridge U.P., London, p. 48, Eq. (2.82.3).

5 Visscher, P.B. and Falicov, L.M. (1970) *Phys. Rev. B*, **2**, 1518.

6 Overhauser, A.W. and Butler, N.R. (1976) *Phys. Rev. B*, **14**, 3371.

7 Overhauser, A.W. and de Graff, A.M. (1968) *Phys. Rev.*, **168**, 763.

8 Bishop, M.F. and Overhauser, A.W. (1977) *Phys. Rev. Lett.*, **39**, 632.

9 Simpson, A.M. (1973) *J. Phys. F*, **3**, 1471; see, also, references listed therein.

10 Holroyd, F.W. and Datars, W.R. (1975) *Can. J. Phys.*, **53**, 2517.

Reprint 16 Phase Excitations of Charge Density Waves[1)]

A.W. Overhauser *

* Department of Physics, Purdue University, 525, Northwestern Avenue, West Lafayette, Indiana 47907, USA

In three-dimensional metals exchange interactions and electron–electron correlations can cause a CDW instability which breaks the translation symmetry of the crystal. Requirements of microscopic charge neutrality require the ions to under-go a compensating periodic displacement, $\vec{A} \sin \vec{Q} \cdot \vec{r}$. Hyperfine fields can be affected in two ways. Electric field gradients will be created or modified, and the Knight shift will become inhomogeneous Incommensurate CDW's have very low frequency excitations, phasons, which cause time-dependent modulation of the phase of the CDW at every site. Consequently hyperfine-field broadening mechanisms can be motionally narrowed, especially in systems for which the CDW amplitude is small or for which the Fermi surface is not highly anisotropic. In this event the hyperfine field broadening will not follow the temperature dependence of the amplitude A. In exceptional cases the residual broadening may be unobservable

16.1 Fermi-Surface Instabilities

The electronic ground state of a metal need not be the mere occupation of the N lowest-energy Block states. Coulomb interactions can cause a modulated, collective deformation of the charge density to have lower total energy [1]. A charge density wave (CDW) is a static deformation for which the conduction-electron density is [2]

$$\rho = \rho_0[1 + p \cos(\vec{Q} \cdot \vec{r} + \varphi)] , \tag{16.1}$$

where ρ_0 is the average density and p is the fractional modulation. The wave vector $\vec{Q}$ is nearly equal to the diameter $2k_F$ of the Fermi surface. The phase φ will be discussed at length below.

A spin density wave (SDW) is a modulation wherein p for up and down-spin electrons are equal in magnitude but opposite in sign. In the Hartree-Fock approximation SDW and CDW deformations are electronically degenerate and have lower energy than the undeformed state. Electron correlations split this degeneracy, with SDW's becoming less stable and CDW's more stable [2].

A CDW instability can occur only if the electronic charge density is locally neutralized by an accompanying lattice distortion. Each positive ion will be displaced from its equilibrium site $\vec{L}$ by

$$\vec{u} = \vec{A} \sin(\vec{Q} \cdot \vec{L} + \varphi) . \tag{16.2}$$

This periodic displacement of the ionic lattice allows a CDW to be detected by Bragg diffraction of X-rays, neutrons, or electrons [3]. The new allowed scattering vectors are $\vec{G} \pm \vec{Q}$, where $\vec{G}$ is any reciprocal lattice vector. These new reflections are generally quite weak since the amplitude A of the displacement is small ($\sim 10^{-2}$ to 10^{-1} Å).

The presence of CDW's in a metal affects the electronic and optical properties as well as the diffraction properties. The electrons will experience a periodic potential, proportional to $\cos \vec{Q} \cdot \vec{r}$, in addition to the potential arising from the lattice of ions. Since $\vec{Q}$ is controlled by

1) Hyperfine Interactions 4, 786–797 (1978) North-Hollan Publishing Germany

Anomalous Effects in Simple Metals. Albert Overhauser

ISBN: 978-3-527-40859-7

Fermi surface dimensions, the wave length of a CDW will generally be unrelated to lattice structure periodicities, i. e., the CDW is generally incommensurate. However $\vec{Q}$ will frequently vary with temperature and may lock into a commensurate value at certain temperatures. Since the discovery of CDW s in the transition-metal dichalcogenides [4], many varieties of behavior have been identified and studied.

If $\vec{Q}$ is commensurate with the reciprocal lattice, say $\vec{Q} = \vec{G}/n$, where $\vec{G}$ is some reciprocal lattice vector and n is an integer, then the CDW causes the volume of the new primitive unit cell to be larger by n. The phase φ of Equations (16.1) and (16.2) will assume a fixed value. The overall result can be regarded merely as a somewhat more complicated lattice structure. In this case the role of a CDW instability need arise only if one inquires why such a structure exists.

If $\vec{Q}$ is incommensurate the resulting structure is multiply periodic and the crystal will no longer have a translation group. No two ions will be equivalent. Such a loss of symmetry means that the energy of the system is independent of the phase φ. One can imagine a situation where φ is slowly time dependent, e. g., $\varphi = vt$. In such a case the CDW would be drifting parallel to $\vec{Q}$ with a speed v. Since v can be arbitrarily small, one perceives that the phase degree of freedom provides a new species of low energy excitation. These excitations are called phasons [3], the analog in a CDW structure of spin waves (or magnons) in a ferromagnet. We will consider the physical consequences of phasons below, after first considering the influence of CDW's on hyperfine fields.

16.2 Hyperfine Effects of CDW's

The change in electronic density around each ion and the displacement of the ions from their positions in the absence of a CDW will alter the electric field gradients and therefore also the hyperfine splittings caused by nuclear quadrupole moments. Furthermore the Knight shift depends on the electron density arising from electrons in states near the Fermi energy. Since these states are the ones that contribute most to a CDW, the Knight shifts will also be affected.

In the case of a commensurate CDW, hyperfine structure will undergo further splitting depending on the number of new inequivalent nuclear sites and the phase of the CDW. However in the case of an incommensurate CDW each hyperfine level will become broadened. We now discuss the incommensurate case in greater detail. For simplicity we illustrate the broadening by considering only isotropic Knight shift distributions.

We suppose that the Knight shift K will deviate from its value K_0 in the absence of a CDW by an amount,

$$\Delta K = w \cos(\vec{Q} \cdot \vec{r} + \varphi) . \tag{16.3}$$

The amplitude w is proportional to p but does not equal $p K_0$ The probability distribution $P(\Delta K)$ for the Knight shift deviation is easily derived and is [5]

$$P(\Delta K) = (\pi w)^{-1}[1 - (\Delta K/w)^2]^{-1/2} , \tag{16.4}$$

for $|\Delta K| < w$, and zero otherwise. This distribution, shown by the dashed curve, 1, in Figure 16.1, has singularities at $\pm w$. Consider next the case of a metal with two CDW's of equal amplitude. Let us assume one $\vec{Q}$ is in the $\hat{x}$ direction and the other in the $\hat{y}$. (The result is independent of these directions as long as they are not parallel.) Then,

$$\Delta K = w(\cos Qx + \cos Qy) . \tag{16.5}$$

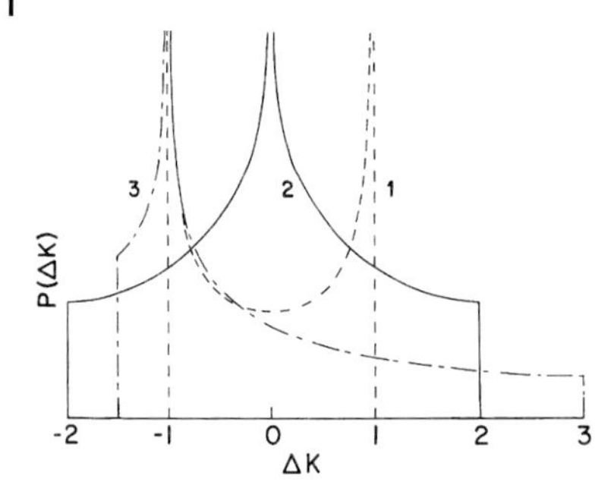

Figure 16.1 Probability distributions for Knight shift deviations ΔK resulting from incommensurate CDW's. Curve 1 is for a single CDW, curve 2 is for two non-colinear CDW's, and curve 3 is for three CDW's with $\vec{Q}_1 + \vec{Q}_2 + \vec{Q}_3 = 0$.

For this situation the probability distribution $P(\Delta K)$ extends from -2 to 2 (in units of w) and has a singularity at 0. It is shown by the solid curve in Figure 16.1. The distributions (16.4) and (16.5) are both symmetric about $\Delta K = 0$.

A more interesting example is one which actually occurs in transition metal dichalcogenides e.g., $NbSe_2$. The distorted phase [6] has three CDW's with $\vec{Q}$'s of equal magnitude lying in the same plane and separated in angle by 120°. For this case the Knight shift deviation is,

$$\Delta K = w[\cos \vec{Q}_1 \cdot \vec{r} + \cos \vec{Q}_2 \cdot \vec{r} + \cos(\vec{Q}_1 + \vec{Q}_2) \cdot \vec{r}] . \tag{16.6}$$

The resulting distribution depends on the choice of sign for the third term, and depends also on whether this term is a sine or cosine. For the particular choice given in Equation (16.6), $P(K)$ is asymmetric, extending from -1.5 to +3 with a singularity at -1, This is shown by curve 3 in Figure 16.1. Had the third term of Equation (16.6) been reversed, $P(K)$ would have been reversed (and would have extended from -3 to +1.5 with a singularity at +1). Had the third term been a sine instead of a cosine, the distribution would have been symmetric and would be almost flat between its limits of -2.6 and 2.6.

Experimental line-shape studies of the ^{93}Nb NMR [7, 8] and the ^{77}Se NMR [9] in 2H-$NbSe_2$ have shown an asymmetric line shape that corresponds to the dependence given in (16.6) and curve 3 of Figure 16.1. This confirms that there are three CDW's throughout the entire volume of the material, instead of a random distribution of $\vec{Q}$ domains, each with one CDW.

We shall now turn to phase excitations of incommensurate CDW's since the time variation of the local charge density can lead to several important effects, including a reduction in hyperfine field broadening as a consequence of time-averaging of ΔK.

16.3 Phasons

Consider a single incommensurate CDW as described by Equations (16.1) and (16.2). Since the energy is independent of phase φ, it follows that there will be low frequency collective excitations corresponding to φ varying slowly in space and time [3]. We express $\varphi(\vec{L}, t)$ by an expansion in running waves,

$$\varphi(\vec{L}, t) = \sum_{\vec{q}} \varphi_{\vec{q}} \sin(\vec{q} \cdot \vec{L} - \omega_{\vec{q}} t) . \tag{16.7}$$

This approach is analogous to treating lattice dynamics in the continuum approximation. The wave vectors $\{\vec{q}\}$ are assumed small compared to the Brillouin zone, and $\{\varphi_{\vec{q}}\}$ are the amplitudes of the normal modes, which we call phasons. Equation (16.7), inserted in Equations (16.1) and (16.2), describes a phase modulation of the CDW. If $\vec{q} \perp \vec{Q}$ the local direction

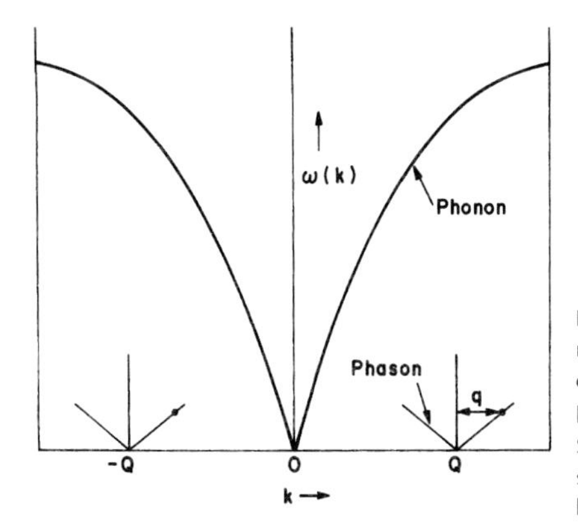

Figure 16.2 Schematic illustration of the vibrational modes in a metal having a CDW structure. The frequency of the phason branch goes to zero at $\vec{Q}$, the location of the CDW satellite reflection in $\vec{k}$-space. Such a diagram has only an approximate meaning since an incommensurate CDW structure does not have a Brillouin zone.

of the CDW wave vector is slightly rotated. If $\vec{q} \parallel \vec{Q}$ the local magnitude of $\vec{Q}$ is periodically modulated. The dynamics of these modes have been studied [3] in the small q limit, assuming negligible damping. The phason frequencies vary linearly with q. The frequency spectrum is plotted schematically in Figure 16.2. Phasons are a type of lattice vibration and are linear combinations of old phonons (in a crystal with no CDW) having wave vectors near $\vec{Q}$ and $-\vec{Q}$, the CDW satellite locations. That is why the phason modes shown in Figure 16.2 approach zero frequency at $\vec{Q}$ and $-\vec{Q}$.

It is expected that ω vs. $\vec{q}$ will be highly anisotropic, because a local rotation of $\vec{Q}$ requires less energy than a change in its magnitude. Consequently the surfaces of constant phason frequency will be very flat (pancake-shaped) ellipsoids. The anisotropy could be as high as 100 to 1, especially if the Fermi surface is nearly spherical.

Emission or absorption of phasons during a diffraction experiment will give rise to pancake-shaped regions of thermal-diffuse scattering surrounding the CDW satellites. Such streaks [10] are observed in the incommensurate phase of IT-TaS_2, and they disappear suddenly at low temperature when the CDW transforms into a commensurate phase, for which phasons are then ordinary optical phonons of higher frequency. This behavior supports the interpretation that the streaks are diffuse scattering by phasons.

16.4 Phason Temperature Factor

Ordinary Bragg reflections are reduced in intensity by the Debye–Waller factor,

$$F_{\mathrm{DW}} \cong \exp\left[-\sum_{\vec{k},i} \frac{G^2 k_{\mathrm{B}} t}{3NM\omega_{\vec{k}i}^2}\right], \tag{16.8}$$

where $\{\omega_{\vec{k},i}\}$ are phonon frequencies, i is the phonon branch, and $\vec{G}$ is the scattering vector of the X-ray, or electron. N is the number of unit cells of mass M in the crystal. This expression is valid only at high temperature. We have shown that the temperature factor for CDW satellites, caused by phason excitations, is quite different [3],

$$F_p \cong \exp\left[-\sum_{\vec{q}} \frac{2k_{\mathrm{B}} t}{NMA^2\omega_{\vec{q}}^2}\right], \tag{16.9}$$

The most important observation is that the exponent in Equation (16.9) depends on A^{-2} where A is the CDW ion displacement, instead of on the square of the scattering vector G^2

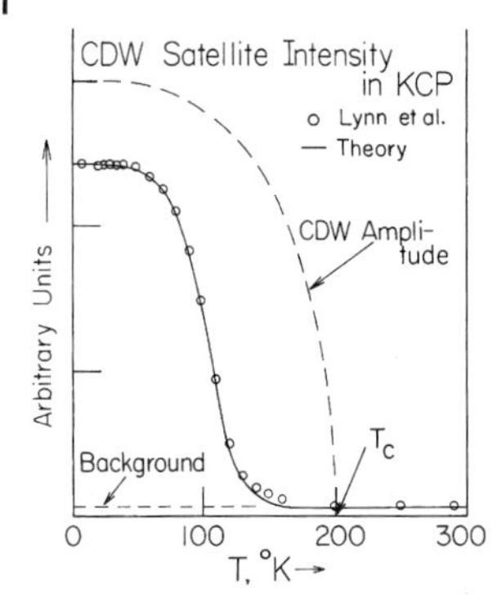

Figure 16.3 Temperature dependence of the CDW satellite in KCP. The circles are the experimental points of Lynn *et al.* The solid line is a theoretical curve which, aside from a scale factor along the ordinate, is the product of the square of the CDW amplitude (dashed curve) and the phason temperature factor.

as in Equation (16.8). The consequences are profound. Since $(GA)^{-2}$ is typically $\sim 10^3$, the phasons are much more effective than phonons in reducing satellite intensity especially as $A \to 0$ near the CDW transition temperature T_c. The phason frequencies also get "soft" near T_c which enhances the reduction.

A striking illustration of the effect of F_p is shown in Figure 16.3. The experimental CDW satellite intensity [11] vs. T for $K_2Pt(CN)_4Br_{0.3}\cdot 3.2D_2O$, (KCP) are the circles. The shape of this curve has not heretofore been explained. The theoretical CDW amplitude vs. T is shown as the dashed curve. Naively one would expect the intensity to be proportional to A^2. Therefore the $A(T)$ curve must be squared and scaled (along both axes) for comparison. The result falls off much too quickly at low temperature, since the experimental data are virtually flat up to 50 °K. The theoretical curve shown in Figure 16.3 is the square of the dashed curve (as shown) multiplied by the phason temperature factor,

$$F_p = \exp\left[\frac{-2\gamma\hbar^2\left(n+\frac{1}{2}\right)}{MA^2 k_B \theta}\right] . \tag{16.10}$$

Equation (16.9) is the high temperature limit of Equation (16.10). Here $\theta(T)$ is the phason "Debye temperature", γ is a numerical factor (~ 3.5) which depends on the anisotropy of the phason frequency spectrum, and n is the Bose–Einstein function,

$$n = [\exp(\theta/T) - 1]^{-1} . \tag{16.11}$$

The important observation is that F_p can cause the CDW satellite intensity to approach zero at a temperature significantly below the CDW transition T_c The CDW still exists, with amplitude $A(T)$ but the phason temperature factor F_p has "washed out" the satellite spots and diffraction intensity has been transferred to multi-phason diffuse background.

16.5 Phason Narrowing of Hyperfine Broadening

Since the hyperfine broadening of an NMR line depends on where each nucleus sits relative to the phase of each CDW present, phase fluctuations can time-average the broadening if they are rapid enough. The residual linewidth $\Delta\omega$ of an interaction of width $\Delta\omega_{int}$, when randomly fluctuating between $+\Delta\omega_{int}$. and $-\Delta\omega_{int}$, is given by [12]

$$\Delta\omega = \Delta\omega_{int}(\Delta\omega_{int}\tau_c) . \tag{16.12}$$

One expects $\Delta\omega_{int}$ to be proportional to the CDW amplitude $A(T)$. The correlation time τ_c is the average time between fluctuations, or the time over which the interaction appears to

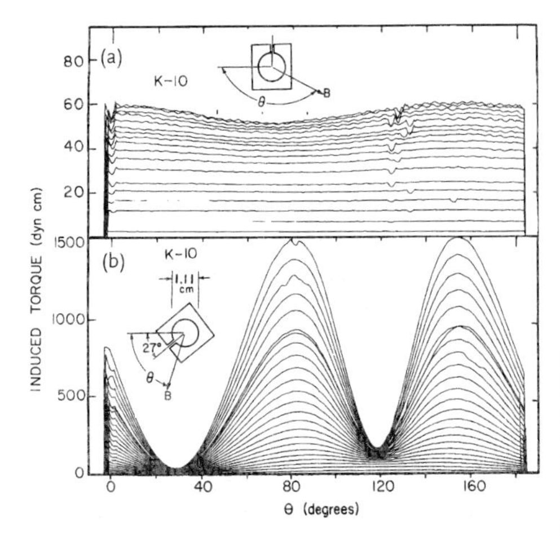

Figure 16.4 Induced torque patterns found by Holroyd and Datars for a sphere of potassium at 1.5 °K in slowly rotating magnetic fields between 1 and 23 kG. θ is the magnet angle in the horizontal plane. In (a) the growth axis was vertical. In (b) it was in the horizontal plane.

be static. Equation (16.12) is valid only when $\Delta\omega_{\text{int}}\tau_c < 1$. For larger values $\Delta\omega$ approaches $\Delta\omega_{\text{int}}$. It follows that line broadening from an incommensurate CDW will approach zero more rapidly than the CDW amplitude if τ_c is sufficiently short. From the foregoing discussions we see that phason excitations can cause a more rapid fall of measured quantities as T_c is approached from below. The extent by which T_c might be underestimated depends on the relevant time scale of the particular measurement.

Phason excitations may even completely mask hyperfine broadening from a CDW. We shall discuss this possibility in the context of metallic potassium, for which there is a wealth of evidence that this "ideal" body-centered-cubic metal has a single, large amplitude CDW. There is striking evidence for this from optical anomalies [13], conduction electron spin resonance [14], magneto-transport properties [15] and others. However diffraction satellites have not been found, a failure which may possibly be attributed to the phason temperature factor discussed above. Furthermore de Haas–van Alphen data indicate (in apparent contradiction to the CDW hypothesis) that the Fermi surface is spherical [16]. We conjecture that this negative evidence may be ascribed to the fact that data have only been taken on poly $\vec{Q}$-domain specimens. The question is still an open one.

The most amazing evidence that potassium is not simple is the recent induced torque study [17] on a spherical sample. This is shown in Figure 16.4. This sample, 1.11 cm in diameter and accurately spherical, was suspended at 1.5 °K in a slowly rotating electromagnet. The induced currents created a torque about the (vertical) axis of suspension, which was then measured as a function of magnet angle (in the horizontal plane). Without a CDW structure, and according to textbook persuasion, all of the curves of Figure 16.4 (taken at various fields from 1 to 23 kG) should be flat. (The conductivity tensor of a cubic crystal is necessarily isotropic.) The observed anisotropy, which reaches values of 45 to 1, together with similar results [18] from prior work, cannot be explained unless the Fermi surface is topologically complicated. The growth axis of the crystal, which was vertical in Figure 16.4a, is obviously unique and indicates the direction of the CDW $\vec{Q}$. In Figure 16.4b the growth axis was in the horizontal plane. (We conjecture that this sample may be the only single-$\vec{Q}$ specimen that has ever been studied.) The low-field torque anisotropies indicate that the residual-resistivity tensor had a cigar-shaped anisotropy of about 5 to 1. This has been quantitatively explained [19] on the basis of a CDW amplitude which simultaneously fits the optical [20] and spin resonance data [14].

The fractional amplitude p of the CDW in potassium is $\sim$ 10%. This could cause a significant broadening of the Knight shift. However, such an effect has been carefully looked for and not found [5]. We shall now show that phase excitations may account for this null result.

The mean square phase fluctuation can be calculated from Equation (16.7).

$$\langle\varphi^2\rangle = \frac{1}{2}\sum_{\vec{q}}\langle\varphi_{\vec{q}}^2\rangle\,. \tag{16.13}$$

The sum is over the allowed phason modes. The mean square phason amplitude $\langle\varphi_{\vec{q}}^2\rangle$ can be calculated by equating the phason kinetic energy, Equation (25) of [3], to $\frac{1}{2}k_B T$.

$$\frac{1}{8}NMA^2\omega_q^2\varphi_q^2 = \frac{1}{2}k_B T\,. \tag{16.14}$$

M is the ionic mass and A is the CDW amplitude. Since $A \sim p/Q$, and $Q \sim 1.33 \times 2\pi/a$ (for potassium $a = 5.2$ Å) we find $A \sim 0.06$ Å. If we now define the phason "Debye temperature" θ by $\hbar(\omega_{\vec{q}})_{\max} = k_B\theta$, and make use of the linear dispersion relation for $\omega_{\vec{q}}$, we find that

$$\langle\varphi^2\rangle \approx \frac{6\hbar^2 T}{MA^2 k_B\theta^2}\,. \tag{16.15}$$

This result is valid for $T \gtrsim \frac{1}{2}\theta$. However because of the isotropy of the "ideal" Fermi surface of potassium we estimate that $\theta \sim 5$°K; so Equation (16.15) is barely applicable at 1.5 °K (the temperature at which the NMR line broadening was investigated).

From Equation (16.15) the root-mean-square phase fluctuation is $\sim$ 10 rad. Since the characteristic phason ω is $\sim 10^{11}$, it follows that for potassium any line broadening from an incommensurate CDW should be time averaged to a negligible value. As discussed in Section 16.2 hyperfine field broadening is observed for incommensurate CDW's in the trans it ion-metal dichalcogenides These materials have very anisotropic Fermi surfaces, so one expects the phason anisotropy to be much less, and θ to be considerably greater than that for potassium. For $\theta \gtrsim 100$°K the root-mean-square phase fluctuation would be a fraction of a radian. Line broadening would then occur.

16.6 Conclusions

It is clear that the most uncertain features of the foregoing quantitative estimates are those related to the phason spectrum. It is extremely important that experimental information on phason frequencies be obtained. In some cases these frequencies may fall in a range accessible to Mössbauer techniques. The small intensity of phason scattering combined with the relative weakness of standard Mössbauer γ ray sources indicates that the experimental difficulties are formidable. In highly anisotropic metals high-resolution inelastic neutron scattering may be able to provide frequency and damping data.

Most theoretical questions related to phase excitation remain unexplored. In some cases these should await the stimulus of experimental results. We have indicated above several areas where phase modulation of CDW's can have important and unusual consequences. Further work is necessary before these suggestions can be confirmed.

A Discussion

E.N. Kaufmann: What mechanism is responsible for forcing $\vec{Q} = \vec{G}/n$ and a commensurate CDW if, as you indicated at the outset, no general restriction on $\vec{Q}$ comes from the direct lattice periodicity, but rather only from dimensions of the Fermi surface?

A. Overhauser: In metals containing more than one atomic species the CDW maxima (or minima) will prefer to be located on one of the species. Therefore if $\vec{Q}$ is near, in magnitude and direction, to some $\vec{G}/n$, the commensurability energy arising from the interaction between the CDW and the ions may allow $\vec{Q}$ to snap into $\vec{G}/n$ at low T. The phason entropy tends to stabilize the incommensurate phase at high T.

K. Andres: Could the effect of zero point phasons be appreciable?

A. Overhauser: There certainly are zero point phasons. I haven't yet thought about their effect, e. g. in narrowing broadened hyperfine lines at $T = 0$.

D.E. Murnick: Would the CDW lead to an electric field-gradient at the nuclear site in potassium metal?

A. Overhauser: Yes, except that the phasons would average the effect to zero, so I can have my cake and eat it too!

P. Peretto: Do phasons occur in spin density waves (SDW) and what are the differences between these and phasons of CDW, e. g. as far as the time ranges are concerned?

A. Overhauser: Phasons occur also in SDW's, but there, they are usually called spin-wave excitations.

M. Eibschutz: PAC measurements are done by substituting an impurity in metal site, e. g. in layer compounds. How would this impurity influence the observation of phasons?

A. Overhauser: The influence of phasons on PAC measurements would depend on the phase correlation time relative to the impurity level lifetime. For example, if the phase correlation time were relatively long, then the PAC spectra would manifest CDW broadening of the hyperfine fields; if it were short, "motional" narrowing could occur.

M. Eibschultz: You said that the broadening of the NMR line in the incommensurate CDW of 2H-$NbSe_2$ is explained by a superposition of three CDW's instead of a random distribution of domains each with one CDW. Another source of the broadening could be due to the temperature dependence of the amplitude of the incommensurate CDW. Could you comment on this?

A. Overhauser: The temperature dependence of the broadening arises primarily from the temperature dependence of the CDW amplitude (as you have surmized). It is the line *shape* (of the broadened line) which distinguishes between single-$\vec{Q}$, double-$\vec{Q}$ and triple-$\vec{Q}$ structures.

References

1 Overhauser, A.W. (1962) *Phys Rev.*, **128**, 1437.
2 Overhauser, A.W. (1968) *Phys. Rev.*, **167**, 691.
3 Overhauser, A.W. (1971) *Phys. Rev. B*, **3**, 3173.
4 Wilson, J.A., DiSalvo, F.J., and Mahajan, S. (1975) *Adv. Phys.*, **24**, 117.
5 Follstaedt, D. and Slichter, C.P. (1976) *Phys. Rev. B*, **13**, 1017.
6 Moncton, D.E., Axe, J.D., and DiSalvo, F.J. (1975) *Phys. Rev. Lett.*, **34**, 734.
7 Berthier, C., Jerome, D., Moline, P., and Rouxel, J. (1976) *Solid State Commun.*, **19**, 131.
8 Stiles, J.A.R. and Williams, D.L. (1976) *J. Phys. C: Solid State Phys.*, **9**, 3941.
9 Borsa, F., Torgeson, D.R., and Shanks, H.R. (1977) *Phys. Rev. B*, **15**, 10.
10 Wilson, J.A., DiSalvo, F.J., and Mahajan, S. (1974) *Phys. Rev. Lett.*, **32**, 882.
11 Lynn, J.W., Iizumi, M., Shirane, G., Werner, S.A., and Saillant, R.B. (1975) *Phys. Rev. B*, **12**, 1154.
12 Pines, D. and Slichter, C.P. (1955) *Phys. Rev.*, **100**, 1014.
13 Overhauser, A.W. and Butler, N.R. (1976) *Phys. Rev. B*, **14**, 3371.

14 Overhauser, A.W. and de Graaf, A.M. (1968) *Phys. Rev.*, **168**, 763.

15 Taub, H., Schmidt, R.L., Maxfield, B.W., and Bowers, R. (1971) *Phys. Rev. B*, **4**, 1134.

16 Shoenberg, D. and Stiles, P.J. (1964) *Proc. Roy. Soc. A*, **281**, 62.

17 Holroyd, F.W. and Datars, W.R. (1975) *Can. J. Phys.*, **53**, 2517.

18 Schaefer, J.A. and Marcus, J.A. (1971) *Phys. Rev. Lett.*, **27**, 935.

19 Bishop, M.F. and Overhauser, A.W. to be published.

20 Overhauser, A.W. (1964) *Phys. Rev. Lett.*, **13**, 190.

Reprint 17 Frictional Force on a Drifting Charge-Density Wave[1)]

M.L. Boriack and A.W. Overhauser**

* Department of Physics, Purdue University, 525, Northwestern Avenue, West Lafayette, Indiana 47907, USA

Received 19 September 1977

We study the effects of impurity scattering on the charge-density-wave (CDW) drift velocity and the electron drift velocity for a three-dimensional system containing an incommensurate unpinned CDW. For the case of isotropic scattering by impurities, we find that the steady-state CDW drift velocity is about 9% of the electron steady-state drift velocity in the limit $G/E_F \ll 1$, where G is the CDW energy gap and E_F is the Fermi energy. For $G/E_F = 0.4$, the steady-state CDW drift velocity is about 22% of the steady-state electron drift velocity.

17.1 Introduction

In previous papers [1, 2] we have discussed the dynamic effects of electric and magnetic fields on the drift velocity of the charge-density wave (CDW) and the electron drift velocity in a three-dimensional system containing an unpinned, incommensurate CDW. In the first of these papers [1], we presented without proof a set of equations of motion for the CDW drift velocity, the electron drift velocity, and the total electric current in the presence of electric and magnetic fields and scattering processes. We have previously presented microscopic discussions of the acceleration of the CDW and the electrons in electric [1] and magnetic fields [2].

Our purpose in this paper is to discuss the effect of scattering processes which cause the CDW and the electrons to acquire finite terminal drift velocities. (In general, the electron drift velocity will be unequal to the CDW drift velocity.) though expressions derived in the paper are for the three-dimensional case, the formalism is applicable also in two dimensions, for which the density of states at the Fermi energy is also finite. This work does not apply to one-dimensional CDW systems, where the density of states at the Fermi energy is zero (at zero temperature).

To facilitate discussion, we present here the results to be derived in this paper so that they may be kept in mind throughout. We take the CDW wave vector $\vec{Q}$ to be in the $\hat{x}$ direction. The rate of change of the CDW drift velocity $\vec{D}$ (symbolic of "drift") due to the frictional force arising from scattering processes is

$$\left(\frac{dD}{dt}\right)_s = -\frac{(D-\beta K_x)}{\tau_D}\,, \tag{17.1}$$

where $\vec{D}$ is along $\vec{Q}$. τ_D is the relaxation time for the CDW drift velocity. K_x is the $\hat{x}$ component of the electron "quasivelocity" $\vec{K}$ which is related to the average wave vector $\langle\vec{k}\rangle_{\text{av}}$ of the electron distribution and the electron mass m by

$$\vec{K} \equiv \hbar\langle\vec{k}\rangle_{\text{av}}/m\,. \tag{17.2}$$

β describes the way a CDW is pulled along by an electric current.

1) Phys. Rev. B 17, 6 (1978)

Anomalous Effects in Simple Metals. Albert Overhauser

ISBN: 978-3-527-40859-7

The rates of change due to scattering of the components of the electron quasivelocity parallel (K_x) and perpendicular (K_y) to are

$$\left(\frac{dK_x}{dt}\right)_s = \frac{-(K_x - \alpha D)}{\tau_x}\,,$$
$$\left(\frac{dK_y}{dt}\right)_s = \frac{-K_y}{\tau_y}\,. \tag{17.3}$$

τ_x and τ_y are the relaxation times for the components of electron quasivelocity parallel (τ_x) and perpendicular (τ_y) to $\vec{Q}$. α describes the way the electrons are pulled along by the drifting CDW. This is the analogy of the acoustoelectric effect for CDW's.

The mean group velocity $\vec{V}$ for the electron distribution is related to $\vec{K}$ and $\vec{D}$ by

$$\vec{V} = (l - \gamma)\vec{K} + \gamma\vec{D}\,, \tag{17.4}$$

where γ is a constant discussed in a previous paper [1]. The total electric current is $\vec{J} = -ne\vec{V}$, where n is the number of electrons per unit volume.

Because our purpose is not to calculate transport coefficients with great accuracy but rather to discuss the physics leading to the form of Equations (17.1) and (17.3), we will treat the case of isotropic scattering. More specifically, we take the scattering cross section for an event $\vec{k} - \vec{k} + \vec{q}$ to be constant for $0 \leq q \lesssim 2k_F$. We believe that most of the essential physics can be discussed in terms of this model, and the mathematical simplicity makes for a much clearer argument.

There is, however, one important result which is greatly underestimated when the impurity potentials are treated in this manner. As mentioned above, isotropic scattering weights all the scattering events with different $\vec{q}$ equally so that the resistivity is very nearly isotropic. A more realistic potential would give much greater weight to the "small-$\vec{q}$" events and can lead to a ratio of resistivity parallel to perpendicular of as much as 5 to 1 [3].

In Section 17.2 we discuss the electron equilibrium distribution in the presence of a moving CDW. The electron relaxation time is derived in Section 17.3. Section 17.4 contains the discussion of the frictional effects of scattering on the CDW drift velocity.

17.2 Equilibrium Electron Distribution

For our jellium model in the presence of a CDW, the one-electron Hamiltonian is of the form

$$\mathcal{H} = p^2/2m + G\cos\vec{Q}\cdot\vec{r}\,, \tag{17.5}$$

where $G\cos\vec{Q}\cdot\vec{r}$ is the totally self-consistent one-electron potential energy. This potential leads to a periodic variation in electron density and gaps in the electron energy spectrum at $\vec{k} = \pm\frac{1}{2}\vec{Q}$. There is an accompanying variation in the density of the positively charged background to preserve charge neutrality.

As discussed in previous work [1], approximate perturbative solutions which are sufficiently accurate for our purposes can be written down by dividing the potential into two parts, one which leads to the gap at $\vec{k} = \frac{1}{2}\vec{Q}$ and the other which leads to the gap at $\vec{k} = -\frac{1}{2}\vec{Q}$. Each part is treated accurately near the gap since the wave functions and energy spectrum are significantly altered there. Away from the gaps the solutions reduce to those found by first-order nondegenerate perturbation theory.

For $\vec{Q}$ in the $\hat{x}$ direction the energies E_+ and E_- for states above and below the energy gaps are

$$E_-(\vec{k}) = \lambda_- + G^2/4(\epsilon_{\vec{k}} - \epsilon_{\vec{k}+\vec{Q}}) ,$$

$$E_+(\vec{k}) = \lambda_+ + G^2/4(\epsilon_{\vec{k}} - \epsilon_{\vec{k}-2\vec{Q}}) , \tag{17.6}$$

for $k_x > 0$. $\lambda_\pm$ is given by

$$\lambda_\pm = \frac{1}{2}(\epsilon_{\vec{k}} - \epsilon_{\vec{k}-\vec{Q}}) \pm \frac{1}{2}\left[(\epsilon_{\vec{k}} - \epsilon_{\vec{k}-\vec{Q}})^2 + G^2\right]^{1/2} , \tag{17.7}$$

for $k_x > 0$. For $k_x < 0$ the energies are

$$E_-(\vec{k}) = \lambda_- + G^2/4(\epsilon_{\vec{k}} - \epsilon_{\vec{k}-\vec{Q}}) ,$$

$$E_+(\vec{k}) = \lambda_+ + G^2/4(\epsilon_{\vec{k}} - \epsilon_{\vec{k}+2\vec{Q}}) , \tag{17.8}$$

where for $k_x < 0$

$$\lambda_\pm = \frac{1}{2}(\epsilon_{\vec{k}} - \epsilon_{\vec{k}+\vec{Q}}) \pm \frac{1}{2}\left[(\epsilon_{\vec{k}} - \epsilon_{\vec{k}+\vec{Q}})^2 + G^2\right]^{1/2} . \tag{17.9}$$

A state $\varphi_{\vec{k}}$ below the gap can be written

$$\varphi_{\vec{k}} = Ae^{i\vec{k}\cdot\vec{r}} + Be^{i(\vec{k}-\vec{Q})\cdot\vec{r}} + Ce^{i(\vec{k}+\vec{Q})\cdot\vec{r}} , \tag{17.10}$$

where for $k_x > 0$,

$$A = \cos\xi , \quad B = -\sin\xi , \quad C = G/2(\epsilon_{\vec{k}} - \epsilon_{\vec{k}+\vec{Q}}) . \tag{17.11}$$

The coefficients $\cos\xi$, $\sin\xi$ are related by

$$\sin 2\xi = G/(\lambda_+ - \lambda_-) = G/W . \tag{17.12}$$

In Equation (17.11) $\lambda_\pm$ is taken from Equation (17.7). For $k_x < 0$ the coefficients are

$$A = \cos\xi , \quad B = G/2(\epsilon_{\vec{k}} - \epsilon_{\vec{k}-\vec{Q}}) , \quad C = -\sin\xi . \tag{17.13}$$

$\cos\xi$ and $\sin\xi$ are given by Equation (17.12) with Equation (17.9) for $\lambda_\pm$ when $k_x < 0$.

A state $\Psi_{\vec{k}}$ above the energy gap is

$$\Psi_{\vec{k}} = Fe^{i\vec{k}\cdot\vec{r}} + He^{i(\vec{k}-\vec{Q})\cdot\vec{r}} + Ie^{i(\vec{k}+\vec{Q})\cdot\vec{r}} , \tag{17.14}$$

where for $k_x > 0$

$$F = \sin\xi , \quad H = \cos\xi , \quad I = 0 . \tag{17.15}$$

For $k_x < 0$ the coefficients are

$$F = \sin\xi , \quad H = 0 , \quad I = \cos\xi . \tag{17.16}$$

In Equation (17.14) we have neglected terms resulting from the mixing of $\vec{k}$ with $\vec{k} \pm 2\vec{Q}$ since these lead to higher-order terms in the final answer. For simplicity we shall discuss here only the case of a simply connected Fermi surface so that only states below the gaps are filled at zero temperature.

In writing down Equations (17.5)–(17.16) we have tacitly assumed that the CDW is at rest. There are two equivalent ways to treat the effects of impurity scattering in a system with a moving CDW. One way is to work in the laboratory frame of reference in which the one-electron potential in Equation (17.4) would become $G\cos[\vec{Q}\cdot(\vec{r}-\vec{D}t)]$ for a CDW with velocity $\vec{D}$. It is possible to find "stationary states" and energies analogous to Equations (17.6)–(17.16) with this time-dependent potential [4]. We could then consider collisions with static impurities. Alternatively, we can work in the reference frame in which the CDW has zero velocity and consider scattering from impurities which are moving with uniform velocity $-\vec{D}$. (Note that we do not make the assumption that the electrons relax to the velocity $-\vec{D}$.) Since the second method is somewhat tidier, it shall be presented here.

The equilibrium distribution for the electrons is that for which the time rate of change of the distribution function $f(\vec{k})$ due to collisions is zero. Since we are interested only in the drift velocity of the equilibrium distribution (which arises due to a uniform shift of the distribution in $\vec{k}$-space), it is sufficient to find the distribution for which the time rate of change of the electron quasi-velocity $\vec{K}$ (or average wave vector) due to collisions with the moving scattering centers is zero. Since $\vec{Q}$ is parallel to $\hat{x}$, $\vec{D}$ is also along $\hat{x}$. By symmetry the resulting shift of the electron distribution must be in the $\hat{x}$ direction. Thus, it suffices to find the shift of the distribution such that the $\hat{x}$ component of the electron quasivelocity $\vec{K}$ does not change with time due to collisions.

These requirements lead to an equation satisfied by $f_{\text{eq}}(\vec{k})$,

$$\int d^3k \int d^3k'[f_{\text{eq}}(\vec{k}) - f_{\text{eq}}(\vec{k}')]k_x\, w(\vec{k},\vec{k}') = 0\,. \tag{17.17}$$

$w(\vec{k},\vec{k}')$ is the golden-rule transition rate for an electron to go from the state with wave vector $\vec{k}$ to one with wave vector $\vec{k}'$.

As discussed above, we have chosen to treat the case of an isotropic scattering cross section. The easiest way to introduce this isotropic scattering is to pretend that the impurities are represented by δ functions in coordinate space. In this case the scattering potential V is given by

$$V = V_0 \sum_{j=1}^{N} \delta(\vec{r} - \vec{r}_j + \vec{D}t)\,, \tag{17.18}$$

where $\vec{r}$, is the position of scattering center j at time $t = 0$. The "strength" of the potential is V_0.

For two states $\varphi_{\vec{k}}$ and $\varphi_{\vec{k}'}$ as in Equation (17.10), the matrix element $\langle\varphi_{\vec{k}'}|V|\varphi_{\vec{k}}\rangle$ is

$$\begin{aligned}\langle\varphi_{\vec{k}'}|V|\varphi_{\vec{k}}\rangle = \sum_{j=1}^{N} e^{i(E'_- - E_-)t/\hbar} e^{i(\vec{k}'-\vec{k})\cdot(\vec{r}_j-\vec{D}t)} \Big[& AA' + BB' + CC' \\ & + (AB' + CA')e^{i\vec{Q}\cdot(\vec{r}_j-\vec{D}t)} + (AC' + BA')e^{-i\vec{Q}\cdot(\vec{r}_j-\vec{D}t)} \\ & + BC'e^{-2i\vec{Q}\cdot(\vec{r}_j-\vec{D}t)} + CB'e^{2i\vec{Q}\cdot(\vec{r}_j-\vec{D}t)}\Big]\,.\end{aligned} \tag{17.19}$$

The golden-rule transition rate $w(\vec{k}, \vec{k}')$ is

$$w(\vec{k}, \vec{k}') = (2\pi N V_0^2/\hbar)\big[(AA' + BB' + CC')^2\delta(\Delta) + (AB' + CA') \times \delta(\Delta - \hbar\vec{Q}\cdot\vec{D}) + (AC' + BA')^2\delta(\Delta + \hbar\vec{Q}\cdot\vec{D}) + (BC')^2 \times \delta(\Delta + 2\hbar\vec{Q}\cdot\vec{D}) + (CB')^2\delta(\Delta - 2\hbar\vec{Q}\cdot\vec{D})\big]\,, \tag{17.20}$$

where

$$\Delta = E'_- - E_- + \hbar(\vec{k}' - \vec{k})\cdot\vec{D}\,. \tag{17.21}$$

For a gas of free electrons the expression in brackets in Equation (17.20) would be simply $\delta(\Delta)$. The additional terms in Equation (17.20) arise since CDW states are mixture of plane-wave states $\vec{k}$ and $\vec{k} \pm \vec{Q}$. Depending on $\vec{k}$ and $\vec{k}'$, the final state can be higher or lower in energy than the initial state by an amount proportional to $\vec{D}$. Since we shall later take the zero-temperature limit and have chosen to discuss the case of a simply connected Fermi surface, we will not carry along the expression similar to Equation (17.20) for states above the gaps.

For $\vec{Q}$ in the $\hat{x}$ direction, any drift velocity associated with the equilibrium distribution will also be along the $\hat{x}$ direction. In the frame of reference in which the scattering centers move with velocity $-D\hat{x}$, the equilibrium distribution for a gas of free electrons would have a drift velocity $-D\hat{x}$ and be centered at $-mD\hat{x}/\hbar$ in $\vec{k}$ space. We expect this result to be modified in the presence of a CDW, and we take the distribution to have a quasivelocity $-(1-\alpha)D\hat{x}$. The problem, then, is to find the value for α which satisfies Equation (17.17).

We write the equilibrium electron distribution function as

$$f_{\text{eq}}(\vec{k}) = (\exp\{[E_- + (1-\alpha)p_x D - E_{\text{F}}]/k_{\text{B}}T\} + 1)^{-1} = f_0(\vec{k}) + \frac{\partial f_0}{\partial E_-}(1-\alpha)p_x D \tag{17.22}$$

to first order in D. $f_0(\vec{k})$ is the Fermi–Dirac distribution function centered at the origin in $\vec{k}$ space,

$$f_0(\vec{k}) = \{\exp[(E_- - E_{\text{F}})/k_{\text{B}}T] + 1\}^{-1}\,. \tag{17.23}$$

k_{B} is the Boltzmann constant. p_x is the $\hat{x}$ component of the one-electron momentum operator $\vec{p} = -i\hbar\vec{\nabla}$.

To perform the integration in Equation (17.17) over the primed variables, it is convenient to use cylindrical coordinates such that

$$\int d^3k' = \int d\varphi' \int dk'_\perp k'_\perp \int dk'_x\,, \tag{17.24}$$

where $k'^2_\perp = k_y^2 + k_g^2$. The integration over φ' is trivial. The integration over $k'_\perp$ involves the δ function by bringing out a factor $m/\hbar^2 k'_\perp$ and limiting the range of the integration over k'_x

It is not possible to solve analytically for these limits of integration using the energy E_- from Equation (17.8). However, using the limits determined with Equation (17.8) rather than the free-electron energy gives corrections to the final answer which are one order higher in G/E_{F} and will be neglected here for the sake of obtaining an analytic expression. Furthermore, keeping terms of order D in the limits of integration gives corrections to α of order D which are negligible in determining the drift velocity of the equilibrium distribution (which is already of order D). Thus, the limits on the $k'_\perp$ integration become $-k$ to k.

For the terms in Equation (17.20) with $\delta(\Delta - \hbar\vec{Q}\cdot\vec{D})$, for example, the expression $[f_{eq}(\vec{k}) - f_{eq}(\vec{k}')]$ becomes

$$f_{eq}(\vec{k}) - f_{eq}(\vec{k}') = \frac{\partial f_0}{\partial E_-}\left[(1-\alpha)(p_x - p_x')D + \hbar(k_x' - k_x - Q)D\right] \tag{17.25}$$

Using this result in Equation (17.17) obtains

$$\begin{aligned}
&\int d^3k \int_{-k}^{k} dk_x k_x \frac{\partial f_0}{\partial E_-}\Big\{\left[(1-\alpha)(p_x - p_x') + \hbar(k_x' - k_x)\right] \\
&\quad \times [(AA' + BB' + CC')^2 + (AC' + BA')^2 + (AB' + CA')^2 + (CB')^2 \\
&\quad + (BC')^2] + \hbar Q[(AC' + BA')^2 - (AB' + CA')^2 - 2(CB')^2 \\
&\quad + 2(BC')^2]\Big\} = 0\,.
\end{aligned} \tag{17.26}$$

It is convenient to make the substitutions $k_x \to k_x$ and $k_x' \to -k_x'$ where appropriate so that all the integrations are over the region with $k_x k_x' > 0$. Note that under $k_x \to -k_x$ the coefficients A, B, and C obey the relations $A \to A$, $B \to C$, $C \to B$. The expression for α becomes

$$\alpha = \frac{\int_{k_x>0} d^3k \int_0^k dk_x' k_x \frac{\partial f_0}{\partial E_-}[(p_x - \hbar k_x)X_1 + \hbar Q X_2]}{\int_{k_x>0} d^3k \int_0^k dk_x' k_x p_x \frac{\partial f_0}{\partial E_-} X_1}\,, \tag{17.27}$$

where

$$\begin{aligned}
X_1 &\equiv (AA' + BB' + CC') + (AA' + BC' + CB')^2 + (AC' + BA')^2 \\
&\quad + (AB' + CA')^2 + (CB')^2 + (BC')^2 + (BB')^2 + (CC')^2 \\
&\quad + (AB' + BA')^2 + (AC' + CA')^2\,, \\
X_2 &\equiv (AC' + BA')^2 + (AB' + BA')^2 - (AB' + CA')^2 - (AC' + CA')^2 \\
&\quad - 2(CB')^2 - 2(CC')^2 + 2(BC')^2 + 2(BB')^2\,.
\end{aligned} \tag{17.28}$$

Using Equations (17.11) and (17.12) and the relation $p_x \cong \hbar k_x - \hbar Q \sin^2\xi$ for the wave function in Equation (17.10), the lowest-order term in the numerator of Equation (17.27) is of order $(G/E_F)^2$ The lowest-order term in the denominator of Equation (17.27) is of order $(G/E_F)^0$. Thus $\alpha \sim (G/E_F)^2$ plus higher-order terms in G/E_F

In the zero-temperature limit, the derivative of the Fermi function can be replaced by a δ function of energy, i. e., $\partial f_0/\partial E_- \to -\delta(E_- - E_F)$. By Equation (17.8) the Fermi surface has cylindrical symmetry about k_x It is convenient to make a change in the variable of integration $d^3k \to dS\,dE_-/|\nabla_k E_-|$, where S is a surface of constant energy. In our case this reduces to $d^3k \to (2\pi m/\hbar^2)dk_x\,dE_-$. Integrating over E_- is trivial with the result that the remaining integrations are over the Fermi surface. The resulting expression for α is

$$\alpha = \frac{\int_0^{k_{xm}} dk_x \int_0^k dk_x' k_x \hbar Q(X_2 - X_1\sin^2\xi)}{\int_0^{k_{xm}} dk_x \int_0^k dk_x' p_x k_x X_1}\,, \tag{17.29}$$

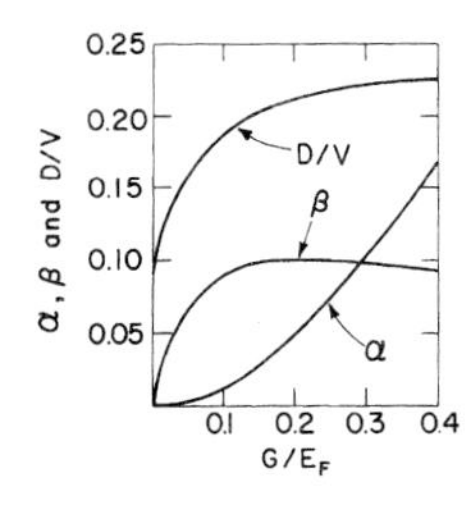

Figure 17.1 α, β, and D/V vs. G/E_F. G is the CDW energy gap and E_F is the Fermi energy, α describes the way in which the electrons are pulled along by a drifting CDW. This is the analogy of the acoustoelectric effect for CDW's. β describes the way a CDW is pulled along by the electric current. D is the steady-state CDW drift velocity and V is the steady-state electron drift velocity.

where k lies on the Fermi surface and k_{xm} is the maximum value for k_x

As in previous work [1, 2] we evaluate α for the Fermi surface in critical contact [5] with the energy gaps. A plot of α vs. G/E_F is shown in Figure 17.1. It is interesting to note from Equations (17.28) and (17.29) that α is a function of G/E_F and does not depend on the strength of the (nonzero) scattering potential. Thus, the strength of the scattering potential does not play a part in determining the (mean wave vector of the) equilibrium distribution of the electrons, but, as will be discovered in Section 17.3, the strength of the scattering potential does appear in the rate at which the electron distribution approaches equilibrium.

17.3 Electron Relaxation Time

In Section 17.2, we determined (the first moment of) the electron equilibrium distribution. We now turn our attention to finding the rate at which the electron distribution comes to equilibrium. As in Section 17.2, we are interested only in the rate of change of the first moment of the electron distribution due to scattering. Therefore, we wish to determine $(d\vec{K}/dt)_s$, where $\vec{K}$ is the quasi-velocity of the electron distribution defined in Equation (17.2).

The rate of change of the electron quasivelocity $\vec{K}$ due to scattering is

$$\left(\frac{d\vec{K}}{dt}\right)_s = \frac{\hbar}{nm}\int \frac{d^3k}{4\pi^3}\int \frac{d^3k'}{4\pi^3}[f(\vec{k}) - f(\vec{k}')]\vec{k}\,w_{\vec{k}\vec{k}'}\,, \tag{17.30}$$

where $w_{\vec{k}\vec{k}}$ is the golden-rule transition rate from Equation (17.21) and n is the number of electrons per unit volume.

In Section 17.2 we demonstrated that for $f_{\text{eq}}(\vec{k})$ given by Equation (17.22), the rate of change of the electron quasi velocity $\vec{K}$ due to scattering was zero. We can add zero the the right-hand side of Equation (17.30) to get

$$\left(\frac{d\vec{K}}{dt}\right)_s = \frac{\hbar}{mn}\int \frac{d^3k}{4\pi^3}\int \frac{d^3k'}{4\pi^3}[f(\vec{k}) - f_{\text{eq}}(\vec{k}) - f(\vec{k}') + f_{\text{eq}}(\vec{k}')]\vec{k}\,w_{\vec{k}\vec{k}'}\,, \tag{17.31}$$

We take the electrons to have quasivelocity $\vec{K}$ so that the electron distribution function $f(\vec{k})$ is given by

$$f(\vec{k}) = \{\exp[(E - \vec{p}\cdot\vec{K} - E_F)/k_B T] + 1\}^{-1}\,. \tag{17.32}$$

This is just the Fermi distribution shifted to center at $m\vec{K}/\hbar$ in $\vec{k}$ space.

The integration over the primed variable can be done in cylindrical coordinates as in Section 17.2. The expression in square brackets in (17.31) becomes

$$\frac{\partial f_0}{\partial E}(\vec{p}' - \vec{p})\cdots[\vec{K} + (1-\alpha)\vec{D}]\,, \tag{17.33}$$

and $(d\vec{K}/dt)$ is

$$\left(\frac{d\vec{K}}{dt}\right)_s = \frac{NV_0^2}{4\pi^4\hbar^2 n}\int d^3k \int_{-k}^{k} dk' \vec{k}\frac{\partial f_0}{\partial E}(\vec{p}' - \vec{p})\cdot[\vec{K} + (1-\alpha)\vec{D}]$$
$$\times [(AA' + BB'CC')^2 + (AB' + CA')^2 + (AC'BA')^2 + (BC')^2 + (CB')^2] , \quad (17.34)$$

where Equation (17.21) for has been used.

Let us first consider the case where only K_x is nonzero. Making appropriate substitutions so that the integration is over $k_x, k_x' > 0$ obtains

$$\left(\frac{dK_x}{dt}\right)_s = \frac{NV_0^2}{2\pi^4\hbar^2 n}[K_x + (1-\alpha)D] \times \int_{k_x>0} d^3k \int_0^k dk_x' k_x p_x \frac{\partial f_0}{\partial E} X_1 . \quad (17.35)$$

X_1 is given in Equation (17.28). Once again we take the low-temperature limit and the case of a simply connected Fermi surface to get

$$\left(\frac{dK_x}{dt}\right)_s = \frac{-[K_x + (1-\alpha)D]}{\tau_x} , \quad (17.36)$$

where

$$\frac{1}{\tau_x} = \frac{NV_0^2}{\pi^3\hbar^4 n}\int_0^{k_{xm}} dk_x \int_0^k dk_x' k_x p_x X_1 . \quad (17.37)$$

k_{xm} is the maximum value for k_x, and k lies on the Fermi surface.

Similarly, by starting with Equation (17.34) and taking only K_y to be nonzero, an expression for $(dK_y/dt)_s$ can be obtained with the result

$$\left(\frac{dK_x}{dt}\right)_s = \frac{K_y}{\tau_y} , \quad (17.38)$$

where

$$\frac{1}{\tau_y} = \frac{NV_0^2 m}{2\pi^3\hbar^3 n}\int_0^{k_{xm}} dk_x \int_0^k dk_x' \left(k_\perp^2 - k_x^2\right) X_1 , \quad (17.39)$$

$k_\perp$ is the radius of a cross section of the cylindrically symmetric Fermi surface measured from the $\hat{k}_x$ axis.

Both τ_x and τ_y can be evaluated analytically in the limit $G \to 0$ with the results $1/\tau_x = 1/\tau_y = 1/\tau_0$, where

$$1/\tau_0 = 2NV_0^2 m^3 E_F^2/3\pi^3\hbar^7 n . \quad (17.40)$$

The variation of τ_x and τ_y with G/E_F is not strong in this model. For G/E_F =0.4, τ_y is about 12% smaller and τ_x about 10% smaller than their $G = 0$ value.

The near equality of τ_x and τ_y for this model is a consequence of the choice of an isotropic scattering cross section which weights all transitions $\vec{k} \to \vec{k} + \vec{q}$ equally for $0 \le q \lesssim 2k_F$. A more realistic scattering potential, weights the smaller values of $\vec{q}$ much more heavily and can lead to anisotropies in the relaxation times as high as $\tau_x/\tau_y \sim \frac{1}{4}$ for $G/E_F = 0.4$ [3].

17.4 Frictional Effects of Scattering on the CDW Drift Velocity

In previous work [1, 2] we have investigated the dynamic effects of electric and magnetic fields in accelerating the CDW. This acceleration arose because the applied electric and magnetic fields polarize the periodic electron density in a manner analogous to the way electric fields polarize atomic wave functions. The polarized electron charge density is out of phase with that of the positively charged background. Thus, there are nonzero net forces on the ions in the background. The ions move in response to these forces in such a way that the CDW accelerates. As long as there is a net "out-of-phase part" to the electron density the CDW will accelerate.

The purpose of this section is to demonstrate that the presence of scattering also introduces an out-of-phase contribution to the electron charge density and that the acceleration of the CDW by frictional forces arising from scattering processes can be described by Equation (17.1).

The electric field $\vec{\mathcal{E}}$ polarized the electron density by mixing in with a state $\varphi_{\vec{k}}$ an amount of $\Psi_{\vec{k}}$ (with the same $\vec{k}$ in a different band) proportional to $\mathcal{E}$. The expression for the electron density contained cross terms like $\mathcal{E}\varphi^*_{\vec{k}}\Psi_{\vec{k}}$ which led to an out-of-phase electron density proportional to $\mathcal{E}$.

Schematically, in the presence of scattering $\varphi_{\vec{k}}$ is replaced by a perturbed wave function $\varphi'_{\vec{k}} = \varphi_k + V_1\varphi_{\vec{k}'} + V_2\Psi_{\vec{k}'} + V_3^2\Psi_{\vec{k}}$, where V_1, V_2, V_3 are appropriate matrix elements and energy denominators. The electron density $|\varphi_{\vec{k}}|^2$ contains out-of-phase terms arising from the cross terms $V_1V_2\varphi^*_{\vec{k}'}\Psi_{\vec{k}'}$, $V_3^2\varphi^*_{\vec{k}}$, $\Psi_{\vec{k}}$ etc. The cross terms, such as $V_1\varphi_{\vec{k}'}\varphi^*_{\vec{k}}$, proportional to odd powers of the scattering potential are zero for a random distribution of impurities.

As in Section 17.2 we choose to work in the reference frame where the CDW is instantaneously at rest. The scattering potential is given by Equation (17.18). It is a lengthy, though not difficult, procedure to use time-dependent perturbation theory to write down the perturbed wave function $\varphi'_{\vec{k}}$ (for a state below the energy gaps) correct to second order in the perturbation.

As discussed schematically above, the out-of-phase electron density $(\Delta n_k)_s \sin\vec{Q}\cdot\vec{r}$ for the state $\varphi_{\vec{k}}$ due to the scattering potential is found from

$$\begin{aligned}(\Delta n_k)_s \sin\vec{Q}\cdot\vec{r} &= |\varphi'_{\vec{k}}|^2 - |\varphi_{\vec{k}}|^2 \\ &= [(\Delta n_k)_1 + (\Delta n_k)_2]\sin\vec{Q}\cdot\vec{r})\,,\end{aligned} \tag{17.41}$$

$(\Delta n_k)_1$ results from the cross term between $\varphi_{\vec{k}}$, and which are each mixed in to first order in V_0. $(\Delta n_k)_2$ results from the second order mixing-in of $\Psi_{\vec{k}}$. Any terms in the electron density linear in V_0 vanish for a random distribution of impurities.

The expressions for $(\Delta n_k)_1$ and $(\Delta n_k)_2$ are

$$\begin{aligned}(\Delta n_{\vec{k}})_1 = -2\pi N V_0^2 &\int \frac{d^3k'}{4\pi^3}\frac{1}{W'}[(AE' + BH' + CI') \\ &\times (AA' + BB' + CC')\delta(\Delta) + (AI' + BF')(AC' + BA') \\ &\times \delta(\Delta - \hbar QD) + (CF' + AH')(CA' + AB')\delta(\Delta + \hbar QD) \\ &+ B^2C'I'\delta(\Delta - 2\hbar QD) + C^2B'H'\delta(\Delta + 2\hbar QD)] \\ &\times (A'I' + B'F' - A'H' - C'F)\,,\end{aligned} \tag{17.42}$$

$$
\begin{aligned}
(\Delta n_{\vec{k}})_2 = {} & -2\pi N V_0^2 (AF - AE + BD - CD) \int \frac{d^3 k'}{4\pi^3} \\
& \times [(AA' + BB' + CC')(FA' + HB' + IC')\delta(\Delta) \\
& + (AC' + BA')(FC' + HA')\delta(\Delta - \hbar QD) + (AB' + CA') \\
& \times (FB' + IA')\delta(\Delta + \hbar QD) + BHC'^2 \delta(\Delta - 2\hbar QD) \\
& + CIB'^2 \delta(\Delta + 2\hbar QD)] \, .
\end{aligned} \tag{17.43}
$$

Δ is given by Equation (17.21). In Equations (17.42) and (17.43) we have assumed that the corresponding δ functions with E'_+ for a state above the gap are never satisfied. This is equivalent to taking $\hbar QD \ll G$.

To determine the total out-of-phase electron density $\Delta N_s \sin \vec{Q} \cdot \vec{r}$ induced by the scattering potential, it is necessary to sum up the $(\Delta n_k)_s$, for all states $\vec{k}$ weighted by the occupation probability. As before we are interested in the case of a simply connected Fermi surface so that only states below the CDW energy gaps are filled at $T = 0\,°\mathrm{K}$. We take the electron distribution to be centered $m K_x / \hbar$ in $\vec{k}$ space so that it is given by Equation (17.32). The integrals over the δ functions can be handled in cylindrical coordinates as in Section 17.2. However, it is convenient here to perform the integration over $k_\perp$ first and then to change variables so that k_x, k'_x integrations are over the regions $k_x, k'_x \geq 0$. The resulting expression for $\Delta N_1 \equiv \sum_{\vec{k}} f(\vec{k})(\Delta n_{\vec{k}})_1$ is

$$
\begin{aligned}
\Delta N_1 = {} & \frac{N V_0^2 m D}{4\pi^4 \hbar} \int\limits_{k'_x > 0} d^3 k' \int\limits_0^{k'} d k_x \frac{1}{W'} \frac{\partial f_0}{\partial E'_-} \\
& \times \left[2k'_z \cos 2\theta' \sin 2\theta - Q(2 \sin 2\theta' \sin 2\theta \cos 2\theta') \right] ,
\end{aligned} \tag{17.44}
$$

where only terms multiplying D which contribute up to order G/E_F have been kept.

Similarly, we find for $\Delta N_2 \equiv \sum_k f(\vec{k})(\Delta n_{\vec{k}})_2$ the expression

$$
\begin{aligned}
\Delta N_2 = {} & -\left(\frac{N V_0^2 m}{4\pi^4 \hbar^2} \right) \int\limits_{k'_x > 0} d^3 k' \int\limits_0^{k'} d k_x \frac{1}{W} \frac{\partial f}{\partial E'_-} \\
& \times \left\{ \hbar D Q \sin 2\theta + [\hbar D (Q - 2k_x) - 2 p_x K] \cos 2\theta \sin 2\theta' \right\} ,
\end{aligned} \tag{17.45}
$$

where terms of order higher than G/E_F have been dropped.

As discussed in detail in previous work [1], there will be a force due to ΔN_s on the positively charged ion background which is easily found using Poisson's equation. This leads to an expression for the time rate of change of the CDW drift velocity D due to the frictional force of scattering processes, where

$$
\left(\frac{dD}{dt} \right)_s = \frac{-(4\pi e^2 \Delta N_s / M P \epsilon_1 Q)}{1 + 8\pi \hbar^2 e^2 G \sum_{\vec{k}} \frac{1/W^3}{M p \epsilon_1}} \equiv -S \Delta N_s \, . \tag{17.46}
$$

ϵ_1 is the electron-gas dielectric function for wave vector Q and p is the fractional modulation of the electron density. The second term in the denominator arises from a self-consistent treatment of the accelerating ions in the positively charged background. (Both of these points are discussed in detail in previous work [1].)

It remains to put these results in the form of Equation (17.1). Remembering that we have been working in the instantaneous rest frame of the CDW we must transform the electron quasivelocity K_x in Equation (17.45) to the laboratory frame by letting $K_x \to K_x - D$. Then

Equation (17.46) can be written

$$\left(\frac{dD}{dt}\right)_s = \frac{-(D - \beta K_x)}{\tau_D}\,, \tag{17.47}$$

where

$$\frac{1}{\tau_D} = \frac{S N V_0^2 m}{4\pi^4 \hbar} \int_{k'_x>0} d^3k' \int_0^{k'} dk_x \frac{\partial f}{\partial E'_-} \times \left\{ \frac{1}{W'}\left[(2k'_x - Q)\cos 2\theta' \sin 2\theta - 2Q \sin 2\theta'\right] - \frac{1}{W}\left[2Q \sin 2\theta + \left(Q - 2k_x + 2\frac{p_x}{\hbar}\right)\cos 2\theta \sin 2\theta'\right]\right\} \tag{17.48}$$

and

$$\beta = -\left(\frac{S\tau_d N V_0^2 m}{2\pi^4\hbar^2}\right) \int_{k'_x>0} d^3k' \int_0^{k'} dk_x \frac{\partial f}{\partial E'_-}\frac{1}{W} P_x \cos 2\theta \sin 2\theta'\,, \tag{17.49}$$

As before we are interested here in the case of a simply connected (cylindrically symmetric) Fermi surface. Treating the $\vec{k}'$ integration as in Section 17.2 obtains

$$\frac{1}{\tau_D} = \frac{S N V_0^2 m}{2\pi^3 \hbar^3} \int_0^{k'_{xm}} dk'_x \int_0^{k'} dk_x \times \left\{ \frac{1}{W}\left[2Q \sin 2\theta + \left(Q - 2k_x + 2\frac{p_x}{\hbar}\right)\cos 2\theta \sin 2\theta'\right] + \frac{1}{W'}\left[2Q \sin 2\theta' + (Q - 2k'_x)\cos 2\theta' \sin 2\theta\right]\right\}, \tag{17.50}$$

and

$$\beta = \frac{S\tau_d N V_0^2 m^2}{\pi^3\hbar^4} \int_0^{k'_{xm}} dk'_x \int_0^{k'} dk_x \frac{P_x \cos 2\theta \sin 2\theta'}{W}\,. \tag{17.51}$$

Equation (17.50) can be readily evaluated analytically in the limit $G \to 0$ for the case of the lemon-shaped Fermi surface in critical contact with the energy gaps. The result in this limit is

$$1/\tau_D^0 = N V_0^2 m Q/\hbar^3 \left(\frac{1}{2}\pi - 1\right) \tag{17.52}$$

where the factor of $(\frac{1}{2}\pi - 1)$ comes from the integration of $1/W^3$ over the volume bounded by the lemon-shaped Fermi surface.

To investigate the behavior of τ_D with G/E_F, it is useful to put Equation (17.50) in the form

$$1/\tau_D = (1/\tau_D^0)\,(\Delta N_s/\Delta N_s^0)\,(m/m^*)\,. \tag{17.53}$$

ΔN_s^0 is the value of ΔN_s evaluated in the limit $G/E_F \to 0$. m^* is the effective mass of the CDW which has been discussed in previous work [1]. The variation of $\Delta N_s/\Delta N_s^0$ with G/E_F is slow and $\Delta N_s/\Delta N_s^0 \cong 1.1$ for $G/E_F = 0.4$. m^* is equal to m at $G/E_F = 0$. The dependence of m^* on G/E_F is roughly linear with a large slope so that $m^* \cong 10^4\, m$ at $G/E_F = 0.1$. Thus, the dependence of τ_D on G/E_F is essentially that of m^*.

It is possible to find an expression for β to first order in G/E_F with the result

$$\beta \cong \frac{2mG}{\pi\hbar^2 Q^2} \ln \frac{\hbar^2 Q^2}{mG} \left(\ln \frac{\hbar^2 Q^2}{mG} - 2 \right) . \tag{17.54}$$

Thus $\beta \to 0$ as $G/E_F \to 0$. A plot of β vs. G/E_F is shown in Figure 17.1 for the lemon-shaped Fermi surface in critical contact with the energy gaps.

It is of interest to compare the terminal drift velocity of the CDW with that of the electrons when there is an applied electric field. From previous work [1, 2] it follows that for an electric field applied parallel to $\vec{Q}$, the ratio of the steady state drift velocity of the CDW to the electron drift velocity V is

$$\frac{D}{V} = \beta + \frac{m\tau_d}{m^* \tau_x} [(1 - \gamma + \beta\gamma) + (m\tau_D/m^* \tau_x)(\alpha + \gamma - \alpha\beta)] . \tag{17.55}$$

m^* is the effective mass of the CDW. γ appears because the drift velocity of the electrons is given by Equation (17.4). Both m^* and γ are discussed in detail in previous work [1].

For $G/E_F \to 0$ we have shown $\alpha \to 0$, $\beta \to 0$ and from previous work [1] $\gamma \to 0$ and $m^* \to m$. Thus, in the limit of $G/E_F \to 0$, $D/V = \tau_D/\tau_s \cong 0.09$. A plot of the variation of D/V with G/E_F is shown in Figure 17.1. $D/V \cong 0.225$ for $G/E_F = 0.4$. Figure 17.1 and Equation (17.4) clearly show that the major variation of D/V with G/E_F is caused by the dependence of β. Since τ_D is proportional to m^*, $m\tau_D/m^*\tau_x$ is almost constant. It must be remembered that the numbers stated here are for the case of isotropic scattering cross sections.

17.5 Conclusions

We have derived equations of motion describing the effects of scattering for the CDW drift velocity and the electron drift velocity in a three-dimensional system containing an unpinned, incommensurate CDW. For the case of scattering from impurities with isotropic cross sections, we found that in the limit $G/E_F \ll 1$, the steady-state CDW drift velocity is about 9% of the electron drift velocity. For $G/E_F = 0.4$, the steady-state drift velocity of the CDW is about 22% of the electron drift velocity.

McMillan [6] has dealt briefly with the problem of drifting CDW's within the framework of a time-dependent Landau theory. In his discussion he assumed that an (unpinned) CDW drifts with the same velocity as the electrons. The results of our microscopic theory clearly indicate that this assumption is not valid. We find, moreover, that in the absence of scattering, there is no coupling between the drift velocity of the CDW and the electron drift velocity.

Acknowledgements

This research was supported by the NSF under Grant Nos. Gh41884 and MRL program DMR 72-03018A04.

References

1 Boriack, M.L. and Overhauser, A.W. (1977) *Phys. Rev. B*, **12**, 5206.
2 Boriack, M.L. and Overhauser, A.W. (1977) *Phys. Rev. B*, **12**, 5256.
3 Bishop, M.F. and Overhauser, A.W. (1977) *Phys. Rev. Lett.*, **39**, 632.
4 Boriack, M.L. and Overhauser, A.W. (1977) *Phys. Rev. B*, **15**, 2847.
5 Overhauser, A.W. and de Graff, A.M. (1968) *Phys. Rev.*, **168**, 763.
6 McMillan, W.L. (1975) *Phys. Rev. B*, **12**, 1197.

Reprint 18 Attenuation of Phase Excitations in Charge-Density Wave Systems[1)]

M.L. Boriack and A.W. Overhauser**

* Department of Physics, Purdue University, 525, Northwestern Avenue, West Lafayette, Indiana 47907, USA

Received 13 February 1978

The attenuation of collective excitations (phasons) corresponding to phase modulation of a charge-density wave (CDW) caused by electron–phason interaction is studied. Phason attenuation is a nonlocal effect and must be treated microscopically because the pertinent length scale is determined by the CDW wave vector $\vec{Q}$ rather than the phason wave vector $\vec{q}$. In three-dimensional jellium, phasons with $\vec{q}$ parallel to $\vec{Q}$ are predominantly attenuated by scattering electrons (in $\vec{k}$ space) near the CDW energy gaps, and the attenuation rate is independent of temperature. The phason attenuation rate is $\gamma \equiv (1/E)dE/dt$ for a phason with energy E. We find $\gamma \sim q\cos^2\theta$, where θ is the angle between $\vec{q}$ and $\vec{Q}$. For $\vec{q}$ parallel to $\vec{Q}$, γ is approximately 0.3 times the phason frequency; i. e., phasons are underdamped. Phasons with $\vec{q}$ perpendicular to $\vec{Q}$ are not attenuated. If heterodyne gaps (caused by potentials of periodicity $\vec{Q} \pm 2\pi\vec{G}$) cut the Fermi surface, the attenuation is increased by a factor of 3.

18.1 Introduction

In three-dimensional metals exchange interactions and electron–electron correlations can give rise to a charge-density wave (CDW) instability [1] which breaks the translation symmetry of the crystal. In the presence of a CDW the spatial density of conduction electrons is of the form

$$\rho = \rho_0[1 + p\cos(\vec{Q}\cdot\vec{r} + \varphi)] , \tag{18.1}$$

where p is the fractional amplitude of the electron-density modulation, ρ_0 is the mean electron density, $\vec{Q}$ is the CDW wave vector, and φ is the CDW phase. This new periodicity arises in addition to the normal periodicity of the crystal lattice; and since Q is nearly equal to the diameter $2k_F$ of the Fermi surface, the CDW periodicity is, in general, incommensurate with that of the lattice. The resulting structure is multiply periodic, and the crystal no longer has a translation group since no two ions are equivalent.

A CDW instability can occur only if the electronic charge density is locally neutralized by an accompanying lattice distortion [1]. Each positive ion will be displaced from its equilibrium lattice site $\vec{L}$ by

$$\vec{u}(\vec{L}) = \vec{A}\sin(\vec{Q}\cdot\vec{L} + \varphi) . \tag{18.2}$$

Because $\vec{Q}$ is incommensurate with the lattice, the energy of the system must be independent of the spatial position of the CDW as determined by the phase φ. It then follows that there will be low-frequency collective excitations corresponding to φ varying slowly in space and time. These elementary excitations are called phasons [2] and have important consequences for experiments which try to detect a CDW with Bragg diffraction of X-rays, neutrons, or electrons or through Knight shift or hyperfine-field effects [3].

1) Phys. Rev. B 17, 12 (1978)

Anomalous Effects in Simple Metals. Albert Overhauser

ISBN: 978-3-527-40859-7

Because phasons have important experimental consequences, it is of interest to determine their rate of attenuation. Our purpose in this paper is to study the attenuation of phasons caused by electron–phason interaction.

We demonstrate below that the length scale of the electron–phason interaction is determined by the CDW wave vector $\vec{Q}$ and not the phason wave vector $\vec{q}$. Because the wavelength of the interaction is the order of a lattice spacing, the electron mean free path will always be long compared to the wavelength of the interaction. Thus, phason attenuation is a nonlocal effect and is correctly described by "Golden Rule" quantum mechanics.

It is certainly incorrect to use Ohm's law or a transport equation with relaxation to a local equilibrium distribution in calculating the phason attenuation rate. Herein lies a major difference between the attenuation of phasons and the attenuation of long-wavelength acoustical phonons, i. e., ultrasonic attenuation. This difference is made manifest in the following argument. Consider a longitudinal acoustical phonon with wavelength $2\pi/q$ equal to, say, 1000 lattice constants and amplitude A. Let us take a length of 100 lattice constants to define a macroscopic chunk of sample. The ion displacements are $u = A \sin \vec{q} \cdot \vec{r}$. The mean displacement of an ion in a typical "chunk" will be $\sim A$ since the displacements are all, in general, in phase. Equivalently, there is a macroscopic motion of the charge density. Now consider a phason with wavelength $2\pi/q$ equal to 1000 lattice constants. As shown below the displacements of the ions are $u \sim A \sin[(\vec{Q} \pm \vec{q}) \cdot \vec{r}]$. That is, the periodicity of the ion displacements is determined by $Q \gg q$ and is the order of one lattice constant. Since $2\pi/Q$ is much smaller than the length of the 100-lattice-constant chunk, the mean displacement of an ion in the chunk is ~ 0. Thus a long-wavelength phason (unlike a phonon) is not accompanied by a macroscopic motion of the charge density.

McMillan [4] has dealt briefly with the problem of phason attenuation within the context of time-dependent Landau theory. The relation of his work to that in this paper is disucssed in the Appendix.

18.2 Phasons and Electron–Phason interaction

In this section we review some of the properties of phasons assuming negligible damping. A more detailed discussion can be found in a previous work by one of the authors [2]. We also derive the electron–phason interaction.

Consider a single incommensurate CDW as described by Equations (18.1) and (18.2). Since the energy of the CDW is independent of phase φ it follows that there will be low-frequency collective excitations corresponding to φ varying slowly in space and time. We express $\varphi(\vec{L}, t)$ as an expansion in running waves,

$$\varphi(\vec{L}, t) = \sum_{\vec{q}} \varphi_{\vec{q}} \sin(\vec{q} \cdot \vec{L} - \omega_{\vec{q}} t) . \tag{18.3}$$

This approach is analogous to treating lattice dynamics in the continuum approximation.

The phason wave vectors $\{\vec{q}\}$ are assumed small compared to the Brillouin zone. $\{\omega_{\vec{q}}\}$ and $\{\varphi_{\vec{q}}\}$ are the frequencies and amplitudes of the normal modes, respectively.

For simplicity consider only a single phason mode to be excited. The atomic displacements relative to the crystal lattice sites are

$$\vec{u}(\vec{L}, t) = \vec{A} \sin[\vec{Q} \cdot \vec{L} + \varphi_{\vec{q}} \sin(\vec{q} \cdot \vec{L} - \omega t)] . \tag{18.4}$$

For the present study we take the amplitude $\vec{A}$ of the atomic displacements to be parallel to $\vec{Q}$. For small $\varphi_{\vec{q}}$ we can express $\vec{u}$ approximately as

$$\vec{u}(\vec{L}, t) \cong \vec{A} \sin \vec{Q} \cdot \vec{L} + \frac{1}{2} \vec{A} \varphi_{\vec{q}} \sin[(\vec{Q} + \vec{q}) \cdot \vec{L} - \omega t] - \frac{1}{2} \vec{A} \varphi_{\vec{q}} \sin[(\vec{Q} - \vec{q}) \cdot \vec{L} + \omega t] . \quad (18.5)$$

If $\vec{q}$ is perpendicular to $\vec{Q}$, the local direction of the CDW wave vector is slightly rotated. If $\vec{q}$ is parallel to $\vec{Q}$, the local magnitude of the CDW wave vector is periodically modulated. Equation (18.5) also shows the relation between phasons and the phonon modes of the undistorted lattice. The last two terms, which represent phason deviations from the equilibrium CDW state, constitute a coherent superposition of phonon modes with wave vectors $\vec{Q} + \vec{q}$ and $\vec{Q} - \vec{q}$.

The energy density of a phason can be written[2]

$$E_{\vec{q}} = \frac{1}{4} n M A^2 \omega^2 \varphi_{\vec{q}}^2 , \quad (18.6)$$

where n is the number of ions per unit volume and M is the ion mass.

It is expected that $\omega_{\vec{q}}$ vs. $\vec{q}$ will be highly anisotropic because a local rotation of $\vec{Q}$ requires less energy than a change in its magnitude. Consequently the surfaces of constant phason energy will be very flat (pancake-shaped) ellipsoids. The anisotropy could be as high as 100 to 1, especially if the Fermi surface is nearly spherical as is the case for the isotropic metals of primary interest in this work. Denoting components of $\vec{q}$ parallel (longitudinal) and perpendicular (transverse) to $\vec{Q}$ as $q_{\parallel}$ and $q_{\perp}$ we can write $\omega_{\vec{q}}$ as

$$\omega_{\vec{q}}^2 = c_{\perp}^2 q_{\perp}^2 + c_{\parallel}^2 q_{\parallel}^2 , \quad (18.7)$$

where $c_{\perp}$ and $c_{\parallel}$ are the corresponding components of phason velocity and $c_{\parallel}/c_{\perp}$ may be as great as 100. The phason frequency spectrum is plotted schematically in Figure 18.1. Note that the phason spectrum is a continuous function of the angle between $\vec{q}$ and $\vec{Q}$. The terms "longitudinal" and "transverse" are used here merely for convenience in specifying the special cases of $\vec{q}$ parallel and perpendicular to $\vec{Q}$ and do not have the same meaning as when they are applied to phonons.

For the electron density to be modulated as in (18.1) requires that the total self-consistent potential in the one-electron Schrödinger equation be of the form [1]

$$V(\vec{r}) = G \cos(\vec{Q} \cdot \vec{r} + \varphi) . \quad (18.8)$$

Taking a single phason mode $\vec{q}$ to be excited, we have for small $\varphi_{\vec{q}}$

$$V(r) = G \cos[\vec{Q} \cdot \vec{r} + \varphi_{\vec{q}} \sin(\vec{q} \cdot \vec{r} - \omega t)] \cong G \cos \vec{Q} \cdot \vec{r} + \frac{1}{2} \varphi_{\vec{q}} G \cos[(\vec{Q} + \vec{q}) \cdot \vec{r} - \omega t] - \frac{1}{2} \varphi_{\vec{q}} G \cos[(\vec{Q} - \vec{q}) \cdot \vec{r} + \omega t] . \quad (18.9)$$

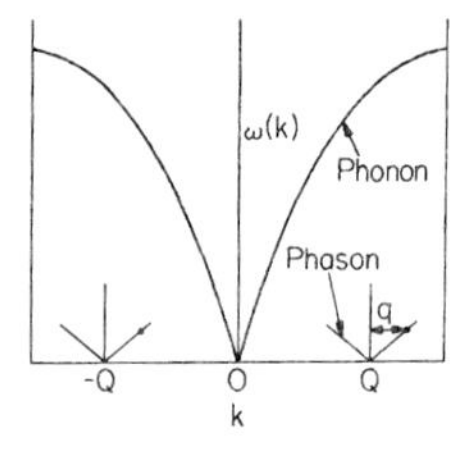

Figure 18.1 Schematic illustration of the vibrational modes in a metal having a CDW structure. The frequency of the phason branch goes to zero at $\vec{Q}$ the location of the CDW satellite reflection in $\vec{k}$ space. Such a diagram has only an approximate meaning since an incommensurate CDW structure does not have a Brillouin zone.

The term $G \cos \vec{Q} \cdot \vec{r}$ is the CDW potential in the absence of phasons. The second and third terms give the electron–phason interaction,

$$V_{e\varphi} = \frac{1}{2}\varphi_{\vec{q}} G \cos[(\vec{Q} + \vec{q}) \cdot \vec{r} - \omega t] - \frac{1}{2}\varphi_{\vec{q}} G \cos[(\vec{Q} - \vec{q}) \cdot \vec{r} + \omega t] \,. \tag{18.10}$$

Note that the wavelengths of the periodic potentials are determined by the CDW wave vector $\vec{Q}$ which is much larger than the phason wave vector $\vec{q}$. Thus, phason attenuation is a nonlocal effect as discussed in Section 18.1.

18.3 Scattering of "Belly" Electrons

The one-electron Schrödinger equation which incorporates the CDW potential is

$$\mathcal{H} = p^2/2m + G \cos \vec{Q} \cdot \vec{r} \,, \tag{18.11}$$

where φ has been chosen equal to zero without loss of generality. This potential deforms the electron wave functions by mixing the plane wave state $\vec{k}$ with $\vec{k} \pm \vec{Q}$ and leads to the modulated electron density of Equation (18.1). The electron energy spectrum is also altered. As shown in Figure 18.2, the CDW potential leads to energy gaps of magnitude G at $\vec{k} = \frac{1}{2}\vec{Q}$. Figure 18.2 also shows the lemon-shaped Fermi surface which results for the case of critical contact, i. e., when the Fermi surface touches the energy gaps at a point.

We shall show that there are two regions in $\vec{k}$ space where electrons are scattered by phasons: electrons near the "belly" of the lemon and "conical point" electrons near the energy gap. For simplicity we begin by considering the "belly" electrons.

Far away from the energy gaps, the electron states which diagonalize the Hamiltonian (18.11) are easily found using first-order perturbation theory. A state $\Psi_{\vec{k}}$ below the energy gap is

$$\Psi_{\vec{k}} = e^{i\vec{k}\cdot\vec{r}} + [G/2(\epsilon_{\vec{k}} - \epsilon_{\vec{k}-\vec{Q}})]e^{i(\vec{k}-\vec{Q})\cdot\vec{r}} + [G/2(\epsilon_{\vec{k}} - \epsilon_{\vec{k}+\vec{Q}})]e^{i(\vec{k}+\vec{Q})\cdot\vec{r}} \,, \tag{18.12}$$

where $\epsilon_k \equiv \hbar^2 k^2/2m$. It is sufficiently accurate to take the energy of $\Psi_{\vec{k}}$ to be the free-electron energy $\epsilon_{\vec{k}}$. We will be considering only the zero-temperature limit so that only states below the gap are needed for the Fermi surface of Figure 18.2. We assume $\hbar\omega \ll G$, i. e., phasons do not cause inter band transitions.

Phasons are attenuated by scattering electrons from states $\Psi_{\vec{k}}$ below the Fermi surface to states $\Psi_{\vec{k}'}$ above the Fermi surface. We now can write down the golden rule transition rate. Confining the region of interest temporarily to "belly" states, we find the transition rate

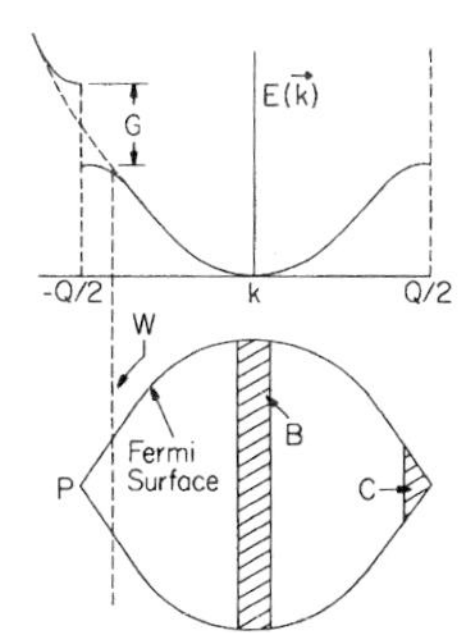

Figure 18.2 One-electron energy spectrum and Fermi surface of a metal with a CDW of wave vector $\vec{Q}$. Energy gap caused by the CDW periodic potential is G. Distortion of the Fermi surface at the conical point P is exaggerated. Electron wave functions most severely deformed are those between the conical point P and the plane w. Electrons which contribute to phason attenuation are those labeled B near the belly and C near the conical point.

using (18.10) for the electron–phason interaction,

$$w_{\vec{k}\to\vec{k}+\vec{q}} = (2\pi/\hbar)\left(\frac{1}{4}\varphi_{\vec{q}}G\right)^2 |\mathfrak{M}|^2 \delta(\epsilon_{\vec{k}-\vec{q}} - \epsilon_{\vec{k}} - \hbar\omega_{\vec{q}}) f_{\vec{k}}(1 - f_{\vec{k}+\vec{q}}) , \tag{18.13}$$

where

$$|\mathfrak{M}|^2 = \left(\frac{1}{2}G\right)^2 [1/(\epsilon_{\vec{k}} - \epsilon_{\vec{k}-\vec{Q}}) - 1/(\epsilon_{\vec{k}+\vec{Q}} - \epsilon_{\vec{k}-\vec{Q}+\vec{q}}) \\ - 1/(\epsilon_{\vec{k}} - \epsilon_{\vec{k}+\vec{Q}}) + 1/(\epsilon_{\vec{k}+\vec{q}} - \epsilon_{\vec{k}+\vec{Q}+\vec{q}})]^2 . \tag{18.14}$$

The Fermi-occupation probabilities $f_{\vec{k}}$ and $f_{\vec{k}+\vec{q}}$ weight the transition rate by the probability that the initial state is filled and the final state is empty.

The phason loses energy to the electrons at the rate

$$\frac{dE_{\vec{q}}}{dt} = -\hbar\omega_{\vec{q}} \sum_{\vec{k}} w_{\vec{k}\to\vec{k}+\vec{q}} . \tag{18.15}$$

Conservation of energy required by the δ function makes $\epsilon_{\vec{k}+\vec{q}} = \epsilon_{\vec{k}} + \hbar\omega$. In the $q \to 0$ limit it follows that the scattered electrons are those with velocity $\vec{v}$ such that the component of $\vec{v}$ along $\vec{q}$ is equal to the phason velocity. This condition is the same as that for ultrasonic attenuation. For longitudinal phasons the energy conserving transitions satisfy $\cos\theta_{\vec{k},\vec{q}} = c_{\parallel}/v_F{\sim}10^{-3}$. The scattered electrons lie on a ring at the belly of the lemon as indicated in Figure 18.2. For these electrons $f_{\vec{k}}(1 - f_{\vec{k}+\vec{q}}) \to \frac{1}{2}\hbar\omega_{\vec{q}}\delta(\epsilon_{\vec{k}} - E_F)$ in the zero-temperature limit. E_F is the Fermi energy.

Furthermore,

$$|\mathfrak{M}|^2 \cong (4mGq_{\parallel}/\hbar^2 Q^3)^2 \tag{18.16}$$

plus higher-order terms in $q_{\parallel}/Q$. Since the matrix element for the transition depends on $q_{\parallel}$ transverse phasons, for which $\vec{q}\cdot\vec{Q} = 0$, are not attenuated. The matrix element for electron–phason scattering depends on the difference of plane-wave coefficients for CDW electrons separated by $\vec{q}$ in $\vec{k}$ space. Since these coefficients are functions only of the component of $\vec{k}$ parallel to $\vec{Q}$, the difference must be zero for $\vec{q}$ perpendicular to $\vec{Q}$.

Changing the sum over $\vec{k}$ in (18.15) to an integral and performing the integration gives

$$dE_{\vec{q}}/dt = -\left(\varphi_{\vec{q}}^2 G^2 m^2 \omega^2/32\pi\hbar^3 q\right) |\mathfrak{M}|^2 . \tag{18.17}$$

The energy attenuation coefficient is $\gamma_{\vec{q}} \equiv -(1/E_{\vec{q}}) \times dE_{\vec{q}}/dt$. Combining Equations (18.6) and (18.17) obtains

$$\gamma_{\vec{q}} = (6\pi\hbar Q^2/MA^2)(mG/\hbar^2 Q^2)^4 q \cos\theta_{\vec{Q},\vec{q}} , \tag{18.18}$$

where $\theta_{\vec{Q},\vec{q}}$ is the angle between $\vec{Q}$ and $\vec{q}$, and we have used the relation $n = k_F^3/3\pi^2$. Note that $\gamma_{\vec{q}}$ is independent of temperature.

Our primary interest in this work is the alkali metals and, in particular, potassium. The CDW hypothesis has been successful in explaining quantitatively several of the anomalous properties of potassium including the anomalous optical absorption [5], splitting of the conduction-electron spin-resonance g factor [6], and the anisotropic residual resistivity [7]. Moreover, the question of phasons is of interest in potassium since the phason Debye–Waller factor would likely greatly reduce the intensity of CDW diffraction satellite peaks [2, 3]. The intensity would not be lost but would be transferred into a pancake-shaped phason cloud about the satellite position. Special techniques that integrate over this diffuse phason pancake may be required to detect the effects of the CDW in a diffraction experiment.

We, therefore, evaluate $\gamma_{\vec{q}}$ for the proposed CDW in potassium. Taking $G/E_F = 0.3$, $A \cong 0.1$ Å, $Q \cong 2k_F$ and longitudinal phasons, for which $\theta_{\vec{q},\vec{Q}} = 0$, we obtain

$$\gamma_{\vec{q}} \cong (20\,\text{cm/s})q\,. \tag{18.19}$$

A useful quantity is the phason quality factor $\mathcal{Q} \equiv \omega_{\vec{q}}/\gamma_{\vec{q}}$ which gives the number of phason cycles during the mean lifetime $1/\gamma_{\vec{q}}$; of the phason. While there is no experimental knowledge of the phason spectrum, a reasonable estimate is that the longitudinal phason velocity is comparable to the longitudinal phonon velocity, i. e., $c_{\|} \sim 3 \times 10^5$ cm/s. Note that since both $\gamma_{\vec{q}}$ and $\omega_{\vec{q}}$ depend linearly on q, $\mathcal{Q}$ is independent of q. For longitudinal phasons $\mathcal{Q} \sim 1.5 \times 10^4$ so that considering only the "belly" electrons leads to negligible damping. Note from (18.7) and (18.18) that $\mathcal{Q}$ is monotonically increasing as $\theta_{\vec{Q},\vec{q}}$ increases from 0 to $\frac{1}{2}\pi$. For transverse phasons $\gamma_{\vec{q}} \equiv 0$.

18.4 Scattering of "Conical Point" Electrons

In the discussion following Equation (18.15) we found that the electrons scattered by a phason in the limit $q \to 0$ are those with velocity $\vec{v}$ such that the component of $\vec{v}$ along $\vec{q}$ is equal to the phason velocity. In addition to the electrons near the belly, discussed in Section 18.3, there exist electrons in $\vec{k}$ space near the energy gaps which satisfy this criteria for energy conserving transitions. This can be seen from Figure 18.2. Owing to Bragg reflection by the periodic CDW potential, electrons at $\vec{k} = \pm\frac{1}{2}Q$ have zero velocity in the $\vec{Q}$ direction. For the Fermi surface in Figure 18.2, the other velocity components are also zero for the states at $+\frac{1}{2}Q$. It follows that electrons near the conical points also contribute to phason attenuation. For each phason only electrons near one of the conical points contribute since the other electrons move in the wrong direction.

The most severely deformed wave functions and energies are for states in the conical point regions. Sufficiently accurate CDW electron wave functions can be found near the energy gap at $\frac{1}{2}\vec{Q}$ by considering only the plane wave states $\vec{k}$ and $\vec{k} - \vec{Q}$. These plane wave states are degenerate at $\vec{k} = \frac{1}{2}\vec{Q}$ in the absence of the CDW potential. Diagonalizing the Hamiltonian (18.11) exactly on this basis leads to the wave functions and energies of states above and below the gap. For the present work only states below the gap need to be considered. It is convenient to move the origin in $\vec{k}$ space to $\frac{1}{2}\vec{Q}$. For the remainder of this section k will be measured from this new origin. A state $\psi_{\vec{k}}$ below the energy gap is

$$\psi_{\vec{k}} = q^{i\vec{Q}\cdot\vec{r}/2}(\cos\xi\, e^{i\vec{k}\cdot\vec{r}} - \sin\xi\, e^{i(\vec{k}-\vec{Q})\cdot\vec{r}})\,, \tag{18.20}$$

and has energy

$$E_{\vec{k}} = \hbar^2\left(k^2 + \frac{1}{4}Q^2\right)/2m - \frac{1}{2}[(\hbar^2\vec{k}\cdot\vec{Q}/m)^2 + G^2]^{1/2}\,. \tag{18.21}$$

The coefficients $\cos\xi$ and $\sin\xi$ are defined by

$$\sin 2\xi = G[(\hbar^2\vec{k}\cdot\vec{Q}/m)^2 + G^2]^{-1/2}\,. \tag{18.22}$$

The golden rule transition rate for a phason to scatter an electron from state $\Psi_{\vec{k}}$ below the Fermi surface to a state $\Psi_{\vec{k}+\vec{q}}$ above is given by (18.13) with $\epsilon_{\vec{k}}$ replaced by $E_{\vec{k}}$ and

$$|\mathfrak{M}|^2 = (\cos\xi \sin\xi' - \sin\xi \cos\xi')^2 . \tag{18.23}$$

Note that $\mathfrak{M}$ depends on cross terms, i. e., the coefficient of $e^{i\vec{k}\cdot\vec{r}}$ is multiplied by the coefficient of $e^{i(\vec{k}-\vec{Q})\cdot\vec{r}}$ etc.

Energy conservation requires $E_{\vec{k}+\vec{q}} = E_{\vec{k}} + \hbar\omega$. For longitudinal phasons the scattered electrons are found using (18.21) to be those which satisfy

$$k_z = -[mc_{\|}(\hbar Q^2/2mG) - 1] - \frac{1}{2}q . \tag{18.24}$$

In the limit $q \to 0$ this reduces to the requirement that the component of electron velocity along $\vec{q}$ is equal to the phason velocity. Near the energy gaps the electrons behave as if they had effective mass $m(\hbar^2 Q^2/2mG - 1)^{-1}$. If $\frac{1}{2}q$ is greater than the first term on the right-hand side in (18.24), the electron will be scattered from a state $\Psi_{\vec{k}}$ one side of the conical point to $\Psi_{\vec{k}+\vec{q}}$ on the other side. For the Fermi surface in Figure 18.2, the electron is scattered from the conical point on the right to that on the left. We shall discuss this point in more detail below.

It is straightforward to evaluate (18.23) and to perform the integration over $\vec{k}$ space in (18.15). The rate at which the phason loses energy to the electrons is

$$d E_{\vec{q}} = \left(\varphi_{\vec{q}}^2 Q^2 \hbar\omega^3 q/128\pi\right) \cos^2\theta_{\vec{Q},\vec{q}} . \tag{18.25}$$

Note that (18.24) is independent of the CDW energy gap. $\mathfrak{M}$ depends on the difference in the coefficients of the wave functions describing states separated by $\vec{q}$ in $\vec{k}$ space. Near the energy gaps the electron effective mass is smaller than the free-electron mass by a factor of order G/E_F. Thus a small change by $\vec{q}$ in $\vec{k}$ space results in a large change in $\mathfrak{M}$. Expression (18.24) is valid for $q/Q < 2mG/\hbar^2 Q^2$. With $G/E_F \cong 0.3$ as proposed for potassium $q \lesssim 10^7\ \text{cm}^{-1}$. Since $d E_{\vec{q}}/dt$ depends on $\cos^2\theta_{\vec{Q},\vec{q}}$, transverse phasons are not attenuated by scattering conical point electrons.

The temperature-independent energy attenuation coefficient for longitudinal phasons in potassium is

$$\gamma_{\vec{q}} \cong (1.1 \times 10^5\ \text{cm/s})q . \tag{18.26}$$

Taking $c_{\|} \sim 3 \times 10^5$ cm/s gives a phason quality factor $\mathcal{Q} \sim 2.7$. Since both $\omega_{\vec{q}}$ and $\gamma_{\vec{q}}$ depend linearly on q, $\mathcal{Q}$ is independent of q. Thus longitudinal phasons are underdamped but, nevertheless, rather strongly attenuated. The $\mathcal{Q}$ obtained here depends on the specific values chosen for A and $c_{\|}$.

This strong scattering by phasons of conical-point electrons should have important experimental consequences. The strong attenuation of longitudinal phasons should effect their contribution to the low-temperature specific heat. Due to phason scattering electrons near a conical point will have a short lifetime. The high rate of scattering of electrons from one conical point to the other may have important consequences for magneto transport. The short lifetime of conical point electrons is also important to the de Haas–van Alphen effect. For an applied magnetic field parallel to $\vec{Q}$, the important cyclotron orbits for the de Haas–van Alphen effect are those at the belly of the lemon where phason scattering is negligible. As the magnetic field is tilted away from $\vec{Q}$, the important cyclotron orbits will intersect the conical

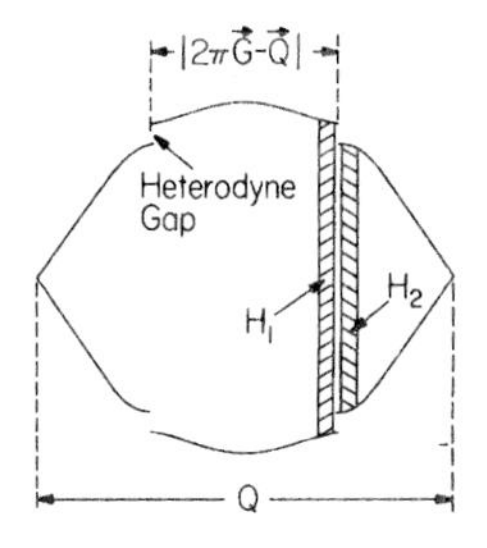

Figure 18.3 If "heterodyne" gaps caused by potentials with periodicity $2\pi\vec{G} \pm \vec{Q}$ cut the Fermi surface as shown, there are two additional sets of electrons, H_1 and H_2 which contribute to attenuation for a longitudinal phason with velocity to the right.

points. An electron on this orbit will be "killed" before it can complete the orbit, and the de Haas–van Alphen signal should disappear. This loss of signal as the orientation of the crystal* changes has been observed in certain samples of potassium [8] and is difficult to explain on the basis of free-electron theory.

With a CDW present, the potential in a metal will have terms with periodicity $\vec{Q}$ and $2\pi\vec{G}$, where $\vec{G}$ is a reciprocal lattice vector. In addition there will be "heterodyne" terms with periodicity $\vec{Q} \pm 2\pi\vec{G}$ [9]. If $\vec{Q} \pm 2\pi\vec{G}$ is less than the diameter of the Fermi surface, there will be additional "heterodyne" gaps in the electron energy spectrum and the Fermi surface will be multiply connected as in Figure 18.3. There will now be two additional places where longitudinal phasons will be strongly attenuated since Equation (18.25) is valid as long as q is small enough as that the scattered electrons are within the strongly deformed regions of the energy spectrum. The quality factor for longitudinal phasons will be $\mathcal{Q} \sim 1$. In the presence of heterodyne gaps as in Figure 18.3, longitudinal phasons will be (approximately) critically damped, but transverse phasons will remain undamped.

We have shown above that phasons with $\vec{q}$ perpendicular to $\vec{Q}$ are not attenuated in jellium. A question which arises is: do the ionic potentials in a real metal lead to finite attenuation of transverse phasons? We have investigated this problem by including the effects of ionic potentials following the method of pseudopotentials. We have found that the first-order matrix element for golden rule transitions is still zero for transverse phasons. However, there is a small but finite attenuation in the next order, i. e., the transition rate depends on the fourth power of the pseudopotential. Thus transverse phasons remain very much underdamped.

18.5 Conclusions

We have derived the electron–phason interaction and have shown that the wavelength of the interaction is determined by the CDW wave vector $\vec{Q}$ and not the wave vector $\vec{q}$ of the phason. Phason attenuation is, therefore, a nonlocal phenomenon and is correctly treated quantum mechanically. Because long-wavelength phasons are not accompanied by macroscopic motion of the charge density, the attenuation of phasons cannot be treated by Ohm's law. Phasons are predominantly attenuated by scattering electrons near the CDW energy gaps. The attenuation rate is independent of temperature. However, transverse phasons are not attenuated. The scattering of conical point electrons leads to phasons which are underdamped; but for longitudinal phasons the damping is strong, and the attenuation rate is ~ 0.3 times the phason frequency for the proposed phasons in potassium. If heterodyne gaps cut the Fermi surface, electrons near these gaps also contribute strongly to phason attenuation (and longitudinal phasons may be nearly critically damped).

A Appendix

McMillan [4] has dealt briefly with the problem of phason attenuation within the context of time-dependent Landau theory. McMillan described the short-wavelength components of charge density by an order parameter $\Psi(\vec{r})$. He treated the long-wavelength components of the conduction-electron charge density as an incompressible fluid with velocity field $\vec{v}(\vec{r})$. He then wrote an "Ohm's law" expression for the power dissipation and found equations of motion for Ψ and $\vec{v}$ which resulted in overdamped phasons.

From the discussion in Section 18.1 of this work, it is clear that McMillan's "Ohm's law" approach is not valid. Phasons are not accompanied by a macroscopic motion of the charge density. The length scale important to phason attenuation is determined by the CDW wave vector $\vec{Q}$ and not the phason wave vector $\vec{q}$. It is incorrect to apply Ohm's law on a 1 – Å scale.

McMillan finds that the lifetime of the phason decreases with increasing conductivity. In other words, the rate of attenuation increases as the conductivity increases. This result is analogous to the (incorrect) result obtained in the theory of ultrasonic attenuation when non-local effects are disregarded. The mechanism for phason attenuation, as we have shown, is scattering of electrons near the CDW energy gaps and is nonlocal. It should also be noted that McMillan's result for the attenuation rate is temperature dependent since it is proportional to the (temperature-dependent) conductivity. This is in direct contrast to our temperature-independent result.

McMillan finds that the phason lifetime $\tau \sim 1/q^2$ (in contrast to the correct result, $\sim 1/q$). An important question, then, is why his phasons are not oscillatory (underdamped) in the $q \to 0$ limit where $\tau \to \infty$. The reason is that McMillan's equations do not allow for oscillatory solutions. The kinetic energy term containing $(\partial\Psi/\partial t)^2$ which would lead to a second-order differential equation for the order parameter was omitted.

References

1 Overhauser, A.W. (1968) *Phys. Rev.*, **167**, 691.
2 Overhauser, A.W. (1971) *Phys. Rev. B*, **3**, 3173.
3 Overhauser, A.W. (1978) *Hyperfine Interactions*, **4**, 786.
4 McMillan, W.L. (1975) *Phys. Rev. B*, **12**, 1197.
5 Overhauser, A.W. and Butler, N.R. (1976) *Phys. Rev. B*, **14**, 3371.
6 Overhauser, A.W. and de Graaf, A.M. (1968) *Phys. Rev.*, **168**, 763.
7 Bishop, M.F. and Overhauser, A.W. (1977) *Phys. Rev. Lett.*, **39**, 632.
8 Shoenberg, D. and Stiles, P.J. (1964) *Proc. R. Soc. A*, **281**, 62.
9 Overhauser, A.W. (1974) *Phys. Rev. B*, **9**, 2441.

Reprint 19 Charge-Density Waves and Isotropic Metals[1]

*A.W. Overhauser**

* Department of Physics, Purdue University, 525, Northwestern Avenue, West Lafayette, Indiana 47907, USA

Received 6 January 1978

Fermi-surface instability leading to charge density waves is a common malady in anisotropic metals, and there is now evidence that it also infects isotropic ones.

19.1 Introduction

The visualization of what is meant by a charge-density-wave (CDW) or spin-density-wave (SDW) structure is made easy by considering the simplest metal of all – "jellium". Jellium is a theorist's idealization which replaces the positive-ion lattice by a uniformly charged jelly. Consequently the quantum problem for the degenerate sea of conduction electrons has translational invariance. It seems natural to assume that the electronic ground state will have a charge density which is spatially uniform, but this assumption was shown to be unsafe when electron–electron interactions are taken into account [14].

Two types of symmetry-breaking instability can occur. A SDW structure has spatially inhomogeneous charge densities for both spin states, but 180° out of phase:

$$\rho\pm = \frac{1}{2}\rho_0(1 \pm p \cos \boldsymbol{Q} \cdot \boldsymbol{r}) . \tag{19.1}$$

The wave-vector Q and the fractional amplitude p are order parameters. Their values vary with temperature in such a way that the free energy is always minimized. As illustrated in Figure 19.1, the total charge density of a SDW is a constant, ρ_0, independent of position. There is a net spin polarization proportional to $p \cos \boldsymbol{Q} \cdot \boldsymbol{r}$.

A CDW structure differs from the SDW case in two important ways [17]. The charge densities for the two spin states are modulated in phase. Consequently the spin density is zero everywhere, but the electronic charge density is $\rho_0(1 + p \cos \boldsymbol{Q} \cdot \boldsymbol{r})$. This does not mean that there will be large, static electric fields in the metal. The positive jelly must undergo a periodic longitudinal displacement given by

$$\boldsymbol{u} = A \sin \boldsymbol{Q} \cdot \boldsymbol{r} . \tag{19.2}$$

Since the local charge density of the jelly is then

$$\rho_+ = -\rho_0(1 - \mathrm{div}\boldsymbol{u}) , \tag{19.3}$$

total charge neutrality is maintained everywhere by choosing the amplitude A of the displacement to be p/Q. If we assume that the jelly has no elastic rigidity, this adjustment will occur automatically with no energy penalty. If the jelly were rigid, the electric-field energy of the CDW would be nonzero, and so large that a CDW instability could never arise. The question of whether a particular real metal has a CDW ground state depends on how nearly its positive-ion lattice approximates the *deformable*-jellium model.

1) Advanced in Physics 27, 3 (1978)

Anomalous Effects in Simple Metals. Albert Overhauser

ISBN: 978-3-527-40859-7

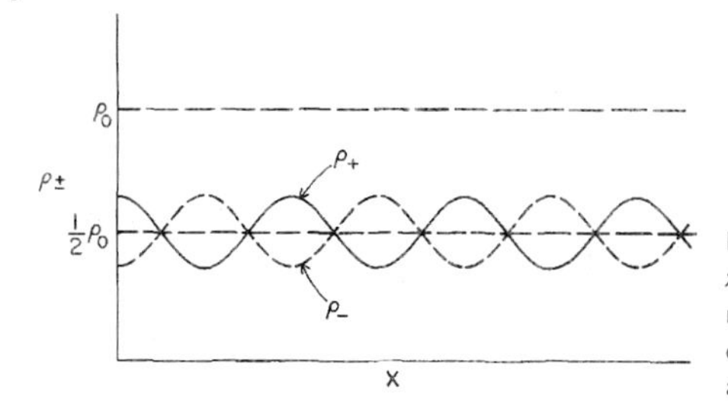

Figure 19.1 Electronic charge density against x of up- and down-spin electrons for an ideal metal with a SDW ground state. The total charge density is a constant, ρ_0. In a CDW, ρ_+ and ρ_- are in phase.

CDWs certainly exist in some anisotropic conducting compounds – in particular, the transition-metal dichalcogenides [39]. This does not establish that such states would be characteristic of electrons in an approximately isotropic and weakly structured background; that would require CDWs to occur in much simpler metals. The orthodox view of the alkali metals is that they closely approach the ideal of a free-electron metal, and are free of CDW instabilities. Several of the experimentally recorded properties of samples of potassium are not consistent with this view, however, and the author has previously argued that these give clear evidence of the existence of CDWs in potassium. Recent experimental work has shown that the significance of these properties is not to be easily discounted. Moreover, other experimentally, recorded properties that have been held to rule out CDWs may now require more careful interpretation, taking into account CDW domain structures and phase excitations.

In this article, following a resume of the theoretical states of CDW instabilities, it is shown that the hypothesis that CDWs form in potassium is well supported by the experimental evidence now available. It gives a more complete explanation of the data than the orthodox assertion that they do not form. This is a very basic question for the physics of the solid state, and the matter should not be pushed to one side until a clear answer is obtained. The directions from which clear-cut experimental evidence will come are identified here.

19.2 Theoretical Summary

19.2.1 Wave-Mechanical Description

A spatially modulated charge density will arise only if each electron experiences a sinusoidal potential, i. e. if the one-electron Schrödinger equation is

$$\left(\frac{p^2}{2m} + V_0 + G\cos \boldsymbol{Q}\cdot\boldsymbol{r}\right)\phi_k = E(\boldsymbol{k})\phi_k\,. \tag{19.4}$$

The potential energy arises from the exchange and correlation hole surrounding each electron. The constant term, V_0, would be the only one present for a ground state with a homogeneous electron density. The cosine term is present only if a CDW distortion has occurred. (The SDW has a Pauli spin operator multiplying the cosine term.) Such instabilities are 'bootstrap' phenomena. A charge modulation occurs only if the periodic part G of the exchange and correlation potential is large enough to sustain a density modulation that could generate the periodic part.

Since the exchange and correlation hole surrounding an electron in a momentum state $\boldsymbol{k}$ depends on $\boldsymbol{k}$, so do the potentials V_0 and G. Therefore the wave-mechanical problem for a CDW ground state involves solving an integral equation for $G(\boldsymbol{k})$. It is much easier to talk

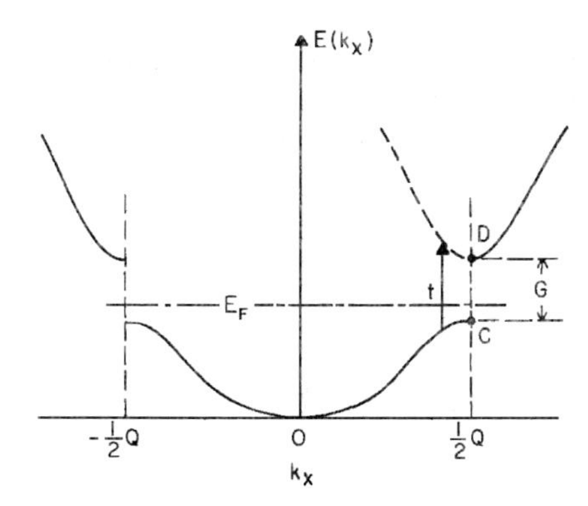

Figure 19.2 Electronic energy $E(k_x)$ against k_x for a CDW structure. The CDW energy gap is G, and the Fermi energy E_F, lies between the gap edges C and D. The optical transition t is allowed for a CDW structure.

about this than to carry it out. The *approximate* solutions of Equation (19.4) have the form

$$\phi_k \cong \exp(i\boldsymbol{k}\cdot\boldsymbol{r}) + \alpha_+ \exp[i(\boldsymbol{k}+\boldsymbol{Q})\cdot\boldsymbol{r}] + \alpha_- \exp[i(\boldsymbol{k}-\boldsymbol{Q})\cdot\boldsymbol{r}]\,, \tag{19.5}$$

where $\alpha_\pm$ are functions of $\boldsymbol{k}$ and differ from zero only if $G(\boldsymbol{k}) \neq 0$. The admixture of these additional terms into the pure-momentum state $\boldsymbol{k}$ causes a charge modulation since

$$|\phi_k|^2 \cong 1 + 2(\alpha_+ + \alpha_-)\cos \boldsymbol{Q}\cdot\boldsymbol{r}\,. \tag{19.6}$$

The fractional modulation p of a CDW is the average value of $2(\alpha_+ + a_-)$ for the occupied momentum states of the degenerate Fermi sea.

The energy spectrum $E(\boldsymbol{k})$ is no longer parabolic in k and isotropic, as is the case for free electrons. Instead there is an energy gap of magnitude G in the direction of $\boldsymbol{Q}$, which we shall take to be along $\hat{\boldsymbol{x}}$. This spectrum, familiar from the weak-binding theory of energy bands in crystals, is shown in Figure 19.2. The most highly perturbed electron states are those at the gap – they are 100% modulated. If the phase of the CDW is chosen so that $G > 0$, then the state C of Figure 19.2 has

$$|\phi_C|^2 \cong 1 - \cos Qx\,. \tag{19.7}$$

The state D just above the gap has

$$|\phi_D|^2 \cong 1 + \cos Qx\,. \tag{19.8}$$

so the charge modulation of D is 180° out of phase with that of C.

A CDW instability will be optimized if all occupied momentum states contribute in-phase to the self-consistent exchange and correlation potential, $G\cos\boldsymbol{Q}\cdot\boldsymbol{r}$. This is assured if the Fermi energy E_F is chosen below that of state D. It should be above that of C so that all highly modulated, in-phase states near C will be occupied. Therefore E_F will lie somewhere in the energy gap, as shown in Figure 19.2.

The foregoing optimization implies that the magnitude Q of the CDW wave-vector is approximately

$$Q \approx 2k_F\,, \tag{19.9}$$

the diameter of the Fermi surface. For this reason SDWs and CDWs are sometimes called Fermi-surface instabilities. The occupied region of $\boldsymbol{k}$ space will be slightly distorted from a Fermi sphere, as shown in Figure 19.3. There may be small necks in the Fermi surface at the energy-gap planes p_1 and p_2. These planes are perpendicular to $\boldsymbol{Q}$. An important pathological consequence of a Fermi-surface instability is that a simply connected surface is converted into a multiply connected one. This has profound effects on the magneto-conductivity properties of a metal.

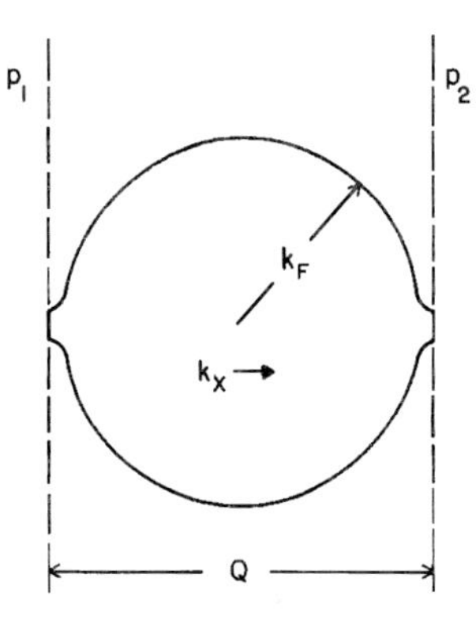

Figure 19.3 Fermi surface of an ideal metal with a CDW ground state. The surface is distorted from that of a sphere of radius k_F near the energy-gap planes p_1 and p_2.

19.2.2 **Detection of CDWs by Diffraction**

A CDW ground state will alter the physical properties of a metal in many ways. The most important is the appearance of Bragg diffraction satellites. Let us return for the moment to the case of a real metal. The positive-ion lattice will have a discrete translation group if a CDW instability has not occurred. The only allowed scattering vectors of an X-ray, a neutron or an electron are the reciprocal lattice vectors $\{\boldsymbol{G}\}$. For the case of a simple-cubic lattice these are

$$\boldsymbol{G}_{hkl} = (h\hat{x} + k\hat{y} + l\hat{z})2\pi/a\ . \tag{19.10}$$

(The lattice constant is a, and h, k and l are integers.) Suppose that the electronic ground state has a single CDW. Each positive ion will then be displaced from its ideal site according to Equation (19.2). A Fourier analysis of the ion density will now have additional Fourier components as well as the usual ones given by Equation (19.10). The new scattering vectors are [20]

$$\boldsymbol{K} = \boldsymbol{G}_{hkl} \pm \boldsymbol{Q}\,. \tag{19.11}$$

There are two CDW satellites associated with each reciprocal lattice vector; their intensity may be two to eight orders of magnitude smaller than the intensities of ordinary Bragg reflections.

Since the magnitude of $\boldsymbol{Q}$ is determined by a diameter (or spanning vector) of the Fermi surface, it will generally be *incommensurate* with the reciprocal lattice; that is, there will *not* exist a (small) integer n together with a reciprocal lattice vector $\boldsymbol{G}$ such that $\boldsymbol{Q} = \boldsymbol{G}/n$. The wavelength of a CDW will be of the same order of magnitude as a lattice spacing, but will be incommensurate with it. In such a case the crystal will no longer have a translation group; instead it is 'multiply periodic'. The primitive unit cell is then the entire sample (in the direction of $\boldsymbol{Q}$).

Suppose that $\boldsymbol{Q}$ is close to a commensurate relationship. If the metal is a compound, then one species of ion will prefer the CDW crests more than others. A slight change in the magnitude (and direction) of $\boldsymbol{Q}$, together with a precise phase adjustment, may allow this preference to result in a lower total free energy, and consequently, 'locked-in' commensurate CDWs can occur. Phase transitions between commensurate and incommensurate CDW structures have been observed in several quasi-two-dimensional metals [39] such as TaS_2 and $TaSe_2$.

The electronic ground state may involve a multiple CDW instability, and in such a case

$$\rho = \rho_0 \left[1 + \sum_i p_i \cos(\boldsymbol{Q}_i \cdot \boldsymbol{r} + \delta_i)\right]\ . \tag{19.12}$$

The two-dimensional layer compounds cited above have three CDWs. Their $\boldsymbol{Q}$s, when projected onto the hexagonal plane, form angles of 120°.

The concept of Q domains is extremely important in cubic crystals if the metal has a single CDW. The preferred direction of Q will then be one of a set of otherwise equivalent crystal directions. A single-crystal sample will in general be divided into domains which differ only by their Q orientation. Physical properties that would be isotropic for cubic symmetry will become uniaxial in an isolated Q domain. The bulk properties of a poly-domain sample will depend critically on the domain distribution. Measurements will not be preproducible from sample to sample (or from run to run on the same sample) unless the domain structure is controlled or eliminated. The non-reproducibility that has plagued research on the alkali metals, discussed in Section 19.3.7, is an example of this difficulty.

19.2.3 Fermi-Surface Instability Theorem

The prediction and understanding of CDW instabilities is, of course, the most important theoretical question. The only rigorous result is the instability theorem in the Hartree–Fock approximation, in which the many-electron wave-function is taken as an antisymmetrized product of one-electron orbitals.

$$\Psi = A \prod_i \phi_i(\mathbf{r}_i) , \tag{19.13}$$

where A is the antisymmetrization operator. For many years it was assumed that the lowest energy state Ψ of this form (for a degenerate electron sea) was the familiar Fermi sphere of occupied plane-wave (PW) states. Such a Ψ is, of course, the ground state of a non-interacting Fermi gas.

It was shown, however [15], that if the one-electron states ϕ_i are taken to have the form given by (19.5), with $Q = 2k_F$ and $\alpha_\pm(\mathbf{k})$ carefully chosen, a lower energy state Ψ can always be found (we call this the instability theorem). The increased kinetic energy of the modulated ϕs is cancelled by a larger decrease in exchange energy, which is the kinematic effect of the Pauli exclusion principle arising from the Coulomb interaction between *parallel-spin*electrons. For this reason the instability theorem applies with equal force to SDW and CDW deformations: the relative phase of the up-spin and down-spin modulations is irrelevant provided that the deformable jellium model is assumed. (For rigid jellium only a SDW instability occurs.) Remarkably, the instability theorem is valid for all electron densities.

This instability and the degeneracy of SDW and CDW states is represented by the energy level diagram to the left of the dashed line in Figure 19.4. The crucial question concerns the effect of electron correlations on the relative energy of the PW, SDW and CDW states.

19.2.4 Role of Electron Correlations

Since parallel-spin electrons are kept apart kinematically by the effect of the antisymmetrization operator A, the dominant correction to a Hartree–Fock energy arises from the dynamical correlation of antiparallel-spin electrons. To account for this, visualize the virtual scattering

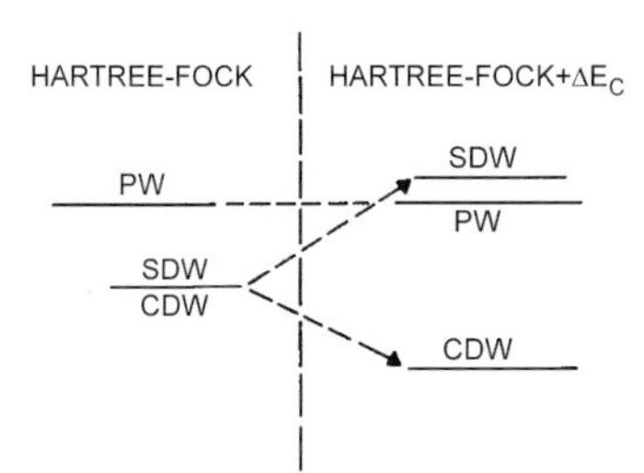

Figure 19.4 Energy-level diagram for the plane-wave (PW), SDW and CDW states of an ideal metal. In the Hartree–Fock approximation the SDW and CDW states are degenerate and have lower energies than the plane-wave state. The degeneracy is split, as shown on the right, by correlation-energy corrections.

of any two (antiparallel-spin) electrons $\boldsymbol{k}$ and $\boldsymbol{k}'$ below E_F to the empty states $\boldsymbol{k}+\boldsymbol{q}$ and $\boldsymbol{k}'-\boldsymbol{q}$ above. Semi-quantitatively, the correlation energy is

$$E_\mathrm{c} \sim \sum_{kk'q} \frac{\mathfrak{M}_q^2}{-w}\,, \tag{19.14}$$

where w is the excitation energy of each virtual transition, and $\mathfrak{M}_q$ the matrix element of a screened Coulomb interaction. It is possible to use a crude expression such as Equation (19.14) for the basis of an analysis because each term is negative.

The change ΔE_c of the correlation energy of a SDW or CDW state, relative to the PW state, arises primarily from the modifications to $\mathfrak{M}_q$, which must be computed with the perturbed one-electron ϕ_ks of Equation (19.5) rather than with plane waves. For most of the low-energy excitations $\mathfrak{M}_q^2$ is, on the average, changed according to [17]

$$\mathfrak{M}_q^2 \to \mathfrak{M}_q^2(1 \pm p^2)\,, \tag{19.15}$$

in which the + sign applies to the CDW case and the − sign to the SDW case – this is easy to understand. For CDWs, up- and down-spin electrons are bunched together in laminar layers (of real space) and consequently their collision 'cross-section' is increased. For SDWs, the opposite-spin electrons are bunched in alternate laminar layers, so their 'cross-section' is reduced. It follows from Equations (19.14) and (19.15) that ΔE_c is negative for CDWs and positive for SDWs; this is shown on the right-hand side of Figure 19.4 by the splitting of the SDW-CDW degeneracy.

The important result is that the correlation correction tends to *enhance* CDW instability and to *suppress* SDW instability; this is the $Q = 2k_\mathrm{F}$ analogue of an important result of many-body theory for $Q = 0$ deformations. Electron correlations *enhance* the compressibility of a metal *over and above the enhancement caused by exchange*, whereas, for the spin susceptibility they cancel part of the exchange enhancement.

We conclude that CDWs should be the prevalent Fermi-surface instability occurring in metals. The rare occurrence of a SDW instability, as in chromium [15], depends on two circumstances: a favourable energy-band structure [10] that allows the Hartree–Fock term to exceed the correlation correction, and a suppression of the CDW instability through the failure of the deformable-jellium model.

19.2.5 Fermi-Surface Shape and Dimensionalty

Most real metals have anisotropic Fermi surfaces. One expects the Fermi-surface shape to be very important in determining both the magnitude of a CDW instability and the direction of $\boldsymbol{Q}$, and this is confirmed by considering the response function of a degenerate Fermi sea of non-interacting electrons in one, two and three dimensions.

The response function $\chi_0(Q)$ is now defined. Suppose that the Fermi sea is perturbed with a small periodic potential

$$V(\boldsymbol{r}) = \epsilon E_\mathrm{F} \cos \boldsymbol{Q}\cdot\boldsymbol{r}\,, \tag{19.16}$$

the strength of the perturbation being the dimensionless parameter ϵ. This potential will cause a small fractional modulation p of the electronic density, given by Equation (19.1). The response function is given by

$$\chi_0(Q) \equiv -p/\epsilon \tag{19.17}$$

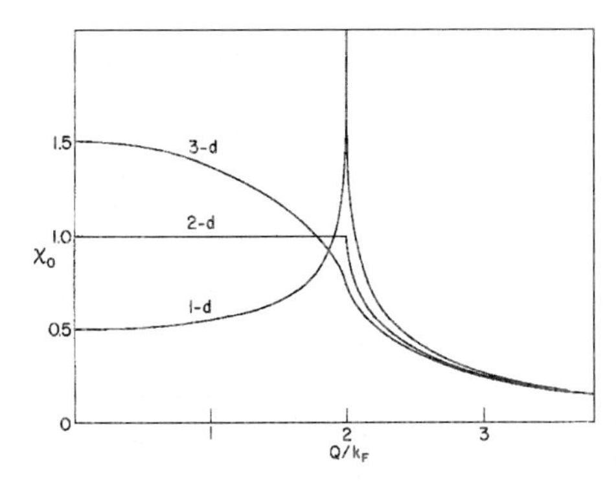

Figure 19.5 The response function $\chi_0(Q)$ of a non-interacting, degenerate Fermi gas for one, two and three dimensions. Q is the wave-vector of a small sinusoidal perturbation.

(in the limit $\epsilon \to 0$); Figure 19.5 shows the response function for each dimensionality. For long-wavelength ($Q \sim 0$) perturbations the response is proportional to the number of dimensions, but for $Q = 2k_F$ the order is reversed.

The relevance of $\chi_0(Q)$ to a consideration of CDW instability is as follows. The one-electron exchange and correlation potential, $G \cos \mathbf{Q} \cdot \boldsymbol{r}$ in Equation (19.4), will be proportional to p (for small p):

$$G = \beta E_F p \,. \tag{19.18}$$

The (dimensionless) coefficient of proportionality β, a measure of the electron–electron interaction strength, is defined by this equation. In general, β is a function of Q. If the potential G generated by the modulation exceeds the value ϵE_F that produced it, an instability will occur, and from Equations (19.16), (19.17) and (19.18) this condition is $\beta\chi_0 > 1$. An alternative way of deriving this condition is to calculate the response function $x(Q)$ which includes the exchange and correlation potential in the perturbation. The result is

$$\chi(Q) = \frac{\chi_0(Q)}{1 - \beta\chi_0(Q)} \,, \tag{19.19}$$

and the criterion for instability is the vanishing of the denominator.

Since $\chi_0(2k_F)$ is smallest for the 3-d curve in Figure 19.5, the exchange and correlation parameter $\beta(2K_F)$ required for instability is largest in this case. An instability at $Q = 2k_F$ is easier for lower dimensionality because the curvature of the Fermi-surface is less. If a Fermi-surface is anisotropic, the CDW $\mathbf{Q}$ will tend to be that vector which connects the flattest regions of the Fermi-surface. However, Fermi-surface 'nesting', as this is called, is not the only factor determining the direction of $\mathbf{Q}$.

The opposition of the positive-ion lattice to the periodic displacement, given by Equation (19.2), is also crucial. In many metals, such as copper, silver and gold, the direct ion–ion interactions caused by overlap of the core-electron wave-functions (the Born–Mayer repulsion) contribute significantly to lattice stiffness. This effect can choke-off a CDW instability completely since it renders invalid the assumption of a perfectly deformable jelly. The contribution of Born–Mayer forces to the self-energy of a periodic lattice displacement depends on the direction of $\mathbf{Q}$, and therefore influences, along with Fermi-surface shape, the optimum value of $\mathbf{Q}$.

The fact that $\chi_0(Q)$ for a 1-d metal has a logarithmic singularity at $Q = 2k_F$, as shown in Figure 19.5, implies that an instability can occur even if there are no electron–electron interactions. This 1-d singularity, discovered by Peierls [25], does not occur in higher dimensions, where exchange and correlation energies are required to stabilize a CDW structure.

19.2.6 CDW Instability in Isotropic Metals

The instability theorem of Hartree–Fock theory [15] and the correlation energy correction [17], which adds stability to a CDW, leads to the conclusion that CDWs can occur in isotropic metals. The major uncertainty for any real metal is whether the positive-ion interactions intervene as described above.

However, from the point of view of response theory, the foregoing conclusion seems unexpected. Consider the denominator of (19.19). For the 3-d case, as shown in Figure 19.5, at $Q = 2k_F$, $\chi_0(Q)$ falls to half its $Q = 0$ value at $Q = 2k_F$. Therefore, if the exchange and correlation potential described by the parameter β is strong enough to cause an instability, a long-wavelength compressional transformation should first occur. This (incorrect) conclusion seems to be well established if the exchange potential is estimated by the most commonly employed approximation – the $\rho^{1/3}$ local potential [30]. For a small-amplitude CDW this potential is, from (19.1),

$$V_x = -\alpha[\rho(1 + p\cos \mathbf{Q} \cdot \mathbf{r})]^{1/3} \tag{19.20}$$

where α is a constant. Comparing the linear term in p from Equation (19.20) with Equation (19.18) leads to

$$\beta = \alpha\rho_0^{1/3}/3E_F , \tag{19.21}$$

which is *independent* of Q, and consequently this theoretical approach does not allow a Fermi-surface instability in an isotropic, jellium model.

The paradox is resolved when it is found that the exchange contribution to $\beta(Q)$ has an (infinite) singularity at $Q = 2k_F$ This follows from the instability theorem, and has been confirmed explicitly by an analysis of the exchange operator [18]. Local-density approximations for exchange and correlation potentials are seriously inadequate for discussing the relative stability of metallic phases.

A further cautionary observation needs to be made: a CDW instability can occur without a divergence in the response function in Equation (19.19). Two important experimental consequences would characterize such a situation: the transition into the CDW phase as temperature is reduced would be first order (rather than second), and the development of a Kohn anomaly in the phonon spectrum as a precursor to the transition would not occur. The theoretical basis for this possibility is an analysis of electron correlations, discussed only briefly above.

The enhancement of $\mathfrak{M}_q^2$ in Equation (19.15) for (virtual) electron-pair excitations applies only to excited states for which the amplitude modulation of $|\phi_{k\pm q}|^2$ is in phase with the CDW, as is the case for the majority of low-lying excited states [17]. However, the amplitude modulations of some of the excited states are out of phase, but their relative number decreases with increasing magnitude of the CDW energy gap G. The resulting anharmonic contribution to the correlation energy correction (vs. amplitude p) could cause a sudden jump (as T is decreased) from $p = 0$ to a finite value.

It is obvious that the occurrence or non-occurrence of CDWs in particular metals cannot be determined by *a priori* theoretical methods. One must consider experimental evidence to discover the characteristics of an exact solution of the many-body problem.

19.3 Experimental Manifestations

19.3.1 The Alkali Metals

The monovalent elemental metals lithium, sodium, potassium, rubidium and caesium, provide important challenges to solid-state theory. Their s-type conduction electrons occupy only half the Brillouin zone in $\boldsymbol{k}$ space. Furthermore, band-structure calculations indicate that energy gaps at the zone boundaries are so small that their Fermi surfaces are almost spherical, and certainly simply connected. They all have a b.c.c. structure at room temperature. Sodium and potassium are particularly suitable for investigation since their electronic energy spectrum is parabolic (with an effective mass $\sim 1 \cdot 2\,\mathrm{m}$) and isotropic to a fraction of 1%. Theoretical metal physics would be in an embarrassing position if the physical properties of these metals could not be explained. But this *is* the current dilemma if the textbook description, just given, is to be believed.

Most investigations which probe Fermi surface symmetry and topology must be carried out at very low temperatures. The significant parameter is usually the product of cyclotron frequency and scattering time, $\omega_c \tau$. For this parameter to be large compared to unity, pure samples are required whose electrical conductivities at 4 K exceed room temperature values by factors of 10^3–10^4. Unfortunately, lithium and sodium undergo a cubic-to-hexagonal transformation near 80 K and 40 K, respectively. Rubidium and caesium are very soft mechanically. For these reasons most research is performed on potassium, but all the alkali metals have been investigated and all are anomalous.

The LAK transport theorems [9] make exact predictions for metals with a single, simply connected Fermi-surface. The electrical resistivity $\rho(B)$ must be independent of the magnetic field B for high fields, i. e. $\omega_c \tau \gg 1$. Nevertheless, the observed $\rho(B)$ continues to increase linearly with respect to B with no evidence of saturation, even at $\omega_c \tau \sim 300$. A number of workers observed this phenomenon using standard four-probe techniques, but when the validity of the technique was questioned, inductive (probeless) methods were developed [26]; the results were the same. According to the LAK theorems a non-saturating $\rho(B)$ can occur (in a monovalent metal) only if the Fermi surface is multiply connected.

The 'Swiss-cheese' model is an attempt to explain away the high-field magnetoresistance by postulating that about 1% of the sample consists of macroscopic voids [33]. This can be ruled out in several ways, principally from the fact that the volumetric density [28] and the lattice constant density [37] agree. Also, in single crystals the high-field effects are very anisotropic, as will be shown below.

A CDW structure provides a means of solving the riddle because the Fermi-surface is then multiply connected. This is caused both by the main CDW energy gaps (of magnitude 67) shown in Figure 19.2, and by small 'heterodyne gaps' associated with periodic potentials whose wave-vectors are given by (19.11).

The alkali metals are good a priori candidates for a CDW because they closely approximate the deformable-jellium model. The pressure variations of the elastic moduli [32] show no evidence of direct ion–ion repulsive forces. The positive ions appear to be weakly confined in a volume kept large by the kinetic pressure of the degenerate sea of conduction electrons.

We shall now focus our attention on the evidence for a CDW structure in potassium. The order of presentation will not be historical because the most convincing data are recent. Many of the anomalous phenomena that characterize potassium have also been observed in one or more of the other four alkali metals. We take it for granted that if potassium is convicted beyond all reasonable doubt of having broken its symmetry, the other four will be immediately suspect by their close association.

19.3.2 Torque Anomalies

The most revealing probeless technique that has been developed for studying the conductivity tensor is the induced-torque method. A spherical, single crystal sample is supported by a vertical rod in a horizontal magnetic field. As the magnet is slowly rotated (in the horizontal plane), induced currents in the sample create a magnetic moment perpendicular to the field B. This causes a torque (on the rod) which is monitored with respect to B and the magnet angle. The theory [35], applicable to a general conductivity tensor, is unambiguous and depends only on Faraday's law of induction. The method is essentially d.c. since typical magnet rotation rates are $\sim 10^{-3}$ Hz.

Early torque studies [27] on potassium were performed on hundreds of specimens. Virtually all showed extremely anomalous high-field torque anisotropies. In a cubic crystal the conductivity tensor is isotropic, so the induced torque must be independent of the magnet angle θ. It was emphasized [8], however, that if the samples were sufficiently distorted from a spherical shape, the observations could be explained, but although torque data at 80 K showed that many of the samples were undistorted, doubt remained until recently.

The anomalous torques were confirmed [5] by measurements on an accurately spherical sample grown in a Kel-F mould; their data are shown in Figure 19.6. In (a) the induced torque as a function of the magnet angle θ is shown for several values of B between 500 G and 17 kG; the growth axis was vertical. In (b) the growth axis was horizontal, and the values of B range from 1 to 23 kG in 1 kG steps. The 'flat' curves in (a) compared to the extreme (45 to 1) anisotropic ones in (b) prove that potassium is electrically uniaxial, not cubic. The ever-increasing torque peaks in (b) prove that the Fermi-surface is multiply connected. The LAK theorems require ρ (and also, as a consequence, torque versus B) to be saturated at $\sim$ 3 kG if the Fermi-surface is simply connected. ($\omega_c\tau \sim 2$ at 1 kG for the Holroyd–Datars sample.)

These striking results are an unmistakable indication of a CDW structure. The spectacular data of Figure 19.6 indicate that (for this sample) Q was the same (almost) everywhere, and parallel to the growth axis. That is, the sample had but one Q domain.

A quantitative theory of the high-field torque peaks is still under development. An attempt to understand them from the topology of heterodyne gaps alone [19] led to the prediction of a magnetically induced low-frequency $1/f$ noise [21], but a careful search for this effect was unsuccessful [4]. Undoubtedly the main CDW gaps, shown in Figures 19.2 and 19.3, play the crucial role.

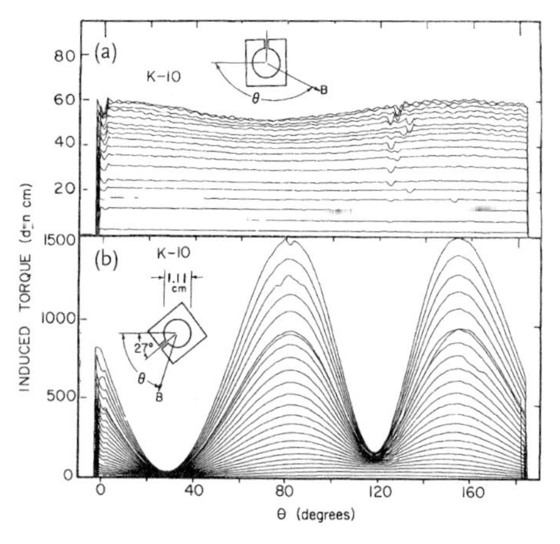

Figure 19.6 Induced-torque against magnetic field direction θ for a potassium sphere 1.11 cm in diameter. In (a) the field B is rotated about the growth axis. The curves shown are for 0.5, 1, 2, 3, ..., kG. In (19.6) the plane of rotation contains the growth axis, and $B = 2, 3, \ldots,$ kG. The data are from Holroyd and Datars [5].

The preferred theoretical direction for Q in potassium is a [110] axis (Overhauser 1971 b). Unfortunately the orientation of the sample K-10 of Figure 19.6 was not determined by a diffraction experiment. However, since potassium films grown on smooth amorphous substrates are known to have a [110] direction perpendicular to the surface [12], it is probable that this was the growth axis. (Nucleation of the crystal was at the bottom of the smooth spherical mould.)

It is clear that a great deal of research on potassium will have to be repeated with samples that are found to be single Q. Induced-torque experiments will become of major importance in investing Q-domain structure.

19.3.3 Optical Anomalies

Free electrons cannot absorb photons. In a solid, the mixing of momentum by periodic potentials, e. g. that given by (19.5), does away with this prohibition and gives rise to inter-band absorption. For potassium the inter-band threshold is 1.3 eV, and the absorption intensity is weak. The discovery by Mayer and El Naby (1963) of an intense optical absorption with a threshold of 0.6 eV created a theoretical crisis. (Their data are shown in Figure 19.7.) A quantitative explanation [16] based on transitions across the energy gap arising from a Fermi-surface instability is indicated by the excellent fit. The transitions involved are shown by the arrow t in Figure 19.2. The only adjustable parameter was the size G of the energy gap.

The inability of others [31] to reproduce the anomaly led many to conclude that it was spurious. However, these unsuccessful repetitions were carried out on evaporated films instead of on a bulk-metal–vacuum interface, as in the original work. These failures actually confirm the CDW-structure interpretation, since the optical properties (as well as the electrical) should be extremely uniaxial. That the original spectrum was not an artifact has been demonstrated by the recent work of Harms, who discovered how to reproduce the anomaly on bulk samples (Overhauser and Butler 1976). Further optical experiments are needed to completely characterize the conditions that control its appearance.

The matrix element for the optical absorption across a CDW energy gap is proportional to the cosine of the angle between Q and the (electric) polarization vector $\hat{\epsilon}$. The latter is (inside the metal) always parallel to the surface for a reflecting infra-red photon. An evaporated potassium film has an allowed Q direction, [110], normal to the surface; interfacial energy will select this direction for Q (in preference to the other five). Consequently the anomalous absorption *should not occur* in evaporated films since e is then perpendicular to Q. That it, in fact, does not occur is a confirmation of the required uniaxial, optical anisotropy. It also verifies that [110] is the optimum direction for Q.

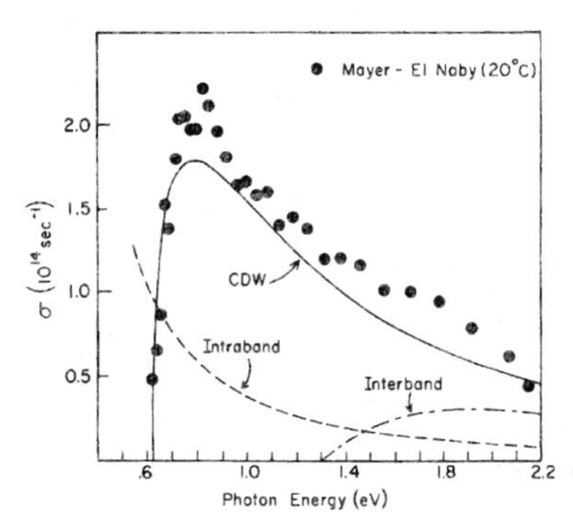

Figure 19.7 Anomalous optical absorption spectrum of potassium. The intra-band conductivity (dashed curve) has been subtracted from the experimental data before being plotted. The solid curve shows the theoretical absorption introduced by a CDW structure. A 'normal' metal would exhibit only the inter-band absorption with a threshold, as shown, at 1.3 eV.

19.3.4 Spin-Resonance Anisotropy

Even if the electron-spin g factor is a function of $\boldsymbol{k}$, conduction-electron spin resonance (CESR) will occur only at a single $\overline{g}$ value, the Fermi-surface average of $g(\boldsymbol{k})$. The reason is that each electron scatters thousands of times during a spin relaxation time. Nevertheless, it was found [36] that when $\boldsymbol{B}$ is oriented so that spin diffusion is restricted to small areas of a thin potassium sample, the CESR splits into well-resolved sharp components. The maximum splitting, 0.5 G at 4200 G, was four times the line-width of the individual components. This is a big effect since the splitting is 10% of the g shift caused by spin–orbit coupling.

The only way to explain such a phenomenon is to suppose the sample to be divided into domains with differing g factors. CDW $\boldsymbol{Q}$ domains provide the solution, since $\overline{g}$ is a function of the angle θ between $\boldsymbol{B}$ and $\boldsymbol{Q}$. A theory of the dependence of $\overline{g}$ on θ led to exact quantitative agreement [24]. The only parameter in the theory was G, and that was fixed by the observed optical anomaly.

19.3.5 Residual-Resistivity Anisotropy

In a cubic metal the resistivity tensor is necessarily isotropic. Nevertheless, the low-field torque data on the potassium shown in Figure 19.6 require that the residual resistivity (T = 1.5 K) of that sample was a cigar-shaped ellipsoid having 4- or 5-to-1 anisotropy. A CDW structure leads to a quantitative explanation of this ratio [2]. The electrical resistivity is proportional to the weighted average of the scattering cross-section $\sigma(\theta)$. The weighting factor, $1 - \cos\theta$, counts 180° scattering heavily and counts small-angle scattering very little.

For impurity scattering, $\sigma(\theta)$ is sharply peaked in the forward direction, i. e. small $\boldsymbol{k}' - \boldsymbol{k}$. The influence of the CDW is easily understood: the wave-function mixing, given by Equation (19.5), caused by the CDW potential allows small-q Fourier components of the impurity potential (which are large in magnitude) to scatter electrons with the conservation rule that

$$\boldsymbol{k}' = \boldsymbol{k} + \boldsymbol{q} \pm \boldsymbol{Q} . \qquad (19.22)$$

This CDW-umklapp effect causes strongly enhanced large-angle scattering, but only for transitions across the Fermi-surface which are nearly parallel to $\pm\boldsymbol{Q}$. Accordingly, the residual resistivity parallel to $\boldsymbol{Q}$ is much larger than it is perpendicularly to $\boldsymbol{Q}$. This effect also accounts for the variation of the residual resistance of potassium wires from run to run [34], which we discuss below.

19.3.6 Hall Coefficient Discrepancy

The LAK magneto-transport theorems required that the *high-field* Hall coefficient R_H is $-1/nec$, where n is the electron density. (This result is for a single simply connected Fermi-surface.) The prediction is precise since n depends only on the lattice constant. Absolute measurements of R_H are difficult using standard four-probe methods, since current distortions near the Hall-field probes cause the sample width at that point to be uncertain. R_H cannot be measured without accurately determining a sample dimension.

Geometric resonances of helicon waves in a flat plate depend only on R_H and the plate thickness [3]. Measurements by this inductive technique on 12 samples at 50 kG showed that R_H is too large by 4 to 8%, depending on crystal orientation. The thickness of the sample was determined by *in situ* ultrasonic time-of-flight measurements. Helicon resonances in a sphere [38] resulted in a discrepancy of 3 to 4%. The diameter of the sphere was determined both mechanically and electrically to a precision of $\frac{1}{2}$%.

These results imply a Fermi-surface that is not simply connected. A quantitative interpretation has not yet been devised.

19.3.7 The Significance of Irreproducibility

Perhaps the most reproducible property of potassium is its irreproducibility. Not only do results vary from sample to sample, they vary drastically from run to run on the same sample. Impurity concentrations are typically $\sim 10^{-4}$ – too small to blame. Less widely appreciated is that the dislocation density is also very small. Primary extinction measurements [20], employing neutron diffraction, imply that the mosaic blocks – the perfect single-crystal regions of a nominal crystal – are about 1 mm in size, much larger than in most other metals. The reason is that the alkali metals continuously recrystallize at room temperature; the debris of deformation anneals out in a fraction of an hour.

Consider some typical difficulties which occur in spite of sample perfection. The residual resistance of a wire will vary by as much as a factor of 2 depending on whether it was cooled to 4 K quickly or slowly, coated with oil or dry, or on how long it was kept at room temperature. One cannot blame the small plastic strains caused by handling since, for potassium, a 1% strain at 4 K causes only a 1.3% increase in resistance [7]. Variations in the linear slope of the high-field magnetoresistance extend over a factor of 6 in single crystals [26] and a factor of 14 in polycrystalline wires [34]. A fivefold change in slope from run to run has been observed in a single specimen.

The most pathological ambivalence of all occurs in the induced-torque patterns of spheres grown in oil [5]. Very little anisotropy is observed in samples which have had the surface oil removed, but if a drop of oil remains at the bottom of a potassium sphere, large high-field torque anomalies appear. The θ dependence is the same as that shown in Figure 19.6, but the peak-to-valley anisotropy is less, e. g. 3 or 4 to 1. The oil itself cannot be the cause since induced torques can result only from bulk currents in the metal.

These and other examples show that there must be some structural parameter of potassium that has escaped experimental control. Since most of these variations involve phenomena that, in themselves, require a multiply connected Fermi-surface, Q domain structure is implicated, which readily explains the residual resistance changes. The anisotropy of the resistivity tensor in a single Q domain, described above, would permit the residual resistance of a sample to change by up to a factor of 4, depending on the distribution of Q domain orientations.

Small potassium spheres grown in oil probably have an isotropic Q domain distribution. A drop of oil (on the bottom) causes, after freezing, a directed differential thermal stress on further cooling to 4 K. This (apparently) permits Q domains to regrow with a preferred orientation in the lower part of the sphere.

19.3.8 The Challenge: Q Domain Control

A new era in research on the alkali metals will begin when it is discovered how to produce and maintain single Q samples. This will be more difficult than the analogous problem in ferromagnets, where applying a large magnetic field produces a single domain. A single SDW domain in chromium was generated [1, 13] by cooling a stress-free crystal through the critical temperature in a 28 kG field. With a 0.6 eV energy gap, the critical temperature of potassium far exceeds the melting temperature. Furthermore, the interaction of a CDW with a magnetic field should be very small.

The available interactions for Q domain orientation are few: uniaxial stress, magnetic field and surface interaction. The stress sensitivity of the magnetoresistance [26] and the oil-drop effect discussed above show that stress can be used. Unfortunately the plasticity of potassium is so high [6] that control by uniaxial stress may be very precarious. Hydroxide layers on the surface may lead to disorientating thermal stress on cooling to 4 K. Perhaps the single Q character of the sphere for which data is shown in Figure 19.6 depended on the fact that it was protected throughout the course of the experiments in the Kel-F mould in which it was grown.

The anisotropy in the diamagnetism of a Q domain is probably $\sim 10^{-8}$ e.m.u. Whether the favoured orientation for Q is parallel or perpendicular to B is not known; however, fields in excess of 100 kG are required to compete with stress effects. Conversion of a poly-domain specimen to a single-domain one may be impeded to some extent by impurities.

The most promising technique at present is the use of the surface interaction at the *face* of nucleation (during crystal growth) to produce a single Q configuration. Subsequent protection from all sources of stress must then be maintained.

19.3.9 Phase Excitations and Satellite Intensity

Although the phenomena we have discussed above show that potassium has a CDW structure, a persistent question is 'where are the CDW diffraction satellites?' Actually a careful search for them has never been reported. Their location in k space is along a [110] direction. However, since $|Q| \sim 1.33(2\pi/a)$, the exact location is very close to a Bragg reflection (at 1.41). This makes the search difficult because the very low phonon frequencies of potassium cause every Bragg reflection to be surrounded by an intense cloud of diffuse phonon scattering. If the direction and magnitude of Q were rigid, the satellite intensity would be about 10^{-3} of a (110) Bragg peak. This is large enough to permit observation at 4 K with a triple-axis neutron spectrometer set at zero energy transfer (to minimize the phonon background).

However every broken symmetry gives rise to new collective excitations, e. g. spin waves in ferromagnets or Goldstone bosons in field theory. For a CDW structure, the new modes are phasons [20]. The ground-state energy of an *incommensurate* CDW is independent of its spatial location relative to the lattice ions, i. e. the energy is independent of the CDW phase ϕ. Therefore the energy will be only slightly increased if ϕ is allowed to vary *slowly* in space and time:

$$\phi(L, t) = \sum_q \phi_q \sin(q \cdot L - \omega_q t) , \tag{19.23}$$

where $\phi_q <$ is the amplitude of each phason mode and ω_q its frequency. The displacement of each ion from its ideal lattice site L is no longer given by Equation (19.2), but by

$$u = A \sin[Q \cdot L + \phi(L, t)] . \tag{19.24}$$

If q is perpendicular to Q there is a periodic, *local* rotation of Q; if q is parallel to Q there is a periodic *local* modulation of the magnitude of Q.

The lattice dynamics of phason excitations leads to a frequency spectrum that one would anticipate to result from a broken symmetry. aω_q approaches zero linearly in $|q|$ and is highly anisotropic. The modes are extremely 'soft' for q perpendicular to Q (the ones associated with local rotations of Q). One consequence is that each CDW satellite is surrounded by a pancake-shaped diffuse cloud caused by phason scattering. Diffraction streaks do surround the CDW satellites in the incommensurate state of 1T-TaS_2 [39]. The diffraction streaks disappear, as expected, in the commensurate state since phasons are then only optical phonons of finite frequency.

Another important property of phase modulation is a reduction in satellite intensity [20, 22]. Total intensity is not lost, however; it is transferred into the pancake-shaped phason cloud. The reduction factor is approximately

$$F_p = \exp\left[\frac{-2\gamma\hbar^2\left(n + \frac{1}{2}\right)}{MA^2 k_B \Theta_D}\right] . \tag{19.25}$$

This is the analogue of a Debye–Waller factor for an ordinary Bragg reflection. However, instead of depending on the square of the scattering vector in Equation (19.11), it depends on $1/A^2$. θ_D is a phason 'Debye temperature', M the ionic mass, n the Bose–Einstein number (for an oscillator of 'frequency' θ_D), and γ a parameter which depends on the allowed $\boldsymbol{k}$ space for phasons and the anisotropy of ω_q

Without experimental knowledge of the phason spectrum, it is difficult to assign a definite value to F_p. Our best estimate is that (for potassium) it is smaller than 10^{-3} at $T = 0$. This implies that finding the CDW satellite in potassium will be extremely difficult. Special techniques that integrate over the diffuse phason "pancake" will no doubt be required.

19.3.10 The de Haas–van Alphen Difficulty

One of the most important and general techniques for probing Fermi surfaces is the de Haas–van Alphen (dHvA) effect. The extremal ($\boldsymbol{k}$ space) areas of cyclotron orbits are directly related to the periodic variations of the diamagnetic susceptibility with $1/B$. A Fermi-surface shape can be mapped out with high precision when the dHvA frequency as a function of crystal orientation (relative to $\boldsymbol{B}$) is known.

The periodic potential in the Schrödinger equation for a CDW structure causes the Fermi surface to be distorted from a spherical shape, as shown in Figure 19.3. However, dHvA studies of potassium [29] indicate that the Fermi-surface is spherical to about 2 parts in 10^3 This result contradicts virtually all the data that have been cited in this article. Is a reconciliation possible?

The samples used in the dHvA studies were spheres grown in oil, which was subsequently removed. Undoubtedly these single crystals contained many Q domains. dHvA signals of a single domain will most likely be observable only when $\boldsymbol{B}$ lies within a conical sector centred along Q, and for these orientations the Fermi-surface cross-section perpendicular to Q is measured. As $\boldsymbol{B}$ is rotated away from Q, the extreme-area orbits will approach the distorted regions of the Fermi-surface, near the necks shown in Figure 19.3. These trajectories will be 'killed' by phason scattering, which is very intense for electrons that approach (in $\boldsymbol{k}$ space) the CDW energy gaps.

Our conjecture is that the reported dHvA isotropy can be attributed to Q domain structure, which provides (for each direction of $\boldsymbol{B}$) a fractional part of the sample with orbit planes nearly perpendicular to Q. One would expect to see interference effects from the dHvA signals of different Q domains, and a fading of the signal during sample rotation has indeed been reported, but was ascribed (without proof) to extrinsic causes. Without question, one of the most important experiments for the future is a dHvA study of a single Q sphere, which exhibits torque patterns similar to those shown in Figure 19.6.

Other experimental phenomena, such as cyclotron resonance, limiting-point resonance and geometric-size-effect resonance, have been interpreted within the standard model – a spherical Fermi surface. We believe that, here also, poly-domain samples have concealed the inherent anisotropy of the CDW structure.

19.4 Prospects for the future

The discovery of CDWs in transition metal dichalcogenides has already led to a broad range of investigations and the developments of phenomenological models to describe the results. Further work along these lines may be expected.

We also look forward to a resurgence of interest in the alkali metals because of their inherent simplicity. For that reason the more venturesome theorist runs a greater risk – he has the opportunity to predict (not merely to explain) and to be found wrong. Many experimental challenges are already clear: direct observation of Q domains, control of domain structure – especially fabrication of single Q specimens, determination of all experimental anisotropies, measurement of the phason spectrum, and of phason–electron interactions, finding evidence of heterodyne gaps and the fields at which they undergo magnetic breakdown, and observation of hyperfine-field effects and the motional narrowing caused by phase excitations.

Finally with experience gained from the studies suggested above, one may investigate whether CDW instability is epidemic in other elemental or nearly isotropic metals.

Acknowledgements

I am indebted to the hospitality of the Aspen Center for Physics during the summer of 1977, which made possible the preparation of this manuscript. The support of the National Science Foundation, including that of the Materials Research Laboratory Program, contributed significantly to this study.

References

1. Arrot, A., Werner, S.A., and Kendrick, H. (1965) *Phys. Rev. Lett.*, **14**, 321.
2. Bishop, M.F. and Overhauser, A.W. (1977) *Phys. Rev. Lett.*, **39**, 632.
3. Chimenti, D.E. and Maxfield, B.W. (1973) *Phys. Rev. B*, **7**, 3501.
4. Hockett, R.S. and Lazarus, D. (1974) *Phys. Rev. B*, **10**, 4100.
5. Holroyd, F.W. and Datars, W.R. (1975) *Can. J. Phys.*, **53**, 2517.
6. Hull, D. and Rosenberg, H.M. (1959) *Phil. Mag.*, **4**, 303.
7. Jones, B.K. (1969) *Phys. Rev.*, **179**, 637.
8. Lass, J.S. (1976) *Phys. Rev. B*, **13**, 2247.
9. Lifshitz, I.M., Azbel, M. Ya, and Kagonov, M.I. (1956) *Zh. Eksp. Teor. Fiz.*, **31**, 63 (English translation: (1957) *Sov. Phys. JETP*, **4**, 41).
10. Lomer, W.M. (1962) *Proc. Phys. Soc.*, **80**, 489.
11. Mayer, H. and El Naby, M.H. (1963) *Z. Phys.*, **174**, 269.
12. Monin, J. and Boutry, G.A. (1974) *Phys. Rev. B*, **9**, 1309.
13. Montalvo, R.A. and Marcus, J.A. (1964) *Phys. Rev. Lett.*, **8**, 151.
14. Overhauser, A.W. (1960) *Phys. Rev. Lett.*, **4**, 462.
15. Overhauser, A.W. (1962) *Phys. Rev.*, **128**, 1437.
16. Overhauser, A.W. (1964) *Phys. Rev. Lett.*, **13**, 190.
17. Overhauser, A.W. (1968) *Phys. Rev.*, **167**, 691.
18. Overhauser, A.W. (1970) *Phys. Rev. B*, **2**, 874.
19. Overhauser, A.W. (1971a) *Phys. Rev. Lett.*, **27**, 938.
20. Overhauser, A.W. (1971b) *Phys. Rev. B*, **3**, 3173.
21. Overhauser, A.W. (1974) *Phys. Rev. B*, **9**, 2441.
22. Overhauser, A.W. (1978) *Hyperfine Interactions*, **4**, 786.
23. Overhauser, A.W. and Butler, N. (1976) *Phys. Rev. B*, **14**, 3371.
24. Overhauser, A.W., and de Graaf, A.M. (1968) *Phys. Rev.*, **168**, 763.
25. Peierls, R.E. (1955) *Quantum Theory of Solids* Oxford University Press, London, p. 108.
26. Penz, P.A. and Bowers, R. (1968) *Phys. Rev.*, **172**, 991.
27. Schaefer, J.A., and Marcus, J.A. (1971) *Phys. Rev. Lett.*, **27**, 935.
28. Schouten, D.R., and Swenson, C.A. (1974) *Phys. Rev. B*, **10**, 2175.
29. Shoenberg, D., and Stiles, P.J. (1964) *Proc.R. Soc. A*, **281**, 62.
30. Slater, J.C. (1951) *Phys. Rev.*, **81**, 385.
31. Smith, N.V. (1969) *Phys. Rev.*, **183**, 634.
32. Smith, P.A. and Smith, C.S. (1965) *J. Phys. Chem. Solids*, **26**, 279.

33 Stroud, D. and Pan, F.P. (1976) *Phys. Rev. B*, **13**, 1434.
34 Taub, H., Schmidt, R.L., Maxfield, B.W., and Bowers, R. (1971) *Phys. Rev. B*, **4**, 1134.
35 Visscher, P.B. and Falicov, L.M. (1970) *Phys. Rev. B*, **2**, 1518.
36 Walsh Jr., W.M., Rupp Jr., L.W., and Schmidt, P.H. (1966) *Phys. Rev.*, **142**, 414.
37 Werner, S.A., Gurmen, E., and Arrott, A. (1969) *Phys. Rev.*, **186**, 705.
38 Werner, S.A., Hunt, T.K., and Ford, G.W. (1974) *Solid State Commun.*, **14**, 1217.
39 Wilson, J.A., Di Salvo, F.J., and Mahajan, S. (1975) *Adv. Phys.*, **24**, 117.

Reprint 20 Residual-Resistivity Anisotropy in Potassium[1)]

Marilyn F. Bishop and A.W. Overhauser**

* Department of Physics, Purdue University, 525, Northwestern Avenue, West Lafayette, Indiana 47907, USA

Received 23 March 1978

We present evidence from induced-torque measurements of Holroyd and Datars supporting an anisotropy in the residual resistivity of potassium of about five to one and argue against explanation of their data by extrinsic mechanisms. We explain how the hypothesis that the conduction electrons are in a static charge-density-wave state leads naturally to the prediction of a large residual resistivity anisotropy. Resistivity calculations are performed using several model impurity-scattering potentials. Numerical results are presented for parameters characteristic of potassium, yielding anisotropics as large as four to one. We suggest that induced-torque experiments might be used to determine whether a given potassium sample has a single charge-density-wave domain and encourage that de Haas–van Alphen experiments be performed on such 'single-domain' samples.

20.1 Introduction

Reports of a giant torque anomaly in potassium have generated considerable controversy in recent years. Potassium is generally considered to be of cubic symmetry and, according to most textbook descriptions, to be an example of a nearly-free electron metal. Within this simple picture, the induced torque in spherical samples of potassium is expected to be independent of magnetic field direction. For this reason, the large anisotropies observed originally by Schaefer and Marcus [1] at 4 K in hundreds of spherical samples of potassium created a severe dilemma. A four-peaked pattern of twofold symmetry that contradicted the cubic symmetry of the metal appeared in the induced torque. Earlier, Lass had measured the induced torque in one large spherical sample and had found the results to be independent of magnetic field direction [2]. In response to the anisotropies, he suggested that the observations could be explained if the samples were sufficiently non-spherical [3], except that torque data at 80 K showed that most samples were essentially undistorted.

In response to these conflicting reports, Holroyd and Datars [4] decided to test the dependence of results on methods of preparation of samples of potassium. They were able to reproduce results of both groups in a controlled manner. In addition, for some samples, they found torque anisotropies larger than those seen by Schaefer and Marcus, and in one particular sample, nearly an order of magnitude larger. No theoretical explanation was given for the observed dependences on sample preparation.

In this paper, we will concentrate on the results, shown in Figure 20.1, of the sample for which the most bizarre induced-torque effects were observed. For this specimen, sample K-10 of Holroyd and Datars, the four-peaked pattern for the largest magnetic fields used exhibited an anisotropy of 45 to one, while this pattern at low fields became two-peaked, with the old minima becoming the new maxima and minima. This anisotropy, extrapolated to zero field, is about three to one, implying that the residual resistivity of potassium has an anisotropy of about five to one, as we shall see in Section 20.2. The purpose of this paper is to present a

1) Phys. Rev. B 18, 6 (1978)

2) A preliminary exposition of this theory appeared in [5].

Anomalous Effects in Simple Metals. Albert Overhauser

ISBN: 978-3-527-40859-7

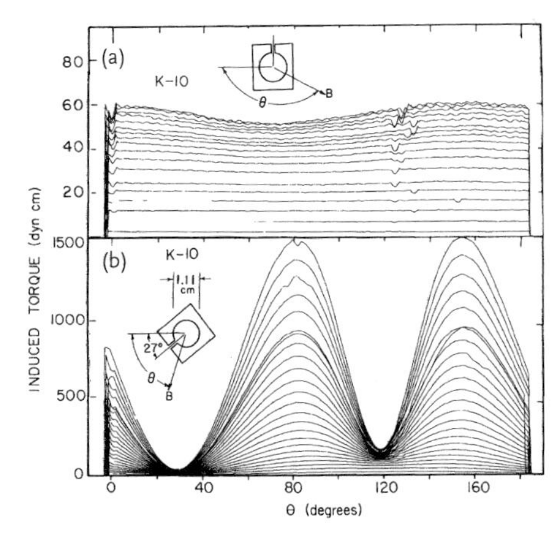

Figure 20.1 Induced-torque vs. magnetic field direction for several field values for sample K-10 of Holroyd and Datars, which was prepared in a mold. The sample was a sphere of diameter 1.11 cm. (a) The field was rotated about the growth axis and the magnitude of the field was 500 G for the lowest curve and ranged from 1 to 17 kG, in steps of 1 kG, for the higher curves, (b) The field was rotated in a plane containing the growth axis, with magnitudes ranging from 1 to 23 kG, in steps of 1 kG.

possible interpretation of this resistivity anisotropy. We will show that the assumption that the conduction electrons in potassium are in a static charge-density-wave state is consistent with the observed residual resistivity anisotropy.[2)] We make no attempt here to explain the exotic high-field behavior.

The organization of the paper is as follows. In Section 20.2, we discuss the induced-torque experiments and include data of Holroyd and Datars. In addition, we consider the possibilities of extrinsic mechanisms explaining the observed anomalies. In Section 20.3, we explicate the relevant properties of charge-density waves. In Section 20.4, we introduce the model impurity scattering potentials to be used for the residual resistivity calculation of Section 20.5 and for the numerical predictions of Section 20.6. Finally, in Section 20.7 we present the conclusions.

20.2 Induced-Torque Experiments

Induced-torque experiments are performed by suspending a sample, usually spherical in shape, in a slowly rotating magnetic field. Because the field varies in time, currents are induced that interact with the magnetic field to exert a torque on the sample. This torque is measured as a function of both the direction and magnitude of the magnetic field. The induced torque in a spherical sample of a simple metal, i.e., a metal with a spherical Fermi surface, should be independent of the orientation of the magnetic field. For a coordinate system in which the axis of rotation is along y and the magnetic field of constant magnitude $\boldsymbol{B}$ is along z, the torque induced is in the y direction and is given by [6–8]

$$N_y = \frac{2\pi R^5}{15}\Omega n^2 e^2 \rho_0 \frac{(\omega_c\tau)^2}{1+\left(\frac{1}{2}\omega_c\tau\right)^2}\,, \tag{20.1}$$

where $\boldsymbol{R}$ is the radius of the sphere, Ω is the rotation speed of the magnet, n is the electron density, $\omega_c(= eB/mc)$ is the cyclotron frequency, τ is the relaxation time, and ρ_0 is the resistivity. While de Haas–van Alphen measurements on potassium seem to indicate that its Fermi surface is spherical to within 0.1% [9, 10], gigantic anisotropics are observed in induced-torque experiments, which are in conflict with this simple model.

In order to study explicitly the results of induced-torque experiments, we refer to the data for sample K-10 of Holroyd and Datars [4] in Figure 20.1. This specimen exhibits an anisotropy that is nearly an order of magnitude larger than that observed on any other sample ever studied and is therefore an excellent example of the effects that we wish to discuss in this paper. The induced torque is plotted versus orientation of the magnetic field for various field strengths, with torque an increasing function of field. In Figure 20.1(a), field values are for 500 G and from 1 to 17 kG in steps of 1 kG. In Figure 20.1(b), field values are for 1–23 kG in steps of 1 kG. The magnetic field was rotated at a speed of 22°/min for axes of rotation along the growth axis (Figure 20.1(a)) and perpendicular to the growth axis (Figure 20.1(b)). The temperature of the sample was kept at 1.5 K during experiments. In order to guarantee a precise shape for the sample, the potassium was grown in a kel-F mold with a spherical cavity of diameter of 1.11 cm, machined to within 10^{-3} in [4]. The angular dependence in Figure 20.1(a), obtained when the magnetic field was rotated in a plane perpendicular to the growth axis, is essentially isotropic and is thus in agreement with the behavior expected from a free-electron metal, as given in Equation (20.1). On the contrary, an anomalous four-peaked anisotropy of 45 to one appears in the torque pattern for high fields for the case in which the growth axis was in the plane of rotation, as in Figure 20.1(b). The magnitudes of these peaks decrease with respect to the minima when the strength of the field is reduced, until at low fields, the four minima become the new maxima and minima of a two-peaked pattern. For this case, the dips occur for the field along the growth axis and the peaks for the field perpendicular to the growth axis. As a consequence of these orientational dependences, the growth axis emerges as a preferred direction in this specimen. Thus potassium must be electrically uniaxial and not cubic Unfortunately, the sample orientation was not determined by a diffraction experiment. However, it is known that K films grown on smooth, amorphous substrates have a [110] direction perpendicular to the surface [11]. Therefore, since the nucleation of the crystal was at the bottom of a smooth spherical mold, it is probable that the growth axis was along some [110] direction.

According to the Lifshitz–Azbel–Kaganov (LAK) transport theorems [12], a metal with a single, simply-connected Fermi surface must have a resistivity $\rho(\boldsymbol{B})$ independent of field $\boldsymbol{B}$ at high fields, i.e., $\omega_c\tau \gg 1$. Since the torque depends on this magnetoresistance, it too must saturate. In fact, for the sample of Figure 20.1, the torque at the peaks should have stopped increasing at about 1 or 2 kG, yet it continues to increase even at 23 kG. Thus, the torque patterns prove in addition that the Fermi surface of K is multiply connected. This is in direct contradiction with de Haas–van Alphen measurements [9, 10].

In this paper we will not attempt to explain the high-field torque patterns and instead turn our attention to the low-field results. The low magnet rotation rate makes the method nearly a dc measurement. In order to examine the relevant data in more detail, we plot with circles in Figure 20.2 the induced torque at 28°, or at the minimum in Figure 20.1(b) (lower curve), and at 118° (middle curve) versus magnetic field. The upper circles at given fields are ratios of the corresponding points in the lower two curves and thus give the torque anisotropy. Solid curves are drawn through the data, with dashed curves possible extensions of the ratio to zero field. Therefore, we -estimate the zero-field limit of the torque anisotropy to be between 2.5 and 3 to 1. If a sample is completely spherical an induced-torque anisotropy can result only if the resistivity tensor ρ of the metal is not isotropic. In general, the expression for the torque N_y induced in a sphere is given by [7, 8]:

$$N_y = \frac{4\pi\left(15c^2\right)^{-1} \boldsymbol{R}^5 \boldsymbol{B}^2 \Omega \lambda}{\lambda\left(\rho_{yy}+\rho_{zz}\right)-\left(\rho_{xx}+\rho_{zz}\right)\rho_{xz}\rho_{zx}-\left(\rho_{xx}+\rho_{yy}\right)\rho_{xy}\rho_{yx}-\rho_{xy}\rho_{yz}\rho_{zx}+\rho_{xz}\rho_{zy}\rho_{yx}}, \tag{20.2a}$$

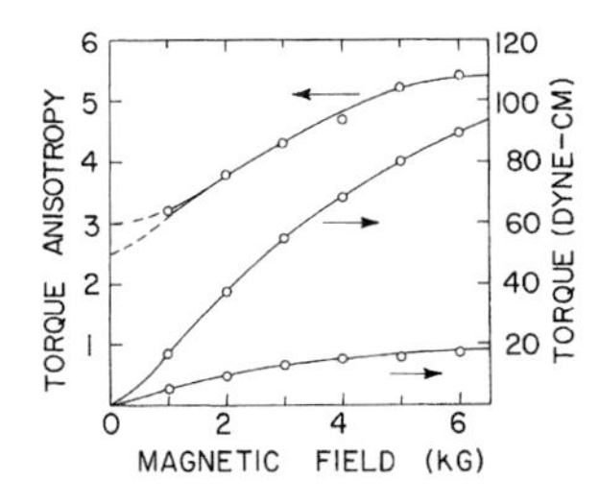

Figure 20.2 Low-magnetic-field values of induced torque shown in Figure 20.1, with data points plotted in circles. The lower curve is the minimum in induced torque taken at an angle $\theta = 28°$ in Figure 20.1, while the middle curve is taken at $\theta = 118°$. The upper curve is the ratio of these limiting values and is the low-field-torque anisotropy. Solid curves are drawn through the experimental points, and the dashed curves are reasonable extrapolations of the anisotropy to zero field.

where

$$\lambda \equiv (\rho_{xx} + \rho_{zz})(\rho_{xx} + \rho_{yy}) - \rho_{yz}\rho_{zy} . \tag{20.2b}$$

Here, the $\{\rho_{ij}\}$ are elements of $\overleftrightarrow{\rho}$. The magnetic field $\vec{B}$ is assumed to be in the z direction with axis of rotation along y.

For sample K-10 of Holroyd and Datars, for which the growth axis was determined from Figures 20.1 and 20.2 to be a preferred direction in the specimen, the zero-field resistivity tensor $\overleftrightarrow{\rho}$ is diagonal in a coordinate system in which the growth axis is along any Cartesian axis. The resistivities parallel and perpendicular to the growth axis are given by $\gamma\rho_0$ and ρ_0, respectively. For low fields, we use the resistivity tensor that includes the Hall effect. If the growth axis is along $\hat{x}$ and $\vec{B}$ along $\hat{z}$, the resistivity tensor for low fields is given by

$$\overleftrightarrow{\rho} = \rho_0 \begin{pmatrix} \gamma & \omega_c\tau & 0 \\ -\omega_c\tau & 1 & 0 \\ 0 & 0 & 1 \end{pmatrix} , \tag{20.3}$$

where $\rho_0\omega_c\tau = B/\text{nec}$ and $\rho_0 = m/ne^2\tau$. From Equation (20.2), we obtain the torque along $\hat{y}$ for this case,

$$N_y^{(1)} = \frac{4\pi R^5 B^2 \Omega}{15c^2\rho_0} \frac{(1+\gamma)^2}{2(1+\gamma)^2 + (1+\gamma)(\omega_c\tau)^2} . \tag{20.4}$$

For a second case with the growth axis and $\vec{B}$ both along $\hat{z}$, the resistivity tensor is

$$\overleftrightarrow{\rho} = \rho_0 \begin{pmatrix} 1 & \omega_c\tau & 0 \\ -\omega_c\tau & 1 & 0 \\ 0 & 0 & \gamma \end{pmatrix} . \tag{20.5}$$

The torque along $\hat{y}$ is then

$$N_y^{(2)} = \frac{4\pi R^5 B^2 \Omega}{15c^2\rho_0} \frac{2(1+\gamma)}{2(1+\gamma)^2 + 2(\omega_c\tau)^2} . \tag{20.6}$$

From Equations (20.4) and (20.6), we may find the ratio $\mathfrak{R}$ between $N_y^{(1)}$ and $N_y^{(2)}$ or the torque anisotropy for low fields,

$$\mathfrak{R} = \frac{(1+\gamma)^2 + (\omega_c\tau)^2}{2(1+\gamma) + (\omega_c\tau)^2} . \tag{20.7}$$

In the limit of zero field, $\omega_c = 0$, and the expression reduces to

$$\mathfrak{R} = \frac{1}{2}(\gamma + 1) . \tag{20.8}$$

Since $\mathfrak{R}$ in this limit is between 2.5 and 3, γ must be between 4 and 5. This is consistent with the value $\gamma = 4$ that was originally estimated from the data of Schaefer and Marcus [8]. From the data in Figures 20.1 and 20.2, we find that $\omega_c\tau \sim 2.5$ at 1 kG, implying that $\tau \sim 1.4 \times 10^{-10}$ s, for the directions perpendicular to the growth axis. Along the growth axis, $\omega_c \sim 0.5$ or $\tau \sim 2.8 \times 10^{-11}$ s.

The anisotropy versus field in Figure 20.2 increases with increasing field, unlike Equation (20.7), which decreases. This is probably a manifestation of high-field effects, which are not explained in this paper. Nevertheless, Equation (20.7) can provide some theoretical basis for the extrapolation procedure. Certainly one cannot easily extrapolate the torque curves themselves to zero, since they must vary according to B^2 as $B \to 0$. The ratio removes this dependence, so that the reasonable curve to extrapolate is that showing the anisotropy. Suppose that one tried to extrapolate this curve to unity, i.e., $\mathfrak{R} \to 1$ as $\boldsymbol{B} \to 0$. This means that $\gamma = 1$ at $\boldsymbol{B} = 0$, as can be seen from Equation (20.8). Since $\mathfrak{R} \approx 3$ at $\boldsymbol{B} = 1$ kG, then it must be true that $\gamma \sim 6$ at 1 kG [Equation (20.7)]. Therefore, this requires a 500% magnetoresistance along the growth axis for 1 kG. However, the observed magnetoresistance in potassium is typically 2% for 1 kG.

The postulation of an oriented array of dislocations arising from unusual strain patterns in a sample could lead to a resistivity that is not isotropic. Strains in samples of interest can be estimated to be on the order of a percent. Jones found the residual resistance ratio in potassium wires to decrease linearly by 1.3% per percent longitudinal strain [13]. Thus, the value of $(\gamma - 1)$ determined from a dislocation model could be at most a few percent, or orders of magnitude smaller than that observed.

As an alternative to the above discussion, Lass [3] has suggested that the data of Schaefer and Marcus could be explained within the model of an isotropic resistivity tensor if the samples had been non-spherical by 10%–15%. The model that he proposed was that of an ellipsoid of revolution with the radius along the axis of the ellipsoid a fraction of the radius R perpendicular to the axis.

The induced torque is then given by [3]

$$N_y = \frac{2\pi}{15c^2}\sigma \boldsymbol{B}^2 \Omega \boldsymbol{R}^5 \eta \left(\frac{1 - E + E\sin^2\beta\sin^2\theta + (\omega_c\tau)^2 E^2 \sin^4\beta \sin^2\theta\cos^2\theta}{1 + \frac{1}{4}(\omega_c\tau)^2[1 - E + E(3+E)\sin^2\beta\cos^2\theta]} \right), \tag{20.9}$$

where β is the angle between the axis of the ellipse and the y direction, and θ is the angle between the magnetic field direction $\boldsymbol{B}$ and the projection of the axis of the ellipse onto the x-z plane. $\sigma = ne^2\tau/m$ is the electrical conductivity and $E = (1-\eta^2)/(1+\eta^2)$ is an asymmetry parameter. With $\eta \approx 0.9$, Lass was able to produce curves similar to those of Schaefer and Marcus. However, the deviations from sphericity were determined experimentally to be 2% or less by performing torque measurements at 80 °K, or for $\omega_c\tau \ll 1$.

An even greater difficulty with this explanation arises in the consideration of the results of sample K-10 of Holroyd and Datars, given here in Figures 20.1 and 20.2. The torque anisotropy at low fields could be reproduced by Lass's model [Equation (20.9)] if $\eta \approx 0.45$, supposing that the mold had been only about half filled.[3] Nevertheless, this could not possibly explain the data. For a magnetic field of 1 kG, the torque calculated from Equation (20.9) with $\beta = \frac{1}{2}\pi$ and $\theta = 0$, corresponding to the minimum in the angular pattern in Figure 20.1(b) (28° in that figure), could never exceed 3.2 dyn cm, regardless of relaxation time

3) However, because the kel-F mold was transparent, both Holroyd and Datars observed that it was completely filled. They also remarked that sample K-10 had a diameter of 1.11 cm (not 1.71), and that the ordinate scale of (their) Figure 20.5(a) was incorrect. Magnet rotation rate for this spectrum was 22°/min (not 35°/min).

τ. The torque of 5.4 dyn cm observed by Holroyd and Datars is therefore much too large to be consistent with this explanation, implying that the mold had to be filled. The exceedingly unlikely possibility that a planar crack developed in the sample perpendicular to the growth axis could remove this objection. The two equal hemispheres thus produced could be represented by two ellipsoids with $\eta = 0.5$, yielding twice the torque of a half-filled mold. On the other hand, even this hypothesis cannot provide a satisfactory explanation of the data. If in Equation (20.9), $\beta = \frac{1}{2}\pi$ and $\theta = 0.5$, the appropriate value of $\omega_c\tau$ is determined by the criterion that the angular pattern in θ remain two peaked through a magnetic field strength of 2 kG and begin the four-peaked behavior by 3 kG. However, this choice, $\omega_c\tau \approx 1.8$ causes the ratio $\mathfrak{R}$ to be more than 50% too large. In addition, an enormous Kohler slope, larger than 0.12, is needed to fit the peaks at 23 kG. Unfortunately, with this value, the dips in the angular pattern no longer agree reasonably with experiment. Therefore, it is impossible to obtain legitimate agreement of the results of this model with the data in Figures 20.1 and 20.2. We conclude, therefore, that the residual resistivity is indeed anisotropic and that Equations (20.2) are the corresponding expressions for the induced torque.

A search for an explanation of the residual resistivity anomaly in potassium leads us to the supposition that the conduction electrons are in a static charge-density-wave state. Other alkali-metal anomalies have been explained as a result of this proposition.[4] As we shall show, this hypothesis provides a quantitative explanation for the anisotropy observed in the zero-field limit in sample K-10 by Holroyd and Datars. Although we have used the high-field induced-torque data in our arguments against proposed explanations by extrinsic mechanisms, we make no attempt here to explain the high-field anomalies.

20.3 Charge-Density Waves

For simplicity, since the effects of a lattice are not important here, we consider a jellium model in which the positive ions are replaced by a uniformly charged deformable jelly. Within the simple free-electron theory of metals, since the positive background is uniform, it is assumed that the electronic ground state will be that of a spatially uniform charge density. However, it has been shown that this is possibly not the case when electron-electron interactions are taken into account [16, 17]. A symmetry-breaking instability can occur, leading to a spin-density-wave (SDW) or charge-density-wave (CDW) state. For alkali metals, theoretical arguments, coupled with experimental evidence, favor the CDW state.

A CDW is characterized by a periodic spatial modulation in electron charge density,

$$\rho(\vec{r}) = \rho_0(1 - p\cos\vec{Q}\cdot\vec{r})\,, \tag{20.10}$$

where $\vec{Q}$ and p are the wave vector and the fractional amplitude of the CDW, respectively. The positive background deforms in order to assure local-charge neutrality, with the static local displacement given by

$$\vec{u}(\vec{r}) = \left(p\vec{Q}/Q^2\right)\sin\vec{Q}\cdot\vec{r}\,. \tag{20.11}$$

In the deformable jellium model, this adjustment incurs no energy penalty. Whether or not a particular real metal has a CDW ground state depends on how closely it resembles this model.

In alkali metals, the Born–Mayer ion–ion interactions are known to be extremely weak [18], enabling the ions to displace easily from their equilibrium positions. The magnitude of $\vec{Q}$

4) For a review, see [14] or [15].

is approximately the diameter of the Fermi surface, as will be discussed below. The lowest energy direction for $\vec{Q}$ is determined by that direction that minimizes the wave vector $\vec{Q}'$ of the static phonon of Equation (20.11), given by [19, 20]

$$\vec{Q}' = (2\pi/a)\,(1,1,0) - \vec{Q}\,, \tag{20.12}$$

where a is the lattice constant of the bcc lattice.

Clearly, the favorable direction for $\vec{Q}$ is along the [110] direction. This is because any deformation energy associated with the neutralization of charge density in Equation (20.10) by displacements in Equation (20.11) should be proportional to the wave vector $\vec{Q}'$.

The spatial modulation of the electron density given in Equation (20.10) will occur only if each electron experiences a sinusoidal potential. Such a potential arises self-consistently through exchange and correlation potentials of the electron gas. The one-electron Schrödinger equation incorporating these effects may be written

$$(p^2 2m + G\cos\vec{Q}\cdot\vec{r})\Psi_{\vec{k}} = E_{\vec{k}}\Psi_{\vec{k}}\,, \tag{20.13}$$

The charge modulation of the electron gas occurs only if the periodic part G of the exchange and correlation potential is sufficiently large. However, the density modulation must be large enough to generate this periodic part of the potential. This process occurs in a self-consistent manner, requiring a $\vec{k}$ dependence of G. Solution of an integral equation for $G(\vec{k})$ is required. Since such analyses overly complicate the problem of interest, we approximate G as a constant, whose value is determined by experiments, assuming that such an assumption does not alter the principal features of our calculations.

Since Equation (20.13) is the Mathieu equation, it does not have compact solutions. On the other hand, we obtain sufficient accuracy by dividing the periodic potential into two parts, one that leads to an energy gap at $k_z = -\frac{1}{2}Q$ and the other that leads to a gap at $k_z = \frac{1}{2}Q$. In the calculation of the residual resistivity, only the states below the gap need be considered. Near each gap, the energies and wave functions are significantly altered, requiring an accurate solution with degenerate first-order perturbation theory. First, we consider the case for $k_z \approx -\frac{1}{2}Q$ and assume that the perturbed state $|\vec{k}\rangle$ mixes with $|\vec{k}+\vec{Q}\rangle$ only. This leads to the secular equation

$$\begin{vmatrix} \epsilon_{\vec{k}} - E_{\vec{k}} & \frac{1}{2}G \\ \frac{1}{2}G & \epsilon_{\vec{k}+\vec{Q}} - E_{\vec{k}} \end{vmatrix} = 0\,, \tag{20.14}$$

within the space defined by the basis set $|\vec{k}\rangle$ and $|\vec{k}+\vec{Q}\rangle$, where $\epsilon_{\vec{k}} = \hbar^2 k^2/2m$, the unperturbed energy of the state $|\vec{k}\rangle$. The energy for the solution below the gap is

$$E_{\vec{k}} = \frac{1}{2}\left(\epsilon_{\vec{k}} + \epsilon_{\vec{k}+\vec{Q}}\right) - \frac{1}{2}\left[\left(\epsilon_{\vec{k}} - \epsilon_{\vec{k}+\vec{Q}}\right)^2 + G^2\right]^{1/2}. \tag{20.15}$$

The corresponding eigenfunction is

$$\Psi_{\vec{k}} = \cos\xi\, e^{i\vec{k}\cdot\vec{r}} - \sin\xi\, e^{i(\vec{k}+\vec{Q})\cdot\vec{r}}\,; \tag{20.16}$$

where

$$\cos\xi(\vec{k}) = G\Big/\left[G^2 + 4\left(\epsilon_{\vec{k}} - E_{\vec{k}}\right)^2\right]^{1/2}. \tag{20.17}$$

Far away from the gap, these solutions reduce to those found by nondegenerate perturbation theory. The CDW instability is optimized if the Fermi energy E_F lies in the CDW energy gap.

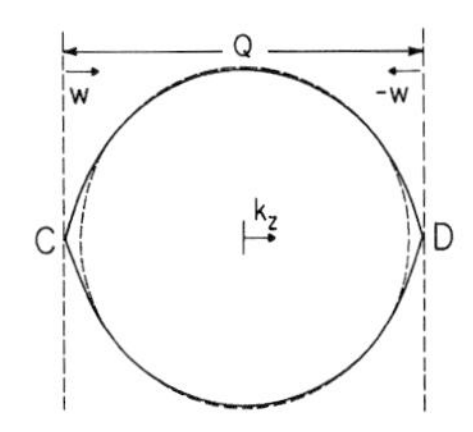

Figure 20.3 Lime-shaped Fermi surface (solid curve) for a jellium model in which conduction electrons are in a CDW ground state with $G/E_F^0 = 0.5$. The dashed curve is that of the free-electron Fermi sphere of radius k_F with the same volume as the lime. Points C and D, the conical points, are the regions of greatest deviation from the sphere.

This is accomplished if one requires critical contact between the Fermi surface and the two energy gaps of magnitude G. This implies that $Q = 2k_F$, or more precisely, that $\vec{Q}$ spans the Fermi surface. For $G = 0$, $Q = 2k_F$. The Fermi surface is distorted into the shape of a lime or lemon, as shown in Figure 20.3. Here Q is the distance between the two conical points at $C(k_z = -\frac{1}{2}Q)$ and $D(k_z = +\frac{1}{2}Q)$.

We obtain the equation for the Fermi surface by setting

$$E_{\vec{k}} = E_F = \hbar^2 \left(\frac{1}{2}Q\right)^2 /2m - \tfrac{1}{2}G \,, \tag{20.18}$$

which makes E_F the energy of the point of critical contact, point C in Figure 20.3, We now choose $\vec{Q}$ in the z direction and introduce dimensionless variables, valid for $-\frac{1}{2}Q < k_z < 0$:

$$u = k_x/Q, \quad v = k_y/Q, \quad w = (k_z + \tfrac{1}{2}Q)/Q \,, \tag{20.19}$$

so that in this coordinate system, the point C is at (0, 0, 0). The equation of the Fermi surface in this system is given by

$$\kappa = (u^2 + v^2)^{1/2} = \left[\left(w^2 + \alpha^2\right)^{1/2} - w^2\right]^{1/2} , \tag{20.20a}$$

where

$$\alpha = mG/\hbar^2 Q^2 \,. \tag{20.20b}$$

The above analysis applies only for the case $k_z < 0$, since for $k_z > 0$, the gap at point $D\,(k_z = \frac{1}{2}Q)$ in Figure 20.3 becomes important. The expressions for $0 < k_z < \frac{1}{2}Q$ may be obtained by replacing $\vec{Q}$ by $-\vec{Q}$ in Equations (20.13)–(20.15). The variables u, v, and κ are defined as before. In addition, if the dimensionless variable w is redefined as

$$w = \left(k_z + \frac{1}{2}Q\right)/Q, \quad k_z < 0 \tag{20.21a}$$

$$w = \left(k_z - \frac{1}{2}Q\right)/Q, \quad k_z > 0 \tag{20.21b}$$

then the equation for the Fermi surface, Equation (20.19), applies for both regions of k_z. The definition (20.21b) corresponds to displacing the right half of the Fermi surface to the left ($-k_z$) such that the two conical points touch. Then w ranges from $-\frac{1}{2}$ to $+\frac{1}{2}$, This is equivalent to considering the Fermi surface in a repeated-zone scheme.

The volume of the Fermi surface determines the number of electrons in the system. The value of Q is obtained by setting this volume equal to the volume of the Fermi surface when $G = 0$, or the volume of the free-electron sphere $\frac{4}{3}\pi k_F^3$. To first order in α, this yields:

$$Q \simeq 2k_F\left(1 + G/4E_F^0\right) \,, \tag{20.22}$$

where $E_F^0 = \hbar^2 k_F^2/2m$ is the free-electron Fermi energy, or the Fermi energy for $G = 0$. However, within this approximation, the volume within the Fermi surface does not remain constant as the value of G is increased to the largest value of G of interest here. In fact, for $G = 0.5 E_F^0$, the volume was about 8% larger than that with $G = 0$, with the consequence that the number of electrons had increased by 8%. Since this is clearly unacceptable, we must include higher-order terms in Q. For consistency, we also include higher-order corrections to the energy and wave functions. In addition, we will write all expressions in a form valid for both $k_z < 0$ and $k_z > 0$ by using the dimensionless variables defined in Equations (20.19)–(20.21).

We include the mixing between the states on the two halves of the Fermi surface, i.e., plane-wave states $|\vec{k} + \vec{Q}\rangle$ with $|\vec{k} - \vec{Q}\rangle$. This is accomplished through the use of nondegenerate perturbative corrections of first order in the wave functions and of second order in the energies. The new expression for $E_{\vec{k}}$ may then be written

$$E_{\vec{k}} = \frac{\hbar^2 Q^2}{2m}\left(\kappa^2 + w^2 + \frac{1}{4} - \left(w^2 + \alpha^2\right)^{1/2} - \frac{\alpha^2}{2(1 - |w|)}\right) . \tag{20.23}$$

The corresponding wave function is

$$\Psi_{\vec{k}} = \left(1 + C_{\vec{k}}^2\right)^{-1/2}\left(A_{\vec{k}} e^{i\vec{k}\cdot\vec{r}} + B_{\vec{k}} e^{i(\vec{k}+\vec{Q})\cdot\vec{r}} + C_{\vec{k}} e^{i(\vec{k}-\vec{Q})\cdot\vec{r}}\right), \quad k_z < 0 \tag{20.24a}$$

$$\Psi_{\vec{k}} = \left(1 + C_{\vec{k}}^2\right)^{-1/2}\left(A_{\vec{k}} e^{i\vec{k}\cdot\vec{r}} + B_{\vec{k}} e^{i(\vec{k}-\vec{Q})\cdot\vec{r}} + C_{\vec{k}} e^{i(\vec{k}+\vec{Q})\cdot\vec{r}}\right), \quad k_z > 0 \tag{20.24b}$$

where $A_{\vec{k}} \cos\varphi_{\vec{k}}$, $B_{\vec{k}} = -\sin\varphi_{\vec{k}}$, and

$$C_{\vec{k}} = \tfrac{1}{2}\alpha/(1 - |w|) . \tag{20.24c}$$

The equation of critical contact is given by:

$$E_{\vec{k}} = E_F = (\hbar^2 Q^2/2m)\left(\tfrac{1}{2} - \alpha - \tfrac{1}{2}\alpha^2\right) . \tag{20.25}$$

Thus, the equation for the Fermi surface may be written as

$$\kappa = \left[\left(w^2 + \alpha^2\right)^{1/2} - \alpha - w^2 + \frac{1}{2}\alpha^2|w|/\left(1 - |w|\right)\right]^{1/2} . \tag{20.26}$$

We obtain the appropriate value of Q by setting the volume within the Fermi surface for a given G equal to the volume of the free-electron sphere,

$$2\pi Q^3 \int_0^{1/2} \kappa^2 dw = \frac{4\pi}{3} k_F^3 = 4\pi n , \tag{20.27}$$

where n is the number of electrons per unit volume. From Equation (20.26) we evaluate this integral and obtain the approximate expression for Q:

$$Q = 2k_F\left\{1 + \frac{\tilde{G}}{4} - \frac{\tilde{G}^2}{32}\left[\frac{\ln(16/\tilde{G}) + \frac{2}{3}\tilde{G}}{1 + \frac{1}{2}\tilde{G}}\right]\right\} , \tag{20.28}$$

where $\tilde{G} = G/E_F^0$. For this value of Q, the difference between the volume within the Fermi surface and that within the free-electron sphere is less than 0.05% for values of $\tilde{G}$ up to 0.5.

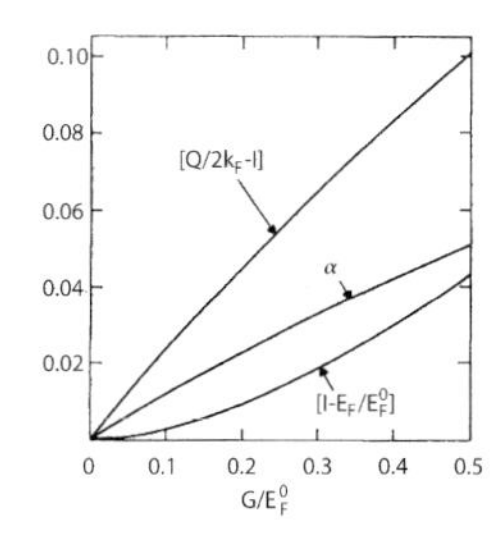

Figure 20.4 Plots of the deviations of Q, α, and E_F from their free-electron values, or values at $G = 0$, vs. G/E_F^0.

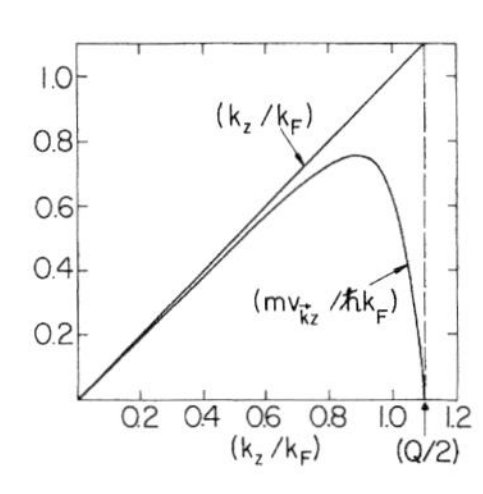

Figure 20.5 Comparison of v_z with k_z as a function of k_z for $G/E_F^0 = 0.5$.

In Figure 20.4, using Equation (20.28), we plot the variation of Q from $2k_F$, $Q/2k_F - 1$, as a function of G/E_F^0. In addition, we plot α from Equation (20.20b). Another consequence of the CDW with its lemon-shaped Fermi surface is to decrease the Fermi energy E_F below its free-electron value E_F^0. This variation, $1 - E_F/E_F^0$, is also plotted in Figure 20.4. Note that Q deviates substantially from the linear approximation given by Equation (20.22).

Another quantity important in following calculations is the velocity of electrons in the z direction along $\vec{Q}$. This is given by

$$v_z = \frac{\partial E_{\vec{k}}}{\partial k_z} = \frac{\hbar Q}{m}\left(w - \frac{w}{2\left(w^2 + \alpha^2\right)^{1/2}} - \frac{\alpha^2 w/|w|}{4\left(1 - |w|\right)^2} \right) . \tag{20.29}$$

In Figure 20.5, for $G/E_F^0 = 0.5$, we compare the plot of v_z with that of k_z as a function of k_z for $k_z > 0$. Note that at the conical point ($k_z = \frac{1}{2}Q$ or $w = 0$) the electron velocity nearly vanishes. This has important consequences, as will be discussed in Section 20.4.

An additional check on the validity of the preceding approximations, we calculate the volume by another method, which involves the electron velocity, to compare with that of the free-electron sphere.

$$4\pi n = \frac{4}{3}\pi k_F^3 = -\frac{1}{4\pi^3}\int v_{\vec{k}_z} k_z \frac{\partial f_{\vec{k}}^0}{\partial E_{\vec{k}}} d^3k$$

$$= \pi Q^3 \int_0^{1/2} \hat{v}_z \hat{k}_z \, dw , \tag{20.30}$$

where $\hat{v}_z = (m/\hbar Q)\, v_z$ and $\hat{k}_z = k_z/Q$, and where $f_{\vec{k}}^0$ is the equilibrium distribution function of the electron gas. For Q given by Equation (20.28), the equality of Equation (20.30) is accurate to within 0.7% for $\tilde{G}$ values up to 0.5.

20.4 Model Scattering Potentials

Since electrons cannot scatter from a perfect lattice, a nonzero resistivity in a metal can arise only if irregularities occur in the lattice. At high temperatures phonons provide this disturbance and a temperature-dependent resistivity results. However, when the temperature approaches absolute zero, the effects of phonons become unimportant, and the remaining temperature-independent resistivity, the residual resistivity, arises through the scattering of conduction electrons from stationary imperfections. If the potential produced by the presence of an impurity is sufficiently weak, the cross section for scattering can be determined within the Born approximation, requiring knowledge only of the Fourier transform of the impurity potential.

In this section we introduce the models for impurity-scattering potentials that will be used in succeeding sections, along with their Fourier transforms. In high-purity alkali metals impurities predominantly enter the crystal substitutionally. Thus, the potential from which electrons scatter is that of the impurity minus that of the missing lattice ion. For mathematical convenience, we will assume a Gaussian form for potentials of lattice and impurity ions, and for convenience in the discussion, we will refer to the host metal as potassium. A potassium ion in the lattice has a potential of the form

$$V_a(\vec{r}) = V_a e^{-r^2/\Gamma_a^2} , \tag{20.31}$$

where $\Gamma_a = a r_s$, with r_2 the Wigner–Seitz radius. The corresponding Fourier transform is

$$\begin{aligned} \tilde{V}_1(\vec{q}) &= \int d^3 r e^{-i\vec{q}\cdot\vec{r}} V_a(\vec{r}) \\ &= V_a \pi^{3/2} \Gamma_a^3 \exp\left(-\tfrac{1}{4}\Gamma_a^2 q^2\right) . \end{aligned} \tag{20.32}$$

If the height V_b of the potential of the impurity ion is different from the height V_a of the potential of the K ion that it replaces, but the width is the same, then the scattering potential will be given by

$$V_1(\vec{r}) = V_1 e^{-r^2/\Gamma_a^2} , \tag{20.33}$$

where $V_1 = V_b - V_a$. The Fourier transform $\tilde{V}_1(\vec{r})$ is the same form as Equation (20.32). If, on the other hand, the width Γ_b of the impurity potential is different from the width Γ_a of the K ion potential, but the height is the same, then the scattering potential is

$$V_2(\vec{r}) = V_b(\vec{r}) - V_a(\vec{r}) , \tag{20.34}$$

where $V_b(\vec{r})$ is obtained from $V_a(\vec{r})$ by replacing Γ_a with Γ_b. The Fourier transform is given by

$$\tilde{V}_2(\vec{q}) = \tilde{V}_b(\vec{q}) - \tilde{V}_a(\vec{q}) . \tag{20.35}$$

The results of the combination of these two cases that might occur will not be considered here, since this would add unnecessary complications without providing additional insight.

In addition to the scattering potential of the impurity ion itself, an additional potential can arise due to a strain field created by the introduction of an impurity into the lattice [21]. The surrounding ions adjust their positions in order to compensate for the misfit of the impurity in the lattice. The strain-field potential is given by the difference between the new positions of the K ions and their positions in the perfect lattice. For a substitutional impurity at the origin, we sum over all lattice points $\vec{L}$ except that of the impurity and write the potential as

$$V_s(\vec{r}) = \sum_{\vec{L}\neq 0} \left[V_a\left(\vec{r} - \vec{L} - \vec{u}_{\vec{L}}\right) - V_a\left(\vec{r} - \vec{L}\right) \right] , \tag{20.36}$$

where $\vec{u}_{\vec{L}}$ is the displacement vector of the ion at site $\vec{L}$. The Fourier transform is given by

$$\tilde{V}_s(\vec{q}) = \tilde{V}_a(\vec{q}) \sum_{\vec{L}\neq 0} e^{-i\vec{q}\cdot\vec{L}} \left(e^{-i\vec{q}\cdot\vec{u}_{\vec{L}}} - 1\right) . \tag{20.37}$$

For small displacements, we may expand the exponential involving $\vec{u}_{\vec{L}}$ and write approximately

$$\tilde{V}_s(\vec{q}) = \tilde{V}_a(\vec{q}) \sum_{\vec{L}\neq 0} \left(-i\vec{q}\cdot\vec{u}_{\vec{L}}\right) e^{-i\vec{q}\cdot\vec{L}} . \tag{20.38}$$

We choose a form for $\vec{u}_{\vec{L}}$ appropriate to that of a spherical center of dilatation in an infinite elastic medium [21–23],

$$\vec{u}_{\vec{L}} = g a_0^3 \vec{L}/L^3 , \tag{20.39}$$

where a_0 is the lattice constant of a bcc lattice and g is a dimensionless constant. We do not use a separate form for the nearest neighbors as for the case of interstitial impurities in [21] since here the nearest neighbors are approximately as distant from the impurity as the next-nearest neighbors in that case. With Equation (20.39), Equation (20.38) reduces to

$$\tilde{V}_s(q) = \tilde{V}_a(\vec{q})\, S(\vec{q}) , \tag{20.40a}$$

where

$$S(\vec{q}) = -i g a_0^3 \sum_{\vec{L}\neq 0} \frac{\vec{q}\cdot\vec{L}}{L^3} e^{-i\vec{q}\cdot\vec{L}} . \tag{20.40b}$$

As we shall see in Section 20.5, the calculation of the residual resistivity with Equation (20.40) requires numerical calculation of a four-dimensional integral. The contribution from each term in the sum of Equation (20.40b) contains many spikes, part of whose contribution is cancelled by higher-order terms. In addition, many sine functions must be computed at each point in the integrand. The integrations are thus difficult to perform and the rate of increase in the cost of computing contributions from each successive term far exceeds the slow rate of convergence of the sum. For this reason, we make a simplifying assumption, as was used in [21], that reduces necessary numerical integrations to three dimensions, reduces the number of sine functions in the integrand, and improves the convergence of the sum. We assume that the N_s sth-nearest neighbors are smeared into a spherical shell of radius L_s, the sth-nearest-neighbor distance. This is accomplished by replacing the sum over the neighbors within a shell by an integral corresponding to an angular average and summing over the shells. This yields

$$\begin{aligned} S(\vec{q}) &= -i g a_0^3 \sum_s \frac{N_s}{4\pi L_s^3} \int q L_s \cos\theta\, e^{-iqL_s\cos\theta}\, d\Omega \\ &= g a_0^3 \sum_s \frac{N_s}{L_s^3} \left(\cos(qL_s) - \frac{\sin(qL_s)}{qL_s}\right) . \end{aligned} \tag{20.41}$$

Since one can, in practice, sum over only a finite number s_f of shells, we compute the remainder to the sum by assuming that the atoms beyond the s_fth shell are spread into a continuum,

$$\begin{aligned} \sum_{s=s_f}^{\infty} \frac{\vec{q}\cdot\vec{L}}{L^3} e^{-i\vec{q}\cdot\vec{L}} &= \frac{2}{a_0} \int_{r_0}^{\infty} L^2 dL \int \frac{Lq\cos\theta}{L^3} e^{-iqL\cos\theta}\, d\Omega \\ &= \frac{-8\pi i}{a_0^3} \frac{\sin(qr_0)}{qr_0} , \end{aligned} \tag{20.42}$$

where $r_0 = N_t^{1/3} r_s$ is the radius of a sphere equal to N_t atomic volumes, with

$$N_t = 1 + \sum_{s=1}^{s_f} N_s \,, \tag{20.43}$$

counting the atoms included in the shells plus the impurity. Combining Equations (20.41) and (20.42), we have

$$S(\vec{q}) \cong g\left[\sum_{s=1}^{s_f} N_s \left(\frac{a_0}{L_s}\right)^3 \left(\cos(qL_s) - \frac{\sin(qL_s)}{qL_s}\right) - 8\pi\frac{\sin(qr_0)}{qr_0}\right] . \tag{20.44}$$

We have computed the necessary integrals for the computation of the residual resistivity and have summed shells out to 17th-nearest neighbors ($s_f = 17$). Good convergence occurs after the inclusion of $s_f = 14$.

Since the total impurity potential arises from a combination of a central-ion potential, $V_1(\vec{r})$ or $V_2(\vec{r})$, and the strain potential $V_s(\vec{r})$, we define two additional potentials that include this. We assume that the central-ion potential contributes a fraction $1-x$ and the strain potential a fraction x to the total potential. In order to simplify this portion of the calculation (and reduce the cost of the integrations), we approximate $S(\vec{q})$ in Equation (20.44) with the continuum term $s_f = 0$, choosing r_0 so that the same resitivities are obtained as with $s_f = 14$. The Fourier transform of the resulting "interference" potential involving $V_1(\vec{r})$ is given by

$$\tilde{V}_{\text{int}}^{(1)}(\vec{q}) = g(V_a/V_1)\beta(1-x)\tilde{V}_1(\vec{q}) + x\tilde{V}_s(\vec{q}) \tag{20.45a}$$

$$= g\tilde{V}_a(\vec{q})\left[\beta(1-x) - 8\pi x(\sin qr_0/qr_0)\right], \tag{20.45b}$$

where β is chosen for each Γ_a such that the resistivity in the absence of a CDW is the same for $x = 0$ as for $x = 1$. The Fourier transform of the corresponding potential involving $V_2(\vec{r})$ is

$$\tilde{V}_{\text{int}}^{(2)}(\vec{q}) = g\beta(1-x)\tilde{V}_2(\vec{q}) + x\tilde{V}_s(\vec{q}) \tag{20.46a}$$

$$= g\left[\beta(x-1)\tilde{V}_b(\vec{q}) - \tilde{V}_a(\vec{q}) \times \left(\beta(x-1) + 8\pi x\frac{\sin qr_0}{qr_0}\right)\right] , \tag{20.46b}$$

where again β is chosen to make resistivities for $x = 0$ and $x = 1$ equal. In both cases [Equations (20.45) and (20.46)], β positive corresponds to g positive (outward displacements of neighbors about the impurity) and β negative to g negative (inward displacements).

We note that the Fourier transforms of the above potentials are effective pseudopotentials for the impurities. Values of $\Gamma \sim r_s$ represent the widths of typical pseudopotential calculations in K [24]. Values within this range cause $\tilde{V}(\vec{q})$ for each of the models presented here to be sharply peaked about $\vec{q} = 0$. This has important consequences in the effects of a CDW on the residual resistivity, which will be explained in Section 20.5.

20.5 Residual-Restivity Calculation

When an electric field $\vec{\mathcal{E}}$ is applied to a metal, the electrons are accelerated in such a way that their equilibrium distribution function $f_{\vec{k}}^0$ is shifted at a constant rate in $\vec{k}$ space, with the distribution function $f_{\vec{k}}$ after a time (Δt) given by[5)]

$$f(\vec{k}) = f_0(\vec{k} - \vec{\delta}) , \tag{20.47a}$$

where

$$\vec{\delta} = -e\vec{\mathcal{E}}(\Delta t)/\hbar . \tag{20.47b}$$

For $\vec{\mathcal{E}}$ in the μ direction, we may write

$$f(\vec{k}) \approx f_0(\vec{k}) - \delta_\mu \frac{\partial f_0(\vec{k})}{\partial k_\mu} = f^0_{\vec{k}} - \delta_\mu v_{\vec{k}\mu} \frac{\partial f^0_{\vec{k}}}{\partial E_{\vec{k}}} , \tag{20.48}$$

where $v_{\vec{k}\mu} = \partial E_{\vec{k}}/\partial k_\mu$ is the velocity of electrons in the state $|\Psi_{\vec{k}}\rangle$ in the μ direction. The current density in the metal is then

$$\vec{J} = e\vec{P}/m , \tag{20.49a}$$

with

$$\vec{P} = \int \frac{d^3k}{4\pi^3} m v_{\vec{k}} f_{\vec{k}} , \tag{20.49b}$$

so that $\vec{J}$ is proportional to the total momentum $\vec{P}$ of the electrons. For the shifted Fermi distribution of Equations (20.47) and (20.48), the current density in the μ direction may be written as

$$\begin{aligned} J_\mu &= e\delta_\nu \int \frac{d^3k}{4\pi^3} v_{\vec{k}\mu} v_{\vec{k}\nu} \frac{\partial f^0_{\vec{k}}}{\partial E_{\vec{k}}} \\ &= e\delta_\mu \int \frac{d^3k}{4\pi^3} v^2_{\vec{k}\mu} \frac{\partial f^0_{\vec{k}}}{\partial E_{\vec{k}}} . \end{aligned} \tag{20.50}$$

For $\mu \neq \nu$, the integral above clearly vanishes for μ, ν Cartesian coordinates and the principal axes in the electron gas described in Section 20.3. The current is thus only along $\vec{\mathcal{E}}$.

In a perfect crystal, this acceleration due to an electric field would take place indefinitely, leading to an infinite current density. However, any real metal contains impurities, so that electrons suffer collisions and the system reaches a steady state. At finite temperatures, electrons also interact with phonons, but here we consider only the limit of zero temperature. In the steady state, the distribution function is displaced by an amount $\vec{\delta}$, where a relaxation time τ replaces (Δt) in Equation (20.47b). A determination of the magnitude of $\vec{\delta}$ leads one to an expression for the resistivity by employing Equation (20.50).

The actual distribution function of an electron gas in the steady-state situation described above is not exactly that of rigidly shifted equilibrium distribution function, but must be obtained from a solution of the equation that equates the rate of change of $f_{\vec{k}}$ due to $\vec{\mathcal{E}}$ with that due to collisions, the linearized Boltzmann transport equation [26],

$$\begin{aligned} -e\left(\vec{v}_{\vec{k}} \cdot \vec{\mathcal{E}}\right) \frac{\partial f^0_{\vec{k}}}{\partial E_{\vec{k}}} = \int \frac{d^3k'}{4\pi^3} \Big[&\left(f_{\vec{k}} - f^0_{\vec{k}}\right) \\ &- \left(f_{\vec{k}'} - f^0_{\vec{k}'}\right)\Big] \mathcal{Q}^{\vec{k}'}_{\vec{k}} , \end{aligned} \tag{20.51}$$

5) See, for instance [25]

where $\mathcal{Q}_{\vec{k}}^{\vec{k}'}$ is the golden-rule transition rate, describing the elastic scattering of electrons from a potential $V_{\text{tot}}(\vec{r})$ of all the impurities in the crystal.

$$\mathcal{Q}_{\vec{k}}^{\vec{k}'} = (2\pi/\hbar)\,|M_{\vec{k}',\vec{k}}|^2 \delta\left(E_{\vec{k}} - E_{\vec{k}'}\right) \tag{20.52a}$$

with

$$M_{\vec{k}',\vec{k}} = \langle \Psi_{\vec{k}'} | V_{\text{tot}}(\vec{r}) | \Psi_{\vec{k}} \rangle \,. \tag{20.52b}$$

For practical purposes, however, it suffices to assume the form (20.48) for $f_{\vec{k}}$ and to impose a slightly weaker condition than Equation (20.51), that is, that the rate of change of the total momentum of the electrons balance in the steady state. Since the rate of change of $\vec{P}$ is given by

$$\frac{d\vec{P}}{dt} = \int \frac{d^3k}{4\pi^3} m\vec{v}_{\vec{k}} \frac{\partial f_{\vec{k}}}{\partial t} \,, \tag{20.53}$$

the required condition is obtained by multiplying both sides of Equation (20.51) by $m\vec{v}_{\vec{k}}$ and integrating over $d^3k/4\pi^3$ to yield

$$\begin{aligned} &- em\mathcal{E}_{\mu} \int \frac{d^3k}{4\pi^3} v_{\vec{k}\mu}^2 \frac{\partial f_{\vec{k}}^0}{\partial E_{\vec{k}}} \\ &= \frac{-m\delta_{\mu}}{2} \int \frac{d^3k}{4\pi^3} \int \frac{d^3k'}{4\pi^3} \left(v_{\vec{k}\mu} - v_{\vec{k}'\mu}\right)^2 \frac{\partial f_{\vec{k}}^0}{\partial E_{\vec{k}}} \mathcal{Q}_{\vec{k}}^{\vec{k}'} \,, \end{aligned} \tag{20.54}$$

which may be solved trivially for δ_{μ}. The resistivity is then obtained by substituting δ_{μ} into Equation (20.50).

The procedure outlined here is equivalent to employing the variational principle to determine the resistivity tensor,[6]

$$\rho_{\mu\mu} = -\frac{1}{2} \int d^3k \int d^3k' \left(\Phi_{\vec{k}\mu} - \Phi_{\vec{k}\mu}\right)^2 \mathcal{Q}_{\vec{k}}^{\vec{k}'} \frac{\partial f_{\vec{k}}^0}{\partial E_{\vec{k}}} \Bigg/ \left| \int d^3k\, ev_{\vec{k}\mu} \Phi_{\vec{k}\mu} \left(\frac{\partial f_{\vec{k}}^0}{\partial E_{\vec{k}}}\right) \right|^2 , \tag{20.55}$$

with a trial function given by $\Phi_{\vec{k}\mu} = v_{\vec{k}\mu}$. This variational expression provides an upper bound on the resistivity, so that the actual resistivity must be less than or equal to that calculated using any given trial function in Equation (20.55).

When the conduction electrons are in a CDW ground state, as described in Section 20.3, the residual resistivity is enhanced more along $\vec{Q}$ than it is perpendicular to $\vec{Q}$, resulting in an anisotropy. The principal factor contributing io this effect is $|M_{\vec{k}',\vec{k}}|^2$, the square of the matrix element in Equation (20.52b), which appears in the golden-rule transition rate in Equation (20.52a). The total potential $V_{\text{tot}}(\vec{r})$, due to all the N_I impurities in the crystal, is the sum over all the single impurity potentials $V(\vec{r})$ located at sites $\vec{R}_i$,

$$V_{\text{tot}}(\vec{r}) = \sum_{i=1}^{N_I} V(\vec{r} - \vec{R}_i) \,, \tag{20.56}$$

6) Reference [24], Sections 7.7 and 7.9.

where $V(\vec{r})$ represents one of the model potentials described in Section 20.4. For $k_z > 0$, $k'_z > 0$, the matrix element becomes

$$M_{\vec{k}'\vec{k}} = \left[\left(1 + C_{\vec{k}}^2\right)\left(1 + C_{\vec{k}'}^2\right)\right]^{-1/2} \sum_{i=1}^{N_I} \Big[\left(A_{\vec{k}} A_{\vec{k}'} + B_{\vec{k}} B_{\vec{k}'} + C_{\vec{k}} C_{\vec{k}'}\right)$$
$$\times\, e^{i\left(\vec{k}-\vec{k}'\right)\cdot\vec{R}_i}\, \tilde{V}\left(\vec{k}-\vec{k}'\right) + \left(A_{\vec{k}} B_{\vec{k}'} + C_{\vec{k}} A_{\vec{k}'}\right) e^{i\left(\vec{k}-\vec{k}'+\vec{Q}\right)\cdot\vec{R}_i}$$
$$\times\, \tilde{V}\left(\vec{k}-\vec{k}'+\vec{Q}\right) + \left(A_{\vec{k}} C_{\vec{k}'} + B_{\vec{k}} A_{\vec{k}'}\right) e^{i\left(\vec{k}-\vec{k}'-\vec{Q}\right)\cdot\vec{R}_i}$$
$$\times\, \tilde{V}\left(\vec{k}-\vec{k}'-\vec{Q}\right) + \left(B_{\vec{k}} C_{\vec{k}'}\right) e^{i\left(\vec{k}-\vec{k}'-2\vec{Q}\right)\cdot\vec{R}_i}$$
$$\times \tilde{V}\left(\vec{k}-\vec{k}'-2\vec{Q}\right) + \left(C_{\vec{k}} B_{\vec{k}'}\right) e^{i\left(\vec{k}-\vec{k}'+2\vec{Q}\right)\cdot\vec{R}_i}\, \tilde{V}\left(\vec{k}-\vec{k}'+2\vec{Q}\right)\Big] . \qquad (20.57)$$

If the impurities are distributed at random [27], the cross terms in $|M_{\vec{k}',\vec{k}}|^2$ cancel, and the result is proportional to $\hat{N}_I$, the number of impurities per unit volume. For $k_z > 0$, $k'_z > 0$,

$$|M_{\vec{k}',\vec{k}}|^2 = \hat{N}_I \left[(1 + C_{\vec{k}}^2)(1 + C_{\vec{k}'}^2)\right]^{-1} \Big[\left(A_{\vec{k}} A_{\vec{k}'} + B_{\vec{k}} B_{\vec{k}'} + C_{\vec{k}} C_{\vec{k}'}\right)^2 \tilde{V}^2\left(\vec{k}-\vec{k}'\right)$$
$$+ \left(A_{\vec{k}} B_{\vec{k}'} + C_{\vec{k}} A_{\vec{k}'}\right)^2 \tilde{V}^2\left(\vec{k}-\vec{k}'+\vec{Q}\right)$$
$$+ \left(A_{\vec{k}} C_{\vec{k}'} + B_{\vec{k}} A_{\vec{k}'}\right)^2 \tilde{V}^2\left(\vec{k}-\vec{k}'-\vec{Q}\right) + \left(B_{\vec{k}} C_{\vec{k}'}\right)^2 \tilde{V}^2\left(\vec{k}-\vec{k}'-2\vec{Q}\right)$$
$$+ \left(C_{\vec{k}} B_{\vec{k}'}\right)^2 \tilde{V}^2\left(\vec{k}-\vec{k}'+2Q\right)\Big] . \qquad (20.58a)$$

Similarly, for $k_z > 0$, $k'_z < 0$,

$$|M_{\vec{k}',\vec{k}}|^2 = \hat{N}_I \left[(1 + C_{\vec{k}}^2)(1 + C_{\vec{k}'}^2)\right]^{-1} \Big[\left(A_{\vec{k}} A_{\vec{k}'} + B_{\vec{k}} C_{\vec{k}'} + C_{\vec{k}} B_{\vec{k}'}\right)^2 \tilde{V}^2\left(\vec{k}-\vec{k}'\right)$$
$$+ \left(A_{\vec{k}} B_{\vec{k}'} + B_{\vec{k}} A_{\vec{k}'}\right)^2 \tilde{V}^2\left(\vec{k}-\vec{k}'-\vec{Q}\right)$$
$$+ \left(A_{\vec{k}} C_{\vec{k}'} + C_{\vec{k}} A_{\vec{k}'}\right)^2 \tilde{V}^2\left(\vec{k}-\vec{k}'+\vec{Q}\right)$$
$$+ \left(B_{\vec{k}} B_{\vec{k}'}\right)^2 \tilde{V}^2\left(\vec{k}-\vec{k}'-2\vec{Q}\right)$$
$$+ \left(C_{\vec{k}} C_{\vec{k}'}\right)^2 \tilde{V}^2\left(\vec{k}-\vec{k}'+2Q\right)\Big] . \qquad (20.58b)$$

The expressions for $k_z < 0$, $k'_z < 0$, and for $k_z < 0$, $k'_z > 0$ may be obtained by replacing $\vec{Q}$ by $-\vec{Q}$ in Equations (20.58a) and (20.58b), respectively.

Examination of Equations (20.58), aided by Figure 20.6, reveals the source of the resistivity anisotropy. Let Θ be the scattering angle, or the angle between $\vec{k}$ and $\vec{k}'$. The factor $(v_{\vec{k}\mu} - v_{\vec{k}\mu})^2$ on the right-hand side of Equation (20.54), which is proportional to $(1-\cos\Theta)$ for a spherical Fermi surface (no CDW), weights 180° scattering heavily and small-angle scattering lightly. As explained in Section 20.4, the Fourier transform of the impurity scattering potential $\tilde{V}(\vec{q})$ is sharply peaked about $\vec{q} = 0$. In the absence of a CDW ($A_{\vec{k}} = 1$, $B_{\vec{k}} = C_{\vec{k}} = 0$), with $\vec{q} = \vec{k}' - \vec{k}$, the scattering cross section thus weights small $\vec{q}$'s the most, which in this case is the same as small angles Θ, as is shown in the right half of Figure 20.6. No preference exists in any particular direction, so the scattering of electrons, and thus the resistivity, is

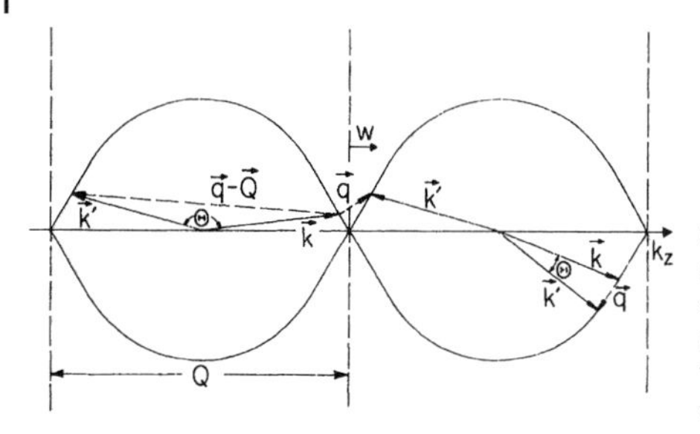

Figure 20.6 Diagram illustrating the umklapp scattering (left portion of figure) vs. the normal scattering (right portion) when the conservation rule of Equation (20.59) is taken into account. The distortion of the conical-point regions is exaggerated for emphasis.

isotropic. In the presence of a CDW, however, the mixing of plane waves, which is responsible for the form of the wave functions $\Psi_{\vec{k}}$ creates a totally different situation. In this case, small-$\vec{q}$ Fourier components of the potential may scatter electrons according to the conservation rule

$$\vec{k}' = \vec{k} + \vec{q} \pm \vec{Q} \,. \tag{20.59}$$

The result is shown in Figure 20.6 on the left, within the reduced-zone scheme, for a wave-vector transfer $\vec{k}$ to $\vec{k}'$ from the conical-point region on one side of the Fermi surface to the other conical-point region on the opposite side. Here Θ is nearly 180°, so that this transition is favored by the weighting factor. Therefore, the scattering along the direction of $\vec{Q}$ (the z direction) is greatly enhanced. This may be viewed alternatively in the repeated zone scheme., as shown in Figure 20.6. The wave-vector transfer $\vec{q}$ between $\vec{k}$ and $\vec{k}'$ scatters an electron from a state on the Fermi surface in one zone to another state in the next, because in the presence of a CDW this-distance is very short in the conical-point region. Since only transitions that are parallel to $\vec{\mathcal{E}}$ contribute to the resistivity, this umklapp scattering does not affect the resistivity perpendicular to the CDW (perpendicular to $\vec{Q}$).

While the factor $(v_{\vec{k}z} - v_{\vec{k}'z})^2$ in the calculation of the component ρ_{zz} of the resistivity tensor enhances umklapp scattering, $v_{\vec{k}z}$ vanishes at the conical point, or at $k_z = \frac{1}{2}Q$, as shown in Figure 20.5. This means that only umklapp transitions between states near the conical points contribute, not transitions between states at the points themselves.

The effect can be seen explicitly as a necessary consequence of Equations (20.58). For $k_z > 0$, $k'_z > 0$ or $k_z < 0$, $k'_z < 0$, the scattering occurs only within one half of the Fermi surface and does not involve umklapp scattering. However, for $k_z > 0$, $k'_z < 0$ or $k_z < 0$, $k'_z > 0$, the CDW-umklapp effect becomes important. In Equation (20.58b), the term involving $\tilde{V}^2\left(\vec{k} - \vec{k}' - \vec{Q}\right)$ is sharply peaked about $\vec{k} - \vec{k}' = \vec{Q}$, and weights the type of transition described by Equation (20.59). In addition, the coefficient $(A_{\vec{k}} B_{\vec{k}'} + A_{\vec{k}'} B_{\vec{k}})^2$ in the region near $k_z \sim \frac{1}{2}Q$ is as large as the coefficient of the first term and larger than the coefficients of the others. For $k_z < 0$, $k'_z > 0$, the corresponding term involves the same coefficient with $\tilde{V}^1\left(\vec{k} - \vec{k}' + \vec{Q}\right)$. Note that if $V(\vec{r})$ were a delta function, $\tilde{V}(\vec{q})$ would be a constant and the CDW-umklapp transitions would no longer be enhanced. Clearly, the resistivity anisotropy depends on the localization of $\tilde{V}(\vec{q})$ about $\vec{q} = 0$. The more sharply peaked $\tilde{V}(\vec{q})$ is, the greater will be the anisotropy. This means that the more extended the potential is in real space, the greater will be the anisotropy.

We now proceed with the calculation of the residual resistivity. In order to perform the integrations indicated in Equation (20.54), we invoke the zero-temperature equality $\partial f^0_{\vec{k}} / \partial E_{\vec{k}} = -\delta\left(E_{\vec{k}} - E_F\right)$ and transform integrations over d^3k to integrations over energy and the dimensionless cylindrical coordinates of Section 20.3,

$$d^3k = \left(m/\hbar^2\right) Q d\theta\, dw\, dE_{\vec{k}} \,, \tag{20.60}$$

where θ is the polar angle. In addition, we define $\psi = \theta - \theta'$ and $\psi' = \frac{1}{2}(\theta + \theta')$, so that $d\theta\, d\theta' = d\psi\, d\psi'$. As an illustration, we consider the first model potential of Section 20.4, given by Equation (20.33). We define

$$T_\mu = \frac{Q}{k_F}\frac{(\Gamma_a k_F)^6}{32\pi^2}\int_{-1/2}^{1/2} dw \int_{-1/2}^{1/2} dw' \int_0^{2\pi} d\psi \int_0^{2\pi} d\psi' \left(\hat{v}_{\vec{k}\mu} - \hat{v}_{\vec{k}'\mu}\right)^2 \left|\hat{M}_{\vec{k}',\vec{k}}\right|^2 , \tag{20.61a}$$

where

$$|M_{\vec{k}',\vec{k}}|^2 \left(\overline{N}_I V_1^2 \pi^3 \Gamma_a^6\right) \left|\hat{M}_{\vec{k}',\vec{k}}\right|^2 , \tag{20.61b}$$

and

$$F = 3\left(\frac{Q}{k_F}\right)^3 \int_0^{1/2} \hat{v}_{\vec{k}\mu}^2\, dw\, . \tag{20.61c}$$

Here, $\overline{N}_I = \hat{N}_I / n$ is the fractional number of impurities in the metal. If we define

$$\delta_\mu = -e\mathcal{E}_\mu \tau_\mu / \hbar\, , \tag{20.62a}$$

then

$$\rho_{\mu\nu} = \left(m/ne^2\tau_\mu\right)\delta_{\mu\nu}\, , \tag{20.62b}$$

where $\delta_{\mu\nu}$ is the Kronecker δ. Combining Equations (20.50), (20.54), (20.61), and (20.62), we have

$$\frac{1}{\tau_z} = \overline{N}_I \frac{V_1^2}{\hbar E_F^0}\frac{T_z}{|F|^2},\quad \frac{1}{\tau_x} = \frac{1}{\tau_y} = \overline{N}_I \frac{V_1^2}{\hbar E_F^0} T_x\, , \tag{20.63}$$

so that the resistivity ratio is given by

$$\gamma = \frac{\rho_{zz}}{\rho_{xx}} = \frac{1}{|F|^2}\frac{T_z}{T_x}\, . \tag{20.64}$$

Note that γ does not depend on the overall strength of the scattering potential or on the number of impurities present.

Equation (20.61) may be written explicitly for the x and z directions, with the substitutions $v_{\vec{k}x} = \hbar k_x / m$ and $v_{\vec{k}z} = (\hbar Q/m)\hat{v}_{\vec{k}z}$ and an integration over ψ', as

$$T_z = \left(\frac{Q}{k_F}\right)^4 \frac{(\Gamma_a k_F)^6}{16\pi}\int_{-1/2}^{1/2} dw \int_{-1/2}^{1/2} dw' \int_0^{2\pi} d\psi \left(\hat{v}_{\vec{k}z} - \hat{v}_{\vec{k}'z}\right)^2 \left|\hat{M}_{\vec{k}',\vec{k}}\right|^2 , \tag{20.65a}$$

$$T_x = \left(\frac{Q}{k_F}\right)^4 \frac{(\Gamma_a k_F)^6}{16\pi}\int_{-1/2}^{1/2} dw \int_{-1/2}^{1/2} dw' \int_0^{2\pi} d\psi$$
$$\times\left[\left(\kappa^2 + \kappa'^2\right) - 2\kappa\kappa' \cos\psi\right]\left|\hat{M}_{\vec{k},\vec{k}'}\right|^2 , \tag{20.65b}$$

where κ is given by Equation (20.26) of Section 20.3. The factor $|\hat{M}_{\vec{k}',\vec{k}}|^2$ does not depend on ψ' and the ψ dependence can be factored out of all the terms in Equations (20.58), so that we may define

$$\left|\hat{M}_{\vec{k}',\vec{k}}\right|^2 = \exp\left\{-\frac{1}{2}\Gamma_a^2 Q^2\left[\left(\kappa^2 + \kappa'^2\right) - 2\kappa\kappa'\cos\psi\right]\right\} |\overline{M}\left(\vec{k}_z', \vec{k}_z\right)|^2\, . \tag{20.66}$$

Following integration over ψ, we have

$$T_z = \left(\frac{Q}{k_F}\right)^4 \frac{(\Gamma_a k_F)^6}{8} \int_{-1/2}^{1/2} dw \int_{-1/2}^{1/2} dw' \left(\hat{v}_{\vec{k}z} - \hat{v}_{\vec{k}'z}\right)^2 \times I_0\left(\Gamma_a^2 Q^2 \kappa\kappa'\right) \left|\overline{M}\left(k_z', k_z\right)\right|^2 , \tag{20.67a}$$

$$T_x = \left(\frac{Q}{k_F}\right)^4 \frac{(\Gamma_a k_F)^6}{8} \int_{-1/2}^{1/2} dw \int_{-1/2}^{1/2} dw' \left[\tfrac{1}{2}\left(\kappa^2 + \kappa'^2\right) I_0\left(\Gamma_a^2 Q^2 \kappa\kappa'\right) - \kappa\kappa' I_1\left(\Gamma_a^2 Q^2 \kappa\kappa'\right)\right] \left|\overline{M}\left(k_z', k_z\right)\right|^2 , \tag{20.67b}$$

where $I_0(z)$ and $I_1(z)$ are modified Bessel functions of the first kind of orders zero and one, respectively. These two-dimensional integrals must be carried out numerically for $G \neq 0$. However, for the case $G = 0$, $T_x = T_z$, $F = 1$, and the solution assumes the simple form, where $\tau_0 = \tau_x = \tau_y = \tau_z$:

$$\frac{1}{\tau_0} = \frac{\overline{N}_I}{6} \frac{V_1^2}{\hbar E_F^0} (\Gamma_a k_F)^2 \left[\left(1 - e^{-2(\Gamma_a k_F)^2}\right) - 2(\Gamma_a k_F)^2 e^{-2(\Gamma_a k_F)^2}\right] . \tag{20.68}$$

If we let $\overline{V}_1 = (\Gamma_a k_F)^3 \pi^{3/2} V_1$ and constrain $\overline{V}_1$ to remain finite as $\Gamma_a \to 0$, then $V_1(\vec{r}) \to \left(\overline{V}_1 / k_F^3\right) \delta(\vec{r})$ and τ_0 reduces to the form

$$\frac{1}{\tau_0} \to \frac{\overline{N}_I}{3\pi^3} \frac{\overline{V}_1^2}{\hbar E_F^0} . \tag{20.69}$$

In order for r to be that of sample K-10 of Holroyd and Datars (Figure 20.1), $\tau \sim 1.4\times 10^{-10}$ s, with $\overline{N}_I = 10^{-5}$ from the reported purity, V_1 must be on the order of $\frac{1}{2}$ eV. This is determined by assuming Equation (20.68) with $\Gamma_a \sim \gamma_s$. This value is consistent with those of typical pseudopotential form factors [24].

In the evaluation of the resistivity with $\tilde{V}_2(\vec{q})$, one obtains three terms, each of which is of the form of $\tilde{V}(\vec{q})$, and the analysis proceeds as above. For $\tilde{V}_s(\vec{q})$ an additional complication arises. Assuming the forms given in Equations (20.40) and (20.44), the steps through Equations (20.65) are unchanged. However, Equation (20.66) no longer applies, since the additional factor of $S(\vec{q})$ enters each of the terms in Equations (20.58), and it is impossible to factor out the dependence on ψ or to perform the ψ integration analytically. Therefore, for this case, three-dimensional numerical integration is needed.

20.6 Numerical Results

Before displaying the results of numerical calculations, we first appeal to experiments for an estimate of the range of values of G/E_F^0 appropriate to K. The CDW model has been successful in the explanations of the Mayer–El Naby optical anomaly [28, 29] and the splitting of the conduction-electron-spin-resonance (CESR) lines [30]. The intense optical absorption with a threshold of 0.6 eV fixed the value of G/E_F^0 at about 0.29, assuming the free-electron mass. This same value produced a quantitative agreement with the CESR data. A maximum

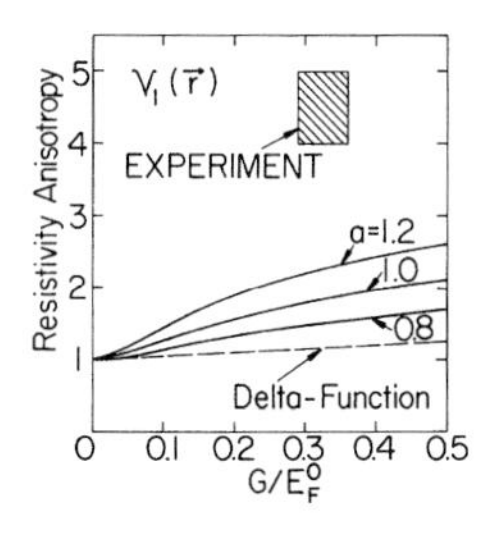

Figure 20.7 Theoretical predictions for the resistivity anisotropy vs. normalized CDW gap G/E_F^0 using the central-ion potential $V_1(\vec{r})$ from Equation (20.33) of the text, for three values of a, where $\Gamma_a = ar_s$ Also shown is the anisotropy expected for a δ-function potential. The shaded rectangle represents the range of values determined by experiments.

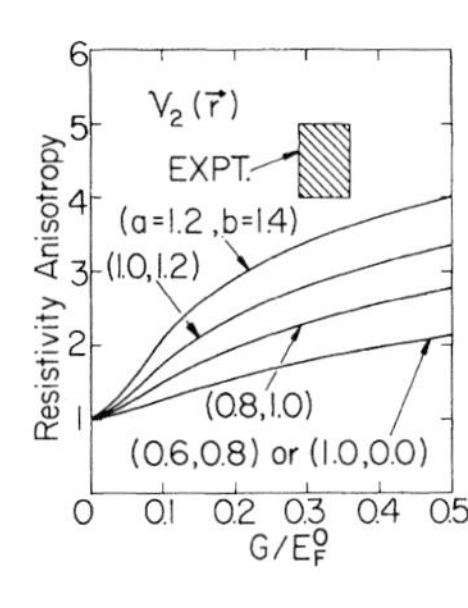

Figure 20.8 Resistivity anisotropy vs. G/E_F^0 using the central-ion potential $V_2(\vec{r})$ from Equation (20.34), for four sets of values (a, b), where $\Gamma_a = ar_s$ and $\Gamma_b = br_s$. Note that the lowest curve is identical to that for $a = 1.0$ in Figure 20.7.

value for G/E_F^0 can be obtained by assuming that $E_F^0 = \hbar^2 k_F^2/2m^*$ where $m^* \approx 1.2m$, the measured cyclotron mass [31], replaces m. This yields $G/E_F^0 \approx 0.36$. Thus, in Figure 20.7 we indicate this range of values, along with the experimental values for the resistivity anisotropy from the torque anisotropy of Figures 20.1 and 20.2, with a shaded rectangle. For K, G/E_F^0 lies between 0.29 and 0.36 and the resistivity anisotropy lies between 4 and 5 at 4 K. For comparison, we also show this shaded rectangle of experimental values in Figures 20.8, 20.9, and 20.11. Clearly, for other alkali metals, a different set of values would be relevant.

We now show the theoretical predictions for the resistivity as a function of G/E_F^0 if one uses the various model scattering potentials described in Section 20.4 and one calculates the resistivity in the manner described in Section 20.5. The first case, shown in Figure 20.7, is that for which the first central-ion potential $V_1(\vec{r})$ from Equation (20.33), is used. We assume that the width Γ_a of the potential is comparable to the Wigner–Seitz radius r_s. In order to show the dependence of the results on this parameter, we let $\Gamma_a = ar_s$ and calculate the anisotropy with the values $a = 0.8, 1.0, 1.2$. Note that the larger Γ_a is, the greater is the spread of the potential in real space, but the more peaked is its Fourier transform in $\vec{q}$ space. If we allow the potential to go to the limit of a δ function, its Fourier transform is a constant, as discussed in Section 20.5. In order to emphasize the importance of a sharply peaked Fourier transform for the enhancement of umklapp scattering, and eventually for a large resistivity anisotropy, we also plot in Figure 20.7 the results of using a δ-function potential. It is clear that the sharp peaking of the Fourier transform about $\vec{q} = 0$ enhances the anisotropy and if no peaking occurs, the anisotropy is very small. The small effect that does occur for the δ-function potential is a consequence of the distorted shape of the Fermi surface (See Figures 20.3 and 20.6.) The elongated shape also produces a slight enhancement of the resistivity perpendicular to the CDW, since it decreases the distance for momentum transfer along that direction. This effect actually suppresses the anisotropy and has been included in these calculations.

The results using the second central-ion potential $V_2(\vec{r})$ of Equation (20.34), are shown in Figure 20.8. The width of the K ion is given by $\Gamma_a = ar_s$ and the width of the impurity ion by $\Gamma_b = br_s$. Since the potential appears squared in the calculations, the roles of these two could equally be reversed. Since this potential is formed by subtracting two Gaussians of the same heights but different widths, in real space the potential vanishes at the origin, so that $V_2(\vec{r})$ is more extended than either $V_a(\vec{r})$ or $V_b(\vec{r})$. Consequently, $\tilde{V}_2(\vec{q})$ is more localized about $\vec{q} = 0$ than either $\tilde{V}_a(\vec{q})$ or $\tilde{V}_b(\vec{q})$. Its shape is nearly that of a Gaussian with a width Γ smaller than Γ_a or Γ_b. Thus this potential produces larger anisotropies than $V_1(\vec{r})$, as can be seen from Figure 20.8. For instance, the lowest curve $(a, b) = (0.6, 0.8)$, is identical to the curve for $a = 1.0$ in Figure 20.7. The difference between the widths of the Fourier transforms $\tilde{V}_a(\vec{q})$ and $\tilde{V}_b(\vec{q})$ can be represented by the quantity $a - b$. The anisotropy varies only slightly as this quantity is varied, as long as the average $\frac{1}{2}(a + b)$ remains constant. That is, the anisotropy produced by scattering from $V_2(\vec{r})$ is principally a function of $\frac{1}{2}(a + b)$.

In Figures 20.7 and 20.8, we have used the value $\Gamma_a = 1.2 r_s$, as the largest acceptable value for the width of the potential of a K ion in the lattice. This value was obtained by assuming that the magnitude of $V_a(\vec{r})$ at a distance r_s from the origin must be less than or equal to half its value at the origin. This implies that $a \lesssim 1/\sqrt{\ln 2} \approx 1.2$. A larger value would result in peaks in the positive background potential halfway between atoms. Values close to this represent a situation that most clearly resembles the jellium model of a uniform positive background. Of course, if one takes into account the actual bcc lattice of K, this maximum value of a should be slightly smaller.

We now turn to the results for the residual resistivity anisotropy for scattering of the electrons only from the strain field created by the presence of the impurity, shown in Figure 20.9. We have used the potential $V_s(\vec{r})$ from Equation (20.36), with the subsequent approximations for $\tilde{V}_s(\vec{q})$ in Equations (20.37)–(20.44). As discussed above, we use $a = 1.2$ as the maximum value of the width of the K-ion potential. In addition, we include the two other values $a = 0.8$, 1.0, as before. Since the strain field begins at the nearest-neighbor distance from the impurity and extends even farther, the Fourier transform $\tilde{V}_s(\vec{q})$ is more sharply peaked than $\tilde{V}_a(\vec{q})$ and thus produces a large anisotropy, considerably larger than that for $\tilde{V}_1(\vec{q})$ in Figure 20.7.

Combining the scattering from the strain-field potential with either of the central-ion potentials can either increase or decrease the anisotropy, depending on the relative strengths of the potentials, which is determined by the parameter x in the expressions in Equations (20.45) and (20.46). The effects of combining the strain-field potential $V_s(\vec{r})$ with the central-ion potential $V_1(\vec{r})$, resulting in $V_{\text{int}}^{(1)}(\vec{r})$ is illustrated in Figure 20.10, for $G/E_F^0 = 0.35$ and $a = 1.0$. When $\beta < 0$, or for contraction of the lattice about the impurity, the anisotropy

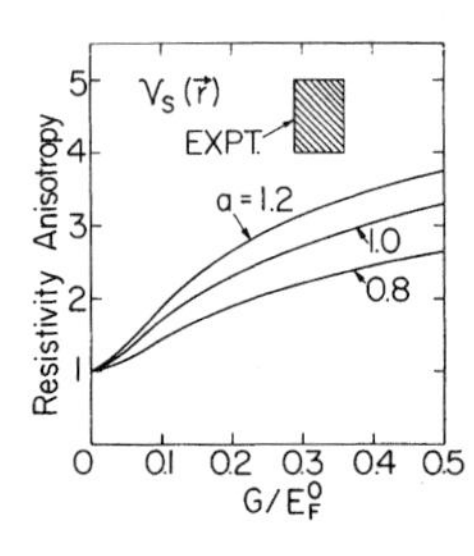

Figure 20.9 Resistivity anisotropy vs. G/E_F^0 using the strain-field potential $V_s(\vec{r})$ from Equations (20.36), (20.40), and (20.44) of the text for three values of a.

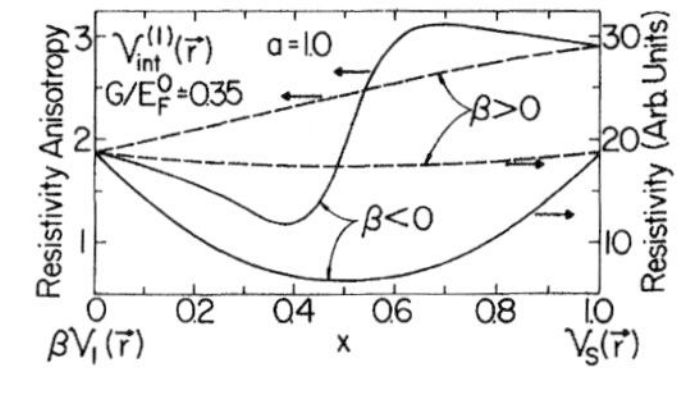

Figure 20.10 Plot of the resistivity anisotropy and the resistivity in arbitrary units for $V_{\text{int}}^{(1)}(\vec{r})$ as a function of x, the "concentration" of the potential $V_1(\vec{r})$, for $G/E_F^0 \sim 0.35$ and $a = 1.0$. Results for both signs of β are shown, allowing for the cases of contraction and expansion of the lattice.

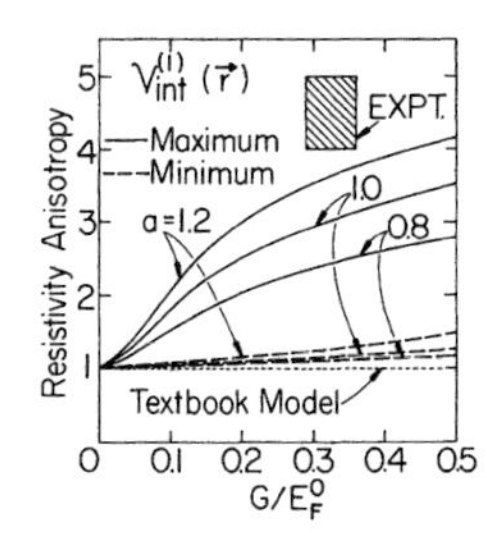

Figure 20.11 Maximum and minimum values for the resistivity anisotropy vs. G/E_F^0 using the potential $v_{\text{int}}^{(1)}(\vec{r})$, determined from Figure 20.10 and similar plots, for three values of a. Also shown is the prediction from the standard "textbook model", for which the electron density is uniform throughout the metal (no CDW).

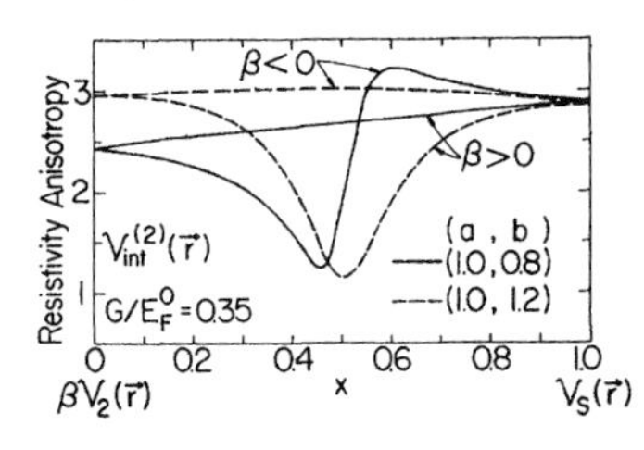

Figure 20.12 Resistivity anisotropy using the potential $v_2(\vec{r})$ as a function of x, with $G/E_F^0 = 0.35$. Here a is associated with the width of a K-ion potential and b with the width of the impurity potential. With $a = 1.0$ fixed, we choose values of b20% larger and smaller than a. We include the possibilities of contraction or expansion of each case by including both signs of β.

reaches a minimum when $x \approx 0.4$ and a maximum when $x \approx 0.65$. The value of the resistivity (in arbitrary units) for $G = 0$ is also shown. For this case, it is clear that the two potentials nearly cancel one another at $x = 0.5$. The minimum occurs when $\tilde{V}_{\text{int}}^{(1)}(\vec{q})$ achieves a cancellation near $\vec{q} = 0$, making the function more extended. Similarly, the maximum occurs when the cancellation occurs in the tails of the potentials, resulting in a more sharply peaked function about $\vec{q} = 0$. The results for $\beta > 0$ are also shown. For this case, since $\tilde{V}_1(\vec{q})$ and $\tilde{V}_s(\vec{q})$ add, no cancellation occurs, and the anisotropy varies nearly linearly with x.

Results for the maximum and minimum in Figure 20.10 and similar plots is shown as a function of G/E_F^0 for three values of a in Figure 20.11. The maximum anisotropies are larger than the curves in Figures 20.7 and 20.9, and the highest curve, for $a = 1.2$, predicts anisotropies nearly as large as have been seen experimentally.

Finally, in Figure 20.12, we show the dependence on x of the resistivity anisotropy that arises with the potential $V_{\text{int}}^{(2)}(\vec{r})$, whose Fourier transform is given by Equation (20.46). As in Figure 20.10, $x = 0$ represents the contribution only from the central-ion potential, in this case $V_2(\vec{r})$, and $x = 1$ from $V_s(\vec{r})$. Here, $G/E_F^0 = 0.35$ and the width of the K-ion potential is fixed by $a = 1.0$. We consider the two cases in which $b = 0.8$ or 1.2, that is, the width of the impurity ion potential is 20% larger or smaller than that of the K ion potential. The case for $b = 0.8$ shown here has a very similar dependence on x as the example illustrated in Figure 20.10. In addition, the shapes of the curves for the resistivity at $G = 0$ (not shown) for $\beta < 0$ and $\beta > 0$ are approximately the same, with minima at $x = 0.5$. However, here for $\beta < 0$ the ratio of the resistivity for $x = 0$ or $x = 1$ with respect to that for $x = 0.5$ is 35 to 1 as opposed to 8 to 1 for the analogous case in Figure 20.10. For $b = 1.2$, the resistivity curve as a function of x is nearly identical to that for $b = 0.8$ except that the curves for $\beta < 0$ and $\beta > 0$ are interchanged. The anisotropy curve, with $G/E_F^0 = 0.35$, shown by the dashed curve, acquires a different shape. The large cancellation occurs for $\beta > 0$ and the function reaches a minimum and not a maximum. The $\beta < 0$ curve achieves a slight maximum. These shapes occur when the anisotropics due to $V_s(\vec{r})$ and $V_2(\vec{r})$ are almost the same. We will not show the variations of the maxima and minima as a function of G/E_F^0, as in Figure 20.11. The shapes of such curves are the same as in Figure 20.11. For all the combinations of parameters considered in Figure 20.8, the maxima are somewhat smaller than those in Figure 20.11 for the same values of a.

Preliminary results to those reported in this section appeared in [5]. The curves displayed there correspond to those in Figures 20.7, 20.9, and 20.11 of this paper with $a = 1.0$. The central-ion potential $V_2(\vec{r})$ was not considered. The earlier curves showed slightly larger anisotropies due to various mathematical approximations in that work.

20.7 Conclusions

We have seen that the results of induced-torque experiments on sample K-10 of Holroyd and Datars prove that K neither has a simply connected Fermi surface nor is of cubic symmetry. In addition, the data of Figures 20.1 and 20.2 forces one to conclude that the residual resistivity in K is anisotropic by as large as a factor of 5 to 1. We have shown that the hypothesis of a CDW structure provides a quantitative explanation of this ratio. Although we do not attempt to explain the high-field torque anomalies here, the CDW clearly contains the features of a preferred axis and a multiply connected Fermi surface.

In this paper, we have not included any of the effects of phasons [14, 15, 32], or phase modulations of the CDW, that might exist in K. It has recently been shown that electrons in the conical-point regions of the Fermi surface shown in Figure 20.3 scatter strongly from phasons [33]. This would produce an additional enhancement of the umklapp scattering discussed here and would therefore increase the residual resistivity anisotropy that one would predict in the presence of a CDW. Although a CDW is expected to lead to weak diffraction satellites, phasons could reduce their intensity in K by several orders of magnitude [14, 15]. As has been suggested in the past, an attempt should be made to search for these satellites at 10^{-7} of the (110) intensity level.

Variation in the results of torque data from sample to sample, which has led to much of the controversy discussed here, implies that a $\vec{Q}$-domain structure exists in most samples. The anisotropy in the resistivity tensor discussed in this paper would permit the residual resistivity to vary up to a factor of 4, depending on the distribution of orientations of $\vec{Q}$ in the domains. It is commonly found that in high-purity samples of K, the resistivity can vary by as much as a factor of 2 from run to run on the same sample, depending on how the sample is handled. As was first suggested in [5], the sample K-10 of Holroyd and Datars may be the first single $\vec{Q}$-domain sample ever studied. Perhaps this could be attributed to the fact that the sample was grown carefully in the smooth spherical kel-F mold and was kept protected by the mold throughout the course of the experiments. Clearly, induced-torque experiments could be used to characterize samples, so that at least one could determine whether a given sample were single $\vec{Q}$.

One of the main difficulties with the CDW model is the apparent contradiction with de Haas–van Alphen experiments [9, 10]. These experiments imply that the Fermi surface is spherical to one part in 10^3. However, it is not clear whether this can be attributed to a multi-$\vec{Q}$-domain structure. On the other hand, results of recent de Haas–van Alphen experiments [34] show some of the effects expected on the basis of a CDW model. We emphasize the importance of performing de Haas–van Alphen experiments on a sample that exhibits the huge induced-torque anisotropies of Figure 20.1 in order to resolve this dilemma.

Acknowledgements

This research was supported in part by the National Science Foundation and the NSF Materials Research Laboratory program. The authors would like to thank F.W. Holroyd and W.R. Datars for many useful discussions and for permission to publish their data.

References

1 Schaefer, J.A. and Marcus, J.A. (1971) *Phys. Rev. Lett. 27*, **935**.
2 Lass, J.S. (1970) *J. Phys. C* **3**, 1926.
3 Lass, J.S. (1976) *Phys. Rev. B* **13**, 2247.
4 Holroyd, F.W. and Datars W.R. (1975) *Can. J. Phys.* **53**, 2517; and private communication.
5 Bishop, M.J. and Overhauser A.W (1977) *Phys. Rev. Lett.* **39**, 632.
6 Landau, L.D. and Lifshitz, E.M. (1960), *Electrodynamics of Continous Media*, Pergamon, New York, p. 209; translated by Sykes, J.B. and Bell, J.S.
7 Visscher, P.B. and Falicov L.M. (1970) *Phys. Rev. B* **2**, 1518.
8 Overhauser, A.W. (1971) *Phys. Rev. Lett.* **27**, 938.
9 Shoenberg, D. and Stiles, P.J. (1964) *Proc. R. Soc. A* **281**, 62
10 Lee, M.J.G. and Falicov, L.M. (1968) *Proc. R. Soc. A* **304**, 319.
11 Monin, J. and Boutry, G.A. (1974) *Phys. Rev. B* **9**, 1309.
12 Lifshitz, I.M., Azbel, M.Y., and Kaganov, M.I. (1956) *Zh. Eksp. Teor. Fiz.* **31**, 63 [(1957) *Sov. Phys.-JETP* **4**, 41].
13 Jones, B.K. (1969) *Phys. Rev.* **179**, 637.
14 Overhauser, A.W. (1971) *Phys. Rev. B* **3**, 3173.
15 Overhauser, A.W. (1978) *Adv. Phys.* **27**, 343.
16 Overhauser, A.W. (1960) *Phys. Rev. Lett.* **4**, 462.
17 Overhauser, A.W. (1962) *Phys. Rev.* **128**, 1437.
18 Smith, P.A. and Smith, C.S. (1965) *J. Phys. Chem. Solids* **26**, 279.
19 Overhauser, A.W. (1971) *Phys. Rev. B* **3**, 3173.
20 Overhauser, A.W. (1974) *Phys. Rev. B* **9**, 2441.
21 Overhauser, A.W. and Gorman, R.L. (1956) *Phys. Rev.* **102**, 676.
22 Eshelby, J.D. (1954) *J. Appl. Phys.* **25**, 255.
23 Love, A.E.H. (1944) *Mathematical Theory of Elasticity*, 4th ed., Dover, New York, p. 187.
24 Rice, T.M. and Sham, L.J. (1970) *Phys. Rev. B* **1**, 4546.
25 Ziman, J.M. (1972) *Principles of the Theory of Solids*, 2nd ed., Cambridge U.P., London, Chap. 7.
26 Ziman, J.M. (1960) *Electrons and Phonons*, Oxford U.P., London, Sec. 7.3.
27 Wilson, A.H. (1953) *The Theory of Metals*, 2nd ed., Cambridge U. P., London, Chap. 9.
28 Overhauser, A.W. (1976) *Phys. Rev. B* **14**, 3371.
29 Overhauser, A.W. (1967) *Phys. Rev.* **156**, 844.
30 Overhauser, A.W. (1968) *Phys. Rev.* **168**, 763.
31 Grimes, C.C. and Kip, A.F. (1963) *Phys. Rev.* **132**, 1991.
32 Overhauser, A.W. (1978) *Hyperfine Interactions* **4**, 786.
33 Boriack, M.L. and Overhauser A.W. (1978) *Phys. Rev. B* **17**, 4549.
34 Altounian, Z., Verge, C., and Datars, W.R. (1978) *J. Phys. F* **8**, 75.

Reprint 21 Detection of a Charge-Density Wave by Angle-Resolved Photoemission[1)]

M.L. Boriack and A.W. Overhauser**

* Department of Physics, Purdue University, 525, Northwestern Avenue, West Lafayette, Indiana 47907, USA

Received 26 June 1978

Angle-resolved photoemission offers a means for determining the presence and the wave vector of an incommensurate charge-density wave in a simple metal.

The purpose of this paper is to show that angle-resolved photoemission offers a means for detecting the presence of an incommensurate charge-density wave (CDW) in a metal. As an example we propose a test for the CDW model of potassium [1, 2]. The abundance of experimental evidence supporting the hypothesis of a CDW ground state in metallic potassium has recently been reviewed [3].

The most conclusive evidence for the presence of a CDW would be a diffraction "picture" of the satellite spots associated with the wave vector $\vec{Q}$ of the CDW. In potassium such spots have not been seen.[2)] However, the surmised location of the spots is very close to the [110] Bragg reflection. Furthermore, a large, expected phason temperature factor [2] may make them nearly unobservable [4]. We show that angle-resolved photoemission offers an alternative means for detecting the presence of a CDW and can be used to determine the CDW wave vector $\vec{Q}$.

Potassium films grown on smooth, amorphous substrates are known to have a [110] direction normal to the surface [5]. (This should, of course, be checked in each experiment.) In addition, there is evidence that the direction of $\vec{Q}$ in such a film is along that [110] direction which is normal to the surface [6]. For this case the geometry is especially simple, and a measurement of the energy distribution of electrons emitted *normal* to the surface can detect the presence of a CDW.

Figure 21.1 illustrates the process of photoemission. An incoming photon with energy $\hbar\omega$ excites an electron from an initial state E_i (measured from the Fermi energy E_F) to a final state. The work function of the metal is φ. The electron escapes from the metal with kinetic energy K. Knowledge of K and $\hbar\omega$ allows one to determine the energy of the initial state,

$$E_i = K + \varphi - \hbar\omega \,. \tag{21.1}$$

The energy of the final state is $E_i + \hbar\omega$. Typically, angle-resolved electron-energy distribution curves (AREDC's) are displayed by plotting the spectral intensity of emitted electrons versus E_i.

AREDC's give information about the energy and direction of electrons emitted from the sample.

In the case under consideration electrons emitted normal to the surface carry information about the energy bands in the [110] direction. For this discussion we shall consider only electrons emitted from states in the bulk (and shall ignore complications arising from electrons in surface states).

1) Phys. Rev. Lett. 41, 15 (1978)

2) No results of a detailed search for the CDW satellite reflections have been published.

Anomalous Effects in Simple Metals. Albert Overhauser

ISBN: 978-3-527-40859-7

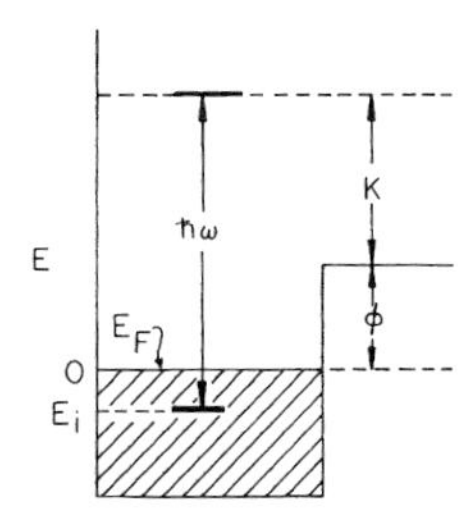

Figure 21.1 Sketch of photoemission. A photon with energy $\hbar\omega$ excites an electron in an initial state with energy E_i below the Fermi energy E_F. The work function of the metal is φ. The emitted electron has kinetic energy K.

As is well known, there is no absorption of photons in a free-electron gas since momentum and energy cannot be conserved. In the presence of the periodic potential of a crystal, *or a CDW*, transitions are allowed. For the case being considered, in which electrons are emitted along [110], it is sufficient to examine the one-electron potential

$$V(\vec{r}) = v_{110} \cos \vec{G}_{110} \cdot \vec{r} + v \cos \vec{Q} \cdot \vec{r} \,. \tag{21.2}$$

The first term is the [110] component of the crystal potential, and the second term is the CDW potential. We neglect here other components of the crystal potential since their magnitude is much smaller. Thus, we neglect the secondary-cone emission of Mahan [7].

In the nearly-free-electron model the effect of $V(\vec{v})$ is to mix the plane-wave state $\vec{k} \pm \vec{G}_{110}$ and $\vec{k} \pm \vec{Q}$. The amount of mixing can be determined by perturbation theory. With these perturbed states there can be absorption as illustrated in Figure 21.2. Because of the multiple periodicity of $V(\vec{r})$it is easiest to think in the extended-zone scheme. If the CDW is absent (there is no gap at point A and) the only transitions allowed are between states separated in k space by $\vec{G}_{110}$ (such as A to C). For $\hbar\omega$ greater than the difference in energy between points C and A there can be no transitions since A is at the Fermi surface in the [110] direction. If a CDW is present, additional transitions are possible (such as A to B) between states separated by $\vec{Q}$ in k space. There are no "Q" transitions for $\hbar\omega$ greater than the energy difference between A and B. As we show below, by varying $\hbar\omega$ it is possible to discriminate between "G" and "Q" transitions and to determine the difference $G_{110} - Q$.

AREDC's for electrons emitted normal to the surface are shown in Figure 21.3 for a number of photon energies. The numerical values, appropriate to potassium, are based on nearly-free-electron energies with $V_{110} = 0.4$ eV [8], a CDW energy gap $V = 0.6$ eV [1], and a Fermi surface in critical contact [1] with the CDW energy gaps.

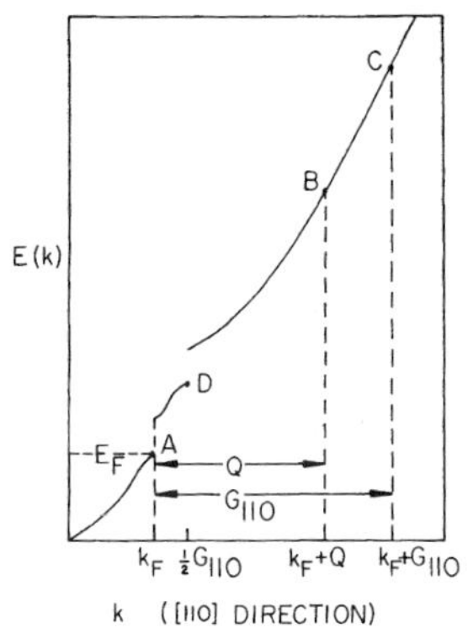

Figure 21.2 Energy versus wave vector in the [110] direction for a nearly-free-electron metal containing a CDW. Gjio is the [110] reciprocal lattice vector and Q is the CDW wave vector. The CDW energy gap is at A and the Brillouin zone gap is at D. With no CDW present only transitions between states separated by G_{110} are allowed, e.g., between points A and C. The periodicity Q of the CDW allows additional transitions such as from A to B.

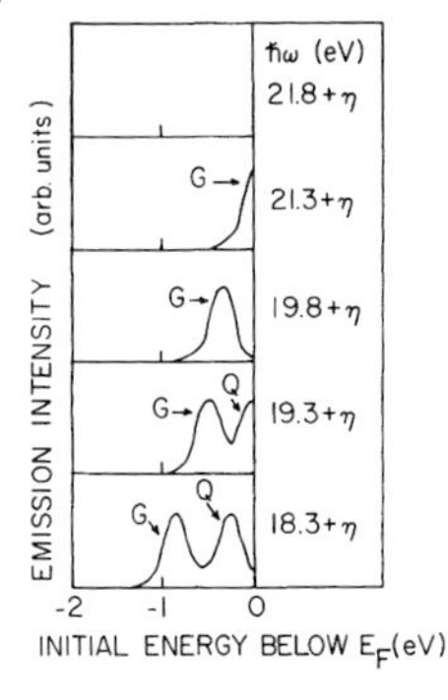

Figure 21.3 A sketch of intensity of emitted electrons versus energy of the initial state below the Fermi energy for a number of photon energies. η is a parameter to account for band-structure and many-electron effects. It is zero for the nearly-free-electron model. Peaks labeled G are allowed by the crystal potential, and peaks labeled Q are allowed by the CDW potential. See text for details.

The photon energies in Figure 21.3 are only meant to be illustrative because of uncertainties in band structure and many-body effects not included in the energies of the electron states. Therefore we have added an unknown energy η to all photon energies. (η would be zero for the nearly-free-electron model under discussion.) The experiment we propose here would be of interest even if the only outcome were a determination of η.

Because of the uncertainties related to band structure and many-body effects we do not estimate peak heights or widths. Since v_{110} and V are of comparable size, the relative heights of the G and Q transitions are expected to be comparable. We choose to represent the distribution of emitted electrons by Gaussians of equal height and full width ~ 0.3 eV. We arbitrarily truncate a Gaussian when it would lead to spectral intensity above $E_i = 0$.

We start first with a high photon energy and construct a series of curves for decreasing values of $\hbar\omega$. As discussed above, for $\hbar\omega$ greater than the energy difference between points A and C in Figure 21.2, there will be no transitions. Such a case (where there are no external photoelectrons) is shown in the top curve of Figure 21.3.

As $\hbar\omega$ is reduced, the G transition first becomes possible for $\hbar\omega = 21.5$ eV, an excitation from A to C in Figure 21.2 (with the initial state A at the Fermi energy). This is illustrated in the second curve of Figure 21.3 for an $\hbar\omega$ slightly below threshold. For smaller $\hbar\omega$ the initial state for the G transition moves to lower energy as shown in the third curve of Figure 21.3.

Below $\hbar\omega = 19.5$ eV the Q transition (A to B in Figure 21.2) becomes possible as shown in the fourth curve of Figure 21.3. As $\hbar\omega$ is decreased further, the energies of initial states for both G and Q transitions decrease. When the Q transition is first allowed, the center of the G peak is ~ 0.4 eV below the Fermi energy in the nearly-free-electron model. The difference in initial energies for the two transitions increases to ~ 0.6 eV as $\hbar\omega$ is lowered.

The calculation is based on a value $Q = 1.33(2\pi/a)$. The separation between the peaks can be used to determine the difference $G_{110} - Q$. [This depends on a knowledge of $E(\vec{k})$ for the initial- and final-state bands.] Thus, angle-resolved photoemission can be used both to detect the presence of a CDW and to measure its periodicity.

We have performed a similar analysis for sodium. Since the Fermi energy in sodium is about 1.5 times that in potassium, the G transition (A to C in Figure 21.2) occurs for $\hbar\omega = 33.4$ eV in the nearly-free-electron model. The Q transition (A to B in Figure 21.2) occurs for $\hbar\omega = 31.4$ eV. The spacing between the two peaks is similar to that in potassium, when a value $Q = 1.35(2\pi/a)$ is assumed for sodium. However, because V_{110} in Na is only ~ 0.2 eV [8], and the CDW energy gap V is thought to be about 1.2 eV, the peak height for the G transition may be much smaller than that for the Q transition. This disparity could lead to experimental difficulties.

We have shown that angle-resolved photoemission offers a means to detect the presence of an incommensurate CDW. The crucial idea is that the periodicity $\vec{Q}$ of the CDW permits new transitions compared to those allowed by the periodicity of the crystal. By plotting AREDC's for various incident photon energies, it should be possible to identify any additional transitions and (if present) to determine $\vec{Q}$.

References

1 Overhauser, A.W. (1968) *Phys. Rev.*, **167**, 691.
2 Overhauser, A.W. (1971) *Phys. Rev. B*, **3**, 3173.
3 Overhauser, A.W. (1978) *Adv. Phys.*, **27**, 343.
4 Overhauser, A.W. (1978) *Hyperfine Interact.*, **4**, 786.
5 Boutry, G.A, and Dormaut, H. (1969) *Philips Tech. Rev.*, **30**, 225.
6 Overhauser, A.W. and Butler, N.R. (1976) *Phys. Rev. B*, **14**, 3371.
7 Mahan, G.D. (1970) *Phys. Rev. B*, **2**, 4334.
8 Ham, F.S. (1962) *Phys. Rev.*, **128**, 82.

Reprint 22 Ultra-low-temperature Anomalies in Heat Capacities of Metals Caused by Charge-density Waves[1)]

M.L. Boriack * *and A.W. Overhauser* *

* Department of Physics, Purdue University, 525, Northwestern Avenue, West Lafayette, Indiana 47907, USA

Received 14 June 1978

A new fruitful area for low-temperature research is discussed. The contribution of phase excitations of incommensurate charge-density waves to the low-temperature heat capacities of metals is calculated, and the characteristic signature of phasons in calorimetric measurements is illustrated. The importance of anisotropy in the phason dispersion relation to the magnitude of the phason heat capacity and to the temperature at which the signature of the phasons can be seen is emphasized. Unless measurements are done at sufficiently low temperatures to *freeze-out* the phasons, it is possible to make significant errors in determining the electronic heat capacity. In particular the possibility of detecting phasons in potassium is discussed.

22.1 Introduction

The electronic ground state of a metal need not be the mere occupation of the N lowest-energy Bloch states. Exchange interactions and electron–electron correlations can cause a modulated, collective deformation of the electronic charge density to have a lower total energy [1]. For such a charge-density-wave (CDW) ground state the conduction-electron density is

$$\rho = \rho_0 \left[1 + p \cos(\vec{Q} \cdot \vec{r} + \varphi) \right] , \tag{22.1}$$

where ρ_0 is the average density and p is the fractional modulation,. The CDW wave vector $\vec{Q}$ is very nearly equal to the diameter $2k_F$ of the Fermi surface. The phase φ will be discussed at length below.

A CDW instability can occur only if the electronic charge density is locally neutralized by an accompanying lattice distortion [1]. Each positive ion will be displaced from its equilibrium lattice site $\vec{L}$ by

$$\vec{u}(\vec{L}) = \vec{A} \sin(\vec{Q} \cdot \vec{L} + \varphi) . \tag{22.2}$$

Since $\vec{Q}$ is controlled by Fermi surface dimensions, the wavelength of a CDW will generally be unrelated to lattice periodicities, i.e., the CDW is incommensurate with the lattice. In this case, the resulting structure is multiply periodic, no two ions are equivalent, and the crystal no longer has a translation group. Such a loss in symmetry means that the energy of the system is independent of phase φ. It then follows that there will be low-frequency collective excitations corresponding to φ varying slowly in space and time. These elementary excitations are called phasons [2, 3] and have important consequences for experiments that try to detect a CDW with diffraction [2] or through Knight-shift or hyperfine-field effects [3].

The purpose of this paper is to discuss the contribution of phasons to the low-temperature heat capacity of metals containing incommensurate CDW's. We illustrate the type of anoma-

1) Phys. Rev. B 18, 12 (1978)

Anomalous Effects in Simple Metals. Albert Overhauser

ISBN: 978-3-527-40859-7

lies which occur in plots of C/T vs. T^2, where C is the total heat capacity. Such anomalies have recently been seen by Sawada and Satoh [4] and led to the identification of CDW's in La-Ge systems. It is expected that the more isotropic the metal, the more anisotropic the phason dispersion relation [3]. We demonstrate that this anisotropy can have profound effects on the size of the phason anomaly and on the temperature at which the anomaly is observed. As an example we discuss the possibility of a phason contribution to the heat capacity of potassium in particular. The search for this type of anomaly in metals provides a new area for ultra-low-temperature research.

22.2 Phason Heat Capacity

Consider an incommensurate CDW described by Equations (22.1) and (22.2). Since the energy of the system is independent of phase φ, it follows that there will be low-frequency collective excitations corresponding to φ varying slowly in space and time [2, 3]. We express $\varphi(\vec{L}, t)$ by an expansion in running waves

$$\varphi(\vec{L}, t) = \sum_{\vec{q}} \varphi_{\vec{q}} \sin(\vec{q} \cdot \vec{L} - \omega_{\vec{q}} t) \,. \tag{22.3}$$

This approach is analogous to treating lattice dynamics in the continuum approximation. The wave vectors $\{\vec{q}\}$ are assumed small compared to the Brillouin zone, and $\{\varphi_{\vec{q}}\}$ are the amplitudes of the phasons. If $\vec{q} \perp \vec{Q}$ the local direction of the CDW vector is slightly rotated. If $\vec{q} \parallel \vec{Q}$ the local magnitude of $\vec{Q}$ is periodically modulated.

The dynamics of these modes have been studied in the small-q limit, assuming negligible damping [2, 3]. The phason frequencies vary linearly with q. The frequency spectrum is plotted schematically in Figure 22.1. Phasons are a type of lattice vibration and are linear combinations of old phonons (in a crystal with no CDW) having wave vectors near $\vec{Q}$ and $-\vec{Q}$. That the phason frequency goes to zero at $\pm\vec{Q}$ is a consequence of the incommensurate nature of the CDW. Since there can be only a given number of vibrational normal modes, phasons and phonons must share the spectral density in the regions near $\pm\vec{Q}$ as is indicated by the dashed region of the phonon dispersion curve in Figure 22.1.

It is expected that ω vs. $\vec{q}$ will be highly anisotropic because a local rotation of $\vec{Q}$ requires less energy than a change in its magnitude. Consequently, the surfaces of constant frequency for the phasons will be flat (pancake-shaped) ellipsoids. Taking $\vec{Q}$ in the z direction and letting q_z and $q_\perp = (q_x^2 + q_y^2)^{1/2}$ be components of $\vec{q}$ parallel and perpendicular to $\vec{Q}$, respectively, we can write

$$\omega_{\vec{q}}^2 = c_z^2 q_z^2 + c_\perp^2 q_\perp^2 \,. \tag{22.4}$$

The ratio of c_z to $c_\perp$ may be as high as 100 to 1, especially if the Fermi surface is nearly spherical. (For the ideal metal "jellium", which has no preferred direction, the ratio would be ∞.) For more anisotropic materials the ratio of c_z to $c_\perp$ may be much smaller.

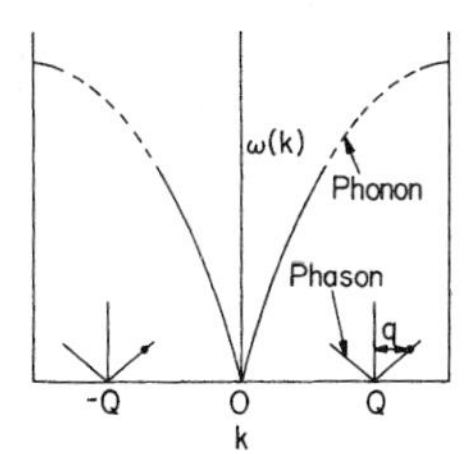

Figure 22.1 Schematic illustration of the vibrational modes of a metal having an incommensurate CDW structure. The frequency of the phason branch goes to zero at $\pm\vec{Q}$, the location of CDW satellite reflections in k space. Such a diagram has only approximate meaning since an incommensurate CDW structure does not have a Brillouin Zone.

Since phasons are (a special type of) lattice vibrations and their frequency varies linearly with q, it is natural to evaluate the phason heat capacity using a Debye model. For phasons the parameters of this model are the radius q_Φ of the phason ("Debye") sphere in k space, the speed c_z of phasons with $\vec{q} \parallel \vec{Q}$ the anisotropy factor $\eta \equiv c_\perp / c_z$. Because of the anisotropy, definition of the phason Debye temperature is somewhat arbitrary. We choose to define the phason Debye temperature in terms of the velocity parallel to $\vec{Q}$, i.e.,

$$\Theta_\Phi \equiv \hbar c_z q_\Phi / k_B \,, \tag{22.5}$$

where k_B is Boltzmann's constant. Thus the free parameters are q_z, c_z, and η.

The total energy E of the phasons is evaluated by summing the contributions $\hbar\omega_{\vec{q}}$ of each normal mode weighted by the Bose–Einstein occupation factor. For a sample with volume Ω,

$$E = \frac{\Omega}{(2\pi)^3} \int_0^{2\pi} d\varphi \int_{-q_\Phi}^{q_\Phi} dq_z \int_0^{\left(q_\Phi^2 - q_z^2\right)^{1/2}} dq_\perp \times q_\perp \hbar\omega_{\vec{q}} \left(e^{\hbar\omega_{\vec{q}}/k_B T} - 1\right)^{-1} . \tag{22.6}$$

In Equation (22.6) a choice of cylindrical coordinates was made.

Note that for phasons there is only one "branch" with the angle between $\vec{q}$ and $\vec{Q}$ varying continuously from $0°$ to $180°$. Thus the factor of 3 present for phonons is absent for phasons.

Changing the variables of integration to dimensionless units, the phason heat capacity at constant volume $C_\Phi = (\partial E/\partial T)_v$ is given by

$$C_\Phi = \frac{\Omega q_\Phi^3 k_B T^3}{2\pi^2 \Theta_\Phi^3 \eta^2} \int_0^{\Theta_\Phi/T} dz \int_0^{\rho_{\max}} d\rho \rho r^2 \times \left[(e^r - 1)(e^{-r} - 1)\right]^{-1} , \tag{22.7}$$

where $r^2 \equiv z^2 + \rho^2$ and

$$\rho_{\max} = \eta(\Theta_\Phi / T)\left[1 - (Tz/\Theta_\Phi)^2\right]^{1/2} . \tag{22.8}$$

It is interesting to evaluate (22.7) in the high- and low-temperature limits. For high temperatures $\Theta_\Phi / T \ll 1$, and (since $\rho, z, r \ll 1$) we can expand the exponentials to lowest order in r. Elementary integrations yield the high temperature Dulong–Petit result $C_\Phi = N_\Phi k_B$, where N_Φ is the number of phason modes contained by the phason "Debye" sphere, i.e., $N_\Phi = \Omega q_\Phi^3 / 6\pi^2$.

To evaluate the phason heat capacity in the $T \to 0$ limit it is convenient to express Equation (22.6) in ellipsoidal coordinates: $q_z = (r/c_z)\cos\theta$; $q_x = (r/c_\perp)\sin\theta\cos\varphi$; $q_y = (r/c_\perp)\sin\theta\sin\varphi$. A change to dimensionless variables then obtains (in the limit $T \to 0$) the usual integral which arises in the Debye theory. It follows that in the low-temperature limit

$$C_\Phi = \left(4\pi^4 N_\Phi k_B / 5\eta^2\right)\left(T/\Theta_\Phi\right)^3 . \tag{22.9}$$

It is important to note the presence of η^2 in the denominator. This factor is not surprising since through Θ_Φ three powers of the phason velocity appear in the denominator of C_Φ. Thus, $C_\Phi \sim c_z^{-1} c_\perp^{-2}$. The anisotropy of the phason velocity and the magnitude of the velocity have a strong effect on the size of C_Φ at low temperatures. In Section 22.3 we discuss the additional and profound effect of η on the temperature at which the presence of phasons can be identified in measurements of the total heat capacity. Both these effects of the phason anisotropy are extremely important to the question of whether phasons can be identified in low-temperature calorimetric measurements on a metal with an incommensurate CDW ground state.

22.3 Total Heat Capacity

In this section we discuss the "signature" of the phasons in a measurement of the total heat capacity. The effects we describe have been seen recently by Sawada and Satoh [4] in a study of heat capacity anomalies in La-Ge systems and analyzed by them in terms of phasons. In their analysis, Sawada and Satoh took the phason velocity to be isotropic. Although in an anisotropic metal the phason dispersion relation is expected to be more isotropic than in an isotropic metal, it is likely that their fit to the experimental data could be made even better with this additional parameter

For our discussion, however, we shall deal with the possibility of detecting the presence of the proposed CDW in potassium [2] by measurements of the low-temperature heat capacity. In this case, phason anisotropy has crucial significance, as will be shown below. The CDW hypothesis has been successful in quantitatively explaining the anomalous optical absorption [5], the anisotropic conduction-electron g factor [6], and the anisotropic residual resistivity [7] of potassium. Although these results and many others [2, 8] present convincing evidence that potassium has a CDW structure, a persistent question is: Where are the CDW diffraction satellites? Actually, a careful search for them has never been reported. It is likely, however that the phason Debye–Waller factor reduces the satellite intensity to a point where the satellites would be extremely difficult to find [2, 3]. Total intensity is not lost, but is transferred into a pancake-shaped phason cloud. Special techniques that integrate over the diffuse pancake may be required to detect CDW's in potassium in diffraction experiments. The existence of phasons also offers an explanation of the "null" result of attempts to detect a CDW in potassium with Knight shift and hyperfine-field effects [3]. Consequently, it is of great interest to investigate whether or not phasons might make their presence known in calorimetric measurements.

To estimate the phason specific heat it is necessary to make reasonable guesses for the three parameters mentioned previously: the anisotropy factor η; the phason-sphere ("Debye") radius q_Φ; and the velocity c_z of phasons with $\vec{q}$ parallel to $\vec{Q}$. To illustrate the importance of η, we shall use the two values $\eta = 0.1$ and $\eta = 0.01$ in our calculations. That is, we discuss phasons with $\vec{q}$ perpendicular to $\vec{Q}$ 10 and 100 times softer than those with $\vec{q}$ parallel to $\vec{Q}$.

Because the number of phason modes N_Φ and, therefore, the magnitude of the phason contribution varies with q_Φ^3, the size of this parameter is of crucial importance. For our discussion we *guess* the radius q_Φ of the phason sphere to be approximately 5% of the phonon Debye radius. The value of $\vec{Q}$ suggested for potassium is very close in k space to the [110] reciprocal-lattice vector. Taking half the difference between Q and the [110] reciprocal lattice vector gives the above estimate for $q_\Phi \sim 4.6 \times 10^6\ \mathrm{cm}^{-1}$. The ratio of the volume of phason and phonon Debye spheres is 1.3×10^{-4} with this estimate.[2] This compares with the Sawada and Satoh experimental result of 8×10^{-5} in La-Ge [4].

The phason velocity c_z (for $\vec{q}$ parallel to $\vec{Q}$) also appears with the third power in the expression for the low-temperature phason heat capacity. Since phasons are a special type of lattice vibration, it might be guessed that phason-vibration velocities are of the same order as phonon velocities.

It is possible to arrive at a rough lower limit for c_z given the choices q_Φ and η above. As will be discussed in detail below, heat-capacity measurements at temperatures above the low-temperature limit for the phasons can lead to errors in the experimentally determined thermal effective mass of the electrons. By requiring that this error be less than, say, 30% of

2) It might seem that the small number of phasons in this sphere would lead to a small phason Debye–Waller factor. However, it can be shown that the extreme anisotropy of the phasons causes the Debye–Waller factor to remain large.

the true thermal effective mass for $\eta = 0.1$, a lower limit on the phason velocity of $c_z \cong 2.7 \times 10^5$ cm/s is obtained. To illustrate the effect of phason velocity we shall compute the phason heat capacity using $c_z = 2.7 \times 10^5$ cm/s and twice this value.

Neglecting phasons for the moment, at low temperatures the heat capacity of a metal is the sum of electronic and lattice contributions. The heat capacity is given by

$$C = \gamma T + AT^3 \,, \tag{22.10}$$

where γT is the electronic contribution and AT^3 is the lattice heat capacity. A plot of C/T vs. T^2 is a straight line. The intercept at $T = 0$ gives γ and the thermal effective mass m_t^* defined by $m_t^*/m_0 = \gamma/\gamma_0$, where m_0 is the free-electron mass and γ_0 is the free-electron value. The slope of the straight line determines the Debye temperature. A CDW leads [1] to a 5%–10% enhancement in γ, which cannot be distinguished experimentally from band-structure, phonon-interaction or many-body effects.

The heat capacity of potassium has been measured between 0.26 and 4.2 °K by Lien and Phillips [9]. They found the heat capacity was well fit by the expression

$$C = 2.08T + 2.57T^3 \tag{22.11}$$

with units of mJ/(mol K).

If phasons are present, the total heat capacity C_T is given by the sum of (22.7) and (22.10) less C_E,

$$C_T = \gamma T + AT^3 + C_\Phi - C_E \,. \tag{22.12}$$

Here C_E is an "Einstein" specific heat subtracted from the phonon specific heat to keep the total number of normal modes equal to $3N$. These phonons correspond to the dashed region of Figure 22.1, where the phonons have lost spectral density at the expense of the phasons. For potassium the Einstein temperature of these phonons is $\sim 20°$K and the contribution of C_E is negligible in the temperature range of interest in this work.

The most convenient way to illustrate the "signature" of phasons in the total heat capacity is to adjust γ and A in (22.12) so that C_T agrees with the measured heat capacity, say, from $T^2 = 0.2$ to $0.4\,°\mathrm{K}^2$. The curves of C_T/T vs. T^2 in Figures 22.2–22.5 have been plotted in this way. It should be noted that for $\eta = 0.01$ the value of A (or, equivalently, the Debye temperature) needed to fit the measured data differs insignificantly from the value found by Lien and Phillips. For $\eta = 0.1$ the change in the Debye temperature needed to fit the experimental data is $\lesssim 2\%$. Thus, phasons have little effect on the slope of the line in this region. As will be discussed in detail below, it was necessary to change γ significantly from the value found by Lien and Phillips.

Plots of C_T/T vs. T^2 are shown in Figure 22.2 for $c_z = 2.7 \times 10^5$ cm/s and $\eta = 0.1, 0.01$ The dashed vertical line indicates the lowest temperatures, $T \sim 0.26°$K, measured by Lien and Phillips. The signature of the phasons is the departure of the curve from a straight line as T decreases.[3] For the lowest temperatures, the curve is once again a straight line with a slope determined by the sum of the phonon and phason contributions. The curve with $\eta = 0.1$ can be seen to depart significantly from a straight line by $T^2 = 0.07°\mathrm{K}^2$. (For smaller q_Φ this departure would be less). The curve for $\eta = 0.01$ does not depart significantly from a straight line until $T^2 = 0.005°\mathrm{K}^2$ or $T \cong 70$ mK.

Similar results for $c_z = 5.4 \times 10^5$ cm/s are shown in Figure 22.3. Note that although the curve for $\eta = 0.1$ does fall below the straight line, the curvature is much less than for the

3) For c_z small enough, the curve can be made to rise slightly above the straight line before falling below.

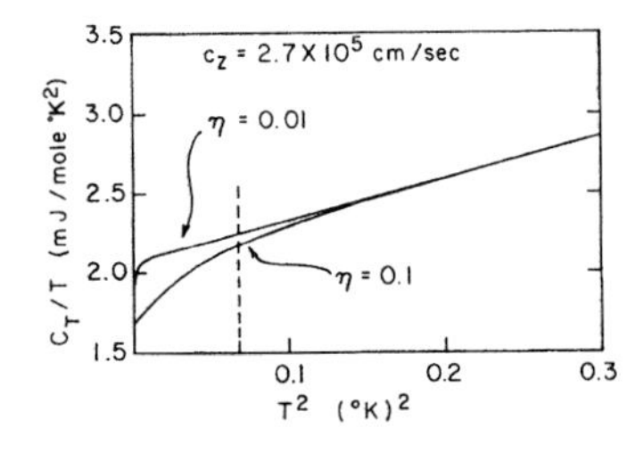

Figure 22.2 Total heat capacity C_T including phasons for potassium plotted as C_T/T vs. T^2. c_z is the velocity of phasons with wave vector $\vec{q}$ parallel to the CDW wave vector $\vec{Q}$. η gives the anisotropy of the phason velocity. See text for details.

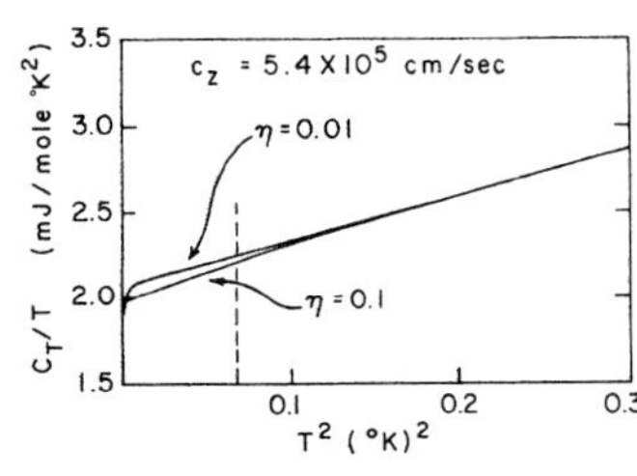

Figure 22.3 C_T/T vs. T^2 with the phason velocity twice that in Figure 22.2.

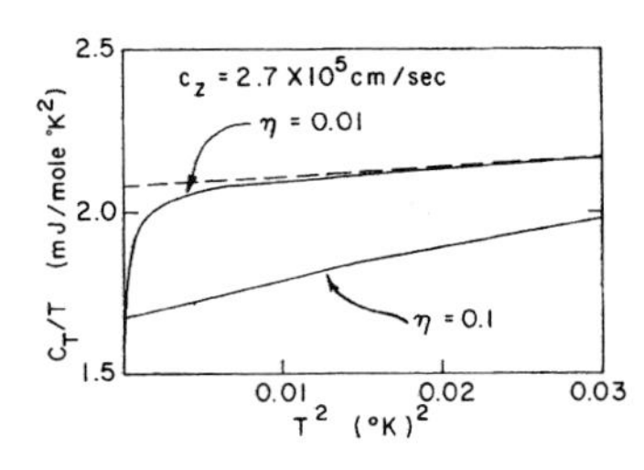

Figure 22.4 Low-temperature range of Figure 22.2.

curves in Figure 22.2 since the phason Debye temperature is greater. Such a curve could be difficult to distinguish experimentally from a straight line.

Figures 22.4 and 22.5 are expanded plots of the low temperature parts of Figures 22.2 and 22.3, respectively. In Figure 22.4 the curve for $\eta = 0.01$ indicates that the heat capacity of the phasons does not reach T^3 behavior until $T \lesssim 20\,\text{mK}$ in this case. Comparison of the straight-line portions for $\eta = 0.01$ and $\eta = 0.1$ clearly indicates the effect of phason anisotropy on the slope of the line, i.e., the magnitude of the phason contribution to the low-temperature heat capacity. It is also evident that phason anisotropy is extremely important in determining where the C/T vs. T^2 curve departs from the higher temperature straight-line behavior. Where this "turn down" occurs is extremely important. Although the Dulong–Petit contribution of the phasons is small and would be easily missed in comparison with the electron and phonon contributions at higher temperatures, the anisotropy of the phasons allows the soft phasons to fill up at much lower temperatures where C_Φ is sizeable in comparison to the other contributions. Thus, for $c_z = 2.7 \times 10^5$ cm/s and $\eta = 0.1$, at $T = 0.07°$K the phason heat capacity is $\sim$ 25% of the total, whereas for $c_z = 5.4 \times 10^5$ cm/s and $\eta = 0.1$, the phason contribution might be missed entirely. This point is crucial for experimental measurements of the phason heat capacity.

The dashed line in Figures 22.4 and 22.5 is the extrapolation of the measurements of Lien and Phillips. The intercept at zero temperatures gives $\gamma = 2.08\,\text{mJ}/(\text{mol K}^2)$. It is evident from the curves in Figures 22.4 and 22.5 that the presence of phasons can lead to a significant error in the determination of γ. Unless calorimetric measurements are performed at temperatures sufficiently low so that the phason heat capacity is in the T^3 regime, extrap-

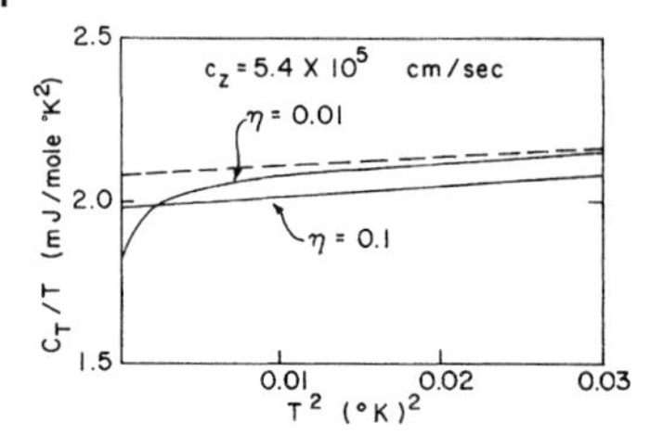

Figure 22.5 Low-temperature range of Figure 22.3.

olation of the C/T vs. T^2 curve to zero temperature can lead to a fictitious and erroneous apparent contribution to γ. In other words, it is necessary to freeze-out the phasons before the electronic heat capacity can be identified with certainty. As discussed above in arriving at a rough lower limit to the phason-velocity, for $\eta = 0.01$ and $c_z = 2.7 \times 10^5$ cm/s the error in γ is $\sim$ 30%. Equivalently, for this case the thermal effective electron mass m_t^* would be $\sim$ 30% too large. From the rest of the curves it can be seen how the "fictitious" γ varies with phason anisotropy and speed.

The possibility that the existence of phasons can lead to an erroneously large thermal effective mass is very interesting in light of the controversy surrounding the theoretical value of m_t^*. It is convenient theoretically to write m_t^* in terms of the free electron mass m_0 and various correction terms so that

$$m_t^* = m_0 \left(1 + \delta_b + \delta_{ep} + \delta_{ee} + \delta_{e\Phi} + \delta_f\right), \tag{22.13}$$

where δ_b arises from energy-band effects, (including that of the CDW), δ_{ep} is the enhancement due to electron–phonon interaction, $\delta_{e\Phi}$ is the analogous (and as yet undetermined) enhancement arising from electron-phason interactions δ_{ee} arises from electron–electron interactions, and δ_f is the "fictitious" contribution from the phasons discussed above. Of course, if the phasons are observed in the T^3 regime it is possible to extrapolate correctly to zero temperature and $\delta_f = 0$. Unless the phasons are frozen out, $\delta_f \neq 0$.

The contribution to m_t^* subject to the most disagreement is δ_{ee}. For an ideal electron gas the deviation of the density-of-states effective mass m^* at the Fermi surface from m is not large, but this is caused by a cancellation of inherently much larger effects [10]. Unfortunately, there is disagreement even as to the sign of δ_{ee}. Rice [11] has shown that Hubbard's theory [12] leads to $m^* > m$ by about 10% for typical metallic densities. In contrast several other calculations [13–15] find $m^* < m$ by a few percent. Clearly, for meaningful comparison between experimental and theoretical values of m_t^*, it is essential that the fictitious phason contribution be known to be zero.

Heat capacity measurements of incommensurate CDW systems at ultralow temperature are, therefore important for two reasons. First, such measurements can yield important evidence of the existence and properties of phasons. Second, by freezing-out the phasons the true electronic contribution to the heat capacity can be measured.

In this study we have assumed that the phason spectrum is that of a pure metal. It is possible for phason distortions of a CDW to be pinned at impurities, and this could alter slightly the phason frequency spectrum. This effect is unlikely for the case of potassium, since the root-mean-square phase fluctuation is estimated to be several radians, even at very low temperatures [3]. A pinned CDW would lead to broadening of the nuclear magnetic resonance, and this has been found to be absent in potassium [16].

22.4 Conclusion

We have discussed the phason contribution to the low-temperature heat capacity of metals containing incommensurate CDW's and illustrated the characteristic signature of phasons in measurements of the total heat capacity. By way of example, we have discussed the possible contribution of phasons to the heat capacity of potassium. It is important to note that the parameters used for the phasons were guessed. Anisotropy of the phasons is extremely important in determining both the magnitude of the low-temperature phason heat capacity and the temperature at which a plot of C/T vs. T^2 departs significantly from a straight line. If heat-capacity measurements are not done at sufficiently low temperatures so that the phasons are frozen out, the electronic heat capacity can be significantly overestimated.

It is likely that a search for phasons in potassium must be performed at temperatures significantly below 0.1 °K. These measurements would be difficult due to the small value of the heat capacity at these temperatures.

Throughout the calculation presented above damping of the phasons has been neglected. This damping has recently been calculated [17]. For potassium phasons with $\vec{q}$ parallel to $\vec{Q}$ are strongly attenuated and nearly critically damped. However, phasons with $\vec{q}$ perpendicular to $\vec{Q}$ are only very weakly attenuated. The effects of damping on phason heat capacity have been investigated. The differences between the finite-damping case and the zero-damping case discussed above are not substantial enough to warrant a separate discussion at this time. Since the characteristic signature of phasons is primarily determined by those phasons with $\vec{q}$ nearly perpendicular to $\vec{Q}$ and these phasons are only very slightly damped it is reasonable that the effects of phason damping on the heat capacity should be relatively small.

References

1 Overhauser, A.W. (1968) *Phys. Rev.*, **167**, 691.
2 Overhauser, A.W. (1971) *Phys. Rev. B*, **3**, 3173.
3 Overhauser, A.W. (1978) *Hyperfine Interact.*, **4**, 786.
4 Sawada, A. and Satoh, T. (1978) *J. Low Temp. Phys.*, **30**, 455.
5 Overhauser, A.W. and Butler, N.R. (1976) *Phys. Rev. B*, **14**, 3371.
6 Overhauser, A.W. and deGraaf, A.M. (1968) *Phys. Rev.*, **168**, 763.
7 Bishop, M.F. and Overhauser, A.W. (1977) *Phys. Rev. Lett.*, **39**, 632.
8 Overhauser, A.W. (1978) *Adv. Phys.*, **27**, 343.
9 Lien, W.H. and Phillips, N.E. (1964) *Phys. Rev.*, **133** A1370.
10 Overhauser, A.W. (1971) *Phys. Rev. B*, **4**, 3318.
11 Rice, T.M. (1965) *Ann. Phys.* (NY), **31**, 100.
12 Hubbard, J. (1957) *Proc. R. Soc London, Ser.*, **A240**, 539.
13 Hedin, L. (1965) *Phys. Rev.*, **139**, A796.
14 Overhauser, A.W. (1971) *Phys. Rev. B*, **3**, 1888.
15 Keiser, G. and Wu, F.Y. (1972) *Phys. Rev. A*, **6**, 2369.
16 Follstaedt, D. and Slichter, C.P. (1976) *Phys. Rev. B*, **13**, 1017.
17 Boriack, M.L. and Overhauser, A.W. (1978) *Phys. Rev. B*, **17**, 4549.

Reprint 23 Analysis of the Anomalous Temperature-dependent Resistivity on Potassium Below 1.6 K[1)]

Marilyn F. Bishop and A.W. Overhauser**

* Department of Physics, Purdue University, 525, Northwestern Avenue, West Lafayette, Indiana 47907, USA

Received 9 March 1979

Recent precision measurements of the resistivity of potassium between 0.38 and 1.6 K revealed a surprising $T^{1.5}$ temperature dependence. We show that scattering of electrons by phasons – the collective excitations of an incommensurate charge-density wave – can provide an explanation.

Recent measurements by Rowlands, Duvvury, and Woods [1], indicate the presence of an anomalous contribution to the resistivity of potassium, which, if fitted to a pure power law, varies as $T^{1.5}$. Previously proposed mechanisms that contribute to the resistivity at low temperatures, including electron–phonon and electron–electron scattering, yield curves of resistivity versus temperature that are the wrong shape to explain this high-precision data. As a resolution of this difficulty, we propose a new mechanism – scattering of electrons with phasons [2], the collective excitations associated with phase modulation of a charge-density wave. We will see that this assumption leads to the prediction of a temperature-dependent resistivity that is in good agreement with the data.

Conventional studies of the resistivity of a metal at low temperatures yield the following contributions:[2)]

$$\rho(T) = \rho_0 + AT^5 + BT^p \exp(-\hbar\omega_0/k_B T) + CT^2 \,, \tag{23.1}$$

where ρ_0 is the residual resistivity The term AT^5 results from normal electron–phonon scattering in the low-temperature limit. It may be greatly reduced in magnitude from the value predicted by the Bloch–Grüneisen formula if phonon-drag effects are important. Electron–phonon umklapp scattering, which is essentially unaffected by phonon drag, contributes a term at low temperatures of the form $BT^p \exp(-\hbar\omega_0/k_B T)$, where ω_0 is the frequency of the phonon with the minimum wave vector that allows participation in an umklapp process. Electron–electron scattering produces the term CT^2 at all temperatures.

The relative importance of each of these mechanisms is revealed by examining the low-temperature resistivity data in potassium. Highly accurate measurements [4–6] exhibit an exponentially decaying resistivity with decreasing temperature below about 4 K, supporting the electron–phonon umklapp scattering mechanism and lending credence to the concept of phonon drag. Below 2 K, however, significant deviations from exponential behavior have been observed, suggesting the presence of an additional scattering mechanism. In order to discover the characteristic behavior of this additional mechanism, one must first subtract the known electron–phonon umklapp contribution from the data Unfortunately, the form of this umklapp term below 2 K is not well determined. While van Kempen *et al.* have fitted their data between 2 and 4 K with the form of the exponential term in Equation (23.1) with $p = 1$ and $\theta = \hbar\omega_0/k_B = 19.9$, Kaveh, Leavens, and Wiser [7], have pointed out that other

1) Phys. Rev. Lett. 42, 26 (1979)
2) For a discussion, see [3].

Anomalous Effects in Simple Metals. Albert Overhauser

ISBN: 978-3-527-40859-7

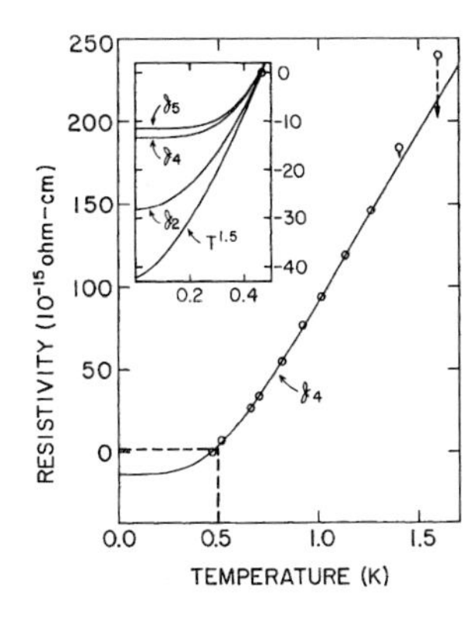

Figure 23.1 Plot of resistivity vs. temperature for the data of simple K2C of Rowlands, Duvvury, and Woods [1], indicted by circles. The dashed box has been enlarged and foreshortened in the inset. The curves and arrows are described in the text.

values of p and θ could allow equally good fits to these data. We avoid this uncertainty by considering only data below which the electron–phonon umklapp term becomes negligible, i.e., below about 1.3 K.

We now turn to a discussion of the new anomalous temperature dependence of the resistivity, as shown by the data of [1], pictured as circles in Figure 23.1, with the lowest point defining the zero of resistivity. The two highest points, above 1.3 K, are displayed to illustrate the residual influence of electron–phonon umklapp scattering on these points. The dashed arrows show the amount that one would subtract if the parameters obtained by van Kempen *et al.* [4] are used. A solid curve has been drawn through all but these two highest data points. For convenience, we will call this curve ρ (phason), with the significance of this designation becoming apparent later. There is no evidence for a T^5 contribution. Electron–electron scattering has been proposed as a possible explanation of the data [1, 4], but this must vary exactly as T^2 and does not allow for the observed dependence. This can be seen most clearly in Figure 23.2, where we show the difference between the shape of the curve defined by the data and attempts to fit it with pure power laws. Because of the scatter in the data, we find it convenient to use ρ (phason) as a reference, or the horizontal line in Figure 23.2, and plot $\rho - \rho$(phason), where ρ is either the data (circles) or T, $T^{1.5}$, and T^2 curves that pass through the first and last data points below 1.3 K. Clearly, the T^2 curve is the wrong shape to fit the data and describes the data as poorly as a straight line (the curve labeled T). If one desires the best-fit power law, that would be $T^{1.5}$ as was first suggested by Rowlands, Duvvury, and Woods [1]. Thus, it seems that electron–electron scattering must be ruled out, at least as the principal mechanism, and a new scattering mechanism must be invoked.

As an explanation of this anomalous temperature dependence of the resistivity, we suggest the mechanism of electrons scattering from phasons, collective excitations corresponding to phase modulation of a charge-density wave (CDW). The proposal of a CDW ground state in potassium has provided successful and consistent explanations of other anomalies, which have recently been reviewed [8]. Phasons have been considered in other contexts and several of their properties have been described [2, 9–11]. Of particular interest here, the electron–phason interaction has been derived to be of the form [10]

$$V_{e\varphi} = \frac{1}{2} G \sum_{\vec{q}} \varphi_{\vec{q}} \left\{ \cos\left[(\vec{Q} + \vec{q}) \cdot \vec{r} - \omega_{\vec{q}} t \right] - \cos\left[(\vec{Q} - \vec{q}) \cdot \vec{r} - \omega_{\vec{q}} t \right] \right\} , \tag{23.2}$$

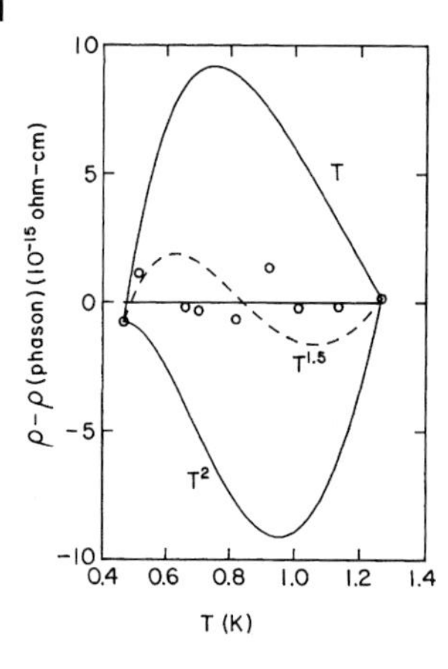

Figure 23.2 Plot of resistivity subtracted from ρ (phason), the smooth curve through the data in Figure 23.1. The data of Figure 23.1 are shown as circles scattered about the horizontal line ρ (phason). Curves labeled T, $T^{1.5}$, and T^2 are the corresponding pure power laws that pass through the first and last data points below 1.3 K.

where $G\cos(\vec{Q}\cdot\vec{r})$ is the total self-consistent potential and $\vec{Q}$ the wave vector of the static CDW. $\varphi_{\vec{q}}$, $\omega_{\vec{q}}$, and $\vec{q}$ are the magnitude, frequency, and wave vector of the phason, respectively.

A phason is actually a normal mode of the lattice whose frequency is zero at the point $\vec{Q}$ in the Brillouin zone and varies linearly with $\vec{q}$ away from that point, with the velocity in the direction of $\vec{Q}$ much greater than that perpendicular to $\vec{Q}$. For simplicity, we assume a Debye-like model and choose a cutoff that reflects the anisotropy of the phason spectrum such that the occupied phason space is small compared to the Brillouin zone. In addition, we assume extreme anisotropy of the phason spectrum, postponing the more general case to a longer, more extended paper. We thus define a phason frequency and temperature as

$$\omega_{\vec{q}} = c_{\parallel} q_{\parallel} \,, \quad \Theta_{\varphi} = \hbar c_{\parallel} q_{\varphi} / k_{\mathrm{B}} \,, \tag{23.3}$$

where $c_{\parallel}$ and $q_{\parallel}$ are the phason velocity and wave vector parallel to $\vec{Q}$. Here, the phason space is approximated by a pillbox of width q_{φ} along $\vec{Q}$.

The temperature-dependent resistivity due to electron–phason scattering along each principal axis was calculated with "golden-rule" perturbation theory using the potential of Equation (23.2). Results were obtained via a variational solution of the Boltzmann equation, assuming the Fermi surface to be rigidly displaced in $\vec{k}$ space by the electric field.[3] For the simple model described above, the resistivity can be expressed in terms of Bloch–Grüneisen functions $\mathcal{J}_r(x)$ as

$$\begin{aligned}\rho_{e\varphi}(T) = A\left[\sqrt{2}T/\Theta_{\varphi}\right]^5 \mathcal{J}_5\left(\Theta_{\varphi}/\sqrt{2}T\right) + B\left(T/\Theta_{\varphi}\right)^4 \mathcal{J}_4\left(\Theta_{\varphi}/T\right) \\ + C\left(T/\Theta_{\varphi}\right)^2 \mathcal{J}_2\left(\Theta_{\varphi}/T\right) \,,\end{aligned} \tag{23.4}$$

where

$$\mathcal{J}_r(x) = \int_0^x \frac{z^n dz}{(e^z - 1)(1 - e^{-z})} \,. \tag{23.5}$$

The expressions for the three coefficients in Equation (23.4) are complicated and their magnitudes depend on the phason anisotropy, as well as on the distribution of $\vec{Q}$ domains

3) A similar calculation was done previously for a CDW system in the study of the residual anisotropy in potassium. This is described in [12].

throughout the sample. In fact, the magnitude of the observed anomalous temperature dependence of the resistivity varies from run to run and sample to sample and seems to be related to the magnitude of the residual resistivity, which also depends on $\vec{Q}$-domain structure.[3] For this reason, we defer discussion of these coefficients to a more extended paper and direct our attention here only to the shape of the resistivity curve as a function of temperature. For this reason, we have fit the data in Figure 23.1 separately with each of the three terms in Equation (23.1), e.g., using the form $\rho_0 + \beta(T/\Theta_\varphi)^4 \mathcal{J}_4(\Theta_\varphi/T)$. Each term individually fits the data as well as the other two, but with different parameters. The excellence of the fit can be seen in Figure 23.1, where the smooth curve through the data can now be identified with the result of fitting with the $\mathcal{J}_4$ term.

Results for the $\mathcal{J}_2$ and $\mathcal{J}_5$ terms would be indistinguishable in Figure 23.1 from the $\mathcal{J}_4$ curve in the regions of the data, but their extrapolations to lower temperatures differ. These are shown in the inset of Figure 23.1, along with the extrapolation of the pure power law $T^{1.5}$ identified in Figure 23.2. The phason temperatures needed for the fit are $\Theta_\varphi = 3.43$ K for $\mathcal{J}_4$, $\Theta_\varphi = 4.58$ K for $\mathcal{J}_5$, and $\Theta_\varphi = 4.85$ K for $\mathcal{J}_2$. The phason temperature and extrapolation below the lowest data point for the true combination of terms in Equation (23.4) would lie between the values for the separate terms. On the other hand, the $T^{1.5}$ curve lies well below these other curves in the ultralow-temperature region. Therefore, it is extremely important that ultralow-temperature resistivity measurements be done in order to determine the precise shape of the curve. If the data would fall on the $T^{1.5}$ curve, this would eliminate electron–phason scattering as the explanation for the anomalous contribution to the resistivity. On the other hand, if the data would lie in the region of the graph between the $\mathcal{J}_5$ and $\mathcal{J}_2$ curves, further information could be obtained about the values of the phason temperature Θ_φ and the phason cutoff q_φ. This information could be coupled with determinations from specific-heat experiments if the predicted signature of phasons [11] is observed there, and a more complete picture of phasons in potassium would be possible.

Acknowledgements

The authors would like to express sincere thanks to J.A. Rowlands for providing the data shown in Figures 23.1 and 23.2 and for many useful discussions. In addition, the authors are grateful to the National Science Foundation Materials Research Laboratory Program for support of this work.

References

1 Rowlands, J.A., Duvvury, C., and Woods, S.B. (1978) *Phys. Rev. Lett.*, **40**, 1201.
2 Overhauser, A.W. (1971) *Phys. Rev. B*, **3**, 3173.
3 Ziman, J.M. (1960) *Electrons and Phonons*. Oxford Univ. Press, London.
4 van Kempen, H., Lass, J.S., Ribot, J.H.J.M., and Wyder, P. (1976) *Phys. Rev. Lett.*, **37**, 1574.
5 Gugan, D. (1971) *Proc. Roy. Soc London, Ser. A*, **325**, 223.
6 Ekin, J.W. and Maxfield, B.W. (1971) *Phys. Rev. B*, **4**, 4215.
7 Kaveh, M., Leavens, C.R. and Wiser, N. (1979) *J. Phys. F*, **9**, 71.
8 Overhauser, A.W. (1978) *Adv. Phys.*, **27**, 343.
9 Overhauser, A.W. (1978) *Hyperfine Int.*, **4**, 786.
10 Boriack, M.L. and Overhauser, A.W. (1978) *Phys. Rev. B*, **17**, 4549.
11 Boriack, M.L. and Overhauser, A.W. (1978) *Phys. Rev. B*, **18**, 6454.
12 Bishop, M.F. and Overhauser, A.W. (1978) *Phys. Rev. B*, **18**, 2447.

Reprint 24 Wave-Vector Orientation of a Charge-Density Wave in Potassium[1)]

G.F. Giuliani and A.W. Overhauser**

* Department of Physics, Purdue University, 525, Northwestern Avenue, West Lafayette, Indiana 47907, USA

Received 23 March 1979

A criterion for the preferred direction of the wave vector $\vec{Q}$ of a charge-density wave (CDW) is obtained by means of a simple theory. Screening of the electric field caused by the CDW is provided by a sinusoidal distortion of the positive-ion lattice. The optimum $\vec{Q}$ direction is that which minimizes the elastic energy of distortion. For potassium the $\vec{Q}$ direction is found to be tilted about 4° away from a [110] direction. The exact value of the tilt depends on the magnitude of $\vec{Q}$.

24.1 Introduction

The concept of a charge-density-wave (CDW) state for electronic systems was introduced many years ago by Overhauser [1], but only recently has this kind of non-normal electronic state been observed. (Spin-density-wave states [2], which are similar, were discovered much sooner.) In suitable quasi-two-dimensional [3] electronic systems direct observation of the typical signature of a CDW state, i.e., the super-lattice structure, has been achieved by means of x-ray electron, and neutron scattering experiments: the existence of satellite spots surrounding the usual Bragg reflections [4].[2)] Very recently evidence has also been found in three-dimensional systems [5]. Furthermore, as has been extensively discussed [6], the alkali metals, and particularly potassium, can for many reasons be expected to suffer a CDW instability. A complete clarification of this important and basic problem is needed.

Ten years ago a neutron scattering experiment on a potassium single-crystal sample was reported [7] in which no evidence for CDW satellites was found. Only high-symmetry directions, and in particular the [110] direction, were scanned. A [110] direction is expected to be the preferred orientation for the CDW wave vector $\vec{Q}$ [4]. The aim of this paper is to present a simple theory of the preferred $\vec{Q}$ orientation and to provide a prediction for the particular case of potassium Unfortunately the previous search for CDW satellites did not include scans along what we now find to be the preferred directions in reciprocal space, which are tilted 4° from [110].[3)]

24.2 Sources of Anisotropy

We consider an ideal jellium model for the interacting gas of conduction electrons in a metal and assume the neutralizing positive background to be deformable without any stiffness. For this system it has been shown [1, 2] that, in Hartree–Fock approximation, the normal state (plane-wave Slater determinant) is unstable with respect to both CDW and spin-density-

1) Phys. Rev. B 20, 4 (1979)

2) Actually many other effects associated with a CDW state can be experimentally observed (see for instance [4] and [6] for an up to date review), but in this paper we will be concerned with the most commonly accepted existence for a CDW i.e., the appearance of satellites.

3) Anything we will say about the [110] direction holds, of course, for all of the other equivalent directions.

Anomalous Effects in Simple Metals. Albert Overhauser

ISBN: 978-3-527-40859-7

wave (SDW) formation. Furthermore an argument has been presented [1] showing how the correlation corrections to this Hartree–Fock result tend to stabilize the CDW with respect to both normal and SDW states.

The electronic charge density in a CDW can be written

$$\rho(\vec{r}) = -en\left[1 - p\cos(\vec{Q}\cdot\vec{r}+\varphi)\right], \tag{24.1}$$

where n is the density of the electron gas, and p, $\vec{Q}$, and φ are the amplitude, wave vector, and phase of the charge distortion.[4] Such a structure for $\rho(\vec{r})$ lowers the exchange and correlation energy with respect to the uniform state. However, it could lead to a macroscopic energy penalty caused by the electrostatic field $\vec{E}_{\text{CDW}}$, proportional to p, which arises from the charge inhomogeneity. If $\rho(\vec{r})$ were not neutralized, then

$$\vec{E}_{\text{CDW}}(\vec{r}) = (4\pi nep\vec{Q}/Q^2)\sin\vec{Q}\cdot\vec{r}\,. \tag{24.2}$$

In this deformable jellium model the crucial point regarding the existence of a CDW state is that this energy penalty is absent because of energetically inexpensive deformations of the positive background. Accordingly, $\vec{E}_{\text{CDW}} \sim 0$. Furthermore the spherical symmetry of the problem allows $\vec{Q}$ to have any orientation.

In more complicated situations the spatial anisotropy of exchange and correlation together with anisotropic band structure (e.g., Fermi-surface nesting) can give rise to a preferred direction of $\vec{Q}$. If we allow for a nonzero stiffness of the positive background, the energy balance will be more complicated. We must take into account the elastic energy cost associated with the screening of $\vec{E}_{\text{CDW}}$, given by Equation (24.2).

At this point we recall that in order to have a CDW ground state, a metal must be as soft as possible (elastically). and that is equivalent to having small stiffness constants c_{ij}. Alkali metals satisfy this condition extremely well [8–10]. Furthermore their practically spherical Fermi surface suggests that elastic behavior may be the principal source of anisotropy in the problem. We shall proceed on the assumption that the $\vec{Q}$ orientation is determined only by the elastic anisotropy.[5]

24.3 Geometrical Factors

Consider now the specific problem of the $\vec{Q}$ orientation in potassium. In this case $\vec{Q}$ lies outside of the Brillouin zone (BZ), and $|\vec{Q}| \approx 1.33(2\pi/a)$ [11], where a is the lattice constant. It is easy to make an approximate choice for the wave vector $\vec{Q}'$ of the acoustic-phonon mode needed to screen the electric field, Equation (24.2), of the CDW. Of course $\vec{Q}'$ must lie within the BZ, so $\vec{Q}' = \vec{G}_{lmn} - \vec{Q}$, where $\vec{G}_{lmn}$ is a reciprocal-lattice vector. From the fee geometry of the reciprocal lattice it is easy to convince oneself [4] that the most energetically favorable condition for the acoustic-mode distortion is with $\vec{Q}$ along [11]. Then

$$|\vec{Q}'| = |\vec{G}_{110} - \vec{Q}| \approx 0.08(2\pi/a)\,.$$

If $\vec{Q}$ were parallel to a cubic axis, then

$$|\vec{Q}'| = |\vec{G}_{200} - \vec{Q}| \approx 0.67(2\pi/a)\,.$$

These acoustic modes have much higher frequencies than the former ones.

4) In the following discussion the phase φ does not play any significant role and we will omit it. We assume that $\vec{Q}$ is incommensurate with the reciprocal lattice.

5) We take here the point of view that in our system only one CDW is present and both the modulus of $\vec{Q}$ and the CDW amplitude parameter p are essentially established by many-body effects. We will assume these are fixed, external data.

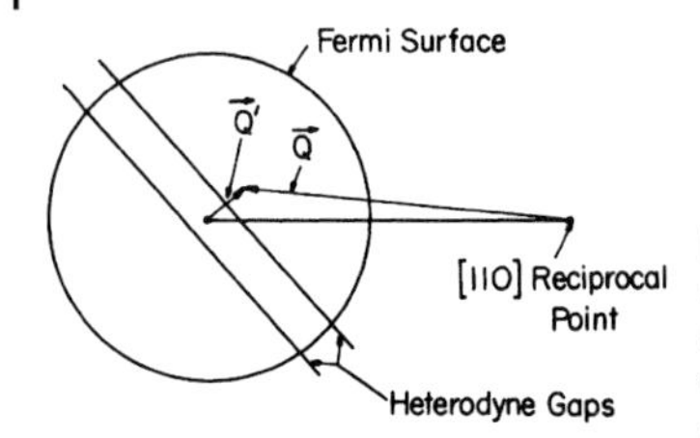

Figure 24.1 Geometrical relation of $\vec{Q}$, $\vec{Q}'$, and $\vec{G}_{110}$. The heterodyne gaps arise from the $\vec{Q}'$ periodicity and can lead to open orbits in large magnetic fields. The main CDW energy gaps, which are perpendicular to $\vec{Q}$ and tangent to the Fermi surface, are not shown.

In Figure 24.1 we show the geometrical relations for $\vec{Q}$ and $\vec{Q}'$ when $\vec{Q}$ is near to [11]: the angles θ and θ' describe the tilt of $\vec{Q}$ and $\vec{Q}'$ from [110]. Also shown are the heterodyne gaps associated with $\vec{Q}'$, which play an important role [12, 13] in the physics of the induced-torque anomalies observed in potassium [14, 15]. The important point is the following: if for some reason $\vec{Q}$ is *slightly* tilted away from [110] (and we will show this is the case) the orientation of $\vec{Q}'$ deviates from [110] by a large angle, as is clear from Figure 24.1. The small value expected for $|\vec{Q}'|$ allows us to work out the theory in the long-wavelength approximation.

24.4 Energy Analysis

The elastic energy associated with the three acoustic modes of wave vector $\vec{Q}'$ is, for a monovalent metal of unit volume,

$$U_E = \frac{1}{4} n M \sum_{i=1}^{3} \omega_i^2 A_i^2 \,, \tag{24.3}$$

where M is the ionic mass, A_i is the amplitude of each (static) excitation, and ω_i is the frequency of the mode (were it allowed to oscillate). Equation (24.3) is the energy penalty which must be paid to screen out $\vec{E}_{\mathrm{CDW}}$. In order to determine how this is partitioned between the three polarizations we have to consider the interaction between the positive ions and the electrostatic potential V_{CDW} caused by the CDW. From Equation (24.2),

$$V_{\mathrm{CDW}}(\vec{r}) = (4\pi n e p / Q^2) \cos \vec{Q} \cdot \vec{r} \,. \tag{24.4}$$

Suppose $e\rho_I(\vec{r})$ is the charge density of each ion. Then the ionic charge density of the deformed lattice is

$$\rho(r) = \sum_{\vec{L}} e\rho_I \left(\vec{r} - \vec{L} - \sum_{i=1}^{3} A_i \vec{\epsilon}_i \sin \vec{Q}' \cdot \vec{L} \right) . \tag{24.5}$$

$\{\vec{L}\}$ are the lattice vectors, and $\{\vec{\epsilon}\}$ re the polarization vectors of the three acoustic modes of wave vector $\vec{Q}'$. The only interaction between Equations (24.4) and (24.5) will involve the Fourier components of wave vector $\pm\vec{Q}$. This is readily calculated from Equation (24.5) if $QA_i \ll 1$,

$$\rho(\pm\vec{Q}) = \frac{1}{2} n e \rho_I(Q) \sum_{i=1}^{3} A_i |\vec{\epsilon}_i \cdot \vec{Q}| \,. \tag{24.6}$$

$\rho_I(Q)$ is the Fourier transform of $\rho_I(\vec{r})$, i.e., it is the ion (or pseudo-ion) form factor. Accordingly the interaction energy (per unit volume) is

$$U_{\mathrm{int}} = 2\pi p n^2 e^2 Q^{-2} \rho_I(Q) \sum_{i=1}^{3} A_i \vec{\epsilon}_i \cdot \vec{Q} \,. \tag{24.7}$$

Observe that the total energy, Equations (24.3) and (24.7), can be written

$$U = \sum_{i=1}^{3} \left(\alpha\omega_i^2 A_i^2 + 2\beta A_i \vec{\epsilon}_i \cdot \vec{Q} \right), \tag{24.8}$$

where α and β are constants. Each of the three amplitudes, A_i is obtained by minimizing U. The minimum energy is

$$U_{\min} = -\frac{\beta^2}{\alpha} \sum_{i=1}^{3} \left(\frac{\vec{\epsilon}_i \cdot \vec{Q}}{\omega_i} \right)^2 . \tag{24.9}$$

$\epsilon_i(\vec{Q}')$ and $\omega_i(\vec{Q}')$ are functions of the direction of $\vec{Q}$ because

$$\vec{Q}' = \vec{G}_{110} - \vec{Q} . \tag{24.10}$$

From Equation (24.9) it is obvious that the optimum direction of $\vec{Q}$ is found by maximizing

$$S(\vec{Q}) \equiv \sum_{i=1}^{3} \left(\frac{\epsilon_i(\vec{Q}') \cdot \vec{Q}}{\omega_i(\vec{Q}')} \right)^2 . \tag{24.11}$$

24.5 Results

As already mentioned in Section 24.3, the smallness of $|\vec{Q}'|$ allows us to determine $\vec{\epsilon}_i$ and ω_i in the long-wavelength, acoustic limit. This means that we need to know only the three elastic (stiffness) modulii: c_{11}, c_{12}, and c_{44}. They are [9, 10]: 4.16, 3.41, and 2.86 × 10^{10} dyn/cm^2 at 4 °K. $nM\omega_i^2$ and $\vec{\epsilon}_i$, are the eigenvalues and eigenvectors of the dynamical matrix

$$\begin{pmatrix} c_{11}x^2 + c_{44}(y^2 + z^2) & (c_{12} + c_{44})xy & (c_{12} + c_{44})xz \\ (c_{12} + c_{44})xy & c_{11}y^2 + c_{44}(z^2 + x^2) & (c_{12} + c_{44})yz \\ (c_{12} + c_{44})xz & (c_{12} + c_{44})yz & c_{11}z^2 + c_{44}(x^2 + y^2) \end{pmatrix}, \tag{24.12}$$

where $\vec{Q}' \equiv (x, y, z)$. We have evaluated $S(\vec{Q})$ numerically for all directions of $\vec{Q}$ near [11]. Contours of $S(\theta, \varphi)/S(0, 0)$ are shown in a polar plot in Figure 24.2. The polar angle θ is the tilt angle of $\vec{Q}$ away from [110]. The azimuthal angle φ (rotation about the [110] polar axis) is 0 when $\vec{Q}$ lies in the (001) plane. The contours were computed from Equations (24.11) and (24.12) for $|\vec{Q}| = 1.33 \times (2\pi/a)$.

It is clear from Figure 24.2 that the maximum value of $S(\theta, \varphi)$ is not at $\theta = \varphi = 0$. Instead, it is at $\theta = 4.1°$, $\varphi = 65.4°$. This implies that the angle between $\vec{Q}'$ and [110] is 47.3°. It is surprising, perhaps, that the point $\theta = \varphi = 0$ is a local *minimum*, even though $|\vec{Q}'|$ here is as small as possible. The reason is that only the high-frequency longitudinal mode contributes to S for this direction. The physical reason why $\vec{Q}$ (and $\vec{Q}'$) are tilted away from [110] is that the low-frequency shear modes then contribute significantly to S, even though their polarization vectors are not closely parallel to $\vec{Q}$. For example, the pronounced "hill" along the $\varphi = 90°$ line in Figure 24.2 arises from the lowest-frequency shear mode. At the absolute maximum S involves all three polarization modes and represents the best compromise between polarization and lattice stiffness.

The location of the maximum S depends on $|\vec{Q}|$, which is not yet known precisely. If $|\vec{Q}| = 1.36(2\pi/a)$, the maximum shifts to $\theta = 2.6°$ and $\varphi = 64.3°$. If $|\vec{Q}| = 1.30(2\pi/a)$, the maximum shifts to $\theta = 5.6°$, $\varphi = 66.9°$. However, as the magnitude of $\vec{Q}$ changes, the

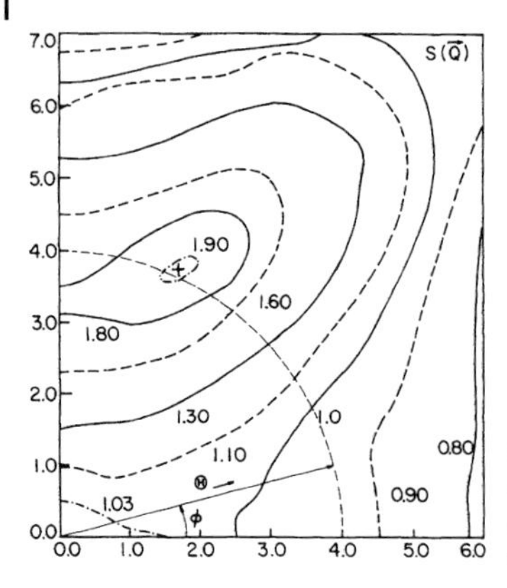

Figure 24.2 Contours of $S(\theta, \varphi)$, Equation (24.11), for potassium with $|\vec{Q}| = 1.33(2\pi/a)$. The contour values are normalized so that $S(0, 0) \equiv 1$. The maximum S, at $\theta = 4.1°$, $\varphi = 65.4°$, is 1.91. The numbers next to the axes measure Θ in degrees.

direction of $\vec{Q}'$ remains practically constant. It is always near $\theta' = 47°$, $\varphi'(= \varphi) = 65°$. In Cartesian notation the directions of $\vec{Q}'$ are the 48 (cubic) equivalents of (1.05, 1.00, 0.40). In high magnetic fields these would be the possible directions for open orbits, caused by the heterodyne gaps, shown in Figure 24.1.

Although the tilt of the CDW wave vector $\vec{Q}$ from [110] is small, $\sim 4°$, it is far enough away that a diffraction scan along a [110] direction would preclude observation of any satellites (if present). Indeed the additional $\vec{Q}$ direction degeneracy (24 instead of 6) implies that satellite intensity will be weaker than otherwise expected, since $\vec{Q}$-domain structure [6] will distribute the CDW satellite intensity among the 48 possible locations surrounding each ordinary reciprocal-lattice vector. Furthermore in a single $\vec{Q}$-domain crystal, one might have to examine all 24 equivalent $\vec{Q}$ axes in order to find a CDW satellite.

Finally, we have carried out equivalent calculations for Na and Rb, based on $T = 78°\mathrm{K}$ elasticity data [16, 17], and with $|\vec{Q}| = 1.35(2\pi/a)$. For Na, $\theta = 3.3°$, $\varphi = 63°$; for Rb, $\theta = 3.2°$, $\varphi = 64°$.

Acknowledgements

We are grateful to the NSF for support of this research.

References

1 Overhauser, A.W. (1968) *Phys. Rev.*, **167**, 691.
2 Overhauser, A.W. (1962) *Phys. Rev.*, **128**, 1437.
3 Wilson, J.A., Di Salvo, F.J. and Mahajan, S. (1975) *Adv. Phys.*, **24**,117.
4 Overhauser, A.W. (1971) *Phys. Rev. B*, **3**, 3173.
5 Sawada, A. and Satoh, T. (1978) *J. Low Temp. Phys.*, **30**, 455.
6 Overhauser, A.W. (1978) *Adv. Phys.*, **27**, 343.
7 Atoji, M. and Werner, S.A. (1969) *Solid State Commun.*, **7**, 1681.
8 See e.g., C Kittel, *Introduction to Solid State Physics*, 4th ed. (Wiley, New York, 1971), p. 149.
9 Marquardt, W.R. and Trivisonno, J. (1964) *J. Phys. Chem. Solids*, **20**, 273.
10 Smith, P.A. and Smith, C.S. (1964) *J. Phys. Solids*, **20**, 279.
11 Overhauser, A.W. (1964) *Phys. Rev. Lett.*, **13**, 190.
12 Overhauser, A.W. (1971) *Phys. Rev. Lett.*, **27**, 938.

13 Overhauser, A.W. (1974) *Phys. Rev. B*, **9**, 2441.

14 Schaefer, J.A. and Marcus, J.A. (1971) *Phys. Rev. Lett.*, **27**, 935.

15 Holroid, F.W. and Datars, W.R. (1975) *Can. J. Phys.*, **53**, 2517.

16 Diedrich, M.E. and Trivisonno, J. (1966) *J. Phys. Chem. Solids*, **27**, 637.

17 Gutman, E.J. and Trivisonno, J. (1967) *J. Phys. Chem. Solids*, **28**, 805.

Reprint 25 Theory of Transverse Phasons in Potassium[1]

G.F. Giuliani and A.W. Overhauser**

* Department of Physics, Purdue University, 525, Northwestern Avenue, West Lafayette, Indiana 47907, USA

Received 6 November 1979

A quantitative theory of the frequency dispersion relation for long-wavelength, transverse phasons of a charge-density-wave state is developed. The emphasis is directed toward alkali metals and in particular potassium. For these the theory is parameter free. The transverse-phason velocity is found to be of the same order of magnitude as the sound velocity. Also, the electric field that would arise from the charge modulation of the conduction electrons is almost completely screened by sinusoidal displacements of the positive ions. The relevance of these results to observable phenomena is discussed.

25.1 Introduction

Phasons [1] are collective modes which describe the lowest-energy excitations of an incommensurate charge-density wave (CDW) [2]. The general properties of this new type of excitation have been extensively investigated [1, 3, 4]. The physical reality of such modes is well established in several CDW systems.[2]

The object of this paper is the development of a quantitative theory for long-wavelength, transverse phasons in a simple metal. We shall assume the reader is familiar with the evidence supporting a CDW ground state for potassium (and other metals), which has been reviewed in a recent paper [5].

The transverse-phason dispersion relation is, of course, interesting in itself. It is also an important input in many theoretical problems related to the CDW state, e.g., the anomalous temperature-dependent resistivity [6], phason attenuation [4], ultra-low-temperature anomalies in heat capacity [7], and the temperature dependence of CDW satellites in neutron scattering [1]. Furthermore the analysis presented here provides a general foundation for predicting the spatial orientation of the CDW wave vector in a simple metal [8].

The structure of the paper is as follows: In Section 25.2 the theory of long-wavelength phasons is reviewed and cast in suitable form. In Section 25.3 (and the appendixes) a theory for the ionic screening in a CDW ground state is developed. In Section 25.4 quantitative results for potassium are obtained. Finally, in Section 25.5 comments on the physical meaning and impact of the results are given.

25.2 Phason Energy Spectrum

Let the conduction-electron gas of a simple metal (at $T = 0\,\mathrm{K}$) be in a CDW ground state. It is characterized by a sinusoidal modulation of the electron charge density [2]

$$\delta\rho_{\mathrm{CDW}}(\vec{r}) = -enp\cos(\vec{Q} - \vec{r} + \varphi)\,, \tag{25.1}$$

where n is the mean electron density. p, $\vec{Q}$, and φ are the amplitude, wave vector, and phase of the CDW. This latter quantity will be of paramount interest in what follows. This charge

2) The most striking evidence in a three-dimensional system is the phason heat capacity in $LaGe_2$, see [12].

Anomalous Effects in Simple Metals. Albert Overhauser

ISBN: 978-3-527-40859-7

inhomogeneity will lead to a macroscopic electric field

$$\vec{E}_{\text{CDW}}(\vec{r}) = \frac{-4\pi e n p}{Q^2}\vec{Q}\sin\left(\vec{Q}\cdot\vec{r}+\varphi\right). \tag{25.2}$$

This will cause a displacement of the ions from their lattice sites

$$\delta\vec{R}_i = \delta\vec{R}(\vec{Q}')\sin\left(\vec{Q}'\cdot\vec{R}_i^0+\varphi\right), \tag{25.3}$$

where $\vec{R}_i^0$ is the lattice site of the ith ion, $\delta\vec{R}(\vec{Q}')$ is a small displacement vector, and $\vec{Q}'$ is the Brillouin-zone reduction of $\vec{Q}$. As discussed elsewhere [1, 8], for the alkali-metal case, $\vec{Q}'$ is related to $\vec{Q}$ by $\vec{Q}' \equiv \vec{G}_{110} - \vec{Q}$. $\vec{Q}$ is expected to lie very nearly along one of the [110] reciprocal lattice directions, e.g., $\vec{G}_{110}$.

The magnitude of $\vec{Q}$ is slightly larger than the diameter $2k_F$ of the Fermi surface [2]. In general it will have no rational relationship to a reciprocal lattice vector. The CDW is then incommensurate. Furthermore, if we consider only very pure samples, it is clear that the energy of the system does not depend on the value of the phase φ. From this φ invariance of the energy, the existence of low-frequency collective excitations (associated with gentle variations of the phase φ in space and time) can be surmised; i.e., the infinite degeneracy of the ground state with respect to the value of φ leads to collective excitations – phasons – having a frequency spectrum that goes to zero [1]. It is clear from Equation (25.3) that phasons are part of the vibrational spectrum of the crystal. Indeed it was shown in [1] that a phason is a coherent linear combination of two "old" phonons (which are no longer normal modes subsequent to the breaking of translational symmetry by the CDW). The orthogonal linear combination describes a modulation of the CDW amplitude, and has a high frequency.

We now assume that φ is a slowly varying function and expand it as follows:

$$\varphi(\vec{R}_i^0, t) = \sum_{\vec{q}} \varphi_{\vec{q}} \sin\left(\vec{q}\cdot\vec{R}_i^0 - \omega_{\vec{q}} t\right), \tag{25.4}$$

where $\varphi_{\vec{q}}$ and $\omega_{\vec{q}}$ are the amplitude and frequency of the phason labeled by the wave vector $\vec{q}$. If only a single phason mode is excited, the ionic displacement Equation (25.3) is,

$$\delta\vec{R}_i = \delta\vec{R}(\vec{Q}')\sin\left[\vec{Q}\cdot\vec{R}_i^0 + \varphi_{\vec{q}}\sin(\vec{q}\cdot\vec{R}_i^0 - \omega_{\vec{q}} t)\right]. \tag{25.5}$$

The meaning of this can be seen by considering a small spatial region near a point where $|\vec{q}\cdot\vec{R}_i^0| \ll 1$. Then, with $\varphi_{\vec{q}} \ll 1$,

$$\varphi\vec{R}_i = \delta\vec{R}(\vec{Q}')\sin\left[(\vec{Q} + \phi_{\vec{q}}\vec{q}\cos\omega_{\vec{q}} t)\cdot\vec{R}_i^0\right]. \tag{25.6}$$

In other words, a phase excitation leads to a local change in $\vec{Q}$ by

$$\delta\vec{Q} = \varphi_{\vec{q}}\vec{q}\cos\omega_{\vec{q}} t\,. \tag{25.7}$$

Long-wavelength phasons cause small changes in the magnitude and orientation of the CDW wave vector $\vec{Q}$. For the special cases where $\vec{q}$ is either parallel or perpendicular to $\vec{Q}$ we say the phasons are purely longitudinal or transverse. A longitudinal phason produces a modulation of the magnitude of $\vec{Q}$, while a transverse phason produces a rotational modulation of $\vec{Q}$.

In order to find the dispersion relation of phasons in the long-wavelength limit it is useful to study both the kinetic and potential energy resulting from the excitation. The kinetic energy density is

$$T = \frac{1}{V}\sum_i \frac{1}{2}M\left(\delta\dot{\vec{R}}_i\right)^2, \tag{25.8}$$

where M is the ionic mass and V the volume of the crystal. The time average of T is easily obtained for small $\varphi_{\vec{q}}$:

$$\langle T \rangle = \frac{1}{8} n M \left[\delta \vec{R}(\vec{Q}') \right]^2 \varphi_{\vec{q}}^2 \omega_{\vec{q}}^2 \,. \tag{25.9}$$

(We have assumed that the number of ions equals the number of conduction electrons.) Now let $\vec{Q}^*$ be the wave vector of the CDW in the ground state. The potential energy of the CDW is a function of $\vec{Q} = \vec{Q}^* + \delta \vec{Q}$. We expand it around $\vec{Q}^*$ to terms of second order in $\delta \vec{Q}$, Equation (25.7). The quantities δQ_α will be the components of $\delta \vec{Q}$ along axes labeled by α ($\alpha = 1, 2, 3$). The potential increment is

$$\delta U(\vec{Q}) = U(\vec{Q}) - U(\vec{Q}^*) = \sum_{\alpha,\beta} \Lambda_{\alpha\beta}(\vec{Q}^*) \delta Q_\alpha \delta Q_\beta \,. \tag{25.10}$$

$\Lambda_{\alpha\beta}$ is a suitable symmetric matrix. The absence of linear terms in δQ_α is guaranteed by the assumption that $\vec{Q}^*$ locates the minimum point in a potential valley. The time average of Equation (25.10) is, with use of Equation (25.7),

$$\langle \delta U(\vec{Q}) \rangle = \frac{1}{2} \varphi_{\vec{q}}^2 \sum_{\alpha,\beta} \Lambda_{\alpha\beta}(\vec{Q}^*) q_\alpha q_\beta \,. \tag{25.11}$$

The virial theorem allows us to equate Equations (25.9) and (25.11). This leads to the (long-wavelength) phason dispersion relation [1]:

$$\omega_{\vec{q}} = \left(\frac{4}{n m \delta R^2(\vec{Q}')} \sum_{\alpha,\beta} \Lambda_{\alpha\beta}(\vec{Q}^*) q_\alpha q_\beta \right)^{1/2} \,. \tag{25.12}$$

As mentioned above, the interesting feature of this spectrum is that $\omega_{\vec{q}}$ varies linearly with $|\vec{q}|$, so phasons can be considered a new branch of the acoustic spectrum. By diagonalization of the matrix $\Lambda_{\alpha\beta}$ (a reduction to principal axes) Equation (25.12) can be written as

$$\omega_{\vec{q}} = \left(C_1^2 q_1^2 + C_2^2 q_2^2 + C_3^2 q_3^2 \right)^{1/2} \,. \tag{25.13}$$

Here, q_1 is the component of $\vec{q}$ along $\vec{Q}^*$, and q_2 and q_3 are perpendicular components corresponding to the principal axes. C_1 is therefore the longitudinal-phason velocity, whereas C_2 and C_3 are the principal values of the transverse-phason velocity. Thus,

$$C_1 = \left(\frac{4}{n m \delta R^2(\vec{Q}')} \lambda_1 \right)^{1/2} \,, \tag{25.14}$$

and for the transverse modes

$$C_2, C_3 = \left(\frac{4}{n m \delta R^2(\vec{Q}')} \lambda_{2,3} \right)^{1/2} \,. \tag{25.15}$$

The members of $\{\lambda_j\}$ are the eigenvalues of $\Lambda_{\alpha\beta}$, Equation (25.12), and are of course positive.

25.3 Energy of the Conduction Electron–Ion System

The object of this section is to find the total energy as a function of the CDW wave vector $\vec{Q}$. This depends on both electronic and ionic degrees of freedom: $\delta\vec{R}(\vec{Q}')$, p, $\vec{Q}$, and φ. In the previous section we showed that the φ degree of freedom led to phason excitations. Consider, now, the energy associated specifically with the electrons. If we assume that exchange and correlation do not depend on the direction of $\vec{Q}$, their contribution to the energy will be a function only of the amplitude p and the magnitude $|\vec{Q}|$ of the CDW.

Let p_0 be the optimum value of p for the case of jellium. The most favorable value for $|\vec{Q}|$ is primarily a theoretical property of the electron gas, and is slightly larger than diameter $2k_F$ of the Fermi sphere [2]. The influence of the ion lattice is primarily in the $\vec{Q}$ orientation problem [8]. For simplicity, we will assume the magnitude of $\vec{Q}$ to be an *a priori*, fixed value.

On the other hand, the ground-state value of p will be established through a balance between the pure electron-gas energy contributions and those arising from electron–ion interaction. The term arising from kinetic, exchange, and correlation energies will be taken to have the following form:

$$U_{kxc}(p\,Q) = E_{kxc}(p_0, Q) + \gamma(P_0, Q)(p - p_0)^2 \,. \tag{25.16}$$

We have assumed here that the final value of p will not be too different from p_0, the jellium value. E_{kxc} and γ can come only from a complete solution of the many-electron problem. The total electronic energy is obtained by adding to U_{kxc} the electrostatic energy of the charge-density modulation. With the help of Equation (25.2) we find

$$U^{\text{el}}(p, Q) = U_{kxc}(p, Q) + \frac{\pi n^2 e^2}{Q^2} p^2 \,. \tag{25.17}$$

Let us consider next the contribution of the ion–ion interactions. In a linear dielectric theory based upon the pseudopotential approach to lattice dynamics [9], the total potential energy of an array of ions embedded in an electron gas can be written as

$$U_{\text{ion}} = 2\pi n^2 e^2 \sum_{\vec{G}} \frac{\bar{\rho}^2_{\vec{G}}}{G^2 \epsilon(\vec{G})} \,, \tag{25.18}$$

where $\bar{\rho}_{\vec{G}}$ is the Fourier transform of the pseudo-ion form factor (see Appendix A), and $\epsilon(G)$ is the dielectric function of the electron gas. If the ions are located on a Bravais lattice, the sum (25.18) is over the reciprocal-lattice vectors $\{\vec{G}\}$.

Suppose the ionic positions have undergone a small sinusoidal displacement described by Equation (25.3). Then the nonzero Fourier components of $\bar{\rho}$ involved in Equation (25.18) have wave vectors $\vec{G}'$ or $\vec{G} \pm \vec{Q}'$. The change in the ionic potential energy up to second order in the small displacements $\delta\vec{R}(\vec{Q}')$ can be shown (by a detailed calculation) to be

$$\Delta U_{\text{ion}} = \pi n^2 e^2 \left(\sum_{\vec{G}} \frac{\bar{\rho}^2_{\vec{G}-\vec{Q}'}}{|\vec{G}-\vec{Q}'|^2} \frac{\left[(\vec{G}-\vec{Q}')\cdot\delta\vec{R}(\vec{Q}')\right]^2}{\epsilon(\vec{G}-\vec{Q}')} - \sum_{\vec{G}\neq 0} \frac{\bar{\rho}^2_{\vec{G}}}{G^2} \frac{\left[\vec{G}\cdot\delta\vec{R}(\vec{Q}')\right]^2}{\epsilon(\vec{G})} \right) . \tag{25.19}$$

The hypothesis underlying Equations (25.18) and (25.19) is that linear-response theory treats adequately the electronic screening of the pseudo-ion interactions. This condition is certainly violated if the electron gas has suffered a CDW instability. Linear-response theory does not take into account the many-body effects which cause a CDW. The Fourier component with wave vector $\vec{Q}(= \vec{G}_{110} - \vec{Q}')$ must be treated in a completely different way.

Our strategy is as follows: We will replace the $\vec{Q}$ term, $\vec{G} = [110]$, in the first sum of Equation (25.19) by an exact electrostatic energy $U^{(2)}_{\text{ion}}(\vec{Q})$ given below. For this particular Fourier component the electrostatic energies of the CDW, the ion–ion interactions, and the coupling between them must be treated consistently.

Since $\vec{Q}$ is only slightly different from $\vec{G}_{110}$, the wave vector $\vec{Q}' \equiv \vec{G}_{110} - \vec{Q}$ is small. It is therefore possible to evaluate the original value of Equation (25.19) in the long-wavelength limit [8]. We find for the *revised* ΔU_{ion},

$$\Delta U_{\text{ion}} - U^{(2)}_{\text{ion}}(\vec{Q}) = \frac{1}{4} n M \sum_{i=1}^{3} A_i^2(\vec{Q}')\omega_i^2(\vec{Q}') - \frac{\pi n^2 e^2 \bar{\rho}_Q^2}{Q^2}\left(\sum_i A_i(\vec{Q}')\vec{e}_i(\vec{Q}')\cdot\vec{Q}\right)^2 \Big/ \epsilon(\vec{Q}) \,. \tag{25.20}$$

The first term on the right-hand side is merely Equation (25.19) expressed in terms of phonon parameters, i.e., the amplitudes A_i frequencies ω_i, and polarization vectors $\vec{e}_i$ of the acoustic modes of wave vector $\vec{Q}'$. The last term of Equation (25.20) is the one which must be subtracted from the first in order to exclude the $\vec{G}_{110}$ term from the sum. The connection with $\delta\vec{R}(\vec{Q}')$ is given by

$$\delta\vec{R}(\vec{Q}') = \sum_{i=1}^{3} A_i(\vec{Q}')\vec{e}_i(\vec{Q}') \,. \tag{25.21}$$

As far as $U^{(2)}_{\text{ion}}(\vec{Q})$ is concerned it is clear that we cannot use a linear-dielectric theory expression. We must generalize Equation (25.1) to include the positive ions in a realistic way. Consider first the electrostatic energy associated with the $\vec{Q}$ component of the static lattice displacement, Equation (25.3):

$$U^{(2)}_{\text{ion}}(\vec{Q}) = \frac{\pi n^2 e^2}{Q^2}\rho^2_{\vec{Q}}\left(\sum_{i=1}^{3} A_i(\vec{Q}')\left[\vec{e}_i(\vec{Q}')\cdot\vec{Q}\right]\right)^2 . \tag{25.22}$$

Here, $\rho_{\vec{q}}$ is the Fourier transform of the real-ion charge (see Appendix A). The final contribution to the energy is that arising from the Coulomb interaction between the CDW and the positive-ion displacements. It is given by [8]

$$U_{\text{int}}(\vec{Q}) = \frac{2\pi n^2 e^2}{Q^2}\rho_{\vec{Q}}\, p \sum_{i=1}^{3} A(\vec{Q}')\left[\vec{e}_i(\vec{Q}')\cdot\vec{Q}\right]. \tag{25.23}$$

The total is obtained from Equations (25.17) (25.20), (25.22), and (25.23).

$$\bar{U}_{\text{tot}} = \frac{\pi n^2 e^2}{Q^2}\rho^2_{\vec{Q}}\left[\sum_i A_i^2\Omega_i^2 - \gamma\left(\sum_i A_i a_i\right)^2 + \frac{1}{x}\left(\frac{p-p_0}{\rho_Q}\right)^2 + \left(\frac{p}{\rho_Q}\right)^2 + \left(\sum_i A_i a_i\right)^2 + 2\frac{p}{\rho_Q}\sum_i A_i a_i\right]. \tag{25.24}$$

We have dropped the constant, $E_{kxc}(p_0, Q)$, and have defined:

$$\Omega_i^2 = M\omega_i^2 Q^2/4\pi n e^2 \rho_{\vec{Q}}^2 , \tag{25.25}$$

$$x = \pi n^2 e^2/\gamma Q^2 , \tag{25.26}$$

$$y = \bar{\rho}_{\vec{Q}}^2/\rho_{\vec{Q}}^2 \epsilon(Q) , \tag{25.27}$$

$$a_i = \vec{e}_i(\vec{Q}') \cdot \vec{Q} . \tag{25.28}$$

It is worthwhile to notice that the last three terms of Equation (25.24) give the total electrostatic energy associated with the Fourier component $\vec{Q}$:

$$U_{\mathrm{ES}} = \frac{\pi n^2 e^2}{Q^2}\left(p + \rho_Q \sum_i A_i a_i\right)^2 . \tag{25.29}$$

One can see that the ion-displacement amplitudes, A_i will acquire nonzero values and thereby screen the electrostatic field Equation (25.2) of the CDW. It is precisely this screening process which allows a CDW ground state to occur [2, 8].

The ground-state energy may now be obtained by minimizing the energy Equation (25.24) with respect to p and $\{A_i\}$. This is carried out in detail in Appendix . We discuss here only a limiting case applicable to the simple metals. The first observation is that the parameter x, Equation (25.26), is very small. This is the ratio between the electrostatic energy change, resulting from an increment Δp, and the corresponding many-body contributions to the energy. For small x the equilibrium value of p differs only slightly from its optimum value p_0 in jellium. Equation (B1) reduces to

$$p \cong p_0 - x\rho_{\vec{Q}} \sum_i A_i a_i . \tag{25.30}$$

A second simplification is that the parameter y, Equation (25.27), is also a small number, of the order of 10^{-2} for potassium. The values of $\{A_i\}$ which correspond to the absolute minimum are then, from Equation (B4),

$$A_i(\vec{Q}') \cong \frac{-a_i p_0}{\rho_{\vec{Q}}\Omega_i^2\left[1 + S(\vec{Q})\right]} , \tag{25.31}$$

where S is

$$S(\vec{Q}) = \sum_{i=1}^{3}\left(\frac{a_i}{\Omega_i}\right)^2 . \tag{25.32}$$

(This S differs from an equivalent one defined earlier [8] only by a constant factor.)

It is interesting at this point to estimate the fractional screening of the CDW resulting from ionic displacements. The electronic charge modulation, Equation (25.1), has a natural counterpart in the positive ion lattice

$$\delta\rho_L(\vec{r}) = -en\rho_{\vec{Q}}\left(\sum_i A_i(\vec{e}_i \cdot \vec{Q})\right)\cos(\vec{Q}\cdot\vec{r} + \varphi) . \tag{25.33}$$

If we define the fractional screening ratio s to be

$$s = -\delta\rho_L/\delta\rho_{\mathrm{CDW}} , \tag{25.34}$$

then from Equation (25.31) we obtain

$$s = \left[1 + S^{-1}(\vec{Q})\right]^{-1} . \tag{25.35}$$

For alkali metals the lattice vibration frequencies are so small that $S(\vec{Q})$ turns out to be of the order of 10^3. Therefore Equation (25.35) indicates that the lattice provides an almost complete screening of the CDW electric field.

The minimum total energy is obtained from Equation (25.24) with the help of Equations (25.30), (25.31), and (25.32). In the limit of small x, y, and $S^{-1}(\vec{Q})$ we obtain, as indicated or from Equation (B5),

$$\bar{U}_{\text{tot}}(\vec{Q}) = \frac{\pi n^2 e^2 p_0^2}{Q^2} \left(-y + \frac{1}{S(\vec{Q})} \right) . \tag{25.36}$$

This energy is a function of the direction of $\vec{Q}$. Clearly the optimum direction is that which maximizes $S(\vec{Q})$. In a previous paper [8] we found that the optimum direction of $\vec{Q}$ in the alkali metals corresponds to a tilt of about 4° from a [110] direction.

The energy depends also on the magnitude of $|\vec{Q}|$. The many-electron contributions to this dependence, Equation (25.16), are not known quantitatively. Consequently it is not possible to calculate the dispersion relation for a longitudinal phason. We will now exploit the assumption that many-electron effects are invariant during roation of $\vec{Q}$ in order to calculate the dispersion relation of transverse phasons.

25.4 Transverse-Phason Velocity in Potassium

In order to carry out a quantitative investigation of the transverse-phason spectrum we have to find the most favorable CDW wave vector $\vec{Q}^*$. This is done by minimizing the total energy Equation (25.36) with respect to $\vec{Q}$ orientation. (For reasons given above the magnitude of $\vec{Q}$ is fixed.) The minimum value of the energy Equation (25.36) corresponds to the maximum value of $S(\vec{Q})$, Equation (25.32). One can see from Equation (25.35) that this also corresponds to the maximum screening fraction s.

The problem of evaluation and maximization of $S(\vec{Q})$ was solved in a previous paper [8]. The most favorable $\vec{Q}$ orientation in potassium is along an axis, [1.00, 0.94, 0.09], which has 24 cubic equivalents. The assumed magnitude of $\vec{Q}$ was [10] $1.33(2\pi/a)$, a being the lattice constant.

Consider now a small deviation $\delta\vec{Q}$ of $\vec{Q}$ in a direction orthogonal to $\vec{Q}$. The potential energy change will be, from Equation (25.36),

$$\begin{aligned} \delta U_{\text{tot}}(\vec{Q}) &= U_{\text{tot}}(\vec{Q}^* + \delta\vec{Q}) - U\text{tot}(\vec{Q}^*) \\ &\cong \frac{\pi n^2 e^2 p_0^2}{Q^2 S(\vec{Q}^*)} \left(\frac{S(\vec{Q}^*)}{S(\vec{Q})} - 1 \right) . \end{aligned} \tag{25.37}$$

The quantity in large parentheses can be expanded to second order, e.g., $\delta Q_\alpha \delta Q_\beta$, where $\alpha, \beta = 2$ or 3 and correspond to orthogonal directions perpendicular to $\vec{Q}^*$. Then

$$\delta U_{\text{tot}} = \frac{\pi n^2 e^2 p_0^2}{Q^2 S(\vec{Q}^*)} \sum_{\alpha,\beta} L_{\alpha\beta} \delta Q_\alpha \delta Q_\beta , \tag{25.38}$$

where $L_{\alpha\beta}$ are the (four) expansion coefficients. The two eigenvalues, l_2 and l_3, of $L_{\alpha\beta}$ correspond to two (λ_2 and λ_3) of the three eigenvalues of $\Lambda_{\alpha\beta}$, Equations (25.12) and (25.15), except for a constant factor. This factor is obtained by comparison of Equation (25.38) with Equation (25.10).

The two principal velocities of the transverse-phason spectrum can now be found from Equation (25.15). However, we must first use Equations (25.21) and (25.31) to evaluate $\delta R^2(\vec{Q}')$:

$$\delta R^2(\vec{Q}') = \left(\frac{p_0}{\rho_{\vec{Q}} S(\vec{Q}^*)}\right)^2 \sum_{i=1}^{3} \left(\frac{a_i}{\Omega_i^2}\right)^2 . \tag{25.39}$$

We have dropped the one in Equation (25.31) since $S(\vec{Q}^*) \sim 10^3$. The two velocities are

$$C_\alpha = \Omega_0 \rho_{\vec{Q}^*} \left(\frac{S(\vec{Q}^*)}{\tilde{S}(\vec{Q}^*)} \Big/ l_\alpha\right)^{1/2} , \tag{25.40}$$

where $\Omega_0 \equiv (4\pi n e^2/M)^{1/2}$ is the ion plasma frequency. Furthermore,

$$\tilde{S}(\vec{Q}^*) \equiv \sum_{i=1}^{3} \left(\frac{a_i Q}{\Omega_i^2}\right)^2 , \tag{25.41}$$

which may be compared with $S(\vec{Q})$, defined by Equation (25.32).

We have evaluated $L_{\alpha\beta}(\vec{Q})$ numerically for $|\vec{Q}^*| = 1.33(2\pi/a)$ and along the directions of type [1.00, 0.94, 0.09], which were determined in our previous work [8] for the case of potassium. The eigenvalues l_2 and l_3, together with the numerical values of $S(\vec{Q}^*)$ and $\tilde{S}(\vec{Q}^*)$ give the following principal values for transverse-phason velocities:

$$C_2 = 2.08 \times 10^5 \text{ cm/s} , \quad C_3 = 0.94 \times 10^5 \text{ cm/s} . \tag{25.42}$$

In many calculations it may be convenient to approximate the anisotropic, *transverse*-phason dispersion with an isotropic one. For this purpose one may use the geometric mean of the two values in Equation (25.42), i.e., $\bar{C} = 1.40 \times 10^5$ cm/s. For the sake of comparison, the velocity of the longitudinal, acoustic mode [11] along [110] at 4 K is 2.71×10^5 cm/s.

It is of interest to remark that the phason velocities Equation (25.42) depend only on the observed elastic moduli C_{11}, C_{12}, and C_{44}, and on the magnitude of $|\vec{Q}^*|$. If we assume $1.36(2\pi/a)$ for this, we obtain $C_2 = 2.13$, $C_3 = 1.00$, and $\bar{C} = 1.46 \times 10^5$ cm/s. The elastic moduli used were the 4 K values [8].

The $\vec{q}$ directions which correspond to the principal phason velocities in Equation (25.42) are, respectively, $[-1.00, 0.86, 2.12]$ and $[0.87, -1.00, 0.81]$.

25.5 Discussion

In the previous section we obtained precise quantitative values for the transverse-phason velocities and for the fractional screening. The main result of the theory can be expressed in the following way: If the electron gas (Fermi surface, exchange and correlation effects, etc.) is isotropic, the velocity of transverse phasons is comparable to the sound velocity.

Since one expects that, in general, it is easier to rotate $\vec{Q}$ than to change its magnitude, the longitudinal-phason velocity is very likely larger than the transverse velocity, perhaps by an order of magnitude or more. The attenuation coefficient of longitudinal phasons was found [4] to be $\gamma_q \cong (1.1 \times 10^5 \text{ cm/s})q$. Comparison of these values indicates that phasons in a simple metal are well-defined excitations having a quality factor > 10 or 100. For phason $\vec{q}$'s making an angle θ with $\vec{Q}$ the attenuation is reduced [4] by $\cos^2\theta$.

The temperature dependence of the electrical resistivity of potassium at low temperature appears to be dominated by electron–phason scattering [6]. The magnitude of this effect depends critically on the phason-frequency spectrum. With the theoretical value of the transverse velocity, derived here, it should be possible to employ the experimental phason-scattering resistivity to gain a quantitative estimate of the longitudinal velocity.

Phason excitations lead to very-low-temperature heat-capacity anomalies [7]. This phenomenon was first reported in the La-Ge system [12]. They also give rise to a reduction of CDW-satellite intensity in diffraction experiments. This temperature factor [1, 3] depends critically on the phason spectrum.

Finally, the result obtained in Equation (25.35), which indicates almost complete screening, shows that the alkali metals are good approximations to the ideal "jellium" model [2, 5]. Therefore one should not be surprised to find in them a CDW structure.

The authors are grateful to the National Science Foundation for support of this research.

A Positive-Ion Form Factors

A.1 Pseudo-Ion Form Factor $\bar{\rho}_{\vec{p}}$

The electrostatic-potential Fourier coefficient caused by the positive-ion lattice is

$$\Phi_{\text{ion}}(\vec{p}) = \frac{4\pi n e}{p^2\epsilon(p)}\rho_{\vec{p}}\,, \tag{A1}$$

where $\rho_{\vec{p}}$ the single-ion form factor, i.e., it would be unity for a single point charge. $\epsilon(P)$ is the electron-gas dielectric function. The pseudopotential seen by an electron is not Equation (A1) times $-e$ because the conduction-electron wave function must be orthogonalized to the occupied ion-core electron states. The $\vec{p}$ component of the pseudopotential may be defined by

$$V_{\text{eff}}(\vec{p}) \equiv \frac{-4\pi n e^2}{p^2\epsilon(p)}\bar{\rho}_{\vec{p}}\,, \tag{A2}$$

where $\bar{\rho}_{\vec{p}}$ is the pseudo-ion form factor.

There are models for the pseudopotential. We have used a simple one consisting of a uniformly charged sphere of radius R and total charge $-S$ together with a uniformly charged spherical shell, also of radius R, but with total charge $S+1$. The unscreened potential of this pseudo-ion is

$$\Phi(r) = \begin{cases} e/r, & r \geq R\,, \\ -\dfrac{e}{2R}\left[S - 2 - S\left(\dfrac{r}{R}\right)^2\right], & r \leq R\,. \end{cases} \tag{A3}$$

The total charge of the pseudo-ion is e, suitable for a monovalent metal. This pseudopotential has the advantage that it is continuous. The pseudo-ion form factor of this model is the Fourier transform of the charge distribution (divided by e):

$$\bar{\rho}_{\vec{p}} = (S+1)\frac{\sin PR}{PR} - 3S\left(\frac{\sin PR - PR\cos PR}{(PR)^3}\right). \tag{A4}$$

The two parameters, R and S, can be adjusted to fit particular data. We have found[10] that the entire phonon spectrum of potassium can be reproduced in a dielectric theory of the lattice dynamics with $R = 1.30/k_F$ and $S = 1.15$. We find for $Q = 1.33(2\pi/a)$,

$$\bar{\rho}_Q \approx -0.20 . \tag{A5}$$

A similar result is obtained from other pseudopotential models, e.g., from that due to Heine and Animalu.

A.2 Real-Ion Form Factor $\rho_{\vec{Q}}$

The form factor of the real positive ion is the Fourier transform of the total ionic charge distribution. For potassium

$$\rho_Q = 19 - \rho_Q^{\mathrm{el}} \approx 2.92 . \tag{A6}$$

The first term is the nuclear charge, and the second is the x-ray form factor [14]. This was evaluated for the Q used above in Equation (A5). With Equations (A5) and (A6) the parameter y, Equation (25.27), can be evaluated. For potassium, and in fact for all the alkali metals, $y \sim 10^{-2}$. The smallness of y allows the CDW minimum energy to be simplified.

B CDW Energy Minimization

The minimization of the function $\tilde{U}_{\mathrm{tot}}$, Equation (25.24), with respect to the amplitudes p and A_i, gives

$$p = \frac{1}{1+x}\left(p_0 - x\rho_{\vec{Q}}\sum_i A_i a_i\right). \tag{B1}$$

The A_i's are determined by the following set of coupled linear equations:

$$\sum_j \left[\delta_{ij}\left(\Omega_i^2 + \zeta a_i^2\right) + (1-\delta_{ij})\zeta a_i a_j\right] A_j = -\frac{p_0}{\rho_{\vec{Q}}}\frac{a_i}{1+x} , \tag{B2}$$

where

$$\zeta \equiv 1 - y - \frac{x}{1+x} . \tag{B3}$$

The system Equation (B2) can be solved exactly:

$$A_i = \frac{-p_0}{\rho\vec{Q}}\frac{1}{1+x}\frac{a_i\Omega_i^{-2}}{1+\zeta S(\vec{Q})} , \tag{B4}$$

where $S(\vec{Q})$ is defined by Equation (25.32) in the text. Finally the minimum total energy is

$$\begin{aligned} U_{\mathrm{tot}}(\vec{Q}) = {} & \frac{\pi n^2 e^2 p_0^2}{Q^2}\left[(1+x)(1+\zeta S)\right]^{-2} \\ & \times \left\{1 + x + S[1 - 2y(1+x)] + S^2[y^2(1+x) - y]\right\} . \end{aligned} \tag{B5}$$

If one expands this to lowest order in the small quantities x, y, and S^{-1}, Equation (25.36) can be obtained.

10) This appendix is based on unpublished lecture notes of A.W. Overhauser. A useful discussion of this topic can be found in the book by [13], where the values of the Abarenkov–Heine pseudopotential are also listed.

References

1. Overhauser, A.W. (1971) *Phys. Rev. B*, **3**, 3173.
2. Overhauser, A.W. (1968) *Phys. Rev.*, **167**, 691.
3. Overhauser, A.W. (1978) *Hyper. Inter.*, **4**, 786.
4. Boriack, M.L. and Overhauser, A.W. (1978) *Phys. Rev. B*, **17**, 4549.
5. Overhauser, A.W. (1978) *Adv. Phys.*, **27**, 343.
6. Bishop, M.F. and Overhauser, A.W. (1979) *Phys. Rev. Lett.*, **42**, 1776.
7. Boriack, M.L. and Overhauser, A.W. (1978) *Phys. Rev. B*, **18**, 6554.
8. Giuliani, G.F. and Overhauser, A.W. (1979) *Phys. Rev. B*, **20**, 1328.
9. Heine, V. and Abarenkov, I.V. (1964) *Philos. Mag.*, **9**, 451.
10. Overhauser, A.W. (1964) *Phys. Rev. Lett.*, **13**, 190.
11. Marquardt, W.R. and Trivisonno, J. (1965) *J. Phys. Chem. Solids*, **26**, 273.
12. Sawada, A. and Satoh, T. (1978) *J. Low Temp. Phys.*, **30**, 455.
13. Harrison, W.A. (1966) *Pseudopotentials in the Theory of Metals*, Benjamin, New York.
14. Cromer, D.T. and Waber, J.T. (1965) *Acta Crystallogr.*, **18**, 104.

Reprint 26 Charge-Density-Wave Satellite Intensity in Potassium[1)]

G.F. Giuliani and A.W. Overhauser***

* Department of Physics, Purdue University, 525, Northwestern Avenue, West Lafayette, Indiana 47907, USA
** Physics Department, Technical University of Munich, Garching8046, West Germany

Received 9 April 1980

The intensity of a charge-density-wave diffraction satellite in potassium is calculated. Velocity dependence of the exchange and correlation potential, which is responsible for the conduction-electron charge modulation, significantly affects the deduced value of the charge-density-wave amplitude. The amplitude of the periodic lattice displacement, which screens the electronic modulation, is reduced to a very small value, 0.03 Å, when the real charge distribution of a positive ion is recognized. A random $\vec{Q}$-domain structure can lead to a reduction by a factor of 24, compared to a single-$\vec{Q}$ specimen, of the satellite intensity. In such a case it is only 1.4×10^{-5} that of a crystallographic Bragg reflection. At temperatures above liquid helium, satellite intensity may be reduced further by phason excitations.

26.1 Introduction

The alkali metals, and in particular potassium, display a wide range of anomalous properties [1, 2] which a normal electron-gas picture cannot account for. In order to find a comprehensive theoretical interpretation of experimental data it has been suggested [3] that the conduction electrons suffer a charge-density-wave (CDW) instability. Exchange and correlation potentials play a role of primary importance in the theory of such an instability. In this paper we will show that the nonlocal velocity dependence of exchange and correlation reduces significantly the observability of a CDW in a diffraction experiment.

The translational invariance of a crystal is broken by a CDW. In order to maintain microscopic charge neutrality the positive-ion lattice undergoes a small sinusoidal displacement. This can be observed directly. Two small diffraction satellites will appear [1, 3] in reciprocal space for each reciprocal-lattice vector. Detection of these is the unequivocal signature of a CDW.

An early estimate of the intensity ratio of a CDW satellite to a Bragg reflection was about 1% [3]. This was based on a jellium model with a charge modulation of 17%, which corresponds to the suggested CDW energy gap of 0.6 eV for potassium. A unique orientation of the CDW wave vector $\vec{Q}$ throughout the sample was assumed.

Atoji and Werner [4] carried out a neutron scattering experiment on potassium at low temperature with a sensitivity of two parts in 10^4. They scanned high-symmetry directions with particular emphasis on the [110], which was expected to be the preferred orientation of the CDW [1]. No satellites were found.

In a recent paper [5] the authors developed a theory for the preferred orientation of the wave vector $\vec{Q}$. In the alkali metals anisotropy of the elastic stiffness is the determining factor. For potassium the optimum direction of $\vec{Q}$ is tilted about 4° away from a [110] direction. Although small, this tilt must be allowed for in a search for the satellites. Furthermore, the

1) Phys. Rev. B 22, 8 (1980)

Anomalous Effects in Simple Metals. Albert Overhauser

ISBN: 978-3-527-40859-7

possibility of 24 different, but equivalent, orientations of $\vec{Q}$ would cut down the intensity of each satellite to 4×10^{-4}.

Motivated by the new information regarding the direction of $\vec{Q}$, Werner, Eckert, and Shirane [6] conducted a new search by neutron diffraction. Their experiment was sufficiently sensitive to detect satellites having an intensity 2×10^{-5} that of an ordinary Bragg reflection. None were found that could be attributed to a CDW.

The aim of this paper is to present an improved estimate for the intensity of a CDW peak in potassium. The result, Equation (26.22), is unfortunately about a factor of 2 smaller than the minimum detectability of the Werner, Eckert, and Shirane experiment.

The satellite intensity depends on the amplitude of the charge-density modulation of the electron gas [3], on the interaction between the electronic CDW and the lattice [7], and also on the excitation spectrum [1] of the system. All of these factors contribute to the revised estimate.

The structure of the paper is as follows. In Section 26.2 the theory of the scattering intensity in a CDW system [1] is reviewed with particular emphasis on the role of CDW collective excitations. In Section 26.3 the ionic lattice distortion is examined. Section 26.4 is devoted to the theory of the CDW fractional amplitude; and a new evaluation of this quantity is reported. Finally, in Section 26.5 the CDW satellite intensity is calculated.

26.2 Neutron-Scattering Elastic Intensity

In a neutron experiment the wave-vector- and frequency-dependent diffracted intensity $I_{\vec{k},\omega}$ is proportional to the following dynamic structure factor:[2)]

$$S_{\vec{k},\omega} = \frac{1}{N} \int_{-\infty}^{+\infty} dt e^{-i\omega t} \sum_{ij} \left\langle e^{i\vec{k}\cdot\vec{R}_i(t)} e^{-i\vec{k}\cdot\vec{R}_j(0)} \right\rangle_T . \qquad (26.1)$$

The indices i and j label the N atoms of the system, $\vec{R}_i(t)$ represents the position of the atom i at the time t, and an equilibrium thermal average is taken of the right-hand side.

In a CDW state the tendency towards microscopic charge neutrality causes the ionic lattice to undergo a distortion with respect to the ideal crystal. The new equilibrium positions of the ions of a single CDW state are given by

$$\vec{R}_i = \vec{R}_i^0 + \delta\vec{R}_0 \cos\left(\vec{Q}\cdot\vec{R}_i^0 + \phi\right), \qquad (26.2)$$

where $\vec{R}_i^0$ are the atomic sites of the original undistorted lattice, $\delta\vec{R}_0$ is the ground-state amplitude, $\vec{Q}$ the wave vector, and ϕ the phase of the distortion. Furthermore, the dynamics of $\vec{R}_i(t)$, the actual ionic positions, are related to the dynamics of the amplitude and the phase of the CDW.

A. Static CDW

We assume for the moment that both phase and amplitude degrees of freedom are frozen. In this case the dynamic structure factor reduces to a simple form:

$$S_{\vec{k},\omega}^{\mathrm{CDW}} = \frac{2\pi\delta(\omega)}{N} \left(\sum_i \exp\left[i\vec{k}\cdot(\vec{R}_i^0 + \delta\vec{R}_0 \cos\vec{Q}\cdot\vec{R}_i^0) \right] \right)^2 , \qquad (26.3)$$

2) See for instance, [8].

where for convenience ϕ has been set equal to zero. In the limit of zero amplitude the usual result for the undistorted lattice is readily recovered:

$$S^{0}_{\vec{k},\omega} = 2\pi N\delta(\omega)\sum_{\vec{G}}\delta(\vec{k}-\vec{G})\,. \tag{26.4}$$

The allowed Bragg reflections are those with scattering vector equal to a reciprocal-lattice vector $\vec{G}$. For finite $\delta\vec{R}_0$ we can use in (26.3) the following formula [1]:

$$e^{iz\sin\alpha} = \sum_{n=-\infty}^{+\infty} e^{in\alpha} J_n(z)\,, \tag{26.5}$$

where n is an integer and $J_n(z)$ a Bessel function of the first kind. $S_{\vec{k},\omega}$ can be rewritten now as

$$S^{\mathrm{CDW}}_{\vec{k},\omega} = 2\pi N\delta(\omega)\sum_{n=-\infty}^{+\infty}\sum_{\vec{G}}\delta\left(\vec{k}-(\vec{G}+n\vec{Q})\right)J_n^2(\vec{k}\cdot\delta\vec{R}_0)\,. \tag{26.6}$$

Notice that the new relevant feature of the diffraction pattern is that satellite spots appear at $\vec{k}=\vec{G}+n\vec{Q}$. As the amplitude $\delta\vec{R}_0$ of the distortion is expected to be small compared to the lattice spacing, and because the limiting behavior of Bessel functions is $J_n(x)\approx\left(\frac{1}{2}x\right)^n/n!$ for small x, each Bragg reflection will be surrounded by a sequence of weak satellite spots. Although in principle all the satellites are present, usually only the set with $n=1$ has sufficient intensity to be easily observed. Even though the magnitude of $\vec{Q}$ may be known [9] from the diameter of the Fermi surface, these first-order satellites may be hard to find if the direction of $\vec{Q}$ is unknown.

B. Dynamic Excitations

For an incommensurate CDW, in the absence of any source of pinning, the phase ϕ is free to assume all possible values. This infinite degeneracy of the ground state results in the existence of a new branch of acoustic, collective modes called phasons [1]. These modes are associated with space and time variations of the variable ϕ of Equation (26.2). Together with phase modulation the CDW may also experience amplitude modulation. Amplitude and phase modes occupy only a small portion of the wave-vector space associated with vibrational excitations of the system. The remaining degrees of freedom are the normal phonons. In the following discussion we will disregard these as they reduce the intensity of all diffraction peaks according to the ordinary Debye–Waller factor.

The inclusion of amplitude modes and phasons in the theory [1, 10] gives for $S^{\mathrm{CDW}}_{\vec{k},\omega}$:

$$S^{\mathrm{CDW}}_{\vec{k},\omega} = 2\pi N\delta(\omega)\sum_{n=-\infty}^{+\infty}\sum_{\vec{G}}\delta\left(\vec{k}-(\vec{G}+n\vec{Q})\right)J_n^2(\vec{k}\cdot\delta\vec{R}_0) \times\exp\left[-2n^2w_\phi+2|n|(|n|-1)w_A\right]+\dots\,, \tag{26.7}$$

where the ellipsis represents inelastic terms, which contain the contribution associated with emission and absorption of phasons and amplitude modes. The factors containing w_A and w_ϕ are the amplitude and the phason temperature factors [1, 10, 11]. They are the analogue of the Debye–Waller factor for an ordinary Bragg reflection. w_A and w_ϕ are proportional to the mean-square fluctuation of the amplitude $|\delta\vec{R}|$ and the phase ϕ of the CDW.

We notice here that the normal Bragg reflection ($n=0$) are unaffected by phase and amplitude fluctuations Furthermore the latter do not alter the intensity of the first satellite spots ($w=\pm1$).

C. Domain Structure

In a recent paper [5] the authors have derived the preferred orientation of the wave vector $\vec{Q}$ of a CDW in alkali metals. Since $\vec{Q}$ is not along a symmetry axis or in a symmetry plane, all 24 cubically equivalent axes will be equally favored. In the absence of uniaxial stress, which could split this 24-fold degeneracy, a single-crystal sample will, in general, be divided into $\vec{Q}$ domains [1, 2] having $\vec{Q}$'s aligned without preference along all allowed directions. This will reduce the intensity of a specific satellite spot by a factor of 24 compared to the intensity it would have if the sample were single $\vec{Q}$.

However, a compensating advantage is that each crystallographic Bragg reflection will be surrounded by 48 CDW satellites. One need not scan along all 24 axes to be sure of finding a satellite.

26.3 Lattice Distortion

The electronic charge of a CDW can be written as [3]

$$\rho(\vec{r}) = -en\left(1 + p\cos\vec{Q}\cdot\vec{r}\right), \tag{26.8}$$

where n is the average electron density, and p is the fractional amplitude of the modulation. Such an electronic state can be energetically more favorable than the usual undistorted homogeneous state provided the underlying positive background is sufficiently deformable to allow microscopic cancellation of the charge modulation. Otherwise a positive electrostatic energy would dominate the energy [3].

An analysis [7] of this phenomenon was carried out It was found that for the alkali metals the screening of the CDW is about 99.9%. The theoretical amplitude $|\delta\vec{R}_0|$ of the periodic lattice distortion is $\simeq$ 0.03 Å. A particularly relevant quantity is $\delta\vec{R}_0 - \vec{Q}$. Equations (3.6), (3.16), and (3.17) of [7] lead to

$$|\delta\vec{R}_0\cdot\vec{Q}| = p/\rho_{\vec{Q}}, \tag{26.9}$$

where p is the CDW fractional amplitude defined above. $\rho_{\vec{Q}}$ is the $\vec{Q}$ component of the ionic charge distribution [5].[3] Since the wavelength of the CDW is small ($\simeq$ 4 Å in potassium), an assumption of pointlike ions is unjustified Incorporation of the ionic form factor takes account of the finite size. For potassium $\rho_{\vec{Q}}$ is easily estimated from available values of the x-ray form factor, $\rho_{el}(\vec{p})$ [12], for a single K^+ ion. We have found

$$\rho_{\vec{Q}} = 19 - \rho_{el}(\vec{Q}) \simeq 2.92\,. \tag{26.10}$$

This result will be employed with Equations (26.6) and (26.9) when the satellite intensity is estimated in Section 26.5.

26.4 CDW Fractional Amplitude in Potassium

An *a priori* theory of the CDW fractional amplitude p could only come from an exact solution of the electronic many-body problem, which is not at hand. So we will rely on a semiempirical calculation to obtain an estimate of this important parameter [3, 13].

3) The ionic form factor is defined as the Fourier transform of the charge distribution of a single ion.

4) In [7] it is shown that, due to the softness of the lattice, this is the case in potassium.

Consider an electron gas embedded in a neutralizing, perfectly deformable jelly and modulated by a single-CDW ground state of wave vector $\vec{Q}$. Each electron is acted upon by a periodic potential which self-consistently sustains the charge modulation. Since any electrostatic (Hartree) field is assumed to be exactly cancelled by an equal modulation of the positive background,[4] this periodic potential originates from exchange and correlation effects [3].

The periodic potential is proportional to cos $\vec{Q} \cdot \vec{r}$, so each one-electron state $\vec{k}$ is mixed with $\vec{k} \pm \vec{Q}$. This causes the charge modulation, and produces a gap Δ (in the single-particle spectrum) on planes perpendicular to $\vec{Q}$ through the points $\pm\vec{Q}/2$. Furthermore, when Δ is small compared to the Fermi energy ϵ_F, the periodic potential can be treated as a perturbation. The free-electron plane waves, which we take as our basis functions, become amplitude modulated. The fractional amplitude p of the CDW is just the Fermi-sea average of this modulation.

A. Local Theory

The simplest choice for the exchange and correlation potential is:

$$V^{\text{xc}}(\vec{r}) = -\Delta \cos \vec{Q} \cdot \vec{r} \,, \tag{26.11}$$

where Δ is independent of $\vec{k}$. The perturbed wave functions are to first order,[5]

$$\Psi_{\vec{k}} \simeq e^{i\vec{k}\cdot\vec{r}} \left(1 - \frac{\Delta}{2\omega_+} e^{i\vec{Q}\cdot\vec{r}} - \frac{\Delta}{2\omega_-} e^{-i\vec{Q}\cdot\vec{r}}\right), \tag{26.12}$$

where the denominators are defined in terms of the free-electron energies, $\vec{E}_k = h^2k^2/2m$.

$$\omega_\pm \equiv E_{\vec{k}} - E_{\vec{k}\pm\vec{Q}} \,. \tag{26.13}$$

We average $|\Psi_{\vec{k}}|^2$ over the occupied states of the Fermi sphere to obtain the charge modulation amplitude defined by Equation (26.8) [3]:

$$p_L = \left(\frac{3\Delta}{4\epsilon_F}\right)\left(1 + \frac{1-u^2}{2u} lu \left|\frac{1+u}{1-u}\right|\right), \tag{26.14}$$

where $u = |\vec{Q}|/2k_F$.

The typical values for potassium are $\Delta/4\epsilon_F \simeq 0.07$ and $|\vec{Q}|/k_F \simeq 1 + \Delta/4\epsilon_F \simeq 1.07$ [3, 9]. In this case Equation (26.14) gives [3]

$$p_L \simeq 0.17 \,. \tag{26.15}$$

B. Nonlocal Theory

As already pointed out by one of the authors [13], in a nonuniform electron gas the periodic part of the exchange operator connecting the electronic states $\vec{k}$ and $\vec{k} + \vec{Q}$, is extremely nonlocal; i.e., it has, for fixed $\vec{Q}$, a dramatic dependence upon $\vec{k}$. When correlation effects are accounted for [14], this dependence is reduced but still remains pronounced. A correct theory of the exchange and correlation periodic potential for a CDW must allow for this nonlocality. A straightforward generalization of the exchange and correlation potential $V^{\text{xc}}(\vec{r})$ which allows for this is,

$$V^{\text{xc}}_{\text{CDW}}(\vec{r}) = \frac{1}{2}\left[\Delta^+(\vec{k})e^{i\vec{Q}\cdot\vec{r}} + \Delta^-(\vec{k})e^{-i\vec{Q}\cdot\vec{r}}\right]. \tag{26.16}$$

5) Strictly speaking this formula loses its validity for electronic states near the energy gaps, it is possible to show that the contribution from these regions of the Fermi sea is negligible for our purpose.

$\Delta^{\pm}(\vec{k})$ is a function of the electronic state. These functions have been calculated by Duff and Overhauser [14] using the "plasmon" model [15, 16].

For the case $|\vec{Q}| \simeq 2k_F$, their results can be summarized as follows: $\Delta(\vec{k})$ is a smooth function of $\vec{k}$; furthermore if $\Delta(\vec{k})$ is equal to Δ at $\vec{k} = -\frac{1}{2}\vec{Q}$, then $\Delta(0) \simeq 0.5\Delta$, $\Delta(\frac{1}{2}\vec{Q}) \simeq 0.36\Delta$. Furthermore $\Delta(\vec{k})$ is practically independent of the component of $\vec{k}$ perpendicular to $\vec{Q}$.

A simple, smooth function with these properties is

$$\Delta^{+}(\vec{k}) = \Delta\left[1 + 0.55\left(\frac{k_z}{Q} + \frac{1}{2}\right) - 0.33\left(\frac{k_z}{Q} + \frac{1}{2}\right)^2\right]^{-1},$$
$$\Delta^{-}(\vec{k}) = \Delta^{+}(-\vec{k}), \tag{26.17}$$

where k_z is measured along $\vec{Q}$. Then the perturbed wave functions are given as before by Equation (26.12), but now Δ is dependent on $\vec{k}$. A numerical calculation gives for the nonlocal value of p, using again a spherical Fermi surface,

$$p_{\mathrm{NL}} \simeq 0.10\,. \tag{26.18}$$

In order to take into account the effect of Fermi-surface distortion caused by the CDW we have repeated the calculation with the distorted Fermi surface given by Bishop and Overhauser,[6] where it is assumed that the Fermi surface touches the gap plane in a single point [9]. The fractional modulation is increased slightly by this improvement. Our final value is

$$p_{\mathrm{NL}} \simeq 0.11\,. \tag{26.19}$$

26.5 Results

In a neutron-diffraction experiment on a CDW system a quantity of interest is the intensity ratio of a satellite spot to a normal Bragg reflection. With the use of Equation (26.7) we obtain the following result:

$$\frac{I_{\vec{G}+n\vec{Q}}}{I_{\vec{G}}} = \frac{1}{D}\left(\frac{J_n(\delta\vec{R}_0\cdot(\vec{G}+n\vec{Q}))}{J_0(\delta(\vec{R}_0))}\right)^2 \times \exp\left[-2n^2 w_\phi + 2|n|(|n|-1)w_A\right]. \tag{26.20}$$

D is the ($\vec{Q}$-domain degeneracy (unity for a single domain sample).

The most intense satellite spots have $n = \pm 1$. The scattering vectors are then $\vec{G} \pm \vec{Q}$. For small CDW amplitude we can use in (26.20) the limiting behavior of the Bessel functions, i.e., $J_1(x) \simeq x/2$ and $J_0(x) \simeq 1$. It follows that a satellite's intensity is proportional to $|\vec{G} \pm \vec{Q}|^2$. This suggests that satellites of large reciprocal-lattice vectors, $\vec{G}$, will be easier to observe. The problem, however, is to find a small satellite in a thermal diffuse background, which also increases in intensity as the square of the scattering vector. For potassium the CDW satellites are expected to be very close (in $\vec{k}$ space) to reciprocal-lattice vectors. The requirements on angular resolution to separate them become severe if satellites of large $\vec{G} \pm \vec{Q}$ are selected for study. The optimum satellite to search for will depend on experimental resolution and available flux.

6) Equation (3.11a) should read $\kappa = \left[(w^2 + \alpha^2)^{1/2} - w^2 - \alpha\right]^{1/2}$ [17].

For purposes of illustration we shall evaluate Equation (26.20) for the CDW satellite at $\vec{k} = \vec{Q}$ for potassium. Equations (26.9) and (26.20) imply:

$$\frac{I_{\vec{G}}}{I_{\vec{G}(110)}} \simeq \frac{1}{24}\left(\frac{p}{2\rho_{\vec{Q}}}\right)^2 e^{-2w_\phi} , \tag{26.21}$$

where p is the fractional modulation given by (26.19). The factor D has been taken to be 24 on account of the degeneracy of equivalent $\vec{Q}$-domain orientations [5]. If we disregard the phason temperature factor as seems justified in a 4 K experiment and use the values (26.10) and (26.19) to obtain $I_{\vec{Q}}$, we find

$$I_{\vec{Q}} \simeq 1.4 \times 10^{-5}\, I_{\vec{Q}(110)} \; . \tag{26.22}$$

We believe that any future experimental search for CDW satellites in potassium should have a sensitivity consistent with this result.

Acknowledgements

We are grateful to the National Science Foundation for support of this research. One of us (A.W.O.) was assisted by an Alexander von Humboldt Award.

References

1 Overhauser, A.W. (1971) *Phys. Rev.*, **3**, 3173.
2 Overhauser, A.W. (1978) *Adv. Phys.*, **27**, 343.
3 Overhauser, A.W. (1968) *Phys. Rev.*, **167**, 691.
4 Atoji, M. and Werner, S.A. (1969) *Solid State Commun.*, **7**, 1681.
5 Giuliani, G.F. and Overhauser, A.W. (1979) *Phys. Rev. B*, **20**, 1328.
6 Werner, S.A., Eckert, J., and Shirane, G. (1980) *Phys. Rev. B*, **21**, 581.
7 Giuliani, G.F. and Overhauser, A.W. (1980) *Phys. Rev. B*, **21**, 5577.
8 Ashcroft, N.W. and Mermin, N.D. (1967) *Solid State Physics*, Holt, Rinehart and Winston, New York, Appendix N.
9 Overhauser, A.W. (1964) *Phys. Rev. Lett.*, **13**, 190.
10 Giuliani, G.F. and Overhauser, A.W. (unpublished).
11 Overhauser, A.W. (1978) *Hyperfine Interactions*, **4**, 786.
12 Cromer, D.T. and Waber, J.T. (1965) *Acta Cryst.*, **18**, 104.
13 Overhauser, A.W. (1970) *Phys. Rev. B*, **2**, 874.
14 Duff, K.J. and Overhauser, A.W. (1972) *Phys. Rev. B*, **5**, 2799.
15 Overhauser, A.W. (1971) *Phys. Rev. B*, **3**, 1888.
16 Lundqvist, B.I. (1967) *Phys. Kondens. Mater.*, **6**, 206.
17 Bishop, M.F. and Overhauser, A.W. (1978) *Phys. Rev. B*, **18**, 2447.

Reprint 27 Theory of Electron–Phason Scattering and the Low-temperature Resistivity of Potassium [1)]

Marilyn F. Bishop [*] *and A.W. Overhauser* [**]

[*] Department of Physics and Atmospheric Science, Drexel University, Philadelphia, Pennsylvania 19104, USA

[**] Department of Physics, Purdue University, West Lafayette, Indiana 47907, USA

Received 24 October 1980

Scattering of electrons by phasons, the collective excitations of an incommensurate charge-density wave (CDW), is presented as a new mechanism in the low-temperature resistivity of potassium. It is shown to provide an explanation for recent precision measurements in potassium between 0.38 and 1.3 K, where conventional mechanisms, such as electron–electron scattering, fail. With this theory, it is possible to explain the shape of the measured resistivity curves, the magnitude of the temperature-dependent part of the resistivity, and the sample dependence. The sample dependence is explained since the measured electron–phason resistivity, which is much larger along the CDW wave vector $\vec{Q}$ than perpendicular to it, depends on the $\vec{Q}$-domain structure of a particular sample. Fitting this theory to these experiments yields the value for the phason temperature $\Theta_\varphi = 3.25$ K and an approximate range of the anisotropy of the phason spectrum, $7.7 \lesssim (1/\eta) \lesssim 9.7$. Further resistivity measurements at ultralow temperatures are needed to test the hypothesis of electron–phason scattering, and, if this mechanism continues to show promise, to provide more accurate estimates for the phason parameters.

27.1 Introduction

Results of recent measurements of the low-temperature electrical resistivity of several simple metals have rejuvenated interest in the basic scattering mechanisms of electrons. One of the motivations for performing these experiments was the desire to find evidence for electron–electron scattering, which was predicted to produce a T^2 term in the resistivity independent of the residual resistivity ρ_0. Therefore, reports of approximate T^2 variations of electrical resistivity below 2 K in potassium [1–4], aluminum [4–6], silver [7], and copper [8] have sparked considerable excitement. However, the interpretation of these data in terms of electron–electron scattering is, in some cases, open to question. Basic discrepancies exist, for example in potassium, between the experiments and the theory in terms of electron–electron scattering.

Only in aluminum has it been shown that the T^2 term is independent of ρ_0 [5, 6]. In fact, in K, the temperature-dependent part of the resistivity increases rapidly with increasing ρ_0. In addition, in K, a T^2 behavior is actually inconsistent with the data. A pure power-law fit [2] produces $T^{1.5}$, but the dependence is probably more complicated. No conventional mechanism can reconcile the observed behavior. For this reason, we suggest that a new scattering mechanism must be invoked [9], namely, the scattering of electrons from phasons [10], excitations associated with phase fluctuations of a charge-density wave (CDW).

The objectives of this paper are to show that all conventional mechanisms fail to explain existing data in K and to demonstrate that the theory of electron–phason scattering can ex-

1) Phys. Rev. B 23, 3638 (1981)

Anomalous Effects in Simple Metals. Albert Overhauser

ISBN: 978-3-527-40859-7

plain both the shape and magnitude of the resistivity anomaly. Previously, in [9], we presented a limiting case of this theory and illustrated its ability to fit the shape of the measured curves. The paper is organized as follows. In Section 27.2, we analyze existing experiments in potassium and discuss conventional resistivity mechanisms. In Section 27.3, we review the properties of charge-density waves and phasons and of the electron–phason interaction. In addition, we calculate the resulting transition matrix elements. We present in Section 27.4 a derivation of the expressions used in calculations of the electron–phason resistivity and in Section 27.5 the results of numerical evaluations of these expressions. Finally, in Section 27.6, we state the conclusions.

27.2 Analysis of Experiments

In order to understand the recent measurements of the lowtemperature resistivity of potassium, let us first consider the conventional theoretical ideas about the contributions to the resistivity $\rho(T)$ of a simple metal at low temperatures [11],

$$\rho(T) = \rho_0 + AT^5 + BT^p \exp(-\hbar\omega_0/k_B T) + CT^2 , \tag{27.1}$$

where ρ_0 is the residual resistivity, a temperature-independent part of the resistivity due to scattering of electrons from impurities and other imperfections in the crystal. The term AT^5 is the low-temperature limiting behavior of the resistivity resulting from normal electron–phonon scattering. If phonon-drag effects are important, i.e, if the phonons drift with the electrons, then the magnitude of this term can be greatly reduced. Electron–phonon umklapp scattering, which is unaffected by phonon drag, contributes a term of the form $BT^p \exp(-\hbar\omega_0/k_B T)$, where ω_0 is the frequency of the phonon with the minimum wave vector that allows the electrons to scatter via an umklapp process. Electron–electron scattering adds a term CT^2 at all temperatures. The assumption that each of these contributions can be added independently is known as Matthiessen's rule.

The determination of the relative importance of each of the above mechanisms has been the object of considerable experimental study. The first of the temperature-dependent terms in Equation (27.1) to be verified was the exponential decay of the electron–phonon umklapp scattering with decreasing temperature below 4 K [12, 13]. Later, theoretical predictions [14–21] of the possibility of detecting the effects of phonon drag at temperatures below 2 K encouraged further low-temperature work. An additional incentive was that if phonon drag diminished the T^5 of normal electron–phonon processes, the only significant remaining temperature-dependent contribution to the resistivity below 2 K might be that of the CT^2 term in Equation (27.1) due to electron–electron scattering [22, 23].

Van Kempen *et al.* [1] measured the temperature-dependent resistivity of potassium between 1.1 and 4.2 K and found evidence for the presence of phonon drag. The magnitude of the resistivity dipped well below the predicted magnitude for the AT^5 that would be present in the absence of phonon drag. Below 2 K, an anomalous sample-dependent component of the resistivity appeared. Since the variation of this new term was consistent with T^2, van Kempen *et al.* suggested that this might be due to electron–electron scattering. Unfortunately, this contribution to the resistivity was sample dependent, in contradiction to conventional theories of electron–electron scattering [22, 23]. In addition, the presence of the exponential tail due to electron–phonon umklapp scattering causes some difficulty in determining the exact temperature dependence of this newly observed feature in the resistivity. Ideally, the umklapp scattering portion of the resistivity should be subtracted from the data before analyzing any new contribution. Unfortunately, difficulties arise because of a lack of knowledge of the exact form of this function below 2 K. While van Kempen *et al.* were able to fit their data well with the exponential term given in Equation (27.1) with $p = 1$ and $\Theta = \hbar\omega_0/k_B = 19.9$, Kaveh, Leavens, and Wiser [24] point out that the validity of extending this form below 2 K is questionable. One may avoid this difficulty by considering only data below which electron–phonon umklapp scattering is negligible compared with the new component of the low-temperature resistivity, i.e., below about 1.3 K. Clearly, then, the remaining temperature range between 1.1 and 1.3 K is too narrow to confirm or deny a T^2 dependence.

This uncertainty prompted Rowlands, Duvvury, and Woods [2] to extend the low-temperature resistivity measurements of potassium down to 0.4 K. Between 4 and 2 K, their data agreed with the previous results. However, they found that, with the extended range of temperatures, the data below 1.3 K were inconsistent with a T^2 dependence, and that the best pure power law that would fit the data was actually $T^{1.5}$. On the other hand, this data need not be fitted to a pure power law. In fact, in a previous paper [9], we showed that the shape of the curve, as well as its sample dependence, could be explained by the scattering of electrons from phasons, the phase fluctuations of a charge-density wave, and this theory yielded Bloch–Grüneisen functions rather than a pure power law.

The sample dependence of the new component in the temperature -dependent resistivity impelled Levy *et al.* [3] to study this aspect further. They measured a variety of samples, with $\rho(T) - \rho_0$ ranging from 8×10^{-14} to 4.2×10^{-13} Ω cm at 1.2 K, where ρ_0 was guessed in order to obtain a T^2 fit. This range of magnitudes was comparable to what was seen in other measurements Although they ascribed a T^2 dependence to the anomalous resistivity, their measurements, which did not extend below 1.1 K, could not determine that behavior. They suggested that the data could be explained by a theory of Kaveh and Wiser [25], in which normal electron–electron processes contribute to the resistivity as the result of anisotropic electron-dislocation scattering, in a breakdown of Matthiessen's rule.

Essentially, the idea is that when electrons scatter from the anisotropic dislocation lines, the steady-state distribution function in the presence of an electric field is distorted in shape from that of the equilibrium distribution function, i.e. a "dimple" forms in the direction of the field. Normal electron–electron scattering can be thought of as a diffusion of electrons on the Fermi surface, and this diffusion "heals" the "dimple" in such a way as to restore the shape of the equilibrium distribution function. In this way, a resistivity arises that is dependent on the density of dislocations and on the normal electron–electron scattering, which varies as T^2.

In order to explain their data in terms of this theory, Levy *et al.* [3] imposed two major assumptions. First, they assumed that the anomalous temperature-dependent resistivity varies as T^2, in contradiction to the measurements of Rowlands *et al.* [2]. Second they supposed that all their samples had a resistivity, $\rho_{0D} = 0.4\,\text{n}\Omega\,\text{cm}$, due to electron-dislocation scattering, independent of impurity concentration or annealing time. (Some samples annealed for more than a month.) No direct measurement was done of the dislocation density or impurity concentration.

The validity of assuming such a large value for ρ_{0D} is questionable. It is known that ρ_{0D} is related, in K, to the dislocation density n_D by [26, 27]

$$\rho_{0D} \simeq 4 \times 10^{-19} n_D\,, \tag{27.2}$$

where ρ_{0D} is in Ω cm and n_D is in cm^{-2}. This relation was extracted from measurements of the resistivity as a function of strain and of strain as a function of applied stress [28]. In obtaining Equation (27.2), the assumption was made that the relation for dislocation density as a function of applied stress, which was determined in the noble metals by electron microscopy and etch pit experiments, may be applied to alkali metals. Thus the value chosen by Levy *et al.* for ρ_{0D} corresponds to a dislocation density of 10^9 cm^{-2}.

Kaveh and Wiser extended the above analysis of low-temperature resistivity measurements in K in terms of their theory by including the data of van Kempen *et al.* [1] and of Rowlands *et al.* [2], in addition to that of Levy *et al.* [3]. They assumed resistivities due to electron-dislocation scattering of $\rho_{0D} \sim 0.53$–$18\,\text{n}\Omega\,\text{cm}$ for van Kempen *et al.* and $\rho_{0D} \sim 1.5$–$4.8\,\text{n}\Omega\,\text{cm}$ for Rowlands *et al.*, corresponding to dislocation densities of $n_D \sim 1.3 \times 10^9$–$4.5 \times 10^9\,\text{cm}^{-2}$ and $n_D \sim 1.8 \times 10^9$–$1.2 \times 10^{10}\,\text{cm}^{-2}$, respectively.

The magnitudes of dislocation densities presumed to exist in unstressed samples of K by Levy *et al.* [3] and Kaveh and Wiser [25] is unreasonably large compared with values obtained from direct measurements. In neutron diffraction experiments of potassium by Overhauser [10] and by Stetter *et al.* [29], it was determined from the rocking curve that the angular mosaic spread was about 0.1 degrees and from primary extinction measurements that the mosaic block was of the order of 1 mm. This corresponds to a dislocation density of less than $10^6\,\text{cm}^{-2}$. In addition, it was found that dislocations produced by thermal stress anneal out in less than an hour [10]. These numbers are then at least 3 orders of magnitude smaller than those assumed by Levy *et al.* [3] and by Kaveh and Wiser [25] in order to explain the data in terms of electron–electron scattering enhanced by electron-dislocation scattering.

In fact, high dislocation densities imply that samples are strained by a significant amount. As given in Equation (27.2), the dislocation density n_D of a sample is related to the resistivity caused by those dislocations. Basinski *et al.* [28] found that large resistivities due to dislocations ρ_{0D} or high dislocation densities n_D were actually difficult to achieve. They found that for small strains, the resistivity increases $\sim 1.3\%$ per percent strain, which was also found by Jones [30]. However, large external stresses were required to produce even small strains. For instance, for the dislocation resistivity p_{0D} suggested by Levy *et al.* [3], $p_{0D} = 0.4\,\text{n}\Omega\,\text{cm}$, the strain required would be about 3.5%, which is attained when the sample is under an external stress of $0.75\,\text{kg/mm}^2$. The largest strain that was achieved in any sample before the wire broke was about 16%, under a stress of nearly $1\,\text{kg/mm}^2$, which corresponds to a dislocation density of $n_D \sim 7 \times 10^9\,\text{cm}^{-2}$. This is considerably smaller than the largest dislocation density assumed by Kaveh and Wiser [25], $\rho_{0D} = 1.2 \times 10^{10}\,\text{cm}^{-2}$, in presumably stress-free samples.

Further difficulties with assuming arrays of oriented dislocations to be responsible for anomalous results in potassium come from other experiments. For example, the resistivity tensor of a single crystal of potassium has been measured to be a cigar-shaped ellipsoid with an anisotropy of 4 or 5 to 1 [31–33]. If oriented arrays of dislocations were responsible for this effect, the resistivity tensor would have to be a pancake-shaped ellipsoid since dislocation lines can be thought of as cylindrically-shaped scattering centers that only scatter electrons that travel perpendicular to the axis of the cylinder.

We now turn our attention to a closer examination of the temperature dependence of the new component in the resistivity. We point out that the best way to analyze the temperature dependence is to employ a procedure that does not depend on the subtraction of the residual resistivity ρ_0 from the data. For instance, a logarithmic plot of the data requires this subtraction, and the extracted temperature dependence of the data can depend on the guess one makes for the magnitude of the residual resistivity. Rowlands *et al.* [2] avoided this difficulty by plotting the total resistivity versus T^n and choosing the value of n that gave a straight line. Here, we plot the data on a linear–linear plot to avoid the problem. In Figure 27.1, we plot with circles the data of sample K2c of Rowlands *et al.* [2], where the zero of resistivity was set at the lowest data point. The two points above 1.3 K are displayed to illustrate the residual contribution of electron–phonon umklapp scattering to these points. The dashed arrows indicate the magnitudes one would subtract from the data if the exponential term in Equa-

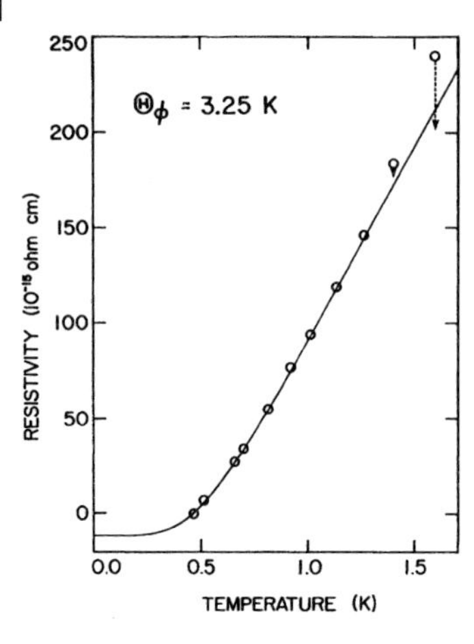

Figure 27.1 Plot of resistivity versus temperature for the data of sample K2c of Rowlands, Duvvury, and Woods [1] indicated by circles. The curve and arrows are described in the text.

tion (27.1) were used with the parameters of van Kempen *et al.* [1] for fitting the data in the region between 2 and 4 K. A solid curve has been drawn through all but the two highest data points. We call this curve ρ(phason), since it is actually the result of the theory of electron–phason scattering, which is presented in this paper.

There is no evidence of a T^5 dependence which is consistent with the existence of phonon drag. A T^2 curve is also the wrong shape to fit the data. This is more obvious in Figure 27.2, where we show directly the difference between a T^2 curve and a smooth curve through the data. The horizontal line is ρ(phason), the smooth curve through the data in Figure 27.1, which we use as a reference, and we plot $\rho - \rho$(phason), where ρ is given either by the data (circles) or by the power laws in T, $T^{1.5}$ and T^2, that pass through the first and last data points below 1.3 K. Clearly the T^2 curve is the wrong shape to describe the data. In fact, that fit is as poor as a straight line through the data (the curve labeled T). If one requires the best-fit pure power law, one obtains $T^{1.5}$, as was first illustrated by Rowlands *et al.* [2]. The estimated error in the data [2] was $\pm 10^{-15}$ Ω cm, which is about the same as the scatter of the data about the horizontal line. The temperature dependence of this data thus rules out electron–electron scattering, at least as the principal mechanism, as an explanation of the data. As we will show in the following sections, the scattering of electrons from phasons, the collective excitations corresponding to phase modulation of a charge-density wave, can explain the shape [9], as

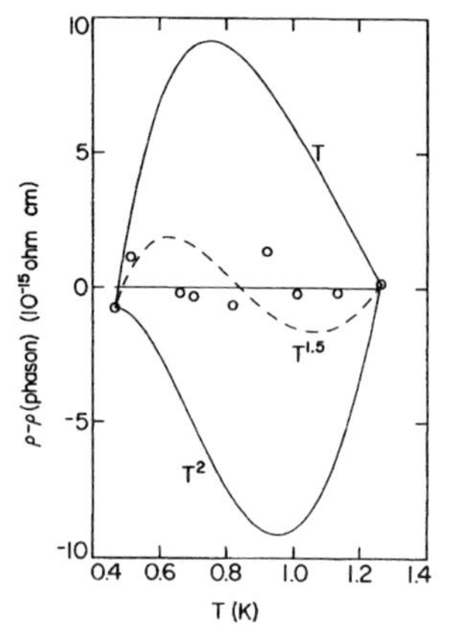

Figure 27.2 Plot of resistivity subtracted from ρ(phason), the smooth curve through the data in Figure 27.1. The data of Figure 27.1 are shown as circles scattered about the horizontal line ρ(phason). Curves labeled T, $T^{1.5}$, and T^2 are the corresponding pure power laws that pass through the first and last data points below 1.3 K.

well as the magnitude, of this new anomalous component in the low-temperature resistivity of potassium.

27.3 Phasons and the Electron–Phason Interaction

A charge-density wave (CDW) is characterized by a static sinusoidal modulation of the electronic charge density,

$$\rho(\vec{r}) = \rho_0 \left[1 + p\cos(\vec{Q}\cdot\vec{r} + \phi)\right], \tag{27.3}$$

where ρ_0 is the average density, p is the fractional modulation, and ϕ is the phase. The magnitude of the CDW wave vector $\vec{Q}$ is approximately the diameter of the Fermi surface. It has been shown that, within the deformable jellium model, such a structure lowers the exchange and correlation energies with respect to those of the state characterized by a uniform charge density [34, 35]. However, large static electric fields will develop in the metal unless the lattice undergoes a compensating distortion in order to ensure macroscopic charge neutrality,

$$\vec{u}(\vec{L}) = \vec{A}\sin(\vec{Q}\cdot\vec{L} + \phi)\,, \tag{27.4}$$

where $\vec{u}(\vec{L})$ is the displacement of the ion from its original site $\vec{L}$, and $\vec{A}$ is the maximum amplitude of that displacement. Theoretically, it has been shown that, in the alkali metals, the lattice provides almost a complete screening of the electric fields produced by the electronic density $\rho(\vec{r})$ [36].

Considerable experimental evidence supports the existence of a CDW ground state in the alkali metals, especially in potassium. The experimental situation in potassium was reviewed recently [37]. Even more recently, the observation of anomalies in the de Haas–van Alphen effect under pressure [38] and of open orbits [39] by the induced torque method have further strengthened the case of the CDW ground state in potassium. Although the predicted CDW diffraction satellites [10, 40] have not yet been observed [41], neutron diffraction experiments have not been performed with sufficient sensitivity to detect the estimated intensities [42].

In the alkali metals, the CDW is incommensurate since the CDW wave vector $\vec{Q}$, which spans the Fermi surface, is not related to any reciprocal-lattice vector by a small integer. For this reason, for very pure samples, the energy of the system does not depend on the value of the phase ϕ. This invariance of the energy leads to low-frequency collective excitations called phasons whose frequency spectrum goes to zero at the point $\vec{Q}$ in $\vec{k}$ space. The phase $\phi(\vec{r}, t)$ of the electronic charge density is modulated slowly in space and time, and the phase $\phi(\vec{L}, t)$ of the lattice follows, so as to screen out the electron density,

$$\phi(\vec{L}, t) = \sum_{\vec{q}} \phi_{\vec{q}} \sin(\vec{q}\cdot\vec{L} - \omega_{\vec{q}} t)\,, \tag{27.5}$$

where $\vec{q}$ and $\omega_{\vec{q}}$ are the wave vector and frequency of the phason, and $\phi_{\vec{q}}$ is its amplitude. Thus a phason is actually a normal mode of the lattice whose frequency vanishes at the point $\vec{Q}$ and varies linearly with $\vec{q}$ away from that point.

Although we are concerned in this paper with the influence of phasons on the electrical resistivity, it is interesting to note that these low-frequency excitations can also affect a number of other physical properties. For example, they can produce an anomaly in the low-temperature heat capacity [43], an effect that was first observed in $LaGe_2$, a three-dimensional system, by Sawada and Satoh [44]. There is even evidence supporting such an anomaly in rubidium, one of the alkali metals [45, 46]. Since extensive theoretical studies have been done

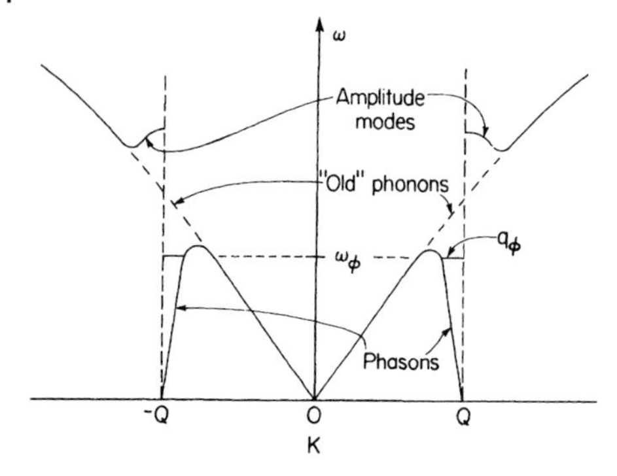

Figure 27.3 Schematic illustration of the vibrational modes in a metal having a CDW structure. The frequency of the phason branch goes to zero at $\pm\vec{Q}$. A phason is a linear superposition of two "old" phonons, and the amplitude modes are the orthogonal linear combination. Phason and amplitude modes quickly merge into the new phonon spectrum, as indicated. ω_ϕ and q_ϕ are the frequency and wave-vector cutoffs for the phason that are used in this paper.

investigating the general properties of phasons [10, 36, 47, 48], we will confine our attention here to those aspects that are required in the calculation of the electron–phason resistivity, referring to earlier work for details.

A phason is a coherent linear combination of two "old" phonons, i.e., phonons of the undistorted lattice, of wave vectors $\vec{q}+\vec{Q}$ and $\vec{q}-\vec{Q}$ [10]. The orthogonal linear combination of these same two "old" phonons is an amplitude mode [49], a collective oscillation of the amplitude $\vec{A}$ in Equation (27.4), which occurs at high frequency. The situation is illustrated schematically in Figure 27.3. Away from the points $\vec{Q}$ and $-\vec{Q}$, the phason and amplitude modes merge quickly into the phonon spectrum, so that these new excitations exist only in a very small volume of phase space. For simplicity, we will assume that at some frequency cutoff ω_ϕ, the phasons transform into phonons, and for the purposes of this paper we will neglect the contributions of the phonons and amplitude modes in the calculation of the resistivity. q_ϕ is the corresponding wave-vector cutoff along $\vec{Q}$. We should comment that Figure 27.3 has only an approximate meaning, since, in the presence of an incommensurate CDW, the system no longer has translational symmetry along $\vec{Q}$. (Actually, $\vec{Q}$ is longer than half the reciprocal-lattice vector along it.)

As illustrated in Figure 27.4, this phason spectrum is expected to be highly anisotropic, such that

$$\omega_{\vec{q}} = \left(c_\parallel^2 q_\parallel^2 + c_\perp^2 q_\perp^2\right)^{1/2} = c_\parallel^2\left(q_\parallel^2 + \eta^2 q_\perp^2\right)^{1/2}, \tag{27.6a}$$

where

$$\eta = c_\perp / c_\parallel , \tag{27.6b}$$

and where $q_\parallel$ and $q_\perp$ are the components of $\vec{q}$ parallel and perpendicular to the CDW wave vector $\vec{Q}$. Correspondingly, $c_\parallel$ and $c_\perp$ are the phason velocities parallel and perpendicular to $\vec{Q}$, or the longitudinal and transverse phason velocities. A characteristic phason temperature Θ_ϕ can now be identified with the frequency and wave vector cutoffs ω_ϕ and q_ϕ as

$$\Theta_\phi \equiv \hbar\omega_\phi / k_B = \hbar c_\parallel q_\phi / k_B . \tag{27.7}$$

This phason temperature can now be determined by the shape of the measured resistivity curve. The value of Θ_ϕ determined from the resistivity will not necessarily be the same as that relevant for the specific heat, since the two measurements probe different properties of the phasons. In the calculation of the specific heat, one must include contributions from all modes (phasons, amplitude modes, phonons). The resistivity, however, depends not only on the modes themselves but on the interaction of electrons with them.

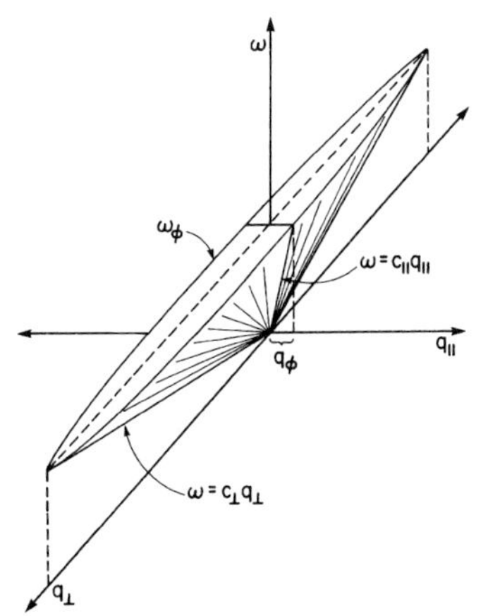

Figure 27.4 Anisotropic cone of the phason spectrum, showing the longitudinal and transverse phason velocities $c_{\parallel}$ and $c_{\perp}$. ω_{φ} is the frequency cutoff of the phason spectrum and q_{φ} is the wave-vector cutoff along $\vec{Q}$.

For small $\vec{q}$, the longitudinal phason represents a lengthening and shortening of $\vec{Q}$, while the transverse phason represents a rotation of $\vec{Q}$ about its static value. In the alkali metals, the magnitude of $\vec{Q}$ is related to the same many-electron effects of exchange and correlation that initially gave rise to the CDW instability. The direction of $\vec{Q}$, on the other hand, depends only on the elastic anisotropy of potassium [40], since the many-electron effects are essentially isotropic. In fact, a calculation of the transverse phason velocity based on this idea has recently been performed [36] and it was found that $c_{\perp}$ is of the order of the acoustic phonon velocities. The magnitude of the longitudinal velocity $c_{\parallel}$ is not known, but it is expected to be much larger than the transverse velocity $c_{\perp}$, since it depends on many-electron interactions. Therefore, η will be much less than unity. In Section 27.5, we will show that the low-temperature resistivity measurements limit the acceptable range of values for η. Although the transverse phason spectrum is also anisotropic, we use here the average transverse phason velocity for $c_{\perp}$.

Since a phason is a harmonic oscillator, its amplitude $\phi_{\vec{q}}$ in Equation (27.5) can be related to its frequency $\omega_{\vec{q}}$. In order to determine that relation, we follow a procedure similar to that for phonons. We first write $\phi(\vec{L}, t)$ of Equation (27.5) in the Heisenberg picture in terms of the creation and annihilation operators $a^*_{\vec{q}}$ and $a_{\vec{q}}$ of the harmonic oscillator,

$$\phi(\vec{L}) = \sum_{\vec{q}} \frac{\phi_{\vec{q}}}{2i} \left(a_{\vec{q}} e^{i\vec{q}\cdot\vec{L}} - a^*_{\vec{q}} e^{-i\vec{q}\cdot\vec{L}} \right). \tag{27.8}$$

For small ϕ, we may rewrite $\vec{u}(\vec{L})$ from Equation (27.4) as

$$\vec{u}(\vec{L}) = \vec{A}\left[\sin \vec{Q}\cdot\vec{L} + \phi(\vec{L}) \cos \vec{Q}\cdot\vec{L}\right]. \tag{27.9}$$

The kinetic energy,

$$T = \frac{1}{2} M \sum_{\vec{L}} \left(\frac{d\vec{u}(\vec{L})}{dt} \right)^2 , \tag{27.10}$$

can be written in terms of $\phi(\vec{L})$ if we recognize that $i\hbar(d\vec{u}/dt) = [\vec{u}, \mathcal{H}]$, where $\mathcal{H} = \hbar\omega_{\vec{q}}(a^*_{\vec{q}} a_{\vec{q}} + \frac{1}{2})$. Thus, we have

$$\frac{d\vec{u}(\vec{L})}{dt} = -\vec{A}\cos(\vec{Q}\cdot\vec{L}) \sum_{\vec{q}} \frac{\omega_{\vec{q}}}{2} \phi_{\vec{q}} \left(a_{\vec{q}} e^{i\vec{q}\cdot\vec{L}} + a^*_{\vec{q}} e^{-i\vec{q}\cdot\vec{L}} \right), \tag{27.11}$$

and

$$T = \frac{1}{4}\rho_m A^2 \sum_{\vec{q}} \frac{\phi_{\vec{q}}^2 \omega_{\vec{q}}^2}{4} \left(a_{\vec{q}} a_{\vec{q}}^* + a_{\vec{q}}^* a_{\vec{q}} \right), \tag{27.12}$$

where ρ_m is the mass density of the crystal, and the volume of a unit cell is set equal to unity. In obtaining this result, Equation (27.12), we have written $\cos^2(\vec{Q}\cdot\vec{L}) = \frac{1}{2} + \frac{1}{2}\cos(2\vec{Q}\cdot\vec{L})$. The second term yields zero, since we average over the long wavelength $\vec{q}$ of the phason.

For a harmonic oscillator, $T = \frac{1}{2}W$, where W is the total energy given by

$$W = \hbar\omega_{\vec{q}} \left(a_{\vec{q}}^* a_{\vec{q}} + \frac{1}{2} \right) = \frac{1}{2}\hbar\omega_{\vec{q}} \left(a_{\vec{q}}^* a_{\vec{q}} + a_{\vec{q}} a_{\vec{q}}^* \right). \tag{27.13}$$

We thus obtain the phason amplitude in terms of $\omega_{\vec{q}}$,

$$\phi_{\vec{q}} = \frac{2}{A}\left(\frac{\hbar}{\rho_m \omega_{\vec{q}}} \right)^{1/2}. \tag{27.14}$$

The property of phasons of most interest in calculating their effect on the low-temperature resistivity is the form of the electron–phason interaction. This may be written formally as

$$V_{e\phi} = G\cos\left(\vec{Q}\cdot\vec{r} + \phi\right) - G\cos\left(\vec{Q}\cdot\vec{r}\right), \tag{27.15}$$

where $G\cos(\vec{Q}\cdot\vec{r})$ is the static self-consistent one-electron potential of the CDW and $G\cos(\vec{Q}\cdot\vec{r} + \phi)$ is the corresponding potential in the presence of weak phase modulation. For small ϕ, we may rewrite this as

$$\begin{aligned} V_{e\phi} &= -G\phi(\vec{r})\sin\vec{Q}\cdot\vec{r} \\ &= \frac{G}{4}\sum_{\vec{q}} \phi_{\vec{q}} \left(a_{\vec{q}} - a_{-\vec{q}}^* \right)\left(e^{i(\vec{q}+\vec{Q})\cdot\vec{r}} - e^{i(\vec{q}-\vec{Q})\cdot\vec{r}} \right), \end{aligned} \tag{27.16}$$

where $\phi(\vec{r})$ is the continuum analog of $\phi(\vec{L})$ in Equation (27.8). Note that the wave vector that appears in this expression is not the small wave vector $\vec{q}$ (long wavelength) but the large wave vector $\vec{q} \pm \vec{Q}$ (short wavelength).

In calculating the electron–phason resistivity, we will employ this interaction $V_{e\phi}$ in Equation (27.16) through the "golden rule" transition rate of an electron scattering from a state labeled $\vec{k}$ to a state labeled $\vec{k}'$:

$$W_{\vec{k}\to\vec{k}'} = \frac{2\pi}{\hbar}|M_{\vec{k},\vec{k}'}|^2 \delta\left(\mathcal{E}_{\vec{k}}^{\text{tot}} - \mathcal{E}_{\vec{k}'}^{\text{tot}} \right) f_{\vec{k}}(1 - f_{\vec{k}'}), \tag{27.17}$$

where $\mathcal{E}_{\vec{k}}^{\text{tot}}$ is the total energy of the initial state (including phasons) and $\mathcal{E}_{\vec{k}'}^{\text{tot}}$ is the total energy of the final state. $f_{\vec{k}}$ is the electron distribution function. Here, the scattering matrix element is

$$M_{\vec{k},\vec{k}'} = \left\langle \Psi_{\vec{k}'}, n_{\vec{q}_1'}, n_{\vec{q}_2'}, \ldots | V_{e\phi} | \Psi_{\vec{k}}, n_{\vec{q}_1}, n_{\vec{q}_2}, \ldots \right\rangle, \tag{27.18}$$

where $n_{\vec{q}_1}$ is the number of phasons of wave vector $\vec{q}_1$, $n_{\vec{q}_2}$ is the number of phasons of wave vector $\vec{q}_2$, etc., and $\Psi_{\vec{k}}$ and $\Psi_{\vec{k}'}$ are wave functions corresponding to solutions of the one-electron Schrödinger equation in the presence of a static CDW,

$$\left(\frac{p^2}{2m} + G\cos\vec{Q}\cdot\vec{r} \right)\Psi_{\vec{k}}(\vec{r}) = E_{\vec{k}}\Psi_{\vec{k}}(\vec{r}). \tag{27.19}$$

The potential $G \cos \vec{Q} \cdot \vec{r}$ deforms both the wave functions and the energy spectrum by mixing the plane wave state $\vec{k}$ with the plane wave state $\vec{k} \pm \vec{Q}$ and produces the modulated electron density given in Equation (27.3). This in turn produces energy gaps of magnitude G at $\vec{k} = \pm \vec{Q}/2$ and distorts the spherical Fermi surface in the same region. In addition, the scattering of electrons by phasons is the most intense in this region. Therefore, for simplicity, we translate our coordinate system in $\vec{k}$ space by $\vec{Q}/2$, so that our new $\vec{k}$ is measured with respect to the point $\vec{Q}/2$.

For small G, the plane wave state $\vec{k} + \frac{1}{2}\vec{Q}$ and $\vec{k} - \frac{1}{2}\vec{Q}$ in the new coordinate system are nearly degenerate. We therefore treat the coupling between these two states exactly by solving the following secular equation, which is written in the basis of these two plane waves:

$$\begin{pmatrix} \left(\frac{\hbar^2}{2m}\left(\vec{k} + \frac{1}{2}\vec{Q}\right)^2 - E_{\vec{k}}\right) & \frac{1}{2}G \\ \frac{1}{2}G & \left(\frac{\hbar^2}{2m}\left(\vec{k} - \frac{1}{2}\vec{Q}\right)^2 - E_{\vec{k}}\right) \end{pmatrix} \begin{pmatrix} \cos \zeta_{\vec{k}} \\ -\sin \zeta_{\vec{k}} \end{pmatrix} = 0\,, \tag{27.20}$$

where $E_{\vec{k}}$ is the energy of a state and $\cos \zeta_{\vec{k}}$ and $\sin \zeta_{\vec{k}}$ are the coefficients of the corresponding wave function. The energy below the gap is given by

$$E_{\vec{k}} = \frac{\hbar^2}{2m}\left(k^2 + \frac{1}{4}Q^2\right)^2 - \frac{1}{2}\left[\left(\frac{\hbar^2}{m}(\vec{k} \cdot \vec{Q})\right)^2 + G^2\right]^{1/2}. \tag{27.21}$$

Above the gap, the energy is given by Equation (27.21), with a plus sign between the two terms. In calculating the resistivity, the electron scatters between states near the Fermi surface, and this, in the present case, includes states only below the gap. We may write Equation (27.21) in a simpler form if we assume that $\vec{Q}$ is along the k_z axis and if we define the following dimensionless units,

$$\kappa = \left(k_x^2 + k_y^2\right)^{1/2} / Q\,, \quad w = k_z / Q\,, \quad \alpha = mG/\hbar^2 Q^2\,. \tag{27.22}$$

With these definitions, Equation (27.21) becomes

$$E_{\vec{k}} = \frac{\hbar^2 Q^2}{2m}\left[\left(\kappa^2 + w^2 + \frac{1}{4}\right) - (w^2 + \alpha^2)^{1/2}\right]. \tag{27.23}$$

The corresponding wave function is given by

$$\Psi_{\vec{k}} = \cos \zeta_{\vec{k}} e^{i[\vec{k} + (\vec{Q}/2)] \cdot \vec{r}} - \sin \zeta_{\vec{k}} e^{i[\vec{k} - (\vec{Q}/2)] \cdot \vec{r}}\,, \tag{27.24}$$

where

$$\cos \zeta_{\vec{k}} = \frac{\alpha/\sqrt{2}}{(w^2 + \alpha^2)^{1/4}[w^2 + (w^2 + \alpha^2)^{1/2}]^{1/2}}\,, \tag{27.25}$$

and the phase is chosen for the state below the gap such that

$$\sin 2\zeta_{\vec{k}} = \frac{\alpha}{(w^2 + \alpha^2)^{1/2}}\,. \tag{27.26}$$

Note that the system is cylindrically symmetric and that the wave-function coefficients are independent of κ. Also, for $w = 0$, $\sin \zeta_{\vec{k}} = \cos \zeta_{\vec{k}} = 1/\sqrt{2}$. In Figure 27.5, we plot in (a) the energy spectrum (below the gap), in (b) the corresponding wave-function coefficients, and in (c) the Fermi surface centered in our new coordinate system with its origin at the gap. For the Fermi surface, we have assumed the case of critical contact at the gap [32, 33]. The equation for this Fermi surface is given by

$$\kappa = \left[(w^2 + \alpha^2)^{1/2} - \alpha - w^2\right]^{1/2} . \qquad (27.27)$$

Only in the conical regions of the Fermi surface, which result from the CDW, are the energy spectrum and wave functions severely distorted from the plane-wave state. For this reason, it is only in this region that electron–phason scattering is intense, as we will see below.

We now turn to a calculation of the matrix element of Equation (27.18). We use the wave functions defined by Equation (27.24) and the interaction $v_{\beta\phi}$ given by Equation (27.16) to obtain

$$\begin{aligned} M_{\vec{k},\vec{k}'} = \left(\frac{G}{4}\right) \sum_{\vec{q}} \phi_{\vec{q}} \Big[\langle \Psi_{\vec{k}'}, n_{\vec{q}} - 1, \ldots | a_{\vec{q}} \left(e^{i(\vec{q}+\vec{Q})\cdot\vec{r}} - e^{i(\vec{q}-\vec{Q})\cdot\vec{r}} \right) \\ \times | \Psi_{\vec{k}'}, n_{\vec{q}}, \ldots \rangle + \langle \Psi_{\vec{k}'}, n_{-\vec{q}} + 1, \ldots | a^*_{-\vec{q}} \\ \times \left(e^{i(\vec{q}+\vec{Q})\cdot\vec{r}} - e^{i(\vec{q}-\vec{Q})\cdot\vec{r}} \right) | \Psi_{\vec{k}'}, n_{\vec{q}}, \ldots \rangle \Big] , \end{aligned} \qquad (27.28)$$

as the only nonvanishing terms. When we evaluate this expression explicitly, we may write the transition rate directly as

$$\begin{aligned} W_{\vec{k}\to\vec{k}'} = H(\vec{k},\vec{k}') f_{\vec{k}} (1 - f_{\vec{k}'}) \Big[n_{\vec{q}} \delta(E_{\vec{k}} - E_{\vec{k}'} + \hbar\omega_{\vec{q}}) \\ + (n_{-\vec{q}} + 1) \delta(E_{\vec{k}} - E_{\vec{k}'} - \hbar\omega_{\vec{q}}) \Big] , \end{aligned} \qquad (27.29)$$

where

$$H(\vec{k},\vec{k}') = \frac{2\pi}{\hbar} \left(\frac{G}{4}\right)^2 \phi^2_{\vec{q}} C^2_{\vec{k},\vec{k}'} \qquad (27.30)$$

and

$$C^2_{\vec{k},\vec{k}'} = (\cos\zeta_{\vec{k}} \sin\zeta_{\vec{k}'} - \cos\zeta_{\vec{k}'} \sin\zeta_{\vec{k}})^2 . \qquad (27.31)$$

with $\vec{q} = \vec{k}' - \vec{k}$. Note that $H(\vec{k},\vec{k}')$ is symmetric with respect to interchange of $\vec{k}$ and $\vec{k}'$.

In writing Equations (27.29)–(27.31), we have neglected the scattering just at the center of the belly of the Fermi surface. For the geometry we have chosen in Figure 27.5, these belly-to-belly transitions would be considered "umklapp" scattering, while the conical-point-to-conical-point transitions would be considered normal transitions. However, in a conventional geometry (centered at the left-hand edge of Figure 27.5), these belly-to-belly transitions would be seen to be no different than transitions within each belly which are negligible in magnitude compared with the scattering within the conical-point regions.

In fcat, the overwhelming contribution to the electron–phason scattering is concentrated in the conical-point regions of the Fermi surface. This can be seen most easily by an examination of Figure 27.5, with the help of Equation (27.31). The wave-function coefficients enter the scattering rate in the form $\cos\zeta_{\vec{k}'} \sin\zeta_{\vec{k}}$. From Figure 27.5(b) and (c), it is clear that $\cos\zeta_{\vec{k}}$ and $\sin\zeta_{\vec{k}}$ are both appreciable only in the conical-point regions of the fermi surface. Since phasons are confined to a very small region of $\vec{k}$ space, as is indicated in Figures 27.3 and 27.4, $\vec{q} = \vec{k}' - \vec{k}$ is always small. Therefore, $\vec{k}$ is nearly the same as $\vec{k}'$, and $\cos\zeta_{\vec{k}'} \sin\zeta_{\vec{k}}$ is large in the same regions as $\cos\zeta_{\vec{k}} \sin\zeta_{\vec{k}}$, which is in the conical-point regions. This can

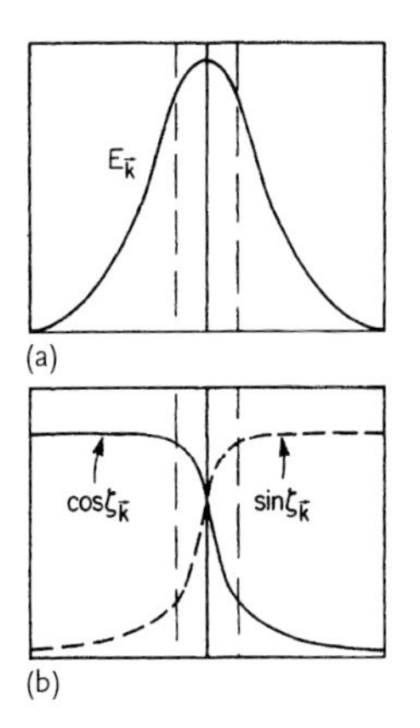

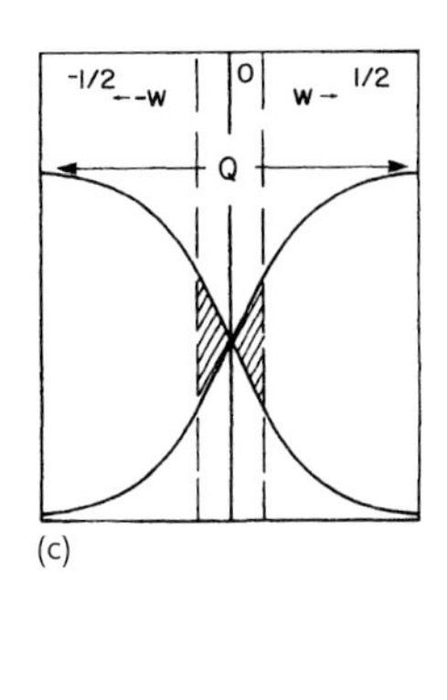

Figure 27.5 (a) Electron energy band $E_{\vec{k}}$ [Equation (27.23)], (b) coefficients of the wave function $\cos\zeta_{\vec{k}}$ and $\sin\zeta_{\vec{k}}$ [Equations (27.24)–(27.26)], and (c) Fermi surface for a CDW system with the origin at the CDW gap.

also be seen from Equations (27.25) and (27.26). For this reason, it is possible to neglect belly scattering compared with conical-point scattering.

Since the only appreciable scattering is in the conical-point regions of the Fermi surface, we can simplify the problem at hand by assuming that the Fermi surface consists of two intersecting cones, which can be obtained from Equation (27.27) by assuming that $|w| \ll 1$. The new Fermi surface is then written as

$$\kappa \simeq \beta |w| , \tag{27.32}$$

where

$$\beta = \left(\frac{1}{2\alpha} - 1 \right)^{1/2} . \tag{27.33}$$

Similarly, we approximate the velocities of the electron to be consistent with this new Fermi surface,

$$v_z \simeq -\left(\frac{\hbar Q}{m} \right) \beta^2 w , \tag{27.34}$$

$$v_x \simeq \left(\frac{\hbar Q}{m} \right) \kappa \cos\theta , \tag{27.35}$$

with κ given by Equation (27.32) and θ the polar angle of cylindrical coordinates. Since the system is cylindrically symmetric, we need only consider the x and z components of velocity.

In Section 27.4, we will employ the transition rate in Equation (27.29) in order to derive expressions for the resistivity. In doing so, we will employ the simplified Fermi surface of two intersecting cones.

27.4 Derivation of the Electron–Phason Resistivity

The residual resistivity in potassium in the temperature region of interest here is at least 3 orders of magnitude larger than the temperature dependent resistivity. That is, the scattering of electrons from impurities and other imperfections in the crystal dominates over the scattering of any temperature-dependent mechanism. For this reason, this scattering will determine the shape of the steady-state electron distribution function. Therefore, we will make the same assumption that we made in a calculation of the residual resistivity for a CDW model of K [32, 33], namely, that the steady-state electron distribution function (relevant for electron–phason scattering) is that of the rigidly shifted equilibrium distribution function. We then balance the total momentum of the electrons arising from an external electric field with the total momentum due to collisions of electrons with phasons. This is equivalent to a variational solution of the Boltzmann transport equation. A detailed discussion of this method is given in [32, 33], where it was used in the calculation of the residual resistivity anisotropy in potassium assuming the electrons to be in a CDW ground state.

The rigidly shifted electron distribution function is given by[2)]

$$f(\vec{k}) = f_0(\vec{k} - \vec{\delta}) \approx f^0_{\vec{k}} - \hbar\delta_\mu v_{\vec{k}\mu} \frac{\partial f^0_{\vec{k}}}{\partial E_{\vec{k}}} , \tag{27.36}$$

where $f^0_{\vec{k}}$ is the equilibrium distribution and the electric field $\vec{\mathcal{E}}$ is assumed to be in the μ direction such that

$$\delta_\mu = -e\mathcal{E}_\mu \tau_\mu / \hbar , \tag{27.37}$$

where $v_{\vec{k}\mu} = (1/\hbar)\partial E_{\vec{k}}/\partial k_\mu$ is the μ component of the velocity of an electron in the state $|\Psi_{\vec{k}}\rangle$. In the absence of collisions, electrons under the influence of $\vec{\mathcal{E}}$ would be accelerated such that $f^0_{\vec{k}}$ would translate at a constant rate in $\vec{k}$ space. Then τ_μ would simply be the length of time after $f^0_{\vec{k}}$ was centered at $\vec{k} = 0$. When collisions are included, δ_μ is independent of time once the system has reached a steady state, and τ_μ is the relaxation time.

We begin the derivation of the electron phason resistivity[3)] with the Boltzmann transport equation, which equates the rate of change of the electron distribution function $f_{\vec{k}}$ due to the electric field $\vec{\mathcal{E}}$ with that due to collisions,

$$-e(\vec{v}_{\vec{k}} \cdot \vec{\mathcal{E}}) \frac{\partial f^0_{\vec{k}}}{\partial E_{\vec{k}}} = \int \frac{d^3k'}{8\pi^3} (W_{\vec{k}' \to \vec{k}} - W_{\vec{k} \to \vec{k}'}) , \tag{27.38}$$

where $W_{\vec{k} \to \vec{k}'}$, given by Equation (27.29) of Section 27.3, is the transition rate for electrons leaving the state $\vec{k}$ by scattering to the state $\vec{k}'$, and $W_{\vec{k}' \to \vec{k}}$ is the rate for electrons entering the state $\vec{k}$ from $\vec{k}'$.

We can transform Equation (27.38) into the equation for the balance of the total momentum of the electrons by multiplying both sides of the equation by $v_{k\nu}$ and integrating over d^3k. Then, if the electric field is in the μ direction, the left-hand side of the equation vanishes unless $\mu = \nu$. Thus,

$$L_\mu \mathcal{E}_\mu = -\hbar\delta_\mu I_\mu , \tag{27.39}$$

where

$$L_\mu = -e \int \frac{d^3k}{4\pi^3} v^2_{\vec{k}\mu} \frac{\partial f^0_{\vec{k}}}{\partial E_{\vec{k}}} , \tag{27.40}$$

and

$$I_\mu = \frac{-1}{\hbar\delta_\mu} \int \frac{d^3k}{4\pi^3} \int \frac{d^3k'}{8\pi^3} \left(W_{\vec{k}' \to \vec{k}} - W_{\vec{k} \to \vec{k}'}\right) . \tag{27.41}$$

2) See, for instance, [50].

3) For a similar derivation of the electron–phonon resistivity, see [51].

L_μ can be evaluated explicitly, yielding,

$$L_x = L_y = ne/m\ , \tag{27.42a}$$

$$L_z = neF/m\ , \tag{27.42b}$$

where n is the electron density, with

$$F = 3\left(\frac{m}{\hbar Q}\right)^2 \left(\frac{Q}{k_F}\right)^3 \int_0^{1/2} v_{\vec{k}z}^2\, dw\ , \tag{27.43}$$

where we have used the variables $E_{\vec{k}}$, w, and Θ of Section 27.3 in Equations (27.22), (27.23), and (27.35) together with the transformation

$$d^3k = \left(\frac{m}{\hbar^2}\right) Q d\theta\, dw\, dE_{\vec{k}} \tag{27.44}$$

to evaluate L_z.

The μ component of the current density in the metal is given by

$$J_\mu = -e\int \frac{d^3k}{4\pi^3} v_{\vec{k}\mu} f_{\vec{k}} = -\hbar\delta_\mu L_\mu = \frac{L_\mu^2}{I_\mu}\mathcal{E}_\mu\ . \tag{27.45}$$

The resistivity tensor is diagonal in this coordinate system and is given formally by

$$\rho_{\mu\mu} = \frac{I_\mu}{L_\mu^2} = \frac{m}{ne^2\tau_\mu}\ . \tag{27.46}$$

The remaining problem is to calculate I_μ. First, we note that if we invoke the identity,

$$n_{\vec{q}} f_{\vec{k}}^0\left(1 - f_{\vec{k}'}^0\right) = \left(n_{\vec{q}} + 1\right) f_{\vec{k}'}^0 \left(1 - f_{\vec{k}}^0\right), \tag{27.47}$$

we may write the transition rates, from (27.29), more simply as

$$\begin{aligned}
(W_{\vec{k}'\to\vec{k}} - W_{\vec{k}\to\vec{k}'}) &= H(\vec{k},\vec{k}')\Big\{\Big[f_{\vec{k}'}(1-f_{\vec{k}}) - f_{\vec{k}'}^0\left(1-f_{\vec{k}}^0\right)\Big] \\
&\times\Big[(n_{\vec{q}}+1)\delta(E_{\vec{k}} - E_{\vec{k}'} + \hbar\omega_{\vec{q}}) + n_{\vec{q}}\delta\left(E_{\vec{k}} - E_{\vec{k}'} - \hbar\omega_{\vec{q}}\right)\Big] \\
&-\Big[f_{\vec{k}}(1-f_{\vec{k}'}) - f_{\vec{k}}^0\left(1-f_{\vec{k}'}^0\right)\Big]\Big[n_{\vec{q}}\delta\left(E_{\vec{k}} - E_{\vec{k}'} + \hbar\omega_{\vec{q}}\right) \\
&+(n_{\vec{q}}+1)\delta(E_{\vec{k}} - E_{\vec{k}'} - \hbar\omega_{\vec{q}})\Big]\Big\}\ .
\end{aligned} \tag{27.48}$$

Then we use the relation Equation (27.36) for the rigidly displaced electron distribution function $f_{\vec{k}}$ together with the identity

$$\frac{\partial f_{\vec{k}}^0}{\partial E_{\vec{k}}} = \frac{-f_{\vec{k}}^0(1-f_{\vec{k}}^0)}{k_B T} \tag{27.49}$$

to obtain

$$\begin{aligned}
(W_{\vec{k}'\to\vec{k}} - W_{\vec{k}\to\vec{k}'}) &= -\frac{\hbar\delta_\mu}{k_B T}(v_{\vec{k}\mu} - v_{\vec{k}'\mu})H(\vec{k},\vec{k}')n_{\vec{q}} \\
&\Big[f_{\vec{k}}^0\left(1-f_{\vec{k}'}^0\right)\delta(E_{\vec{k}} - E_{\vec{k}'} + \hbar\omega_{\vec{q}}) - f_{\vec{k}'}^0\left(1-f_{\vec{k}}^0\right)\delta(E_{\vec{k}} - E_{\vec{k}'} - \hbar\omega_{\vec{q}})\Big]\ .
\end{aligned} \tag{27.50}$$

If we multiply Equation (27.50) by $v_{\vec{k}\mu}$, integrate over d^3k and d^3k', and then interchange $\vec{k}$ and $\vec{k}'$ in the term containing $\delta(E_{\vec{k}} - E_{\vec{k}'} - \hbar\omega_{\vec{q}})$ we can write a simplified form for I_μ

$$I_\mu = \frac{1}{k_B T}\int \frac{d^3k}{4\pi^3}\int \frac{d^3k'}{8\pi^3} H(\vec{k},\vec{k}')(v_{\vec{k}\mu} - v_{\vec{k}'\mu})^2 n_{\vec{q}} f^0_{\vec{k}}\left(1 - f^0_{\vec{k}'}\right)$$
$$\times\, \delta(E_{\vec{k}} - E_{\vec{k}'} + \hbar\omega_{\vec{q}})\,. \quad (27.51)$$

Since the system has cylindrical symmetry, we transform from $\vec{k}$ and $\vec{k}'$ to the coordinates of energies $E_{\vec{k}}$, $E_{\vec{k}'}$ and the cylindrical coordinates w, w', θ, and θ' of Section 27.3. In doing so, we make use of Equation (27.44). We also write explicitly the expression for $H(\vec{k},\vec{k}')$ from Equation (27.30) of Section 27.3. The scattering integral I_μ then becomes

$$I_\mu = \frac{\pi}{(8\pi^3)^2}\left(\frac{mQ}{\hbar^2}\right)^2\left(\frac{G}{A}\right)^2\left(\frac{1}{\rho_m k_B T}\right)$$
$$\times \int dw \int dw' \int d\theta \int d\theta' \int dE_{\vec{k}} \int dE_{\vec{k}'} (v_{\vec{k}\mu} - v_{\vec{k}'\mu})^2$$
$$\times\, C^2_{\vec{k},\vec{k}'}\left(\frac{n_{\vec{q}}}{\omega_{\vec{q}}}\right) f^0_{\vec{k}}(1 - f^0_{\vec{k}'})\delta(E_{\vec{k}} - E_{\vec{k}'} + \hbar\omega_{\vec{q}})\,, \quad (27.52)$$

where, as in Equations (27.29)–(27.31), $\vec{q} = \vec{k} - \vec{k}'$. $f^0_{\vec{k}}(1 - f^0_{\vec{k}'})$ is nearly a delta function of energy at the Fermi surface. By comparison, the quantities $(v_{\vec{k}\mu} - v_{\vec{k}'\mu})^2$ and $C^2_{\vec{k},\vec{k}'}$ are slowly varying in energy in this region, so that it is reasonable to replace them by their values at the Fermi surface in the evaluation of the integrals over energy. With this assumption, the energy-dependent part of I_μ is

$$U = \int dE_{\vec{k}} \int dE_{\vec{k}'} f^0_{\vec{k}}\left(1 - f^0_{\vec{k}'}\right)\delta(E_{\vec{k}} - E_{\vec{k}'} + \hbar\omega_{\vec{q}})$$
$$= \int dE_{\vec{k}} f_0(E_{\vec{k}})\left[1 - f_0(E_{\vec{k}} + \hbar\omega_{\vec{q}})\right]. \quad (27.53)$$

With the definitions,

$$f_0(E_{\vec{k}}) = \frac{1}{(e^\eta + 1)}\,, \quad \eta = \frac{(E_{\vec{k}} - E_F)}{k_B T} \quad (27.54)$$

$$n_{\vec{q}} = \frac{1}{(e^z - 1)}\,, \quad z = \frac{\hbar\omega_{\vec{q}}}{k_B T}\,, \quad (27.55)$$

U may be evaluated as

$$U = (k_B T)\int_{-E_F/k_B T}^{\infty} d\eta \frac{e^{\eta+z}}{(e^\eta + 1)(e^{\eta+z} + 1)} \cong \frac{(k_B T)z}{(1 - e^{-z})} \quad (27.56)$$

and

$$\left(\frac{n_{\vec{q}}}{\omega_{\vec{q}}}\right) U = \left(\frac{\hbar e^{-z}}{(1 - e^{-z})^2}\right). \quad (27.57)$$

In obtaining Equation (27.56), we have used the fact that since $(E_F/k_B T)$ is large and at the lower limit of the integral the integrand is nearly zero, the integral is essentially unchanged if we replace the lower limit by $-\infty$.

In the angular integrations of Equation (27.52) over θ and θ' we first transform to the new angles $\psi = \theta - \theta'$ and $\psi' = \frac{1}{2}(\theta + \theta')$, so that $d\theta d\theta' = d\psi d\psi'$. Then the only factor that depends on ψ' is $(v_{\vec{k}\mu} - v_{\vec{k}'\mu})^2$. If we define

$$(\bar{v}_\mu - \bar{v}'_\mu)^2 = \left(\frac{m}{\hbar Q}\right)^2 \left(\frac{1}{2\pi}\right) \int_{-\pi}^{\pi} d\psi' \left(v_{\vec{k}\mu} - v_{\vec{k}'\mu}\right)^2 , \tag{27.58}$$

then, with the help of Equations (27.34) and (27.35), where the approximate forms of $v_{\vec{k},\mu}$ are given, we may write

$$(\bar{v}_x - \bar{v}'_x)^2 = \frac{1}{2}(\kappa^2 + \kappa'^2) - \kappa\kappa' \cos\psi , \tag{27.59a}$$

$$(\bar{v}_z - \bar{v}'_z)^2 = \beta^4 (w - w')^2 . \tag{27.59b}$$

At this point, since the integrand is even in the variables ψ and w, we agree to integrate only over the positive part of the integration regions of both variables and to multiply the result by 4. (That is, we integrate over the allowed part of ψ between 0 and π it and over the allowed part of w between 0 and $\frac{1}{2}$.)

Next we make a change of variables to center of mass and relative coordinates,

$$\tilde{\xi} = \frac{1}{2}(w + w') , \tag{27.60}$$

$$\tilde{q}_z = q_z/Q = w - w' , \tag{27.61}$$

so that $d\tilde{\xi} d\tilde{q}_z = dw dw'$. We then change from the variable $\tilde{q}_z$ to z, which was defined in Equation (27.55). The relation between $\tilde{q}_z$ and z is

$$\tilde{q}_z^2 = \frac{4\left[\left(\frac{T}{\Theta_\varphi}\right)^2 \left(\frac{q_\phi}{2Q}\right)^2 z^2 - \frac{\eta^2\beta^2}{2}\tilde{\xi}^2 (1 - \cos\psi)\right]}{1 + \frac{\eta^2\beta^2}{2}(1 + \cos\psi)} , \tag{27.62}$$

$$\begin{aligned} dw dw' &= d\tilde{\xi} d\tilde{q}_z \\ &= \frac{d\tilde{\xi}(z dz)}{\tilde{q}_z} \left(\frac{T}{\Theta_\varphi}\right)^2 \left(\frac{q_\phi}{2Q}\right)^2 \frac{1}{\left[1 + \frac{1}{2}\eta^2\beta^2(1 + \cos\psi)\right]} , \end{aligned} \tag{27.63}$$

where the phason temperature $\Theta_\varphi = \hbar\omega_\phi/k_B T$, with $\omega_\phi = c_\parallel q_\phi$, and the wave-vector cutoff q_ϕ of the phason spectrum in the k_z direction were defined in Equation (27.7) and indicated in Figures 27.3 and 27.4. With the assumption of this frequency cutoff for the phason spectrum, the range of integration for z is from 0 to (Θ_φ/T).

The other limits of integration are found simply if we assume the model of two intersecting cones described at the end of Section 27.3 and defined in Equation (27.32). The maximum value $\tilde{\xi}_{max}$ for $\tilde{\xi}$ is actually approximately the average of the maximum values of w and w', which is $\frac{1}{2}$. However, the integrand becomes negligibly small before that, so that for the purposes of numerical integration, this is set at a convenient value. In fact, it is this same localization of the electron–phason scattering near the conical points that allows us to use such a simple approximation for the Fermi surface.

In order to determine the lower limit on $\tilde{\xi}$, we recall that we agreed to integrate only over $w \geq 0$. The integration is then divided into two regions, $w \geq 0$, $w' \geq 0$, or $\tilde{\xi}^2 \geq \frac{1}{4}\tilde{q}_z^2$, or scattering of electrons within a single conical point, and $w \geq 0$, $w' \leq 0$, or $\tilde{\xi}^2 \leq \frac{1}{4}\tilde{q}_z^2$, or scattering of electrons from conical point to conical point. The condition $\tilde{\xi}^2 = \frac{1}{4}\tilde{q}_z^2$ then enables us to determine the lower limit on $\tilde{\xi}$, so that

$$-z\left(\frac{T}{\Theta_\phi}\right)\left(\frac{q_\phi}{2Q}\right)\Big/(1+\eta^2\beta^2)^{1/2} \le \tilde{\xi} \le \tilde{\xi}_{\max}\,, \tag{27.64}$$

where we ensure that $w \ge 0$ by choosing $\tilde{q}_z \ge 0$. The two regions $w \ge 0$, $w' \le 0$ and $w \ge 0$, $w' \ge 0$ are separated by $z(T/\Theta_\phi)(q_\phi/2Q)/(1+\eta^2\beta^2)^{1/2}$. For $\tilde{\xi}$ greater than this value, the scattering is within a single conical region.

The maximum value $\psi_{\max}$ for ψ is determined by requiring that $\tilde{q}_z^2 \ge 0$. Also, $\psi_{\max}$ is never greater than π. The integration is then divided most simply into two regions. The first is

$$-z\left(\frac{T}{\Theta_\phi}\right)\left(\frac{q_\phi}{2Q}\right)/(1+\eta^2\beta^2)^{1/2} \le \tilde{\xi} \le \frac{z}{\eta\beta}\left(\frac{T}{\Theta_\phi}\right)\left(\frac{q_\phi}{2Q}\right),$$
$$0 \le \psi \le \pi\,, \tag{27.65}$$

and the second is

$$\frac{z}{\eta\beta}\left(\frac{T}{\Theta_\phi}\right)\left(\frac{q_\phi}{2Q}\right) \le \tilde{\xi} \le \tilde{\xi}_{\max}\,,$$
$$0 \le \psi \le \cos^{-1}\left[1-\left(\frac{2z^2}{\eta^2\beta^2\tilde{\xi}^2}\right)\left(\frac{T}{\Theta_\phi}\right)^2\left(\frac{q_\phi}{2Q}\right)^2\right]. \tag{27.66}$$

A further approximation that we will make concerns the part of the square of the scattering matrix element $C^2_{\vec{k},\vec{k}'}$, that is defined in Equation (27.31). The cutoff q_ϕ of $\vec{q}$ in the z direction is only about 10^{-3} of the distance from the conical point to the belly of the Fermi surface, so that $\tilde{q}_z$ is always a small quantity. For this reason, we write $C^2_{\vec{k},\vec{k}'}$ in terms of $\tilde{\xi}$ and $\tilde{q}_z$, and expand in powers of $\tilde{q}_z$, keeping only the lowest-order term. Then $C^2_{\vec{k},\vec{k}'}$ becomes

$$C^2_{\vec{k},\vec{k}'} \cong \left(\frac{\tilde{q}_z^2}{4\alpha^2}\right)\hat{C}^2(\tilde{\xi})\,, \tag{27.67}$$

where

$$\hat{C}^2(\tilde{\xi}) = \frac{\alpha^4}{(\tilde{\xi}^2+\alpha^2)^2}\,. \tag{27.68}$$

Combining all these integrations, approximations, and transformations of variables, we can finally write Equation (27.41) in the form

$$I_\mu = \hat{I}_\mu/\bar{\tau}\,, \tag{27.69}$$

where

$$\frac{1}{\bar{\tau}} = \left(\frac{Q^4}{8\pi^4}\right)\left(\frac{m}{n\rho_m}\right)\left(\frac{G}{A}\right)^2\left(\frac{1}{\hbar k_B\Theta_\phi}\right), \tag{27.70}$$

and

$$\hat{I}_\mu = \left(\frac{1}{\alpha^2}\right)\left(\frac{q_\phi}{2Q}\right)^2\left(\frac{T}{\Theta_\phi}\right) \times \int_0^{\Theta_\phi/T} dz \left(\frac{ze^{-z}}{(1-e^{-z})^2}\right) \int_{\tilde{\xi}_{\min}}^{\tilde{\xi}_{\max}} d\tilde{\xi}\, \hat{C}^2(\tilde{\xi}) \times \int_0^{\psi_{\max}} S_\mu(z,\tilde{\xi},\psi) \frac{\left[\left(\frac{T}{\Theta_\phi}\right)^2\left(\frac{q_\phi}{2Q}\right)^2 z^2 - \frac{1}{2}\eta^2\beta^2\tilde{\xi}^2(1-\cos\psi)\right]^{1/2}}{\left[1+\frac{1}{2}\eta^2\beta^2(1+\cos\psi)\right]^{5/2}}, \quad (27.71)$$

where

$$S_z(z,\tilde{\xi},\psi) = 8\beta^4\left[\left(\frac{T}{\Theta_\phi}\right)^2\left(\frac{q_\phi}{2Q}\right)^2 z^2 - \frac{1}{2}\eta^2\beta^2\tilde{\xi}^2(1-\cos\psi)\right], \quad (27.72a)$$

$$S_x(z,\tilde{\xi},\psi) = 2\beta^2\left[\left(\frac{T}{\Theta_\phi}\right)^2\left(\frac{q_\phi}{2Q}\right)^2 z^2(1+\cos\psi) + \tilde{\xi}^2(1-\cos\psi)\right], \quad (27.72b)$$

where the endpoints $\tilde{\xi}_{\min}$, $\tilde{\xi}_{\max}$, $\psi_{\max}$, are determined by Equations (27.65) and (27.66).

The components of the temperature-dependent resistivity tensor due to electron–phason scattering may now be written, from Equations (27.41), (27.46) and (27.69), as

$$\rho_{xx} = \rho_{yy} = \left(\frac{m}{ne^2\bar{\tau}}\right)\hat{I}_x\,, \quad (27.73a)$$

$$\rho_{zz} = \left(\frac{m}{ne^2\bar{\tau}}\right)\frac{\hat{I}_z}{F^2}\,. \quad (27.73b)$$

$\hat{I}_x$ and $\hat{I}_z$ of Equations (27.71) and (27.72) must be obtained in general by three-dimensional numerical integration, which will be discussed in Section 27.5.

The calculation of the contribution to the low-temperature resistivity can be understood in terms of a very simple picture. When one derives the Bloch–Grüneisen formula for the electron–phonon resistivity one assumes that the phonon spectrum is isotropic and that the Fermi surface is spherical. One then pictures the center of a phonon sphere moving along the surface of the Fermi sphere. The position of the center of this phonon sphere marks the initial state $\vec{k}$ of an electron and any point within the phonon sphere but on the Fermi surface, is an allowed scattered state $\vec{k}'$ of the electron. For very low temperatures, as the temperature decreases, the effective size of the phonon sphere decreases, giving rise to a T^5 dependence in the low-temperature limit. For electron–phason scattering, we have an analogous situation, which is shown in Figure 27.6. The phason spectrum is anisotropic, and we assume a frequency cutoff, so that the phonon sphere, for the case of the Bloch–Grüneisen formula, becomes here a phason "pancake", an ellipsoid of revolution, where the ratio of the width to the diameter of the "pancake" is the anisotropy parameter η, defined in Equation (27.6). Since we have assumed the Fermi surface to be two intersecting cones, we can now picture, as in Figure 27.6, the center of the phason "pancake" riding along the surface of the Fermi-surface "cones". An electron in the state $\vec{k}$ at the center of the phason "pancake" may scatter to any state $\vec{k}'$ on the Fermi surface that is within the phason "pancake". As in the case of phonons, for very low temperatures the phason "pancake" decreases in size, leading again to a T^5 dependence in the low-temperature limit, as long as the anisotropy parameter is not too small.

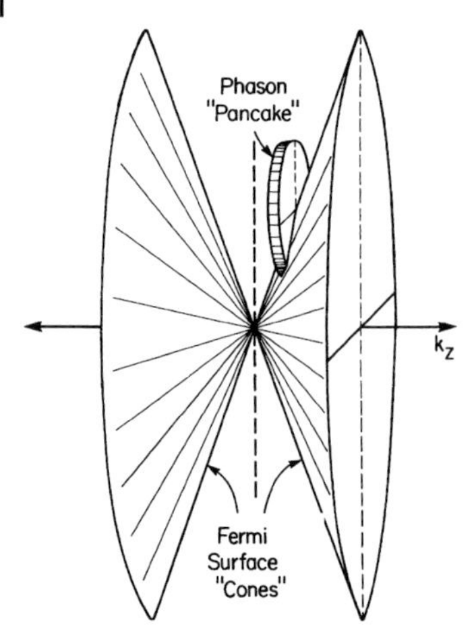

Figure 27.6 Phason "pancake" riding along the surface of the Fermi-surface "cones".

In order to understand what happens in the case of an extremely anisotropic phason spectrum, let us consider the limit of infinite anisotropy, i.e., $\eta \to 0$. In this case, the integrals reduce simply to Bloch–Grüneisen functions, $\mathcal{J}_n(x)$, as was pointed out in [9]. If we keep Θ_ϕ and q_ϕ finite as $\eta \to 0$, then $\hat{I}_x$ and $\hat{I}_z$ become

$$\hat{I}_x = \left(\frac{2\pi\beta^2}{\alpha^2}\right)\left[A_1\left(\frac{q_\phi}{2Q}\right)^3\left(\frac{T}{\Theta_\phi}\right)^2\mathcal{J}_2\left(\frac{\Theta_\phi}{T}\right)\right.$$
$$\left.+A_2\left(\frac{q_\phi}{2Q}\right)^5\left(\frac{T}{\Theta_\phi}\right)^4\mathcal{J}_4\left(\frac{\Theta_\phi}{T}\right)+\left(\frac{q_\phi}{2Q}\right)^6\left(\frac{T}{\Theta_\phi}\right)^5\mathcal{J}_5\left(\frac{\Theta_\phi}{T}\right)\right], \tag{27.74a}$$

$$\hat{I}_z = \left(\frac{8\pi\beta^4}{\alpha^2}\right)\left[A_2\left(\frac{q_\phi}{2Q}\right)^5\left(\frac{T}{\Theta_\phi}\right)^4\mathcal{J}_4\left(\frac{\Theta_\phi}{T}\right)\right.$$
$$\left.+\left(\frac{q_\phi}{2Q}\right)^6\left(\frac{T}{\Theta_\phi}\right)^5\mathcal{J}_5\left(\frac{\Theta_\phi}{T}\right)\right], \tag{27.74b}$$

with

$$A_1 = \int_0^{\tilde{\xi}_{\max}} d\tilde{\xi}\left(\frac{\alpha^4\tilde{\xi}^2}{(\tilde{\xi}^2+\alpha^2)^2}\right), \quad A_2 = \int_0^{\tilde{\xi}_{\max}} d\tilde{\xi}\left(\frac{\alpha^4}{(\tilde{\xi}^2+\alpha^2)^2}\right), \tag{27.75}$$

plus small terms of order$(q_\phi/2Q)^7(T/\Theta_\phi)^6\mathcal{J}_6(\Theta_\phi/T)$. The Bloch–Grüneisen functions are defined as

$$\mathcal{J}_n(x) = \int_0^x \frac{e^{-z}\,dz}{(1-e^{-z})^2}\,. \tag{27.76}$$

The factor of $\sqrt{2}$ that appeared in [9] in the $\mathcal{J}_5$ term was incorrect

We note that, in this limiting case the low-temperature limiting dependence is no longer T^5. In fact, the $\mathcal{J}_2$ term varies as T^2 and the $\mathcal{J}_4$ term as T^4. This is due to the fact that, in the limit $\eta \to 0$, the diameter of the phason "pancake" becomes infinite so that the allowed scattering is from an initial state located on the Fermi surface at w (along the z direction) to a point w' that is located within the region on the Fermi surface between $w \pm q_\phi$. Thus, there is no restriction in the x and y directions. This means that, as the temperature decreases, the effective volume of the phason "pancake" that encloses the Fermi surface does not decrease as rapidly as if the "pancake" were not so anisotropic.

In Section 27.5, we will present the results of numerical integration of $\hat{I}_\mu$ in Equation (27.71) with $\eta \neq 0$. In addition, we will show that comparing the magnitude of the resistivity predicted by this theory with that of existing data enables us to determine an allowed range for the anisotropy parameter η.

27.5 Numerical Results

In this section, we will present the results of the numerical evaluations of the threedimensional integrals $\hat{I}_x$ and $\hat{I}_z$ in Equations (27.71) and (27.72) in the form of the components of the electron–phason resistivity tensor given by Equation (27.73). Since these numerical computations presented considerable difficulties, it is worthwhile to discuss briefly the procedure used to calculate the integrals. The integral over ϕ caused no difficulty. However, in the integral over $\tilde{\xi}$, the integrand varies rapidly in parts of the integration region and must be handled carefully. We used a Gauss–Legendre iterative scheme that divides the regions more finely where the function is rapidly varying and more coarsely where it is slowly varying. Also, in the integration region of Equation (27.66), we employed the transformation

$$x = \tilde{\xi}/(a + \tilde{\xi}) \,, \tag{27.77}$$

and used x as the integration variable. If equal intervals are chosen for x, then intervals for $\tilde{\xi}$ will be close together for small $\tilde{\xi}$ and far apart for large $\tilde{\xi}$. This is useful, since the function decays in this region as $\tilde{\xi}$ increases. a is chosen to optimize the procedure. In the integration over z, we use the same type of transformation as for $\tilde{\xi}$ in Equation (27.77), since the largest part of the integrand is located where z is small. If one tries to perform the integrations over $\tilde{\xi}$ and z with a uniform grid of points a much larger amount of computer time is needed than with the procedures described here.

The magnitude of the calculated electron–phason resistivity agrees very well with the range of experimentally observed values, for reasonable values of the parameters in the theory. In addition, the resistivity tensor due to electron–phason scattering is highly anisotropic which can explain the sample dependence of the data. The resistivity ρ_{zz} along $\vec{Q}$ is much larger than the resistivity ρ_{xx} or ρ_{yy} perpendicular to $\vec{Q}$. Since there are 24 equally preferred directions for $\vec{Q}$ in K [40], one would expect that most samples would not consist of a single domain, i.e., $\vec{Q}$ pointing in the same direction throughout the sample. Rather, many $\vec{Q}$ domains would exist. The domain structure would vary from sample to sample and from run to run on the same sample, depending, perhaps in some uncontrolled way, on the experimental procedures. The effects of these $\vec{Q}$ domains on the residual resistivity were discussed in [32, 33]. Since the residual resistivity is anisotropic by a factor of 4 or 5 to 1, the residual resistivity can change by as much as this factor, with no change in the number of impurities or other imperfections in the sample. Similarly, since the electron–phason resistivity is highly anisotropic, this part of the resistivity can also change if the domain structure changes. In addition, the magnitude of the electron phason resistivity will be correlated with the residual resistivity in such a way as to produce an apparent breakdown of Matthiessen's rule In fact, Matthiessen's rule need not be violated in order to explain the data. That is, one can add the contributions from the residual resistivity and from the electron–phason resistivity independently and still find a correlation between the residual resisitivity and the temperature-dependent electron–phason resistivity, as is observed in experiments.

The parameters in Equations (27.70)–(27.72) that we adjust in order to fit the data are the phason temperature Θ_ϕ, which determines the shape of the resistivity curve, and the anisotropy parameter η, which determines the magnitude of the temperature-dependent part of the data. In order to take into account the $\vec{Q}$-domain structure, we write the total resistivity $\rho(T)$ in the temperature range of the data points in Figure 27.1 below 1.3 K as,

$$\rho(T) = \rho_0 + x\rho_{zz} + (1 - x)\rho_{xx} , \tag{27.78}$$

where x is between 0 and 1 and where the electron–phason resistivities ρ_{zz} and ρ_{xx} are given by Equation (27.73). This means that a fraction x of the domains are along $\vec{Q}$ and a fraction $(1-x)$ are perpendicular to $\vec{Q}$. The procedure is to determine a value of Θ_ϕ [Equation (27.7)] that fits the shape of the data, and for a given value of η [Equation (27.6)], determine x for a particular sample. The result of such a procedure, with $\Theta_\phi = 3.25\,\mathrm{K}$, is the smooth curve through the data in Figure 27.1, or the horizontal line ρ(phason) in Figure 27.2. We recall that, in Figure 27.1, the resistivity was set at 0 for the lowest data point, for experimental convenience, so that the residual resistivity should actually be added to all the data points.

In Figure 27.7, we plot the magnitudes of the components of the electron–phason resistivity tensor, ρ_{xx} and ρ_{zz}, as a function of the anisotropy of the phason spectrum $(1/\eta)$. As can be seen from the figure, the resistivity increases rapidly with increasing isotropy. For the phason temperature, we have used the value $\Theta_\phi = 3.25$ K, which produced the fit to the data in Figure 27.1. The other values of constants used in the calculations displayed in Figure 27.7 are as follows. We employ the average transverse-phason velocity, which was calculated by Giuliani and Overhauser [36] to be $c_\perp = 1.40 \times 10^5$ cm/s, where $c_\perp$ is defined in Equation (27.6a). The phason wave-vector cutoff p_ϕ along $\vec{Q}$ is given, from Equation (27.7), by

$$q_\phi = k_B \Theta_\phi \eta / c_\perp . \tag{27.79}$$

We choose a value for the CDW gap energy $G/E_F^0 = 0.35$ that is within the range determined by previous experiments [32, 33], where the Fermi energy in the absence of a CDW is $E_F^0 = 2.12\,\mathrm{eV}$. For the CDW wave vector $\vec{Q}$, we choose a magnitude consistent with [32, 33], so that for this value of G, $Q/k_F = 2.149$, where $E_F^0 = \hbar^2 k_F^2/2m$. The dimensionless constant $\alpha = 0.0379$ is obtained from Equation (27.22), and then β is given by Equation (27.33). The maximum amplitude A [Equation (27.9)] of the ions from their equilibrium positions has been calculated by Giuliani and Overhauser [40] to be $A \cong 0.03$ Å. The electron density is $n = 1.402 \times 10^{22}\,\mathrm{cm}^{-3}$, and the mass density is $\rho_m = 0.910$.

An examination of Figure 27.7, together with the magnitudes of the temperature-dependent part of the resistivity found in experiments [1–4], permits us to identify a range of values of the anisotropy $1/\eta$ that are consistent with the data. For simplicity, we compare all values for $T = 1$ K. While not all sets of data included this temperature, this is the simplest way of comparing the measurements, since [1] and [3] quote their results in terms of the coefficient a of aT^2 that was used in an attempt to fit the temperature-dependent part of the data. For the data that did not include $T = 1$ K, we simply use the value of that coefficient as the extrapolated value. This gives essentially $\rho - \rho_0$, where ρ_0 is guessed to fit the data. Although this is admittedly a crude procedure, it gives a reasonable idea of the range of magnitudes of the resistivity. This range extends from about 0.05 to 0.29 nΩ cm at $T = 1$ K.

If we assume that the largest value observed, $\rho - \rho_0 \cong 0.29$ nΩ cm, corresponded to all the $\vec{Q}$ domains in the sample lying parallel to the length of the wire, i.e., $x = 1$, then we obtain the maximum possible value of the anisotropy, $1/\eta = 9.7$, that fits the data. For the smallest value observed, $\rho - \rho_0 \cong 0.05$ nΩ cm, we suppose that all the domains were perpendicular to the length of the wire, i.e., $x = 0$. Then we obtain the smallest anisotropy, $1/\eta = 7.7$, that is

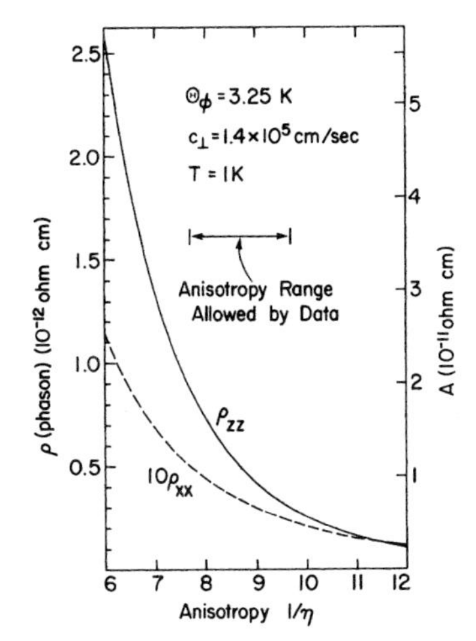

Figure 27.7 Plot of the electron phason resistivities ρ_{xx} and ρ_{zz} [Equation (27.73)] as a function of anisotropy $(1/\eta)$. The range of anisotropies consistent with the data of [1–4] is shown by the double-headed arrow. On the righthand vertical scale, A is the coefficient of the Bloch–Grüneisen function $(T/\Theta_\varphi)^5 \mathcal{J}_5(\Theta_\varphi/T)$ that fits the calculated values of the electron–phason resistivity.

consistent with the data. The actual anisotropy probably lies somewhere between these two extreme values as is indicated in Figure 27.7.

An interesting result of the numerical calculations is that the shape of the resistivity curve resulting from these evaluations is very nearly the same shape as the Bloch–Grüneisen function $\mathcal{J}_5(\Theta/T)$ of Equation (27.76), multiplied by T^5. This is exactly the same function that appears in the simple Bloch–Grüneisen formula for electron–phonon scattering. The shapes of ρ_{xx} and ρ_{zz} agree with $(T/\Theta)^5 \mathcal{J}_5(\Theta_\phi/T)$ to within 1% in the region of the data in Figure 27.1, although the extrapolation at zero temperature is slightly lower for this calculation. The $\mathcal{J}_5$ function extrapolates to -11.5 on the vertical scale of Figure 27.1, while the true function extrapolates to -12.0. For the purposes of analyzing the data, therefore, it might be useful to use the relation

$$\rho(T) \cong \rho_0 + \left[x A_z + (1-x) A_x\right] (T/\Theta)^5 \mathcal{J}_5(\Theta_\phi/T) \tag{27.80}$$

in order to determine parameters of this theory consistent with the data, where the values of A_x and A_z for a given anisotropy can be read from the right-hand vertical scale of Figure 27.7, for $\Theta_\phi = 3.25$ K. This procedure could save considerable computation time. A_z is determined from our calculation by setting $x = 1$, while A_x is determined by setting $x = 0$.

Although we have determined an allowable range of values of the anisotropy that are consistent with the data, as shown in Figure 27.7, there is some flexibility in that range. For instance, we have chosen an average value of the transverse-phason velocity. Giuliani and Overhauser [36] found that the transverse-phason spectrum is actually quite anisotropic, with velocities along the two principal transverse directions given by $c_2 = 2.08 \times 10^5$ cm/s and $c_3 = 0.94 \times 10^5$ cm/s, where the value we have used in Figure 27.7 is the geometric mean, $c_\perp = 1.40 \times 10^5$ cm/s. Assuming that some other average for $c_\perp$ might be more appropriate, we plot in Figure 27.8 the magnitudes of p_{xx} and ρ_{zz} as a function of $c_\perp$ that ranges between these two extreme values. We see that both ρ_{xx} and ρ_{zz} increase rapidly with decreasing $c_\perp$. A smaller value of $c_\perp$ would thus move the allowed range of anisotropics in Figure 27.7 to higher values of $(1/\eta)$. Again, we have labeled the right-hand vertical scale with the magnitude of the coefficient A of the Bloch–Grüneisen function, as in Equation (27.80).

Another factor in the magnitude of the resistivity is the phason temperature Θ_ϕ. Both ρ_{xx} and ρ_{zz} increase rapidly with increasing Θ_ϕ. The dependences of the electron–phason resistivity on the parameters Θ_ϕ, $c_\perp$, and η can be understood qualitatively by considering

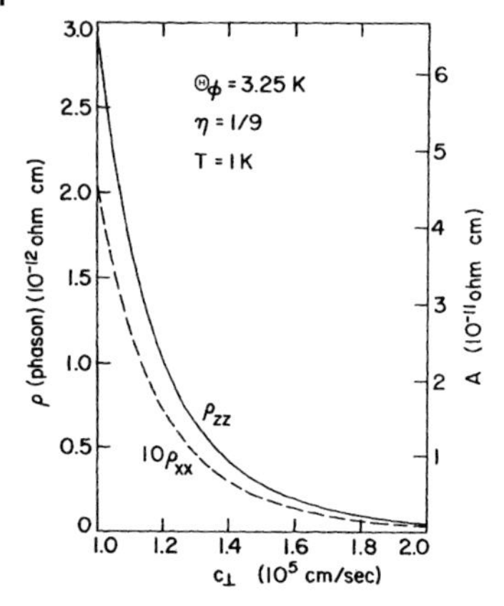

Figure 27.8 Plot of the electron–phason resistivities ρ_{xx} and ρ_{zz} as a function of the transversephason velocity $c_\perp$. A has the same meaning as in Figure 27.7.

the effects of each of these parameters on the total phase space occupied by the phasons. This can be seen by examining Figure 27.4. Since increasing Θ_φ is the same as increasing ω_φ, the frequency cutoff, this enlarges the phase space occupied by phasons and therefore increases the resistivity. As $c_\perp$ decreases, the volume of the anisotropic cone of the phason spectrum increases so that the resistivity increases. If the phason spectrum becomes more isotropic while $c_\perp$ is kept fixed, this is the same as decreasing the longitudinal phason velocity $c_\parallel$, so that the phase space increases and so does the resistivity.

27.6 Conclusions

We have found that the theory of electron–phason scattering can explain the shape, the magnitude, as well as the sample dependence of the anomalous temperature-dependent resistivity in potassium below 1.3 K. In addition, we have ruled out all other explanations that have been proposed to explain the data. A comparison of our theory with the data determines parameters of phasons that enter the theory. The phason temperature that gives the best fit to the data is $\Theta_\varphi = 3.25$ K, and the approximate range of acceptable values of the anisotropy of the phason spectrum is $7.7 \lesssim (1/\eta) \lesssim 9.7$. This is the same order of magnitude as the anisotropy in rubidium, $(1/\eta) \approx 11$, which was determined by attributing to phasons the anomaly in the specific heat of Rb [46]. Another feature is that we find the shape of the resistivity curve to be almost the same as that of the famous Bloch–Gr¨uneisen formula for the temperature-dependent resistivity due to normal electron–phonon scattering. Of course, that formula uses instead the phonon Debye temperature $\Theta_D \sim 90$ K.

In this calculation, we have assumed that the CDW in K produces only one gap, which is located at $\pm\vec{Q}$, at the conical points of the Fermi surface. (See Figure 27.5.) However, many orientations of open orbits have been observed in K [39], suggesting that other gaps, resulting from the multiple periodicity of the lattice and the CDW are also important. Electron–phason scattering would probably be intense near each of these gaps on the Fermi surface and could increase the resistivity above what is calculated here. That would have the result of translating the range of allowed anisotropies to higher values of $(1/\eta)$.

It would be useful to have further sensitive measurements on high-purity samples of the temperaturedependent resistivity of K. Measurements should be done that extend from 1.3 K to the lowest possible temperatures in order to test the theory presented here. However, since the electron–phason resistivity should decrease very rapidly below the temperature range of existing data, with a low-temperature limit of T^5, it is likely that the contribution of electron–electron scattering, with its T^2 temperature dependence, would appear before the electron–phason T^5 dependence is actually achieved. In fact, electron–electron scattering should be enhanced in the presence of a CDW by umklapp scattering from conical point to conical point on the Fermi surface. This would make the electron–electron scattering anisotropic, depending on the direction of the CDW wave vector $\vec{Q}$ with respect to the electric field. Then, if the $\vec{Q}$-domain structure changed from sample to sample, the electron–electron scattering would also be sample dependent.

Acknowledgements

The authors would like to thank J.A. Rowlands for providing the data shown in Figures 1 and 2 and for many useful discussions. In addition, the authors are grateful to the National Science Foundation and the NSF Materials Research Laboratory Program for support of this research.

References

1 van Kempen, H., Lass, J.S., Ribot, J.H.J.M., and Wyder, P. (1976) *Phys. Rev Lett.*, **37**, 1574.
2 Rowlands, J.A., Duvvury, C., and Woods, S.B. (1978) *Phys Rev. Lett*, **40**, 1201, and private communication.
3 Levy, B., Sinvani, M., and Greenfield, A.J. (1979) *Phys. Rev. Lett*, **43**, 1822.
4 Ribot, J.H.J.M. (1979) *Doctoral Dissertation*, Katholieke Universiteit te Nijmegen (unpublished).
5 van Kempen, H., Ribot, J.H.J.M., and Wyder, P. (1978) *J. Phys.* (Paris) **39**, C6 1048.
6 Ribot, J.H.J.M. Bass, J., van Kempen, H., and Wyder, P. (1979) *J. Phys. F*, **9**, L117.
7 Khoshnevisan, M., Pratt, W. P., Schroeder, P. A., Steenwyk, S., and Uher, C. (1979) *J. Phys. F*, **9**, L1.
8 Koshnevisan, M., Pratt, W.P. Jr., Schroeder, P.A., and Steenwyk, S.D. (1979) *Phys, Rev. B*, **19**, 3873.
9 Bishop, M.F. and Overhauser, A.W. (1979) *Phys. Rev. Lett.*, **42**, 1776.
10 Overhauser, A.W. (1971) *Phys. Rev. B*, **3**, 3173.
11 Ziman, J.M. (1960) *Electrons and Phonons*, Oxford University Press, London.
12 Gugan, D. (1971) *Proc. R. Soc. London*, **A325**, 223.
13 Ekin, J.W. and Maxfield, B.W. (1971) *Phys. Rev. B*, **4**, 4215.
14 Kaveh, M. and Wiser, N. (1974) *Phys. Rev. B*, **9**, 4042.
15 Kaveh, M. and Wiser, N. (1974) *Phys. Rev. B*, **9**, 4053.
16 Kaveh, M. and Wiser, N. (1972) *Phys. Rev. Lett.*, **29**, 1374.
17 Leavens, C.R. and Laubitz, M.J. (1974) *Solid State Commun.*, **15**, 1909.
18 Leavens, C.R. and Laubitz, M.J. (1975) *J. Phys. F*, **5**, 1519.
19 Taylor, R., Leavens, C.R., and Shukla, R.C. (1976) *Solid State Commun.*, **19**, 809.
20 Shukla, R.C. and Taylor, R. (1976) *J. Phys. F*, **6**, 531.
21 Froböse, K. (1977) *Z. Phys. B*, **26**, 19.
22 Lawrence, W.E. and Wilkins, J.W. (1973) *Phys. Rev. B*, **7**, 2317.
23 Kukkonen, C.A. and Smith, H. (1973) *ibid*, **8**, 4601.
24 Kaveh, M., Leavens, C.R., and Wiser, N. (1979) *J. Phys. F*, **9**, 71.
25 Kaveh, M. and Wiser, N. (1980) *J. Phys. F*, **10**, L37.
26 Brown, R.A. (1977) *J. Phys. F*, **7**, 1283.
27 Basinski, Z.S., Dugdale, J.S., and Howie, A. (1963) *Philos Mag.*, **8**, 1989.
28 Basinski, Z.S., Dugdale, J.S., and Gugan, D. (1959) *Philos. Mag.*, 4, 880.

29 Stetter, G., Adlhart, W., Fritsch, G., Steichele, E., and Lüscher, E. (1978) *J. Phys. F*, **8**, 2075.

30 Jones, B.K. (1969) *Phys. Rev.*, **179**, 637.

31 Holroyd, F.W. and Datars, W.R. (1975) *Can. J. Phys.*, **53**, 2517.

32 Bishop, M.F. and Overhauser, A.W. (1977) *Phys. Rev. Lett.*, **39**, 632.

33 Bishop, M.F. and Overhauser, A.W. 81978) *Phys. Rev. B*, **18**, 2447.

34 Overhauser, A.W. (1962) *Phys. Rev.*, **128**, 1437.

35 Overhauser, A.W. (1968) *Phys. Rev.*, **167**, 691.

36 Giuliani, G.F. and Overhauser, A.W. (1980) *Phys. Rev. B*, **21**, 5577.

37 Overhauser, A.W. (1978) *Adv. Phys.*, **27**, 343.

38 Altounian, Z., Verge, C., and Datars, W.R. (1978) *J. Phys. F*, **8**, 75.

39 Coulter, P.G. and Datars, W.R. (1980) *Phys. Rev. Lett.*, **45**, 1021.

40 Giuliani, G.F. and Overhauser, A.W. (1979) *Phys. Rev. B*, **20**, 1328.

41 Werner, S.A., Eckert, J., and Shirane, G. (1980) *Phys. Rev. B*, **21**, 581.

42 Giuliani, G.F. and Overhauser, A.W. (1980) *Phys. Rev. B*, **22**, 3639.

43 Boriack, M.L. and Overhauser, A.W. (1978) *Phys. Rev. B*, **18**, 6454.

44 Sawada, A. and Satoh, T. (1978) *J. Low Temp. Phys.*, **30**, 455.

45 Lien, W.H. and Phillips, N.E. (1964) *Phys. Rev.* **133**, A1370.

46 Giuliani, G.F. and Overhauser, A.W. (1980) *Phys. Rev. Lett.*, **45**, 1335.

47 Overhauser, A.W. (1978) *Hyperfine Interactions*, **4**, 786.

48 Boriack, M.L. and Overhauser, A.W. (1978) *Phys. Rev. B*, **17**, 4549.

49 Lee, P.A., Rice, T.M., and Anderson, P.W. (1974) *Solid State Commun.*, **14**, 703.

50 Ziman, J.M. (1972) *Principles of the Theory of Solids*, 2nd ed., Cambridge University Press, London, Chap. 7.

51 Kubo, R. and Nagamiya, T. (1969) *Solid State Physics*, McGraw-Hill, New York, pp. 139–155.

Reprint 28 Structure Factor of a Charge-Density Wave[1]

G.F. Giuliani and A.W. Overhauser**

* Department of Physics, Purdue University, 525, Northwestern Avenue, West Lafayette, Indiana 47907, USA

Received 17 December 1980

The elastic structure factor of an unpinned charge-density wave (CDW) is derived. Both phase and amplitude excitations are studied. Phase modes cause a reduction of the intensity of CDW satellite peaks of any order. Amplitude modes, however, do not alter appreciably the intensity of the firstorder CDW satellite. In fact, they enhance the intensity of peaks of higher order on account of their nonlinear nature. Neither phase nor amplitude excitations affect the usual Bragg reflections. The mean-square fluctuation of the CDW phase and the associated reduction of the satellite peaks are discussed for the case of potassium and are shown to depend critically on the phason spectrum.

28.1 Introduction

Systems containing incommensurate modulations of normal crystalline periodicity have recently attracted much interest. In particular incommensurate spin-density-wave [1] and charge-density-wave [2] (CDW) systems have been studied extensively since their observation in chromium [3] and in layered compounds [4].

An interesting feature of these materials is the existence of a new branch of acoustic, collective modes named phasons [5] associated with fluctuations in space and time of the relative phase between the lattice and the incommensurate modulation. Such extra low-frequency modes, which coexist and merge with the normal acoustic phonons [6], affect many physical properties. Examples are the low-temperature heat capacity [7–9], electrical resistivity [10], NMR spectrum [11], lattice thermal conductivity, and diffraction pattern [5, 11]. The latter phenomenon is the object of the present paper. Our emphasis will be on CDW systems but most of our results are directly applicable to any displacive, incommensurate modulated structure.[2]

The static structure factor of an incommensurate CDW has been thoroughly studied in [5], where it was shown that new collective modes associated with CDW phase modulation provide a peculiar contribution to the Debye–Waller factor of the CDW satellites in a diffraction pattern. In the present paper we extend the analysis of the structure factor to include the amplitude modes [14]. This is necessary for a complete and consistent treatment. The paper is organized as follows: In Section 28.2 we derive the general expression for the structure factor of an incommensurate CDW. In Section 28.3 we propose a theoretical model for the dispersion relation of the lowest-lying collective modes. In Section 28.4 we discuss the temperature dependence of the CDW structure factor and the mean-square fluctuations of the phase and amplitude variables. Finally Section 28.5 contains further discussion and application, especially to some problems involving alkali metals.

1) Phys. Rev. B 23, 8 (1981)
2) See, for instance, [12], and [13].
3) See, for instance, [15].

Anomalous Effects in Simple Metals. Albert Overhauser

ISBN: 978-3-527-40859-7

28.2 Dynamical Structure Factor for a CDW

The dynamical structure factor for an array of N ions is given by[3)]

$$S_{\vec{k},\omega} = \frac{1}{N}\int_{\infty}^{\infty} \frac{dt}{2\pi} e^{-i\omega t} \sum_{i,f} \left\langle e^{i\vec{k}\cdot\vec{R}_i(t)} e^{-\vec{k}\cdot\vec{R}_j(0)} \right\rangle_T , \tag{28.1}$$

where $\vec{R}_i(t)$ represents the position of an ion i at time t. The subscript T denotes that a thermal average is taken over an equilibrium distribution of states. In a CDW system the ionic lattice undergoes a small distortion from the normal crystal situation [2]. The new equilibrium positions are given by

$$\vec{R}_i = \vec{R}_i^0 + A_0 \sin\left(\vec{Q}' \cdot \vec{R}_i^0 + \phi_0\right), \tag{28.2}$$

where $\vec{R}_i^0$ is the equilibrium position of the ion i in the undistorted lattice. $\vec{Q}'$, $\vec{A}_0$, and ϕ_0 are the wave vector, amplitude, and phase of the CDW in the ground state. $\vec{Q}'$ is the Brillouin-zone reduction of $\vec{Q}$, the wave vector of the corresponding electronic-charge-density modulation, which is assumed to be incommensurate with respect to the ionic lattice.

In a CDW system the vibrational modes of the lattice are strongly modified for wave vectors near the CDW wave vector $\vec{Q}'$. In this region the eigenstates of the distorted lattice are phasons [5] and amplitons [14]. These vibrational modes are associated with modulations of the phase ϕ and magnitude $|\vec{A}_0|$ of the CDW ionic displacements.

Away from $\vec{Q}'$ these new modes merge with the phonon modes The effects of phonons on the structure factor are well known. Moreover, phonons are not relevant as far as the typical features of the distorted state are concerned. For instance, they do not affect appreciably the ratio of the CDW satellite intensities to the usual Bragg reflections in a diffraction analysis [5, 11, 16]. Accordingly we will disregard altogether these degrees of freedom in what follows. Allowance for phase and amplitude modulations in the CDW causes the positions of the ions, Equation (28.2), to change in time and in space

$$\vec{R}_i(t) = \vec{R}_i^0 + A_0\left[1 + \delta A(\vec{R}_i^0, t)\right] \times \sin\left[\vec{Q}' \cdot \vec{R}_i^0 + \delta\phi(\vec{R}_i^0, t)\right], \tag{28.3}$$

where $\delta\phi$ and δA are the magnitudes of the phase and amplitude modulations. For convenience ϕ_0 has been taken to be zero. Inserting Equation (28.3) into (28.1) we obtain

$$\begin{aligned} S_{\vec{k},\omega} = {} & \frac{1}{N}\sum_{i,j}\sum_{n,m=-\infty}^{+\infty} e^{i\vec{k}\cdot\left(\vec{R}_i^0 - \vec{R}_j^0\right)} e^{i\vec{Q}'\cdot\left(n\vec{R}_i^0 - m\vec{R}_j^0\right)} J_n\left(\vec{k}\cdot\vec{A}_0\right) J_m\left(\vec{k}\cdot\vec{A}_0\right) \\ & \times \int_{-\infty}^{+\infty} \frac{dt}{2\pi} e^{-i\omega t} \Phi_{nm,ij}(t) A_{nm,ij}(t) , \end{aligned} \tag{28.4}$$

where n and m are integers and J_n is a Bessel function of the first kind. The functions $\Phi_{nm,ij}$ and $A_{nm,ij}$ are defined as follows:

$$\Phi_{nm,ij}(t) = \left\langle e^{in\delta\phi(\vec{R}_i^0,t)} e^{-im\delta\phi(\vec{R}_j^0,t)} \right\rangle_T , \tag{28.5}$$

$$A_{nm,ij}(t) = \frac{\left\langle J_n\{\vec{k}\cdot\vec{A}_0[1+\delta A(\vec{R}_i^0,t)]\} J_m\{\vec{k}\cdot\vec{A}_0[1+\delta A(\vec{R}_j^0,0)]\} \right\rangle_T}{J_n(\vec{k}\cdot\vec{A}_0) J_m(\vec{k}\cdot\vec{A}_0)} . \tag{28.6}$$

These quantities are correlation functions and contain the dynamics of the phase and amplitude modulations. In obtaining Equation (28.4) extensive use has been made of the Jacobi–Anger generating function for the Bessel functions:

$$e^{iz\sin x} = \sum_{n=-\infty}^{+\infty} e^{inx} J_n(z) \,. \tag{28.7}$$

Finally since the eigenvectors of the amplitude and the phase modes are orthogonal, the average in Equation (28.1) can be carried out independently for the two categories of fluctuation.

A. Phase Excitations

We start with the quantization of the phase variable $\delta\varphi$ of the CDW. Accordingly the phase field $\delta\hat{\varphi}(\vec{R}_i^0, t)$ is defined as

$$\delta\hat{\varphi}(\vec{R}_i, t) = \sum_{\vec{q}} \frac{\delta\varphi_{\vec{q}}}{2i}\left(a_{\vec{q}} e^{i\left(\vec{q}\cdot\vec{R}_i^0 - \omega_{\vec{q}} t\right)} - a_{\vec{q}}^{\dagger} e^{-i\left(\vec{q}\cdot\vec{R}_i^0 - \omega_{\vec{q}} t\right)}\right), \tag{28.8}$$

where $\vec{q}$, $\delta\varphi_{\vec{q}}$, and $\omega_{\vec{q}}$ are the wave vector, amplitude, and frequency, respectively, of the phase mode created and destroyed by the operators $a_{\vec{q}}^{\dagger}$ and $a_{\vec{q}}$. These operators satisfy the usual commutation relations for bosons. From Equation (28.8) it follows that

$$\left[\delta\hat{\varphi}\left(\vec{R}_i^0, t\right), \delta\hat{\varphi}\left(\vec{R}_j^0, 0\right)\right] = -2i\sum_{\vec{q}} \left|\frac{\delta\varphi_{\vec{q}}}{2}\right|^2 \sin[\gamma_{ij}(\vec{q}, t)], \tag{28.9}$$

where $\gamma_{ij}(\vec{q}, t)$ is an abbreviation for $\vec{q}\cdot(\vec{R}_i^0 - \vec{R}_j^0) - \omega_{\vec{q}} t$. This commutator is clearly a c number, and both $\delta\hat{\varphi}(\vec{R}_i^0, t)$ and $\delta\hat{\varphi}(\vec{R}_j^0, 0)$ commute with it. Now the function $\Phi_{nm,ij}$ of Equation (28.5) can be expressed as

$$\begin{aligned} \Phi_{nm,ij}(t) = \exp\left(nm\sum_{\vec{q}}\left|\frac{\delta\varphi_{\vec{q}}}{2}\right|^2 \sin[\gamma_{ij}(\vec{q}, t)]\right) \\ \times\left\langle \exp\left\{i\left[n\delta\hat{\varphi}\left(\vec{R}_i^0, t\right) - m\delta\varphi\left(\vec{R}_j^0, 0\right)\right]\right\}\right\rangle_T . \end{aligned} \tag{28.10}$$

Since the exponent in the last factor is a linear form in the boson operators $a_{\vec{q}}$, the thermal average in (28.10) reduces to a simpler result when we take the phasons to be harmonic oscillators.

$$\begin{aligned} &\left\langle \exp i\left[n\delta\hat{\varphi}\left(\vec{R}_i^0, t\right) - m\delta\hat{\varphi}\left(\vec{R}_j, 0\right)\right]\right\rangle_T \\ &= \exp -\frac{1}{2}\left\langle\left|n\delta\varphi\left(\vec{R}_i^0, t\right) - m\delta\varphi\left(\vec{R}_j^0, 0\right)\right|^2\right\rangle_T . \end{aligned} \tag{28.11}$$

The remaining thermal average is then computed in a straightforward manner with the aid of Equation (28.8), i.e.,

$$\begin{aligned} &\left\langle\left|n\delta\hat{\varphi}\left(\vec{R}_i^0, t\right) - m\delta\hat{\varphi}(\vec{R}_j, 0)\right|^2\right\rangle_T \\ &= \sum_{\vec{q}}\left|\frac{\delta\varphi_{\vec{q}}}{2}\right|^2 (2N_{\vec{q}} + 1)\left\{n^2 + m^2 - 2nm\cos\left[\gamma_{ij}(\vec{q}, t)\right]\right\} . \end{aligned} \tag{28.12}$$

$N_{\vec{q}}$ is the temperature-dependent mean occupation number of the phason mode of $\vec{q}$, $\langle a_{\vec{q}}^{\dagger} a_{\vec{q}} \rangle_T$, and is given by the usual expression $[\exp(\hbar\omega_{\vec{q}}/k_B T) - 1]^{-1}$. The mean-square fluctuation of the phase at a given lattice site is defined by

$$\langle \delta\phi^2 \rangle = \left\langle \left| \delta\phi\left(\vec{R}_i^0, t\right) \right|^2 \right\rangle_T . \tag{28.13}$$

By means of Equation (28.8) one can readily show that

$$\langle \delta\phi^2 \rangle = \sum_{\vec{q}} \left| \frac{\delta\phi_{\vec{q}}}{2} \right|^2 (2N_{\vec{q}} + 1) . \tag{28.14}$$

Then with the use of Equations (28.10), (28.11), (28.12), and (28.14) the function $\Phi_{nm,ij}$ can be finally written as

$$\Phi_{nm,ij}(t) = \exp\left[-\frac{1}{2}(n^2 + m^2)\langle \delta\phi^2 \rangle \right] \exp nm \times \sum_{\vec{q}} \left| \frac{\delta\phi_{\vec{q}}}{2} \right|^2 \left\{ (2N_{\vec{q}} + 1) \cos\left[\gamma_{ij}(\vec{q}, t)\right] + i \sin\left[\gamma_{ij}(\vec{q}, t)\right] \right\} . \tag{28.15}$$

If we define Φ_{nm} as the time-independent component of the function $\Phi_{nm,ij}(t)$ we have

$$\Phi_{nm} = e^{-1/2(n^2+m^2)\langle \delta\phi^2 \rangle} . \tag{28.16}$$

B. Amplitude Excitations

Consider the function $J_n[\vec{k} \cdot \vec{A}_0(1 + \delta A)]$ entering Equation (28.6). For small amplitude modulations we can expana this quantity to second order in δA. In the limit of small CDW amplitude, $(\vec{k} \cdot \vec{A}_0 \ll 1)$, we use standard recursion relations for the Bessel functions and write

$$J_n \left\{ \vec{k} \cdot \vec{A}_0 \left[1 + \delta A\left(\vec{R}_i^0, t\right) \right] \right\} \simeq J_n\left(\vec{k} \cdot \vec{A}_0\right) \left(1 + |n| \delta A\left(\vec{R}_i^0, t\right) + \frac{|n|(|n| - 1)}{2} \delta A^2\left(\vec{R}_i^0, t\right) \right) . \tag{28.17}$$

In analogy with Equation (28.8) we introduce the field $\delta\hat{A}(\vec{R}_i^0, t)$ defined as

$$\delta\hat{A}(\vec{R}_i^0, t) = \sum_{\vec{q}} \frac{\delta A_{\vec{q}}}{2i} \left(b_{\vec{q}} e^{i\left(\vec{q} \cdot \vec{R}_i^0 - \Omega_{\vec{q}} t\right)} - b_{\vec{q}}^{\dagger} e^{-i\left(\vec{q} \cdot \vec{R}_i^0 - \Omega_{\vec{q}} t\right)} \right) \tag{28.18}$$

where $\vec{q}$, $\delta A_{\vec{q}}$, and $\Omega_{\vec{q}}$ are the wave vector, amplitude, and frequency, respectively, of the amplitons created or destroyed by the operators $b_{\vec{q}}^{\dagger}$ and $b_{\vec{q}}$. These also satisfy the usual commutation rules for bosons. The function $A_{nm,ij}(t)$ can be evaluated explicitly. With the use of Equations (28.17) and (28.16) in (28.6), we get

$$A_{nm,ij}(t) \simeq 1 + \frac{1}{2}\left[|n|(|n| - 1) + |m|(|m| - 1) \right] \langle \delta A^2 \rangle + |nm| \sum_{\vec{q}} \left| \frac{\delta A_{\vec{q}}}{2} \right|^2 \left\{ (2\tilde{N}_{\vec{q}} + 1) \cos\left[\tilde{\gamma}_{ij}(\vec{q}, t)\right] + i \sin\left[\tilde{\gamma}_{ij}(\vec{q}, t)\right] \right\} . \tag{28.19}$$

The functions $\tilde{N}_{\vec{q}}$ and $\tilde{\gamma}_{ij}(\vec{q}, t)$ differ from $N_{\vec{q}}$ and $\gamma_{ij}(\vec{q}, t)$ only by the substitution of $\Omega_{\vec{q}}$ for $\omega_{\vec{q}}$. In complete analogy with $\langle\delta\phi^2\rangle$, Equation (28.13), the CDW fractional amplitude fluctuation $\langle\delta A^2\rangle$ appearing in Equation (28.19) is defined as

$$\langle\delta A^2\rangle = \left\langle\left|A\left(\vec{R}_i^0, t\right)\right|^2\right\rangle_T = \sum_{\vec{q}} \left|\frac{\delta A_{\vec{q}}}{2}\right|^2 \left(2\tilde{N}_{\vec{q}} + 1\right) . \tag{28.20}$$

With the use of Equations (28.15) and (28.18) in (28.4) the dynamical structure factor of the CDW system can be obtained. Finally notice we use our approximations, $\delta A \ll 1$ and $\vec{k} \cdot \vec{A}_0 \ll 1$, to calculate A_{nm}, the time-independent part of $A_{nm,ij}(t)$. It can be written in the following suggestive way:

$$A_{nm} \simeq \exp\frac{1}{2}\left[|n|(|n|-1) + |m|(|m|-1)\right]\langle\delta A^2\rangle. \tag{28.21}$$

The exponent of Equation (28.21) is positive definite.

C. Elastic Structure Factor

We focus our attention now on the elastic part of the dynamic structure factor, Equation (28.1). From Equation (28.4) it is clear that the elastic contributions to $S_{\vec{k},\omega}$ come only from the time-independent components of both $\Phi_{nm,ij}(t)$, Equation (28.5), and $A_{nm,ij}(t)$, Equation (28.6). By means of the results (28.16) and (28.21) in (28.4), the elastic structure factor of an incommensurate CDW can be expressed as

$$S_{\vec{k}}^0 = \sum_{\vec{G}} \sum_{n=-\infty}^{\infty} \delta\left[\vec{k} - (\vec{G} + n\vec{Q})\right] J_n^2(\vec{k} \cdot \vec{A}_0) F_n^{\phi}(T) F_n^A(T) , \tag{28.22}$$

where $\vec{G}$ is a vector in the reciprocal lattice. The temperature dependence of this quantity is contained in the functions F_n^{ϕ} and F_n^A defined as follows:

$$F_n^{\phi}(T) = e^{-2n^2 W_{\phi}(T)} , \tag{28.23}$$

with

$$W_{\phi}(T) = \frac{1}{2}\langle\delta\phi^2\rangle \tag{28.24}$$

and

$$F_n^A(T) = e^{2|n|(|n|-1) W_A(T)} , \tag{28.25}$$

with

$$W_A(T) = \frac{1}{2}\langle\delta A^2\rangle . \tag{28.26}$$

The functions F_n^{ϕ} and F_n^A are, respectively, the phason [11] and ampliton temperature factors.

The pattern associated with the elastic structure factor $S_{\vec{k}}^0$, Equation (28.22), has been thoroughly discussed in [5] and [16]. We add here just a few remarks The phason and ampliton temperature factors do not affect the normal Bragg reflections, ($n = 0$). As is apparent from

Equation (28.23), F_n^{φ} plays for the phase modes a role similar to that of the usual Debye–Waller factor. The phase oscillations of a CDW reduce the intensity of the satellite peaks.

The amplitude fluctuations, on the other hand, do not have any significant effect on the intensities of both the Bragg reflections and their first, ($n = 1$), satellites which are the most relevant features of the diffraction pattern. Finally a noteworthy result is that the intensity of higher-order satellites, $|n| \gtrsim 2$, is enhanced by amplitude fluctuations.[4)]

28.3 Excitation Spectrum

The strong coupling of phonons with wave vector $\pm\vec{Q}'$ induced by the presence of the electronic CDW gives rise to a qualitative modification of the vibrational dispersion relation As already pointed out, the regions of the spectrum most strikingly modified are those assigned to phase [5] and amplitude modes [14]. The dispersion relation of the lowest-lying modes is shown schematically in Figure 28.1, where the wave vector $\vec{q}$ of a phason (or an ampliton) is also defined. The situation can be described by the following simple model [9]. We start with the assumption that only the lowest acoustic branch of the phonon spectrum of the undistorted lattice is relevant. That is, for $\vec{k} = \pm\vec{Q}'$ the other branchs have much higher frequencies. In this case, for small $\vec{q}$ (i.e., for $\vec{k} \simeq \pm\vec{Q}'$) the vibrational modes of the system can be described by a 2×2 dynamical matrix

$$D_{ij} = \begin{bmatrix} \omega_0^2 & \omega_0^2 F(\vec{q}) \\ \omega_0^2 F(\vec{q}) & \omega_0^2 \end{bmatrix}, \tag{28.27}$$

where ω_0 is the frequency of the unperturbed phonons with $\vec{k} = \pm\vec{Q}'$ (see also Figure 28.1). These phonons are assumed to be dispersionless in the vicinity of $\pm\vec{Q}'$. The off-diagonal coupling $F(\vec{q})$ is associated with new terms appearing in the electronic dielectric-response matrix and caused by the CDW [6, 14]. The spectrum of collective modes described by (28.27) is given by

$$\omega_{\pm}(\vec{q}) = \omega_0 \left[1 \pm F(\vec{q})\right]^{1/2}. \tag{28.28}$$

The frequency ω_+ is Ω (the ampliton frequency), Equation (28.18), whereas ω_- is $\omega_{\vec{q}}$ (the phason frequency), Equation (28.8). The situation here differs significantly from that of a one-dimensional CDW system. In that case the Peierls mechanism [17] is related to a giant Khon anomaly [14, 18] in the phonon spectrum. Such an effect need not occur in the diagonal part of the dynamical matrix, Equation (28.27).

The function $F(\vec{q})$ dictates the crossover region where phase and amplitude mode merge into normal phonons (as $|\vec{q}|$ increases). The knowledge of this function is crucial in the present context since (as discussed above) the two different sets of collective modes contribute in a completely different way to the Debye–Waller factor of the system. At present a theory for $F(\vec{q})$ is not at hand. Nevertheless the problem can be easily solved in the limit of long wavelength ($\vec{q} \to 0$) by taking advantage of the phason dispersion relation which is known in this regime.

4) A more accurate treatment, not reported here for sake of simplicity shows that in addition to the factor of (28.25) the amplitons provide a slight motional reduction of the intensity of both Bragg and satellite spots. Each mode contributes a factor similar to, and of the same order of magnitude as the Debye–Waller factor of a single phonon mode (with the same frequency). As pointed out in [5] such a quantity is much smaller than the corresponding term in W_{ϕ} or W_A.

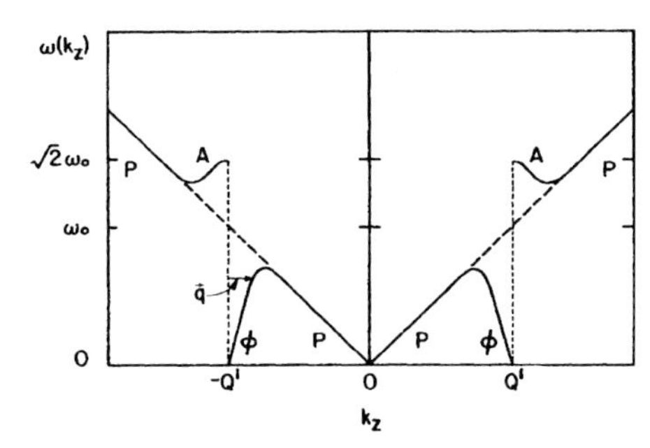

Figure 28.1 Schematic dispersion relation for the lowlying excitations of a CDW (in a three-dimensional metal). The phonon, phason and ampliton regions are denoted by P, ϕ, and A. ω_0 is the frequency of the lowest-energy phonon (of the normal state) at $\vec{k} = \pm\vec{Q}'$. The phason (or ampliton) wave vector $\vec{q}$ is the deviation of $\vec{k}$ from $\pm\vec{Q}'$.

As was first shown in [5], the phason frequency goes to zero linearly with $|\vec{q}|$ and can be expressed as

$$\omega_{\vec{q}} = \left(c_x^2 q_x^2 + c_y^2 q_y^2 + c_z^2 q_z^2\right)^{1/2} , \quad \vec{q} \to 0 . \tag{28.29}$$

c_x, c_y, and c_z are the phason velocities along orthogonal principal axes $\hat{x}$, $\hat{y}$, and $\hat{z}$. Usually $\hat{z}$ can be taken along $\vec{Q}$. $\omega_{\vec{q}}$ is in general very anisotropic In jellium, for instance, the spatial isotropy causes c_x and c_y, the transverse-phason velocities, to be zero. (The longitudinal-phason velocity c_z is finite.) For a CDW in simple metals c_x and c_y are expected to be of the order of magnitude as the sound velocity [19], whereas c_z is thought to be larger by one order of magnitude [9, 20].

If we substitute $\omega_{\vec{q}}$, Equation (28.29), for ω_- in Equation (28.28), we obtain

$$F(\vec{q}) = 1 - \frac{\left(c_x^2 q_x^2 + c_y^2 q_y^2 + c_z^2 q_z^2\right)}{\omega_0^2} , \quad \vec{q} \to 0 . \tag{28.30}$$

The corresponding expression for the ampliton frequency, is obtained from (28.28).

$$\Omega_{\vec{q}} = \left[2\omega_0^2 - \left(c_x^2 q_x^2 + c_y^2 q_y^2 + c_z^2 q_z^2\right)\right]^{1/2} , \quad \vec{q} \to 0 . \tag{28.31}$$

Expressions (28.29), (28.30), and (28.31) provide a correct description of the excitation spectrum only for small $|\vec{q}|$ but fail to account for the details of the dispersion relation when $\omega_{\vec{q}}$, Equation (28.29), becomes comparable to ω_0. A model for $F(\vec{q})$ can be postulated when such details are needed [9].

28.4 Phason and Ampliton Temperature Factors

This section is devoted to the study of the temperature dependence of both the phason and ampliton temperature factors F_n^{ϕ} and F_n^A.

A. Phasons: W_ϕ

The quantity W_ϕ is proportional to the mean-square fluctuation of the CDW phase and is defined in Equation (28.23). With the use of Equation (28.14) we can write

$$W_\phi = \frac{1}{8} \sum_{\vec{q}} |\delta\phi_{\vec{q}}|^2 \coth\left(\frac{\hbar\omega_{\vec{q}}}{2k_B T}\right) , \tag{28.32}$$

where the sum runs over all the N_ϕ phason modes of the system. The virial theorem allows us to obtain the amplitude coefficient of a single phase mode, defined by Equation (28.8).

$$\delta\phi_{\vec{q}} = \left(\frac{4\hbar}{\rho V A_0^2 \omega_{\vec{q}}} \right)^{1/2} , \qquad (28.33)$$

where ρ and V are the mass density and the total volume of the ionic lattice. As discussed in [5] and [11], W_ϕ does not depend explicitly upon the magnitude of the CDW wave vector $\vec{Q}'$. Notice also that the contribution to W_ϕ of a phason compared to an equivalent phonon's contribution to the Debye–Waller factor is a ratio (in the exponent) of $(|\vec{G}|\vec{A}_0)^{-2}$, typically of the order of 10^3. Of course this factor does not carry through to the total sum (28.32) when compared to the total Debye–Waller factor, since N_ϕ is much smaller than $3N$, the total number of vibrational modes of a monatomic lattice.

The qualitative behavior of W_ϕ as a function of temperature can be readily obtained by an explicit evaluation of the sum (28.32). We approximate $\omega_{\vec{q}}$ by Equation (28.29). The phason wave-vector space is then taken as the volume in $\vec{q}$ space contained in a surface of constant frequency ω_0, the frequency at $\vec{Q}'$ of the lowest-energy phonon of the normal lattice.[5] Due to the strong anisotropy of the phason spectrum, such a region of the Brillouin zone has the shape of a pancake with the short axis along $\vec{Q}$. Within this model the quantity $W_\phi(T)$ can be expressed as

$$W_\phi(T) = W_\phi^0 f_\phi(T) , \qquad (28.34)$$

where the temperature-dependent function of $f_\phi(T)$ is

$$f_\phi(T) = 8 \left(\frac{T}{\Theta_\phi} \right)^2 \int_0^{\Theta_\phi/2T} dx\, x \coth x . \qquad (28.35)$$

Θ_ϕ is the phason characteristic temperature defined as $\Theta_\phi = \hbar\omega_0/k_B$. The constant factor W_ϕ^0 in Equation (28.34) is

$$W_\phi^0 = \frac{\eta (k_B \Theta_\phi)^2}{8\pi^2 \hbar \rho A_0^2 C_\perp^3} . \qquad (28.36)$$

An average transverse-phason velocity $c_\perp \equiv \sqrt{c_x c_y}$ has been introduced together with the quantity η, the phason anisotropy ratio, defined as

$$\eta = \frac{c_\perp}{c_z} . \qquad (28.37)$$

If the temperature is much smaller than the cutoff Θ_Φ, $f_\phi(T) \simeq 1$, i.e.,

$$\lim_{T/\Theta_\phi \to 0} W_\phi(T) = W_\phi^0 . \qquad (28.38)$$

W_ϕ^0 represents the contribution of zero-point phase fluctuations of the CDW. In the high-temperature limit, when T is greater than Θ_ϕ, W_ϕ is proportional to T. In this regime we can write

$$W_\phi(T) \simeq 4 \frac{T}{\Theta_\phi} W_\phi^0 , \quad T > \Theta_\phi . \qquad (28.39)$$

5) In the case of a one-dimensional CDW system the off-diagonal coupling $F(\vec{q})$ and the crossover region from CDW-like (phasons and amplitons) to normal phonons is studied in detail in [21]. It is found that such a crossover is continuous but very quick.

The temperature dependence of F_n^ϕ, Equation (28.23), is readily deduced with the use of Equations (28.35), (28.38), and (28.39). As the temperature increases F_n^ϕ goes to zero exponentially with T from its zero-temperature value $\exp(-n^2 W_\phi^0)$.

B. Amplitons: W_A

The mean-square fluctuation of the fractional amplitude of the CDW, $2\,W_A$, Equation (28.26), can be rewritten by means of Equation (28.20) as

$$W_A = \frac{1}{8}\sum_{\vec{q}} |\delta A_{\vec{q}}|^2 \coth(\hbar\Omega_{\vec{q}}/2k_B T)\,. \tag{28.40}$$

The sum in Equation (28.40) runs over the same range of $\vec{q}$ as in Equation (28.32). Then the analysis can be carried out with the use of Equation (28.31) and follows the same procedure outlined in the discussion of W_ϕ. For the sake of brevity, we report here just the most relevant steps and results.

The virial theorem allows us to show that $\delta A_{\vec{q}}$ is also given by Equation (28.33), with $\Omega_{\vec{q}}$ replacing $\omega_{\vec{q}}$. The final result can be cast in the following form:

$$W_A(T) = W_A^0 f_A(T)\,, \tag{28.41}$$

with

$$W_A^0 = \frac{\pi-2}{2} W_\phi^0 \simeq 0.57\,W_\phi^0 \tag{28.42}$$

and

$$f_A(T) = \frac{4}{\pi-2}\int_0^1 dx \frac{x^2}{(2-x^2)^{1/2}} \coth\left(\frac{\Theta_\phi}{2T}(2-x^2)^{1/2}\right). \tag{28.43}$$

Notice that the zero-point, fractional-amplitude fluctuations, Equation (28.42), are roughly half of the corresponding phase fluctuations. In the high-temperature limit $W_A(T)$ increases also linearly with T, with a slope which is approximately one-quarter of the corresponding slope of $W_\phi(T)$, Equation (28.39). The behavior of $F_n^A(T)$, $|n| \geq 2$, is obtained via Equation (28.25) and is of course the inverse of that of $F_n^\phi(T)$, as the signs of the exponents in Equations (28.23) and (28.25) differ. From its zero-temperature value, $\exp[|n|(|n|-1)\,W_A^0]$, $F_n^A(T)$ grows exponentially with T. Nevertheless the product $F_n^A(T)_n^\phi(T)$, entering the expression for the CDW structure factor, Equation (28.22), goes to zero exponentially as the temperature is increased.

28.5 Discussion

In the previous sections we have discussed the theory of the dynamical structure factor in a CDW system. The elastic part of this quantity can be directly analyzed by a neutron-diffraction experiment. The typical signature of a CDW is the presence of extra spots, the satellite reflections in the diffraction pattern [2]. The ratio of the intensity of a satellite peak $\vec{k} = \vec{G} + n\vec{Q}$ to that of a normal Bragg reflection $\vec{k} = \vec{G}$, can be readily expressed with the use of Equation (28.22):

$$\frac{I_{\vec{G}+n\vec{Q}}}{I_{\vec{G}}} = \left(\frac{J_n\left[(\vec{G}+n\vec{Q})\cdot\vec{A}_0\right]}{J_0(\vec{G}\cdot\vec{A}_0)}\right)^2 F_n^\phi(T)F_n^A(T)\,. \tag{28.44}$$

F_n^{φ} and F_n^A are given in Equations (28.23) and (28.25). In particular, for the first-order satellite this ratio is[6)]

$$\frac{I_{\vec{G}+\vec{Q}}}{I_{\vec{G}}} \simeq \left(\frac{(\vec{G}+\vec{Q})\cdot\vec{A}_0}{2}\right)^2 e^{-2W_{\varphi}(T)} . \tag{28.45}$$

This quantity is explicitly evaluated in [16] for the case of a CDW state in metallic potassium. Equation (28.45) gives also a satisfactory description of the temperature dependence of the satellite spots in the quasi-one-dimensional conductor KCP [11]. Notice that for *higher-order* satellites the amplitude fluctuations tend to oppose the reduction of intensity caused by phase fluctuations. This phenomenon may be relevant in the explanation of the anomalously Intense high-order satellites observed in modulated structures such as Na_2CO_3 [22].

Recently the authors discussed some aspects of lattice dynamics in alkali metals assumed to have a CDW ground state [16, 19, 23]. In particular the CDW amplitude $|\vec{A}_0|$ and the average transverse-phason velocity $c_{\perp}$ were calculated. The specific values for potassium are $|\vec{A}_0| \simeq 0.03$ Å [16] and $c_{\perp} \simeq 1.4 \times 10^5$ cm/s [19]. As far as the phason anisotropy ratio η [Equation (28.37)] is concerned, no theory is currently at hand. However an estimate for this quantity can be obtained in an indirect way by fitting experimental data [9, 20]. According to those analyses, η is thought to be roughly 0.1. Furthermore if ω_0, the frequency cutoff for phasons, is chosen as in Section 28.4, its value for potassium is $\omega_0 \simeq -1.2 \times 10^{12}$ Hz. Accordingly $\Theta_{\varphi} \simeq 9$ K. With the use of Equation (28.36) we can now evaluate the zero-point, mean-square phase fluctuation for this CDW model. The result is $W_{\varphi}^0 \simeq 0.85 \times 10^{-2}$, corresponding to a temperature factor which is practically unity at zero temperature. At $T \sim 10$ K, $W_{\varphi}(T) \simeq 0.37$. The reduction of the satellite intensity would be given by $F_1^{\varphi} = \exp[-2W_{\varphi}(T)] \simeq 0.93$.

This indicates that at low temperatures ($T \leq 10$ K), the phase fluctuations in the system are small and do not seriously reduce the satellite intensity. It should be noticed, however, that W_{φ}^0 is extremely sensitive to the value of $c_{\perp}$: a change of a factor of three in $c_{\perp}$, for instance, would completely reverse these conclusions.

The phase fluctuations in a CDW are also relevant in the NMR spectrum. A static CDW theory for the NMR spectrum in potassium [11, 24] leads to an inhomogeneous shift having a full width of ~ 40 G.[7)] What is observed [24] is a single narrow line with a full width of ~ 0.2 G. This result could be explained within a CDW model if motional narrowing of the Knight shift by thermal phase fluctuations [11] were large. The results obtained above seem to make this explanation unlikely. The discrepant NMR linewidth is an important challenge for the CDW theory of alkali metals.

References

1 Overhauser, A.W. (1960) *Phys. Rev. Lett.*, 4, 462.
2 Overhauser, A.W. (1968) *Phys. Rev.*, **167**, 691.
3 Overhauser, A.W. (1962) *Phys Rev.*, **128**, 1437.
4 Wilson, J.A., DiSalvo, F.J., and Mahajan, S. (1975) *Adv. Phys.*, **24**, 117.

6) Equation (28.44) and (28.45) are valid only when the orientation of the CDW wave vector $\vec{Q}'$ is the same everywhere in the crystal (see also [16]).

7) A local theory for the exchange and correlation effects in the CDW would give a full width of 70 G. This value is reduced to ~ 40 G, when nonlocality in the exchange and correlation is properly recognized as indicated in [16].

5 Overhauser, A.W. (1971) *Phys. Rev. B*, **3**, 3173.

6 Giuliani, G.F. and Tosatti, E. (1978) *Nuovo Cimento B*, **47**, 135.

7 Sawada, A. and Satoh, T. (1978) *J. Low Temp. Phys.*, **30**, 455.

8 Boriack, M.L. and Overhauser, A.W. (1978) *Phys. Rev. B*, **18**, 6454.

9 Giuliani, G.F. and Overhauser, A.W. (1980) *Phys. Rev. Lett.*, **45**, 1335.

10 Bishop, M.F. and Overhauser, A.W. (1979) *Phys. Rev. Lett.*, **42**, 1776.

11 Overhauser, A.W. (1978) *Hyper. Inter.*, **4**, 786.

12 Axe, J.D. (1978) in *Neutron Inelastic Scattering 1977*, Vienna, Austria,International Atomic Energy Agency, Vienna, Austria, Pt. II, pp. 101–121.

13 Axe, J.D. (1980) *Phys Rev. B*, **21**, 4181.

14 Lee, P.A., Rice, T.M., and Anderson, P.W. (1974) *Solid State Commun.*, **14**, 703.

15 Ashcroft, N.W. and Mermin, N.D. (1967) *Solid State Physics*, Holt, Rinehart and Winston, New York, Appendix N.

16 Giuliani, G.F. and Overhauser, A.W. (1980) *Phys. Rev. B*, **22**, 3639.

17 Peierls, R.E. (1955) *Quantum Theory of Solids*, Oxford University Press, London, pp. 108–112.

18 Rice, M.J. and Strassler, S. (1973) *Solid State Commun.*, **13**, 1931.

19 Giuliani, G.F. and Overhauser, A.W. (1980) *Phys. Rev.*, **21**, 5577.

20 Bishop, M.F. and Overhauser, A.W. (1981) *Phys. Rev. B*, **23**, 3627.

21 Giuliani, G.F. and Tosatti, E. (1979) in *Lecture Notes in Physics*, (eds S. Barisic, A. Bjelis, J.R. Cooper, and B. Leontic), Springer, Berlin, Vol. 95, p. 191.

22 Hogervorst, A., Peterse, W.J.A.M., and deWolff, P.M. (1979) in *Modulated Structures-1979*, Kailua Kona, Hawaii, Proceedings of the International Conference on Modulated Structures, (eds J.M. Cowley, J.B. Cohen, M.B. Salamon, and J.B. Wuensch, AIP, New York, pp. 217–219.

23 Giuliani, G.F. and Overhauser, A.W. (1979) *Phys. Rev. B*, **20**, 1328.

24 Follstaedt, D. and Slichter, C.P. (1976) *Phys. Rev. B*, **13**, 1017.

Reprint 29 Effective-Medium Theory of Open-Orbit Inclusions[1)]

M. Huberman * *and A.W. Overhauser* *

* Department of Physics, Purdue University, 525, Northwestern Avenue, West Lafayette, Indiana 47907, USA

Received 5 February 1981

We extend the effective-medium approximation for the electrical conductivity of a heterogeneous material to a nonsymmetric effective conductivity tensor. We apply it to a binary mixture of oriented crystallites in which one component has open orbits and the other does not. We calculate the transverse magnetoresistance as a function of field strength and orientation. The open-orbit magnetoresistance peaks that would be observed in a homogeneous sample persist in the mixture, but are broader and smaller. Nevertheless, when the product of the volume fraction f of the open-orbit component and $\omega_c\tau$ is greater than unity, the peak height increases quadratically with field strength.

29.1 Introduction

A commonly used method for calculating the electrical conductivity of a heterogeneous material is the effective-medium approximation.[2)] It was first derived by Bruggeman [2] and Landauer [3] and recently discussed by Stroud [4]. In previous applications, the effective conductivity tensor was assumed to have an axis of symmetry.[3)] In this paper, we extend the effective-medium approximation to a nonsymmetric effective conductivity tensor.

We apply this theory to a binary mixture of oriented crystallites in which one component has open electron orbits and the other does not. In an applied magnetic field, the conductivity tensor of the mixture has no axis of symmetry. We calculate the transverse magnetoresistance as a function of field strength and orientation.

For a homogeneous sample having open orbits, the transverse magnetoresistance has sharp peaks as a function of field orientation [5]. When the magnetic field is perpendicular to an open-orbit direction, the magnetoresistance increases quadratically with the field strength; in all other directions, it saturates. We show that the open-orbit peaks that would be observed in a homogeneous sample persist in the mixture but are broader and smaller. When the product of the volume fraction f of the open-orbit component and $\omega_c\tau$ is greater than unity, the peak height increases quadratically with field strength.

29.2 Approximations for the Effective Conductivity

Following Stroud and Pan [6], we review the derivations of three approximations for the conductivity of a heterogeneous material: the small-fraction approximation, the Maxwell–Garnett approximation, and the effective-medium approximation. We consider a heterogeneous material composed of crystallites or grains of the pure components. For simplicity we consider a two-component material; the generalization to more components is straightforward. We

1) Phys. Rev. B 23, 12 (1981)

2) The theory and application of the effective-medium approximation are reviewed by [1].

3) In an example of [6], the effective conductivity lacks symmetry due to the shape of the inclusions. We are concerned with mixtures which have a nonsymmetric effective conductivity due to the conductivities of the pure components.

Anomalous Effects in Simple Metals. Albert Overhauser

ISBN: 978-3-527-40859-7

assume that the grain size is much larger than an electron mean free path but much smaller than the dimensions of the sample. If the grain size is much larger than the electron mean free path, the current density $\vec{j}$ and electric field $\vec{E}$ at any position $\vec{x}$ are related by the conductivity $\overleftrightarrow{\sigma}$ of the pure component at that position,

$$\vec{j}(\vec{x}) = \overleftrightarrow{\sigma}(\vec{x}) \cdot \vec{E}(\vec{x}) . \tag{29.1}$$

Let $\langle \vec{j} \rangle$ be the spatial average of the current density and $\langle \vec{E} \rangle$ be the spatial average of the electric field. The effective conductivity $\overleftrightarrow{\sigma}_{\text{eff}}$ is defined by

$$\langle \vec{j} \rangle = \overleftrightarrow{\sigma}_{\text{eff}} \cdot \langle \vec{E} \rangle . \tag{29.2}$$

If the grain size is much smaller than the sample dimensions, $\overleftrightarrow{\sigma}_{\text{eff}}$ is independent of the sample geometry or boundary conditions.

The common ingredient of all three approximations for the effective conductivity is the solution of the following problem: Consider a single spherical inhomogeneity of conductivity $\overleftrightarrow{\sigma}$ in an infinite homogeneous medium of conductivity $\overleftrightarrow{\sigma}_{\text{ext}}$. If a uniform electric field $\vec{E}_{\text{ext}}$ is applied, what is the electric field $\vec{E}_{\text{in}}$ inside the inhomogeneity? It can be shown[4] that $\vec{E}_{\text{in}}$ is uniform and proportional to $\vec{E}_{\text{ext}}$,

$$\vec{E}_{\text{in}} = \overleftrightarrow{\alpha}(\overleftrightarrow{\sigma}_{\text{ext}}, \overleftrightarrow{\sigma}) \cdot \vec{E}_{\text{ext}} . \tag{29.3}$$

For a single inhomogeneity in an infinite uniform medium, the applied field is equal to the average field so that

$$\vec{E}_{\text{in}} = \overleftrightarrow{\alpha}(\overleftrightarrow{\sigma}_{\text{ext}}, \overleftrightarrow{\sigma}) \cdot \langle \vec{E}_{\text{ext}} \rangle . \tag{29.4}$$

$\overleftrightarrow{\alpha}$ depends on the conductivities, $\overleftrightarrow{\sigma}_{\text{ext}}$ and $\overleftrightarrow{\sigma}$, of the medium and inhomogeneity.

If the volume fraction f of one of the components is small, we expand the effective conductivity to first order in f, obtaining the small-fraction approximation. Let the conductivity of the small-fraction component be $\overleftrightarrow{\sigma}_2$ and the conductivity of the other component be $\overleftrightarrow{\sigma}_1$. Rewriting Equation (29.1) as

$$\vec{j}(\vec{x}) = \left\{ \overleftrightarrow{\sigma}_1 + \left[\overleftrightarrow{\sigma}(\vec{x}) - \overleftrightarrow{\sigma}_1 \right] \right\} \cdot \vec{E}(\vec{x}) . \tag{29.5}$$

and taking the volume average, we obtain

$$\langle \vec{j} \rangle = \overleftrightarrow{\sigma}_1 \cdot \langle \vec{E} \rangle + f(\overleftrightarrow{\sigma}_2 - \overleftrightarrow{\sigma}_1) \cdot \langle \vec{E}_{\text{in}} \rangle . \tag{29.6}$$

$\langle \vec{E}_{\text{in}} \rangle$ is the average value of the electric field in the small-fraction component. To solve for the effective conductivity to first order in f, it is sufficient to solve for $\vec{E}_{\text{in}}$ to zeroth order, that is, for a single inhomogeneity in an infinite homogeneous medium. Approximating the actual shapes of the grains by spheres and applying Equation (29.4), we obtain

$$\overleftrightarrow{\sigma}_{\text{eff}} = \overleftrightarrow{\sigma}_1 + f(\overleftrightarrow{\sigma}_2 - \overleftrightarrow{\sigma}_1) \cdot \overleftrightarrow{\alpha} \tag{29.7}$$

with $\overleftrightarrow{\alpha} = \alpha(\sigma_1, \sigma_2)$.

4) The proof for the analogous elastic field problem is given by [7].

If the volume fraction of neither component is small such an expansion is not a good approximation. Instead we make a mean-field type of approximation. We replace the actual environment of a grain by a uniform medium of conductivity $\overleftrightarrow{\sigma}_{\text{ext}}$ and calculate the electric field within the grain when a uniform field $\vec{E}_{\text{ext}}$ is applied. Approximating the shape of the grain by a sphere we obtain for the electric field within the grain

$$\vec{E}_{\text{in}}^{(i)} = \overleftrightarrow{\alpha}_i \cdot \vec{E}_{\text{ext}} , \quad i = 1, 2 . \tag{29.8}$$

The current density within the grain is thus

$$\vec{j}_{\text{in}}^{(i)} = \overleftrightarrow{\sigma}_i \cdot \overleftrightarrow{\alpha}_i \cdot \vec{E}_{\text{ext}} , \quad i = 1, 2 . \tag{29.9}$$

Averaging the electric field (29.8) and current density (29.9) over all the grains of the sample and solving for $\langle \vec{j} \rangle$ in terms of $\langle \vec{E} \rangle$ we obtain for the effective conductivity

$$\overleftrightarrow{\sigma}_{\text{eff}} = \left(\sum_i f_i \overleftrightarrow{\sigma}_i \cdot \overleftrightarrow{\alpha}_i \right) \cdot \left(\sum_i f_i \overleftrightarrow{\alpha}_i \right)^{-1} . \tag{29.10}$$

Letting $\delta \overleftrightarrow{\sigma}_i = \overleftrightarrow{\sigma}_i - \overleftrightarrow{\sigma}_{\text{ext}}$, we may rewrite (29.10) as

$$\overleftrightarrow{\sigma}_{\text{eff}} = \overleftrightarrow{\sigma}_{\text{ext}} + \left(\sum_i f_i \delta \overleftrightarrow{\sigma}_i \cdot \overleftrightarrow{\alpha}_i \right) \cdot \left(\sum_i f_i \overleftrightarrow{\alpha}_i \right)^{-1} . \tag{29.11}$$

$\overleftrightarrow{\sigma}_i$ and f_i are the conductivity and volume fraction of the ith component; $\overleftrightarrow{\alpha}_i = \overleftrightarrow{\alpha}(\overleftrightarrow{\sigma}_{\text{ext}}, \overleftrightarrow{\sigma}_i)$ depends on $\overleftrightarrow{\sigma}_{\text{ext}}$, which is unspecified so far.

Different approximations correspond to different choices for $\overleftrightarrow{\sigma}_{\text{ext}}$. If one component is identifiable as the host in which the other component is embedded, a possible choice is the conductivity of the host component. $\vec{E}_{\text{ext}}$ is then equal to the average electric field in the host. This is the Maxwell–Garnett approximation.[5] On the other hand, if the material is a true mixture of different components, an appropriate choice is the self-consistent one,

$$\overleftrightarrow{\sigma}_{\text{ext}} = \overleftrightarrow{\sigma}_{\text{eff}} , \tag{29.12}$$

which is the effective-medium approximation. $\vec{E}_{\text{ext}}$ is then equal to the average electric field $\langle \vec{E} \rangle$ in the mixture.

The effective-medium equations, (29.11) and (29.12), may be derived and expressed in several equivalent ways [3, 4, 6, 10]. An advantage of this formulation is that the equations may be solved by iteration. Starting from an initial choice for $\overleftrightarrow{\sigma}_{\text{eff}}$, we evaluate the right-hand side of (29.11), determining a new solution for $\overleftrightarrow{\sigma}_{\text{eff}}$. This procedure is repeated until a self-consistent solution is obtained. We have found that this is an efficient method of solution. Starting from a free-electron conductivity, the solution typically converged after five iterations to within 1%.

29.3 Electric Field in a Spherical Inhomogeneity

In order to apply the effective-medium approximation, we require the solution for $\overleftrightarrow{\alpha}(\overleftrightarrow{\sigma}_{\text{ext}}, \overleftrightarrow{\sigma})$, which relates the electric field inside a spherical inhomogeneity to the applied field. As shown

5) [8]. An illuminating derivation of the Maxwell–Garnett theory, showing its relationship to the effective-medium theory, is given by [9].

by Stroud and Pan [11],

$$\overleftrightarrow{\alpha}(\overleftrightarrow{\sigma}_{\text{ext}}, \overleftrightarrow{\sigma}) = \left[\overleftrightarrow{1} - \overleftrightarrow{\Gamma}_{\text{ext}} \cdot \left(\overleftrightarrow{\sigma} - \overleftrightarrow{\sigma}_{\text{ext}}\right)\right]^{-1} \tag{29.13}$$

with

$$\overleftrightarrow{\Gamma}_{\text{ext}} = \int_S da \vec{\nabla} G(\vec{x}) \hat{n} \,. \tag{29.14}$$

The integration is over the surface of the inhomogeneity, which is centered at the origin $\vec{x} = 0$. The unit vector $\hat{n}$ is an outward-pointing normal. $G(\vec{x})$ is the Green's function for the medium defined as the solution of the equation

$$\vec{\nabla} \cdot \overleftrightarrow{\sigma}_{\text{ext}} \cdot \vec{\nabla} G(\vec{x}) = -\delta(\vec{x}) \,, \tag{29.15}$$

satisfying the boundary condition $G(\vec{x}) \to 0$ as $|\vec{x}| \to \infty$.

The conductivity tensor $\overleftrightarrow{\sigma}_{\text{ext}}$ may be expressed as a sum of a symmetric tensor $\overleftrightarrow{\sigma}_s$ and an antisymmetric tensor $\overleftrightarrow{\sigma}_a$. Equation (29.15) for the Green's function depends only on the symmetric part $\overleftrightarrow{\sigma}_s$. Since $\overleftrightarrow{\sigma}_s$ is a real symmetric matrix, it has an orthonormal basis of eigenvectors $\hat{v}_i$ and real eigenvalues λ_i,

$$\overleftrightarrow{\sigma}_s \cdot \hat{v}_i = \lambda_i \hat{v}_i \,, \quad i = 1, 2, 3 \,. \tag{29.16}$$

The eigenvectors $\hat{v}_i$ are the principal axes of $\overleftrightarrow{\sigma}_s$. Since the conductivity tensor $\overleftrightarrow{\sigma}_{\text{ext}}$ is positive definite, so is its symmetric part $\overleftrightarrow{\sigma}_s$; therefore the eigenvalues λ_i are positive. In the principal-axes coordinate system, the equation for the Green's function is

$$\lambda_1 \frac{\partial^2 G}{\partial x_1^2} + \lambda_2 \frac{\partial^2 G}{\partial x_2^2} + \lambda_3 \frac{\partial^2 G}{\partial x_3^2} = -\delta(x_1)\delta(x_2)\delta(x_3) \tag{29.17}$$

whose solution is

$$G(x_1, x_2, x_3) = \frac{1}{4\pi} \frac{1}{(\lambda_1 \lambda_2 \lambda_3)^{1/2}} \frac{1}{\left(x_1^2/\lambda_1 + x_2^2/\lambda_2 + x_3^2/\lambda_3\right)^{1/2}} \,. \tag{29.18}$$

We evaluate the components of the tensor $\vec{\Gamma}_{\text{ext}}$ in the principal-axes coordinate system. For a spherical inhomogeneity $\vec{\Gamma}_{\text{ext}}$ is diagonal in this representation. Changing the surface integral in (29.14) into a volume integral, we obtain

$$\vec{\Gamma}_{\text{ext}} = \int_V dv \vec{\nabla} \vec{\nabla} G(\vec{x}) \,. \tag{29.19}$$

Making use of (29.17), we note that the diagonal elements Γ_i of $\vec{\Gamma}_{\text{ext}}$ satisfy

$$\sum_{i=1}^{3} \lambda_i \Gamma_i = 1 \,. \tag{29.20}$$

Thus we need to evaluate only two diagonal elements, the third being determined from (29.20). Substituting (29.18) in (29.14), we obtain

$$\Gamma_i = \frac{-1}{4\pi R} \frac{1}{(\lambda_1 \lambda_2 \lambda_3)^{1/2}} \int_S da \frac{x_i^2/\lambda_i}{\left(x_1^2/\lambda_1 + x_2^2/\lambda_2 + x_3^2/\lambda_3\right)^{3/2}}$$

for a spherical inhomogeneity of radius R. To evaluate this integral, it is sufficient to evaluate

$$\Gamma \equiv \sum_i \Gamma_i = \frac{-1}{4\pi R} \frac{1}{(\lambda_1\lambda_2\lambda_3)^{1/2}} \int_S da \frac{1}{\left(x_1^2/\lambda_1 + x_2^2/\lambda_2 + x_3^2/\lambda_3\right)^{1/2}}$$

since $\Gamma_i = \partial(\sqrt{\lambda_i}\Gamma)/\partial(\sqrt{\lambda_i})$. In spherical polar coordinates, the integration over the polar angle can be done. The remaining integral is related to elliptic integrals [12]. The derivatives of elliptic integrals are expressible in terms of elliptic integrals [13].

Let the eigenvalues λ_i be ordered according to the rule $\lambda_1 \le \lambda_2 \le \lambda_3$. If $\lambda_1 < \lambda_2 < \lambda_3$, then

$$\Gamma_1 = -\frac{1}{\lambda_1} \frac{1}{(\lambda_1\lambda_2\lambda_3)^{1/2}} \frac{1}{(1/\lambda_1 - 1/\lambda_3)^{1/2}(1/\lambda_1 - 1/\lambda_2)}(F - E) ,$$

$$\Gamma_3 = -\frac{1}{\lambda_3} \frac{1}{(\lambda_1\lambda_2\lambda_3)^{1/2}} \frac{1}{(1/\lambda_1 - 1/\lambda_3)^{1/2}(1/\lambda_2 - 1/\lambda_3)} \times \left(\frac{(\lambda_3\lambda_1)^{1/2}}{\lambda_2^{1/2}} (1/\lambda_1 - 1/\lambda_3)^{1/2} - E \right) .$$

F and E are elliptic integrals of the first and second kinds [14], defined by

$$F(\delta, k) = \int_0^\delta (1 - k^2 \sin^2 \phi)^{-1/2} d\phi ,$$

$$E(\delta, k) = \int_0^\delta (1 - k^2 \sin^2 \phi)^{1/2} d\phi ,$$

having amplitude and modulus

$$\delta = \sin^{-1}(1 - \lambda_1/\lambda_3)^{1/2} ,$$
$$k = (1/\lambda_1 - 1/\lambda_2)^{1/2}/(1/\lambda_1 - 1/\lambda_3)^{1/2} .$$

If $\lambda_1 = \lambda_2 < \lambda_3$ then $\Gamma_1 = \Gamma_2$ and

$$\Gamma_3 = -\frac{1}{1 - \lambda_1/\lambda_3} \frac{1}{\lambda_3} \left(1 - \frac{(\lambda_1/\lambda_3)^{1/2} \sin^{-1}(1 - \lambda_1/\lambda_3)^{1/2}}{(1 - \lambda_1/\lambda_3)^{1/2}} \right) .$$

If $\lambda_1 < \lambda_2 = \lambda_3$, then $\Gamma_2 = \Gamma_3$ and

$$\Gamma_1 = \frac{1}{\lambda_3/\lambda_1 - 1} \frac{1}{\lambda_1} \left(1 - \frac{(\lambda_3/\lambda_1)^{1/2} \sinh^{-1}(\lambda_3/\lambda_1 - 1)^{1/2}}{(\lambda_3/\lambda_1 - 1)^{1/2}} \right) .$$

If $\lambda_1 = \lambda_2 = \lambda_3$, then $\Gamma_1 = \Gamma_2 = \Gamma_3 = -1/3\lambda_3$.

29.4 Magnetoresistance of Open-Orbit Inclusions

We apply this theory to a binary mixture of oriented crystallites, in which one component has open electron orbits and the other does not. Let the closed-orbit component have a free-electron conductivity tensor,

$$\overleftrightarrow{\sigma} = \sigma \begin{bmatrix} \frac{1}{1+\beta^2} & \frac{-\beta}{1+\beta^2} & 0 \\ \frac{\beta}{1+\beta^2} & \frac{1}{1+\beta^2} & 0 \\ 0 & 0 & 1 \end{bmatrix} .$$

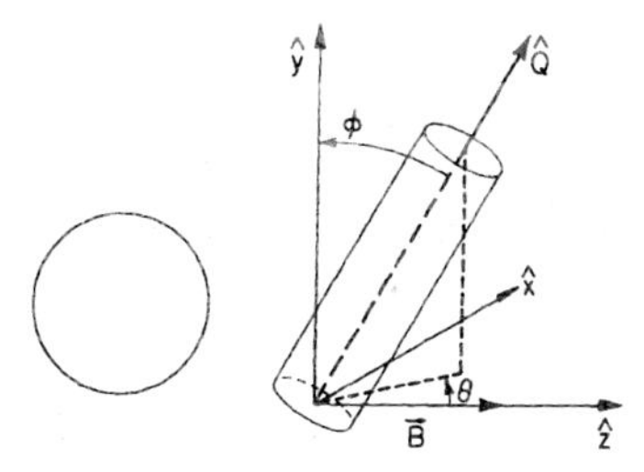

Figure 29.1 Fermi surface of the open-orbit component. The orientation of $\hat{Q}$ with respect to the wire ($\hat{y}$ axis) and the magnetic field ($\hat{z}$ axis) is described by the angles ϕ and θ.

where $\beta = \omega_c \tau$. $\sigma = ne^2\tau/m$ is the conductivity in zero magnetic field, n is the electron density, and τ is the relaxation time. $\omega_c = eB/mc$ is the cyclotron frequency.

Let the open-orbit component have two independent bands, a sphere and a cylinder. In one band, the electrons are free, having spherical constant energy surfaces. In the other band, the electrons move freely perpendicular to a given direction $\hat{Q}$, but cannot move parallel to $\hat{Q}$. Since the electron velocity $\vec{v}$ is proportional to $\vec{\nabla}_{\vec{k}} E(\vec{k})$, the constant energy surfaces are cylinders about the axis $\hat{Q}$. When an applied magnetic field is perpendicular to $\hat{Q}$, the electron orbits in the cylindrical band are open.

For simplicity, let the relaxation times in the two bands be the same. Let η be the fraction of electrons in the cylindrical band. As shown in Figure 29.1, let ϕ be the angle between $\hat{Q}$ and the y axis and let θ be the angle between the projection of $\hat{Q}$ in the x–z plane and $\vec{B} = B\hat{z}$. The conductivity tensor is [15]

$$
\begin{aligned}
\sigma_{xx} &= \frac{(1-\eta)\sigma}{1+\beta^2} + \frac{\eta\sigma}{1+\delta^2}(1-\sin^2\phi\sin^2\theta)\,,\\
\sigma_{xy} &= \frac{-(1-\eta)\beta\sigma}{1+\beta^2} + \frac{\eta\sigma}{1+\delta^2}(-\sin\phi\cos\phi\sin\theta - \delta\sin\theta\cos\theta)\,,\\
\sigma_{xz} &= \frac{\eta\sigma}{1+\delta^2}(-\sin^2\phi\sin\theta\cos\theta + \delta\cos\phi)\,,\\
\sigma_{yx} &= \frac{(1-\eta)\beta\sigma}{1+\beta^2} + \frac{\eta\sigma}{1+\delta^2}(-\sin\phi\cos\phi\sin\theta + \delta\sin\phi\cos\theta)\,,\\
\sigma_{yy} &= \frac{(1-\eta)\sigma}{1+\beta^2} + \frac{\eta\sigma}{1+\delta^2}\sin^2\phi\,,\\
\sigma_{yz} &= \frac{\eta\sigma}{1+\delta^2}(-\sin\phi\cos\phi\cos\theta - \delta\sin\phi\sin\theta)\,,\\
\sigma_{zx} &= \frac{\eta\sigma}{1+\delta^2}(-\sin^2\phi\sin\theta\cos\theta - \delta\cos\phi)\,,\\
\sigma_{zy} &= \frac{\eta\sigma}{1+\delta^2}(-\sin\phi\cos\phi\cos\theta + \delta\sin\phi\sin\theta)\,,\\
\sigma_{zz} &= (1-\eta)\sigma + \frac{\eta\sigma}{1+\delta^2}(1-\sin^2\phi\cos^2\theta)\,,
\end{aligned}
$$

where $\sigma = (ne^2\tau)/m$, $\beta = \omega_c\tau$, and $\delta = \omega_c\tau\sin\phi\cos\theta$. n is the electron density and τ is the relaxation time. $\omega_c = eB/mc$ is the cyclotron frequency.

For a homogeneous sample having open orbits, the transverse magnetoresistance has sharp peaks as a function of field orientation, occurring when $\vec{B}$ is perpendicular to $\hat{Q}$. The peak height increases quadratically with field strength If the relaxation times are the same in the two bands, the Hall coefficient has the free-electron value $R_H = -1/nec$. The effect of open orbits on the longitudinal magnetoresistance is negligible.

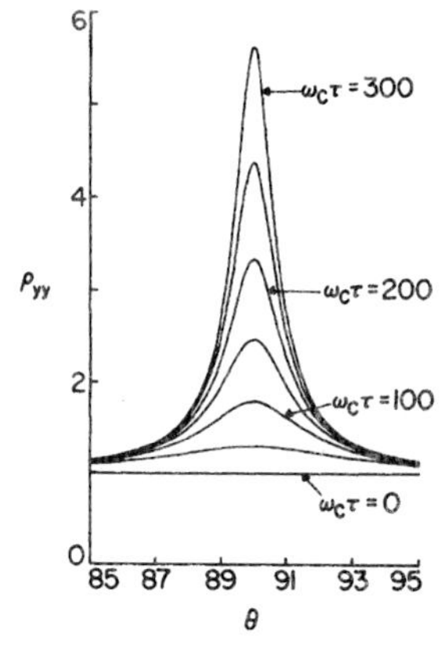

Figure 29.2 Effective magnetoresistance ρ_{yy} as a function of θ for $\omega_c\tau = 0, 50, 100, 150, 200, 250, 300$. The polar angle $\phi = 45°$, the volume fraction $f = 0.01$, and the open-orbit electron fraction $\eta = 0.1$.

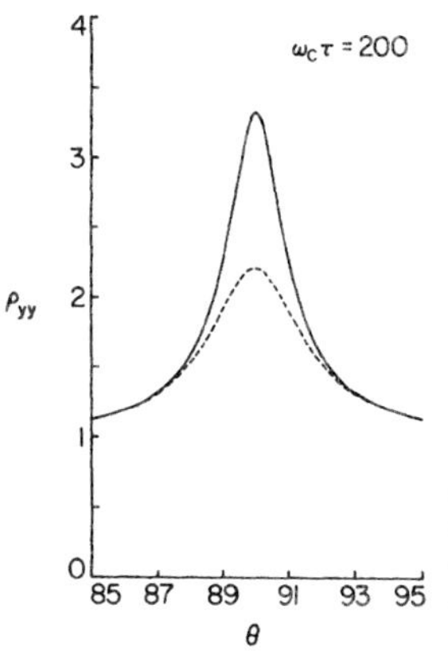

Figure 29.3 Comparison of ρ_{yy} as a function of θ for the effective-medium theory (full curve) and for the Maxwell–Garnett and small-fraction theories (dashed curve). The Maxwell–Garnett and small-fraction theories are the same. $\omega_c\tau = 200$. The polar angle $\phi = 45°$, the volume fraction $f = 0.01$, and the open-orbit electron fraction $\eta = 0.1$.

Let f be the volume fraction of the open-orbit component in the mixture. Let the electron density and relaxation time be the same in both components. We suppose the sample is a long wire in the $\hat{y}$ direction. The magnetic field $\vec{B} = B\hat{z}$ is perpendicular to the wire. Using the effective-medium theory, we have calculated the magnetoresistance ρ_{yy} along the wire as a function of θ for a given value of ϕ. ρ_{yy} is plotted in units of $p = 1/\sigma$, the resistivity of the free-electron component. The polar angle is $\phi = 45°$. Except in Figures 29.5 and 29.6, the volume fraction is $f = 0.01$ and the open-orbit electron fraction is $\eta = 0.1$.

In Figure 29.2, we plot ρ_{yy} as a function of θ about $\theta = 90°$, showing how the peaks grow with field strength. In Figure 29.3, we compare the effective-medium, Maxwell–Garnett, and small-fraction approximations. The Maxwell–Garnett and small-fraction approximations are the same. The effective-medium theory predicts a narrower and larger peak. In Figure 29.4, we compare the peak height at $\theta = 90°$ as a function of field strength for the three approximations. $\Delta\rho_{yy} = \rho_{yy}(B) - \rho_{yy}(0)$ is the change in resistivity at $\theta = 90°$ due to the applied magnetic field. The difference between the effective-medium theory and the other two shows that interactions between the open-orbit inclusions become significant when $(\omega_c\tau) f \gtrsim 1$. In this regime, the effective-medium theory predicts a quadratic increase whereas the other two predict a linear increase.

We have examined the dependence of the rate of increase of the peak height on the volume fraction f of the open-orbit component and on the open-orbit electron fraction η. In Fig-

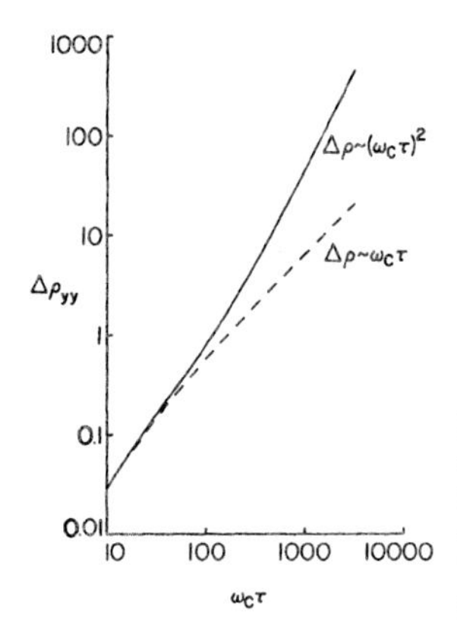

Figure 29.4 Comparison of the peak height $\Delta\rho_{yy} = \rho_{yy}(B) - \rho_{yy}(0)$ at $\theta = 90°$ as a function of $\omega_c\tau$ for the effective-medium theory (full curve) and for the Maxwell–Garnett and small-fraction theories (dashed curve). The Maxwell–Garnett and small-fraction theories are the same. The polar angle $\phi = 45°$, the volume fraction $f = 0.01$, and the open-orbit electron fraction $\eta = 0.1$.

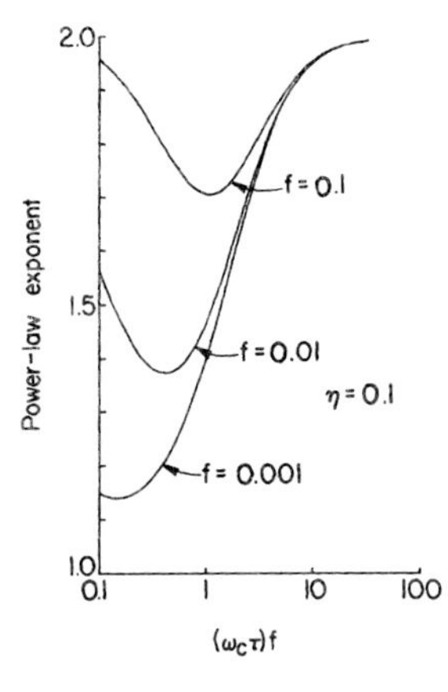

Figure 29.5 Rate of increase of the peak height $\Delta\rho_{yy}$ at $\theta = 90°$ as a function of $(\omega_c\tau) f$ for volume fractions $f = 0.001, 0.01$, and 0.1 The open-orbit electron fraction $\eta = 0.1$ and the polar angle $\phi = 45°$.

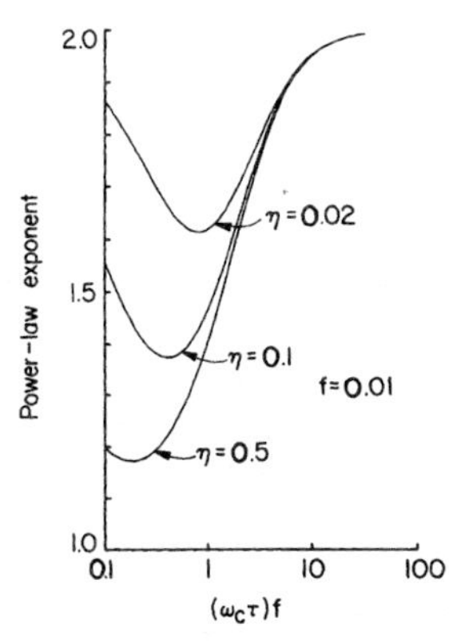

Figure 29.6 Rate of increase of the peak height $\Delta\rho_{yy}$ at $\theta = 90°$ as a function of $(\omega_c\tau) f$ for open-orbit electron fractions $\eta = 0.02, 0.1$, and 0.5. The volume fraction $f = 0.01$ and the polar angle $\phi = 45°$.

ure 29.5, the powerlaw exponent α defined by $\Delta\rho_{yy} \propto (\omega_c\tau)^\alpha$, is plotted for $f = 0.001, 0.01$, and 0.1 with $\eta = 0.1$. In Figure 29.6, it is plotted for $\eta = 0.02, 0.1$, and 0.5 with $f = 0.01$. When $(\omega_c\tau) f \lesssim 1$, the peak height increases at a rate between linear and quadratic, depending on f and η. When $(\omega_c\tau) f \gtrsim 1$, the peak height increases quadratically, independent of f and η.

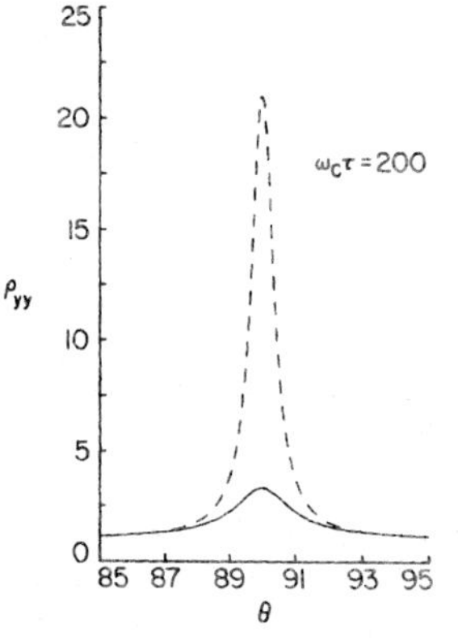

Figure 29.7 Comparison of the magnetoresistance ρ_{yy} as a function of θ for a homogeneous (dashed curve) and inhomogeneous (full curve) sample. The polar angle $\phi = 45°$ and $\omega_c\tau = 200$. In the inhomogeneous sample, the Volume fraction $f = 0.01$ and the open-orbit electron fraction $\eta = 0.1$. The homogeneous sample has the same total fraction of open-orbit electrons ($\eta = 0.001$) as the inhomogeneous sample.

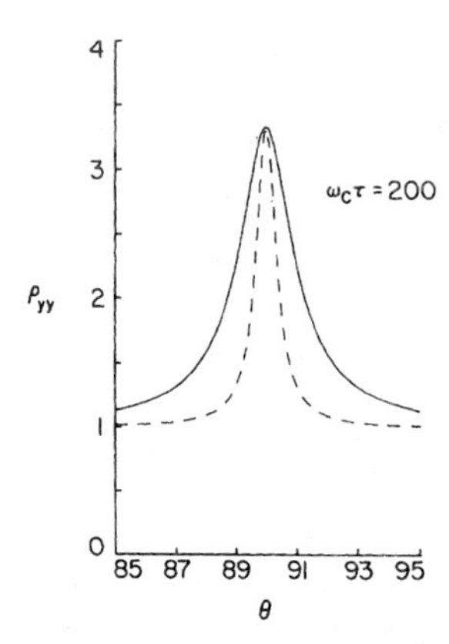

Figure 29.8 Comparison of the magnetoresistance ρ_{yy} as a function of θ for a homogeneous (dashed curve) and inhomogeneous (full curve) sample. The polar angle $\phi = 45°$ and $\omega_c\tau = 200$. In the inhomogeneous sample, the volume fraction $f = 0.01$ and the open-orbit electron fraction $\eta = 0.1$. The open-orbit electron fraction in the homogeneous sample is chosen to give the same peak height as in the inhomogeneous sample.

In Figures 29.7 and 29.8 we compare the magnetoresistance peaks of an inhomogeneous and homogeneous sample. In Figure 29.7, the total fraction of open-orbit electrons is chosen to be the same in the homogeneous and inhomogeneous samples. The peak in the inhomogeneous sample is smaller and broader. The broadening is shown more clearly in Figure 29.8, for which the open-orbit electron fraction in the homogeneous sample is adjusted so that the peak heights are the same.

Finally, we have calculated the Hall coefficient and longitudinal magnetoresistance The Hall coefficient has the free-electron value. The open-orbit component has a negligible effect on the longitudinal magnetoresistance.

29.5 Discussion

When $(\omega_c\tau) f \gtrsim 1$, interactions between the open-orbit inclusions become significant. This is consistent with calculations of the current distribution around an isolated open-orbit inclusion of dimension R, which show that the current distortion extends a distance of the order of $(\omega_c\tau) R$ parallel to the magnetic field [16].

When $(\omega_c\tau) f \gtrsim 1$, the peak magnetoresistance of oriented open-orbit inclusions increases quadratically with field strength. This contrasts with the magnetoresistance of voids [11, 17, 18], which is linear for all $\omega_c\tau > 1$, and the magnetoresistance of randomly oriented open-orbit inclusions [6, 10], which eventually saturates.

The effective-medium theory for a nonsymmetric effective conductivity is required in several practical problems. In polycrystalline wires, the crystallite are not necessarily randomly oriented, but may have a preferred orientation (texture). If the conductivity of a single crystal is anisotropic, the effective magnetoconductivity of the polycrystal has no axis of symmetry. The general theory is necessary in order to treat the effect of texture.

Recent experiments [19] show that potassium has open orbits. A charge-density-wave state has open orbits whose directions are determined by the charge-density-wave vector $\vec{Q}$. There are 24 preferred orientations of $\vec{Q}$ with respect to the crystal axes [20]. A single-crystal sample of potassium which is divided into $\vec{Q}$ domains is electrically equivalent to a 24-component mixture of oriented open-orbit crystallites. The general theory developed here is required to calculate its magnetoresistance as a function of field strength and orientation.

Acknowledgements

This research was supported by the Materials Research Laboratory program of the National Science Foundation.

References

1. Landauer, R. (1978) in *Electrical Transport and Optical Properties of In-homogeneous Media*, Ohio State University, 1977, Proceedings of the First Conference on the Electrical Transport and Optical Properties of Inhomogeneous Media, (eds J.C Garland and D.B. Tanner), AIP, New York, p. 2.
2. Bruggeman, D.A.G. (1935) *Ann. Phys.* (Leipzig), **24**, 636.
3. Landauer, R. (1952) *J. Appl. Phys.*, **23**, 779.
4. Stroud, D. (1975) *Phys. Bev. B*, **12**, 3368.
5. Kittel, C. (1963) *Quantum Theory of Solids*, Wiley, New York.
6. Stroud, D. and Pan, F.P. (1979) *Phys. Eev. B*, **20**, 455.
7. Eshelby, J.D. (1957) *Proc. K. Soc. London Ser. A*, **271**, 376.
8. Maxwell-Gamett, J.C (1904) *Philos. Trans. R. Soc. London Ser. A*, **203**, 385.
9. Smith, G.B. (1977) *J. Phys. D*, **10**, L39.
10. Stachowiak, H. (1970) *Physica*, **45**, 481.
11. Stroud, D. and Pan, F.P. (1976) *Phys. Rev. B*, **13**, 1434.
12. Gradshteyn, I.S. and Ryzhik, I.M. (1980) *Table of Integrals, Series, and Products*, Academic, New York, Sec. 4.577.
13. Gradshteyn, I.S. and Ryzhik, I.M. (1980) *Table of Integrals* Series, and Products*, Academic, New York, Sec. 8.123.
14. Gradshteyn, I.S. and Ryzhik, I.M. (1980) *Table of Integrals, Series, and Products*, Academic, New York, Sec. 8.111.
15. Overhauser, A.W. (1974) *Phys. Rev. B*, **9**, 2441.
16. Martin, P.M., Sampsell, J.B., and Garland, J.C. (1977) *Phys. Rev. B*, **15**, 5598.
17. Sampsell, J.B. and Garland, J.C. (1976) *Phys. Rev. B*, **13**, 583.
18. Schotte, K.D. and Jacob, D. (1976) *Phys. Status Solidi A*, **34**, 593.
19. Coulter, P.G. and Datars, W.R. (1980) *Phys. Rev. Lett.*, **45**, 1021.
20. Giuliani, G.F. and Overhauser, A.W. (1979) *Phys. Rev. B*, **20**, 1328.

Reprint 30 Theory of the Open-Orbit Magnetoresistance of Potassium[1)]

M. Huberman * *and A.W. Overhauser* *

* Department of Physics, Purdue University, 525, Northwestern Avenue, West Lafayette, Indiana 47907, USA

Received 1 May 1981

The discovery of open-orbit magnetoresistance peaks in potassium shows that its Fermi surface is multiply connected. A chargedensity-wave structure (which would have 24 domain orientations and five open-orbit directions per domain) explains the main features. The magnetic field at which open-orbit peaks appear depends on domain size, which we find to be ~ 0.1 mm.

Recently Coulter and Datars [1] discovered open-orbit magnetoresistance peaks in potassium. Their data on two crystals obtained with the induced-torque method, are reproduced in Figure 30.1. In this paper we show that a charge-density-wave (CDW) structure [2, 3] explains the large number of open-orbit peaks, their field dependence, their width, and variations in the data from run to run.

More than twenty anomalous properties reported during the last eighteen years, which are inconsistent with the prevalent view that potassium has a spherical Fermi surface, have been explained quantitatively or qualitatively by the CDW model. (See [3] for highlights.) Therefore the authors make no apology for describing as factual the properties of a CDW derived from extensive theoretical and experimental research. Nevertheless, any reader who feels that the data shown in Figure 30.1 (which are but two examples of approximately 100 runs on twelve specimens [1]) can be reconciled with a horizontal line (demanded by a spherical Fermi surface) is free to substitute the subjunctive mood in what follows.

The CDW wave vector $\vec{Q}$ in potassium has a magnitude [4, 5] $Q = 1.33(2\pi/a)$, 8% larger than the Fermi-surface diameter. Its direction is near a [110] axis [6] but, because of elastic anisotropy, $\vec{Q}$ is tilted 4.1° away and lies in a plane oriented 65.4° from the cubic (001) plane [7]. Accordingly there are 24 symmetry-related, preferred axes for $\vec{Q}$; so any single crystal will likely be divided into $\vec{Q}$ domains, each having its $\vec{Q}$ along one of these 24 axes. The resulting domain structure, which depends on metallurgical history, is responsible for the nonreproducibility of conductivity data from one run to the next, if the temperature has suffered a large excursion [3].

The open orbits are created by three pairs of energy gaps, shown in relation to the Fermi surface in Figure 30.2. The main gaps, associated with the CDW periodic potential, cause the 0.6-eV Mayer–El Naby optical anomaly [4, 5] and distort the Fermi surface nearby to form small necks or points of critical contact. The heterodyne gaps, [8] shown in Figure 30.2, arise from a sinusoidal displacement of the positive ions. The wave vector $\vec{Q} - \vec{G}_{110}$ of this static displacement lies within the Brillouin zone, unlike $\vec{Q}$ which is much larger. Finally, a third pair of energy gaps arises from a phason instability [9] and they are also shown. Other periodicities occur, but we believe their energy gaps to be negligible.

Five open-orbit directions can be identified in Figure 30.2. For example, if there is no magnetic breakdown [10] an electron traveling in k space on a cyclotron orbit from B to C

1) Phys. Rev. Lett. 47, 9 (1981)

Anomalous Effects in Simple Metals. Albert Overhauser

ISBN: 978-3-527-40859-7

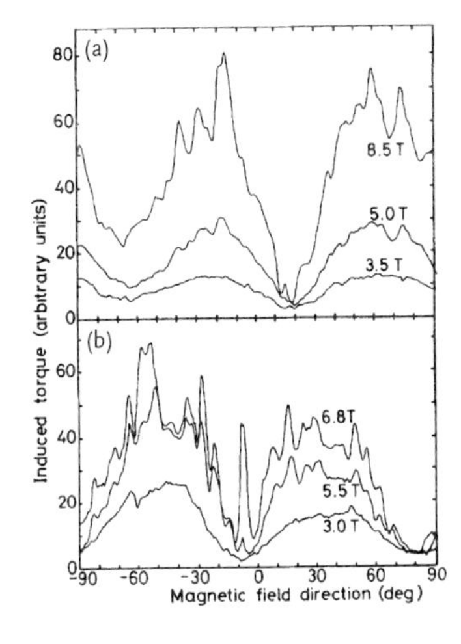

Figure 30.1 Magnetoresistance of potassium vs. $\vec{H}$ (Coulter and Datars, [1]). (a) Single crystal grown in oil, $\vec{H}$ in a (211) plane. (b) Single crystal grown in a Kel-F mold, $\vec{H}$ in a (321) plane.

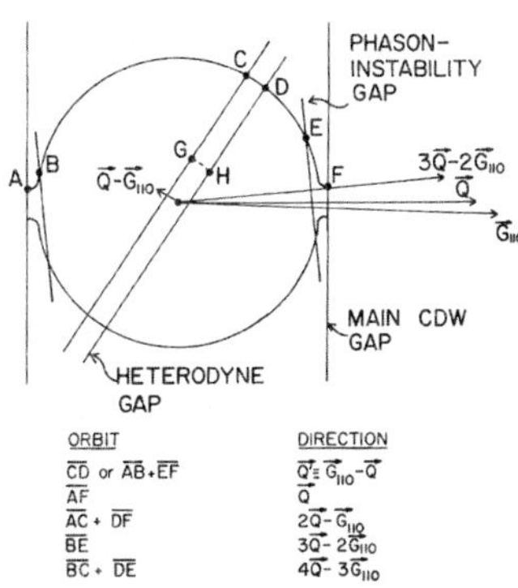

Figure 30.2 Fermi surface, energy gaps, and open orbits (for $\vec{H}$ perpendicular to the plane shown).

will be Bragg reflected to D, and will continue to E, where it will be Bragg reflected back to B. The open-orbit direction is $4\vec{Q} - 3\vec{G}_{110}$. However, if the heterodyne gaps at C and D have suffered magnetic breakdown, the electron will travel continuously from B to E where it will be reflected back to B. The open-orbit direction is then $3\vec{Q} - 2\vec{G}_{110}$. A nonequatorial open-orbit such as G to H (with reflection back to G) can occur even though the same heterodyne gaps have undergone magnetic breakdown on the equatorial orbit at C and D.

It is well known [11, 12] that if a metal has only closed orbits, there can be no high-field magnetoresistance A nonsaturating magnetoresistance, proportional to $\vec{H}^2$ can occur whenever $\vec{H}$ is perpendicular to an open-orbit direction. This effect disappears if $\vec{H}$ is rotated slightly into a nonperpendicular orientation. Sharp magnetoresistance peaks, observed by rotating $\vec{H}$ (or the crystal), are the signature of open orbits. They provide the method for measuring open-orbit directions. The induced-torque technique [1] requiring no electrical contacts, has been so utilized for more than a decade.

Obviously a theory for the angular dependence of the magnetoresistivity of a single crystal containing $24\vec{Q}$ domains, having therefore 120 open-orbit directions, must employ simplifying approximations. We model the magnetoconductivity of a single domain as if it were caused by six Fermi-surface fractions – the dominant one a spherical surface containing $1 - 5\eta \approx 90\%$ of the conduction electrons. The other five are cylindrical surfaces, each having an axis parallel to one of the open-orbit directions shown in Figure 30.2 and containing $\eta \approx 2\%$ of the electrons. The conductivity of a macroscopic single crystal (but broken up into 24$\vec{Q}$-domain varieties) is calculated by means of the effective-medium approximation [13–15].

Accordingly the conductivity tensor for each type, n, of $\vec{Q}$ domain is given by

$$\overleftrightarrow{\sigma}_n = (1 - 5\eta)\overleftrightarrow{\sigma}_s + \sum_{j=1}^{5} \overleftrightarrow{\sigma}_{cnj} \,. \tag{30.1}$$

$\overleftrightarrow{\sigma}_s$ is the magnetoconductivity tensor of a spherical Fermi surface and $\overleftrightarrow{\sigma}_{cnj}$ is the magnetoconductivity tensor of a cylindrical Fermi surface having an axis parallel to the jth open-orbit direction of $\vec{Q}$ domain n [16]. The effective conductivity tensor of the heterogeneous medium is

$$\overleftrightarrow{\sigma}_{\text{eff}} = \overleftrightarrow{\sigma}_{\text{ext}} + \sum_{n=1}^{24} f_n(\overleftrightarrow{\sigma}_n - \overleftrightarrow{\sigma}_{\text{ext}}) \cdot \overleftrightarrow{\beta}_n \,, \tag{30.2}$$

where f_n is the volume fraction of $\vec{Q}$ domain n, and β_n is an explicit tensor, depending on $\overleftrightarrow{\sigma}_n$ and $\overleftrightarrow{\sigma}_{\text{ext}}$ which we have evaluated [16]. The effective-medium approximation is obtained by requiring the conductivity $\overleftrightarrow{\sigma}_{\text{ext}}$ of the "host", surrounding each domain, to be just $\overleftrightarrow{\sigma}_{\text{eff}}$. This self-consistent tensor,

$$\overleftrightarrow{\sigma}_{\text{eff}} = \overleftrightarrow{\sigma}_{\text{ext}} \tag{30.3}$$

can be found quickly by iteration of Equation (30.2).

We took the $\vec{Q}$-domain distribution to be either random $\left(f_n = \frac{1}{24}\right)$ or textured:

$$f_n = \frac{1}{24}\left\{1 + \alpha\left[\frac{3}{2}\left(\hat{Q}\cdot\hat{T}\right)^2 - \frac{1}{2}\right]\right\} \tag{30.4}$$

where $\hat{T}$ is the unit vector of a texture axis. Finally, the electron scattering time τ for the spherical Fermi surface was taken to be 1.5×10^{-10} s, appropriate to potassium at 4 °K if the residual-resistivity ratio is ~4000. Then $H = 2\,\text{T}$ corresponds to $\omega_c\tau = 50$.

Since an open orbit is destroyed when it crosses a domain boundary, the relaxation time τ_{open} used in calculating $\overleftrightarrow{\sigma}_{cnj}$ is shorter than τ, and is given approximately by

$$\frac{1}{\tau_{\text{open}}} = \frac{1}{\tau} + \frac{8v_F}{3D} \,. \tag{30.5}$$

v_F is the Fermi velocity and D is the diameter of a $\vec{Q}$ domain.

The magnetoresistivity ρ_{xx} for $\vec{H}$ in a plane perpendicular to a [211] or a [321] axis was calculated according to the above theory and is shown in Figure 30.3. Induced-torque curves were also calculated [17] and, except for the vertical scale, were found to be indistinguishable from the magnetoresistance curves shown. The magnetic field strength at which open-orbit peaks emerge depends somewhat on domain size; $D = 0.05$ and 0.12 mm, respectively, in Figures 30.3(a) and 30.3(b). A texture parameter $a = -1$, with an axis at $\theta = 0$, was used in Figure 30.3(a); random orientation, $a = 0$, was used in Figure 30.3(b).

The remarkable similarity between Figure 30.3 and the data of Figure 30.1 illustrates the ease with which (otherwise) incomprehensible data can be explained once one recognizes the CDW structure of potassium de Haas–van Alphen data, which are sensitive only to closed

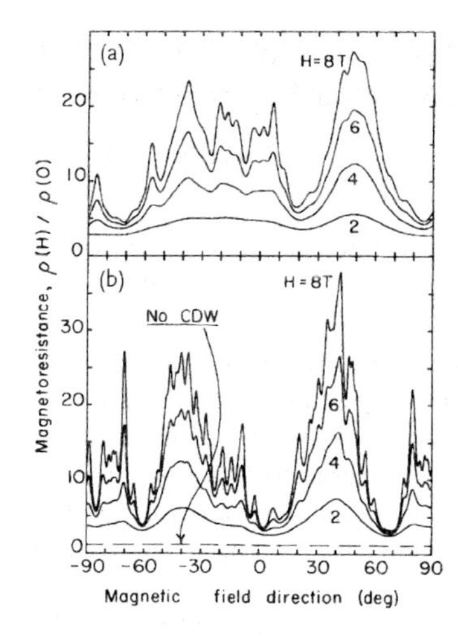

Figure 30.3 Theoretical magnetoresistance vs. $\vec{H}$. (a) $\vec{H}$ in (211) plane, $D = 0.05$ mm; texture: $\alpha = -1$, $\theta = 0$. (b) $\vec{H}$ in (321) plane, $D = 0.12$ mm; no texture ($\alpha = 0$).

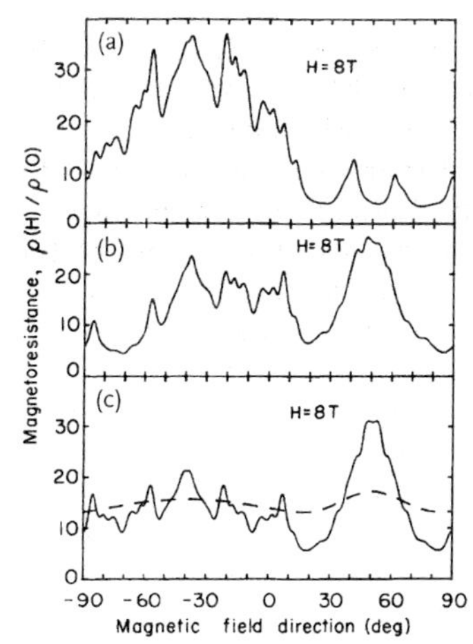

Figure 30.4 Theoretical magnetoresistance for $\vec{H}$ in a (211) plane, $D = 0.05$ mm. (a) $\alpha = 2$, $\theta = 60°$; (b) $\alpha = -1$, $\theta = 0$; (c) $\alpha = 0$. The dashed curve is for $D = 0.005$ mm.

orbits, can no longer be cited as evidence for a spherical Fermi surface. For then, the only allowed behavior of a magnetoresistance spectrum is the horizontal, dashed line shown in Figure 30.3(b).

Until $\vec{Q}$-domain structure can be experimentally controlled, it will not be possible to make detailed comparisons between theory and data. We illustrate this in Figure 30.4, where three spectra are shown for $\vec{H}$ in a (211) plane. Only the texture parameters have been changed. Finally, the dashed curve in Figure 30.4(c) shows how open-orbit spectra can be suppressed if metallurgical preparation has caused too small a domain size.

References

1 Coulter, P.G. and Datars, W.R. (1980) *Phys. Rev. Lett.* **45**, 1021, and private communication.

2 Overhauser, A.W. (1968) *Phys. Rev.* **167**, 691

3 For an experimental and theoretical review, see Overhauser, A.W. (1978) *Adv. Phys.* **27**, 343.

4 Overhauser, A.W. (1964) *Phys. Rev. Lett.* **13**, 190.

5 Overhauser, A.W. and Butler, N.R. (1976) *Phys. Rev. B* **14**, 3371.

6 Overhauser, A.W. (1971) *Phys. Rev. B* **3**, 3173.

7 Giuliani, G.F. and Overhauser, A.W. (1979) *Phys. Rev. B* **20**, 1328.

8 Reitz, J.R. and Overhauser, A.W. (1968) *Phys. Rev.* **171**, 749.

9 Giuliani, G.F. and Overhauser, A.W. (1981) *Bull. Am. Phys. Soc.* **26**, 471.

10 Blount, E.I. (1962) *Phys. Rev.* **126**, 1636.

11 Lifshitz, I.M. Azbel, M.Ia. and Kaganov, M.I. (1956) *Zh. Eksp. Teor. Fiz.* **31**, 63.

12 Lifshitz, I.M. Azbel, M.Ia. and Kaganov, M.I. (1957) *Sov. Phys. JETP* 4, 41.

13 Bruggeman, D.A.G. (1935) *Ann. Phys. (Leipzig)* **24**, 636.

14 Landauer, R. (1952) *J. Appl. Phys.* **23**, 779.

15 Stroud, D. and Pan, F.P. (1979) *Phys. Rev. B* **20**, 455.

16 Huberman, M. and Overhauser, A.W. (1981) *Phys. Rev. B* **23**, 6294.

17 Visscher, P.B. and Falicov, L.M. (1970) *Phys. Rev. B* **2**, 1518.

Reprint 31 Open-Orbit Magnetoresistance Spectra of Potassium[1)]

M. Huberman * *and A.W. Overhauser* *

* Department of Physics, Purdue University, 525, Northwestern Avenue, West Lafayette, Indiana 47907, USA

Received 25 September 1981

A theory of the openorbit magnetoresistance of potassium, discovered by Coulter and Datars, is developed The open orbits caused by the multiple periodicities of a chargedensity-wave (CDW) state are derived. Since a single crystal consists of domains, each having its CDW wave vector along one of 24 preferred axes, we employ effective-medium theory to calculate the magnetoresistance rotation pattern. The effects of domain texture and size are illustrated. The magnetic field at which open-orbit peaks appear depends on the domain size, which we find to be −0.1 mm. The rotation pattern at 24 T is calculated in order to exhibit the detailed information that becomes available at extremely high fields. High-field experiments would aid in the determination of open-orbit directions and allow magnetic-breakdown studies of the energy gaps.

31.1 Introduction

The discovery of open-orbit magnetoresistance peaks in potassium by Coulter and Datars [1] shows that its Fermi surface is multiply connected. A charge-density-wave (CDW) structure [2, 3] explains [4] the main features: the large number of open-orbit peaks, their field dependence, their width, and variations in the data from run to run.

The nonsaturating magnetoresistance of the alkali metals has been a long-standing puzzle [5, 6]. Both the transverse and longitudinal magnetoresistance of potassium increase linearly with magnetic field in fields as high as 10 T [7–10]. Fear that this was an artifact causal by probes led to the development of inductive techniques [10–12], which confirm the results of probe methods. Helicon-resonance experiments on intentionally deformed samples [11] make surface imperfections [13] an unlikely explanation. Voids can cause a linear magnetoresistance [14] but the volume fraction required is known to be too high [15, 16]. The linear magnetoresistance is observed in both poly-crystalline and single-crystal samples. It depends upon sample preparation and history, varying by almost 2 orders of magnitude for differently prepared specimens and by about 50% for nominally the same specimens [7, 8, 11].

The most revealing results have been obtained by the induced torque technique, a probeless method for measuring the magnetoresistance as a function of field orientation [17, 18]. In this experiment (Figure 31.1), a single-crystal sphere is suspended in ja uniform magnetic field $\vec{H}$. Slow rotation of $\vec{H}$ (or the crystal) about the suspension axis induces circulating currents in the sample. The magnetic moment of the induced current interacts with $\vec{H}$, producing a torque about the suspension axis. The torque depends on the induced current, which in turn depends on the magnetoconductivity of the sample in the field $\vec{H}$.

Early experiments by Schaefer and Marcus [19] on potassium single crystals in fields up to 3 T showed a nonsaturating torque, having a twofold anisotropy. The high-field torque had four broad peaks in a 360° rotation pattern. The twofold symmetry contradicted the supposed cubic symmetry, implying a preferred direction in the crystal.

1) Phys. Rev. B 25, 4 (1982)

Anomalous Effects in Simple Metals. Albert Overhauser

ISBN: 978-3-527-40859-7

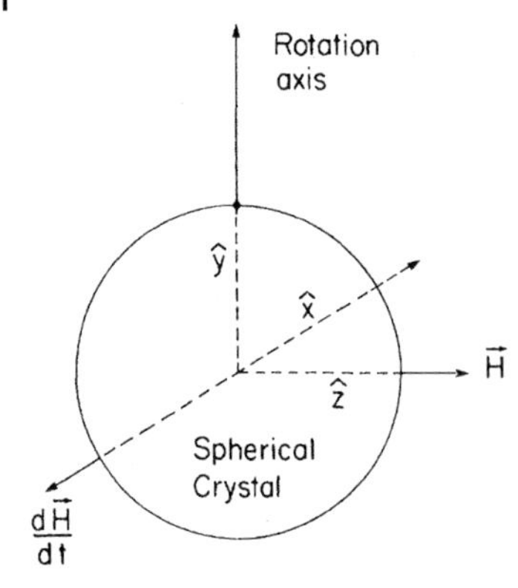

Figure 31.1 Induced-torque experiment. Rotation of $\vec{H}$ induces currents, which interact with $\vec{H}$, producing a torque about the suspension axis.

Holroyd and Datars [20] showed that, although the torque was always nonsaturating, its anisotropy depended on the sample preparation and treatment. The torque was isotropic when all surface oil on the sample was removed, but had a four-peak, twofold anisotropy when there was a nonuniform coating of oil (oil-drop effect). The anisotropy cannot be blamed on sample shape or strain since ellipsoidal and spherical samples had the same anisotropy, and intentionally strained samples showed no significant anisotropy. Voids or surface scratches cannot be the explanation, since deep cylindrical holes drilled into the sample had no significant effect.

In experiments at higher magnetic fields, Coulter and Datars [1] discovered narrow, nonsaturating torque peaks, appearing above about 5 T. The sharp structure was observed in approximately 100 runs on 12 specimens, when the torque at lower fields was either isotropic or anisotropic. The data for two crystals, about 4 mm in diameter, one grown in oil and the other in a mold, at a temperature 1.4 K are reproduced in Figure 31.2. The magnetic field is in a (211) plane and a (321) plane, respectively, for Figures 31.2(a) and 31.2(b).

Semiclassical transport theorems [21, 22], which are experimentally verified for many metals, predict that if a metal has only closed orbits, the high-field magnetoresistance saturates. For example, in copper crystals which are oriented in a magnetic field so that all orbits are closed, the predicted saturation of the magnetoresistance is observed [23]. A nonsaturating transverse magnetoresistance, proportional to H^2, can occur whenever $\vec{H}$ is perpendicular to an open-orbit direction. This effect disappears if $\vec{H}$ is rotated slightly into a nonperpendicular orientation. Sharp, nonsaturating magnetoresistance peaks, observed by rotating $\vec{H}$ (or the crystal), are the signature of open orbits.

An exact solution for the induced torque in terms of the components of the magnetoresistivity tensor has been derived by Visscher and Falicov [24]. Assuming the high-field behavior predicted by semiclassical transport theorems [21, 22], the torque is approximately proportional to the magnetoresistance (ρ_{xx}) in the direction perpendicular to $\vec{H}$ and the rotation axis [24]. The induced-torque technique is thus an open-orbit detector, yielding sharp, nonsaturating torque peaks whenever $\vec{H}$ is perpendicular to an open-orbit direction [17, 18].

Potassium has a body-centered cubic lattice and is monovalent. The first Brillouin zone is only halffilled. The electrons at the Fermi surface are far from the zone boundary and their energy is only weakly perturbed by the lattice potential. The Fermi surface is expected to be nearly spherical and certainly simply connected. The occurrence of open orbits requires an additional periodic potential (besides the lattice potential), producing energy gaps at the Fermi surface.

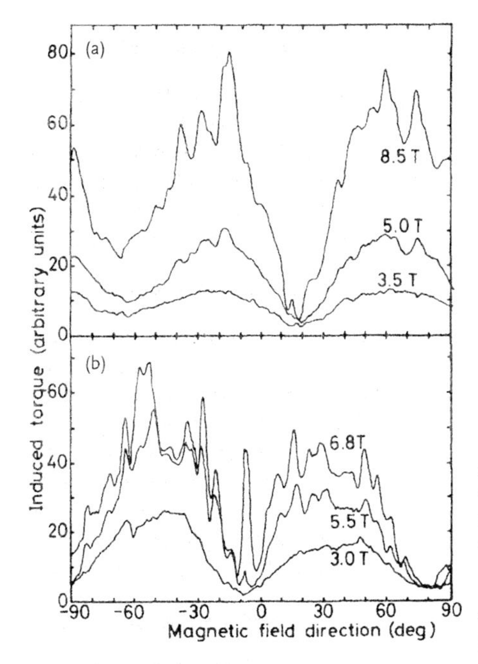

Figure 31.2 Magnetoresistance of potassium versus $\vec{H}$ (Coulter and Datars [1]). (a) Single crystal grown in oil. $\vec{H}$ in the (211) plane; its direction in the plane of rotation is not known. (b) Single crystal grown in a *Kel-F* mold. $\vec{H}$ in the $(\bar{2}\bar{3}1)$ plane; its direction at $\Theta = -90°$ is [128] and its sense of rotation is clockwise about the $[\bar{2}\bar{3}1]$ axis.

In this paper we present a theory of the open-orbit magnetoresistance of potassium, shown in Figure 31.2. We first review the theory of open-orbit magnetoresistance. The open-orbit directions in potassium, caused by a CDW, are then derived. The magnetoresistance of a single crystal, having a domain structure, is then calculated We conclude with a discussion of directions for future research,

31.2 Open-Orbit Magnetoresistance

We illustrate the theory of open-orbit magnetoresistance with a simple model: a Fermi surface consisting of two separate surfaces, a sphere and a cylinder [25]. This model has the virtue that its exact conductivity tensor is derivable.

In a magnetic field $\vec{H} = H\hat{z}$, the conductivity tensor $\overleftrightarrow{\sigma}_s$ of a spherical Fermi surface is

$$\sigma_s = \sigma \begin{pmatrix} \frac{1}{1+(\omega_c\tau)^2} & \frac{-\omega_c\tau}{1+(\omega_c\tau)^2} & 0 \\ \frac{\omega_c\tau}{1+(\omega_c\tau)^2} & \frac{1}{1+(\omega_c\tau)^2} & 0 \\ 0 & 0 & 1 \end{pmatrix} . \tag{31.1}$$

$\sigma = ne^2\tau/m$ is the zero-field conductivity, $\omega_c = eH/mc$ is the cyclotron frequency, and τ is the electron relaxation time.

The conductivity tensor $\overleftrightarrow{\sigma}_c$ of a cylindrical Fermi surface, having an axis $\hat{w}$, is most conveniently expressed in the uvw coordinate system, shown in Figure 31.3, defined by $\hat{u} = \hat{w}x\hat{y}/\sin\phi$ and $\hat{v} = \hat{w}x\hat{u}$. ϕ is the angle between $\hat{w}$ and $\hat{y}$ and θ is the angle between the projection of $\hat{w}$ in the xz plane and $\hat{z}$. In the uvw frame,

$$\sigma_c' = \sigma \begin{pmatrix} \frac{1}{1+\delta^2} & \frac{-\delta}{1+\delta^2} & 0 \\ \frac{\delta}{1+\delta^2} & \frac{1}{1+\delta^2} & 0 \\ 0 & 0 & 0 \end{pmatrix} \tag{31.2}$$

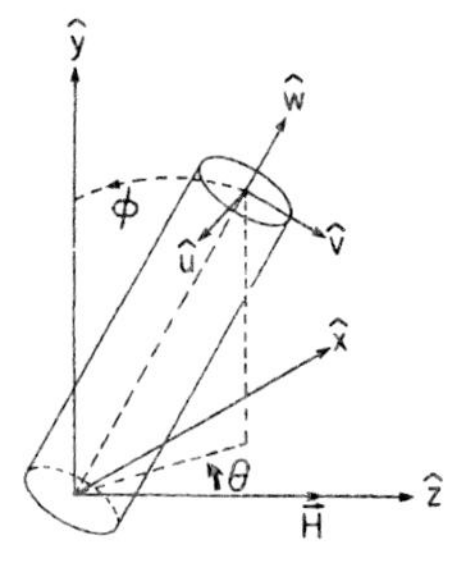

Figure 31.3 Relative orientation of the uvw and xyz axes. $\hat{w}$ is the open-orbit direction and is parallel to $\hat{z}$.

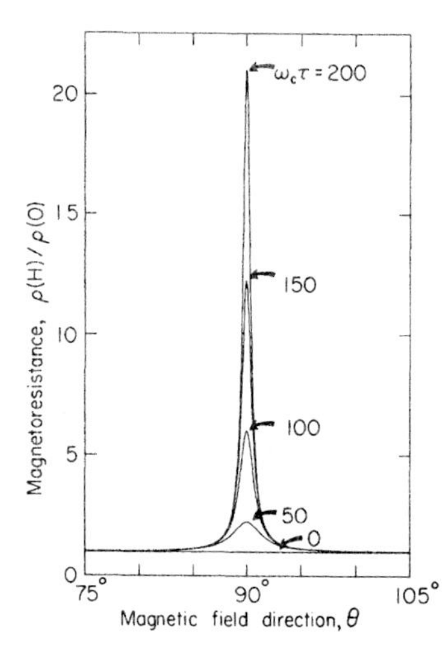

Figure 31.4 Open-orbit magnetoresistance. Sharp peaks occur when $\vec{H}$ is perpendicular to the open-orbit direction $\hat{w}$. Open-orbit electron fraction $\eta = 0.001$.

where $\delta \equiv \omega_c\tau \sin\phi \cos\theta$. Transforming into the xyz frame,

$$\sigma_c = \tilde{S}\sigma_c' S\,, \tag{31.3}$$

where the orthogonal transformation matrix is

$$S \equiv \begin{pmatrix} -\cos\theta & 0 & \sin\theta \\ \cos\phi\sin\theta & -\sin\phi & \cos\phi\cos\theta \\ \sin\phi\sin\theta & \cos\phi & \sin\phi\cos\theta \end{pmatrix} \tag{31.4}$$

Letting η be the electron fraction of the cylindrical Fermi surface, the total conductivity of a sphere and cylinder is

$$\overleftrightarrow{\sigma} = (1-\eta)\overleftrightarrow{\sigma}_s + \eta\overleftrightarrow{\sigma}_c\,. \tag{31.5}$$

The zero-field resistivity is approximately the same as for a spherical Fermi surface, but, in a magnetic field, the open-orbit electrons affect the resistivity dramatically.

Figure 31.4 shows the transverse magnetoresistance ρ_{xx} as a function of field strength and orientation. The open-orbit electron fraction is $\eta = 0.001$ and the angle between the open-orbit direction $\hat{w}$ and the rotation axis $\hat{y}$ is $\phi = 45°$. Sharp, nonsaturating peaks, proportional to H^2 occur when $\vec{H}$ is perpendicular to $\hat{w}$. It can be shown that the peak height is

$$\Delta\rho_{xx} = [\eta/(1-\eta)]\rho(\omega_c\tau)^2\sin^2\phi\,, \tag{31.6}$$

where $\Delta\rho_{xx} \equiv \rho_{xx}(H) - \rho_{xx}(0)$ at $\theta = 90°$ and $\rho \equiv 1/\sigma$. The peak fullwidth at half maximum, inversely proportional to $\vec{H}$, is

$$\Delta\theta = 2/(\omega_c\tau \sin\phi) . \tag{31.7}$$

It can be shown that the Hall coefficient has the free-electron value, $R_H = -1/nec$. Open orbits have a negligible effect on the longitudinal magnetoresistance, which saturates when $\omega_c\tau > 1$.

31.3 Open-Orbit Directions

In a CDW state the conduction electron density has a sinusoidal modulation due to the effects of electron-electron interactions. The CDW wave vector $\vec{Q}$ in potassium has a magnitude [26, 27] $Q = 1.33(27\pi/a)$, 8% larger than the Fermi surface diameter. Its theoretically predicted direction is near a [11] axis [28], but because of elastic anisotropy, is tilted $\sim 4.1°$ away and lies in a plane oriented 65.4° away from the (001) plane [29].

Due to the underlying cubic symmetry, there are 24 symmetry-related preferred axes for $\vec{Q}$. Any single crystal will likely be divided into $\vec{Q}$ domains, each having its $\vec{Q}$ along one of these 24 axes. The domain distribution and size depend upon the metallurgical history, being affected, for example, by sample shape and size, surface orientation and preparation, annealing time, and thermal stress. A directed stress due to a nonuniform oil drop or oxide spot can produce a preferred domain orientation. At temperatures above 15 K, where potassium begins its mechanical recovery [30], domain growth can occur in order to relieve internal strains. The uncontrolled domain structure explains the variability of the open-orbit magnetoresistance for samples prepared by different methods and for the same sample after warming to liquid-nitrogen temperature.

The open orbits are created by three pairs of energy gaps, shown in relation to the Fermi surface in Figure 31.5 [31]. In a CDW state, the conduction electron density is

$$\rho(\vec{r}) = \rho_0(1 - p\cos\vec{Q}\cdot\vec{r}) . \tag{31.8}$$

ρ_0 is the average electron density; p and $\vec{Q}$ are the fractional modulation and wave vector of the CDW. The main gaps, which cause the 0.6 eV Mayer–El Naby optical-absorption anomaly [26, 27], arise from the exchange and correlation potential

$$V(\vec{r}) = G\cos\vec{Q}\cdot\vec{r} \tag{31.9}$$

of the CDW. Since the CDW wave vector $\vec{Q}$ is only slightly larger than the Fermi-surface diameter, these gaps distort the Fermi surface nearby, forming small necks or points of critical contact.

The heterodyne gaps [32], shown in Figure 31.5, arise from a sinusoidal displacement of the positive ions, which occurs in order to maintain charge neutrality. The new ionic positions $\vec{R}_{\vec{L}}$ are related to the old positions $\vec{L}$ by

$$\vec{R}_{\vec{L}} = \vec{L} + \vec{A}\sin\vec{Q}\cdot\vec{L} . \tag{31.10}$$

$\vec{A}$ is the amplitude of the lattice distortion. It's magnitude is 0.03 Å and its direction is parallel to the vector (18.5,54.0,53.1) when $\vec{Q} = (2\pi/a)(0.966, 0.910, 0.0865)$ [33]. The ionic charge density is thus

$$\rho(\vec{r}) = \sum_{\vec{L}} \rho\left(\vec{r} - \vec{L} - \vec{A}\sin\vec{Q}\cdot\vec{L}\right) , \tag{31.11}$$

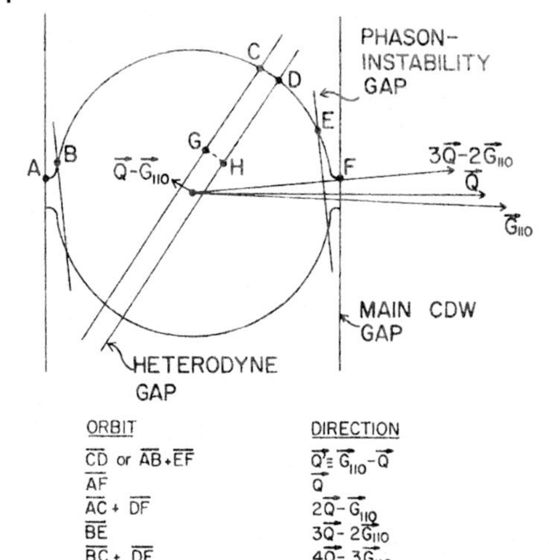

Figure 31.5 Fermi surface, energy gaps, and open orbits of potassium (for $\vec{H}$ perpendicular to the plane shown).

where $\bar{\rho}(\vec{r})$ is the charge density of a single ion at the origin and the sum is over all lattice sites $\vec{L}$.

The Fourier transform of $\rho(\vec{r})$ is

$$\rho_{\vec{q}} = \bar{\rho}(\vec{q}) \sum_{\vec{L}} \exp\left[-i\vec{q}\cdot\left(\vec{L} + \vec{A}\sin\vec{Q}\cdot\vec{L}\right)\right], \tag{31.12}$$

where $\bar{\rho}(\vec{q})$ is the ionic form factor. Expanding to first order in the amplitude $\vec{A}$, we obtain

$$\rho_{\vec{q}} = \bar{\rho}(\vec{q}) \sum_{\vec{L}} e^{-i\vec{q}\cdot\vec{L}}\left(1 - i\vec{q}\cdot\vec{A}\sin\vec{Q}\cdot\vec{L}\right) \tag{31.13}$$

which is nonzero only if $\vec{q} = \vec{G}$ or $\vec{G} \pm \vec{Q}$, where $\vec{G}$ is any reciprocal lattice vector. Of the new periodicities $\vec{G} \pm \vec{Q}$, only $\vec{G}_{110} - \vec{Q}$ is smaller than the Fermi-surface diameter. Its energy gaps intersect the Fermi surface whereas those due to the others are not expected to do so.

The phason-instability gaps, also shown in Figure 31.5, arise from a static phason instability [34], having wave vector $2\vec{Q}$ which lowers the total energy. The ionic positions $\vec{R}_{\vec{L}}$ are now given by

$$\vec{R}_{\vec{L}} = \vec{L} + \vec{A}\sin\left(\vec{Q}\cdot\vec{L} + \beta\sin 2\vec{Q}\cdot\vec{L}\right), \tag{31.14}$$

where β is the amplitude of the static phason. Taking the Fourier transform of the ionic charge density and expanding to first order in $\vec{A}$ and β yields

$$\vec{\rho}_{\vec{q}} = \bar{\rho}(\vec{q}) \sum_{\vec{L}} e^{-i\vec{q}\cdot\vec{L}}\left[1 - i\vec{q}\cdot\vec{A}\left(\sin\vec{Q}\cdot\vec{L} + \beta\cos\vec{Q}\cdot\vec{L}\sin 2\vec{Q}\cdot\vec{L}\right)\right], \tag{31.15}$$

which is nonzero only if $\vec{q} = \vec{G}$, $\vec{G} \pm \vec{Q}$, or $\vec{G} \pm 3\vec{Q}$, where $\vec{G}$ is any reciprocal lattice vector. Of the new periodicities $\vec{G} \pm 3\vec{Q}$, four are smaller than the Fermi-surface diameter, producing energy gaps at the Fermi surface. Of these four, $3\vec{Q} - 2\vec{G}_{110}$ is most longitudinal, i.e., has the largest component parallel to $\vec{A}$. Since $\rho_{\vec{q}}$ is proportional to $\vec{q}\cdot\vec{A}$, it is expected to have the largest energy gap.

Besides these, other energy gaps, which are higher order in the amplitudes $\vec{A}$ and β occur. Except for the main gap, the sizes of these energy gaps are not known. In order to illustrate the theory, we include only the three sets of gaps shown in Figure 31.5.

The main gap, which is 0.6 eV, is too large to be broken down by the magnetic field, but the others, which are smaller, may be. (The breakdown field for the main gap is about 1000 T.) The probability of an energy gap being broken down by a magnetic field $\vec{H}$ is [35, 36]

$$P = \exp\left(\frac{-\pi mcE_g^2}{2\hbar^2 e\left|\vec{K}\cdot\left(\vec{v}\times\vec{H}\right)\right|}\right) \equiv \exp\left(\frac{-H_0}{H}\right). \tag{31.16}$$

$\vec{K}$ is the wave vector of the periodic potential producing the energy gap E_g and $\vec{v}$ is the velocity that an electron would have at the energy gap if it were free. Taking $\vec{H}$ parallel to $\hat{z}$ and $\vec{K}$ parallel to $\hat{x}$, Equation (31.16) reduces to

$$P = \exp\left(\frac{-\tau E_g^2}{4\hbar\omega_c m\left|v_x v_y\right|}\right). \tag{31.17}$$

Since the breakdown field H_0 depends on the electron velocity, an energy gap may simultaneously be broken down by some orbits on the Fermi surface and not by others.

Five open-orbit directions can be identified in Figure 31.5. For example, if there is no magnetic breakdown, an electron traveling in k space on a cyclotron orbit from B to C will be Bragg reflected to D, will continue to E where it will be Bragg reflected back to B. The open-orbit direction is $4\vec{Q} - 3\vec{G}_{110}$. However, if the heterodyne gaps at C and D have suffered magnetic breakdown, the electron will travel continuously from B to E where it will be reflected back to B. The open-orbit direction is then $3\vec{Q} - 2\vec{G}_{110}$. A nonequatorial open orbit such as G to H (with reflection back to G) can occur even though the same heterodyne gaps have undergone magnetic breakdown on the equatorial orbit at C and D. The breakdown field of the orbit G to H shown projected into the plane of Figure 31.5, is greater than that of the orbit C to D, since $v_x v_y$ is less.

Since the magnitudes of the heterodyne and phason-instability gaps are not known, neither are the breakdown fields. Again, in order to illustrate the theory, we include all of the five possible open-orbit directions, indicated in Figure 31.5.

31.4 Open-Orbit Magnetoresistance of Potassium

Obviously a theory for the angular dependence of the magnetoresistivity of a potassium crystal containing 24 types of $\vec{Q}$ domains, having therefore 120 open-orbit directions, must employ simplifying assumptions. We model the magnetoconductivity of a single domain as if it were caused by six Fermi-surface fractions, a sphere and five cylinders. The sphere contains $(1 - 5\eta) \approx 90\%$ of the conduction electrons. The cylinders, whose axes are parallel to the open-orbit directions in Figure 31.5, each contain $\eta \approx 2\%$ of the conduction electrons.

Since the magnetic breakdown probability, given by Equation (31.16), depends on the angle between $\vec{H}$ and the periodicity $\vec{K}$ producing the energy gaps, the open-orbit electron fraction η is different for each of the 120 open-orbit directions. Moreover it is field dependent, either increasing or decreasing as H increases. For example, as H increases, the electron fraction having the open-orbit direction $\vec{Q}$ will increase whereas that having the open-orbit direction

$\vec{Q}' \equiv \vec{G}_{110} - \vec{Q}$ will decrease. In the model, 120 field-dependent values are thus replaced by one constant.

Since the energy gaps and breakdown fields are unknown, η cannot be estimated from the predicted Fermi surface (Figure 31.5). We shall show, however, that the overall scale of the open-orbit magnetoresistance depends on η but the structure is insensitive to its value. Lacking absolute measurements of the induced torque, we assume the value $\eta \approx 2\%$.

Accordingly, the conductivity tensor for each type n of $\vec{Q}$ domain is given by

$$\overleftrightarrow{\sigma}_n = (1 - 5\eta)\overleftrightarrow{\sigma}_s + \eta \sum_{j=1}^{5} \overleftrightarrow{\sigma}_{cnj} . \tag{31.18}$$

$\overleftrightarrow{\sigma}_s$ is the magnetoconductivity tensor of a spherical Fermi surface, given by Equation (31.1), and $\overleftrightarrow{\sigma}_{cnj}$ is the magnetoconductivity tensor of a cylindrical Fermi surface having an axis parallel to the jth open-orbit direction of $\vec{Q}$-domain n, given by Equations (31.2)–(31.4).

The conductivity tensor of a macroscopic single crystal, broken up into 24 $\vec{Q}$-domain varieties, is calculated by means of the effective medium approximation [37–39]. This is a mean-field type of approximation, which is expected to be valid when the domain size is greater than the electron mean free path but smaller than the sample dimensions. The effective conductivity tensor of the heterogeneous medium is [39]

$$\overleftrightarrow{\sigma}_{\text{eff}} = \overleftrightarrow{\sigma}_{\text{ext}} + \left(\sum_{n=1}^{24} f_n \left(\overleftrightarrow{\sigma}_n - \overleftrightarrow{\sigma}_{\text{ext}} \cdot \overleftrightarrow{\alpha}_n\right)\right) \cdot \left(\sum_{n=1}^{24} f_n \overleftrightarrow{\alpha}_n\right)^{-1} . \tag{31.19}$$

f_n and $\overleftrightarrow{\sigma}_n$ are the volume fraction and conductivity of $\vec{Q}$-domain n. $\overleftrightarrow{\sigma}_{\text{ext}}$ is the conductivity of the "host" surrounding each domain. $\overleftrightarrow{\alpha}_n$ is a tensor depending on $\overleftrightarrow{\sigma}_n$ and $\overleftrightarrow{\sigma}_{\text{ext}}$, given by [40]

$$\overleftrightarrow{\alpha}_n = \left[\overleftrightarrow{1} - \overleftrightarrow{\Gamma}_{\text{ext}} \cdot \left(\overleftrightarrow{\sigma}_n - \overleftrightarrow{\sigma}_{\text{ext}}\right)\right]^{-1} , \tag{31.20}$$

where the symmetric tensor $\overleftrightarrow{\Gamma}_{\text{ext}}$ depends only on the symmetric part $\overleftrightarrow{\sigma}_{\text{ext}}^{(s)}$ of $\overleftrightarrow{\sigma}_{\text{ext}}$ having the same principal axes. The eigenvalues Γ_i of $\overleftrightarrow{\Gamma}_{\text{ext}}$, which satisfy the equation

$$\sum_{i=1}^{3} \lambda_i \Gamma_i = -1 , \tag{31.21}$$

are functions of the eigenvalues λ_i of $\overleftrightarrow{\sigma}_{\text{ext}}^{(s)}$ [41]. If $\lambda_1 < \lambda_2 < \lambda_3$ then

$$\Gamma_1 = -\frac{1}{\lambda_1} \frac{1}{(\lambda_1\lambda_2\lambda_3)^{1/2}} \frac{1}{(1/\lambda_1 - 1/\lambda_3)^{1/2}(1/\lambda_1 - 1/\lambda_2)} (F - E) ,$$

$$\Gamma_3 = -\frac{1}{\lambda_3} \frac{1}{(\lambda_1\lambda_2\lambda_3)^{1/2}} \frac{1}{(1/\lambda_1 - 1/\lambda_3)^{1/2}(1/\lambda_2 - 1/\lambda_3)} \times \left[(\lambda_3\lambda_1/\lambda_2)^{1/2} (1/\lambda_1 - 1/\lambda_3)^{1/2} - E\right] .$$

F and E are elliptic integrals of the first and second kinds, defined by

$$F(\delta, k) = \int_0^\delta (1 - k^2 \sin^2 \phi)^{-1/2} d\phi ,$$

$$E(\delta, k) = \int_0^\delta (1 - k^2 \sin^2 \phi)^{1/2} d\phi ,$$

having amplitude and modulus

$$\delta = \sin^{-1}(1 - \lambda_1/\lambda_3)^{1/2} ,$$

$$k = \frac{(1/\lambda_1 - 1/\lambda_2)^{1/2}}{(1/\lambda_1 - 1/\lambda_3)^{1/2}} .$$

If $\lambda_1 = \lambda_2 < \lambda_3$, then $\Gamma_1 = \Gamma_2$ and

$$\Gamma_3 = -\frac{1}{\lambda_3 - \lambda_1} \times \left(1 - \frac{(\lambda_1/\lambda_3)^{1/2} \sin^{-1}(1 - \lambda_1/\lambda_3)^{1/2}}{(1 - \lambda_1/\lambda_3)^{1/2}}\right) .$$

If $\lambda_1 < \lambda_2 = \lambda_3$ then $\Gamma_2 = \Gamma_3$ and

$$\Gamma_1 = \frac{1}{\lambda_3 - \lambda_1} \times \left(1 - \frac{(\lambda_3/\lambda_1)^{1/2} \sinh^{-1}(\lambda_3/\lambda_1 - 1)^{1/2}}{(\lambda_3/\lambda_1 - 1)^{1/2}}\right) .$$

If $\lambda_1 = \lambda_2 = \lambda_3$, then $\Gamma_1 = \Gamma_2 = \Gamma_3 = -\frac{1}{3}\lambda_3$.

The effective-medium approximation is obtained by requiring the conductivity $\overleftrightarrow{\sigma}_{\text{ext}}$ of the host, surrounding each domain, to be just $\overleftrightarrow{\sigma}_{\text{eff}}$. This selfconsistent tensor,

$$\overleftrightarrow{\sigma}_{\text{eff}} = \overleftrightarrow{\sigma}_{\text{ext}} , \tag{31.22}$$

can be found quickly by iteration of Equation (31.19). Starting from a free-electron conductivity tensor, the solution typically converged after five iterations to within 1%.

The 24 parameters f_n which specify the $\vec{Q}$-domain distribution are unknown. We take the $\vec{Q}$-domain distribution to be either random ($f_n = \frac{1}{24}$) or textured:

$$f_n = \frac{1}{24}\left\{1 + \alpha\left[\frac{3}{2}\left(\hat{Q}\cdot\hat{T}\right)^2 - \frac{1}{2}\right]\right\} , \tag{31.23}$$

where $\hat{T}$ is the unit vector of the texture axis and α is a texture parameter lying between -1 and +2. $\alpha = 0$ corresponds to no texture, $\alpha > 0$ to a prolate texture, and $\alpha < 0$ to an oblate texture.

The twofold anisotropy of the induced torque observed at lower magnetic fields, which contradicts the supposed cubic symmetry, shows that the $\vec{Q}$-domain distribution may have a preferred direction. Equation (31.23) is one example of such a distribution. In the model the 24 parameters f_n are thus replaced by three parameters, the texture axis $\hat{T}$ and texture parameter α. We shall show, however, that the main qualitative features of the open-orbit magnetoresistance do not depend on the assumed texture.

The electron scattering time τ at 1.4 K, which is primarily due to impurity scattering, is determined from the zero-field resistivity ρ. In zero magnetic field, the resistivity is approximately the same as for a spherical Fermi surface,

$$\rho = \frac{m}{ne^2\tau} , \tag{31.24}$$

since the cylindrical surfaces contain only a small fraction, $5\eta = 10\%$, of the electrons. The zero-field resistivity of the induced-torque samples was not measured, but the residual resistivity of pure potassium is typically about 4000 times smaller than the room-temperature resistivity $\rho = 7\,\mu\Omega\,\text{cm}$, implying $\tau \approx 1.5\times10^{-10}$ s at 1.4 K. Assuming this value, $\omega_c\tau = 50$ at $H = 2\,\text{T}$.

The relaxation time r applies to the spherical surface, having conductivity tensor $\overleftrightarrow{\sigma}_s$. In high magnetic fields, the open-orbit conductivity is much greater than the closed-orbit conductivity. Since an open-orbit is destroyed when it crosses a domain boundary, the relaxation time τ_{open} to be used in $\overleftrightarrow{\sigma}_{cnj}$ is shorter than τ. An electron in an open orbit travels in a straight line at the Fermi velocity v_F and the average distance it travels before leaving a spherical domain of diameter D is $3D/8$. τ_{open} is thus given approximately by

$$\frac{1}{\tau_{\text{open}}} = \frac{1}{\tau} + \frac{8v_F}{3D} \tag{31.25}$$

which may be rewritten as

$$\frac{\tau}{\tau_{\text{open}}} = 1 + \frac{8\lambda}{3D}, \tag{31.26}$$

where $\lambda = v_F\tau \approx 0.13$ mm is the electron mean free path. In principle, Equation (31.25) applies only when $\vec{H}$ is nearly perpendicular to the open-orbit direction, but this modification has a small effect.

The magnetoresistivity ρ_{xx} for $\vec{H}$ in a plane perpendicular to a [211] or [321] axis was calculated according to the above theory and is shown in Figure 31.6. Induced-torque curves were also calculated using the Visscher–Falicov formula [24] and are shown in Figure 31.7. Except for the vertical scale, they are indistinguishable from the magnetoresistance curves. The crystal orientations in Figure 31.6 correspond to the experimental data in Figure 31.2. The crystallographic direction in the plane of rotation, which was not known for the data in Figure 31.2(a), was adjusted in Figure 31.6(a).

The magnetic field strength at which open-orbit peaks emerge depends somewhat on domain size. $D = 0.05$ mm and $D = 0.12$ mm were used, respectively, in Figures 31.6(a) and 31.6(b); the corresponding open-orbit relaxation times are $\tau_{\text{open}} = \tau/8$ and $\tau_{\text{open}} = \tau/4$. A texture parameter $\alpha = -1$, with an axis at $\theta = 0$, was used in Figure 31.6(a); random orientation, $\alpha = 0$, was used in Figure 31.6(b).

A CDW domain structure is indeed consistent with the main features of the experimental results: the large number of open-orbit peaks, their field dependence, their width, and variations in the data from run to run. Since there is more than one open-orbit direction per domain, the peak locations are not expected to correspond to a single set of 24 symmetry-related directions.

Magnetoresistance spectra for different open-orbit electron fractions η are shown in Figure 31.8. The structure in the magnetoresistance is the same; only the overall scale is affected. Absolute measurements of the induced torque, together with the sample diameter, rotation frequency, and zero-field resistivity, will enable an estimate of η.

Magnetoresistance spectra for different textures are shown in Figure 31.9. Although the qualitative features are the same, the quantitative details are very different. Some peaks present for one texture are absent for another. Until $\vec{Q}$-domain structure can be controlled, a detailed fit of theory and experiment is impossible, since it would require adjusting all 24 values of the domain probabilities f_n. The uncontrolled domain structure, both texture and size, explains the variability of induced-torque data.

Magnetoresistance spectra for different domain sizes are shown in Figure 31.10. $D = 0.1$ mm in Figure 31.10(b) is the approximate domain size of the experimental data in Figure 31.2. If metallurgical preparation has caused too small a domain size, the open-orbit peaks can be suppressed. If the growth of larger domains is stimulated, about twice as many open-orbit peaks are clearly resolved. Induced-torque experiments thus provide a measure of the domain size.

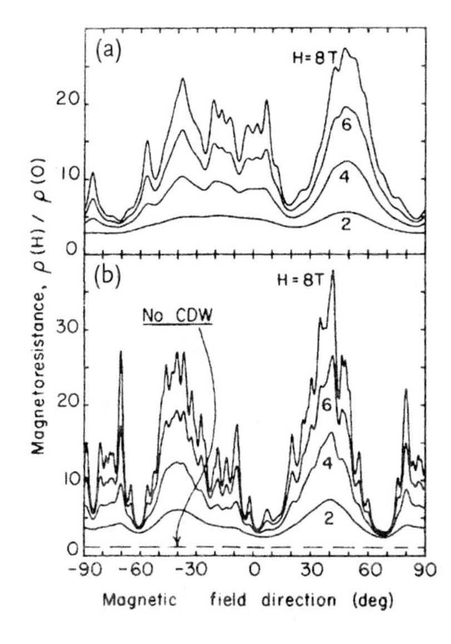

Figure 31.6 Theoretical magnetoresistance versus $\vec{H}$. (a) $\vec{H}$ in the (211) plane; its direction at $\theta = -90^\circ$ is $[\bar{1}20]$ and its sense of rotation is clockwise about the [211] axis. $D = 0.05$ mm; texture: $\alpha = -1$, $\theta = 0$. (b) $\vec{H}$ in the $(\bar{2}\bar{3}1)$ plane; its direction at $\theta = -90^\circ$ is [128] and its sense of rotation is clockwise about the $[\bar{2}\bar{3}1]$ axis. $D = 0.12$ mm; no texture ($\alpha = 0$).

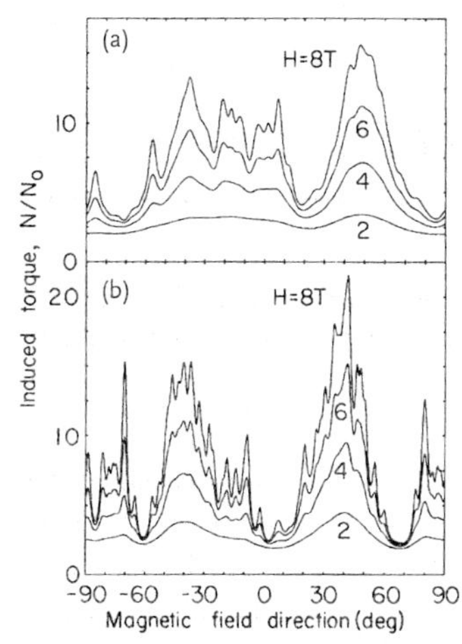

Figure 31.7 Theoretical induced torque versus $\vec{H}$. The curves shown were *really* computed by the Visscher–Falicov formula, using the same parameters as in Figure 31.6. For a spherical Fermi surface, the induced-torque saturates at $N_0 = (8\pi/15)R^5\Omega(ne)^2\rho$, where ρ is the zero-field resistivity, n is the electron density, R is the sample radius, and Ω is the angular rotation frequency.

31.5 Directions for Future Research

The immediate theoretical and experimental challenge is the determination of the open-orbit directions in potassium from its magnetoresistance spectra. Since the location of an open-orbit peak determines only the plane of the open-orbit direction, many experimental runs on accurately oriented crystals are required. If the temperature is allowed to rise above 15 K between runs, some peaks which are present in one run may be absent in the next due to domain distribution changes. Since only about 20 of the 120 predicted open-orbit peaks are present in Figure 31.2, each observed peak contains many unresolved peaks, which further complicates the data analysis.

The additional information that becomes available at extremely high magnetic fields is illustrated in Figures 31.11 and 31.12 by magnetoresistance spectra at 24 T, which we have computed for $\vec{H}$ in a (211) or (321) plane. Approximately twice as many open-orbit peaks are

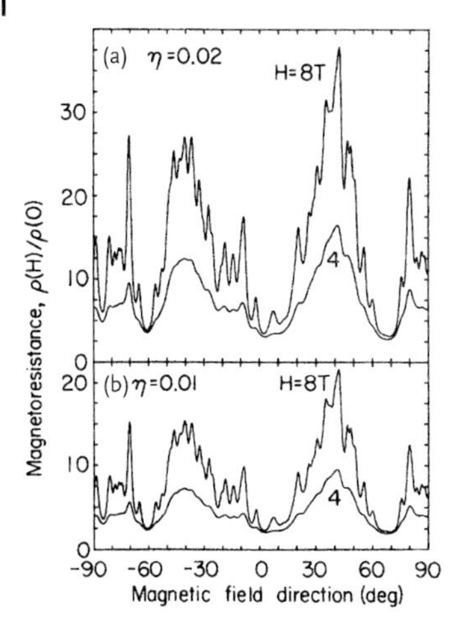

Figure 31.8 Theoretical magnetoresistance for $\vec{H}$ in a (321) plane. (a) $\eta = 0.02$. (b) $\eta = 0.01$.

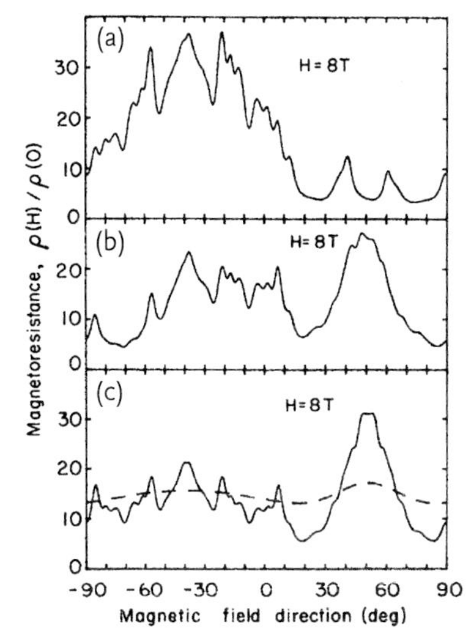

Figure 31.9 Theoretical magnetoresistance for $\vec{H}$ in a (211) plane, $D = 0.05$ mm. (a) $\alpha = 2$, $\theta = 60°$; (b) $\alpha = -1$, $\theta = 0$; (c) $\alpha = 0$. The dashed curve is for $D = 0.005$ mm.

sharply resolved at 24 T as at 8 T. We assumed as before that all five open-orbit directions in each $\vec{Q}$ domain are equally likely, even though magnetic breakdown may require this simple assumption to be modified. For example, if the heterodyne and phason-instability gaps can be neglected at such high fields, each $\vec{Q}$ domain would have only one open-orbit direction $\vec{Q}$, reducing the total number of open-orbit directions from 120 to 24. High-field experiments will thus also allow magnetic-breakdown studies of the energy gaps.

31.6 Conclusion

Potassium, the simplest metal of all, is for solid-state theory the analog of the hydrogen atom. Its properties must have explanations. Only two views are under serious consideration: (1) potassium has a CDW; (2) potassium has a spherical Fermi surface. The open-orbit magnetoresistance observed by Coulter and Datars [1] shown in Figure 31.2, supports

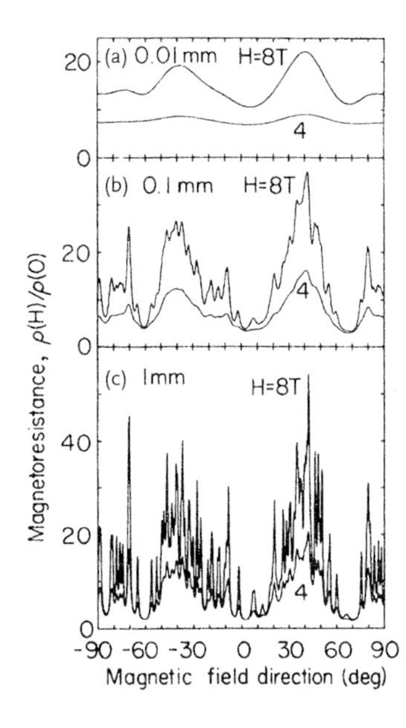

Figure 31.10 Theoretical magnetoresistance for $\vec{H}$ in a (321) plane. (a) $D = 0.01\,\text{mm}$; (b) $D = 0.1\,\text{mm}$; (c) $D = 1\,\text{mm}$.

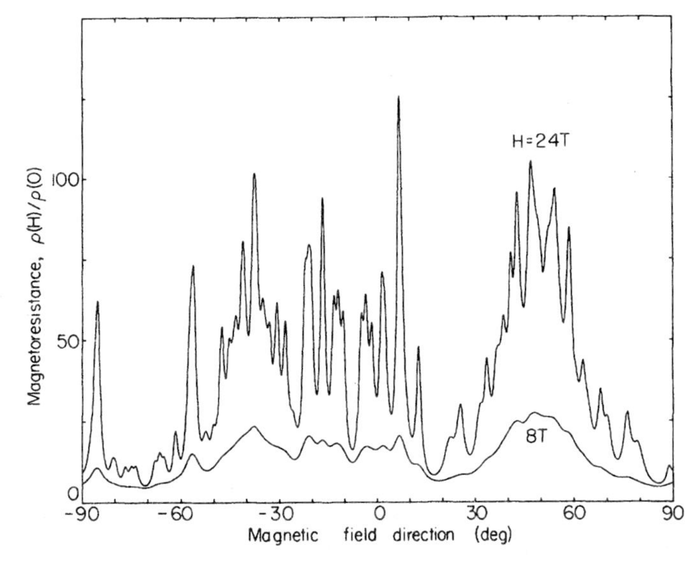

Figure 31.11 Theoretical magnetoresistance at 24 T for $\vec{H}$ in a (211) plane.

the first view, which explains the main features, and rules out the second view, for which the only allowed magnetoresistance spectrum is the horizontal, dashed line shown in Figure 31.6(b).

A question that is often asked is whether dislocations can cause the narrow torque peaks in Figure 31.2. An oriented array of dislocations scatters electrons perpendicular to the dislocation axis. The total relaxation time due to impurity and dislocation scattering is then anisotropic. We have calculated the induced torque for oriented dislocations, assuming

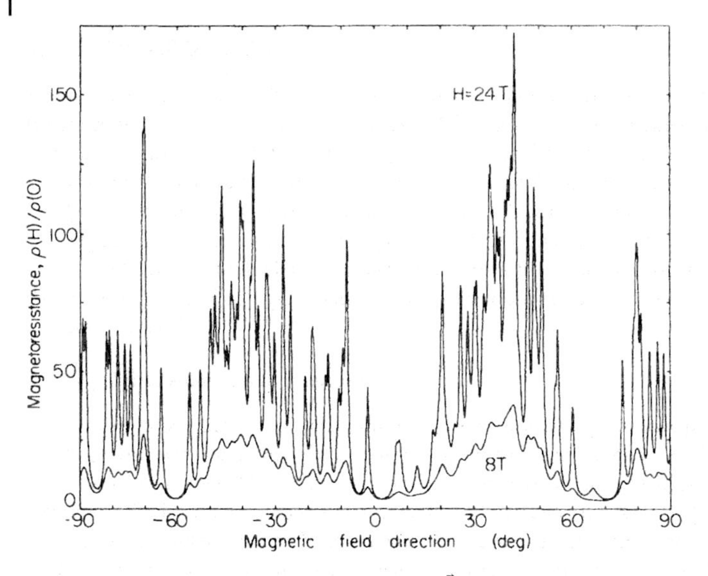

Figure 31.12 Theoretical magnetoresistance at 24 T for $\vec{H}$ in a (321) plane.

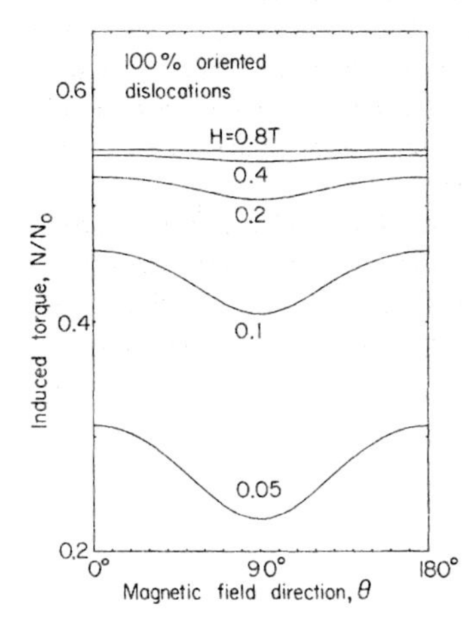

Figure 31.13 Induced torque of oriented dislocations. Resistivity anisotropy = 0.1. The torque saturates at $N/N_0 = 0.55$, where $N_0 = (8\pi/15)R^5\Omega(ne)^2\rho$.

a zero-field resistivity

$$\overleftrightarrow{\rho} = \rho \begin{pmatrix} 1 & 0 & 0 \\ 0 & 1 & 0 \\ 0 & 0 & 0.1 \end{pmatrix} \tag{31.27}$$

which is 10 times less parallel to the dislocation axis than perpendicular to it. Taking the dislocation axis in the rotation plane, we show in Figure 31.13 the dependence of the induced torque on the angle θ between $\vec{H}$ and the dislocation axis. The torque saturates at relatively low fields and is completely isotropic [42].

An outstanding enigma is why the effects of a CDW are not observed in de Haas–van Alphen (dHvA) experiments. The dHvA frequency is proportional to the extremal Fermi-surface cross-sectional area. Although the pressure dependence of the dHvA frequency in potassium is anomalous [43–45], its angular dependence is nearly isotropic [46–48]. The relatively large distortions of the Fermi surface near the energy-gap planes caused by a CDW are not observed.

Since the dHvA effect is sensitive only to closed orbits, it cannot detect the open orbits observed by the induced-torque technique. We expect that in a single domain sample the Fermi-surface anisotropy would be observed and can only conclude that the apparent isotropy is a consequence of the domain structure. We speculate that, if the $\vec{Q}$-domain size is small, the observed dHvA frequency is an average over many domains. A theory of the dHvA effect in a polydomain sample, taking into account electron–electron interactions, remains to be developed.

Deliberate attempts to prepare dHvA effect samples having large domain sizes are encouraged. The factors affecting the domain distribution in order of importance are surface orientation, directed stress, and a magnetic field. Surface faceting may control the domain size, since different surface orientations favor different $\vec{Q}$ directions. Etching, which produces larger facets, may thus stimulate the growth of larger domains. The small samples (about 1 mm in diameter) typically used in dHvA experiments may be "all surface"; their domain size may be limited by the surface facet size. The use of larger samples, having an interior region, may thus permit larger domains to grow. Simultaneous induced-torque measurements, interpreted using the theory developed here, would provide an estimate of the domain size.

Acknowledgements

We are grateful to the NSF-MRL program for support under Grant No. DMR77-23798.

References

1 Coulter, P.G. and Datars, W.R. (1980) *Phys. Rev. Lett.* **45**, 1021; and private communication.
2 Overhauser, A.W. (1968) *Phys. Rev.* **167**, 691.
3 For an experimental and theoretical review, see Overhauser, A.W. (1978) *Adv. Phys.* **27**, 343.
4 A brief exposition of this theory appeared in Huberman, M. and Overhauser, A.W. (1981) *Phys. Rev. Lett.* **47**, 682.
5 Justi, E. (1948) *Ann. Phys. (Leipzig)* **3**, 183.
6 MacDonald, D.K.C (1956) in *Handbuch der Physik*, (ed. S. Flugge), Springer, Berlin, Vol. 14, p. 137.
7 Taub, H., Schmidt, R.L., Maxfield, B.W., and Bowers, R. (1971) *Phys. Rev. B* **4**, 1134.
8 Jones, B.K. (1969) *Phys. Rev.* **179**, 637.
9 Fletcher, R. (1980) *Phys. Rev. Lett.* **45**, 287.
10 Lass, J.S. (1970) *J. Phys. C* **3**, 1926.
11 Penz, P.A. and Bowers, R. (1968) *Phys. Rev.* **172**, 991.
12 Simpson, A. M. (1973) *J. Phys. F* **3**, 1471. The interpretation of this experiment may require revision since an isotropic transverse resistivity was assumed.
13 Bruls, G.J.C.L., Bass, J., van Gelder, A. P., van Kempen, H., and Wyder, P. (1981) *Phys. Rev. Lett.* **46**, 553.
14 Stroud, D. and Pan, F.P. (1976) *Phys. Rev. B* **13**, 1434.
15 Schouten, D.R. and Swenson, C.A. (1974) *Phys. Rev. B* **10**, 2175.
16 Stetter, G., Adlhart, W., Fritsch, G., Steichele, E., and Luscher, E. (1978) *J. Phys. F* **8**, 2075.
17 Moss, J.S. and Datars, W.R. (1967) *Phys. Lett.* **24A**, 630.
18 For an application to copper, see Datars, W.R. (1970) *Can. J. Phys.* 48, 1806.

19 Schaefer, J.A. and Marcus, J.A. (1971) *Phys. Rev. Lett.* **27**, 935.

20 Holroyd, F.W. and Datars, W.R. (1975) *Can. J. Phys.* **53**, 2517.

21 Lifshitz, I.M., Azbel, M.Ia., and Kaganov, M.I. (1956) *Zh. Eksp. Teor. Fiz.* **31**, 63 [Sov. Phys – JETP 4, 41 (1957)].

22 For a theoretical and experimental review, see Fawcett, E. (1964) *Adv. Phys.* **13**, 139.

23 Martin, P.M., Sampsell, J.B., and Garland, J.C. (1977) *Phys. Rev. B* **15**, 5598.

24 Visscher, P.B. and Falicov, L.M. (1970) *Phys. Rev. B* **2**, 1518.

25 Overhauser, A.W. (1974) *Phys. Rev. B* **9**, 2441.

26 Overhauser, A.W. (1964) *Phys. Rev. Lett.* **13**, 190.

27 Overhauser, A.W. and Butler, N.R. (1976) *Phys. Rev. B* **14**, 3371.

28 Overhauser, A.W. (1971) *Phys. Rev. B* **3**, 3173.

29 Giuliani, G.F. and Overhauser, A.W. *Phys. Rev. B* **20**, 1328 (1979).

30 Gurney W.S.C. and Gugan, D. (1971) *Philos. Mag.* **24**, 857.

31 In Figure 31.5 the angle between $\vec{Q}$ and $\vec{G}_{110}$ is 2°. For the predicted tilt angle of 4.1°, the heterodyne and phason-instability energy-gap planes intersect on part of the Fermi surface. This complicates the figure, but does not change the open-orbit directions.

32 Reitz, J.R. and Overhauser, A.W. (1968) *Phys. Rev.* **171**, 749.

33 Giuliani, G.F. and Overhauser, A. W. (1980) *Phys. Rev. B* **22**, 3639; and private communication.

34 Giulani, G.F. and Overhauser, A.W. (1981) *Bull. Am. Phys. Soc.* **26**, 471.

35 Blount, E.I. (1962) *Phys. Rev.* **126**, 1636.

36 Animalu, A.O.E. (1977) *Intermediate Quantum Theory of Crystalline Solids*, Prentice-Hall, New Jersey.

37 Bruggeman, D.A.G. (1935) *Ann. Phys. (Leipzig)* **24**, 636.

38 Landauer, R. (1952) *J. Appl. Phys.* **22**, 779.

39 Stroud, D. and Pan, F.P. (1979) *Phys. Rev. B* **20**, 455.

40 Stroud, D. and Pan, F.P. (1976) *Phys. Rev. B* **13**, 1434.

41 Huberman, M. and Overhauser, A.W. (1981) *Phys. Rev. B* **23**, 6294. In Eq. (20) of Ref. [41] the minus sign in Eq. (21) was inadvertently omitted.

42 Overhauser, A. W. (1971) *Phys. Rev. Lett.* **27**, 938.

43 Altounian, Z., Verge, C., and Datars, W.R. (1978) *J. Phys. F* **8**, 75.

44 Altounian, Z. and Datars, W.R. (1980) *Can. J. Phys.* **58**, 370.

45 Templeton, I.M. (1981) *J. Low Temp. Phys.* **43**, 293.

46 Shoenberg, D. and Stiles, P.J. (1964) *Proc R. Soc. London Ser. A* **281**, 62.

47 Lee, M.J.G. and Falicov, L.M. (1968) *Proc. R. Soc. London Ser A* **104**, 319.

48 O'Shea, M.J. and Springford, M. (1981) *Phys. Rev. Lett.* **46**, 1303.

Reprint 32 The Open Orbits of Potassium[1)]

A.W. Overhauser *

* Department of Physics, Purdue University, 525, Northwestern Avenue, West Lafayette, Indiana 47907, USA

Received 25 November 1981

The high-field magnetoresistance of potassium, which fails to saturate, can be an intrinsic property only if the Fermi surface is multiply connected. Direct observation of open-orbit magnetoresistance peaks by Coulter and Datars confirms this theoretical principle. Since the Brillouin zone of potassium is only half full, and the Fermi surface is nearly spherical, open orbits can occur only if the translational symmetry of the crystal is broken by a charge-density-wave structure. The open-orbit distributions that result explain the observed magnetoresistance patterns.

Le magnétorésistance en champ élevé du potassium, qui ne présente pas de saturation, ne peut être une propriété intrinsèque que si la surface de Fermi est à connexion multiple. L'observation directe par Coulter et Datars de pics de magnétorésistance correspondant à des orbites ouvertes confirme ce principe théorique. Comme la zone de Brillouin du potassium n'est remplie qu'à demi et que la surface de Fermi est presque sphérique, il ne peut y avoir des orbites ouvertes que si la symétrie de translation du cristal est brisée par une structure d'onde de densité de charge. Les distributions d'orbites ouvertes qui en résultent expliquent l'allure observée pour les courbes de magnétorésistance.

32.1 Introduction

A widely held belief is that potassium has a nearly spherical Fermi surface, with deviations of only one part per thousand. This view appears to be substantiated by the observed isotropy of the de Haas–van Alphen frequency [1]. In contrast the high-field magnetoresistance fails to saturate [2] and this behavior can be explained (intrinsically) only if the Fermi surface is multiply connected [3].

It is easy to understand why a spherical Fermi surface cannot lead to a large magnetoresistance. Suppose that **H** is in the $\hat{z}$ direction. Newton's equations for the total electronic momentum **P** are:

$$dP_x/dt = E_x + \omega_c P_y - P_x/\tau = 0 \tag{32.1a}$$

$$dP_y/dt = E_y - \omega_c P_x - P_y/\tau = 0 \tag{32.1b}$$

The first two terms are the accelerations caused by electric and magnetic fields. ω_c is the cyclotron frequency, eH/mc. The last term is the frictional force caused by scattering, the mean collision time being τ. (All units have been chosen so that Equation (32.1) has the indicated simplicity.) $d\mathbf{P}/dt = 0$ when the nonequilibrium steady state is reached. If the current is constrained to flow along $\hat{x}$, then $P_y = 0$. Solution of Equation (32.1a) is then,

$$E_x = P_x/\tau \tag{32.2}$$

1) Can. J. Phys. 60, 687 (1982)

Anomalous Effects in Simple Metals. Albert Overhauser

ISBN: 978-3-527-40859-7

which is Ohm's law. The resistivity, $1/\tau$, is independent of H! The LAK theorems [3] show that this result applies to any simply connected Fermi surface for $\omega_c\tau \gg 1$. They also show that open orbits are necessary to obtain a nonsaturating resistivity.

Since the observed high-field behavior of potassium is quantitatively erratic, some workers are tempted to attribute what they cannot explain to extrinsic causes such as voids or dislocations. I shall discuss this below and show that such possibilities can be discarded.

I have shown that a simple metal should manifest a charge-density-wave (CDW) instability [4]. The conduction-electron density $n(\mathbf{r})$ is then

$$n = n_0(1 + p \cos \mathbf{Q} \cdot \mathbf{r}) \tag{32.3}$$

instead of a constant value n. (The fractional modulation p is small compared to unity.) The wave vector $\mathbf{Q}$ of the CDW is related to the diameter $2k_F$ of the Fermi surface and is incommensurate with the reciprocal lattice. The positive ions will be slightly displaced from their ideal lattice sites $\mathbf{L}$ by

$$\mathrm{U} = \mathbf{A} \sin \mathbf{Q} \cdot \mathbf{L} \tag{32.4}$$

in order to maintain microscopic charge neutrality. Accordingly the translation symmetry of the lattice is broken and extra periodic potentials arise. These lead to new energy gaps truncating the Fermi surface, so that it becomes multiply connected. Open orbits of various types are activated at high fields, but these give rise to only a few of the many anomalous properties caused by a CDW structure. The most important effects have been reviewed elsewhere [5].

32.2 Direct Observation of Open Orbits

The most spectacular property of potassium observed to date is the high-field magnetoresistance spectrum [6]. This appears for $H > 4\,\mathrm{T}$, which corresponds to $\omega_c\tau > 100$. Sharp

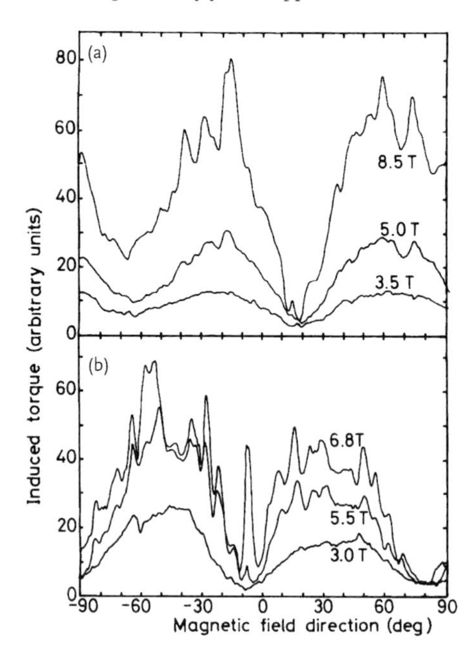

Figure 32.1 Magnetoresistance spectrum of two single crystals of potassium [6]. (a) Crystal grown in oil and **H** in a (211) plane; (b) crystal grown in a Kel-F mold and **H** in a (321) plane.

peaks in resistivity vs. field direction arise and grow rapidly with increasing H. Data obtained by Coulter and Datars on two single crystals are shown in Figure 32.1. The induced torque method was used. They have obtained similar data in all of 100 runs on 12 specimens [7]. Sharp magnetoresistance peaks are the signature of open orbits, and cannot arise from a simply connected Fermi surface. It is clear from Figure 32.1 that there are many open orbits in potassium.

Although the peak value of an open-orbit resistivity increases as H^2 the area of such a peak increases as H. Therefore in polycrystalline samples a (high-field) linear magnetoresistance can be expected, and this is generally observed in potassium [2]. Similar phenomena occur in the other alkali metals.

The CDW **Q** will have a preferred crystal orientation (α, β, γ), which is not along a symmetry axis or in a symmetry plane [8]. Consequently there are 24 cubically equivalent **Q** axes; so a single crystal will be subdivided into 24 **Q**-domain types. Furthermore each **Q** domain may have several open-orbit directions (I shall describe five below). A total of 120 open-orbit peaks could be present in a single crystal, so great a number that fields of $\sim$ 30 T may be necessary to resolve most of them. At present the data of Coulter and Datars are not sufficient to determine the crystal direction of **Q**. The **Q**-domain distribution also is unknown and can vary from run to run, depending on the thermal history of each crystal. Naturally only the major features of the data in Figure 32.1 can be expected to correspond to a theoretical spectrum. This has been achieved [9].

32.3 Open Orbits of a Single Q Domain

An illustration of three pairs of energy-gap planes is shown in Figure 32.2. The magnetic field is assumed to be perpendicular to the plane of the figure, which contains both **Q** and the reciprocal lattice vector $\mathbf{G}_{110}$. The "main" energy gaps are created by the CDW periodic potential, of wavevector **Q**. The "heterodyne gaps" arise from the periodicity $\mathbf{Q}' \equiv \mathbf{Q} - \mathbf{G}_{110}$ which is the "beat periodicity" between the lattice and the CDW. The "phason instability gaps" arise from an (unfulfilled) tendency of the CDW to become commensurate with the lattice [10]. Optical absorption data indicate that the main gaps are large, −0.6 eV [5]. These will not undergo magnetic breakdown. I believe the other gaps are much smaller, perhaps

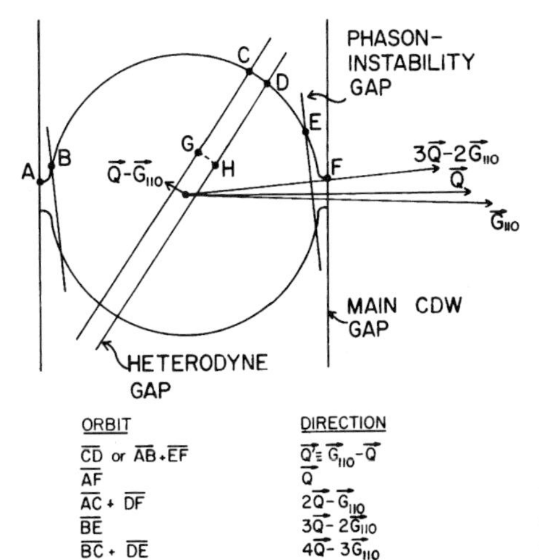

Figure 32.2 Fermi surface, energy gaps, and open orbits of potassium (for **H** perpendicular to the plane shown).

$E_g \sim 10$–100 meV. The probability of magnetic breakdown is given [11] by

$$P = \exp\left(-\pi E_g^2/4\hbar\omega_c m|v_x v_y|\right), \tag{32.5}$$

where v_y and v_y are the velocity components of the electron at the magnetic breakdown points. As a consequence, for example, the heterodyne gaps can be broken down at points C and D of Figure 32.2, but not at points G and H. If one takes advantage of such possibilities, five open-orbit combinations can be constructed from the orbit segments shown in the figure. They are listed along with their **k**-space directions. (It is possible that other energy gaps are also important. However, the three pairs mentioned here are likely the most important.)

The topology of the Fermi surface shown in Figure 32.2 is obviously too complex to treat theoretically without resorting to a model representation. A convenient approximation is to suppose that the Fermi surface is equivalent to a spherical part and five cylindrical parts. The advantage is that this allows one to construct a magnetoconductivity tensor. For the spherical part

$$\sigma_s = \sigma_0 \begin{pmatrix} \frac{1}{1+(\omega_c\tau)^2} & \frac{-\omega_c\tau}{1+(\omega_c\tau)^2} & 0 \\ \frac{\omega_c\tau}{1+(\omega_c\tau)^2} & \frac{1}{1+(\omega_c\tau)^2} & 0 \\ 0 & 0 & 1 \end{pmatrix}, \tag{32.6}$$

where $\sigma_0 \equiv ne^2\tau/m$. Each cylindrical piece is most easily described in a frame of reference defined by its cylinder axis. If ϕ is the angle between **H** and the axis, then

$$\sigma_s = \sigma_0' \begin{pmatrix} \frac{1}{1+\delta^2} & \frac{-\delta}{1+\delta^2} & 0 \\ \frac{\delta}{1+\delta^2} & \frac{1}{1+\delta^2} & 0 \\ 0 & 0 & 0 \end{pmatrix}, \tag{32.7}$$

where $\delta \equiv \omega_c\tau_0 \cos\phi$, τ_0 being the lifetime in an open orbit.

Suppose we consider a conductivity tensor arising from a spherical surface and one cylinder, the latter having a fraction η of the total electron number. Then the conductivity tensor is

$$\sigma = (1-\eta)\sigma_s + \eta\tilde{s}\sigma_c s, \tag{32.8}$$

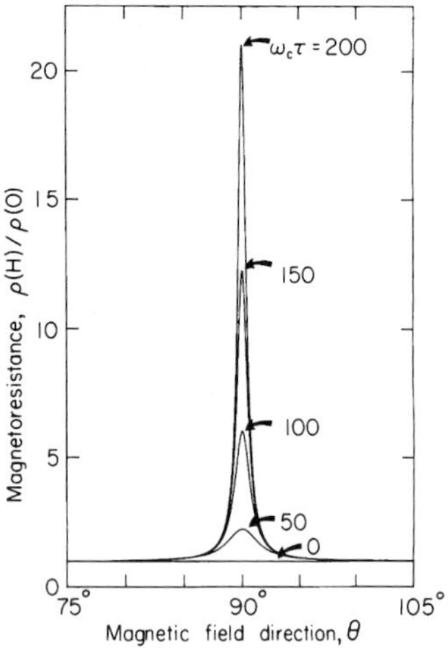

Figure 32.3 Open-orbit magnetoresistance vs. θ, the angle between **H** and the Fermi-surface cylinder axis. The open-orbit electron fraction was taken to be $\eta = 0.001$.

where s is the orthogonal transformation matrix that takes the frame used in Equation (32.7) into the frame used in Equation (32.6) ($\tilde{s}$ is the transpose of s).

It is easy to invert σ to obtain ρ_{xx}, the magneto-resistance vs. angle between **H** and the cylinder axis. Typical results are shown in Figure 32.3 for several values of $\omega_c \tau$. The sharp magnetoresistance peak occurs when **H** is perpendicular to the cylinder axis. (This is also the orientation which leads to open orbits.) It is clear from the figure that rotation patterns of magnetoresistance provide a definitive technique for detecting open orbits.

Equation (32.8) can be generalized to provide a model of the conductivity for a Fermi surface with five open-orbit directions. For simplicity we take all open-orbit fractions equal. Then

$$\sigma = (1 - 5\eta)\sigma_s + \eta \sum_{i=1}^{5} \tilde{s}_i \sigma_c s_i \tag{32.9}$$

A 180° rotation pattern for ρ_{xx}, obtained by inverting Equation (32.9), would show five separate peaks (in general).

32.4 Effective-Medium Theory for Q Domains

Effective-medium theory is the only method available for calculating the macroscopic conductivity of a poly-domain sample [12]. One assumes that the volume fraction f_n of each **Q**-domain type is known. One supposes that σ_{ext} is the conductivity of the medium surrounding a **Q** domain having a conductivity tensor σ_n. The effective conductivity tensor is then given by

$$\sigma_{\text{eff}} = \sigma_{\text{ext}} + \sum_{n=1}^{24} f_n(\sigma_n - \sigma_{\text{ext}}) \cdot \beta_n \tag{32.10}$$

where β_n is an explicit tensor given in [12]. Effective-medium theory is defined by the self-consistent requirement:

$$\sigma_{\text{eff}} = \sigma_{\text{ext}} \tag{32.11}$$

In practice, one starts with a guess for σ_{ext} and iterates Equation (32.10). Convergence to Equation (32.11) is extremely rapid. Usually only seven or eight iterations are needed.

The relaxation time τ_0 for an electron in an open orbit is shorter than that for an electron in a cyclotron orbit. The reason is that an open orbit is interrupted when the electron crosses a **Q**-domain boundary. (This is less likely for a cyclotron orbit when $\eta \ll 1$.) Accordingly,

$$\frac{1}{\tau_0} \cong \frac{1}{\tau} + \frac{8v_F}{3D} \tag{32.12}$$

D is the diameter of a **Q** domain and v_F is the Fermi velocity. Equation (32.12) introduces the only dependence of σ_{eff} on domain size. The theoretical results, shown below, then allow an estimate to be made for D by comparison with the data of Figure 32.1.

The simplest choice for a domain distribution is $f_n = 1/24$ (for all n). One can also employ preferred textures such as

$$f_n = \frac{1}{24}\left\{1 + \alpha\left[\frac{3}{2}(\hat{Q} \cdot \hat{T})^2 - \frac{1}{2}\right]\right\} \tag{32.13}$$

where $\hat{T}$ is the unit vector of a texture axis and α is a parameter between -1 and 2. (The most general domain distribution has 23 parameters.)

Figure 32.4 shows calculated magnetoresistance spectra for single crystals having the rotation axes of the data in Figure 32.1. The striking similarity between theory and experiment provides convincing confirmation of a CDW structure. The only allowed spectrum for a spherical Fermi surface (with no CDW) is the horizontal, dashed line in Figure 32.4b. The traditional model can be ruled out! The domain size D affects the field at which open-orbit peaks emerge; $D = 0.05$ mm in Figure 32.4a and 0.12 mm in Figure 32.4b.

Although the structure in Figures 32.1 and 32.4 is quite detailed at 8 T, it is not sufficient for a measurement of open-orbit directions. This can be appreciated from the theoretical curve in Figure 32.5, computed for $\mathbf{H} = 24$ T [13]. The 8 T curve shown is the same as that

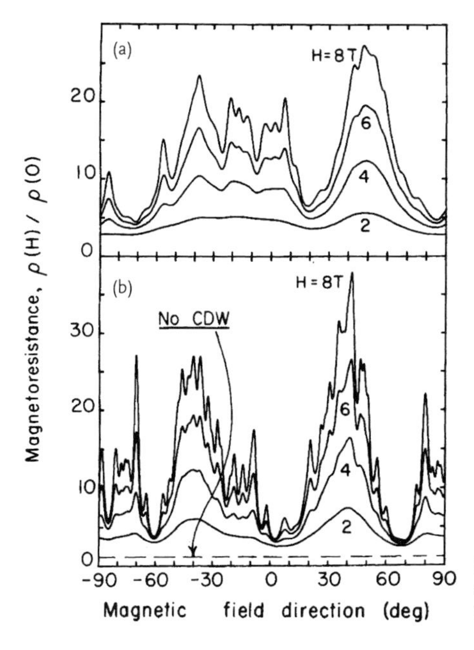

Figure 32.4 Theoretical magnetoresistance spectra of potassium for the orientations shown in Figure 32.1, and $\eta = 0.02$.

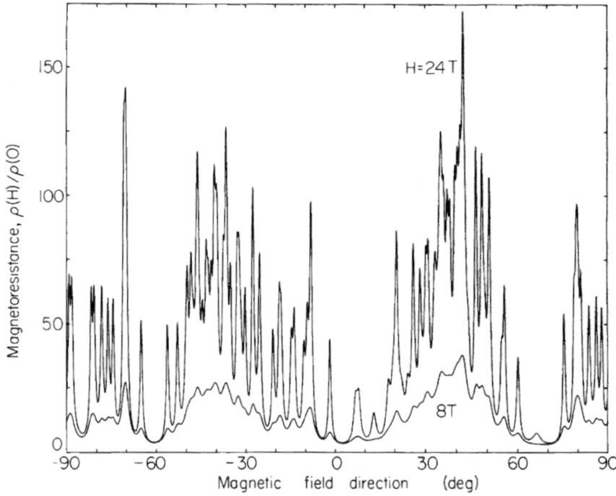

Figure 32.5 Theoretical magnetoresistance spectra at $\mathbf{H} = 24$ T (and 8 T) for $\mathbf{H}$ in a (321) plane.

in Figure 32.4b. Obviously extremely high-field data is needed before any attempt to identify specific open orbits can be made. Very likely, magnetic breakdown effects will be important at high fields and can be studied quantitatively from the field dependence of the magnetoresistance spectrum. (Such effects have been omitted in calculating the spectra shown in Figures 32.4 and 32.5.)

32.5 Discussion

Prior to the discovery of anisotropic magnetoresistance spectra in potassium, some workers had suggested that macroscopic voids could be the cause of a (high-field) linear magnetoresistance. Detailed calculations showed, however, that the volume fraction of voids would have to be $\sim$ 1% [14]. Independent evidence for voids is lacking. In fact precision lattice-parameter measurements [15], when compared with volumetric density data [16], show that any void fraction must be less than a few parts in 10^5. The void enthusiast is stymied when faced with data like that in Figure 32.1, and may postulate (unseen) cracks and scratches on the specimen surface. But this avenue of escape (from a CDW structure) has been ruled out by Holroyd and Datars [17]. They drilled holes in some of their samples, yet they found no observable consequence in their rotation patterns.

It is well known that dislocations can cause an anisotropic resistivity since any momentum transfer (to scattered electrons) is perpendicular to the dislocation line. But even if one has a sample with perfectly oriented dislocations, it is not possible to obtain induced torque patterns like those shown in Figure 32.1. This can be appreciated from calculated rotation patterns [13] of a metal (having a spherical Fermi surface). The residual resistivity was taken to be 10% from impurities and 90% from oriented dislocations. For $\mathbf{H} < 0.1\,\mathrm{T}$ the torque curves have a sinusoidal anisotropy, but they become isotropic above 0.5 T, as shown in Figure 32.6.

The apparent isotropy of the Fermi surface of potassium, suggested by de Haas–van Alphen data, is still a mystery, but not necessarily an insoluble one. The de Haas–van Alphen oscillations measure closed orbits; the magnetoresistance peaks measure open orbits. Both types of orbits occur for a CDW structure. Serious attempts to observe CDW effects in de Haas–van Alphen data [18] do not indicate a spherical Fermi surface, as claimed, for O'Shea

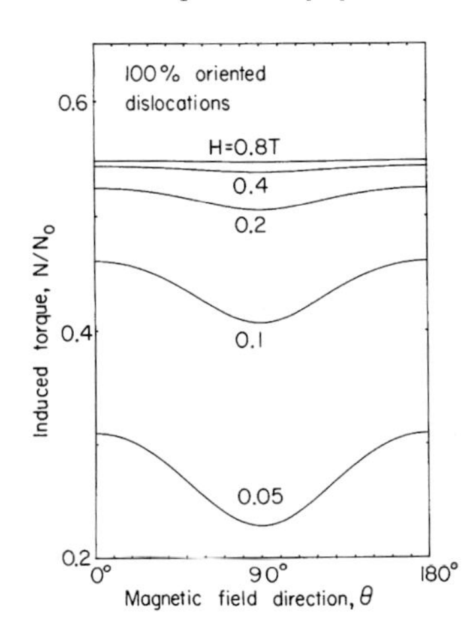

Figure 32.6 Induced torque of an ideal metal with scattering anisotropy of 10:1, assumed to arise from oriented dislocations.

and Springford report that their samples showed a non-saturating magnetoresistivity (which requires a multiply connected Fermi surface [3]).

How can one reconcile isotropic de Haas–van Alphen frequencies with the anisotropic Fermi surface of a CDW [5]? The answer must depend on the **Q**-domain structure of typical specimens. Fermi surface anisotropy would have to appear in a single **Q**-domain sample. (It is not yet known how to produce one.)

Typical de Haas–van Alphen samples are about 1 mm in size, have highly curved surfaces, and are usually not etched. Their domain size D may be only $\sim 10^{-3}$ mm. Since the electronic mean free path is ~ 0.1 mm, and the cyclotron radius at 4 T is $\sim 10^{-3}$ mm, the coherence volume of a Landau level is $(0.1)^2 \times 10^{-3} = 10^{-5}\,\text{mm}^3$. (A Landau level is localized in only one direction.) This "coherence volume" would contain, therefore, $\sim 10^4$**Q** domains. In the absence of a preferred, orientational texture the sample would appear isotropic. It would be unjustified to assume (without proof) that de Haas–van Alphen signals from the many **Q** domains in a coherence volume should be summed independently (and so lead to a "noisy mishmash" for the total signal). Coulomb interactions (between the electrons) might "pull" the many contributions into a "clean" signal appropriate to the *average* extremal area (of the Fermi surface).

The foregoing suggestion is speculative, since theoretical study of Coulomb interaction effects on the Landau levels of a fine-grain, heterogeneous sample has not yet been attempted.

I conclude by emphasizing that the open-orbit magnetoresistance peaks of potassium are facts, just as the isotropic de Haas–van Alphen frequencies are facts. What seems inappropriate is the view that only an isotropic, simply connected Fermi surface can lead to an isotropic de Haas–van Alphen frequency; and that all other phenomena, however spectacular and however reproducible, which contradict that view must be discarded as artifacts.

References

1 Shoenberg, D. and Stiles, P.J. (1964) *Proc R. Soc London, Ser. A* **281**, 62.
2 Taub,H., Schmidt, R.L., Maxfield, B. W., and Bowers, R. (1971) *Phys. Rev. B* **4**, 1134.
3 Lifshitz, I.M., Azbel, M.Ia., and Kayanov, M. I. (1956) *Zh. Eksp. Teor. Fiz.* **31**, 63. Engl. Transl. (1957) *Sov. Phys. JETP* **4**, 41.
4 Overhauser, A.W. (1968) *Phys. Rev.* **167**, 691.
5 Overhauser, A. W. (1978) *Adv. Phys.* **27**, 343.
6 Coulter, P.G. and Datars, W.R. (1980) *Phys. Rev. Lett.* **45**, 1021.
7 Coulter, P.G. and Datars, W.R. Personal communication.
8 Giuliani, G.F. and Overhauser, A.W. (1979) *Phys. Rev. B* **20**, 1328.
9 Huberman, M. and Overhauser, A.W. (1981) *Phys. Rev. Lett.* **47**, 682.
10 Giuliani, G.F. and Overhauser, A.W. To be published.
11 Blount, E. I. (1962) *Phys. Rev.* **126**, 1636.
12 Huberman, M. and Overhauser, A.W. (1981) *Phys. Rev. B* **23**, 6294.
13 Huberman, M. and Overhauser, A.W. (1982) *Phys. Rev. B* **25**.
14 Stroud, D. and Pan, F.P. (1976) *Phys. Rev. B* **13**, 1434.
15 Stetter, G., Adlhart, W., Fritsch, G., Steichele, E., and Luscher, E. (1978) *J. Phys. F* **8**, 2075.
16 Schouten, D.R. and Swenson C.A. (1974) *Phys. Rev. B* **10**, 2175.
17 Holroyd, F.W. and Datars, W.R. (1975) *Can. J. Phys.* **53**, 2517.
18 O'Shea, M.J. and Springford, M. (1981) *Phys. Rev. Lett.* **46**, 1303.

Reprint 33 Open-Orbit Effects in Thermal Magnetoresistance[1)]

M. Huberman and A.W. Overhauser*

* Department of Physics, Purdue University, 525, Northwestern Avenue, West Lafayette, Indiana 47907, USA

Received 22 January 1982

The thermal magnetoresistance rotation pattern of a metal having two Fermi-surface fractions, a sphere and a cylinder, is calculated. An isotropic lattice conductivity κ_g is assumed. Sharp peaks, proportional to H^2 occur when $\vec{H}$ is perpendicular to the open-orbit direction, as in the electrical magnetoresistance. The lattice conductivity, however, causes the peak height to saturate in high fields, providing a means of measuring κ_g. Moreover, the peak becomes inverted. A surprising phenomenon, an open-orbit crevasse, is predicted.

33.1 Introduction

In view of forthcoming experimental results, we have calculated the thermal magnetoresistance rotation pattern expected from a metal having open orbits, in order that the unexpected can be recognized. Several surprising and interesting effects were found, which are reported here.

The thermal conductivity of a metal is a sum of an electronic and a lattice conductivity. If the lattice conductivity is negligible, semiclassical transport theorems [1] predict that the thermal magnetoresistance has the same field dependence as the electrical magnetoresistance. Although the lattice conductivity of a pure metal is much less than the electronic conductivity in zero field, this is no longer true in high fields since a magnetic field reduces the electronic conductivity.

It is well known [2, 3] that if a metal has only closed orbits, there is no high-field electrical magnetoresistance [4]. A nonsaturating magnetoresistance proportional to H^2 occurs whenever $\vec{H}$ is perpendicular to an open-orbit direction. This effect disappears if $\vec{H}$ is rotated slightly into a non-perpendicular orientation. Sharp magnetoresistance peaks, observed by rotating $\vec{H}$ (or the crystal), are the signature of open orbits.

In this paper the effect of lattice conductivity or open-orbit thermal magnetoresistance is investigated. The thermal magnetoconductivity tensor is derived, and rotation patterns are presented and discussed.

33.2 Theory

We begin by reviewing the theory of thermal magnetoresistance for a metal having only closed orbits [5]. We model the Fermi surface by a sphere. In a magnetic field $\vec{H} = H\hat{z}$, the thermal conductivity tensor $\overleftrightarrow{\kappa}_s$ of a sphere is

$$\kappa_s = \kappa_e \begin{pmatrix} \frac{1}{1+(\omega_c\tau)^2} & \frac{-\omega_c\tau}{1+(\omega_c\tau)^2} & 0 \\ \frac{\omega_c\tau}{1+(\omega_c\tau)^2} & \frac{1}{1+(\omega_c\tau)^2} & 0 \\ 0 & 0 & 1 \end{pmatrix} . \tag{33.1}$$

1) Phys. Rev. B 25, 12 (1982)

Anomalous Effects in Simple Metals. Albert Overhauser

ISBN: 978-3-527-40859-7

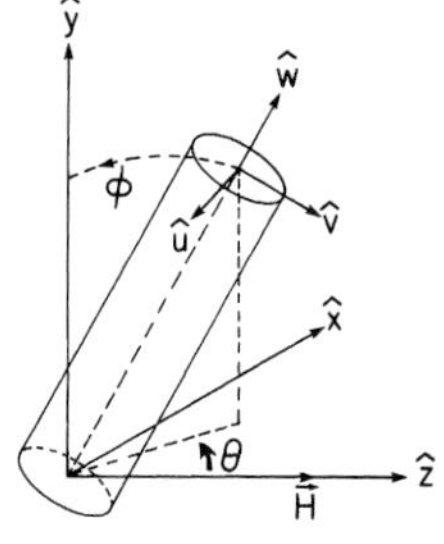

Figure 33.1 Relative orientation of the uvw and xyz axes. $\hat{\omega}$ is the open-orbit direction and $\vec{H}$ is parallel to $\hat{z}$.

κ_e is the zero-field electronic conductivity, $\omega_c \equiv eH/mc$ is the cyclotron frequency, and τ is the electron relaxation time.

The total thermal conductivity $\overleftrightarrow{\kappa}$ is

$$\overleftrightarrow{\kappa} = \overleftrightarrow{\kappa}_s + \overleftrightarrow{\kappa}_g , \tag{33.2}$$

where an isotropic lattice conductivity

$$\kappa_g = \begin{pmatrix} \kappa_g & 0 & 0 \\ 0 & \kappa_g & 0 \\ 0 & 0 & \kappa_g \end{pmatrix} \tag{33.3}$$

is assumed. Inverting Equation (33.2), the transverse magnetoresistance is

$$W = W_e \frac{(1+\gamma) + \gamma(\omega_c\tau)^2}{(1+\gamma)^2 + \gamma^2(\omega_c\tau)^2} , \tag{33.4}$$

where $\gamma \equiv \kappa_g/\kappa_e$ and $W_e \equiv 1/\kappa_e$. The lattice conductivity causes the magnetoresistance to increase initially proportional to H^2 saturating in high fields when $\gamma\omega_c\tau > 1$ to $W_g \equiv 1/K_g$. The quadratic component is proportional to the lattice conductivity κ_g.

For a metal having open orbits, we model the Fermi surface by two separate surfaces, a sphere and a cylinder [6, 7]. Magnetic breakdown is neglected. The thermal conductivity tensor $\overleftrightarrow{\kappa}_c$ of a cylinder, having an axis $\hat{\omega}$, is most conveniently expressed in the uvw coordinate system (Figure 33.1), defined by $\hat{u} = (\hat{w} \times \hat{y})/\sin\phi$ and $\hat{v} = (\hat{w} \times \hat{u})$. ϕ is the angle between $\hat{w}$ and $\hat{y}$, and θ is the angle between the projection of $\hat{w}$ in the xz plane and $\hat{z}$. In the uvw frame,

$$\kappa_c = \kappa_e \begin{pmatrix} \frac{1}{1+\delta^2} & \frac{-\delta}{1+\delta^2} & 0 \\ \frac{\delta}{1+\delta^2} & \frac{1}{1+\delta^2} & 0 \\ 0 & 0 & 0 \end{pmatrix} , \tag{33.5}$$

where $\delta \equiv \omega_c\tau \sin\phi\cos\theta$.

Letting η be the electron fraction of the cylinder, the total thermal conductivity is obtained by summing Equations (33.1), (33.3), and (33.5) after transforming Equation (33.5) into the xyz frame,

$$\overleftrightarrow{\kappa} = (1-\eta)\overleftrightarrow{\kappa}_s + \eta\tilde{S}\overleftrightarrow{\kappa}_c S + \overleftrightarrow{\kappa}_g , \tag{33.6}$$

where the orthogonal transformation matrix is

$$S \equiv \begin{pmatrix} -\cos\theta & 0 & \sin 0 \\ \sin\theta\cos\phi & -\sin\phi & \cos\theta\cos\phi \\ \sin\theta\sin\phi & \cos\phi & \cos\theta\sin\phi \end{pmatrix} . \tag{33.7}$$

33.3 Results

Thermal magnetoresistance rotation patterns are measured by rotating the magnetic field or the crystal about the current direction (Figure 33.2). Taking the current in the y direction, we have calculated the magnetoresistance as a function of field strength and orientation by inverting Equation (33.6). (Thermoelectric effects are again neglected.) Sharp peaks proportional to H^2, shown in Figure 33.3, occur when $\vec{H}$ is perpendicular to the open-orbit direction $\hat{w}$, as in the electrical magnetoresistance [3, 7]. The lattice conductivity, however, causes the peak height to saturate in higher fields, as shown in Figure 33.4. Comparison of the magnetoresistance at and away from the peak shows that it saturates earlier when there is open-orbit current In Figures 33.3 and 33.4, the open-orbit electron fraction η is 0.1, the ratio $\gamma \equiv \kappa_g/\kappa_e$ of the lattice and electronic conductivities is 0.001, and the angle ϕ between the open-orbit and current directions is 45°.

A surprising phenomenon occurs when the open-orbit direction is perpendicular to the current. In this orientation, the open orbits cannot conduct in the direction transverse to the current and magnetic field, and therefore open-orbit peaks do not occur. Instead, sharp minima (open-orbit crevasses), shown in Figure 33.5, are predicted. This phenomenon has the same origin as the open-orbit minima in the electrical magnetoresistance of compensated metals [3]. The resistance increase caused by lattice conduction (or compensation) is reduced by the open-orbit current. This effect is even more startling when the open-orbit direction is tilted out of the rotation plane. Then an open-orbit peak appears but becomes inverted, as shown in Figure 33.6.

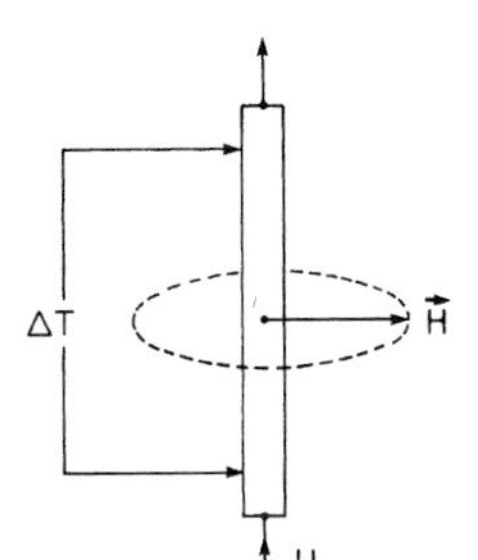

Figure 33.2 Thermal magnetoresistance rotation experiment. U is the heat current and ΔT is the temperature difference.

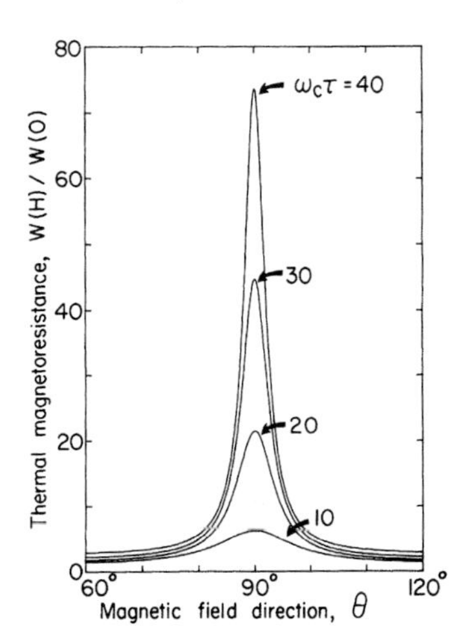

Figure 33.3 Thermal magnetoresistance. $\kappa_g/\kappa_e = 0.001$, $\eta = 0.1$, $\phi = 45°$.

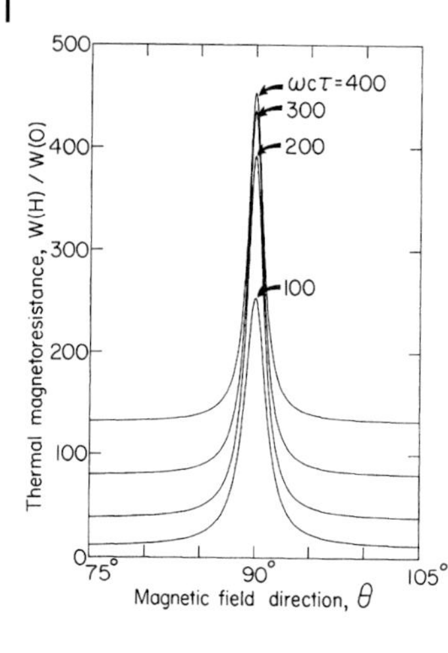

Figure 33.4 Thermal magnetoresistance. $\kappa_g/\kappa_e = 0.001$, $\eta = 0.1$, $\phi = 45°$. Notice that the magnetoresistance at $\theta = 90°$ saturates.

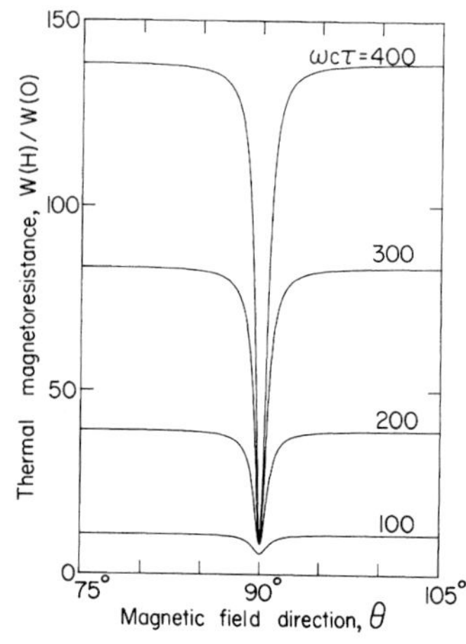

Figure 33.5 Thermal magnetoresistance. $\kappa_g/\kappa_e = 0.001$, $\eta = 0.1$, $\phi = 90°$. An open-orbit crevasse is shown.

From Equation (33.6), we find that the peak magnetoresistance ($\theta = 90°$) is

$$W = W_e \frac{(1 - \eta \sin^2 \phi + \gamma) + (\eta \cos^2 \phi + \gamma)(\omega_c \tau)^2}{(1 - \eta + \gamma)(1 + \gamma) + \gamma(\eta + \gamma)(\omega_c \tau)^2}, \tag{33.8}$$

which increases initially proportional to H^2 but saturates in high fields when $\gamma(\eta + \gamma)(\omega_c \tau)^2 > 1$ to

$$W \to W_g \frac{\eta \cos^2 \phi + \gamma}{\eta + \gamma} \tag{33.9}$$

where $\gamma \equiv \kappa_g/\kappa_e$, $W_e \equiv 1/\kappa_e$, and $W_g \equiv 1/\kappa_g$. If γ is much less than η and $\eta \cos^2 \phi$, the saturation value (33.9) reduces to $W_g \cos^2 \phi$, providing a measurement of the lattice conductivity if the open-orbit direction is known.

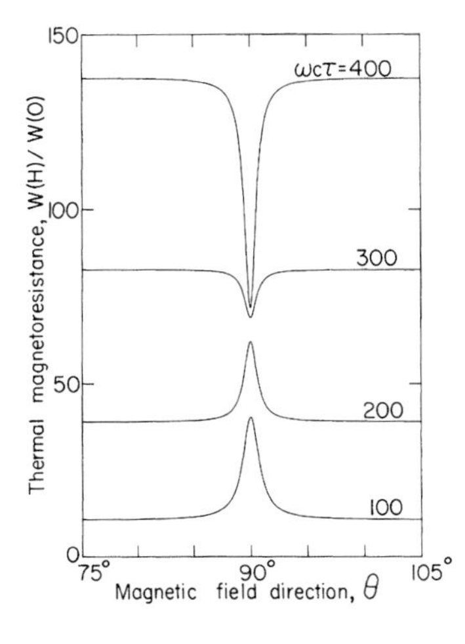

Figure 33.6 Thermal magnetoresistance. $\kappa_g/\kappa_e = 0.001$, $\eta = 0.1$, $\phi = 75°$.

33.4 Discussion

Exploiting the open-orbit magnetoresistance to measure the lattice conductivity requires pure metals having a large open-orbit electron fraction η. Possible candidates are the noble metals Cu, Ag, and Au. For example, in an annealed Cu crystal, having a residual resistivity ratio RRR = 20 000 [8], the low-temperature electron relaxation time is $\tau \approx 5 \times 10^{-10}$ s, implying $\omega_c\tau \approx 1000$ when $H = 100\,\mathrm{kG}$. Using the Wiedemann–Franz law, the zero-field electronic conductivity $\kappa_e = 3.1 \times 10^2\,\mathrm{T\,W/cm\,K^2}$, whereas from alloy measurements [9] the lattice conductivity $\kappa_g = 1.8 \times 10^{-3}\,\mathrm{T^2\,W/cm\,K^3}$, yielding $\gamma \equiv \kappa_g/\kappa_e = 0.6 \times 10^{-5}\,\mathrm{T}$. Assuming $\eta = 0.1$ and $T = 2\,\mathrm{K}$, $\gamma\eta(\omega_c\tau)^2 \sim 1$ so that saturation may be observable.

Other methods for measuring the lattice conductivity include suppression of the electronic conductivity either by alloying [9] or with a magnetic field in the Corbino geometry [10] and measurement of the quadratic component in the thermal magnetoresistance [5]. The latter two methods may give spurious results when there are open orbits. The open-orbit contribution to the quadratic component may be avoided by rotation diagram experiments, since all orbits are closed away from a peak. A theory of the Corbino effect for a metal having open orbits remains to be developed.

Electrical magnetoresistance rotation diagrams have been used for many years to detect and to measure open-orbit directions [3]. According to the theory developed here, thermal rotation diagrams can serve the same purpose. Open-orbit directions lying in the rotation plane, which would not produce open-orbit peaks in an electrical measurement, may be detectable in a thermal measurement as open-orbit crevasses.

Acknowledgements

We are grateful to the NSF Materials Research Laboratory Program for support of this research.

References

1. Azbel, M.Ia., Kaganov, M.I., and Lifshitz, I.M. (1957) *Zh. Eksp. Teor. Fiz.* **32**, 1188 [(1957) Sov. Phys. – JETP 5, 967].
2. Lifshitz, I.M., Azbel, M.Ia., and Kaganov, M.I. (1956) *Zh. Eksp. Teor. Fiz.* **31**, 63 [(1957) Sov. Phys – JETP 4, 41].
3. For a theoretical and experimental review of high-field electrical magnetoresistance, see Fawcett, E. (1964) *Adv. Phys.* **13**, 139.
4. Here and in the rest of this paper, only uncompensated metals, i.e., those having unequal numbers of electrons and holes, are considered.
5. Fletcher, R. (1974) *J. Phys. F* **4**, 1155.
6. Overhauser, A.W. (1974) *Phys. Rev. B* **9**, 2441.
7. Huberman, M. and Overhauser, A.W. (1982) *Phys. Rev. B* **25**, 2211.
8. Martin, P.M., Sampsell, J.B., and Garland, J.C. (1977) *Phys. Rev. B* **15**, 5598.
9. White, G.K. and Woods, S.B. (1955) *Can. J. Phys.* **33**, 58.
10. De Lang, H.N., van Kempen, J., and Wyder, P. (1978) *J. Phys. F* **8**, L39.

Reprint 34 Insights in Many-Electron Theory From the Charge Density Wave Structure of Potassium[1]

*Albert W. Overhauser**

* Department of Physics, Purdue University, 525, Northwestern Avenue, West Lafayette, Indiana 47907, USA

Approximations frequently employed to simplify fundamental treatments of the theory of metals can lead to qualitatively incorrect conclusions. A striking example is the broken translational symmetry of the ground state of metallic potassium. Experimental confirmation of its charge-density-wave structure is reviewed. Such a structure cannot occur if local-density approximations to exchange and correlation are employed, or if Coulomb potentials are replaced by screened interactions.

34.1 Introduction

Many-electron theory becomes a most delicate problem in systems having a degenerate ground state. The Jahn–Teller effect comes to mind as an example. It hardly seems necessary to argue that an infinite degeneracy would provide an even greater challenge to theory. Metals fit this category since their finite density of states at the Fermi surface implies the existence of an infinite number of excited states within ε of the ground state (however small ε might be). Superconductivity, spin density waves, charge density waves, the Kondo effect, and the Mott metal–insulator transition point to a rich spectrum of intriguing phenomena of which, perhaps, many are yet undiscovered. That some have been given names does not mean they are well understood.

Consider the simplest metal of all: "jellium" – a favorite model for theorists, in that the positive-ion background is taken to be without structure. Electrons have charge and spin. So the fundamental response mechanisms are described by a (wave-vector-dependent) polarizability and spin susceptibility. Many theorists have tried to calculate these functions. Bloch [1] was the first to determine the enhancement of the spin susceptibility (over the Pauli value) resulting from exchange interactions. Wigner [2] pointed out that dynamic correlations reduce the magnitude of the enhancement. This latter contribution is a true many-body effect. (The exchange enhancement is a "mean-field" term and is known exactly.) A survey of theoretical "opinion" was compiled by Kushida et al. [3], and is shown in Figure 34.1. The lack of consensus can hardly be exaggerated. Clearly evidence from the real world is necessary to discipline our habits of thought and to dictate limits to our favorite approximations.

Fortunately two of the alkali metals Na and K, come close to the theoretical ideal. The crystalline potential acting on the conduction electrons is so small that the degenerate Fermi sea is free-electron like. A common belief, supported by accurate de Haas–van Alphen measurements [4], is that the Fermi surface deviates by only one or two parts per thousand from a perfect sphere. Nevertheless, many other measurements show that the translation symmetry of Na and K has been broken by a CDW instability, and that their Fermi surfaces are in fact multiply connected.

In the following sections we shall review some of the phenomena which reveal the CDW structure of K. Most research is carried out on K because Na crystals are destroyed by

1) P.-O. Löwdin and B. Pullman (eds.), New Horizons of Quantum Chemistry, 357–372. Copyright © 1983 by D. Reidel Publishing Company.

Anomalous Effects in Simple Metals. Albert Overhauser

ISBN: 978-3-527-40859-7

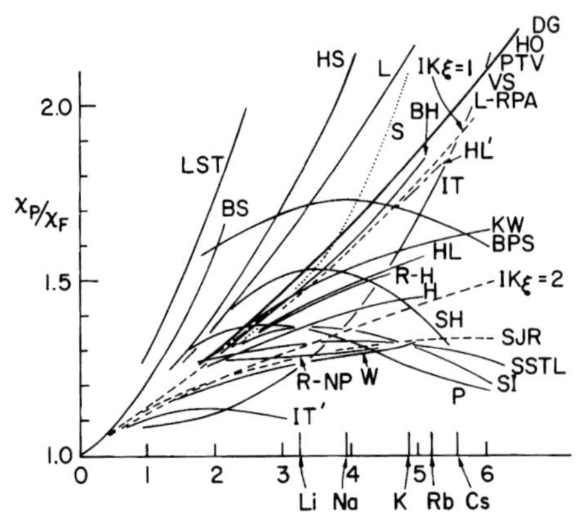

Figure 34.1 Enhancement of the Pauli spin susceptibility for an electron gas, over that for free Fermions, versus electron separation parameter r_s. Values of r_s for the alkali metals are indicated. The labels on the curves refer to the theorists involved and may be interpreted using [3].

a Martensitic transformation near 35 °K. Our attention will focus on recent experimental and theoretical research, which confirms the broken symmetry of the electronic ground state in a spectacular way.

The relevance of these facts to many-electron theory will be discussed in Section 34.5. We note here that broken-symmetry states in simple metals were anticipated theoretically, [5] and that a microscopic theory of the correlation energy [6] shows a CDW structure, rather than a spin density wave, to be favored. Controversy arises from oversimplified accounts of many-electron effects, such as local-density treatments of exchange and correlation, or the use of screened interactions in energy calculations. Such approximations do not permit broken symmetries in simple metals. Appreciation of the ground state of K transcends, therefore, a mere account of the peculiarities of a particular metal.

34.2 Optical Absorption of a CDW

An incommensurate CDW arises from, and causes, an extra periodic potential acting on the conduction electrons.

$$V(\vec{r}) = G \cos \vec{Q} \cdot \vec{r} \,. \tag{34.1}$$

The wavevector $\vec{Q}$ is approximately equal to the diameter $2k_F$ of the Fermi surface and, in general, is not an integral fraction of some reciprocal lattice vector. The presence of this potential in the one-electron Schrödinger equation causes energy gaps of magnitude G to appear in the energy spectrum $E(\vec{k})$ which, for jellium, would have been merely $\hbar^2 k^2/2m$. The modified spectrum is shown in Figure 34.2. The Fermi level E_F is shown to coincide with the lower edge of the CDW energy gap. New optical transitions from filled states (below E_F) to empty states (above) become allowed [7]. An important property of the new absorption is that it is uniaxial; it is proportional to $\cos^2(\hat{\varepsilon} \cdot \hat{Q})$, where $\hat{\varepsilon}$ is the polarization vector of the photon inside the metal. CDW optical transitions are vertical, in Figure 34.2, and have a threshold equal to the gap G.

The first evidence for a CDW structure in K came from the optical absorption measurements of Mayer and El Naby [8]. They found a strong, anomalous absorption having a threshold $\sim$ 0.6 eV, well below the fundamental interband edge at 1.3 eV. Their work has been reproduced by Hietel [9] and by Harms [10], all of whom worked on bulk K surfaces. Other workers, using evaporated films, have not observed the Mayer–El Naby anomaly; and this can be explained, because $\hat{Q}$ should be perpendicular to the surface of an evaporated film [11].

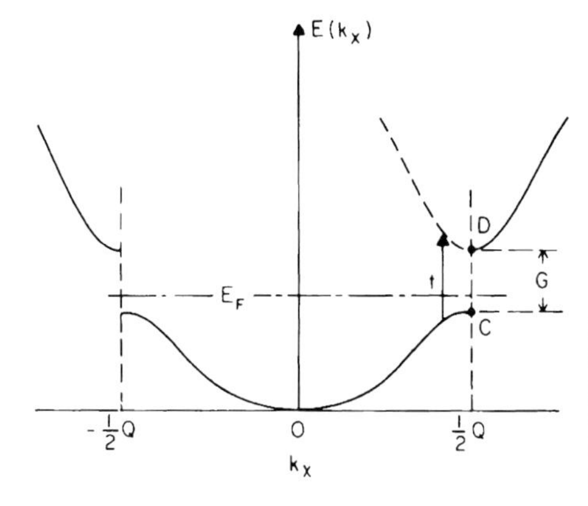

Figure 34.2 Conduction-electron energy spectrum $E(k_x)$ for a CDW with $\vec{Q}$ along $\hat{x}$. The energy G, between states C and D is the magnitude of the CDW potential, Equation (34.1). The optical transitions responsible for the Mayer–El Naby anomaly are indicated by the arrow t.

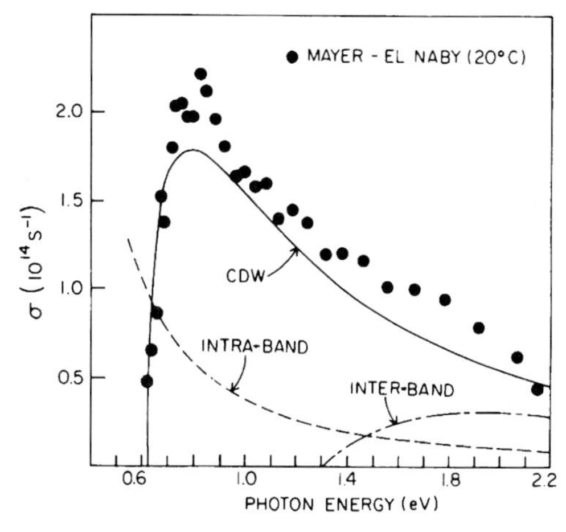

Figure 34.3 Optical absorption of K. The intra-band Drude contribution has been subtracted from the data. Without a CDW structure only the inter-band contribution would remain. The solid curve is theoretical and assumes the presence of a CDW with energy gap $G = 0.62$ eV, [7].

Inside the metal, then, is perpendicular to $\hat{Q}$ and the anomalous absorption strength should be ~ 0.

The Mayer–El Naby anomaly is shown in Figure 34.3. It is independent of temperature and even persists into the liquid state. Its size and spectral shape are in excellent agreement with the theoretical spectrum [7], for which the gap G is the only unknown parameter. A similar anomaly is found [9] in Na, for which the gap is ~ 1.2 eV.

The influence of the potential (34.1) causes the electronic wave-functions to acquire extra momentum components.

$$\psi_{\vec{k}} \cong e^{i\vec{k}\cdot\vec{r}} + \alpha(\vec{k})e^{i(\vec{k}+\vec{Q})\cdot\vec{r}} + \beta(\vec{k})e^{i(\vec{k}-\vec{Q})\cdot\vec{r}} . \qquad (34.2)$$

When one sums $|\psi_{\vec{k}}|^2$ over all occupied states, the total electron density is no longer uniform. It has a small sinusoidal modulation p.

$$n(\vec{r}) = n_0\left[1 - p\cos\vec{Q}\cdot\vec{r}\right]. \qquad (34.3)$$

The Coulomb potential that would result from this charge modulation is compensated by a corresponding modulation in the positive-ion lattice [6]. The ions suffer a periodic displacement from their lattice sites $\vec{L}$:

$$\vec{u}(\vec{L}) = \vec{a}\sin\vec{Q}\cdot\vec{L} . \qquad (34.4)$$

Because of this cancellation, which is almost perfect [12], the CDW potential arises exclusively from exchange and correlation interactions. CDW's provide a critical test of many-electron theory since, for them, exchange and correlation is not a mere correction to a dominant Hartree potential

34.3 Other CDW Phenomena in K

Before discussing the recent, direct observation of open orbits in K, which show that its Fermi surface is multiply connected, we shall give a brief review of a number of anomalous properties studied during the last eighteen years. These also provide direct evidence for the broken symmetry; and give considerable information about other microscopic details.

34.3.1 Conduction-Electron Spin-Resonance Splitting

The g-factor for spin resonance (CESR) in a metal is a measure of the Fermi surface average of $g(\vec{k})$. This averaging occurs because electrons are scattered through $\sim 10^4$ states at the Fermi surface during one spin-lattice relaxation time. Nevertheless the CESR in K was found to be split into well resolved components [13] separated up to $\sim$ 0.5 Oe in an applied field of 4000 Oe. Splittings up to 5 Oe were observed for $H = 40\,000$ Oe [14]; so a g-factor splitting is involved. It is about 10% of the g shift caused by spin-orbit coupling. This splitting was explained quantitatively [15] by the Fermi surface distortion created by the CDW, which leads to a g-factor dependent on the angle between $\vec{H}$ and $\vec{Q}$. The splitting arises for samples subdivided into $\vec{Q}$ domains, and is proportional to the CDW energy gap G, Equation 34.1. The same value, $G = 0.6$ eV, obtained from the optical anomaly is required to fit the observed splitting.

34.3.2 Doppler-Shifted Cyclotron-Resonance Shift

Helicon waves (of wave vector q) generated by radio frequency excitation in a magnetic field $\vec{H}$ (normal to a K surface) can cause cyclotron resonance. This occurs when electrons traveling at the Fermi velocity, $v_F = \hbar k_F / m^*$, satisfy the doppler-shifted frequency condition [16]:

$$q v_F = e H_r / m^* c\,. \qquad (34.5)$$

Since m^* cancels in this equation, the resonance field H_r depends only on k_F which in turn depends only on the lattice constant. For 10 MHz helicons [17], the resonance field is 4% less than the 18 kOe value required by an energy band with spherical symmetry. This shift arises from the effect of the CDW on the $\vec{Q}$ component of conduction-electron velocity. It is explained quantitatively [18] by a CDW energy gap $G = 0.6$ eV.

34.3.3 Residual Resistivity Anisotropy

At helium temperature the electrical resistivity of a metal is dominated by electron scattering from impurities. Since these are point imperfections, the residual resistivity must then be isotropic (if the metal has cubic symmetry). However, measurements on single crystal spheres [19, 20] have shown that, instead, the residual-resistivity tensor is a cigar-shaped ellipsoid with a major to minor axis ratio $\gtrsim 5$. The presence of the CDW energy gaps, Figure 34.2, and the perturbed wavefunctions Equation (34.2), cause the resistivity parallel to $\vec{Q}$ to be about four times larger than perpendicular to $\vec{Q}$ [21] This ratio is reached if the CDW gap is $G = 0.6$ eV.

34.3.4 Linear Magnetoresistance

In large magnetic fields, such that the cyclotron frequency $\omega_c\tau$ and scattering time τ satisfy $\omega_c\tau \gg 1$, a metal (such as Na or K) should not exhibit a magnetoresistance if its Fermi surface is simply connected [22]. All workers have found instead that the resistance increases linearly in H without evidence of saturation, even for $\omega_c\tau \sim 400$. This can be explained (as an intrinsic property) only if the Fermi surface is multiply connected [22]. In Section 34.4 we shall show how the required open orbits arise from a CDW.

34.3.5 Induced Torque Anisotropy

If a spherical single crystal is placed in a rotating magnetic field, the induced currents create a torque; and for $\omega_c\tau \gg 1$ this torque is proportional to the resistivity [23]. For a metal with a spherical Fermi surface the high-field torque should be isotropic. Measurements on ~ 200 specimens [19] of K revealed, however, a striking four-peaked pattern, even when the rotation axis was the three-fold, (111) axis. The anisotropy exhibited a range from 3:1 to 15:1. These observations were verified by Holroyd and Datars [20], who found one sample with an anisotropy of 45:1 (at 23 kOe) This sample was spherical in shape to a precision of 2×10^{-3}. The torque anisotropics show that the symmetry of K (and also of Na) is less than cubic. The broken symmetry of a CDW is indicated.

34.3.6 The Oil Drop Effect

If small, spherical samples are supported in an induced-torque apparatus without being fastened with frozen oil, nearly isotropic torque patterns can be obtained [20]. However, if these same samples are remeasured after a drop of oil has been attached, large torque anisotropies are recovered. This phenomenon provides evidence for variations in $\vec{Q}$-domain texture (analogous to magnetic domains in a ferromagnet). An oil-free specimen may be subdivided into a random distribution of the twenty four possible orientations of $\vec{Q}$. It would then be macroscopically isotropic. A frozen oil drop creates a differential thermal stress which breaks the 24-fold degeneracy of $\vec{Q}$ orientations. This leads to a preferred texture (similar to the alignment of magnetic domains in a ferromagnet) and subsequently to an anisotropic induced-torque pattern.

34.3.7 Variability of the Residual Resistivity

The residual resistivity, caused by impurity scattering, varies from run to run on the same sample [24]. The changes can be in either direction depending on sample history, environment, and the rate of cooling from room to helium temperature. Sometimes the changes are as much as a factor of four. The residual resistivity anisotropy, Section 34.3.3, provides an explanation; since each run, beginning with a fresh cooldown from room temperature, permits a realignment of the Q-domains, stimulated by thermal stress and altered surface conditions. The magnitude of the linear magnetoresistance Section 34.3.4, also varies from run to run.

34.3.8 High-Field Hall Constant

In magnetic fields so high that $\omega_c\tau \gg 1$ the Hall coefficient must approach the constant value, $-1/nec$, where n is the conduction-electron density, known with a precision of 3×10^{-4} from the lattice constant. Different experimental techniques [25–27] all show violations, of the order of 3–8%, depending on crystal orientation. This phenomenon also indicates a multiply-connected Fermi surface.

34.3.9 Phason Heat Capacity Anomaly

Whenever there is a broken symmetry (involving a group of infinite order – e.g., the translation group) new collective excitations arise and have a frequency spectrum that goes to zero. For an incommensurate CDW these modes correspond to a dynamic phase modulation of the CDW [28]. They lead to a heat-capacity anomaly near 1 °K [29, 30]. This effect has been observed [31], and has the expected location in temperature and magnitude.

34.3.10 Direct Observation of Electron-Phason Scattering

Pointcontact spectroscopy is a new technique (similar to that using superconducting tunnel junctions) for measuring inelastic scattering in normal metals [32]. The method has been applied to K [33], and the observed spectrum above 2 meV is in good agreement with the theory for phonon scattering [34]. However a reproducible, sharp anomaly below 1 meV is also found which cannot be ascribed to ordinary acoustic phonons. Inelastic scattering of electrons by the new CDW excitations – phasons – explains the anomaly quantitatively [35], and without an adjustable parameter.

34.3.11 Difficulties

There are other phenomena not listed above – e.g., cyclotron phase resonance, ultrasonic attenuation, low-field Hall effect, etc. which violate the expected behavior for a nearly-free electron model. However, there are also data which agree with the standard model and which have not yet been satisfactorily explained for a CDW structure. One of these is the apparent lack of an inhomogeneously broadened Knight shift in nuclear magnetic resonance [36]. Possibly very-long-range, indirect exchange interactions [37] may have something to do with this. The most important problem, however, is the non-observation of CDW effects in de Haas–van Alphen measurements [4]. A speculative explanation is that samples used have very small $\vec{Q}$-domain size – perhaps so small that the coherence volume of a Landau level contains 10^3–10^4 domains. Interactions may lead then to an average de Haas–van Alphen frequency. Samples with larger $\vec{Q}$-domain size should reveal unusual effects. It should be noted, in this respect, that workers frequently discard data solely for the reason it doesn't behave "properly". In Cs, which also has an optical anomaly and a non-saturating magnetoresistance up to 90% of the de Haas–van Alphen data is rejected [38]. Perhaps the discarded samples were in fact the better ones.

34.4 The Open Orbits of Potassium

The most spectacular evidence for a CDW structure in K is the observation of open-orbit peaks of magnetoresistance versus crystal orientation. Open orbits in momentum space (and in real space) occur whenever energy gaps truncate the Fermi surface so that some electrons cannot complete a closed cyclotron orbit. Instead the energy gaps cause a periodic Bragg reflection which forces those electrons to travel in (essentially) straight lines, even though $\omega_c \tau \gg 1$.

34.4.1 Origin of Open Orbits

The existence of a CDW in K causes its Fermi surface to be cut by several pairs of energy-gap planes, as shown in Figure 34.4. The fact that the positive ions form a discrete lattice, and not a smooth background, is the cause of (at least) two extra pairs, in addition to the main gaps. The heterodyne gaps arise from the beat periodicity between the lattice and the CDW.

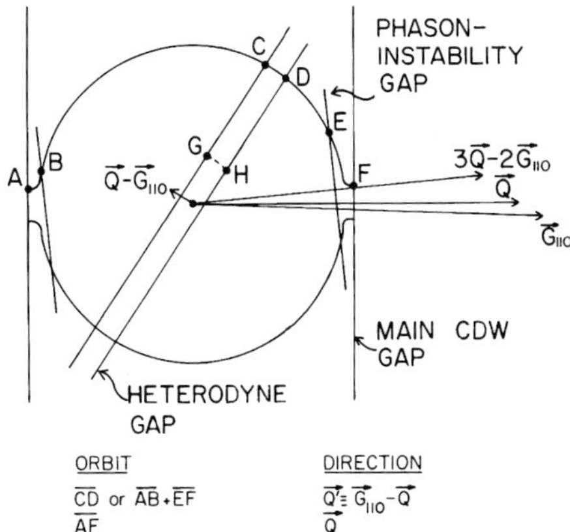

Figure 34.4 Fermi surface, energy-gap planes and open orbits for K with an incommensurate CDW. Magnetic breakdown is possible at the heterodyne gaps and the phason-instability gaps, but not at the main CDW gaps.

A static phason instability [39], caused by the lack of commensuration, will also lead to a pair of small, subsidiary gaps, as shown.

In large magnetic fields the smaller gaps can be "broken down". That means that electrons can jump the gaps as if they were not there. Employment of all such options leads to five possible open orbit directions (in momentum space); and these are shown in Figure 34.4.

34.4.2 Theoretical Open-Orbit Spectrum

The optimum $\vec{Q}$ direction for a CDW does not coincide with a symmetry axis [40]. Consequently in a single crystal there will be twenty four equivalent axes. In general a macroscopic sample will be divided into $\vec{Q}$ domains. So the number of possible open-orbit directions is $5 \times 24 = 120$.

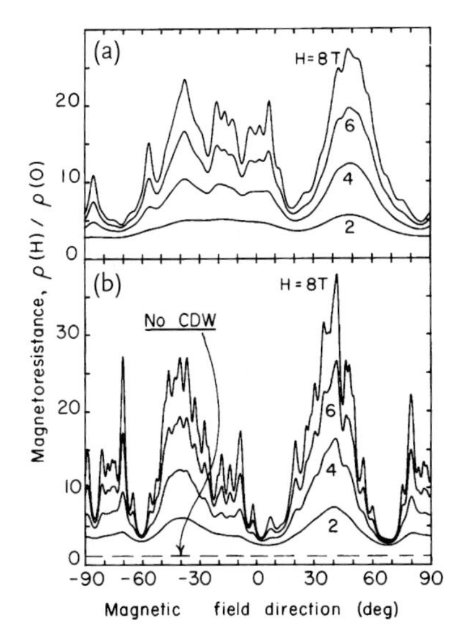

Figure 34.5 Theoretical magnetoresistance rotation patterns calculated using effective medium theory. In (a) the $\vec{Q}$-domain size was taken to be 0.05 mm, and $\vec{H}$ was allowed to rotate in a (2$\bar{1}\bar{1}$) plane. In (b) the $\vec{Q}$-domain size was 0.12 mm, and $\vec{H}$ was in a (321) plane. Note the horizontal, dashed line in (b), which is the only allowed theoretical result for a spherical Fermi surface and no broken symmetry.

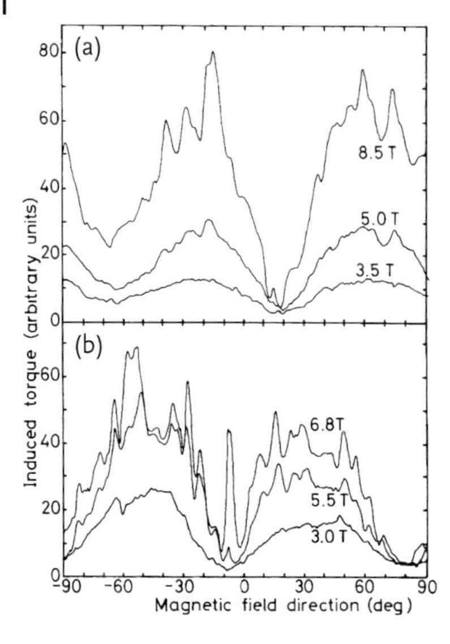

Figure 34.6 Experimental magnetoresistance rotation patterns for single-crystal spheres of K. The data, taken by Coulter and Datars at 1.4 °K, cannot be compared in detail with Figure 34.5 since the volume fractions for each of the twenty four types of $\vec{Q}$-domain are unknown. The rotation planes are the same as in Figure 34.5.

In order to calculate the magnetoresistance rotation spectrum for such a complicated system one must model the resistivity tensor for each $\vec{Q}$-domain, and then use effective medium theory [41] to treat the heterogeneous domain structure. This leads to patterns [42] such as those shown in Figure 34.5. Sharp open-orbit structure appears for fields above 4 T, if the $\vec{Q}$-domain size is taken to be $\sim$ 0.1 mm. The patterns shown are to be compared with the horizontal line in Figure 34.5b, which is the theoretical behavior (for all H) if the metal has a spherical Fermi surface.

34.4.3 Observed Open-Orbit Spectrum

Coulter and Datars have measured [43] magnetoresistance rotation patterns to 8.5 T by means of the induced torque technique. The results of two runs (typical of one hundred experiments) are shown in Figure 34.6. There can be no doubt that the Fermi surface of K is multiply connected. Further work at much higher fields is needed in order to determine the crystallographic directions of the various open orbits. Theoretically, some orbits will remain unresolved even at 24 T [42]. Similar open-orbit patterns are also found in Na.

34.5 Implications for Many-Electron Theory

Since the origin of CDW's in simple metals is understood microscopically [5, 5], the purpose of this section is to stress the limitations of several approximations frequently employed in many-electron theory. The failure (when using these) to foresee CDW structure, or to provide understanding, even after CDW's have been exhibited, should be kept in mind whenever one is tempted to use them.

34.5.1 Local-Density Approximations to Exchange and Correlation are not Predictive

The Slater approximation to the exchange potential [44] is

$$V_{XS} = -3e^2 k_F(\vec{r})/2\pi \,, \tag{34.6}$$

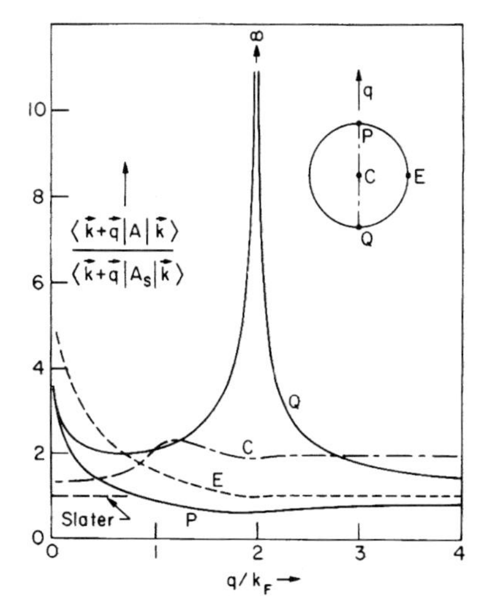

Figure 34.7 Ratio of exact matrix elements of the exchange operator to those given by Slater's local-density approximation. Each curve is for the point $\vec{k}$ of the Fermi sea shown in the inset. The electron gas was taken to have a small sinusoidal density modulation, $\varepsilon \cos \vec{q} \cdot \vec{r}$. The matrix elements are in the (pure) momentum representation.

where $k_F(\vec{r})$ is the Fermi wave vector for a degenerate electron gas of local density $n(\vec{r})$.

$$k_F(\vec{r}) = [3\pi^2 n(\vec{r})]^{1/3} \tag{34.7}$$

It has been argued [45] that Equation (24.7) should be reduced by a factor 2/3. Many introduce a parameter a, so that the exchange (and correlation) potential is:

$$V_{XC} = \alpha V_{XS} \,. \tag{34.8}$$

It is easy to show that a CDW could never occur in K if this approximation were taken seriously.

Consider the periodic potential, Equation (34.1), and let $G = 0.6\,\text{eV}$ and $Q = 2k_F$. A calculation of the fractional density modulation, p of Equation (34.3), leads to $p \approx 0.17$. If this modulation is inserted into Equation (34.7), the result for the amplitude of V_{XC}, Equation (34.8) with $\alpha = 2/3$, is 0.2 eV. A local approximation is inconsistent by a factor of three. G would quickly iterate to zero.

Exchange potentials are highly non-local. This has been shown [46] by calculating the matrix elements of the exact exchange operator A for an electron gas with a small sinusoidal density modulation, $p \cos \vec{q} \cdot \vec{r}$. The off-diagonal matrix elements $\langle \vec{k} + \vec{q}|A|\vec{k}\rangle$, in Slater units defined by Equation (34.6), are illustrated in Figure 34.7 for four values of $\vec{k}$ in the Fermi sea. The enormous disparity between the four curves exhibits a pathologic, non-local behavior. The singularity at $q = 2k_F$ for the $\vec{k}$ value denoted by $\vec{Q}$, in the figure, is responsible for the SDW and CDW instability theorem [5]. When dynamic correlation potentials are included, the non-locality remains, though it is somewhat smaller [47]. A Kohn–Sham local-exchange potential corresponds to a horizontal line in Figure 34.7 (at the ordinate 2/3) for all values of $\vec{k}$.

Na and K are the quantum systems which come closest to the idealization of a nearly-uniform electron gas, upon which all local-density functionals are based. Yet for these cases, where agreement ought to be best, the local-density method fails to reveal the qualitative nature of the ground-state charge density.

34.5.2 Screened Interactions are Dangerous

Since the work of Bohm and Pines [48] and the Fermi liquid theory of Landau [49], it has been common to treat conduction electrons with a Fermion Hamiltonian employing screened Coulomb interactions. This is a valid approximation when electron–electron scattering processes are involved in transport problems. But such a model Hamiltonian cannot be used to investigate alternative ground-state structures. Calculations with dynamically screened interactions [50] have shown that any tendency towards SDW or CDW instability is suppressed. Screened interactions are qualitatively correct for treating spin-density excitations, but are qualitatively wrong for charge-density excitations (or instabilities).

34.5.3 The Coulomb Hole is Important

If one expresses the Coulomb energy for N electrons (in pure momentum states) by means of a Fourier analysis, the result is:

$$E_c^0 = N \sum_{\vec{q}} \frac{2\pi e^2}{q^2} + \tag{34.9a}$$

$$-\frac{1}{2} \sum_{\vec{k},\vec{q}} \frac{4\pi e^2}{q^2} \left(n_{\vec{k}\uparrow} n_{\vec{k}-\vec{q}\uparrow} + n_{\vec{k}\downarrow} n_{\vec{k}-\vec{q}\downarrow} \right), \tag{34.9b}$$

where the n's are occupation numbers. The first term is just another way of writing the total rest mass, Nmc^2. The second term, (34.9b), is the exchange energy. The dielectric response of the electrons to the electric fields associated with the energy E_c^0 can be represented by a dielectric function ε_q. Recall that the energy of a charged capacitor is reduced by ε if a dielectric is moved into the electric field; so also will the Coulomb energy of the N electrons be reduced. This reduction, the Coulomb energy part of the correlation energy, is:

$$\Delta E_c = N \sum_{\vec{q}} \frac{2\pi e^2}{q^2} \left(\frac{1}{\varepsilon_q} - 1 \right) \tag{34.10a}$$

$$-\frac{1}{2} \sum_{\vec{k},\vec{q}} \frac{4\pi e^2}{q^2} \left(\frac{1}{\varepsilon_q} - 1 \right) \left(n_{\vec{k}\uparrow} n_{\vec{k}-\vec{q}\uparrow} + n_{\vec{k}\downarrow} n_{\vec{k}-\vec{q}\downarrow} \right), \tag{34.10b}$$

ΔE_C is generally small compared to the exchange energy, (34.9b). However (34.9b) and (34.10b) can be combined; and their sum is the screened-exchange energy. The term (34.10a) is the Coulomb hole energy. It is much larger than screened exchange. It appears to be an innocuous constant, since it doesn't depend on the occupation numbers $n_{\vec{k}}$, and some workers treat it as such. But that is a serious error. If one cancels off the dynamic contributions of bare exchange (34.9b) by combining it with (34.10b), the dynamic effects of the Coulomb hole, (34.10a), become dominant. The Coulomb hole, from this point of view, is the driving force for a CDW ground state.

The more accurate approach to many-electron effects is to leave the two dynamic corrections (34.10a) and (34.10b) combined, for their sum is small. Then the large kinematic term (34.9b) can be calculated with precision. Unscreened exchange is, with this approach, the major cause of a CDW ground state, and ΔE_C provides a small reinforcement [6].

References

1 Bloch, F. (1929) *Z. Physik* **57**, 545.
2 Wigner, E. (1938), *Trans. Faraday Soc.* **34**, 678.
3 Kushida, T., Murphy, J.C. and Hanabusa, M. (1976) *Phys Rev. B* **13**, 5136.
4 Shoenberg, D. and Stiles, P.J. (1964) *Proc. Roy. Soc. A* **281**, 62.
5 Overhauser, A.W. (1962) *Phys. Rev.* **128**, 1437.
6 Overhauser, A.W. (1968) *Phys. Rev.* **167**, 691.
7 Overhauser, A.W. (1964) *Phys. Rev. Lett.* **13**, 190.
8 Mayer, H. and El Naby, M.H. (1963) *Z. Physik* **174**, 269.
9 Hietel, B. and Mayer, H. (1973) *Z. Physik* **264**, 21; and personal communication.
10 Harms, P. (1972) Dissertation, Technische Universität Clasthal, (unpublished).
11 Overhauser, A.W. and Butler, N.R. (1976) *Phys. Rev. B* **14**, 3371.
12 Giuliani, G.F. and Overhauser, A.W. (1981) *Phys. Rev. B* **23**, 3737.
13 Walsh Jr., W.M., Rupp Jr., L.W. and Schmidt, P.H. (1966) *Phys. Rev.* **142**, 414.
14 Dunifer, G., personal communication.
15 Overhauser, A.W. and de Graaf, A.M. (1968) *Phys. Rev.* **168**, 763.
16 Kjeldaas Jr., T. (1959) *Phys. Rev.* **113**, 1473.
17 Penz, P.A. and Kushida, T. (1968) *Phys. Rev.* **176**, 804.
18 Overhauser, A.W. and Rodriguez, S. (1966) *Phys. Rev.* **141**, 431.
19 Schaefer, J.A. and Marcus, J.A. (1971) *Phys. Rev. Lett.* **27**, 935.
20 Holroyd, F.W. and Datars, W.R. (1975) *Can. J. Phys.* **53**, 2517.
21 Bishop, M.F. and Overhauser, A.W. (1978) *Phys. Rev. B* **18**, 2447.
22 Lifshitz, I.M., Azbel, M.Ya. and Kagonov, M.I. (1956) *Zh. Eksp. Teor. Fiz.* **31**, 63 (English translation: (1957) Sov. Phys.-JETP 4, 41).
23 Lass, J.S. and Pippard, A.B. (1970) *J. Phys. E: J. Sci. Instrum.* **3**, 137.
24 Taub, H., Schmidt, R.L., Maxfield, B.W. and Bowers, R. (1971) *Phys. Rev. B* **4**, 1134.
25 Penz, P.A. (1968) *Phys. Rev. Lett.* **20**, 725.
26 Chimenti, D.E. and Maxfield, B.W. (1973) *Phys. Rev. B* **7**, 3501.
27 Werner, S.A., Hunt, T.K. and Ford, G.W. (1974) *Solid St. Comm.* **14**, 1217.
28 Overhauser, A.W. (1971) *Phys. Rev. B* **3**, 3173.
29 Boriack, M.L. and Overhauser, A.W. (1978) *Phys. Rev. B* **18**, 6454.
30 Giuliani, G.F. and Overhauser, A.W. (1980) *Phys. Rev. Lett.* **45**, 1335.
31 Amarasekara, C.D. and Keesom, P.H. (1981) *Phys. Rev. Lett.* **47**, 1311.
32 Yanson, I.K. (1974) *Zh. Eksp. Teor. Fiz.* **66**, 1035 (English translation (1974) Sov. Phys.-JETP 39, 506).
33 Jansen, A.G.M., van den Bosch, J.H., van Kempen, H., Ribot, J.H.M., Smeets, P.H.H. and Wyder, P. (1980) *J. Phys. F: Metal Phys.* **10**, 265.
34 Ashraf, M. and Swihart, J.C. (1982) *Phys. Rev. B* **25**, 2094.
35 Ashraf, M. and Swihart, J.C. (1982) *Phys. Rev. Lett.*, in press.
36 Follstaedt, D. and Slichter, C.P. (1976) *Phys. Rev. B* **13**, 1017.
37 Giuliani, G.F. and Overhauser, A.W. (1982) *Phys. Rev. B* 26, in press.
38 Gaertner, A.A. and Templeton, I.M. (1977) J. Low Temp. Phys. **29**, 205.
39 Giuliani, G.F. and Overhauser, A.W. (1982) *Phys. Rev. B* **26**, in press.
40 Giuliani, G.F. and Overhauser, A.W. (1979) *Phys. Rev. B* **20**, 1328.
41 Huberman, M. and Overhauser, A.W. (1981) *Phys. Rev. B* **23**, 6294.
42 Huberman, M. and Overhauser, A.W. (1982) *Phys. Rev. B* **25**, 2211.
43 Coulter, P.G. and Datars, W.R. (1980) *Phys. Rev. Lett.* **45**, 1021.
44 Slater, J.C. (1951) *Phys. Rev.* **81**, 385.
45 Kohn, W. and Sham, L.J. (1965) *Phys. Rev.* **140**, A1133.
46 Overhauser, A.W. (1970) *Phys. Rev. B* **2**, 874.
47 Duff, K.J. and Overhauser, A.W. (1972) *Phys. Rev. B* **5**, 2799.
48 Bohm, D. and Pines, D. (1953) *Phys. Rev.* **85**, 338.
49 Landau, L.D. (1956) *Zh. Eksp. Teor. Fiz.* **30**, 1058 (English translation (1956) Sov. Phys.-JETP 3, 920).
50 Hamann, D.R. and Overhauser, A.W. (1966) *Phys. Rev.* **143**, 183.

Reprint 35 Charge Density Wave Phenomena in Potassium[1)]

*A. W. Overhauser**

* Department of Physics, Purdue University, 525, Northwestern Avenue, West Lafayette, Indiana 47907, USA

35.1 The Mysteries of the Simple Metals

35.1.1 Introduction

The importance of many-electron effects in metals was realized in 1929 by Bloch [1]. He showed that exchange interactions cause a significant enhancement of the spin susceptibility χ_s over the value χ_p derived by Pauli for free electrons. The enhancement factor (in the Hartree–Fock approximation) is given by

$$\chi_s/\chi_p = (1 - r_s/6.03)^{-1} . \tag{35.1}$$

r_s is the radius of a sphere containing one electron and is related to the density n by

$$n = \left[\frac{4\pi}{3}(r_s a_B)^3\right]^{-1} , \tag{35.2}$$

where a_B is the Bohr radius. For metals this factor would range from 1.5 to 6.

In 1938 Wigner [2] pointed out that electron-electron correlations caused by the dynamical scattering of opposite-spin electrons, would reduce the enhancement given by (35.1). Since then more than fifty theorists have attempted to calculate the effect of correlations on χ_s. The results have been displayed dramatically by Kushida, et al. [3]. The collection of theoretical curves for χ_s vs. r_s given in their Figure 6 looks like a plate of spaghetti.

The purpose of this presentation is to summarize experimental evidence that exchange and correlation effects lead to a broken-symmetry state in a simple metal. Theoretical arguments, which we feel are convincing, will be presented later. This strategy seems best in view of the widely different approaches current in the study of many-electron effects (with different results as cited above)

The most widely studied, simple metal is potassium. Lithium and sodium experience phase transitions on cooling to 4° K, and this destroys single crystals. Rubidium and cesium are very soft and consequently difficult to handle. De Haas–van Alphen experiments on potassium, e. g. Shoenberg and Stiles [4], indicate an isotropic Fermi surface – to about one part in 10^3. Nevertheless, as we will review, other experiments show its Fermi surface is not even simply connected.

35.1.2 Charge-Density-Wave Structure

A CDW is a broken-symmetry, ground state of a metal in which the conduction-electron density has a small sinusoidal modulation [5]

$$\rho = \rho_0[1 + p\cos(\vec{Q}\cdot\vec{r} + \phi)] , \tag{35.3}$$

(over and above any modulation caused by the positive-ion lattice). In general the wavevector $\vec{Q}$ is incommensurate with the lattice, so the phase ϕ is an arbitrary parameter which

1) Electron Correlations in Solids, Molecules, and Atoms. Edited by Jozef T. Devreese and Fons Brosens (Plenum Publishing Corporation, 1983).

Anomalous Effects in Simple Metals. Albert Overhauser

ISBN: 978-3-527-40859-7

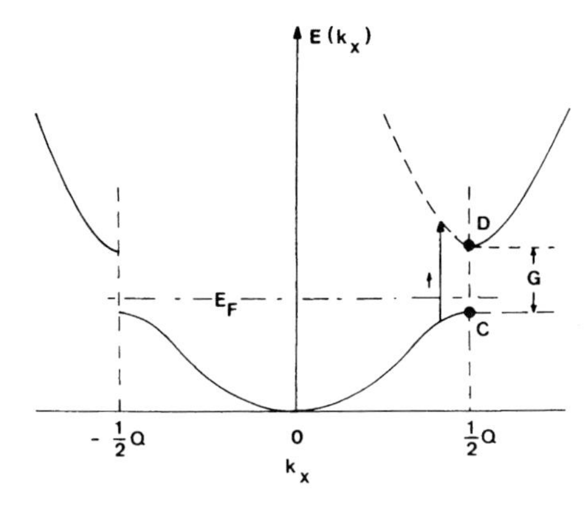

Figure 35.1 Electronic energy $E(k_x)$ against k_x for a CDW structure. The CDW energy gap is G, and the fermi energy E_F lies between the gap edges C and D. The optical transition t is allowed for a CDW structure.

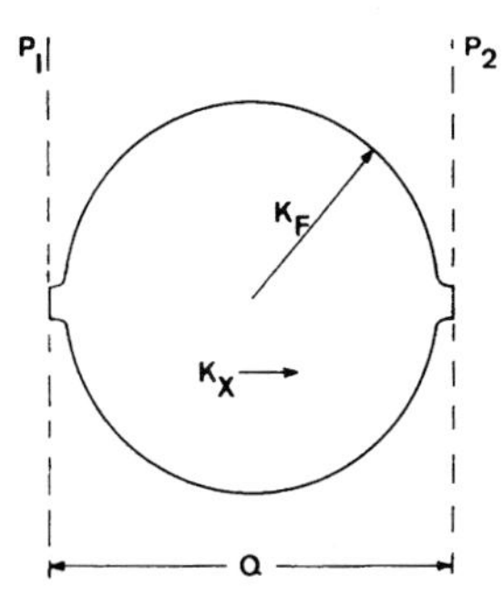

Figure 35.2 Fermi surface of an ideal metal with a CDW ground state. The surface is distorted from that of a sphere of radius k_F near the energy-gap planes p_1 and p_2.

labels the infinite number of degenerate ground states. The one-electron wavefunctions are amplitude modulated by an exchange and correlation potential,

$$V = -G \cos(\vec{Q} \cdot \vec{r} + \phi) , \tag{35.4}$$

in Schrödinger's equation. The theoretical problem addresses the question whether ρ and G are selfconsistent and, if so, whether the CDW state has lower energy than the normal one. Since a modulation arises only when occupied, free-electron states $\vec{k}$ are mixed with empty ones $\vec{k} \pm \vec{Q}$, the optimum $|\vec{Q}|$ is $\sim 2k_F$, the diameter of the Fermi surface.

An obvious consequence of the potential V is the creation of two new energy gaps in k-space along planes passing through $\pm 1/2\vec{Q}$. See Figure 35.1. This leads to a small distortion of the Fermi surface – possibly to the formation of necks, as shown in Figure 35.2. As mentioned above, such effects have not been reported in de Haas–van Alphen studies, a failure which will be discussed below.

35.1.3 Mayer–El Naby Optical Anomaly

The first indication that potassium had a CDW structure was the discovery by Mayer and El Naby [6] of an intense interband, optical transition with a threshold at 0.6 eV, followed by an asymmetric peak at 0.8 eV, as shown in Figure 35.3. The ordinary interband absorption – caused by the Brillouin-zone energy gaps – has a threshold of 1.3 eV. A quantitative account of the anomalous absorption succeeds merely by letting the CDW gap $G = 0.6$ eV and by calculating the optical transition rate across these new energy gaps [7]. The solid curve in Figure 35.3 shows exceptional agreement.

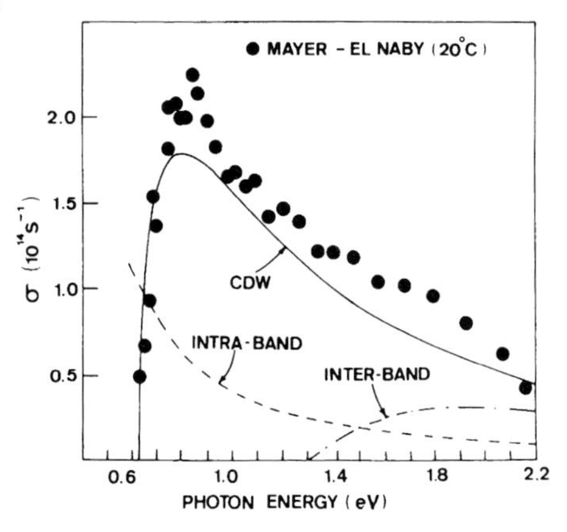

Figure 35.3 Anomalous optical absorption spectrum of potassium. The intra-band conductivity (dashed curve) has been subtracted from the experimental data before being plotted. The solid curve shows the theoretical absorption introduced by a CDW structure. A "normal" metal would exhibit only the inter-band absorption with a threshold, as shown, at 1.3 eV.

Ordinarily potassium, which is body-centered cubic, would have isotropic optical properties. However, the new absorption occurs only from the component of the photon polarization vector $\vec{\varepsilon}$ parallel to $\vec{Q}$. The CDW optical absorption is uniaxial! This is crucial to an explanation of why the Mayer–El Naby anomaly is not obtained in measurements on evaporated films. Such films, deposited on smooth, amorphous substrates, always have a [110] cyrstal direction normal to the surface. It turns out that the optimum direction of Q is then normal to the surface too. Since an infrared photon incident on the film will have its polarization vector ε parallel to the surface, *inside* the metal (even at oblique incidence), anomalous optical absorption *cannot* occur. At bulk–metal vacuum surfaces, the Mayer–El Naby anomaly has been reproduced by three workers. Such anisotropy requires a CDW.

35.1.4 Low-Temperature Magnetoresistance

It is easy to understand that a metal with a spherical Fermi surface will have no magnetoresistance. Suppose that a magnetic field is in the $\vec{z}$ direction, and that the electric field E is in the x–y plane. Let $\vec{P}$ be the total momentum of the electron Fermi sea, and suppose that the units employed allow the following equations for Newton's laws:

$$\frac{dP_x}{dt} = E_x + \omega_c P_y - P_x/\tau \,, \tag{35.5a}$$

$$\frac{dP_y}{dt} = E_y + \omega_c P_x - P_y/\tau \,, \tag{35.5b}$$

The second term on the right hand side is the Lorentz force, with $\omega_c \equiv eH/mc$. The relaxation term was added to take account of scattering by impurities, which allows the electrons to reach equilibrium, $\vec{P} = 0$. In the non-equilibrium steady state $dP/dt = 0$. Furthermore if the wire confines the current to the $\hat{x}$ direction, $P_y = 0$. Equation (35.5a) can be solved immediately for the current.

$$P_x = \tau E_x \,, \tag{35.6}$$

which is Ohm's law. The resistance is independent of the magnetic field! Lifshitz et al. [8] proved that this is true for any simply-connected Fermi surface whenever $\omega_c\tau \gg 1$.

Experimentally all workers have found that $\rho(H)$ for the alkali metals increases with H, without evidence of saturation, even at fields where $\omega_c\tau > 300$. The only *intrinsic* mechanism which allows this behavior is the presence of open orbits. These occur only if the Fermi surface is multiply connected, which would result from a CDW structure. Extrinsic mechanisms for a linear magnetoresistance, such as voids, have been ruled out by measurements which show there are no voids [9].

The size of the magnetoresistance in potassium varies from sample to sample, and from run to run on the same sample. This behavior, found perplexing by the experimentalists, is easily explained by CDW's. The $\vec{Q}$ direction has some optimum axis (α, β, γ) energetically. There are 24 cubically equivalent (α, β, γ) axes. Consequently a macroscopic sample will be divided into $\vec{Q}$ domains, for which the direction of $\vec{Q}$ varies from domain to domain. Each time a sample is cooled to 4°K from room temperature it will have a different $\vec{Q}$-domain distribution. This will cause a different open-orbit distribution and, consequently, a different magneto-resistance.

35.1.5 Induced-Torque Measurements

Inductive techniques were applied to resistance measurements on alkali metals to avoid problems that could arise from placement of the potential probes. The induced-torque method was first used by Schaefer and Marcus [10] to study the orientation dependence of the magnetoresistance of sodium and potassium. Not only did they confirm the non-saturating magnetoresistance but they found that it was highly anisotropic – as much as 15 to 1. There were always four peaks in a 360° rotation, even if a threefold [111] axis were the rotation axis. These results were repeated about 200 times during a two-year period prior to publication. Their results were reproduced by Holroyd and Datars [11], who found one specimen with an anisotropy (at 23 kG) of 45 to 1. See Figure 35.4. These results prove that potassium (and sodium) have neither cubic symmetry nor simply-connected Fermi surfaces.

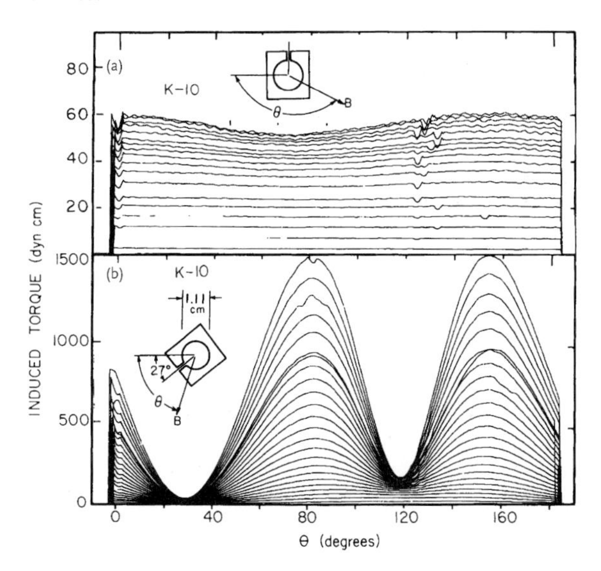

Figure 35.4 Induced-torque against magnetic field direction θ for a potassium sphere 1.11 cm in diameter. In (a) the field B is rotated about the growth axis. The curves shown are for 0.5, 1, 2, 3, ..., kG. In (b) the plane of rotation contains the growth axis, and $B = 1, 2, 3, \ldots$, kG. The data are from Holroyd and Datars (1975).

35.1.6 The Oil Drop Effect

Holroyd and Datars were able to obtain isotropic torque patterns by cleaning off all the oil (in which the spherical samples were grown) The samples were mounted in the apparatus by resting them on cotton. However if these same samples were later fastened to their supports with a drop of oil (which freezes on cooling to 4° K), they then displayed the familiar four-peaked anisotropy. $\vec{Q}$-domain structure, present if a CDW is present, provides a reasonable explanation. Randomly oriented $\vec{Q}$-domains will exhibit macroscopic isotropy (as long as $H < 30\,\text{kG}$) The thermal stress which arises from frozen oil (on cooling) will break the 24-fold degeneracy of the $\vec{Q}$-domains and produce a preferred orientation. This leads to an anisotropic resistivity.

35.1.7 Other Anomalous Phenomena

Many other properties exhibit extraordinary behavior and require that potassium have a CDW ground state. Splitting of the conduction-electron spin resonance [12], anisotropy of the residual resistance [13], direct observation of the phase excitations in point-contact spectroscopy [14], discrepancies in the Hall coefficient from its required value [15], etc. Some of these are discussed in a review by Overhauser [16]. Even de Haas–van Alphen measurements under pressure require anomalous structure (to escape non-conservation of electron number), as reported by Altounian and Datars [17]. The recently observed open-orbit torque peaks by Coulter and Datars [18] at 80 kG are conclusive. They are shown in Figure 35.5. Theoretical magnetoresistance spectra, based on the CDW model, were computed by Huberman and Overhauser [19]. They are shown in Figure 35.6.

I believe the failure to observe anomalous behavior in ordinary de Haas–van Alphen experiments is a consequence of sample preparation leading to $\vec{Q}$-domain sizes smaller than the diameter of a cyclotron orbit. Specimens are typically 1 mm in size, and etching is avoided. Possibly larger specimens (5–10 mm) heavily etched so that [110] surface facets become visible to the naked eye, will lead to larger Q-domains. Some authors have admitted discarding data which did not behave "properly". Perhaps these cases were in fact the important ones.

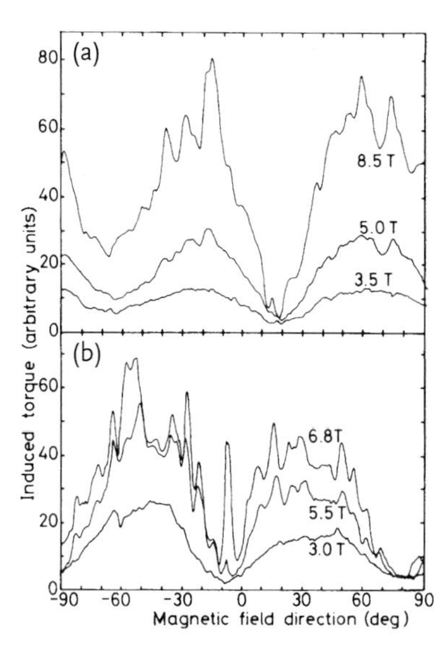

Figure 35.5 Magnetoresistance of K vs. $\vec{H}$ (Coulter and Datars). (a) Single crystal grown in oil, $\vec{H}$ in a (211) plane. (b) Single crystal grown in a mold, $\vec{H}$ in a (321) plane.

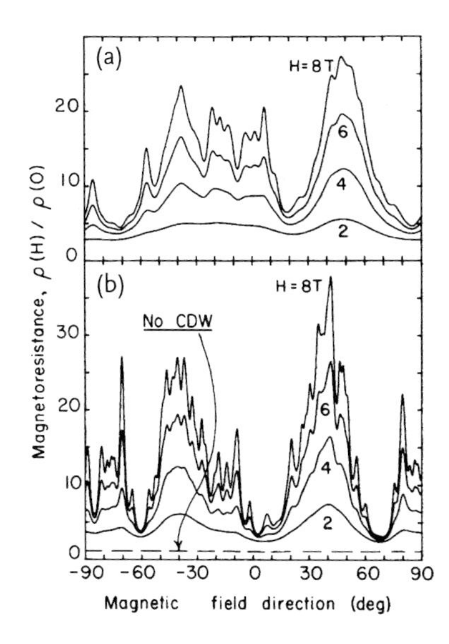

Figure 35.6 Magnetoresistance of potassium vs. $\vec{H}$ computed by Huberman and Overhauser using the CDW model.

35.2 Phasons: What they are and what they do

35.2.1 Introduction

Charge-Density-Wave (CDW) structure in an isotropic metal is a broken-symmetry ground-state caused exclusively by exchange and correlation effects. Since translation symmetry is broken, there will be a continuous spectrum of collective excitations (reaching zero frequency if the CDW is incommensurate) [20]. The occurrence of new excitations when new order parameters arise has been well-known in solid state physics for many decades (e. g. spin waves). In elementary particle theory this has become known as Goldstone's theorem.

A CDW is described by a sinusoidal modulation of the conduction electron density:

$$\rho = \rho_0[1 + p\cos(\vec{Q}_0 \cdot \vec{r} + \phi)] \,. \tag{35.7}$$

The new order parameters are p, $\vec{Q}_0$ and ϕ. The latter can take on any value between 0 and 2π, and provides a label for the infinite number of degenerate ground states. The ground state energy will lie at the bottom of a parabolic, four-dimensional valley, corresponding to the components of $\vec{Q}_0$ and p. The change in energy caused by small deviations from the equilibrium values, $\delta\vec{Q}$ and δp, can be written,

$$E = E_0 + \alpha(\delta Q_{\parallel})^2 + \beta(\delta Q_{\perp})^2 + \gamma(\delta p)^2 \,. \tag{35.8}$$

For simplicity we have assumed axial symmetry along the $\vec{Q}$ axis. I will use this expression as an energy density when $\delta\vec{Q}$ is allowed to be a slowly varying function of $\vec{r}$.

35.2.2 Phase Modulation

One may discover that the local direction of $\vec{Q}$ varies sinus-oidally in space and time if one allows the phase ϕ in Equation (35.7) to be,

$$\phi = \phi_q \sin(\vec{q} \cdot \vec{r} - \omega t) \,, \tag{35.9}$$

instead of a constant. This may be seen readily by taking the amplitude, ϕ_q, of the phase modulation to be small, and by expanding the sine function for small $\vec{q} \cdot \vec{r}$ and $t = 0$. Then Equation (35.7) becomes

$$\rho = \rho_0\{1 + p[\cos(\vec{Q}_0 + \phi_q \vec{q}) \cdot \vec{r}]\} . \tag{35.10}$$

If $\vec{q}$ is parallel to $\vec{Q}_0$ the *local* magnitude of $\vec{Q}$ is changed. If $\vec{q}$ is perpendicular to $\vec{Q}_0$ the *local* direction of $\vec{Q}$ is rotated. In general, the phase modulation given by Equation (35.9) corresponds to a sinusoidal modulation of both the local direction and magnitude of $\vec{Q}$. Actually we will assume that

$$|\vec{q}| \ll |\vec{Q}_0| , \tag{35.11}$$

and will discuss below the theoretical limits of $\vec{q}$.

The collective modes will be harmonic oscillators, so we call them "phasons". The potential energy of a phase excitation can be calculated from Equation (35.8),

$$U = \int \Delta E[\vec{Q}_{\text{local}}(\vec{r})] d^3 r \tag{35.12}$$

The kinetic energy of a phase excitation arises from the motion of the positive ions. It must be noted that a CDW cannot occur unless the metal remains electrically neutral. This means that the positive ions undergo a sinusoidal lattice displacement from their periodic crystal sites $\vec{L}$. Accordingly,

$$\vec{U}(\vec{L}) = \vec{A} \sin[\vec{Q}_0 \cdot \vec{L} + \phi_q \sin(\vec{q} \cdot \vec{L} - \omega t)] . \tag{35.13}$$

The displacement amplitude $\vec{A}$ is proportional to p, and its direction depends on the direction of $\vec{Q}$ and the elastic moduli [21]. The kinetic energy is easily computed from,

$$T = \frac{1}{2} \sum_{\vec{L}} M[\dot{\vec{U}}(\vec{L})]^2 , \tag{35.14}$$

where M is the ionic mass. For an harmonic oscillator the time average of U and T are equal. This requirement leads to the phason dispersion relation [20]:

$$\omega = \left[c_{\|}^2 q_{\|}^2 + c_{\perp}^2 q_{\perp}^2\right]^{\frac{1}{2}} . \tag{35.15}$$

The principal-axis values, $c_{\|}$, and $c_{\perp}$ of the phason velocity are proportional to $\alpha^{1/2}$ and $\beta^{1/2}$, respectively.

35.2.3 Relation Between Phasons and Phonons

When advantage is taken of the infinitesimal value of ϕ_q, Equation (35.13) can be written,

$$\vec{U}(\vec{L}) \cong \vec{A} \sin \vec{Q} \cdot \vec{L} + \frac{1}{2} \vec{A} \phi_q \sin[(\vec{q} + \vec{Q}) \cdot \vec{L} - \omega t] + \frac{1}{2} \vec{A} \phi_q \sin[(\vec{q} - \vec{Q}) \cdot \vec{L} - \omega t] . \tag{35.16}$$

Clearly, a phason is a linear combination of the two "old" phonon modes: $\vec{q} + \vec{Q}$ and $\vec{q} - \vec{Q}$. When these "old" modes are coupled together, one linear combination is pushed down (the phasons) and the orthogonal linear combination is pushed up. One can show that the latter mode corresponds to an amplitude modulation of the CDW [22]. See Figure 35.7.

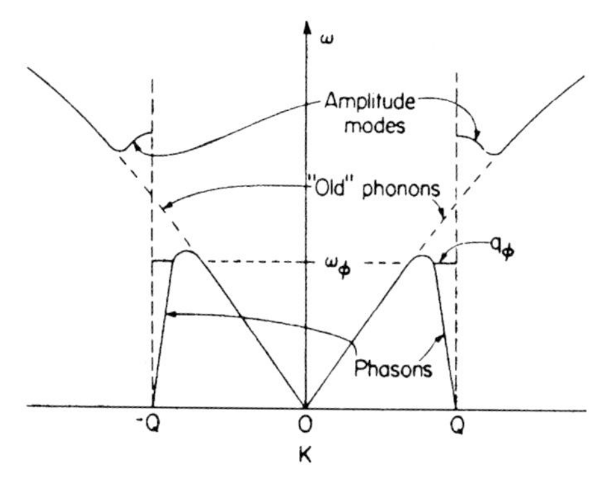

Figure 35.7 Schematic illustration of the vibrational modes in a metal having a CDW structure. The frequency of the phason branch goes to zero at $\pm\vec{Q}$. A phason is a linear superposition of two "old" phonons and the amplitude modes are the orthogonal linear combination. Phason and amplitude modes quickly merge into the old phonon spectrum, as indicated.

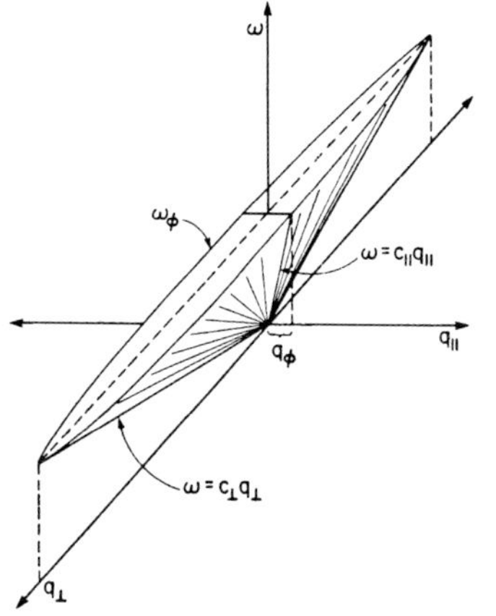

Figure 35.8 Anisotropic cone of the phason spectrum, showing the longitudinal and transverse phason velocities $c_{\|}$ and $c_{\perp}$. ω_ϕ is the frequency cutoff of the phason spectrum and q_ϕ is the wave-vector cutoff along Q.

Phasons have three important physical consequences because of their linear dispersion relation. One may visualize the phason "cone" as the result of "pressing down" on the lowest phonon dispersion surface near $\vec{Q}_0$ (or $\vec{Q}_0 - \vec{G}$, to get into the Brillouin zone) with an anisotropic ice cream cone until the point touches $\omega = 0$. The phason cutoff frequency ω_ϕ is essentially the frequency of the (old) phonon mode at $\vec{Q}_0$. See Figure 35.8.

The three important phason parameters that should be measured are $c_{\|}$, $c_{\perp}$ and ω_ϕ. They enter all calculations of observable properties.

35.2.4 The Phason Heat Capacity

A low temperature anomaly will appear in the heat capacity of a metal having a CDW. It can be calculated only if one knows how the phasons and amplitude modes blend into the phonon branch as $|\vec{q}|$ increases. A simple model employs a 2×2 matrix [23]:

$$\begin{vmatrix} \omega_\phi^2 - \omega^2 & \omega_\phi^2 F(q) \\ \omega_\phi^2 F(q) & \omega_\phi^2 - \omega^2 \end{vmatrix} = 0 \,. \tag{35.17}$$

Is easy to verify that,

$$F(q) \cong 1 - \left(\frac{c_q q}{\omega_q}\right)^2 + \ldots \,, \tag{35.18}$$

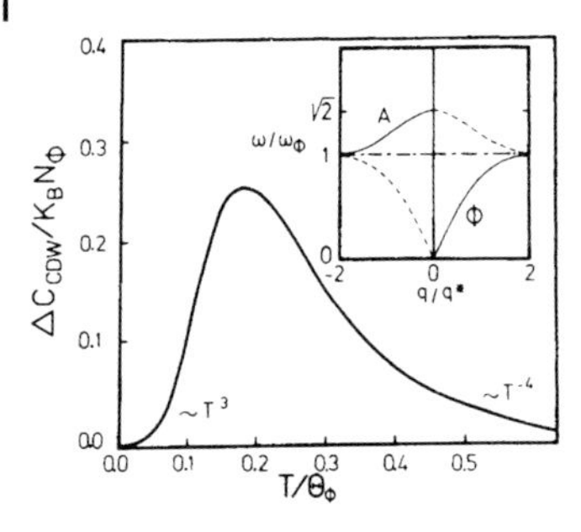

Figure 35.9 Heat-capacity anomaly caused by an incommensurate CDW. The inset shows the vibrational spectrum near $\vec{Q}'$. ϕ is the phason branch and A the amplitude-mode branch. The horizontal line, at $\omega = \omega_\phi$ is the phason cutoff frequency – the phonon frequency at $\vec{Q}'$ with no CDW.

if ω is to approach 0 linearly at $q = 0$. c_q is the phason velocity for the $\vec{q}$ direction. The function $F = \exp\left[-(c_q q/\omega_q)^2\right]$ satisfies this requirement, and is a convenient guess. It leads to the phason (and amplitude mode) spectrum shown in the inset of Figure 35.9. The old phonon frequencies have been set equal to the constant ω_ϕ for simplicity. If $R(\vec{q})$, $S(\vec{q})$ and P are the Einstein heat capacities of a phason, amplitude mode, and a phonon mode (of frequency ω_ϕ), then the change in heat capacity caused by a CDW is:

$$\Delta C_{\text{CDW}} = \sum_{\vec{q}} \left[R(\vec{q}) + S(\vec{q}) - 2P \right] . \tag{35.19}$$

The shape of ΔC_{CDW} is universal and has its peak near $0.18\theta_\phi$, $(\hbar\omega_\phi \equiv k_B \theta_\phi)$. See Figure 35.9.

An anomaly of this type was first seen by Sawada and Satoh [24] in $LaGe_2$ and shown to be associated with a CDW structure. Recently, Amarasekara and Keesom have observed a similar anomaly in potassium [25].

35.2.5 Low Temperature Resistivity

The low temperature resistivity of a metal such as potassium is expected to have four contributions:

$$\rho = \rho_0 + AT^m E^{-(T_0/T)} + BT^2 + CT^5 . \tag{35.20}$$

The residual resistivity ρ_0 is caused by impurities (and other imperfections). The exponential term is caused by the freeze out of umklapp processes. The T^2 term arises from electron-electron scattering, and the T^5 term from acoustic phonon scattering.

A term approximately quadratic in T was observed in potassium by van Kempen et al. [26]. However the coefficient B varied widely from sample to sample, indicating that ordinary electron-electron scattering could not be the cause. Later Rowlands et al. [27] found a similar result. This work extended to 0.4° K and had an experimental scatter of only $10^{-15}\,\Omega$ cm. It fit a power law of $T^{1.5}$. Power law fits proportional to T or T^2 were well outside the experimental scatter. A theory of electron–phason scattering [28] fits extremely well, as can be seen in Figure 35.10. The arrows at the two highest temperatures show where the points would fall if the umklapp tail were subtracted. Phason scattering is important at low temperatures, unlike acoustic phonons, since (as can be seen from Equation (35.16)) they cause large momentum transfers, $\sim \pm\vec{Q}$, even though q is very small [28].

Variability of the data from sample to sample is also explained (as an intrinsic property) since the large momentum transfers occur along $\pm\vec{Q}$. Thus the average resistivity will depend on the orientation distribution of $\vec{Q}$ domains.

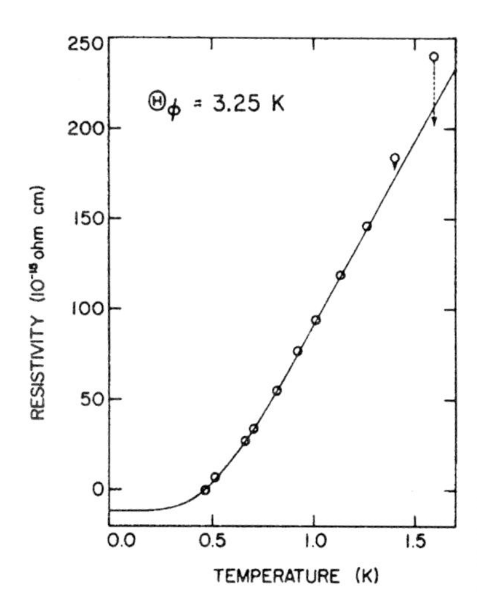

Figure 35.10 Plot of resistivity versus temperature for the data of sample K2c of Rowlands, Duvvury, and Woods indicated by circles. The curve and arrows are described in the text.

35.2.6 Point Contact Spectroscopy

Two metal points touched together in a small contact (size $\ll$ mean free path) create a junction similar to a superconducting tunnel junction. Spectra of d^2V/d^2I vs. voltage V reveal phonon spectra (weighted by a square matrix element and a geometric factor) The geometric factor arises in this case only because an electron must be scattered back through the contact it has just crossed in order for the subsequent phonon event to contribute to the contact resistance. Accordingly low energy acoustic phonons, which scatter electrons at the Fermi surface only through small angles, contribute a vanishing contribution at voltages comparable to their energy.

Nevertheless, data on potassium by Jansen et al. [29] show a striking anomaly below 1 meV, where there should be no contribution at all. Recently Ashraf and Swihart [30] have found that electron–phason scattering explains the anomaly – its shape and size – without any adjustable parameters. Data on the phason spectrum were taken from the fit to the resistivity data mentioned above [28].

35.2.7 Phason Thermal Diffuse Scattering

A CDW structure leads to new diffraction peaks. Each reciprocal lattice vector $\vec{G}$ acquires two satellites, for which the Laue equation is

$$\vec{k}' = \vec{k} + \vec{G} \pm \vec{Q} . \tag{35.21}$$

The phasons cause a thermal diffuse scattering around each satellite:

$$\vec{k}' = \vec{k} + \vec{G} \pm \vec{Q} \pm \vec{q}_{\text{phason}} . \tag{35.22}$$

This diffuse scattering will produce a pancake-shaped cloud because of the phason anisotropy. In electron diffraction it shows up as streaks, as seen in TaS_2 [31].

35.3 Theory of Charge Density Waves

35.3.1 Introduction

The purpose of this section is to show that a CDW is the anticipated electronic ground state of a simple, isotropic metal. We will assume that the positive-ion background has been replaced by a uniformly charged jelly, having *no mechanical* rigidity. The last assumption – a key property of the *deformable* jellium model guarantees that all microscopic electric fields will be zero. (There will be *no* Hartree term in the one-electron Schrödinger equation.) A CDW arises exclusively from exchange and correlation, as we will show. The alkali metals, in the aspects relevant here, correspond well to this model. Their elastic stiffness is two orders of magnitude less than that of copper, and the pressure dependence of their moduli show no evidence of a Born–Mayer ion–ion repulsion.

The fact that potassium has a CDW structure not only confirms the theoretical arguments, but at the same time proves that some treatments of exchange and correlation are fundamentally incomplete, and sometimes dramatically wrong. This will be discussed at length in the concluding section.

35.3.2 SDW-CDW Instability Theorem

The exact ground-state energy of an electron system is,

$$E = E_{\mathrm{HF}} + E_{\mathrm{corr}} \,, \tag{35.23}$$

where E_{HF} is the Hartree–Fock ground state. This equation is a definition of the correlation energy. The spin-density wave instability theorem [32], which I proved in 1962, showed that the normal (paramagnetic) state always has an instability of the SDW type. The theorem is rigorous and applies for all electron densities in the Hartree–Fock approximation. If one allows the jellium to be deformable, then the instability can be any admixture of SDW and CDW [5].

$$\rho_{\pm} = \frac{1}{2}\rho_0[1 + p\cos(\vec{Q}\cdot\vec{r} \pm \phi)] \,, \tag{35.24}$$

where $\rho_{\pm}$ are the electron densities for up and down spin. See Figure 35.11. The HF energy is invariant to the choice of ϕ as long as the jellium background cancels all the charge modulation from the electrons. The kinetic energy and the exchange interactions (up with up and down with down) do not care about the relative phase of the up-spin and down-spin modulations.

Proof of this theorem is intricate. However the strategy is simple. One chooses $Q = 2k_{\mathrm{F}}$ and lets the energy gap G be a variable. This gap is caused by the exchange potential, and is created by the amplitude modulation of the one-electron wavefunctions. One considers only those states near the energy gap at point C in Figure 35.1 – the ones displaced $\sim 1/2G$ in energy from the free electron parabola and $\sim$ 100% amplitude modulated. These are in a

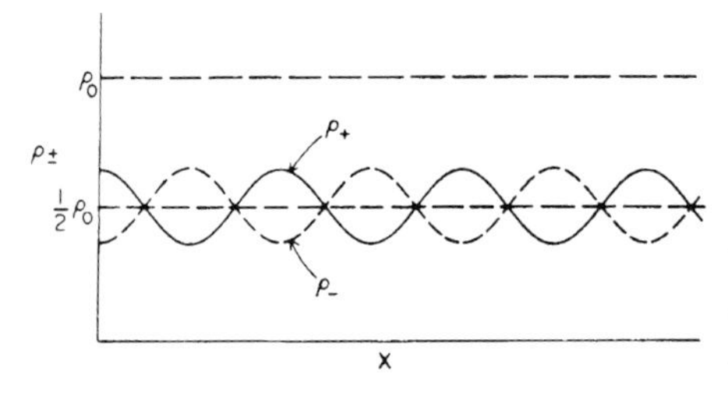

Figure 35.11 Electronic charge density against x of up- and down-spin electrons for an ideal metal with a SDW ground state. The total charge density is a constant, ρ_0. In a CDW, ρ_+ and ρ_- are in phase.

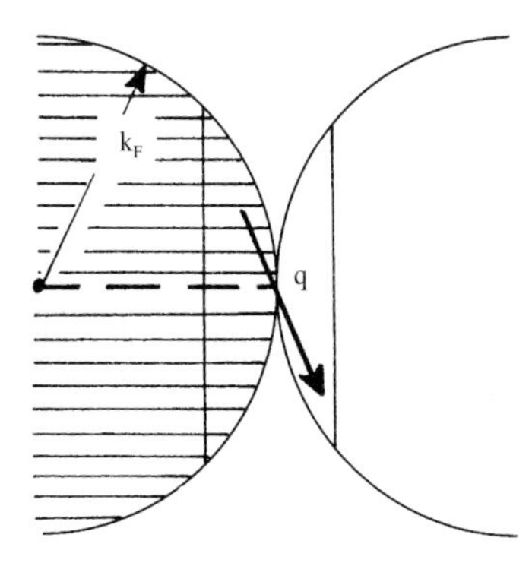

Figure 35.12 The new exchange interactions that cause SDW or CDW instabilities are those between momentum states at the head and tail of the arrow $\vec{q}$. (States $\vec{k}$ in the spherical cap enclosing the head of the arrow, which have energy > E_F, are "virtually" occupied by their interaction with states $\vec{k} - \vec{Q}$ having energy < E_F.)

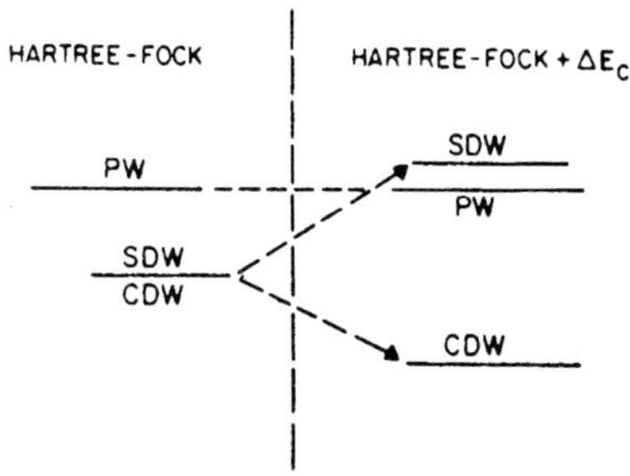

Figure 35.13 Energy-level diagram for the plane-wave (PW) SDW and CDW states of an ideal metal. In the Hartree–Fock approximation the SDW and CDW states are degenerate and have lower energies than the plane-wave state. The degeneracy is split, as shown on the right, by correlation-energy corrections.

spherical cap of height $\sim G$ and radius $\sim G^{1/2}$; see Figure 35.12. Their number N is accordingly, $\sim G^2$. The kinetic energy "investment" to deform these states from plane waves to,

$$\psi \cong e^{i\vec{k}\cdot\vec{r}} + \alpha^+ e^{i(\vec{k}+\vec{Q})r} + \alpha^- e^{i(\vec{k}-\vec{Q})\cdot\vec{r}}\,, \tag{35.25}$$

is their number N times the extra kinetic energy of ψ (compared to a pure plane wave). This is $\sim G$. Therefore,

$$\Delta T \sim G^3 \tag{35.26}$$

New contributions to the exchange energy arise from the interactions of the states just described with virtually excited states associated with the spherical cap on the opposite side of the Fermi sphere. These states lie above k_F in Figure 35.12. They contribute an energy "dividend",

$$\Delta E \sim -\sum_{\vec{k},\vec{k}'} \frac{4\pi e^2}{|\vec{k}' - \vec{k}|^2} \sim -G^3 \ln\left(\frac{E_F}{G}\right)\,. \tag{35.27}$$

This can always be made larger in magnitude than Equation (35.26) by choosing G small enough, which proves the theorem. But one must show too (as it has been) [32] that Equation (35.27) is larger than the loss of exchange energy resulting from the admixtures in (35.25). Consequently, a SDW or CDW state always lies lower than the normal (plane wave) state in the Hartree–Fock approximation, as shown on the left side of Figure 35.13.

35.3.3 **The Correlation Energy Correction**

The algebraic sign of the correlation energy correction for SDW and CDW states is now the crucial issue. We will show that they have opposite sign – in fact a SDW state is destabilized and *a CDW is made more stable.*

The correlation energy can be calculated semi-quantitatively from

$$E_{\text{corr}} \cong -\sum_i \frac{\langle i|V_{\text{eff}}|o\rangle^2}{\Delta E_i} \tag{35.28}$$

where ΔE_i is the two-particle two-hole excitation energy relative to the supposed ground state, $|o\rangle$ What is the meaning of V_{eff}? The dominant contributions to Equation (35.28) are the virtual scattering of opposite-spin electrons from below the Fermi energy to empty states above. This proceeds by a screened interaction V_{scr}.

$$\psi' = C\left\{\psi_o - \sum_i \frac{\langle i|V_{\text{scr}}|o\rangle}{\Delta E_i}\psi_i\right\}, \tag{35.29}$$

where C is a normalizing factor. The new energy is found by taking the expectation value of the *exact* Hamiltonian:

$$E' = \langle\psi'|\sum_\lambda \frac{p_\lambda^2}{2m} + \sum_{\lambda<\mu}\frac{e^2}{r_{\lambda\mu}}|\psi'\rangle . \tag{35.30}$$

If one works this out, one finds [5] Equation (35.28), provided

$$\langle i|V_{\text{eff}}|o\rangle^2 \equiv 2\langle i|V|o\rangle\langle i|V_{\text{scr}}|o\rangle - \langle i|V_{\text{scr}}|o\rangle^2 , \tag{35.31}$$

where $V \equiv e^2/r$. This procedure was first used in correlation energy calculations by Macke [33].

The changes in correlation energy associated with the amplitude-modulated states was worked out in detail by Overhauser [5]. We present here a simplified version which gives accurately the same results. For the SDW case,

$$\phi_\pm \approx e^{i\vec{k}\cdot\vec{r}}(1 \pm p\cos\vec{Q}\cdot\vec{r})^{1/2} , \tag{35.32}$$

where $p \ll 1$. The matrix element for virtual scattering of two opposite-sign electrons is,

$$m = \left\langle(\vec{k}+\vec{q})_+(\vec{k}'-\vec{q})_-\,|V_{\text{eff}}|\,\vec{k}_+\vec{k}'_-\right\rangle . \tag{35.33}$$

Two of the four factors in (35.32) have plus signs, and the other two have minus. It follows that:

$$m \approx m_o\left(1-\frac{1}{2}p^2\right) , \tag{35.34}$$

where m_o is the matrix element for the same event with $p = 0$. As a consequence the correlation energy becomes, to order p^2,

$$E_{\text{corr}}^{\text{SDW}} \cong -\sum_i \frac{m_o^2}{\Delta E_i}(1-p^2) . \tag{35.35}$$

Every term is reduced in magnitude by the factor $(1-p^2)$; so the SDW state is raised *relative to* the (correlated) plane wave state.

For the CDW case all four factors in (32) have a positive sign, say. Accordingly,

$$E_{\text{corr}}^{\text{SDW}} \cong -\sum_i \frac{m_o^2}{\Delta E_i}(1+p^2) . \tag{35.36}$$

Each term in the correlation energy is enhanced in magnitude and this adds stability to the CDW state. (There are high energy virtual excitations – to states above the CDW energy gaps – for which the above remarks are not true; but they are relatively less important.)

It is easy to understand why electron-electron scattering splits the SDW and CDW degeneracy. For the CDW case up-spin and down-spin electrons are stratified in *alternate* laminar layers, and have less probability of scattering from one another. A CDW coaxes both spin types into the *same* layers, and increases their virtual scattering. These effects of correlation are shown on the right side of Figure 35.13.

35.3.4 Analogy with Uniform Deformations

Wigner was the first to point out [2] that electron correlations reduce the enhancement of the spin susceptibility χ_s (compared to the Pauli value χ_p) caused by exchange. The straight line in Figure 35.14 intersecting the horizontal axis near $r_s = 6$ is the Hartree–Fock value. As one approaches ferromagnetism the number of antiparallel-spin pairs approaches zero. So the correlation energy becomes smaller with larger spin polarization, and this cancels part of the exchange enhancement. Therefore the exact value of χ_s/χ_p lies above the Hartree–Fock line (as shown in Figure 35.14).

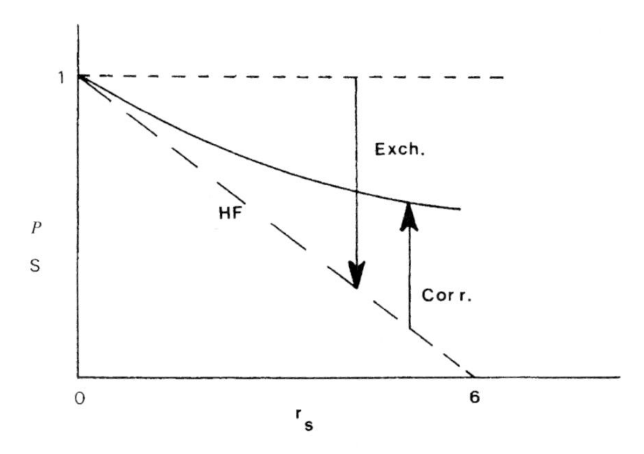

Figure 35.14 Spin susceptibility of an electron gas compared to the Pauli value χ_p versus $r_s \cdot r_s$ is defined by Equation (35.2).

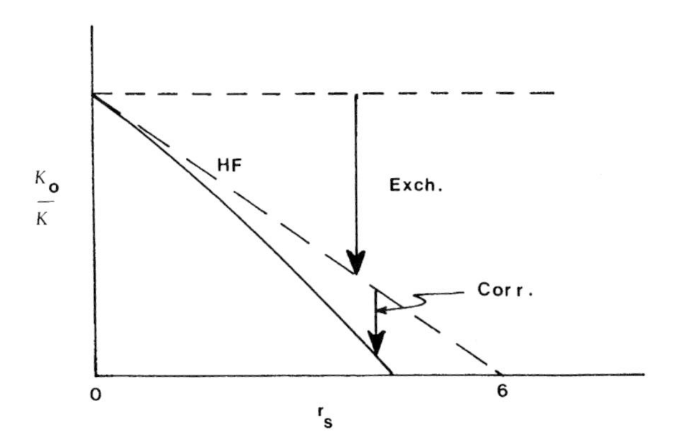

Figure 35.15 Compressibility K of an electron gas compared to K_0 (for a non-interacting Fermi gas) versus r_s.

The compressibility K of an electron gas is also enhanced by exchange – in fact, by the same factor as the spin susceptibility. However, the correlation energy enhances K *even* further. All calculations of E_{corr} vs. r_s show that the second derivative of E_{corr} with respect to density contributes to this. The general conclusion is that exchange and correlation reinforce one another for charge modulations (as shown in Figure 35.15). The results we obtained earlier for the $Q = 2k_F$ case are similar to the uniform, $Q = 0$, examples just discussed.

35.4 Conclusions

A simple metal, with a deformable positive-ion background, should exhibit a CDW instability. Alkali metals are prime candidates for illustrating this phenomenon. Because they are monovalent, their (nearly spherical) Fermi surfaces do not extend to the Brillouin zone boundaries. Anomalous magnetotransport phenomena, clearly ascribable to a multiply connected Fermi surface can then be attributed to a broken-symmetry CDW state. The many striking anomalies of potassium [16], and especially the direct observation of open orbits by Coulter and Datars [18] confirm this behavior.

Several frequently used approximations in many-electron theory are shown to be generally invalid by the CDW structure of potassium. A common view is that electrons in metals can be treated as quasi-particles interacting via short-range, *screened* interactions. This is permitted for scattering processes, but should be forbidden for use in Hamiltonians For example, if screened exchange were used (and it has been) to calculate the electron gas compressibility, one would find a result given by the behavior of the spin susceptibility (shown in Figure 35.14) instead of that given by the second derivative of the total energy with respect to volume (as shown in Figure 35.15). In other words, a screened-exchange treatment leads to a correction (relative to Hartree–Fock) having the *wrong sign*, screened-exchange interactions do not permit a CDW structure in potassium.

Local density approximations are commonly used for exchange and correlation potentials. This may be a satisfactory method when they are used with an adjustable parameter to fit experimental data. But they are not predictive. For example, the CDW structure in potassium requires an exchange potential,

$$V_x = -G \cos \vec{Q} \cdot \vec{r} \, . \tag{35.37}$$

where $G = 0.6$ eV. The resulting charge modulation is

$$\rho = \rho_0 (1 + p \cos \vec{Q} \cdot \vec{r}) \, . \tag{35.38}$$

G and p must be self consistent. If one calculates p from the observed G, assuming V_x is local, $p = 0.17$. If one now uses a Kohn–Sham local-exchange approximation to calculate G from p, one finds $G = 0.2$ eV. The inconsistency is half an order of magnitude. Unfortunately, the Kohn–Sham scheme is supposed to be best when applied to the ground-state problem of a free-electron metal.

The failure of local-exchange approximations to predict the correct ground state of potassium is caused by the fact that CDW states are supported by the non-local, dynamical effects of exchange. This can be easily illustrated by considering a uniform electron gas with an infinitesimal fractional modulation ε. The Slater local-exchange operator is then

$$A_s = -\frac{e^2 k_F}{\pi} (1 + \varepsilon \cos qx)^{1/3} \, . \tag{35.39}$$

Since ε is small, we have:

$$A_s \cong -\frac{e^2 k_F}{3\pi} \varepsilon \cos qx \, . \tag{35.40}$$

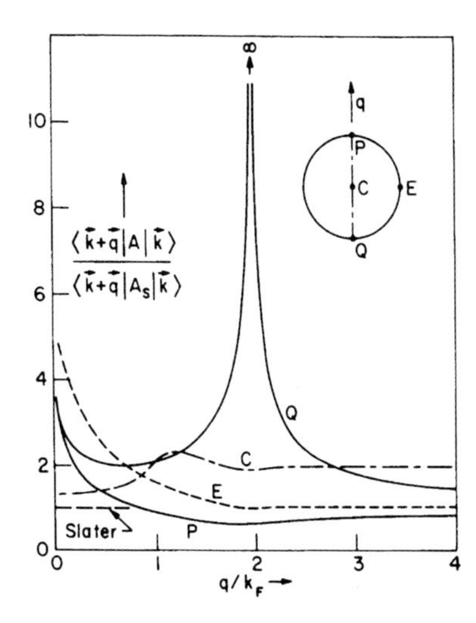

Figure 35.16 Ratio of the off-diagonal matrix elements of the exact exchange operator to the Slater, local-density approximation for a small charge modulation of wave vector $\vec{q}$. Results for four initial states $\vec{k}$ are shown.

The dynamical properties of an operator are displayed by calculating its off-diagonal matrix elements. The only such elements for (35.40) are,

$$\langle \vec{k} \pm \vec{q} | A_s | \vec{k} \rangle = -\frac{e^2 k_F}{6\pi} \,. \tag{35.41}$$

Observe that this result is independent of q and independent of $\vec{k}$. (It is *local* in both senses of the word).

Consider now the exact exchange operator A. It is defined by the equation,

$$A\psi(\vec{r}) = -\sum_i \left[\int \phi_i^+(\vec{s})\psi(\vec{s}) \frac{e^2}{|\vec{r}-\vec{s}|} d^3 s \right] \phi_i(\vec{r}) \,, \tag{35.42}$$

where $\{\phi_i\}$ are the occupied levels. For this illustration these can be taken to be the perturbed electron states associated with a potential such as (35.37), and leading to a fractional polarization $\rho = \varepsilon$. The *exact* off-diagonal matrix elements of A have been calculated [34]. The ratios of these to the local ones from A_s are shown in Figure 35.16 as a function of q, and for four values of $\vec{k}$. The non-local effects are spectacular. Clearly a non-local approximation (constant, horizontal line at unity or 2/3) discards important dynamical properties. The singularity at $q = 2k_F$ is the cause of CDW phenomena.

It seems that progress in many-electron physics may have been impeded during the last fifteen years by the wide use of local, density-functional approximations to exchange and correlation.

References

1 Bloch, F. (1929) *Z. Physik*, **57**, 545.

2 Wigner, E. (1938) *Trans. Faraday Soc.*, **34**, 678.

3 Kushida, T., Murphy, J.C., and Hanabusa, M. (1976) *Phys Rev. B*, **13**, 5136

4 Shoenberg, D. and Stiles, P.J. (1964) *Proc Roy. Soc. A*, **281**, 62.

5 Overhauser, A.W. (1968) *Phys. Rev.*, **167**, 691.
6 Mayer, H. and El Naby, M.H. (1963) *Z. Physik*, **174**, 269.
7 Overhauser, A.W. (1964) *Phys. Rev. Lett.*, **13**, 190.
8 Lifshitz, I.M., Ya Azbel, M., and Kagonov, M. (1956) *Zh. Eksp. Teor. Fiz.*, **31**, 63, English translation: (1957) *Sov. Phys. JETP*, **4**, 41.
9 Stetter, G., Adlhart, W., Fritsch, G., Steichele, E., and Lüscher, E. (1978) *J. Phys. F*, **8**, 2075.
10 Schaefer, J.A. and Marcus, J.A. (1971) *Phys. Rev. Lett.*, **27**, 935.
11 Holroyd, F.W. and Datars, W.R. (1975) *Can. J. Phys.*, **53**, 2517.
12 Overhauser, A.W. and de Graaf, A.M. (1968) *Phys. Rev.*, **168**, 763.
13 Bishop, M.L. and Overhauser, A.W. (1978) *Phys. Rev. B*, **18**, 2447.
14 Ashrof, M. and Swihart, J.C. unpublished.
15 Chimenti, D.E. and Maxfield, B.W. (1973) *Phys. Rev. B*, **7**, 350.
16 Overhauser, A.W. *Adv. Phys.*, **27**, 343 (1978).
17 Altoumian, Z. and Datars, W.R. (1980) *Can. J. Phys.*, **58**, 370.
18 Coulter, P.G. and Datars, W.R. (1980) *Phys. Rev. Lett.*, **45**, 1021.
19 Huberman, M. and Overhauser, A.W. (1981) *Phys. Rev. Lett.*, **47**, 682.
20 Overhauser, A.W. (1971) *Phys. Rev. B*, **3**, 3173.
21 Giuliani, G.F. and Overhauser, A.W. (1979) *Phys. Rev. B*, **20**, 1328.
22 Giuliani, G.F. and Overhauser, A.W. (1981) *Phys. Rev. B*, **23**, 3737.
23 Giuliani, G.F. and Overhauser, A.W. (1980) *Phys. Rev. Lett.*, **45**, 1335.
24 Sawada, A. and Satoh, T. (1978) *J. Low Temp. Phys.*, **30**, 455.
25 Amarasekara, C.D. and Keesomn, P.H. (1981) *Phys. Rev. Lett.*, **47**, 1311.
26 van Kempen, H., Lass, J.S., Ribot, J.H.J.M., and Wyder, P. (1976) *Phys. Rev. Lett.*, **37**, 1574.
27 Rowlands, J.A., Durvury, C., and Woods, S.B. (1978) *Phys. Rev. Lett.*, **40**, 1201.
28 Bishop, M.F. and Overhauser, A.W. (1981) *Phys. Rev. B*, **23**, 3638.
29 Jansen, A.G.M., van den Bosch, J.H., van Kempen, H., Ribot, J.H.J.M., Smeets, P.H.H., and Wyder, P. (1980) *J. Phys. F*, **10**, 265.
30 Ashraf, M. and Swihart, J.C. (1983) *Phys. Rev. Lett.*, **50**, 921.
31 Wilson, J.A., DiSalvo, F.J., and Mahajan, S. (1975) *Adv. Phys.*, **24**, 117.
32 Overhauser, A.W. (1962) *Phys. Rev.*, **128**, 1437.
33 Macke, W. (1950) *Z. Naturf.*, **5a**, 192.
34 Overhauser, A.W. (1970) *Phys. Rev. B*, **2**, 874.

Reprint 36 Energy Spectrum of an Incommensurate Charge-Density Wave: Potassium and Sodium[1)]

F.E. Fragachán * *and A.W. Overhauser* *

* Department of Physics, Purdue University, 525, Northwestern Avenue, West Lafayette, Indiana 47907, USA

Received 24 October 1983

The conduction-electron energy spectrum is calculated for K and Na taking into account the periodic potential $2\alpha\cos(\vec{Q}\cdot\vec{r})$ of the charge-density wave. Accordingly, the Schrödinger equation includes this term in addition to a pseudopotential $2\beta\cos(\vec{G}\cdot\vec{r})$ caused by the ionic lattice. $\vec{Q}$ and $\vec{G}$ are nearly parallel, and differ in magnitude by only a few percent. As a consequence three families of higher-order gaps in $E(\vec{k})$ arise: "minigaps", characterized by wave vectors $(n+1)\vec{Q}-n\vec{G}$; "heterodyne gaps", with periodicities $n(\vec{G}-\vec{Q})$; "second-zone minigaps", with periodicities $(n+1)\vec{G}-n\vec{Q}$. The energy-gap surfaces of the first two families truncate the Fermi surface and cause the open orbits observed by Coulter and Datars in high-field magnetoresistance. They should also lead to new optical-absorption bands in the far infrared.

36.1 Introduction

The conduction-electron spectra of the alkali metals Na and K have long been regarded as examples of nearly-free-electron behavior, i. e., $E(\vec{k}) \approx \hbar^2k^2/2m$. Their Fermi surfaces lie within the Brillouin zone and should be simply connected. During the past two decades, however, considerable evidence has accumulated [1], which indicates that these metals have undergone a spontaneously broken symmetry of the charge-density-wave (CDW) type [2]. An important consequence is that extra energy gaps in $\vec{k}$ space arise. These extra energy gaps lead to anomalous optical-absorption [3] and magnetoresistance behavior. The most spectacular example of the latter is the direct observation [4] of open-orbit magnetoresistance peaks. Theoretical studies [5] based on the multiply connected Fermi surface, "fractured" by the CDW, explain this phenomenon.

The purpose of this investigation is to calculate the size of the new energy gaps introduced by the broken symmetry. The conduction-electron charge density is no longer $\rho_0(\vec{r})$, having the periodicity of a bcc lattice, but is

$$\rho(\vec{r}) = \rho_0[1 - p\cos(\vec{Q}\cdot\vec{r})]\,. \tag{36.1}$$

$\vec{Q}$ is the CDW wave vector and p is its amplitude. The direction of $\vec{Q}$ is known from the observed optical anisotropy [6] to be very near to a (110) axis. Theoretical study [7] indicates that the angular deviation is only 2° or 3°, depending on the magnitude of $\vec{Q}$. This magnitude is somewhat larger than $2k_F = 1.24(2\pi/a)$, since the shape of the CDW optical anomaly indicates [3] that the Fermi surface is in (approximate) critical contact with the main CDW energy gaps. With band structure accounted for, we believe that $|\vec{Q}| \sim (1.36\text{–}1.38)(2\pi/a)$. ($a$ is the lattice constant.)

1) Phys. Rev. B 29, 6 (1984).

Anomalous Effects in Simple Metals. Albert Overhauser

ISBN: 978-3-527-40859-7

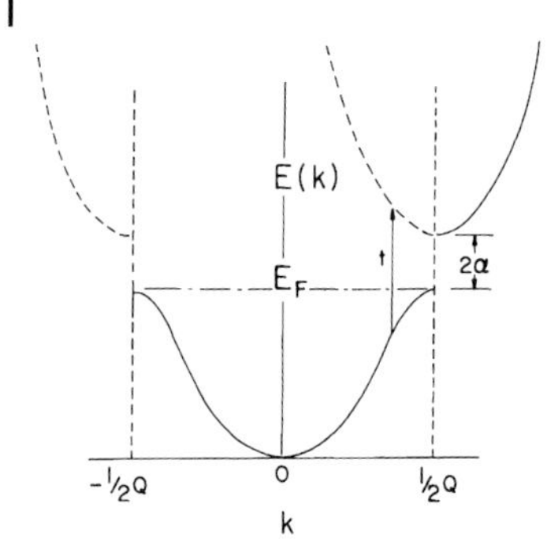

Figure 36.1 Schematic behavior of $E(k)$ for $\vec{k}$ parallel to $\vec{Q}$, when a CDW potential, Equation (36.2), is present. Optical transitions responsible for the Mayer–El Naby anomaly are indicated by the arrow t; the threshold is 2α (if $V_{\text{lat}} = 0$).

The periodic potential which causes the CDW modulation arises from exchange and correlation [2]:

$$V_{\text{CDW}}(\vec{r}) = 2\alpha \cos(\vec{G} \cdot \vec{r}) \,. \tag{36.2}$$

The threshold of the Mayer–El Naby optical anomaly [8] indicates that $2\alpha \sim 0.62\,\text{eV}$. But this identification is accurate only if V_{CDW} is acting alone. Then the CDW energy-gap planes are perpendicular to $\vec{Q}$ and pass through the points $\pm\frac{1}{2}\vec{Q}$ in $\vec{k}$ space (see Figure 36.1). We will find that 2α must be somewhat larger when the pseudopotential,

$$V_{\text{lat}}(\vec{r}) = 2\beta \cos(\vec{G} \cdot \vec{r}) \,, \tag{36.3}$$

of the positive-ion lattice is taken into consideration. $\vec{G}$ is that $\langle 110 \rangle$ reciprocal-lattice vector which is nearly parallel to $\vec{Q}$. Of course, $|\vec{G}| = 1.414(2\pi/a)$, which differs by only a few percent from $|\vec{Q}|$.

The Schrödinger equation which we wish to solve has a potential given by the sum of the incommensurate components (36.2) and (36.3). We have

$$V(\vec{r}) = 2\alpha \cos(\vec{Q} \cdot \vec{r}) + 2\beta \cos(\vec{G} \cdot \vec{r}) \,. \tag{36.4}$$

We shall neglect the other five pseudopotential terms, which are oriented 60° or 90° away from $\vec{G}$, since we are primarily interested in the behavior of $E(\vec{k})$ along $\vec{Q}$ or $\vec{G}$ and far from the ten Brillouin-zone faces defined by the other five periodicities.

It is well known that two incommensurate potentials will lead to new, higher-order energy gaps and that they can have a profound influence on magnetotransport behavior [9]. In principle every periodicity,

$$\vec{K} = m\vec{Q} + n\vec{G} \,, \tag{36.5}$$

with integers, $0 \leq m < \infty$ and $-\infty < n < \infty$, will generate a pair of energy-gap surfaces [10]. For the situation considered here only a few of the m,n combinations lead to gaps which have physical significance.

36.2 Minigaps and Heterodyne Gaps

Solution of the Schrödinger equation having Equation (36.4) for the potential term leads to energy gaps at the Brillouin-zone faces perpendicular to $\vec{G}$ as well as to the main CDW gaps shown in Figure 36.1. In addition there are three important families of higher-order gaps which we name as follows (in accordance with their wave-vector periodicities, $\vec{K}$):

(a) Minigaps,

$$\vec{K} = (n+1)\vec{Q} - n\vec{G} \,, \tag{36.6}$$

(b) second-zone minigaps,

$$\vec{K} = (n+1)\vec{G} - n\vec{Q} \,, \tag{36.7}$$

(c) and heterodyne gaps,

$$\vec{K} = n(\vec{G} - \vec{Q}) \,. \tag{36.8}$$

Here, only positive integers, $n = 1, 2, 3, \ldots$, are intended. The reason why these are the only important families stems from the fact that $\vec{G}$ and $\vec{Q}$ are assumed to differ by only a small wave vector

$$\vec{Q}' \equiv \vec{G} - \vec{Q} \,, \tag{36.9}$$

and the fact (as we shall see below) that the sizes of the gaps fall off rapidly with increasing n. Furthermore, energy gaps having $|m - n| > 1$, with small m,n, in Equation (36.5), do not intersect the Fermi surface.

To understand the origin of the families Equations (36.6)–(36.8), consider the matrix elements of the potential (36.4) in the momentum representation. The only nonzero matrix elements are

$$\langle \vec{k} \pm \vec{Q} | V(\vec{r}) | k \rangle = \alpha \,, \tag{36.10}$$

$$\langle \vec{k} \pm \vec{G} | V(\vec{r}) | k \rangle = \beta \,. \tag{36.11}$$

If one starts from the $\vec{k}$ state, the states which are connected to it by a successively longer chain of couplings α,β are

$$|\vec{k} - \vec{Q}\rangle \,, \quad |\vec{k} - \vec{G}\rangle \,, \tag{36.12}$$

$$|\vec{k} - \vec{Q} + \vec{G}\rangle \,, \quad |\vec{k} - \vec{G} + \vec{Q}\rangle \,, \tag{36.13}$$

$$|\vec{k} - \vec{Q} + \vec{G} - \vec{Q}\rangle \,, \quad |\vec{k} - \vec{G} + \vec{Q} - \vec{G}\rangle \,, \tag{36.14}$$

$$|\vec{k} - \vec{Q} + \vec{G} - \vec{Q} + \vec{G}\rangle \,, |\vec{k} - \vec{G} + \vec{Q} - \vec{G} + \vec{Q}\rangle \,, \tag{36.15}$$

etc. We have assumed here that $\vec{k}$ is near to $\frac{1}{2}\vec{Q}$ or $\frac{1}{2}\vec{G}$. Consequently, the alternation in sign as one proceeds from $|k\rangle$ to (36.12), to (36.13), etc., brings in states that are near in energy to $|\vec{k}\rangle$. Mixing of $|\vec{k}\rangle$ with $|\vec{k} + \vec{G}\rangle$, for example, can be neglected because of the large energy difference. The Hamiltonian matrix obtained from the first seven basis functions is

$$\begin{vmatrix}
k^2 & \alpha & \beta & 0 & 0 & 0 & 0 \\
a & |\vec{k} - \vec{Q}|^2 & 0 & \beta & 0 & 0 & 0 \\
\beta & 0 & |\vec{k} - \vec{G}|^2 & 0 & a & 0 & 0 \\
0 & \beta & 0 & |\vec{k} - \vec{Q} + \vec{G}|^2 & 0 & a & 0 \\
0 & 0 & \alpha & 0 & |\vec{k} - \vec{G} + \vec{Q}|^2 & 0 & \beta \\
0 & 0 & 0 & a & 0 & |\vec{k} - 2\vec{Q} + \vec{G}|^2 & 0 \\
0 & 0 & 0 & 0 & \beta & 0 & |\vec{k} - 2\vec{G} + \vec{Q}|^2
\end{vmatrix} . \tag{36.16}$$

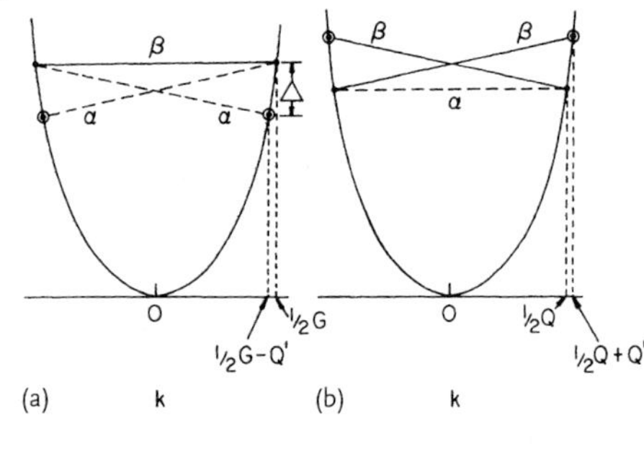

Figure 36.2 First minigaps occur, in (a), at the circled points, which are connected by the coupling chain α,β,α. In (b) the first pair of second-zone minigaps are shown.

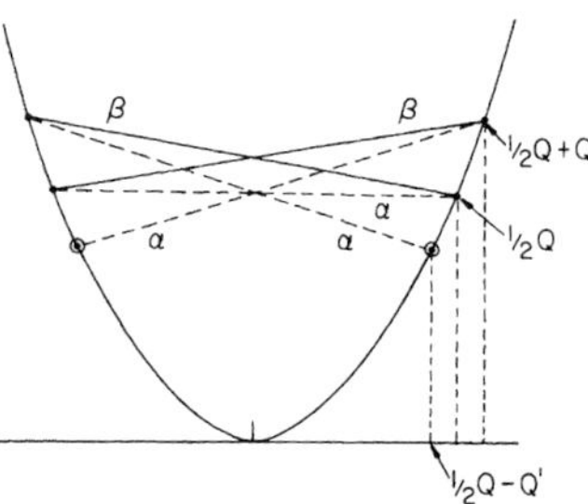

Figure 36.3 Coupling chain $\alpha,\beta,\alpha,\beta,\alpha$ causes the second pair of minigaps at the circled points.

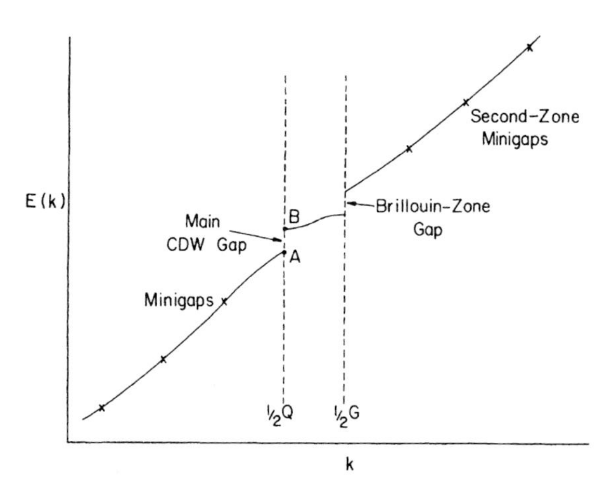

Figure 36.4 Schematic $E(k)$ showing the CDW and Brillouin-zone gaps and the locations of the first three minigaps and the first three second-zone minigaps.

For simplicity we have set $\hbar^2/2m = 1$ in the diagonal elements. One needs to find (with a computer) the eigenvalues (of such a matrix) as a function of $\vec{k}$. An energy gap arises whenever two degenerate basis functions are connected through a chain $\alpha,\beta,\alpha,\beta,\alpha,\ldots$ or $\beta,\alpha,\beta,\alpha,\beta,\ldots$ with a number $(2n+1)$ of links. These chains give, respectively, the minigaps and second-zone minigaps, Equations (36.6) and (36.7). The chains for the first mini-gap and first second-zone minigap are shown in Figure 36.2. The chain for the second minigap is shown in Figure 36.3. The locations of the first three minigaps and second-zone minigaps are indicated on the $E(k)$ curve of Figure 36.4. Since the second-zone minigaps lie beyond the Brillouin zone, they do not cut the Fermi surface of an alkali metal.

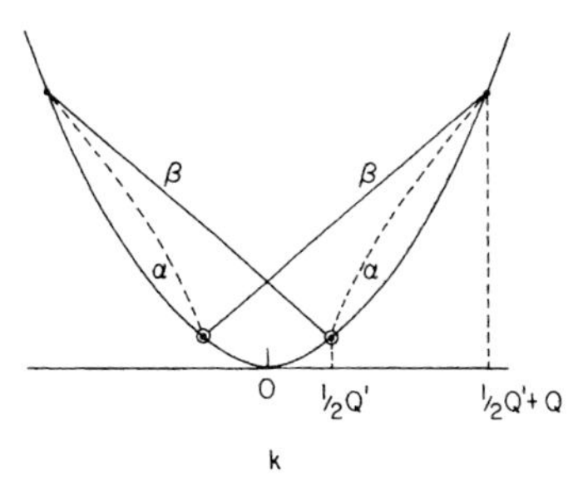

Figure 36.5 Coupling scheme for the first pair of herodyne gaps, located at the circled points.

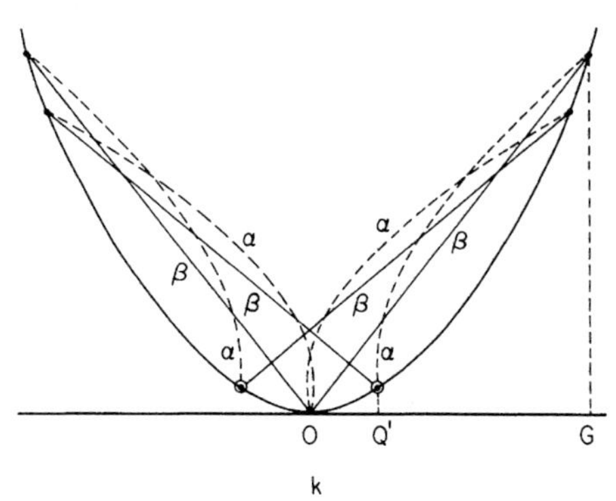

Figure 36.6 Coupling scheme for the second pair of herodyne gaps, located at the circled points.

The third family of higher-order gaps, the heterodyne gaps, lies near $k = 0$. From Equations (36.8) and (36.9) their periodicities can be described by

$$\vec{K} = n\vec{Q}' . \tag{36.17}$$

The minimum number of basis states for the matrix which leads to the first heterodyne gap, $n = 1$, is four:

$$|\vec{k}\rangle , \quad |\vec{k} - \vec{G}\rangle , \quad |\vec{k} + \vec{Q}\rangle , \quad |\vec{k} - \vec{G} + \vec{Q}\rangle . \tag{36.18}$$

The coupling scheme is shown in Figure 36.5. The minimum number of basis states required for the second heterodyne gap is seven – those in (36.18), and

$$|\vec{k} - 2\vec{G} + \vec{Q}\rangle , \quad |\vec{k} - \vec{G} + 2\vec{Q}\rangle , \quad |\vec{k} - 2\vec{G} + 2\vec{Q}\rangle . \tag{36.19}$$

The coupling scheme for this case is shown in Figure 36.6.

A view of $\vec{k}$ space in the plane containing $\vec{Q}$ and $\vec{G}$ is illustrated in Figure 36.7. Six pairs of energy-gap planes are shown together with their approximate location relative to the free-electron Fermi sphere. In this figure we have shown the effect of the small angle θ between $\vec{Q}$ and $\vec{G}$. This arises theoretically [7] by minimizing the elastic strain energy of the periodic lattice displacement (having wave vector $\vec{Q}'$), which tends to neutralize the charge modulation created by the electrons. If $\vec{Q}$ were parallel to $\vec{G}$, only a longitudinal (static) phonon displacement would be involved. The tilt between $\vec{Q}$ and $\vec{G}$ allows a less energetic, transverse phonon mode to provide charge screening. It is a simple exercise to show that the angle θ_n between the nth minigap periodicity $\vec{K}_n$ and $\vec{G}$ is given by

$$\tan \theta_n = \frac{(n+1)Q \sin \theta}{(n+1)Q \cos \theta - nG} . \tag{36.20}$$

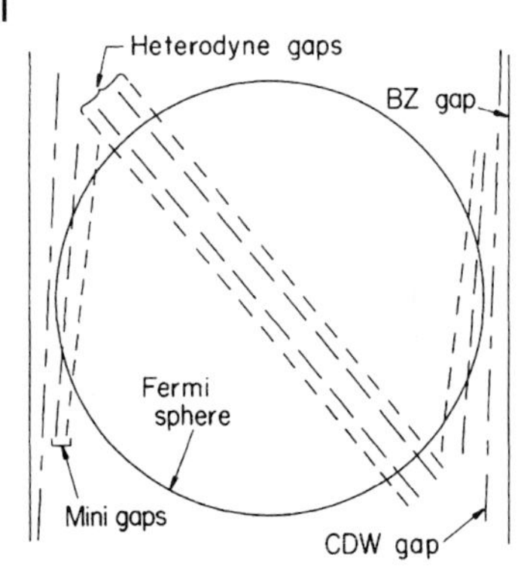

Figure 36.7 The free-electron Fermi sphere and the various energy gaps caused by CDW. Only two (of the twelve) Brillouin-zone gaps (solid vertical lines) are shown.

Since the heterodyne-gap periodicities, Equation (36.17), are all parallel, they have a common angle θ_H relative to $\vec{G}$.

$$\tan\theta_H = \frac{Q\sin\theta}{G - Q\cos\theta} . \tag{36.21}$$

The theory [7] which determines θ and θ_H indicates that θ_H is the angle which is most sharply controlled by the elastic anisotropy of a given metal. That is, θ_H is (virtually) constant when $|\vec{Q}|$ is allowed to take on different values. $\theta_H \approx 48^\circ$ and 47° for Na and K, respectively.

36.3 Results for Na and K

In order to calculate the sizes of the various higher-order gaps discussed in Section 36.2 one must know the parameters α and β. Unfortunately, the pseudopotential parameter β has not been determined in a direct measurement. We shall employ values derived from analyses of de Haas–van Alphen data, similar to those of Ashcroft [11], expressed in eV:

$$\beta_{\text{Na}} = 0.23 , \quad \beta_{\text{K}} = 0.20 . \tag{36.22}$$

The model used to obtain these values employed local pseudopotentials. It is not known how they would change if nonlocal effects of exchange and correlation [12] were incorporated.

As mentioned in Section 36.1, the CDW potential 2α can be estimated from the threshold of the optical anomaly. The observed thresholds, corresponding to transitions from A to B in Figure 36.4, are 0.62 and 1.2 eV for K and Na, respectively. The Mayer–El Naby absorption [8] in K, on bulk samples, has been reproduced by Hietel [13] and by Harms [14]. The anomaly in Na (bulk samples) was studied by Hietel and Mayer [15]. Attempts to observe these optical anomalies using evaporated films have failed [16, 17]. (This is expected if $\vec{Q}$ is epitaxially oriented perpendicular to the film [6].)

When the pseudopotential, $2\beta\cos\vec{G}\cdot\vec{r}$, is included in the Hamiltonian, the threshold energy for the CDW optical anomaly (A to B in Figure 36.4) is smaller than 2α. The reduction depends on the magnitude and tilt angle θ of $\vec{Q}$ relative to $\vec{G}$ and on the pseudopotential β. Accordingly we have adjusted the values of α so that, after the Hamiltonian matrix is diagonalized, the threshold energy agrees with the observed values. The adjusted values of α are given in Table 36.1 along with the corresponding values assumed for $|\vec{Q}|$.

The small (but finite) angle θ between $\vec{Q}$ and $\vec{G}$ leads to several problems in calculating the minigaps and heterodyne gaps. The first is that the gaps no longer lie on planes perpendicular

Table 36.1 Values of α, the CDW potential coefficient, for several values of $|\vec{Q}|$. $a = 5.23$ and 4.23 Å for K and Na, respectively. θ is the angle between $\vec{Q}$ und $\vec{G}$.

$\vert\vec{Q}\vert$ $(2\pi/a)$	θ (deg)	α (eV)
	Potassium	
1.36	2.6	0.36
1.37	2.1	0.38
1.38	1.6	0.40
	Sodium	
1.36	2.2	0.68
1.37	1.7	0.71
1.38	1.2	0.74

Table 36.2 Minigaps (mini) and heterodyne gabs (het.) obtained with several values of $|\vec{Q}|$ for K and Na. All gaps are in meV. $|\vec{Q}|$ is in $2\pi/a$ units.

$\vert\vec{Q}\vert$	First mini	Second mini	Third mini	First het.	Second het.	Third het.
	Potassium					
1.36	97	27	3	27	9	1
1.37	106	43	9	28	12	2
1.38	107	62	25	28	16	5
	Sodium					
1.36	138	57	11	38	14	2
1.37	135	79	31	38	20	5
1.38	116	87	63	37	25	13

to $\vec{K}$ (and which pass through $\pm\frac{1}{2}\vec{K}$). Instead the gaps define curved surfaces in $\vec{k}$ space which are close to the planes described. To find such a surface, consider a coordinate system with $\hat{z}$ parallel to $\vec{K}$ and $\hat{x}$ in the plane containing $\vec{Q}$, $\vec{G}$, and $\vec{K}$. For a fixed k_x, calculate the optical transition energy versus k_z from the eigenvalues of the Hamiltonian matrix. The minimum transition energy is the value of the minigap (for that k_x); and the location k_z of the energy-gap surface is at the value of k_z having that minimum. We find, of course, that the minigap depends on k_x. Usually the minigap varies quadratically with k_X, and its absolute minimum is near $k_x = 0$. The values so obtained for the first three minigaps and heterodyne gaps are given in Table 36.2.

In order to guarantee accuracy for the higher-order gaps, we used Hamiltonian matrices of dimension 21×21 rather than the 7×7 shown in (36.16). Enlarging the matrix beyond that does not change the values given in Table 36.2.

It is evident that the gaps fall off rapidly with increasing order n, defined by Equations (36.6)–(36.8). It is possible to derive a formula for the minigap sequence provided α and β are both small compared to Δ:

$$\Delta \equiv E\left(\frac{1}{2}G\right) - E\left(\frac{1}{2}G - Q'\right) . \tag{36.23}$$

Δ is the energy difference between adjacent basis states, shown in Figure 36.2. The n-th minigap is then

$$g(n) = \frac{2\alpha^{n+1}\beta^n}{(n!)^2\Delta^{2n}} \,. \tag{36.24}$$

A similar formula exists for the second-zone minigaps:

$$\overline{g}(n) = \frac{2\alpha^n\beta^{n+1}}{(n!)^2\Delta^{2n}} \,. \tag{36.25}$$

Derivation of these formulas is left for the reader. They cannot be applied to the case of Na or K because α and β are not sufficiently small. It should also be noted that the formulas are valid only for n which satisfy $n\Delta < E\left(\frac{1}{2}G\right)$. They are of interest because they indicate the rapid rate of decrease of $g(n)$ with increasing n.

36.4 Conclusions

Each minigap or heterodyne gap which cuts the Fermi surface will lead to new optical-absorption bands. From Table 36.2 it is clear that most of these will be in the far-infrared spectrum. The first minigap, near 0.1 eV, should be the easiest to observe. Theoretical study of the shape and intensity of these new absorption bands is underway.

High-order gaps are the ones that lead to the many open-orbit magnetoresistance peaks that have been extensively studied by Coulter and Datars [14, 18]. The size of each gap is of considerable interest because it governs the critical field at which magnetic breakdown of open orbits begins to occur.

We emphasize once again that the values given in Table 36.2 are subject to uncertainty because the ion pseudopotentials are not known from direct experiment. Furthermore, there is likely a strong nonlocal $\vec{k}$ dependence to the CDW potential, since it arises from exchange and correlation. Studies [12] of such nonlocal variations in $\alpha(\vec{k})$ indicate that a significant reduction of the heterodyne-gap magnitudes could be expected. The effect would not be so large for the minigaps, since they are near (in $\vec{k}$ space) to the main CDW gaps. Finally, the directions of $\vec{Q}$ used in the calculations were based on a model [7] which neglects the anisotropy of the conduction electron $E(\vec{k})$, which obtains when a CDW is not present. Refinement of the calculated gaps will become possible if the Bragg reflection satellites [19] of the CDW are found.

Acknowledgements

This research was supported by the National Science Foundation and by a scholarship (to one of us, F.E.F.) from Foninves, Energy and Mine Department, Venezuela.

References

1 Overhauser, A.W. (1978) *Adv. Phys.*, **27**, 343.

2 Overhauser, A.W. (1968) *Phys. Rev.*, **167**, 691.

3 Overhauser, A.W. (1964) *Phys. Rev. Lett.*, **13**, 190.

4 Coulter, P.G. and Datars, W.R. (1980) *Phys. Rev. Lett.*, **45**, 1021.

5 Huberman, M. and Overhauser, A.W. (1982) *Phys. Rev. B*, **25**, 2211.
6 Overhauser, A.W. and Butler, N.R. (1976) *Phys. Rev. B*, **14**, 3371.
7 Giuliani, G.F. and Overhauser, A.W. (1979) *Phys. Rev. B*, **20**, 1328.
8 Mayer, H. and El Naby, M.H. (1963) *Z. Phys.*, **174**, 269.
9 Reitz, J.R. and Overhauser, A.W. (1968) *Phys. Rev.*, **171**, 749.
10 Azbel, M.Y. (1979) *Phys. Rev. Lett.*, **43**, 1954.
11 Ashcroft, N. (1965) *Phys. Rev. A*, **140**, 935.
12 Duff, K.J. and Overhauser, A.W. (1972) *Phys. Rev. B*, **5**, 2799.
13 Hietel, B. Hietel (private communication).
14 Harms, P. (1972) Dissertation, Technische Universität Clausthal; his data are reproduced in [6].
15 Hietel, B. and Mayer, H. (1973) *Z. Phys.*, **264**, 21.
16 Smith, N.V. (1969) **Phys. Rev.**, **183**, 634.
17 Faldt, A. and Wallden, L. (1980) *J. Phys. C*, 13, 6429.
18 Coulter, P.G. and Datars, W.R. (1982) *Solid State Commun.*, **43**, 715.
19 Giuliani, G.F. and Overhauser, A.W. (1980) *Phys. Rev. B*, **22**, 3639.

Reprint 37 Theory of Charge-Density-Wave-Spin-Density-Wave Mixing[1)]

A.W. Overhauser*

* Department of Physics, Purdue University, 525, Northwestern Avenue, West Lafayette, Indiana 47907, USA

Received 5 March 1984

Metals with a charge-density-wave broken symmetry will acquire a spin-density-wave component if the charge-density-wave amplitude is sufficiently large and if the elastic stiffness constants are sufficiently anharmonic. This effect has implications for the compatibility of charge-density waves with superconductivity.

The spin-up and spin-down electron densities of a metal having a broken translational symmetry are [1]

$$\rho^{+}(\vec{r}) = \frac{1}{2}\rho_0[1 + p\cos(\vec{Q}\cdot\vec{r} + \phi)] ,$$
$$\rho^{-}(\vec{r}) = \frac{1}{2}\rho_0[1 + p\cos(\vec{Q}\cdot\vec{r} - \phi)] . \tag{37.1}$$

The wave vector and fractional amplitude of the wave are $\vec{Q}$ and p; $\rho_0(\vec{r})$ is the conduction electron density (having the periodicity of the lattice) which would otherwise prevail. The "spin-split-phase" angle ϕ describes the degree of charge-density-wave-spin-density-wave (CDW-SDW) mixing. If $\phi = 0$, Equation (37.1) describes a pure CDW; if $\phi = \frac{1}{2}\pi$, it describes a pure SDW. In this paper I investigate the conditions which would allow ϕ to have an intermediate value.

Broken translational symmetry in two- or three-dimensional metals is caused by exchange and correlation contributions to the conduction electron energy [1, 2]. The purely electronic energy relative to the normal state ($p = 0$) can be described approximately:

$$\Delta E_e \simeq \varepsilon_0 - \alpha p^2 + \beta p^4 + p^2 D \sin^2\phi . \tag{37.2}$$

The coefficients α and β describe a parabolic minimum which determines p^2 (provided $\Delta E_e < 0$). The constant ϵ_0 has been included to emphasize the fact that Equation (37.2) is not valid for $p \sim 0$. The reasons for this stem from the correlation energy correction [1], and are related to the conclusion that a CDW instability can be a first-order, rather than a second-order, transition [3].

The last term of Equation (37.2) describes the dependence of the correlation energy on the spin-split phase ϕ. Since the correlation energy arises primarily from virtual scattering of antiparallel-spin electrons, it favors a CDW instead of an SDW because the scattering matrix elements are enhanced when the crests in the spin-up and spin-down charge modulations coincide (in real space) [1].

The ordinary Coulomb energy prohibits a CDW instability unless a periodic lattice displacement $\vec{u}(\vec{F})$ of the positive-ion lattice compensates the charge modulation, i. e.,

$$\vec{u} = \vec{A}\sin(\vec{Q}\cdot\vec{r}) . \tag{37.3}$$

1) Phys. Rev. B 29, 12 (1984).

Anomalous Effects in Simple Metals. Albert Overhauser

ISBN: 978-3-527-40859-7

Local charge neutrality requires that the amplitude $\vec{A}$ satisfy

$$z|\vec{A}\cdot\vec{Q}| = p\cos\phi\ , \tag{37.4}$$

where z is the effective charge of the ion for the wave vector $\vec{Q}$. For example, if Z is the nuclear charge of a monovalent metal, then $z = Z - (Z-1)f(Q)$, where $f(Q)$ is the X-ray form factor [4]. The directions of $\vec{Q}$ and $\vec{A}$ depend on the anisotropy of the elastic stiffness constants [5].

Elastic-stress energy is therefore an important energy contribution which impedes a CDW instability. It will be

$$\Delta E_s = \lambda p^2\cos^2\phi + \mu p^4\cos^4\phi\ , \tag{37.5}$$

where λ is proportional to the elastic stiffness (for the directions of $\vec{Q}$ and $\vec{A}$), and μ is a coefficient proportional to the anharmonic correction. The condensation energy of a mixed CDW-SDW is accordingly

$$\Delta = \Delta E_e + \Delta E_s\ . \tag{37.6}$$

It is clear that a large elastic-stress term favors a SDW, whereas the spin-split-phase term of (37.2) favors a CDW. For small p, a pure CDW or SDW occurs depending on whether $D > \lambda$ or $D < \lambda$. The conclusion that Na and K have CDW states [3, 6], whereas Cr has a SDW state [2], should not be thought surprising since the elastic constants of Na and K are two orders of magnitude smaller than those of Cr.

Minimization of the total condensation energy, Equation (37.6), leads to

$$p^2 = \frac{\alpha - D\sin^2\phi - \lambda\cos^2\phi}{2(\beta + \mu\cos^4\phi)}\ . \tag{37.7}$$

This solution to the variational problem applies only if the numerator is positive, otherwise $p = 0$. Further variation of ΔE with respect to ϕ, after substitution of (37.7) for p^2 in ΔE, leads to the optimum amount of CDW-SDW mixing. I first obtain

$$\Delta E = \epsilon_0 = \frac{(\alpha - D\sin^2\phi - \lambda\cos^2\phi)^2}{4(B + \mu\cos^4\phi)}\ . \tag{37.8}$$

The equilibrium value of ϕ is then given by

$$\cos^2\phi = \frac{\beta(D-\lambda)}{\mu(\alpha - D)}\ . \tag{37.9}$$

As noted above, one obtains a pure SDW if $D < \lambda$, so that Equation (37.9) is pertinent only when $D > \lambda$. In this latter case only a pure CDW can occur if $D > \alpha$. Equation (37.9) can be applied only when all of its factors are positive. A SDW component occurs if (37.9) is less than unity, i. e.,

$$\mu > \frac{\beta(D-\lambda)}{\mu(\alpha - D)}\ . \tag{37.10}$$

In order to perceive the trends which lead to CDW-SDW mixing one may neglect λ in comparison to D and D in comparision to α. The criterion, Equation (37.10), reduces to a crude but instructive one:

$$\mu > \frac{\beta D}{\alpha} \quad \text{or} \quad \mu p^2 > \frac{1}{2}D\ . \tag{37.11}$$

The second alternative follows from the relation $p^2 \sim \alpha/2\beta$, obtained from the leading terms of Equation (37.7). Thus the criterion for a SDW admixture is more easily satisfied if the CDW amplitude p and the anharmonic coefficient μ are both large. This is easily understood:

Suppose that α (the driving force of the instability) is slowly increased. The anharmonic term $\mu p^4 \cos^4 \phi$ gradually dictates a preference for further growth in p through a SDW component (which requires no further lattice strain). Only the correlation energy $D p^2 \sin^2 \phi$ intervenes to effect a compromise. Equation (37.10) shows that CDW-SDW mixing can occur only when anharmonic elastic effects are included.

Experimental detection of a mixed CDW-SDW requires the juxtaposition of satellite intensity measurements from X-ray and neutron scattering. X-rays are scattered primarily by the periodic lattice distortion, Equation (37.3). The CDW satellites occur for scattering vectors $\vec{G} \pm \vec{Q}$, where $\{\vec{G}\}$ are the reciprocal-lattice vectors [7]. On the other hand, neutrons are scattered by both the lattice distortion and by the magnetic field of the SDW component which, from magnetic energy considerations, will likely be polarized perpendicular to $\vec{Q}$. Since the SDW exists only in the conduction electrons, the magnetic scattering amplitude will be significant primarily for scattering vectors $\pm\vec{Q}$. Accordingly, the signature of a mixed CDW-SDW will be anomalous neutron scattering amplitudes for the two satellites of $\vec{G} = 0$ in comparison to the satellites of other $\vec{G}$. The difference will be most pronounced for CDW-SDW states of an *s–p* energy band. In this case the $\pm\vec{Q}$ satellites may have intensities so different from those of other $\vec{G}$ that X-ray measurements may not be needed. Interesting polarized-beam effects become possible if the SDW polarization vector can be rotated by external fields [8].

A pure CDW is compatible with superconductivity since the periodic potential caused by the broken symmetry is no different from that of a crystalline periodicity. The electronic density of states at the Fermi energy $N(E_F)$ can be either increased or decreased by the CDW, depending on whether $Q/2k_F$ is greater than or less than unity [1]. The time-reversed partners of amplitude-modulated conduction electrons are still degenerate, so that Cooper pairing is unaffected. However, this is no longer the case for a mixed CDW-SDW state.

The self-consistent periodic potential of a mixed state corresponding to Equation (37.1) is

$$V(\vec{r}) = -W[\cos\phi \cos(\vec{Q}\cdot\vec{r}) - \sigma_z \sin\phi \sin(\vec{Q}\cdot\vec{r})] , \tag{37.12}$$

where σ_z is the Pauli spin operator. This potential does not have time-reversal symmetry. If $\psi_{\vec{k}}$ is an up-spin eigen-state of the Schödinger equation having the potential (37.12), then $T\psi_{\vec{k}}$ is not an eigenstate. ($T \equiv$ time-reversal operator). Indeed, the energy expectation value of $T\psi_{\vec{k}}$ will differ substantially from that of $\psi_{\vec{k}}$. As a consequence, singlet pairing will be suppressed.

A well-known problem in superconductivity is the absence of superconductivity in Li [9], at least down to 6 mK [10]. Calculations of the electron–phonon interaction constant [11, 12] indicate a value $\lambda = 0.4$. This would lead to a transition temperature $T_c \sim 1$ K, based on McMillan's formula with $\mu^* = 0.10$. Optical data [13] of metallic Li indicate an anomalous absorption with a threshold eV, analogous to the Mayer–El Naby anomaly [14] in K ($W \sim 0.6$ eV) and the Hietel–Mayer anomaly [15] in Na ($W \sim 1.2$ eV). The open-orbit magnetoresistance spectra [16] of Na and K show the dramatic effects of CDW structure, which also accounts for the optical anomalies [17, 18]. It is known that Li is very anharmonic, even at liquid-helium temperature [19]. I speculate that the absence of superconductivity in Li may be caused by a mixed CDW-SDW state.

Since a CDW can occur only if a periodic lattice distortion, Equation (37.3), compensates the charge modulation, reduction (or elimination) of the CDW amplitude p may be achieved by interfering with the ability of ions to undergo displacements easily. Interstitial impurities, radiation damage, or freezing-in an amorphous state might accomplish this.[2] The criterion Equation (37.10), for SDW mixing would then be more difficult to satisfy. As a consequence, metals which ought to be superconducting might be enabled to become so.

2) There has been an unconfirmed report that amorphous Li has $T_c = 2.4$ K [20].

References

1 Overhauser, A.W. (1968) *Phys. Rev.*, **167**, 691.
2 Overhauser, A.W. (1962) *Phys. Rev.*, **128**, 1437.
3 Overhauser, A.W. (1978) *Adv. Phys.*, **27**, 343, See 2.6.
4 Giuliani, G.F. and Overhauser, A.W. (1980) *Phys. Rev. B*, **22**, 3639.
5 Giuliani, G.F. and Overhauser, A.W. (1979) *Phys. Rev. B*, **20**, 1328.
6 Overhauser, A.W. (1983) in *Electron Correlations in Solids, Molecules, and Atoms* (eds J.T. Devreese and F. Brosens), Plenum, New York, p. 41.
7 Overhauser, A.W. (1971) *Phys. Rev. B*, **3**, 3173.
8 Werner, S.A., Arrott, A., and Atoji, M. (1969) *J. Appl. Phys.*, **40**, 1447.
9 Grimvall, G. (1981) *The Electron–Phonon Interaction in Metals*, North-Holland, Amsterdam, p. 255.
10 Thorp, T.L., Triplett, B.B., Brewer, W.D. Cohen, M.L., Phillips, N.E., Shirley, D.A., Templeton, J.E., Stark, R.W., and Schmidt, P.H. (1970) *J. Low Temp. Phys.*, **3**, 589.
11 Grimvall, G. (1975) *Phys. Scr.*, **12**, 337.
12 McDonald, A.H. and Leavens, C.R. (1982) *Phys. Rev. B*, **26**, 4293.
13 Inagaki, T., Emerson, L.C., Arakawa, E.T., and Williams, M.W. (1976) *Phys. Rev. B*, **13**, 2305.
14 Mayer, H. and E. Naby, M.H. (1963) *Z. Phys.*, **174**, 269.
15 Hietel, B. and Mayer, H. (1973) *Z. Phys.*, **264**, 21.
16 Coulter, P.G. and Datars, W.R. (1980) *Phys. Rev. Lett.*, **45**, 1021.
17 Overhauser, A.W. (1964) *Phys. Rev. Lett.*, **13**, 190.
18 Overhauser, A.W. and Butler, N.R. (1976) *Phys. Rev. B*, **14**, 3371.
19 McCarthy, C.M., Thompson, C.W., and Werner, S.A. (1980) *Phys. Rev. A*, **22**, 502.
20 Reale, C. *Phys. Lett. A*, **55**, 165 (1975).

Reprint 38 Crystal Structure of Lithium at 4.2 K[1)]

A. W. Overhauser*

* Department of Physics, Purdue University, 525, Northwestern Avenue, West Lafayette, Indiana 47907, USA

Received 26 April 1984

The (heretofore unknown) crystal structure of lithium at low temperature is identified to be close-packed rhombohedral $9R$. The stacking order of the hexagonal layers is $ABCBCACAB$, a nine-layer repeat sequence. There are three atoms per primitive rhombohedral cell, and nine in the (nonprimitive) hexagonal cell The only other element with this structure is samarium.

It seems surprising that the ground-state crystal structure of metallic lithium has remained unknown. In common with the other alkali metals it is body-centered-cubic (bcc) at room temperature. In 1948 Barrett and Trautz [1] discovered that Li undergoes a structural phase transformation on cooling below 70 K. They suggested that the new phase was hexagonal-close-packed (hcp). However, a recent study [2] has shown that the structure is neither hcp nor face-centered-cubic (fcc).

Determination of the atomic arrangement is difficult in this case because single crystals of the low-temperature phase are not available. Only poly crystalline diffraction patterns have been obtained. The best work [2] employed 1.06-Å neutrons. The 4.2-K data of this study show no trace of an hcp (012) reflection; the hcp (Oil) line is also missing. Neither is there an fcc (200) reflection. So the most common close-packed structures have been eliminated as possibilities [2]. The hexagonal-layer stacking sequences of the hcp and fcc lattices are $ABAB\ldots$, and $ABCABC\ldots$ They can be denoted $2H$ and $3R$.

During the following discussion the reader should refer to Figures 1 and 2 of [2], which are "powder" patterns for $\lambda = 1.06$ Å. Perhaps the next stacking sequence to try is the "double hcp" (or AH) structure $ABACABAC\ldots$, with a repeat distance of four. Several elements have this structure, e. g., La, Ce, Pr, Nd, and Am. Such a crystal must have an (014) reflection, at 33.3°, which is equivalent to the (012) line of the hcp lattice. As mentioned above, this line is not present. The stacking sequence ABCACB, which might be called "triple hcp" (or $6H$, does not have a reflection at 33.3°; but its (014) reflection, at 27.9°, coincides with the (200) fcc line, which is also missing.

A somewhat more complex arrangement of hexagonal layers is the $9R$ sequence of samarium: $ABCBCACAB$. The primitive rhombohedral cell has three atoms, but the hexagonal cell has nine.

The scattering angles and structure factors $|F|^2$ for the first 26 diffraction lines are given in Table 38.1. The lattice parameter $a = 3.111$ Å, in the hexagonal plane, has been taken from Barrett [3] since the Bragg angle θ_B of the dominant new line, (110), of the low-temperature Li phase does not depend on the stacking sequence along the c axis. Other θ_B are based on the assumption that ideal close packing obtains, i. e., $c = 9\left(\frac{2}{3}\right)^{1/2} a = 22.86$ Å.

The calculated diffraction pattern of the $9R$ structure appears to agree well with the observed spectrum [2]. Before a discussion of details one should note that approximately 75% of the sample transforms from bcc to the new phase on cooling to 4.2 K. The diffraction pattern continues to change (at 4.2 K) during the first 24 h [2]. The peak heights near 23° and

1) Phys. Rev. Lett. 53, 1 (1984).

Anomalous Effects in Simple Metals. Albert Overhauser

ISBN: 978-3-527-40859-7

Table 38.1 Diffraction lines of the $9R$ structure with $a = 3.111$ Å and $c = 9\left(\frac{2}{3}\right)^{1/2} a$, appropriate to Li at 4.2 K. The structure factors $|F|^2$ for (009) and (00 18) have been divided by three to account for the reduced multiplicity (2 compared to 6) relative to all others. A neutron wavelength $\lambda = 1.06$ Å has been assumed in calculating the scattering angles $2\theta_B$.

(hkl)	$2\theta_B$	$F\|^2$	(hkl)	$2\theta_B$	$F\|^2$
1 0 1	22.85	1.8	2 0 2	46.68	0.8
0 1 2	23.32	0.8	1 1 9	47.10	9
0 0 9	24.09	"3"	0 2 4	47.68	6.4
1 0 4	25.12	6.4	2 0 5	48.43	6.4
0 1 5	26.40	6.4	0 0 18	49.33	"3"
1 0 7	29.55	0.8	1 0 16	49.65	0.8
0 1 8	31.37	1.8	0 2 7	50.38	0.8
1 0 10	35.40	1.8	2 0 8	51.57	1.8
0 1 11	37.58	0.8	0 1 17	52.27	1.8
1 1 0	39.84	9	0 2 10	54.34	1.8
1 0 13	42.19	6.4	2 0 11	55.92	0.8
0 1 14	44.61	6.4	1 0 19	57.69	1.8
0 2 1	46.42	1.8	0 2 13	59.42	6.4

26.5° almost double. This behavior cannot be attributed to further conversion from the bcc phase, since full conversion would lead to only a 33% increase. (Indeed, significant intensity remains in the bcc reflections after 24 h.) Instead, the changes occurring at 4.2 K must be caused (primarily) by annealing of stacking faults in the new phase. Such behavior is expected in any long-period structure.

McCarthy, Thompson, and Werner [2] noted that the first reflection near 23° was too small by a factor of about 8 for an hcp reflection. The first two lines (near 23°) of the $9R$phase have combined intensity, $|F_1|^2 + |F_2|^2$, of 2.6. A full-strength hcp line on this scale would have an intensity of 18, twice the maximum number in Table 38.1 (on account of the extra multiplicity, 12 instead of 6, for hexagonal powder patterns compared to rhombohedral). Thus the unresolved pair at 23° should be 7 times weaker than a full-strength hcp line. In Figure 2 of [2] the shoulder A at 24° and the peak at 26.5° correspond to the (009) and (015) reflections of $9R$. The (104) cannot be separated from the (residual) bcc (110). The two shoulders labeled B in Figure 1 of [2] are the $9R$ (1 0 13) and (0 1 14). A high-resolution study in this region would allow an accurate determination of the lattice parameter c. The reflections in Table I between 29° and 38° are either too small to identify or are masked by the Al (200), from the sample holder, or by the Li bcc (200). The large asymmetric structure (in Figure 1 of [2]) between 47° and 50° is explained very well by the four $9R$ reflections in this region. Here, also a high-resolution study will provide a rigorous test of the proposed $9R$ phase, provided stacking faults are eliminated by annealing.

The low-temperature phase of Li is of particular interest because Li is not superconducting, at least not above 6 mK. Calculations of the electron–phonon interaction parameter [4] show that $\lambda \approx 0.4$, which would imply a superconducting transition temperature $T_c \sim 1$ K. (The calculations apply, of course, to the bcc phase.) One expects that, among all metals, prediction of T_c should be most reliable in one so simple. Therefore failure by a factor of at least 100 is quite disturbing.

A comparison of the theoretical values of λ in the bcc and $9R$ structures becomes a fascinating question. Possibly the "loss" of the slow shear modes (of the bcc structure) can lead to a significant reduction in λ for the $9R$. On the other hand, the Fermi surface in the $9R$ phase is near 38 energy-gap planes, compared to 12 for the bcc (and 14 for an fcc). This could lead to enhanced electron–phonon umklapp coupling associated with low-frequency phonons, and to a *larger* λ.

One may anticipate that the interband optical absorption spectrum of the $9R$ phase will differ substantially from that [5] of the bcc on account of the drastic alteration in energy-gap configuration. Prediction of this is a real challenge since an experimental check is relatively easy.

References

1. Barrett, C.S. and Trautz, O.R. (1948) *Trans. Am. Inst. Min. Metall. Pet. Eng.*, **175**, 579.
2. McCarthy, C.M., Thompson, C.W., and Werner, S.A. (1980) *Phys. Rev. B*, **22**, 574.
3. Barrett, C.S. (1956) *Acta Crystallogr.*, **9**, 671.
4. Grimvall, G. (1975) *Phys. Scr.*, **12**, 337.
5. Inagaki, T., Emerson, L.C., Arakawa, E.T., and Williams, M.W. (1976) *Phys. Rev. B*, **13**, 2305.

Reprint 39 Theory of Induced-Torque Anomalies in Potassium[1)]

Xiaodong Zhu and A.W. Overhauser**

* Department of Physics, Purdue University, 525, Northwestern Avenue, West Lafayette, Indiana 47907, USA

Received 30 January 1984

The nonsaturating, four-peak pattern of induced torque observed in spherical potassium samples has been a challenging puzzle for many years. It is found that this behavior can be quantitatively explained by an anisotropic Hall coefficient. This provides further confirmation of the broken translation symmetry of the ground state of potassium, i. e., a charge-density-wave structure.

39.1 Introduction

The nonsaturating, four-peak induced-torque patterns in potassium spheres [1, 2] indicate that the Fermi surface is multiply connected and lacks cubic symmetry. Such properties are expected if potassium has a charge-density-wave (CDW) structure [3, 4]. In this paper we show that the four-peaked patterns, observed between 10 and 30 kOe, can be quantitatively explained by an anisotropic Hall coefficient.

Potassium is generally considered to have a bcc crystal structure, with its Fermi surface deviating only by about one part in 10^3 from sphericity. For such a metal, the semiclassical galvanomagnetic theory [5, 6] predicts precisely that, in high magnetic fields $\omega_c\tau \gg 1$ its magnetoresistance should saturate, and its Hall coefficient R_H should be exactly that for free electrons, i. e., $R_0 = (-nec)^{-1}$. ω_c is the cyclotron frequency, τ is the electronic scattering time, and n is the electron density. Furthermore, if the induced-torque method [7] is used to measure the magnetoresistance, one expects [7, 8] that the induced torque of a spherical sample will saturate for $\omega_c\tau \gg 1$. However, a large number of experiments have given totally unexpected results.

First, in all experiments the magnetoresistance $\rho(H)$ of potassium is found to increase with H, without evidence of saturation, even at fields for which $\omega_c\tau > 300$. Results obtained by four-terminal [9], helicon-resonance [10], and induced-torque methods [11] substantially agree. These show that $\rho(H) \cong \rho_0(1 + S\omega_c\tau)$, where ρ_0 is the resistance in zero magnetic field. The Kohler slope S is typically 10^4–10^{-2}, depending on sample preparation and metallurgical history. The fact that a linear behavior is found in all samples, regardless of shape, contacts, and quality, rules out explanations invoking extrinsic mechanisms such as voids [12], inhomogeneities [13], or surface imperfections [14]. An explanation based on anisotropic relaxation [15] also falls short by many orders of magnitude.

Second, the Hall coefficient R_H depends on the direction of $\vec{H}$ and sometimes on the magnitude too. Penz and Bowers [10], using a low-frequency helicon, standing-wave technique, observed that $|R_H|$ decreases monotonically with increasing H. The decrease varied from sample to sample and was $\sim$ 7% at 100 kOe. Their measurements of the absolute value of R_H showed a 6–12% enhancement over the free-electron value. However, Chimenti and Maxfield [16] found no field dependence in R_H from 20 to 100 kOe by measuring the resonances of high-frequency helicon waves in a flat plate. Nevertheless, the absolute value of R_H was

1) Phys. Rev. B 30, 2 (1984).

Anomalous Effects in Simple Metals. Albert Overhauser

ISBN: 978-3-527-40859-7

larger than R_0, depending on crystal orientation. For a static magnetic field directed along a ⟨100⟩ direction, R_H was about 4% larger than R_0. For the ⟨010⟩ direction, the enhancement showed the largest deviation about 8%. Helicon resonances in a sphere [17] resulted in an enhanced R_H of 3 to 4%.

Third, the most striking effect of all, induced-torque measurements, shows dramatic angular-dependent anomalies in high magnetic fields. These anomalies will be reviewed below.

39.2 Induced-Torque Anomalies

The induced-torque method has been used to study the orientation dependence of the magnetoresistance of alkali metals. A spherical, single-crystal sample is suspended by a rod in a uniform magnetic field $\vec{H}$. As the magnetic field rotates about the suspension axis at a constant frequency Ω (typically $\Omega \sim 10^{-3}$ Hz), a circulating current is induced in the sample. This leads to an induced torque exerted on the suspension axis.

The experiments of Schaefer and Marcus [1] were the first to show anomalous high-field torque anisotropics (Figure 39.1). They measured induced torques on 200 *spherical*, single crystals of potassium and observed large, twofold torque anisotropics, even when the suspension axis was parallel to a threefold, ⟨111⟩, axis. For fields below $\sim$ 4 kOe, the twofold pattern was sinusoidal. At higher fields, four-peaked patterns emerged, as shown in Figure 39.1. The occurrence of a four-peaked pattern, even when the suspension axis is parallel to a ⟨111⟩ axis, shows that the cubic symmetry has been broken. In samples for which crystal orientation had been determined by X-rays, the low-field torque minima always occurred when the projection of a ⟨110⟩ axis in the horizontal plane was nearly parallel to $\vec{H}$. This indicates that the preferred axis of broken symmetry is most likely near a ⟨110⟩ crystal direction. At high fields, the torque minima increase linearly with field, but the maxima increase at a greater rate. Furthermore, the torque minima perpendicular to the preferred direction described above are always larger than the torque minima parallel to that direction.

These anomalous torque patterns were confirmed by Holroyd and Datars [2]. They also performed experiments on an accurately spherical sample grown in a Kel-F mold. Their data are shown in Figure 39.2. In Figure 39.2a, the induced torque versus magnet angle is shown for several values of H between 500 Oe and 17 kOe. The crystal-growth axis was vertical. In Figure 39.2b, the growth axis was horizontal; and values of H range from 1 to 23 kOe in 1-kOe steps. These results are similar to those of Schaefer and Marcus in every aspect. However, the anisotropy at 23 kOe was 45:1. Schaefer and Marcus found anisotropies which ranged from 3:1 to 15:1. It should also be noted that the four peaks are not evenly spaced in angle. For example, the separation which spans the smallest torque minimum is $\sim$ 100°. Unfortunately, the crystallographic orientation of this sample was never determined.

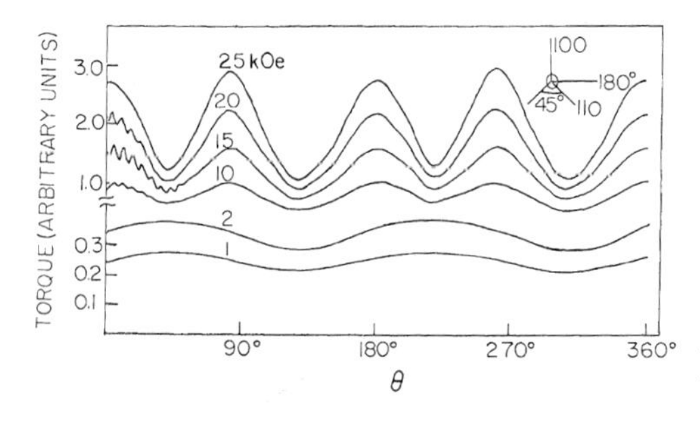

Figure 39.1 Induced torque versus angle θ for a spherical sample of potassium having the ⟨100⟩ axis parallel to the suspension. The data are from Schaefer and Marcus.

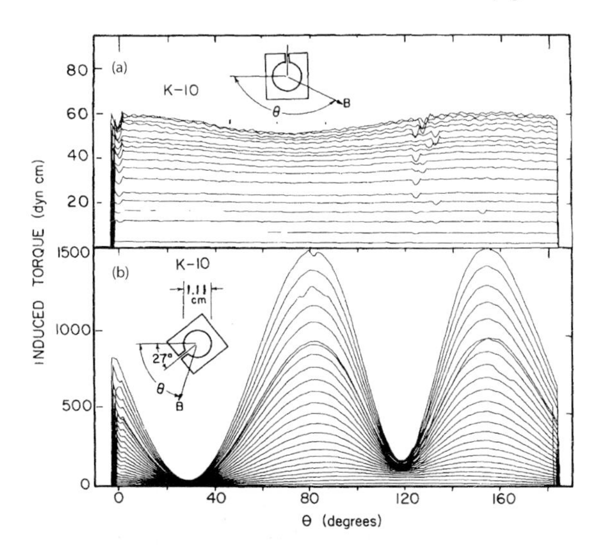

Figure 39.2 Induced torque versus magnetic field direction θ for a potassium sphere 1.11 cm in diameter. In (a) the field $\vec{H}$ is rotated about the growth axis. The curves shown are for 0.5, 1, 2, 3, ... kOe. In (b) the plane of rotation contains the growth axis, and H range from 1 to 23 kOe in 1-kOe steps. The data are from Holroyd and Datars. The crystallographic orientation is unknown.

If potassium had cubic symmetry and a spherical Fermi surface, the resistivity tensor would be isotropic, and the induced torque would be independent of magnet angle θ. To explain the observed anisotropy, several mechanisms have been suggested. They can be divided into three categories.

(a) Anisotropic resistivity. Several possible causes have been considered, such as an oriented array of dislocations or voids [12, 18] or electron–phonon Umklapp processes occurring at well-defined "hot spots" on the Fermi surface [15]. However, the required concentrations of dislocations or voids exceed reasonable values by about three orders of magnitude [2]. The hot-spot model requires variations in $1/\tau$ over the Fermi surface by four orders of magnitude. Measurements [19] of phonon-scattering anisotropy reveal nothing in excess of 10%.

Even if there were a mechanism causing a large residual resistance anisotropy, say 10:1, calculation [20] shows that the induced torque becomes isotropic by 10 kOe, and a four-peak pattern does not develop.

(b) Inhomogeneous residual resistivity. We have solved the induced-torque problem for a sphere having a gradient in impurity concentration. Four-peaked patterns do not arise. This work will be published separately.

(c) Anisotropic sample shape. Lass [21] found that if a sample was sufficiently distorted from a spherical shape, the calculated torque pattern could resemble those of Schaefer and Marcus (Figure 39.1). However, this explanation required deviations from sphericity of 10–15% in order to explain data such as those of Figure 39.1. The measured anisotropy in shape was 2% or less [22]. The data of Figure 39.2 would require a 2:1 shape anisotropy. However, this sample was spherical to a precision of $\frac{1}{2}$%. We shall show that an anisotropic Hall coefficient provides a quantitative explanation of the induced-torque patterns shown in Figures 39.1 and 39.2.

39.3 Magnetoresistivity Tensor of Potassium

We assume that the ground state of potassium has a CDW structure [3]. The conduction-electron density then has a small sinusoidal modulation,

$$\rho_e(\vec{r}) = \rho_e^0[1 - p\cos(\vec{Q}\cdot\vec{r})] , \tag{39.1}$$

where ρ_e^0 is the average electron density, and p and $\vec{Q}$ are the fractional modulation and wave vector of the CDW, respectively. The wave vector $\vec{Q}$ has a magnitude $\sim 1.33(2\pi/a)$, 8% larger than a Fermi sphere's diameter [23, 24]. The direction of $\vec{Q}$ is tilted a few degrees from a $\langle 110\rangle$ axis and is believed to lie in a plane oriented $\sim 65°$ from the $\langle 001\rangle$ plane [25, 26].

The exchange and correlation potential,

$$V(\vec{r}) = G\cos(\vec{Q}\cdot\vec{r}) , \tag{39.2}$$

of the CDW creates two energy gaps in $\vec{k}$ space along planes passing through $\pm\frac{1}{2}Q$. The main gaps, about 0.6 eV [25], distort the Fermi surface nearby, forming small necks or points of critical contact.

To maintain charge neutrality, the positive ions are displaced sinusoidally relative to their ideal bcc lattice sites $\{\vec{L}\}$. The displacements are

$$\vec{u}(\vec{L}) = \vec{A}\sin(\vec{Q}\cdot\vec{L}) , \tag{39.3}$$

where $\vec{A}$ is the amplitude of the lattice distortion. This amplitude is 0.03 Å; the direction is parallel to the vector (18.5, 54.0, 53.1) if $\vec{Q} = (2\pi/a)$ (0.966, 0.910, 0.0865) [27]. The positive-ion density $\rho_i(\vec{r})$ is

$$\rho_i(\vec{r}) = \sum_{\{\vec{L}\}}\delta(\vec{r} - \vec{L} - \vec{A}\sin(\vec{Q}\cdot\vec{L})) , \tag{39.4}$$

where $\delta(\vec{r})$ is the Dirac δ function. We need the Fourier transform of $\rho_i(\vec{r})$,

$$\begin{aligned}\rho_{\vec{q}} &= \int \rho_i(\vec{r})e^{i\vec{q}\cdot\vec{r}}\,d\vec{r} \\ &= \sum_{\{\vec{L}\}} e^{i\vec{q}\cdot[\vec{L}+\vec{A}\sin(\vec{Q}\cdot\vec{L})]} .\end{aligned} \tag{39.5}$$

With the use of the Jacobi–Anger generating function for Bessel functions,

$$e^{iz\sin\phi} = \sum_{n=-\infty}^{\infty} e^{in\phi}J_n(z) , \tag{39.6}$$

we obtain

$$\rho_{\vec{q}} = \sum_{\{\vec{L}\},n} J_n(\vec{q}\cdot\vec{A})e^{i(\vec{q}+n\vec{Q})\cdot\vec{L}} . \tag{39.7}$$

The sum over $\{\vec{L}\}$ yields zero unless $\vec{q} = \vec{G} - n\vec{Q}$, where $\vec{G}$ is any reciprocal-lattice vector. Since we are interested in the energy gaps which truncate the Fermi surface, we only consider those $\vec{q}$ with magnitudes smaller than the Fermi-surface diameter. The so-called "heterodyne gaps" are an important special case, corresponding to $n = 1$ and $\vec{G} = \vec{G}_{110}$. For them, $|\vec{q}| \sim 0.08(2\pi/a)$.

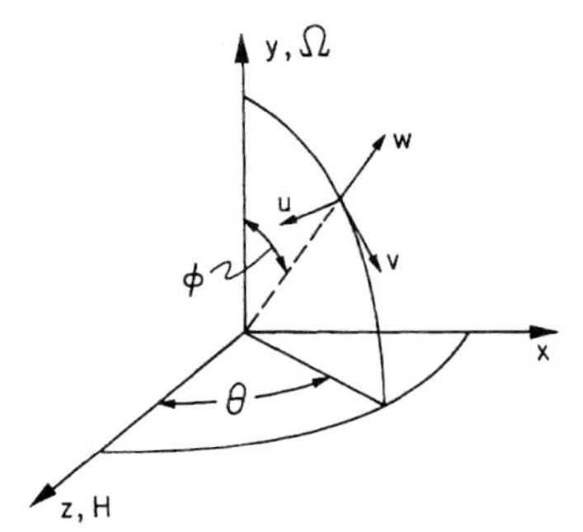

Figure 39.3 Coordinate systems used in the analysis, $\hat{w}$ is parallel to the preferred texture axis for the CDW $\vec{Q}$.

The Fermi surface of potassium is no longer a simply connected, almost-spherical surface. It is multiply connected, with a large number of intersecting energy gaps.

In very high magnetic fields, above 40 kOe, some of the CDW energy gaps can undergo magnetic breakdown. Others will not; and as a consequence, a large assortment of open orbits come into play. The recent very-high-field, induced-torque experiments [28], which reveal open orbits, can be explained by this model [20]. Unfortunately, the Fermi surface in lower fields, even up to 30 kOe, is quite complicated. A microscopic theory taking into account $\vec{Q}$ domains [20] and the effects of a preferred orientational texture has not been completed. Therefore, we shall postulate a phenomenological magnetoresistivity tensor, and shall attempt to justify it by qualitative arguments.

Consider a coordinate system $\hat{u}\hat{v}\hat{w}$ so that the preferred axis of orientational texture (for $\vec{Q}$) is along $\hat{w}$ (Figure 39.3). In the frame shown, we assume that potassium can be described by a magnetoresistivity tensor:

$$\rho' = \rho_0 \begin{pmatrix} 1 + S\omega_c\tau & t_1\omega_c\tau\cos\theta\sin\phi & -t_2\omega_c\tau\cos\theta\cos\phi \\ -t_1\omega_c\tau\cos\theta\sin\phi & 1 + S\omega_c\tau & t_2\omega_c\tau\sin\theta \\ t_2\omega_c\tau\cos\theta\cos\phi & -t_2\omega_c\tau\sin\theta & \gamma(1 + S\omega_c\tau) \end{pmatrix}, \tag{39.8}$$

The meaning of these quantities is as follows: γ is the ratio of zero-field resistivities parallel and perpendicular to $\hat{w}$; t_1 and t_2 describe the anisotropy of the Hall resistivity and its ratio to the free-electron value. For simplicity, we have neglected the anisotropy of the Kohler slope 5, which describes the magnetoresistance. (We have verified that this is unimportant.)

It is well known that the resistivity is proportional to the weighted average of the scattering cross section $\sigma(\theta)$. The weighting factor, $1 - \cos\theta$, counts 180° scattering heavily and small-angle scattering very little. For impurity scattering $\sigma(\theta)$ is sharply peaked in the forward direction, i. e., small $\vec{k}' - \vec{k}$. A CDW mixes plane-wave states $\vec{k}$ with $\vec{k} \pm \vec{Q}$. Thus, the CDW potential allows the usual wave-vector conservation rule to be supplemented by

$$\vec{k}' = \vec{k} + \vec{q} \pm \vec{Q}\,. \tag{39.9}$$

This CDW-Umklapp effect leads to a strongly enhanced large-angle scattering, but only for transitions across the Fermi surface which are nearly parallel to $\pm\vec{Q}$. Accordingly, the resistivity parallel to $\vec{Q}$ is larger than it is perpendicular to $\vec{Q}$, i. e., $\gamma > 1$ [29].

Energy gaps which truncate the Fermi surface cause open orbits for the conduction electrons (in a magnetic field). These lead to an H^2 magnetoresistance when $\vec{H}$ is perpendicular to an open orbit. For $\vec{H} < 25$ kOe the large number of open orbits ($\sim$ 120) prevents individual resolution. A magnetoresistance which is approximately linear in H results when the open-orbit effects are appropriately averaged [20]. Calculation shows that the expected Kohler slope S is indeed about 10^{-4}–10^{-2}.

A CDW also breaks the cubic symmetry of potassium. The combination of closed orbits and open orbits (with a shorter relaxation time [20]) leads to an enhanced Hall coefficient. Indeed it will be anisotropic if the $\vec{Q}$ domains have a preferred orientation. The experiments [17] indicate (from our work) that the Hall coefficient parallel to $\vec{Q}$ is larger than the Hall coefficient perpendicular to $\vec{Q}$. Accordingly, the enhancement of t_1 over t_2 is about 10% or more. We shall assume that $t_2 \approx 1$ since we have verified that only the ratio, t_1/t_2, is relevant to anisotropic torque patterns.

39.4 Calculation of Induced Torque

With the resistivity tensor given by Equation (39.8), and the theory of Visscher and Falicov [8], calculation of the induced torque is straightforward. Here, it is convenient to use a magnetoresistivity tensor $\overleftrightarrow{\rho}$ transformed to the $\hat{x}, \hat{y}, \hat{z}$ coordinate system

$$\overleftrightarrow{\rho} = \tilde{S}\overleftrightarrow{\rho}' S \,. \tag{39.10}$$

The orthogonal transformation matrix is

$$S = \begin{pmatrix} -\cos\theta & 0 & \sin\theta \\ \sin\theta\cos\phi & -\sin\phi & \cos\theta\cos\phi \\ \sin\theta\sin\phi & \cos\phi & \cos\theta\sin\phi \end{pmatrix}, \tag{39.11}$$

where ϕ and θ are the polar and azimuthal angles of the preferred texture axis for $\vec{Q}$. The induced current, in a spherical sample of radius R, satisfying the equations

$$\vec{\nabla} \times \overleftrightarrow{\rho} \cdot \vec{j} = -\frac{1}{c}\dot{\vec{B}} \,, \tag{39.12}$$

$$\vec{\nabla} \cdot \vec{j} = 0 \,, \tag{39.13}$$

and the boundary condition

$$\vec{j} \cdot \vec{r}|_{|\vec{r}|=R} = 0 \,, \tag{39.14}$$

is of the form

$$\vec{j} = \vec{t} \times \vec{r} \,, \tag{39.15}$$

where

$$\vec{t} = -\left(\frac{1}{c}\right)[\mathrm{Tr}(\overleftrightarrow{\rho}) - \tilde{\overleftrightarrow{\rho}}]^{-1}\dot{\vec{B}} \,. \tag{39.16}$$

The torque on the sample is given by the Lorentz force

$$\vec{N} = \frac{1}{c}\int \vec{r} \times (\vec{j} \times \vec{B})d^3r = \frac{1}{c}\int (\vec{t} \times \vec{r})\vec{r} \cdot \vec{B}d^3r$$
$$= \frac{1}{c}\vec{t} \times \left(\int \vec{r}\vec{r}d^3r\right) \cdot \vec{B} \,. \tag{39.17}$$

The tensor in large parentheses is the unit tensor multiplied by a factor $4\pi R^5/15$. Thus,

$$\vec{N} = (4\pi R^5/15c)\vec{t} \times \vec{B} \,. \tag{39.18}$$

Then for our system, with $\vec{B} = B\hat{z}$ and $\dot{\vec{B}} = \Omega B\hat{x}$, the torque about the $\hat{y}$ axis is

$$N_y = -(4\pi R^5/15c)Bt_x .$$

From Equations (39.10), (39.11), and (39.16), we may write N_y explicitly as

$$\begin{aligned} N_y = {} & \frac{4\pi R^5 \Omega}{15c^2 \rho_0}\left(\frac{mc}{\tau}\right)^2 \frac{(\omega_c\tau)^2}{1+S\omega_c\tau} \\ & \times \left[2(1+\gamma)^2 + (1+\gamma)t_2^2\left(\frac{\omega_c\tau}{1+S\omega_c\tau}\right)^2(\cos^2\theta\cos^2\phi + \sin^2\theta)\right. \\ & \left. + 2t_1^2\left(\frac{\omega_c\tau}{1+S\omega_c\tau}\right)^2 \cos^2\theta\sin^2\phi\right]^{-1} \\ & \times \left[(1+\gamma)^2\sin^2\theta\sin^2\phi + 2(1+\gamma)(\sin^2\theta\cos^2\phi + \cos^2\theta)\right. \\ & \left. + (t_1-t_2)^2\sin^4\phi\left(\frac{\omega_c\tau}{1+S\omega_c\tau}\right)^2 \sin^2\theta\cos^2\theta\right] \end{aligned} \tag{39.19}$$

where we have used $\omega_c \equiv eH/mc$. The proportionality factor $(4\pi R^5\Omega/15c^2\rho_0)(mc/e\tau)^2$, which is dependent on the radius of the sample R, the rotation speed Ω, and the relaxation time τ, does not affect the torque pattern. From now on we simply omit it.

The relaxation time τ at 1.4 K is about 1.5×10^{-10} s, which is primarily caused by impurity scattering, and is determined from the zero-field resistivity ρ_0. Accordingly, $\omega_c\tau = 50$ at $H = 20$ kOe.

Now, we are ready to discuss the torque patterns. For simplicity, we take the polar angle describing the orientation of the preferred $\vec{Q}$ direction to be $\phi = \pi/2$. (The results are similar for $\phi < \pi/2$.) In fairly low fields, $\omega_c\tau \sim 1$, the torque shows a twofold anisotropy. The maxima occur when the field is perpendicular to $\vec{Q}$ ($\theta = \pi/2$ or $3\pi/2$), and have the value

$$N_{y\,\max} \approx \frac{1}{2}\frac{(\omega_c\tau)^2}{1+S\omega_c\tau} . \tag{39.20}$$

The minima occur at $\theta = 0$ and $\theta = \pi$, and have the value

$$N_{y\,\max} \approx \frac{1}{1+\gamma}\frac{(\omega_c\tau)^2}{1+S\omega_c\tau} . \tag{39.21}$$

For larger fields, secondary minima develop where the maxima used to be, so one is left with a four-peak pattern. The tourque peaks are

$$N_{y\,\text{peak}} \approx \frac{(\omega_c\tau)^2}{1+S\omega_c\tau}\frac{(t_1-t_2)^2\sin^2\theta\cos^2\theta}{(1+\gamma)t_2^2\sin^2\theta + 2t_1^2\cos^2\theta} , \tag{39.22}$$

which are proportional to the square of $(t_1 - t_2)$. These high-field peaks increase with H faster than linearly. The maxima occur at angles given by

$$\theta = n\pi \pm \sin\left\{\frac{2t_1^2}{(1+\gamma)t_2^2 - 2t_1^2}\left[\frac{t_2}{t_1}\left(\frac{1+\gamma}{2}\right)^{1/2} - 1\right]\right\}^{1/2}$$
$$(n = 0, 1, 2) . \tag{39.23}$$

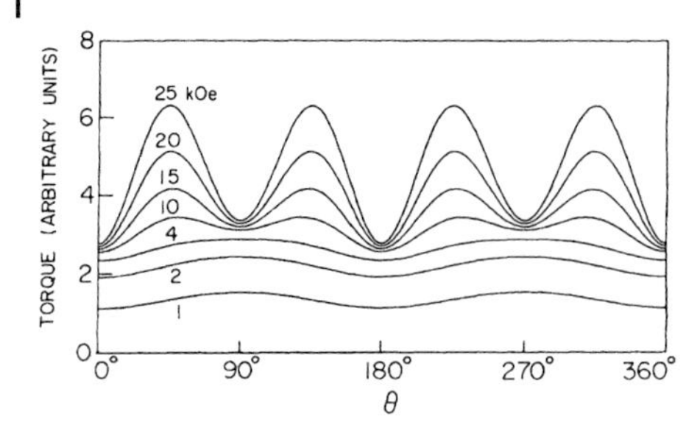

Figure 39.4 Calculated induced torque versus the angle θ between $\vec{H}$ and $\vec{Q}$ for $\gamma = 2.0$, $t_1 = 1.1$, $t_2 = 1.0$, and $S = 0.002$.

There are two pairs of high-field minima. The pair parallel to $\vec{Q}$ is approximately

$$N_\gamma \approx (1 + S\omega_c\tau)\frac{1+\gamma}{t_1^2}\,, \tag{39.24}$$

and the pair perpendicular to Q have a value of

$$N_\gamma \approx (1 + S\omega_c\tau)\frac{1+\gamma}{t_2^2}\,. \tag{39.25}$$

Note that these high-field minima increase linearly with H.

With $\gamma = 2$, $t_1 = 1.1$, $t_2 = 1.0$, and $S = 0.002$ we plot the torque curves for a series of fields. These curves are shown in Figure 39.4. The fit to the curves of Schaefer and Marcus (Figure 39.1) is excellent. At 25 kOe the maxima is $\sim$ 1.8 times the largest minimum and -2.3 times the smallest minimum, just as the data of Figure 39.1. On remembering that the CDW vector $\vec{Q}$ is tilted a few degrees from a $\langle 110\rangle$ axis, one can understand why the high-field minima occur within 5° from parallel or perpendicular to a $\langle 110\rangle$ axis [1].

The enormous high-field torque anisotropy reported by Holroyd and Datars should draw special attention. Since the high-field peak, Equations (39.22), is proportional to the square of $(t_1 - t_2)$, it shows that for this sample the Hall resistivity parallel to the preferred $\vec{Q}$ orientation is about 30% greater than that perpendicular. On taking $\gamma = 2$, $t_1 = 1.3$, $t_2 = 1.0$, and $S = 0.002$, we plot in Figure 39.5 the theoretical curves corresponding to Figure 39.2. Apparently, the preferred texture axis for $\vec{Q}$ orientation was, in this case, the growth axis. Again, we see extraordinary agreement. For the axis of rotation parallel to the growth axis we reproduce the "flat" curves of Figure 39.2. If the plane of rotation contains the axis of preferred $\vec{Q}$, an enormous high-field, four-peak pattern is produced. It is very gratifying to see that our two peaks surrounding the higher minimum (perpendicular to the preferred texture axis) are now closer than 90°; the angle between them is about 80°, which agrees with the data of Figure 39.2.

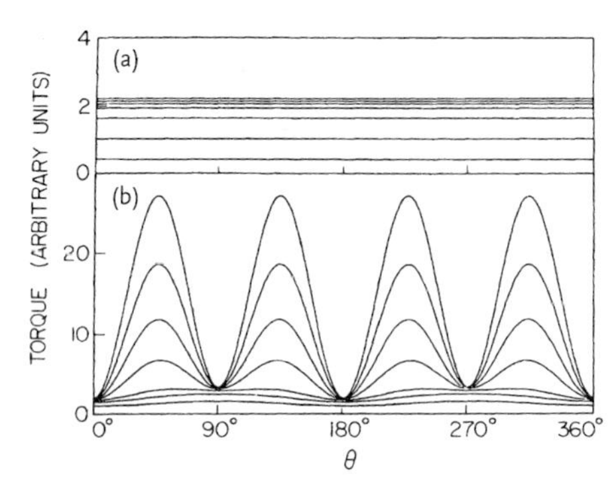

Figure 39.5 Calculated induced torque versus magnetic field direction θ for $\gamma = 2.0$, $t_1 = 1.3$, $t_2 = 1.0$, and $S = 0.002$. (a) corresponds to magnetic fields of 0.5, 1, 2, 4, 8, 14, and 21 kOe, with $\vec{H}$ rotated about the preferred texture axis for $\vec{Q}$. In (b) the plane of rotation contains this axis, the curves shown are 1, 2, 4, 10, 15, 20, and 25 kOe..

39.5 Discussion

The excellent agreement of calculated induced-torque patterns with experiment on accurately spherical samples leaves little doubt about the validity of the magnetoresistivity tensor (39.8). We believe that this confirms the broken symmetry of a CDW structure in potassium.

We have not yet developed a quantitative theory for a 30% anisotropy in the Hall coefficient as is required to explain the data of Holroyd and Datars. We believe that such a microscopic theory will involve *both* the shorter relaxation times for open-orbit electrons [20], resulting from $\vec{Q}$ domain size, and the stochastic occurrence of magnetic breakdown at the many small energy gaps which truncate the Fermi surface [30].

Acknowledgements

We would like to thank Dr. M. Huberman for helpful discussions. We are also grateful to the National Science Foundation Materials Research Laboratories (NSF-MRL) program for support.

References

1 Schaefer, J.A. and Marcus, J.A. (1971) *Phys. Rev. Lett.*, **27**, 935.
2 Holroyd, F.W. and Datars, W.R. (1975) *Can. J. Phys.*, **53**, 2517.
3 Overhauser, A.W. (1968) *Phys. Rev.*, **167**, 691.
4 For an experimental and theoretical review see Overhauser, A.W. (1978) *Adv. Phys.*, **27**, 343; Overhauser, A.W. (1983) in: *Electron Correlations in Solids, Molecules, and Atoms*, (eds. J.T. Devreese and F. Brosens), Plenum, New York, p. 41.
5 Lifshitz, I.M., Azbel, M.Ia., and Kaganov, M.I. (1956) *Zh. Eksp. Teor. Fiz.*, **31**, 63 [Soviet Phys.-JETP 4, 41 (1957)].
6 Fawcett, E. (1964) *Adv. Phys.*, **13**, 139.
7 Lass, J.S. and Pippard, A.B. (1970) *J. Phys. E*, **3**, 137.
8 Visscher, P.B. and Falicov, L.M. (1970) *Phys. Rev. B*, **2**, 1518.
9 Schmidt, R.L., Cornell University Materials Science Center Report No. 1434 (unpublished).
10 Penz, P.A. and Bowers, R. (1968) *Phys. Rev.*, **172**, 991.
11 Lass, J.S. *J. Phys. (Paris) Colloq.*, **31**, C3–1926 (1970).
12 Stroud, D. and Pan, F.P. (1976) *Phys. Rev. B*, **13**, 1434.
13 Herring, C. (1960) *J. Appl. Phys.*, **31**, 1939.
14 Bruls, G.J.C.L. *et al.* (1981) *Phys. Rev. Lett.* **46**, 553.
15 Young, R.A. (1968) *Phys. Rev.*, **175**, 813.
16 Chimenti, D.E. and Maxfield, B.W. (1973) *Phys. Rev. B*, **7**, 3501.
17 Werner, S.A., Hunt, T.K., and Ford, G.W. (1974) *Solid State Commun.*, **14**, 1217.
18 Sampselt, J.B. and Garland, J.C. (1975) *Bull. Am. Phys. Soc.*, **20**, 346.
19 Wagner, D.K. and Bowers, R. (1978) *Adv. Phys.*, **27**, 651.
20 Huberman, M. and Overhuaser, A.W. (1982) *Phys. Rev. B*, **25**, 2211.
21 Lass, J.S. (1976) *Phys. Rev. B*, **13**, 2247.
22 Overhauser, A.W. (1974) *Phys. Rev. B*, **9**, 2441.
23 Overhauser, A.W. (1964) *Phys. Rev. Lett.*, **13**, 190.
24 Overhauser, A.W. and Butler, N.R. (1976) *Phys. Rev. B*, **14**, 3371.
25 Overhauser, A.W. (1971) *Phys. Rev. B*, **3**, 3173.
26 Giuliani, G.F. and Overhauser, A.W. (1979) *Phys. Rev. B*, **20**, 1328.
27 Giuliani, G.F. and Overhauser, A.W. (1980) *Phys. Rev. B*, **22**, 3639.
28 Coulter, P.G. and Datars, W.R. (1980) *Phys. Rev. Lett.*, **45**, 1021.
29 Bishop, M.F. and Overhauser, A.W. (1977) *Phys. Rev. Lett.*, **39**, 632.
30 Fragachan, F.E. and Overhauser, A.W. (1984) *Phys. Rev. B*, **29**, 2912.

Reprint 40 Further Evidence of an Anisotropic Hall Coefficient in Potassium[1]

Xiaodong Zhu and A.W. Overhauser**

* Department of Physics, Purdue University, 525, Northwestern Avenue, West Lafayette, Indiana 47907, USA

Received 6 August 1984

It has been shown previously that an anisotropic Hall coefficient successfully explains the four-peaked induced-torque anomalies observed in potassium spheres. Recently, Elliott and Datars [1], using a modulation method, found an unexpected field dependence of the induced-torque phase (relative to the modulation field). They also reported that a misalignment of the modulation-field direction caused the four-peaked pattern to become twofold. The theory is extended to include these experimental variations, and is shown to account for the new observations.

40.1 Introduction

Elliott and Datars have recently developed an ac technique to measure the induced torque of potassium [1]. A spherical single crystal was suspended in a horizontal, stationary field $\vec{B}_0$. A small oscillatory field $\vec{b}_0 e^{i\omega t}$ was applied perpendicular to both $\vec{B}_0$ and the suspension axis. In this way they measured both the amplitude and phase of the induced torque from 0.005 to 2 T. They also noticed that misalignment of the modulation-field direction caused dramatic changes in the torque patterns (torque versus sample angle).

Their measurements of the amplitude of the induced torque confirmed results published previously by several authors [2, 3]: A twofold torque anisotropy (with peaks separated by approximately 180°) is observed below 0.4 T; a fourfold pattern (with peaks separated by approximately 90°) develops at higher fields. They found that the fourfold pattern at high fields depends sensitively on the direction of the modulation field. A misalignment of this field caused the four-peaked pattern to become twofold.

The phase of the induced torque (relative to the modulation field) was studied in great detail by Coulter and Datars [4] and found to depend on the magnitude and direction of $\vec{B}_0$, as shown in Figure 40.1. For sample orientations corresponding to the minima of the four-peaked pattern, the phase increases monotonically with B_0 and saturates. At orientations corresponding to four-peaked maxima, the phase first increases with B_0 but then decreases dramatically at high fields.

These induced-torque anomalies show that potassium is *not* a simple metal, as some still believe. If potassium had a simply connected, spherical Fermi surface, the torque would not depend on magnetic field direction. The phenomena can occur as a result of a broken symmetry of the charge-density-wave (CDW) type.[2] Orientational texture of the CDW wave vector $\vec{Q}$ leads to a preferred direction in the crystal.

Consider the coordinate system $\hat{u}, \hat{v}, \hat{w}$ of Figure 40.2. the preferred orientation of the CDW wave vector $\vec{Q}$ is along the $\hat{w}$ axis. In our previous paper [7], the magnetoresistivity

1) Phys. Rev. B 30, 12 (1984).

2) For an experimental and theoretical review, see [5, 6].

Anomalous Effects in Simple Metals. Albert Overhauser

ISBN: 978-3-527-40859-7

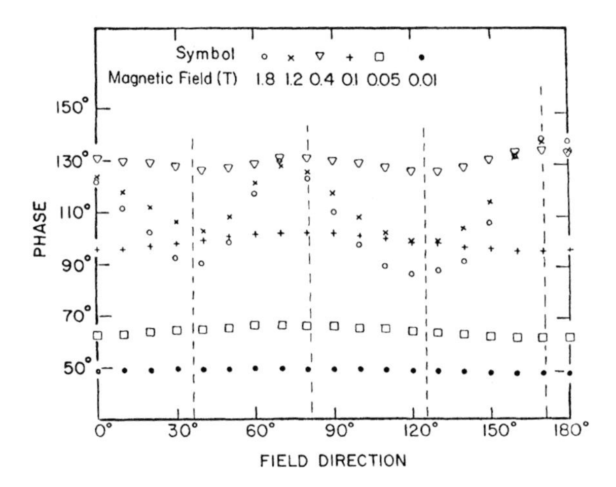

Figure 40.1 Phase of the induced torque versus magnetic field direction for fields of 0.01, 0.05, 0.1, 0.4, 1.2, and 1.8 T. The modulation frequency was 3.2 Hz. The data are from Elliott and Datars.

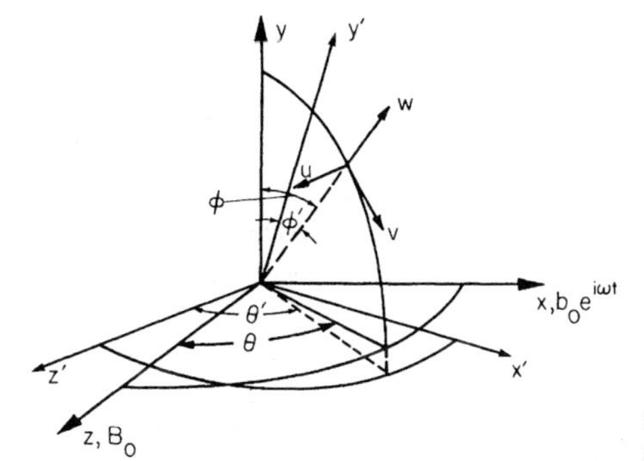

Figure 40.2 Coordinate systems used in the analysis.

tensor (in this coordinate frame) was taken to have the form (appropriate for $T = 0\,\mathrm{K}$)

$$\overleftrightarrow{\rho} = \rho_0 \begin{bmatrix} 1 + S\omega_c\tau & t_1\omega_c\tau\cos\theta\sin\phi & -t_2\omega_c\tau\cos\theta\cos\phi \\ -t_1\omega_c\tau\cos\theta\sin\phi & 1 + S\omega_c\tau & t_2\omega_c\tau\sin\theta \\ t_2\omega_c\tau\cos\theta\cos\phi & -t_2\omega_c\tau\sin\theta & \gamma(1 + S\omega_c\tau) \end{bmatrix} \tag{40.1}$$

ρ_0 is the (zero-field) residual resistivity perpendicular to $\hat{w}$, $\omega_c = eB_0/mc$ is the cyclotron frequency, and τ is the electronic relaxation time. The meaning of the other parameters is as follows: γ is the residual-resistivity anisotropy ($\gamma\rho_0$ is the resistivity parallel to $\hat{w}$), t_1 and t_2 represent the anisotropic enhancement of the Hall coefficient over the free-electron value, $R_0 = -1/nec$, and S is the Kohler slope of the linear magnetoresistance. (For simplicity, we neglect the anisotropy of S.) The crucial feature of the tensor, Equation (40.1), is the anisotropic Hall effect: $t_1 > t_2$.

The anisotropic residual resistivity causes the twofold torque pattern observed below 0.4 T [8]. The anisotropic Hall coefficient causes the four-peak patterns at high fields [7]; the peak-to-valley ratio is proportional to $(t_1 - t_2)^2$. The linear field dependence of the torque minima (at high fields) is proportional to the Kohler slope, $S \sim 10^{-3}$–10^{-2}. The resistivity tensor (40.1) provides an accurate account of induced-torque data from 0.1 to 3 T [7].

In this paper we employ the same magnetoresistivity tensor to explain the phase anomalies and the misalignment effect described above. In order to emphasize the roll of an anisotropic Hall coefficient, we let $\gamma = 1$ and $S = 0$. Were we also to set $t_1 = t_2 = 1$, Equation (40.1) would reduce to the standard magnetoresistivity tensor of an isotropic electron gas. (It may look unusual, because $\vec{B}_0$ is not along the z axis.)

40.2 Misalignment Effect

The applied ac magnetic field $\vec{b}_0 e^{i\omega t}$ induces a current $\vec{j}(\vec{r})$ in a spherical sample (of radius R) according to the equations

$$\vec{\nabla} \times (\overset{\leftrightarrow}{\rho} \cdot \vec{j}) = -\frac{i\omega}{c} \vec{b}_0 e^{i\omega t} , \tag{40.2}$$

$$\vec{\nabla} \cdot \vec{j} = 0 , \tag{40.3}$$

together with the boundary condition

$$(\vec{j} \cdot \vec{r})_{|\vec{r}|=R} = 0 . \tag{40.4}$$

Here we have assumed that ω is sufficiently small to allow the neglect of skin effects, i. e., the ac magnetic field is uniform. One may verify that the solution to Equations (40.2)–(40.4) is

$$\vec{j} = \vec{u} \times \vec{r} e^{i\omega t} , \tag{40.5}$$

where

$$\vec{u} = -\frac{i\omega}{c} \left[\mathrm{Tr}(\overset{\leftrightarrow}{\rho}) - \overset{\leftrightarrow}{\rho}^{\dagger} \right]^{-1} \cdot \vec{b}_0 . \tag{40.6}$$

$\overset{\leftrightarrow}{\rho}^{\dagger}$ is the transpose of $\overset{\leftrightarrow}{\rho}$. The induced magnetic moment, caused by $\vec{j}(\vec{r})$, is, accordingly,

$$\vec{m} = \frac{1}{2c} \int \vec{r} \times \vec{j} d^3 r = \frac{4\pi R^5}{15c} \vec{u} . \tag{40.7}$$

The interaction of $\vec{m}$ with the static field $\vec{B}_0$ creates a torque $\vec{N} = \vec{m} \times \vec{B}$. The component of $\vec{N}$ along the suspension axis of the sphere is

$$N_y = -m_x B_0 = -\frac{4\pi R^5}{15c} B_0 u_x . \tag{40.8}$$

This expression can be evaluated for an arbitrary direction of the ac field, $\vec{b}_0 = (b_x, b_y, b_z)$. For high fields,

$$\begin{aligned} N_y = {} & \frac{4\pi R^5}{15c} \frac{i\omega e^{i\omega t}}{\rho_0} \frac{B_0}{2\left[t_2^2(\cos^2\theta \cos^2\phi + \sin^2\phi) + t_1^2 \cos^2\theta\right]} \\ & \times \left[(t_1 - t_2)^2 \sin^2\theta \cos^2\theta \sin^4\phi\, b_x \right. \\ & + (t_1 - t_2)^2 \cos^2\theta \sin\theta \sin^3\phi \cos\phi\, b_y \\ & + (t_1 - t_2)\left(t_1 \cos^2\theta + t_2 \sin^2\theta\right) \sin\theta \cos\theta \sin^2\phi\, b_z \\ & \left. - (t_1 - t_2)^2 \cos^3\theta \sin\theta \cos^2\phi \sin^2\phi\, b_z \right] . \end{aligned} \tag{40.9}$$

If the modulation field is exactly parallel to the $\hat{x}$ axis, there are four peaks (in 360°), each having a height proportional to $(t_1 - t_2)^2$. This result is a consequence of the first term in square brackets. Usually, the anisotropy $t_1 - t_2 \sim 0.1$ [7]. For the case of a small misalignment $b_y, b_z \ll b_x$ the second term (proportional to b_y) is small compared to the first. On the other hand, the third term (proportional to b_z) is of order $t_1 - t_2$ instead of $(t_1 - t_2)^2$; therefore this "twofold" term can dominate the fourfold one. As a result, the extrema of the high-field *twofold* pattern will be at $\theta = 45°$ and 135° (instead of at 0° and 90°, where they occur for small B_0). This behavior was found experimentally by Elliott and Datars [1].

40.3 Phase Anomalies

In calculating the induced torque, we made an approximation in Equation (40.2): The ac magnetic field was taken to be the applied field. We neglected the field created by the induced currents. In this way, Equation (40.9), which explains the behavior of the torque amplitude, was easily derived. This is a good approximation as far as the torque amplitude is concerned, but it is not sufficient for calculating the phase. The induced current leads to an additional ac field, $\vec{b}_1 e^{i\omega t}$, which is 90° out of phase with the applied field $\vec{b}_0 e^{i\omega t}$. This new field induces, in turn, another field, $\vec{b}_2 e^{i\omega t}$, with phase 90° from $b_1 e^{i\omega t}$ (and 180° from $\vec{b}_0 e^{i\omega t}$), etc. The phase of the total field (and torque) must therefore be derived self-consistently. This requires us to solve Maxwell's equations.

Inside the sphere, Faraday's law,

$$\vec{\nabla} \times (\overleftrightarrow{\rho} \cdot \vec{j}) = (i\omega/t)\vec{b} \ , \tag{40.10}$$

and Ampere's law,

$$\vec{\nabla} \times \vec{b} = (4\pi/c)\vec{j} \ , \tag{40.11}$$

apply. Since the frequency ν is only a few hertz, the radius of the sample R is much smaller than the wavelength, $\lambda = c/\nu$, i. e.,

$$R/\lambda \ll 1 \ . \tag{40.12}$$

Therefore we can neglect the displacement current in Equation (40.11). Similarly, the field outside the sphere obeys

$$\vec{\nabla} \times \vec{b} = 0 \ , \quad \vec{\nabla} \cdot \vec{b} = 0 \ . \tag{40.13}$$

Of course, $\vec{b}$ must be continuous on the surface of the sphere,

$$(\vec{b}_{\text{inside}})_{|\vec{r}|=R} = (\vec{b}_{\text{outside}})_{|\vec{r}|=R} \ . \tag{40.14}$$

It is obvious that the sample can only perturb the field in a finite region, i. e., the resultant magnetic field far away from the sphere is still the applied field,

$$\lim_{|\vec{r}|\to\infty} \vec{b}_{\text{outside}} = \vec{b}_0 \ . \tag{40.15}$$

Equations (40.10), (40.11), and (40.13), together with boundary conditions (40.14) and (40.15), uniquely determine both the total magnetic field and the total induced current, and, thereby, the phase of the induced torque. To solve the equations, however, the coordinate system $\hat{u}, \hat{v}, \hat{w}$ is not a suitable one to work with. Since in this frame the magnetoresistivity tensor,

$$\overleftrightarrow{\rho} = \rho_0 \begin{Bmatrix} 1 & t_1\omega_c\tau\cos\theta\sin\theta & -t_2\omega_c\tau\cos\theta\cos\phi \\ -t_1\omega_c\tau\cos\theta\sin\phi & 1 & t_2\omega_c\tau\sin\theta \\ t_2\omega_c\cos\theta\cos\phi & -t_2\omega_c\tau\sin\theta & 1 \end{Bmatrix}, \quad (40.16)$$

together with Equation (40.10), leads to a complicated relation between $\vec{j}$ and $\vec{b}$,

$$\vec{\nabla}\times\left[\vec{j} - \omega_c\tau(t_2\sin\theta\,\hat{u} + t_2\cos\theta\cos\phi\,\hat{v} + t_1\cos\theta\sin\phi\,\hat{w})\times\vec{j}\right] = (i\omega/\rho_0 c)\vec{b}\,, \quad (40.17)$$

which is difficult to solve directly.

However, Equation (40.17) can be simplified by transforming into the coordinate system $\hat{x}', \hat{y}', \hat{z}'$ shown in Figure 40.2. In this new frame, the magnetoresistivity tensor has a free-electron-like form:

$$\overleftrightarrow{\rho}' = \overleftrightarrow{\tilde{T}}'\cdot\overleftrightarrow{\rho}\cdot\overleftrightarrow{T}' = \rho_0\begin{pmatrix} 1 & \alpha(\theta,\phi) & 0 \\ -\alpha(\theta,\phi) & 1 & 0 \\ 0 & 0 & 1 \end{pmatrix}. \quad (40.18)$$

Accordingly,

$$\vec{\nabla}'\times\left[\vec{j}' - \alpha(\theta,\phi)\hat{z}'\times\vec{j}'\right] = (i\omega/\rho_0 c)\vec{b}'\,, \quad (40.19)$$

The orthogonal transformation matrix which achieves this is

$$\overleftrightarrow{T}' = \begin{pmatrix} -\cos\theta' & 0 & \sin\theta' \\ \sin\theta'\cos\phi' & -\sin\phi' & \cos\theta'\cos\phi' \\ \sin\theta'\sin\phi' & \cos\phi' & \cos\theta'\sin\phi' \end{pmatrix}, \quad (40.20)$$

with

$$\phi' = \tan^{-1}\left(\frac{t_1}{t_2}\tan\phi\right) \quad (40.21)$$

and

$$\theta' = \tan^{-1}\left(\frac{t_2}{t_1\sin\phi\sin\phi' + t_2\cos\phi\cos\phi'}\tan\phi\right). \quad (40.22)$$

Then,

$$\alpha(\theta,\phi) = \omega_c\tau(t_1\cos\theta\sin\phi\cos\theta'\sin\phi' + t_2\cos\theta\cos\phi\cos\theta'\cos\phi' + t_2\sin\theta\sin\theta')\,. \quad (40.23)$$

Inspection of Equations (40.21)–(40.23) shows that if $t_1 = t_2 = t$ then $\phi' = \phi$, $\theta' = \theta$, and $\alpha(\theta,\phi) = \omega_c\tau$, i. e., the coordinate system reduces to the laboratory frame $\hat{x}, \hat{y}, \hat{z}$. This is expected. For the quasi-free-electron case ($t \neq 1$), the $\hat{x}$, $\hat{u}$, $\hat{z}$ frame should be the most convenient one to use. It is the anisotropic Hall effect ($t_2 \neq t_1$) that tilts the solvable frame away from the laboratory frame.

When Equation (40.12) is satisfies, as shown by Ford and Werner [9], Equation (40.19), together with Equations (40.11) and (40.13)–(40.15) in the $\hat{x}', \hat{y}', \hat{z}'$ frame, can be exactly solved, although it requires tedious expansions and lengthy calculations. We merely quote results derived in [9]. The resultant induced magnetic moment is given by

$$\vec{M}' = i(R^3\beta/30)\sum_{m=-1}^{1}\frac{\left(\hat{e'}^{*}_{m}\cdot\vec{b}'_0\right)\hat{e'}_m}{1+i[m\alpha(\theta,\phi)/2-2\beta/21]}\,, \tag{40.24}$$

where

$$\hat{e'} = -(\hat{x'}+i\hat{y'})/2\,,\quad \hat{e'_0} = \hat{z}'\,,\quad \hat{e'_{-1}} = (\hat{x'}-i\hat{y}')/2 \tag{40.25}$$

are the complex orthonormal basis vectors. The dimensionless parameter β is related to the ratio of R to the skin depth δ,

$$\beta = 4\pi\omega R^2/\rho_0 c^2 = 2(R/\delta)^2\,. \tag{40.26}$$

$\vec{b}'_0$ is the applied ac field in the $\hat{x}'$, $\hat{y}'$, $\hat{z}'$ frame. Recalling that $\vec{b}_0 = b_0\hat{x}$ and the orthogonal transformation matrix trom $\hat{x},\hat{y},\hat{z}$ to $\hat{u},\hat{v},\hat{w}$,

$$\overleftrightarrow{T} = \begin{pmatrix} -\cos\theta & 0 & \sin\theta \\ \sin\theta\cos\phi & -\sin\phi & \cos\theta\cos\phi \\ \sin\theta\sin\phi & \cos\phi & \cos\theta\sin\phi \end{pmatrix}\,, \tag{40.27}$$

we have, with Equation (40.20), that

$$\vec{b}'_0 = b_0\overleftrightarrow{\tilde{T}'}\cdot(\overleftrightarrow{T}\cdot\hat{x})\,. \tag{40.28}$$

The desired expression for the induced torque $N_y = -M_x B_0$ can be obtained by transforming Equation ((40.24) into the $\hat{x},\hat{y},\hat{z}$ frame. We obtain

$$M_x = \hat{x}\cdot[\overleftrightarrow{\tilde{T}}\cdot(\overleftrightarrow{\tilde{T}'}\cdot\vec{M}')]\,. \tag{40.29}$$

Finally, the phase of the induced torque relative to $\vec{b}_0 e^{i\omega t}$ is given by

$$\tan^{-1}(\mathrm{Re}\,M_x/\mathrm{Im}\,M_x)+\pi/2\,. \tag{40.30}$$

On the other hand, the induced-torque amplitude is

$$B_0\left[(\mathrm{Re}\,M_x)^2+(\mathrm{Im}\,M_x)^2\right]^{1/2}\,. \tag{40.31}$$

Before we discuss the phase patterns of the induced torque, we must determine the parameters appearing in $\alpha(\theta,\phi)$ and β. Without loss of generality, we consider only the case when the polar angle $\phi = \pi/2$ (the preferred texture axis of $\vec{Q}$ is in the horizontal plane). The results are similar for $\phi < \pi/2$. According to [1, 4], the frequency of the applied ac field was 3.2 Hz, which implies a ratio $R/\delta = 2.2$. It is clear from Equations (40.21)–(40.24) that the induced torque depends only on the ratio t_1/t_2, and thus we shall assume that $t_2 = 1$. The other two parameters, and τ and t_1, are determined by considering the amplitude patterns of the induced torque given by expression (40.31).

Since at $\theta = \pi/2$ (when the preferred axis is perpendicular to $\vec{B}_0$) the magnetoresistivity tensor (40.16) is independent of the parameter t_1 the relaxation time τ can then be found (for this field direction) by plotting the amplitude versus $\omega_c\tau$ as shown in Figure 40.3. The torque grows linearly with $\omega_c\tau$ at low fields, reaches a maximum at $\omega_c\tau = 2.72$, and then

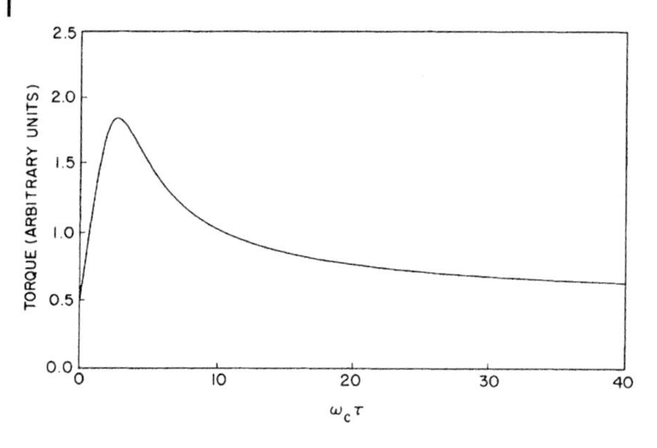

Figure 40.3 Amplitude of the induced torque versus $\omega_c\tau$ at angle $\theta = \pi/2$ ($\vec{B}_0$ perpendicular to the preferred axis of orientation of the CDW, $\vec{Q}$). The calculation was done for $R/\delta = 2.2$.

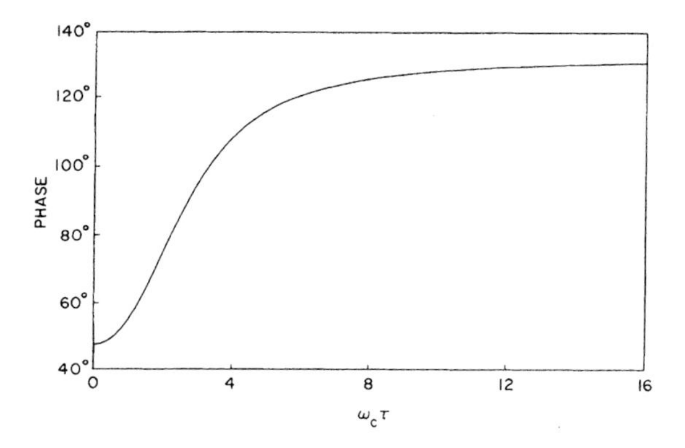

Figure 40.4 Field dependence of the induced-torque phase at $\theta = \pi/2$ for fields ranging from 0 to 0.4 T.

decreases with further increase of the field. Such a curve was found experimentally with a maximum at 0.185 T [1]. However, Coulter and Datars later reported [4] that this measurement was made before room-temperature annealing of the sample took place. When the phase measurements were made later, the maximum occurred at 0.095 T. This implies that $\tau \sim 1.72 \times 10^{-10}$ s, a value which we use in the following.

The field dependence of the induced-torque phase for $\theta = \pi/2$, given by Equation (40.30), was calculated with the parameters determined above and is shown in Figure 40.4. The phase increases with field and saturates. The midpoint in the change of phase is at 90° and corresponds to the field for which $\omega_c\tau = 2.72$. It is in good agreement with the experimental curve shown in Figure 40.5 [1]. Experimentally, the total phase change (in the range from zero field to 0.4 T) is 83°, which equals the change predicted in Figure 40.4 between 0 and $\omega_c\tau = 11.45$. However, the data in Figure 40.5 have a midpoint that occurs at a phase of 102° instead of at the expected value 90°. Fortunately, it was pointed out in [4] that there might have been a constant phase shift of 12° in the electronics.

Elliott and Datars also measured the anisotropy of the fourfold torque amplitude (torque-maximum–to–torque-minimum ratio), which was found to be about 1.5:1 at 1.8 T. Our calculation, from Equation (40.31), shows that such a ratio requires a Hall coefficient with a 10% anisotropy, i. e., $t_1 = 1.1$.

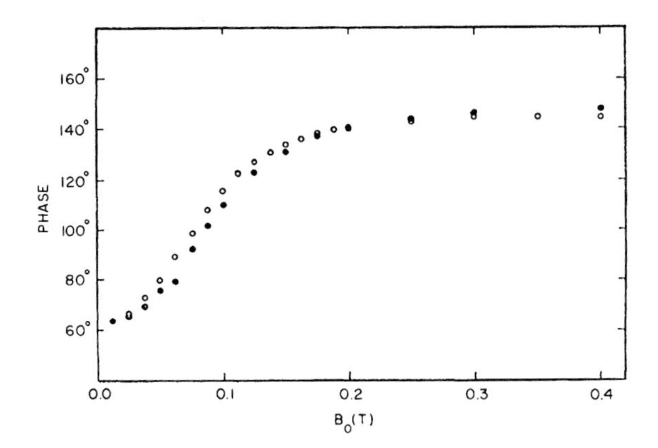

Figure 40.5 Phase of the induced torque as a function of B_0, measured by Elliot and Datars, for $\vec{B}_0$ parallel to the twofold minimum or maximum.

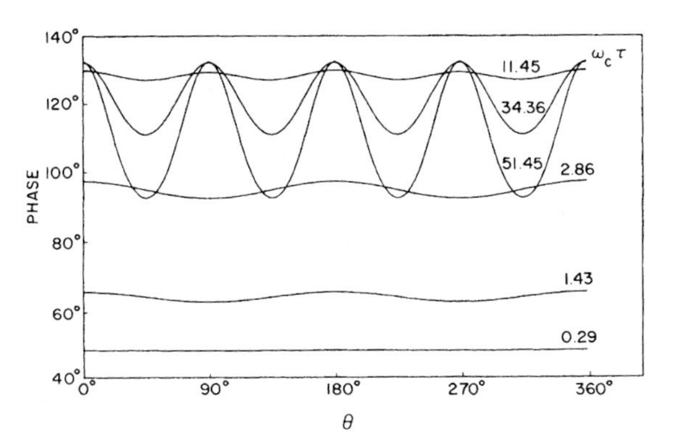

Figure 40.6 Calculated induced-torque phase versus magnetic field direction θ. Values for $\omega_c\tau$ of 0.29, 1.43, 2.86, 11.45, 34.26, and 51.54 were chosen and correspond to the experimental fields shown in Figure 40.1. The parameters determined from the torque amplitude are $t_1 = 1.1$, $t_2 = 1.0$, and $R/\delta = 2.2$.

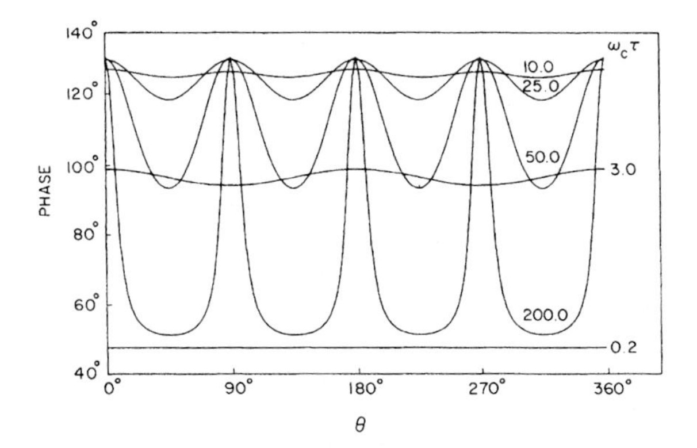

Figure 40.7 Demonstration of the phase spikes at high fields with the same parameters as in Figure 40.6.

We can now calculate the phase patterns and compare them with the experimental results of Figure 40.1. For the same B_0 values shown in Figure 40.1, in Figure 40.6 we plot the phase versus field direction θ. There are no free parameters since $t_1 = 1.1$, $t_2 = 1.0$, $R/\delta = 2.2$, and $\tau = 1.72 \times 10^{-10}$ s were already determined as described above. The agreement with Figure 40.1 is impressive. When $\vec{B}_0$ is parallel or perpendicular to the texture axis (of $\vec{Q}$), the phase increases with B_0 and saturates above 0.4 T. For all other directions the phase reaches a maximum at about 0.45 T and then decreases, giving rise to a fourfold pattern of the induced-torque phase.

At even higher fields we find that there should be four phase "spikes" in a 360° rotation, as shown in Figure 40.7. This behavior was also found experimentally [4].

40.4 Discussion

The induced-torque technique provides an ideal method for studying the low-temperature (below 4.2 K) galvanomagnetic properties of potassium. The anomalous induced-torque amplitude and phase throughout the field region, 0.005–8.5 T, confirm the CDW structure in potassium. It seems appropriate now to present a general review of the theoretical results based on a CDW broken symmetry. For fields below 0.5 T the torque amplitude shows a twofold anisotropy with maxima occurring when $\vec{B}_0$ is perpendicular to the texture axis of preferred $\vec{Q}$ orientation. A recent analysis [10] also gives the correct field dependence of the torque ratio (maximum to minimum). For fields within 0.5–2.5 T, the torque amplitude becomes a four-peak pattern if the modulation field is precisely aligned. On the other hand, the phase of the induced torque tends to a constant value in directions parallel and perpendicular to the texture axis, which are minimum directions of the four-peak torque pattern. The phase minima occur at the four-peak maxima. For fields above 4 T, open-orbit torque peaks become resolved and grow rapidly with B_0, They are caused by the various sets of CDW energy gaps which truncate the Fermi surfaces [11]. Phase dips occur at the peak directions of the amplitude [4, 11].

Anisotropic sample shape, e. g., Lass's ellipsoidal model [12], has sometimes been invoked to explain the induced-torque anomalies for fields below 3 T, although the samples used in the experiments were sufficiently spherical to preclude such an explanation. Curiously, the model is mathematically equivalent to ours. Lass introduced a parameter η to characterize the departure of a sample from a sphere. After a scaling transformation we find that an ellipsoidal sample with a free-electron magnetoresistivity tensor is equivalent to a spherical sample with our anisotropic resistivity tensor, Equation (40.1). The relations between η and the parameters of Equation (40.1) are $\gamma = 1 + \eta$, $t_1 = 1 + \eta$, and $t_2 = 1.0$. The Kohler slope S, which describes the linear magnetoresistance, is introduced manually in either case. It is also caused by (unresolved) open orbits created by the CDW broken symmetry.

Acknowledgements

The authors are grateful to the National Science Foundation Materials Research Laboratories program for support.

References

1 Elliott, M. and Datars, W.R. (1983) *J. Phys. F*, **13**, 1483.
2 Schaefer, J.A. and Marcus, J.A. (1971) *Phys. Rev. Lett.*, **27**, 935.
3 Holroyd, F.W. and Datars, W.R. (1975) *Can. J. Phys.*, **53**, 2517.
4 Coulter, P.G. and Datars, W.R. (1984) *J. Phys. F*, **14**, 911.
5 Overhauser, A.W. (1978) *Adv. Phys.*, **27**, 343.
6 Overhauser, A.W. (1983) in *Electron Correlations in Solids, Molecules, and Atoms*, (eds. J.T. Devreese and F. Brosens, Plenum, New York, p. 41.
7 Zhu, X. and Overhauser, A.W. (1984) *Phys. Rev. B*, **30**, 622.
8 Bishop, M.F. and Overhauser, A.W. (1978) *Phys. Rev. B*, **18**, 2447.
9 Ford, G.W. and Werner, S.A. (1978) *Phys. Rev. B*, **18**, 6752. Also see Ford, G.W., Furdyna, J.K., and Werner, S.A. (1975) *Phys. Rev. B*, **12**, 1452.
10 Huberman, M. and Overhauser, A.W. (1985) *Phys. Rev. B*, **31**, 735.
11 Huberman, M. and Overhauser, A.W. (1982) *Phys. Rev. B*, **25**, 2211.
12 Lass, J.S. (1976) *Phys. Rev. B*, **13**, 2247.

Reprint 41 Field Dependence of the Residual-Resistivity Anisotropy in Sodium and Potassium[1)]

M. Huberman * *and A.W. Overhauser* **

* Department of Physics, Michigan Technological University, 525, Northwestern Avenue, Houghton, Michigan 49931, USA

** Department of Physics, Purdue University, West Lafayette, Indiana 47907, USA

Received 15 August 1984

Recent measurements of the low-field, induced torque in sodium and potassium by Elliott and Datars show that the resistivity anisotropy increases with increasing magnetic field. The zero-field resistivity anisotropy, unexpected for cubic symmetry, is explained by the charge-density-wave (CDW) structure. Due to the wave-function mixing caused by the CDW potential, the momentum transfer (by isotropic impurities) is much larger for electrons near the CDW energy gap. This is modeled by an anisotropic relaxation time in $\vec{k}$ space. The Boltzmann transport equation in a magnetic field can then be solved exactly. The computed resistivity anisotropy is higher for $\omega_c\tau > 1$ compared with its zero-field value. The effect of the magnetic field is to "stir" the electron distribution $f(\vec{k})$; this feeds electrons into the region of rapid relaxation and thereby increases the resistivity anisotropy.

41.1 Introduction

The induced torque of sodium and potassium spheres has different character at low (0–5 kG), intermediate (5–40 kG), and high (> 40 kG) magnetic fields, each stemming from a charge-density-wave (CDW) structure.[2)] At low fields, the torque as a function of magnet rotation angle has a twofold, sinusoidal pattern [2, 3], caused by a residual-resistivity anisotropy [4]. At intermediate fields, it has a smooth, four-peak pattern [2, 3], caused by an anisotropic Hall coefficient [5]. At high fields, it exhibits many sharp peaks [6, 7], caused by open orbits [8].

Recently Elliott and Datars discovered that the magnitude of the twofold, torque anisotropy increases with increasing magnetic field [9, 10]. An example of their data for potassium is shown in Figure 41.1. The deviation of the torque ratio from unity increases by an order of magnitude. Bishop and Overhauser have shown that, if a field-independent resistivity anisotropy is assumed, the torque anisotropy decreases with increasing field [4]. The correct approach, however, is to derive the magnetoresistivity by solving the Boltzmann transport equation. When this is done, the resistivity anisotropy is found to increase with field, explaining the rise of the torque anisotropy.

The zero-field resistivity anisotropy is caused by the CDW potential $V = G\cos(\vec{Q}\cdot\vec{r})$ [4]. The Fermi surface in the repeated-zone scheme is shown in Figure 41.2. Because Q is approximately equal to the Fermi-surface diameter, electron states at the Fermi surface near the conical points $\pm\vec{Q}/2$ are strongly perturbed, the plane waves $\vec{k}$ and $\vec{k}\pm\vec{Q}$ being mixed. In normal electron-impurity scattering, the wave-vector transfer $\vec{q}$ is small, since the impurity potential has small $-q$ Fourier components. Due to the wave-function mixing, however, electrons near the CDW energy gap can also suffer a much larger wave-vector transfer $\vec{q}\pm\vec{Q}$

1) Phys. Rev. B 31, 2 (1985).

2) For an experimental and theoretical review of charge-density waves in simple metals, see [1].

Anomalous Effects in Simple Metals. Albert Overhauser

ISBN: 978-3-527-40859-7

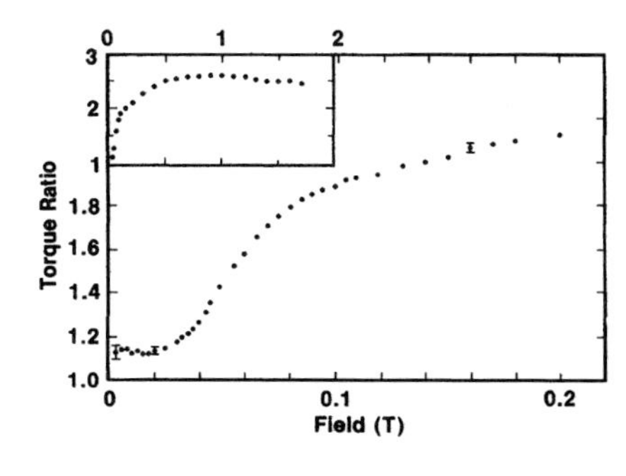

Figure 41.1 The ratio of the twofold torque maximum and minimum for a potassium sphere (Elliott and Datars [10]).

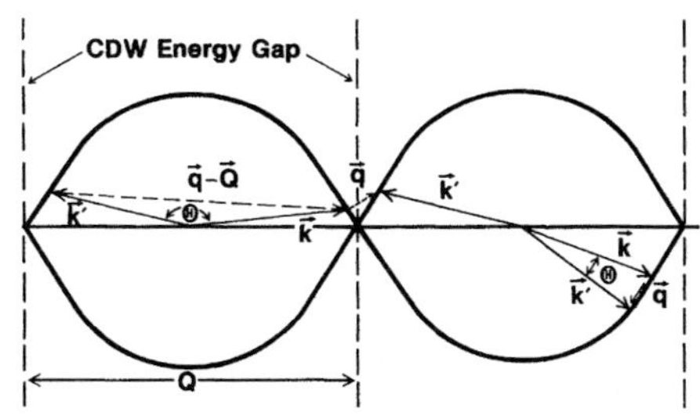

Figure 41.2 Electron-impurity normal scattering ($\Delta\vec{k} = \vec{q}$) and CDW–umklapp scattering ($\Delta\vec{k} = \vec{q} - \vec{Q}$). $\vec{Q}$ is the CDW wave vector.

(CDW–umklapp scattering). Since large-angle scattering contributes more to the resistivity than small-angle scattering, the residual resistivity is higher parallel to $\vec{Q}$ than perpendicular to $\vec{Q}$.

In order to solve the Boltzmann transport equation in a magnetic field, we model the impurity scattering by an anisotropic relaxation time in $\vec{k}$ space. The large momentum transfer near the conical points $\pm\vec{Q}/2$ is mimicked by a rapid relaxation rate. In zero magnetic field, the steady-state distribution $f(\vec{k})$ is nonspherical, being displaced less (from the equilibrium distribution) where the relaxation time is short. The effect of a magnetic field is to "stir" the distribution function, feeding electrons into the region of rapid relaxation. This increases the resistivity anisotropy and thereby the torque anisotropy.

41.2 Anisotropic Relaxation Time

We approximate the Fermi surface by a sphere, neglecting the distortion near the energy-gap planes. The relaxation time of an electron at the Fermi energy has uniaxial symmetry. We model it by

$$\tau(\theta) = \begin{cases} \tau\,, & |\cos\theta| < \cos\eta \\ \tau'\,, & |\cos\theta| > \cos\eta \end{cases} \tag{41.1}$$

where θ is the angle between the electron wave vector $\vec{k}$ and the CDW wave vector $\vec{Q}$ (Figure 41.3). Since the resistivity is higher parallel to $\vec{Q}$, τ' is shorter than τ. The angle η defines the region of umklapp scattering. A microscopic argument, presented in the Appendix, indicates $\eta \approx 25°\text{C}$ for sodium and potassium. The fraction of electrons on the Fermi surface experiencing rapid relaxation is then about 10%.

Semiclassical transport theorems [11] predict for a simply connected Fermi surface that the magnetoresistance saturates and the Hall coefficient is isotropic, when $\omega_c\tau > 1$. Since

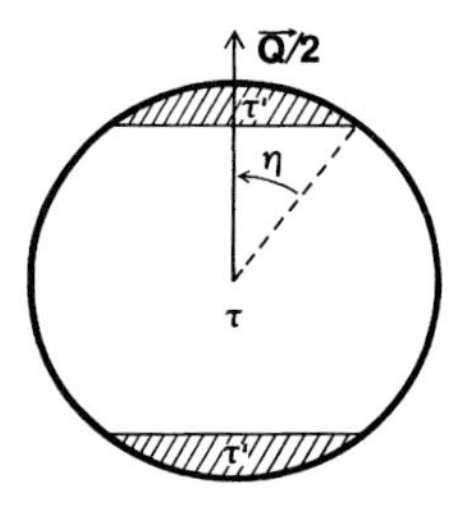

Figure 41.3 Anisotropic relaxation-time on the Fermi surface. $\vec{Q}$ is the CDW wave vector.

the assumed Fermi surface is simply connected, the model omits the nonsaturating magnetoresistance and (high-field) Hall-coefficient anisotropy.

This model has one adjustable parameter, the relaxation-time ratio τ'/τ. The magnitude of the torque-anisotropy increase is approximately explained by choosing $\tau'/\tau \approx 0.05$. Such a large anisotropy of the relaxation time $\tau(\theta)$ is not inconsistent with the small anisotropy (about 10%) of electron-impurity scattering deduced from de Haas–van Alphen effect studies [12]. $\tau(\theta)$ is a momentum-relaxation lifetime, weighting large-angle scattering more heavily than small-angle scattering, whereas the ordinary lifetime determined by the de Haas–van Alphen effect weights all collisions equally.

The model (41.1) focuses on the essential feature causing the torque-anisotropy rise, namely, anisotropic scattering. It has the significant advantage of being tractable (at any magnetic-field strength). Its drawback, of course, is that it is not derived microscopically,

41.3 Zero-Field Resistance

The Boltzmann equation for the steady-state electron distribution $f(\vec{k})$ is

$$-\frac{e}{\hbar}\vec{E}\cdot\vec{\nabla}_{\vec{k}} f - \frac{e}{\hbar c}(\vec{v}\times\vec{B})\cdot\vec{\nabla}_{\vec{k}} f = -\frac{f-f_0}{\tau(\theta)} . \tag{41.2}$$

Letting $g \equiv f - f_0$ be the deviation from equilibrium yields

$$-e\vec{E}\cdot\vec{v}\frac{d f_0}{d\epsilon} - \frac{e}{\hbar c}(\vec{v}\times\vec{B})\cdot\vec{\nabla}_{\vec{k}} g = -\frac{g}{\tau(\theta)} \tag{41.3}$$

to first order in $\vec{E}$. In zero magnetic field, the immediate solution is

$$g = \tau(\theta) e\vec{E}\cdot\vec{v}\frac{d f_0}{d\epsilon} , \tag{41.4}$$

the electron distribution

$$f = f_0 + g \cong f_0(\vec{k} + [\tau(\theta)e/\hbar]\vec{E}) , \tag{41.5}$$

being illustrated (for $\vec{E}$ parallel to $\vec{Q}$) in Figure 41.4.

Evaluating the current density yields the electrical conductivity. For a degenerate equilibrium distribution, the resistivities parallel and perpendicular to $\vec{Q}$ are

$$\rho_{\parallel} = \frac{m}{ne^2\langle\tau(\theta)\rangle_{\parallel}} , \qquad \rho_{\perp} = \frac{m}{ne^2\langle\tau(\theta)\rangle_{\perp}} , \tag{41.6}$$

the weighted angular averages over the Fermi surface being defined by

$$\begin{aligned}\langle\tau(\theta)\rangle_{\parallel} &= \frac{3}{4\pi}\int d\Omega \cos^2\theta\, \tau(\theta) , \\ \langle\tau(\theta)\rangle_{\perp} &= \frac{3}{8\pi}\int d\Omega \sin^2\theta\, \tau(\theta) .\end{aligned} \tag{41.7}$$

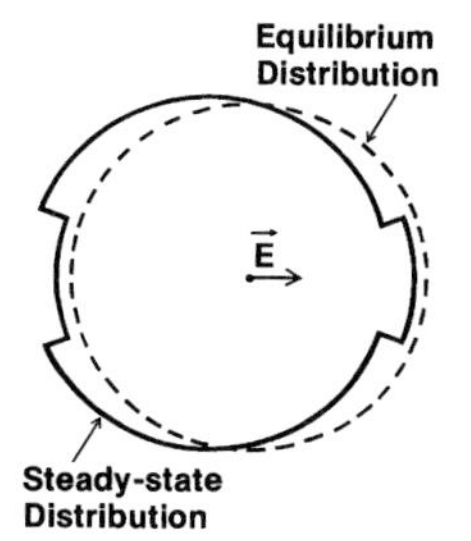

Figure 41.4 Zero-field steady-state distribution $f(\vec{k})$ for $\vec{E} \parallel \vec{Q}$.

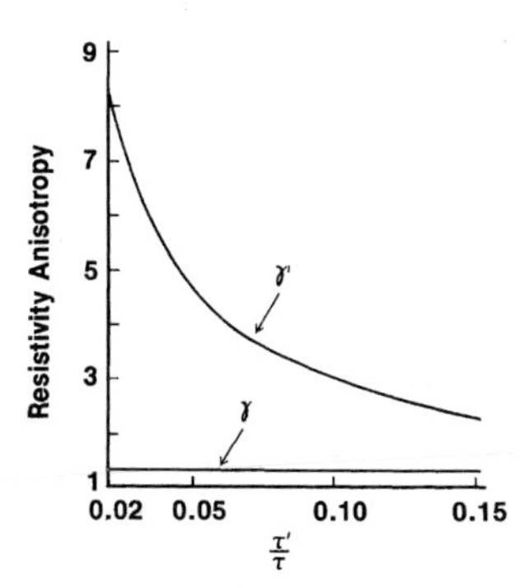

Figure 41.5 Exact (γ) and approximate (γ') zero-field resistivity anisotropy. $\eta = 25°$.

Substituting (41.1) for $\tau(\theta)$, we find

$$\rho_{\parallel} = \rho_0 \left[1 - \left(1 - \frac{\tau'}{\tau}\right)(1 - \cos^3 \eta)\right]^{-1},$$

$$\rho_{\perp} = \rho_0 \left[1 - \left(1 - \frac{\tau'}{\tau}\right)\left(1 - \frac{3}{2}\cos\eta + \frac{1}{2}\cos^3\eta\right)\right]^{-1}, \tag{41.8}$$

where $\rho_0 \equiv m/(ne^2\tau)$. The resistivity anisotropy $\gamma \equiv \rho_{\parallel}/\rho_{\perp}$ is plotted in Figure 41.5. For $\eta = 25°$ and $\tau'/\tau = 0.05$, $\rho_{\parallel}/\rho_0 = 1.32$, $\rho_{\perp}/\rho_0 = 1.01$, and $\gamma = 1.30$.

It is instructive to evaluate the (zero-field) resistivity, assuming a rigidly shifted equilibrium distribution

$$f = f_0(\vec{k} - \vec{\delta}), \tag{41.9}$$

illustrated in Figure 41.6. Evaluating the current density yields

$$\vec{j} = -ne\frac{\hbar\vec{\delta}}{m}. \tag{41.10}$$

The displacement $\vec{\delta}$, which is proportional to $\vec{E}$, is determined by balancing the momentum (per unit time) gained from the applied electric field and lost by collisions. Accordingly, the so-computed resistivities parallel and perpendicular to $\vec{Q}$ are

$$\rho'_{\parallel} = \frac{m}{ne^2}\left\langle\frac{1}{\tau(\theta)}\right\rangle_{\parallel}, \quad \rho'_{\perp} = \frac{m}{ne^2}\left\langle\frac{1}{\tau(\theta)}\right\rangle_{\perp}, \tag{41.11}$$

the angular averages having the same definitions as before.

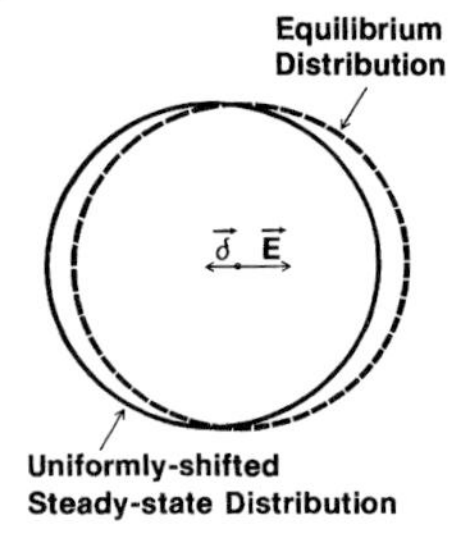

Figure 41.6 Uniformly shifted, steady-state distribution $f(\vec{k})$. The displacement $\vec{\delta}$ depends on the direction of $\vec{E}$ relative to $\vec{Q}$.

By the variational principle [13, 14], the approximate resistivities (41.11) are higher than the exact resistivities (41.6). Substituting (41.1) for $\tau(\theta)$ yields

$$\rho'_{\parallel} = \rho_0 \left[1 + \left(\frac{\tau}{\tau'} - 1 \right) (1 - \cos^3 \eta) \right] ,$$

$$\rho'_{\perp} = \rho_0 \left[1 + \left(\frac{\tau}{\tau'} - 1 \right) \left(1 - \frac{3}{2} \cos \eta + \frac{1}{2} \cos^3 \eta \right) \right] , \tag{41.12}$$

where $\rho_0 \equiv m/(ne^2\tau)$. The so-computed resistivity anisotropy $\gamma' = \rho'_{\parallel}/\rho'_{\perp}$ is plotted in Figure 41.5. For $\eta = 25°$ and $\tau'/\tau = 0.05$, $\rho'_{\parallel}/\rho_0 = 5.86$, $\rho'_{\perp}/\rho_0 = 1.24$, and $\gamma' = 4.71$.

It is easy to understand why γ' is much larger than γ. For small angles η, the rapid relaxation affects primarily the parallel resistivities; the perpendicular resistivities ($\rho_{\perp}$ and $\rho'_{\perp}$) are approximately equal to ρ_0. In the exact distribution function for $\vec{E} \parallel \vec{Q}$, shown in Figure 41.4, the deviation from equilibrium in the region of rapid relaxation is minimized. In the uniformly shifted distribution function, shown in Figure 41.6, this deviation is much larger (for the same total current), many more electrons suffering rapid relaxation. Thus $\rho'_{\parallel}$ is much larger than $\rho_{\parallel}$.

In the next section, we calculate exactly the resistivity in a magnetic field. We shall show that the resistivity anisotropy γ', computed with a uniformly shifted distribution function, becomes exact when $\omega_c\tau > 1$.

41.4 Magnetoresistance

The magnetoresistance depends on the angle between $\vec{B}$ and $\vec{Q}$. We consider two orientations, $\vec{B} \parallel \vec{Q}$ and $\vec{B} \perp \vec{Q}$.

For these orientations, there are five resistivities, depicted in Figure 41.7, and two Hall coefficients. The resistivity for a current $\vec{J}$ is either parallel or perpendicular, depending on whether $\vec{J} \parallel \vec{Q}$ or $\vec{J} \perp \vec{Q}$, and longitudinal or transverse, depending on whether $\vec{J} \parallel \vec{B}$ or

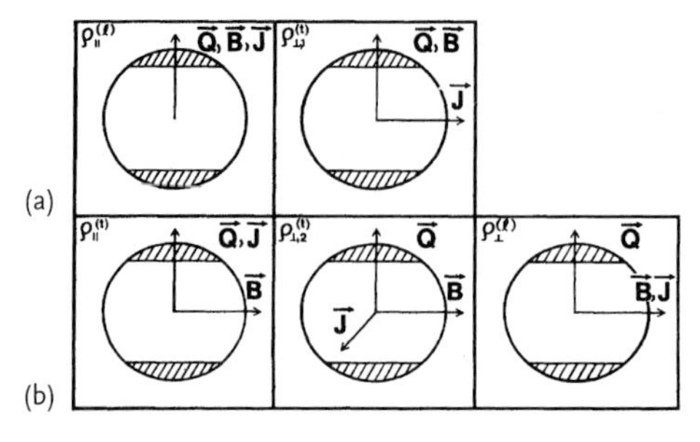

Figure 41.7 Resistivities for (a) $\vec{B} \parallel \vec{Q}$ and (b) $\vec{B} \perp \vec{Q}$. $\vec{J}$ is the current.

$\vec{J} \perp \vec{B}$. There are two transverse, perpendicular resistivities, one with $\vec{Q} \parallel \vec{B}$ and the other with $\vec{Q} \perp \vec{B}$. The two Hall coefficients are $R_{\parallel}$ with $\vec{Q} \parallel \vec{B}$ and $R_{\perp}$ with $\vec{Q} \perp \vec{B}$.

For $\vec{B} \parallel \vec{Q}$, an electron's relaxation time is unchanged by its cyclotron motion. The Boltzmann equation (41.3) is easily solved by the ansatz

$$g = \tau(\theta) e \vec{A} - \vec{v} \frac{d f_0}{d\epsilon} \tag{41.13}$$

with $\vec{A}$ depending on $\cos\theta \equiv \hat{v} \cdot \hat{Q}$. For $\vec{B} = B\hat{z}$, the solution for $\vec{A}$ is

$$\begin{aligned} A_x &= \frac{E_x - \omega_c \tau E_y}{1 + (\omega_c \tau)^2} , \\ A_y &= \frac{E_y + \omega_c \tau E_x}{1 + (\omega_c \tau)^2} , \\ A_z &= E_z . \end{aligned} \tag{41.14}$$

$\omega_c = eB/mc$ is the cyclotron frequency; the θ dependence of τ is understood. Evaluating the current density yields the magnetoconductivity tensor

$$\sigma = \frac{ne^2}{m} \begin{pmatrix} \left\langle \frac{\tau}{1+(\omega_c\tau)^2} \right\rangle_{\perp} & -\left\langle \frac{\omega_c\tau^2}{1+(\omega_c\tau)^2} \right\rangle_{\perp} & 0 \\ \left\langle \frac{\omega_c\tau^2}{1+(\omega_c\tau)^2} \right\rangle_{\perp} & \left\langle \frac{\tau}{1+(\omega_c\tau)^2} \right\rangle_{\perp} & 0 \\ 0 & 0 & \langle\tau\rangle_{\parallel} \end{pmatrix} , \tag{41.15}$$

the angular averages being defined as before.

The resistivity and Hall coefficient are obtained by inverting (41.15). Although the longitudinal resistivity

$$\rho_{\parallel}^{(l)} = \frac{m}{ne^2 \langle\tau\rangle_{\parallel}} \tag{41.16}$$

is field independent, the transverse resistivity increases with field. Its saturation value is

$$\rho_{\perp}^{(t)} \to \frac{m}{ne^2} \left\langle \frac{1}{\tau} \right\rangle_{\perp} , \quad \omega_c \tau > 1 , \tag{41.17}$$

equal to the resistivity $\rho'_{\perp}$, computed with a uniformly shifted distribution function.[3] The Hall coefficient decreases (in absolute value) with increasing field, being equal to the free-electron value $R_0 = -1/(nec)$ when $\omega_c \tau > 1$. The two resistivities ($\rho_{\parallel}^{(l)}$ and $\rho_{\perp,l}^{(t)}$) and the Hall coefficient ($R_{\parallel}$) are plotted in Figures 41.8–41.10 for $\tau(\theta)$ given by (41.1).

For $\vec{B} \perp \vec{Q}$, the magnetoconductivity can be derived by the effective-path method [16], which is equivalent to solving the Boltzmann equation. The steady-state current density

$$\vec{J} = \frac{-2e}{\Omega} \sum_{\vec{k}} f(\vec{k}) \vec{v}(\vec{k}) \tag{41.18}$$

is equal to the rate of change of the dipole moment per unit volume. Only electrons out of equilibrium need be considered, since the contribution from the equilibrium distribution vanishes. The applied electric field excites electrons out of equilibrium at the rate (to first order in $\vec{E}$)

$$\left(\frac{df}{dt} \right)_{\text{field}} = e\vec{E} \cdot \vec{v}(\vec{k}) \frac{d f_0}{d\epsilon} . \tag{41.19}$$

3) It is perhaps surprising that the transverse resistance is field dependent, even though the cyclotron motion does not change an electron's relaxation time. The same effect occurs in the two-band model, see [15].

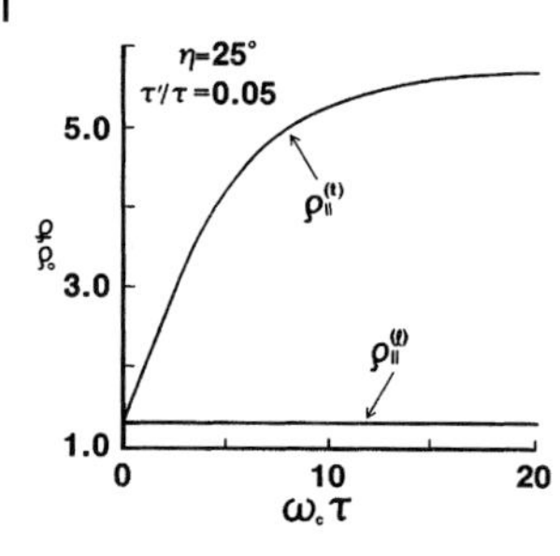

Figure 41.8 Parallel resistivities vs. magnetic-field strength. The saturation value of $\rho_{\parallel}^{(t)}$ is $\rho_{\parallel}' = 5.86\rho_0$, where $\rho_0 = m/(ne^2\tau)$.

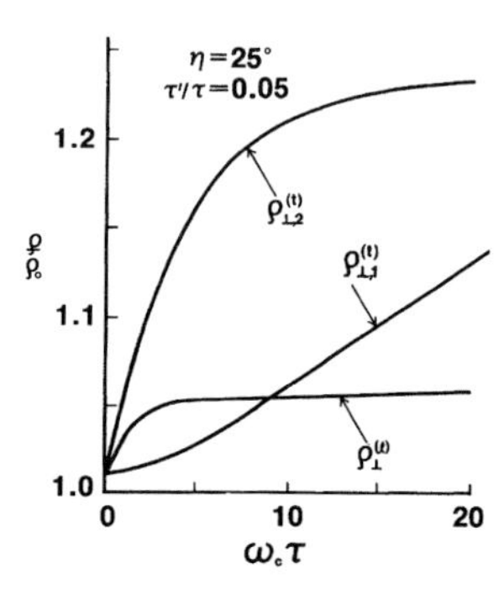

Figure 41.9 Perpendicular resistivities vs. magnetic-field strength. The saturation value of $\rho_{\perp,1}^{(t)}$ and $\rho_{\perp,2}^{(t)}$ is $\rho_{\perp}' = 1.24\rho_0$, where $\rho_0 \equiv m/(ne^2\tau)$. Note the different vertical scales in Figures 41.8 and 41.9.

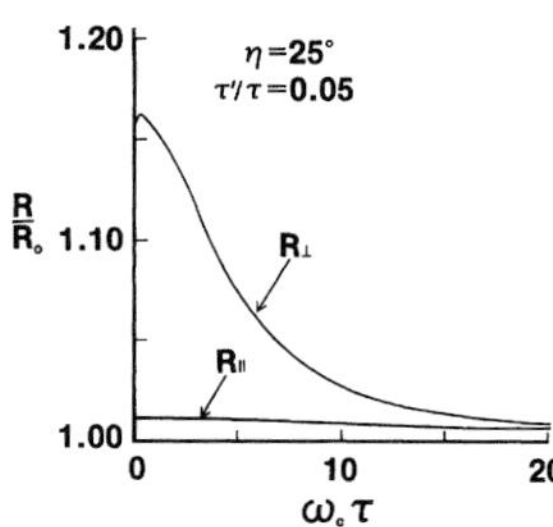

Figure 41.10 Hall coefficients for $\vec{Q} \parallel \vec{B}$ and $\vec{Q} \perp \vec{B}$. For $\omega_c\tau > 1$, $R_{\parallel}$ and $R_{\perp}$ are equal to $R_0 \equiv -1/(nec)$.

The average displacement (effective path) of an electron, created with wave vector $\vec{k}$, until it returns to equilibrium is

$$\vec{L}(\vec{k}) = \int_0^{\infty} dt \vec{v}(\vec{k}(t)) P(\vec{k}, t) , \tag{41.20}$$

where the survival probability $P(\vec{k}, f)$ is

$$P(\vec{k}, t) = \exp\left(- \int_0^t \frac{dt'}{\tau(\vec{k}(t'))} \right) . \tag{41.21}$$

The current density $\vec{J}$ to first order in $\vec{E}$ is thus

$$\vec{J} = \frac{-2e}{\Omega} \sum_{\vec{k}} e\vec{E} \cdot \vec{v}(\vec{k}) \frac{d f_0}{d\epsilon} \int_0^{\infty} dt \vec{v}(\vec{k}(t)) P(\vec{k}, t) , \tag{41.22}$$

which is equivalent to the path-integral solution [17, 18] of the Boltzmann equation.

In a magnetic field $\vec{B} = B\hat{z}$. the electron wave vector $\vec{k}$ rotates about $\vec{B}$ at the cyclotron frequency ω_c,

$$\vec{k}(t) = k\hat{n}(\theta, \phi + \omega_c t) . \tag{41.23}$$

$\hat{n}\theta, \phi = (\sin\theta\cos\phi, \sin\theta\sin\phi, \cos\theta)$ is a unit vector in the direction (θ, ϕ). Substituting into (41.22) and setting $\vec{v}(\vec{k}) = \hbar\vec{k}/m$ for free electrons yields an exact formula for the magnetoconductivity,

$$\sigma_{\alpha\beta} = \frac{ne^2}{m\omega_c}\frac{3}{4\pi}\int d\Omega\, n_\beta(\theta,\phi)\int_0^\infty d\phi' n_\alpha(\theta,\phi+\phi')P(\theta,\phi;\phi') , \tag{41.24}$$

where $P(\theta, \phi; \phi')$ is the survival probability for rotation by an angle ϕ',

$$P(\theta,\phi;\phi') = \exp\left[-\int_0^{\phi'} \frac{d\phi''}{\omega_c\tau(\theta,\phi+\phi'')}\right] . \tag{41.25}$$

Breaking the range of integration of ϕ' into sections of length 2π and using the periodicity of $\hat{n}(\theta, \phi)$ and $\tau(\theta, \phi)$ leads to a (summable) geometric series. The formula for $\sigma_{\alpha\beta}$ reduces to

$$\sigma_{\alpha\beta} = \frac{ne^2}{m\omega_c}\frac{3}{4\pi}\int d\Omega\, n_\beta(\theta,\phi)[1-P(\theta)]^{-1}\int_0^{2\pi} d\phi' n_\alpha(\theta,\phi+\phi')P(\theta,\phi;\phi') \tag{41.26}$$

with $P(\theta) = P(\theta, \phi; 2\pi)$ being the survival probability for a complete revolution.

For $\vec{Q} = Q\hat{x}$, $\sigma_{xz}, \sigma_{zx}, \sigma_{yz}$, and σ_{zy} all vanish by symmetry.[4] In the high-field limit, the exponential in (41.25) can be expanded in powers of $1/H$, yielding an expansion of (41.26). Using the symmetry of $\tau(\theta, \phi)$, it can be shown that the leading terms are

$$\sigma_{xx} \cong \frac{ne^2}{m\omega_c^2}\left\langle\frac{1}{\tau}\right\rangle_\perp ,$$

$$\sigma_{xy} = -\sigma_{yx} \cong -\frac{ne^2}{m\omega_c} ,$$

$$\sigma_{yy} \cong \frac{ne^2}{m\omega_c^2}\left\langle\frac{1}{\tau}\right\rangle_\parallel ,$$

$$\sigma_{zz} \cong \frac{ne^2}{m}\frac{3}{4\pi}\int d\Omega\cos^2\theta\left[\frac{1}{2\pi}\int_0^{2\pi}\frac{d\phi'}{\tau(\theta,\phi')}\right]^{-1} , \tag{41.27}$$

the parallel and perpendicular averages having the same definitions as before.

The resistivity and Hall coefficient are obtained by inverting the magnetoconductivity. From (41.27), the saturation values of the transverse resistivities are

$$\rho_\perp^{(t)} \to \frac{m}{ne^2}\left\langle\frac{1}{\tau}\right\rangle_\perp , \quad \rho_\parallel^{(t)} \to \frac{m}{ne^2}\left\langle\frac{1}{\tau}\right\rangle_\parallel , \quad \omega_c\tau > 1 \tag{41.28}$$

equal to the resistivities $\rho'_\perp$ and $\rho'_\parallel$ computed with a rigidly shifted distribution function. For $\omega_c\tau > 1$, the Hall coefficient has the free-electron value $R_0 = -1/(nec)$.

4) If $\vec{E}$ is parallel to $\vec{B}$, so is $\vec{J}$, implying $\sigma_{zx} = \sigma_{zy}$. By the Onsager symmetry relations, and also vanish.

The three resistivities ($\rho_{\parallel}^{(t)}, \rho_{\perp,2}^{(t)}$, and $\rho_{\perp}^{(l)}$) and the Hall coefficient ($R_{\perp}$), evaluated from (41.26) for $\tau(\theta)$ given by (41.1), are plotted in Figures 41.8–41.10. The low-field magnetoresistance "knees", which have been observed in potassium [19, 20], are linear in B, all the way to $B = 0$.

For $\omega_c\tau > 1$, the transverse resistivities (both for $\vec{Q} \parallel \vec{B}$ and $\vec{Q} \perp \vec{B}$) are equal to the resistivities computed with a rigidly shifted distribution function. For a spherical Fermi surface, this can be proven even without making a relaxation-time approximation [21].

41.5 Induced Torque

In an induced-torque experiment (Figure 41.11), a single-crystal sphere is suspended in a uniform magnetic field B. Slow rotation of $\vec{B}$ (or the crystal) induces circulating currents, giving rise to a torque on the sample. The torque magnitude depends on the induced current, which in turn depends on the magnetoconductivity of the sample in the field $\vec{B}$.

An exact theory of the induced torque, depending only on Faraday's law of induction, has been derived for a general magnetoconductivity tensor [22]. If there is no longitudinal-transverse mixing, i. e., σ_{xz}, σ_{zx}, σ_{yz}, and σ_{zy}, all vanish, the formula for the torque N about the rotation axis reduces to

$$N \propto \frac{\rho_{xx} + \rho_{zz}}{(\rho_{xx} + \rho_{zz})(\rho_{yy} + \rho_{zz}) - \rho_{xy} + \rho_{yx}} . \tag{41.29}$$

The proportionality constant, which is $4\pi a^5 \omega B^2/15c^2$, depends on the sphere radius a and rotation frequency ω. At low fields, when the induced currents circulate in the y–z plane normal to $d\vec{B}/dt$, the Hall resistivity is negligible, yielding

$$N \propto \frac{1}{\rho_{yy} + \rho_{zz}} , \qquad \omega_c\tau < 1 . \tag{41.30}$$

At high fields, when the Lorentz force rotates the current loops into the horizontal x–z plane, the Hall resistivity dominates, yielding

$$N \propto \frac{\rho_{xx} + \rho_{zz}}{-\rho_{xy}\rho_{yx}} , \qquad \omega_c\tau > 1 . \tag{41.31}$$

The $\vec{Q}$-vector orientation of the experimental samples studied in [9, 10] is not known. For simplicity, we suppose $\vec{Q}$ lies in the rotation plane. Then, at the maximum of the twofold torque pattern, $\vec{Q} \perp \vec{B}$; and at the minimum, $\vec{Q} \parallel \vec{B}$. For $\vec{Q} \perp \vec{B}$, $\rho_{yy} = \rho_{\perp,2}^{(t)}$ and $\rho_{zz} = \rho_{\perp}^{(l)}$;

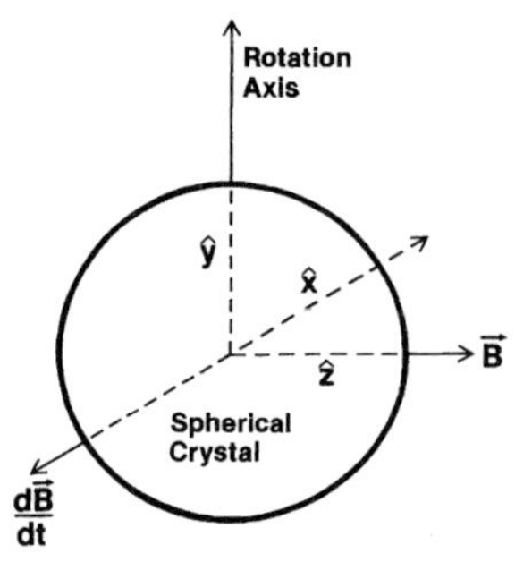

Figure 41.11 Induced-torque experiment. Rotation of $\vec{B}$ induces currents, which interact with $\vec{B}$, producing a torque about the suspension axis.

for $\vec{Q} \parallel \vec{B}$, $\rho_{yy}^{(t)} = \rho_{\perp,1}^{(t)}$ and $\rho_{zz} = \rho_{\parallel}^{(l)}$. The low-field torque anisotropy $\mathcal{R} = N_{max}/N_{min}$ is thus

$$\mathcal{R} \cong \frac{\rho_{\perp,1}^{(t)} + \rho_{\parallel}^{(l)}}{\rho_{\perp,2}^{(t)} + \rho_{\perp}^{(l)}} , \qquad \omega_c\tau < 1 . \tag{41.32}$$

As $B \to 0$, the perpendicular and parallel magnetoresistivities are equal, respectively, to the zero-field resistivities $\rho_\perp$ and $\rho_\parallel$. The zero-field limit of the torque anisotropy is thus

$$\mathcal{R}_0 = \frac{\rho_\perp + \rho_\parallel}{2\rho_\perp} = \frac{1+\gamma}{2} , \tag{41.33}$$

as deduced previously [4]. γ is the zero-field resistivity anisotropy. For the model (41.1) with $\eta = 25°$ and $\tau' = 0.05$, $\gamma = 1.30$, so that $\mathcal{R}_0 = 1.15$.

Since the induced-current loops rotate into the horizontal plane as B increases, it might be expected that the torque would be isotropic at high fields. For $\vec{Q} \perp \vec{B}$, $\rho_{xx} = \rho_{\parallel}^{(t)}$, $\rho_{zz} = \rho_{\perp}^{(l)}$, and $\rho_{yx} = -\rho_{xy} = R_\perp B$; for $\vec{Q} \parallel \vec{B}$, $\rho_{xx} = \rho_{\perp,1}^{(t)}$, $\rho_{zz} = \rho_{\parallel}^{(l)}$, and and $\rho_{yx} = -\rho_{xy} = R_\parallel B$. The high-field torque anisotropy is thus

$$\mathcal{R} \cong \left(\frac{R_\parallel}{R_\perp}\right) \frac{\rho_{\parallel}^{(t)} + \rho_{\perp}^{(l)}}{\rho_{\perp,1}^{(t)} + \rho_{\parallel}^{(l)}} , \qquad \omega_c\tau > 1 . \tag{41.34}$$

For an anisotropic relaxation-time model, the high-field Hall coefficient is isotropic. (In potassium, the high-field Hall-coefficient anisotropy $R_\parallel/R_\perp$ is typically about 1.1, although it can be as large as 1.3 [5].) Setting $R_\parallel/R_\perp = 1$ yields

$$\mathcal{R} \cong \frac{\rho_{\parallel}^{(t)} + \rho_{\perp}^{(l)}}{\rho_{\perp,1}^{(t)} + \rho_{\parallel}^{(l)}} , \qquad \omega_c\tau > 1 . \tag{41.35}$$

If the magnetoresistance were independent of field orientation, i. e., $\rho_{\parallel}^{(t)} = \rho_{\parallel}^{(l)}$ and $\rho_{\perp}^{(t)} = \rho_{\perp}^{(l)}$ the high-field torque ratio $\mathcal{R}$ would indeed be unity, implying a decrease compared to its zero-field value.

The magnetoresistance, however, depends on the field direction. This is illustrated for the model (41.1) in Figures 41.8 and 41.9. For small η, the perpendicular resistivities are only slightly affected by the magnetic field, all being approximately equal. But the high-field, parallel resistivity is much larger for $\vec{B} \perp \vec{Q}$ than for $\vec{B} \parallel \vec{Q}$, thus causing the torque anisotropy to increase.

In high fields, the transverse resistivities $\rho_{\parallel}^{(t)}$ and $\rho_{\perp}^{(t)}$ are equal to $\rho'_\parallel$ and $\rho'_\perp$, the resistivities computed with a rigidly shifted distribution function; the longitudinal resistivity $\rho_{\parallel}^{(l)}$ being field independent, is equal to the zero-field resistivity $\rho_\parallel$. The high-field limit of the torque anisotropy (41.35) is thus

$$\mathcal{R}_\infty = \frac{\delta + \gamma'}{1 + \epsilon\gamma} , \tag{41.36}$$

where $\gamma' = \rho'_\parallel/\rho'_\perp$, $\delta = \rho_{\perp}^{(l)}/\rho_{\perp}^{(t)}$, and $\epsilon \equiv \rho_\perp/\rho'_\perp$. For the model (41.1) with $\eta = 25°$ and $\tau'/\tau = 0.05$, $\delta = 0.85$, $\epsilon = 0.81$, $\gamma = 1.30$, and $\gamma' = 4.71$, so that $\mathcal{R}_\infty = 2.70$.

The torque ratio, evaluated from (41.29) for the model (41.1), is plotted in Figure 41.12.[5] The small dip at low fields is a consequence of the torque formula, the resistivities themselves

5) In order to compare Figure 41.12 with experiment, note that typically $\omega_c\tau \approx 5$ at 0.2 T. (The procedure used in [10] to determine $\omega_c\tau$ may not be valid for an anisotropic relaxation time.)

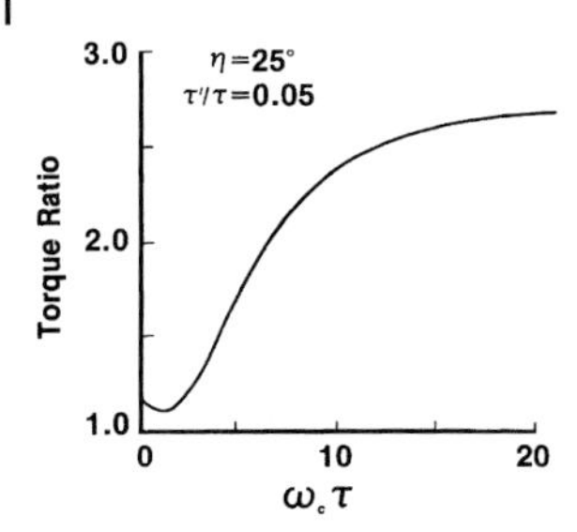

Figure 41.12 Theoretical ratio of the torque maximum and minimum. The zero-field and high-field values are $\mathcal{R}_0 = 1.15$ and $\mathcal{R}_\infty = 2.70$.

increasing monotonically. The initial decrease, which may be masked in a polydomain sample, arises from the linear field dependence of the $(\vec{Q} \perp \vec{B})$ resistivities in the denominator of the torque-ratio formula (41.32).

41.6 Discussion

The low-field, induced-torque experiments by Elliott and Datars [9, 10] show that the residual-resistivity anisotropy of sodium and potassium increases with increasing magnetic field. For potassium, the zero-field torque anisotropy (for the samples studied) was in the range 1.05–1.25. In an applied magnetic field, this anisotropy increased by about an order of magnitude.

If one believes that potassium and sodium have cubic symmetry, these results are triply perplexing. First, how can the zero-field resistivity be anisotropic? Second, how can the resistivity anisotropy increase in a magnetic field? Third, how can the measured resistivity anisotropy vary for nominally identical samples?

The zero-field resistivity anisotropy is explained by anisotropic scattering, caused by the charge-density-wave potential [4]. As we have shown here, this same mechanism also explains the increase of the resistivity anisotropy in a magnetic field. The uncontrolled domain structure, depending on metallurgical history, explains the variability of the data from sample to sample.

Exact numerical calculations of the zero-field resistivity anisotropy in potassium, based on a microscopic theory of electron-impurity scattering for a charge-density-wave state, yield an anisotropy of about 2 [23]. (This value, which applies to a single Q domain, is an upper bound for experiment.) By Equation (41.33), the zero-field torque anisotropy is about 1.5, agreeing with experiment.

The low-field, induced torque of sodium and potassium spheres is expected to be isotropic if the Fermi surface is spherical. Three symmetry-breaking mechanisms have been suggested: (i) nonspherical samples [24]; (ii) oriented, nonspherical scattering centers; (iii) a nonspherical Fermi surface. The first two have been ruled out [4, 10, 25]. Only the third, caused by a charge-density-wave structure, explains the main features of the data.

Acknowledgements

It is a pleasure to thank Xiadong Zhu for helpful discussions. We are grateful to the National Science Foundation and the NSF Materials Research Laboratories Program for financial support.

A Appendix

The angle η in the relaxation-time model (41.1) can be estimated if the region of umklapp scattering is determined by the wave-function mixing (rather than by the Fourier components of the impurity potential). The CDW potential $V = G\cos(\vec{Q}\cdot\vec{r})$ mixes the plane-wave state $\vec{k}$ with $\vec{k}\pm\vec{Q}$. The degree of mixing, which is complete (100%) at an energy-gap plane, decreases with increasing distance of the state $\vec{k}$ from the energy gap plane. This continuous change is modeled in (41.1) by an abrupt transition. The boundary between complete and zero mixing is found by equating the average mixing of the occupied states for the relaxation-time model and a CDW state.

In a CDW state, the electron wave functions satisfy the Schrödinger equation

$$\frac{-\hbar^2}{2m}\nabla^2\psi_{\vec{k}} + G\cos(\vec{Q}\cdot\vec{r})\psi_{\vec{k}} = E_{\vec{k}}\psi_{\vec{k}}\,. \tag{A1}$$

For $\vec{Q} = Q\hat{z}$, the energy denominator for the mixing of the plane-wave state $\vec{k}$ with $\vec{k}\pm\vec{Q}$ is

$$\epsilon_{\vec{k}\pm\vec{Q}} - \epsilon_{\vec{k}} = (\hbar^2/2m)\left(\pm 2k_z Q + Q^2\right)\,. \tag{A2}$$

Thus, if $k_z < 0$, the state $\vec{k}$ is mixed most with $\vec{k}+\vec{Q}$; if $k_z > 0$, the state $\vec{k}$ is mixed most with $\vec{k}-\vec{Q}$.

Because $\vec{Q}$ is slightly greater than the diameter $2k_F$ of the free-electron Fermi sphere, only the states below the energy gap (at $k_z = \pm Q/2$) are occupied. For these states, an approximate solution is derived as follows [4]. For $k_z < 0$, we assume that the plane-wave state $\vec{k}$ mixes only with $\vec{k}+\vec{Q}$. This leads to the secular equation

$$\begin{vmatrix} \epsilon_{\vec{k}} - E_{\vec{k}} & \frac{1}{2}G \\ \frac{1}{2}G & \epsilon_{\vec{k}+\vec{Q}} - E_{\vec{k}} \end{vmatrix} = 0\,, \tag{A3}$$

which has two solutions for the energy eigenvalue $E_{\vec{K}}$. The lower one, belonging to the state below the gap, is

$$E_{\vec{k}} = \frac{1}{2}(\epsilon_{\vec{k}} + \epsilon_{\vec{k}+\vec{Q}}) - \frac{1}{2}\left[(\epsilon_{\vec{k}+\vec{Q}} - \epsilon_{\vec{k}})^2 + G^2\right]^{1/2}, \quad k_z < 0\,. \tag{A4}$$

If the corresponding eigenfunction is denoted by

$$\Psi_{\vec{k}} = \cos(\xi)e^{i\vec{k}\cdot\vec{r}} - \sin(\xi)e^{i(\vec{k}+\vec{Q})\cdot\vec{r}}\,, \tag{A5}$$

the degree of mixing $\zeta(\vec{k}) = 2\sin^2\xi$ is

$$\zeta(\vec{k}) = 2(\epsilon_{\vec{k}} - E_{\vec{k}})^2 \Big/ \left[\left(\frac{1}{2}G\right) + (\epsilon_{\vec{k}} - E_{\vec{k}})^2\right]\,. \tag{A6}$$

For $k_z > 0$, we assume that the plane-wave state $\vec{k}$ mixes only with $\vec{k}-\vec{Q}$, yielding the energy eigenvalue

$$E_{\vec{k}} = \frac{1}{2}(\epsilon_{\vec{k}} + \epsilon_{\vec{k}-\vec{Q}}) - \frac{1}{2}\left[(\epsilon_{\vec{k}-\vec{Q}} - \epsilon_{\vec{k}})^2 + G^2\right]^{1/2}, \quad k_z > 0\,. \tag{A7}$$

The degree of mixing $\zeta(\vec{k})$ is the same as for $k_z < 0$.

It is convenient to introduce the dimensionless variables

$$w = \begin{cases} \left(k_z + \frac{1}{2}Q\right)/Q\,, & k_z < 0 \\ \left(k_z - \frac{1}{2}Q\right)/Q\,, & k_z > 0 \end{cases} . \tag{A8}$$

Making these substitutions yields

$$E_{\vec{k}} = \frac{\hbar^2 Q^2}{2m}\left[\kappa^2 + w^2 + \frac{1}{4} - (w^2 + \alpha^2)^{1/2}\right], \tag{A9}$$

$$\zeta(\vec{k}) = 1 - w(w^2 + \alpha^2)^{-1/2}, \tag{A10}$$

with $\alpha \equiv mG/(\hbar^2 Q^2)$.

We assume that the Fermi surface makes critical contact with the energy-gap planes, as shown in Figure 41.2. For $\vec{k} = \frac{1}{2}\vec{Q}$, $\epsilon_{\vec{k}} = \epsilon_{\vec{k}-\vec{Q}} = \hbar^2\left(\frac{1}{2}Q\right)^1/2m$. Substituting in (A7) yields the Fermi energy

$$E_F = \hbar^2\left(\frac{1}{2}Q\right)\Big/2m - \frac{1}{2}G\,. \tag{A11}$$

From (A9), the equation of the Fermi surface is then

$$\kappa^2 = (w^2 + \alpha^2)^{1/2} - w^2 - \alpha\,. \tag{A12}$$

The magnitude of Q as a function of G is found by equating the volume within the Fermi surface (A12) to the volume $(4\pi/3)k_F^3$ of the free-electron Fermi sphere, yielding

$$k_F^3 = \frac{3}{4}Q^3\left[\frac{1}{2}\left(\frac{1}{4} + \alpha^2\right)^{1/2} + \alpha^2 \sinh^{-1}\left(\frac{1}{2\alpha}\right) - \frac{1}{12} - \alpha\right] \tag{A13}$$

with $\alpha = mG/(\hbar^2 Q^2)$ and $k_F^3 = w3\pi^2 n$. Finally, averaging the mixing (A10) over the occupied states, we obtain

$$\langle\zeta\rangle = 1 - \frac{3}{2}(Q/k_F)^3\left[\frac{1}{8} - \left(\frac{1}{12} + \alpha - \frac{2}{3}\alpha^2\right)\left(\frac{1}{4} + \alpha^2\right)^{1/2} + \left(1 - \frac{2}{3}\alpha\right)\alpha^2\right]. \tag{A14}$$

In the relaxation-time model, the degree of mixing $\xi = 0$ for $|k_z| < k_F \cos\eta$ and $\zeta = 1$ for $k_F \cos\eta < |k_z| < k_F$. The average mixing of the occupied states is then

$$\langle\zeta\rangle = 1 - \frac{3}{2}\cos\eta + \frac{1}{2}\cos^3\eta\,. \tag{A15}$$

For sodium, the free-electron Fermi energy $E_F = \hbar^2 k_F^2/2m$ is 3.24 eV; the CDW-potential amplitude G, equal to the threshold energy of the optical absorption anomaly, is 1.2 eV; and $G/E_F^0 = 0.37$ [26]. For potassium, $E_F^0 = 2.12$ eV, $G = 0.6$ eV, and $G/E_F^0 = 0.28$ [27–29]. Solving numerically Equations (A13)–(A15) yields $\eta = 27°$ for sodium and $\eta = 24°$ for potassium.

References

1 Overhauser, A.W. (1978) *Adv. Phys.*, **27**, 343.
2 Schaefer, J.A. and Marcus, J.A. (1971) *Phys. Rev. Lett.*, **27**, 935.
3 Holroyd, F.W. and Datars, W.R. (1975) *Can. J. Phys.*, **53**, 2517.
4 Bishop, M.F. and Overhauser, A.W. (1978) *Phys. Rev. B*, **18**, 2477.
5 Zhu, X. and Overhauser, A.W. (1984) *Phys. Rev. B*, **30**, 622.
6 Coulter, P.G. and Datars, W.R. (1980) *Phys. Rev. Lett.*, **45**, 1021.
7 Coulter, P.G. and Datars, W.R. (1982) *Solid State Commun.*, **43**, 715.
8 Huberman, M. and Overhauser, A.W. (1982) *Phys. Rev. B*, **25**, 2211.
9 Elliott, M. and Datars, W.R. (1983) *Solid State Commun.*, **46**, 67.
10 Elliott, M. and Datars, W.R. (1983) *J. Phys. F*, **13**, 1483.
11 Lifshitz, I.M., Azbel, M.Ya., and Kaganov, M.I. (1956) *Zh. Eksp. Teor. Fiz.*, **31**, 63 [(1957) Sov. Phys. – JETP, **4**, 41].
12 Llewellyn, B., Paul, D.McK., Randies, D.L., and Springford, M. (1977) *J. Phys. F*, **7**, 2531.
13 Wilson, A.H. (1953) *Theory of Metals*, 2nd edn., Cambridge University Press, Cambridge, England.
14 Ashcroft, N.W. and Mermin, N.D. (1976) *Solid State Physics*, Holt, Rinehart and Winston, New York.
15 Ziman, J.M. (1964) *Principles of the Theory of Solids*, Cambridge University Press, Cambridge, England.
16 Pippard, A.B. (1964) *Proc. R. Soc. London A*, **282**, 464.
17 Reif, F. (1965) *Fundamentals of Statistical and Thermal Physics*, McGraw-Hill, New York.
18 Ashcroft, N.W. and Mermin, N.D. (1976) *Solid State Physics*, Holt, Rinehart and Winston, New York.
19 Babiskin, J. and Siebenmann, P.G. (1969) *Phys. Condens. Mater.*, **9**, 113.
20 Taub, H., Schmidt, R.L., Maxfield, B.W., and Bowers, R. (1971) *Phys. Rev. B*, **4**, 1134.
21 Wagner, D.K. (1972) *Phys. Rev. B* **5**, 336. Wagner, D.K. (1971) Ph.D. thesis, Cornell University. If the Fermi surface is not spherical, the steady-state distribution function for $\omega_c \tau > 1$ is no longer a rigidly shifted equilibrium distribution. It still, however, depends only on the Fermi-surface geometry and not on the scattering mechanisms.
22 Visscher, P.B. and Falicov, L.M. (1970) *Phys. Rev. B*, **2**, 1518.
23 DeGennaro, S. and Borchi, E. (1982) *J. Phys. F*, **12**, 963. DeGennaro, S. and Borchi, E. (1982) *J. Phys. F*, **12**, 2363. We have used the results for $G/E_F = 0.3$, the value appropriate to potassium. The uncertainty of the theoretical calculation stems primarily from the unknown impurity potential. We have disregarded the calculations using a screened impurity pseudopotential, since the many-electron dielectric function $\epsilon(q)$ is not well known near $q = 2k_F$.
24 Lass, J.S. (1976) *Phys. Rev. B*, **13**, 2247.
25 Gugan, D. (1982) *J. Phys. F*, **12**, L173.
26 Hietel, B. and Mayer, H. (1973) *Z. Phys.*, **264**, 21.
27 Mayer, H. and El Naby, M.H. (1963) *Z. Phys.*, **174**, 269.
28 Overhauser, A.W. (1964) *Phys. Rev. Lett.*, **13**, 190.
29 Overhauser, A.W. and Butler, N.R. (1976) *Phys. Rev. B*, **14**, 3371.

Reprint 42 Effect of an Inhomogeneous Resistivity on the Induced-Torque Pattern of a Metal Sphere[1)]

Xiaodong Zhu * *and A.W. Overhauser* *

* Department of Physics, Purdue University, 525, Northwestern Avenue, West Lafayette, Indiana 47907, USA

Received 7 May 1984

We have sought to determine whether an inhomogeneous residual resistivity could cause the low-temperature induced-torque anisotropy observed in single-crystal spheres of potassium, especially the nonsaturating, four-peak patterns observed between 5 and 25 kOe. On the basis of our previous results and the analysis presented herein, we conclude that the residual-resistivity gradient cannot explain induced-torque anomalies.

The induced-torque method is particularly convenient for measuring galvanomagnetic properties of alkali metals [1]. A spherical single-crystal sample is suspended by a rod in a uniform magnetic field $\vec{H}$. Slow rotation of the magnetic field about the suspension axis induces a circulating current which causes a torque on the rod. Only Faraday's law of induction and Ohm's law (in tensor form) is needed to relate theory and experiment.

Schaefer and Marcus [2] measured induced torques on a large number ($\sim$ 200) of spherical single crystals of potassium. For magnetic fields below 3 kOe, they generally observed a twofold, sinusoidal anisotropy (except in a few samples having a [110] axis parallel to the suspension axis). At higher fields, the torque developed four broad peaks in a 360° rotation. The maxima grew almost quadratically with field, while the minima grew linearly with field. The observed patterns violate the supposed cubic symmetry since the four-peak patterns occurred even when a threefold axis was parallel to the axis of rotation. The anomalous torques were confirmed by Holroyd and Datars [3], who found (among their many results) one specimen with an anisotropy (at 23 kOe) of 45 to 1.

These induced-torque anomalies were totally unexpected from galvomagnetic theory [4], which has been remarkably successful in explaining observations in other more complicated metals [5]. Potassium is generally considered to have a body-centered-cubic (bcc) crystal structure. Its Fermi surface is believed to deviate only one or two parts in 10^3 from sphericity. The theory predicts (precisely) that, in high magnetic fields, $\omega_c\tau \gg 1$, its magnetoresistance should saturate and the Hall coefficient should exactly equal that for free electrons $(-1/nec)$. ω_c is the cyclotron frequency, τ is the relaxation time, and n is the electron density. It follows that the induced torque of a spherical sample of potassium would saturate for $\omega_c\tau \gg 1$ and be independent of the magnet angle [6].

Some have feared that the experimental results were an artifact, and several possible causes have been suggested: Anisotropic scattering, such as an oriented array of dislocations. However, calculation [7] shows that, even if one has an anisotropic resistivity of, say, 10 to 1, the induced-torque saturates and becomes isotropic at relatively low fields. An anisotropic sample shape can cause such an effect. Lass [8] has shown that if a sample is sufficiently distorted from a sphere (> 10%), the calculated torque pattern could be similar to those of Schaefer and Marcus. However, the deviations of their samples from sphericity were 2% or less. To fit the data of Holroyd and Datars, sample K-10, a 50% deviation from sphericity is required. But that sample was spherical to a precision of 2×10^{-3} [9, 10]. An inhomogeneous

1) Phys. Rev. B 31, 2 (1985).

Anomalous Effects in Simple Metals. Albert Overhauser

ISBN: 978-3-527-40859-7

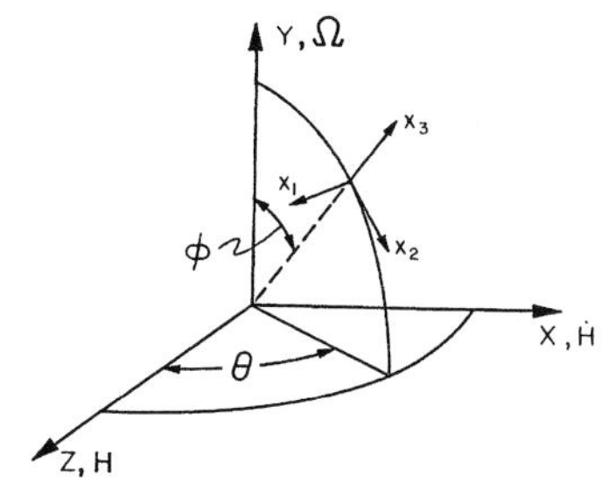

Figure 42.1 The coordinate system used in the analysis. The magnetic field $\vec{H}$ is along $\hat{Z}$, not along x_3.

distribution of impurities has been suggested from time to time. The segregation process on solidification can cause a significant impurity gradient. Typical samples are of high purity potassium (99.95%, which leads to a residual resistivity ratio of $\sim$ 5000). An impurity gradient will cause, of course, an inhomogeneous residual resistivity.

Consider a coordinate system with the origin at the center of a spherical sample of radius R. Let the x_3 axis be the direction of the impurity gradient, as shown in Figure 42.1. A metal conforming to the nearly-free-electron model will have the following magnetoresistivity tensor in the x_1, x_2, x_3 frame (of Figure 42.1):

$$\rho = \rho_0 \begin{pmatrix} q(1+gx_3/R) & \omega_c\tau_0\cos\theta\sin\phi & -\omega_c\tau_0\cos\theta\cos\phi \\ -\omega_c\tau_0\cos\theta\sin\phi & q(1+gx_3/R) & \omega_c\tau_0\sin\theta \\ \omega_c\tau_0\cos\theta\cos\phi & -\omega_c\tau_0\sin\theta & q(1+gx_3/R) \end{pmatrix} . \tag{42.1}$$

$\rho_0 = m/ne^2\tau_0$ is the residual resistivity at the origin. The dimensionless coefficient g represents the gradient in residual resistivity. The coefficient $q(H)$ has been inserted to describe an isotropic magnetoresistance.

It is worthwhile to compare our present hypothesis with Lass's ellipsoidal model. Let us orient the ellipsoid so that its short axis (of length $2bR$) is in the x_3 direction (b is the compression factor). We might view such an ellipsoid as a sphere with an infinite resistivity reaching in a distance $(1-b)R$ from the poles along x_3. Thus, an anisotropic shape appears equivalent to an extreme case of inhomogeneous resistance in a sphere. It is natural to inquire whether a *finite* inhomogeneous resistivity can cause the observed induced-torque anomalies, especially a nonsaturating, four-peak pattern. This line of thought provided the motivation for this paper.

We begin by following Visscher and Falicov [11]. The induced current in a spherical sample of radius R satisfies the equations,

$$\vec{\nabla} \times \overleftrightarrow{\rho} \cdot \vec{j} = -\dot{\vec{H}}/c , \tag{42.2}$$

$$\vec{\nabla} \cdot \vec{j} = 0 , \tag{42.3}$$

and boundary condition,

$$\vec{j} \cdot \vec{r}_{|\vec{r}|=R} = 0 . \tag{42.4}$$

The torque on the sample is given by the Lorentz force:

$$\vec{N} = \frac{1}{c}\int \vec{r} \times (\vec{j} \times \vec{H}) d^3r . \tag{42.5}$$

With the position-dependent magnetoresistivity tensor $\overleftrightarrow{\rho}$ given in Equation (42.1), only when $\dot{\vec{H}}$ is parallel to the concentration gradient and H is infinitesimal can one obtain the exact solution for the induced current $\vec{j}$:

$$\vec{j}(H \to 0, \theta = (n+1)\pi/2, \phi = \pi/2) = \frac{\vec{r} \times \dot{\vec{H}}}{2\rho_0 q c(1 + g x_3/R)} \quad (n = 1, 2) . \tag{42.6}$$

For finite fields and general field directions it seems impossible to find analytic solutions for $\vec{j}$. From the definition of g (the gradient coefficient of residual resistivity), Equation (42.1), g must be less than unity in order that the resistivity remain positive near $x_3 = -R$. In what follows, we shall treat g as a small parameter. Accordingly, we seek solutions for $\vec{j}$ in the form of power series in g:

$$\vec{j} = \sum_{n=0}^{\infty} g^n \vec{j}_n . \tag{42.7}$$

We now define a matrix $\overleftrightarrow{Z}$ by

$$Z = \begin{pmatrix} 1 & (\omega_c \tau_0/q)\cos\theta \sin\phi & -(\omega_c \tau_0/q)\cos\theta\cos\phi \\ -(\omega_c \tau_0/q)\cos\theta\sin\phi & 1 & (\omega_c \tau_0/q)\sin\phi \\ (\omega_c \tau_0/q)\cos\theta\cos\phi & -(\omega_c \tau_0/q)\sin\phi & 1 \end{pmatrix} , \tag{42.8}$$

and rewrite (42.1) as

$$\overleftrightarrow{\rho} = \rho_0 q(\overleftrightarrow{Z} + g x_3 \overleftrightarrow{I}/R) . \tag{42.9}$$

$\overleftrightarrow{I}$ is the unit matrix. We next insert this expression into (42.2) and equate terms having equal powers of g in Equations (42.2)–(42.4). The following relations are found:

$$\begin{aligned} &\vec{\nabla} \times \overleftrightarrow{Z} \cdot \vec{j}_0 = -\dot{\vec{H}}/\rho_0 q c \\ &\vec{\nabla} \cdot \vec{j}_0 = 0 , \\ &\vec{j}_0 \cdot \vec{r}|_{|\vec{r}|=R} = 0 . \end{aligned} \tag{42.10}$$

The equations for $\vec{j}_n$ $(n \neq 0)$ are

$$\begin{aligned} &\vec{\nabla} \times \left(\overleftrightarrow{Z} \cdot \vec{j}_n\right) = -\vec{\nabla} \times (x_3 \vec{j}_{n-1}/R) \\ &\vec{\nabla} \cdot \vec{j}_n = 0 , \\ &\vec{j}_n \cdot \vec{r}|_{|\vec{r}|=R} = 0 . \end{aligned} \tag{42.11}$$

The solution of (42.10) has the form

$$\vec{j}_0 = \vec{t} \times \vec{r} , \tag{42.12}$$

with

$$\vec{t} = -\frac{1}{\rho_0 q c}[\mathrm{Tr}(\overleftrightarrow{Z}) - \overleftrightarrow{Z}]^{-1} \dot{\vec{H}} . \tag{42.13}$$

The Equations (42.11) lead to unique solutions for $\vec{j}_n$; for example, $\vec{j}_1$ are of the form

$$(\vec{j}_1) = A_i + B_{ijk} x_j x_k , \tag{42.14}$$

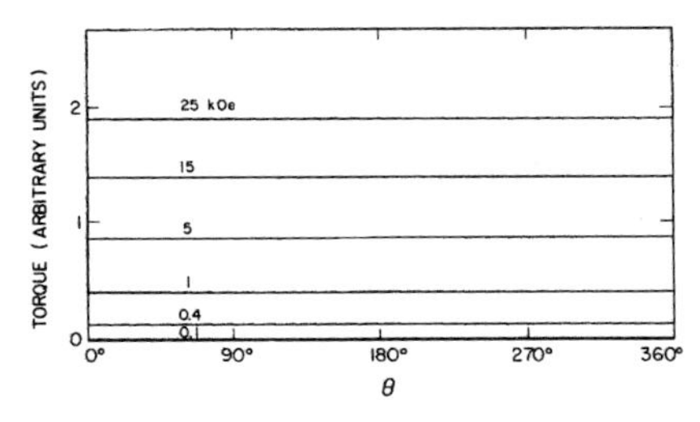

Figure 42.2 Total induced-torque vs. $\vec{H}$, for the case $g = 0.5$. This corresponds to a variation of 3 to 1 in impurity concentration across the sphere. The Kohler slope $S = 0.03$. The curves shown are for 0.1, 0.4, 1, 5, 15, and 25 kOe. The axis of the impurity concentration gradient is in the rotation plane (which corresponds to the maximum effect). θ is the magnet angle in the horizontal plane.

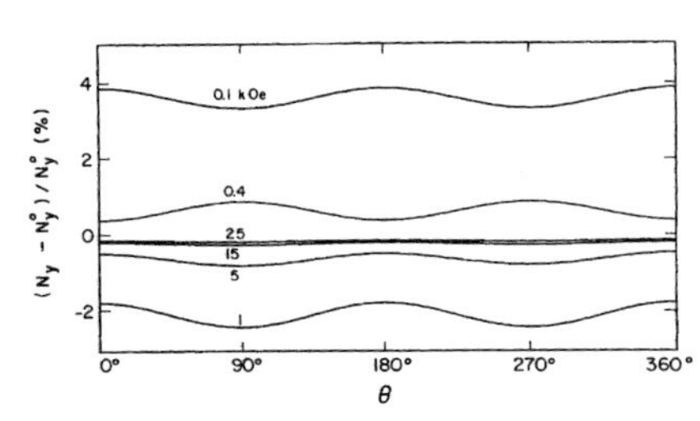

Figure 42.3 Expanded plot of the calculated torque curves shown in Figure 42.2. $(N_y - N_y^0)/N_y^0$ shows the change in torque created by a variation in impurity concentration of 3:1 across the sample.

where the summation convention has been adopted and the subscripts run from 1 to 3. $\{A_i\}$ designate the three components of the constant part of $\vec{j}_i$, whereas $\{B_{ijk}\}$ involve 18 independent coefficients of the quadratic part. They are functions of the field strength as well as the angle between the field and the gradient. They are determined from Equations (42.11) in terms of $\vec{j}_0$. Careful analysis of Equations (42.11) leads to just 21 independent, linear, inhomogeneous equations which can be solved easily with a small computer.

Similarly $\vec{j}_2$, which depends on the solution for $\vec{j}_1$ can be written:

$$(\vec{j}_2)_i = C_{ij} x_j + D_{ijkl} x_j x_k c_l \,. \tag{42.15}$$

There are 39 independent equations to determine the 39 independent coefficients (9 for C_{ij}, and 30 for D_{ijkl}).

From Equations (42.5) and (42.14), one can see that $\vec{j}_1$ has no contribution to the torque since the integrand is an odd function of x_1. As a consequence, the inhomogeneous residual resistivity affects the induced-torque pattern only to second order in the parameter g.

Our calculated induced-torque curves, obtained from Equations (42.5), (42.12), and (42.15), are shown in Figures 42.2 and 42.3, for the case $g = 0.5$. This corresponds to a 3:1 resistivity variation across the sample. In order to match the observed linear increase of the torque with H at the minima, we have taken $q = 1 + S\omega_c\tau_0$, with a Kohler slope $S = 0.03$ [See Equation (42.1)]. Figure 42.2 shows the total induced torque versus $\vec{H}$ for fields between 0.1 and 25 kOe. The axis of the impurity concentration gradient is in the rotation plane ($\phi = \pi/2$). This orientation leads to the maximum anisotropy. Observe that the curves are almost flat.

In Figure 42.3 the results of Figure 42.2 are reproduced in an expanded plot. $(N_y - N_y^0)/N_y^0$, the change in torque caused by the resistivity gradient (divided by the torque with $g = 0$), is shown. Observe that the anisotropy is twofold, and has its maximum value, 1%, at small magnetic fields. At 25 kOe the anistropy is only 3×10^{-5}, even with a 3:1 impurity concentration variation across the sample. Obviously our results show that reasonable resistivity gradients have nothing to do with the four-peak patterns observed experimentally.

So far all *extrinsic* mechanisms which have been considered have failed to explain the induced-torque anomalies. It is appropriate, then, to look for a hidden *intrinsic* mechanism.

Overhauser [12, 13] has pointed out that the ground state of potassium is likely a charge-density-wave (CDW) state. CDW theory has explained many experimental anomalies in potassium. The most striking success is the explanation [7] of the sharp open-orbit torque peaks observed by Coulter and Datars [14] in fields between 40 and 85 kOe.

We have recently found [15] that the four-peak patterns (observed from 5 to 25 kOe), which are the subject of this paper, can be explained quantitatively by an *intrinsic* anisotropic Hall coefficient. Such an intrinsic property stems from the multiply connected Fermi surface (created by a CDW broken symmetry [16]). Nevertheless, it has seemed appropriate to prove that a residual-resistivity gradient cannot be invoked as an alternative explanation.

Acknowledgements

This research was supported by the Materials Research Laboratory Program of the National Science Foundation.

References

1 Delaney, J.A. and Pippard, A.B. (1972) *Rep. Prog. Phys.*, **35**, 677.
2 Schaefer, J.A. and Marcus, J.A. (1971) *Phys. Rev. Lett.*, **27**, 935.
3 Holroyd, F.W. and Datars, W.R. (1975) *Can. J. Phys.*, **53**, 2517.
4 Lifshitz, I.M., Azbel, M.Ya., and Kaganov, M. (1956) *Zh. Eksp. Teor. Fiz.*, **31**, 63 [Soviet Phys. JETP 4, 41 (1957)].
5 Fawcett, E. (1964) *Adv. Phys.*, **13**, 139.
6 Lass, J.S. and Pippard, A.B. (1970) *J. Phys. E*, **3**, 137.
7 Huberman, M. and Overhauser, A.W. (1982) *Phys. Rev B*, **25**, 2211.
8 Lass, J.S. (1976) *Phys. Rev. B*, **13**, 2247.
9 Overhauser, A.W. (1974) *Phys. Rev. B*, **9**, 2441.
10 Overhauser, A.W. (1983) in *New Horizons of Quantum Chemistry*, (eds. P.O. Lowdin and B. Pullman), Reidel, Dordrecht, The Netherlands, p. 357.
11 Visscher, P.B. and Falicov, L.M. (1970) *Phys. Rev. B*, **2**, 1518.
12 Overhauser, A.W. (1968) *Phys. Rev.* 167, 691.
13 For an experimental and theoretical review, see Overhauser, A.W. (1978) *Adv. Phys.*, **27**, 343; Overhauser, A.W. (1978) *Electron Correlations in Solids, Molecules, and Atoms*, (eds. J.T. Devreese and F. Brosens), Plenum, New York, 1983, p. 41.
14 Coulter, P.G. and Datars, W.R. (1980) *Phys. Rev. Lett.*, **45**, 1021.
15 Zhu, X. and Overhauser, A.W. (1984) *Phys. Rev. B*, **30**, 622.
16 Fragachan, F.E. and Overhauser, A.W. (1984) *Phys. Rev. B*, **29**, 2912.

Reprint 43 Infrared-Absorption Spectrum of an Incommensurate Charge-Density Wave: Potassium and Sodium[1)]

F.E. Fragachán* and A.W. Overhauser*

* Department of Physics, Purdue University, 525, Northwestern Avenue, West Lafayette, Indiana 47907, USA

Received 7 May 1984

New optical-absorption edges of a metal having a charge-density-wave ground state arise from transitions across charge-density-wave energy gaps. If the charge-density-wave wave vector Q is incommensurate with the reciprocal-lattice vector **G**, three families of higher-order gaps in $E(\mathbf{k})$ arise: "minigaps", characterized by wave vectors $(n+1)\mathbf{Q}-n\mathbf{G}$; "heterodyne gaps", with periodicities $n(\mathbf{G}-\mathbf{Q})$; and "second-zone minigaps", with periodicities $(n+1)\mathbf{G}-n\mathbf{Q}$. The energy-gap surfaces of the first two families truncate the Fermi surface and lead to additional absorption edges in the far-infrared region. The absorption peaks associated with the first three minigaps are calculated for K and Na, and are found to be an order of magnitude larger than both the interband absorption and the main charge-density-wave peak. However, they are much smaller than the room-temperature Drude absorption. Consequently, a search for far-infrared edges must be carried out at low temperature, and in samples for which the orientation of **Q** allows observation of the Mayer–El Naby anomaly.

43.1 Introduction

The optical properties of the alkali metals have been studied extensively during recent years. Early measurements made by Duncan and Duncan [1], and by Ives and Briggs [2, 3], have been analyzed in terms of the nearly-free-electron theory by Butcher [4], Wilson [5], and Cohen [6]. Measurements extending into the infrared have been reported by Hodgson [7, 8], Mayer and co-workers [9–12], Althoff and Hertz [13] (far infrared), and more recently by Smith [14], Palmer and Schnatterly [15], and by Hietel and Mayer [16].

This work aims to explain the nature of some observed anomalies and to predict new optical-absorption edges, in the far infrared. Our work assumes a charge-density-wave (CDW) ground state and is based on the recent improved understanding of the electronic band structure of such simple metals [17].

In this section, we review the current, rather confused, status of the optical properties of the alkali metals. In Section 43.2 we present the method employed to determine the absorption coefficient, and in Section 43.3 we report the results obtained for sodium and potassium.

The interband optical-absorption threshold for potassium is 1.3 eV, in agreement with the theory of Butcher [4], and the absorption intensity is weak. In 1963, Mayer and El Naby [9–11] discovered a rather intense optical absorption with a threshold of $\simeq 0.6$ eV, well below the normal interband threshold, in the near-infrared reflection spectrum of potassium; see Figure 43.1. Most attempts to explain this anomalous absorption failed. Only one model [18–21], which assumes that potassium has a CDW ground state, has given a quantitative explanation [22]; see Figure 43.1 and notice the excellent agreement, based on vertical transitions across the CDW energy gap 2α created by the CDW potential. See Figure 43.2. The CDW

1) Phys. Rev. B 31, 8 (1985).

Anomalous Effects in Simple Metals. Albert Overhauser

ISBN: 978-3-527-40859-7

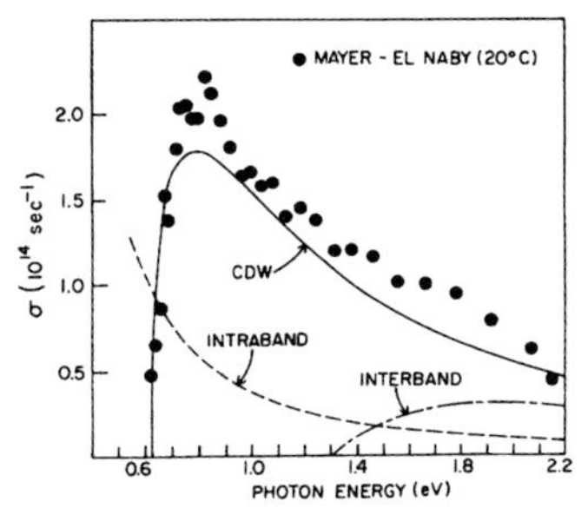

Figure 43.1 Anomalous optical-absorption spectrum of potassium. The intraband conductivity (dashed curve) has been subtracted from the experimental data before being plotted. The solid curve shows the theoretical absorption introduced by a CDW structure. A normal metal would exhibit only the interband absorption with a threshold, as shown, at 1.3 eV. The anomaly is independent of temperature between 80 K and the melting point.

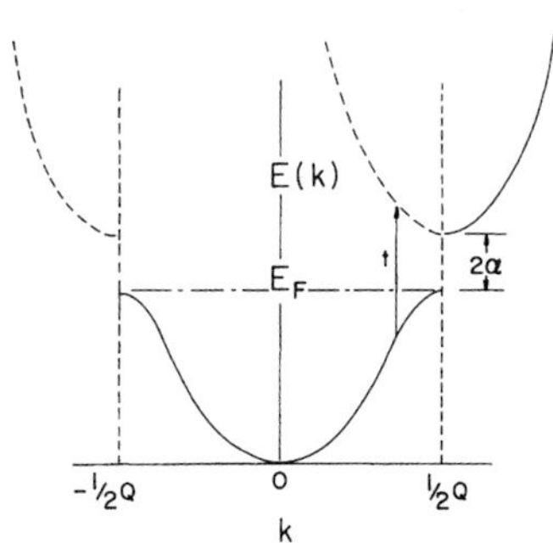

Figure 43.2 Schematic behavior of $E(\mathbf{k})$, for $\mathbf{k}$ parallel to $\mathbf{Q}$, when a CDW potential, Equation (43.1), is present. Optical transitions responsible for the Mayer–El Naby anomaly are indicated by the arrow "t"; the threshold is 2α (if $V_{lat} = 0$).

potential is

$$V_{CDW}(\mathbf{r}) = 2\alpha \cos(\mathbf{Q} \cdot \mathbf{r}) . \tag{43.1}$$

$\mathbf{Q}$ is the CDW wave vector, which in general is incommensurate with the lattice. As a consequence, a new optical-absorption mechanism, which the standard (Butcher) theory cannot account for, will arise having a threshold $\hbar\omega = 2\alpha$. The theoretical absorption coefficient $(2n\kappa/\lambda)$ caused by these transitions is given by [22]

$$\frac{2n\kappa}{\lambda} = \frac{(2\alpha)^2 e^2 Q}{4\pi\hbar c W^2} \left(\frac{W - 2\alpha}{W + 2\alpha}\right)^{1/2} \left(1 - \frac{W + 2\alpha}{2\mu Q}\right) \cos^2\theta , \tag{43.2}$$

where W is the transition energy, n and κ are the optical constants, λ is the vacuum wavelength, $\mu \equiv \hbar^2 Q/2m$, and θ represents the angle between $\mathbf{Q}$ and the photon polarization vector $\hat{\epsilon}$. Therefore, the CDW optical absorption is uniaxial. This is crucial to an explanation of why the Mayer–El Naby anomaly is not seen in measurements on evaporated films. When soft metals are evaporated on amorphous substrates (e. g., glass), the crystal grains have a preferred texture. The close-packed planes of the lattice lie parallel to the surface [23]. For potassium, the normal to a glass-metal interface will be a (110) direction which, in fact, is very near to the preferred $\mathbf{Q}$ direction [24]. When light reflects from a metal surface the polarization vector (inside the metal) is parallel to the surface. As a result, $\mathbf{Q}$ is nearly perpendicular to $\hat{\epsilon}$ $(\theta = \pi/2)$, and the anomalous optical absorption *cannot* occur. At bulk-metal–vacuum surfaces, the Mayer–El Naby anomaly has been reproduced by Hietel [12, 25] and by Harms [26]. The intensity of the CDW absorption is (approximately) independent of temperature between 80 K and temperatures above the melting point, where the anomaly persists [9–11, 25].

Fäldt and Walldén [27] have suggested that the Mayer–El Naby anomaly is due to particles produced as a result of water vapor adsorption. They studied the optical properties of K films, prepared by evaporation onto a liquid-nitrogen-cooled substrate. In the photon energy

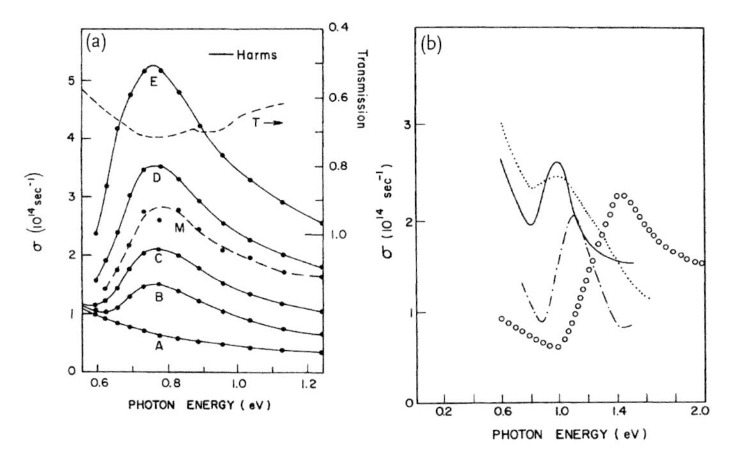

Figure 43.3 Comparison between (a) Harms's experiment, [24, 26], and (b) Fäldt and Walldén, [27]. (a) Curve *A* is the optical conductivity (absorption) of a fresh bulk-potassium–vacuum interface. Curves *B*–*E* were obtained on the same specimen after successive exposure to trace amounts of H_2O. The dashed curve *M* is one obtained by Mayer and El Naby [9–11]. The dashed curve *T* is an inverted plot of the optical transmission through 0.01 cm of KOH. All data were taken at room temperature, (b) The optical conductivity of an evaporated film of potassium after exposure to trace amounts of H_2O. The substrate was at 78 K. The absorption peaks caused by the exposure occur at various photon energies in different experimental runs. All structure disappeared above 110 K.

range 0.6–4.0 eV they found optical anomalies after exposure of the sample to water vapor. However, they never found an anomaly with a 0.6 eV threshold. Furthermore, all new optical peaks disappeared above 110 K, unlike the Mayer–El Naby anomaly which persists even into the liquid state. Taut [28] has followed the misinterpretation of Fäldt and Walldén [27] by claiming that surface states caused by KOH are involved in the Mayer–El Naby anomaly. The absorption peaks observed by Fäldt and Walldén, which have no similarity with Harms's absorption peaks [26] (see Figure 43.3), allow one to conclude that Harms's experiment actually reproduces, in a controlled way, the Mayer–El Naby anomaly. The mechanism suggested by Taut [28] requires that the optical polarization vector $\hat{\epsilon}$ have a component perpendicular to the surface, which would occur for a rough surface. Possibly the absorption peaks observed by Fäldt and Walldén involve surface states since they disappeared at the same temperature as the surface plasmon peak (which also requires a rough surface).

The CDW anomaly in Na has a 1.2-eV threshold [16]. Not only does it persist into the liquid metal (as in K), where the metal–vacuum interface is smooth, but is largest there. The influence of slight amounts of water vapor on the intensity of the Mayer–El Naby peak, but not on its spectral shape or location (unlike the various peaks of Fäldt and Walldén; see Figure 43.3), can be explained by the influence of KOH layers on the crystallographic orientation at the metal surface [24], which alters θ in Equation (43.2).

The existence of this new absorption mechanism indicates that one must consider a Schrödinger equation having a potential

$$V(\mathbf{r}) = 2\alpha\cos(\mathbf{Q}\cdot\mathbf{r}) + V_{\text{lat}}(\mathbf{r})\,, \tag{43.3}$$

i. e., two incommensurate periodic potentials: a CDW potential and a lattice pseudopotential:

$$V_{\text{lat}} = 2\beta\cos(\mathbf{G}\cdot\mathbf{r})\,. \tag{43.4}$$

$|\mathbf{Q}| \simeq 1.36{-}1.38(2\pi/a)$ and $|\mathbf{G}| = 1.414(2\pi/a)$; a is the lattice constant and $|\mathbf{G}|$ is the magnitude of the (110) reciprocal-lattice vector. Solution of the Schrödinger equation having

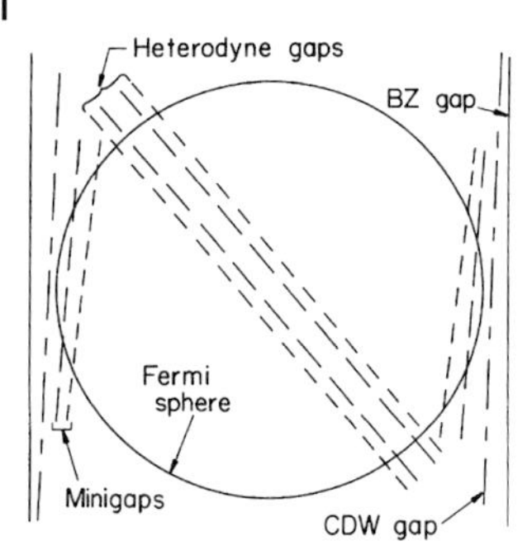

Figure 43.4 Free-electron Fermi sphere and various energy gaps caused by a CDW. Only two (of the 12) Brillouin-zone gaps, solid vertical lines, are shown.

Equation (43.3) for the potential leads to three main families of energy gaps which we name as follows (in accordance with their wave-vector periodicities, $\mathbf{q}$).

(a) For minigaps:

$$\mathbf{q} = (n+1)\mathbf{Q} - n\mathbf{G} \,. \tag{43.5}$$

(b) For heterodyne gaps:

$$\mathbf{q} = n(\mathbf{G} - \mathbf{Q}) \,. \tag{43.6}$$

(c) For second-zone minigaps:

$$\mathbf{q} = (n+1)\mathbf{G} - n\mathbf{Q} \,. \tag{43.7}$$

There remain, of course, the energy gaps at the Brillouin-zone faces perpendicular to $\mathbf{G}$, and the CDW gaps shown in Figure 43.2. For each type, $n = 1, 2, 3, \ldots$ Only the mini-gaps and the heterodyne gaps truncate the Fermi sphere, as shown in Figure 43.4.

43.2 Minigap Absorption

We shall now calculate the optical-absorption coefficient associated with the first three mini-gaps as if each in turn were created by a potential of the form

$$V(\mathbf{r}) = \Delta \cos(\mathbf{q} \cdot \mathbf{r}) \,, \quad |q| < 2k_{\mathrm{F}} \,, \tag{43.8}$$

see Figure 43.5. Δ is the minigap and q is the minigap wave vector. It turns out that independent treatment is not a good approximation for the heterodyne gaps and, therefore, we have left them for a future analysis.

In light of Figure 43.5(b), we can see qualitatively that the optical absorption is caused by vertical transitions from the lower to the upper band, and that the threshold energy is $\hbar\omega = \Delta$. Note that both bands are occupied at $k_\perp = 0$ (near the gap) so no absorption can occur there. However, as we move parallel to the energy-gap plane the two branches of $E(\mathbf{k})$ are parallel and separated by Δ. Hence, absorption can occur on the portion of the plane shown shaded in Figure 43.5(b). For simplicity we neglect any deformation of the Fermi surface caused by the potential in Equation (43.8). (It can easily be included, and we will comment later on what the effect would be.)

The wave function $\Phi_{k'}$ of an electron, in the neighborhood of the gap at A, Figure 43.6, below the gap, is

$$\Phi_{k'} = \sin\gamma\, e^{i\mathbf{k}'\cdot\mathbf{r}} - \cos\gamma\, e^{i(\mathbf{k}'+\mathbf{q})\cdot\mathbf{r}} \,. \tag{43.9}$$

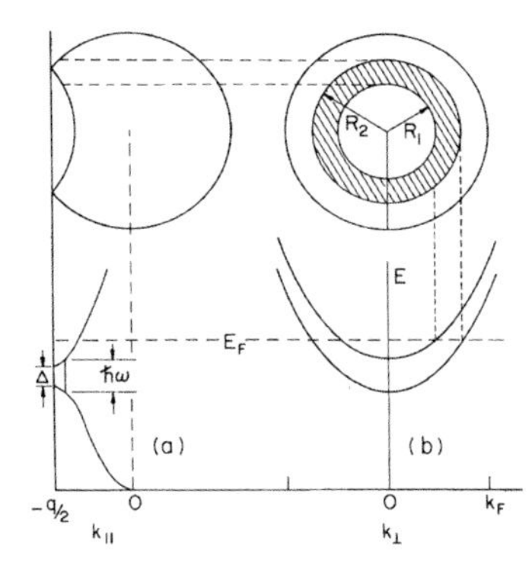

Figure 43.5 Portions of the Fermi sphere responsible for optical absorption caused by a minigap of magnitude Δ. (a) is a side view and (b) is a front view (along a line parallel to $\mathbf{q}$). Relevants plots of $E(\mathbf{k})$ are shown.

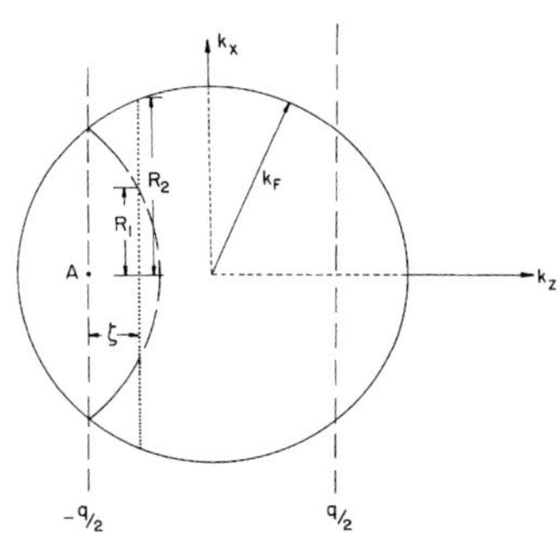

Figure 43.6 Schematic view of the Fermi sphere with parameters employed in the text to derive the minigap absorption from states along the dotted vertical line.

The corresponding state above the gap, which is connected with $\Phi_{k'}$, in an absorption event, is,

$$\Phi_k = \cos\gamma\, e^{i\mathbf{k}\cdot\mathbf{r}} + \sin\gamma\, e^{i(\mathbf{k}+\mathbf{q})\cdot\mathbf{r}} . \tag{43.10}$$

The coefficients obey the relation [22]

$$\sin 2\gamma = \frac{\Delta}{E_+ - E_-} \equiv \frac{\Delta}{W} , \tag{43.11}$$

with

$$E_\pm = \epsilon_k + \mu\zeta \pm \left(\mu^2\zeta^2 + \frac{\Delta^2}{4}\right)^{1/2} , \tag{43.12}$$

where $\epsilon_k = \hbar^2 k^2/2m$, $\mu \equiv \hbar^2 q/2m$, ζ is the perpendicular distance in $\mathbf{k}$ space from the gap at A (Figure 43.6), and $W = E_+ - E_-$ is the vertical transition energy.

The macroscopic photon field in a metal can be described by a vector potential

$$\mathbf{A}(z) = \hat{\epsilon}\, e^{-\kappa\omega z/c} \cos\left(n\frac{\omega}{c}z - \omega t\right) , \tag{43.13}$$

where $\hat{\epsilon}$ is the unit polarization vector of the electric field and n and κ are the optical constants. The interaction Hamiltonian of an electron with this potential is

$$H' = \frac{e}{mc}\mathbf{A}\cdot\mathbf{p} . \tag{43.14}$$

The strategy of the calculation is to calculate the flow of energy at $z = 0^+$ and equate it to the transition rate times $\hbar\omega$ caused by the perturbation H'. The flow of energy at $z = 0^+$ is, from Poynting's vector,

$$P = \frac{c}{4\pi}\langle|\mathbf{E}\times\mathbf{H}|\rangle = \frac{n\omega^2}{8\pi c}\,, \tag{43.15}$$

where the $\langle\rangle$ indicates time averaging. The last equality follows from Equation (43.13), with $\hat{\epsilon}$ along $\hat{\mathbf{s}}$. On the other hand, the transition rate caused by the perturbation is, with $W \equiv \hbar\omega$,

$$\frac{P}{W} = 4\sum_{\mathbf{k}',\mathbf{k}}\left(\frac{2\pi}{\hbar}\right)\langle|V_{\mathbf{k}',\mathbf{k}}|^2\rangle\delta(E_{\mathbf{k}} - E_{\mathbf{k}'} - \hbar\omega)\,, \tag{43.16}$$

where the factor 4 takes account of the spin degeneracy and the two gaps at opposite sides of the Fermi surface. From Equations (43.9), (43.10), (43.13), and (43.14) one finds (considering only the contribution from absorption)

$$\langle|V_{\mathbf{k}',\mathbf{k}}|^2\rangle = \left(\frac{\hbar e\hat{\epsilon}\cdot\mathbf{q}\sin\gamma\cos\gamma}{2mc}\right)\left[\left(k_z' - k_z + n\frac{\omega}{c}\right)^2 + \left(\kappa\frac{\omega}{c}\right)^2\right]^{-1}. \tag{43.17}$$

The $\mathbf{k}$ components transverse to z are conserved. Changing the k sum in Equation (43.16) to an integral (and using properties of the δ function) we obtain

$$\frac{P}{W} = 4\left(\frac{2\pi}{\hbar}\right)\sum_{k_z'}\frac{1}{(2\pi)^3}\int|V_{k_\omega',k_z}|^2\frac{dS}{|\nabla_{\mathbf{k}}(E_{\mathbf{k}} - E_{\mathbf{k}'})|}\,, \tag{43.18}$$

where the integration is over a surface of constant transition energy, $E_{\mathbf{k}} - E_{\mathbf{k}'}$, and for which $\mathbf{k}$ is occupied and $\mathbf{k}'$ is empty. The matrix element will not change as k is swept over this surface, so we take it outside the integral:

$$\frac{P}{W} = 4\left(\frac{2\pi}{\hbar}\right)\sum_{k_z'}|V_{\mathbf{k}',k_z}|^2\frac{1}{(2\pi)^3}\int\frac{dS}{|\nabla_{\mathbf{k}}(E_{\mathbf{k}} - E_{\mathbf{k}'})|}\,. \tag{43.19}$$

The joint density of states (dN/dW) for unit volume is

$$\frac{dN}{dW} = \frac{1}{(2\pi)^3}\int\frac{dS}{|\nabla_{\mathbf{k}}(E_{\mathbf{k}} - E_{\mathbf{k}'})|}\,. \tag{43.20}$$

Hence,

$$\frac{P}{W} = 4\left(\frac{2\pi}{\hbar}\right)\frac{dN}{dW}\sum_{\mathbf{k}_z'}|V_{\mathbf{k}_\omega',\mathbf{k}_z}|^2\,. \tag{43.21}$$

We evaluate the summation over k_z' by converting it to an integral:

$$\frac{P}{W} = 4\left(\frac{2\pi}{\hbar}\right)\frac{dN}{dW}\frac{1}{2\pi}\int_{-\infty}^{\infty}|V_{\mathbf{k}_\omega',\mathbf{k}_z}|^2 dk_z'\,. \tag{43.22}$$

Since Equation (43.17) is a Lorentzian, the integration is trivial We obtain

$$P = \frac{4\pi c}{\kappa}\frac{dN}{dW}\left(\frac{\hbar eq\Delta\cos\phi}{4mcW}\right)^2, \tag{43.23}$$

with ϕ being the angle between $\hat{\epsilon}$ and $\mathbf{q}$. [We have used Equation (43.11) to eliminate $\sin\gamma$ and $\cos\gamma$.] From Equations (43.15) and (43.23) we find the theoretical absorption coefficient:

$$\frac{2n\kappa}{\lambda} = \frac{2\pi\hbar^3}{cW^3}\left(\frac{e}{m}q\Delta\cos\phi\right)^2\frac{dN}{dW}\,. \tag{43.24}$$

The density of states will be determined below.

43.3 Results for K and Na

In order to calculate the optical-absorption spectrum caused by transitions across the minigaps we need to know their sizes [17]. This requires knowledge of the CDW gap, 2α, and the (110) pseudopotential, β. Unfortunately, the value of β is still an unanswered question. Estimates vary by up to a factor of 3. For example, analyses [29, 30] of Fermi-surface anisotropy (from de Haas–van Alphen data) which assume local pseudopotentials lead to values at the higher limit. Much lower values are obtained when nonlocal pseudopotentials [31] are employed. The ranges for the magnitudes of β are

$$\beta_K = 0.07\text{–}0.2\,\text{eV}\,, \tag{43.25a}$$

$$\beta_{Na} = 0.12\text{–}0.23\,\text{eV}\,. \tag{43.25b}$$

The lower values create a problem with regard to the intensity of the ordinary (Wilson–Butcher) interband optical absorption, since this is proportional to β^2. Serious discrepancy with experiment results unless enhancement of the matrix element by the collective effects of exchange and correlation are included [32].

The CDW potential 2α can be estimated from the threshold of the Mayer–El Naby optical anomaly [22]. As mentioned in the Introduction, the observed threshold is 0.62 and 1.2 eV for K and Na, respectively.

As shown in Figure 43.4, there is a small angle θ between **Q** and **G** [17, 33].[2)] This small tilt leads to several problems in calculating the sizes of the various minigaps. The first is that the gaps no longer lie on planes perpendicular to **q** (and which pass through $\pm\mathbf{q}/2$, see Figure 43.6). Instead the gaps define curved surfaces in **k** space which are close to the planes described. To find such a surface, consider a coordinate system with $\hat{\mathbf{z}}$ parallel to **q** and $\hat{\mathbf{x}}$ in the plane containing **Q**, **G**, and **q**. For a fixed k_x, plot the optical transition energy versus k_z. The minimum transition energy is the value of the minigap (for that k_x), and the location k_z of the energy-gap surface is at the value of k_z having that minimum. We found that the gaps vary quadratically with k_x, with minima near $k_x = 0$. The values obtained for the first three minigaps are given in Table 43.1 where, for completeness, we have also included the heterodyne gaps.

In order to determine the joint density of states, allow (as a first approximation) **Q** to be parallel to **G**. For this situation, the joint density of states is obtained by calculating the shaded area in Figure 43.5(b) for a small disc of width $d\zeta$ at a distance ζ from point A (see Figure 43.6), i. e. (R_1 and R_2 are found by simple geometry from Figure 43.6),

$$\frac{dN}{dW} = \frac{1}{(2\pi)^3}\pi\left(R_2^2 - R_1^2\right)\frac{d\zeta}{dW}\,, \tag{43.26}$$

if $0 \le \zeta < (k_F - q/2)$, with $2k_F = 1.24(2\pi/a)$; for $\zeta \ge (k_F - q/2)$, $R_1 = 0$. Therefore (for **Q** parallel to **G**) we obtain

$$\frac{2n\kappa}{\lambda} = \frac{e^2\Delta^2 q}{2\pi c\hbar W^2}\cos^2\phi \quad \text{if} \quad 0 \le \zeta < \left(k_F - \frac{q}{2}\right) \tag{43.27}$$

and

$$\frac{2n\kappa}{\lambda} = \frac{e^2\Delta^2 q}{2\pi c\hbar W^2}\frac{\left(\hbar^2 R_2^2/2m\right)}{(W^2 - \Delta^2)^{1/2}}\cos^2\phi \quad \text{if} \quad \zeta \ge \left(k_F - \frac{q}{2}\right). \tag{43.28}$$

2) For $|\mathbf{Q}| \simeq 1.36 \to 1.38(2\pi/a)$, $\theta \simeq 2.6° \to 1.6°$. See [17].

Table 43.1 Minigaps and heterodyne gaps obtained for several values of **Q** for K and Na. All gaps are in meV. $|\mathbf{Q}|$ is in $2\pi/a$ units. ($\beta \equiv |V_{110}|$.)

$\|\mathbf{Q}\|$	First minigap	Second minigap	Third minigap	First heterodyne	Second heterodyne	Third heterodyne
			Potassium, $\beta = 0.20$ eV			
1.36	97	27	3	27	9	1
1.37	106	43	9	28	12	2
1.38	107	62	25	28	16	5
			Potassium, $\beta = 0.07$ eV			
1.36	33	5.9	0.10	8.4	0.94	0.03
1.37	38	7.1	0.26	8.4	1.40	0.06
1.38	44	11	0.91	8.4	2.3	0.19
			Sodium, $\beta = 0.23$ eV			
1.37	138	57	11	38	14	2
1.38	135	79	31	38	20	5
1.39	116	87	63	37	25	13
			Sodium, $\beta = 0.12$ eV			
1.37	75	18	1.5	18	4	0.25
1.38	80	29	4.5	18	6.1	0.71
1.39	78	46	17	17	9.2	2.6

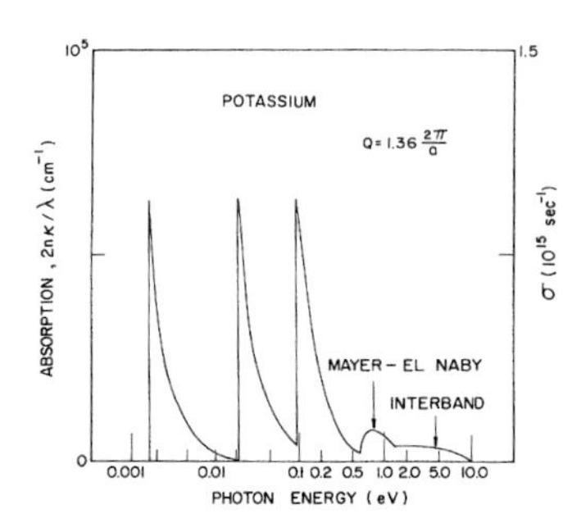

Figure 43.7 Optical-absorption spectrum of potassium for $|\mathbf{Q}| = 1.36(2\pi/a)$ and **Q** parallel to **G**. ($2n\kappa/\lambda = 2\sigma/c$.) The zone boundary energy gap 2β has been taken to be 0.4 eV, and the Drude absorption has been ignored. For room temperature the Drude absorption at 0.1 eV would be 4 times larger than the peak from the minigap at 0.1 eV. For $\beta = 0.01$ eV the first minigap threshold occurs at 0.04 eV (instead of 0.1).

This absorption spectrum is shown in Figures 43.7 and 43.8 for K. We have included the first three minigaps and have used alternative values for **Q**. We have taken $\langle \cos^2 \phi \rangle = \frac{1}{3}$. In Figure 43.8 we have included the contribution from intraband transitions (Drude background) at 4.2 K, assuming a sample with a residual resistivity ratio of $\simeq 6000$. At room temperature the Drude background is too high for these new absorption edges to be observed. This is the reason why any search for such new edges must be done at low temperatures.

Finally we turn to the evaluation of the joint density of states when the small tilt between **Q** and **G** is considered. In this situation, because of the variation of the gap with k_x, we fix k_x, as shown in Figure 43.9, and consider the $k_y - k_z$ plane (Δk_x thick) corresponding to that fixed k_x. The joint density of states in a small "disc" of width $d\zeta$, a distance ζ from point B, Figure 43.9, is given by

$$\frac{dN}{dW} = \frac{1}{(2\pi)^3}(2R_G - 2R_S)\Delta k_x \frac{d\zeta}{dW} . \tag{43.29}$$

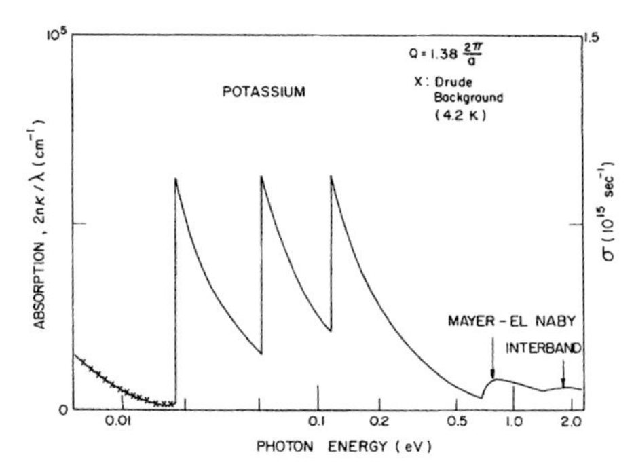

Figure 43.8 Drude background (X) at 4.2 K and the total absorption coefficient (or conductivity) of potassium for $|\mathbf{Q}| = 1.38(2\pi/a)$ and $\mathbf{Q}$ parallel to $\mathbf{G}$. The zone boundary energy gap was assumed to be $2\beta = 0.4$ eV.

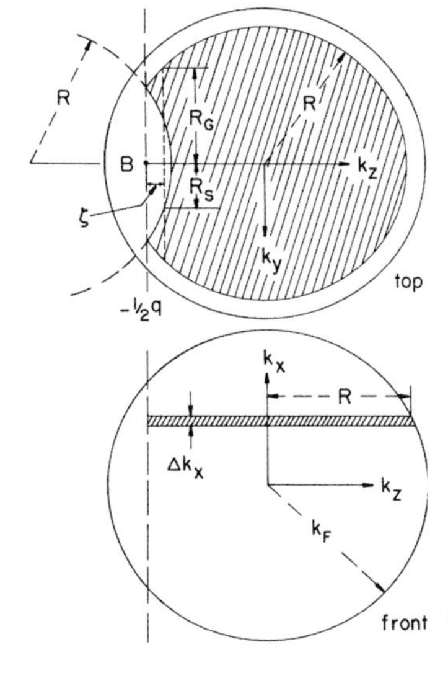

Figure 43.9 Projections of the Fermi sphere used to calculate the absorption when $\mathbf{Q}$ is not parallel to $\mathbf{G}$. Each disc (of thickness Δk_x, shown in the front view) has a slightly different energy gap at $k_z = -q/2$. Only the shaded region of the top view has unoccupied final states.

Therefore,

$$\frac{2n\kappa}{\lambda} = \sum_{k_x} \frac{2\pi\hbar^3}{cW^3} \left(\frac{e}{m} q\Delta \cos\phi \right)^2 \frac{dN}{dW} \,. \tag{43.30}$$

The important dependence on $\cos^2\phi$ should be noted. It implies that the minigap absorption is also uniaxial. The spectrum is obtained by numerical evaluation of the sum over k_x. Examples for the first minigap in K are shown in Figure 43.10 for several values of $|\mathbf{Q}|$. The major difference from the results obtained in Figures 43.7 and 43.8, with $\mathbf{Q}$ parallel to $\mathbf{G}$, is that the sudden rise in absorption at threshold is less abrupt. Had we included the slight distortion of the Fermi surface near the energy gaps, the rise in absorption at threshold would be somewhat steeper.

The absorption profiles of all the minigaps are very much alike. The peaks are about an order of magnitude larger than the interband absorption or the main CDW anomaly. We emphasize once again that the threshold values given in Table I are subject to uncertainty because the pseudopotentials of the periodic lattice are not known from a direct experiment.

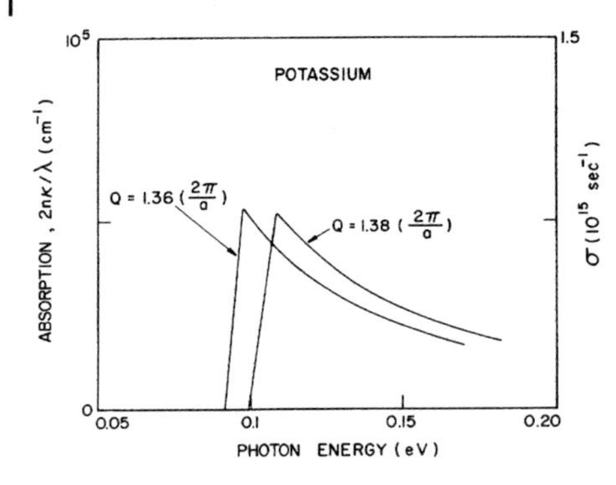

Figure 43.10 Optical-absorption spectrum of the first minigap of potassium showing the gradual rise at threshold caused by the small angle between **Q** and **G**. The zone boundary energy gap 2β was assumed to be 0.4 eV. The threshold is reduced to $\simeq$ 0.04 eV if $2\beta = 0.14$ eV.

43.4 Conclusions

We have found that new optical-absorption edges in the far-infrared spectrum of a metal having a CDW ground state arise from transitions across the minigaps. The absorption peaks associated with the first three minigaps have been calculated for K and Na and found to be about an order of magnitude larger than both the interband and the main CDW peak. However, they are much smaller than the room-temperature Drude background. Therefore any search for such edges must be carried out at low temperatures, where the Drude background is significantly reduced, and in samples for which the orientation of Q allows the observation of the Mayer–El Naby anomaly. The minigap absorption, as well as the main CDW absorption, is *uniaxial*.

Acknowledgements

This research was supported by the National Science Foundation and for one of us (F.E.F.) by the Energy and Mine Department, Venezuela.

References

1 Duncan, R.W. and Duncan, R.C. (1913) *Phys. Rev.*, **1**, 294.
2 Ives, H.E. and Briggs, H.B. (1936) *J. Opt. Soc. Am.*, **26**, 238.
3 Ives, H.E. and Briggs, H.B. (1937) *J. Opt. Soc. Am.*, **27**, 181.
4 Butcher, P.N. (1951) *Proc. Phys. Soc. London A*, **64**, 765.
5 Wilson, A.H. (1958) *The Theory of Metals*, Cambridge University Press, Cambridge.
6 Cohen, M.H. (1958) *Philos. Mag.*, **3**, 762.
7 Hodgson, J.N. (1963) *J. Phys. Chem. Solids*, **24**, 1213.
8 Hodgson, J.N. (1963) *Phys. Lett.*, **7**, 300.
9 Mayer, H. and El Naby, M.H. (1963) *Z. Phys.*, **174**, 269.
10 Mayer, H. and El Naby, M.H. (1963) *Z. Phys.*, **174**, 280.
11 Mayer, H. and El Naby, M.H. (1963) *Z. Phys.*, **174**, 289.
12 Mayer, H. and Hietel, B. Hietel (1966) in *Proceedings of the International Colloquium on Optical Properties and Electronic Structure of Metals and Alloys*, (ed. F. Abeles), Paris, 1965, North-Holland, Amsterdam, p. 47.
13 Althoff, R. and Hertz, J.H. (1967) *Infrared Phys.*, **7**, 11.
14 Smith, N.V. (1969) *Phys. Rev.*, **183**, 634.
15 Palmer, R.E. and Schnatterly, S.E. (1971) *Phys. Rev. B*, **4**, 2329.
16 Hietel, B. and Mayer, H. (1973) *Z. Phys.*, **264**, 21.

17 Fragachán, F.E. and Overhauser, A.W. (1984) *Phys. Rev. B*, **29**, 2912.

18 Overhauser, A.W. (1960) *Phys. Rev. Lett.*, **4**, 462.

19 Overhauser, A.W. (1962) *Phys. Rev.*, **128**, 1437.

20 Overhauser, A.W. (1968) *Phys. Rev.*, **167**, 691.

21 Overhauser, A.W. (1978) *Adv. Phys.*, **27**, 343.

22 Overhauser, A.W. (1964) *Phys. Rev. Lett.*, **13**, 190.

23 Jaklevic, R.C. and Lambe, J. (1975) *Phys. Rev. B*, **12**, 4146.

24 Overhauser, A.W. and Butler, N.R. (1976) *Phys. Rev. B*, **14**, 3371.

25 Hietel, B. (private communication).

26 Harms, P. (1972) Dissertation, Technische Universität Clausthal (unpublished); his data are reproduced in [24].

27 Fäldt, Å. and Walldén, L. (1980) *J. Phys. C*, **13**, 6429.

28 Taut, M. (1982) *J. Phys. F*, **12**, 2019.

29 Overhauser, A.W. (unpublished).

30 Ashcroft, N. (1965) *Phys. Rev. A*, **140**, 935.

31 Lee, M.J.G. (1971) in *Computational Methods in Band Theory*, (eds. P.M. Marcus, J.F. Janak, and A.R. Williams), Plenum, New York.

32 Overhauser, A.W. (1967) *Phys. Rev.*, **156**, 844.

33 Giuliani, G.F. and Overhauser, A.W. (1979) *Phys. Rev. B*, **20**, 1328.

Reprint 44 Dynamic *M*-Shell Effects in the Ultraviolet Absorption Spectrum of Metallic Potassium[1)]

A.C. Tselis and A.W. Overhauser**

* Department of Physics, Purdue University, 525, Northwestern Avenue, West Lafayette, Indiana 47907, USA

Received 20 November 1984

The filled M shell of potassium, set into oscillation by the electric field of a photon, leads to inter-band transitions that explain a large absorption peak at 8 eV. Interference between this collective perturbation and the ordinary $\mathbf{A} \cdot \mathbf{p}$ coupling (between a photon and conduction electron) shows that the sign of the V_{110} pseudopotential is negative.

Twelve years ago Whang, Arakawa, and Callcott [1] discovered a large ultraviolet absorption near 8 eV in metallic potassium. Their data are shown in Figure 44.1, which includes the Drude absorption (below 1 eV) and the Wilson-Butcher interband peak at 2 eV. Explanation of the ultraviolet peak, which also occurs in Rb and Cs [1], has remained elusive. Hermanson [2] showed that it could not be attributed to plasmons; and an attempt [3] with a limited-basis-set, tight-binding model for the conduction band (instead of a nearly free-electron one) led only to a weak and highly structured absorption.

We show here that the absorption in K near 8 eV arises from a collective contribution to the ordinary interband matrix element. It is caused by the interaction of a conduction electron with the dynamic oscillation of the eight M-shell electrons. A nearly free-electron, pseudopotential approximation is the method we employ.

The periodic pseudopotential acting on a conduction electron can be divided into two parts:

$$V(\mathbf{r}) = V_M(\mathbf{r}) + V_I(\mathbf{r}) \,. \tag{44.1}$$

V_I is the inner ion-core potential, caused by the nuclei and their ten tightly bound K- and L-shell electrons. We shall take V_I to be the Coulomb potential of a static bcc array of point ions, each having charge $9e$. V_M is the pseudopotential of the M shells. In the electric field of a photon, described by a vector potential $A\hat{\mathbf{x}} \cos \omega t$, the eight M-shell electrons (on each ion) undergo a coherent oscillation of amplitude a. Accordingly, the static V_M must be replaced by a dynamic one:

$$V_M(\mathbf{r}) \to V_M(\mathbf{r} - a\hat{\mathbf{x}} \sin \omega t) \,. \tag{44.2}$$

The shell model of Dick and Overhauser [4] can be used (with Newton's second law) to calculate the amplitude of the oscillation, $x = a \sin \omega t$:

$$8m\ddot{x} = -8m\omega_M^2 x - \left(\frac{8e\omega}{c}\right) A \sin \omega t \,. \tag{44.3}$$

$8m\omega_M^2$ is the shell-model spring constant; its value will be discussed below. The amplitude which satisfies Equation (44.3) is

$$a = \omega e A / mc\left(\omega^2 - \omega_M^2\right) \,. \tag{44.4}$$

The shell model should be applicable as long as ω is small compared to ω_M.

1) Phys. Rev. Lett. 54, 12 (1985).

Anomalous Effects in Simple Metals. Albert Overhauser

ISBN: 978-3-527-40859-7

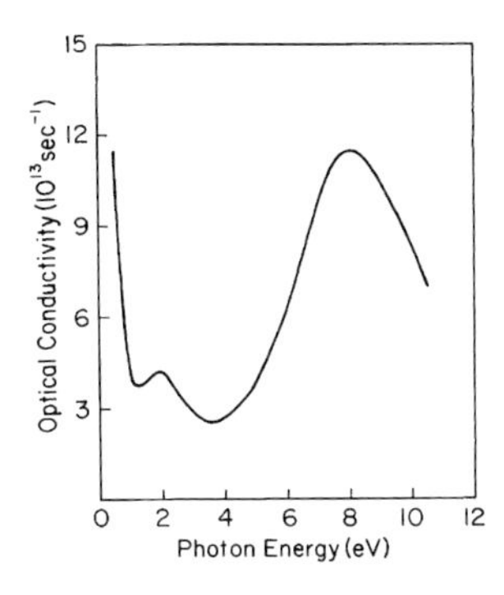

Figure 44.1 Experimental, room-temperature, absorption spectrum of potassium. The rise below 1 eV is the Drude absorption. The Wilson–Butcher interband peak is at 2 eV. Data are from [1].

In the nearly free-electron model, optical absorption is attributed to transitions (in the extended-zone scheme) from $\mathbf{k}$ to $\mathbf{k} + \mathbf{G}$, as shown in Figure 44.2. Consider one of the [110] reciprocal-lattice vectors, $\mathbf{G}$, and take it to be along $\hat{\mathbf{x}}$. The initial-state wave function, corrected to first order in the pseudopotential, is

$$\psi_{\mathbf{k}} \cong e^{i\mathbf{k}\cdot\mathbf{r}} - [V_{110}/W(\mathbf{k})]e^{i(\mathbf{k}+\mathbf{G})\cdot\mathbf{r}} , \tag{44.5}$$

where V_{110} is the $\mathbf{G}$ Fourier coefficient of $V(\mathbf{r})$, Equation (44.1). $W(\mathbf{k})$ is the transition energy:

$$W(\mathbf{k}) = E(\mathbf{k} + \mathbf{G}) - E(\mathbf{k}) . \tag{44.6}$$

The final-state wave function is

$$\psi_{\mathbf{k}+\mathbf{G}} \simeq e^{i(\mathbf{k}+\mathbf{G})\cdot\mathbf{r}} + [V_{110}/W(\mathbf{k})]e^{i\mathbf{k}\cdot\mathbf{r}} . \tag{44.7}$$

The time-dependent perturbation which connects (44.5) to (44.7) arises from the $\mathbf{A}\cdot\mathbf{p}$ coupling with the photon and also from the oscillatory part, ΔV_M, of the M-shell potential (44.2), having Fourier periodicity $\mathbf{G}$:

$$\Delta V_M = V_{MG}\, e^{i\mathbf{G}\cdot\mathbf{r}}(-iaG \sin \omega t) . \tag{44.8}$$

This expression is the leading term in an expansion of $\exp[i\mathbf{G} \cdot (\mathbf{r} - a\hat{\mathbf{x}} \sin \omega t)]$ in powers of a. Higher Fourier components of V_M affect absorption only above 20 eV. Optical transitions arise from (the positive-frequency part of) the total perturbation,

$$H' = (eA/mc)p_x \cos \omega t - iaG\, V_{MG}\, e^{i\mathbf{G}\cdot\mathbf{r}} \sin \omega t . \tag{44.9}$$

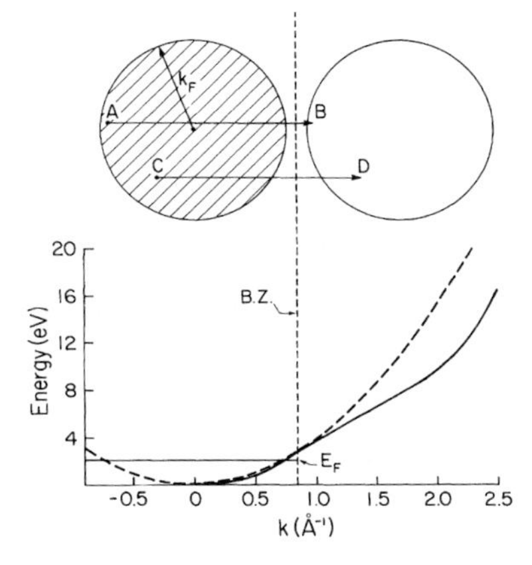

Figure 44.2 Interband transitions from occupied states of the Fermi sphere to empty, excited states. The wave-vector change ($A \to B$ or $C \to D$) is $\mathbf{G}$, a [110] reciprocal-lattice vector. The solid curve is the electronic $E(k)$ given by (44.18); the dashed curve is a free-electron parabola. A Brillouin-zone boundary is shown.

The first term is the only one that is usually considered. Collective effects, analogous to the second term, have been discussed for atomic transitions [5], in a time-dependent Hartree–Fock formalism, and usually lead to only small corrections.

For alkali metals, the first term of (44.9), which causes the Wilson–Butcher interband peak [6], is small because p_x connects the large term of (44.5) with the small term of (44.7), and vice versa. Because of the factor $\exp(i\mathbf{G}\cdot\mathbf{r})$, the M-shell term of H' connects the large terms of both (44.5) and (44.7). That is one reason why the M-shell effect is relatively more important here. The (positive-frequency) matrix element of H' between initial and final states is

$$\langle \mathbf{k}+\mathbf{G}|H'|k\rangle = \frac{-e\hbar G A}{2mc W}\left[V_{110} + \frac{W^2}{(\hbar^2\omega_M) - W^2} V_{MG}\right] , \tag{44.10}$$

when we have used Equation (44.4) and $W = \hbar\omega$ is the transition energy. Note that both terms (in square brackets) are real, so that there will be an interference phenomenon between the two contributions. This effect will permit an experimental determination of the sign of V_{110}, the potential which causes energy gaps at the Brillouin-zone boundary.

We now turn to a discussion of the Fourier coefficients, V_{110} and V_{MG}. The Fourier-transform version of Equation (44.1) for the [110] reciprocal-lattice vector $\mathbf{G}$ is

$$V_{110} = V_{MG} + V_{IG} \; . \tag{44.11}$$

The magnitude of V_{110}, estimated by an analysis of de Haas–van Alphen data [7], is

$$V_{110} \approx -0.2\,\text{eV} \; . \tag{44.12}$$

We have inserted the negative sign as a consequence of our conclusion below. Since $\mathbf{G}$ is here the smallest reciprocal-lattice vector, we may take the inner-ion-core term (charge $= 9e$) to be (Ω is the atomic volume)

$$V_{IG} \approx -36\pi e^2/\Omega G^2 = -7.9\,\text{eV} , \tag{44.13}$$

since the form factor of the ten inner-core electrons is ≈ 10 for wave vector $\mathbf{G}$. It follows from Equations (44.11) to (44.13) that

$$V_{MG} \approx 7.7\,\text{eV} \; . \tag{44.14}$$

Note that V_{MG} is 40 times larger than V_{110}. From Equation (44.10) it is easy to understand why the oscillator strength of the M-shell-mediated peak is 2 orders of magnitude larger than the Wilson–Butcher one.

The only parameter of Equation (44.10) that remains to be determined is ω_M which, from Equation (44.3), is the resonant frequency of the M-shell oscillation. For a free K^+ ion, the shell-model spring constant [4] can be found by fitting the K^+ polarizability [8], $\alpha = 0.9 \times 10^{-24}\,\text{cm}^2$. This approach leads to a free-ion, M-shell resonance at $\hbar\omega = 31\,\text{eV}$. However, in the metal, this resonance will be shifted to a smaller frequency as a result of interaction of each M shell with the ionic potentials of near neighbors. (There is also a small reduction from the dielectric screening of the conduction electrons.) A theoretical estimate of the shift is $\sim -7\,\text{eV}$ [9]. However, it is not necessary to depend on such calculations, since the M-shell resonant frequency has been measured for K by synchrotron-radiation absorption [10] and is

$$\hbar\omega_M \approx 26\,\text{eV} \; . \tag{44.15}$$

We emphasize here that all of the parameters appearing in the interband matrix element, Equation (44.10), are determined from independent experiments. (The vector potential A, of course, disappears when the optical absorption is evaluated.)

The optical conductivity σ can be computed if we equate the mean "Joule heating" to the golden-rule transition-rate times W, i.e.,

$$\frac{1}{2}\sigma E^2 = 8W[(2\pi/\hbar)\langle \mathbf{k}+\mathbf{G}|H'|\mathbf{k}\rangle^2 \rho(W)] , \tag{44.16}$$

where E is the peak electric field ($\omega A/c$) and $\rho(W)$ is the joint density of states (per spin). The factor 8 includes a factor 2 for spin and a factor, $12\langle\cos^2\theta\rangle_{\mathrm{av}} = 4$, to account for the twelve [110] **G**'s and their average polarization factor. (θ is the angle between $\hat{\mathbf{x}}$ and **G**.) Accordingly, the optical conductivity is

$$\sigma = \frac{8\pi e^2\hbar^3 G^2}{m^2 W^3}\left(V_{110}\frac{W^2}{(\hbar\omega_M)^2 - W^2}V_{MG}\right)^2 \rho(W) . \tag{44.17}$$

This result reduces to Butcher's formula [6] if the M shells are rigidly bound ($\omega_M = \infty$), and if $\rho(W)$is calculated from Equation (44.16) with $E = \hbar^2k^2/2m$.

Band calculations [11] show that $E(k)$ is not free-electron-like beyond the (first) Brillouin zone. Therefore we use an $E(k)$ which behaves appropriately beyond $k = G/2$. We accomplish this with the following heuristic function:

$$E = \frac{\hbar^2k^2}{2m}\frac{1+0.08u^4}{1+0.55u^2} , \tag{44.18}$$

where $u = (2k/G)-1$. The u^2 in the denominator provides a "sag" which allows $E(k)$ to "fit" the band calculations in the second zone. The u^4 in the numerator enables $E(k)$ to resume its parabolic course for larger k. The function given by Equation (44.18) is shown in Figure 44.2. The sag at $k = G$, relative to the free-electron parabola (also shown), is ~ 4 eV. Such a value is consistent with the band structure.

Since $E(k)$, given by Equation (44.18), cannot be factored into x, y, and z components, $\rho(\omega)$ must be calculated numerically. The resulting $\sigma(\omega)$ is shown in Figure 44.3. The observed Drude absorption has been added to Equation (44.17) in order to facilitate comparison with the experimental spectrum of Figure 44.1. The M-shell-mediated absorption (taken alone) is the curve having $V_{110} = 0$. Comparison of the other two curves, with $V_{110} = \pm 0.2$ eV, shows the extraordinary interference effect between the $\mathbf{A}\cdot\mathbf{p}$ interaction and the dynamic M-shell potential. It is obvious that V_{110} must be negative. Most pseudopotential form factors, $V(q)$,

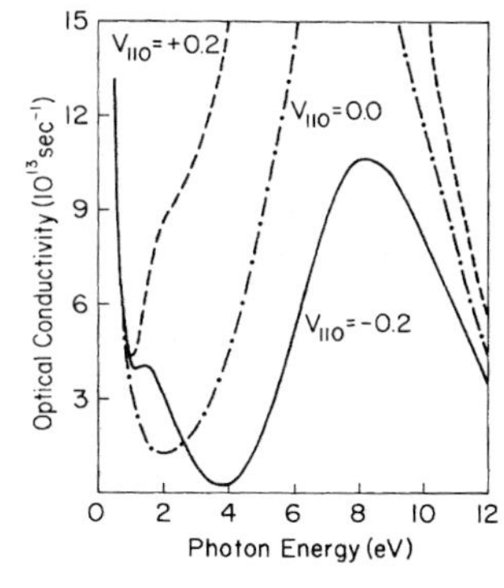

Figure 44.3 Theoretical optical-absorption spectrum of potassium computed from Equation (44.17), for three values of the pseudopotential V_{110}. The experimental Drude absorption has been added. The peak at 8 eV is caused by the M-shell-mediated photon–electron interaction.

found in the literature, though negative for small q, change sign and are positive at $q = G$. One notable exception is the nonlocal potential of Rasolt and Taylor [12], which remains negative.

The small value of σ near 4 eV (for the solid curve of Figure 44.3) is caused by exact cancellation of the two terms in the matrix element (44.10). Only the Drude tail contributes at 4 eV. This is an artifact caused by our neglect of damping. The conduction electrons (undergoing interband transitions) will exert a reaction force. A small phase shift relative to the M-shell oscillation will then prevent the interference dip near 4 eV from being so dramatic [9].

The remarkable agreement in the height and width of the 8-eV peak leaves little doubt that the dynamic M-shell mechanism is responsible. The only adjusted parameters were those in Equation (44.18), which affect somewhat the energy of the ultravolet peak. It seems likely that dynamic polarization will be important in the vacuum-ultraviolet spectra of other materials.

Acknowledgements

The authors are grateful to the National Science Foundation for support of this research.

References

1. Whang, V.S., Arakawa, E.T. and Callcott, T.A. (1972) *Phys. Rev. B*, **6**, 2109.
2. Hermanson, J. (1972) *Phys. Rev. B*, **6**, 400.
3. Ching, W.Y. and Callaway, J. (1973) *Phys. Rev. Lett.*, **30**, 441.
4. Dick Jr., B.G. and Overhauser, A.W. (1958) *Phys. Rev.*, **112**, 90.
5. Dalgarno, A. and Victor, G.A. (1966) *Proc. Roy. Soc. London A*, **291**, 291.
6. Butcher, P.N. (1951) *Proc. Phys. Soc. London A*, **64**, 765.
7. Ashcroft, N. (1965) *Phys. Rev. A*, **140**, 935. A similar analysis [Lee, M.J.G. (1971) in *Computational Methods in Band Theory*, (eds P.M. Marcus, J.F. Janak, and A.R. Williams), Plenum, New York] leads to $|V_{110}| \sim 0.07$ eV, a value which underestimates the Wilson–Butcher peak by an order of magnitude.
8. Dalgarno, A. (1962) *Adv. Phys.*, **11**, 281.
9. Tselis, A.C. and Overhauser, A.W. unpublished.
10. Sato, S., Miyahara, T., Hanyu, T., Yamaguchi, S., and Ishii, T. (1979) *J. Phys. Soc. Jpn.*, **47**, 836.
11. Ham, F.S. (1962) *Phys. Rev.*, **128**, 82.
12. Rasolt, M. and Taylor, R. (1975) *Phys. Rev. B*, **11**, 2717.

Reprint 45 Broken Symmetry in Simple Metals

*A.W. Overhauser**

* Department of Physics, Purdue University, 525, Northwestern Avenue, West Lafayette, Indiana 47907, USA

45.1 Introduction

The purpose of this lecture is to explain why isotropic metals such as Na and K have a broken symmetry of the charge density wave (CDW) type, and to review the many new phenomena that arise as a consequence. It would be logical to begin with a theoretical discussion of the many-electron effects [1, 2] which lead to CDW's in three-dimensional metals.

However, treatment of the many-body problem varies so much among experts that little conviction can result from a purely intellectual discussion. The disarray that has accumulated from theoretical studies of the spin susceptibility of an electron gas is exemplified by the compilation [3] of Figure 45.1. The influence of exchange is known exactly [4]:

$$\frac{\chi_{\mathrm{P}}}{\chi_{\mathrm{F}}} = \frac{1}{1 - r_{\mathrm{s}}/6.03} \,. \tag{45.1}$$

The enhancement of χ_{P} over the (noninteracting) Pauli value χ_{F} diverges when the Wigner–Seitz (electron separation) parameter r_{s} approaches 6.03. This is the Hartree–Fock, mean-field result. Electron correlations, which embody true many-body effects, reduce [5] the enhancement caused by exchange alone. One can see from Figure 45.1 that almost every conceivable behavior for $\chi_{\mathrm{P}}(r_{\mathrm{s}})$ has been espoused during the last thirty years.

The possibility that interacting Fermi systems can have a translationally broken symmetry was discovered in 1960. Attractive interactions lead to density waves [6] and repulsive inter-

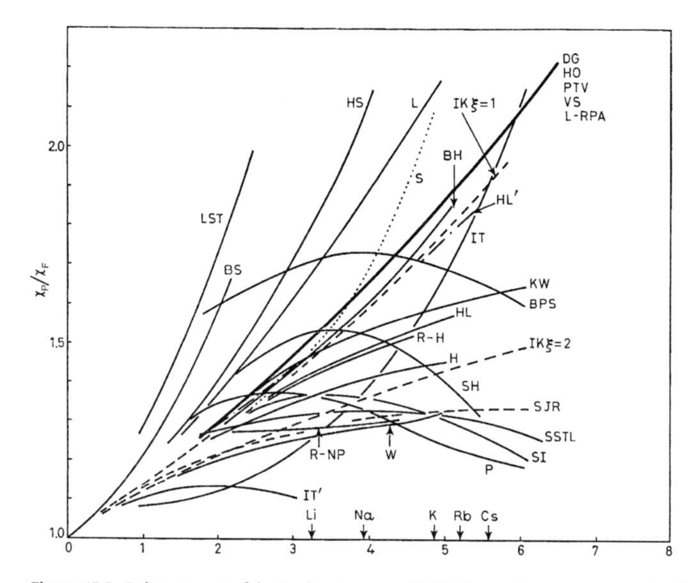

Figure 45.1 Enhancement of the Pauli spin susceptibility for an electron gas, over that for free fermions, *vs.* electron separation parameter r_{s}. Values of r_{s} for the alkali metals are indicated. The labels on the curves refer to the theorists involved and may be interpreted using [3].

Anomalous Effects in Simple Metals. Albert Overhauser

ISBN: 978-3-527-40859-7

actions to spin density waves [7]. In a metal where the positive-ion background can be easily modulated, a CDW becomes possible [2]. (The charge modulation of the conduction electrons and that of the positive ions approximately cancel.)

The phenomenon discussed here is different from a Peierls' instability [8], which results from a singularity in the response function of a one-dimensional, noninteracting Fermi sea. For a two- or three-dimensional, isotropic metal the response function of the electron density to a sinusoidal perturbation, $\cos \mathbf{q} \cdot \mathbf{r}$, is a monotonic decreasing function of q. (Many-electron effects are here ignored.) Only for the one-dimensional case is there a logarithmic singularity at $q = 2k_F$,where k_F is the Fermi wave number.

When exchange interactions are included in the theory – i.e. in the Hartree–Fock approximation – a SDW or CDW can always occur [1]. This theorem is the foundation for broken symmetry in an isotropic metal. nevertheless the role of electron correlations is important. Discussion of these, however, will be postponed to Section 45.3. The overriding significance of CDW structure in simple metals is the lesson provided on many-body effects. But first one must be sure CDW's are there.

45.2 The Evidence

Two of the alkali metals, Na and K, come close to the theoretical ideal. The crystalline potential acting on the conduction electrons is so small that the degenerate Fermi sea is free-electron-like. A common belief, supported by accurate de Haas–van Alphen measurements [9], is that the Fermi surface deviates by only one or two parts per thousand from a perfect sphere. Many other phenomena contradict this conclusion by indicating instead a multiply connected Fermi surface. This requires that the translation symmetry of the b.c.c. lattice be broken by an incommensurate CDW structure. In this section we will not review the historical development which began with the discovery of optical anomalies [10] and their explanation [11] utilizing a broken symmetry. Instead we will begin with recent discoveries in the high-field magnetoresistance of Na and K.

In order to understand the significance of magnetoresistance measurements, it is helpful to examine the consequences of a CDW on the conduction electron energy spectrum $E(\mathbf{k})$. The Brillouin zone of an alkali metal has twelve congruent faces, each perpendicular to a [11] reciprocal-lattice vector. The distance of each face from $k = 0$ is 14% larger than k_F. The energy gaps at the zone faces, caused by the ionic potential, are about 0.4 eV in K [12] and leave the Fermi surface undistorted from its spherical shape. However, the optical anomaly indicates that there is another periodic potential [10] ($\sim$ 0.6 eV) with wave vector $\mathbf{Q}$ nearly parallel to one of the [11] axes [13], for which the reciprocal-lattice vector is $\mathbf{G}$. The angle between $\mathbf{Q}$ and $\mathbf{G}$ is expected to be only a few degrees [14]. It is sufficient for most purposes to consider a Schrödinger equation having a potential term:

$$V(\mathbf{r}) = 0.4 \cos \mathbf{G} \cdot \mathbf{r} + 0.6 \cos \mathbf{Q} \cdot \mathbf{r} , \tag{45.2}$$

$|\mathbf{G}| = 1.414(2\pi/a)$ and $|\mathbf{Q}| = 1.38(2\pi/a)$. A solution of the Schrödinger equation having this potential leads to three main families of energy gaps in addition to the 0.4 eV and 0.6 eV gaps associated with the periodicities $\mathbf{G}$ and $\mathbf{Q}$. We designate these higher-order gaps together with their wave vector periodicities as follows:

$$\begin{cases} a) \text{ minigaps,} & \mathbf{K} = (n+1)\mathbf{Q} - n\mathbf{G} ; \\ b) \text{ heterodyne gaps,} & \mathbf{K} = n(\mathbf{G} - \mathbf{Q}) ; \\ c) \text{ second-zone minigaps,} & \mathbf{K} = (n+1)\mathbf{G} - n\mathbf{Q} . \end{cases} \tag{45.3}$$

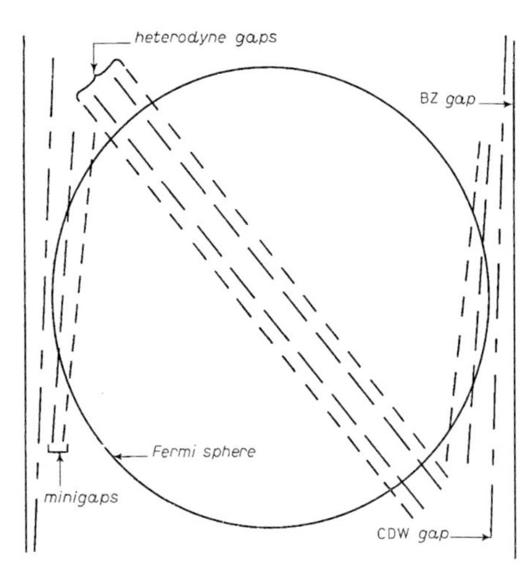

Figure 45.2 The free-electron Fermi sphere and the various energy gaps caused by a CDW. Only two (of the twelve) BZ gaps, solid vertical lines, are shown; they are separated by a reciprocal-lattice vector **G**. The two CDW gaps are separated by **Q**. The periodicities of the other gaps are given by Equation (45.3).

For each type $n = 1, 2, 3, \ldots$ Only the minigaps and the heterodyne gaps truncate the Fermi sphere, as shown Figure 45.2. The angle between the heterodyne gap periodicity and **G** is about 47° [14]. The first minigap is $\sim$ 0.1 eV, and the first heterodyne gap is $\sim$ 0.02 eV [15]. The size of these gaps decreases rapidly with increasing n.

Consider, now, a large magnetic field perpendicular to the plane of Figure 45.2. Open orbits having several directions are possible, especially when magnetic breakdown [16] effects are allowed at the small gaps. If only the first heterodyne gap and the first minigap are taken into account, five open-orbit directions are possible [17].

The "easy" crystallographic axes $(\alpha\beta\gamma)$ for **Q** are determined [14] by the anisotropy of the elastic stiffness constants. Since **Q** does not lie in a symmetry plane, there are 24 equivalent $(\alpha\beta\gamma)$ axes. Accordingly, a macroscopic sample will be subdivided (in general) into **Q** domains, analogous to magnetic domains in a ferromagnet. Such a specimen would exhibit 120 open-orbit directions in magnetotransport phenomena.

At high magnetic fields, that is when $\omega_c\tau \gg 1$ $(\omega_c \equiv eH/mc,\ \tau \equiv$ scattering time), the magnetoresistance is independent of H provided there are no open orbits [18]. On the other hand, if H is rotated through a plane perpendicular to an open orbit, a sharp magnetoresistance peak occurs. This is shown in Figure 45.3. The height of the peak increases as H^2, and its width becomes narrower as $1/H$. Fermi-surface topology can be explored by studying the angular orientations of open-orbit peaks. Inductive techniques employing single-crystal spheres are particularly suitable [19], and the theory for anisotropic resistivity tensors has been developed [20].

A generalization for heterogeneous samples – to account for Qdomain structure – requires an effective medium approximation [21]. This method has been employed to calculate the open-orbit magnetoresistance spectra [22] expected for potassium having a CDW. The theory was based on the Fermi-surface model of Figure 45.2. It was assumed (for simplicity) that all 120 open-orbit directions involve equal electron fractions. Typical results are shown in Figure 45.4. Sharp open-orbit peaks show up at $\sim$ 4 T, for which $\omega_c \sim 100$. The only parameter in the theory is the **Q** domain size, which limits the maximum length of an open-orbit trajectory. This length influences slightly the initial field at which open-orbit peaks appear.

It is particularly important to notice the horizontal dashed line at the bottom of Figure 45.4. This line is the only allowed experimental outcome for a metal having a simply connected Fermi surface. Even if an otherwise isotropic sample were filled with oriented dislocations [22],

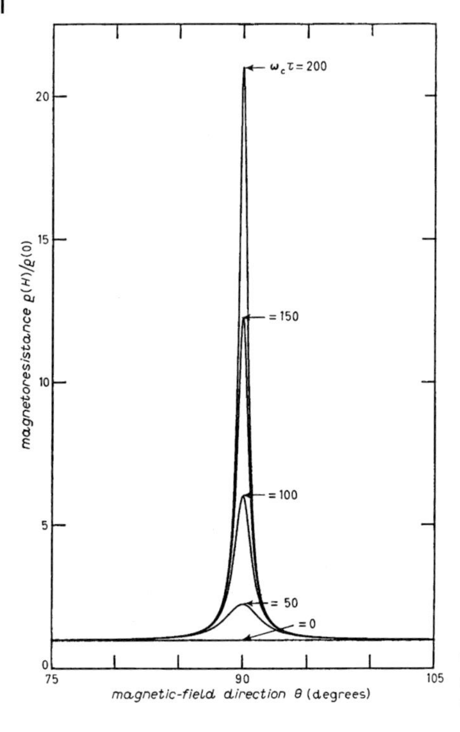

Figure 45.3 Open-orbit magnetoresistance *vs.* θ, the angle between **H** and the open-orbit direction in **k**-space. The open-orbit electron fraction was taken to be 0.001.

or if it were to have a spatial gradient in impurity concentration [23], the horizontal line would be obtained when $H > 1\,\mathrm{T}$. There can be little doubt that high-field magnetoresistance experiments provide decisive discrimination for the presence (or lack) of a CDW. Magnetoresistance is an open-orbit detector.

Coulter and Dataks [24] have carried out induced-torque experiments on Na and K spheres to fields of 8.5 T. Typical data for K are shown in Figure 45.5. The growth of open-orbit peaks, their width, their approximate number and the field where they become prominent are in excellent agreement with the calculations described above and shown in Figure 45.4. Obviously perfect coincidence cannot be expected, since **Q** domain distributions are unknown. The observed threshold field suggests that for the samples of Figure 45.5 (4 mm in diameter) the **Q** domain size was $\sim$ (0.05÷0.1) mm. These experiments have been repeated more than a hundred times on many specimens. Data more spectacular than those of Figure 45.5 have also been reported [25]. The open-orbit peaks are always observed. Many more should become resolved at fields of 24 T [22].

The puzzle remains as to why effects of CDW structure do not appear in de Haas–van Alphen data. A CDW breaks the cubic symmetry (as well as the translation symmetry), so the extremal area of closed orbits should reveal a lower symmetry. Of course, if a sample contains a million **Q** domains, randomly oriented, the expected anisotropy would be averaged out. But one might anticipate a mishmash of slightly different frequencies. Possibly the **Q** domain size in 1 mm samples (that are typically used) is $\sim 10^{-3}$ mm. In such a case there would be $\sim 10^4$ domains in the coherence volume of a Landau level [26]. Many-electron effects might then "pull" the mishmash into a single frequency.[1]

1) Exchange narrowing in ferromagnetic resonance is an example of such a phenomenon.

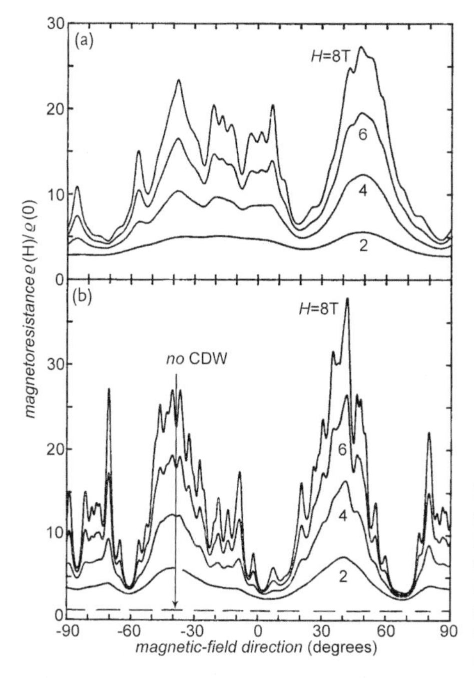

Figure 45.4 Magnetoresistance spectrum of potassium for several values of $|\mathbf{H}|$ computed by Huberman and Overhauser [22]. In (a), $\mathbf{H}$ lies in a (211) plane and the $\mathbf{Q}$ domain size is 0.05 mm. In (b), $\mathbf{H}$ lies in a (321) plane, and the $\mathbf{Q}$ domain size is 0.12 mm. The horizontal dashed line is the only allowed result for an alkali metal without a CDW.

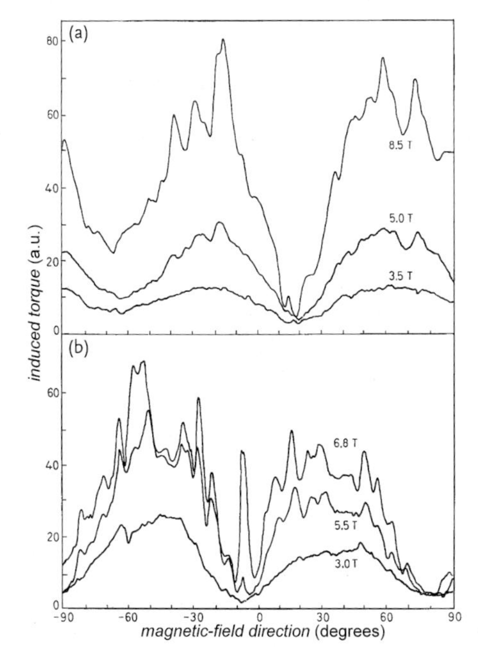

Figure 45.5 Magnetoresistance spectrum of K measured by Coulter and Dataks using the induced-torque method [24]. In (a), $\mathbf{H}$ was in a (211) plane, and the single-crystal sphere was grown in oil. In (b), $\mathbf{H}$ was in a (321) plane, and the sphere was grown in a Kel-F mold.

The foregoing is just a conjecture! However, several experimenters have cautioned against etching alkali metal samples destined for de Haas–van Alphen studies. Usually etching leads to reduction of surface damage and strain, and, therefore, to better samples. This is surely the case here also. But etching causes alkali metal surfaces to become faceted; and such facets can be visible to the naked eye. It is quite likely that large surface facets lead to a large **Q** domain size, which would be deleterious for one who demands that the de Haas–van Alphen signal have cubic symmetry. Many samples are discarded merely because they do not conform to the expected symmetry or signal amplitude. Perhaps these are the ones that should be subjected to exhaustive study.

The sharp magnetoresistance peaks in single-crystal spheres of Na and K are facts. The only known explanation involves open orbits, which can exist in these metals solely as a consequence of a broken symmetry. The isotropic de Haas–van Alphen frequencies of Na and K are also facts. The obvious interpretation is a simply connected, spherical Fermi surface. Clearly the contradiction can only be apparent. Resolution of this paradox is urgently needed.

Section 45.4 includes a review of many other phenomena which require (for an adequate explanation) the existence of the CDW periodic potential – i.e. the second term of Equation (45.2). The size of its coefficient, $\sim 0.6\,\text{eV}$, can be estimated independently from optical, spin resonance and electronic-transport data. Many of these phenomena vary from run to run on the same sample. Such non-reproducibility requires history-dependent structural variations. **Q** domains, which can regrow each time a sample is subjected to thermal strain, provide an immediate interpretation. Control of **Q** domain texture is perhaps the most challenging quest for the experimentalist.

45.3 Theory of Charge Density Waves

The purpose of this section is to show that a CDW is the anticipated electronic ground state of a simple, isotropic metal. We will assume that the positive-ion background has been replaced by a uniformly charged jelly, having *no mechanical* rigidity. The last assumption – a key property of the *deformable* jellium model – guarantees that all microscopic electric fields will be zero. (There will be *no* Hartree term in the one-electron Schrödinger equation.) A CDW arises exclusively from exchange and correlation, as we will show. The alkali metals, in the aspects relevant here, correspond well to this model. Their elastic stiffness is two orders of magnitude less than that of copper, and the pressure dependence of their moduli shows no evidence of a Born–Mayer ion–ion repulsion.

The fact that potassium has a CIW structure not only confirms the theoretical arguments, but at the same time proves that some treatments of exchange and correlation are fundamentally incomplete, and sometimes dramatically wrong. This will be discussed at length in the concluding part of this section.

45.3.1 SDW-CDW instability theorem

The exact ground-state energy of an electron system is

$$E \equiv E_{\text{HF}} + E_{\text{corr}} \,, \qquad (45.4)$$

where E_{HF} is the Hartree–Fock ground state. This equation is a definition of the correlation energy. The spin density wave instability theorem [1], which I proved in 1962, showed that the normal (paramagnetic) state always has an instability of the SDW type. The theorem is rigorous and applies for all electron densities (in the Hartree–Fock approximation). If one allows the jellium to be deformable, then the instability can be any admixture of SDW and CDW [2]

$$\varrho_{\pm} = \tfrac{1}{2}\varrho_0 \left[1 + p\cos\left(\mathbf{Q}\cdot\mathbf{r} \pm \varphi\right)\right] \,, \qquad (45.5)$$

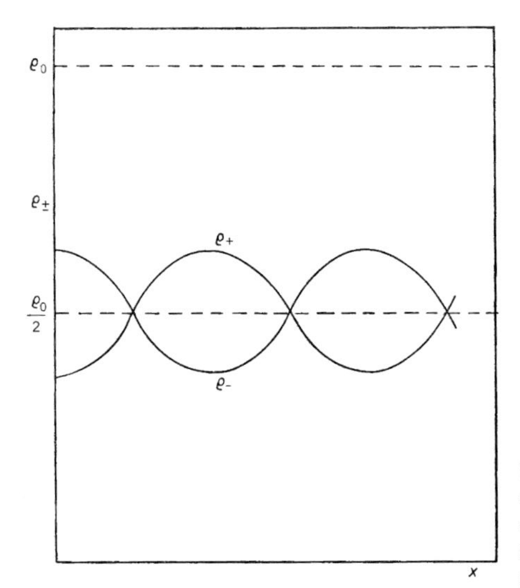

Figure 45.6 Electronic charge density *vs.* x of up- and down-spin electrons for an ideal metal with a SDW ground state. The total charge density in a constant, ϱ_0. For a CDW, ϱ_+ and ϱ_-, are in phase.

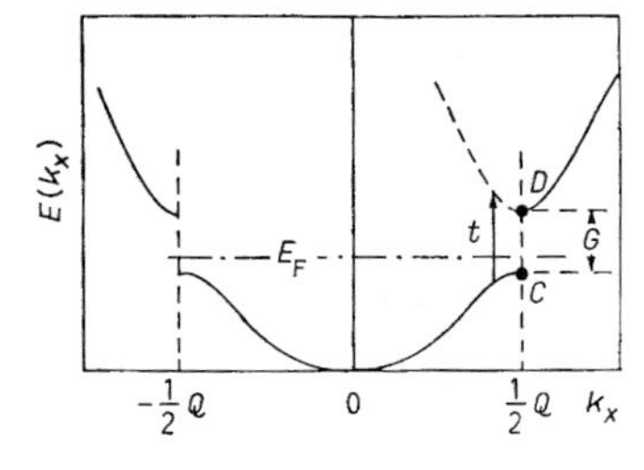

Figure 45.7 Electronic energy $E(k_x)$ *vs.* k_x for a CDW potential, $G \cos Qx$. The Fermi energy E_F lies between the gap edges C and D. The Mayer–El Naby optical absorption is caused by the transitions, t, which become possible as a consequence of the extra momentum components in the wave functions, Equation (45.6).

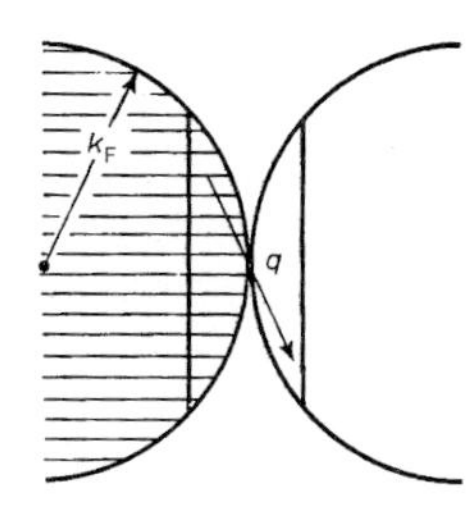

Figure 45.8 The new exchange interactions that cause SDW and CDW instabilities are those between momentum states at the head and tail of the arrow **q**. States **k** in the spherical cap enclosing the head of the arrow, which have energy $> E_F$, are "virtually" occupied by their coupling to states **k** − **Q** (not shown) having energy $< E_F$.

where $\varrho_\pm$ are the electron densities for up and down spin (see Figure 45.6). The HF energy is invariant to the choice of φ as long as the jellium background cancels all the charge modulation from the electrons. The kinetic energy and the exchange interactions (up with up and down with down) do not care about the relative phase of the up-spin and down-spin modulations.

Proof of this theorem is intricate. However, the strategy is simple. One chooses $Q = 2k_F$ and lets the energy gap G be a variable. This gap is caused by the exchange potential and is created by the amplitude modulation of the one-electron wave functions. One considers only those states near the energy gap at point C in Figure 45.7 – the ones displaced $\sim \frac{1}{2}G$ in energy from the free-electron parabola and $\sim$ 100% amplitude modulated. These are in a spherical cap of height $\sim G$ and radius $\sim G^{\frac{1}{2}}$ (see Figure 45.8). Their number N is accordingly $\sim G^2$.

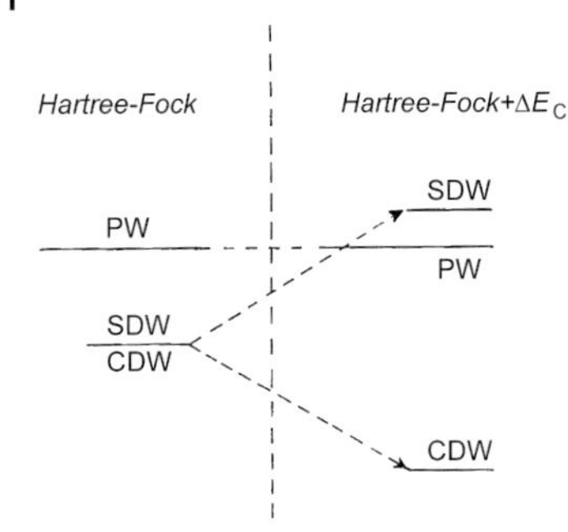

Figure 45.9 Energy level diagram for the plane-wave (PW), SDW and CDW states of an ideal metal. In the Hartree–Fock approximation the SDW and CDW states are degenerate and have lower energy than the PW state. The degeneracy is split, as shown on the right, by correlation energy corrections.

The kinetic-energy "investment" to deform these states from plane waves to

$$\psi \simeq \exp\left[i\mathbf{k}\cdot\mathbf{r}\right] + \alpha^{+}\exp\left[i(\mathbf{k}+\mathbf{Q})\cdot\mathbf{r}\right] + \alpha^{-}\exp\left[i(\mathbf{k}-\mathbf{Q})\cdot\mathbf{r}\right] \tag{45.6}$$

is their number N times the extra kinetic energy of ψ (compared to a pure plane wave). This is $\sim G$. Therefore,

$$\Delta T \sim G^3 . \tag{45.7}$$

New contributions to the exchange energy arise from the interactions of the states just described with virtually excited states associated with the spherical cap on the opposite side of the Fermi sphere. These states lie above k_F in Figure 45.8. They contribute an energy "dividend"

$$\Delta E_\mathrm{ex} \sim -\sum_{\mathbf{k},\mathbf{k}'} \frac{4\pi e^2}{|\mathbf{k}'-\mathbf{k}|^2} \sim -G^3 \ln \frac{E_\mathrm{F}}{G} . \tag{45.8}$$

This can always be made larger in magnitude than Equation (45.7) by choosing G small enough, which proves the theorem. But one must show too (as it has been [1]) that (45.8) is larger than the loss of exchange energy resulting from the admixtures in (45.6). Consequently, a SDW or CDW state always lies lower than the normal (plane wave) state in the Hartree–Fock approximation, as shown on the left side of Figure 45.9.

45.3.2 The correlation energy correction

The algebraic sign of the correlation energy correction for SDW and CDW states is now the crucial issue. We will show that they have opposite sign – in fact, a SDW state is destabilized and a CDW *is made more stable.*

The correlation energy can be calculated semi-quantitatively from

$$E_\mathrm{corr} \simeq -\sum_i \frac{\langle i|V_\mathrm{eff}|0\rangle^2}{\Delta E_i} , \tag{45.9}$$

where ΔE_i is the two-particle two-hole excitation energy relative to the supposed ground state, $|0\rangle$. What is the meaning of V_eff? The dominant contributions to Equation (45.9) are the virtual scattering of opposite-spin electrons from below the Fermi energy to empty states above. This proceeds by a screened interaction V_scr:

$$\psi = C\left\{\psi_0 - \sum_i \frac{\langle i|V_\mathrm{scr}|0\rangle}{\Delta E_i}\psi_i\right\} , \tag{45.10}$$

where C is a normalizing factor. The new energy is found by taking the expectation value of the *exact* Hamiltonian:

$$E' = \langle \psi' | \sum_{\lambda} \frac{p_\lambda^2}{2m} + \sum_{\lambda<\mu} \frac{e^2}{r_{\lambda\mu}} | \psi' \rangle \,. \tag{45.11}$$

If one works this out, one finds [2] Equation (45.9), provided

$$\langle i | V_{\text{eff}} | 0 \rangle^2 \equiv 2 \langle i | V | 0 \rangle \langle i | V_{\text{scr}} | 0 \rangle - \langle i | V_{\text{scr}} | 0 \rangle^2 \,, \tag{45.12}$$

where $V = e^2/r$. This procedure was first used in correlation energy calculations by Macke [27]. Parallel-spin scattering is less important because of the partial cancellation of direct and exchange terms.

The change in correlation energy associated with the amplitude-modulated states was worked out in detail by Overhauser [2]. We present here a simplified version which gives accurately the same results. For the SDW case,

$$\varphi_\pm \approx \exp[i\mathbf{k}\cdot\mathbf{r}]\left(1 \pm p \cos \mathbf{Q}\cdot\mathbf{r}\right)^{\frac{1}{2}} \,, \tag{45.13}$$

where $p \ll 1$. The matrix element for virtual scattering of two opposite-spin electrons is

$$m = \langle (\mathbf{k}+\mathbf{q})_+ (\mathbf{k}'-\mathbf{q})_- | V_{\text{eff}} | \mathbf{k}_+ \mathbf{k}'_- \rangle \,. \tag{45.14}$$

Two of the four φ's in (45.14) have plus signs (see Equation (45.13)) and the other two have minus. It follows that (after integration)

$$m \approx m_0 \left(1 - \tfrac{1}{2} p^2\right) \,, \tag{45.15}$$

where m_0 is the matrix element for the same event with $p = 0$. As a consequence the correlation energy becomes, to order p^2,

$$E_{\text{corr}}^{\text{SDW}} \simeq -\sum_i \frac{m_0^2}{\Delta E_i} (1 - p^2) \,. \tag{45.16}$$

Every term is reduced in magnitude by the factor $1 - p^2$; so the SDW state is raised *relative to* the (correlated) plane-wave state.

For the CDW case all four factors in (45.14) have the same sign. Accordingly,

$$E_{\text{corr}}^{\text{SDW}} \simeq -\sum_i \frac{m_0^2}{\Delta E_i} (1 + p^2) \,. \tag{45.17}$$

Each term in the correlation energy is enhanced in magnitude and this adds stability to the CDW state. (There are high-energy virtual excitations – to states above the CDW energy gaps – for which the above remarks are not true; but they are relatively less important.)

It is easy to understand why electron–electron (virtual) scattering splits the SDW and CDW degeneracy. For the SDW case up-spin and down-spin electrons are stratified in *alternate* laminar layers and have less probability of scattering from one another. A CDW coaxes both spin types into the *same* layers and increases their virtual scattering. These effects of correlation are shown on the right-hand side of Figure 45.9.

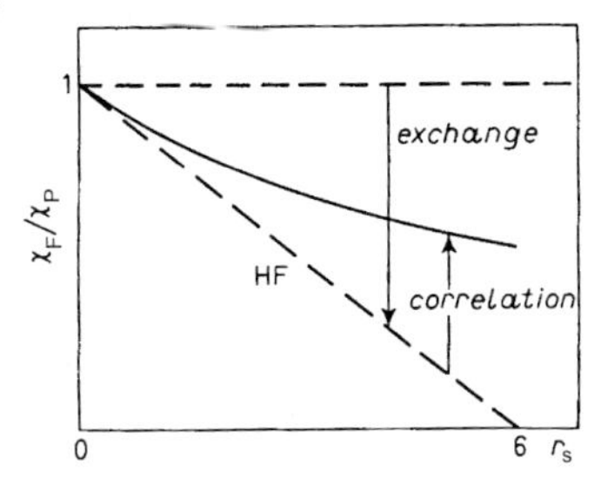

Figure 45.10 Ratio of the (noninteracting) Fermi-gas spin susceptibility, χ_F, to the susceptibility, χ_P, of an electron gas *vs.* electron separation parameter r_s. The dashed line is the HF value, obtained with exchange interactions only.

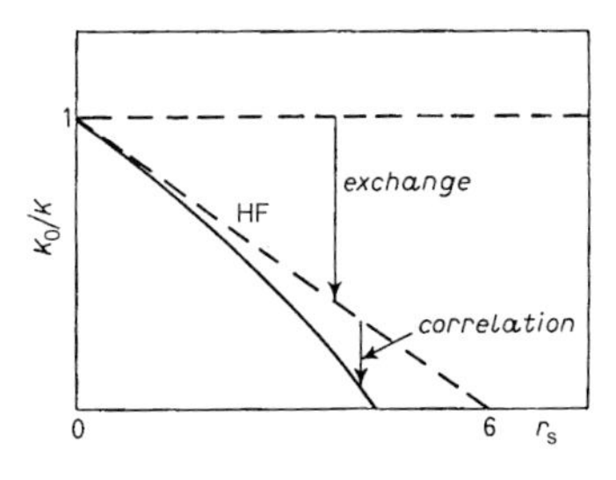

Figure 45.11 Ratio of the (noninteracting) Fermi-gas compressibility, K_0, to K of an electron gas *vs.* r_s. The dashed line is the HF value (exchange interactions only).

45.3.3 Analogy with uniform deformations

Wigner was the first to point out [5] that electron correlations reduce the enhancement of the spin susceptibility caused by exchange. The straight line in Figure 45.10, intersecting the horizontal axis near $r_s = 6$, is the Hartree–Fock result given by Equation (45.1). With increasing spin polarization the number of antiparallel-spin pairs decreases (and reaches zero in the ferromagnetic limit). So the correlation energy becomes smaller in magnitude with increasing polarization. This leads to the reduced spin response. The exact value of χ_F/χ_P lies above the HF line, as shown in Figure 45.10.

The compressibility K of an electron gas is also enhanced by exchange – in fact, by the same factor as the spin susceptibility. However, the correlation energy enhances K *even further.* All calculations of E_{corr} *vs.* r_s show that the second derivative of E_{corr} with respect to density contributes to this. The general conclusion is that exchange and correlation reinforce one another for charge modulations. The fact that the "exact" compressibility in Figure 45.11 lies below the HF line illustrates that the dichotomy in the relative effects of exchange and correlation for the uniform ($Q = 0$) case is the same as that found above for $Q = 2k_F$.

45.3.4 Implications for many-electron theory

I believe that the foregoing arguments provide a microscopic understanding of the origin of CDW's in simple metals (having a deformable positive-ion background). However, several frequently used approximations in many-electron theory fail to anticipate this broken symmetry, which is caused by exchange and correlation.

Local approximations to exchange, first introduced by Slater [28], fail to predict this phenomenon. Consider the periodic potential required for potassium

$$V_x = -G \cos \mathbf{Q} \cdot \mathbf{r} \,, \tag{45.18}$$

where $G = 0.6$ eV. The resulting charge modulation is

$$\varrho = \varrho_0 \left(1 + p \cos \mathbf{Q} \cdot \mathbf{r}\right) \,. \tag{45.19}$$

G and p must be self-consistent. If one calculates p from the observed G, assuming V_x is local, $p = 0.17$. If one now uses a Kohn–Sham local-exchange approximation [29] to calculate G from p, one finds $G = 0.2$ eV. The inconsistency is half an order of magnitude. *Local* exchange (and correlation) is too weak to sustain the broken symmetry. G would quickly iterate to zero.

The failure of local-exchange approximations to predict the correct ground state of potassium is caused by the fact that CDW states are supported by the nonlocal, dynamical effects of exchange. This can be easily illustrated by considering a uniform electron gas with infinitesimal fractional modulation ε. The Slater local-exchange operator is then

$$A_s = -\frac{3e^2 k_F}{2\pi}(1 + \varepsilon \cos qx)^{\frac{1}{2}} . \tag{45.20}$$

Since ε is small, we have

$$A_s \simeq -\frac{e^2 k_F}{2\pi}\varepsilon \cos qx . \tag{45.21}$$

The dynamical properties of an operator are displayed by calculating its offdiagonal matrix elements. The only such elements for (45.21) are

$$\langle \mathbf{k} \pm \mathbf{q} | A_s | \mathbf{k} \rangle = -\frac{e^2 k_F}{4\pi} . \tag{45.22}$$

Observe that this result is independent of q and independent of $\mathbf{k}$. (It is *local* in both senses of the word.)

Consider now the exact operator A. It is defined by the equation

$$A\psi(\mathbf{r}) = -\sum_i \left[\int \varphi_i^+(\mathbf{s})\psi(\mathbf{s}) \frac{e^2}{|\mathbf{r}-\mathbf{s}|} d^3 s \right] \varphi_i(\mathbf{r}) , \tag{45.23}$$

where $\{q_i\}$ are the occupied levels. For this illustration these can be taken to be the perturbed electron states associated with a potential such as (45.18) and leading to a fractional polarization $p = \varepsilon$. The *exact* off-diagonal matrix elements of A have been calculated [30]. The ratios of these to the local ones from A_s are shown in Figure 45.12 as a function of q, and for four values of $\mathbf{k}$. The nonlocal effects are spectacular. Clearly a local approximation (constant, horizontal line at unity or $\frac{2}{3}$) discards important dynamical properties. The singularity at $q = 2k_F$ is the cause of CDW phenomena.

Another common view is that electrons in metals can be treated as quasi-particles interacting via short-range, *screened* interactions. This is permitted for scattering processes, but should be forbidden for use in Hamiltonians.

For example, if screened exchange were used (and it has been) to calculate the electron gas compressibility, one would find a result given by the behavior of the spin susceptibility (shown in Figure 45.10) instead of that given by the second derivative of the total energy with respect to volume (as shown in Figure 45.11). In other words, a screened-exchange treatment leads to a correction (relative to Hartree–Fock) having the *wrong sign*. Similarly, screened-exchange interactions, taken alone, do not permit a CDW structure in potassium.[2)]

The many-electron energy can, of course, be broken down in alternative ways [32]:

$$\text{exchange} + \text{correlation} = \text{screened exchange} + \text{Coulomb hole} . \tag{45.24}$$

2) This follows from the fact that screening of exchange interactions suppresses SDW instabilities at metallic densities [31].

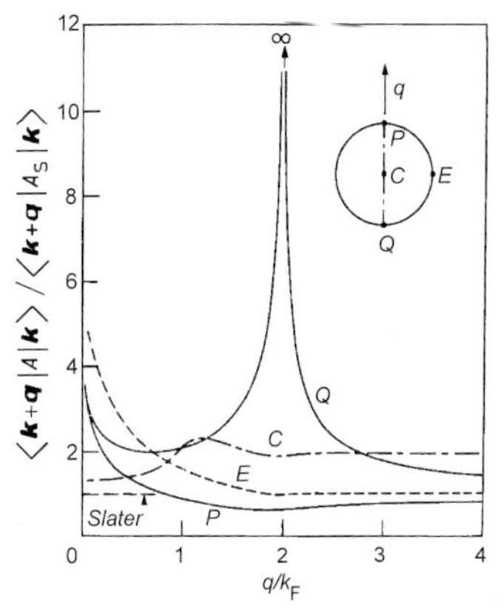

Figure 45.12 Ratio of exact matrix elements of the exchange operator to those given by Slater's local-density approximation. Each curve is for the point **k** of the Fermi sea shown in the inset. The electron gas was taken to have a small sinusoidal density modulation, $\varepsilon \cos \mathbf{q} \cdot \mathbf{r}$. The matrix elements are in the (pure) momentum representation.

This can be taken as a definition of the Coulomb-hole energy. The point that needs to be appreciated is that the Coulomb-hole potential is not an innocuous constant (as is frequently assumed). Consider an expression for the Coulomb energy of N electrons (in pure momentum states) in terms of the usual Fourier components. The result is

$$E_C^0 = N \sum_{\mathrm{q}} \frac{2\pi e^2}{q^2} - \frac{1}{2} \sum_{\mathrm{k,q}} \frac{4\pi e^2}{q^2} \left(n_{\mathrm{k}\uparrow} n_{\mathrm{k-q}\uparrow} + n_{\mathrm{k}\downarrow} n_{\mathrm{k-q}\downarrow} \right) , \tag{45.25}$$

where the n's are occupation numbers. The first term is just another way of writing the total rest mass, Nmc^2. The second term is the exchange energy. The dielectric response of the electrons to the electric fields associated with the energy E_C^0 can be represented by a dielectric function ε_q. Recall that the energy of a charged capacitor is reduced by ε if a dielectric is moved into the electric field; so also will the Coulomb energy of the N electrons be reduced. This reduction, the Coulomb part of the correlation energy, is

$$\Delta E_C = N \sum_{q} \frac{2\pi e^2}{q^2} \left(\frac{1}{\varepsilon_q} - 1 \right) - \frac{1}{2} \sum_{\mathrm{k,q}} \frac{4\pi e^2}{q^2} \left(\frac{1}{\varepsilon_q} - 1 \right)$$
$$\times \left(n_{\mathrm{k}\uparrow} n_{\mathrm{k-q}\uparrow} + n_{\mathrm{k}\downarrow} n_{\mathrm{k-q}\downarrow} \right) , \tag{45.26}$$

ΔE_C is generally small compared to the exchange energy. The second terms of (45.25) and (45.26) can be combined. Their sum is the screened-exchange energy. The first term of (45.26) is the Coulomb-hole energy. It is much larger than screened exchange. If appears to be a constant, since it does not depend on the occupation numbers n_{k}, and some workers treat it as such. But that is a serious error. If one cancels off the dynamic contributions of bare exchange by combining it with the second term of (45.26), the dynamic effects of the Coulomb hole become dominant. The Coulomb hole, from this point of view, is the driving force for a CDW ground state,

The more accurate approach to many-electron effects is to leave the two dynamic corrections of Equation (45.26) combined, for their sum is small. Then the large exchange term of Equation (45.25) can be calculated with precision. Unscreened exchange is, with this approach, the major cause of a CDW ground state, and ΔE_C provides a small reinforcement [2]. The foregoing, over-simplified discussion needs generalization when modulated basis functions, Equation (45.6), appear instead of pure plane waves. Nevertheless, the conclusion remains. Broken symmetry can be attributed to exchange or to the Coulomb-hole term depending on how one assigns the second term of (45.26).

45.4 CDW Phenomena

Most of the physical properties discussed in this section have been studied in potassium. Some have been observed in sodium, and a few in rubidium. There is a good reason why most experimentalists focus their efforts on potassium: It does not undergo a crystallographic transformation on cooling to liquid-helium temperature. Single crystals of Li and Na become poly crystalline after their first warm-up from 4 K. Rb and Cs are extremely soft mechanically and reactive chemically. For practical purposes, then, K is the prototype "ideal" metal.

45.4.1 Optical Anomalies

The first indication that potassium has a CDW structure was the discovery by Mayer and El Naby [10] of an intense inter-band, optical transition with a threshold at 0.6 eV, followed by an asymmetric peak at 0.8 eV, as shown in Figure 45.13. The ordinary interband absorption – caused by the Brillouin-zone energy gaps – has a threshold of 1.3 eV. A quantitative account of the anomalous absorption succeeds merely by letting the CDW gap $G = 0.6$ eV and by calculating the optical-transition rate across these new energy gaps [11]. The transitions are indicated by the arrow t in Figure 45.7. (The threshold transition is from C to D in the figure.)

Ordinarily potassium, which is body-centered cubic, would have isotropic optical properties. However, the new absorption occurs only from the component of the photon polarization vector " parallel to **Q**. The CDW optical absorption is uniaxial! This is crucial to an explanation of why the Mayer–El Naby anomaly is not obtained in measurements on evaporated films. Such films, deposited on smooth, amorphous substrates, always have a [110] crystal direction normal to the surface. It turns out that the optimum direction of **Q** is then normal to the surface too. Since an infra-red photon incident on the film will have its polarization vector " parallel to the surface, *inside* the metal (even at oblique incidence), anomalous optical absorption *cannot* occur. At bulk-metal–vacuum surfaces, the Mayer–El Naby anomaly has been reproduced by Hietel [33][3] and by Harms [34].[4] The intensity of the CDW absorption is (approximately) independent of temperature between 80 K and temperatures *above* the melting point, where the anomaly persists [10, 33].[5]

It has been mistakenly claimed by Fäldt and Walldén [35] that the anomaly is caused by KOH on the surface. They evaporated potassium on an 80 K substrate and found optical anomalies after a few monolayers of H_2O were admitted. However, they never found an anomaly with a 0.6 eV threshold; and all new optical peaks disappeared above 120 K, unlike

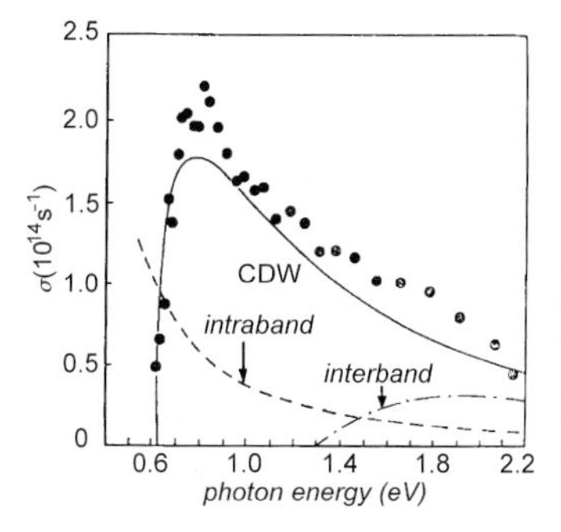

Figure 45.13 Anomalous optical-absorption spectrum of potassium. The intraband conductivity (dashed curve) has been subtracted from the experimental data before being plotted; • experimental data from [10]. The solid curve shows the theoretical absorption introduced by a CDW structure. A "normal" metal would exhibit only the interband absorption with a threshold, as shown, at 1.3 eV.

3) Hietel B.: personal communication
4) His data are reproduced in [13].
5) Hietel B.: personal communication

the Mayer–El Naby anomaly, which persists even into the *liquid* state. Taut [36] has followed this mistake by claiming that surface states caused by KOH are involved in the Mayer–El Naby anomaly. But this mechanism requires that the optical-polarization vector " have a component perpendicular to the surface, which can occur only for a microscopically rough surface. Possibly the absorption peaks observed by Fäldt and Walldén [35] involve surface states, since they disappeared at the same temperature as the surface plasmon peak (which also requires a rough surface).

The CDW anomaly in Na has a 1.2 eV threshold [37]. Not only does it persist into the liquid metal (as in K), where the metal–vacuum interface is smooth, but is largest there. The influence of slight amounts of water vapor on the intensity of the Mayer–El Naby peak [34],[6] which *does not change* its spectral shape or location, can be explained by the influence of KOH layers on the crystallographie orientation at the metal surface [13].

A new, possible confirmation of CDW optical properties awaits experimental verification. Consider the minigaps and heterodyne gaps, Equation (45.3), which are shown in Figure 45.2. They will give rise to optical-absorption thresholds in the infra-red [15]. The first minigap is about 0.1 eV and should cause an absorption peak several times higher than the Mayer–El Naby peak. The first heterodyne gap is $\sim$ 0.2 eV. I am unaware of any optical studies below 0.5 eV on potassium. Hopefully this will be attempted soon.

45.4.2 Conduction Electron Spin Resonance Splitting

The g-factor for spin resonance (CESR) in a metal is a measure of the Fermi-surface average of $g(\mathbf{k})$. This averaging occurs because electrons are scattered through $\sim 10^4$ states at the Fermi surface during one spin-lattice relaxation time. Nevertheless, the CESR in K was found to be split into well-resolved components [38] separated up to $\sim$ 0.5 Oe in an applied field of 4000 Oe. Splittings up to 5 Oe were observed for $H = 40\,000$ Oe [39]; so a g-factor splitting is involved. It is about 40% of the g shift caused by spin-orbit coupling. This splitting was explained quantitatively [40] by the Fermi-surface distortion created by the CDW, winch leads to a g-factor dependent on the angle between **H** and **Q**. The splitting arises for samples subdivided into **Q** domains and is proportional to the CDW energy gap G. The same value, $G = 0.6$ eV, obtained from the optical anomaly is required to fit the observed splitting.

45.4.3 Nonreproducibility, a Consequence of Q Domains

Many fundamental transport properties vary drastically from sample to sample, or from run to ran on the same sample. The CDW structure permits this, since the **Q** direction provides an axis of anisotropy. As already discussed in Section 45.2, a macroscopic sample will generally be subdivided into **Q** domains. The domain distribution will depend on sample history and can change merely by heating or cooling the sample, or by mechanical strain, which stimulates domain regrowth.

Some of the phenomena which exhibit this variability are residual resistivity [41], magnetoresistance [41], Hall effect [42], the electron–electron scattering contribution to the resistivity [43], the induced-torque anisotropy [44] and the CESR splitting [38]. Ability to control **Q** domain texture and size is a prerequisite for future progress. The only successful technique discovered so far is the epitaxial orientation of **Q** normal to the surface of a film evaporated on an amorphous substrate.

6) His data are reproduced in [13].

45.4.4 Linear Magnetoresistance

The LAK magnetotransport theorems [18] require that $\varrho(H)$ saturate when $\omega_c\tau \gg 1$ (for a metal with a simply connected Fermi surface). For typical, pure samples of K, $\omega_c\tau \sim 1$ at 400 Oe. Nevertheless, all workers have found that for $H > 10^3$ Oe the resistivity varies linearly with H:

$$\varrho \equiv \varrho_0(1 + S\omega_c\tau)\,, \quad \omega_c\tau \gg 1\,. \tag{45.27}$$

No evidence of saturation has been found up to fields for which $\omega_c\tau \sim 300$ [45]. The Kohler slope S varies widely – over a range spanning two orders of magnitude – from sample to sample, and by as much as a factor of 5 from run to run on the same sample [45]. Typically, $S \sim 10^{-3} \div 10^{-2}$. Only open orbits can explain the nonsaturating behavior [18]; and the linear dependence arises from the large number of open-orbit directions in a sample having all 24 **Q** domain orientations. Each open-orbit peak increases $\sim H^2$, but its width decreases as $1/H$. Consequently, an angular average leads to a result approximately linear in H [46]. A theoretical histogram [47] for $|\ln S|$ based on random crystal orientation, **H** orientation and **Q** domain texture has the same width as that found experimentally [41].

Prior to the discovery of open-orbit magnetoresistance peaks [24] some workers had attibuted the linear magnetoresistance to the presence of macroscopic voids even though, for some specimens, the volume fraction of voids would have to be ~2%. Comparison of precision lattice parameter measurements [48] with volumetric density data [49] shows that any void fraction could not be larger than a few parts in 10^5.

45.4.5 Hall Coefficient Discrepancy

The LAK magnetotransport theorems [47] require that the *high-field* Hall coefficient R equal $-1/nec$, where n is the electron density. (This result is for a single, simply connected Fermi surface.) The prediction is precise, since n depends only on the lattice constant. Absolute measurements of R are difficult using standard four-probe methods, since current distortions near the Hall-field probes cause the sample width at that point to be uncertain. R cannot be measured without accurate determination of a sample dimension. Geometric resonances of helicon waves in a flat plate depend only on R and the plate thickness [42]. Measurements by this inductive technique on twelve samples at 50 kG showed that R is too large by 4 to 8%, depending on crystal orientation. Sample thickness was determined by *in situ* ultrasonic time of flight. Helicon resonances in a sphere [50] resulted in a $(3 \div 4)$% discrepancy. The diameter of the sphere was determined both mechanically and electrically to a precision of $\frac{1}{2}$%.

The largest discrepancy observed in flat-plate measurements [42] occurred when a [110] crystal axis was normal to the surface. The epitaxial effect mentioned above – that **Q** tends to be normal to a [110] surface – indicates that R for $\mathbf{H} \parallel \mathbf{Q}$ is greater than R for $\mathbf{H} \perp \mathbf{Q}$. (**H** was perpendicular to the flat-plate surfaces.) Undoubtedly, the samples employed had a full spectrum of **Q** domain orientations, with a fractional preferred texture. It may be expected that in a single-**Q** crystal the anisotropy of R will be much larger. I shall cite (in what follows) evidence that it is as large as 30%.

45.4.6 Induced-torque Anisotropy

The most revealing, probeless technique that has been developed for studying the conductivity tensor is the induced-torque method. A spherical, single-crystal sample is supported by a vertical rod in a horizontal magnetic field. As the magnet is slowly rotated (in the horizontal plane), induced currents in the sample create a magnetic moment perpendicular to the

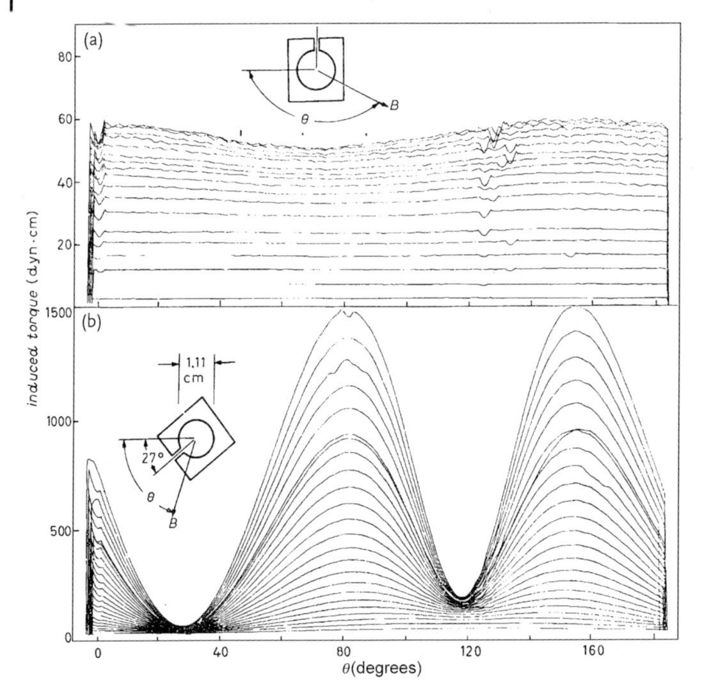

Figure 45.14 Induced torque against magnetic-field direction θ for a potassium sphere 1.11 cm in diameter. In (a) the field B is rotated about the growth axis. The curves shown are for 0.5, 1, 2, 3, ... kG. In (b) the plane of rotation contains the growth, axis, and $B = 1, 2, 3, \ldots$ kG. The data are from [44].

field B. This causes a torque (on the rod) which is monitored *vs.* B and magnet angle. The theory [20], applicable to a general conductivity tensor, is unambiguous and depends only on Faraday's law of induction. The method is practically d.c, since typical magnet rotation rates are $\sim 10^{-3}$ Hz.

Early torque studies [51] on K were done on hundreds of specimens. Virtually all showed extremely anomalous high-field torque anisotropies. In a cubic crystal the conductivity tensor is isotropic, so the induced torque must be independent of magnet angle θ. It was emphasized [52], however, that, if the samples were sufficiently distorted from spherical shape, the observations could be explained. Although torque data at 80 K showed that many of the samples were undistorted, the doubt remained unresolved until 1975.

The anomalous torques were confirmed by Holroyd and Datars with measurements on an accurately spherical sample grown in a Kel-F mold [44]. The data are shown in Figure 45.14. In (a) the induced torque *vs.* magnet angle θ is shown for several values of B between 500 G and 17 kG; the growth axis was vertical. In (b) the growth axis was horizontal and the B values range from 1 to 23 kG in 1 kG steps. The "flat" curves in (a) compared to the extreme (45 to 1) anisotropic ones in (b) prove that K is electrically uniaxial, not cubic, The ever-increasing torque peaks in (b) prove that the Fermi surface is multiply connected. (The high-field torque in a spherical sample must saturate unless there are open orbits.)

A quantitative theory of the torque data, such as that shown in Figure 45.14, has recently been developed [23]. The key ingredient is the existence of an anisotropic Hall coefficient. As pointed out above, an anisotropy in R has been directly observed [42]. The entire family of curves in Figure 45.14 can be accurately fitted if R for $\mathbf{B}$ parallel to the growth axis is

30% larger than R for $\mathbf{B}$ perpendicular to the growth axis. Unfortunately, the crystal axes of this sample were not determined by X-ray diffraction, so one can only speculate that there was a highly preferred $\mathbf{Q}$ domain texture along the growth axis. Typical samples [51] have a four-peaked pattern (in a 360° rotation) with a torque ratio of ~5 M. Such data require a macroscopic (average) anisotropy in the Hall coefficient of about 10%. The *four-peaked* patterns occur even though the rotation axis of $\mathbf{B}$ is parallel to a threefold, [111], symmetry axis. They also occur in single-crystal spheres of Na [51].

45.4.7 The Oil Drop Effect

Some single-crystal spheres do not show any significant four-peaked pattern between 5 and 30 kG. (However, the open-orbit torque peaks always appear above 40 kG [25].) Nevertheless, if a drop of oil is placed on such a sphere and the sample is then cooled to ~ 1 K, the four-peaked rotation pattern occurs. This phenomenon shows that thermal stress (arising after the oil drop freezes) followed by a short period of annealing (during cool-down) creates a preferred $\mathbf{Q}$ domain texture.

This phenomenon suggests a possible approach for systematically altering Qdomain distributions. But the high mechanical plasticity of Na and K poses a severe problem in control.

45.4.8 Residual-resistance Anisotropy

I have already discussed the anisotropic magnetoresistance that occurs in the intermediate-field (5÷30) kOe) and high-field (> 40 kOe) regions, which are characterized, respectively, by smooth, four-peaked patterns and by sharp, manifold open-orbit spectra. The low-field region ((0÷2) kOe) is completely different. It consists of a twofold, sinusoidal anisotropy, as can be seen in the first few traces of Figure 45.14(b)). This behavior corresponds to a residual-resistivity anisotropy.

In a cubic metal the resistivity tensor is necessarily isotropic, Nevertheless, the data from the sample shown in Figure 45.14 require a cigar-shaped tensor with an anisotropy of ~ 4 [53]. A recent analysis [23] of the data taking into account the Hall coefficient anisotropy indicates that the major to minor axis ratio may need to be about 2. This same phenomenon was evident in the original, induced-torque data of Schaefer and Marcus [51] and was recognized as a residual-resistivity anisotropy [54].

A CDW structure leads to a quantitative explanation [55] of this effect. The residual resistance is proportional to the weighted average of the scattering cross-section $\sigma(\theta)$ for impurities. The weighting factor, $1 - \cos 0$, counts 180° scattering heavily and small-angle scattering very little. For impurity scattering, $\sigma(\theta)$ is sharply peaked in the forward direction, i.e. small $\mathbf{k}' - \mathbf{k}$. The influence of the CDW is easily understood: the wave function mixing, given by Equation (45.6) and caused by the CDW potential, allows small-q Fourier components of the impurity potential (which are large in magnitude) to scatter electrons with the conservation rule

$$\mathbf{k}' = \mathbf{k} + \mathbf{q} \pm \mathbf{Q} . \qquad (45.28)$$

This CDW umklapp effect leads to strongly enhanced large-angle scattering, but only for transitions across the Fermi surface in a direction nearly parallel to $\pm\mathbf{Q}$. Accordingly, the residual resistivity parallel to $\mathbf{Q}$ is much larger than that perpendicular to $\mathbf{Q}$. A variational solution of the transport equation [55] leads to anisotropies (depending on the scattering potential) up to 4:1. Values smaller by a factor of two are found from exact, numerical solutions [56] (in the zero-field limit). For magnetic fields at which the measurements are made (~ 1 kOe) an exact theoretical ratio would likely fall between these limits. Precision is not possible because the total scattering potential, which must include effects of elastic strain caused by impurities [55], is unknown.

CDW umklapp scattering leads to a significant anisotropy (in accord with observation). Without a CDW, this effect could not even exist. Dislocations give rise to an oblate, rather than a prolate contribution to the resistivity tensor and are too few in number by at least a factor of 100, even if one assumes they all have a parallel orientation.

45.4.9 Temperature dependence of resistivity near 1 K

The low-temperature resistivity of a metal such as potassium is expected to have four contributions:

$$\varrho = \varrho_0 + AT^m \exp\left[-(T_0/T)\right] + BT^2 + CT^5 \,. \tag{45.29}$$

The residual resistivity ϱ_0 is caused by impurities (and other imperfections). The exponential term is caused by the freeze-out of umklapp processes. The T^2 term arises from electron–electron scattering, and the T^5 term from acoustic-phonon scattering.

Below $\sim$ 1.3 K the phonon umklapp term is negligible and the Bloch T^5 term is (for pure samples) reduced significantly by phonon drag effects. The remaining, temperature-dependent term has been observed by several workers [43, 57–60]. The most striking fact is that the size of this contribution varies from sample to sample by an order of magnitude. If the anticipated mechanism – electron–electron scattering with a reciprocal-lattice vector umklapp (which I call RL umklapp) – were the only one, all samples would have to have the same BT^2 term. Another problem is that for samples with a small coefficient the temperature dependence [57] is more nearly $T^{1.5}$.

A CDW structure causes two new contributions to the low-temperature behavior. The first is electron–phason scattering. Phasons are the new collective modes of an incommensurate CDW [61]; these will be discussed at greater length below. The contribution of phason scattering to the electrical resistivity [62] explains the data [57] for which the power law dependence is close to $T^{1.5}$.

The second mechanism involving the CDW is electron–electron scattering with an umklapp of wave vector $\pm\mathbf{Q}$ (which I call CDW umklapp). Such scattering events satisfy

$$\mathbf{k}_1' + \mathbf{k}_2' = \mathbf{k}_1 + \mathbf{k}_2 \pm \mathbf{Q} \,. \tag{45.30}$$

They lead to a highly anisotropic T^2 term in $\varrho(T)$. The CDW umklapp fraction may be as high as 25% for the resistivity parallel to $\mathbf{Q}$ [63]. It is zero for the resistivity perpendicular to $\mathbf{Q}$. In contrast, the (isotropic) RL umklapp fraction is only a few percent [64].

Both of the foregoing mechanisms will lead to larger resistivity contributions when the preferred $\mathbf{Q}$ domain orientation is parallel to the current. In such a case the second mechanism will likely predominate, and the power law will be T^2. In samples for which $\mathbf{Q}$ tends to be perpendicular to the current – a situation which could occur in very small wires, where $\mathbf{Q}$ domains will tend to orient with $\mathbf{Q}$ normal to the surface – a temperature dependence dominated by phason scattering is more likely. From 0.4 to 1.3 K phason scattering is intermediate between linear and quadratic [62]. These features of CDW structure are consistent with the observations. However, below 0.3 K the observed temperature dependence exhibits a behavior that has not been interpreted [65].

There have been attempts to involve electron scattering by dislocations to explain the variations in T-dependence from sample to sample [66]. But it has been shown [62, 67] that the required dislocation density exceeds reasonable values by at least three orders of magnitude.

45.4.10 Temperature Dependence of the Surface Impedance

Measurements of the surface impedance of potassium between 0.7 and 10 K show a power law behavior (at the low-temperature end) which is *two orders of magnitude* larger than the corresponding behavior discussed immediately above [68]. A theory [69] based on electron–electron scattering (without umklapps) fails completely. It leads to a contribution to the surface resistance proportional to T^2 but having a *negative* coefficient, opposite to the sign of the observed variation.

For pure potassium the surface impedance at 1.4 MHz is in the anomalous skin effect regime. Only electrons having velocity components v_z (normal to the surface), such that $v_z\tau < \delta$, the skin depth, are "effective". Ordinarily the effective electrons (in k-space) are only those in a narrow equatorial strip centered at $k_z = 0$. However, with a CDW there are two additional groups of electrons with small v_z: those near the CDW energy gaps (see Figure 45.1), with $k_z \sim \pm Q/2$. These are the electrons that suffer intense phason scattering [62]. Their contribution to the r.f. surface resistance (because they are "effective") is excessive compared to their contribution to the d.c. resistance [70]. This allows a reconciliation between the r.f. and d.c. behavior.

45.4.11 Deviations from Matthiessen's Rule

At temperatures where impurity arid phonon resistivities have similar magnitude the total resistivity exceeds the sum of the individual contributions. This phenomenon is called DMR and has been studied extensively. The effect is large when there is considerable umklapp scattering, e.g. in polyvalent metals (which have multiply connected Fermi surfaces). However, a study [71] of data on K and Li revealed DMR's comparable to that of polyvalent metals. Theoretical work [72] confirmed that the expected DMR in an alkali metal (without a CDW) should be much smaller. Sharma [73] has pointed out that CDW umklapp scattering, Equation (45.28), explains why the DMR's of alkali metals are much larger than would otherwise be expected.

45.4.12 Doppler-shifted Cyclotron Resonance

Helicons are circularly polarized electromagnetic waves which propagate in a metal in the presence of a magnetic held [74]. (It is necessary that $\omega_c\tau \gg 1$.) The dispersion relation is such that the phase velocity of the wave is typically $\sim 10^4$ cm/s, four orders of magnitude slower than the electron Fermi velocity v_F. Electrons at the Fermi surface, with $\mathbf{v}_F$ parallel to the propagation vector $\mathbf{q}$ of the helicon, will undergo cyclotron resonance when they experience a Dopper-shifted frequency qv_F satisfying

$$q v_F = \omega_c\,. \tag{45.31}$$

A sharp resonance in the surface impedance (the Kjeldaas edge [75]) occurs when Equation (45.31) is satisfied. The fundamental importance of this phenomenon is that the electron's effective mass m^* cancels out of this equation. ($v_F = \hbar k_F/m^*$ and $\omega_c = eH/m^*c$.) It follows that for a given helicon frequency the edge field H_E depends only on the electron density, which is known with precision from the lattice constant.

Fenz and Kushida [76] studied DSCR in single crystals of potassium and found that the resonance appeared $\sim$1 kOe below the value of H_E expected for a free-electron metal. Calculations of this phenomenon for a CDW structure [77]. I predicted this shift when the CDW energy gap was taken to be 0.6 eV, the observed value from optical-absorption data. The shift arises from the reduction in the maximum electron velocity by the CDW periodic potential, Equation (45.18), along an axis parallel to $\mathbf{Q}$.

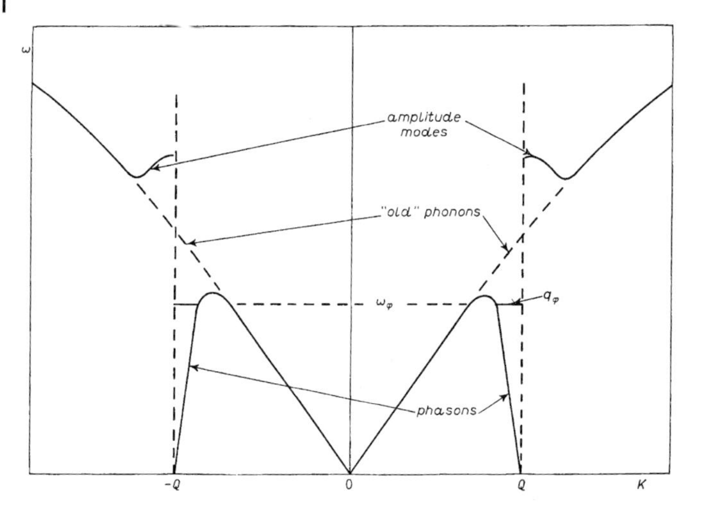

Figure 45.15 Schematic illustration of the vibrational modes in a metal having a CDW structure, The frequency of the phason branch goes to zero at $\pm Q$. A phason is a linear superposition of two "old" phonons, and the amplitude modes are the orthogonal linear combination. Phason and amplitude modes quickly merge into the old phonon spectrum, as indicated.

45.4.13 Phason Anomaly in Point Contact Spectroscopy

Every broken symmetry gives rise to new collective excitations, e.g. spin waves in ferromagnets or Goldstone bosons in field theory. For a CDW structure the new modes are phasons [61]. The ground-state energy of an *incommensurate* CDW is independent of its spatial location relative to the lattice ions, i.e. the energy is independent of the CDW phase φ. Therefore, the energy will be only slightly increased if φ is allowed to vary *slowly* in space and time:

$$\varphi(\mathbf{L}, t) = \sum_{\mathbf{q}} \varphi_{\mathbf{q}} \sin(\mathbf{q} \cdot \mathbf{L} - \omega_{\mathbf{q}} t) \ . \tag{45.32}$$

$\varphi_{\mathbf{q}}$ is the amplitude of each phason mode and $\omega_{\mathbf{q}}$ its frequency. The displacement of each ion from its ideal lattice site $\mathbf{L}$, which occurs in order to maintain microscopic charge neutrality, is

$$\mathbf{u} = \mathbf{A} \sin[\mathbf{Q} \cdot \mathbf{L} + \varphi(\mathbf{L}, t)] \ . \tag{45.33}$$

$\mathbf{q}$ perpendicular to $\mathbf{Q}$ describes a periodic, *local* rotation of $\mathbf{Q}$; $\mathbf{q}$ parallel to $\mathbf{Q}$ describes a periodic, *local* modulation of the magnitude of $\mathbf{Q}$.

The lattice dynamics of phason excitations leads to a frequency spectrum that one would anticipate from a broken symmetry [61]. $\omega_{\mathbf{q}}$ approaches zero linearly in $|\mathbf{q}|$ and is highly anisotropic. The modes are relatively "soft" for $\mathbf{q}$ perpendicular to $\mathbf{Q}$ (the ones associated with local rotations of $\mathbf{Q}$). Their velocity has been calculated from first principles [78]. The relationship between phasons and phonons (before the broken symmetry) is illustrated in Figure 45.15. Two phonon modes with wave vectors differing by $2\mathbf{Q}$ (for potassium, by $2\mathbf{Q} - 2\mathbf{G}_{110}$) are coupled together by off-diagonal elements of the dielectric response matrix [79]. One mode is "pushed" to zero frequency, and the other is pushed to higher frequency. The latter mode can be described as a long-wavelength amplitude modulation of the CDW. It is the low frequency of phasons that leads to their dominant role in low-temperature transport [62]. They also cause heat capacity anomalies [79] at very low temperature. These have

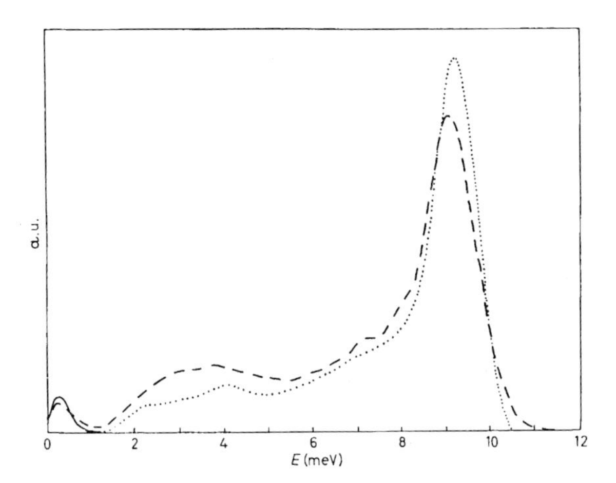

Figure 45.16 Point contact spectrum d^2V/dI^2 for potassium. The long-dashed experimental curve is from [84]. The dotted curve is the theoretical prediction from electron–phonon interactions; it is zero for voltages < 1 meV. The theoretical phason contribution (solid curve) is sensitive to temperature and was calculated by Ashraf and Swihart [85] for 1.2 K, the temperature of the experiment.

been identified in $LaGe_2$ [80]. Observation in K has been reported [81] and contested [82]. Further work is needed.

Point contact spectroscopy is a new technique for measuring inelastic scattering in normal metals [83]. Two metal points touched together in a small contact (size ≪ mean free path) create a junction similar to a superconducting tunnel junction. Spectra of $d^2 V/dI^2$ *vs.* voltage V reveal phonon spectra (weighted by a square matrix element and a geometric factor). The geometric factor arises in this case only because an electron must be scattered back through the contact it has just crossed in order for the phonon event to contribute to the contact resistance. Accordingly low-energy acoustic phonons, which scatter electrons at the Fermi surface only through small angles, contribute a vanishing contribution at voltages comparable to their energy.

Nevertheless, data on potassium by Jansen et al. [84] show a striking anomaly below 1 MeV. where there should be no contribution at all. Recently Ashraf and Swihart [85] have found that electron–phason scattering explains the anomaly – its shape and size – without any adjustable parameters. Data on the phason spectrum were taken from the fit to the resistivity data between 0.1 and 1.3 K [62]. The experimental data and theoretical predictions are compared in Figure 45.16. The scale factor for the ordinate was adjusted to fit the phonon spectrum. Thereafter the phason contribution was completely determined *a priori*. This anomaly is insensitive to the presence of magnetic fields.[7] As expected, it sometimes does not appear, since the **Q** of the point contact region must be approximately parallel to the current to allow backscattering by phasons.

45.5 Conclusion

It is clear that many manifestations of CDW structure – some of which are extraordinary – occur in potassium and sodium. Many other phenomena need to be explored. One of the most important is to locate and measure the Bragg reflection satellites, which occur for scattering

7) Wyder P.: personal communication reported in [85].

vectors

$$\mathbf{K} = \mathbf{G} \pm \mathbf{Q} . \tag{45.34}$$

These reflections arise as a result of the periodic, lattice displacement (45.33). A theory of the satellite intensity [86] indicates that they are five orders of magnitude smaller than an ordinary Bragg reflection. This is below the sensitivity of the best, published experimental search [87]. A more recent, unconfirmed study has located peaks having the appropriate size and location (with $|\mathbf{Q}| \approx 1.38 \times 2\pi/a$), but much effort is needed to rule out artifacts.[8)]

Another interesting area involves nuclear magnetic resonance. The CDW and its accompanying lattice distortion, Equation (45.33), should cause an intrinsic broadening of the NMR line. This has not been observed [88]. Possibly the expected broadening is motionally narrowed by phase fluctuations [89] or narrowed by long-range indirect exchange interactions [90].

Positron annihilation angular correlation can, in principle, measure the momentum distribution in conduction electron wave functions. This is altered in a significant way by the extra wave vector components introduced by a CDW. However, this can masquerade as an incomplete thermalization of the positrons [91]. The effect appears to have been seen, but it too needs detailed study.

The obvious conclusion is that the alkali metals are by no means "simple". The presence of CDW structure introduces a small universe of new phenomena Unit are challenging to theorists and experimentalists alike. The varied behavior transcends the individual effects, no matter how fascinating they might be, since the sole cause of all of them stems from the intricacies of many-electron theory.

References

1 Overhauser A.W. (1962) *Phys. Rev.* **128**, 1437.
2 Overhauser A.W. (1968) *Phys. Rev.* **167**, 691.
3 Kushida T., Murphy J.C. and Hanabusa M. (1976) *Phys. Rev. B* **13**, 5136.
4 Bloch F. (1029) *Z. Phys.* **57**, 545.
5 Wigner E. (1938) *Trans. Faraday Soc.* **34**, 678.
6 Overhauser A.W. (1960) *Phys. Rev. Lett.* **4**, 415.
7 Overhauser A.W. (1960) *Phys. Rev. Lett.* **4**, 462.
8 Peierls R.E. (1955) *Quantum Theory of Solids*, London, p. 108.
9 Shoenberg D. and Stiles P.J. (1964) *Proc. R. Soc. London, Ser. A* **281**, 62.
10 Mayer H. and El Naby M.H. (1963) *Z. Phys.* **174**, 269.
11 Overhauser A.W. (1964) *Phys. Rev. Lett.* **13**, 190.
12 Ashcroft N. (1965) *Phys. Rev.* **140**, A935.
13 Overhauser A.W. and Butler N.R. (1976) *Phys. Rev. B* **14**, 3371.
14 Giuliani G.F. and Overhauser A.W. (1979) *Phys. Rev. B* **20**, 1328.
15 Fragachan F. and Overhauser A.W. (1984) *Phys. Rev. B* **29**, 2912.
16 Blount E.I. (1962) *Phys. Rev.* **126**, 1636.
17 Huberman M. and Overhauser A.W. (1981) *Phys. Rev. Lett.* **47**, 682.
18 Lifshitz I.M., Azbel M.Y. and Kagonov M.I. (1956) *Z. Eksp. Teor. Fiz.*, **31**, 63 (English translation: *Sov. Phys. JFTP* **4**, 41)
19 Moss J.S. and Datars W.R. (1967) *Phys. Lett. A* **24**, 630.
20 Visscher P.B. and Falicov L.M. (1970) *Phys. Rev. B* **2**, 1518.
21 Huberman M. and Overhauser A.W. (1981) *Phys. Rev. B* **23**, 6294.

8) Werner S.A.: personal communication.

22 Huberman M. and Overhauser A.W. (1982) *Phys. Rev. B* **25**, 2211.

23 Zhu X. and Overhauser A.W. (1984) *Phys. Rev. B* **30**, 622.

24 Coulter P.G. and Datars W.R. (1980) *Phys. Rev. Lett.* **45**, 1021.

25 Coulter P.G. and Dataras W.R. (1982) *Solid State Commun.* **43**, 715.

26 Overhauser A.W. (1982) *Can. J. Phys.* **60**, 687.

27 Macke W. (1950) *Z. Naturforsch. Text A* **5**, 192.

28 Slater J.C. (1951) *Phys. Rev.* **81**, 383.

29 Kohn W. and Sham L.J. (1965) *Phys. Rev.* **140**, A1133.

30 Overhauser A.W. (1970) *Phys. Rev. B* **2**, 874.

31 Hamann D.R. and Overhauser A.W. (1966) *Phys. Rev.* **143**, 183.

32 Overhauser A.W. (1971) *Phys. Bev. B* **3**, 1888.

33 Mayer H. and Hietel B. (1960) In: Abeles F. (ed) *Optical Properties and Electronic Structure of Metals and Alloys*, Amsterdam, p. 47.

34 Harms P. (1972) Dissertation. Technische Universität Clausthal, unpublished.

35 Fäldt A. and Walldén L. (1980) *J. Phys. C* **13**, 6429.

36 Taut M. (1982) *J. Phys. F* **12**, 2019.

37 Hietel B. and Mayer H. (1973) *Z. Phys.* **264**, 21.

38 Walsh W.M. Jr., Kupp L.W. Jr. and Schmidt P.H. (1966) *Phys. Per.* **142**, 414.

39 Dunifer G.: personal communication.

40 Overhauser A.W. and de Graaf A.M. (1968) *Phys. Rev.* **168**, 763.

41 Taub H, Schmidt B.L., Maxfield B.W. and Bowers B. (1971) *Phys. Per. B* **4**, 1134.

42 Chimenti D.E. and Maxfield B.W. (1973) *Phys. Per. B* **7**, 3501.

43 van Kempen H., Lass J.S., Ribot J.H.J.M. and Wyder P. (1976) *Phys. Rev. Lett.* **37**, 1574.

44 Holroyd F.W. and Datars W.B. (1975) *Can. J. Phys.* **53**, 2517.

45 Penz P.A. and Bowers R. (1968) *Phys. Rev.* **172**, 991.

46 Overhauser A.W. (1974) *Phys. Rev. B* **9**, 2441.

47 Huberman M. and Overhauser A.W.: unpublished.

48 Stetter G., Adlhart W., Fritsch G., Steichele E. and Luscher E. (1978) *J. Phys. F* **8**, 2075.

49 Schouten D.B. and Swenson C.A. (1974) *Phys. Rev. B* **10**, 2175.

50 Werner S.A. , Hunt T.K. and Ford G.W. (1974) *Solid State Commun.* **14**, 1277.

51 Schaefer J.A. and Marcus J.A. (1971) *Phys. Rev. Lett.* **27**, 935.

52 Lass J.S. (1976) *Phys. Rev. B* **13**, 2247.

53 Bishop M.F. and Overhauser A.W. *Phys. Rev. Lett.* **39**, 632.

54 Overhauser A.W. (1971) *Phys. Rev. Lett.* **27**, 938.

55 Bishop M.F. and Overhauser A.W. (1978) *Phys. Rev. B* **18**, 2447.

56 De Gennaro S. and Borchi E. (1982) *J. Phys. F* **12**, 2363.

57 Bowlands J.A., Durvury C. and Woods S.B. (1978) *Phys. Rev. Lett.* **40**, 1201.

58 Levy B., Sinvani M. and Greenfield A.J. (1979) *Phys. Rev. Lett.* **43**, 1822.

59 van Kempen H., Ribot J.H.J.M. and Wyder P. (1981) *J. Phys. F* **11**, 597.

60 Schroeder P.A. (1982) *Physica B (Utrecht)* **109–110**, 1901.

61 Overhauser A.W. (1971) *Phys. Rev. B* **3**, 3173.

62 Bishop M.F. and Overhauser A.W. (1981) *Phys. Rev. B* **23**, 3638.

63 Bishop M.F. and Lawrence W.E. (1984) *Bull. Am. Phys. Soc.* **29**, 236.

64 Lawrence W.E. and Wilkins J.W. (1973) *Phys. Rev. B* **7**, 2317.

65 Lee C.W., Haerle M.L., Heinen V., Bass J., Pratt W.P. Jr., Bowlands J.A. and Schroeder P.A. (1982) *Phys. Rev. B* **25**, 1411.

66 Kaveh M. and Wiser N. (1981) *J. Phys. F* **11**, 597.

67 Gugan D. (1982) *J. Phys. F* **12**, L173.

68 Love D.P., Van Degrift C.T. and Parker W.H. (1982) *Phys. Rev. B* **26**, 5577.

69 Black J.E. and Mills D.L. (1980) *Phys. Rev. B* **21**, 5860.

70 Danino M. and Overhauser A.W. (1984) *Bull. Am. Phys. Soc.* **29**, 235.

71 Bobel G., Cimberle M.B., Napoli F. and Rizzuto C. (1976) *J. Low Temp. Phys.* **23**, 103.

72 Jumper W.D. and Lawrence W.E. (1980) *Phys. Rev. B* **16**, 3314.

73 Sharma S.M. (1980) *J. Phys. F* **10**, L47.

74 Bowers B., Legendy C. and Bose F. (1961) *Phys. Rev. Lett.* **7**, 339.

75 Kjeldaas T. (1959) *Phys. Rev.* **113**, 1473.

76 Penz P.A. and Kushida T. (1968) *Phys. Rev.*, **176**, 804.

77 Overhauser A.W. and Rodriguez S. (1966) *Phys. Rev.* **141**, 431.

78 Giuliani G.F. and Overhauser A.W. (1989) *Phy. Rev. B* **21**, 5577.

79 Giuliani G.F. and Overhauser A.W. (1980) *Phys. Rev. Lett.* **45**, 1335.

80 Sawada A. and Satoh T. (1978) *J. Low Temp. Phys.* **30**, 455.

81 Amarasekara C.D. and Keesom P.H. (1982) *Phys. Rev. B* **26**, 2720.

82 Van Curen J., Hornung E.W., Lasjaunias J.C. and Phillips X.E. (1982) *Phys. Rev. Lett.* **49**, 1653.

83 Yanson I.K. (1974) *Ž. Ėksp. Teor. Fiz.* **66**, 1035 (English translation: *Sov. Phys. JETP* **39**, 506).

84 Jansen A.G.M., vanden Bosch J.H., van Kempen H., Ribot J.H.J.M, Sheets P.H.H. and Wider P. (1980) *J. Phys. F* **10**, 265.

85 Ashraf M. and Swihart J.C. (1983) *Phys. Rev. Lett.* **50**, 921.

86 Giuliani G.F. and Overhauser A.W. (1980) *Phys. Rev. B* **22**, 3639.

87 Werner S.A., Eckert J. and Shirane G. (1980) *Phys. Rev. B* **21**, 581.

88 Follstaedt D. and Slichter C.P. (1976) *Phys. Rev. B* **13**, 1017.

89 Overhauser A.W. (1978) *Hyperfine Interact.* **4**, 786.

90 Giuliani G.F. and Overhauser A.W. (1982) *Phys. Rev. B* **26**, 1671.

91 Sharma S.M. and Gupta S.C. (1983) *J. Phys. F* **13**, E7.

Reprint 46 Photoemission From the Charge-Density Wave in Na and K[1)]

*A.W. Overhauser**

* Department of Physics, Purdue University, 525, Northwestern Avenue, West Lafayette, Indiana 47907, USA

Received 20 May 1985

Angle-resolved photoemission perpendicular to the surface of a Na or K film (for photon energies near 35 and 26 eV, respectively) can reveal the momentum mixing and band bending caused by the charge-density-wave potential.

Recent photoemission studies of the simple metals [1] Al and [2] Be have demonstrated that fine details in the electronic spectrum $E(\mathbf{k})$ can be measured. It seems timely, therefore, to probe the charge-density-wave (CDW) structure of Na and K by this technique. Angle-resolved photoemission is especially suitable for the study of CDW's in these metals [3] because it is known from optical data that the CDW wave vector $\mathbf{Q}$ is oriented perpendicular to the surface of an evaporated film [4]. This feature, which prevents observation of the CDW optical anomaly in thin films [5], allows one to excite photoelectrons near the CDW energy gap. Wave-function mixing and band bending caused by the CDW potential lead to verifiable predictions about the energy spectrum of photoelectrons emitted perpendicular to the surface. The theory presented here provides a basis for the interpretation of the experimental data on metallic Na given in the preceding paper [6].

The free-electron energy spectrum $E(\mathbf{k})$ for $\mathbf{k}$ along a [110] direction is shown in Figure 46.1 and is the same for all alkali metals. Thin films of Na and K having [110] normal to the surface can be grown on amorphous substrates [4] or on a clean Ni(100) surface [6]. The vertical arrows in Figure 46.1 indicate photoelectric excitations (from occupied states, E_i, below the Fermi level to empty states above). The excited electron, of energy E_F, can escape from the metal perpendicular to the surface. Such photoelectrons cannot arise for all photon energies, $h\nu$, if the Fermi surface is spherical and if there is no CDW broken symmetry. The resulting excitation gap is shown in Figure 46.1.

E_f and $h\nu$ are measured directly, and the initial energy is found from

$$E_i = E_f - h\nu \,. \tag{46.1}$$

A plot of E_i *vs.* $h\nu$ is a valuable way to display photoemission data [1] since no assumptions about $E(\mathbf{k})$ need be made. Such a plot for the free-electron model of Na is shown in Figure 46.2. Note particularly the gap, $\sim$ 6 eV wide, near $h\nu = 35$ eV. It is clear from Figure 46.1 that this gap is caused by the failure of the Fermi sphere to reach the Brillouin-zone boundary. A CDW structure [7] leads to a drastic modification of Figure 46.2.

Conduction electrons near the CDW energy gap are strongly influenced by two periodic potentials:

$$K(\mathbf{r}) = 2\alpha \cos(\mathbf{Q}\cdot\mathbf{r}) + 2\beta \cos(\mathbf{G}\cdot\mathbf{r}) \,. \tag{46.2}$$

$\mathbf{G}$ is the (110) reciprocal-lattice vector and is directed within a few degrees [8] of $\mathbf{Q}$. For the present purpose, $\mathbf{Q}$ and $\mathbf{G}$ can be taken parallel, and both are perpendicular to the surface

1) Phys. Rev. Lett. 55, 18 (1985).

Anomalous Effects in Simple Metals. Albert Overhauser

ISBN: 978-3-527-40859-7

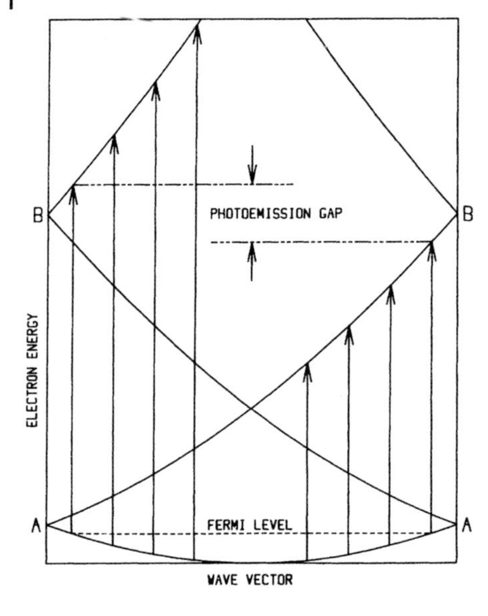

Figure 46.1 Free-electron energy spectrum for a (bcc) alkali metal. The wave vector **k** is parallel to a [110] direction. The arrows denote photoelectric excitations. Small zone-boundary energy gaps at points A and B are not shown.

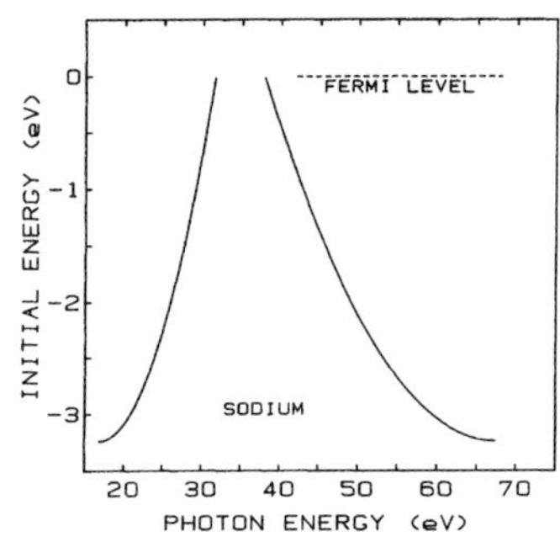

Figure 46.2 Initial photoelectron energy vs. photon energy for the free-electron model of Na (see Figure 46.1).

of the evaporated film. Two sets of energy gaps and the distorted Fermi surface are shown in Figure 46.3. Electrons in a narrow tube surrounding the arrows of **Q** are the ones that (after excitation) leave the metal perpendicular to the surface. Electrons in states near the arrow heads are the ones which (as shown below) will bridge the gap in the spectrum of Figure 46.2.

Solution of the Schrödinger equation having the potentials given in Equation (46.2) is easily accomplished by a plane-wave expansion [9]. The basis functions are

$$\begin{aligned}&|\mathbf{k}\rangle, \quad |\mathbf{k}-\mathbf{Q}\rangle, \quad |\mathbf{k}-\mathbf{G}\rangle, \quad |\mathbf{k}-\mathbf{Q}+\mathbf{G}\rangle,\\ &|\mathbf{k}-\mathbf{G}+\mathbf{Q}\rangle, \quad |\mathbf{k}-2\mathbf{Q}+\mathbf{G}\rangle, \quad |\mathbf{k}-2\mathbf{G}+\mathbf{Q}\rangle,\\ &|\mathbf{k}-2\mathbf{Q}+2\mathbf{G}\rangle\ , \text{ etc.}\end{aligned} \tag{46.3}$$

The off-diagonal matrix elements of the Hamiltonian matrix, from Equation (46.2), have values α, β, or 0. The eigenvalue spectrum is periodic in a narrow strip in k space of thickness $G - Q$. The eigenvalues with the CDW potential $2\alpha = 0$, and the crystal potential $2\beta = 0.5$ eV, are shown in the left panel of Figure 46.4. $Q = 0.96G$ was used. The corresponding solution for $2\alpha = 1.3$ eV, appropriate to the measured CDW optical anomaly in

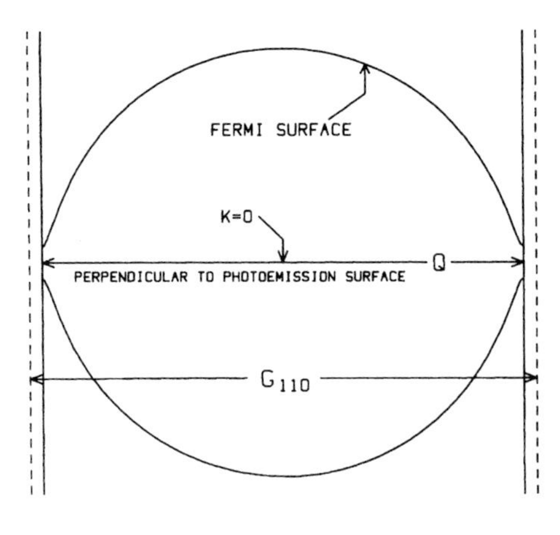

Figure 46.3 Brillouin-zone energy gaps (dashed lines), CDW energy gaps (solid lines), and the distorted Fermi surface of an alkali metal having a CDW of wave vector **Q**. Several sets of higher-order gaps (minigaps and heterodyne gaps, see [9]) are not shown.

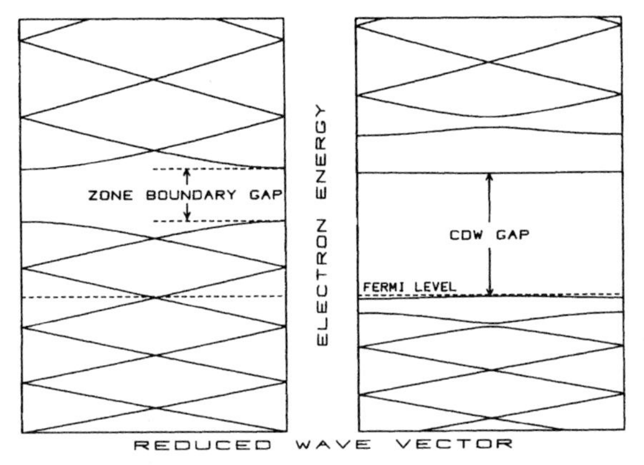

Figure 46.4 Conduction-electron energy spectrum for Na in a reduced zone of width $G - Q = 0.04G$. The left panel applies when the only periodic potential is that of the bcc lattice. The right panel applies when there is also a CDW potential, as given in Equation (46.2) with α and β having values for Na.

Na [10], is shown in the right panel of Figure 46.4. Observe that many of the (formerly) empty states have been pushed below the Fermi level. The strong mixing of the basis functions, (46.3), caused by the CDW, ensures that all "bare"-momentum states up to (and beyond) the zone boundary are included in the occupied eigenstates.

The CDW spectrum of Figure 46.4 was computed by diagonalization of a 15×15 matrix. The eigenvalues were used to calculate a photoelectron plot, analogous to Figure 46.2, and the result is shown in Figure 46.5. Note the appearance of the first and second minigaps [9]. The most important feature is that the 6-eV opening of Figure 46.2 is now bridged by an almost flat top just below the Fermi level. The "bare"-momentum mixing will permit weak extensions of the flat top to appear on either side of the central flat top. For clarity these have been shown as dashed.

An important characteristic of the CDW "bridge" in Figure 46.5 is the energy width w of the external photoelectrons. If the initial-state width is neglected (as is permissible near the Fermi level), the formula [11] for w reduces to

$$w \cong RW/(1 - R) , \tag{46.4}$$

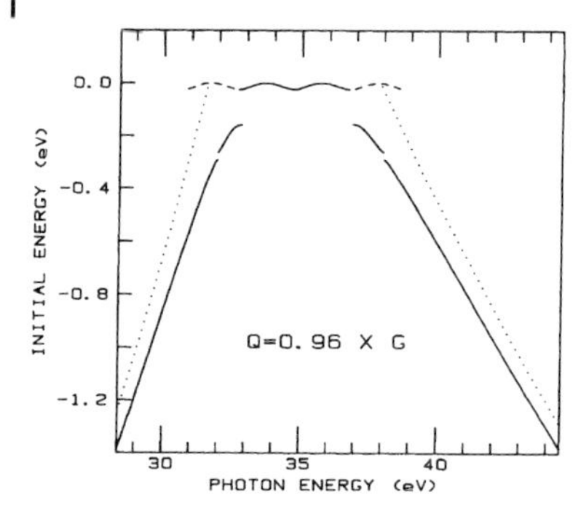

Figure 46.5 Initial photoelectron energy vs. photon energy for Na. The solid curve results when the CDW structure is recognized. The nearly flat "bridge" near 35 eV is the dominant consequence of the CDW. The dotted curve is the free-electron result from Figure 46.2.

where W, $\sim$ 5 eV [1], is the width of the excited states, and R is the ratio of the initial-state slope, $\mathrm{d}E/\mathrm{d}k$, to the final-state slope. For the transitions of Figure 46.5, excluding the bridge, $R \sim \frac{1}{3}$; so that w should be 2–3 eV. However, the initial states that cause the bridge have a very small $\mathrm{d}E/\mathrm{d}k$. (These states come from the band just below the Fermi level in the right half of Figure 46.4.) Consequently, the width of the external photoelectron peak in the bridge region of Figure 46.5 should be *much* smaller than elsewhere.

Confirmation of the foregoing features would be an important addition to the many [5, 12] unusual properties of Na and K caused by their CDW broken symmetry [7]. All that has been described above for Na applies without change to K, except that the bridge is near $h\nu =$ 26 eV. The Fermi surface shown in Figure 46.3 is (in reality) fractured by several higher-order gaps [9], which cause [13] numerous open-orbit peaks in the magnetoresistance [14, 15]. Such minute details cannot be resolved in photoemission.

The writer is grateful to Eric Jensen for several illuminating discussions about photoemission and to the National Science Foundation for research support.

References

1 Levinson H.J., Freuter F., and Plummer E.W. (1983) *Phys. Rev. B* **27**, 727.

2 Jensen E., Bartynski R.A., Gustafsson T., and Plummer E.W. (1984) *Phys. Rev. Lett.* **52**, 2172.

3 Boriack M.L. and Overhauser A.W. (1978) *Phys. Rev. Lett.* **41**, 1066.

4 Overhauser A.W. and Butler N.R. (1976) *Phys. Rev. B* **14**, 3371.

5 Overhauser A.W. (1978) *Adv. Phys.* **27**, 343.

6 Jensen E. and Plummer E.W., preceding Letter [*Phys. Rev. Lett.* **55**, 1912 (1985)].

7 Overhauser A.W. (1968) *Phys. Rev.* **167**, 691.

8 Giuliani G.F. and Overhauser A.W. (1979) *Phys. Rev. B* **20**, 1328.

9 Fragachan F.E. and Overhauser A.W. (1984) *Phys. Rev. B* **29**, 2912.

10 Hietel B. and Mayer H. (1973) *Z. Phys.* **264**, 21.

11 Chiang T.-C., Knapp J.A., Aono M., and Eastman D.E. (1980) *Phys. Rev. B* **21**, 3513 (Equation 14).

12 Overhauser A.W. (1983) In: Devreese J.T. and Brosens F. (eds) *Electron Correlations in Solids, Molecules, and Atoms*, Plenum, New York, p. 41.

13 Huberman M. and Overhauser A.W. (1982) *Phys. Rev. B* **25**, 2211.

14 Coulter P.G. and Datars W.R. (1985) *Phys. Rev. Lett.* **45**, 1021.

15 Coulter P.G. and Datars W.R. (1985) *Can. J. Phys.* **63**, 159.

Reprint 47 Phason Narrowing of the Nuclear Magnetic Resonance in Potassium[1)]

*Y.R. Wang** and *A.W. Overhauser**

* Department of Physics, Purdue University, 525, Northwestern Avenue, West Lafayette, Indiana 47907, USA

Received 15 July 1985

Nuclear magnetic resonance in a metal having an incommensurate charge-density wave (CDW) should exhibit significant broadening because the Knight shift depends on the conduction-electron density surrounding each nucleus. Nevertheless, experiments on metallic potassium at 1.5 K have not revealed such an effect. Thermal excitation of phasons, the low-frequency collective modes of a CDW, cause sufficient motional narrowing to explain this observation. However below 100 mK, as the CDW phase excitations subside, the NMR line should broaden rapidly with decreasing temperature.

47.1 Introduction

Charge-density-wave (CDW) instabilities in simple metals are caused by the many-body effects of exchange and correlation [1]. The conduction electrons have lower energy if they have a sinusoidally modulated charge density,

$$\rho(\mathbf{r}) = \rho_0 \left[1 + p \cos\left(\mathbf{Q}\cdot\mathbf{r} + \phi\right)\right] , \tag{47.1}$$

rather than a uniform value ρ_0. The CDW wave vector is approximately [2]

$$\mathbf{Q} \approx 2k_F \left(1 + G/4E_F\right) , \tag{47.2}$$

where G is the CDW energy gap, and k_F and E_F are the Fermi-surface radius and energy. For an alkali metal the direct of $\mathbf{Q}$ is tilted a few degrees away from a $\langle 110\rangle$ axis [3]. The CDW amplitude p, in the case of potassium, is ~ 0.11 [4]. Since the broken symmetry is incommensurate with the lattice, the total electronic energy is independent of the phase ϕ,

A wealth of experimental evidence [5, 6] has shown the effects of the CDW's in Na and K. The most dramatic example to date is the direct observation of open orbits in high magnetic fields [7]. These orbits arise from higher-order energy gaps, which truncate the Fermi surface [8]. Induced-torque anisotropies [9] below 3 T have also shown that the Fermi surface is multiply connected and lacks cubic symmetry [5, 6]. Both the open-orbit spectra [10] and the induced-torque anisotropy [11] have been explained. These phenomena contradict the simple interpretation of de Haas–van Alphen experiments [12], which suggest a simply connected and isotropic Fermi surface. In light of the open-orbit observations, the most reasonable understanding [13] of de Haas–van Alphen isotropy is that published data have been obtained on samples with a very small **Q**-domain size.[2)]

1) Phys. Rev. B 32, 11 (1985).
2) Some workers have warned that Na and K samples should not be etched (a procedure which generally improves crystalline perfection). It is very likely that etching increases **Q**-domain size through faceting of (otherwise) highly curved surfaces.

Anomalous Effects in Simple Metals. Albert Overhauser

ISBN: 978-3-527-40859-7

A significant theoretical problem[3] which needs study concerns the nuclear-magnetic-resonance (NMR) linewidth in potassium. The CDW broken symmetry will lead to an extremely broadened NMR signal – much larger than the Van Vleck dipolar width [14], $\sim 0.14\,\mathrm{G}$. There are two major sources of additional broadening: quadrupole perturbations and Knight-shift variations.

A CDW in a cubic metal leads to quadrupole broadening because the positive-ion lattice sustains a periodic lattice displacement [1],

$$\mathbf{u}(\mathbf{r}) = \mathbf{A}\sin(\mathbf{Q}\cdot\mathbf{r} + \phi)\,, \tag{47.3}$$

in order to compensate the charge modulation of the conduction electrons. The displacement amplitude has a magnitude [4],

$$A \approx \frac{p}{Q f(Q)} \sim 0.03\,\text{Å}\,, \tag{47.4}$$

where $f(\mathbf{Q})$ is the total-charge form factor of the positive ion for wave-vector $\mathbf{Q}$. ($f \sim 2.9$ rather than unity.) An analysis of quadrupole effects has already been presented in detail [15], but a value $A \sim 0.11$ Å was assumed. The value (47.4) will considerably reduce the estimated quadrupole broadening. Therefore, the present work will focus on the dominant source of CDW broadening: Knight-shift variations.

We shall find that at $T = 0\,\mathrm{K}$ and $H_0 = 6\,\mathrm{T}$ the NMR linewidth caused by Knight-shift variations (from the crests to the troughs of the CDW) is ~ 34 Oe, a value more than 2 orders of magnitude larger than the observed width [15], 0.27 Oe, at $T = 1.5\,\mathrm{K}$. The main thrust of this paper is to show that thermal excitation of phasons [16], the low-frequency collective modes of an incommensurate CDW, provide sufficient motional narrowing [17] to explain the experimental result at $T = 1.5\,\mathrm{K}$, the lowest temperature studied so far. The emphasis will be on the temperature dependence of the phason narrowing. Below about 100 mK the effect of the CDW structure should become apparent, and the NMR line should broaden rapidly as T is further reduced.

47.2 NMR Line Shape at $T = 0$ K

The Knight shift K_0 is defined by the NMR frequency shift in a metal [18],

$$\omega \equiv \omega_d(1 + K_0)\,, \tag{47.5}$$

relative to the frequency in a diamagnetic salt. The shift arises from the spin paramagnetism of the conduction electrons. For potassium, $K_0 = 0.26\%$ [19]. If potassium were truly body-centered cubic (no CDW) each nuclear spin would, for $H_0 = 6\,\mathrm{T}$, experience a hyperfine field of 156 Oe. However, the Knight shift in a CDW state will depend on position in a way similar to the local electron density, Equation (47.1):

$$K(\mathbf{r}) = K_0 + \kappa\cos(\mathbf{Q}\cdot\mathbf{r} + \phi)\,. \tag{47.6}$$

It is a simple exercise to show that the probability distribution $P(\Delta K)$ for a given Knight shift $K_0 + \Delta K$ is

$$P(\Delta K) = \frac{1}{\pi\left[\kappa^2 - (\Delta K)^2\right]^{1/2}}\,. \tag{47.7}$$

3) From an experimental view the most crucial problem is observation of the CDW satellites. These diffraction spots are smaller than Bragg reflections by $\sim 10^5$, and lie very close to the Bragg spots. (See [4].) Sufficiently sensitive experiments have not been reported.

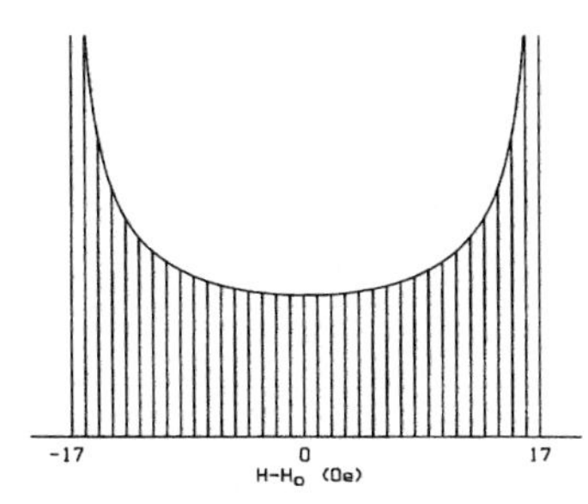

Figure 47.1 NMR line shape at $T = 0$ caused by a CDW in potassium. $H_0 = 6\,\text{T}$.

This distribution is shown in Figure 47.1 and has singularities at the limits $\pm\kappa$. This figure would describe the expected NMR line shape if the CDW phase ϕ did not fluctuate. Analogous line shapes for metals with two or three CDW's have been calculated [20]. In potassium the anisotropy of the CDW-induced optical absorption indicates that there is only one [21].

In order to calibrate the width of the NMR line shape shown in Figure 47.1 it is necessary to determine the parameter κ. It is a reasonable approximation to take

$$\kappa \approx p K_0 \,, \tag{47.8}$$

where p is the fractional-charge modulation, Equation (47.1). This modulation is caused by the exchange and correlation potential of the CDW:

$$V_{\text{CDW}}(x) = G \cos(\mathbf{Q}x) \,. \tag{47.9}$$

(We have chosen the $\mathbf{Q}$ direction along $\hat{x}$.) The CDW gap for potassium is known [2] directly from optical-absorption data [22]; $G = 0.62\,\text{eV}$. A simple calculation of p from Equations (47.9) and (47.2) leads to an incorrect value of 0.17. The reason is that F_{CDW} is nonlocal; it arises entirely from exchange and correlation. The matrix elements of (47.9) are wave-vector dependent:

$$\begin{aligned} \langle k_x + \mathbf{Q}|\, V_{\text{CDW}} |k_x\rangle &\equiv \tfrac{1}{2} G_+(k_x) \,, \\ \langle k_x - \mathbf{Q}|\, V_{\text{CDW}} |k_x\rangle &\equiv \tfrac{1}{2} G_-(k_x) \,. \end{aligned} \tag{47.10}$$

The nonlocal theory [23] of the G_+ and G_- has been fitted numerically [24]:[4)]

$$\mathbf{G}^+(\mathbf{k}_x) \cong \mathbf{G}/\left(1 + 2.2\mu - 0.48\mu^2\right) \,, \tag{47.11}$$

where

$$\mu \equiv (\mathbf{k}_x/\mathbf{Q}) + \tfrac{1}{2} \,. \tag{47.12}$$

The fact that the CDW potential, Equation (47.9), depends on the *axis* of $\mathbf{Q}$, and not on the vector direction, leads to a relation between G^+ and G^-:

$$G^-(\mathbf{K}_x) = G^+(-\mathbf{k}_x) \,. \tag{47.13}$$

Each conduction-electron wave function becomes amplitude modulated when the CDW perturbation (47.9) is included in Schrödinger's equation. Use of the nonlocal matrix elements (47.10) leads to [4]

$$p \approx 0.11 \,. \tag{47.14}$$

4) Equation (17) of [24] contains a transcription error (a factor 4 in the definition of μ) and is corrected in our Equation (47.11). The calculations of [4] were performed with the correct matrix elements.

This value implies, from Equation (47.8), that the total CDW width of the NMR line is (for $H_0 = 6\,\mathrm{T}$)

$$2\kappa H_0 \approx 34\,\mathrm{Oe}\,. \tag{47.15}$$

It must be emphasized that this value is an approximate one because Equation (47.8) is not strictly correct. The charge modulation p corresponds to the Fermi-volume average of the wave-function amplitude modulation. The inhomogeneous Knight shift depends, instead, on the Fermi-surface density-of-states average p' of the wave-function modulation. However, the difference between p' and p is comparable to the uncertainty in p, Equation (47.14), so we shall ignore the distinction.

The NMR line shape described above will be slightly narrowed (a few percent) by zero-point phase fluctuations. The theory of this effect [20] was presented several years ago, before the phason excitation spectrum for potassium was known; and this uncertainty led to an overestimate of the narrowing at $T = 0\,\mathrm{K}$. We now turn to the motional narrowing caused by thermal excitation of phasons.

47.3 Review of Phason Properties

Phasons are the low-frequency collective modes of a CDW broken symmetry [16]. The phase ϕ appearing in Equation (47.1) becomes a dynamic variable. It can be expanded as follows:

$$\phi(\mathbf{r}, t) = \sum_{\mathbf{q}} \phi_q \sin(\mathbf{q}\cdot\mathbf{r} - \omega_q t + \gamma_q)\,, \tag{47.16}$$

where $\mathbf{q}$ and ω_q are the phason wave vector and frequency. γ_q, the phase of the phason, is randomly distributed between 0 and 2π. Phasons have a finite lifetime τ_q, which we discuss below. The randomness of γ_q arises from this fact together with the *assumption* that the CDW is not pinned by lattice imperfections.

The phason spectrum has a linear dispersion relation (for most of its range) and is anisotropic [16]. Suppose we take $\mathbf{Q}$ along $\hat{x}$. Then,

$$\omega_q \cong c_0 \left(q_y^2 + q_z^2 + \gamma^2_{q_x^2}\right)^{1/2}\,. \tag{47.17}$$

c_0 is the velocity of a phason traveling transverse to $\mathbf{Q}$, and γc_0 is the velocity parallel to $\mathbf{Q}$. The phason spectrum extends only to a cutoff frequency ω_ϕ [25]. The allowed $\mathbf{q}$'s lie in a small ellipsoidal volume of $\mathbf{q}$ space.

The three phason-spectrum parameters rae known approximately for potassium:

$$\begin{aligned} c_0 &\approx 1.4\times 10^5\,\mathrm{cm/s}\,,\\ \gamma &\approx 8\,,\\ \hbar\omega_\phi &\approx 3\,\mathrm{K}\,. \end{aligned} \tag{47.18}$$

The estimate for c_0 is theoretical [26], whereas γ and ω_ϕ are based on an analysis of phason contributions to the low-temperature electrical resistivity [27]. A theory of the point-contact spectrum in potassium [28], for which experiments show an anomaly caused by electron–phason interactions, gives excellent agreement with use of the parameter set (47.18).

Another crucial quantity is the phason lifetime τ_q. The theory [29][5] is based on electronic excitations (from below to above E_F) caused by the electron–phason interaction. The result

5) Equation (25) of this reference omitted the factor $\mathbf{G}/4E_F$. We are grateful to G. F. Giuliani for calling this to our attention.

obtained is

$$\frac{1}{\tau_q} = \frac{\hbar \mathbf{Q}^2 q}{32\pi n \mathbf{M} A^2} \left[\frac{G}{4E_F}\right] \cos^2\theta \,, \tag{47.19}$$

where θ is the angle between $\mathbf{q}$ and $\mathbf{Q}$, n is the electron concentration, $\mathbf{M}$ the ionic mass, and A the CDW displacement amplitude, Equation (47.4). G is the CDW energy gap (0.62 eV in potassium). Since Equation (47.19) is proportional to q, the quality factor $\omega_q \tau_q$ for phasons is independent of $|\mathbf{q}|$. For potassium, Equation (47.19) leads to a quality factor

$$\mu \equiv \omega_q \tau_q = \mu_0 / \cos^2\theta \,, \tag{47.20}$$

with $\mu_0 \sim 4$. Another consequence of Equation (47.19) is that $\tau_q \to \infty$ as $q \to 0$. This would imply that a sliding CDW would not experience a fractional force. Such a fractional force, in fact, arises from scattering processes responsible for the electrical resistivity [30]. This mechanism will also contribute to phason damping (but we neglect it here).

We emphasize that the anisotropies of the phason velocity and damping given by Equations (47.17) and (47.19) have not yet been adequately tested. In particular, the theory that leads to Equation (47.19) is based on an electron-lattice interaction for which only longitudinal phonons can scatter electrons. Therefore, the anisotropies (especially for τ_q) discussed above are possibly exaggerated.

In the following section we will find that motional narrowing of the NMR line depends sensitively on the phason anisotropies. Accordingly, we will carry out several calculations depending on whether (or not) anisotropies of ω_q and τ_q are included.

47.4 Motional Narrowing by Phasons

We will employ the theory of Pines and Slichter [17], which determines the resonance width $1/T_2$ caused by frequency shifts $\pm w$ which (on average) continue for a correlation time τ_c. It is assumed that $w\tau_c \ll 1$. The resonance width is the reciprocal of the time it takes for the phase of the transverse magnetization to "diffuse" one radian. The number of steps N for such a random walk, having phase jumps $\pm w\tau_c$, is given by

$$N^{1/2}(w\tau_c) = 1 \,. \tag{47.21}$$

Since $T_2 = N\tau_c$, it follows (on eliminating N) that

$$\frac{1}{T_2} = w^2 \tau_c \,. \tag{47.22}$$

The intrinsic width w is narrowed by the factor $w\tau_c$.

The only difference for the case of phason narrowing is that each phason mode contributes individually to the random walk. Consider a single term of Equation (47.16). The CDW phase change caused by such a mode between $t = 0$ and $t = t$ is, at $\mathbf{r} = 0$,

$$\Delta\Phi_q = \phi_q \left[\sin\left(-\omega_q t + \gamma_q\right) - \sin\gamma_q\right] \,. \tag{47.23}$$

The square of this change must be averaged over the random phase γ_q and over the distribution of phason lifetimes for that mode:

$$P(t) = \frac{1}{\tau_q} \exp\left[\frac{-1}{\tau_q}\right] \,. \tag{47.24}$$

The integrations are elementary. For an average lifetime

$$(\Delta\Phi_q)^2_{\text{av}} = \frac{\omega_q^2\tau_q^2}{1+\omega_q^2\tau_q^2}\phi_q^2 \,. \tag{47.25}$$

The correlation time τ_c will be (approximately) the time for the CDW phase at the point $\mathbf{r} = 0$ to change by one radian as a consequence of the thermal excitation of *all* phason modes. Thus

$$\sum_{\mathbf{q}} \frac{\tau_c}{\tau_q}(\Delta\Phi_{\mathbf{q}})^2_{\text{av}} \approx 1\,. \tag{47.26}$$

The 1 in the denominator of Equation (47.25) may be dropped, because of (47.20), so

$$\frac{1}{\tau_c} \approx \sum_{\mathbf{q}} \frac{1}{\tau_q}\langle\phi_q^2\rangle\,, \tag{47.27}$$

where the angular brackets around ϕ_q^2 indicate the thermal-equilibrium average. Since the kinetic energy of a phason mode (normalized in unit volume) is [31]

$$T_q = \tfrac{1}{8} n M A^2 \omega_{\mathbf{q}}^2 \phi_{\mathbf{q}}^2\,, \tag{47.28}$$

and since the mean thermal energy of a mode is

$$2\langle T_{\mathbf{q}}\rangle = \hbar\omega_{\mathbf{q}}\left[\exp(\hbar\omega_q/k_B T) - 1\right]^{-1}\,, \tag{47.29}$$

we obtain

$$\langle\phi_{\mathbf{q}}^2\rangle = \frac{4\hbar}{nMA^2\omega_{\mathbf{q}}}\left[\exp(\hbar\omega_q/k_B T) - 1\right]^{-1}\,. \tag{47.30}$$

Equations (47.27) and (47.30) determine the correlation time. Integration over the ellipsoidal $\mathbf{q}$ space of the phason spectrum leads to

$$\tau_c = \frac{\pi^2 n M \hbar^2 A^2 c_0^3}{k_B^3 T^3 D F(\Theta/T)}\,. \tag{47.31}$$

k_B is Boltzmann's constant, Θ is the phason cutoff frequency (~ 3 K), and

$$F(\Theta/T) \equiv \int_0^{\Theta/T} \frac{x^2 \mathrm{d}x}{\exp(x) - 1}\,. \tag{47.32}$$

D is a numerical factor which depends on the anisotropies of the phason velocity and damping:

$$D \equiv \int_0^{\pi} \mathrm{d}\theta \frac{\sin\theta}{\mu(\theta)(\sin^2\theta + \lambda^2\cos^2\theta)^{3/2}}\,. \tag{47.33}$$

μ and λ are defined by Equations (47.17) and (47.20).

We have already mentioned that the anisotropies which affect the value of D are not well established. So we shall evaluate D for four models (with $\mu_0 = 4$).

(a) Isotropic velocity ($\lambda = 1$) and isotropic damping ($\mu = \mu_0$): $D = 0.5$.
(b) Isotropic velocity ($\lambda = 1$) and anisotropic damping ($\mu = \mu_0/\cos^2\theta$): $D = 0.167$.

(c) Anisotropic velocity ($\lambda = 8$) and isotropic damping ($\mu = \mu_0$): $D = 0.63$.
(d) Anisotropic velocity ($\lambda = 8$) and anisotropic damping ($\mu = \mu_0 \cos^2 \theta$): $D = 0.0018$.

The value of $F(\Theta / T)$ at $T = 1.5\,\mathrm{K}$ is 0.99. The theoretical correlation times from Equation (47.31) are then as follows:

$$\tau_c = 5.6 \times 10^{-10}\ , \tag{47.34a}$$

$$\tau_c = 1.7 \times 10^{-9}\ , \tag{47.34b}$$

$$\tau_c = 4.4 \times 10^{-9}\ , \tag{47.34c}$$

$$\tau_c = 1.6 \times 10^{-7}\ , \tag{47.34d}$$

in units of seconds. We take the longest of these τ_c's, case (47.34d), and use it to calculate the CDW linewidth from Equation (47.22). For NMR, resonance at 12 MHz ($H_0 \approx 6\,\mathrm{T}$), for which $w = p K_0 \omega_r = 2 \times 10^4\,\mathrm{s}^{-1}$

$$\omega \tau_c \sim 3 \times 10^{-3}\ . \tag{47.35}$$

The motionally narrowed CDW linewidth is *at most*,

$$\Delta H_{\mathrm{CDW}} = 34(w\tau_c) = 0.1\,\mathrm{Oe}\ . \tag{47.36}$$

This value is consistent with the observed linewidth 0.27 Oe at $T = 1.5\,\mathrm{K}$ and $H_0 = 6\,\mathrm{T}$.

Since the motionally narrowed CDW broadening and the dipolar broadening are expected to have Gaussian shapes, the total linewidth involves addition in quadrature:

$$\Delta H = \left[(0.215)^2 + (34 w\tau_c)^2\right]^{1/2}\ . \tag{47.37}$$

(0.215 Oe is the zero-field linewidth extrapolated from measurements at several H_0 [15].) The maximum contribution of the CDW to the linewidth at 1.5 K and 6 T is therefore −0.02 Oe.

47.5 Temperature Dependence of ΔH

The phason excitations will, of course, subside at low temperature. Consequently, the correlation time τ_c will increase rapidly with decreasing T. The CDW broadening should then come in dramatically. The theoretical behavior, based on Equations (47.31) and (47.37), is shown in

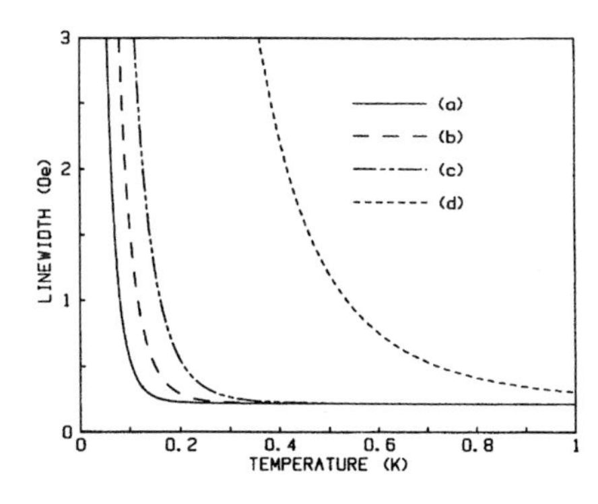

Figure 47.2 NMR linewidth versus temperature for four models of phason anisotropy (described in Section 47.4). $H_0 = 6\,\mathrm{T}$.

Figure 47.2 for the four models described above. It is clear that measurements below 100 mK should be attempted to investigate whether (or not) this interesting phenomenon occurs.

There is also the possibility that at some (low) temperature the CDW may become pinned by lattice imperfections. In such a case there would be a sudden increase in NMR linewidth, and the profile of Figure 47.1 might then emerge.

The spin-lattice relaxation time T_1 should also be affected by CDW phase excitations. The dominant mechanism would probably involve quadrupole coupling, since the Knight-shift fluctuations cause hyperfine fields parallel to H_0. We have not investigated these questions in detail.

Acknowledgements

The authors are grateful to the National Science Foundation for support of this research.

References

1 Overhauser A.W. (1968) *Phys. Rev.* **167**, 691.
2 Overhauser A.W. (1964) *Phys. Rev. Lett.* **13**, 190.
3 Giuliani G.F. and Overhauser A.W. (1979) *Phys. Rev. B* **20**, 1328.
4 Giuliani G.F. and Overhauser A.W. (1980) *Phys. Rev. B* **22**, 3639.
5 Overhauser A.W. (1978) *Adv. Phys.* **27**, 343.
6 Overhauser A.W. (1985) In: Bassani F., Fumi F., and Tosi M.P. (eds) *Proceedings of the International School of Physics "Enrico Fermi"*, Course LXXXIX, North-Holland, Amsterdam, p. 194.
7 Coulter P.G. and Datars W.R. (1985) *Can. J. Phys.* **63**, 159.
8 Fragachan F.E. and Overhauser A.W. (1984) *Phys. Rev. B* **29**, 2912.
9 Schaefer J.A. and Marcus J.A. (1971) *Phys. Rev. Lett.* **27**, 935.
10 Huberman M. and Overhauser A.W. (1982) *Phys. Rev. B* **25**, 2211.
11 Zhu X. and Overhauser A.W. (1984) *Phys. Rev. B* **30**, 622.
12 Shoenberg D. and Stiles P.J. (1964) *Proc. R. Soc. London, Ser. A* **281**, 62.
13 Overhauser A.W. (1982) *Can. J. Phys.* **60**, 687.
14 Van Vleck J.H. (1948) *Phys. Rev.* **74**, 1168.
15 Follstaedt D. and Slichter C.P. (1976) *Phys. Rev. B* **13**, 1017.
16 Overhauser A.W. (1971) *Phys. Rev. B* **3**, 3173.
17 Pines D. and Slichter C.P. (1955) *Phys. Rev.* **100**, 1014.
18 Townes C.H., Herring C., and Knight W.D. (1950) *Phys. Rev.* **77**, 852.
19 Jones W.H. Jr., Graham T.P., and Barnes R.G. (1960) *Acta Metall.* **8**, 663.
20 Overhauser A.W. (1978) *Hyperfine Interact.* **4**, 786.
21 Overhauser A.W. and Butler N. (1976) *Phys. Rev. B* **14**, 3371.
22 Mayer H. and El Naby M.H. (1963) *Z. Phys.* **174**, 269.
23 Duff K.J. and Overhauser A.W. (1972) *Phys. Rev. B* **5**, 2799.
24 Giuliani G.F. and Overhauser A.W. (1980) *Phys. Rev. B* **22**, 3639.
25 Giuliani G.F. and Overhauser A.W. (1980) *Phys. Rev. Lett.* **45**, 1335.
26 Giuliani G.F. and Overhauser A.W. (1980) *Phys. Rev. B* **21**, 5577.
27 Bishop M.F. and Overhauser A.W. (1981) *Phys. Rev. B* **23**, 3638.
28 Ashraf M. and Swihart J.C. (1983) *Phys. Rev. Lett.* **50**, 921.
29 Boriack M.L. and Overhauser A.W. (1978) *Phys. Rev. B* **17**, 4549.
30 Boriack M.L. and Overhauser A.W. (1978) *Phys. Rev. B* **17**, 2395.
31 Overhauser A.W. (1971) *Phys. Rev. B* **3**, 3173 (Equation 25).

Reprint 48 Theory of the Perpendicular-Field Cyclotron-Resonance Anomaly in Potassium[1)]

G. Lacueva* and A.W. Overhauser*

* Department of Physics, Purdue University, 525, Northwestern Avenue, West Lafayette, Indiana 47907, USA

Received 29 October 1985

A simple metal, having a spherical Fermi surface, should not exhibit cyclotron resonance when the magnetic field is perpendicular to the surface. Nevertheless, a sharp resonance was observed by Grimes in potassium. This phenomenon can be explained by a charge-density-wave (CDW) broken symmetry. A small cylindrical piece of Fermi surface, bounded by the CDW gap and the first mini-gap, contains electrons having very small velocity. These electrons provide a mechanism for the anomalous resonance even though their relative concentration is only $\sim 4 \times 10^{-4}$. This same group of electrons is responsible for the sharp photoemission peak (reported by Jensen and Plummer) from (110) surfaces of Na and K.

48.1 Introduction

Cyclotron resonance in metals is ordinarily studied in the Azbel–Kaner configuration [1], for which the dc magnetic field is parallel to the surface. In addition to the fundamental resonance (at H_c), there appears a sequence of subharmonic resonances at H_c/n, $n = 2, 3, \ldots$ This phenomenon was observed in the simple metals Na and K by Grimes and Kip [2]. Theoretical studies [3] of the surface impedance of an ideal metal, having a conduction-electron energy spectrum $E(\boldsymbol{k})$ that is parabolic and spherically symmetric, account for the main features of the data and allow accurate determination of the cyclotron mass ($m^*/m = 1.24$ and 1.21 for Na and K) [2].

Cyclotron resonance is *not* expected when the magnetic field is oriented perpendicular to the surface [4]. The theoretical curve of Figure 48.1 illustrates the variation of the surface impedance with magnetic field for a circularly polarized microwave signal. (The sign convention adopted here corresponds to resonance when $\omega_c/\omega = -1$; $\omega_c \equiv eH/m^*c$.) Structure at resonance should not appear because electrons traveling at the Fermi velocity ($v_F \sim 10^8$ cm/s) spend too short a time in the skin depth ($\delta \sim 10^{-5}$ cm) compared to the period $2\pi/\omega$ of the microwave field. Electrons near the metal surface travel in and out along (dc) magnetic field lines perpendicular to the surface.

Nevertheless, Grimes found[2)] a very sharp *and large* resonance in potassium. This is shown in Figure 48.1. The amplitude of the sharp feature is comparable to the total change in surface impedance, $Z(H)$, between $H = 0$ and $H = H_c$. From our analysis in Section 48.4 the sharpness corresponds to $\omega_r \sim 60$, where τ is the electron relaxation time. The DPPH (diphenylpicrylhydrazyl free radical) marker indicates where cyclotron resonance could occur if $m*/m$ were unity.

The sharp feature shown in Figure 48.1 has been attributed [5] to excitation of a Fermi-liquid mode [6] described by an ellipsoidal deformation of the Fermi surface. However, this interpretation has a serious difficulty. The $l = 2$ Fermi-liquid mode does not couple to an

1) Phys. Rev. B 33, 6 (1986).

2) Grimes C.C. (private communication). The data shown in Figure 1 were taken in 1966. Similar data (due to Grimes) were published in [5] and [8].

Anomalous Effects in Simple Metals. Albert Overhauser

ISBN: 978-3-527-40859-7

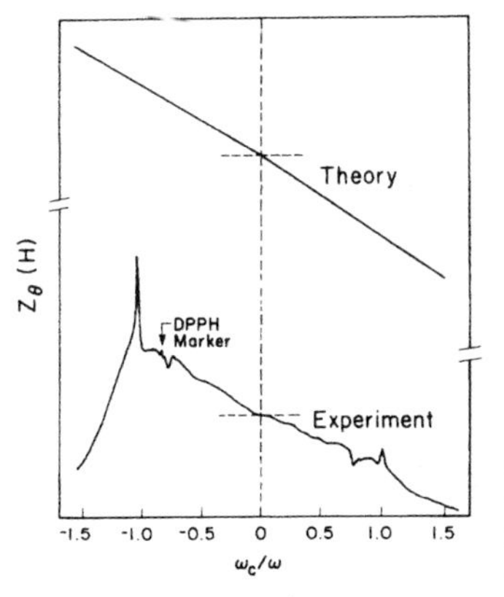

Figure 48.1 Surface impedance of potassium vs magnetic field for circularly polarized radiation at $T = 2.5$ K. $\omega/2\pi = 23.9$ GHz. The theoretical curve is for an ideal metal (spherical Fermi surface) with a specular surface. $\omega_c \equiv eH/m^*c$.

electric field (which stimulates only $l = 1$ Fermi-surface distortions). The $l = 2$ mode can arise only from a steep spatial gradient in the electron distribution function $f(\mathbf{k}, \mathbf{r}, t)$. Variations which occur in the skin depth are too small by 4 orders of magnitude to explain the observed resonance [5]. When one assumes diffuse reflection of electrons at the surface (instead of specular) the shortfall is alleviated by about 1 order of magnitude [7].

The existence of this resonance anomaly remains enigmatic. Effects of surface roughness (on a scale larger than that required for diffuse scattering) were investigated [8]. The model postulated was represented by a very short relaxation time, $\omega\tau \sim 0.1$, within a distance $\sim \frac{1}{3}\delta$ from the surface. At greater depths a bulk value $\omega\tau \sim 10$ was considered typical. The sharp spatial discontinuity of $\omega\tau$ did lead to a significantly larger coupling to the $l = 2$ Fermi-liquid mode. However, there appears to be experimental evidence for discounting this possible explanation.

The samples employed by Grimes were similar[3] to those used in (parallel-field) cyclotron-resonance studies [2]. Well-developed subharmonic resonances were observed, and these required $\omega\tau \sim 10$. Grimes has kindly sent us data which show that the sharp resonance (in perpendicular field) and the Azbel–Kaner signal (in parallel field) occur in *the same specimen.* Subharmonics out to $n = 8$ were visible. Comparison with the theoretical expression[4] for the Azbel–Kaner signal indicates that $\omega\tau$ was between 10 and 15. In fact, the $n = 2$ subharmonic would disappear if $\omega\tau < 3$, and the fundamental would disappear at $\omega\tau = 1$.[5] We believe, therefore, that high-order Azbel–Kaner sub harmonics cannot be expected in samples when $\omega\tau \ll 1$ in a significant fraction of the skin depth. We therefore propose an alternative theory of the perpendicular-field anomaly, an idea that acquires plausibility from recent photoemission studies of Na and K [9].

48.2 Charge-density-wave Structure and the Fermi Surface

The evidence for a broken translational symmetry of the charge-density-wave (CDW) type in Na and K is extensive [10]. One of the most striking observations is the large number of open-orbit magnetoresistance peaks found in potassium [11]. The origin of this phenomenon is the existence of several small energy gaps in $E(\mathbf{k})$, which cut through the Fermi surface,

3) See footnote 7 of Baraff, Grimes, and Platzman [*Phys. Rev. Lett.* **22**, 590 (1969)].
4) See Equation (28) of Mattis and Dresselhaus [*Phys. Rev.* **111**, 403 (1958)].
5) See Figure 48.2 of Mattis and Dresselhaus [*Phys. Rev. Lett.* **111**, 403 (1958)].

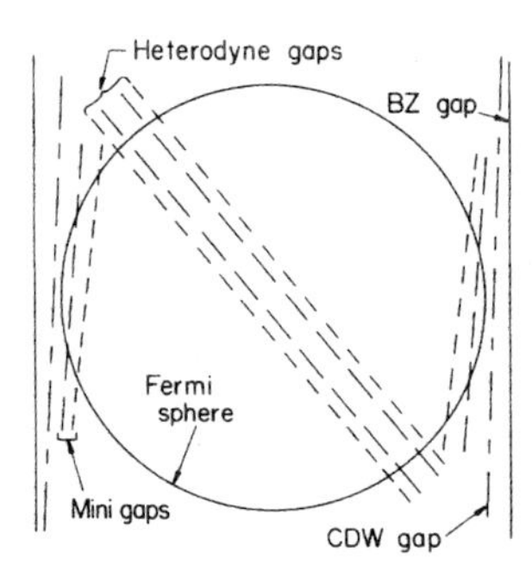

Figure 48.2 Schematic illustration of some of the energy gaps that arise (in an alkali metal) from a CDW broken symmetry. The undistorted Fermi sphere is shown. (Spacings between adjacent high-order gaps have been exaggerated.)

as shown in Figure 48.2. The minigaps and heterodyne gaps arise naturally in the solution of a Schrödinger equation having two incommensurate periodic potentials [12]. Figure 48.2 is schematic; the separations between adjacent high-order gaps have been exaggerated, and the Fermi-surface distortion is not shown. The heterodyne gaps are very small, $\sim$ 1–10 meV, and can suffer magnetic breakdown. The theory of the open-orbit magnetoresistance spectrum [13] provides a rather complete account of the observations [11]. It is worth noting that without a CDW structure the Fermi surface would be simply connected and open orbits could not even occur.

The exact shape of the Fermi surface near the main CDW energy gap cannot be predicted because nonlocal many-body corrections [14] to $E(\mathbf{k})$ are important and are difficult to calculate, especially for a nonuniform electronic charge density. That the Fermi surface comes close to the CDW energy gap is known from the shape (near threshold) of the Mayer–El Naby optical absorption [15].

There is now direct evidence that a small cylindrical Fermi-surface sheet exists between the first minigap ($\sim$ 0.1 eV) and the main CDW gap (0.6 eV), as shown in Figure 48.3. Solution [16] of the Schrödinger equation for $E(k_z)$ in the region *between* these two gaps leads to an $E(k_2)$ that is almost a constant, E_c. The relevant question is whether E_c is above or below E_F, the Fermi energy. If $E_c < E_F$, there will be a small cylindrical volume of occupied states (between the two gaps) and these will give rise to a sharp photo-emission peak (normal to a [110] metal surface) at photon energies for which (otherwise) photo-emission is forbidden [16]. Such a photoemission signal has recently been reported for Na and K [9].

Electrons on the cylindrical Fermi surface will have $v_z \ll v_F$, and will accordingly remain in the skin depth for a much longer time than other Fermi-surface electrons. We should remark here that the $\hat{z}$ direction (the direction of the CDW wave vector $\mathbf{Q}$) is known from optical studies to be perpendicular to the surface of a thin film [17] or of a thin slab grown between smooth, amorphous surfaces (from studies of Doppler-shifted cyclotron resonance [18]).

The purpose of the present work is to show that this "new" group of electrons, even though their number is extremely small, can account for the existence and size of the perpendicular-field cyclotron-resonance anomaly, shown in Figure 48.1.

In Section 48.3 we shall calculate the surface impedance $Z(H)$ for a metal having both a standard spherical Fermi surface *and* a small cylindrical sheet (with its axis parallel to the magnetic field $\mathbf{H}$). It is obvious from Figure 48.2 that the theoretical model used in what follows greatly oversimplifies the full reality of the CDW structure. Not only do we ignore effects of other high-order gaps, but we neglect the small angle between Q and the [110] reciprocal-lattice vector [19]. This deviation will cause the cylinder axis to be tilted away from $\hat{z}$. However, the cyclotron frequency on the cylinder (and for motion perpendicular to $\hat{z}$) is probably not seriously affected. Finally, we remark that the cylinder radius will be taken con-

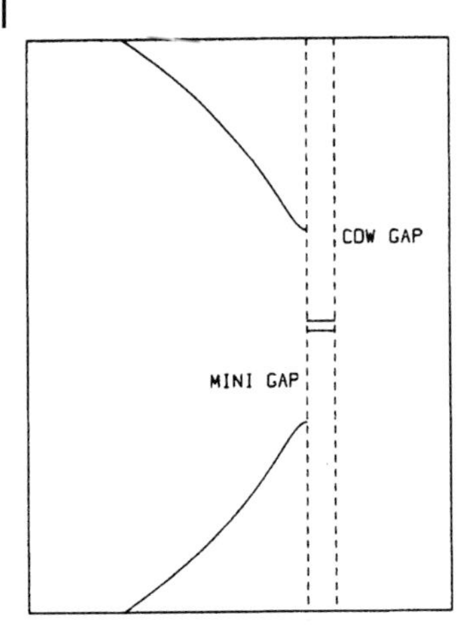

Figure 48.3 Sketch of the Fermi surface near the CDW energy gap of potassium. For simplicity the CDW **Q** and the reciprocal-lattice vector $\mathbf{G}_{110}$ have been taken parallel, although a small tilt ($\sim 1°$) is anticipated. The small cylindrical sheet of Fermi surface between the CDW gap and the (first) minigap is the focus of this study. (The horizontal axis is parallel to $\hat{z}$.)

stant (independent of k_z), which corresponds to $E(k_z) = E_c$. It is clear that the cylinder's radius will actually be slightly larger at the minigap plane compared to its radius at the CDW gap plane. Neglect of this variation is equivalent to taking $v_z = 0$ for electrons in the cylinder. Therefore, if such electrons are in the rf skin depth, they remain there during their relaxation time. It follows that a very small concentration suffices to explain the large, anomalous resonance. There are, of course, two cylinders: one near $k_z = \frac{1}{2}Q$ and another near $-\frac{1}{2}Q$.

48.3 Theory of the Surface Impedance

In accordance with the foregoing discussion, we shall calculate the surface impedance $\mathbf{Z}(\mathbf{H})$ for a metal having both a spherical Fermi surface and a small cylindrical one. The axis of the latter will be parallel to **H** and perpendicular to the surface. The infinite-medium conductivity $\sigma(q, \omega)$ can be written as the sum of two contributions, one from each group,

$$\sigma = \sigma_s + \sigma_c \, . \tag{48.1}$$

We shall follow standard procedures [20] to evaluate the conductivity σ_s of the sphere by solving the Boltzmann equation for $f(\mathbf{k}, \mathbf{r}, r)$,

$$\frac{\partial f}{\partial t} + \mathbf{v} \cdot \nabla_{\mathbf{r}} f - \frac{e}{\hbar}\left[\mathbf{E} + \frac{\mathbf{v}}{c} \times \mathbf{H}\right] \cdot \nabla_{\mathbf{k}} f = \frac{f_0 - f}{\tau} \, , \tag{48.2}$$

where $f_0(\mathbf{k})$ is the equilibrium Fermi–Dirac distribution. We are interested in deviations from f_0 to first order in the electric field,

$$\mathbf{E} \exp(i\mathbf{q} \cdot \mathbf{r} - iwt) \, . \tag{48.3}$$

Accordingly, one looks for a solution having the form

$$f = f_0 + f_1 \exp(i\mathbf{q} \cdot \mathbf{r} - iwt) \, . \tag{48.4}$$

It follows from Equations (48.2)–(48.4) that f_1 must satisfy

$$\left[\frac{1}{\tau} + i(\mathbf{q} \cdot \mathbf{v} - \omega)\right] f_1 + \frac{eH}{\hbar c} (\mathbf{v} \times \nabla_{\mathbf{k}}) \, f_1 = e\mathbf{E} \cdot \mathbf{v} \frac{\partial f_0}{\partial \varepsilon} \, , \tag{48.5}$$

where $\varepsilon = \hbar^2 k^2/2m^*$. It is convenient to change coordinates from $\mathbf{k}$ to ε, k_z, ϕ, where

$$\phi \equiv \tan^{-1}\left(k_y(k_x\right) . \tag{48.6}$$

ϕ is the azimuthal angle of an electron in its (k-space) cyclotron orbit. Equation (48.5) then becomes

$$\left[\frac{1}{\tau} + i(\mathbf{q}\cdot\mathbf{v} - \omega)\right] f_1 + \omega_c \frac{\partial f_1}{\partial \phi} = e\mathbf{E}\cdot\mathbf{v}\frac{\partial f_0}{\partial \varepsilon} . \tag{48.7}$$

Equation (48.7) is easily solved when the propagation vector $\mathbf{q}$ is parallel to $\hat{z}$, which is the case of interest here.

$$f_1 = \frac{e}{\omega_c}\frac{\partial f_0}{\partial \varepsilon} \int_{-\infty}^{\phi} \mathrm{d}\phi' \mathbf{E}\cdot\mathbf{v} \times \exp\left(\frac{[1 + i\tau(qv_z - \omega)]}{\omega_c \tau}(\phi' - \phi)\right) . \tag{48.8}$$

The current density $\mathbf{j}(\mathbf{r}, t)$ can now be evaluated:

$$\mathbf{j} = -\frac{2e}{8\pi^3} \int f_1 \mathbf{v} d^3 k . \tag{48.9}$$

In tensor notation,

$$j_i = \sum_m \sigma_{im} E_m . \tag{48.10}$$

The tensor components of conductivity (from the sphere) are

$$\sigma_{im} = -\frac{2e\hbar m^*}{\omega_c (2\pi)^3} \int \mathrm{d}\varepsilon \frac{\partial f_0}{\partial \varepsilon} \int \mathrm{d}k_z \int_0^{2\pi} \mathrm{d}\phi\, v_i(\varepsilon, k_z, \phi) \int_{-\infty}^{\phi} \mathrm{d}\phi' v_m(\varepsilon, k_z, \phi) \exp\left[\frac{[1 + i\tau(qv_z - \omega)]}{\omega_c \tau}(\phi' - \phi)\right] . \tag{48.11}$$

For circular polarization of $\mathbf{E}$ we define

$$\sigma_s^+ \equiv \sigma_{xx} - i\sigma_{xy} . \tag{48.12}$$

Evaluation of Equation (48.11) then leads to

$$\sigma_s^+ = \frac{3ne^2\tau}{4m^*} \int_0^{\pi} \mathrm{d}\theta \frac{\sin^3\theta}{1 - i(\omega + \omega_c)\tau + iql\cos\theta} , \tag{48.13}$$

where n is the electron density and l is the mean free path, $v_F \tau$. ($\cos\theta = k_z/k_F$.) The integration in Equation (48.13) is elementary and yields

$$\sigma_s^+(q, w) = \frac{3ne^2\tau}{2m^* x^2} \{2ap - 1 + \mathbf{r}\left(x^2 + 1 - a^2\right) + i\left[a + p\left(x^2 + 1 - a^2\right) - 2ar\right]\} , \tag{48.14}$$

where

$$a \equiv (\omega + \omega_c)\tau \,,$$
$$x \equiv ql \,,$$
$$p \equiv \frac{1}{4x} \ln \left[\frac{1 + (x+a)^2}{1 + (x-a)^2} \right] ,$$
$$r \equiv \frac{1}{2x} \left[\tan^{-1}(x+a) + \tan^{-1}(x-a) \right]$$

We next consider the conductivity from electrons in the small cylindrical section of the Fermi sea, shown in Figure 48.3. We shall assume that the total number of "cylindrical" electrons is ηn. For the reasons given at the end of Section 48.2, v_z for these electrons is neglected. It follows that the transverse conductivity is then the same as in a local theory:

$$\sigma_c^+(\omega) = \frac{\eta n e^2 \tau}{m^*} \left[\frac{1}{1 - i\tau(\omega + \omega_c)} \right] . \tag{48.15}$$

(We have assumed that the effective mass m^* is the same for the x, y components of motion on the cylinder as it is for the sphere.)

The surface impedance Z, which is the ratio of the tangential electric field at the surface to the total current density inside the metal, characterizes the microwave-metal interaction [21]. For specular reflection (of the electrons) at the surface, Z^+ is related to the total conductivity [22], the sum of (48.14) and (48.15),

$$\sigma^+(\mathrm{q}, \omega) = \sigma_s^+(\mathrm{q}, \omega) + \sigma_c^+(\omega) \,, \tag{48.16}$$

by

$$Z^+ = -8i\omega \int_0^\infty \frac{dq}{c^2 q^2 - 4\pi i \omega \sigma^+} \tag{48.17}$$

(where c is the velocity of light). The integration in Equation (48.17) can only be done numerically.

There are essentially only two unknown parameters in σ^+: $\omega\tau$ and η. We shall adjust $\omega\tau$ so that the width of the sharp resonance is reproduced. The fraction of electrons in the cylinder determines the (vertical) size of the resonance, compared to the change in Z between $H = 0$ and H_c. The values we find are $\omega\tau \sim 60$ and $\eta \sim 4 \times 10^{-4}$.

48.4 Results and Discussion

Before fitting the theory (just derived) to the data of Figure 48.1 it is necessary to discuss in detail several experimental details. The microwave bridge used by Grimes was one built a number of years before by Galt et al. It did not incorporate feedback tuning of the klystron to the center of the microwave-cavity resonance. As a consequence, the bridge response corresponded to an unknown linear combination of surface resistance and reactance, which we shall call Z_θ:

$$Z_\theta = \cos\theta \, \mathrm{Re}(Z^+) + \sin\theta \, \mathrm{Im}(Z^+) \, . \tag{48.18}$$

The early work on Bi produced such robust signals that the microwave cavity was rhodium plated to reduce the cavity **Q** [17]. This precaution tends to reduce any effect of detuning, so that the bridge output would then be a measure of surface resistance.

Grimes plated the cavity with high-purity copper to maximize **Q** and optimize sensitivity [24]. Effects of thermal expansion, magnetoresistance, etc. can result in an unknown (and nonconstant) value of θ during, for example, sweeping of the magnetic field. Another effect of the plating can be seen in the data of Figure 48.1. The broad resonant dips near $\omega_c/\omega \sim \pm 0.8$ were probably caused [24] by hydrated $CuSO_4$ particles trapped in the copper plating.[6] (Their resonance frequency corresponds to $g \sim 2.15$.) Finally, the relative sizes of the sharp resonances near $\omega_c/\omega \sim \pm 1$ indicate that the circular polarization was $\sim$ 90%. The bridge was balanced at $H = 0$, so the signal shown in Figure 48.1 is the change in Z_θ with H, relative to its *unknown* value at $H = 0$. The vertical scale factor is also unknown.

Theoretical results from Equations (48.16) and (48.17) are shown in Figure 48.4 for three values of θ. The sharp feature (at cyclotron resonance) is reproduced rather well with $0 \sim 47°$, $\omega\tau \sim 60$, and $\eta \sim 4 \times 10^{-4}$.

One aspect of the data still unexplained is the smooth falloff of Z_θ for magnetic fields beyond resonance. It is, of course, possible that other types of orbits, created by the heterodyne gaps shown in Figure 48.2, may be responsible. We have found no way to incorporate such orbits quantitatively.

It is of interest to estimate the radius R of the cylinder having a volume equivalent to the value of η given above. The result depends on the magnitude of the CDW wave vector **Q**, which is now precisely known. We shall take (from unpublished diffraction data) [25]

$$\mathbf{Q} = 0.985\mathbf{G}_{110} , \tag{48.19}$$

where $\mathbf{G}_{110}$ is the (smallest) reciprocal-lattice vector. It follows that the *total* length L of the two cylinders is[7]

$$L = \mathbf{G}_{110} - \mathbf{Q} = 0.015\mathbf{G}_{110} . \tag{48.20}$$

We have, of course,

$$\pi R^2 L = \frac{4\pi}{3}\eta k_F^3 . \tag{48.21}$$

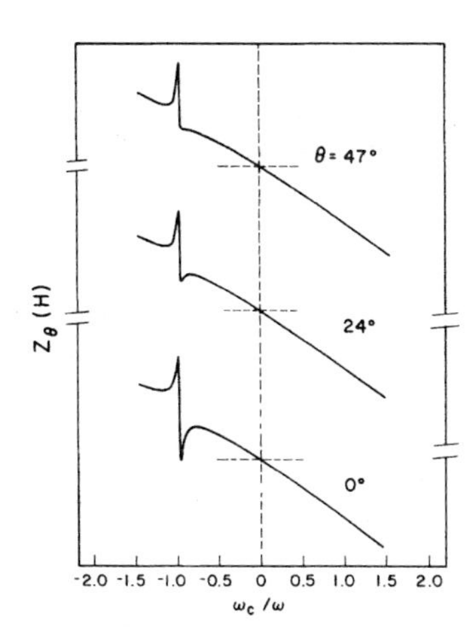

Figure 48.4 Theoretical surface impedance $Z_\theta(H)$ for potassium with a CDW broken symmetry. The mixing angle θ (for surface resistance and reactance) is defined by Equation (48.18).

6) Since the copper in the cavity was off the cylindrical axis, the $CuSO_4$ particles see mostly linearly polarized radiation. Accordingly, the resonances on either side of $H = 0$ are of almost equal size.

7) The spacing between the CDW gap and the first minigap is the same as that between the CDW gap and the Brillouinzone gap (see [12]).

From Equations (48.20) and (48.21) and the fact that $\mathbf{G}_{110} = 2.3k_F$,

$$R \sim 0.1k_F \,. \tag{48.22}$$

Thus the cross-sectional area of the cylinder (perpendicular to $\hat{z}$) is about 100 times smaller than the cross section of the Fermi sphere.

It is of interest to note that Baraff [8] surmised that the sharp cyclotron resonance found by Grimes could be explained by $\sim \frac{1}{20}$ of a monolayer of electrons confined to the surface (and free to move parallel to the surface). He emphasized that it was highly unlikely that such a *specularly* reflecting thin film could be formed on the surface of the sample; only it was interesting that so few electrons were needed to give the observed magnitude of the peak. It is not a coincidence that the number of CDW cylinder electrons within a skin depth δ of the surface would constitute $\sim \frac{1}{20}$ of a monolayer equivalent. In other words, the CDW structure provides a small group of electrons, having local properties, that fulfill the essential requirements of Baraff s imagined specular film.

48.5 Conclusions

A cylindrical sheet of electrons caused by the CDW broken symmetry of potassium explains the sharp perpendicular-field cyclotron-resonance anomaly found by Grimes. This same group of conduction electrons is responsible [16] for the angle-resolved photoemission anomaly observed by Jensen and Plummer.

The foregoing conclusion can be tested by searching for de Haas–van Alphen or Shubnikov–de Haas oscillations having an oscillation period about 100 times larger than the traditional one. Samples with an oriented CDW wave vector are needed and can probably be obtained with evaporated films. In this case, resistance oscillations (versus H) would likely be the easiest method.

Acknowledgements

We are very grateful to Dr. C.C. Grimes for providing detailed information about his experiments and for sending us some of his unpublished data. We appreciate the financial support of the Materials Research Laboratory Program of the National Science Foundation.

References

1 Azbel M.Y. and Kaner E.A. (1957) *Zh. Eksp. Teor. Fiz.* **32**, 896 [*Sov. Phys. JETP* **5**, 730].
2 Grimes C.C. and Kip A.F. (1963) *Phys. Rev.* **132**, 1991.
3 Mattis D.C. and Dresselhaus G. (1958) *Phys. Rev.* **111**, 403.
4 Chambers R.G. (1965) *Philos. Mag.* **1**, 459.
5 Baraff G.A., Grimes C.C., and Platzman P.M. (1969) *Phys. Rev. Lett.* **22**, 590.
6 Cheng Y.C., Clarke J.S., and Mermin N.D. (1968) *Phys. Rev. Lett.* **20**, 1486.
7 Baraff G.A. (1970) *Phys. Rev. B* **1**, 4307.
8 Baraff G.A. (1969) *Phys. Rev.* **187**, 851.
9 Jensen E. and Plummer E.W. (1985) *Phys. Rev. Lett.* **55**, 1919.
10 Overhauser A.W. (1978) *Adv. Phys.* **27**, 343.
11 Coulter P.G. and Datars W.R. (1985) *Can. J. Phys.* **63**, 159.
12 Fragachán F.E. and Overhauser A.W. (1984) *Phys. Rev. B* **29**, 2912.

13 Huberman M. and Overhauser A.W. (1982) *Phys. Rev. B* **25**, 2211.

14 Zhu X. and Overhauser A.W. (1986) *Phys. Rev. B* **33**, 925.

15 Overhauser A.W. (1964) *Phys. Rev. Lett.* **13**, 190.

16 Overhauser A.W. (1985) *Phys. Rev. Lett.* **55**, 1916.

17 Overhauser A,.W. and Butler N.R. (1976) *Phys. Rev. B* **14**, 3371.

18 Penz P.A. and Kushida T. (1968) *Phys. Rev.* **176**, 804.

19 Giuliani G.F. and Overhauser A.W. (1979) *Phys. Rev. B* **20**, 1328.

20 Lifshitz I.M., Azbel M.Y., and Kaganov M.I. (1956) *Zh. Eksp. Teor. Fiz.* **31**, 63. [*Sov. Phys. JETP* **4**, 41 (1957)].

21 Reuter G.E.H. and Sondheimer E.H. (1948) *Proc. R. Soc. London, Ser. A* **195**, 336.

22 Kaner E.A. and Skobov V.G. (1968) *Adv. Phys.* **17**, 605.

23 Galt J.K., Yager W.A., Merritt F.R., Cetlin B.B., and Brailsford A.D. (1959) *Phys. Rev.* **114**, 1396.

24 Grimes C.C. (private communication).

25 Giebultowicz T.M., Overhauser A.W., and Werner S.A., *Phys. Rev. Lett,* (to be published).

Reprint 49 Direct Observation of the Charge-Density Wave in Potassium by Neutron Diffraction[1]

T.M. Giebultowicz, A.W. Overhauser**, and S.A. Werner****

* National Bureau of Standards, 525, Northwestern Avenue, Gaithersburg, Maryland 20899, USA

** Department of Physics, Purdue University, West Lafayette, Indiana 47907, USA

*** Department of Physics and Astronomy, University of Missouri, Columbia, Missouri 65211, USA

Received 13 January 1986

Sharp charge-density-wave satellites in potassium have been located at (0.995, 0.975, 0.015). They are smaller than the {110} reflections by $\sim 10^5$, and each is surrounded by a prolate ellipsoidal cloud of diffuse phason scattering having its major axis along a line through the point (1,1,0).

We announce discovery of the charge-density-wave (CDW) satellites in potassium. CDW instability of conduction electrons in a simple metal is a many-body effect, driven by exchange [1] and correlation [2]. Nevertheless, neutrons can interact with this charge modulation because a parasitic, periodic lattice distortion $A\hat{\varepsilon} \sin \mathbf{Q} \cdot \mathbf{r}$ arises to provide charge neutralization [2]. Theoretical prediction and experimental evidence requires A to be so small [3], ~ 0.03 Å, and $\mathbf{Q}$ to be so near the reciprocal lattice vector $\mathbf{G}_{110}$ [4], that finding the diffraction satellites at $\mathbf{G}_{hkl} \pm \mathbf{Q}$ is extremely difficult. Diffraction searches require a triple-axis spectrometer so that neutron energy transfer can be set equal to zero, providing thereby the needed isolation from inelastic phonon scattering.

The satellite structure we present here was obtained at 4.2 K, near [110], and with 4.08- and 4.70-Å neutrons. The beam was filtered with 6 in. of Be at 78 K. [An additional 6 in. of Be did not change the satellite-to-(110) intensity ratio, which confirmed that order contamination was unimportant.] We have studied the satellites for several years on four triple-axis spectrometers (at the National Bureau of Standards' research reactor), with five wavelengths from 2.35 to 4.70 Å, and near the [110], [220], [200], and [211] Bragg reflections. Highest resolution is obtained near [110] because the pyrolytic-graphite {002} monochromator spacing nearly matches that of potassium {110}.

Although each microscopic region of a potassium crystal has a single $\mathbf{Q}$ [5], there are 24 $\mathbf{Q}$ domains in bulk and, consequently, 48 satellites near every hkl reflection [6]. Most of these are swamped by background (near the Ewald sphere of reflection) caused by mosaic distribution. Others have small intensity because the lattice-displacement vector $\hat{\text{n}}$ is nearly perpendicular to the neutron scattering vector. We define the major satellites to be the family (one of six) for which $\mathbf{Q}_i$, $i = 1, 2, 3, 4$, is nearly parallel to [110]. A satellite map in the $(1\bar{1}0)$ plane near [110] is shown in Figure 49.1. Each satellite is surrounded by a diffuse cloud of phason scattering, which redistributes the anticipated intensity [4]. A previous search [7] for CDW's in potassium did not probe this region, so close to the (110) point.

The single crystal used was a cylinder 2 cm in diameter and 7 cm long. It was sealed in a He-filled Al can and supported only by compressed quartz wool at each end. It was an exquisite specimen, having a mosaic width of less than 0.1° in three mutually perpendicular

1) Phys. Rev. Lett. 56, 14 (1986).

Anomalous Effects in Simple Metals. Albert Overhauser

ISBN: 978-3-527-40859-7

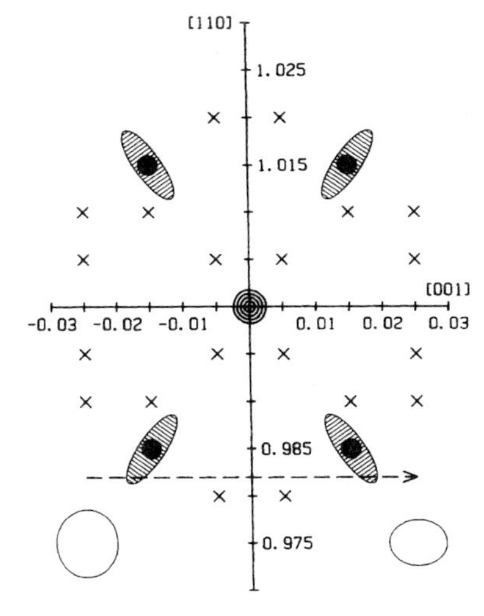

Figure 49.1 Map of the $(1\bar{1}0)$ scattering plane surrounding the [110] Bragg peak. The black dots are the main CDW satellites: The lower two come from the origin, [000], and the upper two from the [220]. The twenty crosses mark satellite locations generated by other nearby reciprocal-lattice vectors. Each point in the figure is the projection of two satellites (located slightly above and below the plane). The shaded ellipses represent diffuse phason scattering. The dashed line represents a transverse hhq scan having $h = 0.982$. The open ellipse in the lower left-hand corner is the experimental resolution for $\lambda = 4.08$ Å, collimation 40′-20′-20′-40′ and energy resolution 0.15 meV. The other ellipse was measured with $\lambda = 4.70$ Å, collimation 20′-20′-20′-20′, and energy resolution 0.07 meV, and pertains to the data in Figure 49.5.

planes. All rocking curves were free of structure. It was grown from potassium having a residual resistance ratio of $\sim$ 8000. (The bcc lattice constant of potassium at 4 K is $a =$ 5.2295 Å [8].)

Transverse hhq scans in the $(1\bar{1}0)$ plane which reveal CDW structure are presented in Figure 49.2. The shaded inset is a similar scan through [110]; it shows that the CDW satellites are extremely sharp, almost equal in sharpness to the Bragg reflections. Inelastic phason scattering suffices to explain any excess width. Observe that for $h = 0.975$ the satellite peaks occur at $q = \pm 0.022$. As h increases this separation decreases and, within experimental error, the maxima lie along lines through [110]. For shorter neutron wavelength (and poorer h resolution), this tilt of the phason ellipsoid is less apparent. (Convolution of the resolution function with the final CDW structure distribution explains such behavior.) Three more transverse hhq scans, with trajectories closer to the [110], are shown in Figure 49.3. The peak heights of the CDW diffraction continue to increase, but the effect of the nearby [110] grows very rapidly, causing the peak centered at $q = 0$.

Each satellite peak in Figures 49.2 and 49.3 arises from a pair, one above and the other below the $(1\bar{1}0)$ scattering plane. The sample was reoriented in order to measure the satellite separations along the $[1\bar{1}0]$ direction. The sample was first rotated by 90° about the original [110] scattering direction. The [001] direction was then perpendicular to the (new) scattering plane. For reasons of beam geometry the sample was then rotated 90° about the (new) vertical direction. A series of twelve scans were made in this (001) plane. Three of these and the final map for the (001) plane are shown in Figure 49.4. The major satellites depicted here correspond to **Q** domains for points $h = 0.990$, $r = \pm 0.015$, and $h = 1.010$, $r = \pm 0.015$ in Figure 49.1. Once again the major axes of phason scattering appear to pass through [110]. The intensity reversal between scans A and B of Figure 49.4 is not surprising. The peaks on the right- and left-hand sides of a given scan need not be equal, since they arise from different **Q** domains. However the **Q** domains which cause the right-hand peak in scan A cause the left-hand peak in scan B. A similar (but less dramatic) asymmetry reversal can be noticed in the upper two scans of Figure 49.2.

Close approach to the [110] requires longer wavelength and tighter collimation. Three higher-resolution scans are shown in Figure 49.5. The price paid was a sixteenfold reduction in counting rate. However, comparison of the $h = 0.985$ scans of Figures 49.3 and

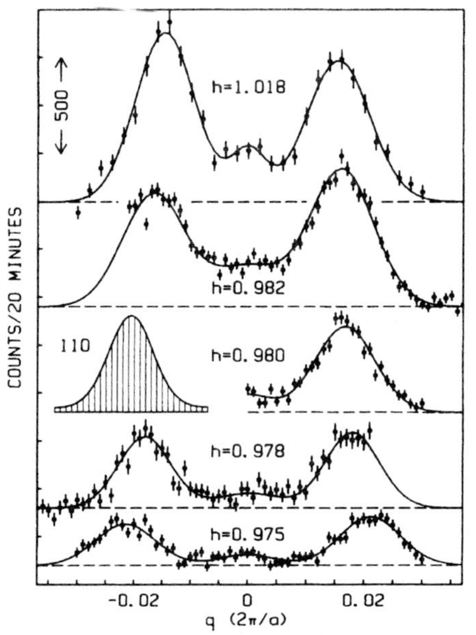

Figure 49.2 Transverse hhq scans through the CDW structure in the $(1\bar{1}0)$ plane. The shaded inset is a scan through the [110] peak, after normalizing the data by 1.5×10^{-5}. (Potassium, 4.2 K, $\lambda = 4.08$ Å.) The curves through the data are three-Gaussian optimum fits. Background lines have been positioned for clarity.

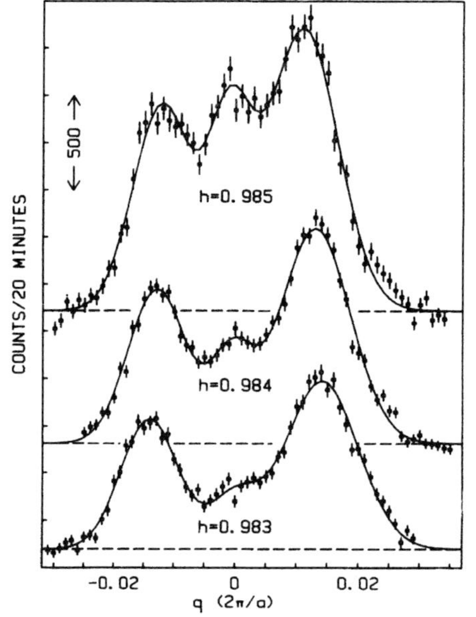

Figure 49.3 Transverse scans in the $(1\bar{1}0)$ plane near the maxima of the CDW structure. The rapidly increasing component (*vs.* h) at $q = 0$ is caused by the intense [110]. (Potassium, 4.2 K, $\lambda = 4.08$ Å.) The curves through the data are three-Gaussian optimum fits. Background lines have been positioned for clarity.

49.5 shows that the needed resolution was achieved. These high-resolution scans allow determination of the CDW satellite coordinate along the [110] scattering direction. A plot of the integrated intensities of all transverse hhq scans versus h is also shown in Figure 49.5. (Intensities from the peaks at $q = 0$ were subtracted.) The coordinate of the "maximum" scan is $h = 0.985$. The curve through the integrated-intensity data was taken to be the sum of a Gaussian and a Lorentzian. Such a decomposition is purely heuristic. At present this is our

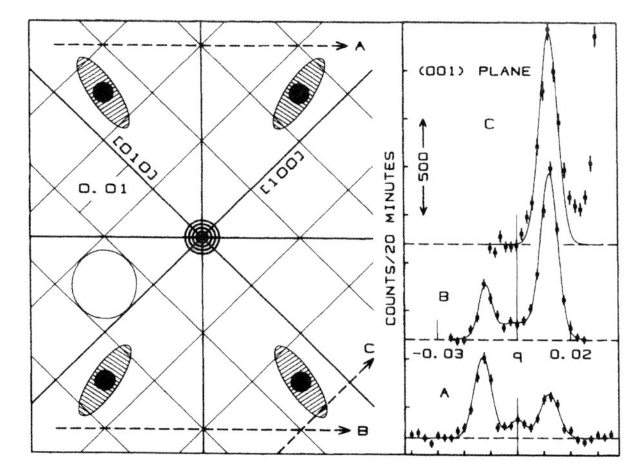

Figure 49.4 Map of the (001) scattering plane near the [110] Bragg peak. On the right are three scans through the CDW structure indicated by the dashed lines on the left. *A* is $(1.02 + q, 1.02 - q, 0)$. *B* is $(0.98 + q, 0.98 - q, 0)$. *C* is $(0.936 + q, 0.97, 0)$. The rapidly rising background at the end of scan *C* is mosaic "splash", which occurs on approaching the Ewald sphere of reflection (at 1.03, 0.97, 0), where the 20-min count reached 4×10^4. (The Bragg-peak count was 3×10^7.) Curves for *A* and *B* are three-Gaussian optimum fits. (Potassium, 4.2 K, $\lambda = 4.08$ Å.)

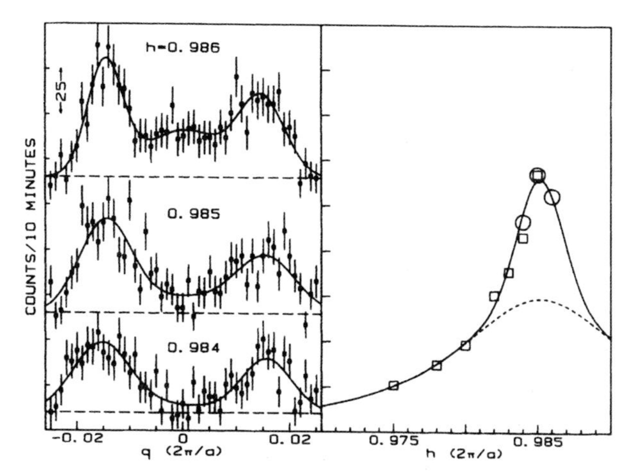

Figure 49.5 High-resolution transverse hhq scans near the maxima of the CDW satellites, $\lambda = 4.70$ Å. The seven squares in the right-hand panel are the integrated intensities of the CDW components from Figures 49.2 and 49.3. The three circles are integrated intensities of the CDW components from the data on the left-hand side. A scale factor of 16 was used to force coincidence of the two points at $h = 0.985$. The solid curves in the left-hand panel are three-Gaussian optimum fits. The solid curve in the right-hand panel is the sum of a Gaussian and a Lorentzian. The sizes of the circles and squares represent our estimate of the relative precision.

best attempt to estimate the ratio of total intensities from elastic and phason contributions. The ratio found was about 1:7.

A series of constant-energy hhq scans were made at $h = 0.970$. These CDW peaks were essentially all of phason origin. Neutron energy transfer was varied between $\Delta E = -0.5$ and 0.5 meV. The (phason) satellites reached their maxima at $\Delta E = 0.3$ meV, i.e., ~ 3.5 K. This value confirms the phason characteristic temperature (the phason analog of θ_D) found from analyses of resistivity [9], point-contact spectroscopy [10], and heat-capacity [11] data. A crucial experiment for the future would be a measurement of how much of the phason scattering is "pulled back" into the elastic peak at $T \sim 0.3$ K.

The structure we report cannot arise from phonons or Huang scattering, i.e., from frozen phonons caused by lattice imperfections. We have convolved the (four-dimensional) experimental resolution with the (known) lattice dynamics of potassium. The calculation was normalized to the measured intensity of the longitudinal phonon at (1.1,1.1,0). The resulting scattering function was too small by an order of magnitude, and had no sharp features resembling our data. Phonon scattering having energies less than 0.07 meV would increase by almost a factor of 20 at 78 K. This temperature independence is expected for CDW satellites since the resolution ellipsoid of the spectrometer (for this particular experiment) was larger than the phason ellipsoids shown in Figure 49.1. The anisotropy of Huang scattering is similar to that from phonons, and cannot explain the sharp satellites we observe.

The CDW wave vector **Q** determined by the data, with a precision of $\sim$ 0.0004 for each component, is

$$\mathbf{Q} = (0.995, 0.975, 0.015) \ . \tag{49.1}$$

The magnitude is $\mathbf{Q} = 1.393(2\pi/a)$, which compares well with [12]

$$\mathbf{Q} \approx 2k_F(1 + g/4E_F) \ , \tag{49.2}$$

where g is the CDW energy gap (0.62 eV) and E_F is the experimental Fermi energy (1.5 eV) [13]. Equation (49.2) leads to $\mathbf{Q} \approx 1.37(2\pi/a)$. The angle between **Q** and [110] is 0.85°. The theory [6] of this tilt predicts 1.0° if the experimental $|\mathbf{Q}|$ is employed. The expected intensity of a satellite is 1.4×10^{-5} of the Bragg reflection [3]. A tight quantitative comparison cannot be made because the [110] reflection is reduced by extinction. The CDW satellites are also reduced by phason scattering [4]. Since these two corrections tend to compensate, the observed intensity ratio has approximately the predicted value.

Conduction electrons in potassium exhibit many anomalous properties attributable to static and dynamic effects of a CDW [14, 15]. Our diffraction data show that neutrons traversing metallic potassium provide direct evidence of the underlying phenomenon.

Acknowledgements

We are grateful to a number of colleagues, particularly to J. A. Gotaas, J. J. Rhyne, and J. M. Rowe, for assistance and counsel during the course of this investigation. We also thank the National Science Foundation for major support provided through the Materials Research Laboratory Program.

References

1 Overhauser A.W. (1962) *Phys. Rev.* **128**, 1437.

2 Overhauser A.W. (1968) *Phys. Rev.* 167, 691 (1968).

3 Giuliani G.F. and Overhauser A.W. (1980) *Phys. Rev. B* **22**, 3639.

4 Overhauser A.W. (1971) *Phys. Rev. B* **3**, 3172.

5 Overhauser A.W. and Butler N.R. (1976) *Phys. Rev. B* **14**, 3371.

6 Giuliani G.F. and Overhauser A.W. (1979) *Phys. Rev. B* **20**, 1328.

7 Werner S.A., Eckert J., and Shirane G. (1980) *Phys. Rev. B* **21**, 581.

8 Werner S.A., Gurmen E., and Arrott A. (1969) *Phys. Rev.* **186**, 705.

9 Bishop M.F. and Overhauser A.W. (1981) *Phys. Rev. B* **23**, 3638.

10 Ashraf M. and Swihart J.C. (1983) *Phys. Rev. Lett.* **50**, 921.

11 Amarasekara C.D. and Keesom P.H. (1982) *Phys. Rev. B* **26**, 2720.

12 Overhauser A.W. (1964) *Phys. Rev. Lett.* **13**, 190.

13 Jensen E. and Plummer E.W. (1985) *Phys. Rev. Lett.* **55**, 1912, and personal communication.

14 Overhauser A.W. (1978) *Adv. Phys.* **27**, 343.

15 Overhauser A.W. (1985) In: Bassani F., Fumi F., and Tosi M.P. (eds) *Highlights in Condensed Matter Theory*, International School of Physics "Enrico Fermi", Course 89, North-Holland, Amsterdam, p. 194.

Reprint 50 Phason Anisotropy and the Nuclear Magnetic Resonance in Potassium[1)]

Y.R. Wang * *and A.W. Overhauser* *

* Department of Physics, Purdue University, 525, Northwestern Avenue, West Lafayette, Indiana 47907, USA

Received 2 June 1986

Recent neutron diffraction data indicate that the spectrum of phase excitations (phasons) of the charge-density wave (CDW) in potassium is very anisotropic. The inelastic scattering profile appears to be a prolate ellipsoid having its major axis along $\mathbf{G}_{110} - \mathbf{Q}$. Phasons propagating parallel to this axis have very low frequency. As a consequence, motional narrowing of the NMR signal (expected to be significantly broadened by the CDW at $T = 0$) could narrow the absorption to its natural width above 40 mK for an experiment at 6 T.

In a previous paper [1], the authors studied the linewidth of the nuclear magnetic resonance (NMR) in potassium. A wealth of experimental evidence [3], especially a recent neutron diffraction experiment [4], shows that metallic potassium has a charge-density-wave (CDW) structure. A CDW broken symmetry will lead to an extremely broadened NMR signal: [1] at $T = 0$ K and $H_0 = 6$ T the NMR linewidth caused by Knight-shift variations should be ~ 34 Oe. However, an experiment [5] at 1.5 K revealed a linewidth of only 0.27 Oe. We have surmised that phasons, the low-frequency collective excitations of an incommensurate CDW, provide sufficient motional narrowing to explain the observed experimental value [1]. Nevertheless, at very low temperatures, as the CDW phase excitations subside, the NMR line can broaden rapidly with decreasing temperature. We found that the temperature at which the NMR line begins to broaden depends on the following parameters: (a) the anisotropy λ of the phason-frequency spectrum, (b) the transverse phason velocity c_t, (c) the phason lifetime τ_q, and (d) the phason characteristic frequency ω_ϕ (an analog of the Debye frequency for phonons). When we employed the following set of parameters,

$$\begin{aligned} \lambda &= 1\,, \\ c_t &= 1.4 \times 10^5\ \text{cm/s}\,, \\ \tau_q &= (16\pi\hbar n M A^2/mG)q^{-1}\,, \\ \hbar\omega_\phi &\approx 3.5\ \text{K}\,, \end{aligned} \tag{50.1}$$

we found that the NMR line should begin to broaden below $T \sim 100$ mK.

Some of the above values were estimated from an analysis of experimental data. For example, the CDW energy gap, ($G = 0.6$ eV, is known from the optical absorption threshold [6], and $\hbar\omega_\phi$ is known from point-contact spectroscopy [7]). On the other hand, the value [8] for c_t and the amplitude, $A = 0.03$ Å, of the CDW periodic-lattice displacement [9], are theoretical estimates. [In (1), M is the ionic mass and n is the conduction-electron density, $\mathbf{q}$ is the phason wave vector.] $\lambda = 1$ corresponds to an isotropic phason velocity. In the past it had seemed reasonable that inelastic phason scattering would be an oblate spheriod, with $\lambda \sim 8$. However, the recent neutron scattering data [4] indicate a prolate-spheriodal shape, requiring $\lambda \sim 0.1$. The CDW satellite intensity agrees with that expected for the theoretical value of

1) Phys. Rev. B 34, 9 (1986).

Anomalous Effects in Simple Metals. Albert Overhauser

ISBN: 978-3-527-40859-7

A given above. The purpose of this paper is to revise our prior estimate of the temperature dependence of phason narrowing by taking into account the new experimental information.

The conduction-electron charge density $\rho_0(\boldsymbol{r})$ acquires (because of exchange [10] and correlation [11]) an additional sinusoidal modulation:

$$\rho = \rho_0 \left\{1 + p \cos\left[\boldsymbol{Q} \cdot \boldsymbol{r} + \phi(\boldsymbol{r}, t)\right]\right\} . \tag{50.2}$$

The fractional CDW modulation is, for potassium [9], $p \cong 0.1$. The wave vector $\boldsymbol{Q}$ was found [4] to be (0.995, 0.975, 0.015) in units of $2\pi/a$, $a = 5.23$ Å at 4.2 K. $\phi(\boldsymbol{r}, t)$ is the phase of the CDW. Its variation (in space and time) gives rise to low-frequency collective modes, i.e., the phasons,

$$\phi(\boldsymbol{r}, t) = \sum_q \phi_q \sin\left(\boldsymbol{q} \cdot \boldsymbol{r} - \omega_q t\right) . \tag{50.3}$$

Neutron diffraction indicates [4] that the contours of thermal diffuse phason scattering are prolate ellipsoids having their major axes parallel to $\boldsymbol{Q}^* \equiv \boldsymbol{G}_{110} - \boldsymbol{Q}$. ($\boldsymbol{G}_{11}$ is the reciprocal-lattice vector nearly parallel to the CDW wave vector $\boldsymbol{Q}$.) We shall define $\hat{z}$ to be parallel to $\boldsymbol{Q}^*$, so the phason spectrum can be approximated [12]:

$$\omega_q \cong c_t \left(q_x^2 + q_y^2 + \lambda^2 q_z^2\right)^{1/2} . \tag{50.4}$$

c_t is the (geometric mean) phason velocity perpendicular to $\hat{z}$.

The effect of phason scattering on the diffraction intensity of a CDW satellite can be analyzed as follows [12]: The diffraction pattern of a monoatomic crystal is obtained by Fourier analysis of the positive-ion density $n(\boldsymbol{r})$. In a CDW state,

$$n(\boldsymbol{r}) = \sum_L \delta\left(\boldsymbol{r} - \boldsymbol{L} - \boldsymbol{u}(\boldsymbol{L})\right) , \tag{50.5}$$

where

$$\boldsymbol{u}(\boldsymbol{L}) = A\hat{\boldsymbol{\varepsilon}} \sin\left[\boldsymbol{Q} \cdot \boldsymbol{r} + \phi(\boldsymbol{r}, t)\right] \tag{50.6}$$

is the lattice displacement of an ion at $\boldsymbol{L}$. The Fourier amplitudes for a scattering vector $\boldsymbol{K}$ is then

$$n(\boldsymbol{K}) = \sum_L \exp\left\{i\boldsymbol{K} \cdot \left[\boldsymbol{L} + \boldsymbol{A} \sin(\boldsymbol{Q} \cdot \boldsymbol{L} + \phi)\right]\right\} . \tag{50.7}$$

With the help of the Jacobi–Anger generating function for Bessel functions, we get

$$n(\boldsymbol{K}) \cong \sum_L e^{i\boldsymbol{K} \cdot \boldsymbol{L}} \left[J_0(\boldsymbol{K} \cdot \boldsymbol{A}) + J_1(\boldsymbol{K} \cdot \boldsymbol{A}) \times \left(\prod_q J_0\left(\phi_q\right)\right) \left(e^{i\boldsymbol{Q} \cdot \boldsymbol{L}} - e^{-i\boldsymbol{Q} \cdot \boldsymbol{L}}\right)\right] . \tag{50.8}$$

The observed integrated intensity of a CDW satellite is, therefore, proportional to $\prod_q J_0^2\left(\phi_q\right)$. Since $\left(\phi_q\right)$ is small, this factor can be expressed as

$$e^{-2W_\phi} = \prod_q J_0^2\left(\phi_q\right) \cong \exp\left(-\frac{1}{2} \sum_q \phi_q^2\right) . \tag{50.9}$$

This factor is the phason temperature factor [12] (an analog of the Debye–Waller factor). Since the diffraction experiments were performed at 4.2 K, which is higher than the phason characteristic temperature, we set the mean kinetic energy of a phason to $kT/2$. (k is Boltzmann's

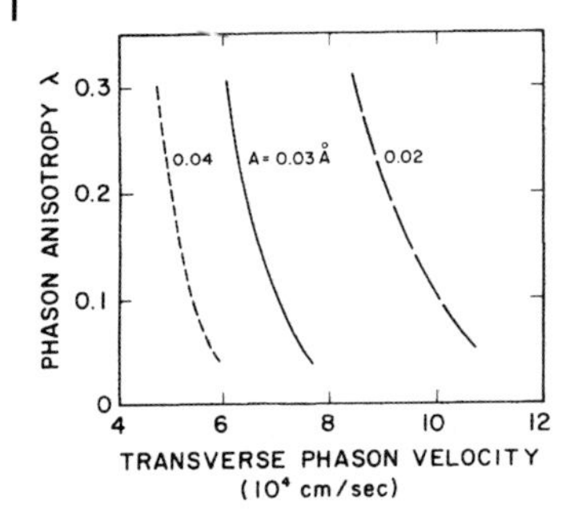

Figure 50.1 Relationship between the phason anisotropy λ and the (transverse) phason velocity c_t required by assuming the phason temperature factor at 4.2 K is $\frac{1}{8}$. The curves are for three values of A, the amplitude of the CDW periodic lattice displacement.

constant.) By direct calculation, the mean kinetic energy of a phason is $\frac{1}{8} n M A^2 \phi_q^2 \omega_q^2$[12] (for a 1-cm^3 sample). Therefore,

$$\frac{1}{8} n M A^2 \phi_q^2 \omega_q^2 = \frac{1}{2} k T \, . \tag{50.10}$$

The phason temperature factor can now be rewritten as

$$e^{-2W_\phi} = \exp\left(-\frac{2kT}{nMA^2} \sum_q \frac{1}{\omega_q^2} \right) . \tag{50.11}$$

An attempt was made [4] to separate the total observed satellite intensity into an elastic (Gaussian-shaped) part and an inelastic (Lorentzian-shaped) part. The ratio found was about 1:7. We shall assume here that this (imputed) reduction of elastic intensity is caused by a phason temperature factor. If true, this interpretation would lead to an overall restriction on the parameters λ, c_t, and A.

A rough estimate of the phason anisotropy can be obtanied by comparing the width of the Lorentzian in Figure 5 of [4] with the excess width of the CDW satellites compared to that of the 110 Bragg peak. (See Figure 2 of [4].) We "guess" $\lambda \sim 0.1$. Such a value implies that phasons propagating along $\boldsymbol{Q}^*$ have very low velocity. In such a case, the wave vector cutoff along $\boldsymbol{Q}^*$ should be $|\boldsymbol{Q}^*|$, the distance to the (110) Bragg point, rather than a q-space distance determined by the frequency cutoff ω_ϕ. However, in a direction perpendicular to $\boldsymbol{Q}^*$ we shall continue to assume that the q-space cutoff is ω_ϕ / c_t. Accordingly, our assumed phason volume (in q space) is a cylinder of radius ω_ϕ / c_t with an axis parallel to $\boldsymbol{Q}^*$. The height of the cylinder is $2|\boldsymbol{Q}^*|$.

We now evaluate the phason temperature factor, Equation (50.11), by integrating over the cylindrical volume just described. ω_q, given by Equation (50.4), is taken to have no dispersion. We find

$$2W_\phi = \frac{kT|\boldsymbol{Q}^*|}{\pi^2 n M A^2 c_t^2} \left[\frac{1}{2} \ln\left(1 + y^2\right) + y \tan^{-1}\left(y^{-1}\right) \right] , \tag{50.12}$$

where $y \equiv \omega_\phi / \lambda c_t |\boldsymbol{Q}^*|$. According to our (speculative) interpretation of the observed satellite shape, the temperature factor at 4.2 K is $\sim \frac{1}{8}$. The relationship between λ, c_t and A required by Equation (50.12), together with the value $\frac{1}{8}$ for $\exp(-2W_\phi)$, is shown in Figure 50.1. We calculate next the NMR linewidth ΔH *vs.* T for several values of the phason anisotropy λ. For each such value we adjust the transverse velocity c_t according to the constraint shown in Figure 50.1.

The relation between the NMR linewidth and the nuclear-spin correlation time τ_c is [1]

$$\Delta H = \left[0.215^2 + (34 w \tau_c)^2 \right]^{1/2} , \tag{50.13}$$

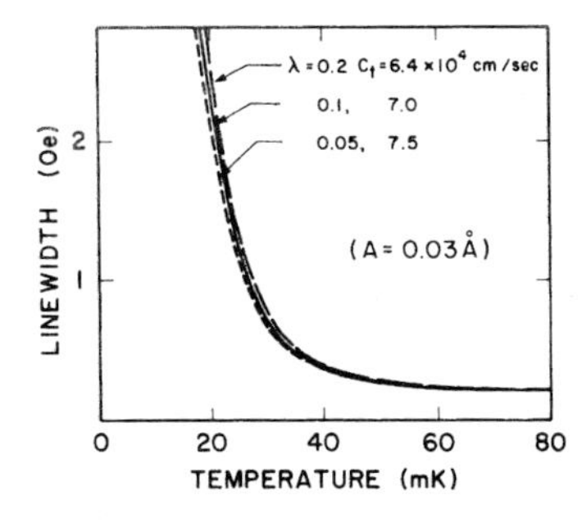

Figure 50.2 NMR linewidth vs temperature for $A = 0.03$ Å and for three values of the phason anisotropy λ. $H = 6$ T.

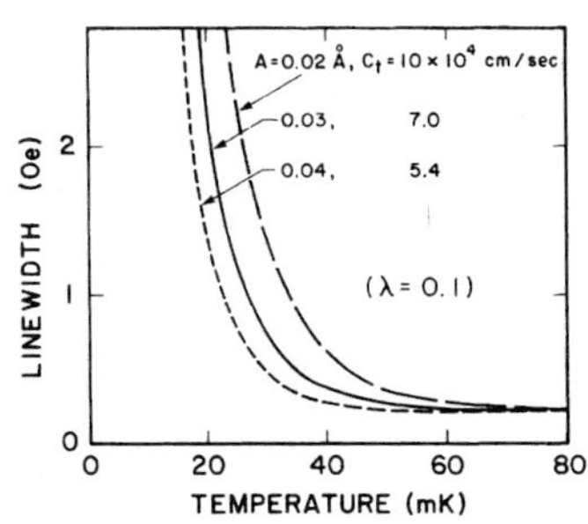

Figure 50.3 NMR linewidth *vs.* temperature for $\lambda = 0.1$ and three values of the amplitude A of the CDW lattice displacement. $H = 6$ T.

where 0.215 Oe is the zero-field linewidth [5] and 34 Oe is the anticipated CDW broadening at $T = 0$ and $H = 6$ T. $w \approx p K_0 \omega_r$, where p is the CDW amplitude, Equation (50.2), and $K_0 = 0.26\%$ is the potassium Knight shift. ($\omega_r/2\pi$ is the NMR frequency.) The correlation time is the time required for the CDW phase (at a given point) to change by one radian (as a consequence of phase excitations). The resulting relation, which one must solve for τ_c, is [1]

$$\sum_q \frac{\tau_c}{\tau_q} \frac{\omega_q^2 \tau_q^2}{1 + \omega_q^2 \tau_q^2} \langle \phi_q^2 \rangle \cong 1 , \qquad (50.14)$$

where τ_q is the phason lifetime, Equation (50.1). The angular bracket indicates the thermal-equilibrium average, so [1]

$$\left\langle \phi_q^2 \right\rangle = \frac{4\hbar}{n M A^2 \omega_q} \left[\exp(\hbar \omega_q / kT) - 1\right]^{-1} . \qquad (50.15)$$

Equation (50.14) is evaluated by integrating over the cylindrical volume described above. $\Delta H(T)$, from Equation (50.13), is shown in Figures 50.2 and 50.3 for several combinations of λ and A. (For each combination c_t was taken from the constraint depicted in Figure 50.1.)

These new results indicate that NMR broadening by the CDW becomes appreciable only below $T \sim 40$ mK. In our previous study [1] the critical temperature was ~ 100 mK (or higher). The changed prediction derives from the significantly smaller phason frequencies seemingly required by the neutron diffraction data [4]. However, we emphasize that this interpretation is tentative. Higher-resolution diffraction studies of the CDW satellites (and their temperature dependence) may alter the present view. Nevertheless it seems appropriate to present the implications of current knowledge as a guide to those contemplating low-temperature NMR studies. The curves shown in Figures 50.2 and 50.3 were calculated for an experiment at $H = 6$ T; their field dependence can be surmised from Equation (50.13).

References

1 Wang Y.R. and Overhauser A.W. (1985) *Phys. Rev. B* **32**, 7103.

2 Overhauser A.W. (1978) *Adv. Phys.* **27**, 343.

3 Overhauser A.W. (1985) In: Bassani F., Fumi F., and Tosi M.P. (eds) *Highlights in Condensed Matter Theory*, Proceedings of the International School of Physics "Enrico Fermi", Course LXXXIX, North-Holland, Amsterdam, p. 194.

4 Giebultowicz T.M., Overhauser A.W., and Werner S.A. (1986) *Phys. Rev. Lett.* **56**, 1485.

5 Follstaedt D. and Slichter C.P. (1976) *Phys. Rev. B* **13**, 1017.

6 Overhauser A.W. (1964) *Phys. Rev. Lett.* **13**, 190.

7 Ashraf M. and Swihart J.C. (1983) *Phys. Rev. Lett.* **50**, 921.

8 Giuliani G.F and Overhauser A.W. (1980) *Phys. Rev. B* **21**, 5577.

9 Giuliani G.F. and Overhauser A.W. (1980) *Phys. Rev. B* **22**, 3639.

10 Overhauser A.W. (1962) *Phys. Rev.* **128**, 1437.

11 Overhauser A.W. (1968) *Phys. Rev.* **167**, 691.

12 Overhauser A.W. (1971) *Phys. Rev. B* **3**, 3173.

Reprint 51 Satellite-Intensity Patterns From the Charge-Density Wave in Potassium[1)]

A.W. Overhauser [*]

[*] Department of Physics, Purdue University, 525, Northwestern Avenue, West Lafayette, Indiana 47907, USA

Received 2 June 1986

The polarization vector $\hat{\epsilon}$ of the lattice displacement which accompanies the charge-density wave (CDW) in potassium is calculated. The angle between $\hat{\epsilon}$ and the CDW wave vector $\mathbf{Q}$ is found to be 43°. Since diffraction intensities of CDW satellites are proportional to $(\hat{\epsilon}\cdot\mathbf{K})^2$, where $\mathbf{K}$ is the scattering vector, relative strengths of the 48 satellites which surround each Bragg point can be predicted for a sample having a random $\mathbf{Q}$-domain distribution.

The theory of diffraction by a charge-density wave (CDW) was elaborated many years ago [1]. For neutron scattering the conduction-electron charge density,

$$\rho(\mathbf{r}) = \rho_0[1 + p\cos(\mathbf{Q}\cdot\mathbf{r})]\,, \tag{51.1}$$

does not contribute to the diffraction amplitude. Any CDW satellites are caused by the periodic lattice distortion [2],

$$\mathbf{u}(\mathbf{L}) = A\hat{\epsilon}\sin(\mathbf{Q}\cdot\mathbf{L})\,, \tag{51.2}$$

which creates a positive-ion density (with Fourier component $\mathbf{Q}$) in order to neutralize the charge modulation of Equation (51.1). For each reciprocal-lattice vector, $\mathbf{G}_{hkl}$, there are two diffraction satellites, at

$$\mathbf{K} = \mathbf{G}_{hkl} \pm \mathbf{Q}\,. \tag{51.3}$$

The intensity of each satellite is proportional to

$$I \sim A^2(\hat{\epsilon}\cdot\mathbf{K})^2\,. \tag{51.4}$$

The polarization vector $\hat{\epsilon}$ of the lattice distortion would be parallel to $\mathbf{Q}$ for an ideal metal (jellium) having no lattice structure. However, in reality $\hat{\epsilon}$ has a different orientation, especially in a metal that is elastically anisotropic [3]. This is the case for alkali metals.

The CDW wave vector $\mathbf{Q}$ in potassium has recently been determined and is [4],

$$\mathbf{Q} = (0.995, 0.975, 0.015)\,, \tag{51.5}$$

in units of $2\pi/a$. The 4.2-K lattice parameter of potassium is $a = 5.2295$ Å [5]. Since $\mathbf{Q}$ is not along a symmetry axis and does not lie in a symmetry plane (of the bcc structure), there are 24 equivalent $\mathbf{Q}$ axes. Accordingly, from Equation (51.3), there will be 48 satellites associated with each G_{hkl}, two from each $\mathbf{Q}$ domain. Since $\mathbf{Q}$ differs only slightly from a [110] reciprocal-lattice vector, it is convenient to define

$$\mathbf{Q}^* \equiv \mathbf{G}_{110} - \mathbf{Q} = (0.005, 0.025, -0.015)\,. \tag{51.6}$$

1) Phys. Rev. B 34, 10 (1986).

Anomalous Effects in Simple Metals. Albert Overhauser

ISBN: 978-3-527-40859-7

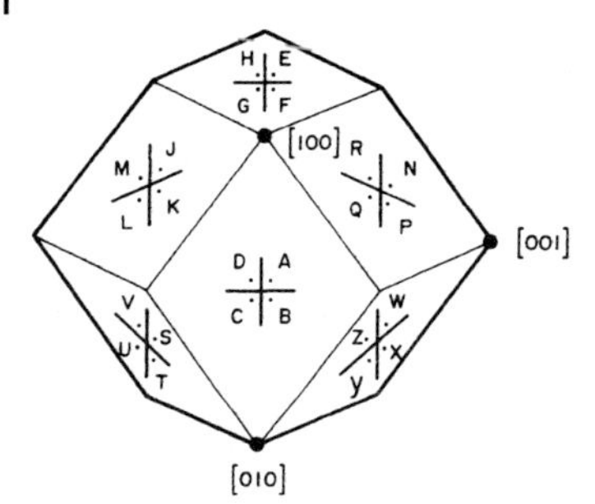

Figure 51.1 The small dots labeled alphabetically represent the **Q**-axis directions of the 24 **Q** domains. The angular deviation of each from the [110] direction passing through the same face (of the dodecahedron) is 0.85°.

The 48 satellites near a given Bragg point are (in reality) satellites of the 12 nearest-neighbor reciprocal points.

It is helpful to *name* the 24 **Q** domains according to the scheme shown in Figure 51.1. The letters label (schematically) the directional deviation of each **Q** from the nearest [110] vector. Each such vector passes through a face center of the regular dodecahedron shown in Figure 51.1. (The angular deviation is [4] only 0.85°.) The 24 satellites near the (110) Bragg point in the ($\bar{1}10$) scattering plane are plotted in Figure 51.2. They are identified by their domain "names". Only satellites lying *above* the scattering plane are shown since, in this projection, each coincides with one (of 24 others) below the plane.

The relative intensities of the satellites shown in Figure 51.2 (in the event all **Q** domains have equal volume fraction within the sample) depend on the directions of their $\hat{\epsilon}$'s relative to the scattering vector **K**. If $\hat{\epsilon}$ were perpendicular to K the intensity, from Equation (51.4), would be zero. The theoretical expression for $\hat{\epsilon}$ is [3],

$$\hat{\epsilon} = \alpha \sum_{i=1}^{3} \frac{\hat{\epsilon}_i \cdot \mathbf{Q}}{\omega_i^2} \hat{\epsilon}_i \,, \tag{51.7}$$

where α is a normalizing constant (so $|\hat{\epsilon}| = 1$). $\{\omega_i\}$ and $\{\hat{\epsilon}_i\}$ are the phonon frequencies and polarizations at $\mathbf{Q}^*$ in the Brillouin zone. (**Q** lies outside the zone, so a reduced wave vector $\mathbf{Q}^*$ must be used.) Equation (51.7) is easy to understand. All three of the acoustic modes play a role in screening the CDW: Each mode contributes a term proportional to the component of its $\hat{\epsilon}_i$ along **Q** and inversely proportional to its elastic stiffness.

Let μ be the density of the metal. For potassium at 4.2 K, $\mu = 0.908\,\mathrm{g/cm^3}$. $\mu\omega_i^2$ are the eigenvalues of

$$\begin{bmatrix} c_{11}q_x^2 + c_{44}\left(q_y^2 + q_z^2\right) & (c_{12}+c_{44})\,q_x q_y & (c_{12}+c_{44})\,q_x q_z \\ (c_{12}+c_{44})\,q_x q_y & c_{11}q_y^2 + c_{44}\left(q_x^2 + q_z^2\right) & (c_{12}+c_{44})\,q_y q_z \\ (c_{12}+c_{44})\,q_x q_z & (c_{12}+c_{44})\,q_y q_z & c_{11}q_z^2 + c_{44}\left(q_x^2 + q_y^2\right) \end{bmatrix} . \tag{51.8}$$

$\{\hat{\epsilon}_i\}$ are the associated eigenvectors. The elastic stiffness constants of potassium at 4.2 K are [6, 7]

$$\begin{aligned} c_{11} &= 4.16 \times 10^{10}\,\mathrm{dyn/cm^2}\,, \\ c_{12} &= 3.41 \times 10^{10}\,\mathrm{dyn/cm^2}\,, \\ c_{44} &= 2.86 \times 10^{10}\,\mathrm{dyn/cm^2}\,. \end{aligned} \tag{51.9}$$

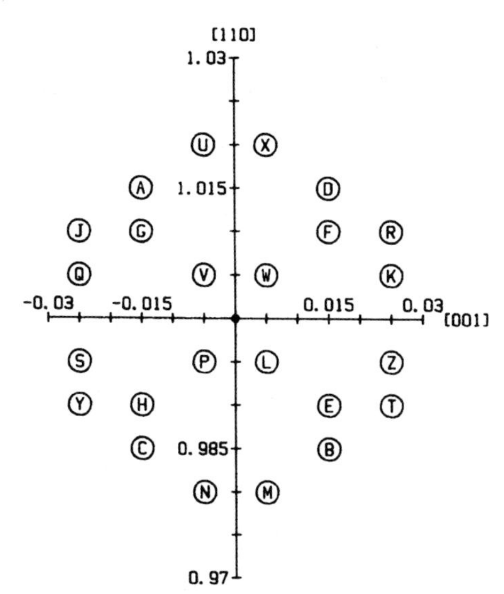

Figure 51.2 Satellite pattern in the $(\bar{1}10)$ scattering plane tor the 24 **Q** domains defined in Figure 51.1. Only satellites lying above the plane (towards [020]) are shown. All satellites are equidistant (in three dimensions) from the (110) Bragg point (located at the intersection of the axes). Their projections on the $(\bar{1}10)$ plane are at (hhq). h is the coordinate along the vertical axis and q along the horizontal.

Elasticity theory can be used here because $|\mathbf{Q}^*|$ is quite small compared to π/a. From Equations (51.7), (51.8), and (51.9), with $\mathbf{q} = \mathbf{Q}^*$

$$\hat{\epsilon} = (0.333, 0.687, 0.646) \,, \tag{51.10}$$

for the domain with **Q** given by Equation (51.5). The angle between $\hat{\epsilon}$ and **Q** is 43°. This result reflects the strong elastic anisotropy of potassium. Much of the CDW screening (by the ions) arises from the two shear modes and can be attributed to their low frequencies and canted polarization vectors.

The relative intensities of the diffraction satellites shown in Figure 51.2 are now easily found. Only the $(\hat{\epsilon} \cdot \mathbf{K})^2$ polarization factors (normalized to 100%) are given in Figure 51.3 Observe that the average intensity is 33% (as it must be from cubic symmetry). Note that the intensity of domain M is expected to be near zero. This explains why a large $q = 0$, on-axis peak was not seen from domains M and N in Figure 51.2, during transverse (hhq) scans [4] near $h = 0.980$. (Only one of four quadrants is shown in Figure 51.3 since the others can be obtained by reflection in the horizontal and vertical axes.) The most intense satellite in Figure 51.3 arises from **Q**-domain L. Unfortunately, this satellite is swamped experimentally by background from the (110) Bragg peak [4]. The intensity patterns near other Bragg reflections [not collinear with (110)] differ from those given in Figure 51.3. Naturally they can be easily computed from Equation (51.4). Had $\hat{\epsilon}$ been purely longitudinal, i.e., parallel to **Q**, the strengths of A, B, C, and D would have been 100%, E, F, G, and H zero, and all others 25%.

An ultimate goal in CDW research is to find techniques that will lead to production of a single-**Q** sample. In such an event, 46 of the 48 possible satellites will have zero intensity. Progress towards that goal – the production of preferred **Q**-domain textures – can be monitored by deviations of observed intensity patterns from those computed here.

Significant progress in studying the spin-density wave (SDW) in chromium [8] occurred following the discovery by Montalvo and Marcus [9] that the SDW wave vector **Q** could be oriented throughout a single crystal by field cooling through the SDW transition. Optical studies show that the energy gap of the CDW in potassium is ~ 0.6 eV [10], which suggests a transition temperature of ~ 2000 K, well above the melting temperature 63.65°C. Accordingly,

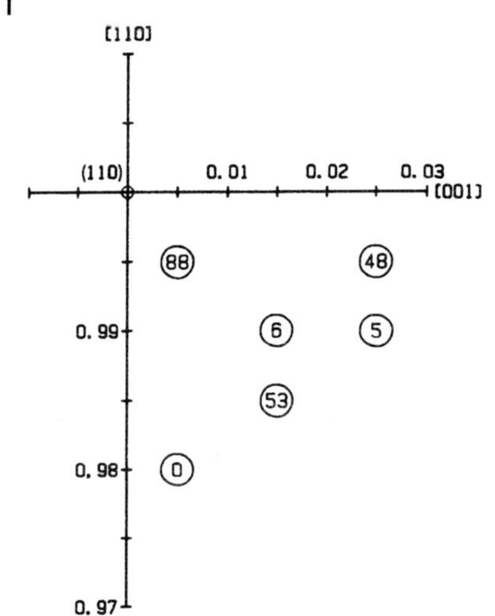

Figure 51.3 Satellite intensities (in percent of maximum value) of the six **Q** domains which appear in the lower-right quadrant of Figure 51.2. These values are derived from Equation (51.7) and are based on the observed **Q** (0.995, 0.975, 0.015) of the CDW in potassium.

attempts to orient the CDW **Q** must focus on mechanisms that can operate during solidification. Success in such an endeavor would lead to a 24-fold increase in satellite intensity. This advantage would facilitate further studies of the phason excitation spectrum [1], its influence on satellite intensity, and the possible occurrence of a static phason instability [11].

A final interesting question concerns the observation of satellites *K*, *Q*, *S* and *Z* of Figure 51.2. Their intensity should be similar to that of the observed satellites *A*, *B*, *C*, and *D*. Nevertheless, they cannot be detected because they are so close to the horizontal axis, which is tangent to the mosaic sphere of reflection. Although the mosaic width of the crystal used in [4] was quite narrow ($\sim 0.1°$), there were long mosaic tails that gave rise to an elastic (mosaic) background of $\sim$ 40 000 counts/20 min near the satellites *K*, *Q*, *S* and *Z*. This background was 1000 times smaller than the (110) Bragg peak, but was still about 40 times larger than a satellite peak. Satellites arising from **Q** domains *K*, *Q*, *S* and *Z* can, of course, be made observable by rotating the sample so that the scattering vector is almost parallel to the appropriate {110} axes. The small *E*, *F*, *G*, and *H* satellites of Figure 51.2 were measured [4] by diffraction near the $[\bar{1}10]$ direction of Figure 51.1.

References

1. Overhauser A.W. (1971) *Phys. Rev. B* **3**, 3173.
2. Overhauser A.W. (1968) *Phys. Rev.* **167**, 691.
3. Guiuliani G.F. and Overhauser A.W. (1979) *Phys. Rev. B* **20**, 1328.
4. Giebultowicz T.M., Overhauser A.W., and Werner S.A. (1986) *Phys. Rev. Lett.* **56**, 1485.
5. Werner S.A., Gurmen E., and Arrott A. (1969) *Phys. Rev.* **186**, 705.
6. Marquardt W.R. and Trivisonno J. (1964) *J. Phys. Chem. Solids* **20**, 273.
7. Smith P.A. and Smith C.S. (1964) *J. Phys. Chem. Solids* **20**, 279.
8. Overhauser A.W. (1962) *Phys. Rev.* **128**, 1437.
9. Montalvo R.A. and Marcus J.A. (1964) *Phys. Lett.* **8**, 151.
10. Overhauser A.W. (1964) *Phys. Rev. Lett.* **13**, 190.
11. Giuliani G.F. and Overhauser A.W. (1982) *Phys. Rev. B* **26**, 1660.

Reprint 52 Magnetoserpentine Effect in Single-Crystal Potassium[1)]

A.W. Overhauser*

* Department of Physics, Purdue University, 525, Northwestern Avenue, West Lafayette, Indiana 47907, USA

Received 10 August 1987

Analysis of four-terminal magnetoresistance data (published recently) from single-crystal cylinders of potassium indicates that open orbits, created by a charge-density-wave structure, are present. Apparent *negative* resistance for some orientations of the magnetic field shows that large charge-density-wave **Q** domains can cause the current to follow a serpentine path having a reversed direction between the voltage contacts.

Twenty years ago Garland and Bowers[2)] studied the transverse magnetoresistance of single-crystal cylinders of Na and K versus magnet angle. Their work was never published because the results were (then) incomprehensible. Recently Soethout et al. [3] have reported high-field data on single crystals of K which deserve attention and explanation. Their research was focused on verifying that equal resistance values are obtained if current and voltage leads are interchanged and **H** is reversed [4].[3)] Confirmation was achieved by measurements on K, In, and Al. The extraordinary data found for K, which I describe and explain below, appeared to be a serendipitous discovery.

The K crystals were 3 cm long and 2 mm in diameter. The Pt current and voltage leads were arranged as shown in Figure 52.1. The residual resistance ratios were $\simeq 3 \times 10^3$, i.e., the resistivities at 4.2 K were $\simeq 2 \times 10^{-9}\ \Omega$ cm. A magnetic field ($H = 7.5$ T) was oriented perpendicular to the cylinder axis, and the angular dependence of the resistance was found for all directions of **H** in the perpendicular plane. Thermal emfs were eliminated by current reversals, and Hall voltages were eliminated by **H** reversals.

The 7.5-T resistances (versus magnet angle) of the two crystals are shown in Figures 52.1 and 52.2. The horizontal dashed lines (for $H = 0$) are the anticipated high-field values for a simple metal having a (simply connected) spherical Fermi surface. Such simplicity is clearly not attributable to K. The large (30–40-fold) resistance enhancement and the peaked structure can only arise from open orbits[6]. Therefore, a multiply connected Fermi surface created by charge-density-wave (CDW) energy gaps [7] is indicated. Ubiquitous open-orbit resonances have been studied extensively by Coulter and Datars using an inductive (probeless) method in both K [8] and Na [9], and have been explained by CDW structure [10]. Many varieties of other anomalous phenomena in K also require a CDW broken symmetry [7, 11]. Nevertheless, some workers appear to have lost confidence in Faraday's law of induction and have expressed a need to observe open-orbit effects in four-terminal measurements. The work of Soethout et al. [3] fulfills that need.

The resistance spectrum shown in Figure 52.1 is bizarre indeed since it is *negative* throughout an angular range of 60°. At 140° the negative value is three times larger than the $H = 0$ (positive) resistance. The data shown are the average of **H** and −**H** values (from Figure 2

1) Phys. Rev. Lett. 59, 17 (1987).
2) Garland J.C. and Bowers R., unpublished. Polar plots of their data were still displayed on the laboratory walls in February, 1972. The results were equivalent to the induced-torque patterns discovered by Schaefer and Marcus [1]. The phenomenon is now understood [2].
3) For a discussion of four-terminal reciprocity see [5].

Anomalous Effects in Simple Metals. Albert Overhauser

ISBN: 978-3-527-40859-7

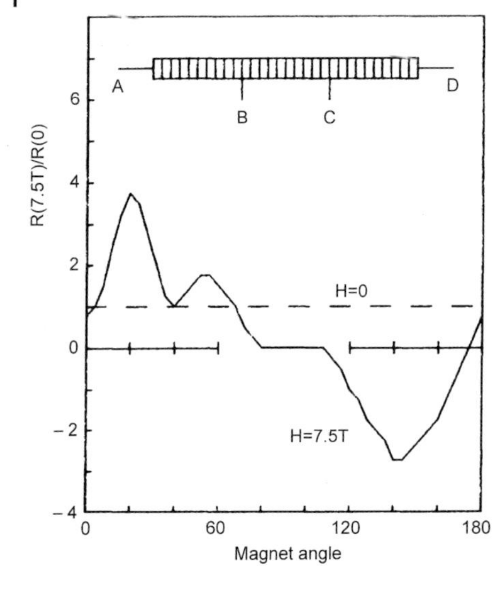

Figure 52.1 Four-terminal resistance of a K single crystal (depicted above) vs. magnet angle in the plane perpendicular to the cylinder axis (AD). The solid curve is derived from the data of Soethout et al. [3], taken at 4.2 K.

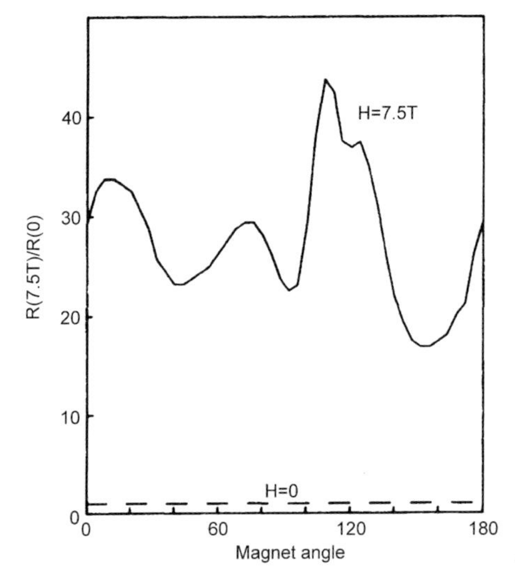

Figure 52.2 Resistance data of a second K single crystal, "identical" to the one used in Figure 52.1.

of [3]), so that Hall voltages have been eliminated. Furthermore, the same curve is obtained when the current and voltage leads are interchanged. With the positive current lead at B (and the negative at C), the voltage at A is *negative*. In other words, for **H** near 140°, lead A seems to be connected to C (and B to D) instead of vice versa! An unusual pattern of CDW **Q** domains can explain this extraordinary observation, as I now show.

Figure 52.3 depicts a network of 170 resistors with current leads A, D, and voltage contacts at B, C. The fixed resistors have resistance R. The variable resistors (ganged together) have resistance R, and are located in such a way as to model two large **Q** domains which exhibit open-orbit resistivity peaks at the same magnet angle. I have calculated the voltage across the contacts B, C as a function of R/r. The resulting behavior of the apparent resistance is

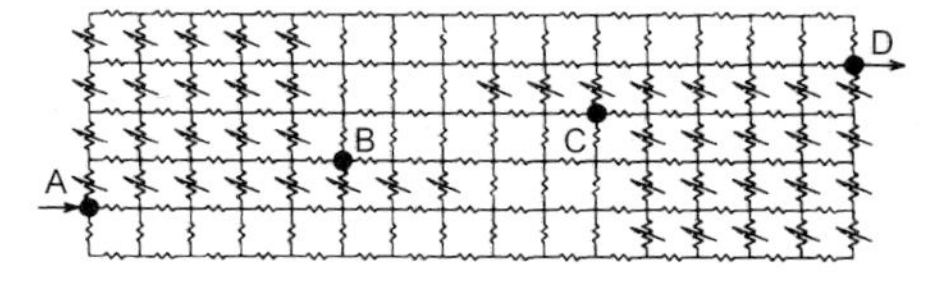

Figure 52.3 Resistor network modeling a K single crystal having three large **Q** domains. The two domains having the variable resistors (ganged together) allow imitation of an open-orbit resistivity peak. The other resistors have fixed resistance r.

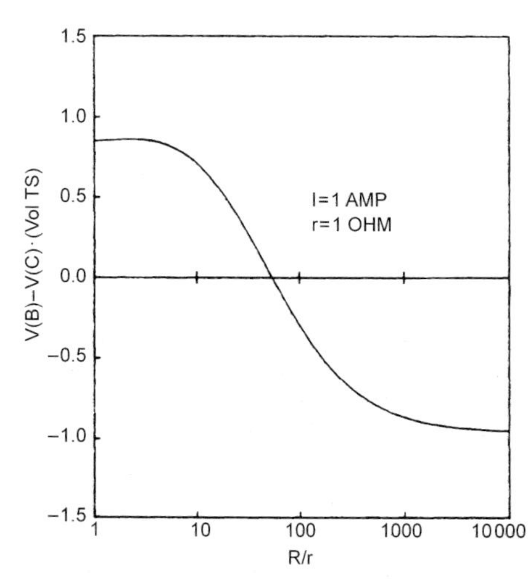

Figure 52.4 Voltage across the contacts B, C of Figure 52.3 vs. R/r, the ratio of resistance values used in the network of Figure 52.3. The curve shown results from an exact calculation.

shown in Figure 52.4. The voltage changes sign near $R/r \simeq 50$. For larger values the current follows a serpentine path from A to C, back to B, and then to D.

The foregoing calculation demonstrates that the "resistance" behavior shown in Figure 52.1 can result from a propitious (but possibly rare) **Q**-domain structure of a single crystal. Open-orbit resistivity peaks have maxima proportional to $(\omega_c\tau)^2$, where ω_c is the cyclotron frequency and τ the electron scattering time. Since $\omega_c\tau \simeq 200$ at $H = 7.5\,\text{T}$ in pure K, the resistivity ratio needed for the serpentine effect is easily achieved.

Juxtaposition of the data of Figures 52.1 and 52.2 proves that two identical crystals of K can exhibit vastly disparate properties. For metals having a CDW broken symmetry, it is necessary to specify the **Q**-domain size and morphology. Techniques do not exist yet for such a determination. A more promising approach is to control **Q**-domain growth by epitaxy, elastic stress, or extreme magnetic fields [7].

Too small a **Q**-domain size can mask CDW effects.[4] A conduction electron, traveling 10^8 cm/s, spends only 10^{-12} s in a domain of micron size. The small, higher-order gaps [12], which are the ones which truncate the Fermi surface, are too weak to cause coherent Bragg-like reflections in so short a time. (Such a failure is similar to magnetic breakdown.) The absence of published de Haas–van Alphen data which exhibit CDW effects is a likely consequence of specimen preparation (and selection) inadvertently favoring extremely small **Q** domains [13].[5] Etched samples are avoided even though such treatment improves metallurgical specimens. In the alkali metals, surface faceting occurs, which very likely promotes growth of large **Q** domains near the surface.

4) See Figure 10 of [10].

5) I have in my file a letter from a de Haas–van Alphen worker who says, "During the course of all this work there were a number of occasions when for one of several reasons we decided to reject the experimental data. The most common such reason was simply that our measurements did not display the expected cubic symmetry... Another worker has reported that up to 90% of his data are discarded for such reasons".

The same physical specimen can exhibit different orientational texture of **Q** domains from one run to the next, depending on conditions of elastic stress during cooldown to 4.2 K. Controlled cycling of the residual resistance by more than a factor of 2 has been achieved [14].[6] Drastic alteration of induced-torque patterns by a drop of oil on a K sphere is another example [16]. Without extreme precautions K samples will acquire a KOH crust. Differential thermal stress can then cause nucleation and growth of a new **Q**-domain pattern for each thermal cycle. Consequently, inadvertent preferred orientations can occur.[7] This is particularly relevant in diffraction studies of CDW satellites [17]. Failure to observe satellites upon scanning of just a few of the 24 possible **Q** orientations does not contravene other indications.

The presentation of data by Soethout et al., which shows that a magnetic field can cause an electric current to follow a serpentine path in a cubic crystal of evident purity, is an event that should not pass without appreciation.

References

1 Schaefer J.A. and Marcus J.A. (1971) *Phys. Rev. Lett.* **27**, 935.

2 Zhu X. and Overhauser A.W. (1984) *Phys. Rev. B* **30**, 622.

3 Soethout L.L., van Kempen H., van Maarseveen J.T.P.W., Schroeder P.A., and Wyder P. (1987) *J. Phys. F* **17**, LI29.

4 Casimir H.B.G. (1945) *Rev. Mod. Phys.* **17**, 343.

5 Büttiker M. (1986) *Phys. Rev. Lett.* **57**, 1761.

6 Fawcett E. (1964) *Adv. Phys.* **13**, 139.

7 Overhauser A.W. (1978) *Adv. Phys.* **27**, 343.

8 Coulter P.G. and Datars W.R. (1985) *Can. J. Phys.* **63**, 159.

9 Coulter P.G. and Datars W.R. (1986) *Phys. Rev. B* **34**, 2963.

10 Huberman M. and Overhauser A.W. (1986) *Phys. Rev. B* **25**, 2963.

11 Overhauser A.W. (1985) In: Bassani F., Fumi F., and Tosi M.P. (eds) *Highlights of Condensed Matter Theory*, North-Holland, Amsterdam, p. 194.

12 Fragachan F.E. and Overhauser A.W. (1984) *Phys. Rev. B* **29**, 2912.

13 Overhauser A.W. (1982) *Can. J. Phys.* **60**, 687.

14 Taub H., Schmidt R.L., Maxfield B.W., and Bowers R. (1978) *Phys. Rev. B* **4**, 1134.

15 Bishop M.F. and Overhauser A.W. (1978) *Phys. Rev. B* **18**, 2447.

16 Holroyd F.W. and Datars W.R. (1975) *Can. J. Phys.* **53**, 2517.

17 Overhauser A.W. (1986) *Phys. Rev. B* **34**, 7362.

6) This effect stems from the CDW-induced residual resistivity anisotropy [15]

7) The oil drop effect occurs also if a spot of KOH, obtained by exposure of a portion of the K sphere to air for a short period, is substituted for the drop of oil; see ([8], Section 5.5).

Reprint 53 Charge Density Wave Satellites in Potassium?[1]

S.A. Werner, T.M. Giebultowicz**, and A.W. Overhauser****

* Department of Physics and Astronomy, University of Missouri-Columbia, 525, Northwestern Avenue, Columbia, Missouri 65211, USA

** National Bureau of Standards, Gaithersburg, Maryland 20899, USA

*** Department of Physics, Purdue University, West Lafayette, Indiana 47907, USA

Received 8 April 1987

Recent neutron diffraction experiments have shown structure at the intensity level 10^{-5} very close to the 110b.c.c. Bragg point in metallic potassium. Three possible explanations of these data have been proposed: (1) Charge Density Waves. (2) Martensitic Embyros. (3) Spectrometer resolution/double-scattering events.

These alternative interpretations are discussed.

A year ago, we published a paper entitled "Direct Observation of the Charge Density Wave in Potassium by Neutron Diffraction" [1]. Since that time we have carried out a number of additional neutron diffraction experiments on potassium, and our view of the possible alternative origins of the structure observed in the region of the 110 b.c.c. Bragg point at a sensitivity level of 10^{-5} to 10^{-6} of its peak intensity, has evolved. During the past year a synchrotron X-ray experiment was carried out at Brookhaven by Axe and collaborators, in which no detectable structure was observed [2]. In addition, an European group led by Pintschovious working at Saclay have carried out another neutron diffraction experiment [3]. They observe the same structure, at approximately the same sensitivity level, as in our experiment, but they attribute it to spectrometer resolution/double scattering events. In the meantime Wilson and dePodesta [4] wrote a paper suggesting that the structure observed in our experiments on potassium might be due to martensitic embryos having the 9R structure described in detail last year by Berliner and Werner as the one appropriate to lithium metal at low temperatures [7]. Thus, if nothing more were to be said, we would be left with three possible explanations for the observed structure near the 110 Bragg point in neutron diffraction experiments: (1) Charge Density Waves, (2) Martensitic Embyros, (3) Spectrometer resolution/double scattering events. But, of course there is a great deal more to say, and the purpose of this paper is to assess the current experimental situation.

Seven years ago Werner, Eckert and Shirane carried out a detailed search for CDW's in potassium at Brookhaven using 2.35 Å neutrons [8]. A number of small peaks were found along symmetry axes, but none of them had the necessary pairwise symmetry relative to Brillouin zone boundaries, characteristic of a sinusoidally modulated structure. Some of them could be accounted for as double scattering events and others due to contaminants (oxides, hydroxides, etc.). The theory of Giuliani and Overhauser had just appeared, in which they had derived a new criteria for the orientation of the CDW (**Q**-vector [9]. Their calculation was based upon the requirement that the electric field caused by the CDW in the electron gas must be screened by a sinusoidal distortion of the positive-ion lattice, which takes full advantage of the elastic symmetry of K, in particular the low-energy shear mode propagating in the [110] direction. The result of this calculation required the **Q**-vector to lie slightly off the [110] axis, in one of four symmetry-related directions, a few degrees apart. Thus, for each [110] axis

1) Physica Scripta T19, 266–272 (1987).

Anomalous Effects in Simple Metals. Albert Overhauser

ISBN: 978-3-527-40859-7

Figure 53.1 A three-dimensional model of the predicted CDW satellite structure surrounding each b.c.c. Bragg point in potassium.

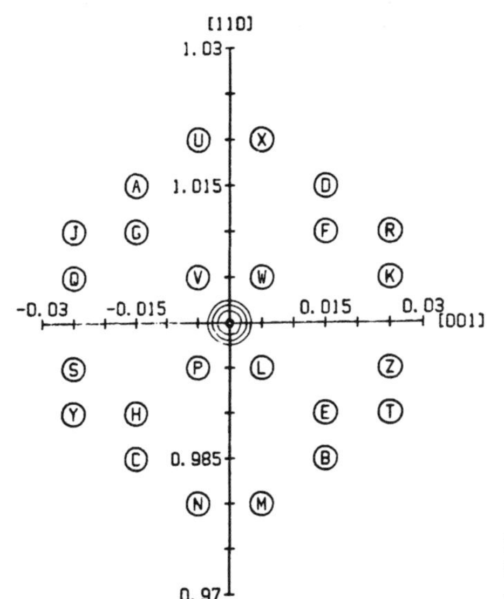

Figure 53.2 Projection of the satellite pattern into the $(1\bar{1}0)$ scattering plane for the 24 **Q** domains. (See [10].)

there should be 4 CDW satellite peaks. Since there are 6 [110] axes, there exists the possibility that a macroscopic sample contains 24 **Q**-vector domains. Since a CDW is a standing wave, there will then be 48 CDW satellite peaks surrounding each b.c.c. reciprocal lattice point. A photograph of a model of this satellite structure surrounding a given reciprocal lattice point is shown in Figure 53.1. From an experimental point of view, this is a nightmare, to say the least. For a macroscopic sample, in which the domain populations are equal, the predicted intensity of the strongest satellites is $\sim 10^{-5}$ of the 110 Bragg reflection. The intensity corresponding to the ith domain is proportional to $A^2(\hat{\varepsilon} \cdot »)^2$, where $\hat{\varepsilon}$ is the polarization of the ion modulation wave of amplitude A, and » is the neutron scattering vector $\kappa = \mathbf{G}_{hkl} \pm \mathbf{Q}_i$. These satellites are predicted to be very close to their associated b.c.c. Bragg point; closer than $\sim 0.03(2\pi/a)$. Thus, not only must one look for a complicated series of small satellite peaks,

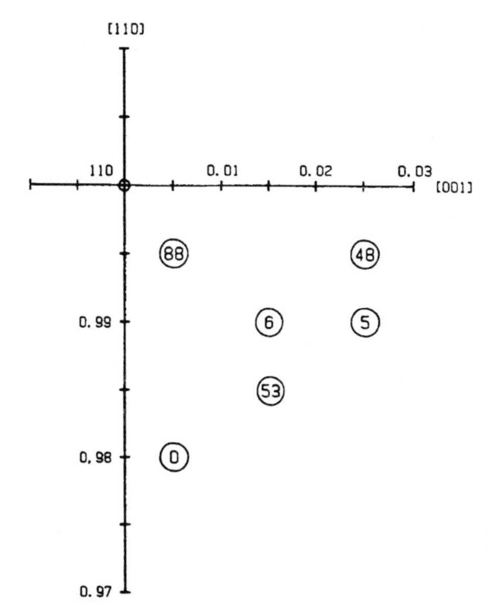

Figure 53.3 Satellite intensities (in percent of maximum value) of the six **Q** domains which appear in the lower right quadrant of Figure 53.2. The intensities and locations are based on $\mathbf{Q} = (0.995, 0.975, 0.015)2\pi/a$. (See [10].)

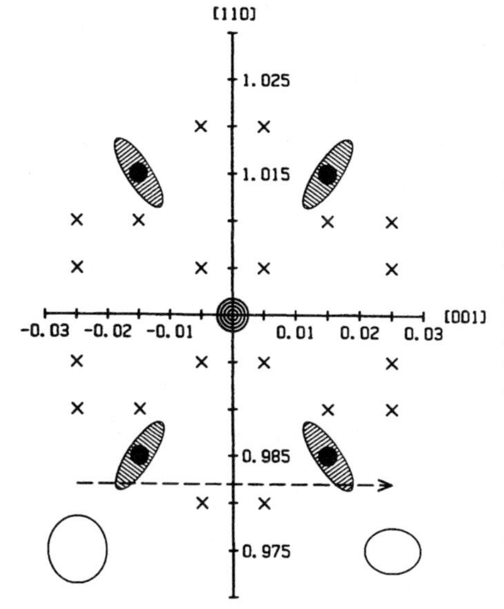

Figure 53.4 Map of the ($\bar{1}10$) scattering plane surrounding the 110 Bragg peak. The black dots are the main CDW satellites: The lower two come from the origin, 000, and the upper two from the 220. The twenty crosses mark satellite locations generated by other nearby reciprocal-lattice vectors. Each point in the figure is the projection of two satellites (located slightly above and below the plane). The shaded ellipses represent diffuse phason scattering. The dashed line represents a transverse hhq scan having $h = 0.982$. The open ellipse in the lower-left corner is the experimental resolution for $\lambda = 4.09$ Å, collimation: 40′-20′-20′-40′; energy resolution was 0.15 meV. The other ellipse was measured with $\lambda = 4.7$ Å, collimation: 20′-20′-20′-20′, energy resolution: 0.08 meV, and pertains to the data in Figure 53.6.

but one must look for them on the sides of a huge mountain. In the Brookhaven experiments scans were not carried out as close to the b.c.c. 110 Bragg point as those in our experiments at the National Bureau of Standards. However, some structure in the region of (0.975, 0.975, 0) was seen, which we later attributed, at least in part, to instrumental origin.

We now turn to a review of the reported results of our experiments at the National Bureau of Standards. We show in Figure 53.2 a projection of the CDW satellite positions onto the ($1\bar{1}0$) reciprocal lattice plane [10]. This projection is essential from the experimental point of view, since our vertical resolution is about 3°. Each of the 24 circles here represents 2 satellites, one above, and one below the scattering plane, both accepted by the vertical apertures of the spectrometer resolution. Each is given a letter name. The satellites A, B, C, D are the ones on which we will concentrate our attention. These are the only ones, near the (110) Bragg point which are predicted to be strong, and not too close to the (110) reflection and its associated mosaic tail to be obscured under conditions of spectrometer resolution available to us. The expected relative intensity of the various satellites is shown in Figure 53.3 [10]. These figures are drawn on the basis of the CDW wave vector $\mathbf{Q} = (0.995, 0.975, 0.015)2\pi/a$.

We carried out scans on the BT-4 triple-axis spectrometer at NBS with incident neutrons of wavelength 4.09 Å and 4.7 Å. Most of the scans were transverse to the [110] axis as shown in Figure 53.4. The 50% resolution ellipsoids appropriate to these two sets of experiments are shown. The data is consistent with the occurrence of CDW satellites located at the positions shown here, but surrounded by an ellipsoidal cloud of low energy phason scattering, and oriented along the lines passing through the 110 Bragg point. The energy resolution (0.15 meV and 0.07 meV for 4.09 Å and 4.7 Å incident neutrons respectively) was adequate to isolate the "elastic" scattering from the ordinary low energy phonons in K, as was verified by inelastic scans, and by detailed resolution calculations.

A reproduction of the original data is shown in Figures 53.5 and 53.6. The intensity of the side peaks continues to rise as one approaches the 110 Bragg point. With this data alone we could not tell whether the structure is due to streaks terminating at the 110 Bragg point, or a peak along these lines, a little closer in. We had numerous discussions on this point. We agreed that in order to interpret this data as due to CDW's we must go up over a peak on closer approach to 110. Subsequently, the 3 scans shown in Figure 53.7 were taken at higher resolution. A fit to the data revealed that we had indeed gone over a peak, as shown by the right-hand panel; but, just barely. We were all very uneasy about this. But we decided to publish in hopes that we or someone else might pursue still higher resolution experiments. That has now been done [3]. At the time we did not have an alternative explanation for the data.

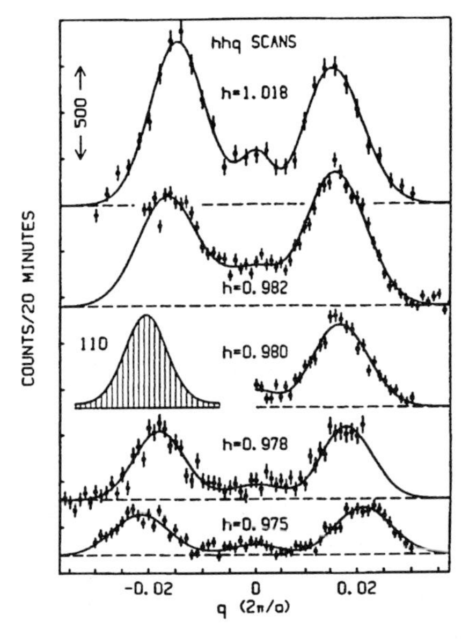

Figure 53.5 Transverse hhq scans through the CDW structure in the $(\bar{1}10)$ plane. The shaded inset is a scan through the 110 peak, after normalizing the data by 1.5×10^{-5}. (Potassium, 4.2 K, $\lambda = 4.09$ Å.) The curves through the data are three-Gaussian, optimum fits. Background lines have been positioned for clarity.

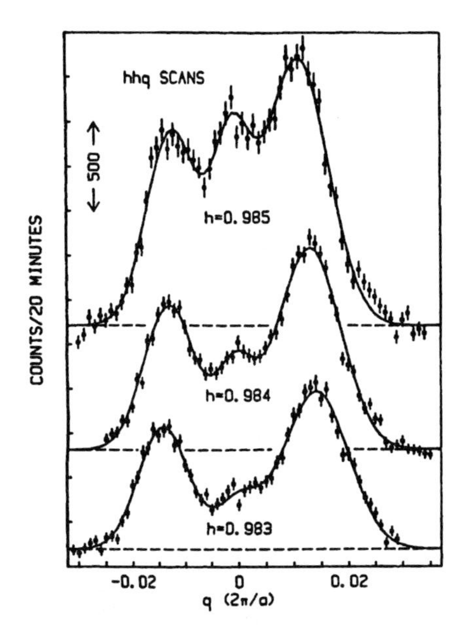

Figure 53.6 Transverse scans in the ($\bar{1}10$) plane near the maxima of the CDW structure. The rapidly increasing component (vs. h) at $q = 0$ is caused by the intense 110. (Potassium, 4.2 K, $\lambda = 4.08$ Å.) The curves through the data are three-Gaussian, optimum fits. Background lines have been positioned for clarity.

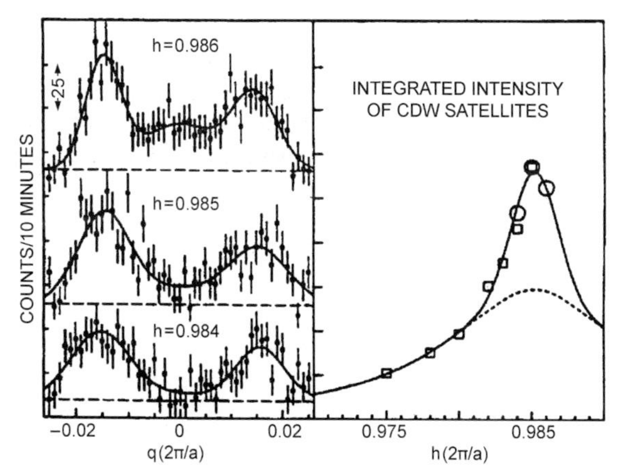

Figure 53.7 High-resolution transverse hhq scans near the maxima of the CDW satellites, $\lambda = 4.70$ Å. The seven squares in the right panel are the integrated intensities of the CDW components from Figures 53.5 and 53.6. The three circles are the integrated intensities of the CDW components from the data on the left. A scale factor of 16 was used to force coincidence of the two points at $h = 0.985$. The solid curves in the left panel are three-Gaussian optimum fits. The solid curve in the right panel is the sum of a Gaussian and a Lorentzian. The sizes of the circles and squares represent our estimate of the relative precision.

Early in March 1986 we came up with an explanation based upon spectrometer resolution effects which could account for the observation of 2 streaks, passing through a Bragg point, and separated by an angle approaching $2\theta_B$. Our paper at Physical Review Letters had already been accepted. We called the PRL office and asked if they could hold the paper – fortunately it hadn't yet gone to press. We arranged to get time on the BT-4 spectrometer to check out this explanation of our data on a narrow Si crystal. Actually, we had earlier checked for spurious

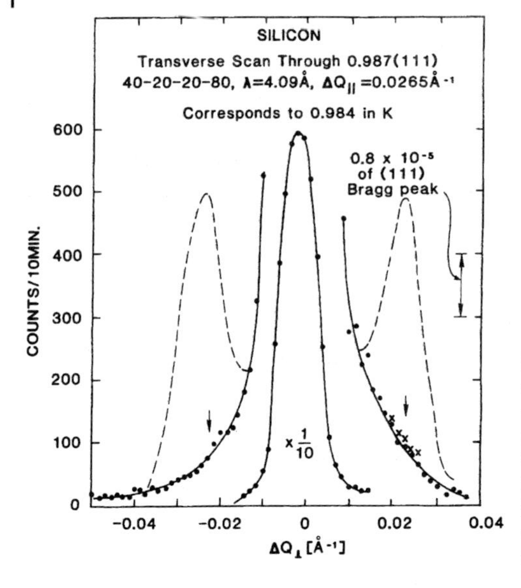

Figure 53.8 A scan transverse to the [111] axis in a narrow (0.03°) Si crystal at 0.987 (111). The collimation is 40-20-20-80. The wavelength is 4.09 Å. The arrows indicate the positions of the expected peaks due to the resolution-generated streaks discussed in the text. The dashed curves indicate the expected peaks based on the potassium data of Figure 53.9.

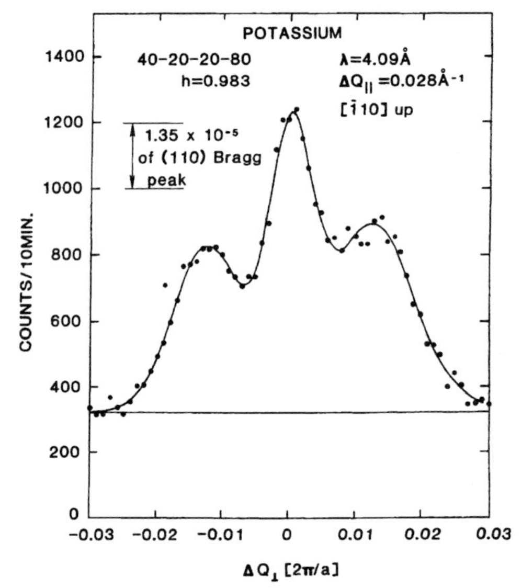

Figure 53.9 Transverse scan at $h = 0.983$ (110) in potassium. The resolution conditions are the same as for the Si scan in Figure 53.8, aside from the fact that the distance from the Bragg point is a little larger. ($\mathbf{G}_{110} = 1.699\,\text{Å}^{-1}$.)

spectrometer structure with a Ge crystal – but the 111 Bragg intensity of that crystal was down by a factor of about 5 from the 110 Bragg peak in K, and this resolution-streaking effect idea depended upon the strength of the nearby Bragg reflection. What we found for a transverse scan in a Si crystal is shown in Figure 53.8. The scan here is transverse to the [111] axis at 0.987. In absolute units, this scan is a distance $\Delta \mathbf{Q}_{\parallel} = 0.0265\,\text{Å}^{-1}$ away from the 111 Bragg point; which would correspond to a transverse scan in K at 0.984 [110]. The arrows indicate where peaks are expected on the basis of instrumental streaking. A similar scan on K, a little further away from the Bragg point is shown in Figure 53.9 where $\Delta \mathbf{Q}_{\parallel} = 0.028\,\text{Å}^{-1}$. The Bragg intensity of the two crystals is the same to within 20%. If the spectrometer were

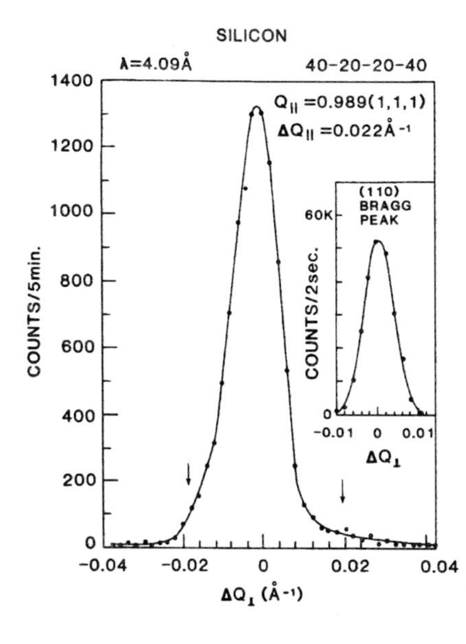

Figure 53.10 Transverse scan in Si at 0.989 (111). The collimation is 40-20-20-40. The wavelength is 4.09 Å. The arrows indicate the expected positions of peaks due to streaks. ($\mathbf{G}_{111}$ = 2.004 Å^{-1}.)

generating the structure in K, it should show up on the Si scan, having a magnitude and location shown by the dashed curves. Nothing that large was observed, so we called PRL and asked them to publish the paper, with an additional sentence saying we had also done this Si crystal check. This fact was published in an erratum.

Since that time we have carried out a number of other neutron diffraction experiments on K. We will not review them all here, because of space limitations. We will concentrate on the crucial ones. We directly measured the extinction factor ε for our crystal at λ = 4.09 Å, by normalizing the (110) integrated intensity (detector wide open) to the incident beam flux on an absolute scale. We find $\varepsilon \approx 1/7$, which requires that the satellite peaks are down by a factor of $\sim 3 \times 10^{-6}$ from the extinction-free 110 reflection. In December, we pursued the Si experiment again, being quite convinced that the resolution-streaking effect must be observable at some level of sensitivity. And here are the results. A scan at 0.989 (111), or at a distance of 0.022 Å^{-1} from the Si (111) Bragg point is shown in Figure 53.10. The arrows indicate where we expected to see structure from the instrumental streaks. Nothing is apparent. The resolution here is somewhat better than in the earlier Si scan of Figure 53.9. However, transverse scans at 0.987 (111) and 0.985 (111), are shown in Figure 53.11. Clearly there is structure. How big is it relative to K data? This is shown in Figure 53.12. The structure in the Si experiment is a factor of about 5 less than observed in the K experiment. But, there was no aluminum cryostat container, heat shield or sample capsule around the Si crystal. Very recently we have repeated these scans with an aluminum tube (with wall thickness of 3 mm) around the Si crystal. This roughly duplicates the amount of aluminum surrounding the K sample. The structure in Si is found to increase by about 60%, but it is still a factor of about 4 less than that observed in K.

We now turn to a brief explanation of the instrumental resolution origin of the streaks mentioned above. It can be shown that the close examination of a Bragg point **G**, in any diffraction experiment will reveal the presence of two streaks, intersecting at **G** and symmetrically situated about the radial line leading from the origin of reciprocal space to the point **G** as shown in Figure 53.13. The angle χ between the two streaks will depend upon the instrumental parameters, but can be expected to approach $2\theta_B$ in most cases.

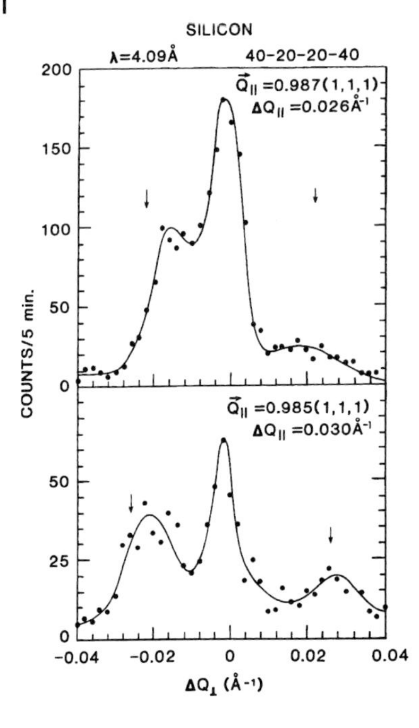

Figure 53.11 Transverse scans in Si at 0.987 (111) and 0.985 (111). The resolution is the same as in Figure 53.10. The arrows indicate the expected position of peaks due to streaks.

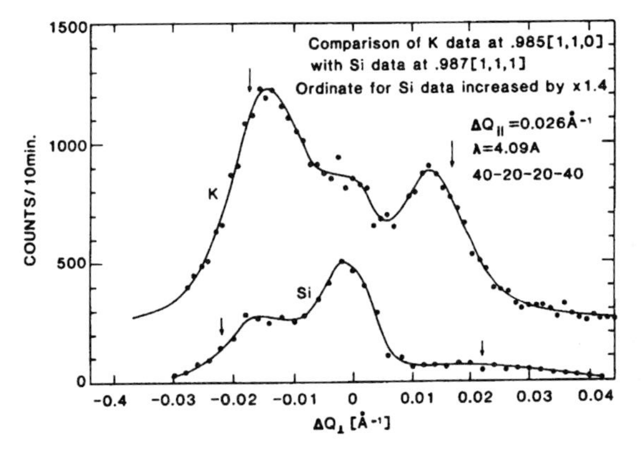

Figure 53.12 Comparison of the structure in transverse scans at $\Delta \mathbf{Q}_{\|} = 0.026\,\text{Å}^{-1}$ observed in Si and K. The intensity axis is scaled to the Bragg peaks, (110) in K and (111) in Si.

The two streaks are labelled type-I and type-II. The occurrence of the type-I streak is well-known, although its physical origin is rarely examined in any detail. It occurs when the Bragg point **G** falls on the sphere of reflection, giving a strong reflected beam at a scattering angle $2\theta = 2\theta_B$, but the detector is situated at an angle γ away from the nominal scattering angle $2\theta_B$ that is required to fully accept this Bragg beam as shown in Figure 53.14. Normally, if γ is several times the resolution width of the detector soller collimator, we assume that few, if any Bragg reflected neutrons will be counted. However, we know that at some level of sensitivity, some neutrons will penetrate through the soller slits, and be reflected by the tail of the analyzer crystal's mosaic distribution. We claim there fore, at least for this reason, that the

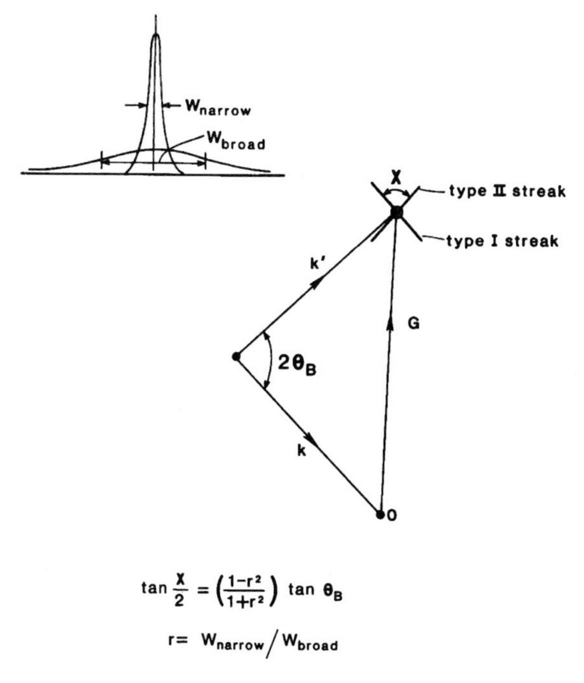

Figure 53.13 Diagram showing the orientation of the streaks in reciprocal space due to instrumental resolution.

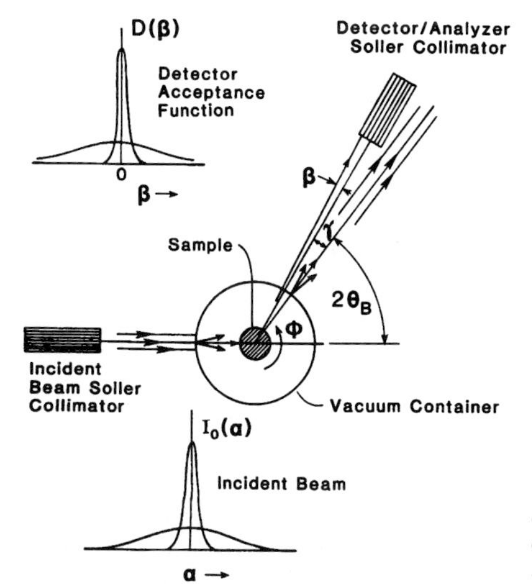

Figure 53.14 Diagram showing the physical origin of the narrow and broad components of the incident beam $I_0(\alpha)$ and the detector acceptance function $D(\beta)$.

detector/analyzer acceptance function, $D(\beta)$, will involve two parts, a very narrow part due to neutrons passing directly through the soller slots, and a much broader past due to neutrons penetrating the soller blades. There are other physical origins of the broad part of $D(\beta)$. For example, the neutron when penetrating the sample capsule, cryostat heat shield, or vacuum container, will suffer small angle scattering, thus broadening the narrow angular distribution of Bragg reflected neutrons. At some level of sensitivity air scattering will also play a role. The neutrons being counted due to the broad part of $D(\beta)$, and the narrow part of $I_0(\alpha)$, appear as a streak I in reciprocal space. The angular distribution of the incident neutron beam, $I_0(\alpha)$, must also have two parts (a narrow component and a broad component) due to the same physical reasons mentioned above. Streak II is due to neutrons in the broad part $I_0(\alpha)$ being accepted by the narrow part of the detector acceptance function $D(\beta)$. If the broad parts of these functions are 10^{-5} of the narrow parts, we get streaks at the level of 10^{-5} of the main Bragg peak.

Can this explain all the structure observed in our neutron diffraction experiments on K? The answer is not clear. We have conflicting pieces of information which would appear to require a combination of alternative interpretations. We had observed earlier that the separation of our peaks in transverse scans was the same for the $(\bar{1}10)$ and the (001) scattering planes. More recently we have tilted the crystal 45° about the [110] axis, and again observe the same structure with approximately the same peak separation. These observations are in agreement with the results of Pintschovious et al. [3]. This result would lead one to conclude that the structure is entirely due to instrumental streaking. However, it may be that the CDW (**Q**-vectors are fluctuating on a conical surface. We also did a series of experiments at reasonably high resolution (10-10-10-20) at 2.35 Å designed to verify the κ^2 dependence of the satellite intensities required by a CDW model. We did transverse scans at $h = 0.978$ (near 110) and at $h = 1.978$ (near 220). According to the κ^2 rule, an integration over the transverse ℓ-index (hhℓ) requires the integrated intensity ratio to be 4.5. The measured ratio, corrected for the Debye–Waller factor 1.1) is 3.2 ± 0.4. The peak separation in these two scans were $0.020(2\pi/a)$ and $0.025(2\pi/a)$ at $h = 0.978$ and $h = 1.978$ respectively. These facts are in "rough" agreement with a CDW model. The instrumental streak model would require the ratio of the peak separations to be 2.46, which is incompatible with the experimental facts. For a sample crystal of mosaic width narrow in comparison to the monochromator and analyzer (which is the case for our K crystal), the instrumental streaking model predicts that the widths of the "satellite" peaks in transverse scans should be a factor $\sqrt{2}$ broader than the b.c.c. 110 rocking curve width. This is also contrary to the experimental facts.

On the other hand, the separation between the peaks in transverse scans near 110 is found to depend on the incident neutron wavelength. The separation between the peaks is roughly proportional to $\tan\theta_B$, as is required by the instrumental streaking interpretation. Our results at $\lambda = 2.35$ Å, 4.09 Å, and 4.7 Å have been extended to include 5.71 Å and 6.28 Å by Pintchovius et al. [3]. The peak separations for a given $\Delta\mathbf{Q}_{\parallel}$ follow the $\tan\theta_B$ rule, approximately. We had earlier interpreted this trend as due to a resolution effect, where the spectrometer resolution ellipsoid passes through the tail of the CDW satellite and its extended phason "cloud" of scattering. This interpretation does not seem to be compatible with the longer wavelength data of Pintchovius et al., unless the center of the CDW satellites is closer to $|\mathbf{G}_{110}|$ than deduced from our earlier data.

Recently, we have modelled the instrumental streaking effect in K [11]. The results are shown in Figure 53.15. It is interesting to note how rapidly, as a function of $\Delta\mathbf{K}_y$ ($= \Delta\mathbf{Q}_{\parallel}$), the structure evolves from a clear two peak distribution through an interval where three peaks arc observed and into a region where only one peak (the tail of the Bragg point) remains. This rapid evolution is the feature that fooled us in our initial Si crystal experiments designed to check for this effect.

We would like to comment briefly on the recent X-ray diffraction search for CDW's in K. These experiments involved exhaustive searches near 110 using 1.05 Å X-rays under three different resolution conditions of the X-ray diffractometer. It is only the low resolution (LR) results which are pertinent, since the resolution volumes under the medium and high reso-

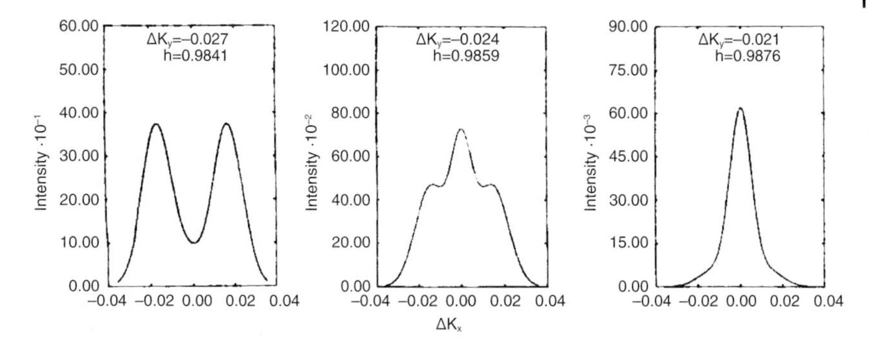

Figure 53.15 Results of model calculations on the instrumental resolution streak effect. The parameters are appropriate to the potassium data taken at $\lambda = 4.09$ Å; 40-20-20-40 collimations. The ratio of the width of the narrow component to the broad component $r = 1/5$, and the ratio of the height of the narrow component to the broad component is taken to be 10^4. The Bragg angle $\theta_B = 33.6°$. Each panel is a transverse scan, where $\Delta\mathbf{K}_x$, $(= \Delta\mathbf{Q}_\perp)$ is in Å^{-1}, and $\Delta\mathbf{K}_y$ $(= \Delta\mathbf{Q}_\parallel)$ is also in Å^{-1}. $\Delta\mathbf{K}_x = \Delta\mathbf{K}_y = 0$ corresponds to the 110 Bragg point, where $h = 1$.

lution conditions are much too small (for the sample temperature, 10 K) to integrate over a sufficiently large fraction of the total (static CDW + phason) scattering to yield a detectable intensity. In the LR scans no feature was detected at a sensitivity level of about 10^{-6} of I_{110}, which is approximately at the same level as the peaks observed in the neutron experiments, when normalized to the extinction-corrected 110. However, the shape, orientation and size of the resolution volumes in the LR X-ray experiment and our neutron experiments are really quite different.

We have not yet discussed the suggestion of Wilson and dePodesta [4], that the structure observed in the neutron diffraction experiments is due to martensitic embryos having a 9*R* hexagonal structure. We do not see how this idea can give rise to satellites on the far side of the 110 b.c.c. Bragg point. We are aware of no other evidence for a martensitic phase transformation in K.

One may now come to the conclusion that the instrumental streaking interpretation of the neutron diffraction data casts doubt on the CDW interpretation of the numerous experimental anomalies in K described by Overhauser over the past 20 years. We are of course disappointed for not having experimentally verified that our data could be interpreted, at least in part, as due to instrumental streaking a year ago. However, the factor of 4 difference in the intensity observed in the K data relative to the Si data, casts doubt on concluding that the structure in K is due solely to instrumental streaking. Nevertheless, we have been concerned about the neutron diffraction data from the beginning. There has always existed the possibility that the CDW **Q**-vector may not be locked rigidly to the b.c.c. lattice of K. We leave the reader with a picture (Figure 53.16) which is appropriate to the paramagnetic phase of the SDW metal, chromium. Incommensurate SDW satellites are of course not observed above the Neel temperature (312 K). Magnetoelastic anomalies extend far into the paramagnetic phase (up to at least 600 K) as discussed by Fawcett et al. [12]. The SDW is still

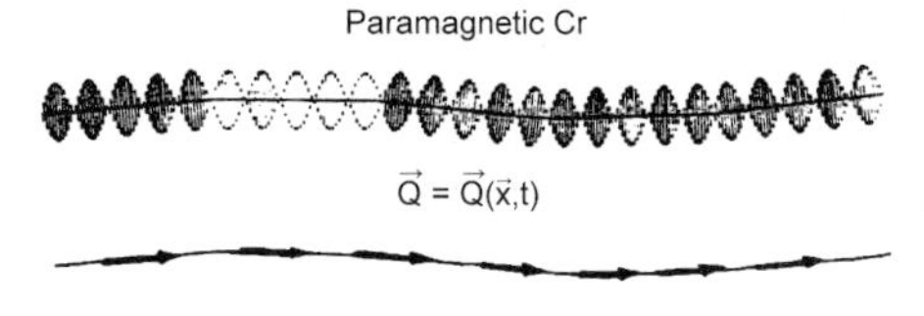

Figure 53.16 Schematic diagram of the phason-fluctuation influence on the SDW in chromium metal in the paramagnetic phase.

there, but phason fluctuations (wobbling of **Q**-vector, etc.) destroy the long-range coherent ordered state. The resulting neutron scattering is all inelastic, and confined to a small region of **k**-space [spherical region with diameter $\approx 0.07(2\pi/a)$] centered at (100) [13]. This means that the angular fluctuations of $\mathbf{Q}(\mathbf{x})$ are only $\sim \pm 2$ degrees. Could this also be the case in K at all temperatures? It is only with high-resolution inelastic neutron scattering measurements that this question can be answered definitively, if at all.

Acknowledgements

We are very appreciative of the considerable assistance, advice and counsel given to us by the members of the National Bureau of Standards neutron scattering group during the course of this work.

References

1 Giebultowicz T.M., Overhauser A.W. and Werner S.A. (1987) *Phys. Rev. Lett.* **56**, 1485 (1987).
2 You H., Axe J.D., Hohlwein D. and Hastings J.B. (1987) *Bull. Am. Phys. Soc.* **32**, 445; preprint of paper.
3 Pintschovius L., Blaschko O., Krexner G., de Podesta M. and Currat R. (1987) *Phys. Rev. B, Rapid Communications* (Submitted March 1987), preprint.
4 Wilson J.A. and de Podesta M. (1986) *J. Phys. F.* **16**, L121.
5 Berliner R. and Werner S.A. (1986) *Phys. Rev. B* **34**, 3586.
6 Overhauser A.W. (1984) *Phys. Rev. Lett.* **53**, 64.
7 Smith H.G. (1987) *Phys. Rev. Lett.* **58**, 1228.
8 Werner S.A., Eckert J. and Shirane G. (1980) *Phys. Rev. B* **21**, 581.
9 Guiuliani G.F. and Overhauser A.W. (1979) *Phys. Rev. B* **20**, 1328.
10 Overhauser A.W. (1986) *Phys. Rev. B* **34**, 7362.
11 Werner S.A., "Streaks of Instrumental Origin Near a Bragg Point" submitted to *Acta Crystallographica*.
12 Fawcett E., Kaiser A.B. and White G.K. (1986) *Phys. Rev. B* **34**, 6248.
13 Grier B.H., Shirane G. and Werner S.A. (1985) *Phys. Rev. B* **31**, 2892.

Reprint 54 Fermi-Surface Structure of Potassium in the Charge-Density-Wave State[1)]

*Yong Gyoo Hwang** and *A.W. Overhauser**

* Department of Physics, Purdue University, 525, Northwestern Avenue, West Lafayette, Indiana 47907, USA

Received 8 March 1988

The neutron-diffraction determination of the charge-density-wave (CDW) wave vector **Q** in potassium is used to calculate the detailed structure of the conduction-electron Fermi surface. A Schrödinger equation having two periodic potentials, $2\alpha\cos(\mathbf{Q}\cdot\mathbf{r})$ from the CDW, and $2\beta\cos(\mathbf{G}\cdot\mathbf{r})$ from the lattice, is solved numerically. Minigaps in $E(\mathbf{k})$, caused by perturbations having periodicity $(n+1)\mathbf{Q}-n\mathbf{G}$, $n=1, 2, \ldots$, lead to small cylindrical sections of Fermi surface. These cylinders give rise to perpendicular-field cyclotron resonance and unexpected angle-resolved photoemission peaks. Heterodyne gaps, having periodicity $n(\mathbf{G}-\mathbf{Q})$, are created in the equatorial region of the Fermi "sphere". Open-orbit magnetoresistance peaks observed by Coulter and Datars (in K and Na) arise from the multiply connected Fermi-surface topology caused by the minigaps and heterodyne gaps.

54.1 Introduction

It has long been held that the energy spectrum $E(\mathbf{k})$ of conduction electrons in alkali metals is nearly parabolic, and that the Fermi surface is nearly spherical (and consequently simply connected). However, there has been a large accumulation of experimental data [1, 2] which suggest a nonspherical and multiply connected structure for the Fermi surface. The microscopic theory of charge-density-wave (CDW) instability [3, 4] explains the existence of such Fermi-surface structure and provides successful explanations of most of the anomalous data [1, 2].

In a CDW state, a sinusoidal modulation of electronic charge density is sustained self-consistently through exchange and correlation interactions, and gives rise to a sinusoidal potential incommensurate with the usual crystal potential. As a result, the Fermi surface suffers a distortion, and is sliced into many pieces by extra energy gaps. The modulated electron charge density is neutralized by a sinusoidal deformation of the positive-ion background. This lattice deformation contributes extra diffraction peaks in a neutron-scattering experiment. These satellite reflections have been observed by Giebultowicz et al. [5, 6] (GOW), and the CDW wave vector **Q** was determined accurately.

Subsequent work by Pintschovius et al. [7, 8] (PB) has suggested that the satellites observed by GOW were instead experimental artifacts caused by double scattering [9, 10]. However, recent experiments [11] by Werner et al. (WOG) have shown that the satellites observed by GOW have the *same* spacing near the (220) Bragg point as they have near the (110) point. Double-scattering artifacts exhibit spacings proportional to $\tan\theta_B$, where θ_B is the Bragg angle, and would have to be 2.5 times further off the [110] axis near the (220) Bragg point compared to their spacing near the (110) point. The samples used by PB showed an incremental increase in mosaic width on each cooldown [7, 8]. Consequently they were severely plastically deformed, and neutron scattering from dislocations exceeded by 2 orders of magnitude the total incoherent scattering observed by GOW and WOG in their stress-free sample. Conse-

1) Phys. Rev. B 39, 5 (1989-I).

Anomalous Effects in Simple Metals. Albert Overhauser

ISBN: 978-3-527-40859-7

quently PB could not have observed CDW satellites. The structure PB reported was indeed caused by double scattering originating from the severe plastic strain.

The purpose of this study is to calculate the detailed structure of the Fermi surface of potassium based on the neutron-diffraction measurement of the CDW wave vector **Q**. In Section 54.2, effects of the CDW potential will be discussed. In Section 54.3, the approximate shape of the Fermi surface is found. The results are summarized in Section 54.4.

54.2 Plane-wave Expansion

To find the energy spectrum for conduction electrons in a CDW state, one has to solve a Schrödinger equation with two sinusoidal potentials, one arising from the crystal structure and the other from the CDW instability:

$$\left[-\frac{\hbar^2}{2m}\nabla^2 + \mathrm{V}(\mathbf{r})\right]\Psi = E\Psi \,, \tag{54.1}$$

where

$$\mathrm{V}(\mathbf{r}) = 2\alpha\cos(\mathbf{Q}\cdot\mathbf{r}) + 2\beta\cos(\mathbf{G}\cdot\mathbf{r}) \,. \tag{54.2}$$

Q is the CDW wave vector, $\mathbf{Q} = (0.995, 0.975, 0.015)$ [5, 6], and **G** is the $\langle 110\rangle$ reciprocal-lattice vector, with $\mathbf{G} = (1, 1, 0)$ in units of $2\pi/a$, where a is the lattice constant of the bcc lattice. $2a$ is the CDW potential and is about 0.87 eV (the CDW energy gap is about 0.62 eV), as indicated by the threshold of the Mayer–El Naby optical anomaly [12, 13]. 2β is the Brillouin-zone energy gap and is approximately 0.40 eV, as derived from an analysis of de Haas–van Alphen data [14]. We will neglect contributions of the other five pseudopotentials, oriented 60° or 90° away from $\mathbf{G}_{110}$, since the coupling through them involves plane-wave states with energies too far removed to contribute significantly.

We choose a coordinate system such that $\mathbf{G} = G\hat{z}$ and $\hat{y}$ is parallel to $\mathbf{G}\times\mathbf{Q}$,

$$\mathbf{Q} = (\mathbf{Q}_x, 0, \mathbf{Q}_z) \,, \tag{54.3a}$$

$$\mathbf{G} = (0, 0, \mathbf{G}) \,. \tag{54.3b}$$

The angle between **Q** and **G** is 0.85°, and $Q/G \cong 0.985$. When **Q** is parallel to **G**, the potential $V(\mathbf{r})$ does not have a translational symmetry. An approximate energy spectrum was obtained for this case by Fragachan and Overhauser [15].

The potential $V(\mathbf{r})$ in Equation (54.2) is periodic under a translation through the vector **R**,

$$\mathbf{R} = m\mathbf{a}_1 + n\mathbf{a}_2 \,, \tag{54.4}$$

where

$$\mathbf{a}_1 = (-\mathbf{Q}_z/\mathbf{Q}_x, 0, 1) \cong (-67.6, 0, 1) \,, \tag{54.5a}$$

$$\mathbf{a}_2 = ((\mathbf{G}-\mathbf{Q}_z)/\mathbf{Q}_x, 0, 1) \cong (1.03, 0, 1) \,. \tag{54.5b}$$

The size of this unit cell is 68.6 times that of the primitive bcc unit cell. The corresponding reciprocal-lattice vectors are

$$\mathbf{b}_1 = \mathbf{G} - \mathbf{Q} = \mathbf{Q}' \,, \tag{54.6a}$$

$$\mathbf{b}_2 = \mathbf{Q} \,. \tag{54.6b}$$

The angle between $\mathbf{Q}'$ and $\mathbf{G}$ is 44.2°, and $\mathbf{Q}'/\mathbf{G} = 0.021$. The size of the new Brillouin zone is about 0.014 times that of the bcc Brillouin zone. As a result, the reciprocal-space structure is much more complicated, especially near the Fermi surface.

According to Bloch's theorem, wave functions have the following form:

$$\Psi_{\mathbf{k}} = e^{i\mathbf{k}\cdot\mathbf{r}} u_{\mathbf{k}}(\mathbf{r}) \,, \tag{54.7}$$

where $u_{\mathbf{k}}(\mathbf{r}+\mathbf{R}) = u_{\mathbf{k}}(\mathbf{r})$ and $\mathbf{k}$ is quantized in accord with the periodic boundary conditions,

$$\Psi(N_1\mathbf{a}_1 + \mathbf{r}) = \Psi(\mathbf{r}) \,, \tag{54.8a}$$

$$\Psi(N_2\mathbf{a}_2 + \mathbf{r}) = \Psi(\mathbf{r}) \,. \tag{54.8b}$$

$N_1\mathbf{a}_1$ and $N_2\mathbf{a}_2$ are the sizes of the sample in the directions $\mathbf{a}_1$ and $\mathbf{a}_2$, respectively. $u_{\mathbf{k}}(\mathbf{r})$ can be expanded in terms of plane-wave states,

$$\begin{aligned} u_{\mathbf{k}}(\mathbf{r}) &= \sum_{m',n'} a_{m',n'} e^{i(m'\mathbf{b}_1+n'\mathbf{b}_2)\cdot\mathbf{r}} \\ &= \sum_{m,n} a_{m,n} e^{i(m\mathbf{Q}+n\mathbf{G})\cdot\mathbf{r}} \,. \end{aligned} \tag{54.9}$$

The Schrödinger equation can be put into matrix form:

$$\begin{aligned} &\sum_{m',n'} \left[\frac{\hbar^2}{2m}(\mathbf{k} + m\mathbf{Q} + n\mathbf{G})^2 \delta_{m,m'}\delta_{n,n'} + \alpha(\delta_{m,m'-1}\delta_{m,m'+1})\delta_{n,n'} \right] \\ &+\beta\delta_{m,m'}(\delta_{n,n'-1} + \delta_{n,n'+1})\big] a_{m',n'} = E(\mathbf{k}) a_{mn} \,. \end{aligned} \tag{54.10}$$

By solving these matrix equations, one finds the energy spectrum and the wave functions. Each state $|\mathbf{k}\rangle$ is coupled to $|\mathbf{k} \pm \mathbf{Q}\rangle$ through the CDW potential and to $|\mathbf{k} \pm \mathbf{G}\rangle$ through the crystal potential. Since the matrix Schrödinger equation cannot be solved exactly, we resort to an approximate scheme presented in the next section.

54.3 Approximate Solutions

Since the potential given by Equation (54.2) does not have any geometrical symmetry other than inversion, the zone structure determined by energy gaps, which in general are curved surfaces, need not be the same as the Brillouin zone structure. But they should not be drastically different because the energy-gap surfaces pass through the points $(m\mathbf{Q} + n\mathbf{G})/2$. The zone structure determined by energy gaps can be shown to be equivalent to the Brillouin-zone structure obtained from cutting and rearranging zones determined by planes bisecting reciprocal-lattice vectors [16].

To find the Fermi surface, we consider a constant energy surface in the extended-Brillouin-zone scheme. Near the Fermi surface, two distinct groups of zone boundaries are important: (a) minigap zone boundaries passing through $\mathbf{K}_n/2 = [(n+1)\mathbf{Q} - n\mathbf{G}]/2$, where $n = 0, 1, 2, \ldots$, and (b) heterodyne-gap zone boundaries passing through $\mathbf{H}_n/2 = n(\mathbf{G} - \mathbf{Q})/2$, where $n = 1, 2, \ldots$ Since the Fermi surface has inversion symmetry, we will consider only $\mathbf{k} \cdot \hat{\mathbf{Q}} \geq 0$.

54.3.1 Near $\mathbf{k} \cdot \hat{\mathbf{Q}} \cong Q/2$ (minigap region)

Because of the small differences in magnitude and direction of $\mathbf{G}$ and $\mathbf{Q}$, heterodyne gaps near $\mathbf{Q}/2$ are of very high order. These gaps can be neglected since the size of an energy gap decreases very rapidly as the order of gap increases, as can be seen in Table 54.1. The sizes of the minigaps in Table 54.1 were calculated parallel to $\hat{\mathbf{Q}}$ starting from $\mathbf{Q}/2$ using the approximation explained in Section 54.3.2.

Approximate solutions of the matrix Schrödinger equation can be found by neglecting couplings between states too far apart in energy. There are two distinct series of minigaps. The minimum number of states needed to calculate the energy spectrum in this region are as follows.

(i) Even-integer minigaps involving states separated by $K_{2m} = (2m+1)Q - 2mG$, where $m = 0, 1, 2, \ldots$

The zeroth minigap (CDW main gap):

$$|\mathbf{k}\rangle, |\mathbf{k}-\mathbf{Q}\rangle \; .$$

The second minigap: the above set and

$$|\mathbf{k}-\mathbf{Q}+\mathbf{G}\rangle, |\mathbf{k}-2\mathbf{Q}+\mathbf{G}\rangle \; ,$$
$$|\mathbf{k}-2\mathbf{Q}+2\mathbf{G}\rangle, |\mathbf{k}-3\mathbf{Q}+2\mathbf{G}\rangle \; .$$

The fourth minigap: the above sets and

$$|\mathbf{k}-3\mathbf{Q}+3\mathbf{G}\rangle, |\mathbf{k}-4\mathbf{Q}+3\mathbf{G}\rangle \; ,$$
$$|\mathbf{k}-4\mathbf{Q}+4\mathbf{G}\rangle, |\mathbf{k}-5\mathbf{Q}+4\mathbf{G}\rangle \; ,$$

etc.

(ii) Odd-integer minigaps involving states separated by $\mathbf{K}_{2m+1} = (2m+2)\mathbf{Q} - (2m+1)\mathbf{G}$, where $m = 0, 1, 2, \ldots$

The first minigap:

$$|\mathbf{k}\rangle, |\mathbf{k}-\mathbf{Q}\rangle, |\mathbf{k}-\mathbf{Q}+\mathbf{G}\rangle, |\mathbf{k}-2\mathbf{Q}+\mathbf{G}\rangle \; .$$

The third minigap: the above set and

$$|\mathbf{k}-2\mathbf{Q}+2\mathbf{G}\rangle, |\mathbf{k}-3\mathbf{Q}+2\mathbf{G}\rangle \; ,$$
$$|\mathbf{k}-3\mathbf{Q}+3\mathbf{G}\rangle, |\mathbf{k}-4\mathbf{Q}+3\mathbf{G}\rangle \; .$$

Table 54.1 Sizes of CDW energy gaps.

Order of gap	Heterodyne gap (meV)	Minigap (meV)
1	16	90
2	14	67
3	12	51
4	8	34
5	3	15
6	0.6	4
7	0.06	0.6
8	0.01	0.06

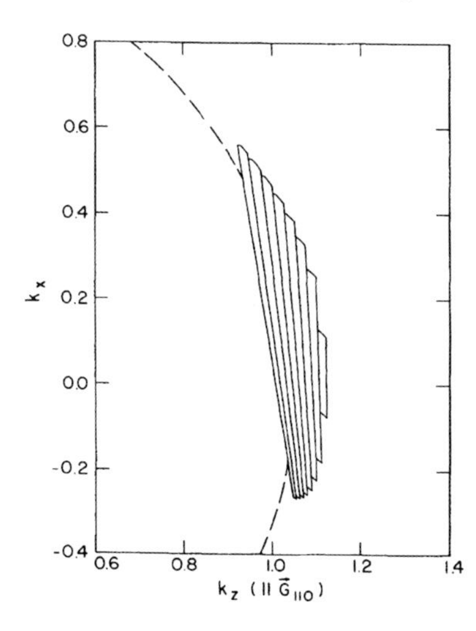

Figure 54.1 Approximate shape of the Fermi surface in the mini-gap region ($\mathbf{k} \cdot \hat{\mathbf{Q}} \cong Q/2$) for $E(\mathbf{k}) = 2.35$ eV. k_z and k_x are in units of k_F. The dashed curve is that of a free-electron Fermi sphere having the same energy.

The fifth minigap: the above sets and

$$|\mathbf{k} - 4\mathbf{Q} + 4\mathbf{G}\rangle, |\mathbf{k} - 5\mathbf{Q} + 4\mathbf{G}\rangle ,$$
$$|\mathbf{k} - 5\mathbf{Q} + 5\mathbf{G}\rangle, |\mathbf{k} - 6\mathbf{Q} + 5\mathbf{G}\rangle ,$$

etc.

Equal numbers of states above and below (in energy) compared to the states required in the minimum sets were included. Dimensions of matrices diagonalized were 30 × 30 near even-integer minigap surfaces and 28 × 28 near odd-integer minigap surfaces. The cross section of the Fermi surface in the $\hat{\mathbf{G}}$–$\hat{\mathbf{Q}}$ plane is shown in Figure 54.1. The energy spectrum is approximately cylindrical. Note the small cylindrical sections near $\mathbf{Q}/2$, which cause the perpendicular-field cyclotron resonance [17] as well as unexpected sharp peaks in recent angle-resolved photoemission data [18, 19].

54.3.2 Near $\mathbf{k} \cdot \hat{\mathbf{Q}}' \cong \mathbf{Q}'$ (heterodyne-gap region)

Near this equatorial region minigaps are of very high order and can be neglected, since the size of an energy gap decreases very rapidly as the order of the gap increases, as can be seen in Table 54.1. The sizes of the hetero-dyne gaps in Table 54.1 were calculated parallel to $\hat{\mathbf{Q}}'$, starting from the origin and with use of the approximation explained in Section 54.3.1.

Approximate solutions can be found by neglecting couplings between states too far apart in energy. There are two distinct series of heterodyne gaps. The minimum number of states needed to calculate the energy spectrum in this region are as follows.

(i) Odd-integer heterodyne gaps involving states separated by $\mathbf{H}_{2m-1} = (2m-1)\mathbf{Q}'$, where $m = 1, 2, \ldots$

The first heterodyne gap:

$$|\mathbf{k}\rangle, |\mathbf{k} + \mathbf{Q}\rangle, |\mathbf{k} + \mathbf{Q} - \mathbf{G}\rangle, |\mathbf{k} - \mathbf{G}\rangle .$$

The third heterodyne gap: the above set and

$$|\mathbf{k} + 2\mathbf{Q} - \mathbf{G}\rangle, |\mathbf{k} + 2\mathbf{Q} - 2\mathbf{G}\rangle, |\mathbf{k} + \mathbf{Q} - 2\mathbf{G}\rangle ,$$
$$|\mathbf{k} + 3\mathbf{Q} - 2\mathbf{G}\rangle, |\mathbf{k} + 3\mathbf{Q} - 3\mathbf{G}\rangle, |\mathbf{k} + 2\mathbf{Q} - 3\mathbf{G}\rangle .$$

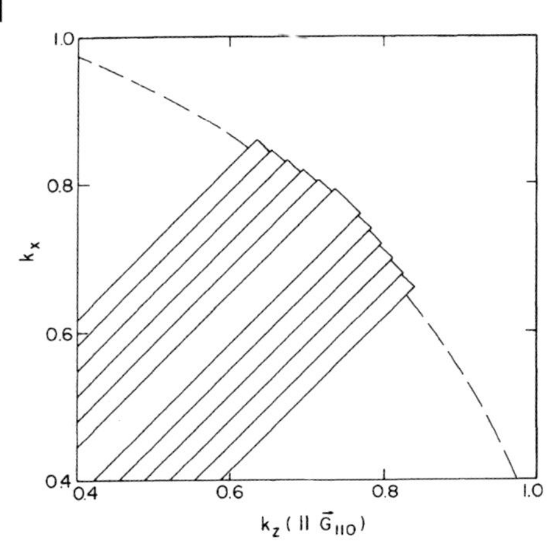

Figure 54.2 Approximate shape of the Fermi surface in the heterodyne gap region ($\mathbf{k} \cdot \hat{\mathbf{Q}}' \cong Q'/2$) for $E(\mathbf{k}) = 2.35$ eV. k_z and k_x are in units of k_F. The dashed curve is that of a free-electron Fermi sphere having the same energy.

The fifth heterodyne gap: the above sets and

$$|\mathbf{k}+4\mathbf{Q}-3\mathbf{G}\rangle, |\mathbf{k}+4\mathbf{Q}-4\mathbf{G}\rangle, |\mathbf{k}+3\mathbf{Q}-4\mathbf{G}\rangle,$$
$$|\mathbf{k}+5\mathbf{Q}-4\mathbf{G}\rangle, |\mathbf{k}+5\mathbf{Q}-5\mathbf{G}\rangle, |\mathbf{k}+4\mathbf{Q}-5\mathbf{G}\rangle \ ,$$

etc.

(ii) Even-integer heterodyne gaps involving states separated by $H_{2m} = 2m\mathbf{Q}'$, where $m = 1, 2, \ldots$

The second heterodyne gap:

$$|\mathbf{k}\rangle, |\mathbf{k}+\mathbf{Q}\rangle, |\mathbf{k}+\mathbf{Q}-\mathbf{G}\rangle, |\mathbf{k}-\mathbf{G}\rangle \ ,$$
$$|\mathbf{k}+2\mathbf{Q}-\mathbf{G}\rangle, |\mathbf{k}+2\mathbf{Q}-2\mathbf{G}\rangle, |\mathbf{k}+\mathbf{Q}-2\mathbf{G}\rangle \ .$$

The fourth heterodyne gap: the above set and

$$|\mathbf{k}+3\mathbf{Q}-2\mathbf{G}\rangle, |\mathbf{k}+3\mathbf{Q}-3\mathbf{G}\rangle, |\mathbf{k}+2\mathbf{Q}-3\mathbf{G}\rangle,$$
$$|\mathbf{k}+4\mathbf{Q}-3\mathbf{G}\rangle, |\mathbf{k}+4\mathbf{Q}-4\mathbf{G}\rangle, |\mathbf{k}+3\mathbf{Q}-4\mathbf{G}\rangle \ .$$

The sixth heterodyne gap: the above sets and

$$|\mathbf{k}-5\mathbf{Q}-4\mathbf{G}\rangle, |\mathbf{k}-5\mathbf{Q}-5\mathbf{G}\rangle, |\mathbf{k}+4\mathbf{Q}-5\mathbf{G}\rangle \ ,$$
$$|\mathbf{k}+6\mathbf{Q}-5\mathbf{G}\rangle, |\mathbf{k}+6\mathbf{Q}-6\mathbf{G}\rangle, |\mathbf{k}+5\mathbf{Q}-6\mathbf{G}\rangle \ ,$$

etc.

Equal numbers of states above and below (in energy) compared to the states required in the minimum sets were included as before. Dimensions of matrices diagonalized were 31 × 31 near even-integer heterodyne-gap surfaces and 28 × 28 near odd-integer heterodyne-gap surfaces. The cross section of the Fermi surface in the $\hat{\mathbf{G}}$–$\hat{\mathbf{Q}}$ plane is shown in Figure 54.2. The energy spectrum is approximately symmetric about $\hat{\mathbf{Q}}'$ in this region. Note the truncations of the Fermi surface, which lead to the presence of open orbits. Open-orbit motion is responsible for magnetoresistance peaks [20–22] and splittings of the spin-wave sidebands in conduction-electron spin-resonance transmission experiments [23].

54.4 Conclusions

The Fermi surface of potassium is shown to be anisotropic and multiply connected by solving a Schrödinger equation having two sinusoidal potentials: one from the CDW and the other from the lattice. An approximate shape of the Fermi surface, calculated numerically, is shown in Figure 54.3. The Fermi energy was chosen to be 2.35 eV so that the radius of the smallest cylindrical section is about $0.1k_F$, a value required to explain the perpendicular-field cyclotron resonance [17]. (The diameters of the cylindrical sections near the main gap depend on the Fermi energy and their positions relative to the CDW gap, which are not known.) Far off the $\hat{\mathbf{G}}$–$\hat{\mathbf{Q}}$ plane, the contributions of other pseudopotentials may not be neglected and the actual Fermi surface may be different from that suggested in Figure 54.3. The Fermi surface is sliced into many pieces by two families of extra energy gaps: minigaps and heterodyne gaps. Minigaps are caused by perturbations having periodicity $\mathbf{K}_n = (n+1)\mathbf{Q} - n\mathbf{G}$, where $n = 0, 1, 2, \ldots$ The presence of cylindrical sections near $|\mathbf{k}| \cong Q/2$ is responsible for the perpendicular-field cyclotron resonance. Heterodyne gaps are caused by perturbations with periodicity $\mathbf{H}_n = n(\mathbf{G} - \mathbf{Q})$, where $n = 1, 2, \ldots$ The occurrence of open orbits caused by these extra energy gaps leads to the open-orbit magnetoresistance peaks observed by Coulter and Datars [20–22]. CDW-domain patterns (arising from the 24 crystallo-graphically equivalent directions of $\mathbf{Q}$) play an important role in explaining the observed splitting of the conduction-electron spin resonance [24] and that of the spin-wave sidebands in transmission experiments [23].

We emphasize as a final remark that the data in Table 54.1, and the details of the Fermi surface shown in Figure 54.3 are merely suggestive. Since the exchange and correlation potentials which cause a CDW are highly nonlocal [25, 26], the CDW potential 2α will be a (possibly rapid) function of $\mathbf{k}$. Such nonlocal effects have been ignored in the exercise presented here. They were estimated in a calculation of the fractional charge modulation of the CDW [27]; and the result was a 40% reduction in the predicted CDW charge modulation. We anticipate that if the $\mathbf{k}$ dependence of 2α were included in the calculations presented here,

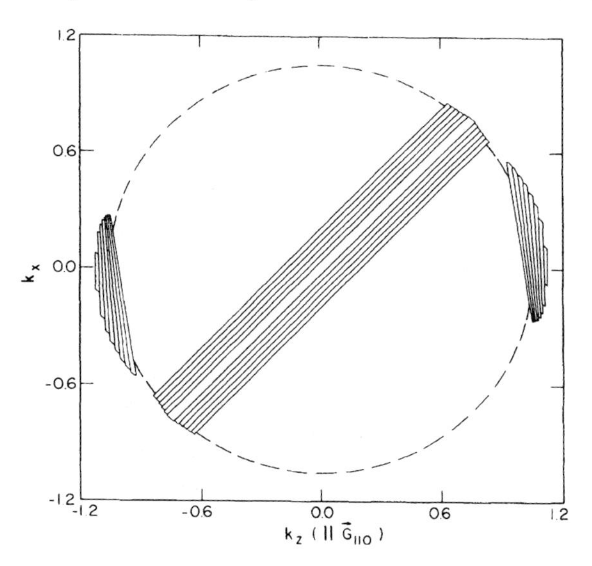

Figure 54.3 Approximate shape of the Fermi surface for $E(\mathbf{k}) = 2.35$ eV. k_z and k_x are in units of k_F. The dashed curve is that of a free-electron Fermi sphere having the same energy.

then the minigaps and heterodyne gaps would fall off more rapidly with increasing order than the sequence shown in Table 54.1.

Acknowledgements

This work was supported by the Materials Research Laboratory Program of the National Science Foundation.

References

1 Overhauser A.W. (1985) In: Bassani F., Fumi F., and Tosi M.P. (eds) *Highlights of Condensed-Matter Theory*, Proceedings of the International School of Physics "Enrico Fermi", Course LXXXIX, North-Holland, Amsterdam.
2 Overhauser A.W. (1978) *Adv. Phys.* **27**, 343.
3 Overhauser A.W. (1962) *Phys. Rev.* **128**, 1437.
4 Overhauser A.W. (1962) *Phys. Rev.* **167**, 691.
5 Giebultowicz T.M., Overhauser A.W. and Werner S.A. (1986) *Phys. Rev. Lett.* **56**, 1485.
6 Werner S.A., Giebultowicz T.M., and Overhauser A.W. (1987) *Phys. Scr.* **T19**, 266.
7 Pintschovius L., Blaschko O., Krexner G., de Podesta M., and Currat R. (1987) *Phys. Rev. B* **35**, 9330.
8 Blaschko O., de Podesta M., and Pintschovius L. (1988) *Phys. Rev. B* **37**, 4258.
9 Giebultowicz T.M., Overhauser A.W., and Werner S.A. (1986) *Phys. Rev. Lett.* **56**, 2228.
10 Werner S.A. and Arif M. (1988) *Acta Crystallogr. Sect. A* **44**, 383.
11 Werner S.A., Overhauser A.W., and Giebultowicz T.M. (unpublished).
12 Mayer H. and El Naby M.H. (1963) *Z. Phys.* **174**, 269.
13 Overhauser A.W. (1964) *Phys. Rev. Lett.* **13**, 190.
14 Ashcroft N. (1965) *Phys. Rev.* **140**, A935.
15 Fragachan F.G. and Overhauser A.W. (1984) *Phys. Rev. B* **29**, 2912.
16 Seitz F. (1940) *The Modern Theory of Solids*, McGraw-Hill, New York, Section 61.
17 Lacueva G. and Overhauser A.W. (1986) *Phys. Rev. B* **33**, 3765.
18 Jensen E. and Plummer E.W. (1985) *Phys. Rev. Lett.* **55**, 1912.
19 Overhauser A.W. (1985) *Phys. Rev. Lett.* **55**, 1916.
20 Coulter P.G. and Datars W.R. (1980) *Phys. Rev. Lett.* **45**, 1021.
21 Coulter P.G. and Datars W.R. (1982) *Solid State Commun.* **43**, 715.
22 Coulter P.G. and Datars W.R. (1985) *Can. J. Phys.* **63**, 159.
23 Hwang Y.G. and Overhauser A.W. (1988) *Phys. Rev. B* **38**, 9011.
24 Overhauser A.W. and de Graaf A.M. (1968) *Phys. Rev.* **168**, 763.
25 Overhauser A.W. (1970) *Phys. Rev. B* **2**, 874.
26 Duff K.J. and Overhauser A.W. (1972) *Phys. Rev. B* **5**, 2779.
27 Giuliani G.F. and Overhauser A.W. (1980) *Phys. Rev. B* **22**, 3639.

Reprint 55 Neutron-Diffraction Structure in Potassium Near the [011] and [022] Bragg Points[1)]

S.A. Werner*, A.W. Overhauser**, and T.M. Giebultowicz***

* Department of Physics and Astronomy, University of Missouri-Columbia, 525, Northwestern Avenue, Columbia, Missouri 65211, USA

** Department of Physics, Purdue University, West Lafayette, Indiana 47907, USA

*** National Bureau of Standards, Gaithersburg, Maryland 20899, USA

Received 20 April 1989

Neutron diffraction from a stress-free crystal of potassium shows no increase in its ($\sim 0.03°$) mosaic width after repeated cycling between 4 and 300 K. Furthermore, no temperature-dependent, Huang-type diffuse scattering near [011] and [022] was found. Properties such as these, indicative of thermally-induced plastic deformation, were reported for the potassium crystals studied by Pintschovius *et al.* and by Blaschko *et al.* Their data also exhibited a sample-caused background as much as 100 times larger than our stress-free crystal. Using 2.35-Å-wavelength neutrons and tight collimation, we find that the satellite structure near [011] retains the same spacing near [022]. The structure cannot be attributed to a double-scattering artifact, which requires a spacing proportional to tan θ_B. The data are consistent with a charge-density-wave interpretation.

We present new neutron-diffraction data on the structure near the [011] and [022] Bragg points in potassium that are consistent with our previous data and conclusions [1]. A critical review is given of other neutron-scattering data on K, recently reported by Pintschovius *et al.* [2] (hereafter called PB1) and by Blaschko *et al.* [3] (hereafter called PB2). We believe that the quality of the K crystals used in PB1 and PB2 was so poor that charge-density-wave (CDW) satellites could probably not have been observed. The structure reported in PB1 was indeed caused by a double-scattering artifact [4], a possibility which we had entertained [5], but were able to exclude for our experiments on the basis of not observing this structure in a Si test crystal [6]. The cause of the artifacts in PB1 was an extraordinarily large diffuse scattering generated by sample imperfections. This spurious scattering caused a background that was 100 times larger in the PB2 experiment than that found in the high-quality, stress-free crystal studied in this work and in [1].

We interpreted our original data [1] as being due to CDW satellites in a $(0\bar{1}1)$ diffraction plane near [011], as shown by the black dots in Figure 55.1. There are other satellite locations which, however, are not expected to be observable because either their intensities are too weak or they lie too close to the mosaic sphere passing through each Bragg point. A CDW satellite is surrounded by an ellipsoidal cloud of phason diffuse scattering [7] indicated by the shaded regions of Figure 55.1. The centroids of these inelastic scattering phason clouds is shifted towards the [011] Bragg point and away from the location of the elastic CDW satellite points. The reason for this is shown schematically in Figure 55.2. The interaction between the phason modes originating from the CDW satellite point Q and the acoustic phasons originating from the Bragg point G gives rise to a branch splitting. The vertical width of these

1) Phys. Rev. B 41, 18 (1990-II).

Anomalous Effects in Simple Metals. Albert Overhauser

ISBN: 978-3-527-40859-7

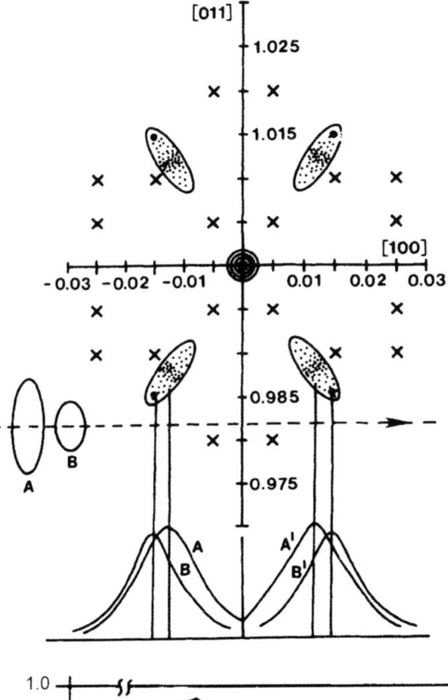

Figure 55.1 Upper half: map of the (0$\bar{1}$1) diffraction plane surrounding the [011] Bragg point. The large dots at $\pm 0.015, 0.985$, and ± 1.015 are the principal CDW satellite locations. The diffuse phason clouds surrounding these satellites are asymmetric (as expected from Figure 55.2). The $\times$'s are the locations of other satellites which are either too weak to be observed or are too near the [100] axis to be isolated from the [011] mosaic tail. Lower half: schematic diffraction profiles for a transverse scan (dashed line) along the trajectory $[h, 0.982, 0.982]$. The apparent shift of the profiles (A, A' to B, B') when the neutron wavelength is changed from 2.35 Å (resolution ellipsoid A) to 4.08 Å (resolution ellipsoid B) arises from the altered convolution with the phason diffuse scattering.

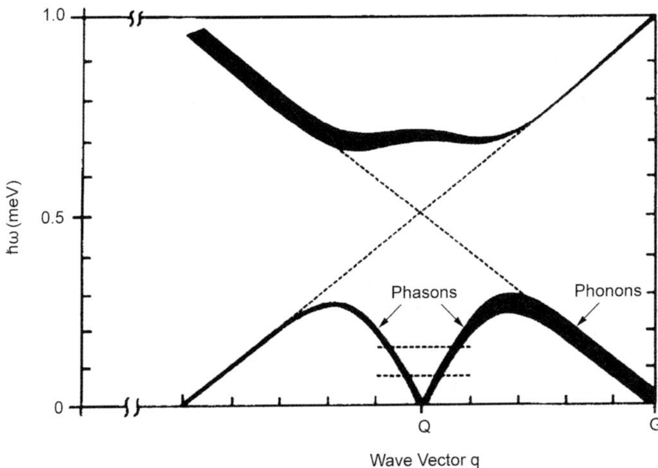

Figure 55.2 Phason and amplitude-mode spectra near a CDW satellite at $q = Q$. The diagonal dashed lines are the phonon dispersion curves (appropriately translated) that would be present if there were no CDW. The vertical widths of the solid curves are proportional to the neutron-scattering cross section for each mode (but do not include the Bose–Einstein thermal factors). The maximum frequency of the phason spectrum is ~ 0.3 meV for K. The horizontal dashed lines depict energy resolutions for two neutron wavelengths and indicate why phason clouds appear to shift with neutron wavelength.

curves is proportional to the square of the amplitudes of these modes. Even at 4 K, about 90% of the integrated intensity resides in the phason cloud [1]. Since phason scattering is inelastic, the apparent location of the satellite will depend on both the energy and momentum resolution of the triple-axis neutron spectrometer. This effect is illustrated schematically by

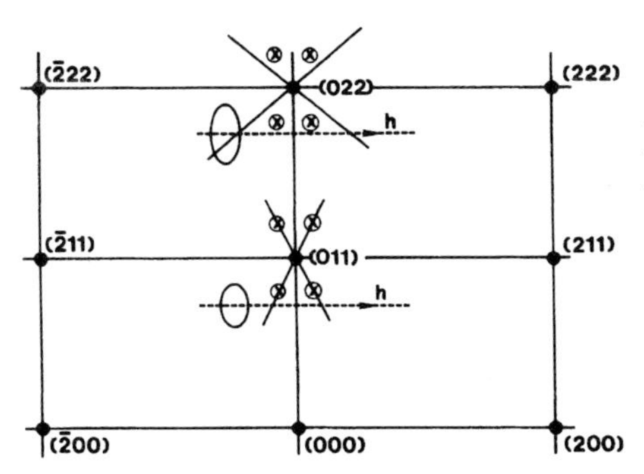

Figure 55.3 Schematic map of the $(0\bar{1}1)$ diffraction plane near the [011] and [022] Bragg points. The diagonal lines through [011] and [022] depict the orientation of streaks due to possible double-scattering artifacts. The angle of these lines, relative to ⟨011⟩, is the Bragg angle θ_B. When the resolution ellipsoids are convolved with the CDW satellites – the four *x*'s surrounding [011] and [022] – the satellite spacings along the indicated transverse scans will be the same near [011] and [022]. The spacings between double-scattering artifacts will be proportional to tan θ_B.

curves *A* and *B* in Figure 55.1. This explains why the satellite locations appear to shift with incident-neutron wavelength [1].

Our new data involve scanning the structures surrounding both the [022] and [011] Bragg points as shown in Figure 55.3. A neutron wavelength, $\lambda = 2.35$ Å, was chosen so that both regions could be studied with comparable resolution. (The [022] point cannot be reached with $\lambda = 4.08$ Å neutrons.) If the observed structure were caused by a double-scattering artifact, then the transverse spacing between the observed peaks (satellites) would be 2.5 times larger near [022] than near [011]. The reason is that the angle between the double-scattering streaks is equal to $2\theta_B$ [4]. However, we find the spacing between the peaks in scans transverse to the ⟨011⟩ axis near [022] and [011] to be equal, which is the expected result for CDW satellites. Representative scans at $k = 0.980$ and 1.980 are shown in Figure 55.4. Unfortunately, the resolution is not quite good enough to separate the structures into three separated peaks. We fitted the data for the two scans to three Lorentzian peaks, each having a width identical to the measured transverse widths of the [011] and [022] Bragg peaks, respectively. The relative integrated intensity of the fitted peaks near [022] to those near [011] is 7.6 For a CDW modulation this ratio should be 4.6, which is obtained from assuming that the CDW satellite cross section, along with its phason cloud, is proportional to the convolution of the scattering cross section with the four-dimensional resolution function. The resolution ellipsoid near [022] overlaps a larger fraction of the intense region of the phason cloud than near [011], This effect is difficult to model accurately, but likely accounts for the larger intensity observed near [022].

We have investigated the temperature dependence of the diffraction structure. For *T* sufficiently high so that the phason temperature factor [8, 9] of the elastic component is small, the integrated intensity of the phason cloud should be independent of *T*. As *T* increases, the shape of the phason cloud will change gradually, since the convolution of multiple-phason events leads to a diffusivelike growth. Accordingly, the apparent amplitude of the CDW-related scattering can either increase, decrease, or remain constant with increasing *T*, depending on the location of the scan line in reciprocal space relative to the satellite center. We have found for a transverse scan $(h, 1.975, 1.975)$ that the structure has about the same amplitude at 4 and 78 K. Further experiments are necessary to explore the scan dependence of this (possibly surprising) result.

Since K undergoes a 5% volume contraction on cooling to 4 K, it is essential to mount the crystal in a manner that prevents plastic deformation. We achieved this requirement by supporting our crystal between compressed-quartz wool at each end, within a He-gas-filled aluminum capsule. Our data have never shown any signs of crystal deformation and the crystal has retained its sharp mosaic width after more than 50 cool-downs to 4 K. A recent study [10] of this crystal with 50 kV x-rays showed that the mosaic width is 0.03°. By contrast, the crystals of PB1 and PB2 suffered incremental increases in mosaic width on each cool-

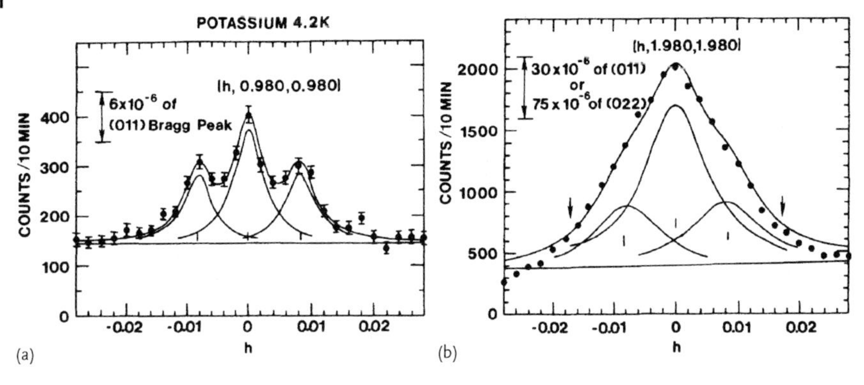

Figure 55.4 (a) Transverse scan of the satellites in K near [011] for $\lambda = 2.35$ and $T = 4.2°$K. The width of the Lorentzian components ($2\Gamma = 0.0056$) was determined by a fit to the [011] rocking-curve profile; (b) transverse scan near [022]. The width of the Lorentzian components ($2\Gamma = 0.0124$) was fixed by the [022] rocking-curve profile. The arrows mark the locations of potential double-scattering artifacts. The central components in (a) and (b) arise from the overlap of the resolution ellipsoid with the nearby Bragg peaks. The monochromator and analyzer are PG (002). The incident beam contains a pyrolytic graphite (PG) filter between the monochromator and the reactor end of the Soller collimator. The collimations are 10′-12′-12′-21′. The energy resolution is 0.28 meV.

down, a behavior which can only be attributed to plastic deformation. The PB1 and PB2 experiments at 5 K were done on a crystal with a mosaic spread of 0.3°–0.5°, more than 10 times the mosaic width of our crystal. The large increases in diffuse scattering near the base of the [011] and [022] Bragg reflections reported by PB2 [and shown in Figure 55.5(a)] during cool-down from 100 to 5 K shows the consequence of additional plastic deformation. This increase, which alone exceeds our total background by a factor of $\sim$ 50, can be ascribed to newly generated dislocations. The cause(s) of the plastic deformation in the crystals of PB1 and PB2 is either the specimen clamp, KOH degradation, or hydrocarbon contamination left from crystal growth. A comparison of longitudinal scans through [022] at 78 and 4 K in our crystal with the data of PB2 is shown in Figures 55.5(b) and 55.5(c). Note the factor-of-100 difference in the background levels between our data and that of PB2.

The extinction factor e of the [011] Bragg peak is an important quantity, and is needed to calibrate satellite intensities and to comp are data from different samples. For $\lambda = 4.08$ Å, we found $\epsilon = 1/7$. This was done by normalizing the scattered intensity to the incident beam intensity using a National Bureau of Standards (NBS) calibrated fission chamber,[2] neutron photographic images of the diffracted beam to determine the effective sample scattering volume, and a wide-open detector to accept the full Bragg-diffracted beam. Since the samples used by PB1 and PB2 had mosaic widths 10 times larger than ours, and since their samples were subjected to severe plastic deformation with each cool-down, the only reasonable hypothesis is that $\epsilon \sim 1$ for those specimens.[3] We show in Figure 55.6 a comparison of the data of PB1 and of Giebultowicz *et al.* [1]. At comparable wavelengths (4.05 and 4.08 Å) the ratio of the background to the [011] Bragg peak intensity is seven times higher in PB1 than for our data [1]. Furthermore, if ϵ is in fact 1 for PB1 as we surmise, their sample-caused background would be 50 times higher than ours, thus obviating the possibility of PB1 observing the same structure reported in our original paper.

2) The help of Dr. D. Gilliam in the incident-beam-intensity calibration is appreciated.

3) PB1 states that $\epsilon \approx 1/40$ for their sample as measured by normalizing to a powder diffraction line. We are not familiar with that normalization procedure, which obviously requires a detailed understanding of the spectrometer vertical resolution function.

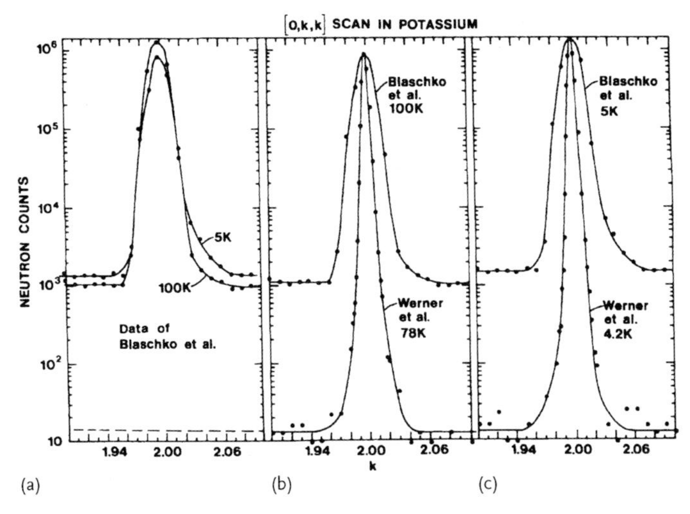

Figure 55.5 Longitudinal scans through the [022] Bragg point of K using $\lambda = 2.35$ Å neutrons. Note the logarithmic vertical scale, (a) Data of Blaschko *et al.* [3] which show a large increase in diffuse scattering near the base of the [022] on cooling from 100 to 5 K; (b) comparison of the $\sim$ 100 K data of Blaschko *et al.* with that of Werner *et al.* (this paper) normalized to a common peak height. Note the 100-fold disparity in diffuse background scattering; (c) comparison of the $\sim$ 5 K data of Blaschko *et al.* with that of Werner *et al.*, normalized to a common peak height. The disparity in diffuse scattering is larger than that in (b), since there is no change in background for our K crystal.

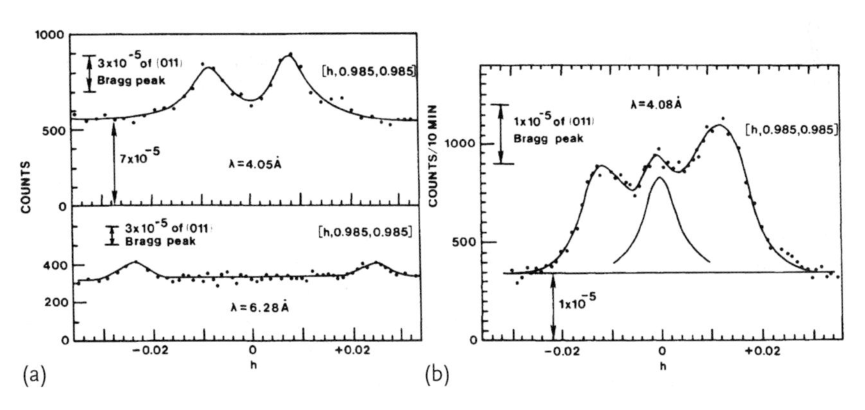

Figure 55.6 Transverse scans near the [011] Bragg peak of K; (a) data of Pintschovius *et al.* [2] for $\lambda = 4.05$ and 6.28 Å, which show structure caused by a double-scattering artifact; (b) data of Giebultowicz *et al.* [1]. Note the factor-of-7 difference in background levels.

It seems appropriate to emphasize again [6] that the symmetry of this structure about the [011] Bragg point is incompatible with scattering caused by martensitic embryos. (Such embryos could lead to two diffuse peaks on the [000] side of [011] but not (also) to two peaks on the [022] side of [011].)

CDW satellites can also be studied with x-ray diffraction. However, an attempt to find satellites in potassium by You *et al.* failed [11]. They employed 1.05 Å synchrotron radiation, which has an absorption length of 200 μm. Since the Bragg angle for K [011] is 8.16°, their diffracting volume was within 10 μm of the sample surface. It has been established that surface elastic strain can dramatically alter the electrical resistivity of potassium. Such effects have been studied in wires [12] and in spheres [13] (the so-called oil-drop effect [14]). The presumption is that strain can alter the CDW Q direction and, as a consequence, CDW satellites could be smeared out. Potassium samples inevitably have a think KOH surface layer which will create thermal stress near the surface. 50 keV x-rays should be used in a search for CDW satellites since a bulk sample 2 cm thick can then be penetrated.

Acknowledgements

We are grateful to the National Science Foundation for financial support.

References

1 Giebultowicz, T.M., Overhauser, A.W. and Werner, S.A. (1986) *Phys. Rev. Lett.* **56**, 1485.
2 Pintschovius, L., Blaschko, O., Krexner, G., de Podesta, M. and Currat, R. (1987) *Phys. Rev. B* **35**, 9330.
3 Blaschko, 0., de Podesta, M. and Pintschovius, L. (1988) *Phys. Rev. B* **37**, 4258.
4 Werner, S.A. and Arif, M. (1988) *Acta Crystallogr. Sect. A* **44**, 383.
5 Giebultowicz, T.M., Overhauser, A.W. and Werner S.A. (1986) *Phys. Rev. Lett.* **56**, 2228.
6 Werner, S.A., Giebultowicz, T.M. and Overhauser, A.W. (1988) *Phys. Scr.* **T19**, 266. The last paragraph of this paper was intended to begin "One may *not* come to the conclusion ... " instead of "one may *now* come to the conclusion... "
7 Overhauser, A.W. (1986) *Phys. Rev. B* **34**, 7362.
8 Overhauser, A.W. (1971) *Phys. Rev. B* **3**, 3172.
9 Giuliani, G.F. and Overhauser, A.W. (1981) *Phys. Rev. B* **23**, 3737.
10 Colella, R. and Quin Zhao (unpublished).
11 Hoydoo You, Axe, J.D., Hohlwein, Dietmar and Hastings, J.B. (1987) *Phys. Rev. B* **35**, 9333.
12 Taub, H., Schmidt, R.L., Maxfield, B.W. and Bowers, R. (1971) *Phys. Rev. B* **4**, 1134.
13 Holroyd, F.W. and Datars, W.R. (1975) *Can. J. Phys.* **53**, 2517.
14 Overhauser, A.W. (1978) *Adv. Phys.* **27**, 343.

Reprint 56 Quantum Oscillations From the Cylindrical Fermi-Surface Sheet of Potassium Created by the Charge-Density Wave[1)]

Graciela Lacueva * *and A.W. Overhauser* **

* Department of Physics, John Carroll University, 525, Northwestern Avenue, University Heights, Ohio 44118, USA

** Department of Physics, Purdue University, West Lafayette, Indiana 47907, USA

Received 6 February 1991

Oscillations reported by Dunifer *et al.* in microwave transmission through thin K layers are found to be periodic in $1/H$. The oscillations arise from conduction-electron Landau levels passing through a small cylindrical sheet of the Fermi surface. This cylinder had been envisioned theoretically after incorporating both charge-density-wave and crystalline potentials in Schrödinger's equation. The cylinder's cross-sectional area is found to be $\pi k_F^2/69$, in agreement with the area inferred from the perpendicular-field cyclotron resonance, discovered by Grimes in the surface impedance.

Microwave transmission through thin K layers in a magnetic field **H** perpendicular to the surface reveals several signals having no conventional explanation [1]. One of these surprising phenomena – the high-frequency oscillation – was seen in nine samples and is the focus of this work. The periodicity turns out to be linear in $1/H$, which is the signature of Landau-level oscillations, but their frequency is 69 times smaller than the well-known de Haas–van Alphen value.

Data from sample K-4, which had the longest and clearest sequence, are shown in Figure 56.1. The conduction-electron spin resonance (together with associated spin-wave sidebands) and the Gantmakher–Kaner oscillations have been understood for many years. The source of the new Landau-level oscillations, as will be explained below, is a small cylindrical sheet of Fermi surface created by the charge-density-wave (CDW) broken symmetry [2]. This cylinder, which contains only a fraction $\eta = 4 \times 10^{-4}$ of the conduction electrons, is also responsible for the dramatic cyclotron resonance near $\omega_c/\omega = 1$ in Figure 56.1 [3].

Before discussing the origin and geometry of the Fermi-surface cylinder, we shall analyze the data of Figure 56.1, for which the horizontal axis, linear in H, is $x = \omega_c/\omega$, where ω_c is the Azbel–Kaner cyclotron frequency and $\omega/2\pi$ is the microwave frequency. $\omega_c = eH/m^*c$. For K, $m^*/m = 1.21$ [4]. Twelve complete oscillations are rather well defined within the two vertical lines of Figure 56.1. The ω_c/ω coordinates of the maxima and minima are given in Table 56.1. (These values apply only for the microwave frequency $f = 79.18$ GHz.) At first glance the oscillations appear to be periodic in x, but we shall test this supposition by finding the best discrete sequence,

$$x_n = sn + b \,, \qquad (56.1)$$

1) Phys. Rev. B Condensed Matter 46, 3 (1992-I).

Anomalous Effects in Simple Metals. Albert Overhauser

ISBN: 978-3-527-40859-7

Table 56.1 ω_c/ω values, X_n, for the maxima and minima of the signal shown in Figure 56.1.

n	Maxima	Minima
1	1.085	1.090
2	1.098	1.105
3	1.112	1.121
4	1.128	1.137
5	1.146	1.154
6	1.163	1.173
7	1.181	1.190
8	1.202	1.210
9	1.220	1.230
10	1.238	1.249
11	1.259	1.267
12	1.277	1.286

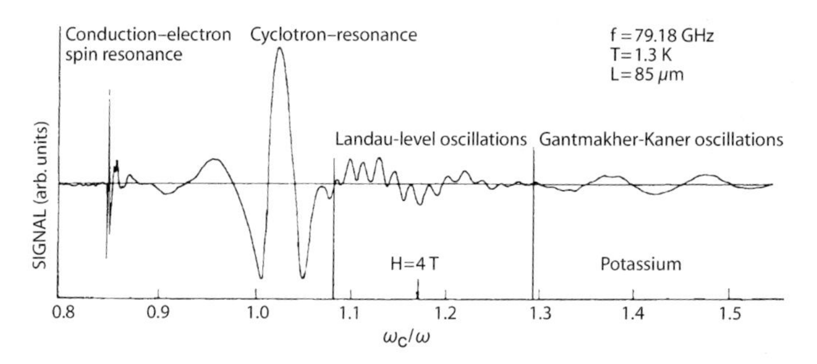

Figure 56.1 Microwave transmission signal vs. H through a K plate in a perpendicular magnetic field. H = 3.42 T at $\omega_c/\omega = 1$. The phase of the microwave reference was adjusted so that the cyclotron resonance was symmetric. The data displayed on this figure were provided by G.L. Dunifer (unpublished).

which fits the 12 maxima (or minima). The values of s and b that minimize the mean-square deviation,

$$D = \frac{1}{12}\sum_{n=1}^{12}[X_n - (sn + b)]^2 \,, \tag{56.2}$$

are $s = 1.779 \times 10^{-2}$, $b = 1.060$ for the maxima and $s = 1.804 \times 10^{-2}$, $b = 1.067$ for the minima. The mean-square *fractional* deviation, $D/(X_6)^2$, is 7.9×10^{-6} for the maxima and 4.4×10^{-6} for the minima.

The foregoing fit should be compared to one obtained by supposing the periodicity to be linear in $1/H$. We then seek the best sequence,

$$\frac{1}{x_n} = b' - s'n \,. \tag{56.3}$$

The mean-square deviation to be minimized is now

$$D' = \frac{1}{12}\sum_{n=1}^{12}\left(\frac{1}{X_n} - (b' - s'n)\right)^2 \,. \tag{56.4}$$

The solutions are $s' = 1.287 \times 10^{-2}$, $b' = 0.936$ and $s' = 1.288 \times 10^{-2}$, $b' = 0.930$ for maxima and minima, respectively. The mean-square *fractional* deviations, $D'(X_6)^2$, are 1.54×10^{-6} and 9.8×10^{-7}. Observe that these χ^2 fitting errors are much smaller, by factors of 5.1 and 4.5, respectively, than those obtained with the linear sequence, Equation (56.1). A detailed analysis of χ^2 versus the periodicity power is given below. It was reported that the periodicity did not depend on sample thickness. Accordingly, we conclude that the high-frequency oscillations are best described by periodicity in $1/H$, and so arise from Landau levels passing through an appropriate sheet of the conduction-electron Fermi surface.

Landau-level quantization causes periodic variation of *any* physical property that depends on conduction-electron response [5]. A property y will then acquire an oscillatory component,

$$\Delta y \sim \cos\left(\frac{2\pi F}{H} + \phi\right) , \tag{56.5}$$

where F is the de Haas–van Alphen frequency,

$$F = \frac{\hbar c A}{2\pi e} . \tag{56.6}$$

A is the external area in **k** space (perpendicular to **H**) of the Fermi surface involved. For **K**, with lattice constant $a = 5.2295$ Å at 4.2 K, a free-electron Fermi sphere has $A_0 = \pi k_F^2$ and $F_0 = 1.828 \times 10^4$ T. ($k_F = 7.45 \times 10^7$ cm^{-1}.)

The frequency F_c that describes the oscillations in Figure 56.1 is H_0/s', where H_0 is the magnetic field corresponding to $\omega_c/\omega = 1$. $\omega/2\pi = 9.18$ GHz, so we obtain $H_0 = 3.42$ T and

$$F_c = 266\ \text{T} . \tag{56.7}$$

If πk_c^2 is the cross section of the Fermi surface causing these oscillations, then

$$\frac{k_F^2}{k_c^2} = \frac{F_0}{F_c} = 69 . \tag{56.8}$$

We now turn our attention to the microscopic origin of the small Fermi-surface structure which explains the ratio (56.8).

The CDW wave vector **Q** in **K** has been measured by neutron diffraction [6], and identification of the satellites as CDW structure was recently confirmed [7] [7] reviews other data, too):

$$Q = (0.995, 0.975, 0.015) \times \frac{2\pi}{a} , \tag{56.9}$$

which differs from the $\mathbf{G}_{110}$ reciprocal-lattice vector by 2%. Electronic band structure near $\frac{1}{2}\mathbf{Q}$ was studied by solving Schrödinger's equation having both a CDW potential and a lattice potential [8]:

$$V(\mathbf{r}) = 2\alpha \cos(\mathbf{Q} \cdot \mathbf{r}) + \cos(\mathbf{G}_{110} \cdot \mathbf{r}) . \tag{56.10}$$

A family of higher-order "minigaps" appears in the region near $\pm\frac{1}{2}\mathbf{Q}$. These new gaps will introduce infrared structure in the optical properties [9], and recent work using the inverse photoelectric effect attributes infrared-emission peaks in Li [10] and in Na and K [11] to these minigaps.

The effect of the CDW on K's Fermi surface was studied [12] after **Q** was measured. One or more small cylindrical sheets of Fermi surface are generated by the minigap structure. The

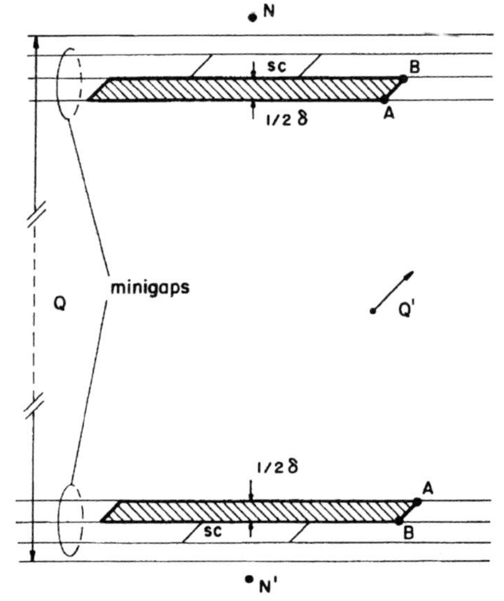

Figure 56.2 Schematic drawing of a Fermi-surface cylinder formed by two CDW minigaps. The two shaded half-cylinders can be joined because the A points are the same quantum state, and so are the B points. (Two points in **k** space can be equivalent if they differ by $\mathbf{Q} - n\mathbf{Q}'$.) There is evidence that indicates the existence of a smaller cylinder (sc) in the adjacent minigap region. For theoretical details see [12]. The points N and N' are the centers of the two Brillouin-zone faces (on opposite sides of the Fermi surface). 99.96% of the occupied volume within the Fermi surface, i.e., the region bounded (in part) by the planes through the A points, is not shown.

axis of each cylinder is parallel to $\mathbf{Q}' \equiv \mathbf{G}_{110} - \mathbf{Q}$. One such cylinder is shown in Figure 56.2 and consists of two half-cylinders pieced together by Bragg reflections having the relevant periodicity. The angle between $\mathbf{Q}'$ and [110] is $\sim 44°$, but the cross section is approximately circular in a plane perpendicular to [110]. The length of the cylinder is $|\mathbf{Q}'|$, but its height (parallel to [110]) is

$$\delta = 0.015 \times \sqrt{2}\left(\frac{2\pi}{a}\right) . \tag{56.11}$$

If k_c is the cylinder radius (in the plane perpendicular to [110]), then the volume (including both halves) is

$$\Omega = \pi k_c^2 \delta . \tag{56.12}$$

The fractional number η of conduction electrons within the cylinder is, of course,

$$\eta = 3\Omega / 4\pi k_F^2 . \tag{56.13}$$

Four years ago we found [13] that an unexpected cyclotron resonance in the perpendicular-field surface impedance [14] could be explained if a small Fermi-surface cylinder was included in the theory. The **k**-space volume of the cylinder was determined by the size of the anomaly relative to the change in surface impedance between $H = 0$ and H at resonance. The value of η needed was

$$\eta = 4 \times 10^{-4} . \tag{56.14}$$

Despite this small value, the cylinder accounts for 2% of the density of states at the Fermi energy. Equations ((56.12)–(56.14) provide an independent determination of the cylinder's radius:

$$\frac{k_F^2}{k_c^2} = 64 . \tag{56.15}$$

The agreement between (56.8), derived from Landau-level quantization, and (56.15), derived from semiclassical magnetotransport, is remarkable.

The small value of $k_c \sim k_F/8$ solves a puzzle that can be noticed in Grime's data. The sharpness of the resonance requires $\omega_c\tau \sim 60$, whereas the measured bulk resistivity (at 2.5 K) would yield $\omega_c\tau \sim 10$. The longer relaxation time, τ, by a factor ~ 6, for the cylinder electrons can be understood, since their velocity is approximately six times smaller than v_F. Consequently, they encounter scattering centers at a reduced rate. A similar factor (i.e., ~ 3) is also necessary to reconcile the sharpness of the cyclotron resonance seen in microwave transmission. (Without the cylinder the resonance should not even occur.)

There are now five phenomena that require a small Fermi-surface cylinder in K.

- The cyclotron resonance in the perpendicular-field surface impedance [13].
- The cyclotron resonance in the perpendicular-field microwave transmission [3].
- The Landau-level oscillations in the microwave transmission.
- Subharmonic cyclotron resonance, i.e., at $H_0/2$, $H_0/3$, etc., in the microwave transmission. This phenomenon was found in 14 samples [1]. The angular tilt of the cylinder axis by $\sim 44°$ causes the elliptical orbits in *real space* to be tilted by 44°. The resulting electron vibrations parallel to **H** lead to subharmonic coupling (as in the Azbel–Kaner effect) even though **H** is perpendicular to the surface.
- A sharp (angle-resolved) photoemission peak originating from electron states near the Fermi level and at photon energies for which no emission should occur. This phenomenon has been observed in Na [15] and K [16]. The sharp Fermi-energy peak is caused by electrons that (before excitation) occupy the cylinder [17]. The sharpness results from the fact that $E(\mathbf{k})$ is flat along the cylinder axis, unlike the steep E versus $|\mathbf{k}|$ that would otherwise occur near E_F. Photoexcitation of the cylinder electrons is ubiquitous because a cylinder wave function $|\mathbf{k}\rangle$ includes many other components, e.g., $|\mathbf{k} \pm n\mathbf{G}'\rangle$.

The common feature of these five striking experimental discoveries is that they all occur in thin samples which may be expected to have epitaxial texture as a result of adhesion to a smooth surface. This causes a [110] axis to be perpendicular to the sample face [18]. **Q** is tilted only 0.8° from [110]. Each crystallite can have one of four **Q**′ directions: $(0.025, 0.005, \pm 0.015)$ or $(0.005, 0.025, \pm 0.015)$. When all crystallites are considered, the cylinder axes will be randomly distributed on a cone having a 44° angle with [110], the surface normal. Nevertheless all cross-sectional areas perpendicular to **H** (and to [110]) and all cyclotron masses would be the same. Naturally, imperfect epitaxial texture will lead instead to an unpredictable distribution in cyclotron mass or extremal area. One should then expect to find structure in the cyclotron resonance, and the field range of the Landau-level oscillations will be limited. These features do indeed occur and vary from sample to sample [1]. (Such effects cannot be attributed to a martensitic transformation, since a new low-temperature phase does not occur in K.)

There are many other phenomena that reveal the complex structure of K's Fermi surface, caused by the CDW [19]. The relevance of the nearly isotropic de Haas–van Alphen effect is, however, frequently interposed. A typical specimen is a mm-size sphere or cylinder that has *not* been etched. **Q**-domain sizes are likely too small to allow coherent CDW effects in magnetic quantization [20]. Furthermore, the cylinder axes in a single crystal would have 24 possible orientations. One should not be surprised, then, if the Landau-level oscillations from the cylinders were obscured by a lack of orientational uniformity. The quantum oscillations in microwave transmissions show that de Haas–van Alphen studies of K have been incomplete and misleading.

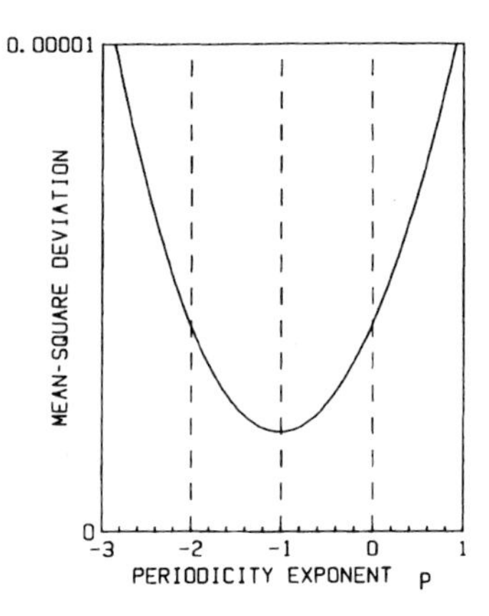

Figure 56.3 Mean-square deviation $[X_n - x_n(p)]^2_{av}$ vs p, the periodicity exponent in Equation ((56.16).

Finally, we search for the periodicity exponent p that best fits the transmission peaks of Table 56.1 (The inquiry above involved only $p = 1$ or -1, values that have physical interpretations.) Suppose the oscillations were described by

$$\Delta y \sim \cos(\lambda H^p + \phi) , \tag{56.16}$$

instead of by Equation (56.5). The parameters λ and ϕ, which optimize the fit, were found for all values of p, and the peak locations $x_n(p)$ were computed. The mean-square deviations from the observed peaks X_n are shown in Figure 56.3. The optimum fit occurs for $p = -1.0$, the precise value appropriate for Landau-level oscillations.

Landau-level oscillations observed by a de Haas–van Alphen effect increase in amplitude with increasing field. However, such monotonic behavior should *not* occur for microwave transmission. These oscillations arise from a periodic variation in the scattering rate $1/\tau$ (caused by an oscillation in the Fermi-surface density of states). An easy exercise to see how τ affects microwave transmission (parallel to a magnetic field) is to calculate the propagation constant q (starting from Maxwell's curl **H** equation). The result for conduction electrons, in a local approximation, is

$$q = \left(\frac{4\pi i (ne^2\tau/m)/c^2}{1 + i(\omega_c - \omega)\tau} \right)^{1/2} . \tag{56.17}$$

($ne^2\tau/m$ is the dc conductivity.) Observe that when $(\omega_c - \omega)\tau$ is near zero (i.e., not far from cyclotron resonance), q depends on τ. Consequently, the transmitted signal will have oscillations. However, when $(\omega_c - \omega)\tau \gg 1$, the τ's in the numerator and denominator of (56.17) cancel. Accordingly, q no longer depends on τ, so the Landau-level transmission oscillations will disappear at high magnetic fields, as exhibited by the data in Figure 56.1.

Acknowledgements

We are grateful to the National Science Foundation, Condensed-Matter Theory Program for financial support. We especially thank G. L. Dunifer for providing the data shown in Figure 56.1.

References

1 Dunifer, G.L., Sambles, J.F. and Mace, D.A.H. (1989) *J. Phys. Condens. Matter* **1**, 875.

2 Overhauser, A.W. (1978) *Adv. Phys.* **27**, 343.

3 Lacueva, Graciela and Overhauser, A.W. (unpublished).

4 Grimes, C.C. and Kip, A.F. (1963) *Phys. Rev.* **132**, 1991.

5 Shoenberg, D. (1984) *Magnetic Oscillations in Metals* (Cambridge University Press, London), Chap. 4.

6 Giebultowicz, T.M., Overhauser, A.W. and Werner, S.A. (1986) *Phys. Rev. Lett.* **56**, 2228.

7 Werner, S.A., Overhauser, A.W. and Giebultowicz, T.M. (1990) *Phys. Rev. B* **41**, 12536.

8 Fragachán, F.E. and Overhauser, A.W. (1984) *Phys. Rev. B* **29**, 2912.

9 Fragachán, F.E. and Overhauser, A.W. (1985) *Phys. Rev. B* **31**, 4802.

10 Kobzar', Yu.M., Bodnar, N.N., Kozych, V.Ya. and Kovtun, V.G. (1989) *Pis'ma Zh. Eksp. Teor. Fiz.* **50**, 323 [*JETP Lett.* **50**, 358].

11 Kobzar', Yu.M. and Bodnar, N.N. (1990) *Pis'ma Zh. Eksp. Teor. Fiz.* **52**, 686 [*JETP Lett.* **52**, 37 (1990)].

12 Hwang, Y.G. and Overhauser, A.W. (1989) *Phys. Rev. B* **39**, 3037.

13 Lacueva, G. and Overhauser, A.W. (1986) *Phys. Rev. B* **33**, 3765.

14 Baraff, G.A., Grimes, C.C. and Platzman, P.M. (1969) *Phys. Rev. Lett.* **22**, 590.

15 Jensen, E. and Plummer, E.W. (1985) *Phys. Rev. Lett.* **55**, 1912.

16 Itchkawitz, B.S., Lyo, In-Whan and Plummer, E.W. (1990) *Phys. Rev. B* **41**, 8075.

17 Overhauser, A.W. (1985) *Phys. Rev. Lett.* **55**, 1916 (1985); **58**, 959 (1987).

18 Overhauser, A.W. and Butler, N.R. (1976) *Phys. Rev. B* **14**, 3371.

19 Overhauser, A.W. (1983) In *Highlights of Condensed-Matter Theory*, Proceedings of the International School of Physics "Enrico Fermi", Course LXXXIX, Varenna on Lake Como, 1983, edited by Bassani, F., Fumi, F. and Tosi, M.P. (North-Holland, Amsterdam, 1985), p. 194.

20 Overhauser, A.W. (1982) *Can. J. Phys.* **60**, 687.

Reprint 57 Magnetotransmission of Microwaves Through Potassium Slabs[1]

Graciela Lacueva * *and A.W. Overhauser* **

* Department of Physics, John Carroll University, 525, Northwestern Avenue, University Heights, Ohio 44118, USA
** Department of Physics, Purdue University, West Lafayette, Indiana 47907, USA

Received 3 May 1993

Five signals which emerge after transmission through potassium slabs (~0.1 mm thick) have been studied by Dunifer, Sambles, and Mace at 1.3 K with 79-GHz microwaves and large magnetic fields, $H\hat{\mathbf{z}}$, perpendicular to the slabs. They are (i) conduction-electron-spin resonance (together with spin-wave sidebands); (ii) Gantmakher–Kaner (GK) oscillations; (iii) cyclotron resonance; (iv) cyclotron-resonance subharmonics; and (v) high-frequency oscillations. Spin-resonance transmission has been understood for many years, but not the other four. The isotropic-Fermi-surface model is used here to calculate the transmitted power versus H by solving Maxwell's equations self-consistently with the Boltzmann transport equation. GK oscillations do emerge, but the (theoretical) transmitted power is too large by a factor ~10 000. In contrast, the remaining three signals should not even exist. It has been shown that a charge-density-wave broken symmetry creates a family of (higher-order) minigaps which cuts through the Fermi surface near its extremities along the slab normal. These gaps diminish the effectiveness of conduction electrons having v_z near the Fermi velocity. Calculation confirms that this ineffectiveness dramatically reduces the predicted amplitude of the GK oscillation, and thereby provides an interpretation of the huge discrepancy. The minigaps also give rise to a small Fermi-surface cylinder which (as recently shown) leads to Landau-level oscillations that quantitatively explain signal (v). Signals (iii) and (iv) can also be attributed to the small Fermi-surface cylinder, but only because its axis is tilted ~ 45° relative to $\hat{\mathbf{z}}$. A nonlocal theory for such a scenario is yet to be formulated.

57.1 Introduction

From a theoretical point of view, potassium is a most important metal. Being monovalent and having Brillouin-zone energy gaps of only ~0.4 eV, its Fermi surface is thought (by some) to be spherical to ~0.1%. Lithium could have been the exemplar, except that below ~78 K it undergoes a martensitic phase transformation (from bcc) to a rhombohedral structure having a nine-layer stacking sequence [1]. This phase change destroys cubic symmetry and thereby precludes a simple characterization of low-temperature electronic behavior. The Fermi surface is no longer simply connected. A similar situation exists for sodium below ~35 K [2]. In contrast, potassium retains its bcc structure to 4 K, as is known from neutron diffraction [3] and ultrasonic attenuation [4] on strain-free crystals. Accordingly, potassium plays the dominant role in the confrontation between theory (for a simple metal) and reality.

The focus of this work is on the transmission of microwaves through potassium slabs in a magnetic field, $H\hat{\mathbf{z}}$, perpendicular to the sample. An extensive study of the five signals that emerge is due to Dunifer, Sambles, and Mace [5], who reported data on 15 potassium samples. Four of the transmission signals for sample K-4 (kindly provided to us by G.L.

1) Phys. Rev. B 48, 23 (1993-I).

Anomalous Effects in Simple Metals. Albert Overhauser

ISBN: 978-3-527-40859-7

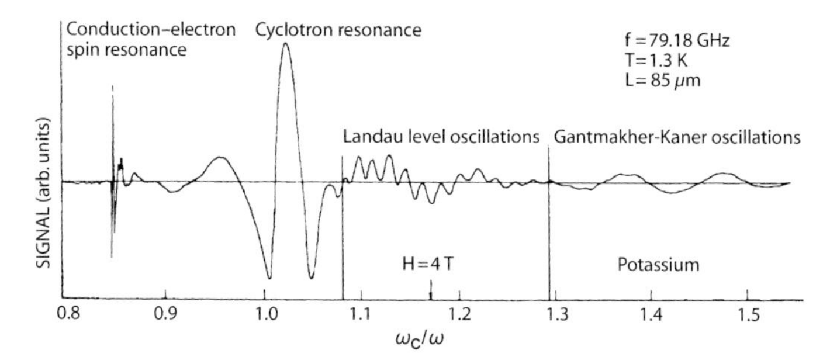

Figure 57.1 Microwave transmission signal vs. H through a potassium slab in a perpendicular magnetic field. ($H = 3.42\,\mathrm{T}$ at $\omega_c/\omega = 1$). The phase of the microwave reference was adjusted so that the cyclotron resonance is symmetric. The data are from sample K-4 of [5]. Not shown is the signal between $\omega_c/\omega = 0$ and 0.8, which exhibits cyclotron-resonance subharmonics at $\frac{1}{2}$, $\frac{1}{3}$, and $\frac{1}{4}$. The Gantmakher–Kaner oscillations in the low-field range are smaller than those near $\omega_c/\omega = 1.5$ by a factor of 5.

Dunifer) are shown versus ω_c/ω in Figure 57.1. $\omega_c \equiv eH/m^*c$, and $m^* = 1.21\,\mathrm{m}$ [6]. Not shown are the subharmonic cyclotron resonances at $\omega_c/\omega = \frac{1}{2}, \frac{1}{3}$, and $\frac{1}{4}$. ($\omega/2\pi$ is the microwave frequency *f*.) Sample K-4, 85 μm thick, was selected by G.L. Dunifer for the overall clarity of its transmission signals. The amplitude of its Gantmakher–Kaner (GK) signal was consistent with those from other samples, whether formed in argon or in vacuum.

Conduction-electron-spin-resonance transmission, and the associated spin-wave sidebands, have been studied extensively [7] and provide the only phenomenon of the five which is correctly accounted for by a simple theoretical model. (Even in this case, however, sideband splittings indicate [8] the presence of a charge-density-wave broken symmetry [9].) Unlike the spin precession signal, which is collective in nature, the four remaining signals are critically dependent on the one-electron energy spectrum and the Fermi-surface topology. Magnetotransmission of microwaves is therefore an important tool for validating or contradicting theoretical points of view.

To gain perspective it is useful to display the field dependence of the transmission signal when the electrical conductivity is local. That is, the current density (in the $\hat{\mathbf{x}}\hat{\mathbf{y}}$ plane) can be obtained from [10]

$$\mathbf{j}(z) = \frac{\sigma_0}{1+(\omega_c\tau)^2}\begin{pmatrix} 1 & -\omega_c\tau & 0 \\ \omega_c\tau & 1 & 0 \\ 0 & 0 & 1+(\omega_c\tau)^2 \end{pmatrix}\mathbf{E}(z)\,, \tag{57.1}$$

where $\mathbf{j}$ and $\mathbf{E}$ are column vectors, and $\sigma_0 = ne^2\tau/m^*$ (τ being the relaxation time, and n the electron density). The sample configuration is here a semi-infinite specimen, and $H\hat{\mathbf{z}}$ is normal to the surface. The (transverse) electric field at $z = 0$ is

$$\mathbf{E} = E_0\hat{\mathbf{x}}e^{-i\omega t}\,. \tag{57.2}$$

On using Maxwell's curl equations, together with Equation (57.1), one finds two propagating microwave fields:

$$E_\pm = (E_x \pm iE_y) = E_0 e^{i(q_\pm z - \omega t)}\,, \tag{57.3}$$

where

$$q_\pm = \left[\frac{4\pi\omega i\sigma_0/c^2}{1-i(\omega\pm\omega_c)\tau}\right]^{1/2}\,. \tag{57.4}$$

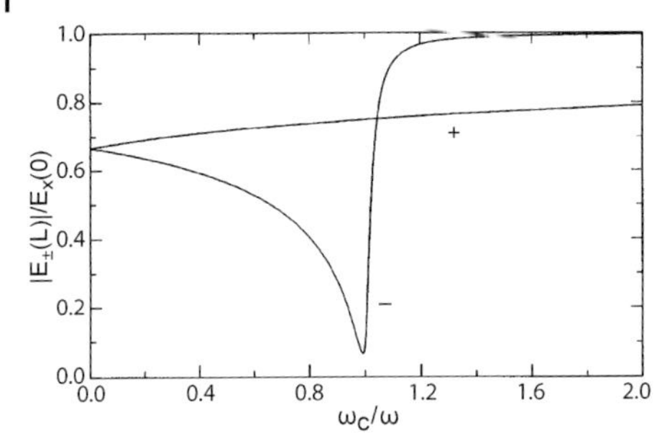

Figure 57.2 Field dependence of the circularly polarized microwave signals at a depth $d = 2 \times 10^{-6}$ cm if the conductivity σ_0 is local, i.e., for propagation constants given in Equation (57.4). The sharp dip near $\omega_c/\omega = 1$ occurs in the wave which rotates with the same sense as an electron's cyclotron motion. ($\omega_c\tau = 70$.)

Figure 57.2 shows the field dependence of the two circularly polarized microwave signals at a depth $d = 2 \times 10^{-6}$ cm. The field variable is represented by ω_c/ω, for $f = 79.18$ GHz. ($\omega_c/\omega = 1$ when $H = 3.42$ T.) The relaxation time τ was chosen so that $\omega_c\tau = 70$, which causes a cyclotron-resonance width equal to that of the resonance in Figure 57.1.

Notice that the E_+ wave increases monotonically with increasing H. In contrast, the E_- wave exhibits a sharp cyclotron-resonance dip at $\omega_c/\omega = 1$. The sharp dip at resonance is the only important transmitted feature. The reason for the dip can be understood from Equation (57.4). At resonance the microwave skin depth for the E_- mode, $1/\mathrm{Im}(q_-)$, reaches its minimum value. Off resonance, the denominator in Equation (57.4) becomes large, and so the skin depth increases, indicating a greater penetration of the microwave field.

Needless to say, the local conductivity Equation (57.1) does not apply to potassium at 1.3 K. The conduction-electron mean free path $l = v_F\tau$ is then much larger than the skin depth. Accordingly a nonlocal theory must be used to determine the transmitted signal. Nevertheless, the most significant intuition that can be gained from Figure 57.2 is that a standard theoretical model leads to a cyclotron-resonance *dip* in the transmitted power. This expected behavior is contradicted by the very large resonance *peak* in the data of Figure 57.1. Such a peak occurs in all samples [5].

Before proceeding to the nonlocal theory, it is necessary to define the transmission efficiency in absolute terms. Consider a slab (of thickness L) which is infinite in the $\hat{\mathbf{x}}\hat{\mathbf{y}}$ plane. Suppose the incident microwave field (on the surface at $z = 0$) has an energy flux P_0 (in ergs/cm^{2}s). Because the reflection coefficient is almost unity, the electric fields of the incident and reflected waves at $z = 0$ will almost cancel. However, the magnetic fields will add. For a linearly polarized incident wave, $E_x' = H_y'$. The reflected wave (at $z = 0$) will have $E_x'' \cong -E_x'$ and $H_y'' \equiv H_y'$. So the total fields at the front surface are $E_x \cong 0$ and $H_y \cong 2H_y'$.

At the rear surface ($z = L$), there is only an outgoing wave. If only the $\hat{\mathbf{x}}$ polarized component were detected, the transmitted field (at $z = L$) will have $E_x(L) \cong H_y(L)$. The ratio of the transmitted power P_t to P_0 is, therefore,

$$\frac{P_t}{P_0} \cong \left| \frac{E_x(L)}{\frac{1}{2}H_y(0)} \right|^2 . \tag{57.5}$$

This definition is general; it applies to the local-conductivity calculation, based on Equation (57.4), as well as to the nonlocal ones which will follow below.

For the (theoretical) local case treated above, with $\tau = 1.4 \times 10^{-10}$ s, $f = 79.18$ GHz, and $n = 1.4 \times 10^{22}$ cm^{-3}, the classical skin depth in the low-field limit is $\approx |q|^{-1}$, which from Equation (57.4) is

$$\delta_0 = 5 \times 10^{-6}\ \mathrm{cm} . \tag{57.6}$$

It is necessary to include both e^{iqz} and e^{-iqz} waves in order to match continuity equations for **E** and **H** at both surfaces. Then the power transmission ratio, from (57.5), (57.6), and Maxwell's curl **E** equation, is, for $L = 85$ μm,

$$\frac{P_t}{P_0} \cong \left[\frac{8\pi f \delta_0}{c} \exp(-L/\delta_0)\right]^2 \sim 10^{-1500} \,, \tag{57.7}$$

which is rather small. Nonlocal effects are necessary to explain the existence of a transmitted signal.

57.2 Nonlocal Theory for an Isotropic Fermi Surface

The conduction-electron mean free path $l = v_F \tau$ in potassium (of typical purity) below 4 K is ~0.1 mm, i.e., ~100 times the (nonlocal) skin depth for microwaves. The transverse conductivity can be found by solving self-consistently the Boltzmann transport equation together with Maxwell's equations. This exercise is a well-known one, so we shall merely quote the result for $\sigma_\pm(q, \omega, H)$ that we derived for a related problem [11]:

$$\sigma_\pm = \frac{3ne^2\tau}{2m^* x^2} \{2ap - 1 + r(x^2 + 1 - a^2) + i[a + p(x^2 + 1 - a^2) - 2ar]\} \,, \tag{57.8}$$

where

$$a = (\omega \pm \omega_c)\tau \,,$$
$$x = ql \,,$$
$$p = \frac{1}{4x} \ln\left[\frac{1 + (x + a)^2}{1 + (x - a)^2}\right] \,,$$
$$r = \frac{1}{2x}[\tan^{-1}(x + a) + \tan^{-1}(x - a)] \,. \tag{57.9}$$

Application of this result to transmission of microwaves through a slab of thickness, L, is facilitated when one assumes that the two surfaces are smooth, so that specular reflection of electrons occurs. Sensitive experiments [12] involving potassium that has crystallized in contact with glass have shown that specularity is readily achieved.

Following the formalism of Urquhart and Cochran [13], we expand the field and current distributions (for $0 < z < L$) in a Fourier cosine series:

$$E_\pm(z) = \sum_{n=0}^{\infty} E_\pm^n \cos(q_n z) \,, \tag{57.10}$$

$$j_\pm(z) = \sum_{n=0}^{\infty} \sigma_\pm E_\pm^n \cos(q_n z) \,, \tag{57.11}$$

where $q_n = n\pi/L$. From Maxwell's curl equations it follows that

$$\frac{\partial^2 E_\pm}{\partial^2 z} = -\frac{4\pi i\omega}{c^2} j_\pm \,. \tag{57.12}$$

(This displacement-current term has been dropped.) The coefficients $E_\pm^n$ can be obtained by using (57.11) and (57.12) and multiplying the resulting equation by $\cos(g_m z)$, followed by an integration from 0 to L. One finds, with $E'_\pm$ denoting $\partial E_\pm/\partial z$,

$$E_\pm^m = \frac{(2 - \delta_{m0})[(-1)^m E'_\pm(L) - E'_\pm(0)]}{L[q_m^2 - (4\pi i\omega/c^2)\sigma_\pm(q_m)]} \,, \tag{57.13}$$

where δ_{m0} is 1 for $m = 0$, and zero otherwise. The integral involving the left-hand side of (57.12) must be carried out by partial integration, without using (57.10). The reason for this caution is that the series for $E'_{\pm}$ converges to 0 at $z = 0$ and L, whereas (in fact) $E'_{\pm}$ is finite at both surfaces.

Suppose, for the time being, that the sample were very thick. Then a solution of interest is a propagating wave in the $\hat{\mathbf{z}}$ direction. $E'_{\pm}(L)$ can be set to zero in Equation (57.13). $E'_{\pm}(0)$ may be evaluated from Maxwell's curl E equation, which requires

$$E'_{\pm} = \pm\frac{\omega}{c}H_{\pm}(0)\,. \tag{57.14}$$

If the incident electromagnetic wave (for $z < 0$) is polarized along $\hat{\mathbf{x}}$, then H(0) is along $\hat{\mathbf{y}}$. We now take the total $H_y(0)$, from incident and reflected waves, to be 1 cgs unit, so that

$$H_{\pm}(0) = \pm i\,. \tag{57.15}$$

Accordingly, the right-hand side of Equation (57.14) is $i\omega/c$. The coefficients (57.13) for the $\hat{\mathbf{z}}$-propagating solution are then

$$E^m_{\pm} = -i\frac{\omega}{cL}(2-\delta_{m0})[q_m^2 - (4\pi i\omega/c^2)\sigma_{\pm}(q_m)]^{-1}\,. \tag{57.16}$$

At the rear surface $z = L$, there will arise a reflected wave traveling along $-\hat{\mathbf{z}}$. The two waves at $z = L$ must combine so that $|H|$ and $|E|$ are equal, because that is required by the boundary conditions appropriate to having only a transmitted wave (for $z > L$). Inside the metal, $|H|$ for a traveling wave is several orders of magnitude larger than $|E|$. Accordingly, the two waves must combine so that (at $z = L$) the magnetic fields almost cancel. The opposite Pointing vectors then require that the two contributions to E are essentially equal and parallel. Recognizing this factor of 2, the electric field at $z = L$ is

$$E_{\pm}(L) = -i\frac{2\omega}{cL}\sum_{m=0}^{\infty}\frac{(2-\delta_{m0})\cos(m\pi)}{q_m^2 - (4\pi i\omega/c^2)\sigma_{\pm}(q_m)}\,. \tag{57.17}$$

A detailed analysis which leads to this factor of 2 has been presented by Cochran [14]. From Equation (57.5) and $H_y(0) \cong 1$, the transmitted power ratio for a detector sensitive to polarization π is

$$P_{\pi}/P_0 = 4|E_{\pi}(L)|^2\,. \tag{57.18}$$

It will be shown below that for the sample of Figure 57.1, the observed ratio is $\sim 10^{-22}$, i.e., 220 decibels below the power incident on the metal surface from the microwave configuration.

Evaluation of Equation (57.17) presents numerical difficulty because the series converges very slowly and, moreover, is an alternating one, since $\cos(m\pi) = (-1)^m$. Nevertheless, a successful calculation results from following the method of [13]. The first 3000 terms can be summed without compromise. For $m > 3000$, one can replace the denominator of Equation (57.16) by q_m^2. Accordingly, one must calculate

$$\sum_{3001}^{\infty}\frac{(-1)^m}{m^2} = \sum_{1}^{\infty}\frac{(-1)^m}{m^2} - \sum_{1}^{3000}\frac{(-1)^m}{m^2} \tag{57.19}$$

(precision to 14 decimals is required.) The first term on the right-hand side is [15] $-\pi^2/12$, and the second causes no numerical problem. We have verified that the results to be presented below are unaltered when the cutoff is changed from 3000 to 5000.

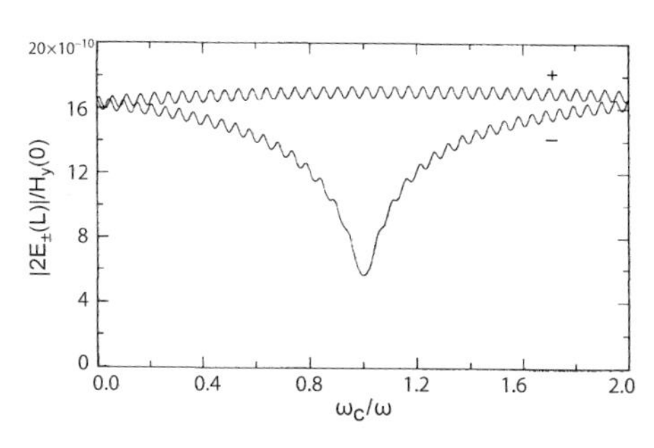

Figure 57.3 Theoretical field dependence of the circularly polarized microwave-transmission amplitudes for a potassium slab of thickness $L = 85\ \mu\text{m}$. The broad, symmetric dip at $\omega_c/\omega = 1$ occurs in the signal rotating with the cyclotron motion. The nonlocal conductivity Equation (57.8) for an isotropic Fermi surface was employed. ($\omega_c\tau = 70$.) The periodicity of the oscillations is half that of Gantmakher–Kaner oscillations (which do not occur in the circularly polarized amplitudes). The small oscillations correspond to ballistic transit times for electrons, having $|v_z| = v_F$, and which travel from $z = 0$ to L and back again, being an integral multiple of the cyclotron period.

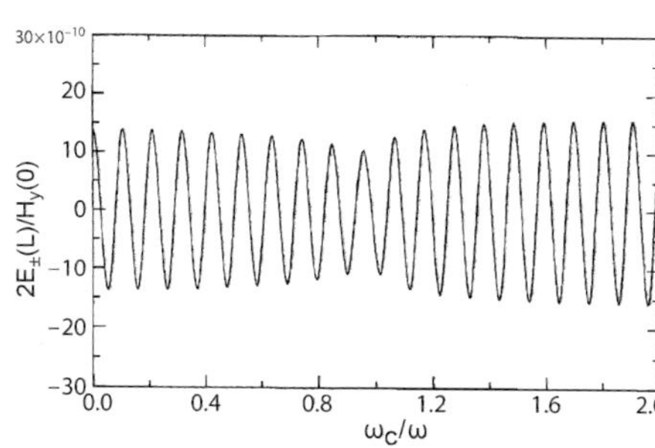

Figure 57.4 Theoretical field dependence of the linearly polarized microwave-transmission amplitude for a potassium slab of thickness $L = 85\ \mu\text{m}$. The nonlocal conductivity Equation (57.8) for an isotropic Fermi surface was employed. There is no cyclotron resonance peak and, of course, no Landau-level oscillations. The power transmission ratio is the square of the ordinate, $2E_x(L)/H_y(0)$, and exceeds that observed in the Gantmakher–Kaner oscillations of Figure 57.1 by 2×10^4.

Figure 57.3 shows the *predicted* amplitude of $E_+(L)$ and $E_-(L)$ for the conditions of Figure 57.1. Observe that the cyclotron resonance is a *moderate* and *broad dip* in E_0 near $\omega_c/\omega = 1$. Both waves exhibit oscillations having half the GK period. Figure 57.4 shows the transmitted signal $E_{\hat{x}}(L)$. Observe the prominent (GK) oscillations, but notice that near $\omega_c/\omega = 1$ there is only a small decrease in the oscillation amplitude. There is no evidence of a sharp cyclotron resonance peak, there are no cyclotron resonance subharmonics, and there are no high-frequency oscillations between $\omega_c/\omega = 1.1$ and 1.3, as in Figure 57.1.

The periodicity of the GK oscillation is related to the Fermi velocity v_F of the conduction electrons, and depends on the sample thickness L. The phenomenon is caused by those electrons having the fastest velocity parallel to **H**. Each oscillation occurs when the number of cyclotron rotations during the transit time, L/v_F, is an integer n, i.e.,

$$n = \omega_c L/2\pi v_F \ . \tag{57.20}$$

On setting $\omega_c = eH_n/m^*c$ and $v_F = \hbar k_F/m^*$, Equation (57.20) leads to GK peaks H_n which are linear in n:

$$H_n = (2\pi\hbar c k_F/eL)n \ . \tag{57.21}$$

Notice that m^* cancels out, so the oscillation period provides a measure of the Fermi-surface radius k_F.

The foregoing characteristics of the GK signal were verified in many specimens [5]. However, what has never been tested until now, to our knowledge, is the absolute magnitude of the GK signal. Equation (57.18) together with the amplitude of $E_{\hat{x}}(L)$ near $\omega_c/\omega = 1.5$ in Figure 57.4 leads to a *theoretical* power transfer ratio at the maximum in the oscillation,

$$P_t/P_0 = 2 \times 10^{-18} \ . \tag{57.22}$$

G.L. Dunifer has kindly provided the calibration of his microwave-transmission instrument. The power transfer ratio was $3s^2 \times 10^{-23}$, where s is the detector output in μv. From the original chart recordings of Figure 57.1, $s \approx 2\mu v$ near $\omega_c/\omega = 1.47$. Accordingly, the *experimental* power transfer ratio is

$$P_t/P_0 \approx 1 \times 10^{-22} \, . \tag{57.23}$$

The discrepancy between (57.22) and (57.23) – a factor exceeding 10 000 – is serious. It cannot be attributed to sample imperfection, e.g., rough or nonparallel surfaces. Such causes would degrade the GK resonances for high values of H_n compared to low ones. Such is not the case. The GK oscillation near $\omega_c/\omega \sim 1.5$, i.e., $H = 5.1\,\text{T}$, which corresponds to $n \sim 14$ in Equation (57.21), is considerably larger than those for $n = 1$–5 (not shown in Figure 57.1). Dunifer, Sambles, and Mace [5] report that larger GK signals for $\omega_c/\omega > 1$ are usually observed. Furthermore, spin-wave sidebands, which are also geometric resonances, are routinely observed out to $n = 30$–50 [7]. Consequently, geometric imperfection of samples cannot be imputed. We have also verified that the prediction (57.22), for which $\omega_c\tau = 70$ was used, is not sensitive to variations in τ by factors of 2. In Section 57.3 we shall show that altered Fermi-surface topology created by a charge-density-wave broken symmetry can account for the extreme weakness of the observed GK signal.

57.3 Suppression of GK Oscillations by a Charge-Density Wave

The existence of a charge-density-wave (CDW) broken symmetry [16] in potassium has been established by many extraordinary electronic properties [9, 17], that now number ~30. The wave vector **Q** of the CDW is tilted about 1° away from a [110] crystal direction; so in a large crystal 24**Q** domains are possible. The theory of the 1° tilt [18] stems from the elastic anisotropy of potassium. Phonon screening of the CDW requires less elastic stress if $\mathbf{Q}^* = \mathbf{G}_{110} - \mathbf{Q}$, i.e., the wave vector of the three phonon modes involved, is tilted about 45° from the [110] axis closest to **Q**.

An important property of potassium is that when thin specimens crystallize in contact with smooth, amorphous silica or sapphire, the (polycrystalline) grains are epitaxially oriented. Each grain (usually) has a [110] direction perpendicular to the surface. (The reason is that {110} planes are the most closely packed.) The CDW charge modulation of the conduction electrons can optimize the interfacial energy if the direction of **Q** is also close to the surface normal. This behavior is indicated, for example, by the anisotropic optical conductivity [9, 19], and has also been noticed with low-energy electron diffraction.

The conduction-electron energy spectrum is profoundly affected by the CDW periodic potential $G\cos(\mathbf{Q}\cdot\mathbf{r})$. In addition to the energy gaps defined by **Q**, there are two families of higher-order gaps which arise when the crystal potential $V\cos(\mathbf{G}_{110}\cdot\mathbf{r})$, is also in the Schrödinger equation [20]. The "heterodyne" gaps are described by the Fourier components

$$\mathbf{Q}_n = n(\mathbf{G}_{110} - \mathbf{Q}) = n\mathbf{Q}^* \, , \tag{57.24}$$

and the "minigaps" correspond to

$$\mathbf{Q}_m = (m+1)\mathbf{Q} - m\mathbf{G}_{110} \tag{57.25}$$

(m and n are integers). Some of the predicted infrared transitions [21] made possible by these gaps have been reported experimentally [22]. The Fermi-surface topology which takes into account the tilt of Q and $\mathbf{Q}^*$ relative to the [110] axis has been elaborated [23]. A schematic

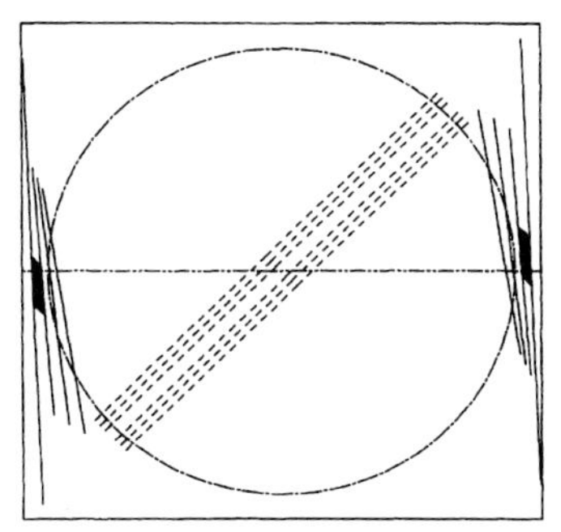

Figure 57.5 The $k_z = 0$ plane of the Brillouin zone of potassium. The horizontal axis is along [110] and is parallel to H. The three sets of (dashed) energy-gap planes passing near the center are heterodyne gaps, having periodicities defined by Equation (57.24). The three sets of (solid and short) energy-gap planes near the extremities (along H) are minigaps having periodicities defined by Equation (57.25). The pair of planes nearest the zone boundary are the main CDW energy gaps. The two black volumes, when joined end to end, describe a Fermi-surface cylinder, having an axis tilted $\sim 45°$ from [110]. The cylinder's volume is $\sim 4 \times 10^{-4}$ times that of the (undistorted) Fermi sphere, which is also shown.

view of the (001) plane in the Brillouin zone is shown in Figure 57.5. For clarity we have exaggerated the disparity between $|\mathbf{Q}|$ and $|\mathbf{G}_{110}|$, which is actually only 1.5%. (The figures in [23] are accurate.)

The horizontal direction in Figure 57.5 is the magnetic-field direction $\hat{\mathbf{z}}$ in the transmission experiments. It is self-evident that fast electrons traveling in the $\hat{\mathbf{z}}$ direction will have their cyclotron rotation interrupted by the minigaps. Consequently their contribution to the GK signal will cease. A possible way to model this effect is to let the scattering time τ be a continuous function of v_z. (Any artificial discontinuity will introduce spurious oscillations in the signal.) A simple option is to replace τ, which we redefine as τ_0, by

$$\tau = \frac{\tau_0}{1 + \gamma |v_z / v_F|} , \tag{57.26}$$

so that scattering times become short for electrons having v_z near the minigap regions (γ is an adjustable parameter). $\sigma_\pm(q, \omega, H)$ must be reevaluated, starting from Equation (57.13) of [11] (but with τ inside the integral). Instead of Equation (57.8),

$$\begin{aligned}\sigma_\pm = \frac{3ne^2\tau_0}{4m^*} \Bigg\{ & \frac{2(\gamma^2 - x^2)}{(\gamma^2 + x^2)^2} - \frac{\gamma}{\gamma^2 + x^2} \\ & + f_1 p_1 - g_1 r_1 + f_2 p_2 - g_2 r_2 \\ & + i \Bigg[f_1 r_1 + g_1 p_1 + f_2 r_2 + g_2 p_2 \\ & - \frac{2a(\gamma^2 - x^2)}{(\gamma^2 + x^2)^2} \Bigg] \Bigg\} . \end{aligned} \tag{57.27}$$

$x = ql$ and $a = (\omega \pm \omega_c)\tau$, as in Section 57.2, and

$$\begin{aligned} f_1 &= [(1 - a^2 - \gamma^2 + x^2)(3\gamma x^2 - \gamma^3) \\ &\quad + 2(a + \gamma x)(3\gamma^2 x - x^3)]/(x^2 + \gamma^2)^3 , \\ g_1 &= [2(a + \gamma x)(\gamma^3 - 3\gamma x^2) \\ &\quad + (1 - a^2 - \gamma^2 + x^2)(3\gamma^2 x - x^3)]/(x^2 + \gamma^2)^3 , \\ p_1 &= \frac{1}{2} \ln \left[\frac{(a - x)^2 + (\gamma + 1)^2}{(1 + a^2)} \right] , \\ r_1 &= \tan^{-1} \left[\frac{a\gamma + x}{\gamma + 1 + a^2 - xa} \right] . \end{aligned} \tag{57.28}$$

Furthermore, $f_2(x) = f_1(-x)$, $g_2(x) = g_1(-x)$, $p_2(x) = p_1(-x)$, and $r_2(x) = r_1(-x)$.

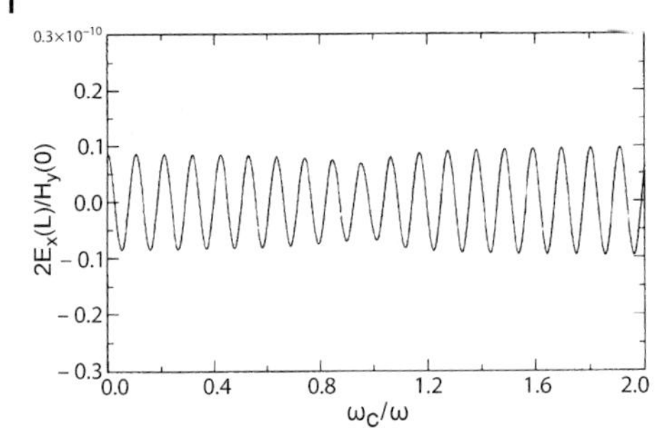

Figure 57.6 Theoretical field dependence of the Gantmakher–Kaner transmission signal based on the nonlocal conductivity Equation (57.27), having a v_z-dependent scattering time given by Equation (57.26) with $\gamma = 6$. The power transmission ratio is the square of the ordinate, and equals the value 1×10^{-22}, observed near $\omega_c/\omega = 1.5$ in Figure 57.1. Magnetic breakdown of minigaps and heterodyne gaps can be simulated by letting γ be a decreasing function of H. Accordingly, the GK amplitude would then increase with H, as is observed.

The transmission signal for $\gamma = 6$ was computed as described above, and is shown in Figure 57.6. As expected, the GK oscillations are substantially reduced, and can be as small as desired by increasing γ. The summation break [see Equation (57.19)] had to be extended to 10 000.

The problem that remains is why, if Figure 57.5 is correct, there are any GK signals at all. Cyclotron rotation of electrons having $v_z \approx v_F$ is thwarted by the minigaps. The answer is that even though most CDW domains will have **Q** perpendicular to the sample face, there can be a few domains for which **Q** is $\sim 60°$ from the [110] surface normal (near another [110] direction). One may surmise from Figure 57.5 that, in this event, "clear" regions of the Fermi surface can occur along the axis parallel to **H**.

The sizes of the minigaps and heterodyne gaps fall off very quickly [20, 23] with the integers n and m in (57.24) and (57.25). Consequently, many of the gaps will lose their effectiveness as a result of magnetic breakdown – a high-field phenomenon. Possibly the larger GK signals (seen at high fields) can be explained by magnetic breakdown of the weaker minigaps and heterodyne gaps which would, in small fields, jeopardize the cyclotron motion of electrons responsible for the GK periodicity.

The physical presence of minigaps and heterodyne gaps has been decisively confirmed by the extensive open-orbit resonance experiments of Coulter and Datars [24], phenomena that are accounted for quantitatively with a CDW structure [25]. There seems to be little risk in concluding that the discrepant (10 000-fold) loss in GK signal has a similar origin.

57.4 Conclusion

The foregoing treatment brings into comprehension one more of the five microwave signals that emerge from potassium. Only two remain to be explained, since the high-frequency oscillations were recently shown to be Landau-level oscillations belonging to a small Fermi-surface cylinder formed by the minigaps [26]. This cylinder, which is pieced together from the two black cylinders shown in Figure 57.5, contains only 4×10^{-4} electrons per atom [11]. This small number was sufficient, however, to explain the sharp cyclotron resonance peak in the microwave surface impedance [27]. The cross-sectional area ($\pi k_F^2/69$) of the cylinder, determined from the Landau-level periodicity (vs. $1/H$) [26], when combined with the cylinder length (determined from CDW neutron-diffraction satellites [28], agrees with the originally surmised fraction 4×10^{-4}. Landau-level oscillations from the main Fermi surface, having a cross section πk_F^2, are too rapid and too small to be seen in a microwave transmission experiment. (They have been studied frequently by the de Haas–van Alphen effect.)

Adding a cylindrical Fermi-surface component to the microwave transmission theory does not (with the cylinder axis parallel to $\hat{\mathbf{z}}$) explain the cyclotron resonance peak (near $\omega_c/\omega = 1$), even if the modified $\sigma_\pm$, Equation (57.27), is employed. We believe that the 45°

tilt of the cylinder axis that is required theoretically [18, 23] plays a crucial role. This tilt is the only possible explanation for the cyclotron-resonance subharmonics because, then, the individual orbits (in real space) oscillate back and forth along $\hat{\mathbf{z}}$ (as well as rotate in the $\hat{\mathbf{x}}\hat{\mathbf{y}}$ plane). Such back and forth motion along the electric field gradient is the behavior which creates subharmonics in the Azbel–Kaner effect [29].

A nonlocal theory which incorporates a cylindrical Fermi surface tilted $\sim 45°$ relative to **H** presents a formidable theoretical endeavor. The reduced symmetry destroys the isolation of the E_+ and E_- fields, and the longitudinal oscillations of the cylinder orbits introduce coupling to plasma modes. A quantitative explanation of the dominant microwave-transmission signal – the large and ubiquitous cyclotron resonance peak – will likely remain pending until a nonlocal study can be brought to completion.

Acknowledgements

We are especially indebted to Professor G.L. Dunifer for sending us the original recorder traces of his microwave transmission data for sample K-4, for developing the absolute calibration of his instruments, and for his advice in helping us to quantify the power transmission ratio reliably. We are grateful to Mi-Ae Park for independently verifying the calculations. Professor J.F. Cochran kindly pointed out the expansion technique for treating transmission through slabs of finite thickness. Finally, we are indebted to the National Science Foundation for financial support.

References

1 Overhauser, A.W. (1984) *Phys. Rev. Lett.* **53**, 64.
2 Berliner, R., Fajen, O., Smith, H.G. and Hitterman, R.L. (1989) *Phys. Rev. B* **40**, 12 086.
3 Werner, S.A., Overhauser, A.W. and Giebultowicz, T.M. (1990) *Phys. Rev. B* **41**, 12 536.
4 Kubinski, D. and Trivisonno, J. (1993) *Phys. Rev. B* **47**, 1069.
5 Dunifer, G.L., Sambles, J.F. and Mace, D.A.H. J. (1989) *Phys. Condens. Matter* **1**, 875.
6 Grimes, C.C. and Kip, A.F. (1963) *Phys. Rev.* **132**, 1991.
7 Mace, D.A.H., Dunifer, G.L. and Sambles, J.F. (1984) *J. Phys. F* **14**, 2105.
8 Hwang, Y.G. and Overhauser, A.W. (1988) *Phys. Rev. B* **38**, 9011; see especially Figs. 7, 8, and 9.
9 Overhauser, A.W. (1978) *Adv. Phys.* **27**, 343.
10 Kittel, C. (1986) *Introduction to Solid State Physics*, 6th ed. (Wiley, New York), p. 155.
11 Lacueva, G. and Overhauser, A.W. (1986) *Phys. Rev. B* **33**, 3765.
12 Penz, P.A. and Kushida, T. (1968) *Phys. Rev.* **176**, 804.
13 Urquhart, K.B. and Cochran, J.F. (1986) *Can. J. Phys.* **64**, 796.
14 Cochran, J.F. (1970) *Can. J. Phys.* **48**, 370.
15 Spiegel, M.R. (1968) *Mathematical Handbook of Formulas and Tables* (McGraw-Hill, New York), formula 19.22, p. 108.
16 Overhauser, A.W. (1968) *Phys. Rev.* **167**, 691.
17 Overhauser, A.W. (1985) in *Highlights of Condensed-Matter Theory, Proceedings of the International School of Physics "Enrico Fermi," Course LXXXIX, Varenna on Lake Como, 1983,* edited by Bassini, F., Fumi, F. and Tosi, M.P. (North-Holland, Amsterdam), p. 194.
18 Giuliani, G.F. and Overhauser, A.W. (1979) *Phys. Rev. B* **20**, 1328.
19 Overhauser, A.W. and Butler, N.R. (1976) *Phys. Rev. B* **14**, 3371.
20 Fragachan, F.E. and Overhauser, A.W. (1984) *Phys. Rev. B* **29**, 2912.

21 Fragachan, F.E. and Overhauser, A.W. (1985) *Phys. Rev. B* **31**, 4802.

22 Kobzar', Y.M. and Bodnar, B.N. (1990) *Pis'ma Zh. Eksp. Teor. Fiz.* **52**, 686 [JETP Lett. **52**, 37 (1990)].

23 Hwang, Y.G. and Overhauser, A.W. (1989) *Phys. Rev. B* **39**, 3037.

24 Coulter, P.G. and Datars, W.R. (1985) *Can. J. Phys.* **63**, 159.

25 Huberman, M. and Overhauser, A.W. (1982) *Phys. Rev. B* **25**, 2211.

26 Lacueva, G. and Overhauser, A.W. (1992) *Phys. Rev. B* **46**, 1273. Equation (17) of this paper is missing a factor ω in the numerator.

27 Baraff, G.A., Grimes, C.C. and Platzman, P.M. (1969) *Phys. Rev. Lett.* **22**, 590.

28 Giebultowicz, T.M., Overhauser, A.W. and Werner, S.A. (1986) *Phys. Rev. Lett.* **56**, 2228.

29 Kittel, C. (1963) *Quantum Theory of Solids* (Wiley, New York), p. 315.

Reprint 58 Microwave Surface Resistance of Potassium in a Perpendicular Magnetic Field: Effects of the Charge-Density Wave[1)]

Mi-Ae Park and A.W. Overhauser**

* Department of Physics, Purdue University, 525, Northwestern Avenue, West Lafayette, Indiana 47907, USA

Received 29 December 1995

The microwave surface resistance of potassium in a perpendicular magnetic field, measured by Baraff, Grimes, and Platzman in 1969, has never been completely explained until now. The sharp cyclotron resonance peak (at a magnetic field H_c) is caused by the small cylindrical section of Fermi surface created by the charge-density-wave (CDW) minigaps, having periodicities $\vec{K}_n = (n+1)\vec{Q} - n\vec{G}_{110}$. The shape of the observed resonance requires a tilt of the CDW vector $\vec{Q}$ away from [110], predicted by Giuliani and Overhauser in 1979. An abrupt drop of the surface resistance for $|H| > |H_c|$ is caused by the heterodyne gaps, which have periodicities $\vec{K}_n = n(\vec{G}_{110} - \vec{Q})$. These very small gaps, which begin to undergo magnetic breakdown for fields $H > 1$ T, interrupt the cyclotron motion of equatorial orbits. The abrupt drop in surface resistance for $|H| > |H_c|$ is caused by the resulting partial loss of carrier effectiveness for electrons having velocities nearly parallel to the surface. [S0163-1829(96)03227-4]

58.1 Introduction

Cyclotron resonance of the conduction electrons in potassium was first observed by Grimes and Kip [1] using the Azbel–Kaner configuration [2], for which the dc magnetic field $\vec{H}$ is parallel to the metal's surface. The effective mass was found to be $m^* = 1.21m$. Resonant peaks in the (microwave) surface resistance also occur at subharmonic values [3], H_c/n, $n = 2, 3, 4, \ldots$, in addition to the fundamental resonance which occurs at $H_c = m^*\omega c/e$. For conduction electrons having an energy spectrum $E(\vec{k})$ that is spherically symmetric, a resonance in the surface resistance should never occur if $\vec{H}$ is perpendicular to the surface [4].

Nevertheless, Baraff, Grimes, and Platzman, using a perpendicular-field configuration, found a sharp fundamental resonance in the surface resistance of potassium [5]. Their data are shown in Figure 58.1 together with the theoretical $R(H)$, which has no resonant structure at all. The magnetic-field sweep, expressed as ω_c/ω (where $\omega/2\pi$ is the microwave frequency, 23.9 GHz, and $\omega_c = eH/m^*c$), includes both positive and negative values because the microwave field was circularly polarized. The sharp cyclotron resonance, at $\omega_c/\omega = -1$, corresponds to $H_c = 1.03$ T. Not only was the existence of the resonance unexpected, but the sharp drop of $R(H)$ for $|H| > H_c$ has remained unexplained for twenty-five years.

The reason why a resonance is not expected in a perpendicular field is easily understood. The skin depth is $\sim 10^{-4}$ cm and the Fermi velocity is $\sim 10^8$ cm/s. Accordingly, the time an electron (having the Fermi velocity) remains in the microwave field ($\sim 10^{-12}$ s) is an order of magnitude shorter than the microwave period. (Electrons do not return periodically to the skin depth in a perpendicular field, as occurs if the parallel-field configuration is employed.)

The resonance cannot be an Azbel–Kaner signal from an oblique surface patch (at the sample's edge) since there are no subharmonics. Neither can the resonance be attributed to

1) Phys. Rev. B 54, 3 (1996-I).

Anomalous Effects in Simple Metals. Albert Overhauser

ISBN: 978-3-527-40859-7

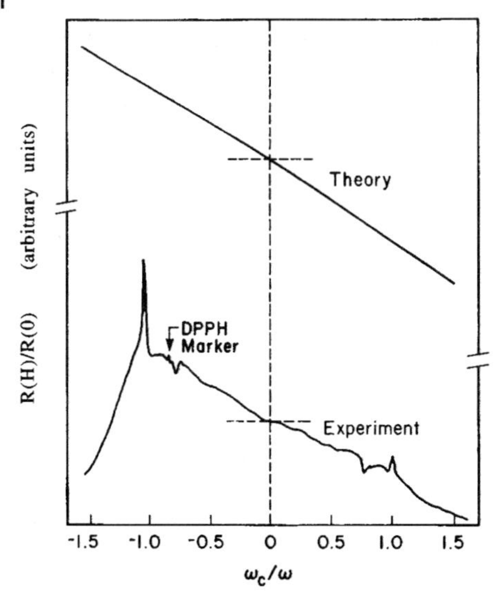

Figure 58.1 Surface resistance of potassium versus magnetic field ($\omega_c = eH/m^*c$). The data, due to Baraff, Grimes, and Platzman [5], for $T = 2.5$ K, and circularly polarized radiation at $\omega/2\pi = 23.9$ GHz. The dips near ± 0.77 are due to particles of $CuSO_4 \cdot 5H_2O$, embedded in the cavity walls during fabrication [13]. The cyclotron resonance, at $\omega_c/\omega = -1$, occurs when $H = 1.03$ T. The small resonance at $\omega_c/\omega = 1$ is caused by a small admixture of the opposite polarization. The theoretical curve is for a purely spherical Fermi surface, which potassium would have in the absence of a CDW broken symmetry.

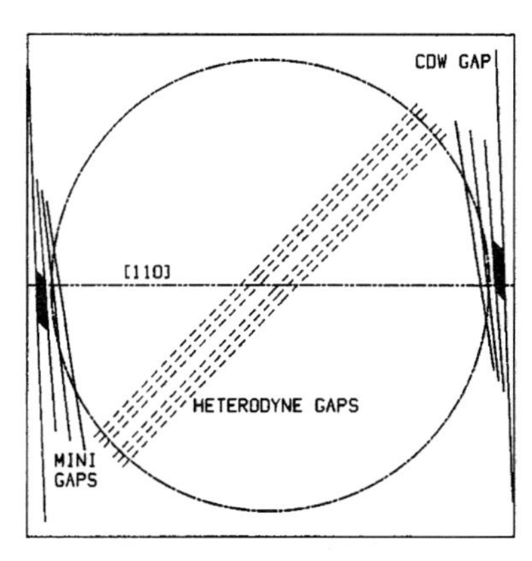

Figure 58.2 The Brillouin zone of potassium on a (001) plane in $\vec{k}$ space. The angular tilt, relative to [110], of the CDW wave vector $\vec{Q}$ has been exaggerated for clarity. The minigaps and heterodyne gaps are associated with the periodicities of Equations (58.1) and (58.2). The shaded areas are the two halves of the Fermi-surface cylinder, which form between the CDW gap and the first minigap. The axis of the cylinder is $\vec{G}_{110} - \vec{Q}$, which is also the direction of the heterodyne-gap vectors. The dc magnetic field $\vec{H}$ is applied parallel to [110], which is the habitual texture direction, perpendicular to smooth potassium surfaces. The (ideal) Fermi sphere is also shown.

electrons in a (110) surface-state band, since the bottom of such a band lies ~ 0.45 eV above the Fermi level.

The only satisfactory explanation of the resonance in a perpendicular field is based on the charge-density-wave (CDW) broken symmetry of potassium [6]. Many anomalous properties (now numbering more than 30) require the presence of a CDW [7], which causes two sequences of small energy gaps to cut the Fermi surface [8], as illustrated in Figure 58.2. The "minigaps" are higher-order gaps created by periodicities:

$$\vec{K}_n = (n+1)\vec{Q} - n\vec{G}_{110} \quad (n = 1, 2, \ldots) , \tag{58.1}$$

where $\vec{Q}$ is the CDW wave vector and $\vec{G}_{110}$ is the (110) reciprocal lattice vector parallel to $\vec{H}$. (It is known from optical properties that $\vec{Q}$ and one of the {110} reciprocal lattice vectors are nearly perpendicular to a smooth potassium surface [9].) The calculated values [8] of the first five mini-gaps are given in Table 58.1.

Table 58.1 Calculated values, from [8], of the first five mini-gaps and heterodyne gaps for K. The main CDW gap was taken to be 0.62 eV and the zone-boundary energy gap was 0.40 eV.

n	Minigap (meV)	Heterodyne gap (meV)
1	90	16
2	67	14
3	51	12
4	34	8
5	15	3

In Figure 58.2 the black regions outline a small cylindrical section of Fermi surface formed by the CDW energy gap and the first minigap. Only a small fraction, $\eta \sim 4\times10^{-4}$, of the conduction electrons are enclosed by this Fermi-surface cylinder. Nevertheless, these electrons are responsible for the cyclotron-resonance structure in the surface resistance [10]. Landau-level oscillations caused by the cylinder have been observed in microwave transmission [11]. The periodicity of the oscillations (versus $1/H$) indicates that the cylinder radius is $k_F/8$ [12]. The small velocities of the cylinder electrons enable them to remain in the microwave skin region and to exhibit a sharp resonance absorption.

The prior treatment of this resonance succeeded in identifying the cylindrical Fermi-surface component as its cause [10]. However, two puzzles remained. The calculated shape of the resonance was antisymmetric rather than (nearly) symmetric. It was possible to "fix" this problem by mixing almost equal amounts of surface reactance and surface resistance. A small amount of such mixing could be tolerated experimentally [13], but the required mixing angle of $\sim 47°$ seems excessive. In Section 58.3 we will show that this problem disappears when one recognizes that the cylinder's axis is $\sim 45°$ from the [110] (and $\vec{H}$) direction. This axis tilt is required theoretically [14], and has been verified experimentally by the location of the CDW diffraction satellites [15]. (The cylinder's axis is parallel to $\vec{G}_{110} - \vec{Q}$, which is tilted $\sim 45°$ when $\vec{Q}$ is only $\sim 1°$ away from [110] [8]. The experimental resonance shape can then be ascribed to the surface resistance alone.

The second puzzle is the sharp drop in $R(H)$ for $|H| > H_c$, mentioned above. In the following section, we will show that this effect arises from the "heterodyne" gaps, created by the periodicities,

$$\vec{K}_n = n(\vec{G}_{110} - \vec{Q}) \quad (n = 1, 2, \ldots) . \tag{58.2}$$

The energy-gap planes of this family are shown by the dashed lines in Figure 58.2, which cut at an angle, $\sim 45°$, through the central region of the Fermi "sphere". The calculated values [8] of the first five heterodyne gaps are given in Table 58.1. Cyclotron orbits for which k_z is near zero can be "Bragg" reflected by the periodic potentials associated with $\{\vec{K}_n\}$, Equation (58.2). When such reflections occur, the electrons become "ineffective" with regard to their cyclotron rotation. A quantitative model for this phenomenon is presented in Section 58.2; and the observed behavior of $R(H)$ when $|H| > |H_c|$ is explained.

58.2 Effect of the Heterodyne Gaps

In this section we will develop a model to account for the disruption of cyclotron motion caused by the heterodyne gaps, which cut through the central section of the Fermi sphere, as shown by the dashed lines in Figure 58.2. (The dc magnetic field $\vec{H}$ is parallel to the horizontal, $\hat{z}$ axis.) The main contribution to the surface resistance $R(H)$ arises from electrons having velocities nearly parallel to the surface; so these electrons (with $k_z \sim 0$) necessarily encounter the heterodyne gaps.

An electron which meets a heterodyne gap during its cyclotron motion can suffer a momentum transfer $\pm\hbar\vec{K}_n$, given by Equation (58.2). The result is a disruption of its cyclotron motion (in the $\hat{x}\hat{y}$ plane); and the change in $\hat{z}$ component of its velocity can cause it to rapidly leave the microwave skin depth, so its cyclotron motion is no longer fully effective. We introduce a factor $f < 1$ which describes the probability that the electron behaves "effectively", i.e., as if there were no gaps.

An electron encountering a small energy gap can also continue on its path in $\vec{k}$ space, as it would if the gap were not present. This phenomenon is called "magnetic breakdown". The breakdown probability P depends exponentially on H [16]:

$$P = e^{-H_0/H} . \tag{58.3}$$

The parameter H_0 depends critically on the energy gap E_g and the orbit geometry:

$$H_0 = \frac{\pi m c E_g^2}{2\hbar^2 e \left|\vec{K}\cdot(\vec{v}\times\hat{H})\right|} , \tag{58.4}$$

where $\hat{H}$ is a unit vector parallel to $\vec{H}$, and $\vec{v}$ is the electron's velocity at the energy-gap plane (if E_g were zero). This invariant form [17] for H_0 is equivalent to the result derived by Blount [16]. It is clear from Figure 58.2 that an electron with $k_z \sim 0$ will encounter several heterodyne gaps. For simplicity, we will still employ Equation (58.3) to describe the net result of all such encounters. The effective fraction, on taking into account magnetic breakdown, is then

$$f_{\text{eff}}(k_z = 0) = f + (1 - f)e^{-H_0/H} . \tag{58.5}$$

At very high fields, when magnetic breakdown is complete, $f_{\text{eff}} = 1$, i.e., the electrons behave as they would without a CDW. For small H, $f_{\text{eff}} = f$, the parameter we introduced above, f, a constant, will be adjusted to fit the data. (f is not zero because electrons with $k_z \sim 0$ sustain part of their cyclotron motion.) On account of the complexity, the breakdown parameter H_0 cannot be calculated reliably; but we have estimated it to be $H_0 \sim 4\,\text{T}$.

Equation (58.5) applies only to orbits for which $k_z \sim 0$; so we must generalize the effective fraction for all k_z. Electrons having a rapid speed along $\hat{z}$ do not remain in the skin layer very long anyway, so the interruption of their $\hat{x}\hat{y}$ motion by the heterodyne gaps is of little consequence. Thus their effectiveness will approach unity as $|k_z|$ increases. This behavior can be described heuristically by

$$f_{\text{eff}}(k_z) = \frac{f + (1 - f)e^{-H_0/H} + \beta\left|k_z/k_F\right|}{1 + \beta\left|k_z/k_F\right|} . \tag{58.6}$$

The constant β will be adjusted to fit the surface-resistance data. The fitted values are $f = 0.8$ and $\beta = 20$. It is clear that f_{eff} approaches unity rapidly as k_z becomes appreciable; and (of course) f_{eff} equals Equation (58.5) when $k_z = 0$.

The foregoing ideas are needed to correct the theoretical electron-gas conductivity, $\sigma_{\alpha\beta}(q, \omega)$, which is obtained by solving the Boltzmann transport equation. For an isotropic, free-electron metal the solution is standard. However, we display σ_{xx} and σ_{xy}, the components derived from Equations (58.12) and (58.13) of [10]:

$$\sigma_{xx} = \frac{3\sigma_0}{8}\int_{-1}^{1} dt(1-t^2)\left[\frac{1}{1-ia_+ + ixt} + \frac{1}{1-ia_- + ixt}\right],$$

$$\sigma_{xy} = \frac{3i\sigma_0}{8}\int_{-1}^{1} dt(1-t^2)\left[\frac{1}{1-ia_+ + ixt} - \frac{1}{1-ia_- + ixt}\right], \tag{58.7}$$

where

$$\begin{aligned} \sigma_0 &= \frac{ne^2\tau}{m^*}, \\ a_+ &\equiv (\omega+\omega_c)\tau, \\ a_- &\equiv (\omega-\omega_c)\tau, \\ x &\equiv ql = qv_F\tau, \\ t &\equiv \frac{k_z}{k_F}. \end{aligned} \tag{58.8}$$

τ is the scattering time, and the magnetic field H (parallel to $\hat{z}$) appears linearly in ω_c, the cyclotron frequency, eH/m^*c. The Cartesian components of σ are displayed here, instead of the circularly polarized ones, to anticipate the requirements of Section 58.3.

Notice that the factor $(1-t^2)$ in the integrand of Equation (58.7) is proportional to the cross-sectional area of the Fermi surface for $t = k_z/k_F$, i.e., to the number of electrons in the slice of width dt. However, as argued above, the heterodyne gaps reduce the effective number by the factor Equation (58.6). Consequently, we must replace

$$(1-t^2) \to (1-t^2)\, f_{\text{eff}}(k_z), \tag{58.9}$$

when the integrals are evaluated. Fortunately, these integrals can be found analytically because, as is evident in what follows, the surface resistance involves a further integration over the wave vector q, which can only be carried out numerically. The analytic expressions for σ_{xx} and σ_{xy} which incorporate the substitution Equation (58.9) are given in Appendix A.

Now, the surface impedance Z for an isotropic metal, having an $\hat{x}\hat{y}$ surface at $z = 0$, is defined by

$$Z = \frac{\mathcal{E}_x(0)}{\int_0^\infty j_x(z)dz}. \tag{58.10}$$

With the use of Stoke's theorem for a circuit in the $\hat{y}\hat{z}$ plane and the two Maxwell curl equations,

$$\mathcal{E}_x'(0) = \frac{4\pi i\omega}{c^2}\int_0^\infty j_x(z)dz. \tag{58.11}$$

The prime indicates $\partial/\partial z$, and the time dependence of the fields is taken as $\exp(-i\omega t)$. It follows that

$$Z = \frac{4\pi i\omega}{c^2}\frac{\mathcal{E}_x(0)}{\mathcal{E}_x'(0)}. \tag{58.12}$$

Solution of Maxwell's equations in the metal with specular boundary conditions at $z = 0$ can be found in [18], which we follow. For the $\alpha = \hat{x}, \hat{y}$ components of polarization,

$$\frac{d^2\mathcal{E}_\alpha(z)}{dz^2} + \frac{\omega^2}{c^2}\mathcal{E}_\alpha(z) = -\frac{4\pi i\omega}{c^2} j_\alpha(z) . \tag{58.13}$$

Solution of this equation may be obtained by Fourier transform. It has been shown experimentally [19] that conduction electrons are specularly reflected from shiny potassium surfaces. Under these conditions, one can treat the metal as infinite, instead of semi-infinite, provided $\mathcal{E}(z)$ is extended symmetrically to the region $z < 0$. This means that at $z = 0$, $\mathcal{E}'$ must undergo a jump from $-\mathcal{E}'(0)$ to $\mathcal{E}'(0)$. Accordingly, integration by parts gives

$$\begin{aligned}\int_{-\infty}^{\infty} \mathcal{E}'' e^{-iqz}\, dz &= \left(\int_{-\infty}^{-0} + \int_{+0}^{\infty}\right) \mathcal{E}'' e^{-iqz}\, dz \\ &= -2\mathcal{E}'(0) - q^2 E(q) .\end{aligned} \tag{58.14}$$

The Fourier transform of Equation (58.13) is then

$$\left(-q^2 + \frac{\omega^2}{c^2}\right) E_\alpha(q) = -\frac{4\pi i\omega}{c^2} J_\alpha(q) + \sqrt{\frac{2}{\pi}}\mathcal{E}'_\alpha(0) , \tag{58.15}$$

where for each component, $\alpha = x, y$,

$$\begin{aligned} E(q) &= \frac{1}{\sqrt{2\pi}} \int_{-\infty}^{\infty} \mathcal{E}(z) e^{-iqz}\, dz , \\ J(q) &= \frac{1}{\sqrt{2\pi}} \int_{-\infty}^{\infty} j(z) e^{-iqz}\, dz , \\ \mathcal{E}(z) &= \frac{1}{\sqrt{2\pi}} \int_{-\infty}^{\infty} E(q) e^{iqz}\, dq , \\ j(z) &= \frac{1}{\sqrt{2\pi}} \int_{-\infty}^{\infty} J(q) e^{iqz}\, dq . \end{aligned} \tag{58.16}$$

Equation (58.15) is actually a pair of coupled equations because the conductivity tensor (58.7) has off-diagonal components. On using $\sigma_{i,j}$ to eliminate $J_\alpha(q)$, Equation (58.15) becomes

$$D_{ij}(q,\omega) E_j(q) = -\sqrt{\frac{2}{\pi}}\mathcal{E}'_i(0) , \tag{58.17}$$

where

$$D_{ij}(q,\omega) \equiv \begin{pmatrix} q^2 - \frac{\omega^2}{c^2} - \frac{4\pi i\omega}{c^2}\sigma_{xx} & -\frac{4\pi i\omega}{c^2}\sigma_{xy} \\ -\frac{4\pi i\omega}{c^2}\sigma_{yx} & q^2 - \frac{\omega^2}{c^2} - \frac{4\pi i\omega}{c^2}\sigma_{yy} \end{pmatrix} . \tag{58.18}$$

For a spherical Fermi surface, $\sigma_{xx} = \sigma_{yy}$ and $\sigma_{yx} = -\sigma_{xy}$. Equation (58.17) can then be solved:

$$E_x(q) = -\sqrt{\frac{2}{\pi}} \frac{[q^2 - \omega^2/c^2 - (4\pi i\omega/c^2)\sigma_{xx}]\mathcal{E}'_x(0) + (4\pi i\omega/c^2)\sigma_{xy}\mathcal{E}'_y(0)}{[q^2 - \omega^2/c^2 - (4\pi i\omega/c^2)\sigma_{xx}]^2 + [(4\pi i\omega/c^2)\sigma_{xy}]^2} . \tag{58.19}$$

We now introduce circularly polarized waves accordingly to the convention

$$\vec{\mathcal{E}}_{\pm}(z) = (\hat{x} \pm i\hat{y})\mathcal{E}_{\pm}(0)e^{i(qz-\omega t)} . \tag{58.20}$$

It follows that

$$\mathcal{E}'_y(0) = \pm i\mathcal{E}'_x(0) . \tag{58.21}$$

This relation allows one to solve Equation (58.19) for $E_x(q)/\mathcal{E}'_x(0)$. Subsequently, the third relation of Equation (58.16), with $z = 0$, can be used to find $\mathcal{E}_x(0)/\mathcal{E}'_x(0)$, which is all one needs to evaluate the surface impedance (58.12). The final result is, after restricting the integration to positive q,

$$Z_{\pm}(H) = -8i\omega \int_0^{\infty} \frac{dq}{c^2q^2 - \omega^2 - 4\pi i\omega(\sigma_{xx} \pm i\sigma_{xy})} . \tag{58.22}$$

[That the integrand is even in q follows from the symmetry of $\mathcal{E}(z)$ mentioned above.] The integration in dq must be carried out numerically with the expressions for σ_{xx} and σ_{xy} from Appendix A. It was found sufficient to sum from $q = 0$ to 500 000 in 50 000 steps. (Doubling the range or reducing the step size by 10 did not alter the output noticeably.)

Inspection of the experimental data of Figure 58.1 reveals that the cavity was not driven in a pure "–" mode. Accordingly, we have calculated the surface resistance given by

$$R(H) = \mathrm{Re}[0.8Z_{-}(H) + 0.2Z_{+}(H)] . \tag{58.23}$$

The residual-resistance ratio of potassium, $\rho(300\,\mathrm{K})/\rho(4\,\mathrm{K})$, is typically $\sim$5000. This value implies a scattering time $\tau \sim 2 \times 10^{-10}$ s. For 23.9 GHz, $\omega\tau = 30$. $R(H)$ calculated from Equation (58.23) is shown in Figure 58.3. The heterodyne gaps cause the surface resistance to decrease when $|H| > H_c$ and to level off near $|\omega_c/\omega| \sim 2$. Not shown is the eventual recovery of $R(H)$ to the ideal Fermi-sphere result for $|\omega_c/\omega| > 3$. The rate of this high-field approach to the ideal $R(H)$ depends on the magnetic-breakdown parameter H_0; so H_0 can in principle be estimated by studying $R(H)$ in the high-field regime. Baraff has reported [20] that unpublished data of Grimes do indeed show the recovery of $R(H)$ just described. (We have not seen these particular data.)

Interruption of the cyclotron motion for electrons having $k_z \sim 0$, caused by the heterodyne gaps, reproduces the observed behavior of $R(H)$ when $|H| > H_c$. The sharp peaks at cyclotron resonance, however, are caused by the cylindrical section of Fermi surface shown in Figure 58.2, and will be explained below. The observed resonance dips near $\omega_c/\omega = \pm 0.77$ have nothing to do with the potassium sample. They are caused by embedded particles of $Cu_2SO_4 \cdot 5H_2O$ in the cavity walls created during fabrication [13].

58.3 Resonance from the Fermi-Surface Cylinder

The minigaps, shown by the short, solid lines in Figure 58.2, correspond to the periodicities of Equation (58.1). The sizes of the first few minigaps [8], tabulated in Table 58.1, are substantial. The two black patches in Figure 58.2 represent a small Fermi-surface cylinder which forms between the first minigap and the main CDW gap (having periodicity $\vec{Q}$). It has already been shown[10] that such a cylinder can explain the occurrence of the sharp cyclotron resonance observed by Baraff, Grimes, and Platzman and reproduced in Figure 58.1.

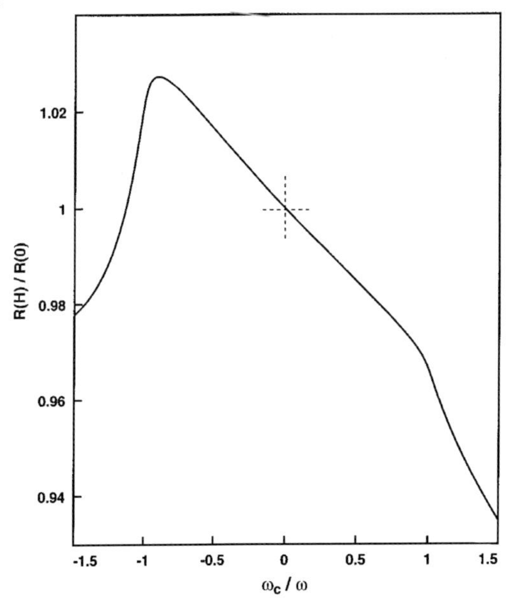

Figure 58.3 Theoretical surface resistance for a Fermi sphere having only heterodyne-gap intersections. The parameters of Equation (58.6), which quantify the loss in effective cyclotron motion on equatorial orbits, are $f = 0.8$ and $\beta = 20$. The drop in R for $|H| > H_c$ increases with decreasing f. The steepness of the decline increases with increasing β. The magnetic-breakdown field is $H_0 = 4\,\mathrm{T}$. The electron scattering time corresponds to $\omega\tau = 30$.

The size of the resonance requires the volume of the cylinder (pieced together from the two halves) to be a very small fraction, $\eta \sim 4 \times 10^{-4}$, of the Fermi-sphere volume. It is noteworthy that this volume fraction agrees with the value calculated from the product of the cylinder's length and its cross-sectional area. The former is obtained from the neutron-diffraction measurement of $\vec{Q}$ [15], and the latter from the periodicity of the Landau-level oscillations [12], observed in microwave transmission [11]. The cylinder's radius is $k_c = k_F/8$, and its length (projected along [110]) is $0.015\,G_{110}$. Although $\vec{Q}$ is tilted from [110] by about 1°, the cylinder's axis, $\vec{Q}' \equiv \vec{G}_{110} - \vec{Q}$, is tilted about 45° from [110],

$$\vec{Q}' \approx (0.025, 0.015, 0.005)\frac{2\pi}{a} \ . \qquad (58.24)$$

Its cross section is approximately circular in a plane perpendicular to [110].

It is of interest to calculate first the surface resistance $R(H)$ caused by a cylinder having its axis parallel to the magnetic field $\vec{H}$. On account of the cylinder's small size, electron velocities on the Fermi surface of the cylinder are also small. Accordingly, we will use a local conductivity tensor for the cylinder. The dc conductivity in the $\hat{x}\hat{y}$ plane is $\eta\sigma_{0c}$ and $\sigma_{zz} = 0$, where σ_{0c} is $ne^2\tau_c/m^*$. The sharpness of the observed resonance corresponds to $\omega\tau_c \sim 150$. That τ_c (on the cylinder) should be ~5 times larger than τ on the main Fermi surface is reasonable because of the smaller velocities of the cylinder electrons. The cylinder's conductivity tensor is then

$$\sigma^{\text{cyl}} = \frac{\eta\sigma_{0c}}{(1 - i\omega\tau_c)^2 + (\omega_c\tau_c)^2}\begin{pmatrix} 1 - i\omega\tau_c & -\omega_c\tau_c & 0 \\ \omega_c\tau_c & 1 - i\omega\tau_c & 0 \\ 0 & 0 & 0 \end{pmatrix} . \qquad (58.25)$$

For this exercise we will neglect the effect of the heterodyne gaps. Consequently, σ^{cyl}, Equation (58.25), is added to the conductivity, Equation (58.7), for an ideal Fermi sphere. The surface impedance is still given by Equation (58.22), and $R(H)$ for 80% circular polarization is obtained from (58.23). The result is shown in Figure 58.4 with $\omega\tau_c = 150$. A sharp cyclotron resonance is obtained but, unlike the data of Figure 58.1, the shape is asymmetric.

The sharp, asymmetric resonance shown in Figure 58.4 was obtained previously [10], but the remedy attempted then involved introduction of a more than 50–50 admixture of surface reactance and surface resistance. However, a remedy not involving such an admixture is possible. Since the cylinder's axis must, theoretically, be tilted $\sim 45°$ from [110] [14], an angle confirmed by neutron diffraction [15], we now study the effect of such a tilt on the resonance shape.

The equation for a cylindrical surface of constant energy $\epsilon = E_F$, having an axis at an angle θ relative to the direction of $\vec{H}$, and with $\vec{k}$ relative to the cylinder's center, is

$$\epsilon - \epsilon_0 = \frac{\hbar^2}{2m^*}\left[(k_x - k_z \tan\theta)^2 + k_y^2\right]. \tag{58.26}$$

This cylinder has a circular cross section in the $\hat{x}\hat{y}$ plane. Consequently, the cyclotron frequency, with $\vec{H}$ along $\hat{z}$, is unchanged. (For the cylinder of interest here, $\epsilon - \epsilon_0 = E_F/64$.) On account of its small size, as already discussed, the electron velocities on this surface are $\sim v_F/8$. We will therefore employ local equations of motion to find the tilted cylinder's conductivity tensor σ^{cyl}. The Lorentz equation for motion in the electric and magnetic fields is

$$\dot{\vec{v}} = -eM^{-1}(\vec{k})\left[\vec{\mathcal{E}} + \frac{1}{c}\vec{v}\times\vec{H}\right] - \frac{\vec{v}}{\tau_c}, \tag{58.27}$$

where $\hbar\vec{v} = \Delta_k\epsilon(\vec{k})$, and the effective mass tensor is

$$[M^{-1}(\vec{k})]_{ij} = \frac{1}{\hbar^2}\frac{\partial^2\epsilon(\vec{k})}{\partial k_i \partial k_j}. \tag{58.28}$$

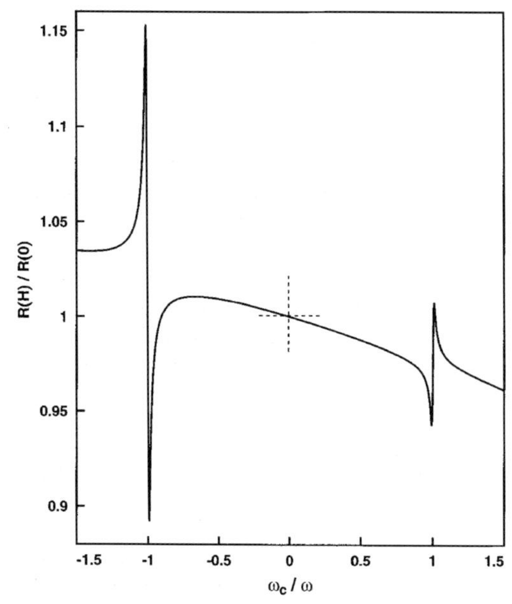

Figure 58.4 Theoretical surface resistance of conduction electrons having $\omega\tau = 30$ on the Fermi sphere and $\omega\tau_c = 150$ on the Fermi-surface cylinder (containing $\sim 4 \times 10^{-4}$ electrons per atom). The axis of the cylinder is, here, parallel to $\vec{H}$, and the heterodyne gaps (intersecting the sphere) are ignored. A 4:1 ratio of left to right circular polarization is assumed.

Then

$$v_x = \frac{\hbar}{m^*}(k_x - k_z \tan\theta) ,$$
$$v_y = \frac{\hbar}{m^*}k_y ,$$
$$v_z = -\frac{\hbar}{m^*}(k_x - k_z \tan\theta)\tan\theta . \tag{58.29}$$

Equations (58.26)–(58.29) can now be used to find the conductivity of the cylinder:

$$\sigma^{\text{cyl}} = \frac{\eta\sigma_{0c}}{(1 - i\omega\tau_c)^2 + (\omega_c\tau_c)^2} \begin{pmatrix} 1 - i\omega\tau_c & -\omega_c\tau_c & -\tan\theta(1 - i\omega\tau_c) \\ \omega_c\tau_c & 1 - i\omega\tau_c & -\tan\theta(\omega_c\tau_c) \\ -\tan\theta(1 - i\omega\tau_c) & \tan\theta(\omega_c\tau_c) & \tan^2\theta(1 - i\omega\tau_c) \end{pmatrix} . \tag{58.30}$$

After comparing this tensor with Equation (58.25), for which $\theta = 0$, it is clear that the electric field may now have a longitudinal, $\hat{z}$ component. J_x and J_y of Equation (58.15) now involve $\mathcal{E}_z$ because σ_{xz} and σ_{yz} are no longer zero. However, we can express $\mathcal{E}_z$ in terms of $\mathcal{E}_x$ and $\mathcal{E}_y$ by using the requirement that the total longitudinal current J_z be zero everywhere. Accordingly,

$$J_z(q) = \sigma_{zx}^{\text{cyl}} E_x + \sigma_{zy}^{\text{cyl}} E_y + \left(\sigma_{zz}^{\text{sph}} + \sigma_{zz}^{\text{cyl}}\right) E_z = 0 . \tag{58.31}$$

The longitudinal conductivity of the spherical portion of the Fermi surface must be calculated nonlocally using the Boltzmann transport equation. For a longitudinal electric field, proportional to $e^{iqz - i\omega t}$,

$$\sigma_{zz}^{\text{sph}} = \frac{3\sigma_0}{2}\int_{-1}^{1} dt \frac{t^2}{1 - i\omega\tau + ixt}$$
$$= \frac{3\sigma_0}{2x^3}[2x - r + 2\omega\tau p + \omega^2\tau^2 r + i(p - 2\omega\tau x$$
$$+ 2\omega\tau r - \omega^2\tau^2 p)] , \tag{58.32}$$

with

$$p = \frac{1}{2}\ln\left[\frac{1 + (x - \omega\tau)^2}{1 + (x + \omega\tau)^2}\right] ,$$
$$r = \tan^{-1}(x - \omega\tau) + \tan^{-1}(x + \omega\tau) . \tag{58.33}$$

Equation (58.31) together with Equation (58.15) changes Equation (58.17) as follows:

$$\left[q^2 - \frac{\omega^2}{c^2} - \frac{4\pi i\omega}{c^2}\left(\sigma_{xx} - \frac{\sigma_{xz}\sigma_{zx}}{\sigma_{zz}}\right)\right] E_x$$
$$- \frac{4\pi i\omega}{c^2}\left(\sigma_{xy} - \frac{\sigma_{xz}\sigma_{zy}}{\sigma_{zz}}\right) E_y = -\sqrt{\frac{2}{\pi}}\mathcal{E}_x'(0) ,$$
$$- \frac{4\pi i\omega}{c^2}\left(\sigma_{yx} - \frac{\sigma_{yz}\sigma_{zx}}{\sigma_{zz}}\right) E_x$$
$$+ \left[q^2 - \frac{\omega^2}{c^2} - \frac{4\pi i\omega}{c^2}\left(\sigma_{yy} - \frac{\sigma_{yz}\sigma_{zy}}{\sigma_{zz}}\right)\right] E_y = -\sqrt{\frac{2}{\pi}}\mathcal{E}_y'(0) , \tag{58.34}$$

where $\sigma_{ij} = \sigma_{ij}^{\text{sph}} + \sigma_{ij}^{\text{cyl}}$. These two equations can be expressed compactly:

$$D_{ij}(q,\omega)E_j(q) = -\sqrt{\frac{2}{\pi}}\mathcal{E}_i'(0) \tag{58.35}$$

with

$$D_{ij}(q,\omega) \equiv \begin{pmatrix} q^2 - \frac{\omega^2}{c^2} - \frac{4\pi i\omega}{c^2}\sigma'_{xx} & -\frac{4\pi i\omega}{c^2}\sigma'_{xy} \\ -\frac{4\pi i\omega}{c^2}\sigma'_{yx} & q^2 - \frac{\omega^2}{c^2} - \frac{4\pi i\omega}{c^2}\sigma'_{yy} \end{pmatrix} . \tag{58.36}$$

Here,

$$\sigma'_{ij} = \sigma_{ij} - \frac{\sigma_{iz}\sigma_{zj}}{\sigma_{zz}} . \tag{58.37}$$

i and j denote x or y. The difference between Equation (58.36) and Equation (58.18) of Section 58.2 is that all transverse conductivities σ_{ij} in Equation (58.18) are replaced by σ'_{ij} in Equation (58.36). For example, σ'_{xx} includes σ_{xz}, σ_{zx}, and σ_{zz} as well as σ_{xx}. The longitudinal motion of electrons in the cylinder leads to creation of an electric field in the $\hat{z}$ direction. The $\mathcal{E}_z$ which arises (to preserve charge neutrality) plays a role in producing the transverse currents j_x and j_y due to the nonzero values of σ_{xz} and σ_{yz}. (The tilted Fermi-surface cylinder mixes the transverse and longitudinal motions.) Even though the number of electrons in the cylinder is small, this mixing causes a large change in the surface impedance. The total conductivity tensor has the following properties:

$$\sigma_{yx} = -\sigma_{xy} , \quad \sigma_{zy} = -\sigma_{yz} , \quad \sigma_{zx} = \sigma_{xz} . \tag{58.38}$$

Accordingly, from Equation (58.37), $\sigma'_{yx} = -\sigma'_{xy}$. Equation (58.35) may now be solved:

$$E_x(q) = -\sqrt{\frac{2}{\pi}}\frac{(q^2c^2 - \omega^2 - 4\pi i\omega\sigma'_{yy})\mathcal{E}_x'(0) + 4\pi i\omega\sigma'_{xy}\mathcal{E}_y'(0)}{(q^2c^2 - \omega^2 - 4\pi i\omega\sigma'_{xx})(q^2c^2 - \omega^2 - 4\pi i\omega\sigma'_{yy}) + (4\pi i\omega\sigma'_{xy})^2} ,$$
$$E_y(q) = -\sqrt{\frac{2}{\pi}}\frac{-4\pi i\omega\sigma'_{xy}\mathcal{E}_x'(0) + (q^2c^2 - \omega^2 - 4\pi i\omega\sigma'_{xx})\mathcal{E}_y'(0)}{(q^2c^2 - \omega^2 - 4\pi i\omega\sigma'_{xx})(q^2c^2 - \omega^2 - 4\pi i\omega\sigma'_{yy}) + (4\pi i\omega\sigma'_{xy})^2} . \tag{58.39}$$

Using Equation (58.11), we express $\mathcal{E}_i'(0)$ in terms of the total current density J_i,

$$J_i = \int_0^\infty j_i(z)dz . \tag{58.40}$$

The third equation of (58.16), together with (58.38)–(58.40), give the electric field at $z = 0$:

$$\mathcal{E}_x(0) = -\frac{8i\omega}{c^2}\int_0^\infty dq \frac{(q^2c^2 - \omega^2 - 4\pi i\omega\sigma'_{yy})J_x + 4\pi i\omega\sigma'_{xy}J_y}{(q^2c^2 - \omega^2 - 4\pi i\omega\sigma'_{xx})(q^2c^2 - \omega^2 - 4\pi i\omega\sigma'_{yy}) + (4\pi i\omega\sigma'_{xy})^2} ,$$
$$\mathcal{E}_y(0) = -\frac{8i\omega}{c^2}\int_0^\infty dq \frac{-4\pi i\omega\sigma'_{xy}J_x + (q^2c^2 - \omega^2 - 4\pi i\omega\sigma'_{xx})J_y}{(q^2c^2 - \omega^2 - 4\pi i\omega\sigma'_{xx})(q^2c^2 - \omega^2 - 4\pi i\omega\sigma'_{yy}) + (4\pi i\omega\sigma'_{xy})^2} . \tag{58.41}$$

These expressions can be written compactly:

$$\mathcal{E}_x(0) = Z_{xx} J_x + Z_{xy} J_y ,$$
$$\mathcal{E}_y(0) = -Z_{xy} J_x + Z_{yy} J_y , \tag{58.42}$$

which by inspection of (58.41) defines the four components of $Z_{\alpha\beta}$, the surface impedance tensor.

It is clear from Equations (58.40) and (58.41) that $J_\alpha(\alpha = x, y)$ depend intricately on the bulk electric fields. Anisotropy caused by the cylinder's tilt causes J_α to be a complicated function of the conductivity components. This asymmetry also prevents the field from having perfect circular polarization. This behavior is studied in Appendix B. Nevertheless, on account of the small size of the cylinder, the electric-field polarization is almost circular. Accordingly,

$$H_y(0) \approx \pm i H_x(0) ,$$
$$J_y \approx \pm i J_x . \tag{58.43}$$

The electric field at the surface will be $\mathcal{E}_x = \mathcal{E}_0 e^{-i\omega t}$ and $\mathcal{E}_y = i\mathcal{E}_x$, which corresponds to right circular polarization. ($\mathcal{E}_0$ is real.) Then from Equation (B25),

$$J_x = \frac{c\mathcal{E}_0}{2\pi} e^{-i\omega t} ,$$
$$J_y = i\frac{c\mathcal{E}_0}{2\pi} e^{-i\omega t} . \tag{58.44}$$

The power absorbed per unit area per unit time is

$$S_z = \frac{c}{4\pi}\{\mathrm{Re}[\vec{\mathcal{E}}(0)] \times \mathrm{Re}[\vec{H}(0)]\}_z$$
$$= \frac{c}{4\pi}\{\mathrm{Re}[\mathcal{E}_x(0)]\,\mathrm{Re}[H_y(0)] - \mathrm{Re}[\mathcal{E}_y(0)]\,\mathrm{Re}[H_x(0)]\}$$
$$= \mathrm{Re}[\mathcal{E}_x(0)]\,\mathrm{Re}[J_x] + \mathrm{Re}[\mathcal{E}_y(0)]\,\mathrm{Re}[J_y] . \tag{58.45}$$

We now separate $Z_{\alpha\beta}$ [defined by (58.41) and (58.42)] into their real and imaginary parts, i.e.,

$$Z_{\alpha\beta} = R_{\alpha\beta} + i I_{\alpha\beta} , \tag{58.46}$$

where $R_{\alpha\beta}$ is the real part of $Z_{\alpha\beta}$ and $I_{\alpha\beta}$ its imaginary part. It follows that

$$\mathrm{Re}(J_x) = \frac{c\mathcal{E}_0}{2\pi}\cos(\omega t) ,$$
$$\mathrm{Re}(J_y) = \frac{c\mathcal{E}_0}{2\pi}\sin(\omega t) , \tag{58.47}$$

$$\mathrm{Re}(\mathcal{E}_x) = \frac{c\mathcal{E}_0}{2\pi}[R_{xx}\cos(\omega t) + I_{xx}\sin(\omega t) + R_{xy}\sin(\omega t) - I_{xy}\cos(\omega t)] ,$$

$$\mathrm{Re}(\mathcal{E}_y) = \frac{c\mathcal{E}_0}{2\pi}[R_{yy}\sin(\omega t) - I_{yy}\cos(\omega t) - R_{xy}\cos(\omega t) - I_{xy}\sin(\omega t)] .$$

By using these expressions in Equation (58.45) and averaging over time, we find the absorbed power.

$$\begin{aligned}\overline{S_z} &= \overline{\mathrm{Re}[\mathcal{E}_x(0)]\,\mathrm{Re}[J_x] + \mathrm{Re}[\mathcal{E}_y(0)]\,\mathrm{Re}[J_y]} \\ &= \frac{c^2\mathcal{E}_0^2}{8\pi^2}(R_{xx} + R_{yy} - 2I_{xy}) \\ &= \frac{c^2\mathcal{E}_0^2}{4\pi^2}\,\mathrm{Re}\left[\frac{1}{2}(Z_{xx} + Z_{yy}) + iZ_{xy}\right] .\end{aligned} \tag{58.48}$$

The effective surface resistance is therefore

$$R = \mathrm{Re}\left[\frac{1}{2}(Z_{xx} + Z_{yy}) + iZ_{xy}\right] . \tag{58.49}$$

From Equations (58.41) and (58.42), and $\mathcal{E}_y = i\mathcal{E}_x$ for right circular polarization, the surface impedance is

$$Z_R = \frac{1}{2}(Z_{xx} + Z_{yy}) + iZ_{xy} = -\frac{8i\omega}{c^2}\int_0^\infty dq \frac{[q^2c^2 - \omega^2 - 2\pi i\omega(\sigma'_{xx} + \sigma'_{yy})] - 4\pi\omega\sigma'_{xy}}{(q^2c^2 - \omega^2 - 4\pi i\omega\sigma'_{xx})(q^2c^2 - \omega^2 - 4\pi i\omega\sigma'_{yy}) + (4\pi i\omega\sigma'_{xy})^2} . \tag{58.50}$$

For a left circularly polarized wave on the front surface, i.e., $\mathcal{E}_y = -i\mathcal{E}_x$, the surface impedance is

$$Z_L = \frac{1}{2}(Z_{xx} + Z_{yy}) - iZ_{xy} = -\frac{8i\omega}{c^2}\int_0^\infty dq \frac{[q^2c^2 - \omega^2 - 2\pi i\omega(\sigma'_{xx} + \sigma'_{yy})] + 4\pi\omega\sigma'_{xy}}{(q^2c^2 - \omega^2 - 4\pi i\omega\sigma'_{xx})(q^2c^2 - \omega^2 - 4\pi i\omega\sigma'_{yy}) + (4\pi i\omega\sigma'_{xy})^2} . \tag{58.51}$$

Equations (58.50) and (58.51) must be evaluated numerically, as in Section 58.2. The effective surface resistance applicable to the experiment, for which the polarization was about a 4:1 admixture of L and R, is now

$$R(H) = \mathrm{Re}[0.8Z_L(H) + 0.2Z_R(H)] . \tag{58.52}$$

The theoretical $R(H)$, which includes effects from both the tilted cylinder and the heterodyne gaps, is shown in Figure 58.5. The agreement with the experimental data of Figure 58.1 is remarkable.

58.4 Conclusion

Inspection of Figures 58.1–58.5 allows one to recognize that the CDW in potassium [6–8] has profound consequences in studies of the perpendicular-field cyclotron resonance. The fact that cyclotron resonance even exists (in the surface resistance, R vs. H) attests to the presence of the small Fermi-surface cylinder (the dark areas of Figure 58.2), created by the CDW gap and the first minigap. A theory based on only a spherical Fermi surface does not allow any structure near $\omega_c = \omega$, as shown by the top curve of Figure 58.1.

The shape of the $R(H)$ resonance (compare Figs. 58.4 and 58.5) reveals that the cylinder's axis is tilted away from [110] (the field direction) by $\sim 45°$, as was found theoretically [8, 14]. (The reason for the tilt is to minimize the elastic-stress energy involved in creating the

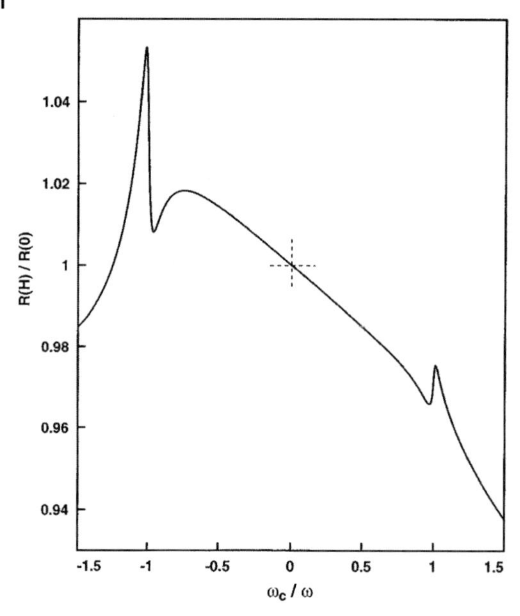

Figure 58.5 Theoretical $R(H)$ for potassium based on the heterodyne-gap parameters of Figure 58.3 and the Fermi-surface cylinder model of Figure 58.4, except that the cylinder's axis is tilted 45° from [110]. [The tilt is required to minimize the elastic stress of the periodic lattice distortion needed to neutralize the electronic CDW [14].] This calculated behavior should be compared with Baraff, Grimes, and Platzman's data in Figure 58.1.

periodic lattice distortion, of wave vector $\vec{G}_{110} - \vec{Q}$, needed to screen the electronic CDW [14].

The drop in R for $|H| > |H_c|$, see Figure 58.3 and the experimental data of Figure 58.1, arises from the heterodyne gaps (Figure 58.2), which interrupt the cyclotron motion of equatorial orbits, and cause a partial loss in carrier effectiveness.

The volume of the Fermi-surface cylinder (corresponding to $\nu = 4 \times 10^{-4}$ electrons/atom) was determined from the size of the resonance relative to $R(H_c) - R(0)$ [10]. The fact that this volume equals the product of the cylinder's length (along [110]), determined from $\vec{Q}$ (observed in neutron diffraction [15]) and the cylinder's cross section (perpendicular to [110]), defined by the periodicity of Landau-level oscillations observed in microwave transmission [12], indicates a compelling consistency among relevant phenomena.

Fracture of potassium's Fermi surface by CDW minigaps and heterodyne gaps, Figure 58.2, is not only evident in the surface resistance anomalies studied here, but is the cause of many other magnetotransport effects, the most spectacular of which are the multitudinous open-orbit resonances [21] created by the minigaps and heterodyne gaps. These open-orbit spectra have been explained within the same framework employed here [17]. Without a broken symmetry, potassium would be the simplest metal of all since, unlike Li [22] or Na, it would retain its cubic symmetry to helium temperature. However, as a consequence of its CDW, potassium has provided (during the last 33 years) a veritable universe of unanticipated behavior – a challenge to all who seek to understand electrons in metals.

Acknowledgements

The authors are indebted to the National Science Foundation, Division of Materials Research for support. We are grateful to Graciela Lacueva and C.C. Grimes for helpful discussions. Dr. Grimes kindly provided original data, including perpendicular-field resonance and Azbel–Kaner (A–K) oscillations on the same specimen. The A–K oscillations usually had five subharmonics, consistent with theory [3] for $\omega\tau \sim 30$, throughout the skin depth.

A Calculation of the Conductivity

Equation (58.6) can be rearranged as follows:

$$f_{\text{eff}} = \frac{f + (1-f)e^{-H_0/H} + 1 - 1 + \beta|t|}{1+\beta|t|}$$

$$= 1 + \frac{(1-f)(e^{-H_0/H} - 1)}{1+\beta|t|} \,. \tag{A1}$$

From Equations (58.7) and (58.9) we have expressions for σ_{xx} and σ_{xy},

$$\sigma_{xx} = \frac{3\sigma_0}{8}\int_{-1}^{1} dt(1-t^2)\left[1 + \frac{f+(1-f)e^{-H_0/H}-1}{1+\beta|t|}\right] \times \left[\frac{1}{1-ia_+ + ixt} + \frac{1}{1-ia_- + ixt}\right]$$
$$= \sigma_{xx}^{\text{orig}} + [f + (1-f)e^{-H_0/H} - 1]\sigma'_{xx} \,, \tag{A2}$$

$$\sigma_{xy} = \frac{3i\sigma_0}{8}\int_{-1}^{1} dt(1-t^2)\left[1 + \frac{f+(1-f)e^{-H_0/H}-1}{1+\beta|t|}\right] \times \left[\frac{1}{1-ia_+ + ixt} - \frac{1}{1-ia_- + ixt}\right]$$
$$= \sigma_{xy}^{\text{orig}} + [f + (1-f)e^{-H_0/H} - 1]\sigma'_{xy} \,, \tag{A3}$$

where

$$\sigma'_{xx} = \frac{3\sigma_0}{8}\int_{-1}^{1} dt\frac{1-t^2}{1+\beta|t|}\left[\frac{1}{1-ia_+ + ixt} + \frac{1}{1-ia_- + ixt}\right],$$

$$\sigma'_{xy} = \frac{3i\sigma_0}{8}\int_{-1}^{1} dt\frac{1-t^2}{1+\beta|t|}\left[\frac{1}{1-ia_+ + ixt} - \frac{1}{1-ia_- + ixt}\right], \tag{A4}$$

and $\sigma_{xx}^{\text{orig}}$ and $\sigma_{xy}^{\text{orig}}$ are the same as σ_{xx} and σ_{xy} in Equation (58.7). These expressions were evaluated previously [10],

$$\sigma_{xx}^{\text{orig}} = \frac{3\sigma_0}{4x^2}\{2a_+p_+ + 2a_-p_- - 2 + r_+(x^2+1-a_+^2) + r_-(x^2+1-a_-^2) + i[a_+ + a_- + p_+(x^2+1-a_+^2) - p_-(x^2+1-a_-^2) - 2a_+r_+ - 2a_-r_-]\} \,,$$

$$\sigma_{xy}^{\text{orig}} = \frac{3\sigma_0}{4x^2}\{a_- - a_+ + p_-(x^2+1-a_-^2) - p_+(x^2+1-a_+^2) + 2a_+r_+ - 2a_-r_- + i[2a_+p_+ - 2a_-p_- + r_+(x^2+1-a_+^2) - r_-(x^2+1-a_-^2)]\} \,, \tag{A5}$$

where

$$a_\pm = (\omega \pm \omega_c)\tau \,,$$

$$x = ql \,,$$

$$p_\pm = \frac{1}{4x}\ln\left[\frac{1+(x+a_\pm)^2}{1+(x-a_\pm)^2}\right] ,$$

$$r_\pm = \frac{1}{2x}\left[\tan^{-1}(x+a_\pm)+\tan^{-1}(x-a_\pm)\right] . \tag{A6}$$

Integration of Equation (A3) is tedious but straightforward. The final forms are

$$\begin{aligned}
\sigma_{xx} = \; & \frac{3\sigma_0}{4x^2}\{2a_+p_+ + 2a_-p_- - 2 + r_+(x^2+1-a_+^2) \\
& + r_-(x^2+1-a_-^2) + [f+(1-f)e^{-H_0/H}-1] \\
& \times [x(f_+ - f'_+ + f_- - f'_-) - \beta a_+(f_+ + f'_+) \\
& - \beta a_-(f_- + f'_-) - \beta(g_+ + g'_+ + g_- + g'_-) \\
& + (s_+ + s'_+ + s_- + s'_-)x^2] \\
& + i[a_+ + a_- + p_+(x^2+1-a_+^2) \\
& + p_-(x^2+1-a_-^2) - 2a_+r_+ - 2a_-r_-] \\
& + i[f+(1-f)e^{-H_0/H}-1] \\
& \times [x(g_+ - g'_+ + g_- - g'_-) - \beta a_+(g_+ + g'_+) - \beta a_-(g_- + g'_-) \\
& + \beta(f_+ + f'_+ + f_- + f'_-) + (t_+ + t'_+ + t_- + t'_-)x^2]\} , \\
\sigma_{xy} = \; & \frac{3\sigma_0}{4x^2}\{a_- - a_+ + p_-(x^2+1-a_-^2) \\
& - p_+(x^2+1-a_+^2) + 2a_+r_+ - 2a_-r_- \\
& + [f+(1-f)e^{-H_0/H}-1][x(g_- - g'_- - g_+ + g'_+) \\
& + \beta a_+(g_+ + g'_+) - \beta a_-(g_- + g'_-) + \beta(f_- + f'_- - f_+ - f'_+) \\
& + (t_- + t'_- - t_+ - t'_+)x^2] + i[2a_+p_+ - 2a_-p_- \\
& - r_-(x^2+1-a_-^2) + r_+(x^2+1-a_+^2)] \\
& + i[f+(1-f)e^{-H_0/H}-1][x(f_+ - f'_+ - f_- + f'_-) \\
& - \beta a_+(f_+ + f'_+) + \beta a_-(f_- + f'_-) - \beta(g_+ + g'_+ - g_- - g'_-) \\
& + (s_+ + s'_+ - s_- - s'_-)x^2]\} ,
\end{aligned} \tag{A7}$$

where

$$u_\pm = \frac{1}{2}\ln\left[\frac{1+a_\pm^2}{1+(x+a_\pm)^2}\right],$$
$$v_\pm = \tan^{-1}(x+a_\pm) - \tan^{-1}(a_\pm),$$
$$f_\pm = \frac{1}{2}\frac{1}{\beta^2+(x-\beta a_\pm)^2}\left[-2a_\pm u_\pm - v_\pm\left(a_\pm^2 - x^2 - 1\right) - x\right],$$
$$g_\pm = \frac{1}{2}\frac{1}{\beta^2+(x-\beta a_\pm)^2}\left[-2a_\pm v_\pm + u_\pm(a_\pm^2 - x^2 - 1) - \frac{x(x-2a_\pm)}{2}\right],$$
$$s_\pm = \frac{1}{2}\frac{\beta^2}{\beta^2+(x-\beta a_\pm)^2}\left[\left(\frac{1}{\beta}-\frac{1}{\beta^3}\right)\ln(1+\beta) + \frac{1}{\beta^2}\left(1-\frac{\beta}{2}\right)\right],$$
$$t_\pm = \frac{1}{2}\frac{-\beta(x-\beta a_\pm)}{\beta^2+(x-\beta a_\pm)^2}\left[\left(\frac{1}{\beta}-\frac{1}{\beta^3}\right)\ln(1+\beta) + \frac{1}{\beta^2}\left(1-\frac{\beta}{2}\right)\right]. \tag{A8}$$

Furthermore, $f'_\pm(x) = f_\pm(-x)$, $g'_\pm(x) = g_\pm(-x)$, $s'_\pm(x) = s_\pm(-x)$, and $t'_\pm(x) = t_\pm(-x)$. The foregoing results are to be used in the integrand of Equation (58.22), which must then be evaluated numerically.

B Polarization of the Field Inside an Anisotropic Metal

Consider a metal in a high-frequency electromagnetic field. To learn how the wave is polarized we shall treat the normal skin effect for which Ohm's law, $\vec{j} = \sigma\vec{E}$, is valid and the conductivity is local. The relevant Maxwell equations are

$$\vec{\nabla}\times\vec{\mathcal{E}} = -\frac{1}{c}\frac{\partial\vec{H}}{\partial t},$$
$$\vec{\nabla}\times\vec{H} = \frac{4\pi}{c}\vec{j}. \tag{B1}$$

We neglect the displacement current. Let us assume that the metal fills the $z > 0$ half space, and that the wave is incident normal to the surface. For a wave propagating in the z direction we shall seek a solution proportional to $\exp(iqz - i\omega t)$. Eliminating the magnetic field $\vec{H}$ from Equation (B1),we can easily find:

$$-\nabla^2\vec{\mathcal{E}} + \vec{\nabla}(\vec{\nabla}\cdot\vec{\mathcal{E}}) + \frac{4\pi}{c^2}\frac{\partial}{\partial t}\vec{j} = 0, \tag{B2}$$

which reduces to

$$\frac{\partial^2}{\partial z^2}\mathcal{E}_\alpha + \frac{4\pi i\omega}{c^2}j_\alpha = 0, \tag{B3}$$

$$j_z = 0, \tag{B4}$$

where $\alpha = x, y$. The conductivity of a nearly-free-electron system in the local approximation is

$$\sigma^s = \frac{ne^2\tau}{m^*}\frac{1}{(1-i\omega\tau)^2+(\omega_c\tau)^2} \times \begin{pmatrix} 1-i\omega\tau & -\omega_c\tau & 0 \\ \omega_c\tau & 1-i\omega\tau & 0 \\ 0 & 0 & \frac{(1-i\omega\tau)^2+(\omega_c\tau)^2}{1-i\omega\tau} \end{pmatrix}. \tag{B5}$$

The conductivity of the Fermi-surface cylinder, as calculated in Section 58.3, is

$$\sigma^c = \frac{\eta n e^2 \tau_c}{m^*} \frac{1}{(1-i\omega\tau_c)^2 + (\omega_c\tau_c)^2} \times \begin{pmatrix} 1-i\omega\tau_c & -\omega_c\tau_c & -\tan\theta\,(1-i\omega\tau_c) \\ \omega_c\tau_c & 1-i\omega\tau_c & -\tan\theta\,(\omega_c\tau_c) \\ -\tan\theta\,(1-i\omega\tau_c) & \tan\theta\,(\omega_c\tau_c) & \tan^2\theta\,(1-i\omega\tau_c) \end{pmatrix} . \tag{B6}$$

The total conductivity is $\sigma^s + \sigma^c$. The usual expression for the conductivity tensor is

$$\sigma = \begin{pmatrix} \sigma_{xx} & \sigma_{xy} & \sigma_{xz} \\ \sigma_{yx} & \sigma_{yy} & \sigma_{yz} \\ \sigma_{zx} & \sigma_{zy} & \sigma_{zz} \end{pmatrix} . \tag{B7}$$

Because the number of electrons enclosed by the cylindrical Fermi surface is only a fraction, $\eta = 4 \times 10^{-4}$, of the total, the following inequalities prevail:

$$\sigma_{xz}, \sigma_{yz}, \sigma_{zx}, \sigma_{zy} \ll \sigma_{xx}, \sigma_{yy}, \sigma_{zz}, \sigma_{xy}, \sigma_{yx} . \tag{B8}$$

Using Ohm's law to express Equations (B3) and (B4), we find a set of homogeneous equations:

$$\begin{aligned} &\left(q^2 - \frac{4\pi i\omega}{c^2}\sigma_{xx}\right)\mathcal{E}_x - \frac{4\pi i\omega}{c^2}\sigma_{xy}\mathcal{E}_y - \frac{4\pi i\omega}{c^2}\sigma_{xz}\mathcal{E}_z = 0 , \\ &-\frac{4\pi i\omega}{c^2}\sigma_{yx}\mathcal{E}_x + \left(q^2 - \frac{4\pi i\omega}{c^2}\sigma_{yy}\right)\mathcal{E}_y - \frac{4\pi i\omega}{c^2}\sigma_{yz}\mathcal{E}_z = 0 , \\ &\sigma_{zx}\mathcal{E}_x + \sigma_{zy}\mathcal{E}_y + \sigma_{zz}\mathcal{E}_z = 0 . \end{aligned} \tag{B9}$$

We next eliminate $\mathcal{E}_z$ in favor of $\mathcal{E}_x$ and $\mathcal{E}_y$, using the third equation of (B9). This allows us to express (B9) with $\mathcal{E}_x$ and $\mathcal{E}_y$ only:

$$\begin{aligned} &\left(q^2 - \frac{4\pi i\omega}{c^2}\sigma'_{xx}\right)\mathcal{E}_x - \frac{4\pi i\omega}{c^2}\sigma'_{xy}\mathcal{E}_y = 0 , \\ &-\frac{4\pi i\omega}{c^2}\sigma'_{yx}\mathcal{E}_x + \left(q^2 - \frac{4\pi i\omega}{c^2}\sigma'_{yy}\right)\mathcal{E}_y = 0 , \end{aligned} \tag{B10}$$

where

$$\sigma'_{\alpha\beta} = \sigma_{\alpha\beta} - \frac{\sigma_{\alpha z}\sigma_{z\beta}}{\sigma_{zz}} . \tag{B11}$$

Here α and β indicate x or y components only. This change of $\sigma_{\alpha\beta}$ to $\sigma'_{\alpha\beta}$ is the main contribution of the electrons in the tilted cylinder. Transverse conductivities are mixed with longitudinal conductivity on account of the longitudinal motion of electrons in the cylinder. The determinant of (B10) must vanish; and this condition leads to the allowed propagation vectors:

$$\begin{aligned} q_1^2 &= \frac{1}{2}\frac{4\pi i\omega}{c^2}\left[\sigma'_{xx} + \sigma'_{yy} + \sqrt{(\sigma'_{xx} - \sigma'_{yy})^2 + 4\sigma'_{xy}\sigma'_{yx}}\right] , \\ q_2^2 &= \frac{1}{2}\frac{4\pi i\omega}{c^2}\left[\sigma'_{xx} + \sigma'_{yy} - \sqrt{(\sigma'_{xx} - \sigma'_{yy})^2 + 4\sigma'_{xy}\sigma'_{yx}}\right] . \end{aligned} \tag{B12}$$

Therefore the two electric-field modes are

$$\vec{\mathcal{E}}_1 = \mathcal{E}_{10}\left\{\hat{x} + \frac{q_1^2c^2 - 4\pi i\omega\sigma'_{xx}}{4\pi i\omega\sigma'_{xy}}\hat{y} - \left[\frac{(q_1^2c^2 - 4\pi i\omega\sigma'_{xx})\sigma_{zy} + 4\pi i\omega\sigma'_{xy}\sigma_{zx}}{4\pi i\omega\sigma'_{xy}\sigma_{zz}}\right]\hat{z}\right\}e^{iq_1z - i\omega t} ,$$

$$\vec{\mathcal{E}}_2 = \mathcal{E}_{20}\left\{\hat{x} + \frac{q_2^2c^2 - 4\pi i\omega\sigma'_{xx}}{4\pi i\omega\sigma'_{xy}}\hat{y} - \left[\frac{(q_2^2c^2 - 4\pi i\omega\sigma'_{xx})\sigma_{zy} + 4\pi i\omega\sigma'_{xy}\sigma_{zx}}{4\pi i\omega\sigma'_{xy}\sigma_{zz}}\right]\hat{z}\right\}e^{iq_2z - i\omega t} . \tag{B13}$$

The amplitudes of the transmitted wave, $\mathcal{E}_{10}$ and $\mathcal{E}_{20}$, can be obtained in terms of the amplitudes of the incident wave $\vec{\mathcal{E}}^I$ by requiring the tangential field components to be continuous at the boundary. There are incident, reflected, and transmitted electric fields on the surface $z = 0$:

$$\begin{aligned}
\vec{\mathcal{E}}^I &= (\mathcal{E}_x^I\hat{x} + \mathcal{E}_y^I\hat{y})e^{iq_0z - i\omega t} , \\
\vec{\mathcal{E}}^R &= (\mathcal{E}_x^R\hat{x} + \mathcal{E}_y^R\hat{y} + \mathcal{E}_z^R\hat{z})e^{-iq_0z - i\omega t} , \\
\vec{\mathcal{E}}^T &= \mathcal{E}_{10}(\hat{x} + \alpha_1\hat{y} + \beta_1\hat{z})e^{iq_1z - i\omega t} \\
&\quad + \mathcal{E}_{20}(\hat{x} + \alpha_2\hat{y} + \beta_2\hat{z})e^{iq_2z - i\omega t} ,
\end{aligned} \tag{B14}$$

where I indicates the incident wave propagating along $\hat{z}$ with wave vector $q_0 = \omega/c$, R indicates the reflected wave traveling along $-\hat{z}$ with wave vector $-q_0$, and T indicates the transmitted wave. Equation (B1) requires the microwave magnetic field to have x and y components only:

$$\begin{aligned}
\vec{H}^I &= (\mathcal{E}_y^I\hat{x} + \mathcal{E}_x^I\hat{y})e^{iq_0z - i\omega t} , \\
\vec{H}^R &= (\mathcal{E}_y^R\hat{x} - \mathcal{E}_x^R\hat{y})e^{-iq_0z - i\omega t} , \\
\vec{H}^T &= \frac{\mathcal{E}_{10}q_1}{q_0}(-\alpha_1\hat{x} + \hat{y})e^{iq_1z - i\omega t} + \frac{\mathcal{E}_{20}q_2}{q_0}(-\alpha_2\hat{x} + \hat{y})e^{iq_2z - i\omega t} ,
\end{aligned} \tag{B15}$$

where

$$\begin{aligned}
\alpha_i &= \frac{q_i^2c^2 - 4\pi i\omega\sigma'_{xx}}{4\pi i\omega\sigma'_{xy}} \\
\beta_i &= -\frac{(q_i^2c^2 - 4\pi i\omega\sigma'_{xx})\sigma_{zy} + 4\pi i\omega\sigma'_{xy}\sigma_{zx}}{4\pi i\omega\sigma'_{xy}\sigma_{zz}} .
\end{aligned} \tag{B16}$$

For the purposes of this appendix we treat potassium as a nearly-free-electron gas characterized by the following parameters: effective mass $m^* = 1.21m$, electron density $n = 1.4\times10^{22}$ cm^{-3}, Fermi radius $k_F = 0.75 \times 10^8$ cm^{-1}, and electron scattering time $\tau = 2.0 \times 10^{-10}$ s (which is appropriate at $T = 2.5$ K). The frequency of the applied microwave field is 23.9 GHz. Accordingly, $\omega\tau = 30$ is used for electrons on the spherical Fermi-surface. On account of the small velocity for electrons in the Fermi-surface cylinder $\omega\tau_c = 150$. (This value is required to fit the observed width of the cyclotron-resonance peak in Figure 58.1.)

The inequalities of Equation (B8) are so extreme that α_1 and α_2 differ from i and $-i$ by $\sim 10^{-6}$. Specifically,

$$\begin{aligned}
\alpha_1 &\approx i , \\
\alpha_2 &\approx -i .
\end{aligned} \tag{B17}$$

The ratio of the x or y component to the z component is about 100, so

$$\mathcal{E}_x^T, \mathcal{E}_y^T \gg \mathcal{E}_z^T \,. \tag{B18}$$

Calculation of $\mathcal{E}_{10}$ and $\mathcal{E}_{20}$ is straightforward by using the continuity of the tangential field at $z = 0$. The final results are

$$\begin{aligned}
\mathcal{E}_{10} &= \frac{-2q_0}{(\alpha_1 - \alpha_2)(q_0 + q_1)}(\alpha_2 \mathcal{E}_x^I - \mathcal{E}_y^I) \,, \\
\mathcal{E}_{20} &= \frac{2q_0}{(\alpha_1 - \alpha_2)(q_0 + q_2)}(\alpha_1 \mathcal{E}_x^I - \mathcal{E}_y^I) \,.
\end{aligned} \tag{B19}$$

The amplitude of the transmitted wave may be found by specifying the incident wave. For right-circular polarization,

$$\mathcal{E}_y^I = i\mathcal{E}_x^I \,. \tag{B20}$$

On account of the extreme inequality (B8), one mode dominates the other by a factor of at least 10^7 for all magnetic fields, i.e.,

$$\mathcal{E}_{10} \gg \mathcal{E}_{20} \,. \tag{B21}$$

For left-circular polarization,

$$\mathcal{E}_y^I = -i\mathcal{E}_x^I \,, \tag{B22}$$

and $\mathcal{E}_{20}$ is much larger than $\mathcal{E}_{10}$.

The magnetic field at $z = 0$ can be found from Equation (B15):

$$\begin{aligned}
H_x(0) &= \frac{2\alpha_1 q_1(\alpha_2 \mathcal{E}_x^I - \mathcal{E}_y^I)}{(\alpha_1 - \alpha_2)(q_0 + q_1)} - \frac{2\alpha_2 q_2(\alpha_1 \mathcal{E}_x^I - \mathcal{E}_y^I)}{(\alpha_1 - \alpha_2)(q_0 + q_2)} \,, \\
H_y(0) &= \frac{-2q_1(\alpha_2 \mathcal{E}_x^I - \mathcal{E}_y^I)}{(\alpha_1 - \alpha_2)(q_0 + q_1)} + \frac{2q_2(\alpha_1 \mathcal{E}_x^I - \mathcal{E}_y^I)}{(\alpha_1 - \alpha_2)(q_0 + q_2)} \,,
\end{aligned} \tag{B23}$$

From Equation (B17) and the fact that $q_1, q_2 \gg q_0$, the magnetic field at the surface is

$$\begin{aligned}
H_x(0) &\approx \pm 2i\mathcal{E}_x^I \quad \text{for } \mathcal{E}_y^I = \pm i\mathcal{E}_x^I \,, \\
H_y(0) &\approx 2\mathcal{E}_x^I \quad \text{for } \mathcal{E}_y^I = \pm i\mathcal{E}_x^I \,.
\end{aligned} \tag{B24}$$

Therefore the total current defined by Equation (58.40) is

$$\begin{aligned}
J_x &= \frac{c}{4\pi} H_y(0) \approx \frac{c}{2\pi}\mathcal{E}_x^I \,, \\
J_y &= -\frac{c}{4\pi} H_x(0) \approx \pm i\frac{c}{2\pi}\mathcal{E}_x^I \,.
\end{aligned} \tag{B25}$$

The foregoing results are incorporated in the calculations of Section 58.3 at Equations (58.43) and (58.44). It must be appreciated that the Fermi-sphere electrons are treated nonlocally in Section 58.3. The purpose of this appendix is to show that the microwave modes in the metal are essentially circularly polarized (despite the broken axial symmetry caused by the tilt of the Fermi-surface cylinder) on account of the small value (4×10^{-4}) of η.

References

1 Grimes, C.C. and Kip, A.F. (1963) *Phys. Rev.* **132**, 1991.
2 Azbel, M.Ya. and Kaner, E.A. (1957) *Zh. Eksp. Teor. Fiz.* **32**, 896 [Sov. Phys. JETP **5**, 730 (1957)].
3 Mattis, D.C. and Dresselhaus, G. (1958) *Phys. Rev.* **111**, 403.
4 Chambers, R.G. (1965) *Philos. Mag.* **1**, 459.
5 Baraff, G.A., Grimes, C.C. and Platzman, P.M. (1969) *Phys. Rev. Lett.* **22**, 590.
6 Overhauser, A.W. (1968) *Phys. Rev.* **167**, 691.
7 Overhauser, A.W. (1978) *Adv. Phys.* **27**, 343; in *Highlights of Condensed-Matter Theory*, Proceedings of the International School of Physics "Enrico Fermi", Course LXXXIX, Varenna on Lake Como, 1983, edited by F. Bassani, F. Fumi, and M. P. Tosi (North-Holland, Amsterdam, 1985), p. 194.
8 Hwang, Y.G. and Overhauser, A.W. (1989) *Phys. Rev. B* **39**, 3037.
9 Overhauser, A.W. and Butler, N.R. (1976) *Phys. Rev. B* **14**, 3371.
10 Lacueva, G. and Overhauser, A.W. (1986) *Phys. Rev. B* **33**, 3765.
11 Dunifer, G.L., Sambles, J.F. and Mace, D.A.H. (1989) *J. Phys. Condens. Matter* **1**, 875.
12 Lacueva, G. and Overhauser, A.W. (1992) *Phys. Rev. B* **46**, 1273.
13 Grimes, C.C. (private communication).
14 Giuliani, G.F. and Overhauser, A.W. (1979) *Phys. Rev. B* **20**, 1328.
15 Giebultowicz, T.M., Overhauser, A.W. and Werner, S.A. (1990) *Phys. Rev. Lett.* **56**,2228 (1986); *Phys. Rev. B* **41**, 12536.
16 Blount, E.I. (1962) *Phys. Rev.* **126**, 1636.
17 Huberman, M. and Overhauser, A.W. (1982) *Phys. Rev. B* **25**, 2211.
18 Kittel, C. (1963) *Quantum Theory of Solids* (John Wiley and Sons, New York, 1963), p. 313.
19 Penz, P.A. and Kushida, T. (1968) *Phys. Rev.* **176**, 804.
20 Baraff, G.A. (1969) *Phys. Rev.* **187**, 851, first paragraph.
21 Coulter, P.G. and Datars, W.R. (1985) *Can. J. Phys.* **63**, 159.
22 Overhauser, A.W. (1984) *Phys. Rev. Lett.* **53**, 64.

Reprint 59 Cyclotron-Resonance Transmission Through Potassium in a Perpendicular Magnetic Field: Effects of the Charge-Density Wave[1)]

Mi-Ae Park and A.W. Overhauser**

* Department of Physics, Purdue University, 525, Northwestern Avenue, West Lafayette, Indiana 47907, USA

Received 24 June 1996

Microwave transmission through potassium by Dunifer, Sambles, and Mace [9] in a perpendicular magnetic field shows five signals. They are Gantmakher–Kaner (GK) oscillations, conduction-electron-spin resonance, high-frequency oscillations, cyclotron resonance, and cyclotron-resonance subharmonics. Only the spin resonance has been successfully explained using a free-electron model. However, such a model predicts GK oscillations which are too large by several orders of magnitude. Lacueva and Overhauser [12] have shown that charge-density-wave (CDW) energy gaps which cut through the Fermi surface reduce the GK signal. CDW gaps also create a small Fermi-surface cylinder. The high-frequency oscillations were shown to result from Landau-level quantization in the cylinder. Recently we found that the anomalous microwave surface resistance, observed by Grimes and Kip [10], can be explained only if the cylinder axis is tilted $\sim 45°$ with respect to the [110] crystal direction perpendicular to the surface. (Such a tilt was predicted by Giuliani and Overhauser [16].) In this study we show that oscillatory motions, parallel to the field, of electrons in the tilted cylinder cause the cyclotron-resonance transmission. This signal and its subharmonics would be completely absent without the tilt. Consequently, four of the five transmission signals require a CDW broken symmetry. [S0163-1829(97)04803-0]

59.1 Introduction

During the last 30 years a variety of magnetoconductivity anomalies have been discovered in potassium, the simplest monovalent metal [1, 2]. Unlike Li and Na, which undergo a crystallographic transformation to the 9R structure [3] when cooled to low temperature, a single crystal of K is not destroyed by cooling. Without a charge-density-wave (CDW) broken symmetry, the bcc lattice of K would support a spherical conduction-electron Fermi surface. Low-temperature transport anomalies could not then arise. Nevertheless, in dc experiments extraordinary phenomena occur which require the Fermi surface to be multiply connected. Examples are the four-peaked induced-torque patterns of single-crystal spheres [4, 5], the many open-orbit resonances [6, 7], and the magnetoserpentine effect [8].

It is not surprising, therefore, that unexpected phenomena also appear in the microwave properties of K. Figure 59.1 shows the microwave transmission signal (at a frequency, $\omega/2\tau\pi = 79.18\,\mathrm{GHz}$) through a K slab in a perpendicular magnetic field. Dunifer, Sambles, and Mace [9] studied this phenomenon in 15 samples at $T = 1.3\,\mathrm{K}$. The data shown (from sample K-4) was kindly selected by G.L. Dunifer, since it revealed clearly all five transmission signals. The horizontal axis of Figure 59.1 is ω_c/ω, which is proportional to the external magnetic field H, since $\omega_c = eH/m^*c$. The cyclotron mass is $m^* = 1.21\,\mathrm{m}$ [10], so the field for the cyclotron resonance, $\omega_c/\omega = 1$, is $H_c = 3.42\,\mathrm{T}$.

The signal at and near the conduction-electron-spin resonance has been explored extensively [11]. It is the only feature in Figure 59.1 that can be explained by a free-electron model.

1) Phys. Rev. B 55, 3 (1997-I).

Anomalous Effects in Simple Metals. Albert Overhauser

ISBN: 978-3-527-40859-7

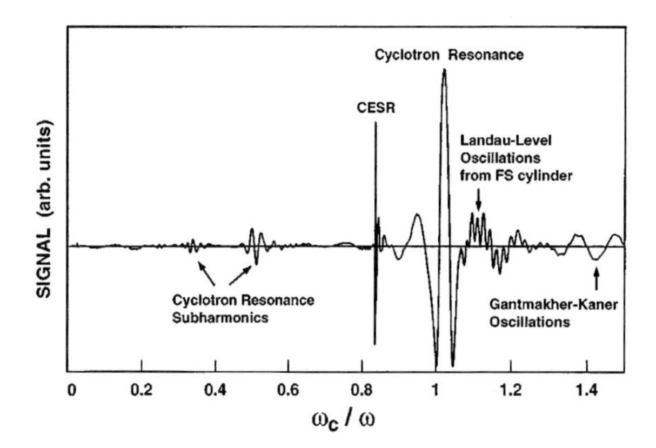

Figure 59.1 Microwave transmission signal vs. H through a potassium slab in a perpendicular magnetic field. ($H = 3.42\,\mathrm{T}$ at $\omega_c/\omega = 1$.) The microwave frequency is 79.18 GHz, and the temperature is 1.3 K. The field at $\omega_c/\omega = 1$ is 3.42 T. The phase of the microwave reference was adjusted so that the cyclotron resonance is symmetric. The slab thickness is $L = 85\,\mu\mathrm{m}$. The data, provided by G.L. Dunifer, were obtained from sample K4, one of 15 samples listed in [9].

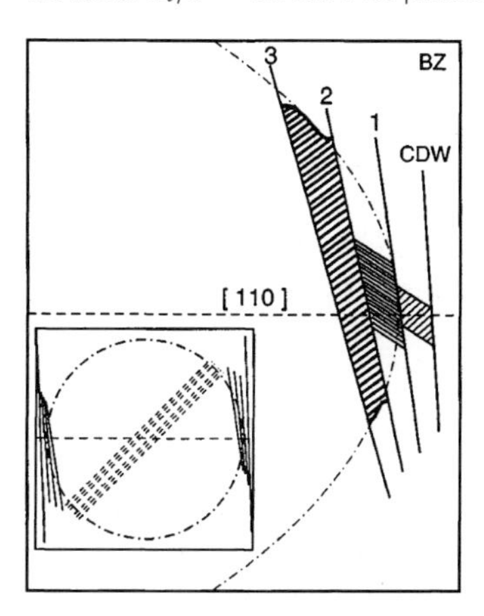

Figure 59.2 Fermi-surface cylinders of potassium. The horizontal axis is parallel to [110] and to the dc magnetic field. The shaded cylinders are created by the CDW gap and the first three minigaps. The thicknesses of the cylinders have been exaggerated by a factor of 10. Each of the half-cylinders shown is joined to a partner on the opposite side of the Fermi surface by Bragg reflection at the energy-gap planes. The complete Brillouin zone is shown in the inset.

Gantmakher–Kaner (GK) oscillations should also appear, but their amplitude should be a hundred times larger [12]. The cyclotron resonance (CR), its subharmonics, and the rapid oscillations shown near $\omega_c/\omega = 1.2$ should not even exist (without a CDW).

The purpose of this study is to show that the CDW broken symmetry of potassium explains CR transmission and CR subharmonics. One must, of course, solve self-consistently a Boltzmann transport equation (for conduction electrons) together with Maxwell's equations. Of crucial importance is the influence of CDW energy gaps on the Fermi-surface topology. A schematic illustration of the complexity introduced by the CDW is shown in Figure 59.2.

It is known from observed optical anisotropy [13] that K has a single CDW. The CDW wave vector $\vec{Q}$ is tilted about a degree from a [110] direction. From detection of neutron diffraction satellites [14, 15], it was found that

$$\vec{Q} = (0.995, 0.975, 0.015)\frac{2\pi}{a} \; . \tag{59.1}$$

Table 59.1 Sizes of CDW minigaps.

Order of gap	Minigap (meV)
1	90
2	67
3	51
4	34
5	15
7	0.6
8	0.06

The magnitude of $\vec{Q}$ is 1.5% smaller than that of the smallest reciprocal-lattice vector $\vec{G}_{110}$. The phonon mode which screens the electronic CDW has wave vector

$$\vec{Q}' = \vec{G}_{110} - \vec{Q} . \tag{59.2}$$

Minimization of the elastic energy required to neutralize the electronic CDW leads to a tilt of $\vec{Q}'$ about 45° away from [110] [16]. The "heterodyne gaps", shown by the dashed lines of the inset in Figure 59.2, are created by the periodicities, $n\vec{Q}'$, $n = 1, 2, \ldots$ They are important in explaining the open-orbit spectra [7] and in understanding the shape of the CR signal observed in perpendicular-field microwave surface resistance [17]. The heterodyne gaps do not significantly affect the microwave transmission, so we shall ignore them in what follows.

The horizontal axis of Figure 59.2 is parallel to [110], parallel to the magnetic field, and perpendicular to the surface of the K slab. Although such samples are polycrystalline, they have recrystallized in contact with the smooth, amorphous-quartz plates used to hold the K slab in the window between the transmit and receive cavities [9]. It is known from low-energy electron diffraction that thin alkali-metal samples, so deposited (or recrystallized), are epitaxially oriented with close-packed (110) planes parallel to the surface. Furthermore, such surfaces are smooth enough for conduction electrons to be specularly reflected [18]. Interfacial energy will be optimized when $\vec{Q}$ is also perpendicular to the surface, and, therefore, nearly parallel to the [110] surface normal.

The important energy gaps for the present study are the main CDW gap and the sequence of "minigaps", shown in Figure 59.2. The wave vectors that describe these small gaps are

$$\vec{K} = (n+1)\vec{Q} - n\vec{G}_{110} . \tag{59.3}$$

The sizes of these gaps have been estimated theoretically [19], and are listed in Table 59.1. The minigaps create several cylindrical sheets of Fermi surface. Each of the three shaded surfaces shown in Figure 59.2, formed by the CDW gap and the first three minigaps are joined (by Bragg reflection) to equivalent surfaces on the opposite side of the Brillouin zone. Accordingly, each cylinder has twice the length that appears in the figure. The cross section of each cylinder in a plane perpendicular to [110], i.e., not perpendicular to the cylinder's axis, is circular.

The rapid oscillations near $\omega_c/\omega = 1.2$, shown in Figure 59.1, have been found to be periodic in $1/H$ to very high precision [20]. The periodicity corresponds to a cross-sectional area 69 times smaller than πk_F^2, the extremal area of the ideal Fermi sphere. Consequently this Landau-level oscillation pattern arises from a cylinder with a radius (in a plane perpen-

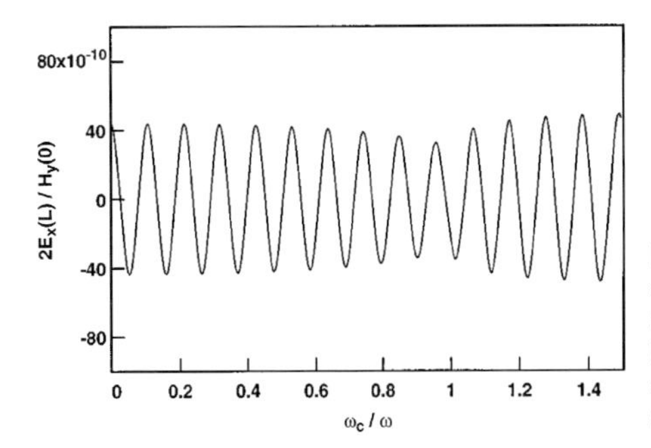

Figure 59.3 Theoretical microwave transmission signal vs. ω_c/ω for potassium if the Fermi surface is spherical. Only Gantmakher–Kaner oscillations appear (since the electron-spin magnetic moment is neglected). The sample parameters are $\omega\tau_0 = 150$ and $L = 85$ μm.

dicular to [110]) equal $k_F/8$. We believe this cylinder is the one formed by the first and second minigaps. The reason for this assignment is that the analogous, rapid oscillations near $\omega_c/\omega = 0.6$ corresponds to a cylinder with a radius, $\sim k_F/16$. This cylinder is likely the one formed by the CDW gap and the first minigap. The third cylindrical surface, shown in Figure 59.2, is the one to which we will attribute the CR transmission and the CR subharmonics. Its radius is estimated in what follows to be $\sim 3k_F/8$.

The influence of the minigaps has also been noticed in a study of the GK oscillations [12]. If a free-electron Fermi-surface sphere is employed, the transmitted power ratio will be 10^{-18} instead of the value, 10^{-22}, observed in sample K-4. Since the microwave transmission signal is carried primarily by electrons having a rapid velocity v_z parallel to [110], they are the ones most affected as they encounter the minigaps while undergoing cyclotron rotation in the $\hat{x}\hat{y}$ plane. The n'th GK oscillation occurs when the time L/v_F to traverse the slab (of thickness L) equals n cyclotron periods. Accordingly, the GK oscillations are periodic in H, as shown by the transmitted signal amplitude in Figure 59.3. This signal was calculated from a free-electron model (without spin). Not only are the features (described above) of the observed signal absent, but the gradual growth from small to large ω_c of the GK oscillations (a factor of 5 in Figure 59.1) does not appear.

The effect of the minigaps on the cyclotron motion of electrons with large $|v_z|$ can be modeled by a v_z-dependent scattering time [12],

$$\tau(v_z) = \frac{\tau_0}{1 + \gamma|v_z/v_F|}, \tag{59.4}$$

where τ_0 is the scattering time attributable to impurities, and γ is an adjustable parameter intended to account for the interruption of cyclotron motion by minigaps. The magnetic breakdown of electron trajectories at the small minigaps implies that γ will be a function of H. The simplest way to model this effect is to let

$$\gamma(H) = \gamma_0[1 - \exp(-H_0/H)], \tag{59.5}$$

where γ_0 is a constant and H_0 is a magnetic-breakdown field. Incorporation of Equations (59.4) and (59.5) into the Boltzmann equation allows one to reduce the GK signal by the required two orders of magnitude, and to fit its observed field dependence. This consequence of the minigaps is treated in Section 59.3.

The major challenge of this study, however, is to include the influence of a tilted Fermi-surface cylinder in the Boltzmann transport theory, solved self-consistently with Maxwell's equations. Complexity arises from the lack of axial symmetry about the magnetic-field direction. The general theory is developed in Section 59.2, and is applied to the microwave transmission of K in Section 59.4.

59.2 Microwave Transmission in an Anisotropic, Nonlocal Medium

Microwave propagation (which we take to be along $\hat{z}$, perpendicular to the metal surface) is governed of course by Maxwell's equations

$$\nabla \times \vec{\mathcal{E}} = -\frac{1}{c}\frac{\partial \vec{H}}{\partial t}\,, \quad \nabla \times \vec{H} = \frac{4\pi}{c}\vec{j} + \frac{1}{c}\frac{\partial \vec{\mathcal{E}}}{\partial t}\,. \tag{59.6}$$

These six equations can be reduced to three by taking the fields proportional to $e^{-i\omega t}$ and eliminating $\vec{H}$:

$$\frac{\partial^2 \mathcal{E}_\alpha(z)}{\partial z^2} + \frac{4\pi i\omega}{c^2} j_\alpha(z) + \frac{\omega^2}{c^2}\mathcal{E}_\alpha(z) = 0\,, \quad \alpha = x, y\,,$$

$$4\pi i\omega c^2 j_z + \frac{\omega^2}{c^2}\mathcal{E}_z = 0\,. \tag{59.7}$$

The current density $\vec{j}$ and the electric field $\vec{\mathcal{E}}$ depend only on z. However, since the conduction-electron mean free path $l = v_F\tau$ in potassium near $T = 0$ is typically $\sim 10^{-2}$ cm, whereas the microwave skin depth is approximately 2×10^{-5} cm, the relation between $\vec{j}$ and $\vec{\mathcal{E}}$ is nonlocal:

$$j_l(z) = \sum_{m=1}^{3} \int_0^L K_{lm}(z, z', \omega)\mathcal{E}_m(z')dz'\,. \tag{59.8}$$

$K_{lm}(z, z', \omega)$ is the (nonlocal) conductivity tensor, and L is the thickness of the metal slab.

Specular reflection of conduction electrons at smooth K surfaces [18] allows a simplification of Equation (59.8) if the Fermi surface has axial symmetry about the surface normal (and the dc field $\vec{H}$). The integration limits can be extended to infinity provided the microwave field is described by a Fourier cosine series (so that $\vec{\mathcal{E}}$ and $\vec{j}$ are symmetric about $z = 0$, L) [18]. Accordingly,

$$j_l(z) = \sum_{m=1}^{3} \int_{-\infty}^{\infty} K_{lm}(z - z', \omega)\mathcal{E}_m(z')dz'\,. \tag{59.9}$$

The Fourier expansions for $\vec{\mathcal{E}}$ and $\vec{j}$ are

$$\mathcal{E}_\alpha(z) = \sum_{n=0}^{\infty} E_\alpha^n \cos(q_n z)\,, \quad j_\alpha(z) = \sum_{n=0}^{\infty} J_\alpha^n \cos(q_n z)\,, \tag{59.10}$$

where

$$q_n = \frac{n\pi}{L}\,, n = \text{integer}\,. \tag{59.11}$$

The fundamental reason Equation (59.9) is allowed arises from the fact that on specular reflection v_z changes sign but not its magnitude. The reflection symmetry of $\mathcal{E}_\alpha(z)$, Equation (59.10), implies that the past history of an electron approaching $z = 0$ from the right is the same as that imputed for an electron approaching from the left along a path which joins that of the original reflection.

In general Equation (59.9) would not be valid for a Fermi surface lacking axial symmetry, e.g., an ellipsoid having its axis tilted from the surface normal. However, for the special case of an anisotropic conductivity caused by a tilted, cylindrical Fermi surface, Equation (59.9) is still valid because v_z retains its magnitude upon reflection. The formalism developed here anticipates that our application in Section 59.4 will be for a Fermi surface having both a spherical and a (tilted) cylindrical piece, created by the CDW broken symmetry. Nevertheless, there is a hidden approximation which is discussed at the end of Section 59.3.

The Fourier coefficients in Equation (59.10) are obtained by multiplying each equation by $\cos q_m z$ and integrating over the interval (0,L),

$$E_\alpha^m = \frac{2-\delta_{0m}}{L}\int_0^L \mathcal{E}_\alpha(z)\cos(q_m z)dz\,,$$

$$J_\alpha^m = \frac{2-\delta_{0m}}{L}\int_0^L j_\alpha(z)\cos(q_m z)dz\,. \tag{59.12}$$

Furthermore, from Equations (59.9) and (59.10), the Fourier components of the conductivity tensor, $\sigma_{lm}(q_n,\omega)$, can be related to the coefficients in Equation (59.12)

$$J_l^n = \sigma_{lm}(q_n,\omega)E_m^n\,. \tag{59.13}$$

All nine components of σ_{lm} will be nonzero when, in Section 59.4, we include the effects of a tilted cylinder.

The electric-field components E_α^n are obtained by taking the Fourier cosine transform of Equation (59.7),

$$\left(q_n^2-\frac{\omega^2}{c^2}\right)E_x^n-\frac{4\pi i\omega}{c^2}J_x^n = \frac{2-\delta_{0n}}{L}\left[\mathcal{E}'_x(L)(-1)^n-\mathcal{E}'_x(0)\right]\,, \tag{59.14}$$

$$\left(q_n^2-\frac{\omega^2}{c^2}\right)E_y^n-\frac{4\pi i\omega}{c^2}J_y^n = \frac{2-\delta_{0n}}{L}\left[\mathcal{E}'_y(L)(-1)^n-\mathcal{E}'_y(0)\right]\,, \tag{59.15}$$

$$\frac{4\pi i\omega}{c^2}J_z^n+\frac{\omega^2}{c^2}E_z^n = 0\,. \tag{59.16}$$

Since $\mathcal{E}'_\alpha(L)$ is ten orders of magnitude smaller than $\mathcal{E}'_\alpha(0)$ in a slab 10 μm thick, we can neglect $\mathcal{E}'_\alpha(L)$. We next use Equation (59.13) to express J_α^n in terms of E_β^n. The longitudinal field can then be related to the transverse field from Equations (59.16) and (59.13),

$$E_z^n = -\frac{\sigma_{zx}E_x^n+\sigma_{zy}E_y^n}{1+\frac{4\pi i}{\omega}\sigma_{zz}}\,. \tag{59.17}$$

With the help of these substitutions Equations (59.14) and (59.15) may be written in a coupled format:

$$\begin{bmatrix} q^2-\frac{\omega^2}{c^2}-\frac{4\pi i\omega}{c^2}\sigma'_{xx} & -\frac{4\pi i\omega}{c^2}\sigma'_{xy} \\ -\frac{4\pi i\omega}{c^2}\sigma'_{yx} & q^2-\frac{\omega^2}{c^2}-\frac{4\pi i\omega}{c^2}\sigma'_{yy}\end{bmatrix}\begin{bmatrix}E_x^n\\E_y^n\end{bmatrix} = -\frac{2-\delta_{0n}}{L}\begin{bmatrix}\mathcal{E}'_x(0)\\ \mathcal{E}'_y(0)\end{bmatrix}, \tag{59.18}$$

where

$$\sigma'_{\alpha\beta} = \sigma_{\alpha\beta}-\frac{\sigma_{\alpha z}\sigma_{z\beta}}{\sigma_{zz}}\,. \tag{59.19}$$

The solution for the x component of the electric field is, for each Fourier component,

$$E_x^n = -\frac{c^2(2-\delta_{0n})}{L} \frac{(c^2 q_n^2 - \omega^2 - 4\pi i\omega\sigma'_{yy})\mathcal{E}'_x(0) + 4\pi i\omega\sigma'_{xy}\mathcal{E}'_y(0)}{(c^2 q_n^2 - \omega^2 - 4\pi i\omega\sigma'_{xx})(c^2 q_n^2 - \omega^2 - 4\pi i\omega\sigma'_{yy}) - (4\pi i\omega)^2\sigma'_{xy}\sigma'_{yx}} . \tag{59.20}$$

At this point we must specify the boundary conditions at the front surface of the sample, so that $\mathcal{E}'_x(0)$ and $\mathcal{E}'_y(0)$ in Equation (59.20) can be determined. Rectangular microwave cavities, excited in a TE_{101} mode, were used [9]. Consequently, from Equation (59.6),

$$\mathcal{E}'_x(0) = \frac{i\omega}{c} H_y(0) . \tag{59.21}$$

The magnetic field $\vec{H}_x(0)$ is zero, so that

$$\mathcal{E}'_y(0) = -\frac{i\omega}{c} H_x(0) = 0 . \tag{59.22}$$

Accordingly the solution, given by Equation (59.20), is completely specified. We need only to evaluate $\mathcal{E}_x(L)$ to find the transmitted signal.

From an analysis given by Cochran [22] and by Lacueva and Overhauser [12], the transmitted electric field just outside the rear surface is twice the incident field at L in the infinite medium. From Equations (59.10) and (59.20), the electric field at $z = L$ is, therefore,

$$\mathcal{E}_x(L) = -\frac{2c\omega i}{L} H_y(0) \times \sum_{n=0}^{\infty} \frac{(2-\delta_{0n})(c^2 q_n^2 - \omega^2 - 4\pi i\omega\sigma'_{yy})\cos(n\pi)}{(c^2 q_n^2 - \omega^2 - 4\pi i\omega\sigma'_{xx})(c^2 q_n^2 - \omega^2 - 4\pi i\omega\sigma'_{yy}) - (4\pi i\omega)^2\sigma'_{xy}\sigma'_{yx}} . \tag{59.23}$$

We now replace $\{q_n\}$ with their values from Equation (59.11), so that the integers $\{n\}$ appear:

$$\mathcal{E}_x(L) = -\frac{2\omega Li}{c\pi^2} H_y(0) \times \sum_{n=0}^{\infty} \frac{(2-\delta_{0n})(-1)^n\left[n^2 - \left(\frac{\omega L}{\pi c}\right)^2 - \left(\frac{4i\omega L^2}{\pi c^2}\right)\sigma'_{yy}\right]}{\left[n^2 - \left(\frac{\omega L}{\pi c}\right)^2 - \left(\frac{4i\omega L^2}{\pi c^2}\right)\sigma'_{xx}\right]\left[n^2 - \left(\frac{\omega L}{\pi c}\right)^2 - \left(\frac{4i\omega L^2}{\pi c^2}\right)\sigma'_{yy}\right] - \left(\frac{4i\omega L^2}{\pi c^2}\right)^2 \sigma'_{xy}\sigma'_{yx}} . \tag{59.24}$$

Since the sum in Equation (59.24) converges very slowly, one needs to include terms to $n = 10^7$ to obtain reliable values for $\mathcal{E}_x(L)$. Now, for large n, the σ'_{lm} terms are much smaller than n^2, so we can take advantage of this disparity as follows [21]:

$$\mathcal{E}_x(L) = -\frac{2\omega Li}{c\pi^2} H_y(0) \left\{ \sum_{n=0}^{M} \frac{(2-\delta_{0n})(-1)^n\left[n^2 - \left(\frac{\omega L}{\pi c}\right)^2 - \left(\frac{4i\omega L^2}{\pi c^2}\right)\sigma'_{yy}\right]}{\left[n^2 - \left(\frac{\omega L}{\pi c}\right)^2 - \left(\frac{4i\omega L^2}{\pi c^2}\right)\sigma'_{xx}\right]\left[n^2 - \left(\frac{\omega L}{\pi c}\right)^2 - \left(\frac{4i\omega L^2}{\pi c^2}\right)\sigma'_{yy}\right] - \left(\frac{4i\omega L^2}{\pi c^2}\right)^2 \sigma'_{xy}\sigma'_{yx}} + 2\sum_{n=M+1}^{\infty} \frac{(-1)^n}{n^2} \right\} . \tag{59.25}$$

The break point M is large. The final sum on the right-hand side can be reexpressed by using the exact value,

$$\sum_{n=1}^{\infty} \frac{(-1)^n}{n^2} = \sum_{n=1}^{M} \frac{(-1)^n}{n^2} + \sum_{n=M+1}^{\infty} \frac{(-1)^n}{n^2} = -\frac{\pi^2}{12} . \tag{59.26}$$

Therefore the transmitted field is

$$\mathcal{E}_x(L) = \frac{2\omega L i}{c\pi^2} H_y(0) \left\{ \frac{\pi^2}{6} + 2\sum_{n=1}^{M} \frac{(-1)^n}{n^2} \right.$$
$$\left. - \sum_{n=0}^{M} \frac{(2-\delta_{0n})(-1)^n \left[n^2 - \left(\frac{\omega L}{\pi c}\right)^2 - \left(\frac{4i\omega L^2}{\pi c^2}\right)\sigma'_{yy}\right]}{\left[n^2 - \left(\frac{\omega L}{\pi c}\right)^2 - \left(\frac{4i\omega L^2}{\pi c^2}\right)\sigma'_{xx}\right]\left[n^2 - \left(\frac{\omega L}{\pi c}\right)^2 - \left(\frac{4i\omega L^2}{\pi c^2}\right)\sigma'_{yy}\right] - \left(\frac{4i\omega L^2}{\pi c^2}\right)^2 \sigma'_{xy}\sigma'_{yx}} \right\} . \tag{59.27}$$

On account of the large reflectivity, the field at the front surface, $H_y(0)$, is twice the incident field H_y^I [12]. In the vacuum, just outside of the metal, $\mathcal{E}_x^I = H_y^I$. Therefore the signal, defined to be the ratio of the transmitted electric field (just beyond $z = L$) to the incident field at $z = 0$, is

$$S = \frac{\mathcal{E}_x(L)}{\mathcal{E}_x^I} = \frac{\mathcal{E}_x(L)}{H_y^I} = \frac{2\mathcal{E}_x(L)}{H_y(0)} . \tag{59.28}$$

From Equations (59.27) and (59.28), the signal is

$$S = \frac{4\omega L i}{c\pi^2} \left\{ \frac{\pi^2}{6} + 2\sum_{n=1}^{M} \frac{(-1)^n}{n^2} \right.$$
$$\left. - \sum_{n=0}^{M} \frac{(2-\delta_{0n})(-1)^n \left[n^2 - \left(\frac{\omega L}{\pi c}\right)^2 - \left(\frac{4i\omega L^2}{\pi c^2}\right)\sigma'_{yy}\right]}{\left[n^2 - \left(\frac{\omega L}{\pi c}\right)^2 - \left(\frac{4i\omega L^2}{\pi c^2}\right)\sigma'_{xx}\right]\left[n^2 - \left(\frac{\omega L}{\pi c}\right)^2 - \left(\frac{4i\omega L^2}{\pi c^2}\right)\sigma'_{yy}\right] - \left(\frac{4i\omega L^2}{\pi c^2}\right)^2 \sigma'_{xy}\sigma'_{yx}} \right\} . \tag{59.29}$$

We have taken the upper limit of n to be $M = 50\,000$, since doubling M does not change S by more than one part in 10^5.

Equation (59.29) can be evaluated once the conductivity tensor $\sigma'_{lm}(q_n, \omega)$, defined by Equations (59.13) and (59.19), is specified. In Section 59.3 we shall determine σ'_{lm} for the main (spherical) part of potassium's Fermi surface, including the influence of the minigaps on the scattering time $\tau(v_z)$, discussed in Section 59.1. In Section 59.4 the contribution of the tilted cylindrical part of the Fermi surface will be incorporated.

59.3 Effect of Minigaps on Microwave Transmission

The conductivity tensor derived in this section will be that for a spherical Fermi surface. However, the effect of the minigaps will be modeled by a v_z-dependent relaxation time described by Equations (59.4) and (59.5). The Boltzmann transport equation for the electron distribution function $f(\vec{k}, \vec{r}, t)$ is

$$\frac{\partial f}{\partial t} + \vec{v} \cdot \nabla_{\vec{r}} f - \frac{e}{\hbar}\left[\vec{\mathcal{E}} + \frac{\vec{v}}{c} \times \vec{H}\right] \cdot \nabla_{\vec{k}} f = \frac{f_0 - f}{\tau} , \tag{59.30}$$

where $f_0(\vec{k})$ is the equilibrium distribution. For the linear response, one takes

$$f(\vec{k}, \vec{r}, t) = f_0(\vec{k}) + f_1(\vec{k}, \vec{r}, t) , \qquad (59.31)$$

with the understanding that f_1 is first order in the microwave field. Accordingly for each Fourier component,

$$\vec{\mathcal{E}}(\vec{r}, t) = \vec{E}(\vec{q}, \omega) \exp(i\vec{q} \cdot \vec{r} - i\omega t) ,$$

$$f_1(\vec{k}, \vec{r}, t) = f_{\vec{q}}(\vec{k}) \exp(i\vec{q} \cdot \vec{r} - i\omega t) . \qquad (59.32)$$

The linearized transport equation is then

$$[1 + i(\vec{q} \cdot \vec{v} - \omega)\tau] f_{\vec{q}} + \frac{eH}{\hbar c} \tau (\vec{v} \times \nabla_{\vec{k}}) f_{\vec{q}} = e\tau \vec{E} \cdot \vec{v} \frac{\partial f_0}{\partial \varepsilon} . \qquad (59.33)$$

It is convenient to change coordinates from $\vec{k}$ to ε, k_z, ϕ, where

$$\varepsilon = \frac{\hbar^2}{2m^*}(k_x^2 + k_y^2 + k_z^2) , \quad \phi = \tan^{-1}(k_y/k_x) ,$$

$$k_x = \left(\frac{2m^* \varepsilon}{\hbar^2} - k_z^2 \right)^{1/2} \cos\phi ,$$

$$k_y = \left(\frac{2m^* \varepsilon}{\hbar^2} - k_z^2 \right)^{1/2} \sin\phi , \quad k_z = k_z . \qquad (59.34)$$

ϕ is the azimuthal angle of an electron in its cyclotron orbit. On introduction of these variables, the last term on the left-hand side of Equation (59.33) reduces to $\omega_c(\partial f_{\vec{q}}/\partial\phi)$, so

$$[1 + i(\vec{q} \cdot \vec{v} - \omega)\tau] f_{\vec{q}} + \omega_c \tau \frac{\partial f_{\vec{q}}}{\partial \phi} = e\tau \vec{E} \cdot \vec{v} \frac{\partial f_0}{\partial \varepsilon} . \qquad (59.35)$$

The velocity in Equation (59.35) is

$$\vec{v}(\vec{k}) = \frac{1}{\hbar} \frac{\partial \varepsilon}{\partial \vec{k}} . \qquad (59.36)$$

Accordingly,

$$v_x = \frac{\hbar}{m^*} k_x = \frac{\hbar}{m^*} \left(\frac{2m^* \varepsilon}{\hbar^2} - k_z^2 \right)^{1/2} \cos\phi ,$$

$$v_y = \frac{\hbar}{m^*} k_y = \frac{\hbar}{m^*} \left(\frac{2m^*}{\hbar^2} - k_z^2 \right)^{1/2} \sin\phi , \quad v_z = \frac{\hbar}{m^*} k_z . \qquad (59.37)$$

Only propagation vectors $\vec{q} = q\hat{z}$ are of interest. Since we shall assume here that v_z is independent of ϕ, Equation (59.35) is easily solved:

$$f_{\vec{q}} = \frac{e}{\omega_c} \frac{\partial f_0}{\partial \varepsilon} \int_{-\infty}^{\phi} d\phi' \vec{E} \cdot \vec{v} \exp \left\{ \frac{[1 + i\tau(qv_z - \omega)]}{\omega_c \tau} (\phi' - \phi) \right\} . \qquad (59.38)$$

The Fourier component of the current density is, naturally,

$$\vec{J}(\vec{q},t) = -\frac{2e}{(2\pi)^3}\int f_{\vec{q}}\vec{v}\mathrm{d}^3k\,, \tag{59.39}$$

and the relation between the current density and electric field is

$$J_i = \sum_{m=1}^{3} \sigma_{im} E_m\,. \tag{59.40}$$

The tensor components of the conductivity are then

$$\begin{aligned}\sigma_{im} &= -\frac{2e^2 m^*}{\hbar^2\omega_c(2\pi)^3}\int d\varepsilon\frac{\partial f_0}{\partial\varepsilon}\int \mathrm{d}k_z\int_0^{2\pi}\mathrm{d}\phi\, v_i(\varepsilon,k_z,\phi)\\ &\times\int_{-\infty}^{\phi}\mathrm{d}\phi' v_m(\varepsilon,k_z,\phi')\\ &\times\exp\left\{\frac{[1+i\tau(qv_z-\omega)]}{\omega_c\tau}(\phi'-\phi)\right\}\,.\end{aligned} \tag{59.41}$$

Here the volume element d^3kin Equation (59.39) has been transformed to $(m^*/\hbar^2)d\varepsilon dk_z d\phi$, appropriate for the cylindrical coordinates of Equation (59.34). Three of the four integrations in Equation (59.41) can be evaluated after use of Equations (59.37) and (59.4). All nine components of the conductivity tensor, $\sigma_{im}(q,\omega)$, can be identified:

$$\begin{aligned}\sigma_{xx} &= \frac{3\sigma_0}{8}\int_{-1}^{1}\mathrm{d}t\left(\frac{1-t^2}{1+\gamma|t|-ia_+ + ixt}+\frac{1-t^2}{1+\gamma|t|-ia_- + ixt}\right)\,,\\ \sigma_{xy} &= \frac{3i\sigma_0}{8}\int_{-1}^{1}\mathrm{d}t\left(\frac{1-t^2}{1+\gamma|t|-ia_+ + ixt}-\frac{1-t^2}{1+\gamma|t|-ia_- + ixt}\right)\,,\\ \sigma_{zz} &= \frac{3\sigma_0}{2}\int_{-1}^{1}\mathrm{d}t\frac{t^2}{1-i\omega\tau_0+ixt}\,,\\ \sigma_{yx} &= -\sigma_{xy}\,,\quad \sigma_{yy}=\sigma_{xx}\,,\sigma_{xz}=\sigma_{zx}=\sigma_{yz}=\sigma_{zy}=0\,,\end{aligned} \tag{59.42}$$

where

$$\begin{aligned} t &= \frac{k_z}{k_F}\,,\quad a_\pm = (\omega\pm\omega_c)\tau_0\,,\\ x &= qv_F\tau_0\,,\quad \sigma_0 = \frac{ne^2\tau_0}{m^*}\,.\end{aligned} \tag{59.43}$$

(The lower limit of the integration in $d\phi'$, which is the first to be executed, contributes nil on account of the exponential factor.) Notice that σ_{zz} depends only on τ_0, the impurity-scattering relaxation time, since we have allowed the minigaps to interrupt only the cyclotron motion, as given by Equation (59.4). The major influence of the minigaps on σ_{zz}, and the nil components of σ in Equation (59.42), will be treated in Section 59.4, where the existence and tilt of a Fermi surface cylinder will be incorporated.

It is fortunate that the remaining integrals in Equation (59.42) can be evaluated analytically since the transmitted signal, Equation (59.29), requires a further sum over the allowed values of q. The integrations are rather tedious; the final expressions are

$$\begin{aligned}
\sigma_{xx} = & \frac{3\sigma_0}{8}\left\{\frac{4(y^2-x^2)}{(y^2+x^2)^2} - \frac{2y}{y^2+x^2} + f_+p_+ \right. \\
& -g_+r_+ + f'_+p'_+ - g'_+r'_+ + f_-p_- - g_-r_- + f'_-p'_- - g'_-r'_- \\
& +i\left[f_+r_+ + g_+p_+ + f'_+r'_+ + g'_+p'_+ + f_-r_- + g_-p_- \right. \\
& \left.\left. + f'_-r'_- + g'_-p'_- - \frac{2(a_+ + a_-)(y^2-x^2)}{(y^2+x^2)^2}\right]\right\}, \\
\sigma_{xx} = & \frac{3\sigma_0}{8}\left\{\frac{2(a_+ - a_-)(y^2-x^2)}{(y^2+x^2)^2} + f_-r_- + g_-p_- + f'_-r'_- \right. \\
& + g'_-p'_- - f_+r_+ - g_+p_+ - f'_+r'_+ - g'_+p'_+ \\
& \left. +i\left[f_+p_+ - g_+r_+ + f'_+p'_+ - g'_+r'_+ - f_-p_- + g_-r_- - f'_-p'_- + g'_-r'_-\right]\right\},
\end{aligned} \tag{59.44}$$

where

$$\begin{aligned}
f_\pm &= \left[(1-a_\pm^2 - y^2 + x^2)(3y^2x - x^3) + 2(a_\pm + yx)(3y^2x - x^3)\right]/(x^2+y^2)^3, \\
g_\pm &= \left[2(a_\pm + yx)(y^3 - 3yx^2) + (1-a_\pm^2 - y^2 + x^2)(3y^2x - x^3)\right]/(x^2+y^2)^3, \\
p_\pm &= \frac{1}{2}\ln\left[\frac{(a_\pm - x)^2 + (y+1)^2}{(1+a_\pm^2)}\right], \\
r_\pm &= \tan^{-1}\left[\frac{a_\pm y + x}{y + 1 + a_\pm^2 - xa_\pm}\right].
\end{aligned} \tag{59.45}$$

Furthermore, $f'_\pm(x) = f_\pm(-x)$, $g'_\pm(x) = g_\pm(-x)$, $p'_\pm(x) = p_\pm(-x)$, and $r'_\pm(x) = r_\pm(-x)$, and

$$\sigma_{zz} = \frac{2\sigma_0}{2x^3}\left[2x - r + 2\omega\tau_0 p + \omega^2\tau_0^2 r + i\left(p - 2\omega\tau_0 x + 2\omega\tau_0 r - \omega^2\tau_0^2 p\right)\right], \tag{59.46}$$

where

$$\begin{aligned}
p &= \frac{1}{2}\ln\left[\frac{1+(x-\omega\tau_0)^2}{1+(x+\omega\tau_0)^2}\right], \\
r &= \tan^{-1}(x-\omega\tau_0) + \tan^{-1}(x+\omega\tau_0).
\end{aligned} \tag{59.47}$$

The results must now be incorporated into Equation (59.29) to obtain S. The transmitted signal shown in Figure 59.3 is obtained by setting $y = 0$, in Equations (59.4), (59.44), and (59.45), i.e., by neglecting all interruptions of the cyclotron motion by the minigaps. The impurity scattering rate corresponds to $\omega\tau_0 = 150$. The observed GK oscillations in Figure 59.1 are smaller in amplitude by 2–3 orders of magnitude [12]. What is perhaps more

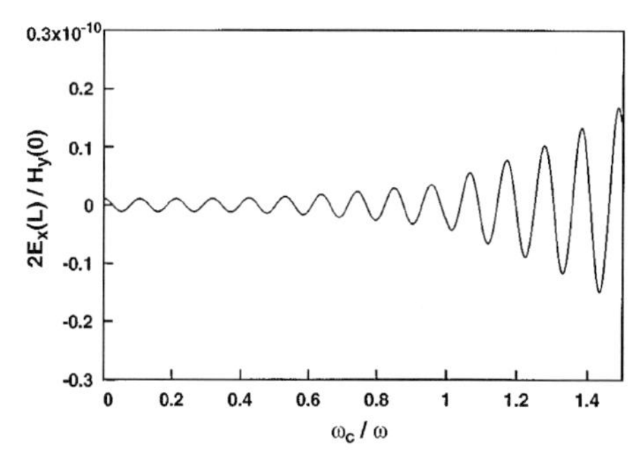

Figure 59.4 Theoretical transmission signal when interruption of cyclotron motion by the CDW minigaps is modeled by the v_z-dependent relaxation time, Equation (59.4), and with magnetic breakdown of the minigaps described by Equation (59.5). The parameters, $y_0 = 21$ and $H_0 = 5.8\,\mathrm{T}$, were adjusted so that the GK amplitude at $\omega_c/\omega = 1.47$ agrees with the (calibrated) data from Figure 59.1, and with its observed growth by a factor of 5 from low to high fields.

puzzling is the growth of the GK amplitude by about a factor of 5 between $\omega_c/\omega = 0.7$ and 1.5. We have adjusted $y_0 = 21$ and $H_0 = 5.8\,\mathrm{T}$ in Equation (59.5) to model the effect of magnetic breakdown of the minigaps. The revised transmission signal is shown in Figure 59.4. The gradual growth of S with increasing ω_c/ω, as shown in the figure, is typical for samples having a thickness $L = 85\,\mu\mathrm{m}$ (or larger). At very high fields the GK oscillations may decrease [9], as might be anticipated if there are slight variations in L across the sample surface.

We have explored the consequences of adding to the conductivity a contribution from a small Fermi-surface cylinder having its axis parallel to the applied field $\vec{H}$. The calculated signal was indistinguishable from the curve shown in Figure 59.4. This result justifies an approximation which we have not yet discussed. When an electron in the cylinder is reflected from the surface, v_x changes sign as well as v_z (but v_y does not). Consequently the $\hat{x}\hat{y}$ cyclotron motion is interrupted, and Equation (59.9) does not strictly apply. However, since the current in the $\hat{x}\hat{y}$ plane arising from the cylinder is too small to affect the transmission signal, as we have just mentioned, the error which has been tolerated is minimal. The CR transmission signal is caused by the $\hat{z}$ component of the cylinder's current, as shown in Section 59.4. Since the sign reversal of v_z on reflection is treated correctly by Equation (59.9), the nonlocal theory for the cylinder's conductivity, given below, should be adequate for microwave transmission (but not for a theory of the surface impedance).

59.4 Conductivity Tensor from a Tilted Fermi-Surface Cylinder

A striking consequence of the minigaps, which correspond to the periodicities of Equation (59.3), is the creation of small Fermi-surface cylinders, shown in Figure 59.2. Giuliani and Overhauser showed that the CDW wave vector $\vec{Q}$ is rotated from the [110] direction by a small angle, $\sim 1°$. The cause of this rotation is the need to minimize the elastic stress energy of the positive-ion lattice distortion which arises to screen the charge of the CDW. The wave vector, $\vec{Q}' = \vec{G}_{110} - \vec{Q}$, of the phonon mode involved is $\sim 45°$ from the [110]. The cylindrical Fermi surfaces created by the minigaps have axes parallel to $\vec{Q}'$ [19]. We will find below that a quantitative fit for the cyclotron resonance signals in Figure 59.1 requires the tilt angle of $\vec{Q}'$ to be $\theta \sim 50°$.

It is necessary to obtain the (nonlocal) conductivity of a tilted cylinder by solving once again the Boltzmann transport equation (59.30). The cross section of a tilted cylinder in a plane perpendicular to [110] is circular [19]. Accordingly the conduction-electron energy spectrum for the cylinder is

$$\varepsilon = \frac{\hbar^2}{2m^*}\left[(k_x - k_z \tan\theta)^2 + k_y^2\right] . \qquad (59.48)$$

It is again appropriate to change notation to cylindrical coordinates ε, k_z, and ϕ:

$$\phi = \tan^{-1}\left(\frac{k_y}{k_x - k_z \tan\theta}\right) ,$$
$$k_x = \left(\frac{2m^*\varepsilon}{\hbar^2}\right)^{1/2} \cos\phi + k_z \tan\theta ,$$
$$k_y = \left(\frac{2m^*\varepsilon}{\hbar^2}\right)^{1/2} \sin\phi . \tag{59.49}$$

The velocity components required for the transport equation are

$$v_x = \frac{\hbar}{m^*}(k_x - k_z \tan\theta) = \frac{\hbar}{m^*}\left(\frac{2m^*\varepsilon}{\hbar^2}\right)^{1/2} \cos\phi ,$$
$$v_y = \frac{\hbar}{m^*}k_y = \frac{\hbar}{m^*}\left(\frac{2m^*\varepsilon}{\hbar^2}\right)^{1/2} \sin\phi ,$$
$$v_z = \frac{\hbar}{m^*}(-\tan\theta)(k_x - k_z \tan\theta)$$
$$= \frac{\hbar}{m^*}(-\tan\theta)\left(\frac{2m^*\varepsilon}{\hbar^2}\right)^{1/2} \cos\phi . \tag{59.50}$$

Since v_z is here a function of ϕ, solution of the differential equation (59.35) for this case is, instead of Equation (59.38),

$$f_{\vec{q}} = \frac{e}{\omega_c}\frac{\partial f_0}{\partial\varepsilon}\int_{-\infty}^{\phi} \mathrm{d}\phi' \vec{E}\cdot\vec{v} \exp[i\alpha(\phi' - \phi) + i\beta(\sin\phi' - \sin\phi)] , \tag{59.51}$$

where

$$\alpha = -i\frac{1 - i\omega\tau_0}{\omega_c\tau_0} , \quad \beta = (-q\hbar\tan\theta/m^*\omega_c)\left(\frac{2m^*\varepsilon}{\hbar^2}\right)^{1/2} . \tag{59.52}$$

The relations (59.39) and (59.40) between the current density and electric field can be used to identify the tensor components of the cylinder's conductivity,

$$\sigma_{im} = -\frac{2e^2 m^*}{\hbar^2\omega_c(2\pi)^3}\int \mathrm{d}\varepsilon \frac{\partial f_0}{\partial\varepsilon}\int \mathrm{d}k_z \int_0^{2\pi} \mathrm{d}\phi\, v_i(\varepsilon, k_z, \phi)$$
$$\times \int_{-\infty}^{\phi} \mathrm{d}\phi' v_m(\varepsilon, k_z, \phi')$$
$$\times \exp[i\alpha(\phi' - \phi) + i\beta(\sin\phi' - \sin\phi)] . \tag{59.53}$$

The integration in $d\phi'$ is enabled by expanding $\exp(i\beta\sin\phi)$ in a Bessel function series [23],

$$\mathrm{e}^{i\beta\sin\phi} = \sum_{m=-\infty}^{\infty} J_m(\beta)\mathrm{e}^{im\beta} . \tag{59.54}$$

We provide details only for σ_{xx},

$$\sigma_{xx} = -\frac{2e^2}{(2\pi)^3}\frac{m^*}{\hbar^2\omega_c}\int \mathrm{d}\varepsilon \frac{\partial f_0}{\partial\varepsilon}\int \mathrm{d}k_z\, I_{xx} , \tag{59.55}$$

where

$$\begin{aligned} I_{xx} &= \frac{2\varepsilon}{m^*} \int_0^{2\pi} d\phi \cos\phi e^{-i\alpha\phi} \\ &\times \sum_{n=-\infty}^{\infty} J_n(-\beta) e^{in\phi} \int_{-\infty}^{\phi} d\phi' \cos\phi' e^{i\alpha\phi'} \\ &\times \sum_{m=-\infty}^{\infty} J_m(\beta) e^{im\phi'} \\ &= \frac{2\varepsilon}{m^*} \sum_{m=-\infty}^{\infty} J_m(\beta) J_{-m}(-\beta) \frac{i(m+\alpha)\pi}{1-(m+\alpha)^2} . \end{aligned} \tag{59.56}$$

Using the relations [23],

$$J_m(-\beta) = (-1)^m J_m(\beta) , \quad J_{-m}(\beta) = (-1)^m J_m(\beta) , \tag{59.57}$$

and letting δ be the length of the cylinder along k_z (i.e., twice the length shown in Figure 59.2), and with k_c the radius of the circular cross section (in the plane perpendicular to $\hat{z}$), we may evaluate Equation (59.55):

$$\sigma_{xx} = \frac{2e^2}{(2\pi)^3} \frac{m^*}{\hbar^2 \omega_c} \sum_{m=-\infty}^{\infty} J_m^2(\beta_F) \frac{i(m+\alpha)\pi}{1-(m+\alpha)^2} k_c^2 \delta , \tag{59.58}$$

where

$$\beta_F = (-q\hbar \tan\theta / m^* \omega_c) k_c . \tag{59.59}$$

Let η be the fraction of the conduction electrons contained in the cylinder, i.e.,

$$\eta = \frac{\pi k_c^2 \delta}{\frac{4}{3}\pi k_F^3} . \tag{59.60}$$

Equations (59.52), (59.58), and (59.60) then determine the component σ_{xx} of the cylinder's conductivity tensor,

$$\sigma_{xx} = \eta \frac{ne^2\tau_0}{m^*} \sum_{m=-\infty}^{\infty} J_m^2(\beta_F) \frac{1 - i(\omega - m\omega_c)\tau_0}{(\omega_c\tau_0)^2 + [1 - i(\omega - m\omega_c)\tau_0]^2} . \tag{59.61}$$

In a similar fashion, the eight remaining components can be found:

$$\begin{aligned} \sigma_{xy} &= \eta \frac{ne^2\tau_0}{m^*} \sum_{m=-\infty}^{\infty} J_m^2(\beta_F) \times \frac{-\omega_c\tau_0}{(\omega_c\tau_0)^2 + [1 - i(\omega - m\omega_c)\tau_0]^2} , \\ \sigma_{xz} &= -\sigma_{xx} \tan\theta , \quad \sigma_{yx} = -\sigma_{xy} , \quad \sigma_{yy} = \sigma_{xx} , \\ \sigma_{yz} &= \sigma_{xy} \tan\theta , \quad \sigma_{zx} = -\sigma_{xx} \tan\theta , \\ \sigma_{zy} &= -\sigma_{xy} \tan\theta , \quad \sigma_{zz} = \sigma_{xx} \tan^2\theta . \end{aligned} \tag{59.62}$$

The transmission signal S, including the effect of the tilted cylinder, can be calculated from Equation (59.29) by adding the conductivity tensor for the main Fermi surface, Equations (59.42)–(59.47), to the tensor just derived for the tilted cylinder, Equations (59.61) and (59.62). The microwave detector is sensitive to the phase of a reference signal from the

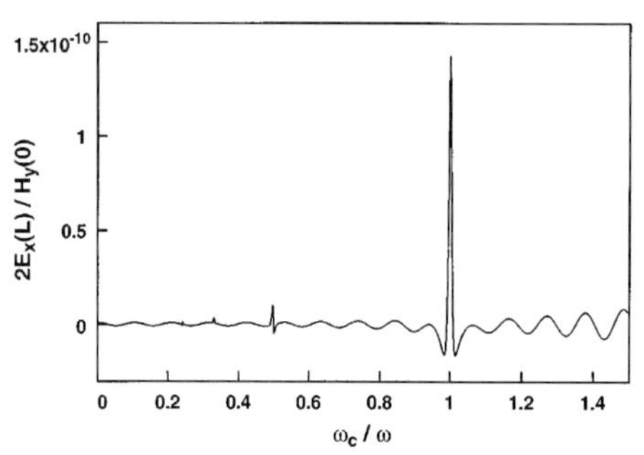

Figure 59.5 Theoretical transmission signal when the nonlocal conductivity of the largest Fermi-surface cylinder, shown in Figure 59.2, is added to the main Fermi-surface conductivity (employed in Figure 59.4). The tilt, $\theta = 50°$, of the cylinder's axis was adjusted so that ratio of the main CR to the first subharmonic (at $\omega_c/\omega = 0.5$) is $\sim$10, consistent with Figure 59.1. The cylinder radius $k_c = 3k_F/8$ in the (110) plane was adjusted so that the ratio of the CR to the high-field GK amplitude agrees with that observed in Figure 59.1. $\omega\tau_0 = 150$ and $L = 85$ μm.

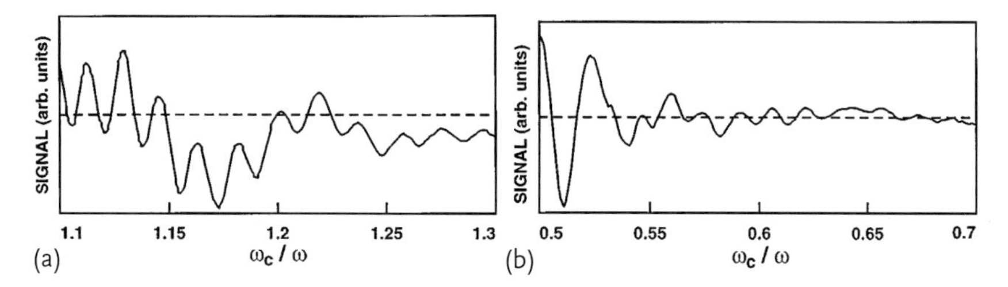

Figure 59.6 Landau-level oscillations near $\omega_c/\omega = 1.2$ and 0.6. The periodicity in (a) requires a cylinder radius, $k_c \sim k_F/8$, corresponding to the middle cylinder of Figure 59.2. The periodicity in (b) requires a cylinder radius, $k_c \sim k_F/16$, appropriate to the smallest cylinder in Figure 59.2.

klystron; and this phase can be adjusted (as was done for the data in Figure 59.1), so that the main cyclotron resonance appears symmetric [9]. Accordingly, the calculated electric field E^T of the transmitted signal will depend on a phase χ:

$$E^T = \mathrm{Re}(S)\cos\chi + \mathrm{Im}(S)\sin\chi\,. \tag{59.63}$$

The calculated cyclotron resonance signal, shown in Figure 59.5, is symmetric with $\chi = 280°$. The experimental [9] value for τ_0 corresponds to $\omega\tau_0 = 150$.

A crucial feature of Equations (59.61) and (59.62) is the resonance denominator, which becomes small whenever $\omega_c = \omega/(m+1)$, $m = 0, 1, 2, \ldots$ This feature is responsible for the occurrence of the cyclotron-resonance subharmonics (as well as the main resonance for $m = 0$), and is similar to the Azbel–Kaner oscillations [24] which occur in parallel-field surface resistance studies [25].

The parameters associated with the cylinder's geometry are δ, k_c, and θ. The cylinder's length is determined by the CDW diffraction satellites [14, 15], and is $0.015|\vec{G}_{110}|$. The ratio of the main cyclotron resonance to the first subharmonic (at $\omega_c/\omega = 0.5$) is sensitive to the tilt angle θ. The observed ratio $\sim$10 in Figure 59.1 is reproduced by the calculated ratio in Figure 59.5 with $\theta \sim 50°$. The absolute size of the main resonance is sensitive to k_c, which we find to be $\sim 3k_F/8$. This value, together with δ, indicates that the conduction-electron fraction of the third and largest cylinder shown in Figure 59.2 is $\eta \sim 0.004$. (The absolute size of the main CR is based on the amplitude of the GK oscillations near $\omega_c/\omega = 1.5$. The GK amplitude was determined in [12] from the original data and instrumental calibrations kindly provided by G.L. Dunifer.) Only the largest cylinder, shown in Figure 59.2, was included in the foregoing calculation. We have found that the influence of the smaller cylinders on the height of the main CR is $\sim$7%. However, their presence is necessary to account for the Landau-level oscillations near $\omega_c/\omega = 0.6$ and 1.2.

59.5 Conclusions

The complex Fermi surface of potassium, illustrated schematically in Figure 59.2, is based on energy-band calculations [19] that incorporate the periodic potential of an incommensurate CDW having the wave vector $\vec{Q}$, Equation (59.1). The presence of the three Fermi-surface cylinders, shown shaded (and with thickness exaggerated by a factor of 10), are manifested by the transmission data of Figure 59.1. The rapid oscillations near $\omega_c/\omega = 0.6$, arise from Landau-level quantization of the smallest cylinder (having a radius $\sim k_F/16$). The Landau-level oscillations near $\omega_c/\omega = 1.2$ arise from the next smallest cylinder, which has a radius $\sim k_F/8$ [20]. Figure 59.6 is an enlarged view of the data in Figure 59.1 showing the Landau-level oscillations near $\omega_c/\omega = 0.6$ and 1.2. The largest cylinder shown in Figure 59.2 has a radius $\sim k_F/8$, a value estimated from the main CR amplitude.

The shape of the main CR varies from sample to sample, as shown in Figure 59.3 of [9]. Resonances from thick samples, $L \geq= 100\,\mu m$ often exhibit structure. This feature is to be expected as a consequence of CDW domains [7]. Four possible CDW $\vec{Q}$'s are nearly parallel to a [110] direction. They can be obtained by interchanging the first two components of Equation (59.1) and by reversing the sign of the third component. It is likely that all four domains occur across the sample area. If the magnetic field is exactly parallel to [110], the cylinders of all four domains will have the same cyclotron frequency. However, if $\vec{H}$ deviates from [110] by even a fraction of a degree, each cylinder will have its own ω_c:

$$\omega_{ci} = \frac{\omega_c \cos\theta_i}{\cos\theta}\,, \quad i = 1, 2, 3, \text{ und } 4\,, \tag{59.64}$$

where Θ_i is the angle between the cylinder's axis $\vec{Q}'_i$ given by Equation (59.2), and the projection of $\vec{H}$ on the plane containing [110] and $\vec{Q}'_i$. (Θ is the angle each axis would have if $\vec{H}$ were exactly parallel to [110].) The CR line can be merely broadened if the deviations of Θ_i from Θ are very small. Splitting of spin-wave sidebands of the spin-resonance signal has also been attributed to CDW domain structure [26].

The major conclusion of this study is, of course, that transmission CR and the associated CR subharmonics arise in potassium from a cylindrical section of Fermi surface created by CDW minigaps. Without the CDW broken symmetry the transmission signal would be the one shown in Figure 59.3.

Acknowledgements

We are grateful to the National Science Foundation, Division of Materials Research for support.

References

1 Overhauser, A.W. (1985) in *Highlights of Condensed-Matter Theory, Proceedings of the International School of Physics "Enrico Fermi," Course LXXXIX, Varenna on Lake Como*, 1983, edited by F. Bassani, F. Fumi, and M. P. Tosi (North-Holland, Amsterdam, 1985), p. 194.

2 Overhauser, A.W. (1978) *Adv. Phys.* **27**, 343.

3 Overhauser, A.W. (1984) *Phys. Rev. Lett.* **53**, 64.

4 Schaefer, J.A. and Marcus, J.A. (1971) *Phys. Rev. Lett.* **27**, 935.

5 Holroyd, F.W. and Datars, W.R. (1975) *Can. J. Phys.* **53**, 2517.

6 Coulter, P.G. and Datars, W.R. (1985) *Can. J. Phys.* **63**, 159.
7 Huberman, M. and Overhauser, A.W. (1982) *Phys. Rev. B* **25**, 2211.
8 Overhauser, A.W. (1987) *Phys. Rev. Lett.* **59**, 1966.
9 Dunifer, G.L., Sambles, J.R. and Mace, D.A.H. (1989) *J. Phys. Condens. Matter* **1**, 875.
10 Grimes, C.C. and Kip, A.F. (1963) *Phys. Rev.* **132**, 1991.
11 Mace, D.A.H., Dunifer, G.L. and Sambles, J.R. (1984) *J. Phys. F* **14**, 2105.
12 Lacueva, G. and Overhauser, A.W. (1993) *Phys. Rev. B* **48**, 16935.
13 Overhauser, A.W. and Butler, N.R. (1976) *Phys. Rev. B* **14**, 3371.
14 Giebultowicz, T.M., Overhauser, A.W. and Werner, S.A. (1986) *Phys. Rev. Lett.* **56**, 2228.
15 Werner, S.A., Overhauser, A.W. and Giebultowicz, T.M. (1990) *Phys. Rev. B* **41**, 12536.
16 Giuliani, G.F. and Overhauser, A.W. (1979) *Phys. Rev. B* **20**, 1328.
17 Park, Mi-Ae and Overhauser, A.W. (1996) *Phys. Rev. B* **54**, 1597.
18 Penz, P.A. and Kushida, T. (1968) *Phys. Rev.* **176**, 804.
19 Hwang, Yong Gyoo and Overhauser, A.W. (1989) *Phys. Rev. B* **39**, 3037.
20 Lacueva, G. and Overhauser, A.W. (1992) *Phys. Rev. B* **46**, 1273.
21 Urquhart, B. and Cochran, J.F. (1986) *Can. J. Phys.* **64**, 796.
22 Cochran, J.F. (1970) *Can. J. Phys.* **48**, 370.
23 McLachlan, N.W. (1941) *Bessel Functions for Engineers* (Oxford University Press, London, 1941), p. 158.
24 Azbel, M.Ya. and Kaner, E.A. (1957) *Zh. Eksp. Teor. Fiz.* **32**, 896 [Sov. Phys. JETP **5**, 730 (1957)].
25 Mattis, D.C. and Dresselhaus, G. (1958) *Phys. Rev.* **111**, 403.
26 Hwang, Yong Gyoo and Overhauser, A.W. (1988) *Phys. Rev. B* **38**, 9011.

Reprint 60 Influence of Electron-Electron Scattering on the Electrical Resistivity Caused by Oriented Line Imperfections[1)]

M.L. Hubermann * *and A.W. Overhauser* *

* Department of Physics, Occidental College, 525, Northwestern Avenue, Los Angeles, California 90041, USA

** Department of Physics, Purdue University, West Lafayette, Indiana 47907, USA

Received 19 August 1996

By means of an exact solution of the Boltzmann transport equation, it is shown for a free-electron metal at low temperatures that electron–electron scattering has no effect (not even a T^2 term) on the electrical resistivity caused by oriented line imperfections. [S0163-1829(97)04007-1]

60.1 Introduction

It is of interest to consider the effect of electron–electron scattering on the electrical resistivity caused by oriented line imperfections. This question is not merely of academic interest, but is also related to a mechanism that has sometimes been invoked to explain anomalies in the low-temperature electrical resistivity of metals, as will be discussed in more detail later.

60.2 Theory

We consider a free-electron metal at low temperatures, supposing that the only scattering mechanisms are isotropic impurity scattering, scattering by oriented line imperfections, and electron–electron scattering. For a homogeneous system in a uniform applied electric field, the Boltzmann equation for the electron distribution $f_{\mathbf{k}}$ is

$$\frac{\partial f_{\mathbf{k}}}{\partial t} - \frac{e\mathbf{E}}{\hbar} \cdot \nabla_{\mathbf{k}} f_{\mathbf{k}} = \left(\frac{\partial f_{\mathbf{k}}}{\partial t} \right)_{\text{coll}} , \tag{60.1}$$

where $f_{\mathbf{k}}$ is the number of electrons having wave vector $\mathbf{k}$ and $(\partial f_{\mathbf{k}}/\partial t)_{\text{coll}}$ is the rate of change of the electron distribution due to scattering processes.[2)] It is convenient to represent the rate of change of the electron distribution due to collisions by a collision operator acting on the electron distribution, $C(f_{\mathbf{k}})$. Making this substitution gives

$$\frac{\partial f_{\mathbf{k}}}{\partial t} - \frac{e\mathbf{E}}{\hbar} \cdot \nabla_{\mathbf{k}} f_{\mathbf{k}} = C(f_{\mathbf{k}}) \, . \tag{60.2}$$

The total collision operator is the sum of the collision operators for scattering by isotropic impurities, scattering by oriented line imperfections, and electron–electron scattering. For scattering by static defects such as impurities and line imperfections, the collision operator in the Boltzmann equation has the form

$$C(f_{\mathbf{k}}) = \sum_{\mathbf{k}'} \left[W_{\mathbf{k}\mathbf{k}'} f_{\mathbf{k}'} \left(1 - f_{\mathbf{k}}\right) - W_{\mathbf{k}'-\mathbf{k}} f_{\mathbf{k}} \left(1 - f_{\mathbf{k}'}\right) \right] , \tag{60.3}$$

1) Phys. Rev. B 55, 7 (1997-I).

2) For an introduction to the Boltzmann transport equation, see, for example [1].

Anomalous Effects in Simple Metals. Albert Overhauser

ISBN: 978-3-527-40859-7

where $W_{\mathbf{k}'\mathbf{k}}$ is the transition rate for electron scattering from the plane-wave state $\mathbf{k}$ to the plane-wave state $\mathbf{k}'$.

Scattering by oriented line imperfections is treated first. For randomly distributed line imperfections, we make the usual approximation that interference terms from scattering by different line imperfections can be neglected. The transition rate $W_{\mathbf{k}'\mathbf{k}}$ for electron scattering by oriented line imperfections is then

$$W_{\mathbf{k}'\mathbf{k}} = N_d w_{\mathbf{k}'\mathbf{k}} \,, \tag{60.4}$$

where N_d is the number of line imperfections and $w_{\mathbf{k}'\mathbf{k}}$ is the transition rate for scattering by a single line imperfection. Because of translation invariance in the z direction, the transition rate $w_{\mathbf{k}'\mathbf{k}}$ for electron scattering by a line imperfection parallel to the z axis conserves the z component of wave vector k_z; moreover, for $k_z = k'$, the transition rate is independent of k_z. Since the scattering is elastic, electron energy is also conserved, i.e., $\varepsilon = \varepsilon'$. Because of cylindrical symmetry, $w_{\mathbf{k}'\mathbf{k}}$ does not depend upon the direction of wave-vector transfer (in the xy plane) $\mathbf{q} = \mathbf{k}' - \mathbf{k}$. It is assumed that $w_{\mathbf{k}'\mathbf{k}}$ is also independent of the magnitude of wave-vector transfer for $q \leq 2k_F$; this assumption corresponds to the leading term for low-energy scattering by a short-range potential, i.e., in an expansion of the scattering rate in powers of the electron energy in the xy plane, $E = \hbar^2(k_x^2 + k_y^2)/2m$.[3] In order to illustrate a scattering rate having these properties, we let $V(\rho)$ with $p = \sqrt{x^2 + y^2}$ be the scattering potential of a line imperfection, treat the scattering by $V(\rho)$ in Born approximation, and assume that the Fourier transform $\Lambda(q)$ of $V(\rho)$ is independent of q for $q \leq 2k_F$, giving

$$w_{\mathbf{k}'\mathbf{k}} = \frac{2\pi}{\hbar}\frac{\Lambda^2}{A^2}\delta_{k_z,k_z'}\,\delta\left(\varepsilon - \varepsilon'\right) \,, \tag{60.5}$$

where A is the cross-sectional area of the sample in the xy plane. Since $\Lambda(q)$ is constant for $q \leq 2k_F$, so is $w_{\mathbf{k}'\mathbf{k}}$. Conservation of k_z in Equation (60.5) implies that a current in the z direction is not degraded by scattering from line imperfections oriented in the z direction; this is the ultimate reason why oriented line imperfections do not contribute to the electrical resistivity in the direction parallel to their orientation.

Substituting Equations (60.4) and (60.5) into the Boltzmann collision operator, Equation (60.3), yields for the scattering by oriented line imperfections,

$$C_d(f_{\mathbf{k}}) = \frac{2\pi}{\hbar}\frac{n_d}{A}\Lambda^2\sum_{\mathbf{k}'}\delta_{k_z,k_z'}\,\delta\left(\varepsilon - \varepsilon'\right)\left(f_{\mathbf{k}'} - f_{\mathbf{k}}\right) \,, \tag{60.6}$$

where C_d is the collision operator for oriented line imperfections and n_d is the number of line imperfections per unit area. If $f_{\mathbf{k}} = g(\varepsilon)k_z$, where g depends only on the electron energy $\varepsilon = \hbar^2k^2/2m$, then it follows from conservation of energy and z component of the wave vector that

$$C_d\left[g(\varepsilon)k_z\right] = 0 \,, \tag{60.7}$$

i.e., $g(\varepsilon)k_z$ is an exact eigenfunction of the collision operator with eigenvalue 0. On the other hand, if $f_{\mathbf{k}} = g(\varepsilon)k_\alpha$ with $\alpha = x$ or y, then because the scattering (in the xy plane) is isotropic, the net scattering into the state $\mathbf{k}$ vanishes, leaving the net scattering out of the state $\mathbf{k}$. Equation (60.6) then gives

$$C_d\left[g(\varepsilon)k_\alpha\right] = -\frac{2\pi}{\hbar}n_d\Lambda^2\rho(E)g(\varepsilon)k_\alpha \,, \quad \alpha = x \text{ or } y \,, \tag{60.8}$$

3) This expansion is given by an effective range theory in two dimensions. For a derivation in three dimensions, see [2].

where $E = \hbar^2\left(k_x^2 + k_y^2\right)/2m$ is the electron energy in the xy plane and $\rho(E)$ is the two-dimensional density of states per unit area (for a single spin direction). But the two-dimensional density of states is a constant, independent of both E and k_z. If this were not the case, then different slices of the Fermi sphere parallel to the xy plane, having different values of E and k_z at the Fermi energy, would relax at different rates due to scattering by line imperfections, and the steady-state electron distribution in an applied electric field would be distorted from a uniformly shifted Fermi sphere. Substituting $\rho(E) = m/2\pi\hbar^2$ in Equation (60.8), we obtain

$$C_d\left[g(\varepsilon)k_\alpha\right] = -\frac{1}{\tau_d}\left[g(\varepsilon)k_\alpha\right], \quad \alpha = x \text{ or } y\,, \tag{60.9}$$

where $1/\tau_d = n_d\Lambda^2 m/\hbar^3$, i.e., $g(\varepsilon)k_\alpha$ with $\alpha = x$ or y is an exact eigenfunction of the collision operator with eigenvalue $-1/\tau_d$.[4)]

We treat next isotropic impurity scattering. Surprisingly, the eigenfunctions of the collision operator C_d for scattering by oriented line imperfections exhibited in Equations (60.7) and (60.9) are also exact eigenfunctions of the collision operator C_i for scattering by isotropic impurities. If $f_{\mathbf{k}} = g(\varepsilon)k_\alpha$ with $\alpha = x$, y, or z, then the effect of scattering by isotropic impurities is

$$C_i\left[g(\varepsilon)k_\alpha\right] = -\frac{1}{\tau_i(\varepsilon)}\left[g(\varepsilon)k_\alpha\right], \quad \alpha = x, y \text{ or } z\,, \tag{60.10}$$

where $1/\tau_i = \int n_i v\sigma(\theta)(1-\cos\theta)d\Omega$, n_i is the density of impurities, $v = \hbar k/m$ is the electron velocity, $\sigma(\theta)$ is the differential cross section for scattering by an impurity, and the integration is over solid angle Ω [4]. Although, in general, the relaxation time τ_i, depends upon electron energy, in a metal only the electron distribution near the Fermi surface is perturbed from equilibrium by an applied electric field. Consequently, for the electrical resistivity of a metal, τ_i is evaluated at the Fermi energy ε_F.

The effects of scattering by isotropic impurities and oriented line imperfections are now combined to derive the electrical resistivity. Expanding the steady-state electron distribution to the first order in the applied electric field,

$$f = f_0 + f_1\,, \tag{60.11}$$

where $f_0(\varepsilon)$ is the equilibrium Fermi distribution, substituting in the Boltzmann equation, Equation (60.2), and noting that the equilibrium distribution f_0 is unchanged by collisions, we obtain to the first order in the applied electric field,

$$-e\mathbf{E}\cdot\mathbf{v} = \frac{df_0}{d\varepsilon} = C_i(f_1) + C_d(f_1)\,. \tag{60.12}$$

Making use of the eigenfunctions of the collision operators in Equations (60.7), (60.9), and (60.10), we find that, if the applied electric field is in the z direction, the solution for f_1 is

$$f_1 = \tau_i e\mathbf{E}\cdot\mathbf{v}\frac{df_0}{d\varepsilon} \tag{60.13}$$

and the electrical resistivity is

$$\rho_{zz} = \frac{m}{ne^2\tau_i}\,; \tag{60.14}$$

4) This agrees with the result given by [3]. More generally, any distribution $f_{\mathbf{k}}$ can be decomposed into an axially symmetric part, obtained by averaging $f_{\mathbf{k}}$ over the azimuthal angle of the wave vector $\mathbf{k}$, and a nonsymmetric part, which is the remainder. Within the approximation that the transition rate is a constant, the symmetric and nonsymmetric parts are exact eigenfunctions of the collision operator with eigenvalues 0 and $-1/\tau_d$, respectively.

whereas, if the applied electric field is in the xy plane,

$$f_1 = \frac{\tau_i \tau_d}{\tau_i + \tau_d} \mathbf{E} \cdot \mathbf{v} \frac{d f_0}{d\varepsilon} \qquad (60.15)$$

and the electrical resistivity is

$$\rho_{xx} = \rho_{yy} = \frac{m}{ne^2} \left(\frac{1}{\tau_i} + \frac{1}{\tau_d} \right) . \qquad (60.16)$$

Two theorems are immediate. First, a comparison of Equations (60.14) and (60.16) proves theorem I: oriented line imperfections do not contribute to the electrical resistivity in the direction parallel to their orientation. Second, the solutions for f_1 in Equations (60.13) and (60.15) show that the steady-state electron distribution is a uniformly shifted Fermi sphere (to the first order in the applied electric field), even in the presence of anisotropic scattering by oriented line imperfections. Although the magnitude of the shift of the electron distribution depends upon the direction of the applied electric field, the shape of the shifted distribution remains spherical. But electron–electron collisions do not affect a uniformly shifted Fermi sphere, as is evident by making a Galilean transformation to the frame of reference in which the drift velocity of the electrons vanishes. This proves theorem II: electron–electron scattering has no effect (to the leading order) on the electrical resistivity caused by oriented line imperfections.

If one were to describe (incorrectly) the anisotropic scattering caused by oriented line imperfections by postulating an anisotropic wave-vector-dependent relaxation time $\tau_d(\theta)$, where $\tau_d(\theta)$ is a (non-negative) function of the angle θ between the wave vector $\mathbf{k}$ and the orientation axis of the line imperfections, then it is easily seen that both theorems proven above are violated. First, such a postulate implies that an electron excited at the Fermi surface with wave vector $\mathbf{k}$ is relaxed to equilibrium uniformly over the entire Fermi surface by collisions with oriented line imperfections. Thus the component of electron velocity parallel to the orientation axis would not be conserved by these collisions, leading to a nonzero contribution by line imperfections to the electrical resistivity in the direction parallel to their orientation, which contradicts theorem I. To show this explicitly, we substitute

$$C_d(f_1) = -\frac{f_1}{\tau_d(\theta)} \qquad (60.17)$$

in Equation (60.12) and solve for f_1, obtaining

$$f_1 = \frac{\tau_i \tau_d(\theta)}{\tau_i + \tau_d(\theta)} \mathbf{E} \cdot \mathbf{v} \frac{d f_0}{d\varepsilon} . \qquad (60.18)$$

Evidently, for $\mathbf{E} = E_z \hat{z}$, the electrical conductivity $\sigma_{zz} = j_z / E_z$ is affected by the scattering from line imperfections oriented in the z direction. Second, it is seen from the solution for f_1 in Equation (60.18) that the steady-state electron distribution is no longer a uniformly shifted sphere (to the first order in the applied electric field), but is distorted from its original spherical shape. Since the effect of electron–electron scattering is to restore the spherical shape of the electron distribution, electron–electron scattering would modify the steady-state electron distribution, thereby increasing the electrical resistivity caused by oriented line imperfections, which contradicts theorem II. (The increase of the resistivity when the distribution is modified follows from the variational theorem [5].) Thus the assumption of a wave-vector-dependent relaxation time, which is often a physically appealing approximation, leads in this instance to qualitatively incorrect conclusions, and therefore the exact eigenfunctions of the collision operator must be used.

60.3 Discussion

At temperatures sufficiently low that electron–phonon scattering can be neglected, the electrical resistivity of a metal is expected to have the form,

$$\rho = \rho_0 + AT^2 , \tag{60.19}$$

where ρ_0 is the temperature-independent residual resistivity due to scattering by impurities, dislocations, and other static defects; and AT^2 is the contribution to the resistivity from electron–electron scattering. This behavior has been observed in potassium metal below about 1.5 K.[5] Contrary to the expectation for a free-electron metal, however, the coefficient of the T^2 term was found to be sample dependent, varying by as much as a factor of 15. Two explanations of this sample dependence have been proposed. One attributes the temperature-dependent term to electron–electron scattering and electron–phason scattering in the presence of a charge-density wave; and explains the sample dependence as a consequence of the uncontrolled orientational texture of the charge-density-wave **Q**-domain distribution [7]. The other postulates a wave-vector-dependent relaxation time $r_d(\mathbf{k})$ due to scattering by oriented dislocations (the scattering being caused primarily by the dislocation cores) and attributes the temperature-dependent term to the effect of electron–electron scattering on the resistivity from oriented dislocations in addition to the direct effect of electron–electron scattering; the sample dependence is then a consequence of variations in the dislocation density [8–10].

Neutron-scattering experiments on potassium show that a well-annealed, carefully grown single crystal has a mosaic block structure, the block size being about 1 mm and the angular spread being about 0.1° [11, 12]. Dislocations occur mainly in the small-angle boundaries between adjacent mosaic blocks, the average distance between dislocations being about 500 lattice parameters and the average dislocation density being less than $10^6\,\mathrm{cm}^{-2}$. Since the distance between dislocations is much larger than the electron Fermi wavelength, the dislocations can be treated as independent scatterers. Since the mosaic block size is about 1 mm, it is likely that the dislocations are oriented over a length scale of an electron mean free path, which is about 0.1 mm in high-purity potassium at low temperatures. (In any case, this assumption is required by the mechanism proposed in [8, 9].) Since large-angle scattering, which is important for the electrical resistivity, is caused mainly by the dislocation core, a dislocation can be approximated by a short-range line potential [13]. Consequently the foregoing theory is applicable to the mechanism proposed in [8, 9]. It has thus been shown here, using the exact eigenfunctions of the Boltzmann collision operator, that this mechanism vanishes in the leading order. (Even if it were present in the leading order, the dislocation density required is at least several orders of magnitude too large, as pointed out by Gugan [14].)

Acknowledgements

We are grateful to the National Science Foundation, Division of Materials Research, for support.

5) For a review, see [6].

References

1. Ashcroft N.W. and Mermin N.D. (1976) *Solid State Physics*, Saunders College, Philadelphia.
2. Bethe H.A. (1949) *Phys. Rev.* **76**, 38.
3. Kaner E.A. and Feldman E.P. (1968) *Fiz. Tverd. Tela* (Leningrad) **10**, 3046. [*Sov. Phys. Solid State* **10**, 2401 (1969)].
4. Peierls R.E. (1955) *Quantum Theory of Solids*, Oxford University Press, London.
5. Ziman J.M. (1960) *Electrons and Phonons*, Oxford University Press, London.
6. Bass J., Pratt W.P., and Schroeder P.A. (1990) *Rev. Mod. Phys.* **62**, 645.
7. Bishop M.F. and Lawrence W.E. (1985) *Phys. Rev. B* **32**, 7009.
8. Kaveh M. and Wiser N. (1980) *J. Phys. F* **10**, L37.
9. Kaveh M. and Wiser N. (1982) *J. Phys. F* **12**, 935.
10. Kaveh M. and Wiser N. (1984) *Adv. Phys.* **33**, 257.
11. Overhauser A.W. (1971) *Phys. Rev. B* **3**, 3173, Sec. III.
12. Adlhart W., Stetter G., Fritsch G., and Luscher E. (1981) *J. Phys. F* **11**, 1347.
13. Brown R.A. (1981) *Can. J. Phys.* **60**, 766.
14. Gugan D. (1982) *J. Phys. F* **12**, L173.

Reprint 61 Theory of the Fourfold Induced-Torque Anisotropy in Potassium[1)]

A.W. Overhauser[*] and G. Lacueva[**]

[*] Department of Physics, Purdue University, 525, Northwestern Avenue, West Lafayette, Indiana 47907, USA

[**] Physics Department, John Carroll University, University Heights, Ohio 44118, USA

Received 15 July 2002

Induced-torque anisotropies observed in single-crystal spheres of potassium prove that the Fermi surface is multiply connected. Cyclotron orbits which intersect heterodyne gaps created by the charge-density-wave broken symmetry lead to an anisotropic Hall effect having longitudinal components. Thereby the theoretical induced torque (in a 360° magnet rotation) has four evenly spaced minima and four maxima with a staggered spacing. The maxima grow almost $\sim H^2$ and can be 30 times higher than the minima. All such features have been observed. (Details depend on crystal growth and orientation.) All are impossible in a spherical crystal with a simply connected Fermi surface.

Potassium, the simplest metal of all, is for solid-state theory the analog of the hydrogen atom.[2)] Virtually all workers, teachers, and authors believe that the conduction electrons are free-electron-like, having a Fermi surface that is essentially a perfect sphere. The experimental properties of K provide therefore a paradigm to test numerous aspects of the quantum theory of metals, including many-electron effects caused by Coulomb interactions.

Experimental data from diverse phenomena which severely contradict the Fermi-sphere model have accumulated during the last forty years. See [2–8] for a partial survey. These "anomalies" (about thirty in number) can be explained if K has a charge-density-wave (CDW) broken symmetry [9]. Such a broken symmetry is theoretically expected for an (otherwise bcc) alkali metal when the extreme nonlocal character of exchange and correlation [2, 10] is recognized. (A CDW cannot arise in a simple metal if commonly used local-density approximations to exchange and correlation are employed [11].)

Some of the energy gaps in $\mathbf{k}$ space that arise from an incommensurate CDW having wave vector $\mathbf{Q}$ are shown in Figure 61.1. $\mathbf{Q}$ is tilted slightly from the [110] direction of the reciprocal-lattice vector $\mathbf{G}_{110}$ [12] and Q is expected to be about 1.33 $(2\pi/a)$, i.e., ~6% smaller than G_{110}. The two families – heterodyne gaps and minigaps – are higher order gaps that arise naturally [13] when the Schrödinger equation for a conduction electron has both potentials included:

$$V(\mathbf{r}) = V_{11}\cos(\mathbf{G}\cdot\mathbf{r}) + V_Q\cos(\mathbf{Q}\cdot\mathbf{r})\ . \tag{61.1}$$

The Brillouin-zone energy gap is $V_{110} \sim 0.4\,\text{eV}$ and the CDW energy gap is $V_Q \sim 0.6\,\text{eV}$ [8]. The heterodyne gaps correspond to periodic potentials with wave vectors

$$\mathbf{G}'_n = n\mathbf{Q}'\,,\quad \mathbf{Q}' \equiv \mathbf{G}_{110} - \mathbf{Q}\,, \tag{61.2}$$

1) Phys. Rev. B 66, 165115 (2002).

2) K remains bcc on cooling to 4 K, unlike Li and Na, which transform to a rhombohedral $9R$ phase: [1].

Anomalous Effects in Simple Metals. Albert Overhauser

ISBN: 978-3-527-40859-7

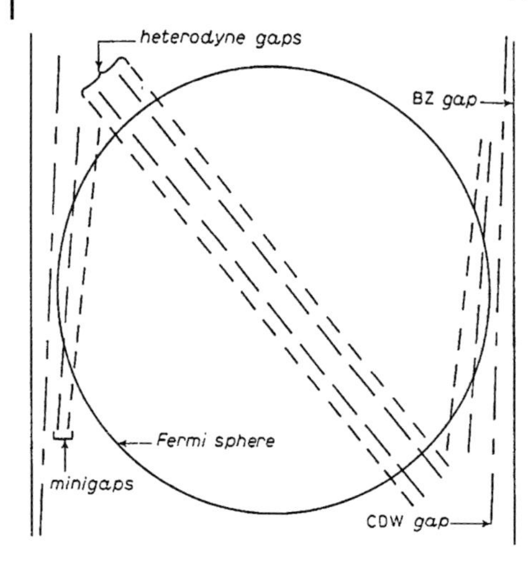

Figure 61.1 Energy-gap planes in the Brillouin zone of K created by the combined influence of the crystal potential $V_{110}\cos(\mathbf{G}_{110}\cdot\mathbf{r})$, and the CDW potential $V_Q\cos(\mathbf{Q}\cdot\mathbf{r})$. The wave vectors associated with the heterodyne gaps and minigaps are given by Equations (61.2) and (61.3). The two solid lines represent faces of the Brillouin zone, and are separated by $\mathbf{G}_{110}$.

$n = 1, 2, 3, \ldots$, and the minigaps have wave vectors

$$\mathbf{G}_n = (n+1)\mathbf{Q} - n\mathbf{G}_{110}\,, \tag{61.3}$$

$n = 1, 2, 3, \ldots$ There are also second-zone minigaps, but they do not intersect the Fermi surface. The minigaps and heterodyne gaps decrease rapidly with increasing n. Heterodyne gaps are likely in the meV range, and minigaps are an order of magnitude larger. Unfortunately, CDW diffraction satellites have not yet been seen in K, so the precise $\mathbf{Q}$ is unknown.

It is clear from Figure 61.1 that the Fermi surface will not only be nonspherical; it will be multiply connected (by virtue of the minigaps and heterodyne gaps). The LAK magnetotransport theorems [14, 15] cannot then apply. Consequently the magnetoresistance need not saturate when $\omega_c\tau \gg 1$, and the high-field Hall coefficient need not be exactly $-1/nec$. ($\omega_c = eH/mc$, is the cyclotron frequency, τ the scattering time, and n the electron density.) Furthermore, open orbits can occur.

The induced-torque technique is a very convenient method to study magnetoresistance since there is then no need for current or voltage leads. A spherical sample is placed in a horizontal magnetic field, which is then slowly rotated about a vertical axis. The induced currents lead to a torque N_y (about this axis) which is, for a metal with an isotropic resistivity ρ_0 (and a simply connected Fermi surface) [16]

$$N_y \frac{2\pi R^5 \Omega n^2 e^2 \rho_0}{15} \frac{(\omega_c\tau)}{1+\left(\frac{1}{2}\omega_c\tau\right)^2}\,. \tag{61.4}$$

R is the radius of the sphere and Ω the rotation rate of the magnet. The torque is isotropic, independent of the angle θ of the magnetic field $\mathbf{H}$ in the horizontal plane. N_y approaches a constant value when $\omega_c\tau \gg 1$. This is the required behavior for K if it has a spherical Fermi surface.

Schaefer and Marcus [17] discovered that the induced torque of single-crystal K spheres is highly anisotropic. In all but seven of two hundred experiments at 4 K, on seventy different samples, $N_y(\theta)$ exhibited four large peaks in a 360° magnet rotation. The anisotropy ratios were typically 3 or 4:1 at $H = 25$ kG. The four peaks appeared irrespective of the crystallographic orientation of the spheres, even if the rotation axis was (the threefold) [111]. Indubitably, K has a broken symmetry that violates its imputed cubic structure. Geometric

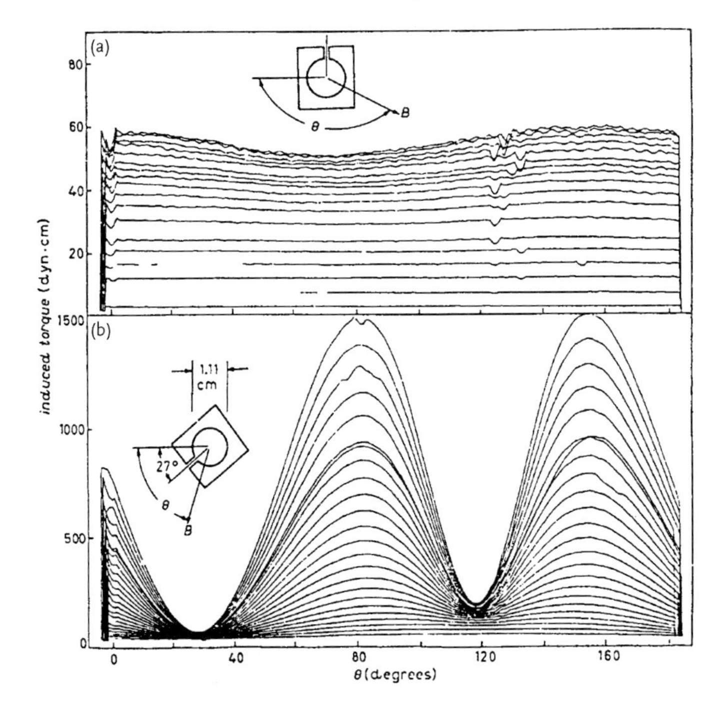

Figure 61.2 Induced torque versus magnet angle θ in the horizontal plane from [18]. The single-crystal sphere was grown (by the Bridgman method) in a spherical Kel-F mold, machined to a precision of 0.002 cm. The magnet rotation rate was 35°/min, i.e., $\Omega = 0.01$ rad/sec. $T = 1.5$ K. Curves are shown for $H = 1$ to 23 kG in 1 kG increments. (There are two traces for 18 kG.) In (a), the growth axis was perpendicular to the (horizontal) plane of rotation. In (b), the growth axis was in the horizontal plane.

distortion from a spherical shape was precluded by requiring torque patterns at 78 K (before further cooling to 4 K) to be accurately isotropic.

Holroyd and Datars [18] confirmed the discoveries of Schaefer and Marcus and extended them in several ways. A single-crystal sphere ($K - 10$), grown in a kel-F mold, exhibits a 30:1 anisotropic torque ($H = 23$ kG) when the growth axis is in the horizontal plane, as shown in Figure 61.2. Another sphere ($K - 4$), grown in oil (but with the oil removed) had an isotropic torque. However, an identical sample ($K - 2$) with surface oil had a 6:1 anisotropic torque ($H = 22$ kG).

The extraordinary variations described above are easy to understand. The optical anisotropy of K requires that there is only a single **Q** for each domain [19]. There are, of course, 24 equivalent axes that **Q** might have. If they are equally represented in a macroscopic sphere, the torque pattern ($H \leq 23$ kG) should be isotropic. However, elastic stress created by cooling (below the freezing point of oil) can lead to an orientational **Q**-domain texture and, consequently, to the fourfold torque anisotropy, as we show below.

At higher fields, $40 < H \leq 85$ kG, 20–30 sharp, open-orbit resonances appear in the induced-torque patterns [3]. This truly spectacular phenomenon, observed in all eighteen K samples studied, has been explained [5], based on the multiply connected Fermi surface of Figure 61.1.

The purpose of the present work is to derive the torque pattern shown in Figure 61.2. Suppose K has a spherical Fermi surface. Then, with **H** along the $\hat{z}$ axis, the resistivity tensor

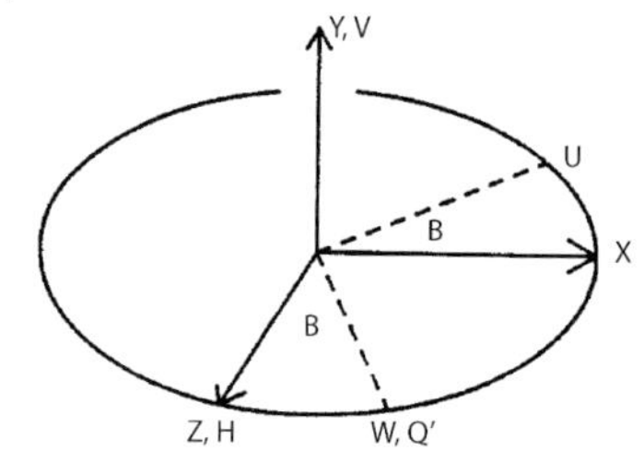

Figure 61.3 Coordinate systems employed to describe the resistivity tensor ρ_{xyz} or ρ_{uvw}. The heterodyne-gap wave vector $\mathbf{Q}'$ is taken to be in the xz plane, at an angle θ from the magnetic field $\mathbf{H}$. (A positive increment in θ here, and in Figure 61.6, corresponds to a negative increment in Figure 61.2.)

in the xyz frame is

$$\rho_{xyz} = \rho_0 \begin{pmatrix} 1 & \omega_c\tau & 0 \\ -\omega_c\tau & 1 & 0 \\ 0 & 0 & 1 \end{pmatrix} . \tag{61.5}$$

We shall examine the effect of one pair of heterodyne gaps and assume that $\mathbf{Q}'$ lies in the $\hat{x}\hat{z}$ plane, an angle θ from $\mathbf{H}$ and parallel to $\hat{w}$, as shown in Figure 61.3. The resistivity tensor in the $\hat{u}\hat{v}\hat{w}$ frame is

$$\rho_{uvw} = S^{-1}\rho_{xyz}S \, , \tag{61.6}$$

where

$$S = \begin{pmatrix} \cos\theta & 0 & \sin\theta \\ 0 & 1 & 0 \\ -\sin\theta & 0 & \cos\theta \end{pmatrix} . \tag{61.7}$$

S^{-1} is the transpose of S. Accordingly,

$$\rho_{uvw} = \rho_0 \begin{pmatrix} 1 & \omega_c\tau\cos\theta & 0 \\ -\omega_c\tau\cos\theta & 1 & -\omega_c\tau\sin\theta \\ 0 & \omega_c\tau\sin\theta & 1 \end{pmatrix} . \tag{61.8}$$

Consider Figure 61.4 and the two (vertical) heterodyne-gap planes. A magnetic field $H\cos\theta$ parallel to $\mathbf{Q}'$ would support ordinary cyclotron orbits, unaffected by the heterodyne gaps. Consequently the uv and vu elements of the tensor (61.8) remain unchanged. However, for a magnetic field, $H\sin\theta$, parallel to a vertical line in Figure 61.4, a typical cyclotron orbit S would obtain if the heterodyne gap is $E_g = 0$. If $E_g \neq 0$, then on orbit S' (in Figure 61.4) the points P and P' have the same wavefunction. So the time it would ordinarily take to travel from P to P' is not required. This effect corresponds to a Bragg-like reflection in real space; but in momentum space, to an instantaneous "Bragg advance". Consequently the effective cyclotron frequency for the vw and wv elements of the tensor (61.8) must be increased from ω_c to $\omega_c(1+\gamma)$:

$$\rho_{uvw} = \rho_0 \begin{pmatrix} 1 & \omega_c\tau\cos\theta & 0 \\ -\omega_c\tau\cos\theta & 1 & -\omega_c\tau(1+\gamma)\sin\theta \\ 0 & \omega_c\tau(1+\gamma)\sin\theta & 1 \end{pmatrix} . \tag{61.9}$$

One can show that, $\gamma \equiv 3Q'/4k_F$, for the simple case just considered. (We will modify γ below to treat magnetic breakdown effects and additional heterodyne gaps.)

The corrected resistivity tensor in the xyz frame is obtained by the inverse transformation

$$\rho_{xyz} = S\rho_{uvw}S^{-1} \, . \tag{61.10}$$

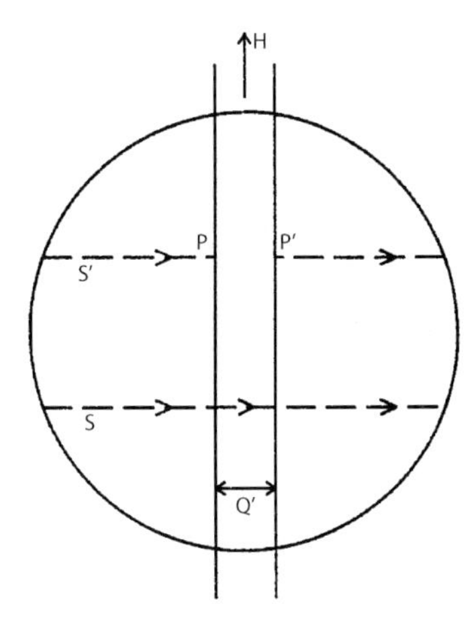

Figure 61.4 Cyclotron orbits S and S' on the Fermi surface (in **k** space). The heterodyne-gap planes are the two vertical lines. If the heterodyne gap E_g is very small, the orbit Swill "pass through" the energy gaps, i.e., magnetic breakdown. If E_g is sufficiently large, the electron on orbit S' will proceed instantly from P to P' (because they are in fact the same quantum state).

Accordingly, from Equations (61.7) and (61.9)

$$\rho_{uvw} = \rho_0 \begin{pmatrix} 1 & \omega_c\tau(1+\gamma\sin^2\theta) & 0 \\ -\omega_c\tau(1+\gamma\sin^2\theta) & 1 & -\omega_c\tau\gamma\sin\theta\cos\theta \\ 0 & \omega_c\tau\gamma\sin\theta\cos\theta & 1 \end{pmatrix}. \tag{61.11}$$

The xy and yx components require the Hall coefficient to be larger in magnitude than, $1/nec$ (the rigorous prediction for a simply connected Fermi surface). Significant Hall enhancements ~4–8%, depending on single-crystal orientation, have been observed (by helicon resonances at high fields) in both slab [20] and spherical [21] geometries.

The most remarkable feature of the tensor (11) is the appearance of yz and zy components. These can be described as a "longitudinal Hall effect", i.e., a current in the $\hat{y}$ direction creates an electric field parallel to **H**. The existence of this longitudinal Hall effect was postulated and shown to cause a four-peaked, induced-torque anisotropy [22], but a derivation (as given above) was unknown.

Of course, the torque N_y is no longer given by Equation (61.4). Visscher and Falicov [23] have derived an expression for N_y, applicable to a general resistivity tensor

$$N_y = \frac{4\pi R^5 \Omega B^2}{15c^2}\frac{\lambda}{\lambda(\rho_{yy}+\rho_{zz})-\mu}, \tag{61.12}$$

where

$$\lambda = (\rho_{xx}+\rho_{zz})(\rho_{xx}+\rho_{yy}) - \rho_{yz}\rho_{zy} \tag{61.13}$$

and

$$\mu = (\rho_{xx}+\rho_{zz})\rho_{xz}\rho_{zx} + (\rho_{xx}+\rho_{yy})\rho_{xy}\rho_{yx} + \rho_{xy}\rho_{yz}\rho_{zx} + \rho_{xz}\rho_{zy}\rho_{yx}. \tag{61.14}$$

For the tensor (61.11) there is considerable simplification:

$$N_y = \frac{2\pi R^5 \Omega n^2 e^2 \rho_0}{15}\frac{(\omega_c\tau)\left[4+(\omega_c\tau\gamma\sin\theta\cos\theta)^2\right]}{4+(\omega_c\tau\gamma\sin\theta\cos\theta)^2+\left[\omega_c\tau(1+\gamma\sin^2\theta)\right]^2}, \tag{61.15}$$

which reduces to Equation (61.4) if $\gamma = 0$. It is clear from the θ dependence that a four-peaked anisotropy occurs for $\omega_c\tau \ll 1$, and that the torque peaks grow roughly as H^2 in the high-field limit.

In order to fit Equation (61.15) to the data of Figure 61.2 it is necessary to introduce magnetic breakdown of the heterodyne gaps. If an energy gap E_g is small and if an electron is approaching (in momentum space) rapidly, there is a probability P_B that the electron will not be "Bragg advanced" at the energy gap. Instead, it will continue on its initial orbit in k space as if the energy gap were not there. An invariant expression for P_B based on original examples [24, 25] is[2]

$$P_B = \exp\left[\frac{-\pi mcE_g^2}{2\hbar^2 e|\mathbf{v}\cdot(\mathbf{K}\times\mathbf{H})|}\right], \tag{61.16}$$

where $\mathbf{K}$ is the wave vector of the periodic potential producing the energy gap E_g and $\mathbf{v}$ is the velocity that the electron would have at the gap if it were free. For the present purpose the breakdown probability reduces to

$$P_B = \exp\left[\frac{-H_0}{H|\sin\theta|}\right], \tag{61.17}$$

where θ is, as in Figure 61.3, the angle between $\mathbf{Q}'$ and $\mathbf{H}$. The appearance of $|\sin\theta|$ in the denominator, was first emphasized by Reitz.[3] H_0 combines all the other factors in Equation (61.16) and is frequently called the breakdown field; it will be an adjustable parameter.

Since the heterodyne gaps decrease rapidly with n, in Equation (61.2), and since H_0 is proportional to E_g, we will take (as an approximate model) the two-channel option depicted in Figure 61.5. We assume that all heterodyne gaps for $n > 2$ are always broken down, and that the heterodyne gap for $n = 1$ is sufficiently large to not break down when $H \leq 23\,\text{kG}$. Accordingly, orbit S' in Figure 61.5 will occur with probability $1 - P_B$, so the remaining channel S will be assigned probability P_B. The value of $\gamma(H)$, which appears in Equations (61.9), (61.11), and (61.15) will then be

$$\gamma(H) = \gamma_0\left[2 - \exp\left(\frac{-H_0}{H|\sin\theta|}\right)\right]. \tag{61.18}$$

Figure 61.6 illustrates the fit of Equations (61.15) and (61.18) to the data of Figure 61.2. (The data for $180° \leq \theta \leq 360°$ are identical to $0 \leq \theta \leq 180°$.) The torque minima are evenly spaced: 90°, 90°, 90°, 90°. However, the torque maxima are staggered: 75°, 105°, 75°, 105°, as observed. The peak height *vs.* H for $\theta = 155°$ in Figure 61.2 is shown in Figure 61.7 along with the (fitted) theoretical curve. The dashed line ($\sim H^2$) indicates that the theoretical torque maxima deviate slightly from an H^2 dependence.

It is known from low-field ($H < 3\,\text{kG}$) induced torque data [3, 17, 26] that the residual resistivity ρ_0 is very anisotropic. This feature is caused by impurity-induced, CDW-umklapp scattering [27]. For simplicity we have ignored this feature, which has an interesting field dependence [28], in order that the behavior of Equation (61.15) can be intelligible and uncomplicated. The comparative data near the torque minima of Figures 61.2 and 61.6 indicate the relevance of an anisotropic residual resistivity.

The extraordinary anisotropy of sample K-10, shown in Figure 61.2, requires a highly oriented $\mathbf{Q}$-domain texture. The theoretical treatment given above was tractable because we considered only a single $\mathbf{Q}$ domain and, even then, neglected many of the gaps shown in

3) Reference [5], Equation 16.

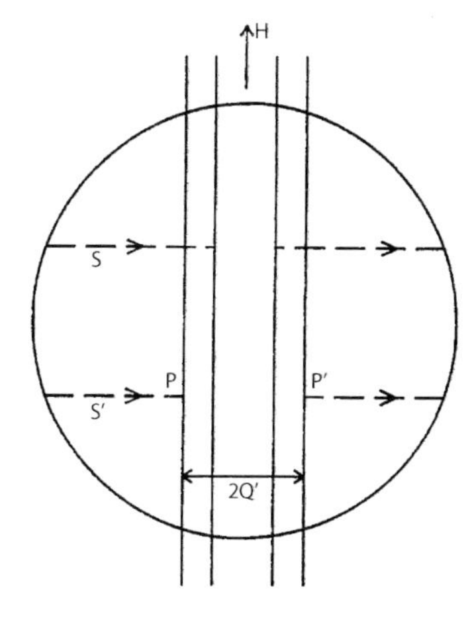

Figure 61.5 Cyclotron orbits S and S' on the Fermi surface, intersected by two pairs of heterodyne gaps. The orbit S is broken down at the outer pair, but is "Bragg advanced" by the inner pair. The orbit S' advances from P to P' if the outer pair is not broken down.

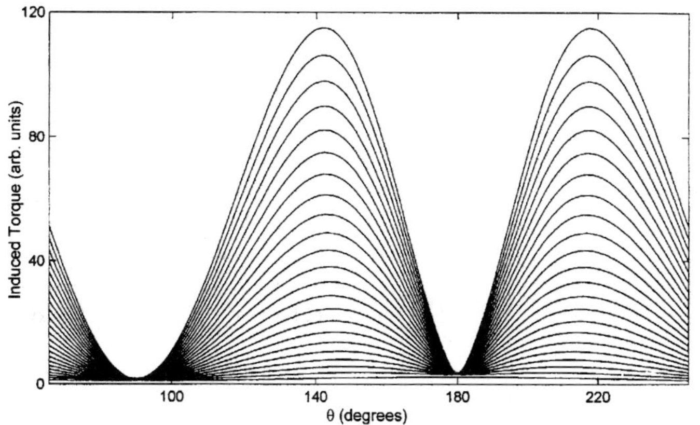

Figure 61.6 Theoretical induced-torque anisotropy based on the resistivity tensor (61.11), calculated from Equation (61.15) and incorporating the breakdown phenomenon modeled by Equation (61.18), in accordance with Figure 61.5. The torque peaks have staggered separations 75°, 105°, 75°, 105°, in agreement with the data of Figure 61.2. The parameters used in Equations (61.15) and (61.18): $\gamma_0 = 0.4$, $H_0 = 2.12\,\text{kG}$, $\omega_c\tau = 2.5\,\text{H}$, with H in kG.

Figure 61.1. Very likely the growth axis of K-10 was [110]. In such a case, a **Q**-domain texture for which the [110] direction would be prominent in the data could comprise four **Q** domains (in approximately equal volumes) with hypothetical **Q**'s:

$$(1.0, 0.9, 0.1), (1.0, 0.9, -0.1), (0.9, 1.0, 0.1), (0.9, 1.0, -0.1)\ , \tag{61.19}$$

(together with minor populations from the twenty other **Q** domains). Since most K single-crystal spheres have a torque anisotropy of 3 or 4:1, their **Q**-domain textures must be more nearly random (compared to K-10) but with a preferred bias created by elastic stress or other metallurgical history.

The anomalous terms, proportional to γ, in Equation (61.11) can also be derived by computing the conductivity tensor σ with the help of the Chambers path-integral method [29], and then inverting σ to obtain ρ. (This calculation is extremely tedious.) One does indeed reproduce the anomalous Hall terms of Equation (61.11). Terms which create an open-orbit

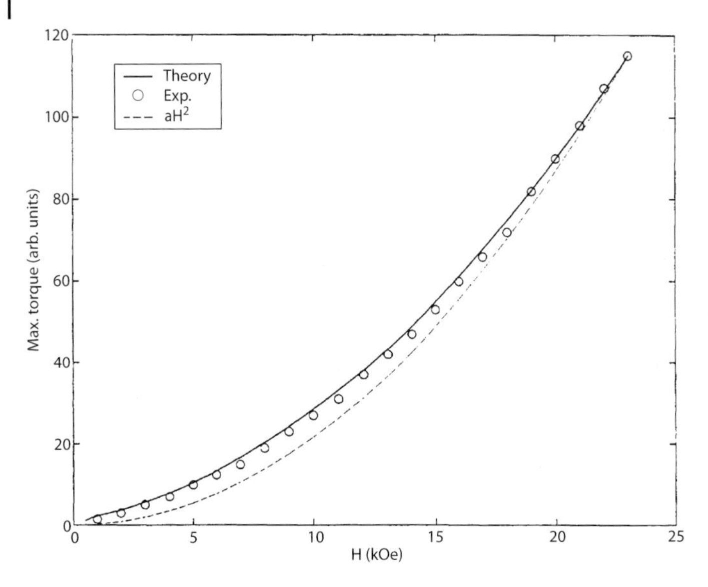

Figure 61.7 Magnetic field dependence of the torque peaks (at $\theta = 155°$ in Figure 61.2). The solid curve is the theoretical behavior from Figure 61.6. The dashed curve is a pure H^2 behavior, shown only to provide a comparison with the theoretical shape.

resonance also appear. We disregard them here because they are unimportant for $H < 30\,\mathrm{kG}$. They are treated in full elsewhere [5].

In conclusion the induced-torque anisotropy of single-crystal K spheres shows that K does not have a simply connected Fermi surface. The data requires an anisotropic Hall effect, indeed one with longitudinal components. Heterodyne gaps, which are a natural consequence of a CDW broken symmetry, explain the four-peaked patterns, including the staggered angular intervals between maxima as well as their field dependence. They also explain the anomalous Hall coefficients observed with helicons.

References

1 Overhauser A.W. (1984) *Phys. Rev. Lett.* **53**, 64.
2 Overhauser A.W. (1985) In: Bassani F., Fumi F., and Tosi M.P. (eds) *Highlights of Condensed-Matter Theory*, Proceedings of the International School of Physics "Enrico Fermi", Course LXXXIX, Varenna on Lake Como, North-Holland, Amsterdam, p. 194.
3 Coulter P.G. and Datars W.R. (1984) *Can. J. Phys.* **63**, 159.
4 Dunifer G.L., Sambles J.F., and Mace D.A.H. (1989) *J. Phys. Condens. Matter* **1**, 875.
5 Huberman M. and Overhauser A.W. (1982) *Phys. Rev. B* **25**, 2211.
6 Lacueva G. and Overhauser A.W. (1992) *Phys. Rev. B* **46**, 1273.
7 Park M.-A. and Overhauser A.W. (1998) *Phys. Rev. B* **55**, 1398.
8 Overhauser A.W. (1978) *Adv. Phys.* **27**, 343.
9 Overhauser A.W. (1968) *Phys. Rev.* **167**, 691.
10 Overhauser A.W. (1970) *Phys. Rev. B* **2**, 874.
11 Overhauser A.W. (1983) In: Devreese J.T. and Brosens F. (eds) *Electron Correlations*

in *Solids, Molecules and Atoms*, Plenum, New York, p. 41.
12 Giuliani G.F. and Overhauser A.W. (1979) *Phys. Rev. B* **20**, 1328.
13 Fragachan F.E. and Overhauser A.W. (1984) *Phys. Rev. B* **29**, 2912.
14 Lifshitz I.M., Azbel M.I., and Kaganov M.I. (1956) *Zh. Eksp. Teor. Fiz.* **31**, 63 [*Sov. Phys. JETP* **4**, 41]
15 Fawcett E. (1964) *Adv. Phys.* **13**, 139.
16 Lass J.S. and Pippard A.B. (1970) *J. Phys. E* **3**, 137.
17 Schaefer J.A. and Marcus J.A. (1971) *Phys. Rev. Lett.* **27**, 935.
18 Holroyd F.W. and Datars W.R. (1975) *Can. J. Phys.* **53**, 2517.
19 Overhauser A.W. and Butler N.R. (1976) *Phys. Rev. B* **14**, 3371.
20 Chimenti D.E. and Maxfield B.W. (1973) *Phys. Rev. B* **7**, 3501.
21 Werner S.A., Hunt T.K., and Ford G.W. (1974) *Solid State Commun.* **14**, 1217.
22 Zhu X. and Overhauser A.W. (1984) *Phys. Rev. B* **30**, 622.
23 Visscher P.B. and Falicov L.M. (1970) *Phys. Rev. B* **2**, 1518.
24 Blount E.I. (1962) *Phys. Rev.* **126**, 1636.
25 Reitz J.R. (1964) *J. Phys. Chem. Solids* **25**, 53.
26 Elliott M. and Datars W.R. (1983) *J. Phys. F* **13**, 1483.
27 Bishop M.F. and Overhauser A.W. (1978) *Phys. Rev. B* **18**, 2447.
28 Huberman M. and Overhauser A.W. (1985) *Phys. Rev. B* **31**, 735.
29 Chambers R.G. (1950) *Proc. R. Soc. London, Ser. A* **202**, 378.

Reprint 62 Observation of Phasons in Metallic Rubidium[1)]

G.F. Giuliani and A.W. Overhauser***

* Physik-Department, Technische Universität München, 525, Northwestern Avenue, Garching, West Germany D-8046, D

** Department of Physics, Purdue University, West Lafayette, Indiana 47907, USA

Received 29 May 1980

The heat-capacity anomaly caused by phasons – the collective excitations of an incommensurate charge-density wave – is identified in rubidium metal at 0.8 °K. Data analysis leads to a value $\Theta_\varphi = 4.5°$K for the phason cutoff frequency. The velocity ratio of a purely longitudinal to a purely transverse phason is found to be ~11.

Heat-capacity data of Lien and Phillips [1] reveal a low-temperature anomaly having the expected size, shape, and location for the phase excitations [2] of a charge-density wave [3] (CDW). Analysis of their data permits the first determination of the phason spectrum in a metal having a CDW ground state.

Alkali metals have many anomalous properties that require a broken translational symmetry [4]. Most studies have been of potassium, but the optical anomaly [5] and the high-field magnetoresistance [6] have been reported also in rubidium. Phasons are collective excitations that arise from the broken symmetry, and they have a frequency spectrum that goes to zero. They can influence a number of physical properties [7], but an anomaly in the low-temperature heat capacity [8] is perhaps their most fundamental effect. The first observation of such an anomaly was in $LaGe_2$ by Sawada and Satoh [9].

A conduction-electron CDW requires a sinusoidal lattice displacement to maintain charge neutrality [3]. Accordingly, the lattice displacement of a phase-modulated CDW is

$$\vec{U} = \vec{A}\cos\left[\vec{Q}\cdot\vec{r} + \beta\sin\left(\vec{q}\cdot\vec{r} - \omega t\right)\right] . \tag{62.1}$$

The phason amplitude, wave vector, and frequency are β, $\vec{q}$, and ω. It has been shown [3] that a phason is a coherent linear combination of two phonon modes having wave vectors $\vec{q} + \vec{Q}$ and $\vec{q} - \vec{Q}$. Since the CDW wave vector $\vec{Q}$ lies outside the Brillouin zone, the phonon mode which screens the CDW has wave vector $\vec{Q}' = \vec{G}_{110} - \vec{Q}$, where $\vec{G}_{110}$ is the (110) reciprocal-lattice vector of the bcc lattice. It is the soft transverse phonon at $\vec{Q}'$ which screens the CDW [10]. Its frequency, $\omega_\varphi \equiv k_B\Theta_\varphi/\hbar$, defines the phason cutoff frequency and characteristic temperature. For rubidium, with $Q = 1.35(2\pi/a)$ and tilted 3.2° from a [110] direction [10], $\Theta_\varphi = 4.2°$K.

When two phonon modes are coupled together, a 2 × 2 dynamical matrix results. The eigenvalue equation is

$$\begin{vmatrix} \omega_\varphi^2 - \omega^2 & \omega_\varphi^2 F(\vec{q}) \\ \omega_\varphi^2 F(\vec{q}) & \omega_\varphi^2 - \omega^2 \end{vmatrix} = 0 . \tag{62.2}$$

The lowest root of this equation is the phason frequency $\omega(\vec{q})$. The highest root is an amplitude mode [11] – a collective oscillation of the amplitude A of Equation (62.1). The off-diagonal coupling arises from new terms in the electronic dielectric matrix caused by the

1) Phys. Rev. B 45, 16 (1980).

ISBN: 978-3-527-40859-7

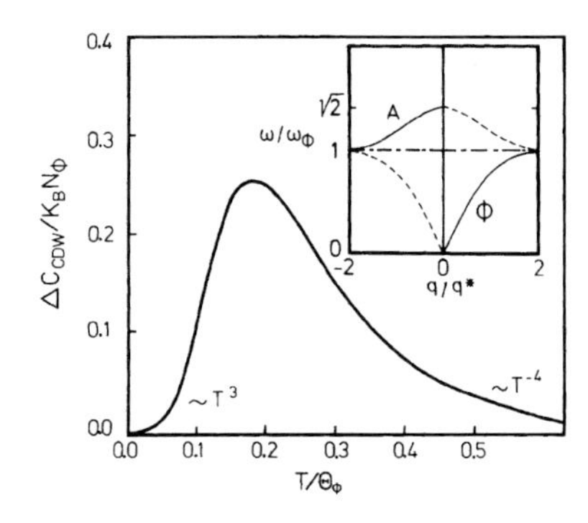

Figure 62.1 Heat-capacity anomaly caused by an incommensurate CDW. The inset shows the vibrational spectrum near $\vec{Q}'$. φ is the phason branch and A the amplitude-mode branch. The horizontal line, at $\omega = \omega_\varphi$, is the phason cutoff frequency – the phonon frequency at $\vec{Q}'$ with no CDW.

CDW [12]. It is easy to show from Equation (62.2) that

$$F(\vec{q}) \approx 1 - \left(c_{\vec{q}}/\omega_\varphi\right)^2 + \dots , \tag{62.3}$$

for small q, in order that the phason dispersion relation have the required linearity and zero frequency [2] at $\vec{q} = 0$. $c_{\vec{q}}$ is the phason velocity in the $\vec{q}$ direction and is expected to be highly anisotropic [2]. A purely longitudinal phason, $\vec{q}$ parallel to $\vec{Q}$, has a much higher velocity than a purely transverse one, $\vec{q}$ perpendicular to $\vec{Q}$. Since $F(\vec{q})$ is unknown for large $\vec{q}$, we choose two arbitrary functions satisfying Equation (62.3) for illustration.

$$F_1 = \exp\left[-\left(q/q^*\right)^2\right] ,$$
$$F_2 = \left[1 + \tfrac{1}{2}\left(q/q^*\right)^2\right]^{-2} , \tag{62.4}$$

where $q^* \equiv \omega_\varphi/c_{\vec{q}}$. Eigenvalues of the dynamical matrix for the case F_1 are shown in the inset of Figure 62.1. Phason and amplitude modes merge quickly into the phonon spectrum. The number of phason modes N_φ is characterized by the ellipsoidal volume of $\vec{q}$ space defined by $|\vec{q}| = q^*$.

The heat-capacity anomaly caused by the CDW structure is

$$\Delta C_{\text{CDW}} = \sum \vec{q} \left(C_{\text{phason}} + C_{\text{amp. mode}} - 2C_{\text{phonon}}\right) . \tag{62.5}$$

Each term is, of course, the heat capacity of an harmonic oscillator. The result for the case F_2 is shown in Figure 62.1. The peak in ΔC_{CDW} occurs at $T = 0.18\Theta_\varphi$ for both F_1 and F_2. For the case F_2 the peak is higher by a factor of 1.67. In all cases $\Delta C_{\text{CDW}} \sim T^3$ for low T and $\sim T^{-4}$ for high T.

The low-temperature heat capacity of a normal metal is

$$C_0 \cong \gamma T + AT^3 + BT^5 + \dots \tag{62.6}$$

The T^5 term is caused by dispersion of the acoustic waves. The coefficient B is determined by a plot of $(C - \gamma T)/T^3$ *vs.* T^2, which should be a straight line. The best fit of Lien and Phillips is shown by the dashed line in Figure 62.2(a) and depends mainly on higher-temperature points not shown. The expanded plot shown here emphasizes the downward singularity of the data points near $T = 0$. This is caused by too large a value for the Sommerfeld constant γ. A slightly smaller value leads to an upward singularity shown in Figure 62.2(c). An intermediate value, $\gamma = 2.37$ mJ/mole °K, eliminates the singularity, as shown in Figure 62.2(b).

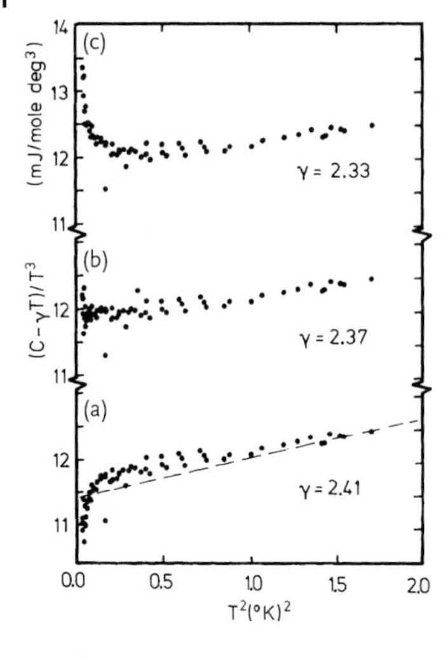

Figure 62.2 Plots of Lien and Phillips' data on Rb for three values of the Sommerfeld constant γ (in mJ/mole units). The dashed line in (a) is Lien and Phillips' best straight-line fit, from Equation (62.6), which depends mostly on higher-T points not shown.

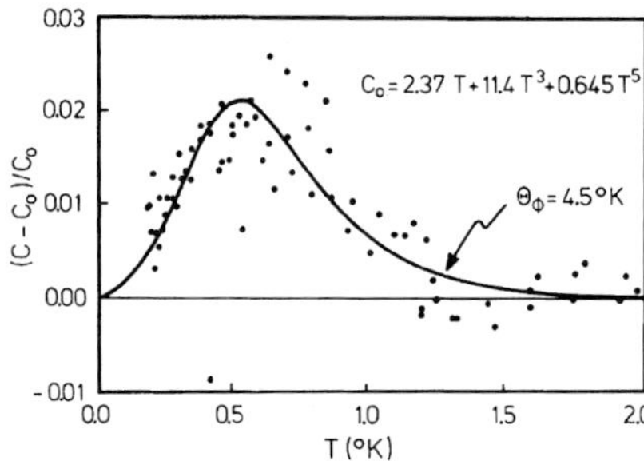

Figure 62.3 Deviations of Lien and Phillips' data from the heat capacity C_0 of a normal metal (no CDW). C_0 incorporates the revised value of γ from Figure 62.2(b). The curve is the theoretical phason anomaly, calculated from Equation (62.5).

A new plot of Lien and Phillips' data is shown in Figure 62.3, where C_0 incorporates the revised value of γ.[2] The deviations of the data from C_0 show a significant anomaly near 0.8 °K. (This peak appears shifted to lower T in Figure 62.3 because percentage deviations are plotted.) The smooth curve through the data is the theoretical anomaly computed from Equation (62.5). The best fit was obtained with $\Theta_\varphi = 4.5°$K, for both choices in Equation (62.4). The excellent agreement with the theoretical value quoted above is noteworthy. The height of the anomaly is 5 standard deviations, computed from point scatter about the theoretical curve.

Mathematical scaling of Equation (62.5) readily shows that the size of ΔC_{CDW} is proportional to $c_t^{-2} c_l^{-1}$, where c_t is the velocity of a purely transverse phason and c_t is that of a longitudinal phason. The observed anomaly implies that this product has a value 1.6×10^{-16} or 1.0×10^{-16}, if F_1 or F_2 is used in Equation (62.4).

The nearly spherical Fermi surfaces in alkali metals allow one to compute transverse phason velocities [13]. The theoretical value for potassium was found to be 1.4×10^5 cm/s. For rubidium it is 0.9×10^5 cm/s. We believe these velocities to be reliable because they depend only on observed elastic moduli. Transverse phase modulation corresponds to a small periodic rotation of $\vec{Q}$ from its optimum direction [2], which does not alter the many-electron effects [3] responsible for a CDW instability.

2) The coefficient B of Equation (62.6) was increased from 0.636 to 0.645 to maintain the fit near 2 °K.

If the theoretical c_t is combined with the values given above for the velocity product, we obtain $c_l = (8-12) \times 10^5$ cm/s. The longitudinal- to transverse-velocity ratio is, accordingly,

$$c_l/c_t \sim 11 \,. \tag{62.7}$$

This result is the first indication of the anticipated phason anisotropy from experimental evidence.

The anomaly shown in Figure 62.3 cannot be made to disappear by adjusting the coefficients A and B of Equation (62.6). We emphasize that the electronic term γ should be determined first by elimination of a low-T singularity, as we have shown in Figure 62.2. Of course, it is always possible to ascribe an anomaly to experimental uncertainties (e.g., the temperature scale).[3] We hope that the analysis given here will stimulate further work to eliminate such problems. In particular high-precision calorimetry on potassium, where the effects of CDW structure are so numerous, provides an important challenge.

Acknowledgements

The authors are grateful to the National Science Foundation and the Alexander von Humboldt Foundation for support of this research.

References

1 Lien W.H. and Phillips N.E. (1964) *Phys. Rev.* **133**, A1370.
2 Overhauser A.W. (1971) *Phys. Rev. B* **3**, 3173.
3 Overhauser A.W. (1968) *Phys. Rev.* **167**, 691.
4 Overhauser A.W. (1978) *Adv. Phys.* **27**, 343.
5 Mayer H. and Hietel B. (1966) In: Abeles F. (ed) *Proceedings of the International Colloquium on Optical Properties and Electronic Structure of Metals and Alloys, Paris, 1965*, North-Holland, Amsterdam, p. 47.
6 MacDonald D.K.C. (1957) *Philos. Mag.* **2**, 97.
7 Overhauser A.W. (1978) *Hyperfine Interact.* **4**, 786.
8 Boriack M.L. and Overhauser A.W. (1978) *Phys. Rev. B* **18**, 6454.
9 Sawada A. and Satoh T. (1978) *J. Low Temp. Phys.* **30**, 455.
10 Giuliani G.F. and Overhauser A.W. (1979) *Phys. Rev. B* **20**, 1328.
11 Lee P.A., Rice T.M., and Anderson P.W. (1974) *Solid State Commun.* **14**, 703.
12 Giuliani G.F. and Tosatti E. (1978) *Nuovo Cimento B* **47**, 135.
13 Giuliani G.F. and Overhauser A.W. (1980) *Phys. Rev. B* **21**, 5577.
14 Martin D.L. (1970) *Can. J. Phys.* **48**, 1327.

3) More recent data [14], which extends only to 0.4 °K, does not at first seem anomalous. However, an anomaly perfectly consistent with our Figure 62.3 appears when one corrects the Rb data for a thermometry error evident in his Cu calibration run.

Reprint 63 Theory of Induced-Torque Anomalies in Potassium[1)]

*A.W. Overhauser**

* Scientific Research Staff, Ford Motor Company, 525, Northwestern Avenue, Dearborn, Michigan 48121, USA

Received 17 August 1971

The bizarre torque anisotropies observed by Schaefer and Marcus in single-crystal spheres of potassium show that the Fermi surface is neither simply connected nor of cubic symmetry. We show further that a charge-density – wave structure provides a complete account of the observed behavior.

According to conventional views the induced torque exerted on a sphere of K by a rotating magnetic field H should be [1]

$$N_y = \frac{2\pi R^5 \Omega n^2 e^2 \rho_0}{15c^2} \frac{(\omega_c \tau)^2}{1 + \left(\frac{1}{2}\omega_c \tau\right)^2}, \tag{63.1}$$

where R is the radius of the sphere; Ω, the rotation speed of the magnet; n, the electron density; ω_c, the cyclotron frequency; τ, the relaxation time; and ρ_0, the resistivity. This result is independent of the crystal orientation and approaches a limiting value at high fields. The extraordinary torque anisotropies observed by Schaefer and Marcus [2] in K (and Na) require attention.

It is possible to obtain torque anisotropies by postulating a parallel array of dislocations, which would cause anisotropy of the resistivity. The torque is then given by the general expression [3]

$$N_y = \frac{4\pi(15c^2)^{-1} R^5 H^2 \Omega \lambda}{\lambda\left(\rho_{yy} + \rho_{zz}\right) - \left(\rho_{xx} + \rho_{zz}\right)\rho_{xz}\rho_{zx} - \left(\rho_{xx} + \rho_{yy}\right)\rho_{xy}\rho_{yx} - \rho_{xy}\rho_{yz}\rho_{zx} - \rho_{xz}\rho_{zy}\rho_{yx}}, \tag{63.2}$$

where

$$\lambda \equiv \left(\rho_{xx} + \rho_{zz}\right)\left(\rho_{xx} + \rho_{yy}\right) - \rho_{yz}\rho_{zy} .$$

$\{\rho_{ij}\}$ are the elements of the resistivity tensor (including the Hall terms). At low fields the torque would have a twofold anisotropy; but at high fields it would again saturate and the anisotropy would vanish. The latter behavior can be understood: In the high-field limit, induced currents circulate in a horizontal plane [1], so the azimuthal angle of the tensor axis is irrelevant. This holds as long as all cyclotron orbits are closed [3]. Therefore a dislocation model cannot explain the high-field torque anisotropy. Even at low fields the observed anisotropy is several orders of magnitude larger than what one might reasonably estimate.

Only open orbits can cause a high-field torque which does not saturate [3]. Since a nonsaturating four-peaked pattern is observed [2] even when a threefold axis of K is vertical, one must conclude that the Fermi surface is neither simply connected nor of cubic symmetry.

1) Phys. Rev. Lett. 27, 14 (1971).

Anomalous Effects in Simple Metals. Albert Overhauser

ISBN: 978-3-527-40859-7

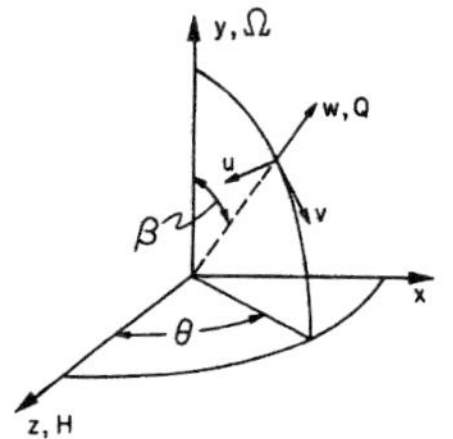

Figure 63.1 Coordinate systems used in the analysis.

A charge-density-wave (CDW) modulation [4] of the conduction electrons has provided successful explanations of other alkali-metal anomalies. These have been summarized [5]. CDW properties which have been previously supposed [5] and are pertinent here are the following: (a) The positive ions are displaced from their bcc lattice sites $\{\vec{L}\}$ by $\vec{A}\sin(\vec{Q}\cdot\vec{L})$. (b) These displacements cause a mixing of the CDW and reciprocal-lattice periodicities, and give rise to "heterodyne" gaps with periodicity $\vec{Q}' = 2\pi\vec{G} \pm \vec{Q}$. (c) The $\langle 110 \rangle$ crystal axes are the preferred $\vec{Q}$ directions. (d) $\vec{Q}$ tends to align parallel to $\vec{H}$ if at 4 °K in a stress-free crystal. (e) Alignment of $\vec{Q}$ is inhibited by elastic stress. (f) There are *low-frequency* vibrational modes – "phasons" – which are describable as a phase modulation of the CDW, and which cause the *local* direction of $\vec{Q}$ to fluctuate slightly about the [110] axis chosen in each $\vec{Q}$ domain. (g) Phason and umklapp scattering between the two conical points of the distorted Fermi surface cause significant zero-field anisotropy of the low-temperature resistivity.

Suppose $\vec{Q}$ is in the $\vec{w}$ direction (Figure 63.1) and that the resistivities parallel and perpendicular to $\vec{Q}$ are $\gamma\rho_0$ and ρ_0. With $\vec{H}$ in the $\vec{z}$ direction the resistivity tensor in the *uvw* frame is

$$(\rho_L) = \rho_0 \begin{pmatrix} 1 & \omega_c\tau\cos\theta\sin\beta & -\omega_c\tau\cos\theta\cos\beta \\ -\omega_c\tau\cos\theta\sin\beta & 1 & \omega_c\tau\sin\theta \\ \omega_c\tau\cos\theta\cos\beta & -\omega_c\tau\sin\theta & \gamma \end{pmatrix} . \tag{63.3}$$

This applies to the lemon-shaped Fermi surface (Figure 63.2) if we neglect the heterodyne gaps. The off-diagonal elements are just the Hall resistivities, τ is the relaxation time appropriate to ρ_0. (ρ_0 will be less than the bulk resistivity when $\gamma > 1$.) For the tiny crystals used experimentally [2], we suppose that the residual thermal stress, caused by cooling to 4 °K after being glued to a support, locks $\vec{Q}$ within some [110] easy-axis valley. Equation (63.2) gives the θ dependence of the torque after we transform (63.3) to the *xyz* reference frame:

$$(\rho_L)_{xyz} = \tilde{S}\,(\rho_L)_{uvw}\,S\,, \tag{63.4}$$

where $\tilde{S}$ is the transpose of S, and

$$S = \begin{pmatrix} -\cos\theta & 0 & \sin\theta \\ \sin\theta\cos\beta & -\sin\beta & \cos\theta\cos\beta \\ \sin\theta\sin\beta & \cos\beta & \cos\theta\sin\beta \end{pmatrix} . \tag{63.5}$$

The torque anisotropy at low field is twofold and nearly sinusoidal. The torque minima occur when $\theta = 0$ and 180°, that is, when (for $\gamma > 1$) $\vec{\Omega}$, $\vec{Q}$, and $\vec{H}$ are coplanar. The observed [2] magnitude of the low-field anisotropy requires $\gamma \sim 4$. As indicated previously, the torque computed from (63.2) and (63.4) becomes isotropic for $\omega_c\tau \gg 1$.

Consider the heterodyne gaps shown in Figure 63.2. If $\vec{Q}$ is parallel to [110], the gaps (dashed lines) are perpendicular to $\vec{Q}$ and cut out a fraction $f \sim 10\%$ of the electrons. If $\vec{Q}$ deviates $\sim 3°$ from [110], $\vec{Q}'$ deviates $\sim 40°$, and the gaps cut out a fraction $f \sim 15\%$.

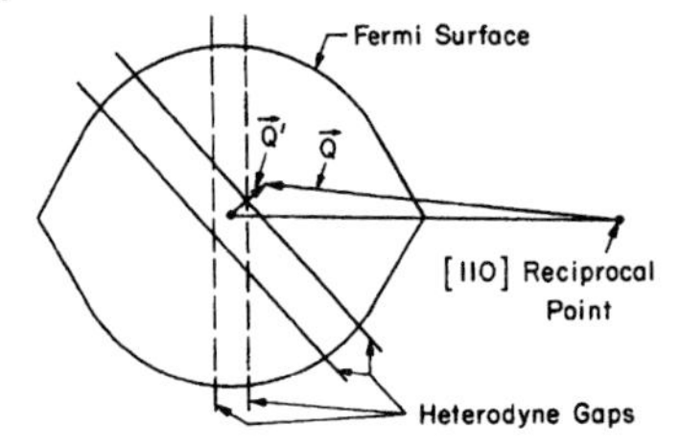

Figure 63.2 Fermi surface and heterodyne gaps of a CDW structure. The main CDW energy gaps, touching the conical points and normal to $\vec{Q}$, are not shown.

The large rotational leverage between $\vec{Q}'$ and $\vec{Q}$ results from the geometry, since $Q/Q' \sim 1.33/0.084$ when they are parallel.

The transport properties of the cut-out electrons are those of a cylindrical Fermi surface. Their conductivity tensor can easily be shown to be

$$(\sigma_C)_{uvw} = \frac{\rho_0^{-1}}{1+\delta^2}\begin{pmatrix} 1 & -\delta & 0 \\ \delta & 1 & 0 \\ 0 & 0 & 0 \end{pmatrix} , \tag{63.6}$$

where $\delta \equiv \omega_c \tau \cos\theta' \sin\beta'$, θ' and β' being the direction angles of $\vec{Q}$. Equation (63.6) applies for a cylinder having n electrons per unit volume.

The total conductivity tensor can now be derived. The two lemon ends can be pieced together to form a simply connected surface containing a fraction $1-f$ of the electrons. Accordingly,

$$(\sigma)_{xyz} \cong (1-f)\,(\rho_L)_{xyz}^{-1} + f\,\tilde{S}'\,(\sigma_c)_{uvw}\,S' \,. \tag{63.7}$$

$(\rho_L)_{xyz}$ is given by (63.4) and σ_C is here transformed to the xyz frame by $S' \equiv S\,(\theta',\beta')$, Equation (63.5). The induced torque is again obtained from (63.2) after (63.7) is inverted.

The torque N_y is a function of θ, β, θ', and β'. Even though we have assumed that residual stress prevents $\vec{H}$ from switching $\vec{Q}$ to another easy-axis valley, $\vec{Q}$ may still *align* (direction sense is irrelevant) slightly. Accordingly, for $0 < \theta < 90°$, θ will lag slightly the projection of the relevant [110] axis in the xz plane. From Figure 63.2 θ' will lead that projection by a much larger angle, say or. For $90° < \theta < 180°$, lead and lag will be interchanged. For simplicity, we take

$$\begin{aligned} \theta' &\approx \theta + \alpha, \quad 0 < \theta < 90° \text{ or } 180° < \theta < 270° \,; \\ \theta' &\approx \theta - \alpha, \quad 90° < \theta < 180° \text{ or } 270° < \theta < 360° \,. \end{aligned} \tag{63.8}$$

If N_y is calculated using (63.8), the high-field pattern becomes four sharp spikes separated in pairs by 2α. Each spike grows as H^2, but its width decreases as H^{-1}.

Phason fluctuations of the $\vec{Q}$ direction must now be accounted for. These will cause $\vec{Q}'$ to fluctuate throughout a much larger solid angle. In particular the angle α will have a Gaussian distribution,

$$P(\alpha) \sim \exp\left[-4(\ln 2)(\alpha-\alpha_0)^2/W^2\right] , \tag{63.9}$$

with a half-width W. Figure 63.3 shows the torque patterns computed using (63.2), (63.7), and (63.8) with a final average of N_y over the distribution (63.9). The high-field torque peaks grow linearly with H as observed [2]. All of the striking features are present, including the transition from a twofold to a four-peaked pattern near $\omega_c \tau \sim 10$.[2] The break in slope at

2) In Na the transition occurs at higher $\omega_c \tau$, which may in part reflect a smaller f, expected for the larger $\vec{Q}$.

3) It seems reasonable to expect that the last three parameters could have some dependence on H, θ, and β.

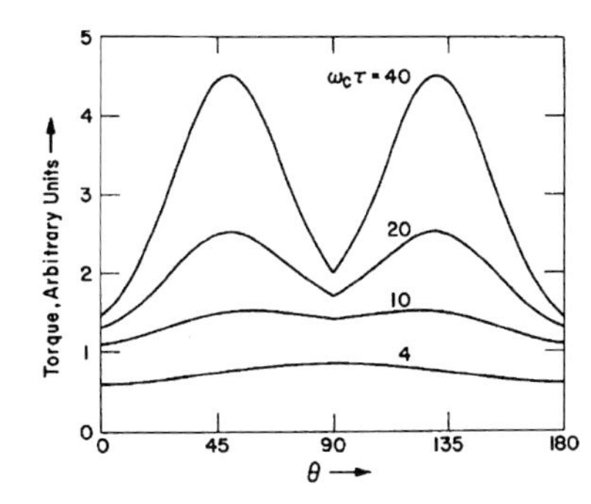

Figure 63.3 Computed torque patterns for a CDW structure. See text for details. The pattern repeats for $180° < \theta < 360°$.

$\theta = 0, 90°$, etc. is caused by the abrupt flopping of $\vec{Q}$ in its easy-axis valley, assumed in (63.8). Experimentally these minima are rounded off.

The parameters used in computing the torque patterns of Figure 63.3 were $\gamma = 4$, $f = 0.15$, $\theta = 35°$, $\beta' = 20°$, $\alpha_0 = 40°$, and $W = 45°$. No other set was tried.[3] The most questionable step in the foregoing derivation involves the phason fluctuations of $\vec{Q}$. These modes cause $\vec{Q}$ to vary in time *and space*. However Equation (63.2) applies only to a homogeneous resistivity tensor. In averaging over $P(\alpha)$ we have replaced what should have been done with a time-only average.

Large torque anisotropies need not always be seen. Three ways to miss them are possible: (i) $\vec{Q}$ is parallel to the axis of rotation. (ii) The crystal is stress free, so that new $\vec{Q}$ domains nucleate and grow as θ increases. (iii) The crystal has many $\vec{Q}$ domains caused by inhomogeneous residual stress. The absence of anisotropy reported previously [1] in a 22-mm specimen may be an example of case (ii).

Reproducibility of the Schaefer–Marcus anomalies in hundreds of experiments [2] can likely be ascribed to their choice of specimen size. Each crystal was comparable in size to the blob of grease used to attach it to its support. This is conducive to obtaining a relatively homogeneous thermal stress. It is important to emphasize that one must not attribute the effects to stress. The stress merely stabilizes an inherent anisotropy of the Fermi surface, allowing it to show forth. (This is reminiscent of the spin-density-wave state in Cr.) Residual stress cannot be accurately controlled; it may even depend on metallurgical history. Some samples will surely be bi-$\vec{Q}$ or poly-$\vec{Q}$ and will have torque patterns of greater complexity.

In recent years many Fermi-surface and cyclotron-orbit experiments have been carried out on Na and K. These are not in obvious conflict with the complexities of the CDW model. Indeed some of them [5] seem also to require it. The major question would appear to be this: Which ones are associated with the pieced-together lemon ends and which with the fluctuating cylindrical surface?

The physical requirements imposed by the experimental anomalies can be summarized: (a) The low-field torque patterns require a highly anisotropic (zero-field) resistivity tensor. A CDW structure provides this by conical-point to conical-point scattering. (b) The nonsaturating high-field torque peaks require open orbits. CDW heterodyne gaps provide these. (c) The constant width of the high-field peaks and their linear increase with H require a broad angular distribution in open-orbit directions. Phason fluctuations of $\vec{Q}$ together with the amplified response of $\vec{Q}'$ provide this. (d) The four-peaked patterns require a biaxial open-orbit configuration unrelated to the crystal axes and in some way controlled by $\vec{H}$. The latter feature seems necessary, since, otherwise, why would not the two axes sometimes have near-coincident projections in the xz plane, resulting in a two-peaked high-field pattern? CDW behavior mimics the former and satisfies the latter by (partial) field alignment of $\vec{Q}$ within an easy-axis potential valley.

It seems fair to conclude that either K has a CDW or it has something else with almost identical properties. Needless to say, certitude depends ultimately on direct observation by neutron diffraction [5].

References

1 Lass J.S. and Pippard A.B. (1970) *J. Phys. E: J. Sci. Instrum.* **73**, 137.

2 Schaefer J.A. and Marcus J.A. (1971) prededing Letter [*Phys. Rev. Lett.* **27**, 935].

3 Visscher P.B. and Falicov L.M. (1970) *Phys. Rev. B* **2**, 1518.

4 Overhauser A.W. (1968) *Phys. Rev.* **167**, 691.

5 Overhauser A.W. (1971) *Phys. Rev. B* **3**, 3173.

Reprint 64 Magnetoflicker Noise in Na and K[1)]

*A.W. Overhauser**

* Department of Physics, Purdue University, 525, Northwestern Avenue, West Lafayette, Indiana 47907, USA

Received 25 October 1973

It is shown that a physical model put forward to explain the induced-torque anomalies observed by Schaefer and Marcus in single-crystal spheres of Na and K requires the existence of a new type of $1/f$ flicker noise in large magnetic fields. A quantitative theory of this effect is developed. The mean-square noise voltage is proportional to $I^2 R_0^2 B^3 \Delta f/f$, where I is the current, R_0 is the zero-field resistance, B is the magnetic field, and Δf is the bandwidth. For a convenient experiment the noise could be as large as $(1 \times 10^{-6}\,\text{V})^2$, 13 orders of magnitude greater than Johnson noise. Experimental study is recommended to illuminate further the perplexing magnetoconductivity properties of Na and K.

64.1 Background

The failure of the magnetoresistance of Na and K to saturate at high fields has been a challenging puzzle for many years [1]. Fear that this anomalous behavior might be an artifice caused by electrode geometry has been eliminated by the advent of inductive techniques [2]. Data obtained by these newer methods substantiate those obtained from well-executed four-terminal measurements [3].

The most spectacular results obtained so far are the torque measurements of Schaefer and Marcus [4] on single-crystal spheres. A torque about the vertical axis arises when a sample experiences a magnetic field rotating in the horizontal plane [5]. The torque is proportional to the induced current, which depends on the magnetoconductivity tensor of the sample. The theoretical treatment [6] for a spherical sample with an isotropic conductivity (and quite generally [7] for tensors of arbitrary symmetry) requires that the high-field torque saturate if the magnetoresistance does. Not only do the observed torques increase linearly at high field, but they are highly anisotropic (sometimes by factors exceeding 5). Data from oriented single crystals [4] indicate that the anisotropy is in conflict with the presumed cubic symmetry of Na and K.

Lass [8] has calculated the torque pattern for a sample having a nonspherical shape, and has shown that it is possible to predict curves similar to those of Schaefer and Marcus. There are, however, serious difficulties with this explanation. A 10% deviation from spherical shape is required to explain the observations. Experimentally, the deviations from sphericity were about 2% or less [9]. This was determined by torque measurements (e.g., at 80 °K) when $\omega_0 \tau \ll 1$. Under these conditions a 1% shape anisotropy gives rise to a 1% torque anisotropy. On the other hand the high-field torque anisotropy is proportional to the square of the shape deviation [8]. Consequently, Lass's explanation appears inadequate by about a factor of 25.

Another problem is that the high-field anisotropies were correlated with [110] crystal axes, whereas shape anisotropies would likely have random orientation. Furthermore, if the degree of shape anisotropy had a reasonable statistical distribution (~ 70 samples were studied) a large fraction should have had quite small high-field anisotropies in view of the quadratic

1) Phys. Rev. B 9, 6 (1974).

Anomalous Effects in Simple Metals. Albert Overhauser

ISBN: 978-3-527-40859-7

dependence mentioned above. This was not the case [4]. A further difficulty is that an intrinsic high-field linear magnetoresistance with a Kohler slope $S \sim 0.025$ had also to be assumed. This value is an order of magnitude larger than that generally observed in single crystals [2, 3]. In this connection, however, a recent study [10] suggests that the longitudinal magnetoresistance may be significantly larger than the transverse. This latter work was carried out on spheres that were accurately round to better than $\frac{1}{2}$%. Nevertheless anisotropies up to a factor 2.4 were observed for $\omega_c \tau = 150$.

It is remotely conceivable that a spherical sample, electrically isotropic at 80 °K (where phonon scattering dominates the resistivity), could be electrically anisotropic at 4 °K from an unusual impurity concentration at one end of the specimen. However, the large diffusion coefficients obtained [11] for solutes in Na and K should provide effective homogenization during the time a sample awaits its destined application.

We shall now turn to the other proposed [12] explanation of the Schaefer–Marcus experiments. This model is based on the hypothesis that the electronic ground state of Na or K has a (almost) static charge-density-wave (CDW) structure [13]. This model had been invoked to explain other alkali-metal anomalies [14]. It is noteworthy that no new embellishments of the CDW model were required to provide a detailed account of all the Schaefer–Marcus results. Our purpose is to observe that this success requires the existence of a large magnetoflicker noise in K wires. Experimental test of this prediction is crucial, since it may show that success can be a measure of ingenuity without being a measure of truth.

64.2 CDW Structure

The possibility for a simple metal to have a CDW ground state rests on a firm theoretical foundation. It has been shown [13] that exchange and correlation act constructively to cause such an instability. The only uncertain contribution is whether the positive-ion background is sufficiently deformable to allow neutralization of the electronic charge density

$$p(\vec{r}) = \rho_0 \left[1 - p\cos(\vec{Q}\cdot\vec{r} + \varphi)\right] . \tag{64.1}$$

$\vec{Q}$ and p are the wave vector and amplitude of the CDW. Neutralization will be optimum if the ion displacements are

$$\vec{u}(\vec{L}) = \left(p\vec{Q}/Q^2\right)\sin\left(\vec{Q}\cdot\vec{L} + \varphi\right) , \tag{64.2}$$

relative to their ideal lattice sites $\vec{L}$. Na and K are prime candidates for such a phenomenon since their direct ion-ion interaction is very weak [15].

The magnitude of $\vec{Q}$ is very nearly the diameter $2k_F$ of the Fermi surface, since that choice optimizes both the exchange [16] and correlation [13] energy. Q is larger than the radius of the Brillouin zone. Accordingly, the wave vector $\vec{Q}'$ of the static phonon equivalent to Equation (64.2) is

$$\vec{Q}' = (2\pi/a)(1,1,0) - \vec{Q} . \tag{64.3}$$

We have concluded [14] that the lowest-energy direction for $\vec{Q}$ is along the [110] direction of the bcc lattice (having lattice constant a because that minimizes the magnitude of $\vec{Q}'$. This is desirable because any deformation energy associated with the neutralization of Equation (64.1) by (64.2) should have some proportional relationship to the energy of a phonon of wave vector $\vec{Q}'$. Since $Q = 1.33(2\pi/a)$ [17],

$$Q' \approx (2\pi/a)(\sqrt{2} - 1.33) \approx 0.08(2\pi/a) . \tag{64.4}$$

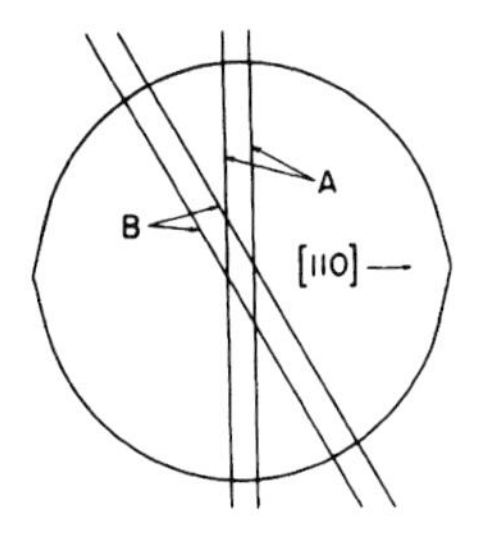

Figure 64.1 Brillouin zone and Fermi surface of K, assuming a CDW ground state. The heterodyne gaps *A* occur when $\vec{Q}$ is parallel to [110], The heterodyne gaps *B* occur when $\vec{Q}$ is tilted 2° from [110].

The periodicity $\vec{Q}'$ gives rise to two small energy gaps (heterodyne gaps [18]) which cut through the Fermi surface very near $k = 0$. Although these energy gaps are quite small [19] ($E_g \sim 0.015$ eV) cyclotron orbits which intersect them will on the average not suffer magnetic breakdown except in fields exceeding 100 kG. The geometric relationship of the heterodyne gaps to the Fermi surface is shown in Figure 64.1 for 5 parallel to [110] and for $\vec{Q}$ having a 2° angular deviation from [110].

The effect of the heterodyne gaps is to divide the Fermi surface into two parts (not three). Two "hemispherical" ends can be pieced together to form one simply connected surface that is almost spherical. (It is somewhat distorted near the poles and the equator.) The second piece is the cylindrical slice, which we shall call the "wedding band". It will give rise to open orbits whenever $\vec{Q}'$ is perpendicular to $\vec{B}$. This allows one to explain the anomalous high-field magnetoresistance, as we show in Section 64.4, and to account [12] for the bizarre torque patterns of Shaefer and Marcus without a separate assumption.[2)] The predicted correlation of the torque patterns with one of the [110] crystal axes was verified in the 15 oriented samples that were studied [4], The fluctuation (in time) of the axis $\vec{Q}'$ of the wedding band plays a crucial role in all of these phenomena. For example, the angular width of the high-field torque peaks is directly related to the range of the fluctuation. In Section 64.5 we show that these fluctuations must give rise to a flicker noise in the resistivity.

The major physical difficulty of the CDW model is a reconciliation with the isotropy [20] of the de Haas–van Alphen effect. Prior to the work of Schaefer and Marcus, one could postulate [14] that $\vec{Q}$ aligns parallel to a strong magnetic field, thereby preventing observation of a CDW anisotropy in the Fermi surface. The torque anomalies "show" that $\vec{Q}$ is confined (within a few degrees) to a [110] axis. An alternative reconciliation is necessary, and this is currently under study. It should be mentioned that the conduction-electron-spin-resonance (CESR) splitting [21] provides indirect evidence of a Fermi-surface distortion ($\sim$ 7%) that is in quantitative agreement [22] with the CDW model. The splitting arises when a sample has two or more $\vec{Q}$ domains, since [22] the CESR g factor depends on the angle between $\vec{B}$ and $\vec{Q}$. Recent experiments [23] at 40 kG show that the splitting is proportional to B, and that the CESR can have as many as five well-resolved components.

64.3 Fluctuations of $\vec{Q}$ and $\vec{Q}'$

A CDW state is an ordered phase with a four-dimensional order parameter. The four components are the amplitude p [Equation (64.1)], the magnitude of $\vec{Q}$, and the two angles needed to specify the directional deviation of $\vec{Q}$ from a [110] axis. Since the charge density amplitude is large [13] ($p \sim 0.17$), we shall neglect its variation and consider only the three re-

2) The magnetic-breakdown model for the magnetoresistance [18] became inapplicable once it was realized [14] that $\vec{Q}$ would align along a [110] direction. (The breakdown model requires hole orbits, which can occur only if $\vec{Q}$ is along a [100] or [111] direction.)

maining components. A dynamical theory of these (fluctuations in $\vec{Q}$) has been given [14]. It was prompted by a necessity to explain why diffraction satellites associated with the new periodicity $\vec{Q}$ could not be observed in a feasible experiment. (Analogous fluctuations in a one-dimensional metal have recently become of interest [24].)

Long-wavelength fluctuations in $\vec{Q}$ can be most easily described by letting the phase angle φ in Equations (64.1) and (64.2) be a slowly varying function of position and time [14]:

$$\varphi\left(\vec{L}, t\right) = \sum_{\vec{q}} \varphi_{\vec{q}} \sin\left(\vec{q} \cdot \vec{L} - \omega_{\vec{q}} t\right) , \tag{64.5}$$

where $\varphi_{\vec{q}}$ is the amplitude of an excitation having wave vector $\vec{q}$ and frequency $\omega_{\vec{q}}$. We have shown that $\omega_{\vec{q}}$ has a linear dispersion relation near $q = 0$. These new modes are, of course, part of the vibrational spectrum of the crystal. In view of their description as a phase modulation of the CDW we have called them phason modes. Relative to the ordinary $\vec{k}$ space of the Brillouin zone the phason $q = 0$ is located a distance Q' from $k = 0$ along one of the [110] directions. The fact that this mode can have $\omega = 0$ (in addition to the three acoustic phonon modes for $k = 0$) depends on $\vec{Q}$ being incommensurate with the reciprocal lattice.

Let q_1, q_2, q_3 be the three components of $\vec{q}$, corresponding to the [110] ($\vec{Q}$ direction), $[\bar{1}10]$, and [001] directions. For small q,

$$\omega_{\vec{q}} = \left[u_1^2 q_1^2 + u_2^2 q_2^2 + i_3^2 q_3^2\right]^{1/2} , \tag{64.6}$$

where u_1, u_2, u_3 are the three principle components of the phason velocity tensor. There has been no attempt to calculate the magnitude of these velocities. However, they are probably smaller than phonon velocities since the condensation energy of a CDW state is only $\sim 2 \times 10^{-4}$ eV per atom [14]. A phason (q_1 0, 0) describes a periodic variation of the magnitude of $\vec{Q}$, whereas (0, q_2, q_3) describes a periodic variation of the direction of $\vec{Q}$. These attributes are based on the assumption that phason modes are underdamped. Damping caused by electron–phason interactions has not been studied.[3)]

The fact that Q' [Equation (64.4)] is 16 times smaller than Q, together with the geometric constraint on these vectors given by Equation (64.3), means that a 1° rotation of $\vec{Q}'$ causes a 16° rotation of $\vec{Q}'$. It is clear that zero-point and thermal excitation of phasons will cause profound time-dependent fluctuations in the Fermi surface and all physical properties which depend on it. We now turn to its influence on the magnetoconductivity tensor. The dominant effect arises from fluctuations of the axis of the wedding band.

Consider a coordinate system u, v, w so that the axis $\vec{Q}'$ of the wedding band is in the w direction. The angular relationship of u, v, w to an x, y, z coordinate system is shown in Figure 64.2, The conductivity tensor of the wedding band in the u, v, w frame is [12]

$$\sigma_{wb} = \frac{\eta \rho_0^{-1}}{1 + \delta^2} \begin{pmatrix} 1 & -\delta & 0 \\ \delta & 1 & 0 \\ 0 & 0 & 0 \end{pmatrix} , \tag{64.7}$$

where $\delta \equiv \omega_c \tau \sin\beta \cos\theta$, ρ_0 is the $B = 0$ resistivity of the metal, and η is the fraction of the conduction electrons between the heterodyne gaps:

$$\eta \approx \pi k_F^2 Q' / \left(\tfrac{4}{3}\pi\right) k_F^3 \sim 0.10 . \tag{64.8}$$

3) Another detail in need of study is the possibility that the minimum energy for $\vec{Q}$ might occur when $\vec{Q}$ has a small angular tilt ($\sim$ 1° or 2°) from a [110] direction. Equation (64.3) would then correspond to a static phonon having a transverse component. It is well known that the [110] transverse mode of a [110] phonon is much softer than the longitudinal mode. Such a tilt might reduce slightly the charge-neutralization energy.

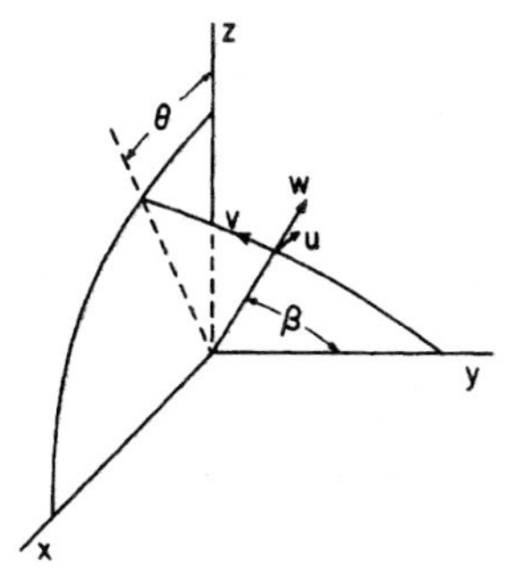

Figure 64.2 Relative orientation of the uvw axes and the xyz axes. The axis $\vec{Q}'$ of the wedding band is parallel to w. The magnetic field $\vec{B}$ is parallel to z.

When $\vec{Q}'$ is tilted the fraction η is somewhat larger. We have taken the field $\vec{B}$ in the z direction.

The quasispherical Fermi surface, which contains the remaining fraction, $1 - \eta$, of electrons will have a conductivity tensor in the xyz frame given approximately by

$$\sigma_{qs} \approx \frac{(1-\eta)\rho_0^{-1}}{1+(\omega_c\tau)^2}\begin{pmatrix} 1 & -\omega_c\tau & 0 \\ \omega_c\tau & 1 & 0 \\ 0 & 0 & 1+(\omega_c\tau)^2 \end{pmatrix} \tag{64.9}$$

We have here neglected any anisotropy in σ_{qs} for $B = 0$ since it is not germane to a calculation of the magnetophason noise. (It was relevant to the low-field torque patterns [12].) The total conductivity in the xyz frame is obtained by summing (64.7) and (64.9) after transforming (64.7) into the xyz frame.

$$\sigma = \tilde{S}\sigma_{wb}S + \sigma_{qs}\,. \tag{64.10}$$

The (orthogonal) transformation matrix is [12]

$$S \equiv \begin{pmatrix} -\cos\theta & 0 & \sin\theta \\ \sin\theta\cos\beta & -\sin\beta & \cos\theta\cos\beta \\ \sin\theta\sin\beta & \cos\beta & \cos\theta\sin\beta \end{pmatrix}. \tag{64.11}$$

It is of course the phason-induced fluctuations of the first term of Equation (64.10) that generate a high-field linear magnetoresistance. This was already shown by the theory of the torque anomalies [12], since the high-field torques are a direct measure of the magnetoresistance [6].

In the following section we shall compute the transverse magnetoresistance in a long, thin wire using Equation (64.10). Even if the wire were a single crystal, it will have many $\vec{Q}$ domains. A $\vec{Q}$ domain is a region within a single crystal where $\vec{Q}$ is aligned parallel to just one of the six [110] axes. There will be $\vec{Q}$-domain boundaries across which the direction of $\vec{Q}$ changes abruptly to another [110]-type axis. In a macroscopic sample the size and distribution of $\vec{Q}$ domains will play an important role in determining electrical transport. One would have to embark on studies of even greater difficulty than those considered by Herring [25]. Very little can be inferred about $\vec{Q}$-domain structure from experiment. If the CDW hypothesis is correct, the variability of $\vec{Q}$-domain structure would certainly be the explanation of the nonrepeatability of experiments from sample to sample, or from run to run on the same sample, which has baffled and frustrated all workers who have studied alkali metals. Slight deformation, thermal or mechanical, may cause nucleation and growth of new $\vec{Q}$ domains and change the transport coefficients by large factors.

The 200 experimental runs reported by Schaefer and Marcus [4] are perhaps the set with the least overall variability to date. Their samples were small, 2–7-mm-diam spheres. It is reasonable to regard the qualitative uniformity of their results, together with the fact that a single z-domain model could explain them, as indicating that $\vec{Q}$ domains can be comparable in size to their specimens. There is also indirect support. We have shown [26] by neutron diffraction studies of primary extinction that the mosaic blocks in single crystal K are about 1 mm in size.

Extreme variability appears to be an intrinsic property of Na and K. We now turn to one effect of $\vec{Q}$-domain structure and phason fluctuations which account, at least in part, for that.

64.4 Magnetoresistance of a Thin Wire

Consider a wire of length L, diameter d, and let it be oriented along the y axis. We shall assume that d is sufficiently small so each $\vec{Q}$ domain extends across the cross section of the wire. The approximation we are attempting to justify is one for which the total resistance is the sum of the resistances of each $\vec{Q}$ domain. With $\vec{B}$ in the z direction the magneto resistivity of a 5 domain will be ρ_{yy}, where the matrix ρ is the inverse of σ [Equation (64.10)]. Calculating this inverse is a strenuous algebraic exercise leading to expressions of great complexity. Our interest is only in the leading term for the high-field limit, $\omega_c\tau \gg 1$:

$$\rho_{yy} \approx \rho_0 + \eta\rho_0 \frac{(\omega_c\tau)^2 \cos^2\beta \sin^2\theta}{1 + (\omega_c\tau)^2 \sin^2\beta \cos^2\theta} . \tag{64.12}$$

We have neglected terms of order η and all terms important only near $\omega_c\tau \sim 1$. Accordingly Equation (64.12) is correct only for the low-field and high-field limits. Furthermore it is valid in the latter case only when the numerator of the second term is large compared to the denominator. From Figure 64.2 one can see that $\epsilon \equiv \sin\beta\cos\theta$ is the angular deviation of $\vec{Q}'$ away from the xy plane, which is perpendicular to $\vec{B}$. The variation of $\Delta\rho_{yy}$ with ϵ is shown in Figure 64.3. It has a Lorentzian shape with a height proportional to $(\omega_c\tau)^2$ and a half-width equal to $2/\omega_c\tau$, The area under this curve is proportional to $\omega_c\tau$.

We now assume that the wire has many $\vec{Q}$ domains of random orientation and that the phason fluctuations of $\vec{Q}'$ cause the directional distribution of $\vec{Q}'$ to be continuous and isotropic. The mean high-field resistivity increase is then

$$\left\langle \frac{\Delta(\omega_c\tau)}{\rho_0} \right\rangle \approx \frac{\eta(\omega_c\tau)^2}{4\pi} \int_0^{\pi} \cos^2\beta \sin\beta\, d\beta \times \int_0^{2\pi} \frac{d\theta}{1 + (\omega_c\tau)^2 \sin^2\beta \cos^2\theta} . \tag{64.13}$$

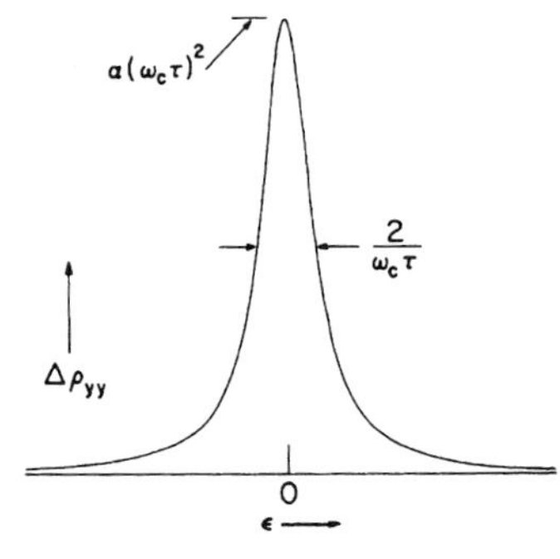

Figure 64.3 Variation of ρ_{yy} with ϵ, the angle between the axis $\vec{Q}'$ of the wedding band and the xy plane. This curve also depicts the voltage pulse that a $\vec{Q}$ domain will contribute if $\epsilon = \nu t$.

We have set $\sin\theta = 1$ when inserting the second term of Equation (64.12) in the integrand of Equation (64.13), since the important contribution to the integral arises only when $\theta \approx \frac{1}{2}\pi$. The θ integration yields $2\pi\left[1 + (\omega_c\tau)^2 \sin^2\beta\right]^{-1/2}$, which for $\omega_c\tau \gg 1$ is just $2\pi/\omega_c\tau\sin\beta$. The β integration is then trivial and contributes a factor $\frac{1}{2}\pi$. We obtain, finally,

$$\langle \Delta\rho_{yy}/\rho_0 \rangle \approx \tfrac{1}{4}\pi\eta\omega_c\tau\,, \tag{64.14}$$

which is the (desired) high-field linear magnetoresistivity. The Kohler slope S is Equation (64.14) divided by $\omega_c\tau$. From Equation (64.8) the expected value of S is 0.08. Observed values of S vary about this value by two orders of magnitude. We therefore define a factor F so that

$$S = \tfrac{1}{4}\pi\eta F\,. \tag{64.15}$$

This factor is intended to account for the sample variation of $\vec{Q}$-domain distribution, as well as one additional effect discussed below.

It is clear from the derivation that the high-field resistance results from resistance pulses $(\omega_c\tau)^2$ in magnitude and lasting a fraction $(\omega_c\tau)^{-1}$ of the time. Suppose a $\vec{Q}$ domain does not occupy the entire cross section of the wire. Its resistance pulses may be "shorted out" by the surrounding material. This is a likely occurrence in large specimens, as is frequently the case with single crystals, which generally show values ~ 0.003 for S. A very interesting theoretical question is the extent to which inductive effects may inhibit this shorting process during the very brief time interval of a resistance pulse.

One can perhaps predict in advance whether mechanical deformation will cause S to increase or decrease. The shorting effect would be enhanced if the $\vec{Q}$ domains become smaller in size. Deformation (which should initially decrease the size) can initiate a strain-anneal process that may then allow $\vec{Q}$ domains to grow larger. Deformation of single-crystal and polycrystal specimens [2] has been found to give an increase in S; but these data were obtained on samples strained at 77 °K, well above 15 °K where K begins its mechanical recovery [27]. In contrast, polycrystalline wires strained at 4 °K show a decrease in S with strain [28]. If the strain temperature is the difference, (and this should be given further study) the contrasting behavior can be understood in terms of the anticipated change in $\vec{Q}$-domain size outlined above.

64.5 Magnetophason Noise

The proposed origin of the high-field linear magneto resistance given previously [12] and in the foregoing section must necessarily cause a large flicker noise, the sum total of all pulses from every $\vec{Q}$ domain delivered randomly in time. We calculate here the noise spectrum and its absolute magnitude without attempting to be exact at every step.

Suppose that the angle ϵ by which $\vec{Q}'$ deviates from the xy plane varies with time as follows:

$$\epsilon \sim \zeta \sin 2\pi\nu t\,, \tag{64.16}$$

where ζ is a characteristic amplitude of the $\vec{Q}'$ oscillation angle, and ν is a characteristic frequency. We expect that $\zeta \sim 0.4$ rad, since 2ζ must equal the width of the Schaefer–Marcus torque peaks ($\sim 45°$). From Figure 64.3 one can see that the resistivity pulse is large only for $|\epsilon| \lesssim 1/\omega_c\tau$. So we may approximate the time dependence of ϵ during one such pulse by

$\epsilon = \nu t$, where ν is an angular velocity $\sim 2\pi\nu\zeta$. If r_0 is the zero-field resistance of the $\vec{Q}$ domain, then from (64.12),

$$\Delta r \approx \eta r_0 (\omega_c \tau)^2 \cos^2\beta / \left[1 + (\omega_c \tau \nu t)^2\right] . \tag{64.17}$$

This resistance pulse has the shape depicted in Figure 64.3. The Fourier transform of (64.17) is

$$\Delta r_f = \int_{-\infty}^{\infty} \Delta r e^{-2\pi i f t} dt . \tag{64.18}$$

where f is the frequency. The integration is elementary. The one-sided spectral density for one such pulse per second is obtained by squaring Equation (64.18) and multiplying by 2:

$$(\Delta r_f)^2 \approx 2 \left(\pi \eta \nu^{-1} r_0 \omega_c \tau \cos^2\beta\right)^2 e^{-4\pi f/\nu\omega_c\tau} . \tag{64.19}$$

The factor 2 is inserted so that Equation (64.19) includes contributions from f and $-f$.

We must next average (64.19) over a suitable angular velocity distribution $P(\nu)$. The maximum velocity associated with a perpendicular traversal of $\vec{Q}'$ through the xy plane is $\nu_m \sim 2\pi\nu\zeta$. The characteristic frequency ν is unknown since the phason velocities [Equation (64.6)] are unknown. A reasonable guess for an upper limit might be u/d, where d is the wire diameter (and $\vec{Q}$-domain size.) For $u \sim 10^5$ cm/s and $d \sim 0.1$ cm, $\nu < 10^6$ Hz.

The traversal of $\vec{Q}'$ through the xy plane may occur, however, at any angle, say α. Equation (64.16) corresponds to $\alpha = \frac{1}{2}\pi$; so in general the angular velocity will be

$$\nu = \nu_m \sin\alpha , \tag{64.20}$$

The probability distribution $P(\nu)$ is related to that for α,

$$P(\nu) = \left(\frac{d\nu}{d\alpha}\right)^{-1} P(\alpha) , \tag{64.21}$$

where $P(\alpha) = 2/\pi$, $0 \le \alpha \le \frac{1}{2}\pi$. One calculates the needed derivative from Equation (64.20), whereupon

$$P(\nu) = 2/\pi \left(\nu_m^2 - \nu^2\right)^{1/2} . \tag{64.22}$$

We must integrate the ν dependent terms of Equation (64.19) with this distribution. The factor of interest is

$$\int_0^{\nu_m} P(\nu) e^{-4\pi f/\nu\omega_c\tau} \nu^{-2} d\nu . \tag{64.23}$$

This integral can be approximated very well by observing that most of the contribution comes from small ν as long as $f \ll \omega_c \tau \nu$. For small ν, Equation (64.22) is just $2/\pi\nu_m$, and the integration becomes elementary. Accordingly expression (64.23) has the approximate value:

$$\omega_c\tau/2\pi^2 f \nu_m = \omega_c\tau/4\pi^3\zeta\nu f . \tag{64.24}$$

At this point we see that the spectral density has a $1/f$ frequency dependence. Magnetophason noise is a $1/f$ noise, in common with many other types of flicker noise. We combine

(64.24) with the remaining terms of (64.19), multiply by 2ν (a mean number of plane crossings per second for an active $\vec{Q}$ domain), and obtain the spectral density per active $\vec{Q}$ domain:

$$\left(\Delta r_f\right)^2 \approx \eta^2 r_0^2 (\omega_c \tau)^3 \cos^4 \beta / \pi \zeta f . \quad (64.25)$$

We note that the characteristic fluctuation frequency ν has dropped out. Our multiplication of (64.19) by 2ν to obtain (64.25) is justified only if pulses from the same $\vec{Q}$ domain are uncorrelated in time. Of course this cannot be completely true. The continuous distribution of phason frequencies between 0 and ν prevents long-time correlations. Nevertheless this step must be regarded as an approximation.

The resistance pulses from different $\vec{Q}$ domains will be uncorrelated. We need only count how many $\vec{Q}$ domains in the sample (of length L and diameter d) will participate. If the average length of a $\vec{Q}$ domain is γd, there will be $L/\gamma d$ domains in the sample. (By definition γ is a shape parameter for the typical $\vec{Q}$ domain, and should be near unity.) Not all of the domains can contribute to the flicker noise, since some will have $\vec{Q}$ oriented too far from the xy plane to allow $\vec{Q}'$ much chance of fluctuating into the plane. Only those domains having $\vec{Q}$ within an equatorial strip 2ζ wide will contribute fully to the noise. The fractional solid angle of this strip is ζ. We must modify the total expected number by the same factor F that corrects the linear magneto-resistance [Equation (64.15)]. Consequently, the total sample spectral density will be (64.25) multiplied by the factor

$$\zeta F L/\gamma d . \quad (64.26)$$

Within the effective (equatorial) strip, $\langle\cos^4 \beta\rangle \cong \frac{3}{8}$. The mean-square noise voltage is the spectral density times the bandwidth Δf, times the square of the measuring current I. We combine all these factors and obtain for the noise voltage V,

$$\langle V^2 \rangle = \frac{3\eta^2 r_0^2 (\omega_c \tau)^3 L F I^2}{8\pi \gamma d} \left(\frac{\Delta f}{f}\right) . \quad (64.27)$$

The sample resistivity for $B = 0$ is $R_0 = L r_0/\gamma d$. We can therefore eliminate r_0 from (64.27), and also F with the help of Equation (64.15). Our final result is

$$\langle V^2 \rangle = \frac{3\gamma \eta d S I^2 R_0^2 (\omega_c \tau)^3}{2\pi^2 L} \left(\frac{\Delta f}{f}\right) . \quad (64.28)$$

It is noteworthy that except for γ (the $\vec{Q}$-domain shape parameter which should be close to unity), none of the factors in (64.28) are adjustable. The fractional number η of electrons enclosed by the heterodyne gaps is given by Equation (64.8). The theoretical uncertainty in η cannot exceed about 50%. (For Na, $\eta \sim 0.07$.) The $B = 0$ resistance and the Kohler slope S must be measured for each sample. Similarly the residual resistance ratio determines $\omega_c \tau$ (which is proportional to B). Equation (64.28) predicts a spectacularly large magneto flicker noise in Na and K. Such a phenomenon is an inescapable conclusion from the premises that Na and K have CDW ground states and that the high-field torque patterns of Schaefer and Marcus are correctly explained by the model. This latter premise is needed so one can assume that $\vec{Q}$ domains are large enough in size to span the cross section of a thin wire.

64.6 Conclusion

Let us evaluate the noise voltage [Equation (64.28)] for a typical experiment. The residual resistivity of K is $\sim 1 \times 10^{-9}\ \Omega\,\text{cm}$. Consequently a specimen, $d = 0.1\,\text{cm}$, $L = 50\,\text{cm}$, will have a $4\,^\circ\text{K}$ resistance $6 \times 10^{-6}\ \Omega$. $\omega_c \tau \sim 10^2$ for $B = 40\,\text{kG}$. Take $I = 1\,\text{A}$, $\Delta f = 100\,\text{Hz}$,

$f = 10^3$ Hz, $S = 0.01$, $\eta = 0.1$, and $\gamma = 1$. The noise power is proportional to

$$\langle V^2 \rangle \sim \left(1 \times 10^{-6}\,\mathrm{V}\right)^2 . \tag{64.29}$$

Let us compare this with Johnson noise under the same conditions:

$$\langle V^2 \rangle_J = 4 k_B T R_0 \Delta f . \tag{64.30}$$

Accordingly, the Johnson noise power will be proportional to

$$\langle V^2 \rangle_J \sim \left(4 \times 10^{-13}\,\mathrm{V}\right)^2 . \tag{64.31}$$

Although Equation (64.29) is 13 orders of magnitude greater than Equation (64.31), carrying out such a measurement is not trivial. There will be serious microphonic noise caused by vibration of the sample and leads in the magnetic field. This can be minimized, of course, by a compensating winding in series opposition. Microphonic noise can be distinguished from magnetophason noise since the former is proportional to B^2 only, whereas the latter is proportional to $I^2 B^3$.

The predicted flicker noise is so unambiguous in magnitude that an experimental measurement will provide important clarification of the mechanisms contributing to numerous anomalies in alkali-metal behavior. One hopes that such a study will be forthcoming soon.

Note added in proof. R.S. Hockett and D. Lazarus [Bull. Am. Phys. Soc. (to be published)]. have carried out the experiment proposed above. They found no noise greater than 4 nV. This result shows that the direction of $\vec{Q}'$ cannot be assumed constant within a $\vec{Q}$ domain. A modified theory of the four-peaked torque pattern is therefore required and takes into account the finite coherence length a for the direction of $\vec{Q}'$. (This will be submitted later.) The magnitude of the magnetoflicker noise is then given by

$$\langle V^2 \rangle \approx \frac{a^3 \omega_c \tau S I^2 R_0^2}{d^2 L} \frac{\Delta f}{f} . \tag{64.32}$$

This result is much smaller than that given by Equation (64.28) and leads to values $\sim (10^{-9}\,\mathrm{V})^2$. The coherence length a is unknown, but it must be $\gtrsim 2 \times 10^{-4}$ cm, the cyclotron radius at 30 kG. Otherwise the de Haas–van Alphen effect would be incompatible with the CDW model.

Acknowledgements

Work supported by the National Science Foundation, Material Research Laboratory Program No. GH 33574.

References

1 Gugan D. (1972) *Nature Phys. Sci.* **235**, 61.

2 Penz P.A. and Bowers R. (1968) *Phys. Rev.* **172**, 991.

3 Taub H., Schmidt R.L., Maxfield B.W., and Bowers R. (1971) *Phys. Rev. B* **4**, 1134.

4 Schaefer J.A. and Marcus J.A. (1971) *Phys. Rev. Lett.* **27**, 935.

5 Datars W.R. and Cook J.R. (1969) *Phys, Rev.* **187**, 769.
6 Lass J.S. and Pippard A.B. (1970) *J. Phys. E* **3**, 137.
7 Visscher P.B. and Falicov L.M. (1970) *Phys. Rev. B* **2**, 1518.
8 Lass J.S. (1972) *Phys. Lett. A* **39**, 343.
9 Schaefer J.A. (private communication).
10 Simpson A.M. (1973) *J. Phys. F* **3**, 1471.
11 Barr L.W., Mundy J.N., and Smith F.A. (1967) *Philos. Mag.* **16**, 1139.
12 Overhauser A.W. (1971) *Phys. Rev. Lett.* **27**, 938.
13 Overhauser A.W. (1968) *Phys. Rev.* **167**, 691.
14 Overhauser A.W. (1971) *Phys. Rev. B* **3**, 3173.
15 Wallace D.C. (1968) *Phys. Rev.* **176**, 832.
16 Overhauser A.W. (1962) *Phys. Rev.* **128**, 1437.
17 Overhauser A.W. (1964) *Phys. Rev. Lett.* **13**, 190.
18 Reitz J.R. and Overhauser A.W. (1968) *Phys. Rev.* **171**, 749.
19 Overhauser A.W. (1972) *Bull. Am. Phys. Soc. 17*, 40.
20 Shoenberg D. and Stiles P.J. (1964) *Proc. R. Soc. A* **281**, 62.
21 Walsh W.M. Jr., Rupp L.W. Jr., and Schmidt P.H. (1966) *Phys. Rev.* **142**, 414.
22 Overhauser A.W. and de Graaf A.M. (1968) *Phys. Rev.* **168**, 763.
23 Dunifer G.L. and Phillips T.G. (private communication).
24 Lee P.A., Rice T.M., and Anderson P.W. (1973) *Phys. Rev. Lett.* **31**, 462.
25 Herring C. (1960) *J. Appl. Phys.* **31**, 1939.
26 Overhauser A.W. (unpublished).
27 Gurney W.S.C. and Gugan D. (1071) *Philos. Mag.* **24**, 857.
28 Jones B.K. (1969) *Phys. Rev.* **179**, 637.

Reprint 65 Influence of Charge-Density-Wave Structure on Paramagnetic Spin Waves in Alkali Metals[1)]

*Yong Gyoo Hwang** and *A.W. Overhauser**

* Department of Physics, Purdue University, 525, Northwestern Avenue, West Lafayette, Indiana 47907, USA

Received 8 January 1988

A spin-wave theory is developed for alkali metals having a charge-density-wave (CDW) structure. For simplicity the spin-dependent part of the many-body interaction between quasiparticles is taken as a constant, thereby neglecting Landau parameters B_n with $n \geq 1$. In a CDW state the velocity distribution is not only anisotropic, but open-orbit motion becomes possible. As a result, the motion of an electron is significantly modified. Since the known properties of alkali metals indicate charge-density-wave structure, the interpretation of experimental spin-wave data is reconsidered. The revised values of the Landau parameter B_0 and the fraction of electrons in open orbits η for potassium are found to be $B_0 = -0.252(\pm 0.003)$ and $\eta = 0.048(\pm 0.015)$ when only the open-orbit effect is included. $B_0 = -0.221(\pm 0.002)$ and $\eta = 0.042(\pm 0.013)$ if Fermi surface distortion is also included. It is also shown that observed splittings of spin-wave side bands can be explained when the effects of the CDW domain structure are recognized.

65.1 Introduction

Since conduction-electron spin resonance (CESR) was first observed in 1952 [1], extensive work has been done to yield valuable information on the spin-dependent part of the many-body interaction. As was first noticed by Dyson [2], CESR becomes possible from the fact that an electron returns many times to the rf skin layer where the external oscillating magnetic field is applied, before it diffuses further into the sample. A theory of CESR was developed by Dyson [2] for a noninteracting electron gas in a static magnetic field normal to the surface, and later extended for an arbitrary angle by Lampe and Platzman [3]. When the many-body interaction is taken into account in the framework of Landau Fermi-liquid theory, it was shown by Silin [4, 5] that an interacting Fermi system in a static magnetic field can have collective excitations with nonzero wave number analogous to spin waves in ferromagnetic materials. A transmission-electron spin resonance (TESR) experiment [6, 7] is one the most powerful tools in investigating the many-body interaction in a Fermi system, since the existence of spin waves exclusively depends on the presence of the many-body interaction. Theories of paramagnetic spin waves have been developed by Platzman and Wolff [6, 7] (PW) and others [8, 9]. In those theories, it has been assumed that the electron energy spectrum is free-electron-like and that the Fermi surface is very close to a perfect sphere. But there is a large amount of experimental data which cannot be explained by free-electron-like theories [10–12].[2)] TESR data in particular show several anomalous features which have not been interpreted by conventional theories. The main CESR sometimes shows a splitting of about 0.5 G [13]. This phenomenon was explained quantitatively as the result of an anisotropic g factor in the CDW state by Overhauser and de Graaf [14]. Spin-wave side bands sometimes show splittings into two or more components [15].

1) Phys. Rev. B 38, 13 (1988).
2) For a review of these phenomena, see [12].

Anomalous Effects in Simple Metals. Albert Overhauser

ISBN: 978-3-527-40859-7

In this study, an extension of the PW theory has been made to incorporate several charge-density-wave (CDW) effects: distortion of the Fermi surface, the presence of open-orbit motion, and domain structure. It is also shown that several anomalous experimental observations can be explained using this theory. In Section 65.2 relevant aspects of the CDW theory are reviewed. In Section 65.3 essentials of the Landau Fermi-liquid theory are recalled. In Section 65.4 a simplified model is developed to incorporate CDW effects, and transmission of rf signals through a finite slab is obtained. In Section 65.5 the Platzman–Wolff theory, its CDW modifications, and experimental data for potassium are compared. In Section 65.6 our results are summarized.

65.2 Brief Review of Charge-Density-Wave Theory

It has been shown by Overhauser [16, 17] that an interacting electron gas always suffers a spin-density-wave (SDW) or a charge-density-wave instability in the Hartree–Fock approximation, and that when correlation effects are taken into account the CDW instability is enhanced while the SDW instability is reduced. Alkali metals have very weak elastic stiffness [18, 19] and very small Born–Mayer ion–ion repulsion; [20] they are expected to suffer (CDW) instabilities. Many otherwise anomalous data have been successfully explained by CDW theory [10–12].[3] Recently neutron diffraction satellites were observed [21],[4] so the direction and magnitude of **Q** are directly determined for potassium.

CDW structure can be described by including an extra sinusoidally varying potential with wave vector **Q** in addition to the usual crystal potential:

$$U^{\text{CDW}} = G\cos(\mathbf{Q}\cdot\mathbf{r})\,, \tag{65.1}$$

where **Q** is the CDW wave vector and G is the periodic part of the exchange and correlation interaction. This periodic potential is sustained self-consistently by the resulting modulation of electron charge density $\rho(\mathbf{r})$:

$$\rho(\mathbf{r}) = \rho_0\left[1 - p\cos(\mathbf{Q}\cdot\mathbf{r}')\right]\,, \tag{65.2}$$

where p is the fractional amplitude of the CDW. The Hartree potential of the electron charge density is neutralized by deformation of the positive-ion background. The one-electron Schrödinger equation can be approximated by

$$\left[\frac{p^2}{2m} + G\cos(\mathbf{Q}\cdot\mathbf{r})\right]\Psi_{\mathbf{k}} = \varepsilon^0_{\mathbf{k}}\Psi_{\mathbf{k}}\,. \tag{65.3}$$

Self-consistency between the CDW potential and the charge modulation requires one to solve an integral equation for $G(\mathbf{k})$ [16, 17]. Since this is a very complicated problem, G will be assumed to be constant throughout our analysis. Although an analytical solution of the above Schrödinger equation is not available, a sufficiently accurate solution can be found in the following way. The CDW potential has off-diagonal matrix elements

$$\langle\mathbf{k}\pm\mathbf{Q}|\,U^{\text{CDW}}|\mathbf{k}\rangle = G/2\,. \tag{65.4}$$

When $k_z \cong -Q/2$, the mixing of $|\mathbf{k}\rangle$ and $|\mathbf{k}-\mathbf{Q}\rangle$ can be treated using nondegenerate perturbation theory, but the energy difference between these two states is so large that the mixing

3) For a review of these phenomena, see [12].

4) Apparently conflicting data: [22], can be intepreted by differing CDW domain structure in different samples.

can be neglected. However, the energy difference between $|\mathbf{k}\rangle$ and $|\mathbf{k}+\mathbf{Q}\rangle$ is a very small, so the mixing between these must be treated using degenerate perturbation theory. This leads, of course, to an energy gap at $k_z = -\mathbf{Q}/2$. In the following analysis, only those states below the gap need be considered. The secular equation for Equation (65.3) becomes

$$\begin{vmatrix} \varepsilon_{\mathbf{k}}^{f} - \varepsilon_{\mathbf{k}}^{0} & \frac{G}{2} \\ \frac{G}{2} & \varepsilon_{\mathbf{k}+\mathbf{Q}}^{f} - \varepsilon_{\mathbf{k}}^{0} \end{vmatrix} = 0 , \tag{65.5}$$

where $\varepsilon_{\mathbf{k}}^{f} = \hbar^2 k^2/2m$, the energy of a free electron with wave vector $\mathbf{k}$. The energy for a state below the energy gap is found from the forgoing secular equation:

$$\varepsilon_{\mathbf{k}}^{0} = \tfrac{1}{2}\left(\varepsilon_{\mathbf{k}}^{f} + \varepsilon_{\mathbf{k}+\mathbf{Q}}^{f}\right) - \tfrac{1}{2}\left[\left(\varepsilon_{\mathbf{k}}^{f} - \varepsilon_{\mathbf{k}+\mathbf{Q}}^{f}\right)^2 + G^2\right]^{1/2} . \tag{65.6}$$

The corresponding eigenfunction is

$$\Psi_{\mathbf{k}} = \cos\phi_{\mathbf{k}} e^{i\mathbf{k}\cdot\mathbf{r}} - \sin\phi_{\mathbf{k}} e^{i(\mathbf{k}+\mathbf{Q})\cdot\mathbf{r}} , \tag{65.7}$$

where $\cos\phi_{\mathbf{k}} = G/[G^2 + 4(\varepsilon_{\mathbf{k}}^{f} - \varepsilon_{\mathbf{k}}^{0})^2]^{1/2}$. These solutions will reduce to those found using nondegenerate perturbation theory when k_z is far away from $-\mathbf{Q}/2$.

A similar solution can be found for $k_z \cong \mathbf{Q}/2$;

$$\varepsilon_{\mathbf{k}}^{0} = \tfrac{1}{2}\left(\varepsilon_{\mathbf{k}}^{f} + \varepsilon_{\mathbf{k}-\mathbf{Q}}^{f}\right) - \frac{1}{2}\left[\left(\varepsilon_{\mathbf{k}}^{f} - \varepsilon_{\mathbf{k}-\mathbf{Q}}^{f}\right)^2 + G^2\right]^{1/2} , \tag{65.8}$$

$$\Psi_{\mathbf{k}} = \cos\phi_{\mathbf{k}} e^{i\mathbf{k}\cdot\mathbf{r}} - \sin\phi_{\mathbf{k}} e^{i(\mathbf{k}-\mathbf{Q})\cdot\mathbf{r}} . \tag{65.9}$$

If dimensionless quantities u, v, and w are defined

$$u = k_x/\mathbf{Q}, \quad v = k_y/\mathbf{Q}, \quad w = k_z/\mathbf{Q} - \frac{1}{2}\frac{k_z}{|k_z|} , \tag{65.10}$$

the above two solutions for $k_z \lessgtr 0$ can be put in a very simple form:

$$\varepsilon_{\mathbf{k}}^{0} = \frac{\hbar^2\mathbf{Q}^2}{2m}\left[u^2 + v^2 + \tfrac{1}{4} - \left(w^2 + \zeta^2\right)^{1/2}\right] , \tag{65.11}$$

$$\Psi_{\mathbf{k}} = \cos\phi_{\mathbf{k}} e^{i\mathbf{k}\cdot\mathbf{r}} - \sin\phi_{\mathbf{k}} e^{i(\mathbf{k}\pm\mathbf{Q})\cdot\mathbf{r}} , \tag{65.12}$$

$$\zeta = \frac{mG}{\hbar^2\mathbf{Q}^2} . \tag{65.13}$$

The Fermi-surface velocity in the ζ direction along $\mathbf{Q}$ is given by

$$v_z = \frac{1}{\hbar}\frac{\partial\varepsilon_{\mathbf{k}}^{0}}{\partial k_z} = \frac{\hbar\mathbf{Q}}{m}\left[w - \frac{w}{2\left(w^2 + \zeta^2\right)^{1/2}}\right] . \tag{65.14}$$

When $w \cong 0$, i.e., near $k_z \cong \pm\mathbf{Q}/2$, the electron velocity nearly vanishes. An approximate shape of the Fermi surface can be obtained by requiring critical contact between the Fermi surface and two CDW gaps (of magnitude G) [23]. It follows from this assumption that

$$\varepsilon_{\mathbf{k}}^{0} = \varepsilon_F = \frac{\hbar^2\left(\mathbf{Q}/2\right)^2}{2m} - \frac{G}{2} , \tag{65.15}$$

$$\kappa \equiv \left(u^2 + v^2\right)^{1/2} = \left[\left(w^2 + \zeta^2\right)^{1/2} - \zeta - w^2\right]^{1/2} . \tag{65.16}$$

The Fermi surface is then distorted into the shape of a lemon, as shown in Figure 65.1.

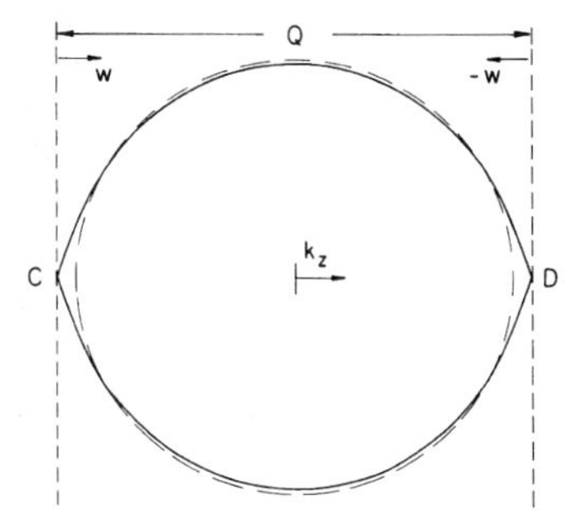

Figure 65.1 Lemon-shaped Fermi surface for $G/E_F = 0.5$. The dashed curve is that of a free-electron Fermi sphere, of radius k_F, having the same volume as the lemon-shaped surface. The dashed curve is drawn $\frac{4}{3}$% larger than actual.

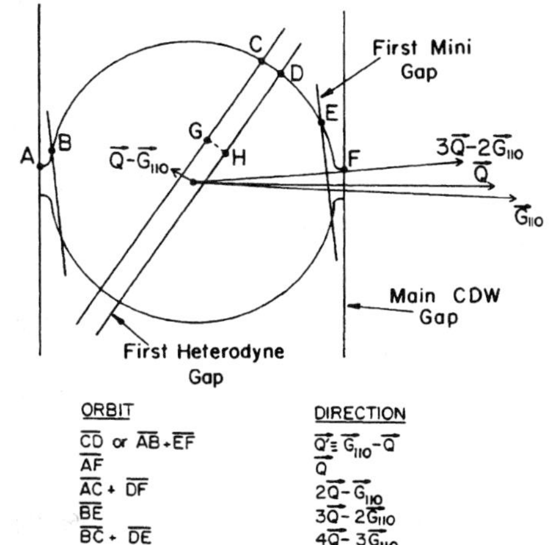

Figure 65.2 Fermi surface, energy gaps, and open orbits (for **H** perpendicular to the plane shown).

When the crystal potential is also included, the Schrödinger equation becomes

$$\left[\frac{p^2}{2m} + G\cos\mathbf{Q}\cdot\mathbf{r} + V\cos\mathbf{G}\cdot\mathbf{r}\right]\Psi_{\mathrm{k}} = \varepsilon_{\mathrm{k}}^0\Psi_{\mathrm{k}}\,. \tag{65.17}$$

Solution of the above Schrödinger equation leads to energy gaps at the Brillouin-zone faces perpendicular to **G** as well as the main CDW gaps. In addition, there are new gaps corresponding to $\mathbf{K}_{m,n} = m\mathbf{Q} + n\mathbf{G}$ with $-\infty < m < \infty$ and $-\infty \le n \le \infty$. Among these only the following three groups of lower-order gaps are important. Since $\mathbf{Q}' \equiv \mathbf{G} - \mathbf{Q}$ is very small, the size of the gaps falls off rapidly with increasing n [24]. The three groups of lower-order gaps are (a) first-zone minigaps; $\mathbf{K}_{n+1,-n} = (n+1)\,\mathbf{Q} - n\mathbf{G}$, (b) second-zone minigaps; $\mathbf{K}_{-n,n+1} = (n+1)\mathbf{G} - n\mathbf{Q}$, and (c) heterodyne gaps; $\mathbf{K}_{-n,n} = n(\mathbf{G} - \mathbf{Q})$, where $n = 1, 2, 3, \ldots$ Minigaps and heterodyne gaps truncate the Fermi surface so that open orbits become possible. Five possible open orbits (depending on magnetic breakdown) are depicted in Figure 65.2.

Since the optimum direction of **Q** does not coincide with a symmetry axis, there are 24 equivalent directions in a single crystal [25]. In general, a macroscopic sample will be divided into **Q** domains, and this plays a very important role in explaining spin-wave data. In the analysis that follows, only consequences of distortion of the Fermi surface, the presence of open orbits, and domain structure will be considered.

65.3 Brief Review of Landau Fermi-Liquid Theory

Landau Fermi-liquid theory [26–29] is based on the assumption that energy levels of an interacting Fermi system can be classified in the same way as the corresponding noninteracting system. That is, a state in the noninteracting system characterized by momentum **p** is assumed to evolve in some way to a corresponding state in the interacting system as the interaction is gradually turned on, and to remain characterized by the same momentum **p**. A state in the interacting system is called a quasiparticle or an elementary excitation, and can be considered as a single particle surrounded by a self-consistent distribution of other particles. A quasiparticle has the same charge as the corresponding noninteracting particle, and they obey Fermi statistics.

Since the energy of a particle depends on the states of the surrounding particles, the total energy of the system becomes a functional of the distribution function $f(\mathbf{p})$, which is envisioned as a statistical matrix with respect to spin. It is assumed that the distribution $f(\mathbf{p})$ characterizes the system completely. For an infinitesimal change in the distribution function $\delta f(\mathbf{p})$ the change in the total energy is expanded to second order in $\delta f(\mathbf{p})$. We shall take the volume of the system to be unity.

$$\begin{aligned}\delta E &= \mathrm{Tr}\int \varepsilon_0(\mathbf{p})\frac{d\mathbf{p}}{(2\pi\hbar)^3} \\ &\quad + \tfrac{1}{2}\mathrm{Tr}\,\mathrm{Tr}'\int F(\mathbf{p},\mathbf{p}')\delta f(\mathbf{p})\delta f(\mathbf{p}')\frac{d\mathbf{p}}{(2\pi\hbar)^3}\frac{d\mathbf{p}'}{(2\pi\hbar)^3} \\ &= \mathrm{Tr}\int \varepsilon(\mathbf{p})\delta f(\mathbf{p})\frac{d\mathbf{p}}{(2\pi\hbar)^3}\,. \end{aligned} \tag{65.18}$$

where

$$\varepsilon(\mathbf{p}) = \varepsilon_0(\mathbf{p}) + \mathrm{Tr}'\int F(\mathbf{p},\mathbf{p}')\delta f(\mathbf{p}')\frac{d\mathbf{p}'}{(2\pi\hbar)^3}\,. \tag{65.19}$$

$\varepsilon(\mathbf{p})$ is the functional derivative of the total energy E with respect to $\delta f(\mathbf{p})$ and corresponds to a change in the energy of the system upon the addition of a single quasiparticle with momentum $\mathbf{p}$; $\varepsilon(\mathbf{p})$ is the energy of the quasiparticle and $\varepsilon_0(\mathbf{p})$ is the energy in the equilibrium distribution. $F(\mathbf{p},\mathbf{p}')$ is the second-order functional derivative of the total energy with respect to the distribution function, and can be considered as the interaction function between quasiparticles.

Since there is a one-to-one correspondence between a state in the interacting system and the corresponding state in the noninteracting system, the entropy of the interacting system can be obtained in the same way as in the noninteracting system.

$$S = -\mathrm{Tr}\int \left[f\ln f + (1-f)\ln(1-f)\right]\frac{d\mathbf{p}}{(2\pi\hbar)^3}\,. \tag{65.20}$$

Using the thermodynamic law

$$\delta E = T\delta S + \mu\delta N\,, \tag{65.21}$$

one can get the distribution function for the quasiparticles, where N is the number of quasiparticles and μ is the chemical potential,

$$f(\mathbf{p}) = \frac{1}{e^{[\varepsilon(\mathbf{p})-\mu]/kT}+1}\,, \tag{65.22}$$

where μ is the chemical potential and is equal to Fermi energy $\varepsilon_F = \varepsilon(p_F)$ at zero temperature. The assignment of a definite momentum to each quasiparticle is possible only when the uncertainty in the momentum due to the finite mean free path is small compared with the momentum and the width of the "transition zone" of the distribution. This leads to the following condition [30].

$$kT \ll \varepsilon_F . \tag{65.23}$$

Nonequilibrium states of a Fermi liquid are described by a distribution function $f(\mathbf{p}, \mathbf{r}, t)$ which depends on both position $\mathbf{r}$ and momentum $\mathbf{p}$, which gives the distribution in a unit volume centered at $\mathbf{r}$. This description is valid as long as the quasiparticle de Broglie wavelength $\hbar/p_F$ is small compared to the wavelength of the inhomogeneity λ,

$$\frac{\hbar}{p_F} \ll \lambda, \quad \text{or } \hbar q \ll p_F . \tag{65.24}$$

Since the frequency ω of the inhomogeneity is of order $v_F q$, the above criterion is equivalent to [30]

$$\hbar\omega \ll \varepsilon_F . \tag{65.25}$$

Therefore Landau theory is applicable only to macroscopic disturbances. Now the change in the total energy becomes

$$\begin{aligned}
\delta E &= \mathrm{Tr} \iint \varepsilon_0(\mathbf{p})\delta f(\mathbf{p}, \mathbf{r}, t)\frac{d\mathbf{p}}{(2\pi\hbar)^3} d\mathbf{r} \\
&\quad + \tfrac{1}{2}\mathrm{Tr}\,\mathrm{Tr}' \iint F(\mathbf{p}, \mathbf{r}, \mathbf{p}', \mathbf{r}')\delta f(\mathbf{p}, \mathbf{r}, t) \\
&\quad \times \delta f(\mathbf{p}', \mathbf{r}', t)\frac{d\mathbf{p}}{(2\pi\hbar)^3}\frac{d\mathbf{p}'}{(2\pi\hbar)^3} d\mathbf{r} d\mathbf{r}' \\
&= \mathrm{Tr} \int \varepsilon(\mathbf{p}, \mathbf{r}, t)\delta f(\mathbf{p}, \mathbf{r}, t)\frac{d\mathbf{p}}{(2\pi\hbar)^3} d\mathbf{r} ,
\end{aligned} \tag{65.26}$$

where

$$\varepsilon(\mathbf{p}, \mathbf{r}, t) = \varepsilon_0(\mathbf{p}) + \delta\varepsilon(\mathbf{p}, \mathbf{r}, t) , \tag{65.27}$$

and

$$\delta\varepsilon(\mathbf{p}, \mathbf{r}, t) = \mathrm{Tr}' \int F(\mathbf{p}, \mathbf{r}, \mathbf{p}', \mathbf{r}')\delta f(\mathbf{p}', \mathbf{r}, t)\frac{d\mathbf{p}'}{(2\pi\hbar)^3} d\mathbf{r}' . \tag{65.28}$$

When the system is assumed to be invariant under spatial translation, the interaction function can only depend on $(\mathbf{r} - \mathbf{r}')$:

$$F(\mathbf{p}, \mathbf{r}, \mathbf{p}', \mathbf{r}') = F(\mathbf{p}, \mathbf{p}'\mathbf{r} - \mathbf{r}') . \tag{65.29}$$

When the interaction is of short range,

$$\delta\varepsilon(\mathbf{p}, \mathbf{r}, t) \cong \mathrm{Tr} \int F(\mathbf{p}, \mathbf{p}')\delta f(\mathbf{p}', \mathbf{r}, t)\frac{d\mathbf{p}'}{(2\pi\hbar)^3} , \tag{65.30}$$

where

$$F(\mathbf{p}, \mathbf{p}') = \int F(\mathbf{p}, \mathbf{r}, \mathbf{p}', \mathbf{r}')d\mathbf{r}' . \tag{65.31}$$

This procedure is not directly applicable to an electron system in a metal, since the Coulomb interaction is a long-range interaction. As shown by Silin [31], this difficulty can be removed if one includes dynamic screening of the particle motion self-consistently. The electrostatic interaction between the average charge distribution of an excited quasiparticle can be described by a space-charge electrostatic field $\mathbf{E}_p(\mathbf{r}, t)$ given by

$$\nabla \cdot \mathbf{E}_p(\mathbf{r}, t) = 4\pi e \mathrm{Tr} \int \delta f(\mathbf{p}, \mathbf{r}, t) \frac{d\mathbf{p}}{(2\pi\hbar)^3} . \tag{65.32}$$

This field can be regarded as an additional applied field. As a result, each excited quasiparticle is surrounded by a polarization cloud of other quasiparticles. The residual interaction $F(\mathbf{p}, \mathbf{r}, \mathbf{p}', \mathbf{r}')$ is short ranged and can be described by the above procedure. The local quasiparticle energy becomes

$$\varepsilon(\mathbf{p}, \mathbf{r}, t) = \varepsilon_0(\mathbf{p}) + \mathrm{Tr}' \int F(\mathbf{p}, \mathbf{p}') \delta f(\mathbf{p}', \mathbf{r}, t) \frac{d\mathbf{p}'}{(2\pi\hbar)^3} . \tag{65.33}$$

In Landau theory, $\varepsilon(\mathbf{p}, \mathbf{r}, t)$ is considered as the Hamiltonian function of the quasiparticle.

The distribution function satisfies a transport equation

$$\frac{df}{dt} = I[f] , \tag{65.34}$$

where $I[f]$ is the collision integral, giving the rate of change in the distribution due to collisions. The explicit time dependence of f contributes a term $\delta f/\delta t$. The dependence on the coordinates and momenta gives terms

$$\frac{\partial f}{\partial \mathbf{r}} \cdot \frac{d\mathbf{r}}{dt} + \frac{\partial f}{\partial \mathbf{p}} \cdot \frac{d\mathbf{p}}{dt} = \frac{\partial f}{\partial \mathbf{r}} \cdot \frac{\partial \varepsilon}{\partial \mathbf{p}} - \frac{\partial f}{\partial \mathbf{p}} \cdot \frac{\partial \varepsilon}{\partial \mathbf{r}} = \{f, \varepsilon\}^{\mathrm{PB}}_{\mathbf{r},\mathbf{p}} ,$$

where Hamilton's equations have been used and PB stands for Poisson bracket. Finally, the time variation of the function as an operator with respect to the spin variables is given by $(\iota/\hbar)[\varepsilon, f] = (\iota/\hbar)[\varepsilon f - f \varepsilon]$. Collecting all the terms, one gets the Landau–Silin equation,

$$\frac{\partial f}{\partial t} + \{f, \varepsilon\}^{\mathrm{PB}}_{\mathbf{r},\mathbf{p}} + \frac{\iota}{\hbar}[\varepsilon, f] = I[f] . \tag{65.35}$$

The distribution function and quasiparticle energy can be decomposed into spin-dependent and spin-independent parts using the Pauli matrices τ:

$$f = n_1 + \tau \cdot \mathbf{n}_2 , \tag{65.36}$$

$$\varepsilon = \varepsilon_1 + \tau \varepsilon_2 . \tag{65.37}$$

By inserting Equations (65.36) and (65.37) into Equation (65.35) and taking a trace, one gets

$$\frac{\partial n_1}{\partial t} + \{n_1, \varepsilon_1\}^{\mathrm{PB}}_{\mathbf{r},\mathbf{p}} + \left\{n_2^j \varepsilon_2^j\right\}^{\mathrm{PB}}_{\mathbf{r},\mathbf{p}} = I[n_1] . \tag{65.38}$$

By inserting Equations (65.36) and (65.37) into Equation (65.35), multiplying by τ, and taking a trace, one gets

$$\frac{\partial \mathbf{n}_2}{\partial t} \{n_1, \varepsilon_2\}^{\mathrm{PB}}_{\mathbf{r},\mathbf{p}} + \{\mathbf{n}_2, \varepsilon_1\}^{\mathrm{PB}}_{\mathbf{r},\mathbf{p}} - \frac{2}{\hbar} \varepsilon_2 \times \mathbf{n}_2 = I[\mathbf{n}]_2 . \tag{65.39}$$

In the presence of external fields, the conjugate momentum **p** is different from the usual momentum **k**:

$$\mathbf{p} = \mathbf{k} + \frac{e}{c}\mathbf{A}\,, \tag{65.40}$$

where one can take the gauge, $\phi = 0$, without loss of generality; i.e.,

$$\mathbf{B} = \nabla \times \mathbf{A}, \quad \mathbf{E} = -\frac{1}{c}\frac{\partial \mathbf{A}}{\partial t}\,. \tag{65.41}$$

It is more convenient to express the transport equations in terms of **k**, which has physical meaning in the absence of external fields [29]. The effect of the vector potential **A** is to shift the origin in **p** space by an amount $e\mathbf{A}/c$. **p** is measured from this shifted origin, and the distribution function $f(\mathbf{p}, \mathbf{r}, t)$ is the same as the distribution $f(\mathbf{k}, \mathbf{r}, t)$ measured from the true origin $\mathbf{k} = 0$. If the Hamiltonian is expressed in terms of **k** and **r**, it has the same form as when $\mathbf{A} = 0$ and $\mathbf{p} = \mathbf{k}$. Now express everything in terms of **k** and **r**;

$$\mathbf{k} = \mathbf{p} - \frac{e}{c}\mathbf{A}\,. \tag{65.42}$$

For operators M and N which depend on **p** and **r**,

$$\left[\frac{\partial m}{\partial t}\right]_{\mathrm{p,r}} = \frac{\partial M}{\partial t} + e\mathbf{E}\cdot\frac{\partial M}{\partial \mathbf{k}}\,, \tag{65.43}$$

$$\{M, N\}^{\mathrm{PB}}_{\mathrm{r,p}}\{M, N\}^{\mathrm{PB}}_{\mathrm{r,k}} + \frac{e}{c}\frac{\partial M}{\partial \mathbf{k}} \times \frac{\partial N}{\partial \mathbf{k}}\cdot\mathbf{B}\,. \tag{65.44}$$

The transport equations become

$$\begin{aligned}&\frac{\partial n_1}{\partial t} + \{n_1, \varepsilon_1\}^{\mathrm{PB}}_{\mathrm{r,p}} = \left\{n_2^j, \varepsilon_2^j\right\}^{\mathrm{PB}}_{\mathrm{r,k}}\\ &+ e\left[\mathbf{E} + \frac{1}{c}\frac{\partial \varepsilon_1}{\partial \mathbf{k}} \times \mathbf{B}\right]\cdot\frac{\partial n_1}{\partial \mathbf{k}} + \frac{e}{c}\frac{\partial n_2^j}{\partial \mathbf{k}} \times \frac{\partial \varepsilon_2^j}{\partial \mathbf{k}}\cdot\mathbf{B} = I[n_1]\,,\end{aligned} \tag{65.45}$$

where the summation over repeated indices is assumed;

$$\begin{aligned}&\frac{\partial \mathbf{n}_2}{\partial t} + \{\mathbf{n}_2, \varepsilon_1\}^{\mathrm{PB}}_{\mathrm{r,p}} + \{n_1, \varepsilon_2\}^{\mathrm{PB}}_{\mathrm{r,p}} - \frac{2}{\hbar}\varepsilon_2 \times \mathbf{n}_2\\ &+ e\left[\mathbf{E} + \frac{1}{c}\frac{\partial \varepsilon_1}{\partial \mathbf{k}} \times \mathbf{B}\right]\cdot\nabla_{\mathbf{k}}\mathbf{n}_2 - \frac{e}{c}\frac{\partial n_1}{\partial \mathbf{k}} \times \mathbf{B}\cdot\frac{\partial \varepsilon_2}{\partial \mathbf{k}} = I[\mathbf{n}_2]\,.\end{aligned} \tag{65.46}$$

When the system is invariant under translation and there is no spin-orbit coupling, the interaction function $F(\mathbf{k}, \mathbf{k}')$ can be decomposed into spin-independent and spin-dependent parts in terms of Pauli spin atrices [6, 7, 29]

$$F(\mathbf{k}, \mathbf{k}') = \eta(\mathbf{k}, \mathbf{k}') + \tau\cdot\tau'\xi(\mathbf{k}, \mathbf{k}')\,. \tag{65.47}$$

Under the presence of a static field $\mathbf{H}_0$ and an rf field **h**,

$$\varepsilon_1 = \varepsilon^0_{\mathbf{k}} + \int \frac{2d\mathbf{k}'}{(2\pi\hbar)^3}\eta(\mathbf{k}, \mathbf{k}')n_1(\mathbf{k}', \mathbf{r}, t)\,, \tag{65.48}$$

$$\varepsilon_2 = -\frac{\gamma_0\hbar}{2}\mathbf{H} + \int \frac{2d\mathbf{k}'}{(2\pi\hbar)^3}\xi(\mathbf{k}, \mathbf{k}')\mathbf{n}_2(\mathbf{k}', \mathbf{r}, t) \tag{65.49}$$

$$= \varepsilon^0_2 + \delta\varepsilon_2\,, \tag{65.50}$$

where

$$\varepsilon_2^0 = -\frac{\gamma\hbar}{2}\mathbf{H}_0 , \tag{65.51}$$

$$\gamma = \gamma_0/\gamma_0^B , \tag{65.52}$$

$$\gamma_0^B = 1 + B_0 , \tag{65.53}$$

$$B_0 = \int \frac{2d\mathbf{k}'}{(2\pi\hbar)^3}\xi(\mathbf{k},\mathbf{k}')\frac{\partial n_{\mathbf{k}}^0}{\partial \varepsilon_{\mathbf{k}'}^0} , \tag{65.54}$$

and γ_0 is the gyromagnetic ratio of an electron. In terms of these quantities, one can linearize the transport equation for spin polarizations m_α ($\alpha = +, -, 0$), where $m_\pm = (m_x \pm \iota m_y)/\sqrt{2}$ and $m_0 = m_z$. The linearized Landau–Silin equation is

$$\begin{aligned} &\frac{\partial m_\alpha}{\partial t} + \frac{e}{c}\mathbf{v}_\mathbf{k} \times \mathbf{H}_0 \cdot \left[\frac{\partial m_\alpha}{\partial \mathbf{k}} - \frac{\partial n_\mathbf{k}^0}{\partial \varepsilon_\mathbf{k}^0}\frac{\partial \delta\varepsilon_{2\alpha}}{\partial \mathbf{k}}\right] + \mathbf{v}_\mathbf{k} \cdot \nabla \left[m_\alpha - \frac{\partial n_\mathbf{k}^0}{\partial \varepsilon_\mathbf{k}^0}\delta\varepsilon_{2\alpha}\right] \\ &- \iota\alpha\Omega_0 \left[m_\alpha - \frac{\partial n_\mathbf{k}^0}{\partial \varepsilon_\mathbf{k}^0}\delta\varepsilon_{2\alpha}\right] = I[m_\alpha] \end{aligned} \tag{65.55}$$

where

$$\Omega_0 = -\gamma\mathbf{H}_0 , \tag{65.56}$$

and

$$\delta\varepsilon_{2\alpha} = -\frac{\gamma_0\hbar}{2}h_\alpha + \int \frac{2d\mathbf{k}'}{(2\pi\hbar)^3}\xi(\mathbf{k},\mathbf{k}')m_\alpha(\mathbf{k}',\mathbf{r},t) . \tag{65.57}$$

65.4 Simplified Model for Charge-Density-Wave Effects

In the following analysis, several CDW effects [10–12][5] will be incorporated into the theory of spin waves in alkali metals; mainly, the anisotropic velocity distribution caused by the distortion of the Fermi surface, the open-orbit motion resulting from truncation of the Fermi surface by the extra energy gaps, and CDW **Q**-domain structure. Among many possible open orbits, the ones due to the heterodyne gaps are believed to be most important, since its direction has the largest angle with respect to **Q**. Also it is believed that $\mathbf{G}_{110}$ is usually perpendicular to the surface of a thin sample [32, 33].

The above three effects will be analyzed using the following simplified model, shown in Figure 65.3. The Fermi surface can be thought of as being composed of two parts: a lemon-shaped surface and a cylindrical surface. The lemon-shaped surface describes the motion

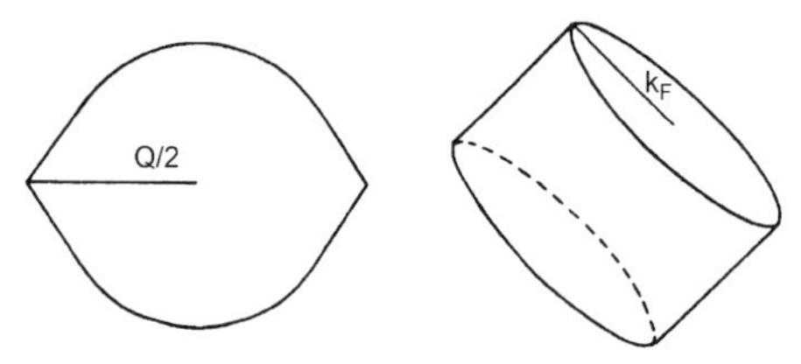

Figure 65.3 Simplified model of the Fermi surface.

5) For a review of these phenomena, see [12].
6) For a review of these phenomena, see [12].

of electrons in closed orbits and the cylindrical surface describes the motion of electrons in open orbits. The axis of the cylindrical surface is taken parallel to $\mathbf{Q}' = \mathbf{G}_{110} - \mathbf{Q}$.

In the following analysis, B_n with $n \geq 1$ will be neglected since they are believed to be quite small compared to B_0 [15]. Instead, we will introduce 77, the fraction of electrons in the cylindrical part, as a new parameter. It should be emphasized that G, the CDW energy gap, and **Q**, the CDW wave vector, are not parameters to be used in fitting the spin-wave data. They are determined from other experiments [10–12].[6] So we have B_0 and η as fitting parameters instead of B_0 and B_1 as in conventional theories. If η were to become known from another experiment, we could include B_1 as an adjustable parameter. We emphasize that B_1 and η play compensatory roles. Determinations of B_1 from spin-wave data have significance only if a CDW structure is not present.

65.4.1 Closed Orbits

For the closed-orbit part, the energy of an electron is given by

$$\varepsilon_{\mathbf{k}}^{0,c} = \frac{1}{2m}\left(k_x^2 + k_y^2\right) + \frac{\hbar^2 Q^2}{2m} f_e(w) , \tag{65.58}$$

where (x, y, z) refers to the c-coordinate system defined in Figure 65.4 with $\hat{z}$ parallel to $\mathbf{Q}$, the superscript c stands for closed orbit, m is the band mass ($m = 1.211\ m_0$), and

$$f_e(w) = w^2 + \tfrac{1}{4} - (w^2 + \xi^2)^{1/2} , \tag{65.59}$$

$$w = \frac{k_z}{\hbar Q} \mp \tfrac{1}{2}\mathrm{sgn}(k_z) , \tag{65.60}$$

$$\xi = \frac{mG}{\hbar^2 Q^2} = 0.0035\ 25 . \tag{65.61}$$

(This value of ξ falls a little short of critical contact of the Fermi surface with the CDW gap.) The coordinate systems used and angles are defined in Figure 65.4, where $\mathbf{Q}$ is parallel to $\hat{z}$.

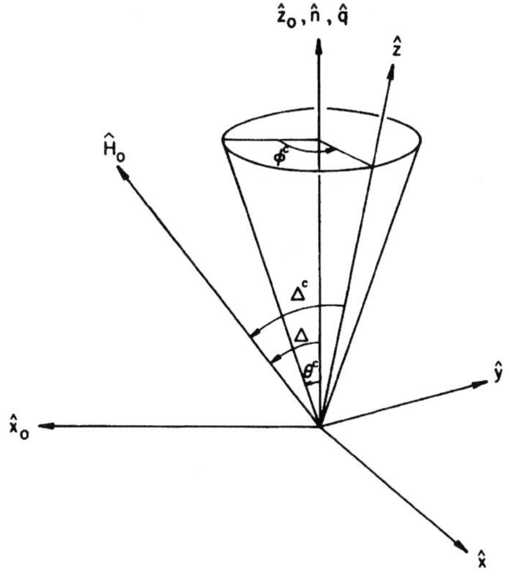

Figure 65.4 Definitions of coordinate systems and angles.

Since m^c_α in Equation (65.55) is proportional to $\partial n^{c,0}_{\mathbf{k}}/\partial \varepsilon^{c,0}_{\mathbf{k}}$, it is convenient to define

$$m^c_\alpha = -\frac{\partial n^{c,0}_{\mathbf{k}}}{\partial \varepsilon^{c,0}_{\mathbf{k}}} g^c_\alpha \,. \tag{65.62}$$

The Landau–Silin equation becomes

$$\delta\left(\varepsilon^{0,c}_{\mathbf{k}} - \varepsilon_F\right)\left[\omega g^c_\alpha - \Omega_0 \overline{g}^c_\alpha + \frac{\iota}{\tau}\left(\overline{g}^c_\alpha - \langle \overline{g}^c_\alpha \rangle^c\right) + \frac{\iota}{T_2}\overline{g}^c_\alpha + \iota \mathbf{v}\cdot\nabla \overline{g}^c_\alpha\right]$$
$$+\iota\frac{e}{c}\left[\mathbf{v}\times\mathbf{H}_0\cdot\frac{\partial}{\partial \mathbf{k}}\right]\delta\left(\varepsilon^{0,c}_{\mathbf{k}} - \varepsilon_F\right)\overline{g}^c_\alpha = 0\,, \tag{65.63}$$

where ⟨⟩ denotes the average over the Fermi surface,

$$\overline{g}^c_\alpha = g^c_\alpha - \frac{\gamma_0 \hbar}{2} h_\alpha + \delta g^c_\alpha\,, \tag{65.64}$$

and

$$\delta g^c_\alpha = \int \frac{2d\mathbf{k}'}{(2\pi\hbar)^3}\xi(\mathbf{k},\mathbf{k}')\delta\left(\varepsilon^{0,c}_{\mathbf{k}} - \varepsilon_F\right) g^c_\alpha(\mathbf{k}')\,. \tag{65.65}$$

We will look for a solution of the form

$$g^c_\alpha = g^c_{0\alpha} + \mathbf{g}^c_{1\alpha}\cdot\mathbf{v}\,, \tag{65.66}$$

which is the simplest function with a term linear in the velocity. The rf magnetization $\mathcal{M}^c_\alpha$ is obtained by integrating m^c_α over $\mathbf{k}$:

$$\mathcal{M}^c_\alpha = \int \frac{2d\mathbf{k}}{(2\pi\hbar)^3}\frac{\gamma_0\hbar}{2} m^c_\alpha = \frac{\gamma_0\hbar}{2}\nu^c_F g^c_{0\alpha}\,, \tag{65.67}$$

where ν^c_F is the density of states at the Fermi energy.

It will be assumed that all fields and spin polarization depend on the coordinate $z_0 = -x\sin\theta^c + z\cos\theta^c$, where $\hat{z}_0$ is perpendicular to the surface of the sample. By integrating Equation (65.63) over $\mathbf{k}$, after multiplying by 1, v_x, v_y, and v_z, we obtain four coupled equations for $g^c_{0\alpha}$, $g^c_{1\alpha x}$, $g^c_{1\alpha y}$, and $g^c_{1\alpha z}$:

$$\omega g^c_{0\alpha} + \left[\alpha\omega_s + \frac{\iota}{T}\right] g^c_{0\alpha'} - \iota\langle v^2_x\rangle^c \frac{\partial}{\partial x} g^c_{1\alpha x}$$
$$+\iota\langle v^2_z\rangle^c\frac{\partial}{\partial y} g^c_{1\alpha z} = 0\,, \tag{65.68}$$

$$\left(\overline{\omega}^c + \alpha\Omega_0\right)\langle v^2_x\rangle^c g^c_{1\alpha x} - \iota\omega_c\Lambda^c_z\langle v^2_y\rangle^c g^c_{1\alpha y}$$
$$+\iota\omega_c\Lambda^c_z\langle v^2_z\rangle^c g^c_{1\alpha z} = \iota\gamma^B_0\langle v^2_x\rangle^c\frac{\partial}{\partial x} g^{c'}_{0\alpha}\,, \tag{65.69}$$

$$\iota\omega_c\Lambda^c_z\langle v^2_x\rangle^c g^c_{1\alpha x} + \left(\overline{\omega}^c + \alpha\Omega_0\right)\langle v^2_y\rangle^c g^c_{1\alpha y}$$
$$-\iota\omega_c\Lambda^c_x\langle v^2_z\rangle^c g^c_{1\alpha z} = 0\,, \tag{65.70}$$

$$-\iota\omega_c\Lambda^c_y\left\langle v^2_x\frac{m\partial v_z}{\partial k_z}\right\rangle^c g^c_{1\alpha x} + \iota\omega_c\Lambda^c_x\left\langle v^2_y\frac{m\partial v_z}{\partial k_z}\right\rangle^c g^c_{1\alpha y}$$
$$+\left(\overline{\omega}^c + \alpha\Omega_0\right)\langle v^2_z\rangle^c g^c_{1\alpha z} = \iota\gamma^B_0\langle v^2_z\rangle^c\frac{\partial}{\partial z} g^{c'}_{0\alpha}\,, \tag{65.71}$$

where

$$g_{0\alpha}^{c'} = g_{0\alpha}^{c} - \frac{\gamma_0 \hbar}{2\gamma_0^B} h_\alpha \,, \tag{65.72}$$

$$T = T_2/\gamma_0^B \,, \tag{65.73}$$

and

$$\overline{\omega}^c = \omega + i/\tau \,. \tag{65.74}$$

The Λ^c's specify the direction of $\mathbf{H}_0$ in the c-coordinate system:

$$\begin{aligned}
\Lambda_x^c &= \sin\Delta \cos\theta^c \cos\phi^c - \cos\Delta\cos\theta^c \,, \\
\Lambda_y^c &= -\sin\Delta\sin\phi^c \,, \\
\Lambda_z^c &= \sin\Delta\sin\theta^c\cos\phi^c + \cos\Delta\cos\theta^c \,,
\end{aligned} \tag{65.75}$$

where Δ is the angle between $\mathbf{H}_0$ and $\hat{z}_0$ (perpendicular to the surface of the sample), and θ^c and ϕ^c are the polar and azimuthal angle of $\mathbf{Q}$ relative to the coordinate system (x_0, x_0, z_0).

Various averages in the above equations can be evaluated to first order in ξ;

$$\begin{aligned}
\langle v_x^c \rangle^c &= \langle v_y^c \rangle^c = \tfrac{1}{3} v_F^2 (1 - 2\xi) \,, \\
\langle v_z^c \rangle^c &= \tfrac{1}{3} v_F^2 \left[1 - 2\xi \left[3\tan^{-1}\frac{1}{2\xi} - 2 \right] \right] , \\
\left\langle v_x^2 \frac{m\partial v_z}{\partial k_z} \right\rangle^c &= \left\langle v_y^2 \frac{m\partial v_z}{\partial k_z} \right\rangle^c = \tfrac{1}{3} v_F^c \left[1 - \left(3\pi + \tfrac{1}{2} \right) \xi \right] .
\end{aligned} \tag{65.76}$$

A solution can be found for $g_{0\alpha}^c$ by solving Equations (65.68)–(65.71):

$$\omega g_{0\alpha}^c + \left(\alpha\omega_s + \iota/T - \gamma_0^B \beta^c \nabla_0^2 \right) g_{0\alpha}^{c'} = 0 \,, \tag{65.77}$$

where

$$\begin{aligned}
\beta^c &= \langle v_x^2 \rangle^c \Big\{ \sin^2\theta^c \left[(\overline{\omega}^c + \alpha\Omega_0) - \mu\omega_c^2 \left(\Lambda_x^c\right)^2 \right] + \lambda\cos^2\theta^c \\
&\quad \times \left[(\overline{\omega}^c + \alpha\Omega_0)^2 - \omega_c^2 \left(\Lambda_z^c\right)^2 \right] + \sin\theta^c\cos\theta^c \left[(\lambda+\mu)\omega_c^2 \Lambda_x^c \Lambda_z^c \right. \\
&\quad \left. - \iota(\lambda-\mu)\omega_c \Lambda_y^c (\overline{\omega}^c + \alpha\Omega_0) \right] \Big\} \Big/ \left\{ (\overline{\omega}^c + \alpha\Omega_0) \left[(\mu\sin^2\Delta^c \right. \right. \\
&\quad \left. \left. + \cos^2\Delta^2)\, \omega_c^2 - (\overline{\omega}^c + \alpha\Omega_0)^2 \right] \right\} , \\
\mu &= \left\langle v_x^2 \frac{m\partial v_z}{\partial k_z} \right\rangle^c \Big/ \langle v_x^2 \rangle^c \,,
\end{aligned} \tag{65.78}$$

By using Equations (65.67) and (65.77), one obtains a modified Bloch equation,

$$\omega \mathcal{M}_\alpha^c + \left(\alpha\omega_s + \iota/T - \gamma_0^B \beta^c \nabla_0^2 \right) \left(\mathcal{M}_\alpha^c - \chi h_\alpha \right) = 0 \,, \tag{65.79a}$$

or equivalently

$$\frac{\partial \mathbf{M}}{\partial t} = \gamma_0 \mathbf{M} \times \mathbf{M} - \frac{\mathbf{M} - \mathbf{M}_{\mathrm{eq}}}{T} + D^c \nabla_0^2 (\mathbf{M} - \mathbf{M}_{\mathrm{eq}}) \,, \tag{65.79b}$$

where $\mathbf{M}_{\mathrm{eq}} = \chi\mathbf{H} = \chi(\mathbf{H}_0 + \mathbf{h})$, $\mathbf{M} = \mathbf{M}_{\mathrm{eq}} + \mathbf{M}^c$, $D^c = \gamma_0^B \beta^c/\iota$, and $\chi = \chi_0/\gamma_0^B$ is the susceptibility of an interacting electron gas. Equation (65.79), originally suggested by

Torrey [34], was first obtained by Walker [9]. When fields and spin polarizations are assumed to be proportional to $e^{\iota q z_0}$, Equation (65.77) is reduced to

$$g_{0\alpha}^{c} = \frac{\Omega_0 \frac{\gamma_0 \hbar}{2}\left[\alpha + \frac{\iota}{\Omega_0 T} + \frac{\beta^c q^2}{\Omega_0}\right] h_\alpha}{\omega + \alpha\omega_s + \iota/T + \gamma_0^B \beta^c q^2} . \tag{65.80}$$

By neglecting terms of the order $1/\Omega_0 T \ll 1$ and $\beta^c q^2/\Omega_0 \ll 1$, in the small-$q$ limit, one gets

$$g_{0\alpha}^{c} = \frac{\Omega_0 \frac{\gamma_0 \hbar}{2} h_\alpha}{\omega + \alpha\omega_s + \iota/T + \gamma_0^B \beta^c q^2} , \tag{65.81}$$

which was first obtained by Platzman and Wolff [6, 7]. The Bloch equation in this limit is

$$\omega \mathcal{M}_\alpha^c + \left(\alpha\omega_s + \iota/T - \gamma_0^B \beta^c \nabla_0^2\right)\mathcal{M}_\alpha^c = \alpha\omega_s \chi h_\alpha , \tag{65.82a}$$

or equivalently

$$\frac{\partial \mathbf{M}}{\partial t} = \gamma_0 \mathbf{M} \times \mathbf{H} - \frac{\mathbf{M} - \mathbf{M}_0}{T} + D^c \nabla_0^2 \mathbf{M} , \tag{65.82b}$$

where $\mathbf{M}_0 = \chi \mathbf{H}_0$ and $\mathbf{M} = \mathbf{M}_0 + \mathbf{M}^c$. Equation (65.79) is more accurate than Equation (65.82) in the sense that no specific spatial dependence of fields and polarizations is assumed. It has the nice property that the magnetization relaxes to the local equilibrium value $\chi \mathbf{H}$, and it is valid for all α. But Equation (65.82), originally suggested by Kaplan [35], would be equally satisfactory for our purpose, since only small-g excitations are of interest and only the $\alpha = -1$ component is resonant. Equation (65.79a) can be put in a slightly different form in terms of $\mathcal{M}_\alpha^{c'}$:

$$\omega \mathcal{M}_\alpha^c + (\alpha\omega_s + \iota/T - \gamma_0^B \beta^c \nabla_0^2)\mathcal{M}_\alpha^{c'} = -\left[\frac{\omega}{\omega_s}\right]\omega_s \chi h_\alpha , \tag{65.83}$$

where

$$\mathcal{M}_\alpha^{c'} = \mathcal{M}_\alpha^c - \chi h_\alpha . \tag{65.84}$$

Since $\omega \cong \omega_s$, one can see that the Platzman–Wolff solution for M_- is valid without the small-g approximation if the rf magnetization in Equation (65.82) is interpreted as the deviation from a local equilibrium value χh_- and h_- is replaced by $(\omega/\omega_s)h_-(\cong h_-)$.

When $\xi = 0$, i.e., for a spherical Fermi surface, Equation (65.78) reduces to that of the PW theory:

$$\beta^{\mathrm{PW}} = \beta^c \gamma_0^B = \tfrac{1}{3} v_F^2 \gamma_0^B (\varpi^c - \Omega_0) \times \left[\frac{\sin\Delta}{\omega_c^2 - (\varpi^c - \Omega_0)^2} - \frac{\cos^2\Delta}{(\varpi^c - \Omega_0)^2}\right] . \tag{65.85}$$

If $|B_0 \omega \tau/(1 + B_0) \gg 1$, β approaches a pure (real) number and g_0 exhibits a branch of singularities along the curve $\omega - \omega_s = \beta q^2 = 0$ in ω–q space. When $B_0 \to 0$, β approaches a pure imaginary number and there are no singularities. Therefore paramagnetic spin waves exist only when the interaction between electrons is strong enough [6, 7].

65.4.2 Open Orbits

Similar procedures to those just described can be followed for the cylindrical Fermi-energy surface defined by

$$\varepsilon_{\mathbf{k}}^{0,o} = \frac{1}{2m}\left(k_x^2 + k_y^2\right) , \tag{65.86}$$

where the superscript o stands for open orbit. The coordinate systems and angles used are defined in Figure 65.4, where $\mathbf{Q}'$ is parallel to $\hat{z}$.

The Landau–Silin equation becomes

$$\delta\left(\varepsilon_{\mathbf{k}}^{0,o} - \varepsilon_F\right)\left[\omega g_0^o - \Omega_0 \overline{g}_\alpha^o + \frac{\iota}{\tau_{op}}\left(\overline{g}_\alpha^o - \langle \overline{g}_\alpha^o \rangle^c\right) + \frac{\iota}{T_2}\overline{g}_\alpha^o - \iota \mathbf{v}\cdot\nabla\overline{g}_\alpha^o\right]$$
$$+ \iota\frac{e}{c}\left[\mathbf{v}\times\mathbf{H}_0\cdot\frac{\partial}{\partial\mathbf{k}}\right]\delta\left(\varepsilon_{\mathbf{k}}^{0,o} - \varepsilon_F\right)\overline{g}_\alpha^o = 0 \tag{65.87}$$

where electrons in open orbits are assumed to have a different momentum relaxation time from that of electrons in closed orbits. We look for a solution having the form

$$g_\alpha^o = g_{1\alpha}^o + \mathbf{g}_{1\alpha}^o\cdot\mathbf{v} , \tag{65.88}$$

as before. The rf magnetization $\mathcal{M}_\alpha^o$ is obtained by integrating m_α^o over $\mathbf{k}$;

$$\mathcal{M}_\alpha^o = \frac{\gamma_0\hbar}{2} v_F^o g_{0\alpha}^o . \tag{65.89}$$

It will be assumed that all fields and the spin polarization depend on the coordinate $z_0 = x\sin\theta^o + z\cos\theta^o$. (x, y, z) refers to the c-coordinate system defined in Figure 65.4 with $\hat{z}$ parallel to $\mathbf{Q}'$.

By integrating Equation (65.87), after multiplying by $1, v_x, v_y$, we obtain three coupled equations in $g_{0\alpha}^o, g_{1\alpha x}^o, g_{1\alpha y}^o$;

$$\omega g_{0\alpha}^o + \left(\alpha\omega_s + \iota/T\right) g_{0\alpha}^{o'} - \iota\langle v_x^2\rangle^o \frac{\partial}{\partial x} g_{1\alpha x}^o = 0 , \tag{65.90}$$

$$\left(\overline{\omega}^o + \alpha\Omega_0\right)\langle v_x^2\rangle^o g_{1\alpha x}^o - \iota\omega_c\Lambda_z^o\langle v_y^2\rangle^o g_{1\alpha y}^o = \tfrac{1}{2}\iota\gamma_0^B\langle v_x^2\rangle^o\frac{\partial}{\partial x} g_{0\alpha}^{o'} , \tag{65.91}$$

$$\iota\omega_c\Lambda_z^o\langle v_x^2\rangle^o g_{1\alpha x}^o + \left(\overline{\omega}^o + \alpha\Omega_0\right)\langle v_y^2\rangle^o g_{1\alpha y}^o = 0 , \tag{65.92}$$

where $\overline{\omega}^o = \omega + \iota/\tau_{op}$. The Λ^o's, defined by equations similar to Equation (65.75), specify the direction of $\mathbf{H}_0$ in the o-coordinate system and θ^o and ϕ^o are the polar and azimuthal angle of $\hat{\mathbf{Q}}'$ relative to the coordinate system (x_0, y_0, z_0). A solution can be found for $g_{0\alpha}^o$ by using Equations (65.90)–(65.92):

$$\omega g_{0\alpha}^o + \left(\alpha\omega_s + \iota/T - \gamma_0^B\beta^o\nabla_0^2\right) g_{0\alpha}^{o'} = 0 , \tag{65.93}$$

$$\beta^o = \tfrac{1}{2}v_F^2\left(\overline{\omega}^o - \Omega_0\right)\frac{\sin^2\theta^o}{\omega_p^2 - \left(\overline{\omega}^o - \Omega_0\right)^2} , \tag{65.94}$$

where

$$\omega_p = \omega_c\cos\left(\hat{\mathbf{Q}}'\cdot\hat{\mathbf{H}}_0\right) \tag{65.95}$$

and

$$\overline{\omega}^{o} = \omega + i/\tau_{op} . \tag{65.96}$$

By using Equations (65.89) and (65.93), one gets Bloch equations corresponding to Equations (65.79) and (65.82).

The main difference between the solution here and the one in Section 65.4.1 is the following: The factor 3 or 2 corresponds to the dimensional freedom of the motion; an average of one component of **v** is reduced from $\frac{1}{3}v_F^2$ to $\frac{1}{2}v_F^2 \sin^2\theta^{o}$. The electrons on the cylindrical part of the Fermi surface cannot move perpendicular to $\mathbf{Q}'$, and can only move in open orbits if $\mathbf{H}_0$ is perpendicular to $\mathbf{Q}'$. ω_c^{o}, the effective cyclotron frequency, becomes 0 in that case.

65.4.3 Mixed Orbits

When we consider both Fermi surfaces at the same time, the cross relaxation of quasiparticles from one surface to another becomes important. When a quasiparticle is destroyed at a point on the Fermi surface, it can end up at a point on either surface with equal probability per unit Fermi-surface area. The collision integral becomes the following:

$$\left[\frac{\partial g_{\alpha}^{c}}{\partial t}\right]_{\text{coll}} = -\frac{\overline{g}_{\alpha}^{c}}{\tau} - \frac{\overline{g}_{\alpha}^{c}}{T_2} + (1-\eta)\frac{\langle \overline{g}_{\alpha}^{c}\rangle^{c}}{\tau} + \eta\frac{\langle \overline{g}_{\alpha}^{o}\rangle^{o}}{\tau_{op}} , \tag{65.97}$$

$$\left[\frac{\partial g_{\alpha}^{o}}{\partial t}\right]_{\text{coll}} = -\frac{\overline{g}_{\alpha}^{o}}{\tau_{op}} - \frac{\overline{g}_{\alpha}^{o}}{T_2} + (1-\eta)\frac{\langle \overline{g}_{\alpha}^{c}\rangle^{c}}{\tau} + \eta\frac{\langle \overline{g}_{\alpha}^{o}\rangle^{o}}{\tau_{op}} , \tag{65.98}$$

where η is the fraction of electrons on the open-orbit Fermi surface and τ_{op} is their momentum relaxation time. It has been shown that τ_{op} is shorter than τ on account of **Q**-domain boundaries [36]:

$$\frac{1}{\tau_{op}} = \frac{1}{\tau} + \frac{8v_F}{3D} , \tag{65.99}$$

where D is the **Q**-domain size, η and D play crucial roles in explaining the splittings of spin-wave side bands as will be shown later. In addition to the coupling through Equations (65.97) and (65.98), g^{c} and g^{o} get coupled through the interaction between electrons:

$$\delta g_{\alpha}^{c} = \delta g_{\alpha}^{o} = B_0\left[(1-\eta)g_{0\alpha}^{c} + \eta g_{0\alpha}^{o}\right] . \tag{65.100}$$

The rf magnetization $\mathcal{M}_{\alpha}$ is obtained by integrating m_{α} over **k**:

$$\mathcal{M}_{\alpha} = \frac{\gamma_0\hbar}{2}v_F\left[(1-\eta)\,g_{0\alpha}^{c} + \eta g_{0\alpha}^{o}\right] . \tag{65.101}$$

Now we will have seven coupled equations, similar to Equations (65.68)–(65.70) and (65.90)–(65.92) with modified collision terms (65.97) and (65.98), and modified interaction terms Equation (65.100), after multiplying by appropriate factors and taking averages over the Fermi surface. Solving these seven equations for g_0^{c} and g_0^{o}, one gets the following two coupled equations:

$$\begin{aligned}
&\omega g_{0\alpha}^{c} + \Bigg[[1+(1-\eta)B_0]\left[\alpha\Omega_0 + \frac{\iota}{T_2}\right] + \iota\eta\left[\frac{1+(1-\eta)B_0}{\tau}\right. \\
&\left.\left. - \frac{(1-\eta)B_0}{\tau_{op}}\right]\right] g_{0\alpha}^{c'} - \beta^{c}\nabla_0^2\left\{[1+(1-\eta)B_0]\,g_{0\alpha}^{c'} + \eta B_0 g_{0\alpha}^{o'}\right\} \\
&+ \left[\eta B_0\left[\alpha\Omega_0 + \frac{\iota}{T_2}\right] - \iota\eta\left[\frac{\eta B_0}{\tau} - \frac{1+\eta B_0}{\tau_{op}}\right]\right] g_{0\alpha}^{o'} = 0 ,
\end{aligned} \tag{65.102}$$

$$\omega g^{c}_{0\alpha} + \left[(1+\eta B_0)\left[\alpha\Omega_0 + \frac{\iota}{T_2}\right] - \iota(1-\eta)\left[\frac{\eta B_0}{\tau} - \frac{1+\eta B_0}{\tau_{op}}\right]\right] g^{o'}_{0\alpha}$$
$$- \beta^{o}\nabla_0^2\left[(1+\eta B_0)g^{o'}_{0\alpha} + (1-\eta)B_0 g^{c'}_{0\alpha}\right] + \left[(1-\eta)B_0\left[\alpha\Omega_0 + \frac{\iota}{T_2}\right]\right.$$
$$\left. -\iota(1-\eta)\left[\frac{1+(1-\eta B_0)}{\tau} - \frac{(1-\eta)B_0}{\tau_{op}}\right]\right] g^{c'}_{0\alpha} = 0 \,. \tag{65.103}$$

Unfortunately the solution for $g_{0\alpha} = (1-\eta)g^{c}_{0\alpha} + \eta g^{0}_{0\alpha}$ cannot be found in a closed form and one has to resort to the approximation that fields and spin polarizations are proportional to $e^{\iota q z_0}$ with small q. The solutions for $g^{c}_{0\alpha}$ and $g^{o}_{0\alpha}$, correct to first order in q^2, can be found by neglecting terms of the order of $\eta(\beta q^2/\omega)^2$;

$$g_{0\alpha'} = (1-\eta)g^{c}_{0\alpha'} + \eta g^{o}_{0\alpha'}$$
$$= \frac{-\omega\gamma\hbar/2}{\omega + \alpha\omega_s + \iota/T + \gamma_0^B\beta q^2} h_\alpha \,, \tag{65.104}$$

where

$$\beta = \gamma_0^B\left[(1-\eta)\beta^c + \eta\beta^0\right] \,, \tag{65.105}$$

and $T = T_2/\gamma_0^B$. By using Equations (65.101) and (65.104), one gets Bloch equations corresponding to Equations (65.79) and (65.82):

$$\omega\mathcal{M}_\alpha + \left(\alpha\omega_s + \iota/T - \iota D^*\nabla_0^2\right)\left(\mathcal{M}_\alpha - \chi h_\alpha\right) = 0 \,, \tag{65.106a}$$

or equivalently

$$\frac{\partial \mathbf{M}}{\partial t} = \gamma_0 \mathbf{M}\times\mathbf{H} - \frac{\mathbf{M}-\mathbf{M}_{\text{eq}}}{T} + D^*\nabla_0^2\left(\mathbf{M}-\mathbf{M}_{\text{eq}}\right) \,, \tag{65.106b}$$

where $\mathbf{M}_{\text{eq}} = \chi\mathbf{H} = \chi(\mathbf{H}_0+\mathbf{h})$, $\mathbf{M} = \mathbf{M}_{\text{eq}} + \mathcal{M}$, $D^* = \gamma_0^B\beta/\iota$, and $\chi = \chi_0/\gamma_0^B$ is the susceptibility of an interacting electron gas.

All the modifications caused by CDW effects incorporated in this work, i.e., the anisotropic velocity distribution, the presence of open orbits, and the CDW **Q**-domain structure, are contained in Equation (65.105). β^o depends on the **Q**-domain size D through Equations (65.96) and (65.99).

65.4.4 Transmitted Signals

A standard way of finding a transmitted signal through a finite slab is to utilize the Bloch equation (65.106) with an appropriate boundary condition. From Equation (65.106), one can see that the magnetization current $\mathbf{J}_{\mathcal{M}}$ is given by $-\nabla_0(\mathbf{M}-\mathbf{M}_{\text{eq}})$, or equivalently $-\nabla_0(\mathcal{M}-\chi h)$ [6, 7]. The subscript α is dropped since only the $\alpha = -1$ component is resonant. The appropriate boundary condition in the absence of surface spin relaxation is

$$\hat{\mathbf{n}}\cdot\mathbf{J}_{\mathcal{M}} = \hat{\mathbf{n}}\cdot\nabla_0(\mathcal{M}-\chi h) = 0 \,. \tag{65.107}$$

A simpler way is to solve the Bloch equation with the above boundary condition for $\mathcal{M}'$;

$$\omega\mathcal{M}' + (-\omega_s + \iota/T - \iota D\nabla_0^2)\mathcal{M}' = -\omega\chi h \,. \tag{65.108}$$

The boundary condition (65.107) becomes

$$\hat{\mathbf{n}}\cdot\nabla_0\mathcal{M}' = 0 \,. \tag{65.109}$$

When fields and magnetization are assumed to depend only on z_0, Equation (65.108) becomes

$$\frac{d^2}{dz_0^2}\mathcal{M}' + k^2\mathcal{M}' = \frac{\omega\chi}{\iota D}h\,, \tag{65.110}$$

where

$$k^2 = \frac{\iota(\omega-\omega_s)T - 1}{D^* T}\,, \tag{65.111}$$

and

$$\left[\frac{d\mathcal{M}'}{dz_0}\right]_{z_0=\pm L/2} = 0\,. \tag{65.112}$$

The solution for $\mathcal{M}$ can be readily found to be

$$\mathcal{M} = \chi h + \frac{\omega\chi}{\iota D^*}\frac{1}{k\sin^2 W}\left\{\cos\left[k\left[z-\frac{L}{2}\right]\right]\int_{-L/2}^{z} dz'\cos\left[k\left[\frac{L}{2}+z'\right]\right]h(z')\right.$$
$$\left. + \cos\left[k\left[z+\frac{L}{2}\right]\right]\int_{z}^{L/2} dz'\cos\left[k\left[\frac{L}{2}-z'\right]\right]h(z')\right\}, \tag{65.113}$$

where $W = kL/2$. Equation (65.113) was obtained by Walker [37], and the second term describes the spin-wave excitation. The transmitted signal can be found to be proportional to the magnetization just inside the surface at

$$H\left(L_+/2\right) = cZ_0 M\left(L_-/2\right)\,, \tag{65.114}$$

where Z_0 is the surface impedance for spinless electrons.

$$H_T \propto \frac{\iota}{[(\omega-\omega_s)T_2 + \iota]}\left[\frac{2W}{\sin 2W}\right]\,, \tag{65.115}$$

which was first obtained by Platzman and Wolff using a Green's-function technique [6, 7].

65.5 Comparison with the Platzman–Wolff Theory and Experimental Data

Although the Wilson-Fredkin theory [8] was shown to explain experimental data better than the Platzman–Wolff theory, the two theories are equally satisfactory [15] if one neglects B_n and $\eta \geq 2$ and anomalous experimental features reproduced in Figure 65.7. Hence we will compare CDW theory only with Platzman–Wolff theory for potassium, for which the direction and magnitude of $\mathbf{Q}$ were recently determined [21].[7] In Figure 65.5, comparisons are made for two different geometries with only open-orbit effects included, assuming that the direction of $\mathbf{G}_{110}$ is perpendicular to the surface of the sample (which is indicated by optical experiments [32, 33]), that the azimuthal angle of $\mathbf{H}_0$ about $\hat{\mathbf{Q}}'$ is zero, defined in Figure 65.4 with $\hat{\mathbf{Q}}'$ parallel to $\hat{z}$, and that the CDW domain size is 0.001 cm. $\mathbf{Q}$ is (0.995,0.975,0.015)

7) Apparently conflicting data: [22], can be intepreted by differing CDW domain structure in different samples.

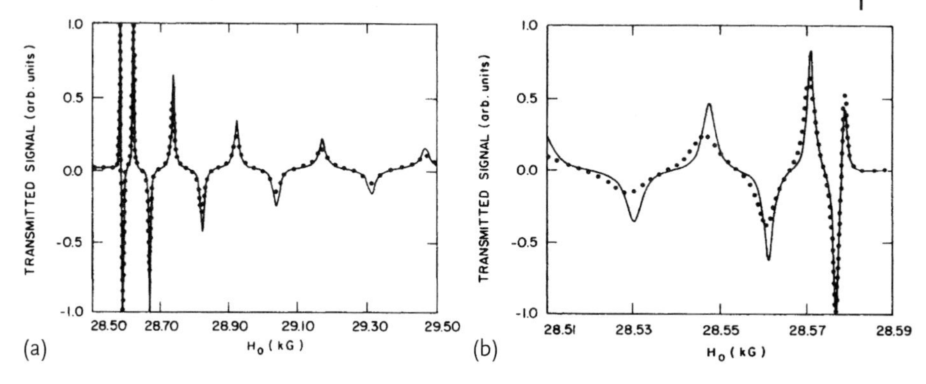

Figure 65.5 Comparison between the CDW theory and PW theory: the dotted curve for the CDW theory with only open-orbit effects included; the solid curve for the PW theory. No attempt was made to make the amplitudes and widths agree with experimental data, by readjusting τ and T_2 (for the CDW case). (a) For the field-perpendicular geometry. (b) For the field-parallel geometry.

in units of $2\pi/a$ and the angle between $\mathbf{Q}$ and $\mathbf{G}_{110}$ is 0.85°. $\mathbf{Q}' = (0.005, 0.025, -0.015)$ in units of $2\pi/a$ and the angle between $\mathbf{Q}'$ and $\mathbf{G}_{110}$ is 44.18°. All modifications caused by CDW effects are contained in the effective diffusion constant D^* in Equation (65.106), or equivalently β in Equation (65.105) through Equations (65.94), (65.96), and (65.99). Since the inclusion of distortion effects does not change the following results qualitatively, all comparisons with experimental data will be made with only open-orbit effects included. The positions of the side bands agree very well, while the different signal amplitudes and widths can be made to agree with experiment by adjusting the relaxation times τ and T_2. T_2 determines the decay rate of the amplitudes of the side band peaks, and τ determines their widths. Our theory does not explain the variation of the width of the main CESR peak versus Δ, and τ and T_2 would be adjusted following the procedure taken by Mace, Dunifer, and Sambles [15]. For the PW theory, $B_0 = -0.292$ and $B_1 = -0.073$ were used [15], whereas for the CDW theory it was found that $B_0 = -0.252\pm0.003$ and $\eta = 0.048\pm0.015$ with the open-orbit effect included and $B_0 = -0.221\pm0.002$ and $\eta = 0.042\pm0.013$ with both the open-orbit effect and the distortion effect included. "$\pm$" indicates the upper and lower bounds for the range of azimuthal angle of $\mathbf{H}_0$ about $\hat{\mathbf{Q}}$, defined in Figure 65.4 with $\hat{\mathbf{Q}}$ parallel to $\hat{\mathbf{q}}$. The variation of the fitting parameters B_0 and η over the azimuthal angle is shown in Figure 65.6. When the $\mathbf{Q}$-domain size is larger than 0.002 cm, B_0 and η vary much more rapidly and spin-wave side bands show splittings into two or more components as will be discussed later. This variation can be attributed to the fact that the magnetic field along $\mathbf{Q}'$ vanishes for some azimuthal angles and the effective diffusion constant near those angles changes rapidly. One can see that the magnitude of B_0 depends critically on the various CDW effects, and that there will be a systematic change in the positions of side band peaks as the azimuthal angle is changed for samples having large CDW domain size.

Although extreme care was taken in preparing samples, it was reported [15] that even the best samples displayed some imperfections: a faint milky appearance, numerous small bubbles at the surface, and microscopic protrusions due to shallow scratches on the quartz windows. When samples showed splitting of either the main CESR or side bands, they were simply thought unacceptable and discarded. It was reported that there were extra features having the characteristic of spin-wave signals in the vicinity of the first two spin waves, reproduced in Figure 65.7(a) from Figure 14(a) of [15], and side bands sometimes split into two or more components, reproduced in Figure 65.7(b) from Figure 13(a) of [15]. Figure 7(a) was

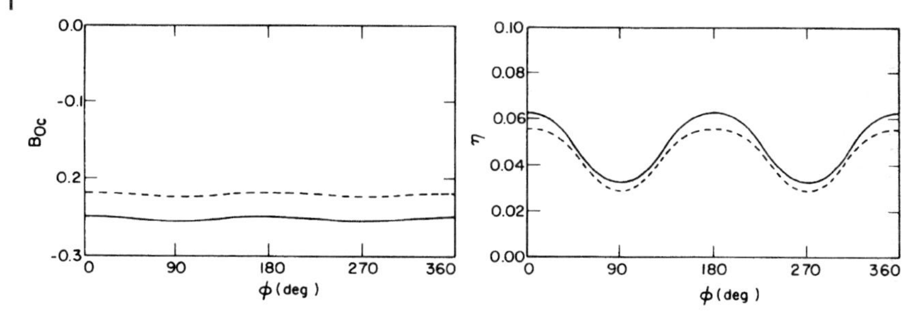

Figure 65.6 Variation of fitting parameters as a function of the azimuthal angle between **Q** and $\mathbf{H}_0$ for domain size $D = 0.001$ cm. The solid curve obtains when only open-orbit effects are included, and the dashed curve applies when the Fermi surface distortion is also included.

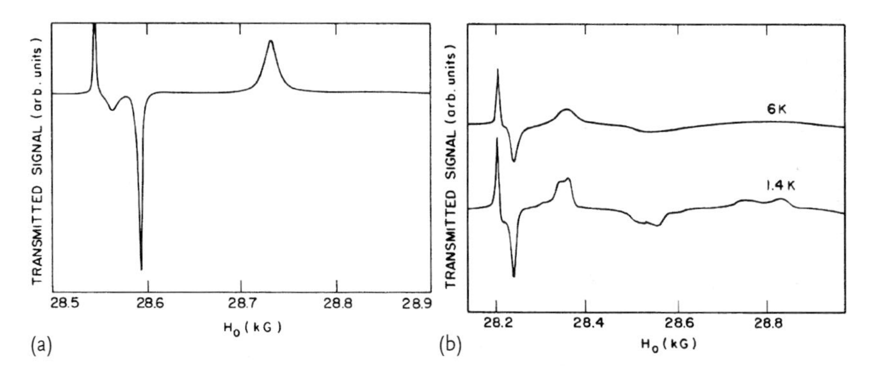

Figure 65.7 Anomalous experimental data. (a) Extra feature between the main CESR and the first spin wave in Na. (b) Splitting of side band peaks in K.

obtained in Na and Figure 65.7(b) was obtained in K, but these features were observed in both Na and K. The anomalous features in Figure 65.7(a) were left unexplained and the splittings in Figure 65.7(b) were attributed in the report to different thickness within the sample, although the variation in thickness was typically less than 1%. But we believe that unless there is a systematic, much larger difference in thickness, splitting of a side band cannot occur, since what is measured is the transmitted microwave from the whole surface of the sample. Also it was reported that the main CESR peak sometimes splits into two or more peaks. This was attributed to inhomogeneity of the static magnetic field, since the signal changes when the inhomogeneity is adjusted. But it is not clear how a small inhomogeneity of 0.6 G, possible in a superconducting magnet, can cause the CESR signal to split into two or more peaks. This phenomena was explained by Overhauser and de Graaf [14] using CDW theory. It is natural for the CESR signal to change with field inhomogeneity since different CDW domains have different locations in the solenoid, and the change does not necessarily indicate that the splitting is due to the field inhomogeneity.

For the field-parallel geometry, $\Delta = 90°$, the domain structure is assumed to be such that any azimuthal angle of the open-orbit direction $\hat{\mathbf{Q}}'$ about the normal to the surface, defined in Figure 65.4 with $\hat{\mathbf{Q}}'$ parallel to $\hat{z}$, is equally probable and the average of the transmitted signal H_t is taken over the azimuthal angle. In Figure 65.8, the splitting of side bands appears naturally for samples with large **Q**-domain size whereas the splitting does not appear for samples with small domain size. For the field-perpendicular geometry, $\Delta = 0°$, different orientations

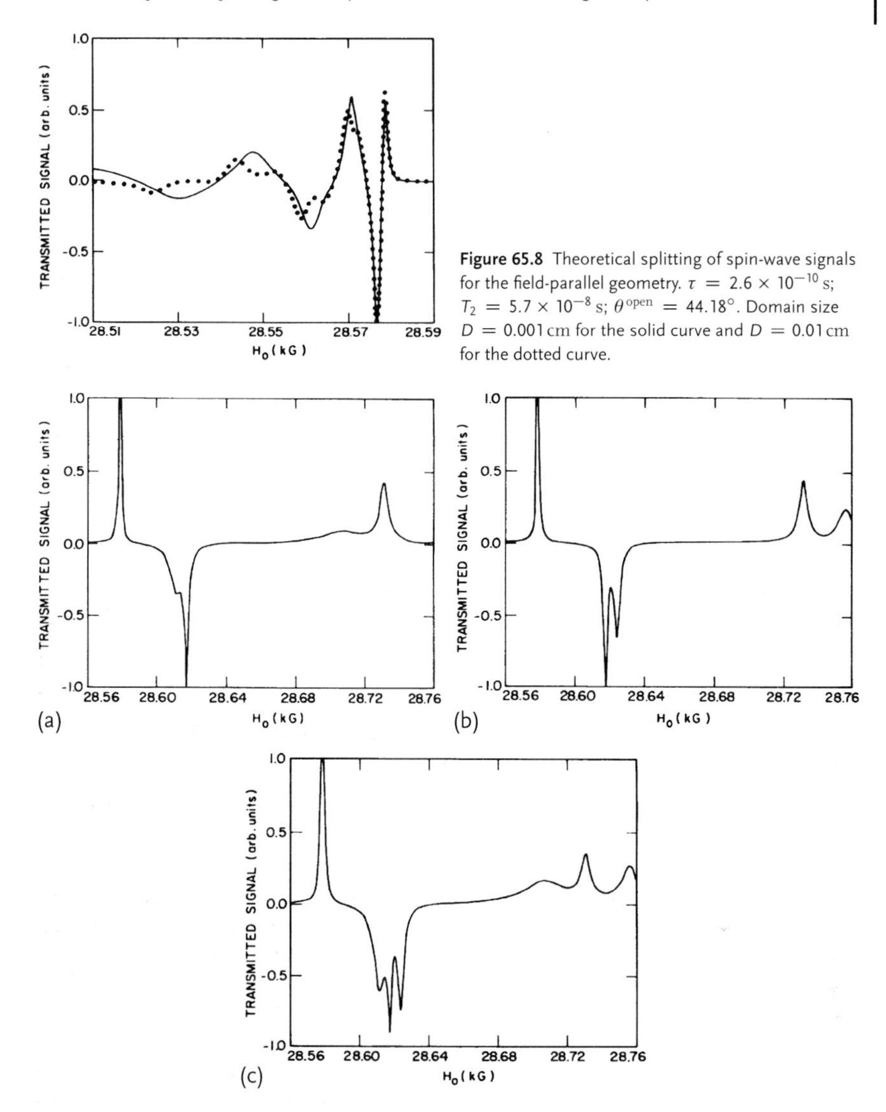

Figure 65.8 Theoretical splitting of spin-wave signals for the field-parallel geometry. $\tau = 2.6 \times 10^{-10}$ s; $T_2 = 5.7 \times 10^{-8}$ s; $\theta^{\text{open}} = 44.18°$. Domain size $D = 0.001$ cm for the solid curve and $D = 0.01$ cm for the dotted curve.

Figure 65.9 Theoretical splitting of spin-wave signals for the field-perpendicular geometry. $\tau = 5.2 \times 10^{-10}$ s, $T_2 = 5.7 \times 10^{-8}$ s, $D = 0.02$ cm. (a) $\theta_1^{\text{open}} = 44.18°$, $\theta_2^{\text{open}} = 61.44°$ (0.5 + 0.5). (b) $\theta_1^{\text{open}} = 44.18°$, $\theta_2^{\text{open}} = 76.17°$ (0.5 + 0.5). (c) $\theta_1^{\text{open}} = 44.18°$, $\theta_2^{\text{open}} = 61.44°$, $\theta_3^{\text{open}} = 76.17°$ (0.2 + 0.5 + 0.3).

of CDW wave vector **Q** which are crystallographically equivalent are included. Other possible open-orbit directions are 17.02°, 61.44°, 76.17°, 103.83° (equivalent to 76.17°), and 118.56° (equivalent to 61.44°). The directions 61.44° and 76.17° were chosen since they result in larger splittings, and were included to obtain Figure 65.9. In Figure 65.9, the splitting of side bands appears naturally for samples with large domain size. These splittings are due to extra spin waves from different CDW domains. Positions of these extra signals depend crucially on the CDW wave vector Q, and the extra features observed for Na in Figure 65.7(a) could be interpreted as extra spin-wave signals from different CDW domains, similar to Figure 65.9(a).

These anomalous features will become less prominent as the temperature of the samples is raised, as observed, since individual components will become broader (as the momentum relaxation time decreases) and will eventually merge into smooth peaks. Lattice imperfections mentioned above are expected to contribute to the formation of CDW domains having **Q** tilted with respect to the normal to the sample surface. Hence elimination of small pockets of argon gas between the metal and quartz window will tend to eliminate such splitting, as observed. Also the adjustments on the homogeneity control (of the static magnetic field) could make anomalous features weaker, since different domains would then be probed with magnetic fields of different strengths.

CDW domain structure is rather sensitive to the past history of a sample and similarly prepared samples can have different CDW domain structure. This explains why spin-wave signals were split for some samples and not for others and why no systematic behavior of the above-mentioned anomalous features was observed for different samples with different thicknesses. As discussed above, the observed anomalous data can be interpreted without recourse to field inhomogeneity or variation in the sample thickness. We suggest that splitting of spin-wave signals results from CDW **Q**-domain structure.

65.6 Conclusion

We have shown that the occurrence of open orbits and distortion of the Fermi surface (caused by a CDW state) modify significantly an electron's motion and influence thereby paramagnetic spin-wave signals. The revised values of the Landau parameter B_0 and the fraction of electrons in open orbits η for potassium, with CDW-domain size $D = 0.001$ cm, are found to be $B_0 = -0.252(\pm 0.008)$ and $\eta = 0.048(\pm 0.015)$ when only open-orbit effects are included. $B_0 = -0.221(\pm 0.002)$ and $\eta = 0.042(\pm 0.013)$ when Fermi-surface distortion effects are also included. We also observed that the splitting of spin-wave signals into two or more components can be naturally explained within the framework of CDW theory.

Acknowledgements

This research was made possible by support from the National Science Foundation, Materials Research Laboratory Program.

References

1 Griswold T.W., Kip A.F., and Kittel C. (1952) *Phys. Rev. Lett.* **88**, 951.
2 Dyson F.J. (1955) *Phys. Rev.* **98**, 349.
3 Lampe M. and Platzman P.M. (1966) *Phys. Rev.* **150**, 340.
4 Silin V.P. (1957) *Zh. Eksp. Teor. Fiz.* **33**, 1227 [*Sov. Phys. JETP* **6**, 945 (1958)].
5 Silin V.P. (1958) *Zh. Eksp. Teor. Fiz.* **35**, 1243 [*Sov. Phys. JETP* **8**, 870 (1959)].
6 Platzman P.M. and Wolff P.A. (1967) *Phys. Rev. Lett.* **18**, 280
7 Platzman P.M. and Wolff P.A. (1973) *Waves and Interactions in Solid State Plasmas*, Academic, New York.
8 Wilson A.R. and Fredkin D.R. (1970) *Phys. Rev. B* **2**, 4656.
9 Walker M.B. (1971) *Phys. Rev. B* **3**, 30.
10 Mayer H. and El Naby M.H. (1963) *Z. Phys.* **174**, 269.
11 Coulter P.G. and Datars W.R. (1980) *Phys. Rev. Lett.* **45**, 1021.
12 Overhauser A.W. (1985) In: Bassani F., Fumi F., and Tosi M.P. (eds) *Highlights of Condensed-Matter Theory*, Proceedings

of the International School of Physics "Enrico Fermi", Course LXXXIX, Varenna on Lake Como, 1983. North-Holland, Amsterdam.
13 Walsch W.M. Jr., Rupp L.W. Jr., and Schmidt P.H. (1966) *Phys. Rev.* **142**, 414.
14 Overhauser A.W. and de Graaf A.M. (1968) *Phys. Rev.* **168**, 763.
15 Mace D.A.H., Dunifer G.L. and Sambles J.R. (1984) *J. Phys. F* **14**, 2105.
16 Overhauser A.W. (1962) *Phys. Rev. 128*, 1437.
17 Overhauser A.W. (1968) *Phys. Rev.* **167**, 691.
18 Kittel C. (1971) In: *Introduction to Solid State Physics*, 4th ed. Wiley, New York.
19 Marquardt W.R. and Trivisonno J. (1965) *J. Phys. Chem. Solids* **26**, 273.
20 Smith P.A. and Smith C.S. (1965) *J. Phys. Chem. Solids* **26**, 279.
21 Giebultowicz T.M., Overhauser A.W., and Werner S.A. (1986) *Phys. Rev. Lett.* **56**, 1485.
22 Pintschovius L., Blaschko O., Krexner G., de Podesta M., and Currat R. (1987) *Phys. Rev. B* **35**, 9930.
23 Overhauser A.W. (1964) *Phys. Rev. Lett.* **13**, 190.
24 Fragachan F.E. and Overhauser A.W. (1984) *Phys. Rev. B* **29**, 2912.
25 Overhauser A.W. (1986) *Phys. Rev. B* **34**, 7632.
26 Landau L.D. (1956) *Zh. Eksp. Teor. Fiz.* **30**, 1058 [*Sov. Phys. JETP* **3**, 920 (1957)].
27 Landau L.D. (1957) *Zh. Eksp. Teor. Fiz.* **32**, 59 [*Sov. Phys. JETP* **5**, 101 (1957)]. 32, 59 (1957) [5, 101 (1957)];
28 Landau L.D. (1958) *Zh. Eksp. Teor. Fiz.* **35**, 97 [*Sov. Phys. JETP* **8**, 70 (1959)].
29 Pines D. and Nozieres P. (1966) *The Theory of Quantum Liquid.* Benjamin, New York.
30 Lifshitz E.M. and Pitaevskii L.P. (1980) *Statistical Physics*, Vol. 2. Pergamon, New York.
31 Silin V.P. (1957) *Zh. Eksp. Teor. Fiz.* **33**, 495 [*Sov. Phys. JETP* **6**, 387 (1958)].
32 Overhauser A.W. and Burtler N.R. (1976) *Phys. Rev. B* **14**, 3371.
33 Giuliani G.F. and Overhauser A.W. (1979) *Phys. Rev. B* **20**, 1328.
34 Torrey H.C. (1956) *Phys. Rev.* **104**, 563.
35 Kaplan J.I. (1959) *Phys. Rev.* **115**, 575.
36 Huberman M. and Overhauser A.W. (1981) *Phys. Rev. Lett.* **47**, 682.
37 Walker M.B. (1970) *Can. J. Phys.* **48**, 111.

Part III Thirty Unexpected Phenomena Exhibited by Metallic Potassium

The thirty anomalous phenomena exhibited by metallic potassium, and their approximate discovery dates, are listed below. (Some are also observed in other alkali metals.) For reasons discussed in the opening section of the Introduction the major focus has been potassium. The sixty-five research papers, [R1] to [R65], reprinted in Part II, document the efforts and success in showing that a CDW broken symmetry can account for all thirty unexpected phenomena. The relevant reprint(s) which explain each puzzle are identified.

1.	1963	Mayer-El Naby optical anomaly. [R4]; [Section R19.3.3] [R11]; [Section R34.2]; [R35]; [Section R45.4.1]
2.	1963	Optical anisotropy. [R11]; [R35]; [R45]; [Section R45.4.1]
3.	1966	Conducton-electron spin-resonance splitting. [R7]; [Section R19.3.4]; [Section R34.3.1]; [Section R45.4.2]
4.	1968	Non-saturating transverse magnetoresistance. [R8]; [Section R19.3.2]; [R35]; [Section R45.4.4]
5.	1968	Kohler-slope variability. [Section R19.3.2]; [R35]; [Section R45.4.4]
6.	1968	Doppler-shifted cyclotron-resonance discrepancy. [R5]; [Section R34.3.2]; [Section R45.4.12]
7.	1969	Perpendicular-field cyclotron resonance. [R59]; [R48]; [R58]
8.	1969	Longitudinal magnetoresistance. [R61]
9.	1971	Four-peaked induced-torque anisotropy. [Section R19.3.2]; [R61] [Section R34.3.5]; [R35]; [R39]; [R40]; [Section R45.4.6]
10.	1972	Residual-resistance variability. [Section R19.3.7]; [Section R19.3.5] [Section R34.3.7]; [R41]; [Section R45.4.3]
11.	1973	Hall-coefficient discrepancy. [Section R19.3.6]; [Section R34.3.8] [R35]; [R40]; [Section R45.4.5]
12.	1974	Cyclotron-resonance transmission. [R59]; [Section R45.4.12]
13.	1975	Oil drop effect. [Section R34.3.6]; [R35] [Section R45.4.7]
14.	1976	Deviations from Matthiessen's rule. [R23]; [Section R45.4.11]
15.	1977	Residual-resistivity anisotropy. [R12]; [Section R19.3.5]; [R41]; [Section R45.4.8]

16.	1978	Low-temperature phason resistivity. [R16]; [R22]; [R25]; [R23]; [R35]; [Section R45.4.9]
17.	1980	Open-orbit magnetoresistance resonances. [R29]; [R30]; [R31]; [R32]; [Section R34.4.2]; [Section R34.4.3]; [R35]
18.	1980	Phason peak in point-contact spectrocopy. [R16]; [Section R34.3.10]; [Section R45.4.13]
19.	1981	Variability of electron-electron scattering resistivity. [R35]; [Section R45.4.9]
20.	1982	Phason heat capacity peak. [R16]; [R22]; [R62]; [R35]
21.	1982	Temperature-dependence of the surface impedance. [Section R45.4.10]
22.	1983	Field-dependence of the residual-resistance anisotropy. [R35]; [R41]
23.	1983	Four-peaked phase anomalies. [R40]
24.	1985	Fermi-energy photoemission peak. [R21]; [R46]
25.	1987	Magneto-serpentine effect. [R52]
26.	1988	Splitting of paramagnetic-spin-wave sidebands. [R65]
27.	1989	Subharmonic cyclotron-resonance transmissions. [R57]
28.	1990	Neutron-diffraction satellites. [R10]; [R49]; [R53]; [R55]
29.	1992	Landau-level oscillations from the cylindrical Fermi surface. [R56]
30.	1993	Gantmakher-Kaner oscillations (too small by 10^4). [Section R57.3]